RARE LAB

# Fundamentals of
# MICROFABRICATION

# Fundamentals of
# MICROFABRICATION

## Marc Madou

**CRC Press**
Boca Raton   London   New York   Washington, D.C.

Development Editor: Marleen Madou
Publisher: Ron Powers
Project Editor: Paul Gottehrer
Pre Press: Gary Bennett, Kevin Luong, Carlos Esser, Walt Cerny, Greg Cuciak
Cover design: Denise Craig

**Library of Congress Cataloging-in-Publication Data**

Madou, Marc J.
    Fundamentals of microfabrication / Marc Madou.
       p.  cm.
    Includes bibliographical references and index.
    ISBN 0-8493-9451-1 (alk. paper)
    1. Microelectronics—Design and construction.   2. Machining.
3. Microelectronic packaging.   4. Lasers—Industrial applications.
I. Title.
TK7836.M33   1997
621.3815′2—dc20
                                      96-43344
                                        CIP

No claim to original U.S. Government works
International Standard Book Number 0-8493-9451-1
Library of Congress Card Number 96-43344
Printed in the United States of America      3  4  5  6  7  8  9  0
Printed on acid-free paper

# Dedication

For my wife, Marleen Madou, thanks for urging me to write a second book and for all your love and support. For my children, Ramses and Maura, sorry for all the time I could have been with you but worked on the book instead. For my late parents to whom I wish I had been able to show this second book.

# Preface

*Fundamentals of Microfabrication* explores the science of miniaturization. Miniaturization methods and materials surveyed include micromachining in single crystal and polycrystalline Si and other micromachining methods and materials based on lithography as well as more traditional non-lithography miniaturization options and materials. In dealing with micromachining techniques borrowed from the electronics industry, we emphasize the processes that differ the most from the standard processes encountered in regular IC and hybrid manufacturing.

New developments in lithography largely determine which direction the IC industry and Si-based microfabrication will take in the coming years. Therefore, Chapter 1 appropriately introduces the book by discussing different lithography techniques and tries to define the optimum lithography for the future which may be different for micromachining than for IC technology. Whereas finer linewidths and standarized materials are the main quest in the IC industry, microfabrication seeks high features and high aspect ratios and the introduction of new materials. Resist patterns created by lithography on a substrate can be transferred to the substrate by subtractive (etching) or by additive (deposition) techniques. Dry etching, discussed in Chapter 2, is an important subtractive pattern transfer method in IC fabrication. Recent progress in deep directional etching as well as environmental concerns helped push dry etching to the foreground in micromachining applications as well. Chapter 3 covers all types of additive pattern transfer techniques. A limited description of thin film deposition technologies and doping methods sufficed as these techniques are, in most cases, the same for both IC and microfabrication. Thick film deposition technologies proved important in the manufacture of all types of new chemical and biological sensors. More effort is dedicated to this subject as many newly pioneered thick film materials and processes are foreign to IC production. Chapter 4 is dedicated to wet bulk micromachining, a key process in sensor fabrication but less common in IC manufacturing. Surface micromachining, a method involving thin film additive techniques as well as wet and dry etching, is covered in Chapter 5. The rapid commercial acceptance of surface micromachining is explained in terms of its compatibility with existing IC equipment and processes. LIGA, the newest microfabrication tool based on deep etch x-ray lithography, electrodeposition, and molding gets extensive coverage in Chapter 6.

Chapter 7 compares traditional and non-traditional microfabrication tools. The arsenal of microfabrication tools has increased dramatically over the last 20 years and different applications require different fabrication means. Several micromachining methods, especially the more traditional precision engineering grounded methods, bear little relevance on today's IC Industry but they are viable microfabrication tools. It is one of the objectives of the book to broaden the perspective of the reader on all of the different options available to manufacture small things. Applying microfabrication tools correctly to the problem at hand might generate a lot more commercial successes. Chapter 7 also speculates on the future of micromachining. We believe in the merging of micromachining and nanomachining. Micromachining often provides the tools (e.g. the scanning tunneling microscope) to enable nanomachining. In nanomachining we distinguish between nanofabrication: heir to micromachining, using the same subtractive and additive processes to build devices in the sub 0.1 μm regime, and molecular engineering where one mimics nature's way to build nanomachinery. Nature has a different way of 'building' things than micromachinists. The elemental building blocks of everything we know are atoms and molecules. In molecular engineering one attempts to build functional structures by building up from the atomic or molecular level in the hope of constructing the same diversity in shapes, functions and memory size offered by nature. In comparison, the micromachining world represents a crude construction site with Si wafers as building blocks as thick as 500 μm, insulating layers of up to a micron thick, Al and Au metal layers between a few tens to thousands of angstroms, and in general a very limited choice of other materials. In micromachining we are building down towards smaller and smaller structures whereas in molecular engineering we are building with a plethora of different atoms and molecules towards bigger and bigger molecular entities. The synergy between nanofabrication, molecular engineering and microfabrication may prove the most fruitful research domain for decades to come.

Once all currently available microfabrication tools have been explored, we turn our attention to new device development and packaging in Chapter 8. Given that no standard design rules yet permeate micromachining and because of the dilemma of partitioning sensing and electronics functions correctly, early attention to design and packaging of micromachines is even more important than in the IC industry. We cannot stress enough the importance of starting the micromachine design from a good understanding of the application and from the application specific package and real world interface and only then applying the preferred micromachine inside the package. Merging of IC design software with micromechanical design code is helping a slow but certain introduction of micromachining with even the most conservative manufacturing firms. We also will see how micromachining itself provides many excellent solutions for future packaging strategies.

After fabrication, design and packaging technologies are exposed, Chapter 9 explores the importance of miniaturization in general with an emphasis on the most difficult to miniaturize components, i.e., actuators and power sources. A good understanding of scaling laws will help the reader develop 'micro-intuition' and assist him/her in making mature decisions about the optimum micromachining approach and design. We will see the emergence of the most exciting opportunities in micromachining in those areas where the macro continuum models break down in the microdomain.

In Chapter 10 on 'Microfabrication Applications' we present a list of current and potential applications and discuss market opportunities.

Because of the rapidly changing nature of the micromachining field a homepage (http://www.crcpress.com/microfab) dedicated to 'Fundamentals of Micromachining' was set up to transform this book into a living, hyperlinked document with frequent updates, questions and input from readers all around the world. Finally, a few Appendices give useful information for aspiring micromachinists ranging from a list of metrology methods (Appendix A), further reading on the internet (Appendix B), detailed Si and $SiO_2$ KOH etching data (Appendix C), important further reading (Appendix D) and a glossary (Appendix E).

The goal of this book is to familiarize the reader with the micromachinists tools, directions and jargon in order to facilitate a confident choice of fabrication method for a particular miniaturization problem.

# Acknowledgments

I would like to thank Mr. John Hines, manager of Sensors 2000! at NASA Ames, for his support and for sharing my vision on micromachines in biomedical applications. From the Center for Advanced Microdevices (CAMD) in Baton Rouge, I acknowledge Dr. Volker Saile, Director of CAMD, for giving me the opportunity to participate in LIGA development work. My thanks also go to the Miller Institute for Basic Research in Science for the Visiting Miller Professorship awarded for my stay at the Berkeley Sensors and Actuators Center (BSAC) at UC Berkeley; my gratitude to Professor Richard White for many stimulating discussions and his open attitude to cross disciplinary work.

My sincere thanks go to all of the colleagues who helped review different chapters of the book. In alphabetical order they are: Mr. Rashid Bashir (National Semiconductor), Dr. Barry Block (consultant), Dr. Luc Bousse (Caliper), Mr. Jim Bustillo (UCB), Mr. Sean Cahill (MicroScape), Mr. Michael Cohn (UCB), Dr. Ben Costello (Berkeley Microinstruments), Mr. Peter Hillen (Congruity), Dr. Keith Jackson (LBNL), Mr. Jack Judy (UCB), Mr. Chris Keller (UCB), Professor Chantal Khan Malek (CAMD/LSU), Dr. Kim Kinoshita (LBNL), Dr. Peter Krulevitch (LLNL), Dr. Adolfo Lopez-Otero (Stanford University), Professor Roy Morrison (Simon Fraser University), Professor Richard Muller (UCB), Professor Michael Murphy (LSU), Dr. Armand Neukermans (Adagio), Dr. Seajin Oh (SRI International), Dr. Tony Ricco (Sandia), Dr Angel Sanjurjo (SRI International), Mr. Tim Slater (consultant), Dr. Michael Thierny (Cygnus), Dr. Volker Saile (CAMD), Dr. Stuart Wenzel (Berkeley Microinstruments), Professor Richard White (UCB)

Dr. Adolfo Lopez-Otero (Stanford University) assisted with reference research, the internet, Appendix A and the index. My son Ramses helped with some of the artwork. Finally, I would like to thank my wife, Marleen for spending countless hours converting my Flemish-English into her Flemish-English.

# Table of Contents

# Chapter 3
# Pattern Transfer with Additive Techniques

# Chapter 4
# Wet Bulk Micromachining

# Chapter 5
# Surface Micromachining

# Chapter 6
# LIGA

# Chapter 10
# Microfabrication Applications

# Lithography

*Microfabrication or micromachining (also micro-manufacturing) in the narrow sense comprises the use of a set of manufacturing tools based on batch thin and thick film fabrication techniques commonly used in the electronics industry. In the broader sense, microfabrication describes one of many precision engineering disciplines which takes advantage of serial direct write technologies, as well as of more traditional precision machining methods, enhanced or modified for creating small three-dimensional (3D) structures with dimensions ranging from subcentimeters to submicrometers, involving sensors, actuators, or other microcomponents and microsystems*

## Introduction

Both micromachining and microelectronic fabrication start with lithography, the technique used to transfer copies of a master pattern onto the surface of a solid material, such as a silicon wafer. In this chapter we review different forms of lithography, adding the most details where the lithography for building micromachines differs from the process used to fashion integrated circuits (ICs).

A short historical note about the origins of lithography is followed by a description of photolithography, including the expansions that enable the method to print the ever-decreasing features of modern ICs. Details on alternative lithographies, including X-ray and charged particle (electrons and ions) lithographies, are given next, followed by promising lithography techniques in the early research and development (R&D) stage.

## Historical Note: Lithography's Origins

After experimenting with various resins in sunlight, Nicéphore Niépce managed to copy an etched print on oiled paper by placing it over a glass plate coated with bitumen dissolved in lavender oil (France, 1822). After 2 or 3 hours of sunlight the unshaded areas in the bitumen became hard compared to the shaded areas which remained soluble and could be washed away with a mixture of turpentine and lavender oil. By etching a plate,

developed by Niépce, in strong acid, the Parisian engraver Lemaître made an etched copy of an **engraving of Cardinal d'Amboise** in 1827 (talking about fast turnaround time!). The latter copy represents the earliest example of pattern transfer by photolithography and chemical milling. The accuracy of the technique was 0.5 to 1 mm.[1]

The word 'lithography' itself (Greek for the words 'stone' [*lithos*] and 'to write' [*gráphein*]) refers to the process invented in 1796 by Aloys Senefelder. Senefelder found that stone (he used Bavarian limestone), when properly inked and treated with chemicals, could transfer a carved image onto paper. Due to the chemical treatment of the stone, image and non-image areas became oil-receptive (water-repellent) and oil-repellent (water-receptive), respectively, attracting ink onto the image area and attracting water on non-image areas.[2]

The Niépce process heralded the advent of photography. Much later, photomasking followed by chemical processing lead to the photolithography now used in fabricating ICs and micromachines. Not until World War II, more than 100 years after Niépce and Lemaître, did the first applications of the printed circuit board come about. Interconnections were made by soldering separate electronic components to a pattern of 'wires' produced by photo-etching a copper foil laminated to a plastic board. By 1961, methods were devised whereby a photo-etching process produced large numbers of transistors on a thin slice of Si. At that time patterns had a resolution of 5 μm.[1] Today in

## *Engraving of Cardinal d'Amboise*

The earliest example of photolithography followed by wet etching. (Photograph from the Science Museum. Courtesy of the Royal Photographic Society.)

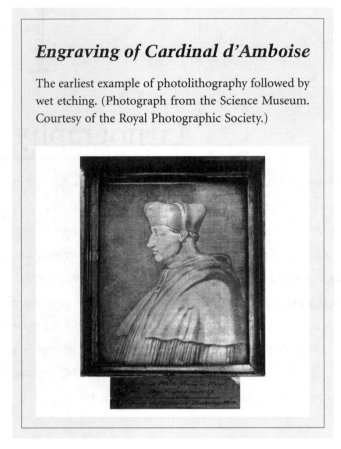

## *Light field and dark field*

Mask-reticle polarities for a field effect transistor showing the gate and contacts.

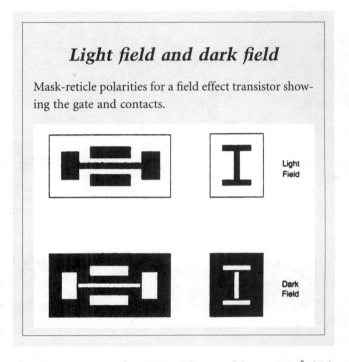

photolithography and X-ray and charged particle lithography submicron accuracy in printing has been achieved.

# Photolithography

## Introduction

The most widely used form of lithography is photolithography. In the IC industry, pattern transfer from masks onto thin films is accomplished almost exclusively via photolithography. Accurate registration and exposure of a series of successive patterns lead to complex multilayered ICs. This essentially two-dimensional process has a limited tolerance for nonplanar topography, creating a major constraint for building microstructures which often exhibit extreme topographies. Photolithography matured rapidly and has become better and better at resolving smaller and smaller features. For the IC industry, this continued improvement in resolution has pushed the adaptation (insertion point) of alternative, higher resolution lithography techniques, such as X-ray lithography, back in time. Research in high aspect ratio resist features, to satisfy the needs of microfabrication, is finally also improving photolithography's capacity to cover wide ranges of topography.

## Masks

The stencil used to generate a desired pattern in resist-coated wafers over and over again is called a mask. In use, a photomask, a nearly optically flat glass (transparent to near UV) or quartz

plate (transparent to deep UV) with a metal (e.g., a 800-Å thick chromium layer) absorber pattern, is placed into direct contact with the photoresist coated surface, and the wafer is exposed to the ultraviolet radiation. The absorber pattern on the photomask is opaque to ultraviolet light, whereas glass or quartz is transparent. A *light field or dark field image* (mask polarity) is then transferred to the semiconductor surface. This procedure results in a 1:1 image of the entire mask onto the silicon wafer. The described masks, making direct physical contact to the substrate (also referred to as 'hard contact'), are called 'contact masks'. Unfortunately, these masks degrade faster through wear than do non-contact, proximity masks (also referred to as 'soft contact' masks) which are slightly raised, say 10 to 20 μm, above the wafer. The defects resulting from hard contact masks on both the wafer and the mask make this method of optical pattern transfer unsuitable for Very Large Scale Integration (VLSI) manufacturing. We are reviewing hard contact masks because they are still in use in R&D, in mask making itself, and for sensor prototyping. Contact mask and proximity mask printing are collectively known as shadow printing. If, instead of placing a mask in direct contact or in proximity to a wafer, the photomask is imaged by a high-resolution lens system onto the resist-coated wafer, one defines projection printing. In the latter case, the only limit to the mask lifetime results from operator handling. The mask pattern can be reduced by the imaging lens 1:5 or 1:10, making the mask fabrication less challenging.

## Spinning Resist and Soft Baking

A common step, before spinning on a resist, with Si as a substrate, is to grow a thin layer of oxide on the wafer surface by heating it to between 900 and 1150°C in steam or in a humidified oxygen stream. Dry oxygen also works, but wet oxygen and steam produce faster results. The oxide can serve as a mask for a subsequent wet etch or boron implant. As the first step in the

## Resist spinner and on-line film thickness monitor

Photoresist is dispensed on a wafer laying on a wafer platen. The wafer is spun at high rates to make a uniform thin resist coating. With uniformities of ±10Å being specified for wafer-to-wafer coatings, each coating parameter (resist dispense rate, dispense volume, spin speed, back temperature, ambient humidity, and temperature, etc.) must be optimized. A real-time, *in situ* thickness monitor can provide full optimization possibilities of the photoresist coating process. The system shown is a multiwavelength reflection spectrometer. Reflected light containing the interference profile is analyzed and the resist thickness is deduced. (From Metz et al., *Semicond. Int.*, 15, 68–69, 1992. With permission.)

lithography process itself, a thin layer of an organic polymer, a photoresist, sensitive to ultraviolet radiation, is deposited on the oxide surface. The photoresist is dispensed from a viscous solution of the polymer onto the wafer laying on a wafer plate in a *resist spinner*.[5] A vacuum chuck holds the wafer in place. The wafer is then spun at a high speed, between 1500 and 8000 rpm, depending on the viscosity and the required film thickness, to make a uniform film. At these speeds, centrifugal forces cause the solution to flow to the edges where it builds up until expelled when the surface tension is exceeded. The resulting polymer thickness, T, is a function of spin speed, solution concentration, and molecular weight (measured by intrinsic viscosity). The empirical expression for T is given by

$$T = \frac{KC^{\beta}\eta^{\gamma}}{\omega^{\alpha}} \qquad 1.1$$

with K an overall calibration constant, C the polymer concentration in grams per 100 ml solution, η intrinsic viscosity, and ω the number of rotations per minute (rpm). Once the various exponential factors (α, β, and γ) have been determined, Equation 1.1 can be used to predict the thickness of the film that can be spun for various molecular weights and solution concentrations of a given polymer and solvent system.[3]

The spinning process is of primary importance to the effectiveness of pattern transfer. The quality of the resist coating determines the density of defects transferred to the device under construction. The resist film uniformity across a single substrate and from substrate to substrate must be ±5 nm for a 1.5-μm film that is ±0.3%) in order to ensure reproducible line-widths and development times in subsequent steps. The coating thickness of the thin glassy resist film depends on the chemical resistance required for image transfer and the fineness of the lines and spaces to be resolved. The application of too much resist results in edge covering or run-out, hillocks, and ridges, reducing

manufacturing yield. For silicon integrated circuits, the resist thickness, after baking (see below), typically ranges between 0.5 and 2 μm; for microfabricated structures, thicknesses of 1 cm and above have been attempted.[4] In the latter case, techniques such as casting or plasma polymerizing of the resist replace the ineffective resist spinners. The challenges involved in making those thicker resist coats needed to make high aspect ratio micromachines will be discussed in detail in Chapter 6 on LIGA. Optimization of the 'regular' photoresist coating process, in terms of resist dispense rate, dispense volume, spin speed, ambient temperature, and humidity presents a growing challenge as a 1-μm resist thickness with a repeatability of ±10 Å (±0.1%) might become the norm in the microelectronic industry dealing more and more with submicron minimum feature sizes. An *on-line film thickness* monitor, possibly a technique based on reflection spectroscopy, will become essential for statistical process control of such demanding photoresist coatings.[5]

After spin coating, the resist still contains up to 15% solvent and may contain built-in stresses. The wafers are therefore soft baked (also prebaked) at 75 to 100°C for 10 minutes to remove solvents and stress and to promote adhesion of the resist layer to the wafer.

Two excellent reference books detailing the resist coating process and other aspects of resists are Moreau's *Semiconductor Lithography*[6] and Thompson et al.'s *Introduction to Microlithography*.[7] Other good, but older, references on lithography are by Sze,[8] Ghandi,[9] and Colclaser.[10]

## Exposure and Post-Exposure Treatment

After soft baking, the resist-coated wafers are transferred to some type of illumination or exposure system where they are aligned with a precision better than 5 μm, using the flat of the wafer or patterns already existing on the wafer. In the simplest case, an exposure system consists of a UV lamp illuminating the resist-coated wafer through a mask without lenses between the mask and the wafer. The purpose of the illumination system is to deliver light to the wafer with the proper intensity, directionality, spectral characteristics, and uniformity across the wafer, allowing a nearly perfect transfer (also 'print') of the mask image onto the resist in the form of a latent image.

In photolithography, wavelengths of the light source used for exposure of the resist-coated wafer range from deep ultraviolet (DUV), i.e., 150 to 300 nm, to near UV, i.e., 350 to 500 nm. In the near UV, one typically uses the g-line (436 nm) or i-line (365 nm) of a mercury lamp. The brightness of most shorter wavelength sources is severely reduced compared to longer wavelength sources, and the addition of lenses further reduces the efficiency of the exposure system (for example, the total collected DUV power for a 1 kW mercury-xenon lamp is only 10 to 20 mW). The additional optics absorb more energy of the short wavelengths passing through them. As a consequence, with shorter wavelengths, a higher resist sensitivity is required. Alternatively, unconventional DUV sources producing a high flux of DUV radiation must be used, e.g., a KrF excimer laser with a short wavelength of 249 nm and a power of 10 to 20 watts at that wavelength.

The incident light intensity (in $W/cm^2$) multiplied by the exposure time (in sec) gives the incident energy ($J/cm^2$) or dose, D, across the surface of the resist film. The radiation induces a chemical reaction in the exposed areas of the photoresist, altering the solubility of the resist in a solvent either directly or indirectly via a sensitizer.

At times the sensitizer in the resist bleaches during the latent image-forming reaction; in other words, exposed resist is rendered transparent to the incoming wavelength. This bleaching allows the use of thick films with high absorbency, since light will reach the substrate through the bleached resist. The absorbency of the unexposed resist should not reach 40% in order to avoid degradation of the image profile through the resist depth, as too large a percentage of the light is absorbed in the top layer. On the other hand, with the absorbency far below 40%, exposure times required to form the image become too long. The smaller the dose needed to 'write' or 'print' the mask features with good resolution onto the resist layer, the better the lithographic sensitivity of the resist.

Post-exposure treatment is sometimes desired as the reactions initiated during exposure might not have run to completion. To bring the reactions to a halt or to induce new reactions, several post-exposure treatments are in use: post-exposure baking, flood exposure with other types of radiation, treatment with reactive gas, and vacuum treatment. Post-exposure baking (sometimes *in vacuum*) and treatment with reactive gas define the image reversal and dry development systems. In the case of a chemically amplified resist, the post-exposure bake is most critical. Although the catalysis reaction induced by the catalyst formed during exposure will take place at room temperature, the timing could be highly improved to a few seconds by baking at 100°C. The precise control of this type of post-exposure bake critically determines the subsequent development itself. Dry developed resists, imaged reversal, and amplified resists will be treated in more detail in subsequent sections, as their importance for creating tall three-dimensional structures is considerable.

## Development

Development transforms the latent resist image formed during exposure into a relief image which will serve as a mask for further subtractive and additive steps (see Chapters 2 and 3, respectively). During the development process, selective dissolving of resist takes place. Two main technologies are available for development: wet development, widely used in circuit and micromachine manufacture, and dry development, still in the exploratory stage.

Wet development by solvents can be based on three different types of exposure-induced changes: variation in molecular weight of the polymers (by cross-linking or by chain scission), reactivity change, and polarity change.[11] Two main types of wet development set-ups are used: immersion and *spray developers*. During immersion developing, cassette-loaded wafers are batch-immersed for a timed period in a bath of developer and agitated at a specific temperature. During spray development, fresh developing solution is directed across wafer surfaces by fan-type sprayers.

The use of solvents leads to some swelling of the resist and a loss of adhesion of the resist to the substrate. Dry development could overcome these problems, as it is either based on a vapor phase process or a plasma.[6] In the latter case, oxygen-reactive

### *Spray developer*

Fresh developing solution is directed across wafer surfaces by a fan-type spraying nozzle. The renewal of developer allows a uniform bath strength to be maintained.

ion etching ($O_2$-RIE) is used to develop the latent image. The latent image formed during exposure exhibits a differential etch rate to $O_2$-RIE rather than differential solubility to a solvent.[3]

With the continued pressure by the U.S. Environmental Protection Agency (EPA) for a cleaner environment, dry development as well as dry etching (see Chapter 2) are bound to get more and more attention. Dry-developed resists, especially the so-called 'DESIRE' process where the surface of the exposed resist is treated with a silicon-containing reagent, will be discussed in more detail further below.

## De-Scumming and Post-Baking

A mild oxygen plasma treatment, so-called de-scumming, removes unwanted resist left behind after development. Especially negative resists but also positive resists leave thin polymer films at the resist-substrate interface. The problem is most severe in small (<1 µm) high-aspect-ratio structures where the mass transfer of developer is poor. Patterned resist areas are also thinned in the de-scumming process, but this is usually of little consequence.

Finally, before etching the substrate or adding a material, the wafer needs post-baking. Post-baking or hard baking removes residual developing solvents and anneals the film to promote interfacial adhesion of the resist weakened by developer penetration along the resist-substrate interface or by swelling of the resist (mainly for negative resists). Hard baking also improves the hardness of the film. Improved hardness increases the resistance of the resist to subsequent etching steps. The post-bake frequently occurs at higher temperatures (120°C) and for longer times (20 min) than the prebaking step. The major limitation for heat application is excessive flow or melt which degrades wall profile angles and makes it more difficult to remove the resist. Special care needs to be taken with the baking temperature above the glass transition temperature, $T_g$, when impurities easily incorporate in the resist (see next section).

Resist does not withstand long exposure to etchants well. As a consequence, with 1:7 buffered HF (BHF), a mixture of 1 part 49% aqueous HF-solution and 7 parts $NH_4F$ that is used to strip $SiO_2$, the post-bake sometimes is repeated after 5 min of etching to prolong the lifetime of the resist layer.

## Glass Transition Temperature of a Resist ($T_g$)

Resists must meet several rigorous requirements: high sensitivity, high contrast, good etching resistance, good resolution, easy processing, high purity, long shelf life, minimal solvent use, low cost, and a high glass transition temperature, $T_g$. Most resists are amorphous polymers. At temperatures above the glass transition temperature, polymers exhibit viscous flow with considerable molecular motion of the polymer chain segments. At a temperature below $T_g$, the motion of the segments is halted and the polymer behaves as a glass rather than as a rubber. If $T_g$ is at or below room temperature, the resist is considered a rubber, while if it lies above room temperature it is considered a glass. Since above $T_g$ the polymer flows easily, heating the resist film above its glass transition temperature for a reasonable amount

of time enables the film to anneal into its most stable energetic state. In the rubber state the solvent easily can be removed from the polymer matrix, i.e., soft-bake the resist. Extreme attention needs to be given to the cleanliness of the working environment with the resist in this state. When softening the resist at or above $T_g$, it may be easier to remove solvent, but the resist tends to pick up all types of impurities. The importance of resist reflow, as we will learn later, also lies in planarizing topography.

In general, polymers that crystallize are not useful as resists because the formation of crystalline segments prevents the formation of uniform isotropic films.[3]

## Resist Tone

The principal components of photoresists are a polymer (base resin), a sensitizer, and a casting solvent. The polymer changes structure when exposed to radiation. The solvent allows spin application and formation of thin layers on the wafer surface. Sensitizers control the chemical reactions in the polymeric phase. In the case of resists without sensitizers, one talks about single-component or one-component systems; in the case of a sensitizer-based resist, one speaks of two-component systems. Solvent and other potential additives do not count as components as they do not directly relate to the photo-activity of the resist.

If the photoresist is of the type called positive (also 'positive tone'), the photochemical reaction during exposure of a resist typically weakens the polymer by rupture or scission of the main and side polymer chains, and the exposed resist becomes more soluble in developing solutions (say ten times more soluble). In other words, the development rate, R, for the exposed resist is about ten times larger than the development rate, $R_0$, for the unexposed resist. If the photoresist is of the type called negative (also 'negative tone'), the reaction strengthens the polymer by random cross-linkage of main chains or pendant side chains, becoming less soluble (slower dissolving). Exposure and development sequences for negative and positive resists are shown in Figure 1.1A.[23]

### Positive Resists

Two well known families of positive photoresists form the single component *poly(methylmethacrylate) (PMMA)* resists and the two-component *DQN* resists comprised of a photoactive *diazoquinone ester (DQ)* (20 to 50 wt%) and a *phenolic novolak resin (N)*.

PMMA becomes soluble through chain scission under deep UV illumination; the maximum sensitivity of PMMA lies at 220 nm; above 240 nm the resist becomes insensitive. PMMA resin by itself constitutes a rather insensitive or slow DUV photoresist requiring doses >250 mJ/cm². Exposure times of tens of minutes were required with the earliest DUV PMMAs available.[6] By adding a photosensitizer such as t-butyl benzoic acid, the UV spectral absorbency of PMMA is increased, and a 150 mJ/cm² lithographic sensitivity can be obtained. PMMA is also used in electron beam, ion beam, and X-ray lithography.

The DQN system is a frequently used, near-UV, two-component positive resist, photochemically transformed into a polar,

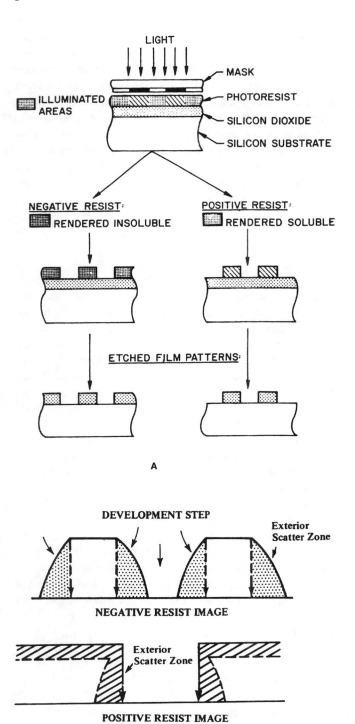

## Poly(methylmethacrylate) or PMMA

Photo-induced chain scission of PMMA resist.

**A**

**Figure 1.1** Positive and negative resist exposure, development, and edge-scattered radiation. (A) Positive and negative resists, exposure, and development. Positive resists develop in the exposed region and usually remain soluble for lift-off. Negative resists remain in the exposed region but are insoluble and not suitable for lift-off (see text). (B) Edge-scattered radiation profile for negative and positive resists. Time-independent development of cross-linked negative resist fails to remove light scatter zone. Development of positive resist rapidly removes exposed region and can be quenched to inhibit removal of lateral scattered exposed resist region. (From Brodie, I. and J. J. Muray, The Physics of Microfabrication, Plenum Press, New York, 1982. With permission.)

## Diazoquinone ester (DQ) and phenolic novolak resin (N), i.e., DQN

The novolak (Novolak) matrix resin (N) is prepared by acid copolymerization of cresol and formaldehyde. The base insoluble sensitizer, a diazoquinone (DQ), undergoes photolysis to produce a carbene which then undergoes a rearrangement to form a ketene. The ketene reacting with water present in the film forms a base-soluble, indenecarboxylic acid photoproduct.

base-soluble product. The hydrophilic novolak resin (N) is in itself alkali soluble but rendered insoluble by the addition of 20- to 50-wt% DQ which forms a complex with the phenol groups of the novolak resin. The resist is rendered soluble again through the photochemical reaction of DQ. The matrix novolak resin is a copolymer of a phenol and formaldehyde. A novolak resin absorbs light below 300 nm, and the DQ addition adds an absorption region around 400 nm. The 365-, 405-, and 435-nm mercury lines can all be used for exposure of DNQ.

Most positive resists are soluble in strongly alkaline solutions and develop in mildly alkaline solutions. Some typical industrially used developers are KOH (e.g., a 0.05- to 0.5-N aqueous solution + a surfactant), tetramethylammonium hydroxide (TMAH), ketones, or acetates. Radiation-induced reactions alter the exposed resist to exhibit hydrophobic and hydrophilic regions. Surfactants and other wetting agents ensure uniform wetting. Solutions also are often buffered to provide a more stable operating window and a longer lifetime. The dissolution rate and the pH depend on the temperature, which needs to be controlled to within ±0.5°C.[3] Typical casting solvents for positive resists are Cellosolve* acetate, methyl Cellosolve, and aromatic hydrocarbons. The semiconductor industry tries hard to move away from organic-based solvents because of health concerns. Consequently, aqueous-based positive resists have become more and more popular.

**Negative Resist**

The prevalent negative photoresists are based on cross-linking of main chains or pendant side chains rendering the exposed parts of the resist insoluble. Commonly used negative-acting, two-component resists are *bis(aryl)azide rubber resists*, whose matrix resin is *cyclized poly(cis-isoprene)*, a synthetic rubber. The bis(aryl)azide sensitizers lose nitrogen and generate a highly reactive nitrene upon photolysis. The nitrene intermediate undergoes a series of reactions that result in the cross-linking of the resin. Often oxidation, involving oxygen from the ambient or dissolved in the polymer, is a competing reaction for polymerization. Consequently, polymerization can be prevented by the quenching of the cross-linking reactions through scavenging of the nitrene photoproduct by oxygen. This competing reaction represents a disadvantage, as exposure might have to be carried out under a nitrogen blanket. Another disadvantage of negative resists is that the resolution is limited by film thickness, as the cross-linking process starts topside where the light hits the resist first. Consequently, overexposure is needed to render the resist insoluble at the substrate interface. The thicker the resist wanted, the greater the overdose needed for complete polymerization and the larger the scattered radiation. Scattered radiation in turn reduces the obtainable resolution. In practice, the obtainable resolution with negative resists is limited to about 2 to 3 μm because of swelling in the development stage, whereas with positive resist the resolution is better than 0.5 μm. To improve the resolution of a negative

resist one can use thinner resist layers; however, when using thin layers of negative resist, pinholes become problematic.

A practical example of a commercial, two-component negative photoresist is the Kodak KTFR (an *azide-sensitized poly(isoprene) rubber*) with a lithographic sensitivity (also photospeed) of 75 to 125 mJ/cm². Negative photoresists, in general, adhere very well to the substrate, and a vast amount of compositions are available (stemming from R&D work in paints, UV curing inks, and adhesives all based on polymerization hardening). Also the negative resists are highly resistant to acid and alkaline aqueous solutions as well as to oxidizing agents. As a consequence, a given thickness of negative resist is more resistant than a corresponding thickness of positive resist. This chemical resistance ensures better retention of resist features even during a long, aggressive wet etch. Negative resists also are more sensitive than positive resists but exhibit a lower contrast (γ smaller) (see below under Contrast and Experimental Determination of Lithographic Sensitivity). More recently, better resolution negative resists have been developed using non-swelling polymers (see O'Brien et al.[12]).

Xylene is the most commonly used solvent for negative resists, although almost any organic solvent can be used. Because negative resists have a line-width limit of only about 2 to 3 μm and because the industry is moving away from organic solvent-based systems in favor of less toxic, water-based developers, positive resists gained in popularity (see above). However, negative resists continue to be used in the production of low-cost, high-volume chips as they require only small amounts of sensitizers and therefore are substantially less expensive than positive resists.

A comparison of negative and positive photoresist features is presented in Table 1.1. This table is not exhaustive and is meant only as a practical guide for selection of a resist tone. The choice depends on a variety of considerations: cost, speed, resolution, etc. The choice of resist tone will even depend on the specific intended pattern geometry known as the optical proximity effect. For example, an isolated single line most easily resolves in a negative resist (higher resolution line), whereas an isolated hole or trench is most easily defined in a positive resist.

In Table 1.2 some common positive and negative resists employed in various lithography strategies are listed together with their lithographic sensitivity. For charged particles (e-beam lithography and ion beam lithography) the sensitivity is expressed in Coulombs (C)/cm²; for photons (optical and X-ray) joule (J)/cm² is used. Ideally, in charged-particle lithography one should select a resist with a sensitivity in the range of $10^{-5}$ to $10^{-7}$ C/cm², and 10 to 100 mJ/cm² in photon lithography in order to minimize the exposure duration.

In reality, the complex resist chemistry contrasts with the simple picture conveyed above. Additives such as plasticizers, adhesion promoters, speed enhancers, and non-ionic surfactants further promote improvements in resist performance.[6] The resist does not adhere well to the wafer when the humidity is too high or if the wafer has been immersed previously in water. Good humidity control (at 40%) and an anneal are requirements to prepare the wafer for resist coating. After oxidation, Si wafers may be vapor primed with reactive silicone

---

\* Cellosolve is a trade name for solvents based on esters of ethylene glycol; these solvents have been identified as possible carcinogens.

## Bis(aryl)azide-sensitized rubber resists with cyclized poly(cis-isoprene) as matrix resin

The primary photoevent in bisarylazide-rubber resists is the production of nitrene which then undergoes a variety of reactions that result in covalent, polymer-polymer linkages. A typical structure of one commonly employed sensitizer is shown. The reaction involved in the synthesis of the cyclized rubber matrix and the bisazide sensitizer is also shown.

**TABLE 1.1** Comparison of Negative and Positive Photoresist

| Characteristic | Resist Type | |
|---|---|---|
| | Positive Resist | Negative Resist |
| Adhesion to Si | Fair | Excellent |
| Available compositions | Many | Vast |
| Contrast $\gamma$ | Higher (e.g., 2.2) | Lower (e.g., 1.5) |
| Cost | More expensive | Less expensive |
| Developer | Aqueous based (ecologically sound) | Organic solvent |
| Developer process window | Small | Very wide, insensitive to overdeveloping |
| Image width to resist thickness | 1:1 | 3:1 |
| Influence of oxygen | No | Yes |
| Lift-off | Yes | No |
| Minimum feature | 0.5 $\mu$m and below | $\pm 2$ $\mu$m |
| Opaque dirt on clear portion of mask | Not very sensitive to it | Causes printing of pinholes |
| Photospeed | Slower | Faster |
| Pinhole count | Higher | Lower |
| Pinholes in mask | Prints mask pinholes | Not so sensitive to mask pinholes |
| Plasma etch resistance | Very good | Not very good |
| Proximity effect | Prints isolated holes or trenches better | Prints isolated lines better |
| Residue after development | Mostly at <1 $\mu$m and high aspect ratio | Often a problem |
| Sensitizer quantum yield $\Phi$ | 0.2–0.3 | 0.5–1 |
| Step coverage | Better | Lower |
| Strippers of resist over | | |
|   Oxide steps | Acid | Acid |
|   Metal steps | Simple solvents | Chlorinated solvent compounds |
| Swelling in developer | No | Yes |
| Thermal stability | Good | Fair |
| Wet chemical resistance | Fair | Excellent |

**TABLE 1.2** Negative and Positive Resist Examples

| Class of Resist | Resist Name | Tone (Polarity) | Lithographic Sensitivity |
|---|---|---|---|
| Optical | Kodak 747 | Negative | 9 mJ/cm$^2$ |
| | AZ-1350J | Positive | 90 mJ/cm$^2$ |
| | Kodak KTFR | Negative | 9 mJ/cm$^2$ |
| | PR 102 | Positive | 140 mJ/cm$^2$ |
| e-Beam | COP (copolymer-[$\alpha$-cyano ethyl acrylate-$\alpha$-amido ethyl acrylate]) | Negative | 0.3 $\mu$C/cm$^2$ |
| | GeSe (germanium selenide) | Negative | 80 $\mu$C/cm$^2$ |
| | PBS (poly-[butene-1-sulfone]) | Positive | 1 $\mu$C/cm$^2$ |
| | PMMA (poly-[methyl-metacrylate]) | Positive | 50 $\mu$C/cm$^2$ |
| X-ray | COP | Negative | 175 mJ/cm$^2$ |
| | DCOPA | Negative | 10 mJ/cm$^2$ |
| | PBS | Positive | 95 mJ/cm$^2$ |
| | PMMA | Positive | 1000 mJ/cm$^2$ |

primers before spin coating to further improve resist adhesion. A typical adhesion promoter is hexamethyldisilazane (HMDS). Baking the SiO$_2$ surface at 250°C for 30 min removes adsorbed water from the silanol groups at the silicon surface which then can react with the amino groups of the HMDS vapor. Sputtering of the surface (see Chapter 2) sometimes is an attractive alternative to vapor priming.

## Wafer Cleaning and Contaminants

An important step even before wafer priming is wafer cleaning. Contaminants include solvent stains (methyl alcohol, acetone, trichloroethylene, isopropyl alcohol, xylene, etc.), dust from operators and equipment, smoke particles, etc. In Table 1.3, some common sources of contaminants are listed, and in Figure 1.2 the cleanroom classification system is elucidated. The allowable contamination particle size in IC manufacture has been decreasing hand in hand with the ever-decreasing minimum feature size. With a 64-Kilobyte dynamic memory chip (DRAM), for example, one can tolerate 0.25 $\mu$m particles, but for a 4-Megabyte DRAM one can only tolerate 0.05 $\mu$m particles.

The smallest feature sizes in these two cases are 2.5 and 0.5 μm, respectively. As a reference point: a human hair has a diameter of 75 to 100 μm (depending on age), tobacco smoke contains particles ranging from 0.01 to 1 μm, and a red blood cell ranges from 4 to 9 μm. Solvent stains and other contaminants on a Si wafer come into clear view via dark field microscopy (special off-axis microscopy).

At least seven different methods for wafer cleaning are in use: wet treatment, the RCA1 and RCA2 cleaning procedures with

**TABLE 1.3   Some Common Contaminant Sources**

Wafer transfer box, cassette

Wafer handling

Process equipment

Residual photoresist or organic coating

Metal corrosion

Solvents, chemicals

Atmosphere

Clothing, lint

Electrostatic charge (cleanroom must have a conductive floor), the electrostatic charge is influenced by humidity and temperature

Unclean room furniture, stationary, etc.

Operator (e.g., smoker vs. nonsmoker)

mixtures of hydrogen peroxide with various acids or bases followed by deionized water rinses; vapor cleaning; thermal treatment, baking at 1000°C in vacuum or in oxygen; plasma or glow discharge techniques, e.g., in Freons with or without oxygen; ultrasonic agitation; polishing with abrasive compounds; and supercritical cleaning. Ultrasonic cleaning, excellent in removing particulate matter from the substrate, is a process prone to contamination and mechanical failure of deposited films. Attributes of wet vs. dry cleaning techniques are compared in Table 1.4. Except for environmental concerns, wet etching still outranks other cleaning procedures.

The prevalent RCA1 and RCA2 wet cleaning procedures are as follows:

- RCA1: Add 1 part of $NH_3$ (25% aqueous solution) to 5 parts of DI water; heat up to boiling and add 1 part $H_2O_2$. Immerse the wafer for 10 min. This procedure removes organic dirt (resist).

- RCA2: Add 1 part HCl to 6 parts of deionized (DI) water; heat up to boiling and add 1 part of $H_2O_2$. Immerse the wafer for 10 min. This procedure removes metal ions.

The second RCA cleaning process is required to keep the oxidation and diffusion furnaces free of metal contamination. Both cleaning processes leave a thin oxide on the wafers. Before a further etch of the underlying Si is attempted, oxide must be stripped off by dipping the wafer in a 1% aqueous HF-solution for a very short time. Water spreads on an oxide surface (hydrophilic) and beads up on a bare Si surface (hydrophobic). This behavior can be used to establish if any oxide remains.

In most IC labs, processing a wafer previously exposed to KOH is prohibited as it is feared that the potassium will spoil the IC fabrication process. In more lenient environments, carefully cleaned wafers using RCA1 and RCA2 are allowed.

*Supercritical cleaning with $CO_2$* is especially suited for microstructure cleaning.[15] Supercritical fluids possess liquid-like solvating properties and gas-like diffusion and viscosity that enable rapid penetration into crevices with complete removal of organic

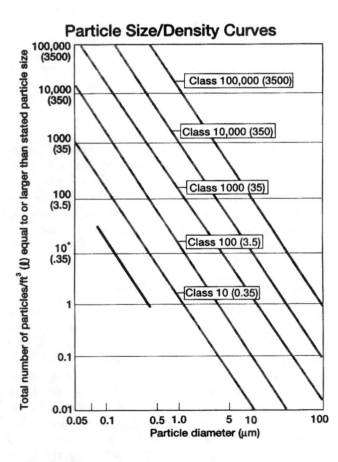

**Figure 1.2**   U.S. Federal Standard 209b for clean room classification. The bottom curve shows the definition of Class 1.[13] (From Cunningham, J. A., *Semicond. Int.*,15, 86–90, 1992. With permission.)

**TABLE 1.4   Wet vs. Dry Cleaning Attributes**

| Attribute | Wet | Dry |
|---|---|---|
| Particle removal | + | − |
| Metal removal | + | − |
| Heavy organics, i.e., photoresist | + | − |
| Light organics, i.e., outgassed hydrocarbon residues | + | + |
| Throughput | + | − |
| Process repeatability | + | + |
| Water usage | − | + |
| Process chemical cleanliness | − | + |
| Environmental impact, purchase, and disposal cost | − | + |
| Single wafer use applicability | − | + |

*Note:*   Dry usually requires wet follow-up. UV-ozone can effectively remove light organic contamination.[14]

*Source:*   Iscoff, R., *Semicond. Int.*, 14, 48–54, 1991. With permission.

and inorganic contaminants contained therein.[15] During wet cleaning of surface micromachined structures, thin, mechanical microstructures tend to stick to each other through surface tension (stiction). Consequently, dry vapor phase and supercritical cleaning with low or no surface tension are preferred (Chapter 5). Vapor phase cleaning also uses significantly less chemicals than wet immersion cleaning.[16]

Wafer cleaning has become a scientific discipline in its own right with journals (e.g., *Microcontamination, The Magazine for Ultraclean Manufacturing Technology*), books (e.g., *Handbook of Contamination Control in Microelectronics — Principles, Applications and Technology*),[17] and dedicated conferences (e.g., the Microcontamination Conference).

## Critical Dimension (CD), Overall Resolution, Line-Width

The absolute size of a minimum feature in an IC or a micromachine, whether it involves a line-width, spacing, or contact dimension, is called the critical dimension (CD). The overall resolution of a process describes the consistent ability to print a minimum size image, a critical dimension, under conditions of reasonable manufacturing variation.[18] The resolution of a lithography can be limited by many aspects of the process, including hardware, materials, and processing considerations. Hardware limitations include diffraction of light or scattering of charged particles (in the case of charged particle lithography or hard X-rays), lens aberrations, mechanical stability of the system, etc. Material properties that impact resolution are contrast, swelling behavior, thermal flow, and chemical etch resistance, etc. The most important process-related resist variables include swelling (during development) and stability (during etching and baking steps). Resolution frequently is measured by line-width measurements using either transmitted or reflected light or other metrology techniques (see Appendix A). Optical techniques perform satisfactorily for features of 1 μm and larger providing a precision of ±0.1 μm (at 2 σ, i.e. all data points within plus and minus two standard deviations). By 1998, devices with features as small as 0.25 μm will have launched equipment requirements

for line-width measurement with a precision of at least ±.02 μm (at 3 σ). Scanning electron microscopes or atomic force microscopes come forward as the methods to reach that goal. *A line-width, L,* is defined as the horizontal distance between the two resist-air boundaries in a given cross-section of the line, at a specified height above the resist-substrate interface.[18] Since different measurements may measure the line-width of the same line at different heights of the cross-section, the measuring technology used always needs identifying. The successful performance of devices depends upon the control of the size of critical structures across the entire wafer and from one wafer to another, referred to as line-width control. A rule of thumb is that the dimensions must be controlled to tolerances of at least ±1/5 the minimum feature size. Typically, a series of features with known sizes across a substrate are measured and then plotted as a function of position on the wafer. The standard deviation at the 1 or 2 σ level is adopted as the line-width control capability of the particular exposure/resist technology. Plotting these data as a function of time enables line-managers to maintain optimum performance on a manufacturing line.[18]

## Lithographic Sensitivity and Intrinsic Resist Sensitivity (Photochemical Quantum Efficiency)

### Lithographic Sensitivity

A distinction must be made between the intrinsic sensitivity of a resist, i.e., the resist's response to radiation, and the lithographic sensitivity defining the measurement of the efficiency that translates resist exposure into a sharp image. In the literature the values given for the lithographic sensitivity of one and the same resist show a tremendous spread as a result of the complex relationship between the intrinsic resist sensitivity and the dose required to successfully process that resist. This relation involves the intrinsic resist sensitivity as well as the bandwidth of the optical exposure system, baking conditions, resist thickness, developer composition, and development conditions. In order to reproduce a reported lithographic sensitivity, all these parameters need exact duplication.

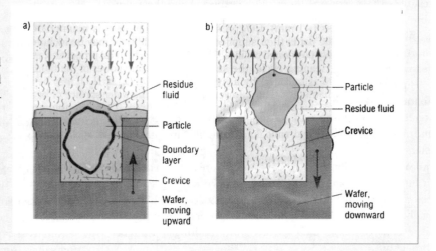

## *Supercritical cleaning*

Dislodging of a particle by supercritical fluid pressure pulsation. (a) 1200 psi, and (b) 800 psi. (From Bok et al., *Solid State Technol.*, 35, 117–119, 1992. With permission.)

a)
Residue fluid
Particle
Boundary layer
Crevice
Wafer, moving upward

b)
Particle
Residue fluid
Crevice
Wafer, moving downward

## Line-width L

The accuracy and precision of line-width measurement techniques are not simple to determine. This figure illustrates that there is a host of significant factors inherent in the characteristics of both line-width sample and the components of the measurement that impact the line-width value. The example here is an optical measurement technique, but SEM or mechanical measurements have their own set of significant factors. (From Wolf, S. and R. N. Tauber, *Silicon Processing for the VLSI Era*, Lattice Press, Sunset Beach, 1987. With permission.)

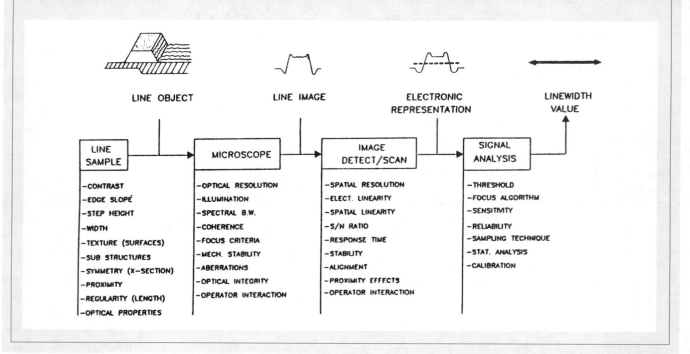

### Intrinsic Sensitivity of a Resist (Photochemical Quantum Efficiency)

A first indication of the intrinsic sensitivity of a resist to a certain wavelength can be deducted from the spectral-response curve of the resist. If the resist strongly absorbs in the ranges where the radiation source shows strong emission lines, relatively short exposure times can be expected and the actinic absorbency accordingly is high. Practical limits confine resist sensitivity: too sensitive a resist might mean an unacceptably short shelf life, and clearly the resist should be insensitive to the yellow and green light of the clean room.

High intrinsic resist sensitivity is a sought after characteristic. To increase resolution of photolithography, shorter and shorter wavelengths must be used. Exposure sources become less bright and optics absorb more at those wavelengths. Since the total energy incident on a resist is a function of light source intensity, time and absorption efficiency of the exposure optics, a decrease in intensity, and an increase in light absorption require compensation through longer exposure times. This results in a smaller throughput of wafers per hour; conversely, a more sensitive resist decreases the exposure time resulting in a higher throughput.

The intrinsic sensitivity or photochemical quantum efficiency, $\Phi$, of a resist is defined as the number of photo-induced events divided by the number of photons required to accomplish that number of events:

$$\Phi = \frac{\text{Number of photo-induced events}}{\text{Number of photons absorbed}} \qquad 1.2$$

As the photochemical event leading to a latent image differs depending on the nature of the resist, Equation 1.2 takes on slightly different forms for different resist systems. For resists with a sensitizer in a polymer matrix, i.e., two-component resists, $\Phi$ corresponds to the number of molecules of sensitizer converted to photo-product divided by the number of absorbed photons required to accomplish that conversion. For polymer resins where the polymer undergoes scission or cross-linking without the need for light-absorbing sensitizers, i.e., one-component resists, a G-value is introduced. The G-value corresponds with the number of scissions or cross-links produced per 100 eV of absorbed energy. For scission reactions the symbol G(s) is used, for cross-linking one uses G(x). In contrast to the lithographic sensitivity, the measurement of intrinsic radiation sensitivity as expressed through $\Phi$, G(s), or G(x) is quite reliable, and values from different sources agree relatively well.

The experimental determination of the quantum efficiency of a one-component resist is a complex undertaking. Samples of the polymer must be exposed to a known dose of gamma radiation and the molecular weight of the irradiated samples must be measured either by membrane osmometry or gel permeation chromatography. Quantitative analysis of the molecular weight vs. dose in polymers that undergo scission leads to an important relationship for a better understanding of resist exposure. We will use this relationship when exploring X-ray lithography for the creation of 'high-rise' PMMA resist structures ($\geq 10$ μm) (see Chapter 6). For a positive resist sample of weight w (in grams) containing $N_0$ molecules, the definition of average molecular weight $M_n^0$ is given by:[6,19]

$$M_n^0 = \frac{wN_A}{N_o} \qquad 1.3$$

where $N_A$ is Avogadro's number. Rearranging Equation 1.3 yields

$$N_o = \frac{wN_A}{M_n^0} \qquad 1.4$$

for the total number of molecules in the sample prior to exposure. Expressing the dose, D, in eV/g, the total dose absorbed by the sample is Dw (in eV). The total number of scissions produced in the sample, $N^*$, is proportional to the absorbed dose, or:

$$N^* = KDw \qquad 1.5$$

where K is a constant dependent on the polymer structure, generally expressed in terms of a G-value. G(s) (for positive resists) and G(x) (for negative resists) like $\Phi$, are figures of merit used to compare one resist material with another. With K expressed in terms of G(s), Equation 1.5 can be rewritten as:

$$N^* = \left(\frac{G(s)}{100}\right)Dw \qquad 1.6$$

in which we divide by 100 to express the number of events per 100 eV. Each time a scission occurs, the number of molecules is increased by one and the new average molecular weight after exposure to dose D is then given by:

$$M_n^* = \frac{wN_A}{N_o + N^*} \qquad 1.7$$

where the total mass of the polymer is assumed to remain constant during exposure. By substituting Equations 1.4 and 1.6 into Equation 1.7, we obtain:

$$M_n^* = \frac{N_A}{\dfrac{N_A}{M_n^0} + \left(\dfrac{G(s)}{100N}\right)D} \qquad 1.8$$

which is independent of the sample mass. Rearranging Equation 1.3 we obtain:

$$\frac{1}{M_n^*} = \frac{1}{M_n^0} + \left(\frac{G(s)}{100N_A}\right)D \qquad 1.9$$

From Equation 1.9 we conclude that a linear relationship exists between the inverse of the molecular weight after exposure and dose. The intercept on the y-axis gives $1/M_n^0$ and the slope allows one to calculate G(s). There is a very high correlation between G(s) values for gamma radiation (the radiation commonly used to determine G(s)) and sensitivity for electrons, ions, and X-rays.

The G(s) of polymers commonly used as one-component, positive resist systems ranges from 1.3 for some PMMAs to approximately 10 for certain poly(olefin sulfones). A PMMA with a G(s) value of 1 has a corresponding photochemical quantum-yield for scission, $\Phi$ (Equation 1.2), of 0.02.[6] PMMA exhibits a rather low cross-linking propensity. For some polymers, both scissioning and cross-linking events occur simultaneously upon exposure. It is possible, even in the latter case, to uniquely determine both scission efficiency G(s) and cross-linking efficiency G(x).[19]

For one-component negative resists the figure of merit for intrinsic sensitivity, G(x), expressed as number of cross-links per 100 eV absorbed dose, ranges from 0.1 for poly(ethylene) to approximately 10 for polymers containing oxirane groups (epoxy groups) in their side chains.

For a two-component positive system such as DNQ, $\Phi$ in Equation 1.2 corresponds to the quantum efficiency, i.e., the number of sensitizer molecules converted to photo-product, divided by the number of absorbed photons required to accomplish that conversion. The quantum efficiency can easily be measured by using a narrow-bandwidth radiation source and a UV-visible spectrophotometer. The quantum efficiency, $\Phi$, of typical diazonaphtoquinone sensitizers ranges from 0.2 to 0.3 (compared to 0.02 for PMMA). Because of the high opacity (i.e., high non-bleachable absorption) of novolak resins in the deep UV (200 to 300 nm) region, other resists like PMMA are used for shorter wavelength exposures (e.g., for DUV, e-beam, and X-ray lithographies). The quantum efficiency of the bis-arylazide sensitizers in negative resist systems ranges from 0.5 to 1, making negative resists more sensitive than positive resists.

## Radiation and Resist Profile

In an ultraviolet exposure, not all photons strike the resist film in an orthogonal fashion. Scattering at the substrate/resist interface causes a broadening of the exposed region. The scattered radiation profiles for positive and negative resists are shown in Figure 1.1B. Figure 1.3 represents the comparison of the prevalent photoresist profiles resulting from those radiation profiles. Here, R is defined as the development rate of the exposed region and $R_0$ as the development rate of the unexposed region. The scattered radiation profile for a positive resist (especially pronounced with overexposure) can, depending on the development mode, lead to a lip or overcut in the case of a fast developer

**A**

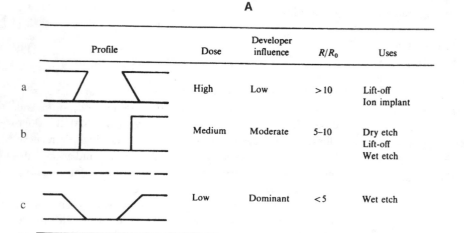

| | Profile | Dose | Developer influence | $R/R_0$ | Uses |
|---|---|---|---|---|---|
| a | | High | Low | >10 | Lift-off<br>Ion implant |
| b | | Medium | Moderate | 5–10 | Dry etch<br>Lift-off<br>Wet etch |
| c | | Low | Dominant | <5 | Wet etch |

**B**

Undercut

**Figure 1.3** Photoresist profiles. (A) Positive resist. (a) Desired resist profile for lift-off, i.e., exposure-controlled profile also called overcut. (b) Perfect image transfer by applying a normal exposure dose and relying moderately on the developer. (c) Receding photoresist structure with thinning of the resist layer, i.e., developer control also called undercut. For a good understanding of these different resist profiles, see also Figure 1.1B where scattered radiation profiles are shown. (B) Negative resist. Profile is mainly determined by the exposure. Development slightly swells the resist but otherwise has no influence on the wall profile. (From Moreau, W. M., *Semiconductor Lithography*, Plenum Press, New York, 1988. With permission.)

($R/R_0 > 10$), or, with a quenched developer ($R/R_0 = 5$ to 10), to a straight resist profile. In the latter case, the removal of the laterally exposed region has been inhibited and one obtains a perfect pattern transfer of the mask features into the resist. In a developer-dominated process ('force' developed with $R/R_0 < 5$), the resist profile (undercut) recedes and thinning of the entire resist layer occurs.

The creation of a lip or overcut can be taken advantage of in the so-called lift-off processes. Lift-off is an important process, for example, for patterning polymeric membranes and catalytic metals such as Pt. Membranes and Pt are frequently used in chemical sensors and do not lend themselves well to direct wet etching, but they can be patterned with relative ease thanks to the lift-off process. In the lift-off process sequence, shown in Figure 1.4, a solvent dissolves the remaining soluble positive photoresist underneath the metal, starting at the edge or lip (also called retrograde wall-angle profile) of the unexposed photoresist, and lifts off the metal in the process. A good photoresist profile for lift-off, of the type shown in Figure 1.3A(a) and Figure 1.4, is realized by using a high-exposure dose, so that the profile is dominated by the absorption of radiation, and the photons are reflected at the substrate-resist interface. The disadvantages of this technique are the rounded profile associated with deposited features and temperature limitations. The rounding is a result of shadowing, leading to features with a rounded top (see Figure 1.4). A more desirable profile for a conductor line has a rectangular cross section minimizing electrical resistance. The latter is one reason why lift-off in IC fabrication, where contact resistance is of prime concern, is used with discretion. With lift-off, the metal deposition technique is limited to temperatures below 200 to 300°C where resist begins to degrade.[20]

Development of a positive resist is time dependent and enables the operator to tailor resist profiles to his needs. The exterior scattering zone in the case of a negative resist becomes

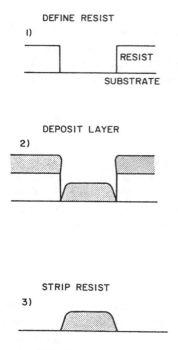

**Figure 1.4** Example of lift-off sequence for the construction of a Pt-based electrochemical sensor electrode. Rounding of deposited features through shadowing is observed (see text).

insoluble and, aside from some swelling, is independent of development (see also Figure 1.1B). With negative resists (Figure 1.3B), the exposed regions remain as they are rendered insoluble, whereas with positive resists the exposed region develops and the unexposed regions usually remain soluble. The swelling in negative resists is one of the reasons they are limited to the manufacture of devices with minimum feature size >3 μm. Scattered radiation and swelling result in a broadening of the remaining resist features. Positive resists do not exhibit this swelling due to a different dissolution mechanism.

a series of positive resist pads of known area are subjected to varying doses and developed in a solvent that does not attack the unexposed film; then the thickness of the remaining film in the exposed area is measured and normalized to the original thickness. The contrast, $\gamma_p$, is determined from the slope of the response curve as:

$$\gamma_p = \frac{1}{\left(\log D_p - \log D_p^o\right)} = \left[\log \frac{D_p}{D_p^o}\right]^{-1} \qquad 1.10$$

and the lithographic sensitivity, $D_p$, is the x-axis intersection. For a given developer, $D_p$ corresponds to the dose required to produce complete solubility in the exposed region, while not affecting the unexposed resist. And $D_p^0$ is the dose at which the developer first begins to attack the irradiated film. For a dose less than $D_p$ but higher than $D_p^0$ another developer could 'force develop' the resist. In force developing, the developer attacks or thins the original, unexposed resist [see also Figure 1.3A (c)]. This describes how a positive resist's profile can be manipulated by the operator.

For a negative resist, the contrast relates to the rate of cross-linked network formation at a constant input dose. This is simpler than in the case of a positive resist where contrast is also very solvent dependent. Consequently, if one negative resist has a higher cross-linking rate compared to another, it also possesses the higher contrast of the two. With negative resists, the onset of cross-linking, as evidenced by gel formation, is not observed until a critical dose, $D_g^i$ (also called the interface gel dose), has been reached (see Figure 1.5B). In Figure 1.5B we show the response or sensitivity curve, i.e., the normalized developed film thickness vs. log dose. Below the interface gel dose, no image can form as the film thickness is insufficient to serve as an etching mask. At higher doses, the image thickness increases until the thickness of the image equals that of the resist prior to exposure (in reality it remains thinner as the film shrinks due to cross-linking). The latter dose is shown in Figure 1.5B as $D_g^0$, the dose required to reach 100% polymerization of initial film thickness (prior to exposure). The contrast, $\gamma_n$, is obtained from the slope of this curve as:

$$\gamma_n = \frac{1}{\left(\log D_g^o - \log D_g^i\right)} = \left[\log \frac{D_g^o}{D_g^i}\right]^{-1} \qquad 1.11$$

The lithography sensitivity ($D_g^x$) defines the dose cross-linking the film to the required thickness for optimal resolution. That required dose sometimes is defined as the dose resulting in dimensional equality of clear and opaque features (corresponding to nominally equal structures on the mask) imaged in the resist. The so-defined lithographic sensitivity can be determined separately from a plot of feature size vs. dose for an opaque and a clear feature of equal size (Figure 1.6). This dose, ($D_g^x$), corresponding to the lithographic sensitivity, transposed on the X-axis of Figure 1.5B, fixes the required crosslinked film thickness after development on the Y-axis (usually 0.5 to 0.7 times the normalized thickness). The lithographic sensitivity also may be taken as the dose $D_{0.7}^x$ at which 70% of the original film is retained after development.[11]

## *Lightly scattered zones do not insolubilize*

Insolubilization of radiated negative tone resist is prevented by oxygen ingress from the unexposed resist areas. The oxygen interdiffusion enhances the resolution. (From Moreau, W. M., *Semiconductor Lithography*, Plenum Press, New York, 1988. With permission.)

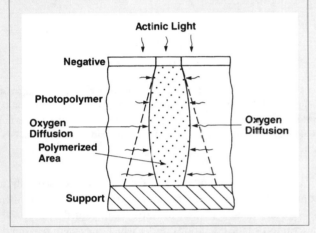

The oxygen effect quenching the cross-linking, as discussed earlier, usually means a disadvantage for negative resists, but sometimes it can turn into an advantage, for example, to improve resolution. We already know that oxygen can scavenge the photo-generated reactive nitrene species and that this reaction eliminates the precursors for the cross-linking and insolubilization. If oxygen is excluded from the top surface of the resist by flushing with an inert gas or by blocking with a polymer topcoat, oxygen, dissolved in the polymer film, moves laterally from the unexposed dark areas into the light zone, which, under illumination, has become an oxygen sink and the *lightly scattered zones* do not insolubilize, leading to a better resolution.[6]

A more rigorous, mathematical treatment of resist profiles is presented further under Mathematical Expression for Resist Profiles.

## Contrast and Experimental Determination of Lithographic Sensitivity

The resolution capability of a resist, defined as the smallest linewidth to be consistently patterned (see above), is directly related to resist contrast, $\gamma$. For positive resists, the contrast is related to the rate of chain scission and the rate of change of solubility with molecular weight. The latter is very solvent dependent. The exposed resist layer thickness after development decreases until, at a critical dose, $D_p$, the film is completely removed. The lithographic sensitivity, $D_p$, and contrast can be obtained from the response curve; a plot of normalized film thickness vs. log D (dose) (Figure 1.5A). To construct an image as in Figure 1.5A

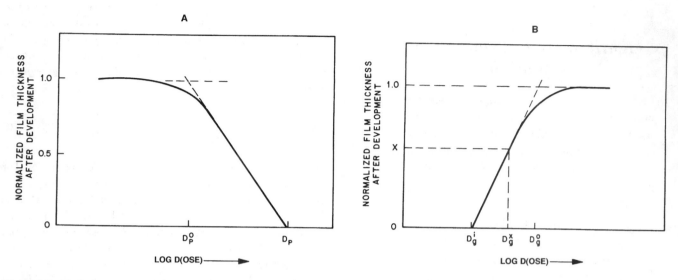

**Figure 1.5** Typical response curves or sensitivity curves. (A) For a positive resist. Contrast, $\gamma_p$, is determined from the slope. The contrast for a positive resist is markedly solvent dependent. A typical contrast value for a positive optical resist is 2.2. (B) For a negative resist. The value of $D_g^x$ usually occurs at 0.5 to 0.7 normalized thickness as shown in Figure 1.6. The slope determines the contrast, $\gamma_n$. A typical value for the contrast of a negative optical resist is 1.5. (From Willson, C. G., in *Introduction to Microlithography*, Thompson, L. F., Willson, C. G., and Bowden, M. J., Eds., American Chemical Society, Washington, D.C., 1994. With permission.)

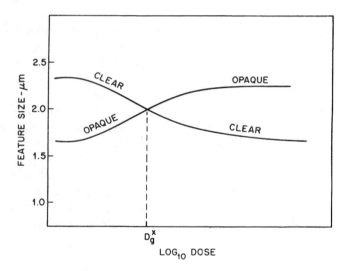

**Figure 1.6** Size of a clear and opaque 2.0-μm feature (on the mask) as a function of the exposure dose for a negative resist. The dose ($D_g^x$) resulting in the correct feature size (same size as on the mask) is called the 'lithographic sensitivity'. (From Willson, C. G., in *Introduction to Microlithography*, Thompson, L. F., Willson, C. G., and Bowden, M. J., Eds., American Chemical Society, Washington, D.C., 1994. With permission.)

Resists with higher contrast result in better resolution than those with lower contrast. This can be explained as follows. In an exposure, energy is delivered in a diffused manner due to diffraction and scattering effects. Some areas outside the mask-defined pattern will receive an unintended dose higher than $D_g^i$ but lower than ($D_g^x$). The resultant resist profile will exhibit some slope after development. The higher the contrast of the resist, the more vertical the resist profile. Since line-width is measured at a specified height above the resist-substrate surface, as the resist profile is less vertical, the less accurately the resist line-width represents the original mask dimension.

The values of $D_p$ and $D_g^x$ are figures of merit used only to compare different resists. For the lithographic sensitivity numbers to have any value at all they must be accompanied by a detailed description of the conditions under which they were measured.

## Photolithography Resolution

### Introduction

The resolution of a lithographic system can be determined by a line-width measurement such as in a scanning electron microscope (SEM). Correct feature size must be maintained within a wafer and from wafer to wafer, as device performance depends on the absolute size of the patterned structures. The term critical dimension (CD) refers to a specific feature size and is a measure of the resolution of a lithographic process. The image profile slope at the mask edge can measure CD control capability (see section above on Critical Dimension (CD), Overall Resolution, and Line-Width). In what follows we are considering the theoretical resolution of different photolithography printing techniques.

### Resolution (R) in Contact and Proximity Printing (Shadow Printing)

In the shadow printing mode (contact and proximity), optical lithography has a resolution with limits set by such factors as diffraction of light at the edge of an opaque feature in the mask as the light passes through an adjacent clear area, alignment of wafer to mask, nonuniformities in wafer flatness, and debris between mask and wafer. Figure 1.7 illustrates a typical intensity distribution of light incident on a photoresist surface after passing through a mask containing a periodic grating consisting of opaque and transparent spaces of equal width, b.[19] Diffraction causes the image of a perfectly delineated edge to become blurred or diffused. The theoretical resolution, R, i.e.,

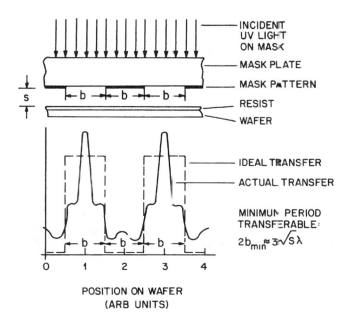

**Figure 1.7**   Light distribution profiles on a photoresist surface after light passed through a mask containing an equal line and space grating. (From Willson, C. G., in *Introduction to Microlithography*, Thompson, L. F., Willson, C. G., and Bowden, M. J., Eds., American Chemical Society, Washington, D.C., 1994. With permission.)

minimum resolved feature size ($2b_{min}$) with a mask (as illustrated in Figure 1.7) employing a conventional resist is given by:

$$2b_{min} = 3\sqrt{\lambda\left(s + \frac{z}{2}\right)} \qquad 1.12$$

where $b_{min}$ stands for half the grating period, s for the gap between the mask and the photoresist surface, $\lambda$ for the wavelength of the exposing radiation, and z for the photoresist thickness.

*Contact Printing*

For contact printing s = 0. From Equation 1.12 with $\lambda$ = 400 nm and a 1-$\mu$m thick resist, we conclude that the maximum resolution is slightly less than 1 $\mu$m; for very thin resist layers, small z, the resolution capability of contact printing is higher than any other optical lithography process. Equation 1.12 clarifies the need to use shorter wavelength in order to achieve higher resolution. The theoretical maximum resolution is seldom achieved; only diffraction effects were taken into account to derive Equation 1.12. The other factors mentioned above usually conspire to make the resolution worse. The required contact between the mask and the wafer causes mask damage and contamination that render the method unsuitable for most modern microcircuit fabrication.

*Proximity Printing*

In proximity printing, spacing of the mask away from the substrate minimizes defects that result from contact. On the other hand, diffraction of the transmitted light reduces the resolution. The degree of reduction in resolution and image distortion depends on the wafer-to-substrate distance which may

vary across the wafer. For proximity printing, Equation 1.12, with s $\gg$ z can be rewritten as:

$$2b_{min} \approx 3\sqrt{\lambda s} \qquad 1.13$$

On the basis of Equation 1.13 for a gap of 10 $\mu$m, using 400 nm exposing radiation, the resolution limit is about 3 $\mu$m.[21]

The smallest features resolvable in a practical UV proximity exposure measure about 2 to 3 $\mu$m for most processes. During proximity printing, wafer-to-mask contact during alignment still poses problems; for feature sizes below 2 $\mu$m, optical projection methods are used.

**Resolution with Self-Aligned Masks**

The most desirable fabrication processes involve *in situ* deposited masks, also called self-aligned or conformable masks. These masks, forming a molecular contact with the substrate, offer a superior resolution as light has no chance to diffract between mask and substrate. Conformable masks comprise sacrificial layers produced in intermediate process steps. Figure 1.8 exemplifies a LIGA microfabrication process involving a self-aligned mask, i.e., an array of mushroom hats on nickel (Ni) posts.[22] The mask is made *in situ* by overplating of Ni from a Ni-plating solution. The set-up illustrates the making of a gold (Au) gate electrode around each Ni post without contacting the Ni posts. The mushroom hats protect (mask) the base of the Ni posts during Au deposition (Figure 1.8E). The gold in a directional evaporation (see Chapter 3) only deposits on top and around the Ni hats. The resulting Au gate electrode is used to electrochemically sharpen the Ni posts they surround.[22] It should be noted that the process illustrated in Figure 1.8 also shows a lift-off sequence where the Ni mushroom mask enables the selective lift-off of the Au layer on the Ni mushroom hats (see also Figure 1.4).

**Projection Printing**

*Types of Projection Methods*

Since 1973, when Perkin-Elmer first introduced its scanning projection system for lithography, **optical projection** of mask patterns became a standard lithography method. In projection printing, wafer contact is completely avoided; a high-resolution lens projects an image of the photomask onto the photoresist-covered wafer. There are three popular projection printer types: projection scanners (e.g., Perkin-Elmer's Micralign), 1 $\times$ step-and-repeat projection systems (e.g., Ultratech), and reduction (e.g., 10$\times$, 5$\times$) step-and-repeat projection systems (e.g., Electromask). A scanning projection system exposes the wafer in a succession of scans without the benefit of the opportunity to realign to local alignment marks in mid-scan. With a deep UV light source, a resolution of 1 $\mu$m can be obtained, a depth of focus (see below) of $\pm$6 $\mu$m is possible, and an overlay accuracy of $\pm$0.25 $\mu$m (1 $\sigma$) has been reached. In practice, these scanners are mainly used for alignment of patterns with critical dimensions in the 3-$\mu$m range and for high throughput applications (e.g., 100, 150 mm wafers per hour). A stepper system exposes one small part of the wafer followed by a new exposure through stepping to the next position. By

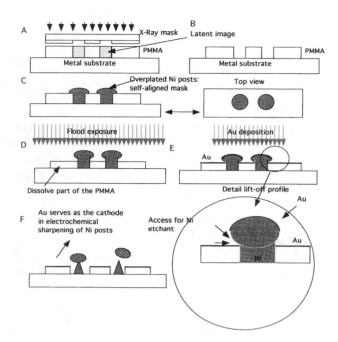

**Figure 1.8** Example of a self-aligned mask. (A) PMMA radiation using an X-ray mask; (B) development; (C) overplating of Ni in PMMA holes to create mushrooms, i.e., making of the actual self aligned mask; (D) Flood exposure of PMMA and dissolution of the thin top layer of PMMA; and (E) Au evaporation (directional); and (F) electrochemical sharpening.[22]

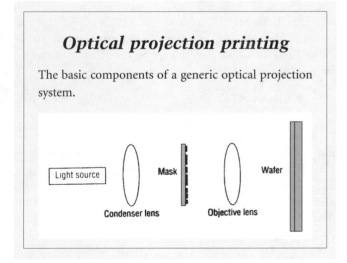

stepping and repeating, the entire wafer is covered with the reticle pattern. Steppers have the ability to align to each field and make adjustments to X, Y, rotation and focus, and tilt. Steppers offer higher alignment accuracies but are slower than scanners. Most steppers use reduction lenses (for example, with a 10:1 or a 5:1 reduction) rather than one-to-one projection printing. The reduction printer exposes part of the wafer to a pattern from a mask 5 to 10 times larger than the projected image. The reduction process makes reticle inaccuracies insignificant, consequently improving the resolution and resulting in easier mask making. The only drawback pertains to the size

of the image field: the higher the reduction, the smaller the image field. Because of lens imperfections and diffraction considerations, projection techniques have a lower resolution for pattern transfer than that provided by a contact or self-aligned mask; however, CDs below 0.5 μm are possible with projection lithography.

*Modulation Transfer Function*

Before we analyze the mathematical expressions governing resolution in projection printing, the meaning of image modulation requires explaining. In projection printing of a grating, a series of undulating maxima and minima are produced, as demonstrated in Figure 1.9.[23] Because of mutual interference, the dark region in an image never becomes completely dark, and the maximum brightness does not correspond to 100% transmission. The quality of the image transfer can be conveniently indicated by the modulation, M, defined as:

$$M = \frac{I_{max} - I_{min}}{I_{max} + I_{min}} \qquad 1.14$$

In this equation, $I_{max}$ and $I_{min}$ are the peak and trough intensities, respectively. As shown in Figure 1.9A, M reveals the degree to which diffraction effects cause incident radiation to fall between the images on a screen of two slits in a mask. Ideal optics would give a modulation equal to 1. All practical exposure systems behave less ideally with an M < 1. The ratio $I_{max}/I_{min}$ also is called the contrast C.[18]

The optical imaging quality, i.e., the capability to reproduce a mask feature on a wafer surface, for a projection system can then be characterized in terms of the modulation transfer function (MTF) curve (Figure 1.9B). The MTF of an exposure system is defined as the ratio of the modulation in the image plane to that in the object or mask plane, or $M_{im}/M_{mask}$. Since the intensity in the mask plane at the center of an opaque feature is essentially zero we can equate $M_{mask} = 1$. Consequently, we can also equate MTF and $M_{im}$. The modulation in the image plane, $M_{im}$ can be measured by scanning a very small photodetector across an image of the grating.[23]

*Mathematical Expressions Governing Resolution in Projection Printing*

The practical limiting resolution R of a lens in projection printing is given by:

$$R = \frac{k_1 \lambda}{NA} \qquad 1.15$$

where $k_1$ is an experimentally determined parameter that depends on resist parameters as well as process conditions and usually fluctuates between 0.5 and 1.0 (production engineers demand that $k_1 > 0.7$ for single layer resists); λ represents the wavelength of the light used for the pattern transfer, and NA is the numerical aperture of the imaging lens system.

In what follows we will explain step by step how the mathematical form of Equation 1.15 comes about. The numerical aperture of the lens in Figure 1.10 in a medium of refractive

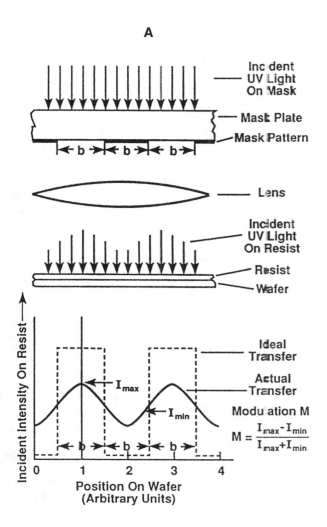

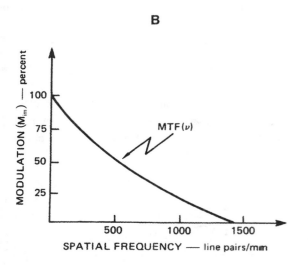

**Figure 1.9** Modulation and modulation transfer function (MTF). (A) Modulation, M, and (B) modulation transfer function; MTF = $M_{im}$. (From Brodie, I. and J. J. Muray, *The Physics of Microfabrication*, Plenum Press, New York, 1982. With permission.)

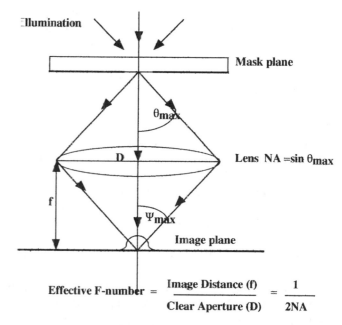

**Figure 1.10** Relationship between the object, image, and focal length and the diameter of a lens to define the numerical aperture.

index n (typically 1.0 in air) defines the angle of acceptance, $2\theta_{max}$, of the cone of diffracted light from an object (the photomask) that the lens can accept. The angle of acceptance and NA are linked via the expression:

$$NA = n\sin\theta_{max} \qquad 1.16$$

Hence, the numerical aperture characterizes the ability of a lens to transmit light and is proportional to the size of the lens. Equation 1.16 is further related to the familiar lens parameters, D, lens diameter, and f, the focal length (see also Figure 1.10), as

$$NA = n\sin\theta_{max} = \frac{D}{2F} \qquad 1.17$$

The effective F number of a projection system is given by F = 1/2NA, and it can be shown that F = (1 + M)f, where f is the F number of the system with the object at infinity and M is the operating magnification. The angles, $\theta_{max}$ and $\psi_{max}$ in Figure 1.10, are equal for unit magnification (M = 1). The larger NA of the projection lens in an exposure system is, the greater the amount of light (containing diffraction information of the mask) collected and subsequently imaged. Because the image is constructed from diffracted light, and the collection of higher orders of diffracted light enhances the resolution of the image, a larger NA, allowing a larger acceptance angle, results in a better resolution. For a set wavelength, the resolution has traditionally been improved by increasing the NA of the optical system (see Equation 1.15). Unfortunately, this accomplishment comes at the expense of increased complexity in the lens system and reduction of the image field (20 mm² is standard for 5× reduction steppers), which in turn places rigorous demands on the required exactness of the mechanical system used to accurately step small image fields over the surface of the wafer.

The lens resolution, as given by Equation 1.15, not only depends on the wavelength and NA of the lens, but it also is a function of the degree of the spatial coherence of the light source (one of the factors contributing to $k_1$).

The *spatial coherence* is a measure of degree to which the light emitted from the light source stays in phase at all points along the emitted wave fronts. Coherence ($\sigma$) varies from $\sigma = 0$ for coherent radiation to $\sigma = \infty$ for fully incoherent illumination. A point source would give an all-coherent light exposure. In reality, we deal with light sources with finite size. The degree of coherence of the light on the resist plane is dependent on that size, and the resulting light is only partially coherent. When imaging a grating with coherent light, the direction of the diffracted rays is given by the grating formula:

$$2bn\sin\theta = N\lambda \qquad 1.18$$

where n is the refractive index in image space (assumed to be 1), b the grating spacing making the spatial frequency, $v = 1/2b$ (see Figures 1.7 and 1.9), $\theta$ the angle of the ray of order N emerging from the grating, and $\lambda$ the wavelength of the exposing radiation. The grating frequency corresponding to the first order (N = 1) diffracted peak is then derived as:

$$v = \frac{1}{2b} = \frac{n\sin\theta}{\lambda} \qquad 1.19$$

For imaging it is required that $\theta \le \theta_{max}$, where $\theta_{max}$ is defined by the numerical aperture of the projection optics (Equation 1.16). At larger angles than $\theta_{max}$, light is no longer captured by the imaging lens. Therefore, the highest grating spatial frequency that can be imaged by a coherent illumination system is given by:

$$v_{max} = \frac{n\sin\theta_{max}}{\lambda} = \frac{NA}{\lambda} = \frac{1}{2\lambda F} = \frac{1}{2b_{min}} \qquad 1.20$$

with F = 1/2 NA. Consequently, the resolution R in the case of coherent light is given by:

$$R = 2b_{min} = \lambda/NA \qquad 1.21$$

In the case of incoherent illumination of a grating, each ray is diffracted by the grating and forms its own image in the wafer plane. The direction of the first diffraction peak (N = 1) for incoherent rays incident at an angle i is given by a more general grating equation:

$$2bn(\sin i + \sin\theta) = \lambda \qquad 1.22$$

representing the path difference for light passing through adjacent slits. For light incident normal to the grating (sin i = 0), Equation 1.22 reduces to Equation 1.18. For image formation it is required that both i and $\theta \le \theta_{max}$. Therefore,

$$v_{max} = \frac{2N\sin\theta_{max}}{\lambda} = \frac{2NA}{\lambda} = \frac{1}{F\lambda} = \frac{1}{2b_{min}} \qquad 1.23$$

or

$$R = 2b_{min} = \frac{\lambda}{2NA} \qquad 1.24$$

so that the maximum resolution for incoherent light measures twice the resolution of coherent light.

The modulation transfer function, the MTF, of an exposure system depends on NA, $\lambda$, mask feature size, and the degree of

---

## Spatial coherence

Spatial coherency is a measure of the degree to which the light emitted from a source is in phase at all points along the emitted wave fronts. A point source, of infinitely small dimension, represents the ideally coherent source (left). Because all lithography systems have radiation sources of finite size, the degree of coherence exhibited by light incident on a plane is dependent on the source size (right).

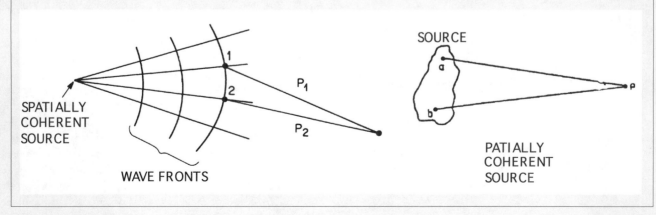

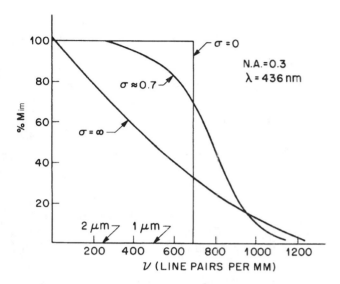

**Figure 1.11** Modulation of an image as a function of spatial frequency, ν, for different coherency factors. (A) σ = 0 (coherent); (B) σ = 0.7 (partially coherent) and σ = infinite (incoherent). (From Brodie, I. and J. J. Muray, *The Physics of Microfabrication*, Plenum Press, New York, 1982. With permission.)

spatial coherency of the illuminating system. Since NA and λ are fixed by system hardware design, a plot of MTF vs. feature size, parametrically changing the spatial coherency, often is used to compare a system's imaging capability. MTF curves (also the modulation $M_{im}$ in the image plane) for coherent, partially coherent, and incoherent illumination are shown in Figure 1.11.[23] Contrary to the coherent case in which MTF remains constant up to the cutoff frequency, MTF for the incoherent cases decreases monotonically as the frequency increases. Completely incoherent light produces an MTF curve that has a greater resolution limit, but for which the MTF value only slowly increases as the feature size increases. For the above reasons, partial coherent light often is preferred.[21] Based on various tradeoffs, a coherence value of ~0.7 typically is selected for a practical exposure. The cutoff frequency shown in Figure 1.11 defines the resolution limit of the exposure system. It is proportional to the numerical aperture of the lens and inversely proportional to the wavelength. Comparing the calculated Equations 1.21 and 1.24 to the empirical Equation 1.15, $k_1$ equals 0.5 for coherent light and 1 for coherent light. A good source for more detailed derivations of Equations 1.21 and 1.24 is Bowden.[22]

### Practical Resolution Limits in Projection Printing

The image of a small point produced by an aberration-free lens in the focal plane is not a sharp point but is spread about into a diffraction pattern known as the Airy disk. The image of two point sources can be resolved only if the corresponding Airy disks are far enough apart. Resolution R in Equation 1.24 actually corresponds to the half-intensity point of the Airy disks of line images. For a first estimate about resolution, one might also use the Rayleigh criterion. Rayleigh spells out his criterion for resolution as two images considered resolved when the intensity between the Airy disks drops to 80% of the image intensity.

Applying that criterion to two images separated by a distance $2b_{min}$, as in the grating in Figure 1.7, leads to a slightly more conservative expression:

$$R = 2b_{min} = \frac{0.6\lambda}{NA} \qquad 1.25$$

For the practical resolution of a resist to equal the Rayleigh limit of the optical system predicted by Equation 1.24 or 1.25, the resist would have to have an infinite contrast. Because a resist always has a finite contrast value, a greater degree of modulation is needed before an adequate image can be formed. The minimum MTF of an optical system to adequately define an image in a resist also depends on the resist contrast value and is defined as the critical MTF, or $CMTF_{resist}$. The relationship between the contrast and CMTF is given by:

$$CMTF_{resist} = \frac{10^{\frac{1}{\gamma}} - 1}{10^{\frac{1}{\gamma}} + 1} \qquad 1.26$$

For an exposure system to adequately print a given feature size, its MTF for that feature size must be larger than or equal to the $CMTF_{resist}$ of the resist used. If the MTF of an exposure system is known for various feature sizes, knowledge of the resist contrast and Equation 1.26 will allow prediction of the smallest features printable when applying that system. A rule of thumb is that for positive resists the actual workable resolution is 2.5 times the Rayleigh limit of the projection printer.

### Depth of Focus

To obtain good line-width control, the latent image must remain in focus through the depth of the resist layer. A certain amount of defocus tolerance wherein the image still remains within specifications is allowed. The defocus tolerance or depth of focus (DOF) or δ of an optical system is given by:

$$DOF = \pm \frac{k_2\lambda}{(NA)^2} \qquad 1.27$$

and by using Equation 1.15:

$$DOF = \pm \frac{k_2 R^2}{k_1^2 \lambda} \qquad 1.28$$

In Equations 1.27 and 1.28, $k_2$ again is a process-dependent constant hovering around 0.5. For an elegant mathematical derivation of Equations 1.27 and 1.28 we again refer to Bowden.[22] As deduced from these equations, a high numerical aperture can result in a depth of focus too small to produce a focused image through the total depth of a typical 1.0- to 1.5-μm thick photoresist. A good NA of a lens for a g-line (436 nm) lithography system is 0.54. With a $k_1$ factor (depending on the resist) of 0.8, this leads to a resolution of 0.65 μm. With i-line (365 nm) lithography, a resolution of 0.65 μm can be achieved with a 0.45-NA lens while exhibiting a superior DOF of 0.9 μm compared to 0.7 μm for the g-line. Widefield i-line *steppers* with *variable numeric aperture* are becoming available with a 17.96 × 25.2-mm field, a 5× reduction ratio, and a resolution of better than 0.45 μm. This type of equipment allows one to balance resolution, DOF, and wafer throughput for different applications. Since the patterns on wafers have their own topology and wafers may not be perfectly flat, a large DOF is needed to ensure that the image stays in focus over the entire field. If even a small part is out of focus, the final product will be ruined.

The microlithographic (or practical) DOF is defined as the total defocus allowable for a desired tolerance on a minimum feature size. It may be quite different from the values estimated from Equations 1.27 and 1.28. Indeed the practical DOF must encompass device topography, resist thickness, wafer flatness, and focus tilt errors. It can easily be understood that for sub-0.5-μm lithography, only a small amount of residual nonplanarity in device topography can be tolerated without negatively affecting the critical dimension control. Small DOF values require expensive planarization processes to bring all IC features in focus within the DOF of the optical system. Some of those planarization processes will be reviewed in Chapter 3 on Pattern Transfer with Additive Processes.

Micromachined structures possess a lot more extreme topologies than ICs, making planarization a bigger challenge yet. Over the last 20 years, progress in expanding the depth of focus of semiconductor equipment did not keep pace with the progress in decreasing the critical dimensions. For micromachining the former is more important. We will learn how topographical masks and using X-ray and e-beam lithography enable the higher DOFs needed for high aspect ratio micromachines.

### Misregistration in Projection Printing

So far we did not dwell much on the alignment of the wafer to the mask with so-called aligners or printers. Almost every type of mask aligner-contact/proximity, projection scanner, and step-and-repeat is used for some of the lithography steps in the fabrication of state-of-the-art integrated circuitry and micromachines today. For a detailed description of mask alignment equipment we refer to the literature.*

Product failure during production is caused mainly by poor alignment between the image being projected and the preexisting patterns on the wafer. The type of errors one encounters in projection printing are illustrated in Figure 1.12. They range from misalignment to mask error, optical distortion, wafer or mask expansion, and magnification change. The pattern registration capability is the degree to which the pattern being printed can 'fit' relative to mask alignment marks and to the previously printed patterns. With an optical stepper, the step-and-repeat operation is performed by laser-interferometer-controlled stages to a positioning accuracy of <0.1 μm, thereby allowing the registration of successive layers of a semiconductor device to a similar precision. Alignment or registration marks such as *optical vernier patterns* are created on the different levels to be aligned.[6] Generally, the misalignment from all causes should not exceed a quarter of the minimum feature size. Alignment errors (A) depict only one part of the total device ground-rule tolerances. Tolerance (T) or the ability to overlay (align) spatially one masking layer with the previously etched or deposited pattern is given by the standard deviations of the independent variables of the mask (M), etch (E), and A errors:[6,18]

$$T = \left( A^2 + M^2 + E^2 \right)^{1/2} \qquad\qquad 1.29$$

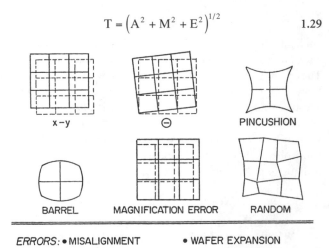

**Figure 1.12**   Misregistration in projection printing.

---

### Variable numeric aperture stepper

With a variable numeric aperture, one can balance the resolution, DOF, and throughput for different applications. (Courtesy of Nikon Precision, Inc.)

Typical mask errors range from 0.1 to 0.3 μm, etch errors from 0.2 to 0.4 μm, and alignment errors from 0.2 to 0.5 μm. Thus, the tolerances of the current run-of-the-mill lithography equipment used in manufacturing lie in the range of 0.5 to 0.8 μm.[18] For the next generation of submicrometer ground rules, a four-fold reduction in alignment and other errors will be necessary. Typically, a tolerance of feature size ±10% will be required. These types of aligners already penetrate the field. For example, commercially available aligners for X-ray lithography, developed in Japan, make possible alignment compatible with feature sizes of 0.5 to 0.35 μm. Since interferometric schemes enable sensitivities to below 10 nm, it appears feasible that aligners compatible with line-widths of 0.1 μm or even 50 nm can be developed.[6,24]

In micromachines, alignment poses more complexity than in IC manufacture. Not only does one deal with high aspect ratio 3D features causing problems for alignment systems with low DOF, one also frequently needs to align 3D features on both sides of the wafer. The object is to position alignment marks opposite each other on the two surfaces of the same wafer, allowing accurate positioning of all later feature-defining photolithographic patterns. There are several options for front to back alignment. One way to accomplish double-sided alignment, not too elegant nor too desirable, is by etching holes through the Si wafer. A better way is to use infrared to see through the silicon or, with visible light, to employ a double-sided mask alignment system defining marks on opposite sides of the wafer with a set of mirrors and two light sources. The latter double-sided mask aligner is illustrated in Figure 1.13. First mask 2 is aligned to mask 1, and then the wafer is inserted and aligned and UV exposure of one or both sides can begin. The alignment accuracy thus procured is claimed to be <1 μm.

Since two-sided mask aligners and infrared microscopes are quite costly, researchers have been looking into less costly but still accurate alternatives. In the simplest, least accurate approach, wafer flats can be used as reference for the double-sided alignment. In a bit more sophisticated procedure, White and Wenzel[25] use a simple jig as shown in Figure 1.14. In this approach, mask 1, containing only alignment marks, is contact printed onto photoresist-coated mask 2 while both are positioned snugly against the three pins on the jig. After developing and etching the alignment marks on mask 2, the individual alignment patterns from the two masks are transferred onto the opposite faces of a semiconductor wafer coated on both sides with photoresist. This is accomplished by sandwiching the resist-coated Si wafer between the two alignment masks (again set snugly against the three pins of the jig), exposing each wafer surface (directly for one side and through the large hole in the jig for the other side). The alignment patterns then are etched into the wafer and used in a conventional one-sided mask aligner. The authors estimate the predictable alignment errors to be less than 1 μm across a 250-μm thick 2-in. wafer. Kim et al.[26] described a different front-to-back alignment technique involving the visual alignment of stepper cross bars (alignment keys) to the center of a free diaphragm etched from the back of the wafer. These authors claim a less than 1-μm alignment error, but no experimental evidence was presented. Tatic-Lucic et al.[27] improved the latter type of double-side alignment method and demonstrated less than 2-μm error across a 4-in. wafer. In this upgrade, only alignment marks on the front side of the wafer are required and formed in a low-stress insulating layer. This insulating layer with the alignment marks is later bulk-micromachined into a free-standing diaphragm with a cavity underneath it. This cavity makes the front-side alignment marks visible from the back side. This allows visual alignment from both sides of the wafer with a traditional GCA 4800 stepper. Yet another alternative to facilitate double-sided alignment is the

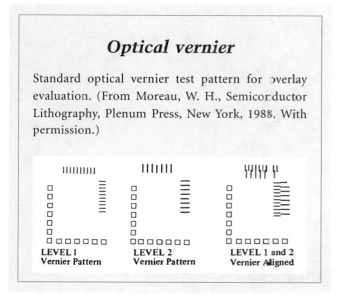

## *Optical vernier*

Standard optical vernier test pattern for overlay evaluation. (From Moreau, W. H., Semiconductor Lithography, Plenum Press, New York, 1988. With permission.)

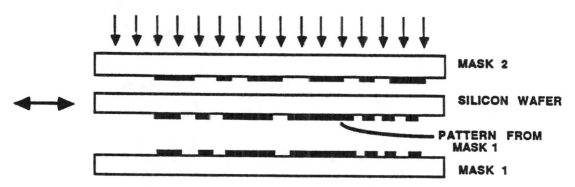

**Figure 1.13**   Double-sided mask aligner. (A) Align mask 2 to 1; (B) insert and align wafer; and (C) expose one or both sides.

pick-up of capacitive signals between conductive metal fingers on the mask and ridges on a small area of the Si wafer. The latter technique, requiring close proximity between mask and wafer, is only applicable for 1:1 proximity lithography as used, for example, in LIGA (see below under X-ray lithography and Chapter 6).

Because of the problems involved with classical exposure tools for highly nonplanar micromachines, maskless exposures to photons (in air) or electrons (in vacuum) with high-precision linear and rotary positioning stages and numerically controlled beam direction and stage position are gaining popularity with micromachinists (see Chapter 7 on Comparison of Micromachining Techniques).

## Mathematical Expression for Resist Profiles

Writing the differential expression for the slope (dZ/dX) of the resist edge of an image of a mask feature of width, W, (see Figure 1.15) for a case where the dose is low and the developer influence dominant) we obtain

$$dZ/dX = (dZ/dD)(dD/dX) \qquad 1.30$$

with Z the thickness dimension and X the lateral dimension in the resist, the term dZ/dD (developer elution term) only resist and resist processing dependent and dD/dX (the derivative of the energy or dose absorbed), also called the intensity profile, a term only dependent on the exposure system and the object. The response of a positive resist to the dose D is given by $\gamma_p$, the contrast of the resist and the developer-process term in Equation 1.30 can be approximated by[28]

$$dZ/dD = -\gamma_p / D_p \qquad 1.31$$

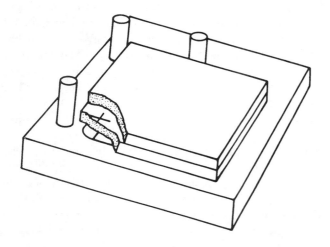

**Figure 1.14** Sketch of two-sided alignment jig with the two alignment masks in place for contact printing marks onto mask 2 (upper mask). The cross represents an alignment mark on mask. (Courtesy of Dr. Richard White, UC, Berkeley, CA.)[25]

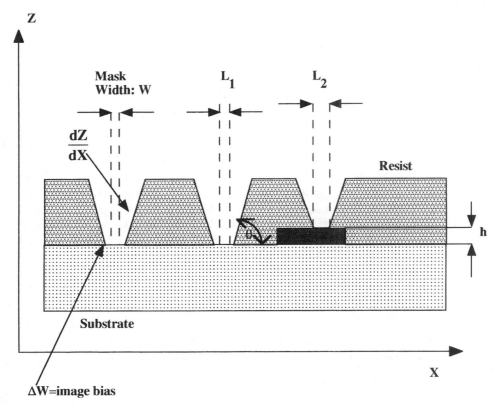

**Figure 1.15** Edge slope of a positive resist, $\dfrac{dZ}{dX}$ primarily determines the image bias, $\Delta W$. The change of line-width of a feature developed in resist with a topographic feature of height h is given by $L_2 - L_1 = \Delta CD$.

In Equation 1.10 for the contrast of a positive resist, $D_p^0$ represents the exposure energy below which no resist removal by the solvent takes place and $D_p$ embodies the sensitivity of the resist or the exposure energy at which no resist remains after development. The exposure energy, $D_p^0$, is independent of the thickness, t, of the resist. The resist sensitivity, $D_p$, on the other hand, will vary with the reciprocal of the film absorbency ($D_p \approx 10^{\alpha t}$). This means that we can rewrite Equation 1.31 as

$$\frac{dZ}{dD} \approx \frac{1}{(a + \alpha t)D_p} \qquad 1.32$$

where a is a constant and $\alpha$ is the resist absorption.[25] The second term in Equation 1.30, dD/dX, is the intensity profile of the image and is affected by the wavelength of exposure, $\lambda$, the numerical aperture NA, the depth of focus of the exposure tool (DOF) and the uniformity of illumination and can be written as[1]

$$\frac{dD}{dX} \approx \frac{2NA}{\lambda\left[1 - k\left(\frac{DOF(NA)^2}{\lambda}\right)\right]^2} \qquad 1.33$$

where k is a parameter depending on the coherence of the light source. Deriving from Equations 1.32 and 1.33 we can rewrite Equation 1.30 describing the resist profile in identifiable parameters,

$$\frac{dZ}{dX} \approx \frac{2NA}{\lambda(a + \alpha t)D_p}\left[1 - k\left(\frac{DOF(NA)^2}{\lambda}\right)\right]^{-2} \qquad 1.34$$

It follows that for a steeper edge slope, dZ/dX, the resist contrast, $\gamma_p$, should be as high as possible, and the shorter the exposure wavelength $\lambda$ and the higher NA the sharper the image profile. Since a higher NA means a lower DOF, deteriorating image profiles at larger depths, there is an optimum NA for each exposure/resist system. The image bias, $\Delta W$, shown in Figure 1.15, is primarily a function of the resist edge slope and becomes smaller as dZ/dX gets steeper. Equation 1.34 also implies that a high-contrast resist with a low absorbency will have a wider exposure latitude and can tolerate a larger intensity variation in exposure system output ($D_p$). At high radiation dose ($>100$ mJ/cm$^2$), the profile of a positive resist is dominated by the absorption of radiation, the reflected photons, and the quantum yield ($\Phi$ or G(s)) of the photochemical reaction (see Figure 1.3A(a)). The absorbed energy per depth Z dominates the rate of dissolution. In a developer dominated (also 'force developed') resist profile (also undercut), the profile recedes and thinning of the whole resist layer occurs (see Figure 1.3A(c)). This thinning does not occur in the case of insolubilized negative resists, rendering those resists less susceptible to overdevelopment (see Figure 1.3B). For negative resists, $\gamma_n$ is not influenced by the solvent and, apart from some uncontrollable swelling of negative resists during

development, the resist profile cannot be manipulated in the development step. By using a normal exposure dose and with moderate developer influence a 'perfect' image transfer can be accomplished as shown in Figure 1.3A(b). At those moderate doses, both the developer elution term and the energy absorption term contribute to the formation of the resist profile.

From Figure 1.15 we can derive how the exposure gradient results in the developed positive resist profile having a larger opening on top, forming a so-called "over-cut" profile. This overcut profile results in a difference of dimension in subsequent patterning steps over features with different topography. The change in critical dimensions, $\Delta CD$, relates to the profile angle, $\theta$, and the topography height, h, and can be expressed as:[2]

$$\Delta CD = L_1 - L_2 = 2h(\tan\theta)^{-1} \qquad 1.35$$

## Improving Resist Sensitivity-Chemical Amplification

Quantum yields for typical positive-tone resists are 0.2 to 0.3; thus, three or four electrons are required for each transformed sensitizer molecule. This places a fundamental limit on the photo-sensitivity of such systems. The situation is worse for DUV lithography. Typical near-UV positive resists (e.g., DNQ sensitized novolak resins) fail to be very useful for deep-UV lithography because of the unacceptable strong unbleachable absorption below 300 nm. Bleaching is the decrease in optical density during exposure enabling further penetration of the resist as the top layers are being cleared. The short wavelength absorption coefficients of the base resin and the added sensitizers often are too high to allow uniform imaging through practical resist thicknesses (about 1 µm for ICs). In general, commercially available resists perform poorly at short wavelengths. Those with a good plasma etching resistance, frequently based on novolak resins, have excessive unbleachable absorption, while those with an acceptable transparency have insufficient etching resistance. This is of concern particularly in micromachining where much thicker resist layers are employed. Short wavelengths do not penetrate the film completely and very long exposure to etchants, needed to create the desired features, degrades the resist. An additional problem, touched upon before, is that DUV optics absorb more light themselves making less light available at the resist plane. To boost quantum efficiency a 100-fold sensitivity improvement is desired, which does not happen through structural tailoring of classical resists alone.

One approach to improve resist sensitivity involves the concept of chemical amplification. In this approach, a single photon initiates a cascade of chemical reactions of the sort that characterizes a silver halide photographic emulsion system. The amplification in resists is based on the photogeneration of a catalytic photoproduct, e.g., an acidic species that catalyzes the scission for a positive-tone resist or the cross-linking of a resin for a negative-tone resist. The overall quantum efficiency of the catalyzed reaction is a lot higher than the efficiency for the initial acid generation.[29] Within a resist, sensitized with an onium salt such as diphenyliodonium hexafluoroarsenate, a Lewis acid is released upon photolysis. In the case of a positive resist, this released acid upon baking at 100°C, catalyzes the cleavage or scission of the resist, making it more soluble. This type of imaging involves the usual formation of a latent image during exposure. The latent image, a three-dimensional distribution of the catalytic photoproduct, does not immediately generate a concomitant change in dissolution rate as with regular resists.

Image formation only takes place after an activating thermal step in a post-exposure bake at 100°C. Typical turnover rates for one acid catalyst molecule are in the range of 800 to 1200 cleavages. Resists thus amplified may attain a photosensitivity of 5 mJ/cm² or better. Tenfold sensitivity increases are common. Both the acid catalyzed scission and cross-linking reactions are very dependent on the post-exposure bake temperature, time, and method. The control of these parameters represents a significant difference between conventional resists and chemically amplified resists. In Figure 1.16, a schematic representation of a generalized chemically amplified resist process (positive and negative) is shown as well as typical chemicals. Resist amplification might play a role in device manufacture with sub-0.5-μm dimensions where the optics of the exposure system force one to work with less light intensity and where the sensitivity of traditional resists would require too long an exposure time.[30]

Although chemically amplified resists provide reasonable throughput for e-beam and deep UV lithographies (short wavelengths!), they often suffer from poor environmental stability and sensitivity to process conditions. The environmental effects cause surface inhibition by neutralization of photogenerated acid in the top resist layer. A resist topcoat can be used to minimize the environmental effect. Unfortunately, such a multilayer resist (MLR) further complicates the photoresist process.

## Image Reversal

Another chemical modification of traditional photoresists involves image reversal. Because swelling does not take place during the development of positive photoresists, several process variations aim at reversing the tone of the image so that the resist can act as a high resolution (!) negative resist. Figure 1.17A demonstrates the process sequence for the image reversal of a positive resist. After the resist has been patterned, an amine vapor (such as imidazole or triethanolamine, or, more general, a base) is diffused into the exposed areas. The amine neutralizes the byproduct of the photodecomposition (a carboxylic acid in this case) and makes these exposed areas highly resistant to further change by exposure to light and highly insensitive to further development. A subsequent blanket or flood UV exposure makes the areas adjacent to the neutralized image areas soluble in conventional positive photoresist developers. The net result gives a negative image of the mask with improved resolution since the scattered forward radiation is not reproduced.

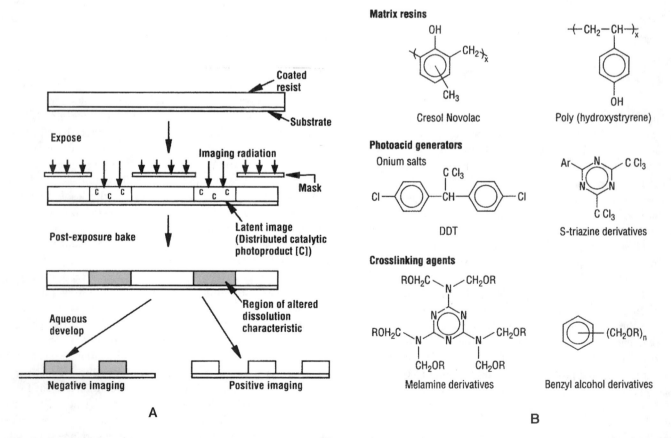

**Figure 1.16** (A) Schematic representation of a generalized chemically amplified resist process. (B) Typical chemicals used in amplification type resists.[29,31] (Part A from Reichmanis, E., et al., *Microlithogr. World*, 7–14 (1992). Part B from Lamola, A. A., et al., *Solid State Technol.*, 53, 53–60 (1991). With permission.)

Image reversal with a negative resist such as KTFR is possible as well (see Figure 1.17B). We already discussed that remaining oxygen in a resist can improve resolution by preventing insolubilization in light scattered zones outside the exposed area. Image reversal is an extension of this insolubilization. By working at low intensity, oxygen flooding can scavenge all photogenerated azide polymerization precursors, making the exposed areas soluble in a developer as they are prevented from polymerizing. A subsequent flood exposure under a nitrogen blanket, at higher intensity, initiates the polymerization in the previously nonexposed areas.[6]

Feely and coworkers[33] developed a novolak-based reversal system with diazonaphtoquinone sensitizer as the acid generator and hydroxymethyl-melanine as the hardening agent. The process, because of its micromachining appeal, will be discussed separately in the New Types of Resist section.

## Multilayer Resist

Variations in resist thickness over thin-film steps lead to linewidth control problems (see Figure 1.15). This lack of linewidth control and the limitations of low source intensity and shallow depth of focus can be overcome (to some degree) by using multilayer resists. A typical multilevel resist as shown in Figure 1.18 consists of a thick, planarizing bottom layer, and a thin top imaging overlayer of resist. These layers function synergistically to achieve good resolution otherwise impossible to obtain with thick, planarizing single-layer resists. After exposure and development of the thin imaging overlayer, the pattern is transferred to the planarizing underlayer. An intermediate isolation layer sometimes separates the planarizing and imaging

layer, preventing their mixing. Multilayer resist processes often are referred to as top-surface imaging techniques. In development (wet etching) of the planarizing resist layer, an undercut (see Figure 1.18) or overcut may be generated. Alternatively, if good fidelity of the mask pattern is desired, dry etching (RIE) may be used (see Figure 1.18).

The commercially available DESIRE (Dry Etching of Silylated Image REsist) process by UCB (Belgium), illustrated in Figure 1.19, exemplifies a multilevel resist scheme. In this case, a short exposure of a thick layer of planarizing DQN-novolak resist produces latent images of indenecarboxylic acid. In a post-exposure bake, in vacuum, the layer is converted into a film enabling selective silylating. In the silylating process, Si atoms only react on the surface of the exposed resist, locally forming a thin Si containing resist layer. The polymer film is typically exposed and silylated to a depth of at least 1000 Å to provide sufficient differential plasma etch resistance for pattern transfer. An oxygen plasma etch subsequently removes the unexposed, unreacted resist and the silylated mask remains due to the formation of a protecting silicon oxide layer, and a negative-tone image of the mask results. The DESIRE process illustrates both dry resist development and image reversal. A major advantage of surface imaging techniques is that only the surface of the resist requires exposure which, coupled with the anisotropic nature of plasma etching, allows for the use of thick resist layers that planarize the underlying topography to a higher degree than resists with more typical thicknesses could. The lithography process also remains unaffected by substrate reflection and resist thickness variation, as light does not have to penetrate the thickness of the resist. Image quality with DESIRE is excellent. The process may present the first dry development system used in full-scale

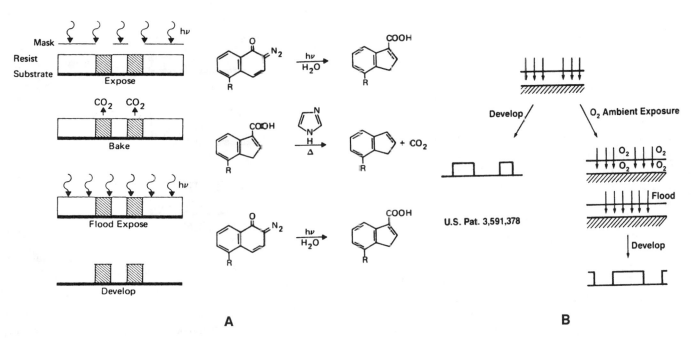

**Figure 1.17** Image reversal process sequence. (A) Positive photoresist. (B) negative photoresist.[32,6] (Part A from Alling, E. and Stauffer, C., *Solid State Technol.*, 37–43 (1988). Part B from Moreau, W. H., *Semiconductor Lithography* (Plenum Press, New York, 1988). With permission.)

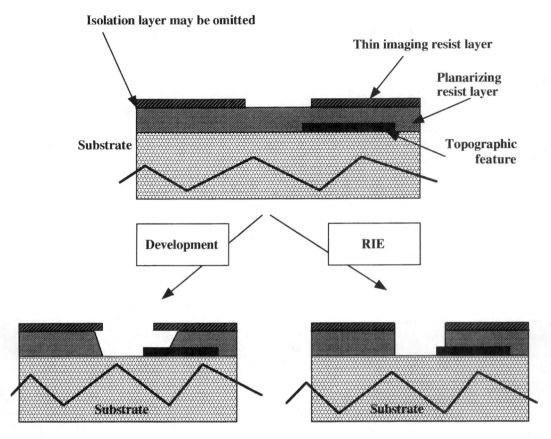

**Figure 1.18**  Multilevel resist scheme. Thick planarizing underlayer for planarizing of wafer topography, optional intermediate isolation layer, and thin imaging overlayer for optimum resolution. Wet etching (development) and dry etching (RIE) produce different resist profiles.

manufacturing.[19] Interested readers can catch up on the DESIRE process and other novel surface imaging techniques by reading, for example, Calvert et al. and references therein.[34]

## Thin Film Interference Effects

When using single layer resists (SLR), the deleterious effects of thin film interference effects need to be overcome for good CD control, especially of submicron features. The effects manifest themselves in nonvertical resist profiles, line-width variations, reflective notching, scumming (underexposed resist leaving organics behind after development), and alignment inaccuracies. The most important thin film interference effects are schematically illustrated in Figure 1.20.[35] Solutions to thin film interference effects involve the use of anti-reflective coatings (ARCs), increasing the absorption of the photoresist, and the use of multilayer resist (see above). Antireflective coatings overcome interference effects caused by reflections from either the top or bottom of the resist (Figure 1.20 a and b). Also, by adding absorbing dyes, the reflection at the substrate/resist interface can be minimized. When a resist is exposed to monochromatic radiation, *standing waves* are formed in the resist as a consequence of coherent interference from reflecting substrates creating a periodic intensity distribution in the direction perpendicular to the surface (see Figure 1.20c). The standing wave effect is a strong function of resist thickness, and exposure

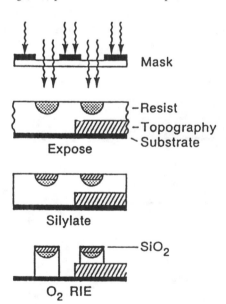

**Figure 1.19**  The DESIRE process developed by UCB Chemical Company. Exposure and silylation are confined to the upper layers of the resist coating.

variations resulting from variation in resist thickness in the vicinity of steps result in changes in line-width. The standing waves are the result of coherent interference of monochromatic light, and the effect can be decreased by using broadband illumination. To minimize the standing wave effects, one also can use thinner resists (<0.3 μm) but the thinner resists cause a degradation in line-width control over varying topography. Reflective notching (Figure 1.20d) comes about when light is reflected from the more reflective topological features buried in the resist.

## New Types of Resists

Some relatively new photoresists with promising sensor and microfabrication applications are introduced below. Several of these new resists enable high aspect ratio photofabrication in polymers.

### Polyimides

*Overview*

In use as photoresist, photosensitive polyimide precursors, called polyamic acids, are spun on the wafer and upon exposure to UV light, cross-linking results. During development, the unexposed regions are dissolved, and final curing by further heat treatment leads to a chemical transformation (known as imidization) of the remaining cross-linked material which yields polyimide. At 275°C, more than 99% of the polyimide precursor is

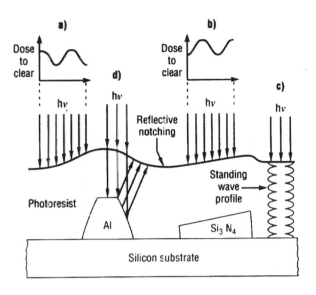

**Figure 1.20** Thin film interference effects. Light reflected from (a) resist/Si or (b) Si₃N₄/Si interfaces through various resist or nitride thicknesses and changes the dose necessary to clear the resist. Also, standing-wave profile (c) and reflective notching effects (d) are shown. (From Horn, M. W., *Solid State Technol.*, November, 57–62, 1991. With permission.)

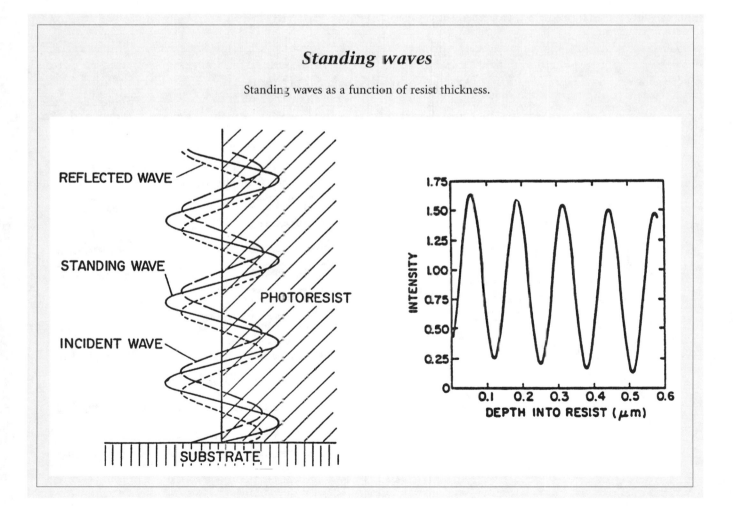

## Standing waves

Standing waves as a function of resist thickness.

converted to polyimide. Photosensitive polyimide is negative tone. As with other negative resists, an oxygen effect is noted in the exposure. The actinic sensitivity of the resist usually is confined to 365 nm.

One often uses polyimides as permanent photoresists: photosensitive polymers that can be left behind as a dielectric, especially when the uniquely low dielectric constant of the polyimide gives devices a decreased capacitance resulting in increased speed for electronic applications.[36] In multiple spin-coats, very thick polyimide films can be deposited (>100 μm). Since the films are so transparent, UV lithography permits the fabrication of high aspect ratio features. This is of particular interest to micromachinists who develop LIGA-like processes based on UV-sensitive polyimides[37] (see Chapter 5 for surface micromachining with polyimides).

### Polyimide Properties

Polyimides are part of a group of high-temperature polymers that have come into commercial use in the 1970s. Due to their complex structure involving very strong carbon ring bonds, they do not melt and flow as do most thermosets and thermoplasts. Polyimide films feature excellent thermal stability (up to ~450°C for short periods of time), good dielectric properties (ε = 3.3 and a resistivity of ~$10^{16}$ ohm-cm), superior chemical resistance, toughness, wear resistance, flame retardance, and (due to their flexibility) interesting mechanical properties. Current 'macro'

applications of polyimides include ball-bearing separators and mechanical seals. Polyimides, photosensitive and nonphotosensitive, have also found many applications in both integrated circuitry and microfabrication. In the IC and micromachining areas, these materials are being used as passivating and interlayer dielectrics, planarizing compounds, reactive ion masks, alpha-particle barriers, micromachined structural elements, humidity-sensitive materials,[38,39] liquid crystal display, color filters[40] (by incorporating dyes) and other optical elements such as waveguides. In planarization, polyimides smooth the undulation caused by topographic features on the wafer so that the top imaging layer has a much smaller thickness variation. An important commercial application of polyimides is in *multiple chip modules* (MCMs).[41] In the multiple chip module application, the use of photosensitive polyimides dramatically simplifies the manufacturing procedure.[41] In a finished device, up to five layers of metallization are insulated by 15- to 20-μm thick polyimide films on a ceramic wiring board. On the negative side, their short shelf-life stability still needs to be addressed by

---

\*   MCMs increase the interconnect density on printed circuit boards, making them smaller and speeding up the processing time. Multichip modules might become an important supporting technology for sensors, as most sensors are better partitioned into separate sensing and electronic parts rather than being built monolithically in Si.

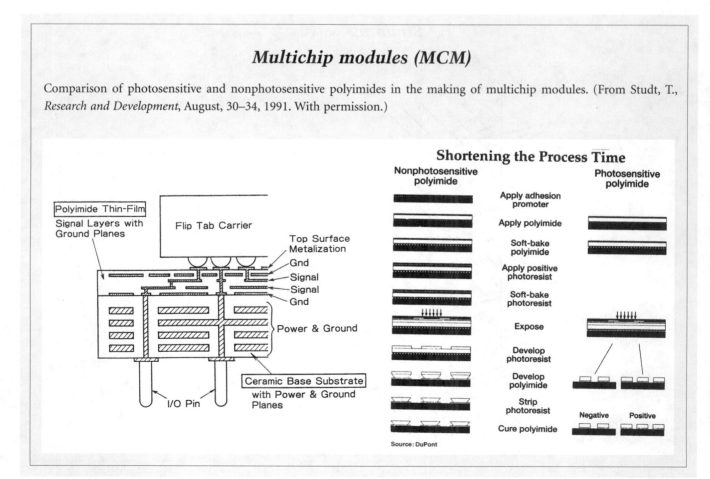

## *Multichip modules (MCM)*

Comparison of photosensitive and nonphotosensitive polyimides in the making of multichip modules. (From Studt, T., *Research and Development*, August, 30–34, 1991. With permission.)

organic chemists. Exploration of micromachining applications of modified polyimides has started as well. For example, special perfluorinated polyimides were synthesized at the NTT laboratories in Japan. These remarkable materials are resistant to soldering temperatures (260°C) and are highly transparent at the wavelengths of optical communications (1.0 to 1.77 μm). In addition, their low dielectric constants and refractive indices match the conventional fluorinated polyimides, whereas their birefringence is lower.[42] As we will see in Examples of Dry Etching in Chapter 2, dry plasma etching of these modified polyimides results in very high aspect ratio microstructures with interesting optical properties.

### SU-8

IBM has started experimenting with SU-8, an epoxy-based, transparent photoresist. High aspect ratio (>10:1) features and straight sidewalled images in thick layers (>200 μm) were obtained by applying standard lithography. No detailed investigation of the sidewall roughness and sidewall run-out of this negative acting material has been undertaken yet, but plating stencils for micromachine fabrication and RIE masks for deep silicon etching look extremely promising. The resist also is chemically inert and temperature stable, and stripping of the material is currently the most problematic aspect.[43]

This type of new resist development will further catapult the micromachining community into making LIGA-like structures inexpensively. The details of resist profile and wall roughness of new UV resists like SU-8 will determine the relative merits of X-ray lithography vs. photolithography (see Chapter 6).

### Positive Resist of the AZ-4000 Series

Engelmann et al.[44] use conventional ultraviolet lithography to process thick AZ-4000 positive resist layers in the 15- to 80-μm range. This resist type is very viscous and like SU-8 has a high transparency. Multi-spin coats result in the desired thick layers. To obtain steep resist profiles, a very good contact between the mask and the photoresist surface is essential. Edge bead removal and vacuum printing promote such a good contact. With a 33-μm thick resist, patterns with an aspect ratio of 7.3 were obtained.[44]

### Polarity Changing Resists

Resists may find their base in illumination-induced solubility changes, etch resistance, or polarity changing. In this section we will explore polarity changing resists. When using polarity changing resists, one takes advantage of the alteration of the hydrophobic nature of resists through illumination. For example, DUV radiation of onium salts such as triarylsulfonium or diaryliodonium results in the generation of an extremely strong protonic acid, leaving unprotected polar phenol groups behind in the polymer backbone. Development in nonpolar solvents selectively removes unreacted resist (nonpolar), leading to a negative image, whereas development in an aqueous base selectively removes the polyvinylphenol (polar) resist to result in a positive image. Such dual-tone, amplifying resists obviously give the engineer an additional degree of freedom.

Polarity changes also are at the heart of the newest ultrathin film (UTF) resist strategies. Ultrathin film resists represent the next step in surface imaging technologies. Whereas in a more traditional surface imaging technique such as the DESIRE process (Figure 1.19) a 1000-Å of the top layer of the resist is converted through silylation, in UTF only the top monolayer of the resist changes, opening up the potential for further resolution improvement. Calvert et al.[45] use a few monolayers (<10 Å) of organosilanes which they chemisorb onto a substrate. These chemisorbed films are easier to prepare and more robust than the physisorbed Langmuir-Blodgett (LB) films. Deep UV irradiation of these films cleaves organo functional groups from the film and produces an extremely hydrophilic surface (water contact angle <10°) to which colloidal Pd/Sn catalyst does not adhere. Subsequent electroless deposition occurs only on those surface regions that were not irradiated. Patterns of electroless Ni, Co, and Cu were thus produced on a variety of substrates including silicon, quartz, alumina, metals, and polymers. These metal patterns serve as an efficient plasma etching mask for pattern transfer. Fabrication of 0.3-μm line-widths has been demonstrated in this way.[34,45]

Polarity changing resists also have demonstrated their usefulness in biosensor applications. In a variation on the Calvert patterning procedures of hydrophobic monolayers, explained above, a hydrophobic silane pattern on a glass slide is produced by 'lift-off' where the hydrophobic silane (dimethyl-trichlorosilane) replaces the more familiar metallization step (see Figure 1.4). Hydrophobic patterning is followed by the deposition of an amino silane (3-aminoethylaminopropyltrimethoxysilane) in the remaining areas; the amino group forming an attachment site for proteins.[46,47] Successful chemical patterning of glass slide surfaces is demonstrated by the growth pattern of cells. Some cells, such as fibroblasts, will grow in the hydrophilic areas but will have difficulty adhering and spreading in hydrophobic areas. Sharp edge definition is thus obtained along edges of hydrophilic-hydrophobic interfaces. Using this technique, neuronal growth can be controlled and neuronal processes may be induced to follow the straight lines of the original mask pattern. It has been suggested that this type of approach could be applied to nerve repair techniques. The method and its range of applications are described in Britland et al.[48]

Protein patterning evolved into an important application for lithography. It is easy to imagine arrays with several different enzymes, antibodies, or DNA-probes immobilized precisely onto a small transducer surface as a diagnostic panel for clinical applications. Polarity changing materials only outline one way to photolithographically pattern chemical sensor arrays. In Chapter 10 on Microfabrication Applications, breakthroughs in the chemical and biosensor field based on innovative patterning of protein and other chemical and biological 'selector' materials are discussed.

### Langmuir-Blodgett Resists

Scattering of radiation during exposure of a resist limits resolution to no better than the resist thickness. Using ever thinner resist layers could further improve the resolution. Employing thinner imaging layers may render a low DOF, associated with high numerical aperture (NA) lenses and short wavelength lithography, less critical. It is in this context that ultrathin, physisorbed Langmuir-Blodgett resist films have been considered for some of the finest lithography applications. Langmuir-Blodgett (LB) films are prepared by transferring organic monolayers floating on a water surface onto solid substrates (see Figure 3.36). Multilayer films of poly(methyl methacrylate) deposited by the LB technique have been investigated as a potential e-beam resist for nanolithography. Major problems concerning the LB films need addressing. Research in this area focuses on speeding up the coating process, simplifying the surface cleaning procedures, and improving etching resistance. Despite a lot of work in the early 1980s to automate LB resist deposition systems, the technology is currently not envisioned as a viable manufacturing option.[34,49]

### Rohm and Haas 3D Resist

A potentially important dual tone resist chemistry, based on a phenolic resin diazonaphthoquinone as acid generator and hydroxymethylmelanine as the hardening agent, was developed by Feely[50] at Rohm and Haas. Initially the Rohm and Haas process drew a lot of attention because the process promised to have microfabrication applications. Figure 1.21A illustrates the process flow options for this resist.[33] Exposure produces the latent image, which, when developed at this stage, results in the 'normal' positive resist relief image (right-hand path in Figure 1.21A). If, on the other hand, the film is heated after exposure to induce the acid-catalyzed reaction of the melanine with the phenolic resin, the exposed areas become insoluble and image reversal is initiated. The reversal is completed by a subsequent flood exposure, solubilizing the nonexposed areas and rendering a negative tone image of the mask (left-hand path in Figure 1.21A). The cross-linked nature of the image imparts resistance to swelling and dissolution by both aqueous and organic solvents as well as dimensional and thermal stability to temperatures >300°C. By using three-tone masks that have opaque, transparent, and partial transmission (50%), stepped (positive tone process) and cantilevered (negative tone process) structures are produced (Figure 1.21B). By appropriate processing of image reversal systems, walls with positive, negative, and vertical walls become possible as well. Figure 1.21C showcases an example of the types of microstructures that can be printed with the Rohm and Haas system. In principle, this resist could be the basis of a complete microstructural universe.

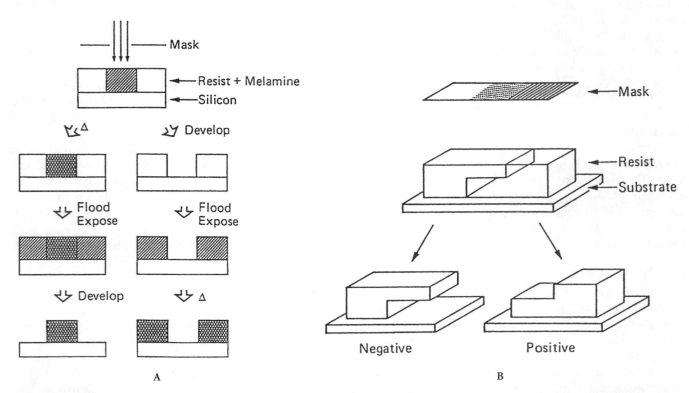

**Figure 1.21**   Rohm and Haas dual tone resist. (A) Process options for Rohm and Haas dual tone resist. The right-hand path produces positive-tone images in which the remaining resist is cross-linked and does not flow when exposed to temperatures above the $T_g$ of the novolak base resin. The left-hand path provides thermally stable negative-tone images. (B) Exposure of the Rohm and Haas resist with a mask exhibiting nominally 0% transmission, 50% transmission, and 100% transmission allows generations of patterns with controlled variations in thickness and/or overhangs and cantilevered structures. (C) Example of the types of features that can be printed using the Rohm and Haas resist. (Courtesy of Dr. Feely, Rohm and Haas.)

# Resist Stripping

## Wet Stripping

Photoresist stripping means, in slightly oversimplified terms, organic polymer etching. The primary consideration is complete removal of the photoresist without damaging the device under construction. To place stripping in its proper context in a typical process flow we are recapping in Figure 1.22 some of the fundamental processes introduced so far.[23] For reasons of simplicity, baking steps of the photoresist are not detailed in this figure. An oxidized wafer (a) is first coated with a 1-µm thick photoresist coat (b). After exposure (c), the wafer is rinsed in a developing solution or sprayed with a spray developer, removing either the exposed areas (positive tone) or the unexposed areas of photoresist (negative tone), leaving a pattern of bare and photoresist-coated oxide on the wafer surface (d). The photoresist pattern is either the positive or negative image of the pattern on the photomask. In a typical next process step after the development, the wafer is placed in a solution of HF or HF + NH$_4$F, meant to attack the oxide but not the photoresist

or the underlying silicon (e). The photoresist protects the oxide areas it covers. Once the exposed oxide has been etched away, the remaining photoresist can be stripped off with a strong acid such as H$_2$SO$_4$ or an acid-oxidant combination such as H$_2$SO$_4$ – Cr$_2$O$_3$ attacking the photoresist but not the oxide or the silicon (f). Other liquid strippers are organic solvent strippers and alkaline strippers (with or without oxicants). Acetone can be used if the postbake was not too long or happened at not too high a temperature. With a post-bake of 20 min at 120°C acetone is still fine. But with a post-bake at 140°C, the resist develops a tough 'skin' and has to be burned away in an oxygen plasma.

The oxidized Si wafer with the etched windows in the oxide (f) now awaits further processing. This might entail a wet anisotropic etch of the Si in the oxide windows with SiO$_2$ as the etch mask.

## Dry Stripping

Dry stripping or oxygen plasma stripping (also ashing) has become more and more popular as it poses fewer disposal

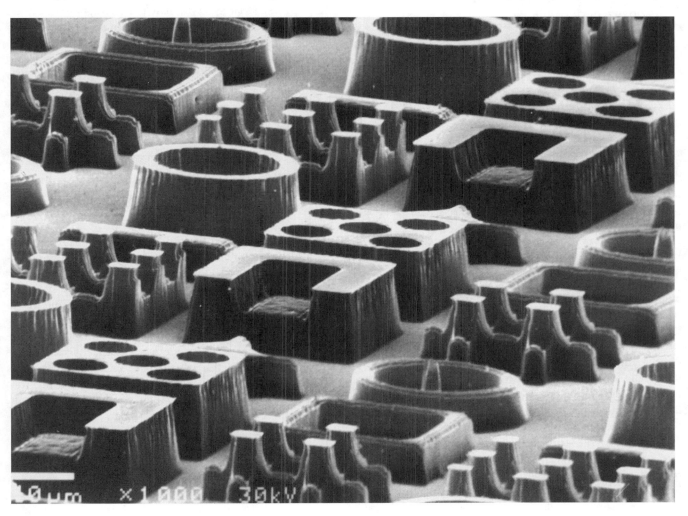

c

**Figure 1.21** (Continued)

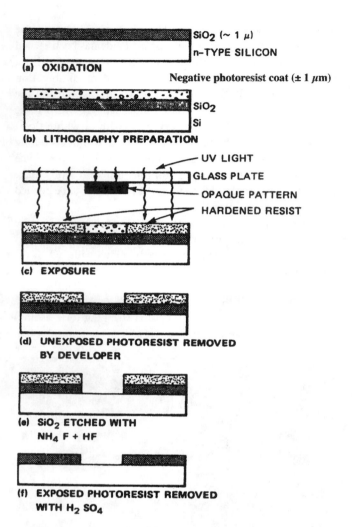

**(a) OXIDATION** — SiO₂ (~ 1 μ), n-TYPE SILICON

**(b) LITHOGRAPHY PREPARATION** — Negative photoresist coat (± 1 μm), SiO₂, Si

**(c) EXPOSURE** — UV LIGHT, GLASS PLATE, OPAQUE PATTERN, HARDENED RESIST

**(d) UNEXPOSED PHOTORESIST REMOVED BY DEVELOPER**

**(e) SiO₂ ETCHED WITH NH₄ F + HF**

**(f) EXPOSED PHOTORESIST REMOVED WITH H₂ SO₄**

**Figure 1.22**  Basic IC process steps on an oxidized Si wafer; photolithography (with a negative-tone resist), including exposure, development, oxide etching, and resist stripping. (From Brodie. I. and J. J. Muray, The Physics of Microfabrication, Plenum Press, New York, 1991. With permission.)

problems with toxic, flammable, and dangerous chemicals. Wet stripping solutions also lose potency in use, causing stripping rates to change with time. Accumulated contamination in solutions can be a source of particles, and liquid phase surface tension and mass transport tend to make photoresist removal difficult and uneven. Also dry stripping is more controllable than liquid stripping, less corrosive with respect to metal features on the wafer, and, importantly, under the right conditions it leaves a cleaner surface. Finally, it causes no undercutting and broadening of photoresist features as wet strippers do.

In solid-gas resist stripping, a volatile product forms either through reactive plasma stripping (e.g., with oxygen), gaseous chemical reactants (e.g., ozone), and radiation (UV) or a combination thereof (e.g., UV/ozone-assisted). Plasma stripping employs a low pressure electrical discharge to split molecular oxygen ($O_2$) into its more reactive atomic form (O). This atomic oxygen converts an organic photoresist into a gaseous product that may be pumped away. This type of plasma stripping

belongs to the category of chemical dry stripping and is isotropic in nature (see Chapter 2 on Pattern Transfer with Dry Etching). In ozone strippers, ozone, at atmospheric pressure, attacks the resist. In UV/ozone stripping, the UV helps to break bonds in the resist, paving the way for a more efficient attack by the ozone. Ozone strippers have the advantage that no plasma damage can occur on the devices in the process. Reactive plasma stripping currently is the predominant commercial technology due to its high removal rate and throughput. Some different stripper configurations* are barrel reactors, downstream strippers, and parallel plate systems. These prevalent stripping systems are reproduced in Figure 1.23.[51] In Chapter 2 more details about these different plasma strippers will be provided.

Dry stripping of resist has been so successful that it accelerated the use of plasmas in other lithography steps such as etching, development, and even deposition of resist. For example, the study of the attack of oxygen on polymers led to the development of dry resists as used in the DESIRE process (see Figure 1.19). Such dry resists are vital for future submicron lithography where underetching and broadening of features are most critical. So far dry resists have not been applied to the field of micromatching, as submicron features have not yet gained enough importance.

For more detailed information on multilayer resists, resist monitoring, dry resists, surface imaging resists (SIR), and other resist aspects not touched upon here or mentioned only briefly, we refer to Moreau.[6] Also *Solid State Technology*, *Semiconductor International*, and *Microlithography World* carry excellent tutorials on these topics.

## Expanding the Limits of Photolithography Through Improved Mask Technology

### Background

Control of the deposition of films in the z-direction can be achieved with remarkable precision, down to 20 Å. The x,y dimensions are more difficult to control and depend on the lithography as described in the preceding sections. The minimum x,y feature size feasible on a substrate in IC and microfabrication technology is determined primarily by the precise focus of the energy source that allows discrimination between areas of exposed resist and nonexposed resist. Other key factors are the tolerances on the mask itself, the ability to align the mask to the wafer and to align subsequent layers of masks to create the proper vertical geometry, and the ability to control the rate

---

\*    For a further comparison of the stripping systems shown in Figure 1.23 we refer to the August 1992 issue of *Solid State Technology*, pp. 37–39.[51] A listing of leading manufacturers of plasma stripping equipment can be found in the October 1992 issue of *Solid State Technology*, pp. 43–48.[52] Finally, in *Semiconductor International*, February 1992, pp. 58–64, a review on how to strip resists toughened through processing and difficult to remove with regular techniques is presented.[53] A lot of the proposed techniques are combinations of wet and dry etching techniques.

and direction of etching and deposition. Sometime after the year 2000, the smallest feature on an integrated circuit that advanced lithographies will be able to write will be about 0.1 μm. Historically, dynamic random access memory chips

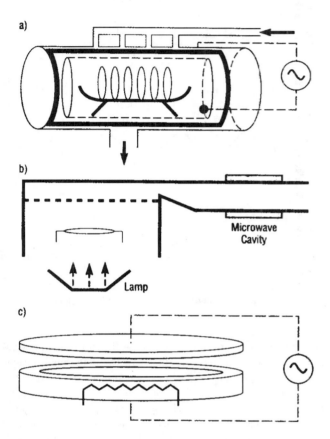

a)

b)

Microwave
Cavity

Lamp

c)

**Figure 1.23** Various dry stripping reactors: (a) barrel reactor; (b) downstream etchers; (c) parallel plate systems. (From Flamm, D. L., *Solid State Technol.*, 35(B), 37–39, 1992. With permission.)

(DRAMs) have driven microlithography more than microprocessors which exhibit more relaxed design rules (see Figure 1.24A).[54] The current DRAMs manifest a smallest feature size of less than 0.5 μm. The smallest structural dimensions in sophisticated commercial memory chips measure 0.35 μm. This evolution in feature size is illustrated again in Figure 1.24B including reference to the required operation modes for the active devices.[55] Transistors with feature sizes between 0.5 and 0.25 μm operated at low voltages still exhibit good gain and low enough leakage currents. Below 0.25 μm, cooling becomes necessary to obtain the same good characteristics. More significant hurdles need to be overcome to extend minimum geometry beyond about 0.2 μm:

- The resistance of the minute connections between and within transistors must be lowered even further.
- The tendency of "leakage" with smaller transistors requires solving.
- High-volume manufacturing equipment capable of creating 0.2-μm features with an accuracy of ±10% still must be developed.

A 0.1-μm feature size will enable DRAMs of a gigabit. The scientific literature lists many examples of electronic devices with 0.1-μm features and smaller, written, for example, with an electron beam or a scanning tunneling microscope (see below). Such technology far outdistances manufacturing. Attempting transistors with minimum feature size below 0.1 μm will also require other strategies such as reliance on quantum devices and molecular electronics (see Chapter 7).

The current approach to increase the resolution of industrial lithography techniques still focuses on reducing the wavelength of the light source and increasing the numerical aperture of the imaging lens (see above). Most commercial lithography work uses UV (such as a Mercury vapor lamp) at a wavelength of

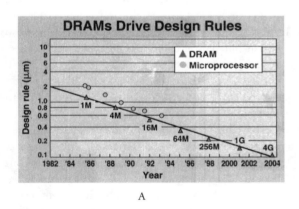

A

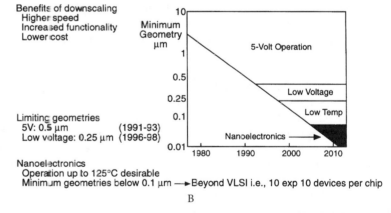

B

**Figure 1.24** Evolution in feature size of ICs vs. year of expected implementation. (A) DRAMs drive design rules (last traditional device: 0.1 μm — 1 to 4 G bit DRAM (2002)) (From an Editorial in *Semicond. Int.*, February, 17, 1993. With permission.); (B) operation modes of active devices with shrinking minimum feature size. (From Bate, T. R., *Solid State Technol.*, November, 101–108, 1989. With permission.)

436 nm (g-line) or 365 nm (i-line), and less common, UV light at 248 and 193 nm with excimer lasers.

By combining shorter wavelengths with improved mask technology, better optics, and more sophisticated resist chemistries, photolithography will be usable up to 0.25 μm, and even features slightly below 0.2 μm will become possible. Innovative resist strategies leading beyond these lithography limits, for example by using ultra-thin films, were discussed earlier.

Our understanding about mask technology enabling improved photolithographies in terms of resolution and depth of focus in the sub-0.5-μm range may need some illuminating. In micromachining, the differences with traditional IC manufacture start at the mask fabrication level. For example, to print a high aspect ratio micromachine, masks with an expanded DOF are favored, but masks with increased resolution remain of greater importance in IC manufacture. Phase-shifting and topographical masks, discussed below, lead, respectively, to improved resolution and improved DOF.

### Phase-Shifting Masks

One method allowing further improvement of the photolithography resolution at a given wavelength ($\lambda$) and numerical aperture is by carefully controlling light diffraction using constructive and destructive interference to help create a circuit pattern. The method is based on building phase-shifting masks. Using a phase-shifting mask, one controls both amplitude and phase of the light and, in particular, one arranges the mask so that light with opposite phase emerges from adjoining mask features. In that case, destructive interference can be used to cancel some of the image-spreading effects of diffraction.

In Figure 1.25 we illustrate the effect for three nearby mask apertures; a classical transmission mask is added for comparison.[56] The amplitude profile at the plane of the classical transmission mask consists of three square-cornered features with positive amplitude. The amplitude at the wafer level is broadened and rounded but does not change sign. With a phase-shifting mask, a shifter layer results in a reversal of the phase of the light at the mask. As the photoresist only shows sensitivity towards the intensity of the light and not to the sign, the three bright features develop identically and the areas where the light of opposite phases causes destructive interference form very dark contrast lines.

How narrow a feature can one make this way? Oki Electric Industries presently sells a MESFET (metal semiconductor field effect transistor) device with a 0.18-μm gate patterned by i-line lithography ordinarily considered capable of only 0.5-μm resolution![56] Phase-shifting masks can thus easily improve resolution by 50 to 100%.

### Topographical Masks

With the introduction of the phase-shifting masks discussed above, a new field of 'mask engineering' emerged. The topographical mask as proposed by op de Beeck[57] exemplifies another engineered mask. The basic idea here is simply to realize

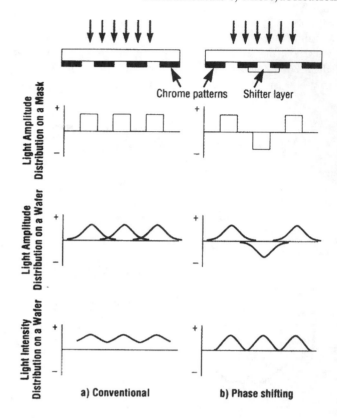

**Figure 1.25** Comparison of a conventional transmission mask (a) and a "Levenson-type" phase-shifting mask (b). Both contain opaque chrome (Cr) regions, but the light passing through some of the apertures of the phase-shifting mask also passes through the transparent material of the phase shifter. This reverses the phase of the light at the mask, giving rise to destructive interference at the photoresist plane. (From Levenson, M. D., *Microlithogr. World*, 6, 6–12, 1992. With permission.)

a mask with metal patterns located at different optical distance from the projection lens. These patterns are focused at a different image distance from the lens, as shown in Figure 1.26. By designing the mask according to the wafer topography, the actual DOF of the total exposure field is enlarged. To create two metal layers at different levels on one mask plate, one uses a conventional transmission mask partially covered by a thick transparent layer. Since the refractive index of this layer differs from that of air, the optical distance between the metal patterns and the projection lens differs for metal patterns with or without the thick layer. This type of mask is called 'thick shifter mask'. Although the topographical mask is somewhat simpler to make than a phase-shifting mask, both pose considerably more fabrication difficulties than a regular mask, hence their slow acceptance in industry. Microfabrication might find an important application in fabricating all types of innovative masks to improve DOF and line resolution, not only for X-ray lithography (see below) but also for topographical masks and novel types of phase shifting masks.

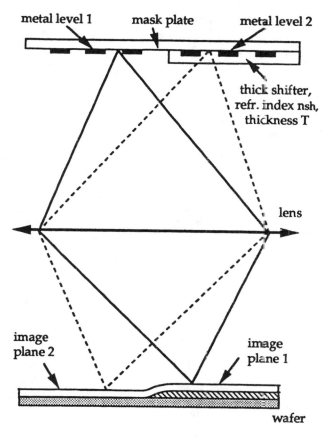

**Figure 1.26** Thick shifter mask: imaging principle. (Courtesy of op de Beeck, IMEC, Belgium.)

# Alternative Lithographies

## Introduction

Besides DUV, we also need to consider X-rays, electron-beams, and ion-beams as options for industrial submicron lithography systems of the future. Figure 1.27 illustrates these four lithography techniques. We believe that IC and micromachining developments will take more and more separate paths in adopting preferred lithography strategies and that alternative lithographies will be introduced faster in micromachining.

In the IC industry, continuous improvements in optical lithography have postponed the industrial adoption of alternative lithographies because of the huge financial investment needed for photolithography equipment. Critics believe photolithography will never be displaced by another lithography technology, since, as they say, cost and technical obstacles will make it pointless to go smaller than 0.1 μm.[58] More likely, at a 0.1-μm feature size (somewhere after the year 2000), where i-line and g-line light ceases to be used, plenty of commercial incentives will beckon to look at other lithographies such as X-ray lithography, ion beam lithography, or even scanning tunneling microscope based lithography.

With ICs one needs finer and finer geometries. Batch processes are a prerequisite. With micromachines, extending the z-direction, i.e., the height of features (skyscraper type structures) and incorporating nontraditional materials (e.g., gas-sensitive ceramic layers) catch the spotlight; batch fabrication is not always a prerequisite. Depending on allowable cost of the micromachine, serial processes such as electron-beam writing, laser machining, and even traditional mechanical precision engineering may become viable manufacturing tools. Taking these facts into consideration, alternative lithography techniques such as X-ray lithography, electron-beam and ion-beam lithography, and laser-based processes may gain importance faster in micromachining rather than in the IC industry. The key factor to spur this development will be the identification of a "killer-application" for a micromachine which might command a much larger price than an IC; say, a S200 microspectrometer to analyze a set of gases or a better magnetic read-write head.

The most likely technology to push beyond photolithography is X-ray lithography. For example, the large depth of focus (virtually infinite), the insensitivity to organic dust, and the capability to draw parallel lines even into very thick resist (up to several cm) gives X-ray lithography a technological edge over photolithography in micromachining. Moreover, whereas UV photons are scattered at all interfaces, leading to standing waves and proximity effects, X-rays are absorbed and do not scatter less.[24] The use of molding to replicate the original X-ray exposed template in a wide variety of materials holds the additional promise of making X-ray lithography much less expensive once the master molds are made (see below and Chapter 6 on LIGA). It seems possible that nontraditional microstructure fabrication might lead the way to a wider acceptance of X-ray lithography. In the next section we will look into some of these alternative lithography techniques, keeping in mind that the decision for an optimum technique might be quite different for micromachines compared to ICs.

## X-Ray Lithography

### Introduction

In contrast with electron lithography and ion-beam lithography, no charged particles are directly involved in X-ray exposures, which eliminates the need for vacuum (see Figure 1.27).[23] Another advantage of X-rays is that one can use flood exposure of resist-coated wafers, ensuring higher throughput than when writing with a thin electron or ion beam. The three main classes of sources for X-ray lithography are electron impact tubes, laser-based plasmas, and synchrotrons. For micromachining, we mainly will be concerned with synchrotron radiation, for reasons explained below.

### X-Ray Lithography in Micromachining

X-ray lithography is superior to optical lithography because of the use of a shorter wavelength and a very large DOF

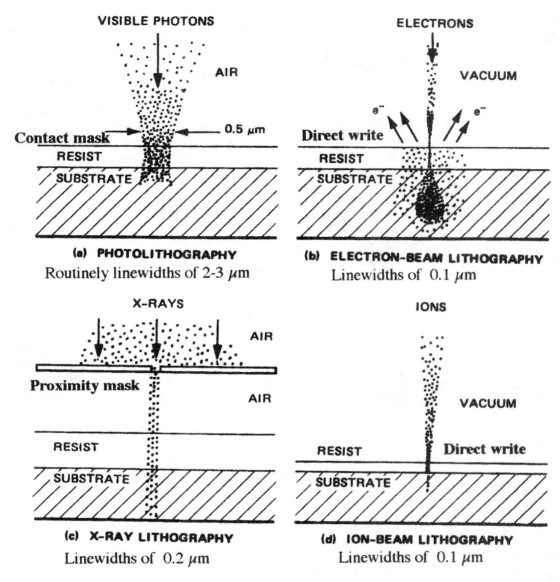

**Figure 1.27**    A comparison of photolithography, electron-beam, ion-beam, and X-ray lithography. (From Brodie, I. and J. J. Muray, *The Physics of Microfabrication*, Plenum Press, New York, 1982. With permission.)

(depth of focus) and due to the fact that exposure time and development conditions are not as stringent. Reproducibility is high as results are independent of substrate type, surface reflections, and wafer topography.* Count as another important benefit the fact that the lithography is immune to low atomic number (Z) particle contamination (dust). With an X-ray wavelength on the order of 10 Å or less, diffraction effects generally are negligible and proximity masking can be used, increasing the lifetime of the mask. With a standard 50-μm proximity gap and using synchrotron X-rays one can print 0.25-μm patterns; by decreasing the proximity gap to 25-μm, patterns of 0.15 μm can be resolved.[59] The obtainable aspect ratio, defined as the

structural height or depth to the minimum lateral dimension, reaches more than 100. (With UV photolithography, under special conditions, an aspect ratio of about ten is possible.) An aspect ratio of 100 corresponds to the aspect ratio attainable by wet-chemical anisotropic etching of monocrystalline Si (see Chapter 4 on Wet Bulk Micromachining).

LIGA exploits all of the above advantages of X-ray lithography (see Chapter 6 and References 60 and 61, for example). The process is schematically illustrated in Figure 6.1. It involves a thick layer of resist (from micrometers to centimeters), high-energy X-ray radiation, and development to make a resist mold. By applying galvanizing techniques the mold is filled with a metal. The resist structure is removed and metal products result. Alternatively, the metal part can serve as a mold itself for precision plastic injection molding. Several types of plastic molding processes have been tested, including reaction injection molding, thermoplastic injection molding, and hot embossing. The so-formed plastic part, just as the original resist structure, can

---

\*    A word of caution about the substrate independence of X-ray lithography: as the energy of the X-rays increases, secondary electron emission from the substrate might cause scattering and the same problem limiting electron lithography also becomes a limiting factor.

serve as a mold again for fast and cheap production since one does not rely on a new X-ray exposure. LIGA enables new building materials and a wider dynamic range of dimensions and possible shapes. A show-piece structure for the LIGA technology is pictured in Figure 1.28* in which an ant holds a Ni gear. As many people do, the ant wonders what to do next with that gear.

### X-Ray Resists

An X-ray resist should have high sensitivity to X-rays; high resolution and resistance to chemical, ion, and/or plasma etching; thermal stability of >140°C; and a matrix or resin absorption of less than 0.35 $\mu m^{-1}$ at the wavelength of interest. No present resist meets all those requirements. One material predominantly used in X-ray lithography is poly (methylmethacrylate) or PMMA, a material better known by its tradenames Plexiglas™ or Lucite™. Another name commonly used for polymers based on polymethyl methacrylate is acrylics. Clear sheets of the material are used, for example to fabricate 'unbreakable' windows, inexpensive lenses, machine guards, clear lacquers on decorative parts, etc. At a wavelength of 8.34 Å, the lithographic sensitivity of PMMA typically hovers about 2 J/cm²; a rather low sensitivity implying a small throughput. A possible approach to make PMMA more X-ray sensitive is the incorporation of X-ray absorbing high-atomic-number atoms. Another approach, discussed earlier, involves the use of chemically amplified photoresists. More recent X-ray resists explored for LIGA applications are poly(lactides). These resists show a considerably enhanced sensitivity and reduced stress corrosion compared to PMMA.[62]

Negative X-ray resists exhibit inherently higher sensitivities compared to positive X-ray resists, although their resolution capability is limited by swelling. Poly(glycidyl methacrylate-co-ethyl acrylate) (PGMA), a negative e-beam resist, has been used in X-ray lithography. In general, resists materials sensitive to e-beam exposure are also sensitive to X-rays and function in the same way; that is, materials that are positive in tone for electron beam radiation typically also are positive in tone for X-ray radiation. A strong correlation exists between the resist sensitivities observed with these two radiation sources, suggesting that the reaction mechanisms might be similar for both types of irradiation (see O'Brien et al. in Reference 12).

The IC industry requires a typical resist layer not more than 1 to 2 μm thick. Thicker layers, between 10 and 1000 μm, are dictated by the need for high aspect ratio micromachines. The technology of applying thicker layers of photoresist still remains in its infancy. Spin coating of multiple resist layers (for relatively thin coats), resist casting with *in situ* polymerization of mildly cross-linked PMMA (for layers above 500 μm) and plasma polymerization of PMMA with a possibility of subsequent diamond grinding of the resulting layers are some of the techniques currently being explored (see Chapter 6 for more details).

In the case of exposures with very high energy X-rays (hard X-rays), the associated wavelength is now measured in angstroms and not in nanometers. At those energies, almost every type of polymer becomes a 'resist' and even 'resistless' lithography becomes possible as thin films can be etched, vaporized, or ion-implanted directly.

### X-Ray Masks

Another challenge in X-ray lithography, besides the low sensitivity of the resists and the high cost of sufficiently bright X-ray sources, is the mask making; already complex for producing DRAMs, but even more complex for 3D structures with high aspect ratios. In Table 1.5, the procedure for making an optical mask is compared to the procedure involved in making an X-ray mask. Technologies developed for manufacturing masks for submicron circuitry with X-ray lithography do not directly transfer to the fabrication of X-ray masks for building micromachines. An X-ray mask basically consists of a pattern of X-ray-absorbing material (a material with a high atomic number, Z, such as gold) on a membrane substrate transparent to X-rays (a low Z material, e.g., Ti, Si, SiC, $Si_3N_4$, BN, Be). The membrane must in certain areas have windows or other features for alignment purposes. Yet the material also needs to be thermomechanically stable to a few parts in $10^6$.[61] Mechanical stress in the absorber pattern can cause in-plane distortion of the supporting thin membrane, requiring a high Young's modulus material. Also, humidity or the X-ray exposure itself might distort the membrane. A fabrication sequence for an X-ray mask is shown in Figure 1.29.[23] In this case the X-ray absorber pattern is ion-beam etched in an Au film after the pattern has been first written in a 300-Å PMMA layer with an e-beam. The absorber film typically consists of two metal layers: a thin layer of chromium for adhesion to the substrate, topped by a thicker layer of gold. The higher the required aspect ratio of the exposed resist, the thicker the gold layer of the mask absorber pattern must be in order to maintain a good contrast. The X-ray mask shown in Figure 1.29 only has a 400-Å-thick gold layer, which adequately covers DRAM manufacture but not high aspect ratio 3D structures where 5 to 15 μm of Au absorber might be required for a 500-μm-thick resist

TABLE 1.5 Optical vs. X-Ray Mask

| Optical Mask | X-Ray Mask |
|---|---|
| Mask design: CAD | Mask design: CAD |
| Substrate preparation | Substrate preparation |
|     Quartz |     Thin membrane substrate (Si, Be, Ti, …) |
|     Thin metal film deposition |     Deposit plating base (50 Å Cr, then 300 Å Au) |
| Pattern delineation | Pattern delineation |
|     Coat substrate with resist |     Coat with resist |
|     Expose pattern (optical, e-beam) |     Expose pattern (optical, e-beam) |
|     Develop pattern etch Cr layer |     Develop pattern |
|     Strip resist | Absorber definition: |
| |     Electroplate Au (~15 μm for hard X-rays) |
| |     Strip resist |
| Cost: $1K–$3K | Cost: $4K–$12K |
| Duration: 3 days | Duration: 10 days |

---

* Figure 1.28 appears as a color plate after page 144.

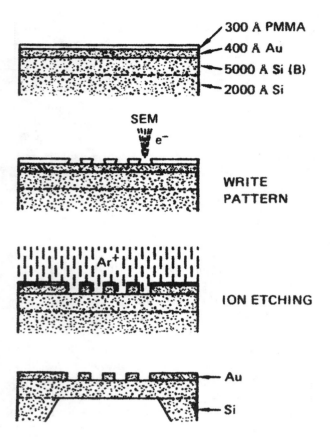

**Figure 1.29** Fabrication of a silicon membrane-based X-ray mask with a gold absorber pattern. For use in high aspect ratio micromachining, the gold absorber layer must be between 5 and 15 μm. (From Brodie, I. and J. J. Muray, *The Physics of Microfabrication*, Plenum Press, New York, 1982. With permission.)

and up to 50 μm for 10-cm-thick resists. A proximity X-ray mask as shown in Figure 1.29 is placed close to the substrate and its pattern is reproduced by exposure to X-rays.

Various schemes to make LIGA X-ray masks are explored in Chapter 6.

### Why Use a Synchrotron to Generate X-Rays?

The full power of X-ray lithography for micromachining only materializes when using hard, collimated *synchrotron radiation.* To appreciate this we will explore the procedure using a less intense, less collimated beam from an electron-beam bombardment source. A schematic diagram of an X-ray exposure system using an electron-beam bombardment-based X-ray source is shown in Figure 1.30.[23] The mask typically is offset above the wafer by about 10 μm, after alignment. A proximity scheme rather than a contact mask is a useful feature, given the fact that an X-ray mask can cost up to $13,000. Since the X-ray source is finite in size and separated by a distance, D, the edge of the mask does not cast a sharp shadow but rather has a region associated with it known as penumbral blur, δ. Image blurring limits the ultimate resolution power of an X-ray exposure system, as shown in Figure 1.30. As diffraction effects can be ignored, simple geometric considerations can be used for

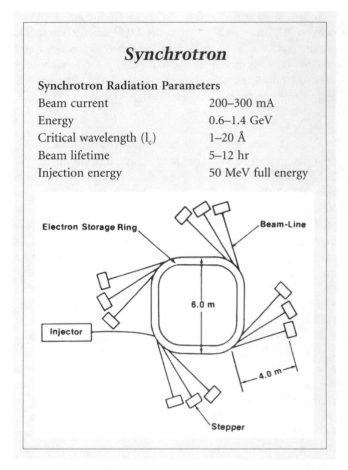

relating the image to the pattern on the mask. From Figure 1.30 we estimate that the blurring, δ, at the resist plane is given by:

$$\delta = s\left(\frac{d}{D}\right) \qquad 1.36$$

where s represents the mask-to-wafer gap, d the source diameter, and D the source-to-substrate distance. In a high-resolution system, δ should be controlled to within 0.1 μm. The spacing, s, should allow the accommodation of large-diameter masks without high risk of contacting the resist and greatly increasing the occurrence of defects. The X-ray source must be sufficiently collimated in order to reduce this so-called penumbral blurring. In practice, this translates into a small source diameter (e.g., a few millimeters) and a large source-to-mask distance. Conventional e-beam-generated X-ray sources have sizes of a few millimeters and are about 40 cm away from the mask. Unfortunately, a large distance required for adequate collimation results in prohibitively long exposure times (e.g., hours) due to the weak intensity of these sources. With synchrotron radiation, on the other hand, penumbral blurring does not limit the spatial resolution. Because of the high collimation of synchrotron radiation, rather large distances between the mask and the wafer can be tolerated (about 1 mm for 1-μm line-width patterns). In the electron storage ring or synchrotron, a magnetic field constrains electrons to follow a circular orbit and the radial acceleration of the electrons causes electromagnetic radiation to be emitted in the forward direction. The radiation is thus strongly collimated in the forward direction and can be

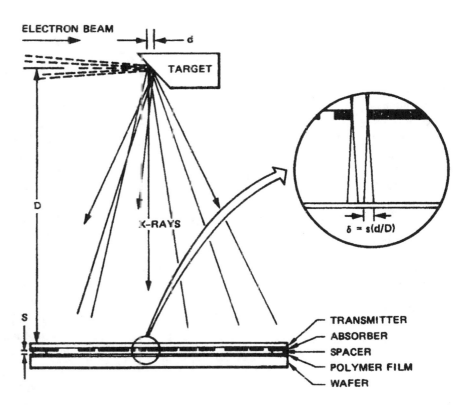

**ELECTRON BEAM**

**TARGET**

**X-RAYS**

TRANSMITTER
ABSORBER
SPACER
POLYMER FILM
WAFER

$\delta = s(d/D)$

**Figure 1.30** X-ray lithography with an electron-beam X-ray source. Inset: extent of penumbral effect calculated from geometric considerations. (From Brodie, I. and J. J. Muray, *The Physics of Microfabrication*, Plenum Press, New York, 1982. With permission.)

assumed to be parallel for lithographic applications. Moreover, because of the much higher flux of usable collimated X-rays, shorter exposure times become possible. The Advanced Light Source (ALS) synchrotron in Berkeley, for example, can deliver a flux of 0.4 W/cm$^2$ at 30 m for 3 to 9 keV radiation. Especially if one wants to generate a highly collimated photon flux in the spectral range required for precise deep-etch X-ray lithography in thick resist layers, synchrotron radiation comes close to being the ideal source because of its intensity, tunability, small source size, and small divergence.[63]

As mentioned before, the IC industry favors improved photolithography over X-rays. The prohibitive cost of introducing a new industrial lithography remains a strong deterrent, pushing existing photolithography to its absolute resolution limit. For example, by using UV phase-shifting mask lithography, planar IC features approaching X-ray lithography resolution have been made. Even in the micromachining field there are continued attempts to squeeze more out of classical photolithography. Using techniques such as modified near UV lithography and cryogenic dry etching, LIGA-like, high aspect ratio features have been produced, hemming in the potential for X-ray lithography even for building micromachines.

## Charged Particle Beam Lithography

### Introduction

Following is a brief introduction to lithographies based on charged particle beams. The mask fabrication process is significantly simpler in the case of narrow beam lithography compared to the flood exposure-based photo- and X-ray lithographies: the computer-stored pattern is directly converted to address the writing particle beam, enabling the pattern to be exposed sequentially, point by point, over the whole wafer. In

other words, the mask is really a 'software mask'. The two manufacturing modes, charged particle vs. noncharged particle, were illustrated in Figure 1.27. Electron-beam (e-beam) and ion-beam (i-beam) lithographies involve high current density in narrow electron or ion beams. The smaller the beam size, the better the resolution, but the more time is spent writing the pattern. This sequential (scanning) type system exposes one pattern element or pixel at a time. Within that area the charged particle beam can deliver some maximum current (i) which is limited primarily by the source brightness and column design. The experimental set-up imposes a limit to the speed at which the writing beam can be moved and modulated, resulting in a 'flash' time in seconds (t). The maximum dose (in Coulombs per cm$^2$) deliverable by a particular beam is given by:

$$D_{max} = \frac{it}{a} \qquad 1.37$$

with 'a' the pixel area in cm$^2$. It will then be necessary to work with resists that react sufficiently fast at $D_{max}$ to produce a lithographically useful, three-dimensional image (latent or direct image). The e-beam method displays a large depth of focus, as active focusing over various topographies is possible.

In principle, flood exposure of a mask in a projection system (i.e., parallel exposure of all pattern elements at the same time) is possible with ion and electron beams as well. The exposure masks are fabricated from heavy metals on semitransparent organic or inorganic membranes. The high cost of mask fabrication, the instability of the mask due to heating, and the difficult implementation of flood exposure systems have postponed the commercial acceptance of high energy exposure systems. Moreover, with ion and electron beams, flood exposure is limited to chip-size fields due to difficulties in obtaining broad, collimated, charged-particle beams. From the high energy sources, only X-ray flood exposure, applicable over a large wafer area, approaches the commercial application stage. The most prevalent use of charged particle beams remains the scanning mode.

The continued development of better charged particle beam sources keeps on widening the many possibilities for nanoscale engineering through lithography, etching, depositing, analyzing, and modifying of a wide range of materials, well beyond any capability of classical photolithography. Table 1.6 lists some ion-beam and electron-beam applications.

**TABLE 1.6**  Electron- and Ion-Beam Applications

| Electron-Beam Applications | Ion-Beam Applications |
| --- | --- |
| Nanoscale lithography | Micromachining and ion milling |
| Low-voltage scanning electron microscopy | Microdeposition of metals |
| Critical dimension measurements | Maskless ion implantation |
| Electron-beam-induced metal deposition | Microstructure failure analysis |
| Reflection high-energy electron diffraction (RHEED) | Secondary ion mass spectroscopy |
| Scanning auger microscopy | |

## Electron-Beam Lithography

### Overview

The e-beam lithography method, like X-ray lithography, does not limit the obtainable feature resolution by diffraction, because the quantum mechanical wavelengths of high energy electrons are exceedingly small. E-beam lithography exhibits some other attractive attributes such as precise control of the energy and dose delivered to a resist-coated wafer, deflection and modulation of electron beams with speed and precision by electrostatic or magnetic fields, and imaging of electrons to form a small point of <100 Å as opposed to a spot of 5000 Å for light. Two distinct ways stand out in the use of scanning electron beams for lithography, i.e., direct writing on a resist-coated substrate or using electrons to create a mask whose pattern can subsequently be transferred onto a wafer.

As major advantages of electron beam lithography we can count: the ability to register accurately over small areas of a wafer, lower defect densities, and a large depth of focus because of continuous focusing over topography. The latter advantage results from the lack of need for intermediate masks.

Some of the disadvantages of electron-beam lithography are

1. Electrons scatter quickly in solids, limiting practical resolution to dimensions greater than 10 nm (although resolutions as small as 2 nm have been obtained in a few materials). The resolution of e-beam lithography tools is not simply the spot size of the focused beam. It also is affected by scattering of the e-beam inside the resist and substrate and by backscattering from the substrate exposing the resist over a greater area than the beam spot size. Collision with the substrate not only causes random

scattering and back-scattered electron generation, but also produces secondary electrons. So-called 'proximity effects' are created by scattered electrons partially exposing the resist up to several micrometers from the point of impact. As a result, serious variations of exposure over the pattern occur when pattern geometries fall in the micrometer and submicrometer ranges. The use of a very thin support substrate (200 to 500 Å) reduces the effect of back-scattered electrons. Applying a very thin (30 to 100 Å) resist layer significantly reduces the electron scattering in the resist itself. Earlier we mentioned this can be accomplished by the Langmuir-Blodgett method. Lithography in the nanometer range can also be accomplished by the use of a thin layer of alkali halides or aliphatic amino acids.[23] Reducing the electron energy voltage to below 5 keV also reduces scattering effects (see also below about low energy electrons from a micro-scanning tunneling microscope developed at Cornell).

2. Electrons, being charged particles, also need to be held in a vacuum, making the apparatus more complex than for photolithography.

3. Other major difficulties with scanning electron-beam lithography include the relatively slow exposure speed (an electron beam must be scanned across the entire wafer) and high system cost. As a result the use of electron-beam lithography has been limited to mask making and direct writing on wafers for specialized applications, e.g., small batches of custom ICs. (As a research solution several groups have been "Rube Goldberging" their standard scanning electron microscope (SEM) to create customized electron-beam writing systems.)

As mentioned earlier an e-beam can also be used as an alternative way to build microstructures directly without the use of a mask, i.e., in a direct write type microfabrication method. The writing can be additive or subtractive. As an example of additive e-beam writing, e-beam-induced metal deposition from a metal organic gas (e.g., W deposition from $W(CO)_6$) has been used for the formation of microstructures of various geometries. These devices are made one by one rather than in a large batch. Usually, this type of slow, expensive fabrication technique prohibits commercial acceptance. But some microstructures, especially intricate microsystems, might be worth the bigger price tag. In that case, serial microfabrication techniques may not necessarily be as prohibitive as they would be in the case of ICs. In Chapter 7 on Comparison of Micromachining Techniques, direct write microfabrication methods are reviewed in more detail.

### Electron-Beam Resists

Numerous commercial e-beam resists are produced for mask making and direct write applications. Bombardment of polymers by electrons causes bond breakage. In principle, any material can function as a resist. However, the important considerations include sensitivity, tone, resolution, and etching resistance. PMMA exemplifies an inexpensive positive e-beam resist with a high resolution capability and a moderate glass transition temperature $T_g$ (114°C). We already know that the same material acts as an X-ray resist as well. This is not coincidental as there is a strong relation between X-ray and e-beam sensitivity. A co-polymer of glycidyl methacrylate and ethyl acrylate (COP) is a frequently used negative resist in mask manufacture. This material, although exhibiting good thermal stability, has poor plasma-etching resistance. The measured G(x) value of representative polymers in the COP family is about 10.

### Electron Emission Sources

**Electron emission (Field emission, thermionic, and photoemission)** underwrites the principle for electron source construction.[64] Schottky emission (SE) and cold field emission (CFE) have been in common use, especially for nanometer-size beams for electron focusing systems. Emission of electrons from a metal under the influence of a field occurs in both CFE and SE. During SE emission, a blunt tungsten emitter tip coated with a low work function material (ZrO) is heated to 1800K and thermionic emission takes place, i.e., heat thermally excites the electrons enough to bring them out of the material. In other words, the field emission is helped along by thermal excitation of the electrons. During cold field emission a much smaller Tungsten wire (radius of <0.1 μm) is used, and a very high field causes electrons to tunnel out of the material. In CFE sources, electrons tunnel from various energies below the Fermi level. With SE cathodes, thermally excited electrons (nontunneling electrons) escape over a field-lowered potential energy barrier. Both SE and CFE sources display similar energy spreads, but their *energy distributions* are mirror images.[64]

Attaining high-current levels in a submicron electron beam at low voltages (500 eV to 1 keV) is of interest for e-beam lithography, as well as in scanning electron microscopy (SEM). When sensitive biological samples or electron beam sensitive resists are involved, SEM pictures must be made at voltages below 1 keV. At these low voltages, high currents are required to attain the needed detail and to minimize edge effects. Traditional SEM cathodes using tungsten hairpin filaments are very limited at low voltages; they cannot supply enough current in submicron beams at the low-voltage end. The newer high brightness CFE and SE sources including LaB₆ and

## *Field and thermionic emission and photoemission*

(From Lindquist et al., *Research and Development*, June, 91–98, 1990. With permission.)

### Electron emission in a water bucket

THE THREE MECHANISMS used by field emission sources all basically involve emitting electrons and ions from a metal surface under the influence of a strong electric field.

Understanding these mechanisms is where the water bucket comes in.

In this analogy, the water level in a bucket represents the Fermi level— the highest occupied energy level in a cathode material. The work function is the energy required to get the water droplets (electrons) from the top of the liquid out of the bucket. This is the distance equivalent to the potential energy barrier.

In photoemission, photon energy excites electrons at the Fermi level of the cathode material and can impart enough kinetic energy to allow the electrons to escape from the bucket.

In thermionic emission, heat thermally excites the electrons, providing enough energy to boil the electrons off and out of the bucket.

In field emission a high electric field can thin the side of the bucket enough so that the electrons can tunnel through it.

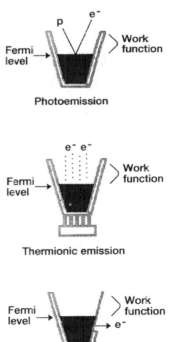

$CeB_6$, can do the job (see inset below). SEM instruments with beam diameters of 1 nm have been made using such cathodes. For further reading about e-beam lithography, refer to Reference 65.

### Ion-Beam Lithography

Ion beam lithography consists in part of work in flood exposure with $H^+$, $He^{2+}$, or $Ar^+$ through a mask of Au on a silicon membrane, but mainly consists of point-by-point exposures with a scanning source of liquid gallium metal. For ion-beam construction, liquid metal ion (LMI) sources are becoming the choice for producing high-current-density submicrometer ion beams. With an LMI source, liquid metal (typically gallium) migrates along a needle substrate. A jet-like protrusion of liquid metal forms at the source tip under influence of an electrical field. The gallium-gallium bonds are broken under the influence of the extraction field and are uniformly ionized without droplet or cluster formation. LMI sources hold extremely high brightness levels ($10^6$ A/cm$^2$ sr) and a very small energy spread, making them ideal for producing high-current-density submicrometer ion beams. Beam diameters of less than 50 nm and current densities up to 8 A/cm$^2$ are the norm. Besides Ga, other pure element sources are available such as indium and gold. By adopting alloy sources, the list expands to dopant materials such as boron, arsenic, phosphorus, silicon, and beryllium.

As in the case of e-beam systems, ion-beam lithography offers direct write and mask fabrication opportunities. Compared to photons (X-rays and light) or electrons, ions chemically react with the substrate, allowing a greater variety of surface modifications such as patterned doping. The resolution of ion-beam lithography is better than for electrons because the secondary electrons produced by an ion-beam are of lower energy and have a short diffusion range so that hardly any back-scattering occurs. The ion-beam spot size has the smallest possible size, smaller than UV, X-ray, or electron-beam spots. The smallest focused ion beam (FIB) spot currently reached is about 8 nm, accomplished by using a two-lens microprobe system and a single-isotope

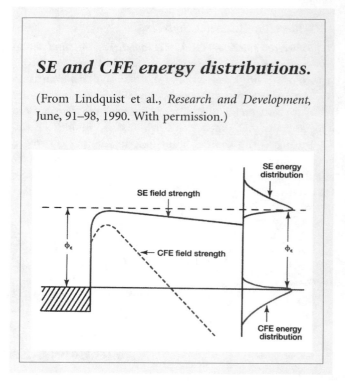

## SE and CFE energy distributions.

(From Lindquist et al., *Research and Development*, June, 91–98, 1990. With permission.)

## Comparison of electron sources

(From Lindquist et al., *Research and Development*, June, 91–98, 1990. With permission.)

| Parameter | Field Emission (Shottky ZrO/W) | Field Emission (Cold) | LaB$_6$ | CeB$_6$ | Tungsten Filament |
|---|---|---|---|---|---|
| Brightness (A/cm$^2$ sr) | $5\cdot10^8$ | $10^9$ | $10^7$ | $10^7$ | $10^6$ |
| Emitting surface area (μm$^2$) | >0.3 | 0.03 | $\gg$1 | $\gg$1 | $\gg$1 |
| Typical service life (hr) | 5000 | 2000 | 1000 | 1500+ | 100 |
| Operating vacuum | $10^{-8}$ | $10^{-10}$ | $10^{-7}$ | $10^{-7}$ | $10^{-5}$ |
| Energy spread (eV) | 0.3–1.0 | 0.2–0.3 | 1.0 | 1.0 | 1.0 |
| Evaporation rate (g/cm$^2$sec) | | | $2.9\cdot10^{-9}$ | $2.1\cdot10^{-9}$ | NA |
| Work function (eV) | | | 2.6 | 2.4 | 4.5 |
| Short-term beam stability (% RMS) | <1 | 4 to 6 | <1 | <1 | <1 |
| Standard cost | | | $775 | $850 | $20 |

gallium ion source. With this set-up arrays of dots were produced in a 60-nm thick PMMA layer with dot dimensions ranging from 10 to 20 nm.[66] Ion-beam lithography experiences the same drawbacks as an electron-beam system, as it requires a serially scanned beam and a vacuum. Research suggests that parallel projection and multiple-beam lithographic equipment may eventually overcome the limitations of serially scanned ion and electron-beam lithography.[23]

Focused ion beams can be used to perform maskless implantation and metal patterning with submicrometer dimensions. Focused ion beam also has been applied to milling in IC repair, maskless implantation, circuit fault isolation, and failure analysis (see Table 1.6). Some micromachining applications of the ion-beam technology will be reviewed in Chapter 2 (Pattern Transfer with Dry Etching) and Chapter 7 (Comparison of Micromachining Techniques). As a machining tool, FIB is very slow. Except for research, it may take a long time for FIB to become an accepted 'micromachining tool'. For additional reading on ion beam lithography in general, refer to Selinger[67] and Brodie,[23] for more specific reading on focused ion-beam-induced deposition.

# Emerging Lithography Technologies

## Proximal Probe Lithography

### Background

The scanning tunneling microscope (STM) can image the surface of conducting materials with atomic-scale detail. STM was invented by Gerd Binnig and Heinrich Rohrer of IBM's Zurich Laboratory in 1985 (they received the 1986 Nobel Prize for their discovery).[69] In general, STM works by bringing a small conducting probe tip up to a conducting surface. When the probe comes very close to the surface (less than 10 Å), very small amounts of currents are produced, because the electrons in the probe and the surface have wave functions extending beyond the physical contact boundaries. To the extent that these spill-over wave functions overlap, a measurable current results. The interesting part about this current is that it depends exponentially on the spacing between the two conductors (as well as the voltage). The z-axis distance variation between tip and sample is accomplished by a piezoelectric transducer. By changing the distance over 1 Å, the current changes by a factor of ten. In practice, the current is kept constant through a feed-back mechanism and the probe moves up and down over the surface following the atomic contours it 'sees'. The images produced by the STM come from the electronic structure as well as from the geometry of the sample. Up to 100 times more powerful than scanning electron microscopes (SEM), scanning tunneling microscopes measure objects in the angstrom range. STM now belongs to a large new family of very local, proximal probes, such as atomic force microscopes (AFM), scanning thermal microscopes, scanning capacitance microscopes, magnetic force microscopes, etc., enabling microscopy of almost any type of material. The common feature of these instruments is that their resolution

is not determined by the wavelength used for the interaction with the probed object, as in conventional microscopy.[70]

## Nanolithography and Nanomachining

STM equipment locally modifies surfaces[71] and thus can be used to perform nanolithography and direct 'nanomachining' (see inset on page 47 **Silicon surface engraved by an STM**). The electric field strength in the vicinity of a probe tip is very strong and inhomogeneous (say, a field of 2 V Å$^{-1}$ concentrated around the probe tip). This field can manipulate atoms, including sliding of atoms over surfaces and transferring atoms by pick (erase) and place (write). Drexel calls this type of machining mechano-synthesis.[72] In Figure 1.31 (top), four Pt atoms have been herded together in a linear array.[71] Since one can manipulate individual atoms with these techniques, the theoretical resolution of a lithography technique based on these atomic probes is a single atom. In practice, lines of 100 Å in width have been written using STM. For reference: a single memory bit can be stored in an area that measures 100 Å on a side. This enables bit storage of $10^{12}$ bits/cm$^2$ as compared to $10^9$ bits/cm$^2$ with conventional technology.[61]

One major negative to bear in mind with atom placing or removing techniques for micromachining is the time involved in generating even the simplest of features. If one wanted to deposit a metal line 10 μm long, 1 μm wide, and 0.5 μm high, $10^{16}$ atoms would need to be manipulated. Even at a deposition rate of $10^9$ atoms per second, this would take more than 100 days. Along this line, in the late 1980s, it took an IBM team a week to spell out the IBM logo with an STM, putting Xe atoms in place on a Ni surface. In 1993, Kamerzki et al.[73] succeeded in shortening the time considerably. Using an *atomic processing microscope (APM)* they imaged selected atoms on a surface, stripped them off, and replaced them with other atoms all in a matter of minutes. The APM relies on several technical innovations such as photon biasing whereby a precise voltage

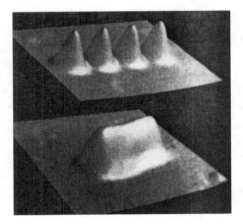

**Figure 1.31** Atom manipulation for nanomatching. (Top) A 60 × 48-Å STM image of four Pt adatoms assembled into a linear array on a Pt(111) surface. Pt atoms were herded four unit cells apart along a close-packed direction of the Pt(111) surface. (Bottom) A 40 × 40-Å STM image of a compact array of seven Pt adatoms. (From Stroscio, J. A. and D. M. Eigler, *Science,* 254, 1319–1326, 1991. With permission.) For more pictures of STM atom manipulation, visit http://www.almaden.ibm.com:80/vis/stm/gallery.html.

and photon pulse is 'tuned' to a specific atom to strip it from the surface. In both cases, though, the number of atoms manipulated is minuscule, and depositing or removing clusters of atoms and parallel processing will be essential for these approaches to become viable. Nature, working with similarly small building blocks (amino acids and proteins) to circumvent the time problem, has a lot of redundancy and parallel processing built in.

Lithography with an STM on a typical photoresist surface is more likely to succeed sooner than direct atom manipulation. The Naval Research Laboratory worked with STM on chemically amplified negative e-beam resist (SAL-601 from Shipley) but also carried out direct surface modification in which thin oxides were induced on Si and GaAs surfaces by slightly increased tip voltages.[74] Resist films from 30 to 70 nm were patterned with typical tip-sample voltages from −15 to −35 V, resulting in sub-25-nm resolution. With the thin oxide approach, 50-nm line-width patterns were achieved. The resulting patterns in each case were sufficiently robust to act as a mask for a reactive ion etching of the substrate with boron trichloride. Speeds close to 1 μm/sec for this type of lithography require major improvement. By using piezoslabs with resonant frequencies in the MHz range, STMs capable of scanning at video rates have been built. But even with tip scan rates of 1 mm/sec writing in a fast resist, it is obvious that the niche for this technology belongs in the nanolithography domain (i.e., <100 nm).[75]

It is conceivable that micromachining will provide a solution to make 'nanolithography' a cost-effective and adequately fast lithography proposition in the future. At Cornell's National Nanofabrication Facility (NNF), for example, a research group has been working on arrays of microfabricated, miniaturized electron lithography systems based on scanning tunneling microscopes. In this STM aligned field emission (SAFE) system, the physical dimension of the electron beam column (length and diameter) are in the orders of millimeters. A field emission tip is mounted onto an STM; the STM feedback principle is used for precision x, y, and z piezoelectric alignment of the tip to a miniaturized electron lens to form a focused probe of electrons (see Figure 1.32).[76] Since many electron-optic aberrations scale with size, microfabrication techniques enable lenses with negligible aberrations, resulting in exceptionally high brightness and resolution. The STM controls also allow the stability of the emission to be controlled by automatically adjusting the z-position through a piezo-element. An array of these microcolumns, each with a field emission tip as the source with individual STM sensors and controls, can generate patterns in parallel, one or more columns per chip. The low voltage of operation of these tips might obviate the need for proximity effect corrections, as low voltage operation proved to eliminate proximity effects.[75,76] For example, low energy electrons (15 to 50 eV) from an STM have been used to write patterns with 23-nm features size,[74] more than four times smaller than can be written on the same substrate and in the same resist with a tightly focused 50-kV e-beam. Attempts to make 10-nm e-beams in the low-kV range are underway as well. An STM maintains the feature size regardless of pattern geometry, indicating the absence of proximity effects. Using a similar approach to the Cornell team, Wada et al.[77] propose an array of 1000 micromachined STMs on a nanolithographic subsystem. About 25 of these subsystems would form a 50 wafers/hour nano-lithography system.

Today the above described microprobe techniques mainly act as fundamental research tools, but it can be envisioned how they might revolutionize the way we think about micromachining in the coming years. Already many studies report how to scratch, melt, erode, indent, and otherwise modify surfaces on a nanometer scale and IBM is already exploring proximal probes for memory storage.

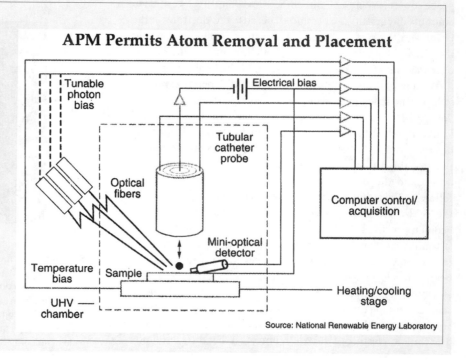

### Atomic processing microscope

Schematic shows tunable pulsed lasers for photon biasing; catheter probe for imaging, processing, and analysis; and computer for device control and data acquisition. (From Editorial in *Research and Development*, April, 17, 1993. With permission.)

**APM Permits Atom Removal and Placement**

Tunable photon bias

Electrical bias

Tubular catheter probe

Optical fibers

Computer control/ acquisition

Mini-optical detector

Temperature bias

Sample

Heating/cooling stage

UHV chamber

Source: National Renewable Energy Laboratory

## Holographic Lithography

Some other lithography technologies an aspiring micromachinist should know about are holographic and stereo lithography. In holographic lithography[78,79] (Figure 1.33), a holographically constructed photomask replaces the standard photomask. In the holographic recording or construction phase of the hologram, one uses the interference of two mutually coherent beams. A well collimated, flood laser (object beam) passes through the photomask and is diffracted by the mask features. This signal beam contains the amplitude and phase information of the photomask. When this diffracted object beam passes through the holographic recording layer, it interacts with another beam n1. the reference beam to create the interference pattern. The reference beam converts the amplitude and phase information into intensity information which is stored in the photosensitive holographic medium. The light-sensitive recording layer stores the holographic image data as variations in the refractive index of the photopolymer. Pattern printing or image reconstruction is accomplished by scanning a collimated laser illumination beam — the phase conjugate of the reference beam — to create the hologram. By interaction with the recorded hologram, the latter beam generates an image of the original photomask at precisely its original position in space. When a photoresist-coated substrate resides in this plane, a copy of the original mask can be printed. The image is only diffraction limited, and because the holographic mask and the wafer can lie very close, a high NA and thus a very good resolution are possible (0.3 $\mu$m has been reported). Only the size of the photomask itself restricts the image field size. The holographic image of the original mask exactly overlays the wafer surface with a full field.[69,70]

## Stereolithography/Micro-Photoforming Process

In stereolithography light exposure solidifies a special liquid resin into a desired 3D shape. Besides producing industrial 3D mock-ups, micromachinists are exploring the same technology

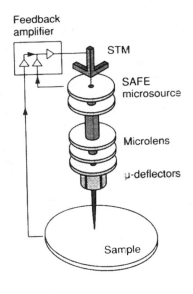

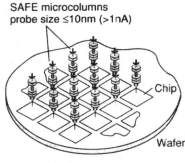

- Maskless
- Sub-100nm lithography
- ≥1 col/chip

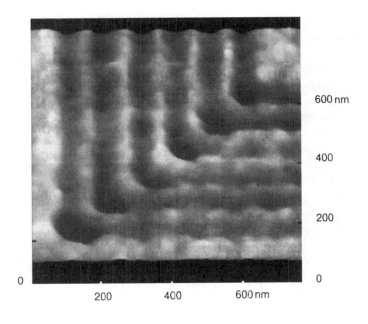

**Figure 1.32** A microcolumn based on STM aligned field emission (SAFE) and arrayed microcolumn lithography. (From an Editorial in *Solid State Technol.*, 25–26, 1993. With permission.)

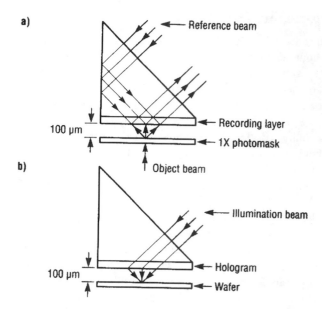

**Figure 1.33** Basic arrangement for total internal reflection holographic lithography. (a) The photomask pattern forms a hologram in the polymer recording layer. (b) Using the illumination beam, the high resolution holographic mask image is reconstructed into a printable masking layer in the resist coated substrate.[78,79]

to produce micromachines. Some concepts illustrating the technology are shown in Figure 1.34. A liquid resin is kept either in the free surface mode (Figure 1.34A) or in the fixed surface mode (Figure 1.34B). The latter has a resin container with a transparent window plate for exposure. The solidification always happens at the stable window/resin interface. An elevator is pulled up over the thickness of one additional layer above the window for each new exposure (Figure 1.34C). In the case of free surface, solidification occurs at the resin/air interface and more care needs to be taken to avoid waves or a slant of the liquid surface. A 3D structure can be made of a UV polymer by exposing the polymer with a set of two-dimensional cross-sectional shapes (masks) of the final structure. These two-dimensional shapes are a set of photomasks used to subsequently expose the work (photomask method, Figure 1.34D), or the sliced shapes can be written directly from a computerized design of the cross-sectional shapes by a beam in the liquid (Figure 1.34E). The scanning method has the advantage of point by point controllability, avoiding unevenness of solidification leading to nonuniform shrinking of the works. When applying the scanning technique, a laser beam (e.g., a He-Cd laser) is used to solidify one microscopic polymer area at a time to arrive at complicated 3D shapes by stacking thin films of hardened polymer layer upon layer. Process control in this case simply is directed from a CAD system containing the 'slice data'. The laser beam is focused down to 5-μm spot size and typical machining time ranges from 30 minutes to an hour. The position accuracy for the laser beam spot is 1 μm in the z-axis, and 0.25 μm in the x and y directions. Takagi et al.[80] obtained a 8-μm resolution with their photoforming set-up, and Ikuta et al.[81,82] report a minimum solidification unit size of 5 by 5 by 3 μm and a maximum size of fabricated structures of 10 by 10 by 10 mm. No physical contact occurs between tools and works. Very complex shapes, including curved surfaces, can be made with this type of desktop microfabrication method. The objects realized this way include liquid chromatography systems and electrostatic

micro-actuators. Takagi et al.[80,83] introduced the combination of this type of plastic micromachining with more traditional Si micromachined substrates. In Figure 1.35 a schematic for a photoformed plastic clamp anchored to a Si substrate is represented. The fabricated plastic clamps measure about 2 mm long and 2 mm high and are 250 μm thick.

To further improve the lithography — 3D photoforming in this case — it is necessary to better understand the shape of a "solidified cell" which depends on both the characteristics of the beam and the resin. Summarizing, micro-photoforming offers several advantages over more classical photolithography based micromachining processes:

1. The turn-around from CAD to prototype takes only an hour or less.
2. Photoforming is an additive process accommodating virtually any shape.
3. Fully automatic process.
4. Small capital investment (less than $30,000).
5. No need for a clean room.

This direct write lithography technique might well mean an alternative to LIGA in cases where 3D shape versatility outweighs accuracy. As with the LIGA technique, the plastic shapes made by stereo lithography may be used as a cast for electroplating metals or for other materials which can be molded into the polymer structures.[73]

## Lithography on Nonplanar Substrates

Jacobsen et al.[84] used a numerically controlled e-beam for nonplanar lithography. They equipped an SEM with high-precision linear and rotary positioning stages inside the vacuum chamber and numerically controlled beam direction and stage position. Cylindrically shaped metals were patterned in this set-up.

**See Appendix B for links to the WWW on lithography.**

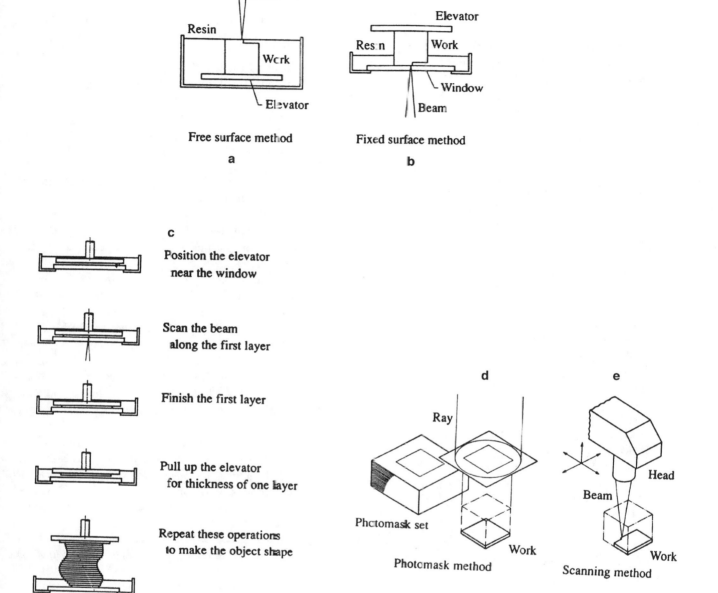

**Figure 1.34** Stereolithography or photoforming: (a) free surface method, (b) fixed surface method, (c) forming process with the fixed surface method, (d) exposure with photomask set, and (e) exposure with a scanning beam. (From Ikuta et al., References 81, 82.)

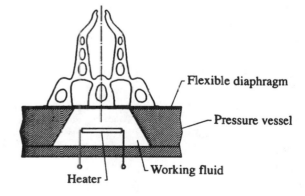

**Figure 1.35** Thermally driven microclamping tool. The clamping tool is made by photoforming, and the Si substrate by Si bulk micromachining. (From Takagi, T. and N. Nakajima, MEMS '94, Oiso, Japan, 1994, pp. 211–216. With permission.)

# References

1. Harris, T. W., *Chemical Milling*, Clarendon Press, Oxford, 1976.
2. Compton, *Compton's Interactive Encyclopedia (Interactive Multimedia). Computer Data and Program*, Compton's New Media, Carlsbad, 1994.
3. Thompson, L. F., "Resist Processing", in *Introduction to Microlithography*, Thompson, L. F., Wilson, C. G. and Bowden, M. J., Eds., American Chemical Society, Washington, DC, 1994, p. 1-17.
4. Guckel, H., T. R. Christenson, T. Earles, J. Klein, J. D. Zook, T. Ohnstein, and M. Karnowski, "Laterally Driven Electromagnetic Actuators", Technical Digest of the 1994 Solid State Sensor and Actuator Workshop, Hilton Head Island, SC, 1994, p. 49-52.
5. Metz, T. E., R. N. Savage, and H. O. Simmons, "In Situ Control of Photoresist Coating Processes", *Semicond. Int.*, 15, 68-69, 1992.
6. Moreau, W. M., *Semiconductor Lithography*, Plenum Press, New York, 1988.
7. Thompson, L. F., C. G. Willson, and M. J. Bowden, *Introduction to Microlithography*, American Chemical Society, Washington, DC, 1994.
8. Sze, S. M., *Semiconductor Devices: Physics and Technology*, John Wiley & Sons, New York, 1985.
9. Ghandhi, S. K., *VLSI Fabrication Principles*, John Wiley & Sons, New York, 1983.
10. Colclaser, R. A., *Microelectronics: Processing and Device Design*, John Wiley & Sons, New York, 1980.
11. Le Barny, P., "Chemistry of Polymer Molecules for Ultrathin Films", in *Molecular Engineering of Ultrathin Polymeric Films*, Stroeve, P. and Franses, E., Eds., Elsevier, New York, 1987, p. 99-150.
12. O'Brien, M. J. and D. S. Soane, "Resists in Microlithography", in *Microelectronics Processing*, Hess, D. W. and Jensen, K. F., Eds., American Chemical Society, Washington, DC, 1989, p. 325-376.
13. Cunningham, J. A., "The Remarkable Trend in Defect Densities and Chip Yield", *Semicond. Int.*, 15, 86-90, 1992.
14. Iscoff, R., "Wafer Cleaning: Can Dry Systems Compete?", *Semicond. Int.*, 14, 48-54, 1991.
15. Bok, E., D. Kelch, and K. S. Schumacher, "Supercritical Fluids for Single Wafer Cleaning", *Solid State Technol.*, 35, 117-119, 1992.
16. Singer, P. H., "Trends in Wafer Cleaning", *Semicond. Int.*, 15, 34-39, 1992.
17. Tolliver, D. L., *Handbook of Contamination Control in Microelectronics-Principles, Applications and Technology*, Noyes Publications, Park Ridge, 1988.
18. Wolf, S. and R. N. Tauber, *Silicon Processing for the VLSI Era*, Lattice Press, Sunset Beach, 1987.
19. Willson, C. G., "Organic Resist Materials", in *Introduction to Microlithography*, Thompson, L. F., Willson, C. G. and Bowden, M. J., Eds., American Chemical Society, Washington, DC, 1994.
20. Sze, S. M., *VLSI Technology*, McGraw-Hill, New York, 1988.
21. Bowden, M. J., "The Lithography Process: The Physics", in *Introduction to Microlithography*, Thompson, L. F., Willson, C. G. and Bowden, M. J., Eds., American Chemical Society, Washington, DC, 1994, p. 19-138.
22. Madou, M. J., *Unpublished work*, CAMD, LSU, 1994.
23. Brodie, I. and J. J. Muray, *The Physics of Microfabrication*, Plenum Press, New York, 1982.
24. Smith, H. I. and M. L. Schattenburg, "Why Bother with X-Ray Lithography ?", *SPIE*, 1671, 282-298, 1992.
25. White, R. M. and S. W. Wenzel, "Inexpensive and Accurate Two-Sided Semiconductor Wafer Alignment", *Sensors Actuators*, 13, 391-395, 1988.
26. Kim, E. S., R. S. Muller, and R. S. Hijab, "Front-to-Backside Alignment Using Resist-Patterned Etch Control and one Etching Step", *J. Microelectromech. Syst.*, 1, 95-99, 1992.
27. Tatic-Lucic, S. and Y.-C. Tai, "Novel Extra-Accurate Method for Two-Sided Alignment on Silicon Wafers", *Sensors and Actuators A*, A41-42, 573-577, 1994.
28. Bruning, J., H., "Optical Imaging for Microfabrication", *J. Vac. Sci. Technol.*, 17, 1148-1155, 1980.
29. Reichmanis, E., L. F. Thompson, O. Nalamasu, A. Blakeney, and S. Slater, "Chemically Amplified Resists for Deep-UV Lithography: A New Processing Paradigm", *Microlithogr. World*, November/December, 7-14, 1992.
30. Reichmanis, E., F. M. Houlihan, O. Nalamasu, and T. X. Neena, "Chemical Amplification Mechanisms for Microlithography", in *Polymers for Microelectronics-Resists and Dielectrics*, Thompson, L. F., Willson, C. G. and Tagawa, S., Eds., American Chemical Society, Washington, DC, 1994.
31. Lamola, A. A., C. R. Szmanda, and J. W. Thackeray, "Chemically Amplified Resists", *Solid State Technol.*, 53, 53-60, 1991.
32. Alling, E. and C. Stauffer, "Image Reversal Photoresist", *Solid State Technol.*, 31, 37-43, 1988.
33. Feely, W. E., J. C. Imhof, and C. M. Stein, "The Role of the Latent Image in a New Dual Image, Aqueous Developable, Thermally Stable Photoresist", *Polym. Eng. Sci.*, 26, 1101-1104, 1986.
34. Calvert, J. M., C. S. Dulcey, M. C. Peckerar, J. M. Schnur, J. H. J. Georger, G. S. Calabrese, and P. Sricharo-enchaikit, "New Surface Imaging Techniques for Sub-0.5 Micrometer Optical Lithography", *Solid State Technol.*, 34, 77-82, 1991.
35. Horn, M. W., "Antireflection Layers and Planarization for Microlithography", *Solid State Technol.*, November, 57-62, 1991.
36. Makino, D., "Recent Progress of The Application of Polyimides to Microelectronics", in *Polymers for Microelectronics-Resists and Dielectrics*, Thompson, L. F., Willson, C. G. and Tagawa, S., Eds., American Chemical Society, Washington, DC, 1994.

37. Frazier, A. B. and M. G. Allen, "Metallic Microstructures Fabricated Using Photosensitive Polyimide Electroplating Molds", *J. Microelectromech. Syst.*, 2, 87-94, 1993.

38. Schubert, P. J. and J. H. Nevin, "A Polyimide-Based Capacitive Humidity Sensor", *IEEE Trans. Electron Devices*, ED-32, 1220-1224, 1985.

39. Ralston, A. R. K., C. F. Klein, P. E. Thoma, and D. D. Denton, "A Model for the Relative Stability of a Series of Polyimide Capacitance Humidity Sensors", 8th International Conference on Solid-State Sensors and Actuators (Transducers '95), Stockholm, Sweden June, 1995, p. 821-824.

40. Latham, W. J. and D. W. Hawley, "Color Filters from Dyed Polyimides", *Solid State Technol.*, 31, 223-226, 1988.

41. Studt, T., "Polyimides: Hot Stuff For the '90's", *Res. & Dev.*, August, 30-31, 1992.

42. Ando, S., T. Matsuura, and S. Sasaki, "Synthesis of Perfluorinated Polyimides for Optical Applications", in *Polymers for Microelectronics-Resists and Dielectrics*, Thompson, L. F., Willson, C. G. and Tagawa, S., Eds., American Chemical Society, Washington, DC, 1994.

43. LaBianca, N. C., J. D. Gelorme, E. Cooper, E. O'Sullivan, and J. Shaw, "High Aspect Ratio Optical Resist Chemistry for MEMS Applications", JECS 188th Meeting, Chicago, IL, 1995, p. 500-501.

44. Engelmann, G., O. Ehrmann, J. Simon, and H. Reichl, "Fabrication of High Depth-to-Width Aspect Ratio Microstructures", Proceedings. IEEE Micro Electro Mechanical Systems, (MEMS '92), Travemunde, Germany, 1992, p. 93-98.

45. Calvert, J. M., W. J. Dressick, C. S. Dulcey, M. S. Chen, J. H. Georger, D. A. Stenger, T. S. Koloski, and G. S. Calabrese, "Top-Surface Imaging Using Selective Electroless Metallization of Patterned Monolayer Films", in *Polymers for Microelectronics*, Thompson, L. F., Willson, C. G. and Tagawa, S., Eds., American Chemical Society, Washington, DC, 1994.

46. Cooper, J. M., J. R. Barker, J. V. Magill, W. Monaghan, M. Robertson, C. D. W. Wilkinson, A. S. G. Curties, and G. R. Moores, "A Review of Research in Bioelectronics at Glasgow University", *Biosens. Bioelectron.*, 8, R22-R30, 1993.

47. Connolly, P., G. R. Moores, W. Monoghan, J. Shen, S. Britland, and P. Clark, "Microelectronic and Nanoelectronic Interfacing Techniques for Biological Systems", *Sensors and Actuators B*, B6, 113-121, 1992.

48. Britland, S. T., G. R. Moores, P. Clark, and P. Connolly, "Patterning and Cell Adhesion and Movement on Artificial Substrate, a Simple Method", *J. Anat.*, 170, 235-236, 1990.

49. Stroeve, P. and E. Franses, Eds. *Molecular Engineering of Ultrathin Polymeric Films*, Elsevier, London, 1987.

50. Feely, W. E., "Micro-Structures", Technical Digest of the 1988 Solid State Sensor and Actuator Workshop., Hilton Head Island, SC, 1988, p. 13-15.

51. Flamm, D. L., "Dry Plasma Resist Stripping Part I: Overview of Equipment", *Solid State Technol.*, 35, 37-39, 1992.

52. Flamm, D. L., "Dry Plasma Resist Stripping Part II: Physical Processes", *Solid State Technol.*, 35, 43-48, 1992.

53. Peters, L., "Stripping Today's Toughest Resists", *Semicond. Int.*, 15, 58-64, 1992.

54. Editorial, "DRAMs, Microprocessors to Drive Technology in the 90s", *Semicond. Int.*, 17, 1993.

55. Bate, T. R., "Nanoelectronics", *Solid State Technol.*, November, 101-108, 1989.

56. Levenson, M. D., "Phase-Shifting Mask Strategies: Isolated Dark Lines", *Microlithogr. World*, 6, 6-12, 1992.

57. op de Beeck, M., "Strategies to Extend the Limits of Optical Lithographies", Ph.D. Thesis, Leuven, Belgium, 1993.

58. Stix, G., "Toward "Point One"", *Sci. Am.*, 272, 90-95, 1995.

59. Arden, W. and K. H. Muller, "Light vs. X-rays: How Fine Can We Get?", *Semicond. Int.*, 12, 128-131, 1989.

60. Becker, E. W., W. Ehrfeld, P. Hagmann, A. Maner, and D. Munchmeyer, "Fabrication of Microstructures with High Aspect Ratios and Great Structural Heights by Synchrotron Radiation Lithography, Galvanoforming, and Plastic Molding (LIGA process)", *Microelectron. Eng.*, 4, 35-56, 1986.

61. Bley, P., W. Menz, W. Bacher, K. Feit, M. Harmening, H. Hein, J. Mohr, W. K. Schomburg, and W. Stark, "Application of the LIGA Process in Fabrication of Three-Dimensional Mechanical Microstructures", 4th International Symposium on MicroProcess Conference, Kanazawa, Japan, 1991, p. 384-389.

62. Wollersheim, O., H. Zumaque, J. Hormes, J. Langen, P. Hoessel, L. Haussling, and G. Hoffman, "Radiation Chemistry of Poly(lactides) as New Polymer Resists for the LIGA Process", *J. Micromech. Microeng.*, 4, 84-93, 1994.

63. Hunter, S., "Opportunities for High Aspect Ratio Micro-Electro-Magnetic-Mechanical Systems (HAR-MEMMS) at Lawrence Berkeley Laboratory", Report No. LBL-34767-UC-411, 1993.

64. Lindquist, J., Ratkey, D. and Fischer, P., "Do You Have The Right Beam For The Job ?", *Res. & Dev.*, June, 91-98, 1990.

65. Brewer, G., Ed. *Electron-Beam Technology in Microelectronic Fabrication*, Academic Press, 1980.

66. Editorial, "Ion Beam Focused to 8-nm Width", *Res. & Dev.*, September, 23, 1991.

67. Seliger, R. L., J. W. Ward, V. Wang, and R. L. Kubena, "A High-Intensity Scanning Ion Probe with Submicrometer Spot Size", *Appl. Phys. Lett.*, 34, 310-312, 1979.

68. Melngailis, J., "Focused Ion Beam Deposition-A Review", *Proc. SPIE - Int. Soc. Opt. Eng.*, 1465, 36-49, 1991.

69. Behm, R. J., N. Garcia, and H. Rohrer, Eds. *Scanning Tunneling Microscopy and Related Methods*, Proc. NATO ASI, Vol. 184, Kluwer, Dordrecht, 1989.

70. Stroscio, J. and W. Kaiser, Eds. *Scanning Tunneling Microscopy*, Academic Press, Boston, 1993.

71. Stroscio, J. A. and D. M. Eigler, "Atomic and Molecular Manipulation with the Scanning Tunneling Microscope", *Science*, 254, 1319-1326, 1991.

72. Drexel, K. E., *Nanosystems: Molecular Machinery, Manufacturing, and Computation*, John Wiley & Sons, New York, 1992.

73. Editorial, "Ingenious STM Puts Atoms Right Where You Want Them", *Res. & Dev.*, April, 71, 1993.

74. Marrian, R. K., A. A. Dobisz, and O. J. Glembocki, *Res. & Dev.*, February, 123-125, 1992.

75. Marrian, C. R. K. and E. A. Dobisz, "Scanning Tunneling Microscope Lithography: A Viable Lithographic Technology?", *SPIE*, 1671, 166-175, 1992.

76. Editorial, "Novel Electron-Beam Lithography System Being Explored at Cornell's NNF", *Solid State Technol.*, 36, 25-26, 1993.

77. Wada, Y., M. I. Lutwyche, and M. Ishibashi, "Micro-Machine Scanning Tunneling Microscope for Nanoscale Characterization and Fabrication", Micromachining and Microfabrication Process Technology II, Austin, Texas, USA, 1996, p. 327-331.

78. Brook, J. and R. Dandliker, "Submicrometer Holographic Photolithography", *Solid State Technol.*, 32, 91-94, 1989.

79. Omar, B., S. Clube, F. Hamidi, M. Struchen, D. and S. Gray, "Advances in Holographic Lithography", *Solid State Technol.*, September, 89-94, 1991.

80. Takagi, T. and N. Nakajima, "Photoforming Applied to Fine Machining", Proceedings. IEEE Micro Electro Mechanical Systems, (MEMS '93), Fort Lauderdale, FL, 1993, p. 173-178.

81. Ikuta, K. and K. Hirowatari, "Real Three Dimensional Micro Fabrication Using Stereo Lithography", Proceedings. IEEE Micro Electro Mechanical Systems, (MEMS '93), Fort Lauderdale, CA, 1993, p. 42-47.

82. Ikuta, K., K. Hirowatari, and T. Ogata, "Three Dimensional Integrated Fluid Systems (MIFS) Fabricated by Stereo Lithography", IEEE International Workshop on Micro Electro Mechanical Systems, MEMS '94, Oiso, Japan, 1994, p. 1-6.

83. Takagi, T. and N. Nakajima, "Architecture Combination by Micro Photoforming Process", IEEE International Workshop on Micro Electro Mechanical Systems, MEMS '94, Oiso, Japan, 1994, p. 211-216.

84. Jacobsen, S. C., D. L. Wells, C. C. Davis, and J. E. Wood, "Fabrication of Microstructures Using Non-Planar Lithography (NPL)", Proceedings. IEEE Micro Electro Mechanical Systems, (MEMS '91), Nara, Japan, 1991, p. 45-50.

# 2

# Pattern Transfer with Dry Etching Techniques

## Introduction

Lithography steps precede a number of subtractive and additive processes. Materials either are removed from or added to a device, usually in a selective manner, using thin and/or thick film manufacturing processes, transferring the lithography patterns into integrated circuits (ICs) or three-dimensional micromachines. This chapter deals with material removal by dry etching processes. Chapter 3, Pattern Transfer with Additive Techniques, focuses on additive technologies.

Table 2.1 lists the most important subtractive processes encountered in micromachining, including mask-based wet and dry etching, and maskless processes such as focused ion-beam etching (FIB), laser machining, ultrasonic drilling, electrochemical discharge machining (EDM), and traditional precision machining. Examples (mostly involving Si), typical material removal rates, relevant reference(s), and some remarks on the techniques supplement the list. One of the most important subtractive techniques for micromachining listed in Table 2.1 is etching. Etching can be described as pattern transfer by chemical/physical removal of a material from a substrate, often in a pattern defined by a protective maskant layer (e.g., a resist or an oxide). From the etching processes listed in Table 2.1, the mask-based dry etching processes including chemical, physical, and physical-chemical etching will be analyzed in the current chapter. Etching in the liquid phase is analyzed separately in Chapter 4, Wet Bulk Micromachining. Maskless material removing technologies, such as focused ion beam (FIB), precision machining, and laser and ultrasonic drilling, will be discussed in Chapter 7, Comparison of Micromachining Techniques.

When introducing resist stripping in Chapter 1, we learned that the popularity of dry stripping of resists (ashing) and the need for better control of critical dimension (CD) precipitated a major research and development effort in all types of dry etching processes based on plasmas. The preference of dry stripping over wet stripping methods is based on a variety of advantages: fewer disposal problems, less corrosion problems for metal features in the structure, and less undercutting and broadening of photoresist features, i.e., better CD control (nanometer dimensions!) and, under the right circumstances, a cleaner resulting surface. Also, with the current trend in IC manufacture leading towards sub-half-micron geometries, surface tension might preclude a wet etchant from

reaching down between photoresist features, whereas dry etching, characterized as selective stripping, precludes any problem of that nature. As in IC manufacture, dry etching evolved into an indispensable technique in micromachining.

## Dry Etching: Definitions and Jargon

Dry etching covers a family of methods by which a solid state surface is etched in the gas phase, physically by ion bombardment, chemically by a chemical reaction with a reactive species at the surface, or combined physical and chemical mechanisms. The plasma-assisted dry etching techniques are categorized according to the specific set-up as either glow discharge techniques (diode set-up) or ion-beam techniques (triode set-up). Using the glow discharge techniques, the plasma is generated in the same vacuum chamber where the substrate is located, whereas when using ion-beam techniques the plasma is generated in a separate chamber from which ions are extracted and directed towards the substrate by a number of grids. Figure 2.1 represents the relation between the various dry etching diode and triode techniques.[14] In sputter/ion etching and ion-beam milling or ion-beam etching, the etching occurs as a consequence of a physical effect, namely momentum transfer between energetic $Ar^+$ ions and the substrate surface. Some chemical reaction takes place in all the other dry etching methods. In the physical/chemical case, impacting ions, electrons, or photons either induce chemical reactions or, in sidewall-protected ion-assisted etching, a passivating layer is cleared by the particle bombardment from horizontal surfaces only. As a result, etching occurs almost exclusively on the cleared planar surfaces. In Table 2.2, the most common dry etching methods are reviewed and the associated jargon clarified. From the methods listed, plasma etching and reactive ion etching are the most widely used techniques in IC manufacture and micromachining.

In selecting a dry etching process, the desired shape of the etch profile and the selectivity of the etching process require careful consideration. In Figure 2.2, different possible etch profiles are shown. Depending on the etching mechanism, isotropic, directional, or vertical etch profiles are obtained. In dry etching anisotropic etch profiles, directional or vertical, can be generated

**TABLE 2.1**  Partial List of Subtractive Processes Important in Micromachining (For More Details See Chapter 7)

| Subtractive Technique | Applications | Typical Etch Rate | Remark | Ref. |
|---|---|---|---|---|
| Wet chem. etch. (iso. and anis.) | Iso.: Si spheres, domes, grooves Anis.: Si angled mesas, nozzles, diaphragms, cantilevers, bridges | Iso.: Si polishing at 50 μm/min with stirring (RT, acid) Anis.: etching at 1 μm/min on a (100) plane (90°C, alkaline) | Iso.: Little control, simple Anis.: With etch-stop more control, simple. | 1, Chapter 4 |
| Electrochem. etch. | Etches p-Si and stops at n-Si (in n-p junction), etches n-Si of highest doping (in n/n⁺). | p-Si etching 1.25–1.75 μm/min. on a (100) plane, 105–115°C (alkaline) | Complex, requires electrodes | 2,3, Chapter 4 |
| Wet photoetch. | Etches p-type layers in p-n junctions | Etches p-Si up to 5 μm/min (acid) | No electrodes required | 4, Chapter 4 |
| Photoelectrochem. etch. | Etches n-Si in p-n junctions, production of porous Si | Typical Si etch rate: 5 μm/min (acid) | Complex, requires electrodes and light | 5, Chapter 4 |
| **Dry chem. etch.** | Resist stripping, isotropic features | Typical Si etch rate: 0.1 μm/min (but with more recent methods up to 6 μm/min) | Resolution better than 0.1 μm, loading effects | 6, 7, Chapter 2 |
| **Phys./Chem. etch.** | Very precise pattern transfer | Typical Si etch rate: 0.1 to 1 μm/min (but with more recent methods up to 6 μm/min) | Most important of dry etching techniques | 7, 8, Chapter 2 |
| **Phys. dry etch., sputter etch., and ion milling** | Si surface cleaning, unselective thin film removal | Typical Si etch rate: 300 Å/ min | Unselective and slow, plasma damage | 8, Chapter 2 |
| Focused ion-beam (FIB) milling | Microholes, circuit repair, microstructures in arbitrary materials | Typical Si etch rate: 1 μm/min | Long fabrication time: >2 h including set-up | 9, Chapter 7 |
| Laser machining (with and without reactive gases) | Circuit repair, resistor trimming, hole drilling, labeling of Si wafers | Typical rate for drilling a hole in Si with a Nd: YAG(400W laser): 1 mm/sec (3.5 mm deep and 0.25 mm dia.) | Laser beams can focus to a 1-μm spot, etch with a resolution of 1 μm³ | 10, Chapter 7 |
| Ultrasonic drilling | Holes in quartz, silicon nitride bearing race rings | Typical removal rate of Si: 1.77 mm/min | Especially useful for hard, brittle materials | 11, Chapter 7 |
| Electrostatic discharge machining (EDM) | Drilling holes and channels in hard brittle metals | Typical removal rate for metals: 0.3 cm³/min | Poor resolution (>50 μm), only conductors, simple, wire discharge machining resolution much better | 12, Chapter 7 |
| Mechanical turning, drilling and milling, grinding, honing, lapping, polishing, and sawing | Almost all machined objects surrounding us | Removal rates of turning and milling of most metals: 1 to 50 cm³/min; for drilling: 0.001 to 0.01 cm³/min | Prevalent machining technique | 13, Chapter 7 |

*Note:* Boldface type indicates techniques reviewed in Chapter 2, RT = room temperature, chem. etch. = chemical etching, iso. = isotropic, anis. = anistropic.

in single crystalline as well as in polycrystalline and amorphous materials. The anisotropy is not a result of the anisotropy of single crystals as in the case of anisotropic wet chemical etching (Chapter 4); rather, the degree of anisotropy is controlled by the plasma conditions. Selectivity of a dry etch refers to the difference in etch rate between the mask and the film to be etched, again controllable by the plasma conditions.

At low pressures, in the $10^{-3}$- to $10^{-4}$-Torr range (see Figure 2.1), obtaining anisotropic etching is easy, but accomplishing selectivity is difficult. At higher pressures (1 Torr) in plasma etching, chemical effects dominate, leading to better selectivity than in the case of ion beams, but etched features are isotropic. In the extreme case of anisotropic etching, vertical sidewalls result (no lateral undercut) with perfect retention of the CD (see Figure 2.2).

To ensure a uniform etch rate over the whole wafer and between wafers, overetching may be required. The drawing in Figure 2.3 illustrates the etch profile for a 25% overetch. Making

the initial lithography smaller compensates for the loss of dimensionality incurred by overetching.

A sharp vertical sidewall does not always agree with the desired edge profile. For example, with line-of-site deposition methods such as resistive evaporation (see Chapter 3), a tapered sidewall is easier to cover than a vertical wall. Another example where a nonvertical process has the advantage is illustrated in Figure 2.4. To take away the layer on the vertical walls shown, an anisotropic etchant would require extensive overetching, whereas an isotropic etchant will remove the material quickly. The resulting etching patterns associated with all the possible different dry etching mechanisms are illustrated in Figure 2.5.[6]

## Plasmas or Discharges

Most dry etching systems find their common base in plasmas or discharges, areas of high energy electric and magnetic fields that will rapidly dissociate any gases present to form energetic

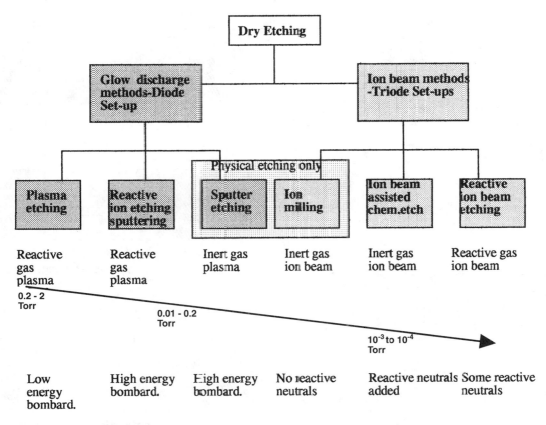

**FIGURE 2.1** Relationship between the various dry etching techniques. (Adapted from Lehmann, H. W., in *Thin Film Processes II*, Vossen, J. L. and Kern, W., Eds. Academic Press, Boston, 1991.)

**TABLE 2.2** Some Popular Dry Etching Systems[a]

|  | CAIBE | RIBE | IBE | MIE | MERIE | RIE | Barrel Etching | PE |
|---|---|---|---|---|---|---|---|---|
| Pressure (Torr) | $\sim10^{-4}$ | $\sim10^{-4}$ | $\sim10^{-4}$ | $10^{-3}–10^{-2}$ | $10^{-3}–10^{-2}$ | $10^{-3}–10^{-1}$ | $10^{-1}–10^{0}$ | $10^{-1}–10^{1}$ |
| Etch mechanism | Chem./phys. | Chem./phys. | Phys. | Phys. | Chem./phys. | Chem./phys. | Chem. | Chem. |
| Selectivity | Good | Good | Poor | Poor | Good | Good | Excellent | Good |
| Profile | Anis. or iso. | Anis. | Anis. | Anis. | Anis. | Iso. or anis. | Iso. | Iso. or anis. |

*Note:* CAIBE = Chemically assisted ion beam etching; MERIE = Magnetically enhanced reactive ion etching; MIE = Magnetically enhanced ion etching; PE = Plasma etching; RIBE = Reactive ion beam etching; RIE = Reactive ion etching.[15,16]

[a] Many of the more recent dry etching techniques are not listed here. For more information on inductively coupled plasmas (ICP), electron cyclotron plasmas (ECR), microwave multipolar plasmas (MMP), and introductory material on helicon plasma sources, helical resonators, rotating field magnetrons, hollow cathode reactors, etc., refer to References 16–22.

ions, photons, electrons, and highly reactive radicals and molecules. We will start our foray into the physics and chemistry of plasmas by looking at the simplest plasma set-up, i.e., a DC-diode glow discharge.

## Physics of DC Plasmas

The simplest plasma reactor may consist of opposed parallel plate electrodes in a chamber maintainable at low pressure, typically ranging from 0.001 to 1 Torr. The electrical potentials established in the reaction chamber filled with an inert gas such as argon at a reduced pressure determine the energy of ions and electrons striking the surfaces immersed in the discharge. Using the set-up shown in Figure 2.6A and applying 1.5 kV between the anode and cathode, separated by 15 cm, results in a 100-V/cm field. Electrical breakdown of the argon gas in this reactor will occur when electrons, accelerated in the existing field, transfer an amount of kinetic energy greater than the argon ionization potential (i.e., 15.7 eV) to the argon neutrals. Such energetic collisions generate a second free electron and a positive ion for each successful strike. Both free electrons reenergize, creating an avalanche of ions and electrons that results in a gas breakdown emitting a characteristic glow. Avalanching requires the ionization of 10 to 20 gas molecules by

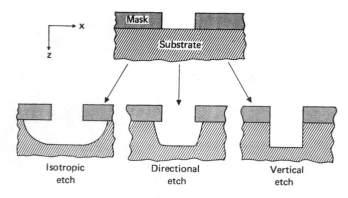

FIGURE 2.2   Directionality of etching processes.

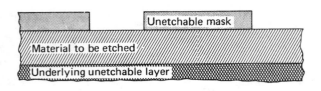

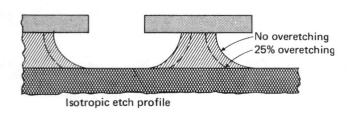

FIGURE 2.3   Example illustrating the loss of dimensionality incurred by an isotropic etch when the etch depth is of the order of the lateral dimensions involved. Making the initial lithography smaller compensates for the loss of dimensionality by overetching.

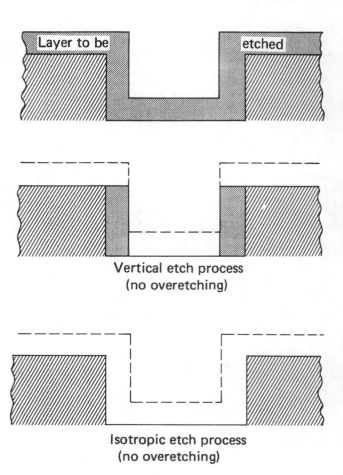

FIGURE 2.4   An example of dry etching illustrating a case where an isotropic etch is preferred to a vertical etch process. To clear the layer on the vertical walls, an isotropic etch only requires a minimal overetch while an anisotropic etch requires a substantial amount of overetching.

one secondary electron. At the start of a sustained gas breakdown, a current begins flowing and the voltage between the two electrodes drops from 1.5 kV to about 150 V. The discharge current builds up to the point where the voltage drop across a current limiting resistor is equal to the difference between the supply voltage (1.5 kV) and the electrode potential difference (150 V). To sustain a plasma, a mechanism to generate additional free electrons must exist after the plasma-generating electrons have been captured at the positively charged anode. Plasma-sustaining electrons are generated at the cathode which emits secondary electrons (Auger electrons) when struck by ions. The continuous generation of those 'new' electrons prompts a sustained current and a stable plasma glow. It is easy to understand why plate (electrode) separation and gas pressure are critical. Plates positioned too closely prevent ionizing collisions, but plates separated too far cause too many inelastic collisions of ions which lose energy. Once equilibrium is reached, the glow region of the plasma, being a good electrical conductor, hardly sustains a field and its potential is almost constant. The potential drop resides at electrode surfaces where electrical double layers are formed in so-called sheath fields, counteracting the loss of electrons from the

plasma (Figure 2.6B). The plasma sheaths coincide with the plasma dark spaces (the dark space in front of the cathode is called the Crookes dark space). The dark spaces develop because the higher energy electrons in those spaces are more likely to cause ionization than light-generating excitation.

The permanent positive charge of a plasma with respect to the electrodes is a striking characteristic and is a result of the random motion of the electrons and ions. The positive charge of a plasma can be understood from kinetic theory which predicts that for a random velocity distribution the flux of ions, $j_i$, and electrons, $j_e$, upon a surface is given by:

$$j_{i,e} = \frac{n_{i,e} \langle v_{i,e} \rangle}{4}$$  2.1

where n and $\langle v \rangle$ are the densities and average velocities, respectively. Because ions are heavier than electrons (typically 4000 to 100,000 times as heavy), the average velocity of electrons is larger. Consequently, the electron flux (according to Equation 2.1) is larger than the ion flux, and the plasma loses electrons to the walls, thereby acquiring a positive charge. The bombarding

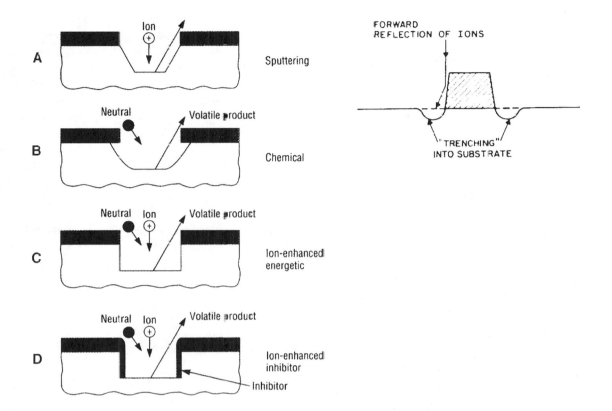

**FIGURE 2.5** Important dry etching profiles associated with the different dry etching techniques. (A) Sputtering and the formation of trenches (ditches) in physical ion etching. The mask is etched most rapidly in the neighborhood of the mask corner. The slope becomes less steep and not all ions are arriving at the etch bottom parallel to the sides any more. Some of them collide with the sides of the mask or substrate before arriving at the etch surface. Consequently, an increase in the number of incident ions close to the edges occurs, locally increasing the etch rate forming ditches. (B) Chemical etching in a plasma at low voltage and relative high pressure leads to isotropic features and lateral undercuts. (C) Ion-enhanced etching: physical-chemical, is the most perfect image transfer, as the undercutting is limited by the combined action of physical and chemical etching. The low pressure and high voltage lead to directional anisotropy. (D) Ion-enhanced inhibitor. Sidewalls are protected from undercutting by a surface species (e.g., a polymer) which starts etching when hit by a particle such as an ion, a photon, or an electron. Since very few particles hit the sidewalls (high voltage and high pressure), undercutting is suppressed. (Left side of figure is from Flamm, D. L., *Solid State Technol.*, October 49–54 (1993). With permission.)

energy of ions is proportional to the potential difference between the plasma potential and the surface being struck by ions. In equilibrium, the plasma potential $V_p$ (see Figure 2.6) averages a few times the electron energy $\langle v_e \rangle$ and typically reaches 2 to 10 volts. The rationale behind the asymmetric voltage distribution at the anode and cathode as seen in Figure 2.6 is as follows: electrons near the cathode rapidly accelerate away from it due to their relatively light mass. The ions, being more massive, accelerate towards the cathode in a slower tempo. Thus, on average, ions spend more time in the Crookes dark space, and at any instant their concentration is greater than that of electrons. The net effect is a very large field in front of the cathode in comparison with the field in front of the anode or in the glow region itself. Consequently, the greatest part of the voltage between the anode and the cathode, $V_e$, is dropped across the Crookes dark space where charged particles (ions and electrons) experience their largest acceleration. We will analyze the voltage distribution in a glow discharge in more detail when introducing RF discharges below.

The degree of ionization in a plasma depends on a balance between the rate of ionization and the rate at which particles are lost by volume recombination and by losses to the walls of the apparatus. Wall losses generally dominate over volume recombination. Accordingly, the occurrence of a breakdown in a given apparatus depends on the gas pressure (particle density), the type of gas, electric field strength (electron velocity), and on surface-to-volume ratio of the plasma. Figure 2.7 represents the Paschen curve and illustrates the gas breakdown voltage required to initiate discharge as a function of the product of the pressure, P, and the electrode spacing, d. The sharp rise in the breakdown voltage at the left side in Figure 2.7 (low Pxd side) occurs because the electrode spacing is too small, or the gas density is so low that electrons are lost to walls without colliding with gas atoms to produce ionization. On the right hand side of this plot (high Pxd side), the slow rise in the required voltage occurs because the electron energy is too low to cause ionization. This occurs at high pressures, because electron collisions with gas atoms become so frequent that the electrons cannot gather sufficient energy to overcome the ionization potential of the inert gas. Different gases exhibit similar 'Paschen' behavior, with the curves more or less shifted from the air-curve depending on the mean free path of the gas molecules involved.

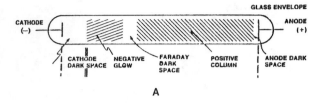

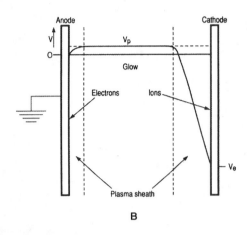

**FIGURE 2.6**  Glow discharge in a DC diode system. (A) Structure of the glow discharge in a DC diode system. (B) Voltage distribution in a DC diode discharge in equilibrium.

To better understand the implication of a Paschen curve, another way of introducing this plot might be of use. At constant pressure (say at 1 atmosphere), on the right side of the minimum in the Paschen curve, the breakdown voltage increases as the distance between the two parallel plane electrodes increases. Based on intuition about electrical breakdown, we predict higher voltages for larger electrode distances. Interestingly and counter intuitively, on the left side of the Paschen curve, i.e., at very short distances between anode and cathode (below 5 μm at atmospheric pressure in air), the breakdown voltage suddenly increases for shorter electrode distances because the electrode distances are too short for avalanche to take effect. We will see in Chapter 9 that a wide variety of electrostatic micromachines such as motors, switches, and gas sensors can operate in air without sparking as they operate on the left side of the Paschen curve. Microstructures, with dimensions of a few microns, operate in air as if surrounded by a reduced pressure environment. This typifies how linear scaling might be misleading when predicting the behavior of microdevices.

All glow discharges are nonequilibrium as the average electron energy ($\langle v_e \rangle = kT_e$, see Equation 2.2) is considerably higher than the average ion energy ($\langle v_i \rangle = kT_i$, see Equation 2.3), so a discharge or plasma cannot be described adequately by one single temperature ($T_e$ (electrons)/$T_i$(ions) = 10 to 100). High temperature electrons in a low temperature gas occur due to the small mass of electrons ($m_e$) compared to ions and neutrals. In collisions between electrons and argon atoms ($M_A$), the ratio $m_e/M_A$ is only $1.3 \cdot 10^{-5}$. Thus, in collisions, electrons have a poor energy transfer and stay warmer longer than the heavier ions and neutrals. Electrons can attain a high average energy, often many electron volts

(equivalent to tens of thousands of degrees above the gas temperature), permitting electron-molecule collisions to excite high temperature type reactions which form free radicals in a low temperature neutral gas. Generating the same reactive species without a plasma would require temperatures in the $\sim 10^3$ to $10^4$ K range, destroying resists and damaging most inorganic films.

Also important to note is that a plasma typically is weakly ionized: the number of ions is very small compared to the number of reactive neutrals such as radicals. The ratio between ionized and neutral gas species in a glow discharge plasma is of the order of $10^{-6}$ to $10^{-4}$. This fact is crucial in understanding which entities are responsible for the actual etching of a substrate placed in a glow discharge.

Of the different dry etching techniques, the choice of technique depends on the efficiency or 'strength' of a particular plasma evaluated by parameters such as the average electron energy, $\langle v_e \rangle$, also called electron temperature, i.e.,

$$\langle v_e \rangle = kT_e \left( \text{e.g., } 1 - 10 \, eV \right) \qquad 2.2$$

the average ion energy, i.e.,

$$\langle v_i \rangle = kT_i \left( \text{e.g., } 0.04 \, eV \right) \qquad 2.3$$

the electron density (e.g., between $10^9$ and $10^{12}$ cm$^{-3}$), plasma ion density (e.g., $10^8$ to $10^{12}$ cm$^{-3}$), neutral species density (e.g., $10^{15}$ to $10^{16}$ cm$^{-3}$), and the ion current density (e.g., 1 to 10 mA/cm$^2$). A quantity of particular use in characterizing a plasma's average electron or ion energy is the ratio of the electrical field to the pressure:

$$kT_{i,e} \sim \frac{E}{P} \qquad 2.4$$

With increasing field strength, the velocity of free electrons or ions increases because of acceleration by the field ($\sim E$) but velocity is lost by inelastic collisions. Since an increase in pressure decreases the electron or ion mean free path, there being more collisions, the electron or ion energy decreases with increasing pressure ($\sim 1/P$).

How do we use the described DC plasmas for dry etching? In one of the arrangements the substrate to be etched is placed on the cathode (target) in an argon plasma, while sufficiently energetic ions (between 200 and 1000 eV) induce physical etching, i.e., ion etching or sputtering in which atoms are billiard ball-wise ejected from the bombarded substrate (see Figure 2.8). Externally applied voltages concentrate across the cathode plasma sheath and ions are accelerated in that field before hitting the cathode. Each ion will collide numerous times with other gas species before transversing the plasma sheath, as the sheath thickness is larger than the mean free path. As a result of these collisions, the ions lose a lot of their energy and move across the sheath with a drift velocity that is less than the 'free fall' velocity. But these vertical velocities are still very large compared to the random thermal velocities discussed above. The resulting bombarding ion flux, $j_i$, is given by:

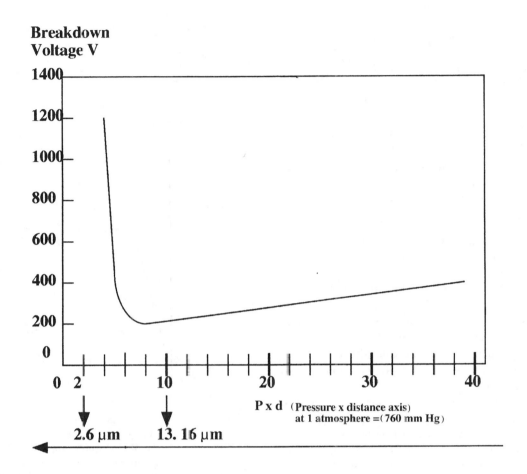

**FIGURE 2.7** The DC breakdown voltage as a function of the gas pressure P and the electrode spacing d for plane parallel electrodes in air. Such curves are determined experimentally and are known as Paschen curves.

$$j_i = qn_i\mu_i E \qquad\qquad 2.5$$

with E, the electric field, $n_i$, the ion density, $\mu_i$, the ion mobility, and q the charge. Reducing the pressure in the reactor increases the mean free path. Consequently, ions accelerated towards the cathode at lower pressures can gain more energy before a collision takes place. In an alternate mode, reactive species generated by the DC plasma may combine with the substrate to form volatile products that evaporate, chemically etching the substrate.

## Physics of RF Plasmas

In an RF-generated plasma, a radio frequency voltage applied between the two electrodes causes the free electrons to oscillate and to collide with gas molecules leading to a sustainable plasma. RF-excited discharges can be sustained without relying on the emission of secondary electrons from the target. Electrons pick up enough energy during oscillation in an RF field to cause ionization, thus sustaining the plasma at lower pressures than in a DC plasma (e.g., 10 vs. 40 mTorr). Another asset

RF has over DC sputtering is that RF allows etching of dielectrics as well as metals. The RF breakdown voltage of a plasma shows the Paschen behavior of a DC plasma, i.e., a minimum in the required voltage as a function of the pressure with the mean free path of the electrons substituting the spacing, d, between the electrodes.

In the simplest case of RF ion sputtering, the substrates to be etched are laid on the cathode (target) of a discharge reactor, for example, a planar parallel plate reactor as shown in Figure 2.8. The reactor consists of a grounded anode and powered cathode or target, enclosed in a low-pressure gas atmosphere (e.g., $10^{-1}$ to $10^{-2}$ Torr of argon). An RF plasma, formed at low gas pressures, again consists of positive cations, negative anions, radicals, vibrationally excited polyatomic species, and photons (the UV photons create the familiar plasma glow). As with a DC plasma, the neutral species greatly outnumber the electrons and ions; the degree of ionization only being on the order of $10^{-4}$ to $10^{-6}$ for parallel plate gas discharges. The RF frequency typically employed is 13.56 MHz (a frequency chosen because of its non-interference with radio-transmitted signals). The RF power

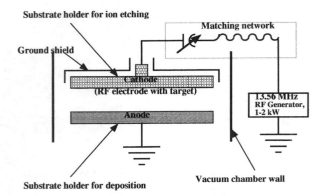

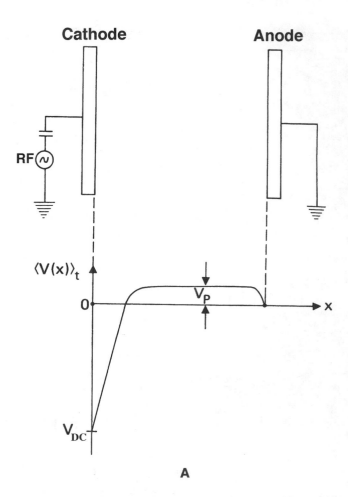

FIGURE 2.8   Two electrode set-up (diode) for RF ion sputtering or sputter deposition. For ion sputtering, the substrates are put on the cathode (target); for sputter deposition, the substrates to be coated are put on the anode.

supply rates between 1 and 2 kW. With one of the two electrodes capacitively coupled to the RF generator, the capacitively coupled electrode automatically develops a negative DC bias and becomes the cathode with respect to the other electrode (see Figure 2.9A). This DC bias (also called 'self-bias' $V_{DC}$) is induced by the plasma itself and is established as follows. When initiating an AC plasma arc, electrons, being more mobile than ions, charge up the capacitively coupled electrode; since no charge can be transferred over the capacitor, the electrode surface retains a negative DC bias.

The energy of charged particles bombarding the surface in a glow discharge is determined by three different potentials established in the reaction chamber: the plasma potential, $V_p$, i.e., the potential of the glow region; the self bias, $V_{DC}$; and the bias on the capacitively coupled electrode, $(V_{RF})_{pp}$ (see also Figure 2.9B). The following analysis clarifies how these potentials relate to one another and how they contribute to dry etching.

The voltage build-up, $V_{DC}$, on an insulating electrode by the electron flux (Equation 2.1) is given by:

$$V_{DC} = \frac{kT_e}{2e} \ln \frac{T_e m_i}{T_i m_e} \qquad \qquad 2.6$$

where $T_e$ and $T_i$ are the electron and ion temperatures defined by Equations 2.2 and 2.3 and $m_e$ and $m_i$ are the electron and ion masses, respectively.[22]

The electron loss creates an electric sheath field in front of any surface immersed in the plasma, counteracting further electron losses. This sheath or dark space also forms a narrow region (typically 0.01 to 1 cm depending on pressure, power, and frequency) between the conductive glow region and the cathode (Crookes dark space) where most of the voltage ($V_{DC}$ in Figure 2.9A) is dropped as in the DC plasma case (the cathode being capacitively coupled acts effectively as an insulator for DC currents). The other electrode is grounded and conductive (no charge build-up and no voltage build-up) and automatically becomes the anode with respect to the capacitively coupled electrode.

FIGURE 2.9   RF plasma: (A) Approximate time-averaged potential distribution for a capacitively coupled planar rf discharge system. (B) Potential distribution in glow discharge reactors. $V_p$: plasma potential, $V_{DC}$: self-bias of cathode electrode, $(V_{RF})_{pp}$: peak-to-peak RF voltage applied to the cathode. (C) Equivalent electrical circuit of an RF plasma.

The time-average of the plasma potential, $V_p$, the DC cathode potential (self-bias potential,) $V_{DC}$, and the peak-to-peak RF voltage $(V_{RF})_{pp}$, applied to the cathode are approximately related as (see Figure 2.9B):

$$2V_p \sim \frac{(V_{RF})_{pp}}{2} - |V_{DC}| \qquad \qquad 2.7$$

Clearly the magnitude of the self-bias depends on the amplitude of the RF signal applied to the electrodes. In the RF discharge the time-averaged RF potential of the glow region, referred to as the plasma potential, $V_p$, is more significantly positive with respect to the grounded electrode than in the case of a DC plasma.

Positive argon ions from the plasma are extracted by the large field at the cathode and are sputtering that electrode at near-normal incidence with energies ranging from a few to several electronvolts, depending on the plasma conditions and the chamber construction. One of the most important parameters determining the plasma condition is the total reactor pressure. As pressure is lowered below 0.05 to 0.1 Torr, the total ion energy, $E_{max}$, rises as both the self-bias voltage and the mean

$$2V_P \sim \frac{(V_{RF})_{pp} - |V_{DC}|}{2}$$

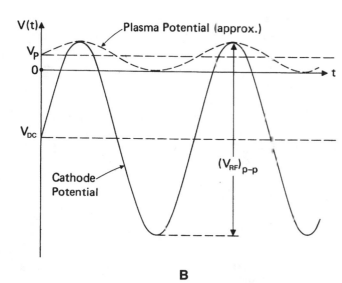

**B**

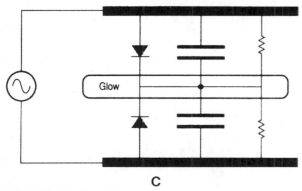

**C**

FIGURE 2.9 (Continued)

free path of the bombarding ions increase. Subsequently, ion-substrate bombardment energy rises sharply with decreasing pressure. The maximum energy of positive ions striking a substrate placed on the cathode is proportional to:

$$E_{max} = e\left(|V_{DC}| + V_p\right) = eV_T \qquad 2.8$$

with $V_T = |V_{DC}| + V_p$, whereas the maximum energy for a substrate on the grounded electrode (i.e., the anode) is proportional to:

$$E_{max} = eV_p \qquad 2.9$$

Equations 2.8 and 2.9 can be deduced from an inspection of Figure 2.9A. The situation where the wafers are put on the cathode is referred to as 'reactive ion etching' or 'reactive sputter etching'.

The chamber construction, especially the ratio of anode to cathode area, influences the rates of $V_T/V_p$ and, consequently, as calculated from Equations 2.8 and 2.9, the energy of the sputtering ions on these respective electrodes. Usually one wants to avoid etching of the anode, i.e., one tries to make $V_p$ small. To deduce the influence of the geometry of the anode and cathode on dry etching, we compare the sheath voltages in front of the anode and cathode, using the Child-Langmuir equation,

expressing the relationship between the ion-current flux, $j_i$, the voltage drop, V, over the sheath thickness of the dark-space, d; and the mass of the current carrying ions, $m_i$. The relation can be deduced from Equation 2.5 assuming the presence of a space-charge limited current:

$$j_i = \frac{KV^{3/2}}{\sqrt{m_i}\,d^2} \qquad 2.10$$

in which K is a constant.[23] The current density of the positive ions must be equal on both the anode and cathode, resulting in the following relation for the sheath-voltages:

$$\text{(cathode)}\ \ \frac{V_T^{3/2}}{d_T^2} = \frac{V_p^{3/2}}{d_p^2}\ \ \text{(anode)} \qquad 2.11$$

The plasma behaves electrically as a diode (large blocking voltage drop towards the capacitively coupled cathode and small voltage drop on the anode/plasma interface) in parallel with the sheath capacitance. As soon as any electrode tends to become positive relative to the plasma, the current rises dramatically, causing the plasma to behave as if a diode were present in the equivalent electrical circuit of the plasma. Hence, a planar, parallel set-up also is called a diode set-up. The equivalent electrical circuit representing an RF plasma is represented in Figure 2.9C. The dark spaces in a plasma are areas of limited conductivity and can be modeled as capacitors, i.e.,

$$C \sim \frac{A}{d} \qquad 2.12$$

The plasma potential is determined by the relative magnitudes of the sheath capacitances which in turn depend on the relative areas of anode and cathode. An RF voltage will split between two capacitances in series according to:

$$\frac{V_T}{V_p} = \frac{C_p}{C_T} \qquad 2.13$$

Using Equations 2.12 and 2.13, we can write:

$$\frac{V_T}{V_p} = \left(\frac{A_p}{d_p}\right)\left(\frac{d_T}{A_T}\right) \qquad 2.14$$

and substituting into Equation 2.11, we obtain:

$$\frac{V_T}{V_p} = \left(\frac{A_p}{A_T}\right)^4 = R^4 \qquad 2.15$$

where $A_p$ is the anode area and $A_T$ the cathode area. If there were two symmetric electrodes, both blocked capacitively, sputtering would occur on both surfaces. If the area of the cathode were significantly smaller than the other areas in contact with the discharge, the plasma potential would be small and little sputtering would occur on the anode, whereas the cathode would sputter very effectively. Since the cathode, in a set-up such as that represented in Figure 2.8, usually is quite large (> 1 m$^2$), allowing many silicon wafers or other substrates to be etched simultaneously, the grounded area needs to be larger yet. In actual sputtering systems, the larger grounded electrode consists of the entire sputtering chamber, creating a very small dark space, where there is hardly any sputtering taking place from these dark spaces. The exponential in Equation 2.15 in practical systems usually is less than 2, rather than 4.[23]

Higher ion energies ($V_T$ large), translate into lower etch selectivity and can be a cause of *device damage.* Consequently, a key feature for good *etch performance* is effective ionization to turn out very high quantities of low energy ions and radicals at low pressures. In answer to this need, equipment builders have come up with low-energy, high-density plasmas, for example, magnetrons, ICPs (inductively coupled plasmas)[19] and ECRs (electron cyclotron resonance)[16] (see below).

# Physical Etching: Ion Etching or Sputtering and Ion-Beam Milling

## Introduction

Bombarding a surface with inert ions, such as argon ions, in a set-up as shown in Figure 2.8, translates into ion etching or sputter etching (DC or RF). With ions of sufficient energy, impinging vertically on a surface, momentum transfer (sputtering) causes bond breakage and ballistic material ejection, throwing the bombarded material across the reactor to deposit on an opposing collecting surface, provided the surrounding pressure is low enough. The kinetic energy of the incoming particles largely dictates which events are most likely to take place at the bombarded surface, i.e., physisorption, surface damage, substrate heating, reflection, sputtering, or ion implantation. At energies below 3 to 5 eV, incoming particles are either reflected or physisorbed. At energies between 4 and 10 eV, surface migration and surface damage results. At energies >10 eV (say, from 5 to 5000 eV), substrate heating, surface damage, and material ejection, i.e., sputtering or ion etching, takes place. At yet higher energies >10,000 eV, ion-implantation, i.e., doping, takes place. The energy requirements for these various processes are summarized in Table 2.3.

## Device damage

Device damage comes in many "flavors" including:
- alkali (sodium) and heavy metal contamination
- catastrophic dielectric breakdown
- current-induced oxide aging
- particulate contamination
- UV damage
- temperature excursions which can activate metallurgical reactions
- "rogue" stripping processes which simply do not remove all the residue
- plasma-induced charges, surface damage, ion implantation

## Etch performance

### Etch Performance

Etch performance is being judged in terms of etch rate, selectivity, uniformity (evenness across one wafer and from wafer to wafer), surface quality, reproducibility, residue, microloading effects, device damage, particle control, post-etch corrosion, CD, and profile control. Selectivity as high as 40:1 might be required in the future for ICs and even more for micromachines. It is generally believed that 'radiation' damage of a plasma can be minimized by keeping ion energies low. It is also generally believed that a lot of the radiation damage can be annealed out. In reality, very little is understood of the damage a plasma can do. The higher the etch rate, the better the wafer throughput. Good selectivity, uniformity, and profile control are more easily achieved at lower etch rates. Trenches of various depths are made in Si in the manufacture of MOS devices. Shallow trenches (0.5 to 1.5 μm) are used to aid in reducing the effects of lateral diffusion during processing and to make flat structures (planarization). Deeper trenches (1.5 to 10 μm) are used to create structures that become capacitors and isolation regions in integrated circuits. For the shallowest trenches, photoresist is adequate as a mask. For deeper etching, a silicon dioxide mask may be needed. In micromachining one would like to obtain aspect ratios of 100 and beyond, so the degree of difficulty in masking keeps mounting. With aspect ratios above 2:1 or 2.5:1, the etching action at the bottom of a trench tends to slow down or stop altogether.

TABLE 2.3 Energy Requirements Associated with Various Physical Processes

| Ion Energy (eV) | Reaction |
|---|---|
| <3 | Physical adsorption |
| 4–10 | Some surface sputtering |
| 10–5000 | Sputtering |
| 10–20 K | Implantation |

The deposition phenomenon mentioned above can be used to deposit materials in a process called sputter deposition (Chapter 3). A low pressure and a long mean free path are required for material to leave the vicinity of the sputtered surface without being backscattered and redeposited.

In what follows, we shall consider two purely physical dry etching techniques at energies >10 eV, i.e., sputtering and ion-beam milling.

## Sputtering or Ion-Etching

The gradient in the potential distribution around the target in Figure 2.8 accelerates the ions prompting them to impinge on the substrate in a direction normal to the surface, and the etch rate in the direction of the impinging ions ($V_z$) becomes a strong function of $E_{max}$ (Equation 2.8). The impinging ions erode or sputter etch the surface by momentum transfer. This offers some advantages: volatility of the etch products is not critical as for dry chemical etching (only a billiard ball effect plays a role in physical etching), consequently the method does not lead to large differences in etch rates for different materials (sputter yields for most materials are within a factor of three of each other), and the method entails directional anisotropy. Physical etching is inherently material nonselective because the ion energy required to eject material is large compared to differences in chemical bond energy and chemical reactivity. When no reactive etching processes are available for a given material, physical etching always offers an option. The directional anisotropy remains as long as the dimensions of the surface topographical structures are small compared to the thickness of the sheath between the bulk plasma and the etched surface. However, etch rates are slow, typically a hundred to a few hundred angstroms per minute. The use of a magnetron can help improve upon the speed of the etch rate. In both DC and RF diode sputtering, most electrons do not cause ionization events with Ar atoms. They end up being collected by the anode, substrates, etc., where they cause unwanted heating. A magnetron adjusts this situation by confining the electrons with magnetic fields near the target surface; consequently, current densities at the target can increase from 1 mA/cm² to 10–100 mA/cm².

As sputter etching is nonselective, it introduces a masking problem. Another hurdle to overcome is the need for rather high gas pressure to obtain large enough ion currents, resulting in a short mean free path, $\lambda_i$, of the ions. With a mean free path smaller than the interelectrode spacing, considerable redeposition of sputtered atoms on the etching substrate laying on the cathode can occur. Electrical damage from ion bombardment also needs watching when critical electronic components reside on the substrate.

## Ion-Beam Etching or Ion-Beam Milling

When the plasma source for ion etching is decoupled from the substrate which is placed on a third electrode, one practices ion beam milling (IBM). The equipment needed is called a triode set-up. In a typical DC triode set-up, as shown in Figure 2.10, control of the energy and flux of the ions to the substrate happens independently. The sample being etched can be rendered neutral by extracting low-energy electrons from an auxiliary thermionic cathode (i.e., a hot filament neutralizer), thus making this DC equipment usable for sputtering insulators as well as conductors. In ion-beam milling, as in ion etching, one generally uses noble gases as they exhibit higher sputtering yields (heavy ions!) and avoid chemical reactions. The argon pressure in the upper portion of the chamber can be quite low, 10⁻⁴ Torr, resulting in a large mean free ion path. Electrons are emitted by a hot filament (typically tantalum or tungsten) and accelerated by a potential difference between the cathode filament and an anode. The discharge voltage must be larger than the gas ionization potential (15.7 eV for argon) and typically is operated at several times this value, about 40 to 50 V, in order to establish a glow discharge. Ions are extracted from the upper chamber by the sieve-like electrode, formed into a beam, accelerated, and fired into the lower chamber where they strike the substrate. Typical etch rates with argon ions of 1-keV energy and an ion current density of 1.0 mA/cm² are in the range of 100 to 3000 Å/min for most materials such as silicon, polysilicon, oxides, nitrides, photo-resists, metals, etc. Inert ion-beam etching (IBE) is, in principle, capable of very high resolution (<100 Å) but aspect ratios are usually less than or equal to unity.

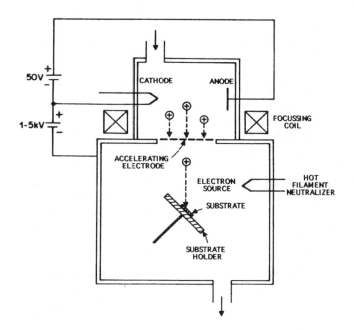

FIGURE 2.10 Ion-beam etching (IBE) apparatus (triode). In an ion beam apparatus, the beam diameter is approximately 8 cm. The substrates are mounted on a moveable holder allowing etching of large substrates. The coils focus the ion beam and densify the ion flux in magnetically enhanced confined ion etching (MIE).

In magnetically enhanced ion etching (MIE) the plasma is guided and made dense by an applied magnetic field (see coils in Figure 2.10). A magnetic field is applied to the extent that electrons cannot pass directly from the anode to the cathode, but follow helical paths between collisions, greatly increasing their path length and ionization efficiency. The cyclotron radius of a 100-eV electron in a field of 0.01 Tesla (T) (100 Gauss (G)) is 3.2 mm, making relatively low field strengths adequate. This results in an increased etch rate due to a denser ion plasma. When plasma density is increased with magnetic confinement, the degree of ionization is between $10^{-2}$ and $10^{-4}$, compared to $10^{-4}$ to $10^{-6}$ for simple plate discharges.

When the inert argon ions, in a triode set-up as shown in Figure 2.10, are replaced with more reactive ions, one refers to reactive ion beam etching (RIBE). The ions not only transfer momentum to the surface, but they also react directly with the surface, i.e., a chemical and physical mechanism is involved. Direct reactive ion etching of a substrate is the exception rather than the rule. We will learn further that the chemical reactions at the surface usually are dominated by radicals.

Ion-beam etching typically involves an ion source energy of up to 1.5 keV and a current density of 25 mA/cm² covering a diameter of 3 to 8 cm. This also is referred to as showered ion etching. A second type of ion beam etching, a maskless technique called focused ion beam (FIB), in which the beam is made extremely narrow and used as a direct writing tool, will be discussed in more detail in Chapter 7.

## Etching Profiles in Physical Etching

The ideal result in dry or wet etching is usually the exact transfer of the mask pattern to the substrate, with no distortion of the critical dimensions. Isotropic etching (dry or wet) always enlarges features and thus distorts the critical dimensions. Chemical anisotropic wet etching is crystallographic. As a consequence, critical dimensions can be maintained only if features are strategically aligned along certain lattice planes (see Chapter 4, Wet Bulk Micromachining). With sputtering the anisotropy is controllable by the plasma conditions.

As can be deduced from Figure 2.5A, ion etching and ion milling do not lead to undercutting of the mask but the walls of an etched cut are not necessarily vertical. A variety of factors contribute to this loss of fidelity in pattern transfer, and they are either caused by involatile sputtering reaction products or by special ion-surface interactions. We shall briefly review these dry physical etching problem areas.

### Faceting Due to Angle-Dependent Sputter Rate

Even when starting out with a vertical mask side wall, ion sputtering exhibits a tendency to develop a facet on the mask edge at the angle of maximum etch rate. This corner faceting is detailed in Figure 2.11A. The corner of the mask, always a little rounded even when the mask walls are very vertical, etches faster than the rest and is worn off. Faceting at the mask corner arises because the sputter yield for materials usually is a function of the angle at which ions are directed at the surface. The sputter-etch rate of resist, for example, reaches a maximum at an

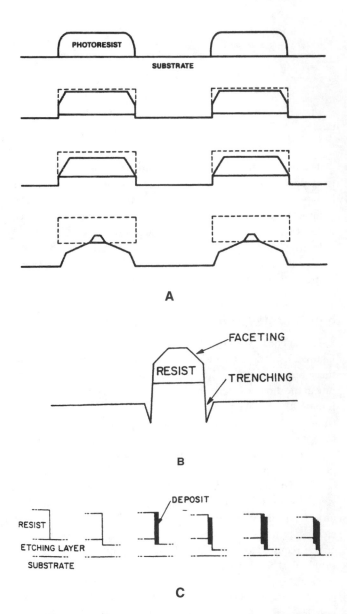

**FIGURE 2.11** Limitations of dry physical etching. (A) Faceting. Sputtering creates angled features. An angled facet (~60°) in the resist propagates as the mask is eroded away. Sloped walls may also be created in the underlying substrate. (B) Ditching due to glancing incidence of ions. (C) Redeposition of material sputtered from the bottom of a trench. (B and C from Shutz, R. J. in *VLSI Technology*, Sze, S. M., Ed., McGraw-Hill, New York, 184–232, 1988. With permission.)

incidence angle of about 60°, more than twice the rate at normal incidence.[8] Sloped mask sidewalls may eventually be followed by sloped etch steps in the substrate. The faceting of the substrate itself will proceed along its own preferred sputtering direction angle. The faceting is more pronounced with an applied bias due to the increased electric field at corners. Usually the faceting affects only the masking pattern, and its influence on the fidelity of the pattern transfer process can be minimized by making the mask sufficiently thick. The faceting can also be minimized or eliminated by a more ideal resist profile, with very little rounding of the mask corners. In the case of a resist mask, post-bake temperatures must be controlled so the reflow does not induce rounding of the resist features, leading to faceting.

It should be noted, though, that some of the disadvantages of physical etching, such as resist corner faceting, can sometimes be exploited. For example, a gently sloping edge is advantageous to facilitate metal coverage or planarization because a tapered sidewall is easier to cover than a vertical wall, especially when using line-of-site deposition methods. The method most often used to obtain such a taper is a controlled resist failure, i.e., *erodible or sacrificial masks*. In other words, the 'negative effect' described in connection with Figure 2.11A is put to good use.

### Ditching

When the slope of the side of the mask is no longer vertical, some ions will collide at a glancing angle with the sloping edges before they arrive at the etch surface. This gives a local increase in etch rate leading to ditches (Figures 2.5A and 2.11B). For this mechanism to be active there must be a sizable fraction of ions with at least slightly off-vertical trajectories, or the sidewall must have a slight taper as shown in Figure 2.5A and 2.11B.[24] The taper of the sidewall could result, for example, from redeposition (see next section) or faceting (see above). Since ditching is a small effect (say, 5%), it often goes unnoticed unless thick layers are etched.

Off-vertical ion trajectories could also be caused by sheath scattering and field nonuniformities. Most ions impinge perpendicularly on the substrate. Scattering of a small fraction of ions in the sheath causes a distribution of impinging angles. This scattering in the sheath is responsible for hourglass-shaped etch profiles that have been observed in trench etching of silicon.[14] Besides sheath scattering, there is a second mechanism active which may lead to skewing of ion directionality; namely, inhomogeneities in the electrical field at the substrate surface. When etching conductors, the bending of electric field lines due to surface topography has the effect of enhancing ion flux at feature edges and leads to ditching. On the other hand, when etching insulators, charging effects may cause appreciable ion fluxes to the sidewalls of a trench and contribute to lateral etching. The latter will again lead to hourglass-shaped trenches. The importance of the effect is very much related to the electrical conductivity of the masking and etching surfaces, with the greatest significance for strongly insulating materials.[25]

### Redeposition

Another sputtering limitation, already alluded to, is the *redeposition* of involatile products on step edges (Figure 2.11C). Redeposition involves sputtered involatile species from the bottom of the trench settling on the sidewalls of the mask and etched trench. The phenomenon manifests itself mainly on sloped sidewalls.

By *tilting and rotating* the substrate during etching, etch profiles can be improved. The reasons for tilting and rotating improvements are a combination of shadowing the bottom of the step (to reduce trenching), partially etching the sidewalls of the mask (to reduce redeposition), and gaining more nearly vertical edges on the etched profiles in the substrate.[8] Especially with aspect ratios exceeding unity, redeposition becomes problematic, and reactive gas additives are necessary to generate volatile etch products. With aspect ratios above 2:1 or 2.5:1, the etching action at the bottom of too fine features tends to slow down or stop altogether. Reactive additives bring us into the realm of chemical-physical etching (discussed further below). For micromachining, tilting and rotating of substrates is a recurring topic. It is one of the desirable modifications of standard

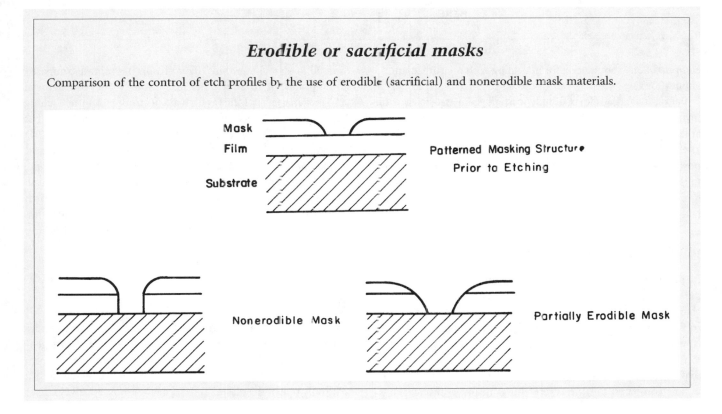

## *Erodible or sacrificial masks*

Comparison of the control of etch profiles by the use of erodible (sacrificial) and nonerodible mask materials.

Mask
Film
Substrate

Patterned Masking Structure
Prior to Etching

Nonerodible Mask

Partially Erodible Mask

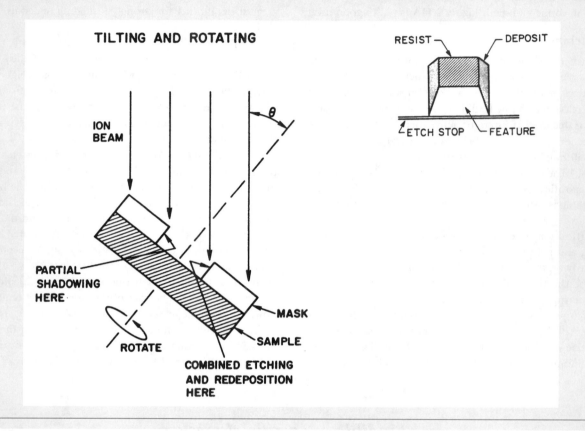

### Rotating and redeposition

Typical configuration for tilting and rotating, as well as an example of redeposition in a high aspect ratio device. By tilting and rotating a sample during ion beam etching, better etching profiles can be obtained. The walls exposed to ion beam etching exhibit combined etching and redeposition while the unexposed walls are partially shadowed; rotation averages these effects out.

equipment one should look for in dealing with micromachining applications.

Backscattering is a form of redeposition again associated with involatile etch products. A fraction of the sputtered and involatile species from the surface is backscattered onto the substrate after several collisions with gas-phase species. This indirect redeposition may involve contaminants from the walls and fixtures in the vacuum chamber. Backscattering fixes the upper pressure limit for ion-enhanced etching. Significant redeposition can take place at pressures as low as 10 mTorr.

### Physical Etching Summary

In physical etching, ion-etching or sputtering, and ion-beam milling, argon or other inert ions extracted from the glow discharge region are accelerated in an electrical field towards the substrate, where etching is purely impact controlled. Sputtering is inherently nonselective because large ion energies compared to the differences in surface bond energies and chemical reactivities are involved in ejecting substrate material. The method is slow compared to other dry etching means, with etch rates limited to several hundreds of angstroms per minute compared to thousands of angstroms per minute and higher for chemical and ion-assisted etching (as high as 6 μm/min!). Sputter etching tends to form facets, ditches, and hourglass-shaped trenches and redeposits material frequently in high aspect ratio (>2:1) features. Electrical damage from ion bombardment and implantation to the substrate can be problematic. Some of the reversible plasma damage can be removed by a thermal anneal. With the continuing increase in device complexity, which includes layers of different chemical composition, inert ion etching and ion-beam sputtering continues to find applications.

## Dry Chemical Etching

### Introduction

In reactive plasma etching, reactive neutral chemical species such as chlorine or fluorine atoms and molecular species generated in the plasma diffuse to the substrate where they form volatile products with the layer to be removed (Figure 2.5B). The only role of the plasma is to supply gaseous, reactive etchant species. Consequently, if the feed gas were reactive enough, no plasma would be needed. At pressures of $>10^{-3}$ Torr, the neutrals strike the surface at random angles, leading to isotropic, rounded features. A dry chemical etching regime can be established by operating at low voltages, eliminating impingement of high-energy ions on the sample, and facilitating surface etching almost exclusively by chemically active, neutral species formed in the plasma. The reaction products, volatile gases, are removed by the vacuum system. The volatility of the formed reaction product introduces a major difference with sputtering, where involatile fragments are ejected billiard ball wise and may be redeposited close by.

## Reactor Configurations

Reactive plasma etching follows the same process we encountered when discussing dry stripping. The different popular configurations for plasma etching (barrel reactor, downstream etcher, and parallel plate system) are shown in Figure 1.23. In the case of resist etching, one refers to the process as ashing. Depending on the configuration, high energy ion bombardment of the substrate can be prevented and plasma-induced device damage avoided. For example, in a barrel reactor the substrates are shielded by a perforated metal shield to reduce substrate exposure to charged high energetic species in the plasma. In a downstream stripper, the geometry of the reactor allows reactant generation and stripping to take place in two physically separated zones (triode-type configuration with a remote plasma). In parallel plate strippers, the substrates are placed inside the plasma source which leads to higher ion damage compared to the two previous methods. The diode-like, parallel plate stripper is called a reactive ion etcher (RIE), although ions rarely reactively 'eat' away the substrate.

## Reaction Mechanism

The plasma etching process can be broken down into as many as six primary steps as illustrated in Figure 2.12. The first step is the production of the reactive species in the gas-phase (1). In a glow discharge a gas such as $CF_4$ dissociates to some degree by impact with energetic particles such as plasma electrons with an average energy distribution between 1 and 10 eV. In the dissociation reactive species such as $CF_3^+$, $CF_3$, and F are formed. This step is vital, because most of the gases used to etch thin films do not react spontaneously with the film, e.g., $CF_4$ does not etch silicon. In a second step, the reactive species diffuse to the solid (2) where they become adsorbed (3), diffuse over the surface, and react with the surface (4). Finally, the reaction products leave the surface by desorption (5) and diffusion (6). As in a parallel resistor combination, the total resistance is determined by the smallest resistance, the reaction with the smallest rate constant determines the overall reaction rate.

Some of the six reaction steps listed occur in the gas-phase and are termed homogeneous reactions, while others occur at the surface and are called heterogeneous reactions. For the homogeneous reactions, the plasma only plays the role of creating highly reactive species from the plasma gas. Radicals are more abundant in a glow discharge than ions because they are generated at a lower threshold energy (e.g., <8 eV), which leads to a higher generation rate. Moreover the uncharged radicals have a longer lifetime. The low-energy ions rarely act as the reactant themselves; instead, neutrals are responsible for most reactive etching (chemical etching) at pressures above 0.001 to 0.005 Torr. Radicals and molecules formed in the plasma are not inherently more chemically reactive than ions, but they are present in significantly higher concentrations. Heterogeneous reactions display even more complexity than the homogeneous reactions just described. In

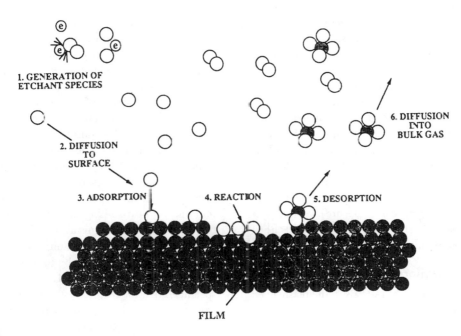

**FIGURE 2.12** Primary process occurring in a plasma etch process.

principle, all the species generated in the plasma may influence the reaction rate; nonreactive species may decrease the reaction rate by blocking surface sites, and adsorption of radicals may enhance the reaction rate. Radicals and other neutrals reach the surface by diffusion, whereas ions are accelerated towards the surface by the negative potential on the substrate electrode. Besides chemically active species and ions, the effect of electron bombardment and irradiation by visible and UV radiation emanating from the plasma requires consideration. Chemical etching occurs at low bias and since, in principle, no highly energetic ions bombard the surface, sputtering itself is not an important surface-removal mechanism and radiation damage to the substrate is reduced. The term 'reactive ion etching' unfortunately is used indiscriminately for all chemical dry etching, even when ions themselves are not the major reactive species. Radicals and molecules also serve as the primary depositing species for all types of films in plasma-enhanced chemical vapor deposition (PECVD) (see Chapter 3). Ions directly participate in chemical etching only in reactive ion beam etching (RIBE), where ion reactions at very low pressures can etch at modest rates of less than 400 Å/min. RIBE is an example of physical-chemical dry etching where the same ion has both a physical component (ion impact) and a chemical component (reactive etching). Working with a remote plasma in a triode-type system (e.g., a downstream stripper) facilitates further ion bombardment reduction of and current flux to the wafer. The steps of reactant generation and actual etching are separated efficiently because charged species (mainly electrons and positive $O_2^+$ ions in case of a pure oxygen plasma used for stripping resist) suffer a much higher loss than reactive neutrals such as atomic oxygen due to the presence of plasma excitation.

In the absence of crystallographic effects (typically seen with III-V compounds but not with Si), chemical dry etching leads to isotropic profiles only when ions do not assist the reaction. In the case where ions do assist the chemical reaction, anisotropy is induced. Isotropic etching of the reactive species leads to mask undercutting. In some cases mask undercutting is required, for example in device fabrication involving 'lift-off' or when layers on sidewalls must be cleared.

Because of the chemical nature of the etching process, a high degree of control over the relative etch rates of different materials (i.e., selectivity) can be obtained by choosing suitable reactive gases. During photoresist stripping, an oxygen plasma removes photoresists by oxidizing the hydrocarbon material to volatile products. Fluorine compounds are used for silicon etching; many materials are susceptible to chlorine etching. For example, aluminum is etched in chlorine but not in fluorine, as aluminum chlorides are volatile and aluminum fluorides are involatile. Mixtures of gaseous compounds such as $CF_4$ (fluorocarbon)-$O_2$ assist in patterning silicon, silicon dioxide, silicon nitride, etc.

### Loading Effects/Uniformity and Nonuniformity

In dry etching the number of radicals in the plasma is in proportion to the number of atoms to be removed, contrary to wet etching where the number of etchant molecules might be $10^5$ times higher than the number of atoms. A so-called loading effect occurs as the result of gas phase etchant being depleted by reaction with the substrate material. The more purely chemical the etching, the bigger the loading effect. With lower pressures the loading effect becomes smaller. Conventional plasmas can sustain enough radicals to etch at a rate of 1000 Å/min. With more effective power sources such as a cyclotron or magnetron $10^4$ Å/min can be achieved.

The loading effect brings an important limitation of dry etching to light; as the etch rate becomes dependent on wafer loading, uniformity is jeopardized. If the supply of reactant limits the etch rate, small variations in flow rate or gas distribution uniformity may indeed lead to etch rate nonuniformities. The gas flow is the most important parameter to control in this regard. The symmetry of the gas flow, i.e., the relative position of gas inlet and pumping port, has to be optimized for the given reactor configuration. Hence, a minimum flow rate of the reactant gas prevents the process from being limited by reactant supply. A utilization factor, U, i.e., the ratio of rate of formation of etch product to the rate of etch gas flow, may be defined and $U > 0.1$ is suggested for uniform etching. To illustrate: a 500-Å/min etch rate of a 3-in Si wafer in a $CF_4$ plasma corresponds to a removal rate of $5 \times 10^{19}$ Si atoms per min, or a $SiF_4$ evolution rate of approximately 2 sccm. This means that the $CF_4$ flow rate should reach at least 20 sccm.[26]

The loading effect is a function of the number of wafers in the chamber and may also change while etching different features on one wafer (local loading or microloading effect). For example, after clearing an etching film from the planar regions of a surface, less residual material remains, and more etchant is available to etch the remaining material residue, say on the sidewalls of a trench.

Etching uniformity also is impacted by the relative reactivity of the wafer surface with respect to the cathode material used. Resulting nonuniformities in this case are referred to as bull's-eyes because circular interference patterns show on an etched wafer. If an aluminum electrode is used for Si or poly-Si etching in an $SF_6$ plasma, a very pronounced bull's-eye effect is evident. The striking nonuniformity in etching pattern results from a lower consumption of reactant species above the aluminum electrode, as aluminum only mildly reacts with reactive species formed by $SF_6$. By applying an electrode material that consumes the fluorine reactive species as fast as Si itself (e.g., a Si cathode), concentration gradients of reactant species at the edge of the wafer are avoided, resulting in uniform etching (see Figure 2.13).[26]

To further avoid nonuniformities, wafers should be positioned away from the edges of electrodes to eliminate edge effects caused by changing sheath thickness as well as varying angles of incidence. A good thermal contact between wafers and cathode also is important. Local temperature variations due to nonuniform heat-sinking can lead to large etch nonuniformities, particularly in chemically dominated processes.[26]

Another example of nonuniformity is the 'grass' structure sometimes observed at the bottom of a Si trench. The 'grass' consists of pillars of silicon also called 'black silicon'; Hasper speculates that the reason for the 'grass' is $Al_2O_3$ contamination from the mask and/or reactor walls.[27] In more recent work 'grass' formation during RIE was explained as a consequence of the sharpening of the ion angular distribution with the increasing aspect ratio of the trench during etching.[24]

## Ion Energy vs. Pressure Relationship in a Plasma

In general, three factors control the etch rate in a plasma reactor: the neutral atom and free radical concentration, the ion concentration, and the ion energy. The ion and radical concentrations control the reaction rate, while the ion energy provides the necessary activation and controls the degree of anisotropy. The respective contribution to chemical and physical action of a plasma can be manipulated by varying voltage and gas pressure. For good critical dimension control, anisotropy is required and positive ions need to be formed at low pressures ($<10^{-2}$ Torr) to strike the surface at normal incidence. Etching at low pressures, with a long mean free path length of the ions, $\lambda_i$, is inherently more anisotropic, i.e., directional and less contaminating because etch reaction by-products show more volatility at lower pressures and are easier to remove, but at these lower pressures the ion density drops off quickly, causing a lower etch rate and lower wafer throughput. Increasing the power or wafer bias will increase the etch rate as the remaining ions will become more energetic. Higher ion energies can cause additional problems in terms of device damage. Really needed is a plasma source operating at low pressure with very high ion density. Operating at low pressures also reduces the effects of chemically induced loading (i.e., the total amount of material to be removed depletes the reactant chemicals) and microloading (e.g., etch-rate dependence on feature size). High plasma ion densities are created, for example, by magnetic coils increasing the electron path length in a magnetron sputtering machine. Just as the ionospheric van Allen belts are confined by the Earth's dipole field, the plasma remains confined within the magnetic envelope. This magnetic envelope also prevents electron loss to plasma exposed surfaces. With a high pressure, short $\lambda_i$, and a low voltage one gets isotropic chemical etching. The pressure-voltage relationship for a plasma is schematically represented in Figure 2.14[29] and summarized in Table 2.4. Sputtering and dry chemical etching thus represent the two extremes of a *continuous dry-etching spectrum* with physical etching by sputtering with inert argon ions at one end and chemical etching with reactive neutral species at the other end.

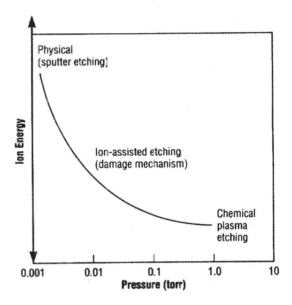

FIGURE 2.14  Ion energy vs. pressure for a plasma. (From Flamm, D. L., *Solid State Technol.*, 35, 37–39, 1991. With permission.)

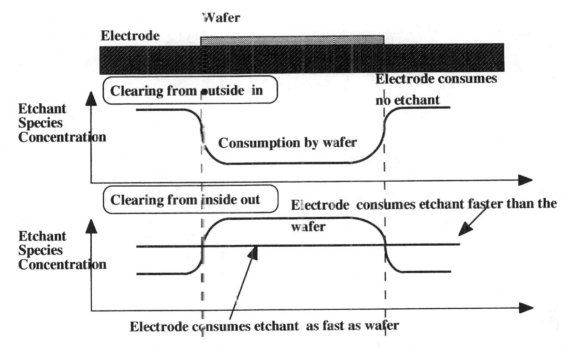

FIGURE 2.13  Chemical and physical effects leading to nonuniform etching across a wafer (bull's-eye-effect). (After Elwenspoek, et al., Universiteit Twente, *Micromechanics*, 1994.)

## Gas Phase Etching without Plasma ($XeF_2$)

Isotropic etching of Si with xenon difluoride ($XeF_2$) does not require a plasma to generate the etching species. Xenon difluoride ($XeF_2$) is a white solid with a room temperature vapor pressure of about 4 Torr which reacts readily with Si. The Si etch occurs in the vapor phase at room temperature and at pressures between 1 and 4 Torr established by a vacuum pump throttled to the right pressure. Hoffman et al.[30] observed silicon etch rates as high as 10 $\mu$m per minute but worked at more typical rates of 1 to 3 $\mu$m. The extreme selectivity of $XeF_2$ to silicon over silicon dioxide and silicon nitride are well documented, but Hoffman et al.[30] have shown that $XeF_2$ also displays extreme etching selectivity over aluminum and photoresist. A 50-nm Al mask or a single layer of hardbaked photoresist suffice as excellent deep etch masks. The simplicity of the process, the fast etch rate as well as the resistance of even very thin layers of oxide, nitride, and Al metal make this etching process a possible choice for etching micromachined structures in the presence of CMOS electronics (in a so-called post-CMOS procedure). In Figure 2.15A, an aluminum hinge (5 $\mu$m wide and 1.1. $\mu$m thick) holding an oxide plate (200 $\mu$m square) suspended over a Si etch pit is shown. The etch pit underneath the suspended plate (not shown in Figure 2.15A) is nearly isotropic and exhibits a surface roughness of several microns. The conductivity of the Al metal was unaffected by the $XeF_2$ etch. The hinge contacts a polysilicon strain gauge, sandwiched between oxide layers, to protect it from the $XeF_2$ etchant (see Figure 2.15B). The flexible hinge enables rotation of the oxide plate out of the plane of the wafer. Hoffman et al. suggest that this contraption might function as a piezoresistive accelerometer, with the hinges providing mechanical support and electrical connectivity between the strain gauges embedded in the oxide plate proof mass and the wafer (see Figure 2.15C).

## Plasma Jet Etching

Plasma jet etching is an atmospheric variant of plasma etching capable of producing very high etch rates without substrate damage. In plasma jet etching reactive chemical species are generated in a DC arc between two electrodes in a noble gas such

TABLE 2.4   Plasma Reactions

|  | Energy of Ions (Pressure, Torr) | |
| --- | --- | --- |
| Reactive Gas | Low (0.1–10) | High (0.001–0.1) |
| Volatile | Plasma etching | Reactive sputtering |
|  | Plasma ashing | Reactive ion etching |
| Involatile | Plasma anodization | Reactive sputtering |

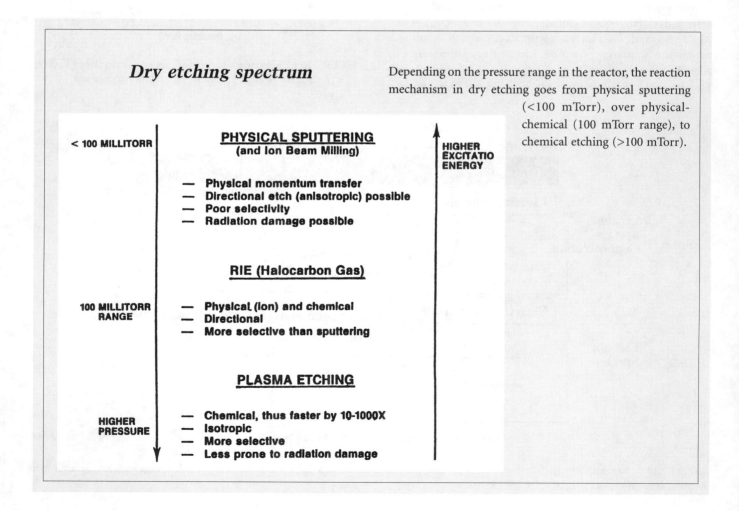

### Dry etching spectrum

Depending on the pressure range in the reactor, the reaction mechanism in dry etching goes from physical sputtering (<100 mTorr), over physical-chemical (100 mTorr range), to chemical etching (>100 mTorr).

**< 100 MILLITORR**

**PHYSICAL SPUTTERING**
**(and Ion Beam Milling)**

— Physical momentum transfer
— Directional etch (anisotropic) possible
— Poor selectivity
— Radiation damage possible

**RIE (Halocarbon Gas)**

**100 MILLITORR RANGE**

— Physical (ion) and chemical
— Directional
— More selective than sputtering

**PLASMA ETCHING**

**HIGHER PRESSURE**

— Chemical, thus faster by 10-1000X
— Isotropic
— More selective
— Less prone to radiation damage

**HIGHER EXCITATIO ENERGY**

**A**

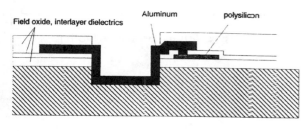

Field oxide, interlayer dielectrics    Aluminum    polysilicon

**B**

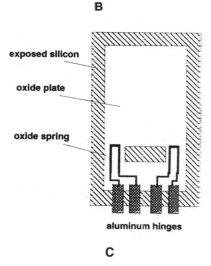

exposed silicon

oxide plate

oxide spring

aluminum hinges

**C**

**FIGURE 2.15** (A) Al hinge (5 μm wide and 1.1 μm thick) holding a suspended oxide plate (plate is on the right side of the photograph). The Si is etched from underneath the Al hinge and oxide plate by XeF₂. The suspended oxide plate is 200 μm square. (Courtesy of Dr. Kris Pister.) (B) Al hinge contacts a poly-Si piezoresistor embedded in the oxide plate to protect it from the XeF₂ etchant. (C) The piezoresistive accelerometer design shown enables rotation of the oxide plate out of the plane to detect orthogonal acceleration. (From Hoffman et al., MEMS '95, 288–293, 1995. With permission.)

as argon at atmospheric pressure. A stream of reactant gas (the gas jet) such as $CF_4$ is then flowed into the arc and onto the wafer which is located up to 12 inches away to avoid damage.[31] The technology may improve the performance and reduce the cost of bulk material removal operations such as mechanical grinding, lapping, and wet chemical etching.

## Physical-Chemical Etching

### Introduction

The most useful plasma etching is neither entirely chemical nor physical. By adding a physical component to a purely chemical etching mechanism, the shortcomings of both sputter-based and purely chemical dry etching processes can be surmounted.

In physical-chemical techniques, the following four types of ion-surface interactions may promote dry etching. A first is found in reactive ion beam etching (RIBE), a rather exceptional case where ions are reactive and etch the surface directly (Figure 2.16). The more general case presents itself in three types of ion-assisted etching. In one type, ion bombardment induces a reaction by making the surface more reactive for the neutral plasma species, for example, by creating surface damage (i.e., energy driven anisotropy). In another type, ions clear the surface of film-forming reaction products, allowing etching with reactive neutrals to proceed on the cleared areas (inhibitor-driven anisotropy). Finally, ion bombardment may supply the energy to drive surface reactions. Defining which step primarily causes enhanced etching is often hard to determine.

### Energy-Driven Anisotropy

During energy-driven anisotropy or ion-assisted etching, bombardment by ions (<1000 eV) disrupts an unreactive substrate and causes damage such as dangling bonds and dislocations, resulting in a substrate more reactive towards etchant species (electrons or photons also can induce surface activation) (see Figure 2.5C). Vacuum pumping removes the volatile reaction products. This type of etching is referred to as reactive ion etching (RIE), when it involves reactive chemicals in a diode type reactor, and as chemically assisted ion-beam etching (CAIBE) when it involves chemical reactants (e.g., $Cl_2$) introduced over the substrate surface in a triode type set-up. Figure 2.8 shows the preferred set-up for RIE. Figure 2.16 represents the ideal set-up for CAIBE. We already indicated that the label RIE can cause confusion since ions are not really the dominant reactive species. In a magnetically enhanced RIE system (MERIE), a higher ion flux is obtained at lower energy through magnetic confinement of the plasma.

The etch rate reached by RIE is substantially higher than in ion etching. For example, the etch rate of Si in Ar sputtering hovers around 100 Å/min compared to 2000 Å/min for a reactive gas such as $CCl_2F_2$. Chemically assisted ion etching can lead to accurate transfers of the mask pattern to the substrate and to a fair selectivity in etching different materials. The directional anisotropy ensues from operating at low pressures and high voltages. Under these conditions, the mean free path of the

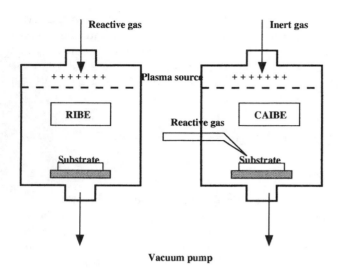

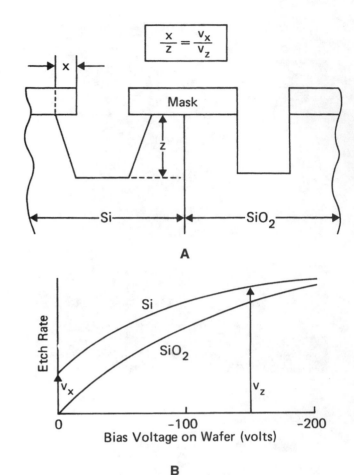

FIGURE 2.16 A chemically assisted ion beam etching system (CAIBE) compared with reactive ion beam etching system (RIBE).

reacting molecules typically grows larger than the depth to be etched, resulting in the horizontal surfaces being hit and etched by the reactive neutrals a lot more than the sidewalls. The need for energetic impinging ions, as in the case of physical etching, reduces as the complex on the surface activates more easily.

A hypothetical etch profile for Si and $SiO_2$ etching with a reactive gas and positive ion bombardment is shown in Figure 2.17A.[8] In the absence of ionic bombardment, the etch rate is assumed to be zero for $SiO_2$; consequently, the $SiO_2$ film etches with perfect anisotropy (vertical sidewalls). The silicon etch rate in the absence of ionic bombardment, on the other hand, is assumed finite but small; consequently, the etched feature ends up having a profile with slanted sidewalls. Mathematically, the above can be expressed in terms of etch depth, Z, undercut, X, the etch rate under bias, $V_Z$, and the etch rate without bias, $V_X$, as:

$$\frac{V_X}{V_Z} = \frac{X}{Z} \qquad 2.16$$

(See Figure 2.17B.) The anisotropy of RIE can further be enhanced by the use of helium backside cooling (as low as –120°C might be required). The low temperature suppression of chemical attack of silicon by fluorine atoms exemplifies how lower temperature further improves anisotropy and critical dimension control. While suppressing the isotropic fluorine reaction the same low temperature only slightly influences ion-assisted reactions and hence improves the anisotropy of profiles; the etch rate at $V_X = 0$ lowers or becomes zero.

With RIE, etched depths of 10 to 100 μm and beyond become within reach in a variety of materials. For example, using RIE equipment, *through-the-wafer vias* have been made through *100-μm thick slices of GaAs.* For this experiment, a GaAs wafer thinned to 100 μm was patterned with an AZ-4000 series photoresist mask 13 to 16 μm thick. The initial open area of the vias spanned 60 by 60 μm. Using a mixture of 15% $Cl_2$ in $SiCl_4$,

FIGURE 2.17 Relationship between the shape of the etched wall profile and the dependence of the etch rate on the wafer potential. (A) Etch profiles; (B) etch rate bias dependence. The etch rate of $SiO_2$ in the Z direction, $V_Z$, increases with increasing wafer bias. Assuming no chemical etching component is present in the $SiO_2$ reaction, i.e., $V_X = 0$ (the etch rate equals 0 at bias = 0), an ideal vertical profile results. The etch rate of Si also is bias dependent but some reaction occurs at $V_X = 0$. In other words, a chemical component to the reaction does not need ion bombardment resulting in nonvertical walls (see text).[8]

vias were etched in the GaAs slabs.[32] For a review on etching high aspect ratio trenches in silicon by RIE, we refer to Jansen et al.[28] This article features an excellent review of possible trench shapes.

## Reactive Ion-Beam Etching and Chemically Assisted Ion-Beam Etching Compared

Reactive ion-beam etching (RIBE), like CAIBE, involves chemical reactants in a triode set-up. We need to point out an important difference, though. In the case of RIBE, the reactive ions are introduced through the ion source itself; in CAIBE, the reactive gas is fed over the substrate to be etched and unreactive ions are generated in the ion source. At very low pressures in RIBE systems, reactive ions, substituting Ar ions, can sustain a modest etch rate of below 400 Å/min. In CAIBE, ion bombardment of a substrate in the presence of a reactive etchant species leads to a synergism where fast directional material removal rates greatly exceed the separate sum of chemical attack and sputtering rates. At one point this type of etching was thought to be caused by direct chemical reactions between the ions and the surface material (as with RIBE). However, in most practical situations for etch rates ranging from 1000 to 10,000 Å/min, that possibility disappears as the ion flux is much lower than the actual surface removal rates.[8]

The distinction between RIBE and CAIBE is not absolute since the presence of a reactive gas in CAIBE will produce some beam ions from species that back-diffuse into the broad-beam ion source. Figure 2.16 compares a CAIBE system with a RIBE setup.

## Inhibitor-Driven Anisotropy

Inhibitor-driven anisotropy embodies another example of a physical-chemical etching technique (Figure 2.5D). In this case, etching leads to the production of a surface-covering agent. Ion bombardment clears the 'passivation' from horizontal surfaces, and reaction with neutrals proceeds on these cleared surfaces only. The protective film may originate from involatile etching products or from film-forming precursors that adsorb during the etching process. Passivating gases, such as $BCl_3$ and some halocarbons (freons such as $CCl_4$ and $CF_2Cl_2$), are sources of inhibitor-forming species. In the latter case, the reactive gas component appears to be adsorbed on the surface, e.g., a polymer is formed, where it is subsequently dissociated by electron, ion, or photon bombardment, clearing the surface for reaction with the reactive neutrals.

An example of etch profile manipulation with inhibitor-driven chemistry using an idealized Si sample is illustrated in Figure 2.18.[8] At the top of this figure we show the Si etch rate as a function of percentage of $H_2$ in $CF_4$ for a biased and an unbiased wafer. From Figure 2.17B we remember that the Si etch rate increases as the wafer bias is increased so the etch rate curve for the biased wafer lays above the one for the unbiased wafer. If $H_2$ is added to a $CF_4$ feed gas, the Si etch rate decreases, and at some value of $H_2$ concentration, the nonbombarded surface etch rate decreases to zero ($V_X = 0$ at 10% $H_2$ in Figure 2.18) while the bombarded surface continues to etch. The decrease in etch rate stems from an increase in the amount of

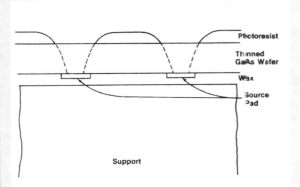

### Via formation in GaAs slab

Through-the-wafer connections are attractive as they improve device gain and packing density. To prepare for the via etch, the front of the wafer with the manufactured devices in place is attached to a glass substrate. The wafer is then thinned to 100 µm, and the back of the wafer is patterned with an AZ-4000 series photoresist mask that is 13 to 16 µm thick. The initial open area of the vias is 60 by 60 µm. To increase the etch resistance of the photoresist and to flow the resist to produce a sloped profile a high-temperature postbake is used. A slope sidewall is desirable to facilitate subsequent metallization. A mixture of 15% $Cl_2$ in $SiCl_4$ at a pressure between 150 and 300 mTorr was used.[32] (From Cooper, III, C. B., Salimian, S., and Day, M. E., *Solid State Technol.*, January, 109–112 (1989). With permission.)

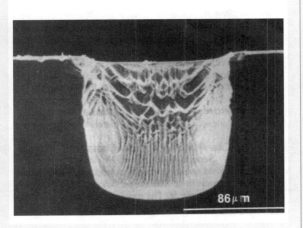

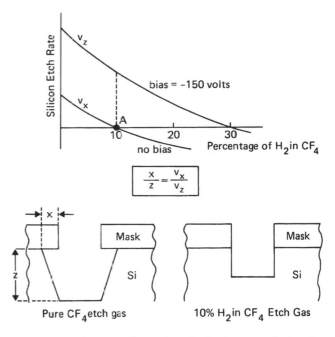

FIGURE 2.18 Trench profile manipulation by decreasing the fluorine-to-carbon ratio (through hydrogen introduction).[8]

passivating polymerization. Aggressive fluorine reacts with the hydrogen so that carbon compounds polymerize more readily. As before, Equation 2.16 can be used to calculate the ratio of underetch, X, to etch depth, Z. At $V_X = 0$, that ratio, of course, is zero.

Similarly, aluminum etching in $CCl_4 + Cl_2$ or $CHCl_3 + Cl_2$ plasmas presents a good example of an inhibitor system. Even though $Cl_2$ and Cl atoms formed in these plasmas are rapid chemical etchants for clean aluminum, aluminum in these plasma mixtures can afford near-vertical profiles with excellent line-width control.

Another typical example of a sidewall mechanism is the etching of phosphorous-doped Si which etches isotropically in a $Cl_2$ plasma but etches anisotropically when $C_2F_6$ is added to the source gas. Qualitatively the effect is accounted for by assuming that the two gases dissociated in the plasma are

$$C_2F_6 + e \rightarrow 2CF_3 + e$$
$$Cl_2 + e \rightarrow 2Cl + e$$

**Reaction 2.1**

The possible surface reactions

$$xCl + Si \rightarrow SiCl_x \text{ (etching)}$$

**Reaction 2.2**

$$CF_3 + Cl \rightarrow CF_3Cl \text{ (recombination)}$$

**Reaction 2.3**

Reaction 2.2 results in etching of the silicon surface, and Reaction 2.3 results in recombination without material removal. Ion bombardment enhances Reaction 2.2, and Reaction 2.3 is assumed to be dominant on the sidewalls. The recombination reaction acts as a sidewall passivant.

In practice, the anisotropy brought about by the directed ion flux in a plasma allows the final etched feature to be within 10% of its dimensions in the mask and submicron resolutions become feasible. A desire to move away from sidewall passivating chemistries to avoid CD loss is prominent, especially in the IC industry. Thick sidewall coatings not only reduce CD control, they also prove difficult to strip. Processes involving little or no sidewall thickness increase are sought for tight CD control in submicron devices. In this respect wafer cooling (to as low as −120°C) presents an attractive way of obtaining sidewall protection without the need for additional chemistries. At low substrate temperatures, reaction products are involatile and can serve as very thin sidewall inhibitors. Quite generally, at lower temperatures lateral etch rate can be suppressed while using simpler chemistries. In micromachining applications where feature size is less important and aspect ratios are extreme, applying passivating chemistries is almost a necessity, but wafer cooling can help the process.

## Gas Composition

Table 2.5 lists frequently used reactive gases and typical applications.[6] Oxidizing additives are added to the plasma to increase etchant concentration and to suppress polymerization. The addition of oxygen to $CF_4$ in Si etching typifies the procedure.[33] Oxygen, at concentrations below 16%, reacts with $CF_X$ radicals

to enhance F atom formation (increase of fluorine-to-carbon ratio F/C) and eliminate polymerization. At yet higher concentration, adsorbed oxygen on the surface depresses the etch rate.[33] Radical scavengers such as hydrogen increase the concentration of inhibitor former and reduce etchant concentration of fluorine in the etchant (decrease of F/C) (see also example in Figure 2.18). Heinecke[34] realized that the etch ratio of $SiO_2/Si$ can be increased either by adding $H_2$ to the $CF_4$ feed gas or by employing $CHF_3$ or, in general, fluorocarbons displaying a smaller F/C ratio than $CF_4$. Adding hydrogen to fluorocarbon gases helps promote $CF_X$ film growth (such as $CF_2$, $C_2F_4$) for selective $SiO_2$ etching. The carbon accumulates less on oxide surfaces than on Si surfaces, as $SiO_2$ surfaces directly react with the hydrogen. By adding hydrogen, the carbon blocking increases (especially on the Si) because hydrogen scavenges fluorine forming HF and preventing reaction with carbonaceous species on the surface.

The same approach helps to optimize the selectivity on other oxide/nonoxide systems (e.g., $SiO_2$/resist, $SiO_2/Si_3N_4$, $Ta_2O_5$/Ta, etc.). The F/C ratio should not be made smaller than two or contamination of the whole system with $C_XF_Y$ polymer sets in. Factors that tend to control polymerization and selectivity are temperature, hydrogen concentration, pressure, and ion bombardment. Some etchants, such as Cl atoms, do not readily etch through thin native oxide films on, for instance, Si, Nb, and Al. These native oxides prevent the onset of etching unless small amounts of native oxide etchants, such as $BCl_3$, are added. Inert gases such as argon or helium help stabilize the plasma, enhance anisotropy, improve uniformity, or reduce the etching rate by dilution. Since helium has a high thermal conductivity, it also improves heat transfer between wafers and the supporting electrodes.[26]

**TABLE 2.5**  Frequently Used Reactive Plasma Gases

| Etchant: Purpose | Composition (Additive-Etchant): Application |
|---|---|
| Oxide etchant: Etches through oxide to initiate etching | $C_2F_6$-$Cl_2$: $SiO_2$<br>$BCl_3$-$Cl_2$: $Al_2O_3$<br>$CCl_4$-$Cl_2$: $Al_2O_3$ |
| Oxidant: Increases etchant concentration and suppresses polymerization | $O_2$-$CF_4$: Si<br>$N_2O$-$CHF_3$: $SiO_2$<br>$O_2$-$CCl_4$: GaAs, InP |
| Inert gas: Stabilizes plasma, dilutes etchant, improves heat transfer | Ar-$O_2$: organic material removal<br>He-$CF_3Br$: Ti, Nb |
| Inhibitor former: improves selectivity, induces anisotropy | $C_2F_6$-$Cl_2$: Si<br>$BCl_3$-$Cl_2$: GaAs, Al<br>$H_2$-$CF_4$: $SiO_2$<br>$CHF_3$-$SF_6$: Si<br>$O_2$(50%)-$SF_6$: Si |
| Water and oxygen scavenger: prevents inhibition, improves selectivity | $BCl_3$-$Cl_2$: Al<br>$H_2$-$CF_4$: $SiO_2$ |
| Radical scavenger: increases film formation and improves selectivity | $H_2$-$CF_4$: $SiO_2$<br>$CHF_3$-$SF_6$: Si<br>$CF_3Br$-$SF_6$: Si<br>$CF_2Cl_2$-$SF_6$: Si |

(Most data from Flamm, D. L., *Solid State Tech.*, 49–54, 1993. With permission.)

The need to eliminate the use of chlorofluorocarbons (CFCs) is changing the type of gases used for dry etching purposes. The evolution in gas mixtures is captured in Table 2.6.[17] The primary new etchant gases are $SF_6$, $NF_3$, and $SiF_4$ (fluorine-based etch); $Cl_2$, $BCl_3$, and $SiCl_4$ (chlorine-based etch); and $Br_2$ (bromine-based etch). For very aggressive etch chemistry at low pressures (e.g., to etch deep trenches in Si) HBr also gained popularity. Other new etchant gases include $SiBr_4$, HI, $I_2$, and even nonhalogenated gases such as $CH_4$ and $H_2$. Table 2.7 lists etchants appropriate for etching common electronic materials. Table 2.8 presents an overview of the etch rates or etch ratios for a variety of important microfabrication materials in some popular reactive gases and gas mixtures.[38] Examples are etching of photoresist on Si, $SiO_2$, Al, etc. in $O_2$ plasma and etching of Si with a metal mask (e.g., Al) in a $CF_4$ plasma. Metal masks influence the etch rate of Si or even $SiO_2$ by catalytic action of fluorinated metal surfaces, leading to excess production of free radicals. And finally, Table 2.9 reviews mask materials listing their suitability for a number of gases.[26] A plus sign in this table corresponds with a low etch rate and high selectivity. For further reading on gas composition for dry etching and stripping, see References 6, 17, 35–38.

## Simplifying Rules

Some simple rules help interpret Tables 2.5 to 2.9 when specifying a choice of dry etchant and mask for a specific application. These rules are a set of semi-empirical observations; they should not be looked upon independently, and some state the same phenomenon in slightly different ways.[39]

1. *Fluorine-to-carbon (F/C) ratio:* During etching, polymerization occurs simultaneously. Etching stems from the fluorine and polymerization from the hydrocarbons. The dominant process will depend on the gas stoichiometry, reactive-gas additions, the amount of material to be etched, and the electrode potential. Adding hydrogen causes HF to form and the F/C ratio to drop, leading to more polymerization and less etching (see example in Figure 2.18). Decreasing the fluorine concentration can also occur by overloading the

**TABLE 2.6** The Evolution in Gas Mixtures for Dry Etching

| Material being etched | Conventional chemistry | New chemistry | Benefits |
|---|---|---|---|
| PolySi | $Cl_2$ or $BCl_3$ / $CCl_4$ ⎫ sidewall<br>/ $CF_4$ ⎬ passivating<br>/ $CHCl_3$ ⎪ gases<br>/ $CHF_3$ ⎭ | $SiCl_4$/$Cl_2$<br>$BCl_3$/$Cl_2$<br>$HBr$/$Cl_2$/$O_2$<br>$HBr$/$O_2$<br>$Br_2$/$SF_6$<br>$SF_6$<br>$CF_4$ | No carbon contamination<br><br>Increased selectivity to $SiO_2$ and resist<br>No carbon contamination<br><br>Higher etch rate |
| Al | $Cl_2$<br>$BCl_3$ + sidewall passivating gases<br>$SiCl_4$ | $SiCl_4$/$Cl_2$<br>$BCl_3$/$Cl_2$<br>$HBr$/$Cl_2$ | Better profile control<br>No carbon contamination |
| Al-Si (1%)-Cu (0.5%) | Same as Al | $BCl_3$/$Cl_2$ + $N_2$ | $N_2$ accelerates Cu etch rate |
| Al-Cu (2%) | $BCl_3$/$Cl_2$/$CHF_3$ | $BCl_3$/$Cl_2$ + $N_2$ + Al | Additional aluminum helps etch copper |
| W | $SF_6$/$Cl_2$/$CCl_4$ | $SF_6$ only<br>$NF_3$/$Cl_2$ | No carbon contamination<br>Etch stop over TiW and TiN<br>No carbon contamination |
| TiW | $SF_6$/$Cl_2$/$O_2$ | $SF_6$ only | |
| $WSi_2$, $TiSi_2$, $CoSi_2$ | $CCl_2F_2$ | $CCl_2F_2$/$NF_3$<br>$CF_4$/$Cl_2$ | Controlled etch profile<br>No carbon contamination |
| Single crystal Si | $Cl_2$ or $BCl_3$ + sidewall passivating gases | $CF_3Br$<br>$HBr$/$NF_3$ | Higher selectivity trench etch |
| $SiO_2$ (BPSG) | $CCl_2F_2$<br>$CF_4$<br>$C_2F_6$<br>$C_3F_8$ | $CCl_2F_2$<br>$CHF_3$/$CF_4$<br>$CHF_3$/$O_2$<br>$CH_3CHF_2$ | CFC alternatives |
| $Si_3N_4$ | $CCl_2F_2$<br>$CHF_3$ | $CF_4$/$O_2$<br>$CF_4$/$H_2$<br>$CHF_3$<br>$CH_3CHF_2$ | CFC alternatives |
| GaAs | $CCl_2F_2$ | $SiCl_4$/$SF_6$<br>/$NF_3$<br>/$CF_4$ | Florine provides etch stop on AlGaAs |
| InP | None | $CH_4$/$H_2$<br>HI | Clean etch<br>Higher etch rate than with $CH_4$/$H_2$ |

(From Peters, L., *Semicon. Intl.*, 66–70, 1992. With permission.)

**TABLE 2.7** Plasma Etchants for Microelectronic Materials

| Material | Common Etch Gases[a] | Dominant Reactive Species | Product | Comment | Vapor Pressure (Torr at 25°C) |
|---|---|---|---|---|---|
| Aluminum | Chlorine-based | Cl, $Cl_2$ | $AlCl_3$ | Toxic gas and corrosive gases | $7 \times 10^{-5}$ |
| Copper | Forms low pressure compounds | Cl, $Cl_2$ | $CuCl_2$ | Toxic gas and corrosive gases | $5 \times 10^{-2}$ |
| Molybdenum | Fluorine-based | F | $MoF_6$ | — | 530 |
| Polymers of carbon and photoresists (PMMA and polystyrene) | Oxygen | O | $H_2O$, CO, $CO_2$ | Explosive hazard | $H_2O = 26$, CO, $CO_2 > 1$ atm |
| III-V and II-VI compounds | Alkanes | — | — | Flammable gas | — |
| Silicon | Fluorine- or chlorine-based | F, Cl, $Cl_2$ | $SiF_4$, $SiCl_4$ | Toxic gas | $SiF_4 > 1$ atm, $SiCl_4 = 240$ |
| $SiO_2$ | $CF_4$, $CHF_3$, $C_2F_6$, and $C_3F_6$ | $CF_X$ | $SiF_4$, CO, $CO_2$ | — | $SiF_4 > 1$ atm, CO, $CO_2 > 1$ atm |
| Tantalum | Fluorine-based | F | $TaF_3$ | — | 3 |
| Titanium | Fluorine- or chlorine-based | F, Cl, $Cl_2$ | $TiF_4$, $TiF_3$, $TiCl_4$ | — | $TiF_4 = 2.10^{-4}$, $TiCl_4 = 16$ |
| Tungsten | Fluorine-containing | F | $WF_6$ | — | 1000 |

[a] Common chlorine containing gases: $BCl_3$, $CCl_4$, $Cl_2$, and $SiCl_4$. Common fluorine containing gases: $CF_4$, $SF_4$, and $SF_6$. (After References 6, 26, 38, and 39.)

**TABLE 2.8** Frequently Used Reactive Plasma Gases, Reported Etch Rates, and Etch Ratios

| Film/Underlayer (F/U) | Etch Ratios-F/U |
|---|---|
| $Si_3N_4$ over AZ-2400 resist | $CF_4$-10 |
| $SiO_2$ over AZ-1350 resist | $CF_4/H_2$-10 |
| Poly Si over PBS resist | $CF_4$-15 |
| PSG over AZ-1350 | $SF_6$-10 |
| $SiO_2$ over Si underlayer | $CF_3$ Cl-30 |
| $Si_3N_4$ over $SiO_2$ underlayer | $NF_3$-50 |
| Poly-Si over $SiO_2$ underlayer | $CCl_4$-10 |
| Si over $SiO_2$ underlayer | $SF_6$-30 |

| Etchant | Material and Etch Rate (Å/min) |
|---|---|
| Ar | Si-124, Al-166, resist-185, quartz-159 |
| $CCl_2F_2$ | Si-2200, Al-1624, resist-410, quartz-533 |
| $CF_4$ | Si-900, PSG-200, thermal $SiO_2$-50, CVD $SiO_2$-75 |
| $C_2F_6$-$Cl_2$ | Undoped Si-600, thermal $SiO_2$-100 |
| $CCl_3F$ | Si-1670 |

*Source:* Moreau, *Semiconductor Lithography* (Plenum Press, New York, 1988). With permission.

**TABLE 2.9** Mask Materials in Dry Etching

| Mask Material | $SF_6$ | $CHF_3$ | $CF_4$ | $O_2$ | $N_2$ |
|---|---|---|---|---|---|
| Si | − | +/− | − | + | + |
| $SiO_2$ | +/− | − | +/− | + | + |
| $Si_3N_4$ | +/− | − | +/− | + | + |
| $Al/Al_2O_3$ | + | + | + | + | + |
| W | − | − | − | + | + |
| Au | + | + | + | + | + |
| Ti | − | − | − | + | + |
| Resist | +/− | +/− | +/− | − | + |
| CFs | + | + | + | + | − |

*Source:* After Elwenspoek et al., Universiteit Twente, *Micromechanics,* (1994).

leads to aggressive resist etching. Gases such as $NF_3$ and $ClF_3$ allow high fluorine concentrations (for aggressive Si etching) without the addition of oxygen, thus avoiding resist attack.

2. *Selective vs. unselective dry etching:* The polymerizing point of the gas (i.e., the composition of the gas where polymerization takes over) primarily determines selectivity. The closer one works to the polymerization point the better the selectivity. Factors that have a tendency to increase polymerization rate increase selectivity: decreased temperature, high hydrogen concentration, low power, high pressure, and high monomer concentration all increase polymerization and thus selectivity. Typical unselective etchants for Si and polysilicon, used for noncritical etching are $CF_4$ and $CF_4$-$O_2$. Less toxic gases such as $CF_4$ and $SF_6$ get preference over fluorine and are more selective, leading to polymerization. The most commonly used gases for selective Si etching are $Cl_2$, $CCl_4$, $CF_2Cl_2$, $CF_3Cl$, $Br_2$, and $CF_3Br$, along with

reactor, leading to overconsumption of fluorine and favoring of polymerization. Adding oxygen to a fluorine carbon mixture leads to formation of CO and $CO_2$ reaction products, increasing F/C and thus the etch rate while decreasing the polymerization tendency. In other words, the addition of oxygen to a gas mixture reduces the tendency of Freons™ to polymerize and increases the concentration of halogen etchants ensuing from these gases. Adding oxygen to improve the F/C ratio mixtures such as $Cl_2$-$C_2F_6$. Small additions of halogens significantly increase the selectivity of fluorine-based etchants, especially the selectivity of Si over silicon dioxide. For pure chlorine, the etch rate of $SiO_2$ is so low that even a native oxide can prevent the Si from etching. The high selectivity of $SiO_2$/Si is a major objective of many plasma processes to mimic the wet HF etch.

3. *Substrate bias:* A negative bias on a surface exposed to the plasma increases, at a constant F/C ratio, the etching behavior over polymerization tendency. This effect is caused by the enhanced energy of the ions striking the surface, resulting in polymer sputtering.

4. and 5. *Dry etching of III-V compounds:* Group III halides, particularly fluorides, tend to be involatile. As a result, F plasmas are usually substituted for chlorine-containing plasmas and elevated substrate temperatures are used to further help volatize the chlorides (fourth rule). The chemical composition (Ga/As ratio) of the different atomic planes in GaAs vary, and crystallographic etch patterns are observed under etch conditions in which chemical processes dominate (fifth rule). The latter establishes a major difference with dry Si etching where crystallography does not play a role.

6. *Metal etching:* Chlorocarbons and fluorocarbons typically are used to etch metal films since they can reduce native metal oxides chemically. Oxygen and water vapor must be rigorously excluded during etching because of the high stability of the metal-oxide bond. Also, because of the stability of that bond, ion bombardment is essential. Since $AlF_3$ is not volatile, chlorine-containing gases are preferred for Al etching, while tungsten can be etched in fluorine ($SF_6$ or $NF_3$/chlorine).

7. *Organic films:* To retain an organic mask while etching away materials such as $SiO_2$ and $Si_3N_4$ one must work in etching conditions close to the polymerization point so that some loss of resist is compensated by condensation of reaction product, e.g., $CF_4$-$C_2H_4$ and $CF_4$-$H_2$. We already discussed that $CF_4$-$O_2$ plasmas severely degrade resist materials.

8. *Carbon-containing additives:* Carbon-containing additives generally degrade the selectivity of Cl- and Br-based inorganic chemistries for polysilicon etching over $SiO_2$. The Br atom attack on $SiO_2$ is thermodynamically unfavorable, whereas reactions between carbon halogen bonded species and $SiO_2$ are exothermic. It was shown that with HBr a polysilicon over silicon dioxide selectivity of 300:1 can be achieved when completely avoiding carbon containing gases or photoresists; the presence of resists or carbon traces in gases severely diminishes this ratio.

## New Plasma Sources

New plasma sources keep emerging. More recently introduced plasma sources are based on *microwave electron cyclotron resonance* (ECR)[40] sources and *inductively coupled plasma (ICP) technology,*[7] both producing very high-density plasmas.

In ECR reactors, a magnetically enhanced source is driven by microwave excitation resonating at the orbital frequency where electrons circle in the magnetic field. The 2.45-GHz microwave frequency is generated by a magnetron and injected into an etching chamber, enclosed by a quartz bell-jar, through a waveguide. A magnetic field generated by solenoid coils is applied to the chamber. The interaction of the magnetic field and the microwaves result in an intense, high-density plasma maintainable at low pressure (usually $10^{-5}$ to $10^{-2}$ Torr). In many ECR systems the wafers are placed downstream of the plasma to further limit their exposure to this intense discharge. Wafers can also be biased to control ion bombardment energy. Like MERIE, ECR produces higher plasma densities with low energy density, reducing charge-up damage. Using ECR, Juan et al.[41] demonstrated that the $n^{++}$ Si etch increases, but the $p^{++}$ Si etch rate decreases with respect to lightly doped Si. It was thus established that with a high microwave power, high RF power, and an increased temperature a large difference in etch rates of $p^{++}$ and $n^{+-}$ Si results.

In an ICP source, a plasma is driven inductively with a power source operating at the standard 13.56 MHz. Such ICP sources create high-density, low-pressure, low-energy plasmas by coupling ion-producing electrons to the magnetic field arising from the RF voltage. In ICP sources, one shields the plasmas from the electric field of the RF to avoid capacitive coupling which tends to create high-energy ions. The Alcatel radio frequency (RF) inductively coupled plasma source has recently been advanced as an ideal source for deep anisotropic Si etching.[7] This equipment, using fluorine based gases, etches silicon at rates of 6 μm/min with an etch uniformity better than +/– 5% while maintaining a Si:$SiO_2$ selectivity of more than 150:1 (wafer at –100°C). Etch depths greater than 250 μm and profile angles of +/– 1 degree were demonstrated while aspect ratios of 9:1 were achieved (e.g. a 4 μm wide and 35 μm deep trench). The electron density in this plasma source reaches >$10^{12}$/$cm^3$ in argon.

The ideal dry etching technology of the future will have high plasma densities (>$10^{11}$/$cm^3$) in order to achieve a high etch rate while operating at a low pressure (1 to 20 m Torr). Low pressure increases ion directionality and discourages microloading. The source also needs to produce ions uniformly in energy and distribution while keeping the ions' energy low. For building high aspect ratio micromachines, cryogenic cooling of the substrate will gain importance and might, for certain applications, become a less expensive means to build molds than the LIGA technique (see Chapter 6).

## Dry Etching Models — *In Situ* Monitoring

A serious impediment to more effective use of plasma etching is the large number of parameters that affect the process.[42] Just consider pressure and temperature. Pressure influences the ion-to-neutral ratio and fluxes of these particles, the sheath potentials and energy of the ions bombarding surfaces, the electron energy, and the relative rate of chemical kinetics. Temperatures, both gas and surface, have a profound influence on discharge chemistry as well. Only the surface temperature is really controllable and influences selectivity, etch rates, degradation of

### Microwave electron cyclotron resonance (ECR)

(From Editorial, *Semicon. Intl.*, 72–76, 1992. With permission.)

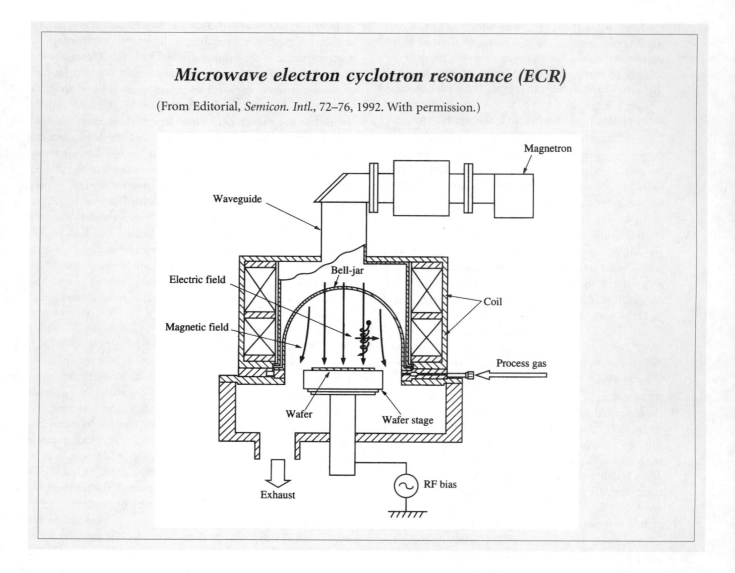

resist masks, and surface roughness (higher temperature leads to a rougher surface).

Also, electrode area ratio, RF frequency, and power will alter plasma and electrode potentials, thereby changing the ion energies. At lower frequencies, where ions can respond directly to the oscillating field, ions can attain the maximum energy corresponding to the maximum field across the sheath. As a result, for a constant sheath potential, ion bombardment energies are higher at lower frequencies (lower than 50 kHZ). Nonuniformity can be caused by gradients in etchant concentration. Such gradients are responsible for the dreaded bull's-eye etching pattern, where the etch rate decreases monotonically from the wafer periphery to its center (see above). Pressure, power, and gas flow must be adapted until the specified uniformity is obtained.

Predicting the exact relation between the flux and the energy of particles striking a wafer surface and the actual etch rate still is not state of the art, largely because no equilibrium is attained in these dry etching reactors and the kinetics are not known for most reactions.[8] Consequently, *in situ* monitoring of the plasma to improve etching uniformity is an important area for further research.[43,44] Besides measuring these controlling parameters, techniques such as optical emission spectroscopy (OES) to map *in situ* etchant concentration are coming to the foreground as sensitive and effective endpoint detectors. Emissions can emanate from etchants, etch products, or their fragments. An optical spectrometer is set to the line of a reaction product of interest and one follows its intensity during an etch cycle. Laser interferometry (for thin transparent films such as polysilicon, silicon dioxide, and silicon nitride) and laser spot reflectance (for opaque, highly reflecting metal films) are used to monitor film thickness and film disappearance, respectively. Also, the ***temperature of the wafer*** during reactive ion etching affects process quality in terms of photoresist integrity, selectivity, etch rate, and etch residues. An *in situ* temperature monitor enables one to improve the performance during the plasma etch.[45]

## Comparing Wet and Dry Etching

In bulk micromachining, requiring more extreme topologies (more z-axis) than classical IC devices, wet etching of crystalline Si still dominates the state of the art. Wet anisotropic etching of Si results in atomically smooth planes and atomically

## Inductively-coupled plasma

This electrostatic, shielded, inductive-coupled plasma source produces electric field lines from helical resonator combined with an electrostatic shield to produce electric field lines that are circumferential in response to the axial RF magnetic field.

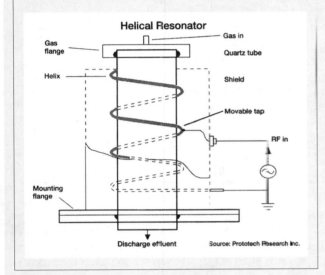

Source: Prototech Research Inc.

## Wafer temperature

Noncontact temperature monitoring of wafer temperature during plasma etch. The monitor relies on the temperature dependence of intrinsic optical properties of silicon, as detected from the back surface of a wafer. (From Duek et al., *Semicon. Intl.*, 208–210, 1993. With permission.)

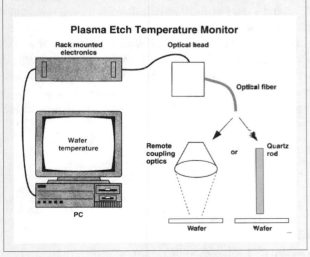

sharp edges, properties which are hard to obtain with dry etching.

Dry etching already is predominately used in the IC manufacture, and, with higher aspect ratio devices and better materials selectivity being attained the same is happening now in micromachining. Besides selectivity and aspect ratio, problems with and concerns about the slow dry etch rate and high sensitivity to operating parameters are being addrressed. Etch rates as high as 6 μm/min have become possible and operation is much more automated. A more recent concern is the search for alternative chemistries since production of chlorofluorocarbons (CFCs) were banned after 1995. According to the Montreal Protocol, all ozone-depleting CFCs must be eliminated by the year 2000. This includes etchants such as $CFCl_3$, $CCl_2F_2$, $CF_3Cl$, $C_2F_5Cl$, and $CF_3Br$.[46] Environmental issues dictated a switch from wet to dry etching for most IC applications. Another more decisive concern was the need for better CD control. A lot more research and development are needed to establish new dry etching schemes that outperform the wet anisotropic etch performances.

Surface micromachining relies on both wet and dry etchants, with processes similar to the ones used in the IC industry. For example, wet etchants are employed to etch away sacrificial layers (e.g., HF to etch CVD phosphosilicate glass) while dry etching helps to define polysilicon structural elements (e.g., in a $NF_3$-$O_2$ plasma). A combination of wet and dry techniques presently makes up the majority of Si micromachining work. Table 2.10 compares the characteristics of dry vs. wet etching techniques.

See Appendix B for Dry Etching URL's.

## Dry Etching Micromachining Examples

Examples of dry etched micromachines are presented below. They involve dry etching in single crystal Si (SCS), single crystal GaAs and polyimide, and combined wet/dry etching. In the IC industry deep trenches for device isolation and narrow contact holes also require higher and higher aspect ratios. Dry etching requirements for future IC development are presented, as well.

### Example 1. Silicon Single Crystal Reactive Etching and Metallization (SCREAM)

As a first example of dry etching in micromachining we will follow some of the work by the Cornell Nanofabrication Laboratory in making laterally driven micromachines.[47,48] In laterally driven micromachines, beams and structures of various shapes should be of a construction often narrow and thick, that can move freely and in parallel to the supporting substrate while rigidly held above it. Increased thickness of beams permits greater driving power for a given voltage and increases stiffness of the device perpendicular to the plane of motion. Thin suspended beams about 2 μm thick can be made with combinations of dry and wet etching in poly-Silicon (surface micromachining, see Chapter 5), but thick single crystal beams, >20 μm, in materials such as quartz, Si, and GaAs, can best be

**TABLE 2.10**   Comparison of Dry vs. Wet Etching Techniques[a]

| Parameter | Dry Etching | Wet Etching |
|---|---|---|
| Directionality | Can be highly directional with most materials (aspect ratio of 25 and higher) | Only directional with single crystal materials (aspect ratio of 100 and higher). |
| Production-line automation | Good | Poor |
| Environmental impact | Low | High |
| Masking film adherence | Not as critical | Very critical |
| Cost chemicals | Low | High |
| Selectivity | Poor | Can be very good |
| Materials that can be etched | Only certain materials can be etched (not, e.g., Fe, Ni, Co) | All |
| Radiation damage | Can be severe | None |
| Process scale-up | Difficult | Easy |
| Cleanliness | Good under the right operational conditions | Good to very good |
| Critical dimension control | Very good ($<0.1$ μm) | Poor |
| Equipment cost | Expensive | Inexpensive |
| Submicron features | Applicable | Not applicable |
| Typical etch rate | Slow (0.1 μm/min) to fast (6 μm/min) | Fast (1μm/min, anis.) |
| Theory | Very complex, not well understood | Better understood (see Chapter 4) |
| Operating parameters | Many | Few |
| Control of etch rate | Good in case of slow etching | Difficult |

[a]   See also Chapter 4.

made with dry etching the single crystalline materials, especially since thick lateral moving microstructures, given the crystallographic limitations, are very difficult to realize using wet bulk micromachining (Chapter 4). A Cornell dry etching process that obtains such laterally driven, single crystal micromachines is called SCREAM (single crystal reactive etching and metallization). This technology uses reactive ion etching both to define and release structures. SCREAM portrays a relatively new micromachining approach and represents an important new technique from several points of view. It is a self-aligned, single mask process, run at low-temperatures ($<300°C$), and completed in less than 8 hours that can be carried out in the presence of integrated circuitry on the same chip. The SCREAM example given here is the fabrication of single crystal, silicon-released beam structures (see Figure 2.19).[49]

As starting substrate an arsenic-doped, 0.005-Ωcm, n-type (100) silicon wafer is used. A masking layer of plasma-enhanced chemical vapor-deposited silicon dioxide (PECVD), 1 to 2 μm thick, is coated with resist (step 1 and 2). Photolithography on the photoresist layer creates the desired pattern (step 2) to be transferred onto the silicon dioxide using $CHF_3$-based, magnetron reactive ion etching (MERIE) (step 3). The photoresist is then stripped in an oxygen barrel etcher (step 4). The silicon dioxide mask pattern subsequently is transferred to the silicon substrate using a $Cl_2/BCl_3$ plasma etch. The etch depth, depending on the intended micromachine, ranges between 4 and 20 μm (step 5). Following the silicon etch, silicon dioxide is conformably deposited using plasma-enhanced chemical vapor deposition (PECVD). This oxide will serve as a protection for the sidewalls during release at a thickness of 0.3 μm (step 6). The next step is very interesting and crucial to the process. The objective is to remove the oxide from the bottom and the top

of the structure without removing the oxide from the sidewalls. This selective clearing of oxide materializes in an anisotropic $CF_4/O_2$ etch at 10 mT. It removes 0.3 μm from the mesas and the trench bottoms while leaving the sidewalls oxides intact (step 7). In the next step, a second deep Si etch ($BCl_3/Cl_2$) deepens the trench further (3 to 5 μm deeper) so as to expose some Si sidewall underneath the oxide-covered part (step 8). This exposed Si will be removed during release. The release is accomplished with a $SF_6$ isotropic etch at 90 mT. The RIE step etches the Si out from underneath the beams, releasing them in the process (step 9). An etchant such as $SF_6$ hardly etches the oxide and gives good selectivity to the process. A layer of aluminum is then sputter-deposited using DC magnetron sputter deposition (see Chapter 3) (step 10). The Al can be contacted and used as drive electrodes in micromechanical resonator structures. Resonator elements with beam elements of 0.5 to 5 μm and aspect ratios >10 have been achieved using this Si SCREAM process.

The measurement of the resonant frequency variation of small resonator devices made this way (see Figure 2.19B) can be used to measure monolayer mass changes, as well as pressure, temperature, acceleration, etc. In the measurement of mass, for example, the sensitivity of the device is a function of resonant frequency and mass of the resonating structure:

$$\text{Sensitivity} \sim \frac{\text{Resonant frequency}}{\text{Mass}} \qquad 2.17$$

To build the most sensitive detector, the mass must be minimized and the resonant frequency maximized. The mechanical resonant frequency of an ideal resonator is given by:

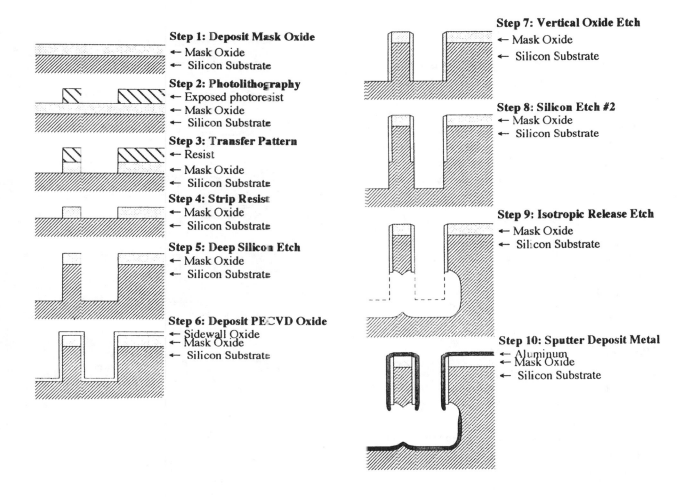

**Step 1: Deposit Mask Oxide**
← Mask Oxide
← Silicon Substrate

**Step 2: Photolithography**
← Exposed photoresist
← Mask Oxide
← Silicon Substrate

**Step 3: Transfer Pattern**
← Resist
← Mask Oxide
← Silicon Substrate

**Step 4: Strip Resist**
← Mask Oxide
← Silicon Substrate

**Step 5: Deep Silicon Etch**
← Mask Oxide
← Silicon Substrate

**Step 6: Deposit PECVD Oxide**
← Sidewall Oxide
← Mask Oxide
← Silicon Substrate

**Step 7: Vertical Oxide Etch**
← Mask Oxide
← Silicon Substrate

**Step 8: Silicon Etch #2**
← Mask Oxide
← Silicon Substrate

**Step 9: Isotropic Release Etch**
← Mask Oxide
← Silicon Substrate

**Step 10: Sputter Deposit Metal**
← Aluminum
← Mask Oxide
← Silicon Substrate

**A**

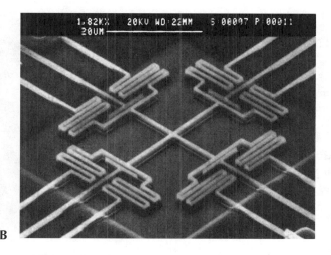

**B**

**FIGURE 2.19** (A) Si-SCREAM process. Cross-section of a typical beam made using the SCREAM process. These figures show a beam and its associated parallel plate capacitor. The released beam is free to move while the plate on the right is static and can be used to measure the motion of the beam. (From Shaw, K. A. et al., *Sensors and Actuators A*, 40, 63–70 (1994). With permission.) (B) A representative SCREAM structure: SEM micrograph of a Si x,y stage with parallel plate capacitive drives and sinuous springs. (From Yao, J. J. et al., *J. Microelectromech. Syst.*, 1, 14–22 (1992). With permission.)

$$f_r = \left(\frac{k}{m}\right)^{\frac{1}{2}} \qquad \textbf{2.18}$$

with k the force constant and m its mass (stiffness/mass)$^{1/2}$. A large spring stiffness must be designed and m minimized to make $f_r$ as high as possible. For some of the very stiff, low mass structures made at Cornell, $f_r$ was 5 MHz, with k = 55 N/m and m = $2.0 \times 10^{-13}$ kg. The above simple considerations illustrate the power of micromachining in enabling more sensitive detectors.

Lateral resonant structures as shown in Figure 2.17B can also be used to make sets of comb electrodes, for example to drive precision x-y microstages for positioning or as comb-drives in accelerometers. Scanning tunneling microscopes (STM), 40 × 40 μm in size, have been manufactured using the SCREAM technology.[49] In the latter case, field emitter tips were integrated on an x-y stage with moving beams with nominal cross-sectional dimensions of 250 nm in width and 1000 nm in height. Displacements in the x-y directions through parallel plate capacitive drives measured ±200 nm for an applied voltage in the 50 V range.

## Example 2. GaAs-SCREAM

Figure 2.20A outlines the SCREAM process steps in the case of a GaAs-based device.[48] Specifically, the process sequence for the fabrication of a cantilever beam structure identical to the Si one in Figure 2.19A is shown here again.[50] Considering two identical structures enables us to better compare the degree of complexity of micromachining in those two materials. Single crystal GaAs, in some respects, is an attractive material for micromechanics because of its optoelectronic properties. With this kind of material one can envision the possibility of vertically etched mirrors, lenses, detectors, waveguides, and semiconductor lasers, all adding up to the possibility of a totally integrated optical bench on GaAs. On the other hand, in comparison to silicon, GaAs possesses a low fracture strength (its yield strength is considerably lower: 2 vs. 7 GPa) and is more difficult to passivate than Si as its own oxides are either leaky or chemically unstable. The process example illustrates the difficulty in passivating GaAs vs. passivating Si (a dual layer dielectric is required to passivate GaAs). A more complete comparison of Si vs. GaAs in micromachining is presented in Chapter 8.

For the simple movable beam in Figure 2.20A, chemically assisted ion beam etching (CAIBE) produces deep vertical trenches, and reactive ion etching (RIE) laterally undercuts and releases the vertically etched structures from the GaAs substrate.[50] This process produces aspect ratios of 25:1 with trenches of 20 μm vertical depth to 400 nm lateral width. Plasma-enhanced chemically vapor-deposited SiN$_x$:H (200 nm) and a second layer of SiO$_x$:H (150 nm) are used to mask the GaAs substrate (Figure 2.20A,a). PECVD-SiN$_x$:H on GaAs easily results in structural buckling and layer cracking, but nevertheless is preferred as a conformable layer because the surface of PECVD-SiO$_x$:H on GaAs is rough due to the diffusion of gallium into PECVD-SiO$_x$:H. The PECVD-SiN$_x$:H deposition temperature of 100°C is selected to obtain simultaneously good dielectric properties and low residual stress in the film. To further protect the GaAs during etching, an additional sacrificial

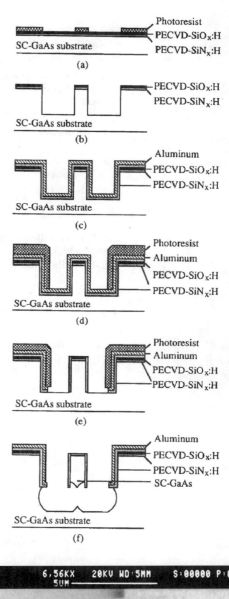

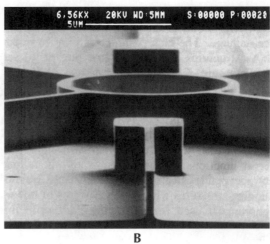

**FIGURE 2.20** (A) Formation of a SC-GaAs cantilever beam with aluminum electrodes adjacent to each side of the beam using PECVD-SIN$_x$: H for insulation and for the top and sidewall etch mask. (From Zhang, et al., *MEMS '92*, 72–77. With permission.) (B) SEM micrography showing SC-GaAs circular and angled straight-line features after the CAIBE step. (From Zhang, Z. L., *J. Microelectromechanical Systems*, Vol. 2, No. 2, June 1993, 66–73. With permission.)

PECVD-SiO$_X$:H is deposited. The thermal expansion mismatch between the two dielectrics is less than between either dielectric and the GaAs. The pattern producing freestanding beams in GaAs is then printed in a photoresist layer on top of the bilayer dielectric. Plasma etching with CHF$_3$/O$_2$ transfers this image into the dielectric layer. The pattern in the dielectric stack subsequently is transferred to the single crystal GaAs by CAIBE (Figure 2.20A,b) using a 500-eV, 0.1-mA/cm$^2$ Ar ion beam and a Cl$_2$ flow of 10 ml/min. After the CAIBE process, the remaining photoresist is stripped in an oxygen plasma and a 300 nm PECVD layer of SiN$_X$:H is conformably deposited. Subsequently, using DC magnetron sputtering, a 400-nm layer of aluminum is deposited (Figure 2.20A,c). Photoresist is spun on the Al layer and the aluminum side-electrode pattern is printed in the resist, developed (Figure 2.20A,d), and then transferred to the Al by a Cl$_2$/BCl$_3$ etch. The latter step removes the Al from the areas that have been cleared of resist. Prior to the aluminum deposition, contact windows are formed to allow electrical contact to both the GaAs substrate and the movable GaAs beams. To clear the PECVD-SiN$_X$:H (from the second deposition of this compound) from the bottom of the trenches and to retain the dielectric elsewhere, a highly anisotropic CHF$_3$/O$_2$ is applied (Figure 2.20A,e). Finally, the GaAs structures are released from the substrate using a chlorine-based RIE (Figure 2.20A,f). The top and sidewalls of the GaAs structure remain protected by the PECVD-SiN$_X$:H and PECVD-SiO$_X$:H mask. At the end of the beam-releasing etch, the PECVD-SiO$_X$:H is consumed, i.e., it acts as a sacrificial mask. Finally, the photoresist for the Al patterning is stripped in oxygen. In a simplification of the process described in Figure 2.20A, Zhang et al.[48] recently discovered an improved PECVD-SiN$_x$H$_y$ conformal coating enabling a single dielectric mask for the GaAs SCREAM process. Figure 2.20B demonstrates circular and angled straight-line features in SC-GaAs produced with the simplified process.

## Example 3. Dry Etching of Polymeric Materials

### Electron Cyclotron Resonance to Etch Polyimide

Juan et al.[51] use electron cyclotron resonance (ECR) to etch polyimide molds for the fabrication of electroplated microstructures in one of several reported 'pseudo-LIGA' processes (see also Chapter 6). A fast polyimide etch rate of 0.91 μm/min and a high selectivity over a Ti etch mask of 3150:1 was reached in an oxygen plasma. Polyimide (Dupont Pyralin® PI-2611) was spun on in multiple coatings and cured at 380°C for one hour. According to these authors, etching with an ECR source is 10 times faster and six times more selective than conventional RIE. The combination of magnetic confinement and microwave power, leading to an order of magnitude greater dissociation efficiency at much lower pressures than RIE, reduces scattering of reactive species and promotes vertical profiles and smooth surface morphology. Cooling down to −130°C helps maintain a vertical etch profile by suppressing spontaneous chemical reactions. This way, lateral resonator elements, 32 μm thick with a 1-μm gap, have been fabricated in polyimide, and electroplated Ni accelerometers, with an aspect ratio of 11:1 were formed

from these polyimide molds. Compared to LIGA, where the average wall roughness is below 50 nm, the walls produced in this process still appear rough, but for many applications they might be good enough.

### MERIE to Etch Polyimide

To obtain the optimum anisotropy in high aspect ratio devices, one cools the substrate, uses high RF power density, works at the lowest possible pressure, utilizes high flow rate, and relies on sidewall-passivating additives. We quote more examples of the anisotropic power of magnetically enhanced reactive ion etching, very similar to the preceding example except for the microwave power, by Murakami et al.[52] and Furuya et al.[53]

Murakami et al.[52] investigated the etching of silicon and polyimide (Kapton H film) with an RIE cathode stage where the temperature could be controlled between 0 and −140°C. To offset the cooling effect on etching, the etch rate was increased by magnetic plasma confinement, a narrow 1-cm gap between the electrodes, and high flow rates. With Si a maximum etch rate of 1.6 μm/min was achieved with normalized side etching (defined as the ratio of the lateral etch to the vertical etch depth) of less than 0.02 at a temperature of −120°C. Etching selectivity for Si over a mask of 0.2-μm Al and 2 μm of Ni went beyond 900 with SF$_6$ at a power density of 4 W/cm$^2$. This work also nicely illustrates the so-called micro-loading effect discussed above. Figure 2.21 shows the relationship between the etch depth in Si and the opening width of the mask pattern.[52] High aspect ratio grooves prevent active species from reaching the bottom of the grooves while reaction products cannot diffuse out from the grooves. For polyimide, the normalized side etch reached less than 0.01 at an etch rate of 0.7 μm/min at −100°C in an oxygen plasma. A 0.3-μm sputtered Al mask was hardly

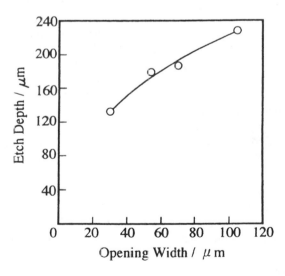

FIGURE 2.21 Micro-loading. Relationship between the etch depth and the opening width of the mask pattern (pressure 2.7 Pa, temperature −120°C, RF-power density 2.0 W/cm$^2$). (From Murakami, K. et al., *Cryogenic Dry Etching for High Aspect Ratio Microstructures, Micro Electromechanical Systems* (Institute of Electrical and Electronics Engineers, Hilton Head, SC 1994), pp. 82–85. With permission.)

etched by the oxygen plasma. The cryogenic temperatures suppress radical reactions on sidewalls while the bottom parts are cleared by ion bombardment.

Furuya et al. at NTT[53] made a fluorinated polyimide microgrid (Figure 2.22) by using MERIE in an oxygen plasma. The grid elements consist of 18 pole-shaped, gold-covered electrodes connected at the base by a common electrode. Individual poles measure 15 µm in diameter and 100 µm high. The micro-grid is formed by pairs of grid elements facing each other. A voltage can be applied between opposing electrodes of the grid-element pairs. A micro-grid like this, the authors suggest, can be used in an electrostatic ion drag micro-pump and in micro-electrophoresis devices. The fluorinated polyimide used in this study etches rapidly (3 to 5 µm/min) and with high selectivity (2600 over a Ti mask) because MERIE produces a high-density, low energy plasma and because the fluorine in the polimide produces a high concentration of oxygen radicals in the plasma. From the F/C rule, presented above, we predict that a fluorocarbon/oxygen atmosphere leads to a more aggressive and less selective etch. In the current case, by adding the fluorine to the polyimide instead of to the etchant gas, a faster etch rate occurs without losing the high selectivity.

The morphology of the fluorinated polyimide surface, as revealed by atomic force microscopy, is smooth at high etching rate, i.e., at high magnetic fields. The NTT fluorinated polyimide is chemically stable and the fluorine content makes the material extremely transparent, so it can be used to make optical components such as optical waveguides and lenses. The optical quality (and the high aspect ratio of the produced polyimide structures) could make this fabrication technology a competitive technology for many X-ray lithography-based LIGA structures.

## Example 4. Combination Wet and Dry Etching

Brugger et al.[54] developed an elegant micromachining process, combining wet and dry etching, to fabricate a meander type cantilever with an integrated high aspect ratio tip for microprobe applications. The crucial part of an atomic force microscope (AFM) or scanning tunneling microscope (STM) microprobe is the tip, which for some applications (e.g., IC trench probing) should have a curvature radius ideally in the range of 50 Å or less, an angle of aperture of about 5°, and a height of 10 µm. The meander structure and integrated tip are presented in Figure 2.23 and the process sequence in Figure 2.24.

First a 30-µm thin Si membrane is anisotropically etched in a double-sided polished 280-µm thick Si wafer (Figure 2.24A). The etching occurs at the back side with 40% KOH at 60°C through an opening etched with BHF in a 1.5 µm thermal oxide. During the KOH etch, the front side is protected by a mechanical chuck. The front side oxide is patterned into a two-step profile using two lithographic and two BHF etching steps. The 0.75-µm thick oxide in the two-step profile acts as the mask for the cantilever and a 1.5-µm thick oxide serves as the mask for the tip (Figure 2.24A). The topside is etched with 15-µm deep RIE using chlorine/fluorine gas mixtures ($C_2ClF_5/SF_6$) to preshape

the cantilevers (Figure 2.23B). By adjusting the RIE parameters, vertical sidewalls of the cantilever can be obtained. The 0.75-µm thick oxide that covered the cantilever is completely removed afterwards and the remaining oxide cap, formerly 1.5 µm and now 0.75 µm thick, serves as a mask for the tip etching. Then a $C_2ClF_5/SF_6$ gas mixture is applied again to RIE the tip (Figure 2.24C). In the final step (Figure 2.24D) an isotropic $HNO_3$:HF:$CH_3$COOH etch lasting 2 minutes forms the tips (using the oxide cap as mask), releases the cantilever by etching through the remaining Si membrane, smoothes the dry-etched rough surface and cleans the remainder of organic photoresist. Mixtures of $HNO_3$, HF, with $CH_3$COOH and/or water etch the silicon isotropically and exhibit high etch rates (10 to 300 µm/min). The $SiO_2$ mask withstands the three successive etch steps: cantilever RIE etching, tip RIE etching, and 1- to 2-min wet etching with HF:$HNO_3$:$CH_3$COOH sharpening the tip. Tip heights up to 20 µm with opening angles of approximately 5 to 10° and tip radii estimated to be 40 nm can be obtained this way.

A postoxidation process yielding Si tips with curvature radius of less than 10 nm by several consecutive oxidation and HF etching steps, exploiting an anomaly of the oxidation behavior at regions with high geometric curvature, has been demonstrated by Marcus et al.[55,56] and by our own group at TSDC.[57] By applying this oxidation sharpening technique to the tips produced in the current example, further sharpening would result. A rigid cantilever structure rather than a meander structure as shown in Figure 2.23A would also be of more use in a practical STM.[57]

## Example 5. Dry Etching Applications In IC Technology

One area where requirements for high aspect ratio in the IC industry are quite extreme is in the etching of deep Si trenches. Trenches in Si can be used to isolate devices in CMOS and bipolar circuitry. Vertical capacitors and transistors are actually constructed inside these trenches. The latter enables denser packing of devices.[24] A width of 1 µm and a depth of 6 to 15 µm might be required. Care needs to be taken that the mask will survive the etching; for example, a 10-µm trench will need 2 µm of $SiO_2$ if the selectivity is 5:1. In micromachining, we might want to etch to a depth of several hundreds of microns, requiring even more selectivity. In Table 2.11 the typical requirements for trench etching and selectivity in the IC industry are provided.

**TABLE 2.11** Typical Process Specifications for IC Etch Requirements[18]

|                    | 16 Mb          | 256 Mb          |
|--------------------|----------------|-----------------|
| Poly-Si vs. $SiO_2$ | >25:1 (must)   | 120:1 (target)  |
| Gate CD            | 0.6 µm         | 0.22 µm         |
| Trench size        | 0.6 × 3 µm     | 0.3 × 5 µm      |
| Aspect ratio       | 3:1            | >5:1            |
| $SiO_2$ vs. poly-Si | >18:1 (must)   | >35:1 (target)  |
| Profile control    | 86°            | 88°             |

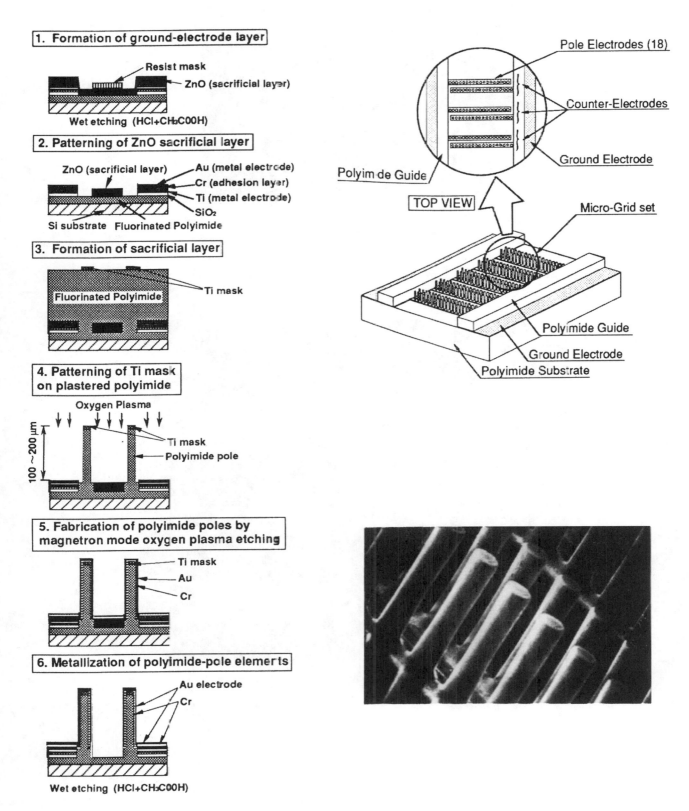

FIGURE 2.22   An array of polyimide posts made by cryogenic MERIE. (From Furuya, A. et al., Micro-Grid Fabrication of Fluorinated Polyimide by using magnetically enhanced reactive ion etching (MERIE), *Micro Electromechanical Systems* (Institute of Electrical and Electronics Engineers, Ft. Lauderdale, FL, 1993, pp. 59–64. With permission.)

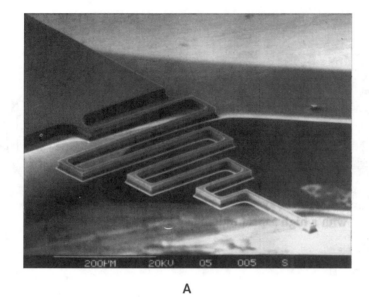

A

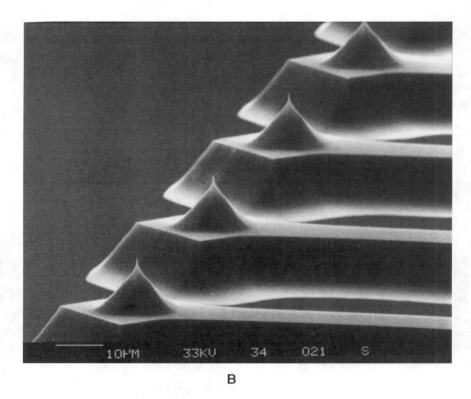

B

**FIGURE 2.23**    Si-cantilever with square cross-section, designed as a meander shape. (From Brugger, et al., *Sensors & Actuators*, A34, 193–200, 1992. With permission.)

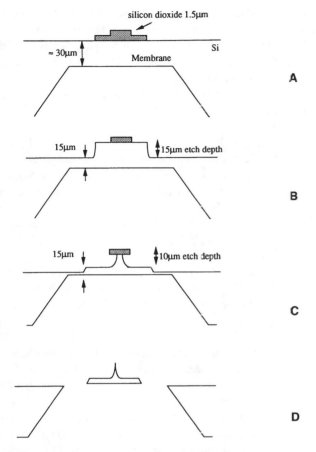

**FIGURE 2.24** Fabrication process for a silicon meander beam with integrated tip. (A) Patterning of oxide mask onto pre-processed 30-μm thin membrane. (B) Dry etching of rough cantilever shape. (C) Dry etching the tip. (D) Final wet etching step releasing the cantilever, forming the tip and smoothing the surface. (From Brugger, J. et al., *Sensors & Actuators*, A34, 193–220 (1992). With permission.)

# References

1. Kern, W. and C. A. Deckert, "Chemical Etching", in *Thin Film Processes*, Vossen, J. L. and Kern, W., Eds., Academic Press, Orlando, 1978.

2. Jackson, T. N., M. A. Tischler, and K. Wise, D., "An Electrochemical P-N Junction Etch-Stop for The Formation of Silicon Microstructures", *IEEE Electron Device Lett.*, EDL-2, 44-45, 1981.

3. Meek, R. L., "Electrochemically Thinned N/N+ Epitaxial Silicon-Method and Applications", *J. Electrochem. Soc.*, 118, 1240-1246, 1971.

4. Yoshida, T., T. Kudo, and K. Ikeda, "Photo-Induced Preferential Anodization for Fabrication of Monocrystalline Micromechanical Structures", *Proceedings. IEEE Micro Electro Mechanical Systems*, (MEMS '92), Travemunde, Germany, 1992, p. 56-61.

5. Watanabe, Y., Y. Arita, T. Yokoyama, and Y. Igarashi, "Formation and Properties of Porous Silicon and Its Applications", *J. Electrochem. Soc.*, 122, 1351-1355, 1975.

6. Flamm, D., L., "Feed Gas Purity and Environmental Concerns in Plasma Etching-Part 1", *Solid State Technol.*, October, 49-54, 1993.

7. Pandhumsoporn, T., M. Feldbaum, P. Gadgil, M. Puech, and P. Maquin, "High Etch Rate, Anisotropic Deep Silicon Plasma Etching for the Fabrication of Microsensors", *Micromachining and Microfabrication Process Technology II*, Austin, Texas, USA, 1996, p. 94-102.

8. Manos, D. M. and D. L. Flamm, Eds., *Plasma Etching, an Introduction*, Academic Press, Boston, 1989.

9. Vasile, M. J., C. Biddick, and S. Schwalm, A., "Microfabrication by Ion Milling : The Lathe Technique", *J. Vac. Sci. Technol.*, B12, 2388-2393, 1994.

10. Chryssolouris, G., *Laser Machining*, Springer Verlag, New York, 1991.

11. Moreland, M. A., "Ultrasonic Machining", in *Engineered Materials Handbook*, Schneider, S. J., Ed., ASM International, Metals Park, OH, 1992, p. 359-362.

12. Kalpajian, S., *Manufacturing Processes for Engineering Materials*, Addison-Wesley, Reading, MA, 1984.

13. DeVries, W. R., *Analysis of Material Removal Processes*, Springer-Verlag, New York, 1992.

14. Lehmann, H. W., "Plasma-Assisted Etching", in *Thin Film Processes II*, Vossen, J. L. and Kern, W., Eds., Academic Press, Inc., Boston, 1991, p. 673-748.

15. Singer, P., "Trends in Plasma Sources: The Search Continues", *Semicond. Int.*, July, 52-56, 1992.

16. Singer, H., P., "ECR: Is the Magic Gone?", *Semicond. Int.*, July, 46-48, 1991.

17. Peters, L., "Plasma Etch Chemistry: The Untold Story", *Semicond. Int.*, May, 66-70, 1992.

18. Singer, P., "Meeting Oxide, Poly and Metal Etch Requirements", *Semicond. Int.*, April, 50-54, 1993.

19. Comello, V., "ICP Etchers Start to Challenge ECR", *R&D Magazine*, 79-80, 1993.

20. Flamm, D. L., "Trends in Plasma Sources and Etching", *Solid State Technol.*, March, 47-50, 1991.

21. Burke, R. R. and C. Pomot, "Microwave Multipolar Plasma for Etching and Deposition", *Solid State Technol.*, February, 67-71, 1988.

22. Brodie, I. and J. J. Muray, *The Physics of Microfabrication*, Plenum Press, New York, 1982.

23. Wolf, S. and R. N. Tauber, *Silicon Processing for the VLSI Era*, Lattice Press, Sunset Beach, 1987.

24. Schutz, R. J., "Reactive Plasma Etching", in *VLSI Technology*, Sze, S. M., Ed., McGraw-Hill Book Company, New York, 1988, p. 184-232.

25. Arnold, J. C. and H. H. Sawin, "Charging of Pattern Features During Plasma Etching", *J. Appl. Phys.*, 70, 5314, 1991.

26. Elwenspoek, M., H. Gardeniers, M. de Boer, and A. Prak, "Micromechanics", University of Twente, Report No. 122830, Twente, Netherlands, 1994.

27. Hasper, A., Ph.D. Thesis, University of Twente, 1987.

28. Jansen, H., M. de Boer, and M. Elwenspoek, "The Black Silicon Method VI: High Aspect Ratio Trench Etching for MEMS Applications", *Proceedings. IEEE Micro Electro Mechanical Systems (MEMS '96)*, San Diego, California, USA, 1996, p. 250-257.

29. Flamm, D. L., "Dry Plasma Resist Stripping Part I: Overview of Equipment", *Solid State Technol.*, 35, 37-39, 1992.

30. Hoffman, E., B. Warneke, E. Kruglick, J. Weigold, and K. S. J. Pister, "3D Structures with Piezoresistive Sensors in Standard CMOS", *Proceedings. IEEE Micro Electro Mechanical Systems, (MEMS '95)*, Amsterdam, Netherlands, 1995, p. 288-93.

31. Editorial, Introducing Plasma Jet Etching, IPEC Precision, Inc., 1996.

32. Cooper III, C. B., S. Salimian, and M. E. Day, "Dry Etching for the Fabrication of Integrated Circuits in III-V Compound Semiconductors", *Solid State Technol.*, January, 109-112, 1989.

33. Mogab, C. J., A. C. Adams, and D. L. Flamm, "Plasma Etching of Si and SiO$_2$-the Effect of Oxygen Additions to CF4 Plasmas", J. Appl. Phys., 49, 3796-803, 1979.

34. Heinecke, R., H., "Control of Relative Etch Rates of SiO2 and Si in Plasma Etching", *Solid State Electron.*, 18, 1146-7, 1975.

35. Flamm, D. L., "Dry Plasma Resist Stripping Part II: Physical Processes", *Solid State Technol.*, 35, 43-48, 1992.

36. Flamm, D., L., "Feed Gas Purity and Environmental Concerns in Plasma Etching-Part 2", *Solid State Technol.*, November, 43-50, 1993.

37. Flamm, D., L., "Dry Plasma Resist Stripping Part III: Production Economics", *Solid State Technol.*, 35, 43-48, 1992.

38. Moreau, W. M., *Semiconductor Lithography*, Plenum Press, New York, 1988.

39. Hess, D. W. and D. B. Graves, "Plasma-Enhanced Etching and Deposition", in *Microelectronics Processing: Chemical Engineering Aspects*, Hess, D. W. and Jensen, K. F., Eds., American Chemical Society, Washington, DC, USA, 1989, p. 377-440.

40. Editorial, "Microwave Plasma Etching System", *Semiconductor International*, 1992, 72-76.

41. Juan, W. H., J. W. Weigold, and S. W. Pang, "Dry Etching and Boron Diffusion of Heavily Doped, High Aspect Ratio Si Trenches", *Proceedings. SPIE-The International Society for Optical Engineering*, Austin, Texas, USA, 1996, p. 45-55.

42. Mucha, J., A. and D. W. Hess, "Plasma Etching", in Introduction to Microlithography, Thompson, L. F., Willson, C. G. and Bowden, M. J., Eds., *ACS Symposium Series*, Washington D.C., 1983.

43. Economou, D., E. S. Aydil, and G. Barna, "In Situ Monitoring of Etching Uniformity in Plasma Reactors", *Solid State Technol.*, 34, 107-111, 1991.

44. Elta, M., "Developing 'Smart' Controllers for Semiconductor Processes", *Res. & Dev.*, February, 69-70, 1993.

45. Duek, R., N. Vofsi, M. Haemek, S. Mangan, and M. Adel, "Improving Plasma Etch With Wafer Temperature Readings", *Semicond. Int.*, July, 208-210, 1993.

46. Mocella, M. T., "The CFC-Ozone Issue in Dry Etch Process Development", *Solid State Technol.*, April, 63-64, 1991.

47. Zhang, L. Z. and N. C. MacDonald, "An RIE Process for Submicron, Silicon Electromechanical Structures", 6th International Conference on Solid-State Sensors and Actuators (Transducers '91), San Francisco, CA, 1991, p. 520-523.

48. Zhang, Z. L. and N. C. MacDonald, "Fabrication of Submicron High-Aspect-Ratio GaAs Actuators", *J. Microelectromech. Syst.*, 2, 66-73, 1993.

49. Yao, J. J., S. C. Arney, and N. C. MacDonald, "Fabrication of High Frequency Two-Dimensional Nanoactuators for Scanned Probe Devices", *J. Microelectromech. Syst.*, 1, 14-22, 1992.

50. Zhang, Z. L., G. A. Porkolab, and N. C. MacDonald, "Submicron, Movable Gallium Arsenide Mechanical Structures and Actuators", *Proceedings. IEEE Micro Electro Mechanical Systems (MEMS '92)*, Travemunde, Germany, 1992, p. 72-77.

51. Juan, W., H., S. W. Pang, A. Selvakumar, M. W. Putty, and K. Najafi, "Using Electron Resonance (ECR) Source to Etch Polyimide Molds For Fabrication of Electroplated Microstructures", Technical Digest of the 1994 Solid State Sensor and Actuator Workshop., Hilton Head Island, SC, 1994, p. 82-85.

52. Murakami, K., Y. Wakabayashi, K. Minami, and M. Esashi, "Cryogenic Dry Etching for High Aspect Ratio Microstructures", *Proceedings. IEEE Micro Electro Mechanical Systems, (MEMS '93)*, Fort Lauderdale, FL, 1993, p. 65-70.

53. Furuya, A., F. Shimokawa, T. Matsuura, and R. Sawada, "Micro-grid Fabrication of Fluorinated Polyimide by Using Magnetically Controlled Reactive Ion Etching (MC-RIE)", *Proceedings. IEEE Micro Electro Mechanical Systems, (MEMS '93)*, Fort Lauderdale, FL, 1993, p. 59-64.

54. Brugger, J., R. A. Buser, and N. F. de Rooij, "Silicon Cantilevers and Tips for Scanning Force Microscopy", *Sensors and Actuators A*, A34, 193-200, 1992.

55. Marcus, B. R. and T. S. Ravi, "Method for Making Tapered Microminiature Silicon Structures", in US Patent 5,201,992, Bell Communications Research, Inc., 1993.

56. Marcus, R. B., T. S. Ravi, T. Gmitter, K. Chin, D. Liu, W. J. Orvis, D. R. Ciarlo, C. E. Hunt, and J. Trujillo, "Formation of Silicon Tips With <1 nm Radius", *Appl. Phys. Lett.*, 56, 236-238, 1990.

57. Editorial, Tips for STM, TSDC (Teknekron), Menlo Park, CA, USA, 1991.

# 3

# Pattern Transfer with Additive Techniques

## Introduction

Adding to a substrate such as an exposed and developed area on a silicon wafer involves surface modification such as ion implantation, annealing, or deposition. Solids can be deposited from a liquid, a plasma, a gas, or the solid state. Deposition processes are additive and especially in the thin film arena often are identical for integrated circuit (IC) applications and micromachines. During growth of a Si single crystal no addition is involved; only a phase change takes place. Growth of a film, as in oxidation of Si in an oxygen atmosphere and nitridation of Si in ammonia, also differs from straight deposition due to the consumption of the substrate surface while forming the film. Silicon oxidation, like Si crystal growth, is a primary process. An understanding of the physical nature of the $Si/SiO_2$ interface is essential for IC manufacture as well as for certain sensors.

Additive processes in microsensor development span a wide range of materials from inorganic to organic. Besides the typical Si microelectronic elements, Al, Au, Ti, W, Cu, Cr, and Ni-Fe alloys, these processes involve deposition of several nontypical elements such as Zr, Ta, Ir, Pt, Pd, Ag, Zn, In, Nb, and Sn. A plethora of exotic compounds ranging from enzyme layers to shape memory alloys (e.g., NiTi) and piezoelectrics (e.g., ZnO) are used to construct micromachines. The number of materials involved in IC fabrication, in comparison, is very limited. Moreover, for sensor construction, particularly chemical sensors, thick film technologies become prevalent. In Table 3.1 additive processes are listed and in Table 3.2 some sensor materials are listed together with their deposition techniques and a typical resulting function of the deposit.

In the deposition methods from the gas phase listed in Table 3.1, two major categories can be distinguished: direct line-of-sight impingement deposition techniques called physical vapor deposition (PVD) and diffusive-convective mass transfer techniques, i.e., chemical vapor deposition (CVD). At the lower pressures employed in a PVD reactor, the vaporized material encounters few intermolecular collisions while traveling to the substrate. Modeling of deposition rates is a relative straightforward exercise in geometry in this case. Evaporation, sputtering, molecular beam epitaxy (MBE), laser ablation deposition, ion-plating, and cluster deposition represent the PVD techniques discussed. In the case of a CVD reactor, the diffusive-convective transport to the substrate involves many intermolecular

TABLE 3.1   Additive Processes

| Additive | Application | Ref. |
|---|---|---|
| Bonding techniques | 7740 glass to silicon | 1 |
| Casting | Thick resist (10–1000 μm) | |
| Chemical vapor deposition | Tungsten on metal | 2 |
| Dip coating | Wire type ion selective electrodes | |
| Droplet delivery systems | Epoxy, chemical sensor membranes | |
| Electrochemical deposition | Copper on steel | 3 |
| Electroless deposition | Vias | 4 |
| Electrophoresis | Coating of insulation on heater wires | |
| Electrostatic toning | Xerography | |
| Ion cluster deposition | | 5 |
| Ion implantation and diffusion of dopants | Boron into silicon | 6 |
| Ion plating | | 5 |
| Laser deposition | Superconductor compounds | 7 |
| Liquid phase epitaxy (CVD) | GaAs | 8 |
| Material transformation (oxidation, nitridation, etc.) | Growth of $SiO_2$ on silicon | 9 |
| Molecular beam epitaxy (PVD) | GaAs | 10 |
| Plastic coatings | Electronics packages | 11 |
| Screen printing | Planar ion selective electrodes (ISEs) | 12 |
| Silicon crystal growth | Primary process | 13 |
| Spin-on | Thin resist (0.1–2 μm) | 13 |
| Spray pyrolysis (CVD) | CdS on metal | 14 |
| Sputter deposition (PVD) | Gold on silicon | 15 |
| Thermal evaporation (PVD) | Aluminum on glass | 15 |
| Thermal spray deposition from plasmas or flames | Coatings for aircraft engine parts and $ZrO_2$ sensors | 16 |
| Thermomigration | Aluminum contacts through silicon | 17 |

collisions. Accordingly, mass and heat transfer modeling of deposition rates becomes more complex. When a molecule has reached the surface, the required reaction analysis is the same, regardless of deposition method. The molecular phenomena at the surface to be considered include sticking coefficient, surface adsorption, surface diffusion, surface reaction, desorption, and

89

TABLE 3.2    IC and Micromachine Deposits, Deposit Method and Typical Application

| Material | Deposition Technique | Function |
|---|---|---|
| **Organic thin/thick films** | | |
| Hydrogel | Silk screen | Internal electrolyte in chemical sensors |
| Photoresist | Spin-on | Masking, planarization |
| Polyimide | Spin-on | Electrical isolation, planarization, microstructures |
| **Metal oxides** | | |
| Aluminum oxide | CVD, sputtering, anodization | Electrical isolation |
| Indium oxide | Sputtering | Semiconductor |
| Tantalum oxide | CVD, sputtering, anodization | Electrical isolation |
| Tin oxide | Sputtering | Semiconductor in gas sensors |
| Zinc oxide | Sputtering | Electrical isolation, piezoelectric |
| **Noncrystalline silicon compounds** | | |
| $\alpha$-Si-H | CVD, sputtering, plasma CVD | Semiconductors |
| Polysilicon | CVD, sputtering, plasma CVD | Conductor, microstructures |
| Silicides | CVD, sputtering, plasma CVD, alloying of metal and silicon film | Conductors |
| **Metals (thin films)** | | |
| Aluminum | Evaporation, sputtering, plasma CVD | Electrical conduction |
| Chrome | Evaporation, sputtering, electroplating | Electrical conduction, adhesion layer |
| Gold | Evaporation, sputtering, electroplating | Electrical conduction |
| Molybdenum | Sputtering | Electrical conduction |
| Palladium | Sputtering | Electrical conduction, adhesion layer |
| Platinum | Sputtering | Electrical conduction, interdiffusion barrier |
| **Alloys** | | |
| Al-Si-Cu | Evaporation, sputtering | Electrical conduction |
| Nichrome | Evaporation, sputtering | Resistors |
| Permalloy | Sputtering | Magnetoresistor, thermistor |
| TiNi | Sputtering | Shape memory alloy |
| **Chemically/physically modified silicon** | | |
| n/p type silicon | Implantation, diffusion, incorporation in the melt | Conduction modulation, etch stop |
| Porous silicon | Anodization | Electrical isolation, light-emitting structures, porous junctions |
| Silicon dioxide | Thermal oxidation, sputtering, anodization, implantation, CVD | Electrical and thermal isolation, masking, encapsulation |
| Silicon nitride | Plasma CVD | Electrical and thermal isolation, masking, encapsulation |

film or crystal growth. Plasma-enhanced CVD (PECVD), atmospheric pressure (APCVD), low pressure CVD (LPCVD), very low pressure (VLPCVD), metallorganic chemical vapor deposition (MOCVD), and spray pyrolysis are the CVD techniques covered.

We believe that more and more micromachines will exploit epitaxial Si. For example, in silicon on insulator (SOI), crystalline epilayers of a predetermined thickness on a thin $SiO_2$ layer afford well-controlled suspended membranes. We will, therefore, analyze epitaxial methods based on physical vapor, chemical vapor, and liquid phase epitaxial deposition methods.

In general, for the deposition of thin films used in microelectronics, low-pressure processes, such as sputtering and low pressure chemical vapor deposition are preferred over the older,

conventional chemical methods such as electroplating, because deposition from an aqueous solution often produces poorer quality films. On the other hand, electrochemical and electroless metal deposition from solution are gaining renewed interest with micromachinists because of their emerging importance to replication of photoresist molds (see Chapter 6). We will review metal deposition from solution as its importance in creating novel functional metal microstructures increases.

Deposition techniques involving thin and thick organic materials are of extreme importance to the construction of chemical and biological sensors. Techniques such as spin-coating, dip-coating, casting, Langmuir-Blodgett, ink-jet printing and drop delivery systems, silk-screening, and lithographically patterning of membranes are highlighted. These methods embody the manufacturing options of choice for chemical and biological sensors such as glucose sensors and ion selective electrodes.

In this chapter we also include a section on planarization technologies. Planarization involves additive deposition processes and poses a significant challenge for the lithographic definition of 'high rise' structures typical in micromachining. This will complement the section in Chapter 1 on planarizing with multilayer resists (MLR).

A particulate deposition technique, such as plasma spraying, is treated here because of its interesting chemical sensor and micromachining applications. Like spray-pyrolysis, plasma spraying does not have IC industry application but might well propel planar thick film chemical sensors into 'batch' manufacturability.

The scope of this book does not warrant a detailed explanation of all additive techniques listed in Tables 3.1 and 3.2. A short description of each of the listed processes emphasizing those most important for building sensors and micromachines should suffice. As in the case of subtractive processes in Chapter 2, only the underlying physical and chemical principles are explained. The references in Table 3.1 are meant to form a basis for further study. A limited number of examples of deposition processes in micromachining is given at the end of this chapter; more examples and further discussion of properties of resulting films are dealt with in subsequent chapters.

# Growth

## Silicon Growth

Silicon crystal growth comprises the primary process toward the construction of a Si IC or micromachine. In the *Czochralski crystal pulling method,*[8] a Si crystal seed is grown into a Si single crystal by pulling it slowly upwards, at about 2 to 5 cm/hr, from a molten and ultra-pure silicon melt. The melt must not be stirred; otherwise, oxygen is transported from the $SiO_2$-Si (liquid) interphase to the Si (liquid)-Si (solid) interphase. The molten silicon slowly rotates, nested in a silica crucible while pulling the growing crystal up. To avoid all contact with the crucible, eliminating, for example, oxygen impurities, *float-zone crystal growth*[18] is used instead. This method requires a slowly rotating polycrystalline silicon rod locally melted and recrystallized by a scanning RF heating coil. Very pure silicon results. The silicon is ideal for devices requiring a low doping level to produce, for example, low leakage diodes as required for detectors and power devices.

Routine evaluation of the grown crystalline material involves the measurement of resistivity and crystal perfection. The first is measured with a four-point probe while X-ray or electron beam diffraction helps evaluate the crystal perfection. To prepare wafers from the grown Si boule, first the diameter is sized by grinding on a lathe; X-rays establish the orientation with an accuracy of ± 0.5° (in the very best case). Diameter grinding is followed by grinding one or more flats along the length of the ingot. Flat areas help orientation determination and placement of slices in cassettes and fabrication equipment (large primary

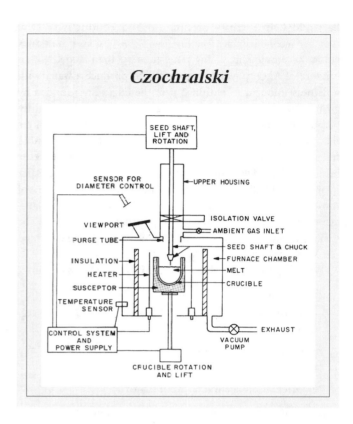

*Czochralski*

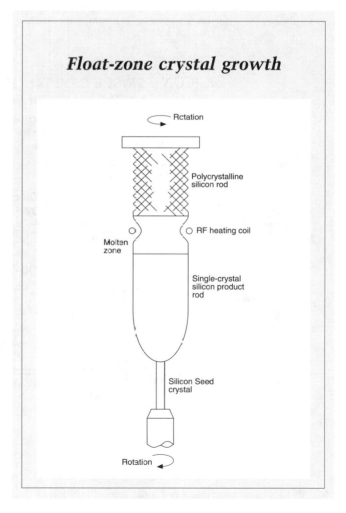

*Float-zone crystal growth*

flat) and help identify orientation and conductivity type (smaller secondary flat) (see Figure 4.5). Argon laser markings guide the subsequent slicing. The slices are then lapped with a mixture of $Al_2O_3$ powder in glycerine to produce a wafer with a flatness uniform to within 2 μm; the edges are rounded by grounding. Wafer edge rounding cannot be taken lightly because it removes microcracks at the wafer edge, minimizes the source for cracks in silicon chips, makes the wafers more fracture-resistant, increases the mask life, and minimizes resist beading. The edge rounding process also makes for a safer quartz boat loading by increasing the number of transfers of those wafers into quartz boats without chipping.[19] After a damage-removing etch, the wafers are polished and cleaned. The slicing, lapping, edge treatment, damage removal, and polishing result in a material loss of up to 50%.

Typical specifications for a 100-mm diameter Si wafer are[8]

- Ø = 100 mm ± 0.5 mm
- Thickness T = 500 μm ± 10 μm
- Primary = flat 30 to 35 mm
- Secondary flat = 16 to 20 mm
- Bow = 40 μm
- Orientation accuracy within 0.5 to 1°

Notice that the thickness control is very limited which has important consequences for fabricating thin Si membranes. Figure 3.1 illustrates wafer flatness and defect parameters such as bow, warp, chips, etc. Brown provides an excellent further introduction to Si crystal growth.[18]

## Oxidation of Silicon

### Kinetics

Silicon dioxide growth involves the heating of a Si wafer in a stream of steam at 1 atm or *wet or dry oxygen/nitrogen mixtures* at elevated temperatures (between 600 and 1250°C). Silicon readily oxidizes, even at room temperature, forming a thin native oxide approximately 20 Å thick. The high temperature aids diffusion of oxidant through the surface oxide layer to the silicon interface to form thick oxides in a short amount of time:

$$Si + 2H_2O \rightarrow SiO_2 + H_2 \text{ (wet)} \qquad \textbf{Reaction 3.1}$$

$$Si + O_2 \rightarrow SiO_2 \text{ (dry)} \qquad \textbf{Reaction 3.2}$$

Another gas phase oxidation method of Si is the *pyrogenic method*. Besides oxygen and nitrogen the gas also contains hydrogen. The ratio of *silicon thickness converted, $X_S$, to resulting oxide thickness, $X_{OX}$*, is proportional to their respective densities:

$$X_s = 0.46 X_{ox} \qquad 3.1$$

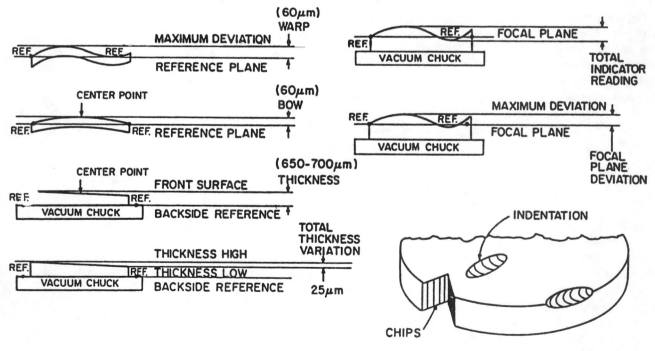

FIGURE 3.1   Wafer flatness and defect parameters such as bow, warp, and chips.[19]

Due to molecular volume mismatch and thermal expansion differences a compressed $SiO_2$ film results. Wet oxidation relaxes the compressive stress somewhat speeding up the $SiO_2$ growth. When 10,000 Å of oxide has grown, 4600 Å of Si will be consumed; in other words, the amount of silicon consumed is 46% of the final oxide thickness. This relationship holds importance for calculating step heights that form in silicon microstructures.

All gas phase oxidation processes involve gas phase transport of oxidant to the surface ($F_1$), diffusion through the existing oxide ($F_2$) and the oxidation reaction itself ($F_3$). Equilibrating the fluxes $F_1 = F_2 = F_3$ in Figure 3.2 yields the concentration of oxidant at the interface oxide/silicon, $C_i$:

$$C_i\left(X_{ox}, k_s, D, P_G\right) \qquad 3.2$$

with $X_{OX}$ = oxide thickness; $k_s$ = Si oxidation rate constant, which is a function of temperature, oxidant, crystal orientation, and doping; $D$ = oxidant diffusivity which is a function of temperature and oxidant; and $P_G$ = partial pressure of the oxidant in the gas phase.

The oxide growth rate is given by:

$$\frac{dX_{ox}}{dt} = \frac{C_i k_s}{N} \qquad 3.3$$

where N stands for the number of molecules of oxidant per unit volume of oxide ($2.2 \times 10^{22}$ cm$^{-3}$ for dry oxygen). Solution of

this differential equation in the Deal-Grove model for oxidation (1965) yields:[20,21,22]

$$X_{ox}(t) = \frac{A}{2}\left\{\left[1 + \frac{(t+\tau)}{A^2}4B\right]^{\frac{1}{2}} - 1\right\} \qquad 3.4$$

where $A = 2D(1/k_s + 1/h)$, $B = 2DC^*/N$, $\tau = (X_i^2 + AX_i)/B$, h = gas phase mass transport coefficient in terms of concentration in the solid, $C^*$ = equilibrium oxidant concentration in the oxide $= HP_G$, with H being Henry's law constant, $X_i$ = initial oxide thickness.

For thin oxides (short oxidation times $(t + \tau) \ll A^2/4B$), the process is reaction rate limited (see Figure 3.3):

$$X_{ox}(t) = \frac{B}{A}(t+\tau) = \frac{C^*}{N}\left(\frac{1}{k_s} + \frac{1}{h}\right)^{-1}(t+\tau) \qquad 3.5$$

where B/A is the linear rate constant (in μm/h).

For thick oxides (long oxidation times with $t \gg \tau$ and $t \gg A^2/4B$), the process is diffusion-limited and a parabolic law results (see Figure 3.3):

$$X_{ox}^2 = B(t+\tau) = 2D\left(\frac{C^*}{N}\right)(t+\tau) \qquad 3.6$$

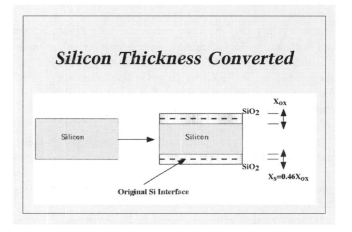

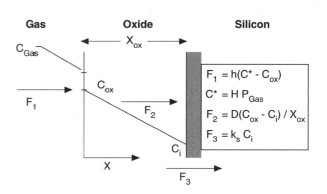

FIGURE 3.2  Fluxes in the silicon oxidation reaction.

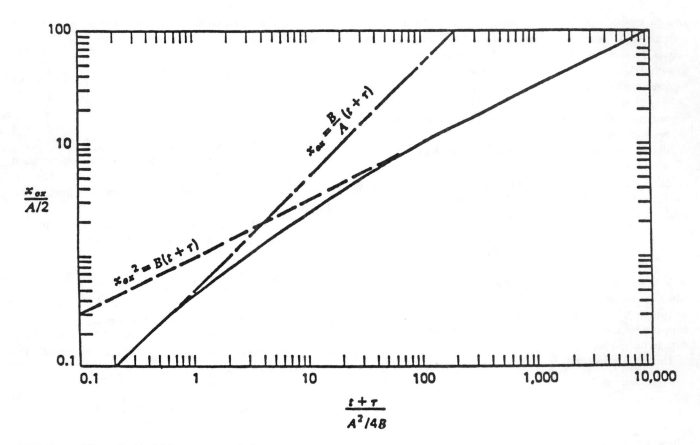

**FIGURE 3.3**   Silicon dioxide thickness vs. growth time.

where B is the parabolic rate constant. From Figure 3.3 we conclude that oxide thickness vs. growth time initially is linear when the oxidation reaction at the surface dominates, but parabolic when the diffusion to the oxide/silicon interface dominates the reaction rate. When the surface reaction rate dominates, the linear rate constant B/A depends on surface orientation (see below).

The oxidation rate depends on such parameters as crystallographic orientation of the Si, Si doping, the presence in the oxidizing gas of halogen impurities, water, or hydrogen; the oxidizing gas pressure, and the use of a plasma or a photon flux during oxide growth.

Silicon dioxide formed by thermal oxidation of silicon is used as a common insulating layer, as a mask, and as a sacrificial material. $SiO_2$ performs excellently as a mask against diffusion of the common dopants in Si; the diffusion coefficient, D, of boron at 900°C in $SiO_2$ is $2.2 \times 10^{-19}$ cm²/sec compared to $4.4 \times 10^{-16}$ in Si; for phosphorous, D equals $9.3 \times 10^{-19}$ cm²/sec compared to $7.7 \times 10^{-15}$ in Si (from the common Si dopants, only Ga diffuses fast through the oxide with D = $1.3 \times 10^{-13}$). For many other elements, $SiO_2$ forms a poor diffusion barrier. The amorphous oxide has a more open structure than crystalline quartz — only 43% of the space is occupied. Consequently, a wide variety of impurities (especially alkali ions such as sodium and potassium) can readily diffuse through amorphous $SiO_2$. Diffusion through the open $SiO_2$ structure happens especially fast when the oxide is hydrated. One of the reasons for the poor performance of silicon dioxide-based ion sensitive field effect transistors (ISFETS) in aqueous solutions can be traced

back to the simple observation that $SiO_2$ in water almost behaves like a sponge for ions.[23] This is why, in such devices, $SiO_2$ is often topped off with the excellent ionic barrier material $Si_3N_4$. The use of $SiO_2$ as a diffusion mask often stems from convenience; $SiO_2$ is easy to grow, whereas one cannot put $Si_3N_4$ directly onto Si.

The quality of silicon dioxide depends heavily on its growth method. Dry oxidation at high temperature (between 900 and 1150°C) in pure oxygen produces a better quality oxide than steam oxidation. Such a thermal oxide is stoichiometric, has a high density, and is basically pinhole free. Wet oxidation in steam occurs much faster but produces a lesser quality oxide, while water causes a loosening effect on the $SiO_2$, making it more prone to impurity diffusion. Both types of oxidation are carried out in a quartz tube. Oxide thicknesses of a few tenths of a micron are used most frequently, with 1 to 2 μm being the upper limit for conventional thermal oxides. Addition of chlorine-containing chemicals during oxidation increases the dielectric breakdown strength and the rate of oxidation and improves the threshold voltage of many electronic devices;[24] however, too high concentrations of halogens at high temperatures can pit the silicon surface. The influence of chlorine, hydrogen, and other gases on the $SiO_2$ growth rate and the resulting interface and oxide quality has been a very fertile research field. A short summary of this field can be found in Fair.[22] It should be noted that in general the electronic quality of the Si/$SiO_2$ interface as expressed through low concentrations of *interface trap states, low fixed oxide surface charges, low bulk oxide trapped charges*, and *mobile charges*[25] (impurity ions such as $Li^+$, $Na^+$, $K^+$) fail to be of importance in micromachining unlike in the IC industry.

## Location of oxide charges in thermally oxidized silicon structures

(From Brodie, I. and J. J. Muray, *The Physics of Microfabrication*, Plenum Press, New York, 1989. With permission.)

In micromachining, the oxide only proves useful as a structural element, a sacrificial layer, or a dielectric in a passive device. One notable exception is the aforementioned ISFET, where the electronic properties of the gate oxide are key to its functioning; a perfect oxide and oxide/semiconductor interface hold as much importance in this case as in the IC industry.

Short of using an ellipsometer to determine the oxide thickness, the silicon oxide thickness color table below comes in handy (Table 3.3). This table also lists some typical applications for oxides of various thicknesses. In Table 3.4, we review some of the more significant properties of thermally grown SiO$_2$.

### Orientation Dependence of Oxidation Rates

The thermal oxidation rate is influenced by the orientation of the Si substrate. The effect involves the linear oxidation rate constant used in the Deal-Grove model[20] in the regime where the reaction is surface reaction rate-limited. This constant was given as B/A (in μm/hr) in Equation 3.5. The ratio of this constant, for a (111) Si plane to that for a (100) Si plane, is given by:

$$\frac{\frac{B}{A}(111)}{\frac{B}{A}(100)} = \frac{C_1(111)\exp\left(-\frac{2.0\mathrm{eV}}{kT}\right)}{C_1(100)\exp\left(-\frac{2.0\mathrm{eV}}{kT}\right)} = 1.7 \qquad 3.7$$

Thus, a (100) surface oxidizes about 1.7 times more slowly than a (111) surface. The lower oxidation rate of (100) surfaces might be due to the fewer silicon bonds with which oxygen can react on such a surface. The available bond density at a (111) plane equals $11.76 \times 10^{14}$ cm$^{-2}$ and $6.77 \times 10^{14}$ cm$^{-2}$ for the (100) plane. The linear oxidation rate for Si follows the sequence (110) > (111) > (311) > (511) > (100), corresponding to an increasing activation energy incorporating a term for the plane bond

**TABLE 3.3**   Oxide Color Table and Applications

| Color | Thickness (Å) | | | | Application |
|---|---|---|---|---|---|
| Grey | 100 | | | | Tunneling oxides |
| Tan | 300 | | | | Gate oxides, capacitor dielectrics |
| Brown | 500 | | | | LOCOS[a] pad oxide |
| Blue | 800 | | | | |
| Violet | 1000 | 2800 | 4600 | 6500 | |
| Blue | 1500 | 3000 | 4900 | 6800 | |
| Green | 1800 | 3300 | 5200 | 7200 | |
| Yellow | 2100 | 3700 | 5600 | 7500 | |
| Orange | 2100 | 3700 | 5600 | 7500 | |
| Orange | 2200 | 4000 | 6000 | | Masking oxides, surface passivation |
| Red | 2500 | 4400 | 6200 | | Field oxides |

a    LOCOS = local oxidation of silicon.

**TABLE 3.4**   Properties of Thermal $SiO_2$

| Property | Value |
|---|---|
| Resistivity at 25°C (Ω cm) | $3 \times 10^{15}$ (dry); $3$–$5 \times 10^{15}$ (wet) |
| Density (g/cm³) | 2.24–2.27 (dry); 2.18–2.20 (wet) |
| Dielectric constant | 3.9 |
| Dielectric strength (V/cm) | $2 \times 10^6$ (dry); $3 \times 10^6$ (wet) |
| Energy gap (eV) | 9 |
| Etch rate in buffered HF (Å/min) | 1000 |
| Refractive index | 1.46 |
| Thermal conductivity (W/cm°C) | 0.014 |
| Specific heat (J/g°C) | 1.0 |
| Stress in film on Si (dyne/cm²) | $2$–$4 \times 10^9$ (compressive) |
| Thermal conductivity (W/cmK) | 0.014 |
| Thermal linear expansion coefficient (°C$^{-1}$) | $5 \times 10^{-7}$ (0.5 ppm/°C) |

density as well as one for the bond orientation. As might be expected, steric hindrance results in higher activation energy.[21]

In anisotropic wet etching, the sequence of the etch rate is mostly reversed. For example, a (100) plane etches up to 100 times faster than a (111) plane. Actually, the slow etching of a (111) Si plane relative to a (100) plane may be due to its more efficient oxidation, protecting it better against etching.[26] In this interpretation one assumes that anodic oxidation, just like thermal oxidation, will occur faster at a (111) Si surface. To our knowledge, no evidence to support this theory has been presented as yet. Seidel et al.[27] pointed out that the huge anisotropy in wet etching of Si could not be credited to the number of bonds available on different Si planes, as that number barely differs by a factor of 2. From the numbers observed such an explanation could be valid for the anisotropy in thermal oxidation. There is no agreement, however, about the exact mechanism, and an understanding of the orientation dependence of the oxidation rate is still lacking. It has been attributed at various times to the number of Si-Si bonds available for reaction and the orientation of these bonds (see above), the presence of surface steps, mechanical effects such as stress in the oxide film, and the attainment of maximum coherence across the Si-$SiO_2$

interface.[21] It should also be noted that the oxidation rate sequence depends on temperature and oxide thickness — one more reason to take any model with a grain of salt.[21]

Electrochemical oxidation of Si has been studied extensively (see, for example, Schmidt[28]) but has never lead to commercial exploitation. Even though these oxides are pinhole free, they exhibit a lot of interface states. Mechanisms for anodic oxidation of Si and introduction of oxide 'dopants' into anodically formed oxides were investigated by Madou et al.[29,30] Such dopants can be used afterwards as a source for doping of the underlying Si at elevated temperatures. Further study material on silicon oxidation can be found in Fair[22] and Katz.[21]

In conclusion, thermal oxidation involving the consumption of a surface results in excellent adhesion and very good electrical and mechanical properties. It is an excellent technique when available.

# Physical Vapor Deposition

## Introduction

Different kinds of thin films in IC and micromachining are deposited by evaporation and sputtering. Both are examples of physical vapor deposition (PVD). PVD reactors may use a solid, liquid, or vapor raw material in a variety of source configurations. Other PVD techniques, very useful in the deposition of complex compound materials, are molecular beam epitaxy (MBE) and laser ablation deposition. Ion-plating and cluster deposition are based on a combination of evaporation and plasma ionization and offer some of the advantages inherent to both techniques. The key distinguishing attribute to a PVD reactor is that the deposition of the material onto the substrate is a line-of-site impingement type deposition.

## Thermal Evaporation

Thermal evaporation represents one of the oldest thin film deposition techniques. Evaporation is based on the boiling off (or sublimating) of a heated material onto a substrate in a vacuum.

From thermodynamic considerations, the number of molecules leaving a unit area of evaporant per second is given by:

$$N = N_o \exp-\left(\frac{\Phi_e}{kT}\right) \qquad 3.8$$

where $N_0$ may be a slowly varying function of temperature, T, and $\Phi_e$ is the activation energy (in eV) required to evaporate one molecule of the material. The activation energy for evaporation is related to the enthalpy of formation of the evaporant, H, as $H = \Phi \times e \times N$ (Avogadro's number) J/mol.

Table 3.5 suggests the need for a good vacuum during evaporation; even at a pressure of $10^{-5}$ Torr, 4.4 contaminating monolayers per second redeposit on the substrate[25] For reference purposes, the number of atoms per unit area corresponding to a monolayer for a metal is about $10^{15}$ atoms/cm². Moreover, to avoid reactions at the source (for example, oxide impurities being formed), the oxygen partial pressure needs to be less than $10^{-8}$ Torr.

In laboratory settings a metal usually is evaporated by passing a high current through a highly refractory metal containment structure (e.g., a tungsten boat or filament). This method is called 'resistive heating' (Figure 3.4A). In industrial applications resistive heating has been surpassed by electron-beam (e-beam) and RF induction evaporation. In the e-beam mode of operation a high-intensity electron beam gun (3 to 20 keV) is focused on the target material that is placed in a recess in a water-cooled copper hearth. As shown in Figure 3.4B, the electron beam is magnetically directed onto the evaporant which melts locally. In this manner, the metal forms its own crucible and the contact with the hearth is too cool for chemical reactions, resulting in fewer source-contamination problems than in the case of resistive heating. One disadvantage of e-beam evaporation is that the process might induce X-ray damage and possibly even some ion damage on the substrate (at voltages > 10 kV the incident electron beam will cause X-ray emission). The X-ray damage may be avoided by using a focused, high-power laser beam instead of an electron beam. However, this technique did not yet penetrate commercial applications. In RF induction heating, a water-cooled RF coupling coil surrounds a crucible with the material to be evaporated. Since about two thirds of the RF energy is absorbed within one skin depth of the surface, the

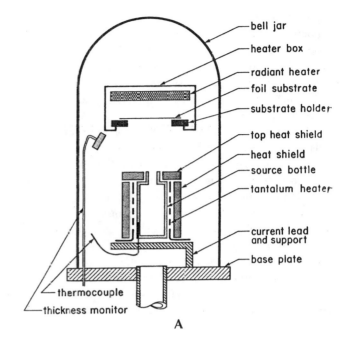

A

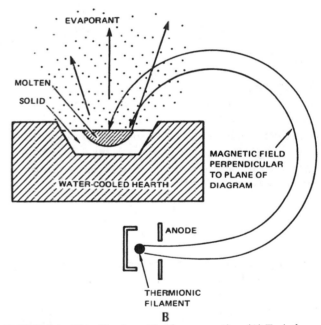

B

**FIGURE 3.4** Thin film deposition by evaporation. (A) Typical evaporation set-up. (B) Diagram of a magnetized deflection electron-beam evaporation system. (From Brodie, I. and J. J. Muray, *The Physics of Microfabrication*, Plenum Press, New York, 1982. With permission.)

TABLE 3.5 Kinetic Data for Air as a Function of Pressure

| Pressure (Torr) | Mean Free Path (cm) | Number Impingement Rate (s⁻¹ · cm⁻²) | Monolayer Impingement Rate (s⁻¹) |
|---|---|---|---|
| $10^1$ | 0.5 | $3.8 \times 10^{18}$ | 4400 |
| $10^{-4}$ | 51 | $3.8 \times 10^{16}$ | 44 |
| $10^{-5}$ | 510 | $3.8 \times 10^{15}$ | 4.4 |
| $10^{-7}$ | $5.1 \times 10^4$ | $3.8 \times 10^{13}$ | $4.4 \times 10^{-2}$ |
| $10^{-9}$ | $5.1 \times 10^6$ | $3.8 \times 10^{11}$ | $4.4 \times 10^{-4}$ |

*Source:* Brodie, I. and J. J. Muray, *The Physics of Microfabrication*, Plenum Press, New York, 1982. With permission.)

TABLE 3.6 Comparison of Heat Sources for Evaporation

| Heat Sources | Advantages | Disadvantages |
|---|---|---|
| Resistance | No radiation | Contamination |
| Electron-beam | Low contamination | Radiation |
| RF | No radiation | Contamination |
| Laser | No radiation, low contamination | Expensive |

frequency of the RF supply must decrease as the size of the evaporant charge increases. With a charge of a few grams of evaporant, frequencies of several hundred kilohertz are sufficient.

In Table 3.6 we compare the different heat sources available for thermal evaporation. The substances used most frequently for thin-film formation by evaporation are elements or simple compounds whose vapor pressures range from 1 to $10^{-2}$ Torr in the temperature interval 600 to 1200°C. Refractory metals such as platinum, molybdenum, tantalum, and tungsten do not easily heat to the temperatures required to reach that vapor pressure range. In order to obtain a deposition rate high enough for practical applications, the vapor pressure at the source must reach at least $10^{-2}$ Torr. An evaporation source operating at a vapor pressure of $10^{-1}$ Torr, for example, can deliver rates of about 1000 atomic layers per second. The rate of atoms or molecules lost from the source as a result of evaporation, N, in molecules/unit area/unit time (Equation 3.8), can also be expressed in the following relationship, derived from kinetic considerations on how the vapor pressure of the evaporant relates to the evaporation rate:

$$N = \frac{P_v(T)}{(2\mu MkT)^{1/2}} = 3.513 \times 10^{22} \frac{P_v(T)}{(MT)^{\frac{1}{2}}}$$

$$3.9$$

where M stands for the molecular weight of the evaporant, T for the source temperature in Kelvin, and $P_V(T)$ for the vapor pressure of the evaporant in Torr. If a high-vacuum is established, most atoms/molecules will deposit on the substrate without suffering intervening collisions with other gas molecules. The fraction of particles scattered by collisions with atoms of residual gas is proportional to:

$$1 - \exp\left(-\frac{d}{\lambda}\right) \qquad 3.10$$

where d is the distance between source and substrate (see Figure 3.5A) and $\lambda$ is the mean free path of the particles. At $10^{-5}$ Torr the mean free path in air hovers about 5 m and about 0.5 m at $10^{-4}$ Torr (see Table 3.5). The source-to-wafer distance must be smaller than the mean free path of the residual gas. Typically, the source-to-wafer distance is 25 to 70 cm.

In a configuration where the evaporant is held in a container with a small hole (as shown in Figure 3.5A) rather than in an open containment structure, the flow through the exit orifice of the evaporant source can range from free molecular to viscous flow. If the Knudsen number (Kn) (ratio of the mean free path, $\lambda$, to the orifice diameter, D) is greater than 1:

$$Kn = \frac{\lambda}{D} > 1 \qquad 3.11$$

i.e., an atom or molecule passes the orifice in a single, straight track. The flow is free molecular. If, on the other hand, the Knudsen number is <0.01, the flow is viscous, i.e., the atom or

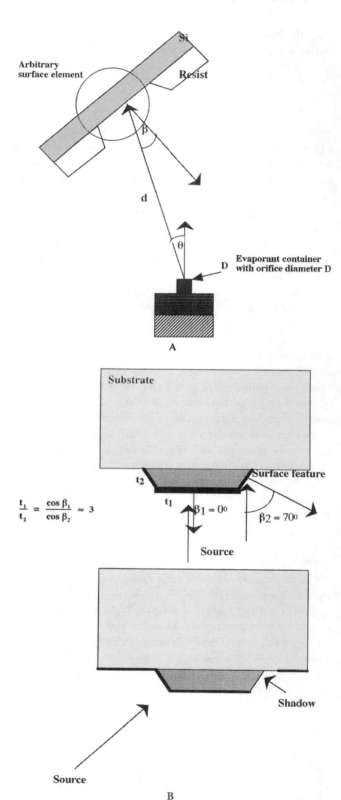

FIGURE 3.5 Geometric considerations in evaporation. (A) Geometric considerations of arrival rate A at an arbitrary surface element in an evaporation experiment. (B) Nonuniform thickness of deposits over varying topography with $\theta = 0^0$ (top) and $\theta \neq 0^0$ C (bottom).

molecule, before emerging from the orifice, bounces several times into the orifice's sidewalls. A transition region exists between the two.

By kinetic molecular theory the mean free path relates to the total pressure as:

$$\lambda = \left(\frac{\pi RT}{2M}\right)^{\frac{1}{2}} \frac{\eta}{P_T} \qquad 3.12$$

where $P_T$ is the total pressure, R the ideal gas constant, and $\eta$ the gas viscosity. Low pressure, low molecular weight, and large viscosity translate into a large mean free path. A high vacuum also ensures a purer material deposit with very little gas inclusions. The arrival rate, A, at distance d from a small evaporation source follows the cosine law of deposition:

$$A \sim \frac{\cos\beta\cos\theta}{d^2} \qquad 3.13$$

where $\beta$ is the angle between the normal to the substrate and the radial vector joining the source to the arrival point being considered, and $\theta$ is the angle between the normal to the evaporation source and the same radial vector (see Figure 3.5A). The deposition, according to Equation 3.13, is not spherical; maximum deposition occurs in directions normal to the evaporation source, where $\cos\theta$ is maximum, that is, $\theta = 0$.

The straight line deposition of evaporant leads to difficulties in obtaining a continuous coating over topographical steps on a wafer, a problem known as shadowing. This is of particular concern in micromachining where high aspect ratio features are common. Figure 3.5B illustrates the problem. Since the thickness of the deposited film, t, is proportional to the $\cos\beta$, the ratio of the film thickness shown in Figure 3.5B, with $\theta = 0°$ is given as:

$$\frac{t_1}{t_2} = \frac{\cos\beta_1}{\cos\beta_2} \qquad 3.14$$

In case of a steep wall, $\beta_2 \sim 70°$ and $\beta_1 \sim 0°$, the ratio of $t_1/t_2$ is approximately 3. The thinner section, $t_2$, is susceptible to cracking at the extreme ends of the interval. Two methods can help overcome this problem. The first method requires heating of the substrate during deposition to 300 to 400°C to increase the surface mobility of the metal atoms. The second method relies on rotating the wafers in planetary wafer holders so that the angle $\beta$ varies during deposition.[31]

Theoretical metal film thickness profiles for deposition into an arbitrary shaped cavity or trench have been calculated for the case of wafers rotating on a planetary system. A computer program named SAMPLE gives the metal profile for particular input parameters.[32,33]

Evaporation is fast (e.g., 0.5 μm/min for Al) and comparatively simple; it registers a low energy impact on the substrate (~ 0.1 eV), i.e., no surface damage results except when using e-beam evaporation. Under proper experimental conditions, it can provide films of extreme purity and known structure. In cases where the purity of the deposited film is of prime importance, evaporation is the preferred technique. For example, for electrochemical sensors, where the electrocatalytic activity of the top monolayers of the sensing electrode can determine the proper operation of such sensors, material purity is a priority consideration. In chemical sensors in general the surface purity of the film is more important than the bulk resistivity of the film (a criterion often used to evaluate IC films). As we just learned, evaporators emit material from a point source (e.g., a small tungsten filament), resulting in shadowing and sometimes causing problems with metal deposition, especially on very small structures. Difficulties also arise for large areas where highly homogeneous films are required, unless special set-ups are chosen. A different problem to overcome arises from the source materials decomposing at high evaporation temperatures. While this risk does not exist when evaporating pure elements, it becomes a problem in the evaporation of compounds and substance mixtures. The e-beam heating systems have an advantage since only a small part of the metal source is evaporating. The initial evaporant stream is rich in the higher-vapor pressure component; however, the melt depletes that constituent locally and eventually an equilibrium rate is established. To deposit complex metal alloys, one uses evaporation of the metallic elements from different sources because of the excellent control in alloy composition possible with modern quartz-crystal deposition rate monitors. When one wants to form the oxides of the deposited metals, evaporation is performed in a low pressure oxygen atmosphere. When oxygen is added one refers to reactive evaporation. The oxygen supply comes from a jet directed at the substrate during deposition. To obtain the correct stoichiometry the deposition needs to take place on a heated substrate.

## Sputtering

Sputtering is preferred over evaporation in many applications due to a wider choice of materials to work with, better step coverage, and better adhesion to the substrate. Actually, sputtering is employed in laboratories and production settings, whereas evaporation mainly remains a laboratory technique. Other reasons to choose sputtering over evaporation can be concluded from a comparison of the two techniques in Table 3.7. Nowadays, sputtering equipment applies films to compact discs, computer disks, large area active-matrix liquid crystal displays, and magneto-optic disks.[34] Also, bearing gears, saw blades, etc. can be coated with a number of hard, wear-resistant coatings such as TiN, TiC, TiAlN, NbN, CrN, TiNbN, or CrAlN.[35]

During sputtering, the target (a disc of the material to be deposited), at a high negative potential, is bombarded with positive argon ions (other inert gases such as Xe can be used as well) created in a plasma (also glow discharge). As most aspects pertaining to the physics and chemistries of DC and RF plasmas were discussed in Chapter 2, we will only amplify the material here as it applies to deposition. The target material is sputtered away mainly as neutral atoms by momentum transfer and ejected surface atoms are deposited (condensed) onto the substrate placed on the anode (Figure 2.8). During ion bombardment, the source is not heated to high temperature and the vapor pressure of the source is not a consideration as it is in vacuum-evaporation. The amount of material, W, sputtered

**TABLE 3.7**  Comparison of Evaporation and Sputtering Technology

|  | Evaporation | Sputtering |
|---|---|---|
| Rate | Thousand atomic layers per second (e.g., 0.5 μm/min for Al) | One atomic layer per second |
| Choice of materials | Limited | Almost unlimited |
| Purity | Better (no gas inclusions, very high vacuum) | Possibility of incorporating impurities (low to medium vacuum range) |
| Substrate heating | Very low | Unless magnetron is used, substrate heating can be substantial |
| Surface damage | Very low; with e-beam, X-ray damage is possible | Ionic bombardment damage |
| *In situ* cleaning | Not an option | Easily done with a sputter etch |
| Alloy compositions, stoichiometry | Little or no control | Alloy composition can be tightly controlled |
| X-ray damage | Only with e-beam evaporation | Radiation and particle damage is possible |
| Changes in source material | Easy | Expensive |
| Decomposition of material | High | Low |
| Scaling-up | Difficult | Good |
| Uniformity | Difficult | Easy over large areas |
| Capital equipment | Low cost | More expensive |
| Number of depositions | Only one deposition per charge | Many depositions can be carried out per target |
| Thickness control | Not easy to control | Several controls possible |
| Adhesion | Often poor | Excellent |
| Shadowing effect | Large | Small |
| Film properties (e.g., grain size and step coverage) | Difficult to control | Control by bias, pressure, substrate heat |

**TABLE 3.8**  Sputter Yields of Several Commonly Used Metals with 500-eV Argon

| Element | Symbol | Sputter Yield |
|---|---|---|
| Aluminum | Al | 1.05 |
| Chrome | Cr | 1.18 |
| Gold | Au | 2.4 |
| Nickel | Ni | 1.33 |
| Platinum | Pt | 1.4 |
| Titanium | Ti | 0.51 |

*Source:*  Vossen, J. L. and W. Kern, *Thin Film Processes*, Academic Press, New York, 1978. With permission.)

from the cathode is inversely proportional to the gas pressure, $P_T$, and the anode-cathode distance, d:[25]

$$W = \frac{kVi}{P_T d} \qquad 3.15$$

with V being the working voltage, i the discharge current, and k a proportionality constant. Other energetic particles such as secondary electrons, secondary ions and photons, and X-rays are created at the target and can be incorporated in the growing film and/or influence its properties through heating, radiation, or chemical reactions. The sputter yield, S, stands for the number of atoms removed per incident ion and is a function of the bombarding species, the ion energy of the bombarding species, the target material, the incident angle of the bombarding species, and its electronic charge. Deposition rate is roughly

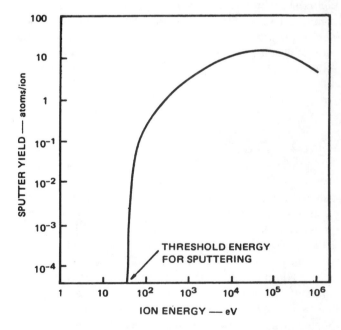

**FIGURE 3.6**  Typical sputter yield characteristic. (From Brodie, I. and J. J. Muray, *The Physics of Microfabrication*, Plenum Press, New York, 1982. With permission.)

proportional to yield for a given plasma energy. Some yield figures are tabulated according to material and ion energy in Table 3.8.[15]

A typical sputter yield as a function of ion energy is shown in Figure 3.6.[25] In the low-energy region the yield increases rapidly from a threshold energy. The sputter threshold, usually in the range of 10 to 130 eV, is essentially independent of the bombarding ion species used and equals about four times the activation energy for evaporation. This threshold voltage decreases with increasing atomic number within each group of the periodic table. Above a few hundred volts, a change-over region takes hold. Its value depends on the ion species and target used, but after its occurrence the sputter yield increases more slowly. Ion energies in the range of 0.5 to 3 kV typically are used for sputter deposition as nuclear collisions are predominant in this range. In this region, the sputter yields typically range from 0.1 to 20 atoms per ion and the yield of most metals is about 1 (see table 8.3). An ion can only penetrate up to a certain depth into the target and still effect a recoil collision that can reach the surface with sufficient energy to eject atoms (e.g., 1 nm/kV penetration for argon ions in copper).

Consequently, at ion bombardment energies in the range of 10 keV to 1 MeV, the sputter yield reaches a maximum and then gradually declines in value as a result of deep ion implantation. Average ejection energies of ions from the target range between 10 and 100 eV. At those energies, the incident ion can penetrate a substrate one to two atomic layers into the surface on which it lands. As a result, the adhesion of sputtered films is superior to films deposited by other methods.[36] Although sputtering is basically a low-temperature process, considerable amounts of energy are dissipated at the target surface. To liberate one atom requires up to 100 to 1000 times the activation energy needed for evaporation, limiting useful sputtering rates to about one atomic layer per second (1000 times less than evaporation from a source operating at $10^{-1}$ Torr). The target needs water cooling as most of the energy, estimated at about 75%, dissipates in the target and excessive heating might result. Only 1 percent of the remaining energy goes into sputtering, and the remainder is dissipated by secondary electrons that bombard and heat the substrate. Figure 2.8 exemplifies a typical sputtering set-up. Sputtering systems often have a load-lock connected to the sputtering chamber used for loading and unloading the substrates. With the use of a load-lock system, the sputtering station can be kept under continuous high vacuum; the target and chamber can be kept free of contamination from gaseous atmospheric components. Instead of pumping the large sputtering chamber, only the small load-lock chamber is pumped, resulting in a significant shorter cycle time. To prevent sputtering of the structural elements of the cathode assembly represented in Figure 3.7, a shield of metal at anode potential is placed around all of the surfaces to be protected at a distance less than that of the dark space at the cathode.[36] No discharge will take place between two surfaces that have a separation less than the Crooks cathode dark space (see Chapter 2).

Crucial to the formation of a sustained DC plasma, as discussed in Chapter 2, is the production of ionizing collisions between secondary electrons released from the cathode and the gas in the sputtering chamber. In effect, this requires a relatively high working pressure of greater than $1 \times 10^{-2}$ Torr. However, if the pressure is too high, significant numbers of the sputtered atoms cannot pass through the sputtering gas and are reflected back to the cathode by collisions. At $10^{-1}$ Torr, the mean free path of the sputtered metal atoms is about 1 mm, which is near

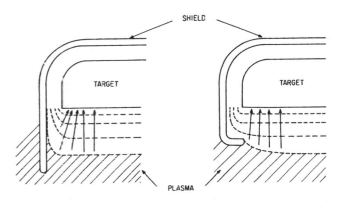

**FIGURE 3.7** Sputtering: use of cathode shield. (Left) Reducing rim effect by extending cathode shield. (Right) Reducing rim effect by wrapping shield around the cathode.[36]

the practical limit. Because of the multiple collisions of the metal atoms in transversing the path between the cathode and anode, the metal atoms arrive at the anode at random incident angles. Any point on the anode surface sees the metal ions impinging from all directions within the hemispherical field of view. This leads to a good step coverage compared to evaporation. Furthermore, as the sputtering target is very broad compared to an evaporation point source, evidence of a shadowing effect decreases drastically. Sputtering at low gas pressure leads to improved film adhesion because the sputtered atoms have a higher energy. Reduced contamination of the film by trapped gas molecules results in films of higher density and purity. The higher material density aspect bears itself out in a slower chemical etch of the deposited film.

For conductors, a DC sputtering set-up can be used. Insulating materials require an RF power supply. Before sputter deposition the substrate may be sputter etched by connecting it to the negative pole of the power supply. Sputter cleaning further promotes adhesion of subsequent metallizations. *In situ* cleaning can also be done efficiently with an ion gun. As in dry etching, the plasma can be confined and densified by using electron cyclotron resonance. Higher ion densities permit higher sputter rates. In a magnetron sputtering apparatus, the crossed electric and magnetic fields contain the electrons and force them into long, helical paths, thus increasing the probability of an ionizing collision with an argon atom. Furthermore, secondary electrons emitted by the cathode due to ion bombardment are bent by the crossed fields and are collected by ground shields. This eliminates the secondary electron bombardment of the substrates, which is one of the main sources of unplanned substrate heating. The deposition rate reaches hundreds of angstrom per minute. Magnetron sputter enhancement can be used in high-frequency sputtering systems as well as in DC sputtering systems. The substrate is also often heated to promote film adhesion. In general, it is preferable to have ions strike the substrate while growing a film. The plasma ions add energy to the film and keep the surface atoms mobile enough to fill virtually every void.

Sputtering often occurs in the presence of a reactive gas, ensuring control or modification of the properties of the deposited film in so-called reactive sputtering. A sensor material produced this way is $IrO_x$, a low-impedance pH sensing material arrived at by sputtering from a pure Ir target with argon ions in the presence of oxygen.[37] The exact mechanism for compound formation in reactive sputtering still eludes researchers. It is conjectured that at low pressures the reaction takes place at the substrate as the film is being deposited. At high pressure, the reaction is believed to occur at the cathode with the compound being transported to the substrate. Composite films can also be made by co-sputtering or by sputtering from a single composite target. Since the various elements in a target have different sputtering rates and sticking probabilities, the composition of the target can differ from that on the substrate. The decomposition is less significant than in evaporation and can be compensated automatically by the changes in target surface composition resulting from preferential sputtering of one of the components until a new balance is reached. High substrate temperatures are needed to obtain the right composition and to promote adhesion.

Commercial in-line sputtering equipment is available, enabling, in a series of evacuated chambers and load locks, depositions of insulators such as $SiO_2$, $Si_3N_4$, $Ta_2O_5$, metals, and semiconductors such as indium tin oxide (ITO) on substrate with sizes of $830 \times 1500$ mm. In one particular set-up one can actually perform *continuous deposition* on substrates, including foils and fabrics. In a high vacuum chamber, the substrate is unwound and passes over the cathode cylinder where thin layers of target material are deposited. The substrate is continuously rewound on the other side of the deposition site. The sputter target is a long cylinder and reactive and nonreactive processes are available.[38] Innovations like these might help revive the potential of micromachined chemical sensors. When manufacturing microfabricated chemical sensors, one often competes with continuous fabrication processes (e.g., continuous printing of glucose sensor paper strips). Classical Si micromachining cannot beat the cost advantages offered by such continuous processes. A process as described here, complemented by continuous lithography, might promote such chemical microsensors.

The most negative aspects in sputtering are the complexity of the process compared to an evaporation process, the excessive substrate heating due to secondary electron bombardment, and, finally, the slow deposition rate. In a regular sputtering process the rate is one atomic layer per second vs. thousand atomic layers per second available from a typical evaporation source at a vapor pressure of $10^{-1}$ Torr. Controls in a sputtering set-up include argon pressure, flow rate, substrate temperature, sputter power, bias voltage, and electrode distance.

## Molecular Beam Epitaxy

Epitaxial techniques arrange atoms in single-crystal fashion upon a crystalline substrate acting as a seed crystal so that the lattice of the newly grown film duplicates that of the substrate. If the film is of the same material as the substrate, the process is called homoepitaxy, epitaxy, or simply epi. Epi deposition represents one of the cornerstone techniques for building micromachines. Of special importance is the fact that Si plates of a predetermined thickness and doping level can be fashioned. The growth rate of an epi layer depends on the substrate crystal orientation. Si (111) planes have the highest density of atoms on the surface, and the film grows most easily on these planes. Important epi applications are Si on Si substrates and GaAs on GaAs substrates. If the deposit is made on a chemically different substrate, usually of closely matched lattice spacing and thermal

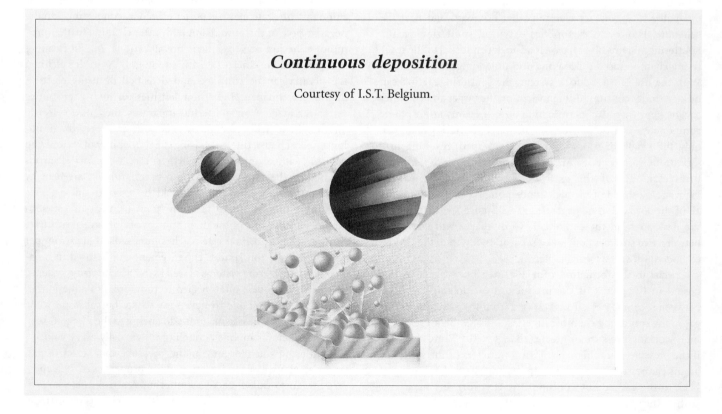

*Continuous deposition*

Courtesy of I.S.T. Belgium.

expansion, the process is termed heteroepitaxy. One important heteroepitaxy application is the deposition of silicon on insulator (SOI), e.g., Si on $SiO_2$ or Si on sapphire ($Al_2O_3$). Various sapphire orientations such as $<011\bar{2}>$, $<10\bar{1}2>$, and $<1\bar{1}0\bar{2}>$ have been used to grow $<100>$-oriented Si layers. Another example of heteroepitaxy is that of gallium phosphide on gallium arsenide.

Various types of epitaxy techniques exist. Chemical vapor phase and liquid epitaxy will be described further below. First, we will briefly discuss PVD epitaxy, i.e., molecular beam epitaxy. In molecular beam epitaxy (MBE), the heated single-crystal sample (say, between 400 and 800°C) is placed in an ultra-high vacuum ($10^{-11}$ Torr) in the path of streams of atoms from heated cells that contain the materials of interest. These atomic streams impinge, in a line-of-sight fashion, on the surface-creating layers with a structure controlled by the crystal structure of the surface, the thermodynamics of the constituents, and the sample temperature. This technique is the most sophisticated form of PVD. The deposition rate of MBE is very low (i.e., about 1 μm/hr or 1 monolayer per second), and considerable attention is devoted to *in situ* material characterization to obtain high-purity epitaxial layers. Fast-acting shutters control the deposition. One or two atomic layers of material lie between the shutter action. This becomes important when an ultra-sharp profile is called for. The low deposition rate gives the operator better control over the film thickness. Ultra-sharp profiles are needed, for example, when making quantum well devices. A quantum well might be 40 Å thick. To uniquely define its energy levels it must be 40 Å ± 2 Å. Figure 3.8 represents a schematic of a molecular beam deposition set-up.[25] The technique has several potential advantages over CVD epitaxy (see below): for example, the relatively low growth temperatures reduce diffusion and autodoping effects. Precise control of layer thickness and doping profile on an atomic layer level is possible. Novel structures such as quantum devices, silicon/insulator/metal sandwiches, and superlattices can be made. Figure 3.8 shows the deposition of n- and p-type GaAs on a GaAs single-crystal surface.[10] With molecular beam epitaxy, virtually any device structure can be made. The limitations to consider lie in volume manufacturing and cost. The ultrahigh vacuum requirements make operation very expensive. From the available epi techniques, MBE still exemplifies the least production-ready technique.[39]

## Laser Sputter Deposition or Ablation Deposition

Laser ablation deposition uses intense laser radiation to erode a target and deposit the eroded material onto a substrate. A high-energy focused laser beam avoids the X-ray damage to the substrate encountered with e-beam evaporation. A high-energy excimer laser pulse coming from, for example, a KrF laser at 248 nm with a pulse energy in the focus of 2 J/cm$^2$ is directed onto the material to be deposited. The energy of the very short wavelength radiation is absorbed in the upper surface of the target, resulting in an extreme temperature flash, evaporating a small amount of material. This material, partially ionized in the laser-induced plasma, is deposited onto a substrate almost without decomposition. This technique is particularly useful when dealing with complex compounds, as

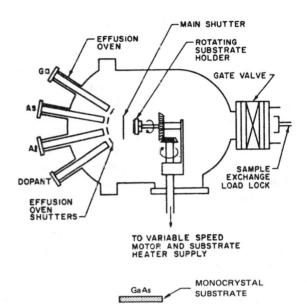

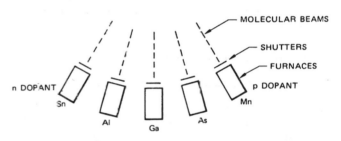

FIGURE 3.8 Schematic of molecular-beam epitaxy growth chamber. Example is the growth of GaAlAs epi on GaAs single crystal. (From Brodie, I. and J. J. Muray, *The Physics of Microfabrication*, Plenum Press, New York, 1989. With permission.)

in the case of the deposition of high temperature superconductor films (HTSC), e.g., $YBa_2Cu_3O_{7-x}$. Pulsed laser deposition faithfully replicates the atomic ratios present in the hot isostatically pressed target disc onto the thin film coating. Achieving complex stoichiometries presents more difficulties with any other deposition technology. Approximately 10,000 pulses (pulse length of 20 ns and a repetition rate of 15 pulses per second) are needed to achieve a film thickness of 0.1 μm on the substrate. Normally, the laser deposited films are amorphous. The energy necessary to crystallize the film comes from heating the substrate (700 to 900°C) and from the energy transferred from the intense laser beam to the substrate via atomic clusters. In Figure 3.9A a schematic set-up of a laser deposition system is shown.[40] Here, a pulsed excimer laser is used to deposit superconductor materials such as those based on YBaCuO.[40] In Figure 3.9B, an improved laser deposition set-up is shown where substrate heating can be replaced in part or completely by additional laser radiation, in this case a cw-$CO_2$ infrared laser(cw stands for continuous wave). With this set-up, using two crossed laser beams, it is possible to deposit and induce the correct crystalline structure in the growing HTSC film while it is being deposited.[41]

A new sensor-related application of pulsed laser deposition is the protection of components in contact with body fluids using the biocompatible calcium phosphate-based ceramic, calcium hydroxylapatite, or $Ca_{10}(PO_4)_6(OH)_2$. This material exemplifies the most stable calcium phosphate in contact with body fluids. Deposition methods of this material include: sputtering, plasma-spraying, electrophoretic deposition, and combinations of these techniques. In all cases, a post-deposition treatment is needed to crystallize the partially or wholly amorphous and/or dehydrated films. With the laser pulse technique, in water vapor-enriched inert gas environments, deposition of the right hydroxyl apatite was observed at temperatures between 400 and 700°C. Adhesion of this material to Si and Ti-6Al-4V was found to be excellent.[7] Laser sputtering also could help prepare complex and stoichiometry-sensitive coatings such as electrochromic devices.

To sum up, it can be said that laser ablation is a good technique for preparing thin films of any desired stoichiometry but, because of the small source size, it is not useful for large-scale coatings. A small source size might make it essential, especially when working with high aspect ratio micromachined structures, to rotate the sample at an angle with respect to the material flux so as to obtain a coverage of the vertical features in the device. For further reading on laser sputtering, refer to References 7, 40, 41, 42, and 43.

## Ion Plating

In ion plating, evaporation of a material is combined with ionization of the atom flux by an electron filament or a plasma. The principle of ion-plating in an argon plasma is illustrated in Figure 3.10.[5] As shown, the addition of a gas (nitrogen in this case) to the reactor enables one to make new compounds (such as TiN) on the substrate surface with the gas reacting with the ionized atoms from the evaporation source (Ti). Because of the high kinetic energy of the impacting ions, a very well-adhering, dense TiN film with extraordinarily low friction coefficient and high hardness coefficient (Vickers hardness of 50000) forms. Because of the thermal nature of the process, very high deposition rates can be achieved.

## Cluster-Beam Technology

When applying cluster-beam technology, ionized atom clusters (say, 100 to 1000 atoms) are deposited on a substrate in a high vacuum ($10^{-5}$ to $10^{-7}$ mbar). Those atom clusters typically carry one elementary charge per cluster and therefore achieve the same energy in an electrical field as a single ion would.

To make these atom clusters, a special evaporation cell must be used, as shown in Figure 3.11.[5] The heating of the evaporant in an evaporation cell with a small opening causes an adiabatic expansion, from more than 100 to $10^{-5}$ or $10^{-7}$ mbar, of the vapor upon exiting that cell. The expansion causes a sudden cooling, inducing the formation of atom clusters. These clusters are then partially ionized by an electron bombardment from a heated filament. Low energy neutral clusters (0.1 eV) and somewhat higher energy ionized clusters (a few eV) arrive at the surface where they flatten and form a film (Figure 3.12)[5] of excellent adhesion and purity, with a relatively low number of defects. Cluster-beam epitaxy is possible at temperatures as low as 250°C, and no charge build-up occurs when depositing on an insulator.

With the wide variety of selective layers needed in chemical sensors and the frequent need for low temperature deposition of thick, well-adhering layers, ion plating, laser sputtering, and

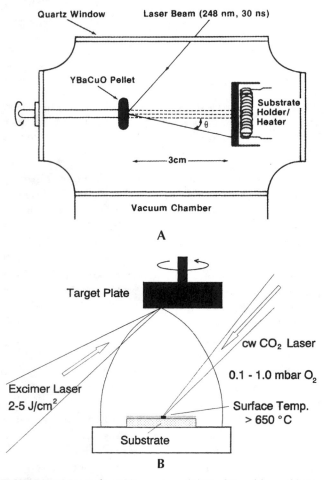

**FIGURE 3.9** Laser deposition system. (A) Traditional laser ablation system. (From Dutta et al., *Solid State Tech.*, 106–110, 1989. With permission.) (B) Laser ablation with additional laser for surface heating.

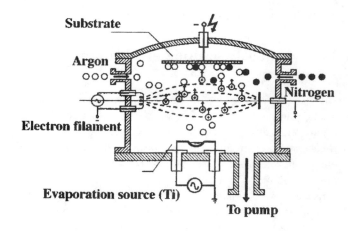

**FIGURE 3.10** Principle of ion plating. (After Menz, W. and P. Bley, *Mikrosystemtechnik für Ingenieure*, VCH, Germany, 1993.)

cluster-beam technology seem destined to play a more crucial role in chemical sensor development in the future. Also, building up devices from atomic constituents, in nanomachining with proximal probes (see Chapter 7), is too time consuming. To speed up the process, one would need too massive an amount of coordinated proximal probes. Cluster-beam technology better enables nanomachining since the building blocks are of a more appropriate size.

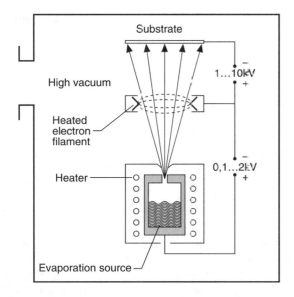

**FIGURE 3.11**  Set-up for ion-cluster beam deposition. (After Menz, W. and P. Bley, *Mikrosystemtechnik für Ingenieure*, VCH, Germany, 1993.)

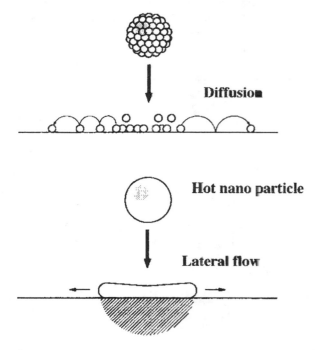

**FIGURE 3.12**  Film forming with atom-clusters. (Top) Old model based on atom diffusion. (Bottom) New model based on particle flow. (After Menz, W. and P. Bley, *Mikrosystemtechnik für Ingenieure*, VCH, Germany, 1993.)

# Chemical Vapor Deposition

## Introduction

During chemical vapor deposition (CVD), the constituents of a vapor phase, often diluted with an inert carrier gas, react at a hot surface to deposit a solid film. In CVD, the diffusive-convective transport to the substrate involves intermolecular collisions. Mass and heat transfer modeling of deposition rates is much more complex than in PVD. In the reaction chamber, the reactants are adsorbed on the heated substrate surface and the adatoms undergo migration and film-forming reactions. Gaseous by-products are desorbed and removed from the reaction chamber. The reactions forming a solid material do not always occur on or close to the heated substrate (heterogeneous reactions) but can also occur in the gas phase (homogeneous reactions). As homogeneous reactions lead to gas phase cluster deposition and result in poor adhesion, low density, and high defect films, heterogeneous reactions are preferred. The slowest of any of the CVD steps mentioned, gas-phase process or surface-process, determines the rate of deposition. The sample surface chemistry, its temperature, and thermodynamics determine the compounds deposited. The most favorable end-product of the physical and chemical interactions on the substrate surface is a stochiometric-correct film. Several activation barriers need surmounting to arrive at this end-product. Some energy source, such as thermal, photons, or ion bombardment, is required to achieve this.

The CVD method is very versatile and works at low or atmospheric pressure and at relatively low temperatures. Amorphous, polycrystalline, epitaxial, and uniaxially oriented polycrystalline layers can be deposited with a high degree of purity, control, and economy. CVD is used extensively in the semiconductor industry and has played an important role in transistor miniaturization by introducing very thin film deposition of silicon. CVD embodies the principal building technique in surface micromachining (Chapter 5).

## Reaction Mechanisms

Figure 3.13 illustrates the various transport and reaction processes underlying CVD schematically:[44]

- Mass transport of reactant and diluent gases (if present) in the bulk gas flow region from the reactor inlet to the deposition zone
- Gas-phase reactions (homogeneous) leading to film precursors and by-products (often unselective and undesirable)
- Mass transport of film precursors and reactants to the growth surface
- Adsorption of film precursors and reactants on the growth surface
- Surface reactions (heterogeneous) of adatoms occurring selectively on the heated surface
- Surface migration of film formers to the growth sites

- Incorporation of film constituents into the growing film, i.e., nucleation (island formation)
- Desorption of by-products of the surface reactions
- Mass transport of by-products in the bulk gas flow region away from the deposition zone towards the reactor exit

Energy to drive reactions can be supplied by several methods (e.g., thermal, photons, or electrons), but thermal energy is the most widely used. In the case of a thermally driven CVD reaction, a temperature gradient is imposed on the reactor, the gas-phase species (e.g., $SiH_4$) forms in a hot region, and the equilibrium shifts towards the desired solid (e.g., Si) in a slightly colder region. Either the gas-phase or the surface processes can determine rate.

The transport in the gas phase takes place through diffusion proportional to the diffusivity of the gas, D, and the concentration gradient across the boundary layer that separates the bulk flow (source) and substrate surface (sink). The flux of depositing material is given by:

$$Fl = D\frac{dc}{dx} \qquad \textbf{3.16}$$

(Fick's first law). The boundary layer thickness, $\delta$ (x), as a function of distance along the substrate, x, can be calculated from (see Figure 3.14):

$$\delta(x) = \left(\frac{\eta x}{\rho U}\right)^{\frac{1}{2}} \qquad \textbf{3.17}$$

where $\eta$ is the gas viscosity, $\rho$ the gas density, and U the gas stream velocity parallel to the substrate. The average boundary layer thickness, in the boundary layer model from Prandtl,[45] over the whole plate can then be calculated as follows:

$$\delta = \frac{1}{L}\int_0^L \delta(x)\,dx = \frac{2}{3}L\left(\frac{\eta}{\rho UL}\right)^{\frac{1}{2}} \qquad \textbf{3.18}$$

where L stands for the length of the plate being deposited on. The Reynolds number for the gas is given by

$$R_e = \frac{\rho UL}{\eta} \qquad \textbf{3.19}$$

$R_e$ is a dimensionless number used in fluid dynamics representing the ratio of the magnitude of inertial effects to viscous effects in fluid motion (see also Chapter 9). For low values (<2000), the gas flow regime is called laminar, while for larger values the regime is turbulent. Substituting Equation 3.19 in Equation 3.18, we obtain:

$$\delta = \frac{2L}{3\sqrt{R_e}} \qquad \textbf{3.20}$$

By substituting this value for the average boundary layer thickness in Equation 3.16 (i.e., with dx = $\delta$) the following expression for the materials flux to the substrate results:

$$Fl = D\frac{\Delta c}{2L}3\sqrt{R_e} \qquad \textbf{3.21}$$

According to Equation 3.21, the film growth rate in the mass flow controlled regime should depend on the square root of the gas velocity, U (the Reynolds number is proportional to U). In Figure 3.15A, a plot of silicon growth rate as a function of the gas flow rate (proportional to the gas velocity in a fixed volume reaction chamber) is shown, as well as the predicted square root dependence.[46] At high flow rates, the growth rate reaches a maximum and them becomes independent of flow. In this regime the reaction rate controls the deposition, evidenced by the exponential dependence of the growth rate on temperature observed at those flow rates (see Figure 3.15B).[46]

Surface reactions can be modeled by a thermally activated phenomenon proceeding at a rate, R, given by:

$$R = R_o e^{-E_a/kT} \qquad \textbf{3.22}$$

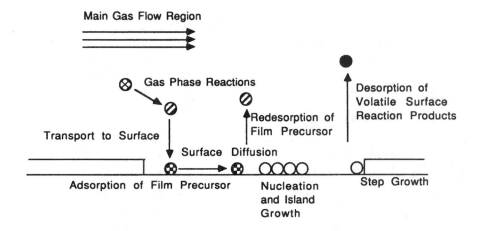

**FIGURE 3.13**  Schematic of transport and reaction processes underlying CVD. (From Jensen, K. F., in *Microelectronic Processing: Chemical Engineering Aspects*, ACS, 1989. With permission.)

where $R_0$ is the frequency factor, $E_a$ is the activation energy in eV, and T is the temperature in degrees K. From the slope of an Arhenius plot, as demonstrated in Figure 3.15B, the activation energy of the rate determining surface process can be deduced. For a certain rate-limiting reaction, the temperature may rise high enough for the reaction rate to exceed the rate at which reactant species arrive at the surface. In such a case, the reaction rate cannot proceed any faster than the rate at which the reactant gases are supplied to the substrate by mass transport, no matter how high the temperature is raised (see plateau in Figure 3.15B). This situation is referred to as a mass-transport-limited deposition process. In this regime, temperature carries less importance than for the reaction-rate limited regime (Arhenius regime). In the latter case, the arrival rate of reactants is less important, since their concentration does not limit the growth rate. A direct practical application of these two possible rate-limiting processes is the way substrates are stacked in low pressure CVD (LPCVD) vs. atmospheric pressure CVD (APCVD) reactors. In a LPCVD reactor, (~ 1 Torr), the diffusivity of the gas species is increased by a factor of 1000 over that at atmospheric pressure, resulting in one order of magnitude increase in the transport of reactants to the substrate. The rate-limiting step becomes the surface reaction. LPCVD reactors enable wafers to be *stacked vertically* at very close spacings as the rate of arrivals of reactants is less important. On the other hand, APCVD, operating in the mass-transport-limited regime, must be designed such that all locations on the wafer and all wafers are supplied with an equal flux of reactant species. In this case the wafers often are *placed horizontally.*

The stability of the flow in a CVD reactor is crucial in achieving uniform deposition. The criterion for flow stability depends on whether the flow is fully developed, i.e., laminar, before it reaches the susceptor. A gas flow is fully developed at a distance $l_F$ from a flat entrance given by:

$$l_F = 0.04 H R_e \qquad 3.23$$

where H stands for the height of the flow channel and Re is based on the channel width. The thermal entrance length $l_T$ for a fully developed radial profile, however, is seven times the velocity entrance length:

$$l_T = 0.28 H R_e \qquad 3.24$$

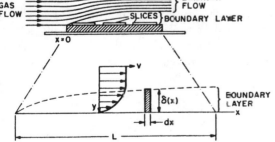

**FIGURE 3.14** Development of a boundary layer in gas flowing over a flat plate. The inset shows an expanded view of the boundary layer.[45]

The Reynolds number in a typical LPCVD reactor is smaller than that for APCVD. Consequently, the thermal entrance length for APCVD is longer than for LPCVD reactors.

The particular flow characteristics are given by the value of the Knudsen number (Kn), defined in this case as the ratio of the mean-free path of the molecules ($\lambda$) to the characteristic dimension of the structure to be coated (e.g., the width of the cevice). The flow around such substrates typically is in the transition regime ($0.1 < Kn = \lambda/w < 10$) or in the free molecular regime (Kn > 10). For typical CVD growth temperatures, $\lambda$ may range from ~0.1 μm at atmospheric to >100 μm at 1 Torr. It is anticipated that the thickness of deposited films in the IC

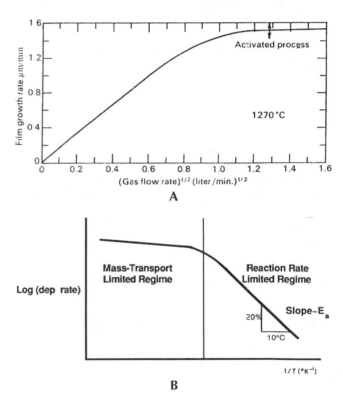

**FIGURE 3.15** Growth rate dependence in a Si CVD process as a function of gas flow rate (A) and temperature (B).[46]

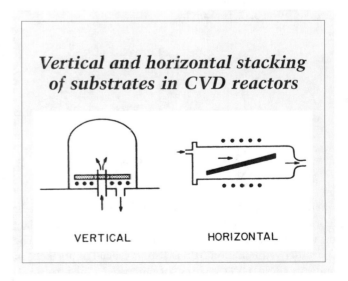

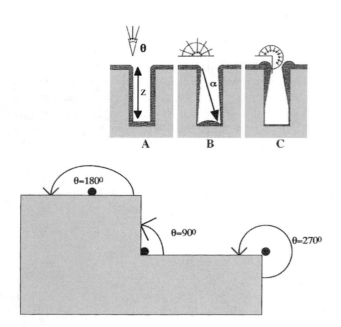

**FIGURE 3.16**  Step coverage cases of deposited film: (A) Uniform coverage resulting from rapid surface migration. (B) Nonconformal step coverage for long mean free path and no surface migration. Distance $\alpha$ is the longest path a molecule travels to reach the corner of the trench. (C) Nonconformal step coverage for short mean free path and no surface migration. (Inset) Different arrival angles in two dimensions.

industry will reach the order of 10 nm, rather than a few microns, and thickness uniformity will be harder and harder to maintain. The tendency is toward large single wafers which are transferred from station to station in so-called cluster tools; considering that no thickness uniformity problem arises when the flow is molecular, it looks like CVD reactors are going to be operated in very low pressure regime (VLPCVD) where molecular flow still dominates.

## Step Coverage

The mean free path of a molecule, $\lambda$, based on slightly modified Equation 3.12, is given by:

$$\lambda = \frac{kT}{2^{1/2}P_T\pi a^2}$$

$$\left(\lambda = \frac{5\times 10^{-3}}{P_T(torr)}\text{ cm at 300 K}\right) \qquad 3.25$$

$$\left(\lambda = \frac{10^{-2}}{P_T(torr)}\text{ cm at 600 K}\right)$$

where a is the molecular diameter.[47] The expressions in brackets are easily memorized and come in handy to correlate the total pressure of the system with the mean free path. The above-derived Equation 3.25 is also of crucial importance in understanding CVD coating of micromachined features.

CVD films have the capacity to passivate or isolate underlying surface features against subsequent layers or the atmosphere, determined by the degree at which edges and pits can be covered uniformly. As demonstrated in Figure 3.16, three cases can be distinguished. Ideally, a uniform, dense coating should form (Figure 3.16A). This can occur in instances where the reactants, after first hitting the solid, have enough energy left for surface migration before a bond is established with the underlying substrate. Coatings in which equal film thickness exists over all substrate topography, regardless of its slope, provide conformal coverage. In a second case, Figure 3.16B, the mean free path of the molecules is large enough to reach the bottom of the trench but little energy remains for surface migration. Finally, in the third case, Figure 3.16C, the mean free path length is too short to reach the bottom and there is little surface migration.

The value of the integral of the material flux in Equation 3.21 ($\int Fld\theta$) and, thus, the CVD film thickness, are directly proportional to the range of feasible angles of arrival, $\theta$, of the depositing species (in the absence of surface migration). Different arrival angles in two dimensions are illustrated in the inset in Figure 3.16. The arrival angle at a planar surface is 180°. At the top of a vertical step, the arrival rate is non-zero over a range of 270°; the resultant film thickness is 270/180, or 1.5 times greater than for the planar case. At the bottom corner of a trench the arrival angle is only 90°, and the film thickness is 90/180, or one half that of the planar case. The CVD profile in Figure 3.16C, where the mean free path is short compared to the trench dimensions and there is no surface migration, reflects the 180, 90, and 270° arrival angles. The thick cusp at the top of the step and the thin crevice at the bottom combine to give a re-entrant shape that is particularly difficult to cover with evaporated or sputtered metal. Gas depletion effects also are observed along the trench walls. Along the vertical walls the arrival angle, $\theta$, is determined by the width of the opening and the distance from the top and can be calculated from

$$\theta = \arctan\frac{w}{z} \qquad 3.26$$

where w is the width of the opening and z the distance from the top surface (Figure 3.16A). This type of step coverage is thinning along the vertical walls and may have a crack at the bottom of the step. For uniformity of deposition, in case of Figure 3.16B, where the mean free path is longer than the distance $\alpha$ (the longest path a molecule travels to reach the corner of the trench), the rate of surface migration of adspecies should exceed the rate of adsorption of adspecies. The condition of $\lambda > \alpha$ can be met by working at low pressures (based on Equation 3.25):

$$\frac{kT}{2^{1/2}P_T\pi a^2} > \alpha \qquad 3.27$$

Equation 3.27 gives us the maximum pressure at which $\lambda > \alpha$. The only variable influencing the requirement for large surface migration is the reactor temperature.

Evaporated and sputtered metal films often have trench profiles as shown in Figure 3.16B, whereas CVD deposited polysilicon and silicon nitride are often uniform and conformal as demonstrated in Figure 3.16A. Plasma deposited $SiO_2$ and $Si_3N_4$ are similar to Figure 3.16B.[49]

A simulation model for nonplanar CVD was recently presented by Coronell et al.[48] The model provides a picture of the evolution of the depositing film profile. The parameters investigated include the sticking coefficient, the surface mobility of the adsorbed reactants, the Knudsen number (the ratio of the mean-free path to the feature size), the feature aspect ratio, and feature geometry.

## Energy Sources for the CVD Processes

The thermal energy is the sole driving force in high temperature CVD reactors; for lower temperature deposition an additional energy source is needed. Radio frequency (RF), photo radiation, or laser radiation can be used to enhance the process, known as plasma-enhanced CVD (PECVD), photon-assisted CVD,[50] or laser-assisted CVD (LCVD),[51] respectively.

With photon- and laser-assisted CVD systems, part of the energy needed for deposition is provided by photons. This method fills the need for an extremely low temperature deposition process. With a laser source, it is possible to write a pattern on a surface directly by scanning the micron-size light beam over the substrate in the presence of the suitable reactive gases. By adjusting the focal point of the laser continuously, it is even possible to grow three-dimensional microstructures such as fibers and springs in a wide variety of materials such as boron, carbon, tungsten, silicon, SiC, $Si_3N_4$, etc. (see Chapter 7).[51]

In PECVD, plasma activation provides the radicals that result in the deposited films, and ion bombardment of the substrate provides the energy required to arrive at the stable desired end-products. The operational temperatures are lower, as part of the activation energy needed for deposition now comes from the plasma.

## Overview of CVD Process Types

In Table 3.9[44,49] we review some important CVD processes, listing applications as well as operational pressures and temperatures for the different types of CVD.

### Plasma-Enhanced Chemical Vapor Deposition (PECVD)

Two simple PECVD set-ups are illustrated in Figure 3.17. The top image demonstrates a rotating susceptor, while the lower one shows an arrangement for heating the susceptor from the back with lamps. The latter set-up also exemplifies the use of a showerhead plate where gases enter and that acts as an electrode. In these set-ups, an RF-induced plasma transfers energy into the reactant gases, allowing the substrate to remain at lower temperatures than in APCVD and LPCVD processes. Clearly, all of the dry etching equipment discussed in Chapter 2 can be used for PECVD as well.

With a simple parallel plate reactor, substrates can be placed horizontally, or, in case the pressure is low enough, vertically (see inset page 107. The vertical position is being

TABLE 3.9    Review of CVD Processes[44,49]

| Process | Advantages | Disadvantages | Applications | Remarks | Pressure/Temperature |
|---|---|---|---|---|---|
| APCVD | Simple, high deposition rate, low temperature | Poor step coverage, particle contamination | Doped and undoped low-temperature oxides | Mass-transport controlled | 100–10 kPa/350–400°C |
| LPCVD | Excellent purity and uniformity, conformable step coverage, large wafer capacity | High temperature and low deposition rate | Doped and undoped high-temperature oxides, silicon nitride, polysilicon, W, $WSi_2$ | Surface-reaction controlled | 100 Pa/550–600°C |
| VLPCVD | | | Single-crystalline silicon and compound semiconductor superlattices | Surface-reaction controlled | 1.3 Pa |
| MOCVD | Excellent for epi on large surface areas | Safety concerns | Compound semiconductors for solar-cells, laser, photocathodes, LEDs, HEMTs, and quantum wells | High volume, large surface area production | |
| PECVD | Lower substrate temperatures; fast, good adhesion; good step coverage; low pinhole density | Chemical (e.g., hydrogen) and particulate contamination | Low-temperature insulators over metals, passivation nitride) | | 200–600 Pa/–300–400°C |
| Spray pyrolysis | Inexpensive | Difficult to control, not compatible with IC | Gas sensors, solar cells, ITO, large area | | Atmospheric (~100kPa)/100–180°C |

used to increase throughput. No loss of film thickness uniformity occurs in the latter case because the PECVD method is surface-reaction limited. Adequate substrate temperature control ensures uniformity. Wafers are placed on the grounded electrode and are subjected to a less energetic bombardment than wafers placed on the powered electrode. In most PECVD systems, the reactor configuration is actually changed so that the potential of both the powered and the grounded electrode, relative to the plasma, become equal.[46] Compared to sputter deposition, PECVD offers several advantages. The lower power densities, higher pressures, and higher substrate temperatures (say, >200°C) all lead to less severe radiation damage than in sputter deposition. Moreover, for radiation-sensitive substrates such as compound semiconductors, afterglow or downstream deposition systems can be used where the radicals are formed in the glow discharge and then transported out of the region to a downstream deposition system. Thus, selective activation of reactants becomes possible without damaging the surface of the substrate.[52] PECVD films are not stoichiometric because the

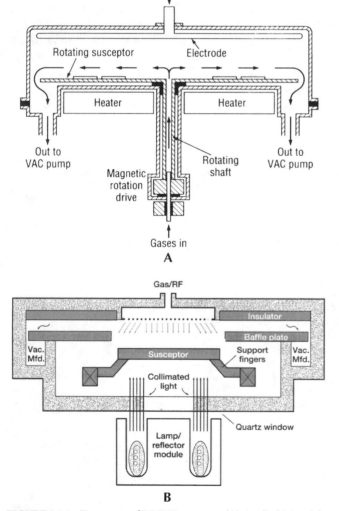

**FIGURE 3.17** Two types of PECVD reactors. (A) Applied Materials Plasma 1 cross-section. (B) Susceptor electrode (grounded electrode) is radiantly heated to provide rapid thermal processing.

deposition reactions vary widely, and particle bombardment during growth of a multicomponent system changes the composition according to the ratios of sputtering yields of the component materials. Despite this negative consequence of particle bombardment, in general, the more ion bombardment the better the film quality. Microstructure, stress, density, and other film properties show marked response, mostly for the better, to ion bombardment during deposition. Good adhesion, low pinhole density, good step coverage, adequate electrical properties, and compatibility with fine line-width pattern transfer processes have led to wide use of PECVD in VLSI. PECVD enables dielectric films such as oxides, nitrides, and oxynitrides to be deposited on wafers with small feature sizes and line-widths at low temperatures and on devices unable to withstand the high temperatures of a thermally activated reaction. Planarization represents only one of the many applications of this versatile technology. Another application is deposition of amorphous-silicon thin films, as used in flat-panel displays, spectacles, and photovoltaic panels. The most significant application is probably the deposition of $SiO_2$ or $Si_3N_4$ over metal lines.[52]

When reviewing dry etching, we discussed plasma chemistry and physics. The following listing further completes that information. Specific attention is paid to PECVD parameter settings influencing thin film properties important in building microstructures, such as film stress and density, sidewall coverage, and gap filling capabilities for planarization. This section also touches upon the field of surface micromachining (Chapter 5). Unlike bulk micromachining, where substrate silicon comprises the sensing element, surface micromachining utilizes deposited thin films, such as polysilicon, silicon nitride, and silicon dioxide. One of the challenges of any surface-micromachined process is to control the intrinsic stresses in those deposited films. Plasma settings act as critical controlling parameters of intrinsic stress in CVD films. Recall that the plasmas of interest are low pressure glow discharges consisting of ions, electrons, and neutral species. The neutral species, molecules, and radicals greatly outnumber the electrons and ions and are not influenced by external electrical fields. Just as etching occurs mainly through neutral species, deposition also almost exclusively involves neutrals. In etching, the ions impart anisotropy to the etching process, while in deposition they alter the properties of the deposited films. The different reactor parameter settings have the following influence on the CVD deposited films:

1. Total Reactor Pressure: since gas density varies with pressure, the mean free path is longer at lower pressures. Consequently, ions accelerated towards the cathode at lower pressure can gain more energy before a collision takes place. Therefore, the effect of ion bombardment is more pronounced at lower pressures and better quality CVD films ensue, characterized as films with a low wet etch rate (high film density) and low *compressive stress*. Experimental results show that as the reactor pressure is lowered film stress goes from *tensile* to compressive and wet etch rates decrease. Often, stress dynamically changes when the film is exposed to the atmosphere and subsequent heating. In

the case of oxide films, for example, tensile films take up water, swell, and become more compressive (the substrate bends down). Compressive oxide films have a greater moisture resistance. A nitride film, on the other hand, absorbs no water and shows no tendency toward compressive stress with time. In the latter case, the amount of Si-H bonds, controlled by annealing at 490°, seems to dominate the stress behavior.[53] Too high pressures promote gas phase polymerization, increasing defect density in the deposited material. In the other extreme, too low pressures (alternatively, the reactant gas is too diluted) change the process from CVD-like to PVD-like, giving way to a columnar film morphology with more defects.

2. Frequency of the RF excitation: at low frequencies, ions experience the full amplitude of the RF voltage, whereas above the ion-transition frequency (>3 MHz) where ions cannot follow anymore, the ion energy is determined by the time average of the RF amplitude. Consequently, lower frequency shifts the ion-energy upward. At the lower frequencies, wet etch rates lower and compressive film stress results. Again, higher energy bombardment yields better films. Higher ion bombardment also improves the film quality on sidewalls. Multiple frequency plasma is emerging rapidly as it allows the user precise control over film properties (particularly stress) over a wide range of process conditions and it can also improve step coverage.

3. RF power effects: an increase in RF power leads to more intense ion bombardment due to the increase in ion current. With a higher ion current, the film deposition rate goes up. In order to separate the effect of RF power on growth rate and film quality, the ratio of power density to deposition rate must be evaluated. This ratio represents a rough measure of the ion bombardment per deposited molecule. In a plot of wet etch rate vs. power density divided by the deposition rate, the maximum ion bombardment, corresponding to highest energy density, leads to the lowest wet etch rate.[54]

4. Growth temperature: the growth temperature has a strong influence on the structure of the film. At low temperatures (and high growth rates), the surface diffusion is slow relative to the arrival rate of film precursors. In this situation, the adsorbed precursor molecule is likely to interact with an impinging precursor molecule before it has a chance to diffuse away on the surface, and an amorphous film is formed. At high temperatures (and low growth rates), the surface diffusion is fast relative to the incoming flux. The adsorbed species can diffuse to step growth sites forming single crystalline materials.

Summarizing, low RF excitation frequency, low reactor pressures, and low deposition rate at high temperature contribute to improved film quality as evidenced by stress and (wet) etch rate measurements. More information on CVD thin films and how they are influenced by the above deposition parameters as well as other parameters (e.g., moisture, flow rate, gas composition, etc.) will be given in Chapter 5 on Surface Micromachining.

**Atmospheric Pressure CVD (APCVD)**

Atmospheric to slightly reduced pressure CVD (±100 to 10 kPa) is used primarily to grow epitaxial (i.e., single-crystalline) films of Si and compound semiconductors such as GaAs, InP, and HgCdTe and to deposit, at high rates, $SiO_2$, e.g., from the reaction of $SiH_4$ and oxygen, at low temperatures of 300 to 450°C (low temperature oxide, LTO).[44] The epitaxy processes (also vapor phase epitaxy, or VPE) involve high growth temperatures (>850°C for Si and 400 to 800°C for compound semiconductors). **Reactor walls typically are cooled** to minimize particulate and impurity problems caused by deposition on the walls. Impurity atoms can be introduced in the gas stream to grow *doped epitaxial layers* (e.g., arsine-doped epitaxial silicon). The APCVD is susceptible to gas phase reactions, and step

## Compressive and tensile stress

Tensile stress causes concave bending (a), and compressive stress causes convex bending (b) of a thin substrate.

DEPOSITED FILM

a)   b)

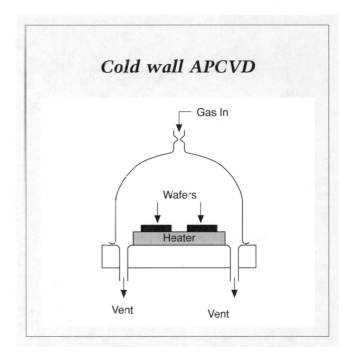

*Cold wall APCVD*

Gas In

Wafers

Heater

Vent   Vent

coverage is often poor. High gas dilutions help avoid gas-phase nucleation. As we saw above, an APCVD reactor operates in the mass-transport regime so that wafer access, in contrast to the PECVD process, becomes more important and temperature control loses importance.

As an example, consider silicon epitaxy at 1200°C:

$$SiCl_4 \text{ (gas)} + 2H_2 \rightarrow Si \text{ (solid)} + 4HCl \text{ (gas)} \qquad \textbf{Reaction 3.3}$$

An intermediate species, $SiCl_2$, is necessary for silicon formation, and below 1000°C no $SiCl_2$ forms. Lower temperature epitaxy can be performed by starting with $SiH_2Cl_2$ or $SiHCl_3$ decomposing more readily to $SiCl_2$. In Figure 3.18 we show the growth rate vs. temperature for the above reaction; the plot delineates a CVD polysilicon deposition region and an epitaxial monocrystalline Si deposition region.[46]

### Low Pressure CVD (LPCVD)

Low pressure CVD (LPCVD) at below 10 Pa allows large numbers of wafers to be coated simultaneously without

detrimental effects to film uniformity. This is the result of the large diffusion coefficient at low pressures leading to a growth limited by the rate of surface reactions rather than by the rate of mass transfer to the substrate. The surface reaction rate is very sensitive to temperature, but temperature is relatively easy to control. Typically, reactants can be used without dilution; therefore, growth rates are only an order of magnitude less than operation at atmospheric pressure allows. LPCVD in some cases can overcome the uniformity, step coverage, and particulate contamination limitations of early APCVD systems. LPCVD polysilicon is used for structural layers in surface micromachines, and LPCVD $SiO_2$ and phosphosilicate glass (PSG) are used as sacrificial layers (see Chapter 5 on Surface Micromachining). Two disadvantages of LPCVD are the low deposition rate and the relatively high operating temperatures.

Horizontal tube, *hot wall reactors* are the most widely used LPCVD reactors. In this case, not only the wafers but also the reaction chamber walls get coated. Such systems require frequent cleaning to avoid serious particulate contamination. They find wide application due to their economy, throughput, and uniformity.

### Very Low Pressure CVD (VLPCVD)

Very low pressure processes (about 1Pa) have been used for the growth of single-crystalline Si at relatively low temperatures. Low pressure operation is also advantageous for the growth of III–V compound superlattices by reducing flow recirculations and improving interface abruptness.[44]

### Metallorganic Chemical Vapor Deposition (MOCVD)

Metallorganic chemical vapor deposition sometimes called organo-metallic vapor-phase epitaxy (OMVPE), relies on the flow of gases (hydrides such as arsine and phosphine or organo-metallics such as trimethyl gallium and trimethyl aluminum) past samples placed in the stream. MOCVD provides thickness control within one atomic layer. Metal organic chemical vapor deposition (MOCVD) has become the preferred epitaxial process, a cost-effective manufacturing process for a variety of compound semiconductor devices. Foremost,

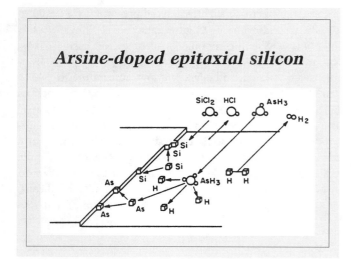

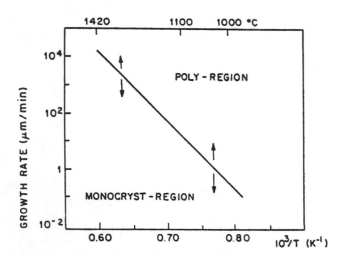

FIGURE 3.18   Growth rate vs. temperature. Epitaxy to CVD transition in CVD polysilicon deposition and epitaxial monocrystalline Si deposition. (From Wolf, R. N., *Silicon Processing for the VLSI Era*, Vol. 1, *Process Technology* (Lattice Press, 1987). With permission.)

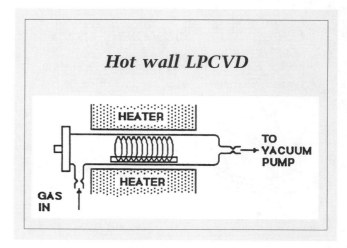

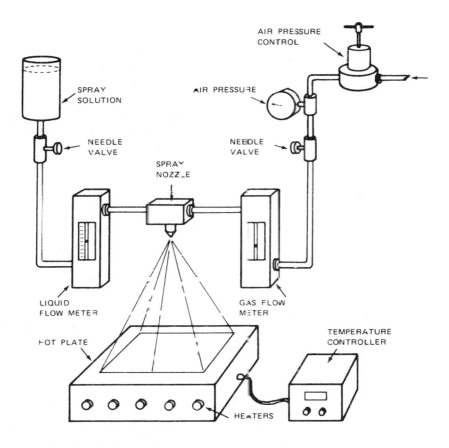

**FIGURE 3.19**   Spray pyrolysis set-up. Courtesy of Mr. Jack Mooney, SRI International, Menlo Park, CA.

MOCVD plays a key role in the manufacture of many optoelectronic devices with III–V compounds for solar cells, lasers, photocathodes, LEDs, HEMTs, and quantum wells.[55]

### Spray Pyrolysis

The CVD technologies discussed so far all revolve around the IC industry. Spray pyrolysis has never been a contender in that arena but is a viable technology for large area devices such as solar cells and antireflective window coatings. In spray pyrolysis, the simplest form of CVD, a reagent dissolved in a carrier liquid is sprayed on a hot surface in the form of tiny droplets. The spray in spraying systems is formed from a liquid pressurized by compressed air, or the liquid is mechanically compressed through a tiny nozzle orifice, or an inert dry vapor carrier is used.

The reagent decomposes or reacts with oxygen on the hot surface to deposit a stable residue. A simple spray-pyrolysis set-up is shown in Figure 3.19.[14] All spraying processes display the same significant variables: the substrate temperature, ambient temperature, chemical composition of the carrier gas and/or environment, carrier gas flow rate, nozzle-to-substrate distance, droplet radius, solution concentration, solution flow rate, and — for continuous processes — substrate motion.[14] Because individual droplets evaporate and react quickly, grain sizes are very small, usually less than 0.1 μm. The small grains pose a disadvantage for most semiconductor applications but not necessarily for sensor applications, e.g., in gas sensors where surface area is important.

The process produces relatively thick films, is difficult to control, and is not compatible with IC processing. Spray pyrolysis is a very simple, inexpensive, useful technique used to produce several compound semiconductors with utility in various devices such as solar cells, gas sensors, antireflection coatings, ion selective electrodes, etc.[56]

## Epitaxy

### Introduction

In surface micromachining, where polycrystalline silicon functional layers are built up on a substrate rather than etched in the bulk of a single-crystal substrate as in bulk micromachining, epi-grown silicon-on-insulator (SOI) wafers are a very attractive alternative. Surface micromachines built from layers of epi silicon and isolated from the substrate by a SiO₂ layer combine the most attractive features of surface micromachining (i.e., CMOS compatibility and freedom in types of structural shapes) with the superior single crystal properties of the epi layer (see Chapter 5). For ICs, SOI is compatible with existing wafer processes, provides 50% faster circuit speed than bulk silicon, allows easier scaling to finer line-widths, and can reduce the number of required mask levels for a given design by ~30%. SOI may very well represent the wave of the future in the IC industry rather than GaAs.[57] Silicon micromachining, being a small industry today, is hostage to progress in the IC industry, and SOI becoming more prevalent in that industry is a good indicator for coming trends in micromachining. SOI, we believe, will simplify building micromachines by reducing the number of masks, reducing packaging concerns, and making integration of electronics easier. In the IC industry, many device parameters such as transistor breakdown voltage, junction capacitance, transistor gain, and AC performance depend on the epi-layer thickness, necessitating precise control of the epi-layer thickness. A precise control of the epi-layer also enables more reproducible and predictable mechanical micromachines.

## Liquid and Solid Phase Epitaxy

In the section on physical vapor deposition (PVD) techniques we discussed molecular beam epitaxy (MBE) as the most advanced PVD technique. In the chemical vapor deposition (CVD) section we encountered two types of epi techniques, i.e., atmospheric pressure CVD (APCVD) and metallorganic CVD (MOCVD). For silicon processing, vapor phase epitaxy (VPE) has met with the widest acceptance since excellent control of impurity concentration and crystalline perfection can be achieved. Here, we briefly mention liquid and solid phase epitaxy. For depositing multilayer structures of different materials on the same substrate, liquid phase epitaxy (LPE) is used. In liquid phase epitaxy, films grow from a liquid solution very near the equilibrium state, making the technique reproducible and resulting in films with low concentrations of growth-induced defects. A schematic for a typical liquid phase epitaxy set-up is presented in Figure 3.20.[25] In operation, a graphite slider plate moves relative to a multiple-well assembly to bring the substrate in a recess in the slider plate in contact with the different solutes. Liquid phase epitaxy has found its widest application in producing epitaxial layers of III–V compounds (e.g., InP, GaAs). Anderson gives a good review of the LPE technique.[58]

Solid phase epitaxy describes the crystalline regrowth of amorphous layers that extend continuously to the underlying single-crystal substrate. At temperatures between 500 and 600°C, in the case of silicon, a recrystallization process occurs on the underlying crystalline substrate and regrowth proceeds toward the surface. Regrowth is faster on (100) than on (111) Si, and impurities such as B, P, and As enhance the regrowth, while O, C, N, or Ar retard regrowth.[46]

## Selective Epitaxy

The incorporation of selective epi in the micromachining arsenal will make yet more versatile microstructures possible. Under the correct growth conditions and/or surface treatment, it is possible to initiate Si growth in selected areas.[46,59] Selective epitaxial growth allows the formation of closely spaced silicon features isolated by $SiO_2$ (see Figure 3.21). Besides increased density over other insulation techniques (important mainly for the IC industry) structures as shown in Figure 3.21A and B might enable a host of interesting mechanical microstructures. For example, imagine that the $SiO_2$ in Figure 3.21B is selectively etched away, resulting in an epi anchor with suspended poly beams. In selective epitaxy of the type shown in Figure 3.21A, silicon atoms possessing high surface mobility migrate to sites on the single crystal where nucleation is favored. In the ideal case, all of the epi grows exclusively in the oxide openings. The silicon mobility improves with the presence of halides. The higher the number of chlorine atoms in the silicon source, the better the degree of selectivity (e.g., $SiHCl_3$ is a better source than $SiH_2Cl_2$). The selective deposition shown in Figure 3.21B is accomplished by simultaneous deposition of epitaxial silicon in the oxide openings and polysilicon on the oxide surfaces. The

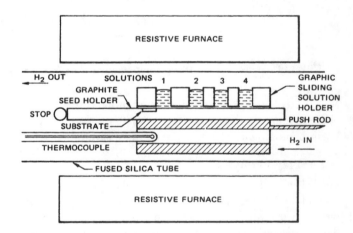

FIGURE 3.20   Liquid phase epitaxy set-up. (From Brodie, I. and J. J. Muray, *The Physics of Microfabrication*, Plenum Press, New York, 1989. With permission.

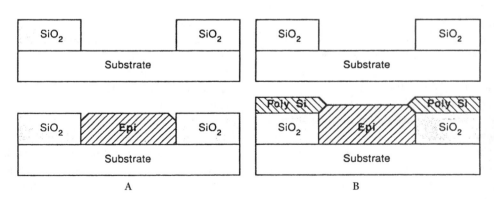

FIGURE 3.21   Selective deposition of epitaxial silicon. (A) Selective deposition of epi Si on Si in $SiO_2$ windows. (B) Simultaneous deposition of epi-Si on Si, and poly Si on $SiO_2$.[46]

angle between the epi and the poly in Figure 3.21B depends on the crystallographic orientation of the substrate. The angle is 90° for <110> orientations, 72° toward the polysilicon for <100> orientations, or tapered towards the single-crystal silicon at 70° for the <111> orientation.[46] These figures represent interesting angles from which to construct micromechanical structures. This selective deposition process also lends itself to CVD, e.g., for selective deposition of W. For more information on selective epitaxy, see Chapter 5.

## Epi-Layer Thickness

The epitaxial layer thickness is a critical parameter both in IC applications and micromachines and therefore must be accurately measured and controlled. Many devices, from discrete transistors and 16-megabyte DRAMs to membrane-based micromechanical structures, use silicon epitaxial layer wafers as their starting material. The thickness of an epitaxial layer forms an integral part in the design of many micromachined devices. For example, in a piezoresistive pressure sensor, the epi-layer thickness control ultimately determines the pressure sensitivity control.

Epi-layer thickness can be measured from infrared reflectance, angle-lap and stain, tapered groove, weighing, capacitance-voltage measurements, and profilometry. The most widely used, nondestructive method of measuring epi thickness is with infrared (IR) instruments. Fourier transform infrared offers automated epilayer thickness measurements.[60] Commercial epitaxial services offer layers ranging from 0.5 to 150 μm, and N, P, N+, and P+ type with a uniformity better than ±5%.

## Recent Trends in Vacuum Equipment

Vacuum equipment to clean, deposit, and etch is being combined more and more in so-called *cluster tools*.[61] Another recent

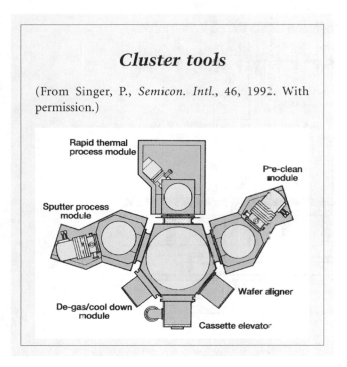

### Cluster tools

(From Singer, P., *Semicon. Intl.*, 46, 1992. With permission.)

Rapid thermal process module

Pre-clean module

Sputter process module

Wafer aligner

De-gas/cool down module

Cassette elevator

trend in vacuum equipment, along the same line, is the development of continuous in-line stations where raw material goes in at one end and a finished product comes out at the other end. The latter goes hand-in-hand with the concept of a *minienvironment* loosely defined as an ultraclean space containing only wafers, one or more process chambers, a robot arm for wafer transport, and a few additional accessories. Potential contaminants are thus tightly controlled at the process level itself, while the surrounding area operates under relaxed cleanliness requirements. Minienvironments offer the advantage of keeping humans out of the very clean areas.[62]

## Electrochemical Deposition

### Introduction

Electrochemical deposition may be of yet greater importance in micromachining than in the IC industry. In micromachining, electrochemical deposition enables the metal replication of high-aspect-ratio resist molds while maintaining the highest fidelity. In the IC industry one tends to avoid wet chemistry when a dry deposition method presents itself, but micromachining needs are forcing reconsideration of electrochemical techniques as a viable solution.[63] The fundamentals of electrochemical deposition are covered below; use of electrochemical techniques in LIGA is treated in Chapter 6, and in Chapter 7 we discuss the technology in comparison with other machining technologies.

### Electroless Metal Deposition

#### Metal Displacement and Electroless Deposition

Simple metal displacement reactions where the surface of a less noble metal such as Zn immersed in a copper sulfate solution gets replaced by a copper surface features the simplest example of electroless deposition. Zinc gives off two electrons and goes into solution while Cu ions receive two electrons and deposit as a metal. The surface of the zinc substrate becomes a mosaic of anodic (zinc) and cathodic (copper) sites. The displacement process continues until almost the entire surface is covered with copper. At this point, oxidation (dissolution) of the zinc anode virtually stops and copper deposition ceases. Deposits range from only 1 to 3 μm thick; hence, plating via this displacement process has limited application.[4]

In order to continuously build thick deposits by chemical means without consuming the substrate, it is essential that a sustainable oxidation reaction be employed as an alternative to the dissolution of the substrate. In Figure 3.22, the difference between immersion plating and electroless deposition is illustrated by comparing deposit thickness vs. time.[4] The oxidation reaction replacing substrate consumption might be that of hypophosphite, resulting in an overall reaction for Ni deposition given by:

Reduction:

$$Ni^{+2} + 2e^- \rightarrow Ni$$

Reaction 3.4

## *Mini-environments*

(From Glanz, J., *Res. & Devel.*, 97–101, 1992. With permission.)

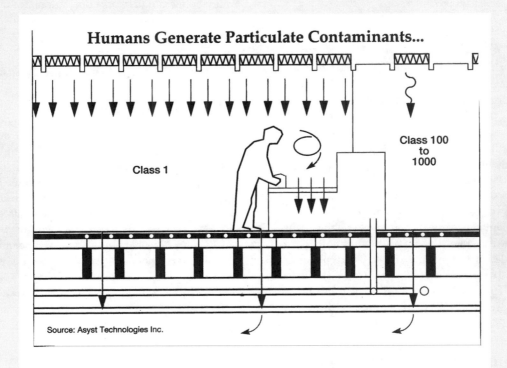

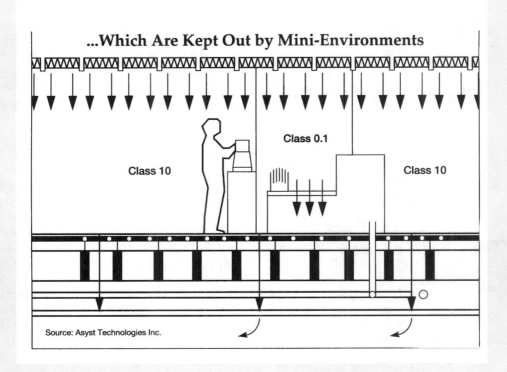

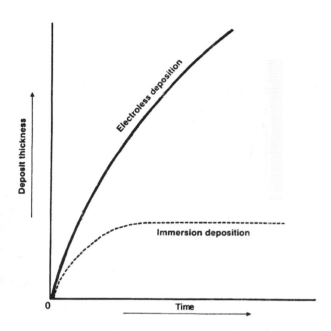

**FIGURE 3.22** Thickness vs. time comparison between electroless and immersion deposition. (From Mallory, J. B., *Electroless Plating, Fundamentals and Applications*, AESF, 1990. With permission.)

Oxidation:

$$H_2PO_2^- + H_2O \rightarrow H_2PO_3^- + 2H^+ + 2e^-$$ Reaction 3.5

$$Ni^{+2} + H_2PO_2^- + H_2O \rightarrow Ni + H_2PO_3^- + 2H^+$$ Reaction 3.6

In parallel to Reaction 3.4 more or less severe hydrogen reduction goes on. Copious hydrogen evolution might upset the quality of the depositing metal film and should be avoided. The hydrogen evolution rate is not directly related to that of the metal deposition, and mainly originates from the reductant molecules. Stabilizers (i.e., catalytic poisons) are needed in electroless deposition baths as the solutions are thermodynamically unstable; deposition might start spontaneously onto the container walls. Poisons for hydrogenation catalysts such as thiourea, $Pb^{2+}$, and mercaptobenzothiazole function as stabilizers in

such electroless baths. Besides stabilizers, the metal salt, and a reducing agent, electroless solutions may contain other additives such as complexing agents, buffers, and accelerators. Complexing agents exert a buffering action and prevent the pH from decreasing too fast. They also prevent the precipitation of metal salts and reduce the concentration of free metal ions. Buffers keep the deposition reaction in the desired pH range. Accelerators, also termed exaltants, increase the rate of deposition to an acceptable level without causing bath instability. These exaltants are anions, such as $CN^-$, thought to function by making the anodic oxidation process easier. In electroless copper, for example, compounds derived from imidazole, pyrimidine, and pyridine can increase the deposition rate to 40 $\mu m$ $hr^{-1}$. The electroless deposition must occur initially and exclusively on the surface of an active substrate and subsequently continue to deposit on the initial deposit through the catalytic action of the deposit itself. Since the deposit catalyzes the reduction reaction, the term autocatalytic is often used to describe the plating process. Electroless plating is an inexpensive technique enabling plating of conductors and nonconductors alike (plastics such as ABS, polypropylene, Teflon, polycarbonate, etc. are plated in huge quantities). A catalyzing procedure is necessary for electroless deposition on nonactive surfaces such as plastics and ceramics. The most common method for sensitizing those surfaces is by dipping into $SnCl_2/HCl$ or immersion in $PdCl_2/HCl$.[4] This chemical treatment produces sites that provide a chemical path for the initiation of the plating process.

Metal alloys such as nickel-phosphorous, nickel-boron, cobalt-phosphorus, cobalt-boron, nickel-tungsten, copper-tin-boron, palladium-nickel, etc. can be produced by codeposition. In the case of Ni deposits, with or without phosphorous and boron incorporated, different electroless solutions are used to obtain optimum hardness, effective corrosion protection, or optimum magnetic properties. Recent experimental results show that composite material also can be produced by codeposition. Finely divided, solid particulate material is added and dispersed in the plating bath. Electroless Ni with alumina particles, diamond, silicon carbide, and PTFE have reached the commercial market. Table 3.10 presents a list of electroless plating baths.[4]

**TABLE 3.10** Typical Electroless Plating Baths

| Component | Concentration (per L) | Application/Remark | pH | Temp (°C) |
|---|---|---|---|---|
| Au | 1.44 g $KAu(CN_2$, 6.5 g KCN, 8 g NaOH, 10.4 g $KBH_4$ | Plate beam leads on silicon ICs, ohmic contacts on n-GaAs | 13.31 | 70 |
| Co-P | 30 g $CoSO_4 \cdot 7H_2O$, 20 g $NaH_2PO_2 \cdot H_2O$, 30 g $Na_3$citrate $\cdot 2H_2O$, 60 g $NH_4Cl$, 60 g $NH_4OH$ | Magnetic properties | 9.0 | 80 |
| Cu | 10 g $CuSO_4 \cdot 5H_2O$, 50 g Rochelle salt, 10 g NaOH, 25 ml conc. HCHO (37%) | Printed circuit boards | 13.4 | 25 |
| Ni-Co | 3 g $NiSO_4 \cdot 6H_2O$, 30 g $CoSO_4 \cdot 7H_2O$, 30 g $Na_2$malate $\cdot 1/2H_2O$, 180 g $Na_3$citrate $\cdot 2H_2O$, 50 g $NaH_2PO_2 \cdot H_2O$ | | 10 | 30 |
| Ni-P | 30 g $NiCl_2 \cdot 6H_2O$, 10 g $NaH_2PO_2 \cdot H_2O$, 30 g glycine | Corrosion and wear resistance on steel | 3.8 | 95 |
| Pd | 5 g $PdCl_2$, 20 g $Na_2EDTA$, 30 g $Na_2CO_3$, 100 mL $NH_4OH$ (28% NH ), 0.0006 g thiourea, 0.3 g hydrazine | Plating rate is 0.26 $\mu m$/min | | 80 |
| Pt | 10 g $Na_2Pt(OH)_6$, 5 g NaOH, 10 g ethylamine, 1 g hydrazine hydrate (added now and then to maintain this concentration) | Plating rate is 12.7 $\mu m$/hr | | 35 |

*Source:* Based on Mallory[4] and Romankiw.[59]

## Mixed Potential and Evan's Diagram

Electrochemically speaking, an electroless deposition reaction is the combined result of two independent electrode reactions: the cathodic partial reaction (Reaction 3.4) and the anodic partial reaction (Reaction 3.5). A close look at an electrolytic cell as shown in Figure 3.23 will clarify some of the terminology used here. In an electrolytic cell a potential is applied over two electrodes submerged in a solution made conductive by adding salts, i.e., the electrolyte solution (e.g., KCl in water). The positive ions ($M^{z+}$) in this solution are called 'cations' and the negative ions ($X^{w-}$) 'anions'. When applying a bias in the presence of an electro-active species (e.g., $NiCl_2$), the positive cations are attracted to the negatively charged cathode which gives off electrons in a cathode reaction (forming Ni); the negative anions are attracted to the positive anode which picks up electrons in an anode reaction (forming $Cl_2$). In electrolysis, electrons flow from the anode to the cathode via the power supply. In electroless deposition, as considered here, no external potential is applied, and both anodic and cathodic partial reactions occur on the same electrode surface. The amount of electroless deposition can be predicted from the polarization curves of the partial anodic and cathodic processes as they take place on two separate electrodes in a cell as shown in Figure 3.23. Figure 3.24 represents the polarization curves in

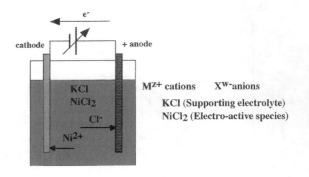

$M^{z+}$ cations      $X^{w-}$ anions

KCl (Supporting electrolyte)
$NiCl_2$ (Electro-active species)

$Ni^{2+} + 2e^- = Ni$ (Cathode reaction)

$2Cl^- \quad = Cl_2 + 2e^-$ (Anode reaction)

---

$NiCl_2 \quad = Ni + Cl_2$

**FIGURE 3.23**  Electrolytic cell with anions and cations in an electrolyte solution.

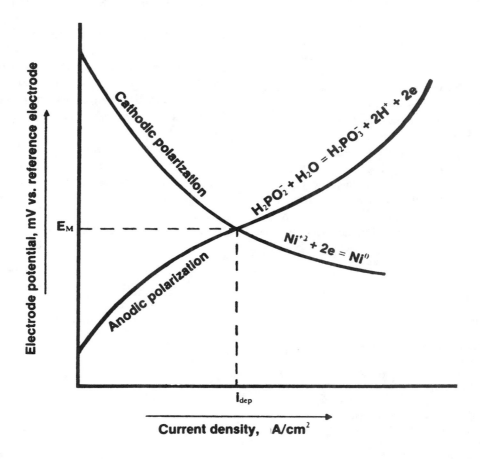

**FIGURE 3.24**  Schematic representation of anodic and cathodic polarization curves combined in an Evan's diagram to determine deposition potential ($E_M$) and deposition current ($i_{dep}$).

an Evan's diagram. The mixed potential, $E_M$, and the deposition current density, $i_{dep}$, are obtained by the intersection of the partial anodic and cathodic polarization curves. At the steady state equilibrium potential (mixed potential), the rate of deposition equals the rate of oxidation of hypophosphite (anodic current density, $i_a$), as well as the rate of the cathodic reaction (cathodic current density, $i_c$). That is:

$$i_{deposition} = i_a = i_c \qquad 3.28$$

The expression in Equation 3.28 and Faraday's laws help to determine the amount of Ni deposited. Faraday's first law states that the amount of any substance discharged is proportional to the amount of electricity passed across the interface where the deposition takes place. The second Faraday law states that if the same amount of electricity passes through a variety of electrolytes, the amount of the different substances discharged is proportional to the chemical equivalent weights of those substances. The quantity of electricity required to deposit one gram equivalent of a substance (the equivalent weight of the substance expressed in grams) is called the Faraday constant F which equals 96,500 coulombs. If an ion of valence z is being discharged electrolytically, each gram ion will contain z gram equivalents and will require zF coulombs to cause its deposition. The total amount of deposited material, $m_{max}$, as a function of the total current I (= $i_{deposition} \times A$, where A is the surface area), is

$$m_{max} = \frac{ItM}{Fz} \qquad 3.29$$

with t the deposition time and M the molecular weight of the depositing material (see Equation 6.5). It should be noted that the above represents an idealized picture. We ignored for example that part of the cathodic current might be attributed to hydrogen evolution, diminishing the metal deposition rate.

### Electroless Plating in the IC Industry and in Micromachining

The IC industry applies electroless metal deposition for a wide variety of applications as incorporated in Table 7.1 and 7.2. In Figure 3.25 we singled out the schematic for electroless deposition of a buried conductor. In micromachining applications one uses electroless deposition for the same purposes as in the IC industry as well as to make structural micro-elements from a wide variety of metals, metal alloys, and even from composite materials. A key advantage of electroless deposition is that a metal can be deposited without the need for electrical contact with a voltage source. This feature makes it suitable for depositing metals on a wafer with CMOS circuitry which might get damaged in a conventional plating process. Plating of high aspect ratio features is of key interest to micromachinists in general and in particular for those interested in LIGA and LIGA-like processes (see Chapter 6).

### Measuring Thickness of Plated Layers

The methods to determine the electroless deposition rate can be split into two categories, namely electrochemical and non-electrochemical techniques. The electrochemical techniques are: Tafel extrapolation method (see Equation 3.45 below), dc polarization method (see Evan's diagram in Figure 3.24), ac impedance method, and anodic stripping method. The non-electrochemical methods are: weight-gain method, optical absorption method, resistance probe method, and acoustic wave method. For more details on each of these techniques see Ohno, 1988 and references therein.[64]

## Electrodeposition — Electroplating and Anodization

### Electroplating

Electroplating takes place in an electrolytic cell as illustrated in Figure 3.23. The reactions involve current flow under an imposed bias. As an example, let's consider the deposition of Ni from $NiCl_2$ in a KCl solution with a graphite anode (not readily attacked by $Cl_2$) and a Au cathode (inert surface for Ni deposition). With the cathode sufficiently negative and the anode sufficiently positive with respect to the solution, the following reactions initiate:

| | |
|---|---|
| $Ni^{2+} + 2e^- \rightarrow$ Ni (Cathode reaction) | **Reaction 3.7** |
| $2Cl^- \rightarrow Cl_2 + 2e^-$ (Anode reaction) | **Reaction 3.8** |
| $NiCl_2 \rightarrow Ni + Cl_2$ | **Reaction 3.9** |

The amount of Ni deposited can be calculated from Equation 3.29. The process differs from the electroless Ni deposition in that the anodic and cathodic processes occur on separate electrodes and that the reduction is affected by the imposed bias

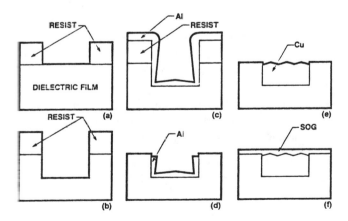

**FIGURE 3.25** Schematic of a buried conductor process. (a) Photoresist pattern over a dielectric. (b) Pattern transfer by anisotropic dry etching. (c) Deposition of a thin Al seed layer. (d) Base metal patterning with lift-off. (e) Trench filling with electroless Cu. (f) Spin-on glass (SOG) dielectric layer for planarization.

rather than a chemical reductant. Important process parameters are pH, current density, temperature, agitation, and solution composition.

## Thermodynamics of Electrochemical Reactions

Electrochemical deposition dates back further than dry physical and chemical vapor deposition and the theory is very well established.[3] Let's analyze in some detail the reaction at an electrode, where a metal ion, $M^{z+}$, is reduced to pure metal (Reaction 3.7). For simplicity we will assume that the metal ion, $M^{z+}$, deposits on a metal electrode M. In this process, the metal ion in the solution leaves its position in the electrolyte and takes up a position in the structure of the metal electrode. The free energy change accompanying this process can be written as:

$$\Delta G = G_m - G_e \qquad 3.30$$

where $G_m$ represents the free energy of the pure metal and $G_e$ that of the ion in the electrolyte. We may also write the free energy change in terms of ion concentration, or more correctly in terms of ion activity,

$$\Delta G = \Delta G^0 - RT \ln a_{M^{z+}} = \Delta G^0 - RT \ln C_{M^{z+}} v_{M^{z+}} \qquad 3.31$$

where $C_M{}^{z+}$ is the ionic concentration of the metal ions, $a_M{}^{z+}$ the ionic activity, and $v_{M^{z+}}$ the ionic activity coefficient. The activity of the metal, M (a pure solid), is 1 and does not change during reaction. The electrical work, w, performed at constant pressure and constant temperature in electrodeposition, may be related to the free energy change as:

$$\Delta G = -w + P\Delta V \qquad 3.32$$

No measurable volume change accompanies the electrolysis process ($\Delta V = 0$). The electrical work performed in the transfer of a charge of zF coulombs through a potential difference, E, is $EzF(= w)$ joules. Hence, we have:

$$\Delta G = -EzF \qquad 3.33$$

Substituting Equation 3.33 into Equation 3.31, one obtains an expression for the potential of the electrode (the cathode here) as a function of the ion concentration

$$E = E^0 + \frac{RT}{zF} \ln a_{M^{z+}} \qquad 3.34$$

i.e., the so-called Nernst equation, in which $E^0$ is called the standard electrode potential, equivalent to the potential at which $\ln a_{M^{z+}} = 0$. A similar equation can be written down for the anodic reaction (Reaction 3.8). The overall reaction in the electrolytic cell is given by the sum of the free energy changes on both electrodes:

$$\Delta G = \Delta G_2 - \Delta G_1 \qquad 3.35$$

or

$$\Delta G = -(E_2 - E_1)zF = -E_{cell}zF \qquad 3.36$$

where $E_{cell}$ is defined as an electromotive force (EMF) equal to the potential differences at the electrodes that make up the cell. If $E_2 > E_1$, the sum of the free energies for the overall reaction is negative and the reactions occur spontaneously (e.g., Cu ions on one electrode and Zn ions on the other electrode in a Daniel cell type battery or 'galvanic cell'). In the electrolysis of $NiCl_2$, on the other hand, no spontaneous reaction takes place and an external potential, $E_{ext} > E_{cell}$, must be applied to force the desired Reactions 3.7 and 3.8. The chemicals in cathodic and anodic compartments of a battery must be kept from reacting directly with each other. To that end, anodic and cathodic compartments are separated by a membrane forcing electrons through the external circuit where they can do useful work. In electrodeposition, work must be delivered by an external power source in order to affect metal deposition.

It must be stressed that the reactions described so far assume thermodynamic reversibility in which $E_{cell}$ is a measure of the reversible EMF of the cell and assumes that no current flows in the circuit. It is thus only a measure of the tendency of a cell to produce a given EMF, but not of the actual occurring process which also involves kinetic effects.

## Kinetics of Electrochemical Reactions
### Activation Control

Considering the preceding section we can make the distinction between possible or impossible thermodynamical processes. A thermodynamic possible reaction may not necessarily occur if the kinetics of the reaction are not favorable.

If the electrode is part of a system in which a finite current passes, it no longer exhibits perfect thermodynamical reversibility, and the electrode potential deviates from its reversible potential. This deviation from the ideal potential is called polarization.

Figure 3.26 depicts the electrode processes in terms of the flow of electrons and ions. The net current, i, equals the sum of the partial current densities $i_a$ and $i_c$ (see also Equation 3.28). If $|i_c| > |i_a|$, then i is negative and a cathodic net process results. On the other hand, if $|i_c| < |i_a|$, i is positive and an anodic net process results. Hence, for a cathode reaction, E is less than $E°$ and for an anode reaction E is greater than $E°$. The difference between E and $E°$ is defined as overpotential at the electrode, $\eta$. Thus, $\eta$ is positive for an anodic current and negative for a cathodic current. A net current density of zero does not imply zero current density but merely represents the equilibrium condition where the partial current densities are equal, that is, $E = E°$ and $\eta = 0$.

We can start the analysis of the polarization process by observing ion transport at the electrode/electrolyte interface without imposed electric field. At equilibrium, anodic and cathodic reactions take place but the sum of these reactions gives a zero net current flow.

Consider the case shown in Figure 3.27 where a positive ion is transported through a double layer to be incorporated into

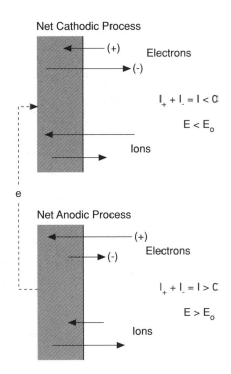

FIGURE 3.26   Flow of electrons and ions between electrode and the electrolyte for net cathodic and net anodic processes. I = i × A (with A the surface area); $I_+ = i_a A$; and $I_- = i_c A$.

an electrode. In accordance with the kinetic theory of quantum chemistry, the ion has to surpass an energy barrier or activation energy before being incorporated. The frequency with which the successful ion jumps over the energy barrier may be expressed according to the Boltzmann distribution theory:

$$\vec{k}_c = \frac{kT}{h} e^{-\frac{\Delta \overline{G}^{\#}}{RT}} \qquad 3.37$$

where $\Delta G^{\#}$ represents the standard free energy of activation or the activation energy without an applied electric field present and with k the Boltzmann constant and h the Planck constant. When an electric field is imposed over the electrode-electrolyte interface, as shown in Figure 3.28, the transport of the ions across the double layer experiences an influence. The electric field, as drawn, will hinder the movement of positive ions toward the electrode as the ions attempt to move against the direction of the electric field (varying $X_2$ in Figure 3.27). Consequently, the energy barrier for the ion movement toward the electrode becomes higher. The electric field affects the movement of ions toward the electrode differently than it affects their movement from electrode to solution (varying $X_1$ in Figure 3.27). We assume that a fraction of the electric work, $\beta F \Delta \phi$, has been applied to hinder the positive ion transport to the electrode and the other portion, $(1-\beta)F \Delta \phi$, has been affecting the ion to solution transport portion. Thus, the activation energy for the ion transport changes to $\Delta G^* = \Delta G^{\#} + \beta F \Delta \phi$ (in case the charge on the ion z=1) and the modified jumping frequency translates into

$$\vec{k} = \vec{k}_c \frac{kT}{h} e^{-\frac{\beta zF \Delta \phi}{RT}} \qquad 3.38$$

for the more general case where the ion carries a charge z.

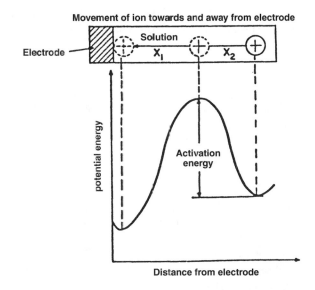

FIGURE 3.27   Construction of a potential energy-distance profile by consideration of the potential-energy changes produced by varying $X_1$ and $X_2$.

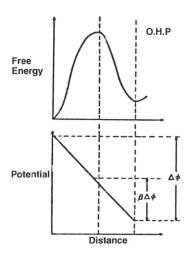

FIGURE 3.28   The electrical work of activating the ion is determined by the potential difference across which the ion has to be moved to reach the top of the free energy-distance relation. O.H.P. = outer Helmholtz plane.

The current density ($\vec{i} = ic = I_-/A$) associated with this process results from the velocity by which ions incorporate into the electrode ($\vec{k}$) and the charges per mole of positive ions (zF) and their concentration $C_m^{z+}$:

$$\vec{i} = \vec{k} zFC_{M^{z+}} = \vec{k}_c zFC_{M^{2+}} \frac{kT}{h} e^{-\frac{\beta zF \Delta \phi}{RT}} \qquad 3.39$$

Similarly, the electric field helps the positive ions jump from the metal to the electrolyte or it lowers the activation barrier by $(1-\beta)F \Delta \phi$, or $\Delta G^* = \Delta G^{\#} - (1-\beta)F \Delta \phi$ (in the case of z = 1). Thus, the associated current density ($i = i_a = I_+/A$) may be written as:

$$\vec{i} = \vec{k} zFC_{M^{z+}} = \vec{k}_c zFC_{M^{2+}} \frac{kT}{h} e^{\frac{(1-\beta)zF \Delta \phi}{RT}} \qquad 3.40$$

for the more general case where the ion carries a charge z.

The measured current density over the interface equals the difference between the two opposite contributions:

$$i = \vec{i} - \overleftarrow{i} \qquad 3.41$$

At thermodynamic equilibrium, the above forward and reverse reactions produce equal current densities, resulting in a zero current density across the interface of this electrode:

$$i_e = \overleftarrow{i} = \overleftarrow{k}\,zFC_{M^{2+}} = \overleftarrow{k}_c\,zFC_{M^{2+}}\frac{kT}{h}e^{\frac{(1-\beta)F\Delta\phi_e}{RT}} = \vec{i} = \overleftarrow{k}_c\,zFC_{M_2^+}\frac{kT}{h}e^{\frac{-\beta zF\Delta\phi_e}{RT}} \qquad 3.42$$

$i_e$ refers to the equilibrium exchange-current density, and $\Delta\phi_e$ represents the equilibrium potential across the electrode-electrolyte interface ($\vec{i} = i_c = I_-/A$ and $\overleftarrow{i} = i_a = i_+/A$). The difference between the nonequilibrium potential and the equilibrium potential, as stated earlier, is defined as the reaction overpotential (reaction polarization):

$$\eta = \Delta\phi - \Delta\phi_e \qquad 3.43$$

Deriving from this definition, the measurable current density i is given by the Butler-Volmer equation, one of the fundamental equations in electrochemical kinetics:

$$i = i_e\left(e^{\frac{(1-\beta)zF\eta}{RT}} - e^{\frac{-\beta zF\eta}{RT}}\right) \qquad 3.44$$

The graphic representation of the above equation is given in Figure 3.29. Note that when η is large enough (large positive or large negative), the Butler-Volmer equation can be simplified as

$$\eta = a + b\log(i) \qquad 3.45$$

which indicates that at large $|\eta|$ a linear relationship holds between the overpotential and log(i). Equation 3.45 is referred to as the Tafel law. Reactions exhibiting the Tafel law are said to be activation controlled. In this case, the rate-limiting step is the reaction at the electrode surface. When the transport of reactive species becomes rate-limiting, one refers to diffusion-limited reactions.

### Diffusion-Limited Reactions

Above, we discussed the relation between current density and overpotential for reactions occurring at an electrode surface. Species in the electrolyte must be transported to and from the electrode before electrode reactions can occur. As species are being consumed or generated at the electrode surface via electrochemical reactions, the concentration of these species at the electrode surface will become smaller, or larger, respectively, than in the bulk of the electrolyte. Suppose we are dealing with an ionic species of concentration $C^0$ in the bulk of the electrolyte ($x = \infty$) being consumed at the electrode, the concentration gradient at the electrode is then given as:

$$\frac{dc}{dx} = \frac{C^0_{x=\infty} - C_{x=0}}{\delta} \qquad 3.46$$

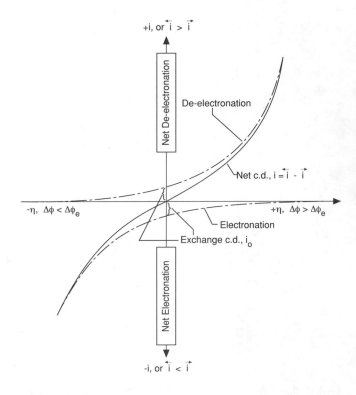

**FIGURE 3.29** The graphical presentation of the Butler-Volmer Equation 3.44. Also indicated are the equilibrium current exchange density, $i_e$ and the approach to the high-field approximation as η becomes large. Net current density (c.d.): $i = \overleftarrow{i} - \vec{i}$

where $C_{x=0}$ is the concentration of the ionic species at the electrode surface and δ represents the boundary layer thickness. From the thermodynamic relationship between the potential and concentration as described by Equation 3.34 (to simplify we will use concentration instead of activity and n for the number of electrons transferred rather than z for the number of charges on the ion) we conclude that the concentration difference described by Equation 3.46 leads to another overpotential:

$$\eta_c = \frac{RT}{nF}\ln\frac{C_{x=0}}{C^0_\infty} \qquad 3.47$$

the expression for the concentration polarization, $\eta_c$ and n is the number of electrons involved in the reaction. On the basis of Faraday's law, we can rewrite Equation 3.46 in terms of the current density as:

$$i = nFD_0\frac{C^0_\infty - C_{x=0}}{\delta} \qquad 3.48$$

with $D_0$ ($cm^2\ sec^{-1}$) representing the diffusion coefficient of the electro-active species and n the number of electrons transferred. At a certain potential $\eta_c$, all of the ionic species arriving at the electrode are immediately consumed and from that potential on the concentration of the electro-active ion at the surface falls to 0, i.e., $C_{x=0} = 0$, and we reach the limiting current density $i_l$:

$$i_l = nFD_0\frac{C^0_\infty}{\delta} \qquad 3.49$$

This equation shows that the limiting current is proportional to the bulk concentration of the ionic species. On this basis, classical amperometric (current based) sensors are used as analytical devices. We can equate

$$\frac{C_x = 0}{C_\infty^0}$$

in Equation 3.47 with $1 - i/i_l$ or:

$$i = i_l \left( 1 - e^{\frac{nF\eta_c}{RT}} \right) \qquad 3.50$$

This expression is illustrated for a cathodically diffusion limited reaction in Figure 3.30.

As shown in Figure 3.30, stirring of the electrolyte strongly affects the limiting current. Higher stirring rates promote convective transport of the ions toward the electrode and result in a smaller boundary layer $\delta$.

The concentration overpotential and the activation overpotential have different origins. In the case of activation overpotential, the slope of i vs. $\eta$ increases with $\eta$; whereas, in the transport limiting case, the slope of i vs. $\eta$ decreases with increasing $\eta$ and effectively becomes zero for sufficiently large values of the overpotential.

### Nonlinear Diffusion Effects on Micro-Electrodes

Electrochemical reactions such as electrodeposition and electrodissolution are influenced by the electrode size relative to the thickness of the diffusion layer. With microstructures of the dimensions of the diffusion layer macroscale electrochemistry theory breaks down and we can exploit some unexpected

beneficial effects afforded us through miniaturization. In Chapter 9 many more such phenomena will be explored in fields ranging from electrostatics to fluidics.

The total diffusion-limited current, $I_l$, on a large substrate of area A is based on Equation 3.49 and given by:

$$I_l = nFAD_0 \frac{C_\infty^0}{\delta}$$

For a small spatially isolated electrode, with its size reduced to the range comparable with the thickness of the diffusion layer, nonlinear diffusion caused by curvature effects must be taken into account. Figure 3.31 illustrates diffusion effects for various types of microelectrode shapes. The analysis shows that as the curvature effects become more pronounced, more diffusion of ions from all directions takes place, thus increasing the supply of the ions to the electrode.

The diffusion layer thickness arising from linear diffusion is time dependent and given by:[23,65]

$$\delta = \left( \pi D_0 t \right)^{\frac{1}{2}} \qquad 3.51$$

Substituting Equation 3.51 in Equation 3.49, we obtain the so-called Cottrell equation:

$$I_l = nFAC_\infty^0 \left( \frac{D_0}{\pi t} \right)^{\frac{1}{2}} \qquad 3.52$$

This equation represents the current-vs.-time response on an electrode after application of a potential step sufficient to cause the surface concentration of electro-active species to reach zero. This equation, at short times after the potential step application, is appropriate regardless of electrode geometry and rate of solution stirring, as long as the diffusion layer thickness is much less

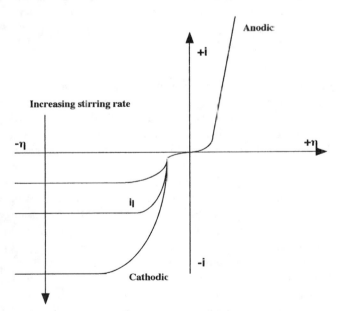

FIGURE 3.30 The cathodic limiting current is indicated for different stirring rates. The cathodic limiting current appears as a horizontal straight line limiting the current that can be achieved at any large negative value of the overpotential.

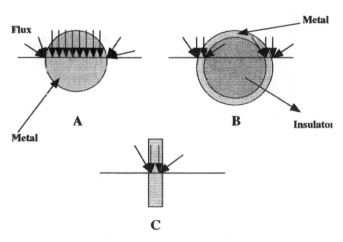

FIGURE 3.31 Convergent flux to small circular (A), ring (B), and band (C) electrodes. Top: side view; bottom: plane view of the electrodes.

than the hydrodynamic boundary layer thickness.* Nonlinear diffusion at the edges of microelectrodes results in deviation from the simple Cottrell equation at longer times. The total current time relation with correction terms becomes:

$$I_l = nFAC_\infty^0 \left(\frac{D_0}{\pi t}\right)^{\frac{1}{2}} + AnFD_0 \frac{C_\infty^0}{r} \qquad \textbf{3.53}$$

At longer times and for small electrodes, Equation 3.53 predicts that the correction term can become significant. In practice, a study on ultra-small electrodes often starts with a linear regression of the measured current vs. $t^{-1/2}$ after application of a potential step. The intercept gives the steady state term. This term, the diffusion-limited current, $I_{l,m}$, at sufficiently long times on ultra-small microelectrodes for some important electrode shapes, is given by:[65]

$$I_{l,m} = \pi rnFD_0 C_\infty^0 \quad \text{(disc)}$$

$$I_{l,m} = 2\pi rnFD_0 C_\infty^0 \quad \text{(hemisphere)} \qquad \textbf{3.54}$$

$$I_{l,m} = 4\pi rnFD_0 C_\infty^0 \quad \text{(sphere)}$$

In the correction term of Equation 3.53, the electrode surface area A is divided by the radius r. Hence, the principal location of charge transfer appears to be on the outer edge of the electrode. This constitutes a very favorable scaling. In Chapter 9 we will learn that phenomena scaling with the linear dimension to the power one becomes important in the microdomain. Surface tension exemplifies another law scaling with the power of one as a very dominant force in the microdomain. In the current case, since $I_{l,m}$ is proportional to the electrode radius ($\sim r^1$) while the background current $I_c$ (associated with the charging current of the Helmholtz capacitance) is proportional to the area ($\sim r^2$), the ratio of the Faradaic current (the charging current) to back-

ground currents should increase with decreasing electrode radius (l/r). For electrodeposition this means that smaller features will be plated faster; for sensor applications, this translates into a higher S/N ratio or an improved sensitivity. The latter makes amperometric sensing with micro-electrodes possible at unprecedented sensitivities.

With a single micro-electrode, the analytical current remains small and the analytical gain is quite relative. However, using an *array of micro-electrodes* with all micro-electrodes connected in parallel and separated by an insulating layer provides an elegant solution to this problem. In arrays of ultra-small electrodes, the electrodes behave as an equivalent number of individual micro-electrodes when sufficiently large spacing occurs between them. This configuration enables analytical currents in a higher range compared to the single-electrode case. This gain is temporary since the diffusion layers around the individual micro-electrodes expand across the insulating surface at a rate given by Equation 3.51. The time at which the overlap occurs is both a function of electrode size and spacing in the array, resulting in a decrease in the steady state current. In an analytical experiment, short compared to the time frame of overlap, more sensitive measurements at reasonably high currents become possible. Another gain, besides higher currents and improved S/N ratio through higher mass transport, results from using an array, because the signal can be averaged over many electrodes in parallel, (an $\sqrt{n}$ improvement in S/N with n the number of electrodes). Other advantages of using micro-electrodes include:

- High mass transfer rates at ultra-small electrodes making it possible to experiment with shorter time scales (faster kinetics)
- An array of closely spaced ultra-small electrodes can collect electro-generated species with very high efficiency
- Electrochemical measurements in high-resistivity media, including possibly in air, become feasible.[66]

Micromachining small reaction chambers in which L, the outer boundary of the electrochemical cell, is made smaller than the diffusion layer thickness, $\delta$, also increases the sensitivity of an electrochemical sensor as in Equation 3.49. $\delta$ is now replaced by a smaller L resulting in a higher diffusion limited current $I_l$. The latter finds its application in the design of thin-layer, channel-type flow cells.

When considering a small metal disc embedded in a photoresist, nonlinear diffusion can be neglected for large substrates (r large or $I_{l,m}$ small). If r becomes small (i.e., of the dimension of and smaller than $\delta$), the contribution of $I_{l,m}$ to the total current becomes larger than $I_l$; i.e., an enhanced mass-transport to small electrodes occurs compared to large area electrodes. Consequently, we can expect that nonlinear behavior increases the plating rate of small features over large features.

Another factor influencing the deposition rate on a small substrate area is the position of the conductor with respect to surrounding insulating surfaces. The correction term in Equation 3.53 was derived for an inlaid microdisc electrode for which the electrochemically active substrate coincides with the

---

*    In diffusion the parameter D is analogous to the kinematic viscosity $v_k$ in convection; these parameters are important in determining the thickness of the diffusion layer and the hydrodynamic boundary layer, respectively. For aqueous solutions, the diffusion coefficient typically is a thousand times smaller than the kinematic viscosity ($D = 10^{-5}$ cm$^2$ s$^{-1}$ and $v_k \approx 10^{-2}$ cm$^2$ s$^{-1}$). It can be shown that the relation between $\delta$, the thickness of a steady state diffusion layer, and $\delta_h$, the thickness of the stagnant layer on an electrode in a stirred solution, is

$$\delta \approx \left(\frac{D}{v_k}\right)^{\frac{1}{3}} \delta_h$$

For $D \approx 10^{-5}$ cm$^2$ s$^{-1}$ and $v_k \approx 10^{-2}$ cm$^2$ s$^{-1}$, $\delta \approx 0.1\ \delta_h$. Thus, the thickness of the diffusion boundary $\delta$ is considerably smaller than that of the hydrodynamic boundary layer. In an unstirred solution, $\delta$ is not well defined and all types of disturbances can affect the transport. To prevent random convective motions from affecting transport to and from the sensing electrode, we want the diffusion layer to be smaller than the hydrodynamic boundary layer and $\delta_h$ to be regular.[23]

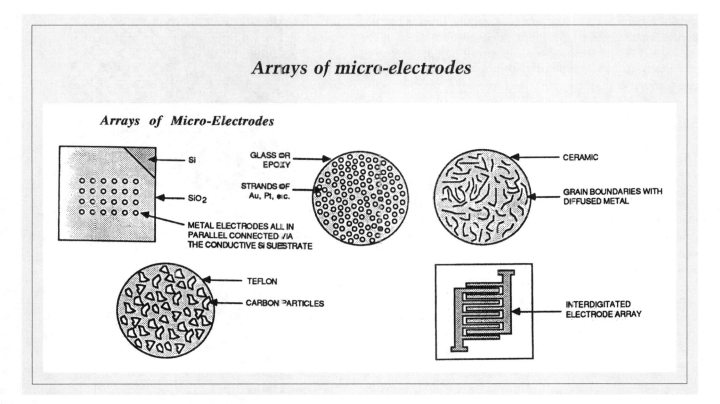

dielectric insulation level (Figure 3.32A). If the conductor disc is recessed within an insulating medium (Figure 3.32B), the diffusion inside the hole must be considered as well, leading to a slight modification of the correction term $I_{l,m}$

$$I_{l,m} = AnFD_0 \frac{C_\infty^0}{r+L} \qquad 3.55$$

in which L stands for the depth of the recession. The mass transport is still enhanced compared to large area substrates, although this advantage may be lost for large L. Van der Putten et al.[67] found that the metal deposition rate at smaller recessed areas (with $r \leq L$) was larger than on large such areas, since overfilling was obtained for the smaller recessed features on the same time scale. When plating features in LIGA (see Chapter 6), we may deal with situations where a metal layer is deeply recessed ($r \ll L$) in a layer of insulating poly(methylmethacrylate) (PMMA). As very high aspect ratio features (L/r large) can be made with LIGA, the thickness of the diffusion layers is increased artificially by the microstructured polymer layer, since it is more difficult for convective flow to reach the bottom of the deep gaps.

The treatise of electroplating in micropatterned resists is of importance for both micromachining and the IC industry. Additional reading on this topic is suggested (e.g., References 68–70). See also Chapter 6 on LIGA and Chapter 7 on the comparison of micromachining techniques.

## Anodization

Anodization is an oxidation process performed in an electrolytic cell (Figure 3.23). The material to be anodized becomes the

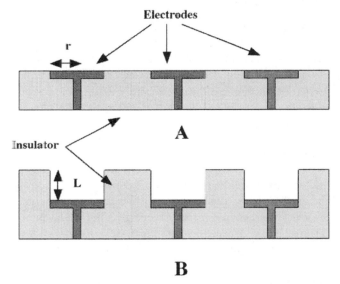

**FIGURE 3.32** Nonlinear diffusion on inlaid and recessed electrodes. (A) Inlaid microelectrodes. (B) Recessed microelectrodes.

anode (+) while a noble metal is made the cathode (–). Depending on the solubility of the anodic reaction products, an insoluble layer (e.g., an oxide) results, or in the case of a soluble reaction product, the electrode etches. If the primary oxidizing agent is water, the resulting oxides generally are porous, whereas organic electrolytes may lead to very dense oxides providing excellent passivation.[30] In the section on thermal oxidation we indicated that anodic oxidation of Si never led to a commercially acceptable process, mainly because the interface state density at the $SiO_2/Si$ interface is prohibitively high for IC applications. Anodization of Si in a highly concentrated HF solution (excellent etchant for the anodic oxidation product $SiO_2$) may lead to porous Si and very high aspect ratio pores, with diameters ranging from 20 Å to several microns. The growth rate and degree of porosity of the Si can be controlled by the current density (see Chapter 4).

# Silk-Screening or Screen Printing

## Introduction

Screen printing presents a more cost-effective means of depositing a wide variety of films on planar substrates than does integrated circuit technology, especially when fabricating devices at relatively low production volumes. The technique constitutes one of several thick-film or hybrid technologies used for selective coating of flat surfaces (e.g., a ceramic substrate). The technology was originally developed for the production of miniature, robust, and, above all, cheap electronic circuits. The up-front investment in a thick-film facility is low compared to that of integrated circuit manufacturing. For disposable chemical sensors, recent industrial experience indicates that screen printing thick films is a viable alternative to Si thin-film technologies.

## How It Works

A paste or ink is pressed onto a substrate through openings in the emulsion on a stainless steel screen (see Figure 3.33A).[71] The paste consists of a mixture of the material of interest, an organic binder, and a solvent. The organic vehicle determines the flow properties (rheology) of the paste. The bonding agent provides adhesion of particles to one another and to the substrate. The active particles make the ink a conductor, a resistor, or an insulator. The lithographic pattern in the screen emulsion is transferred onto a substrate by forcing the paste through the mask openings with a squeegee (Figure 3.33B). In a first step, paste is put down on the screen (Figure 3.34A), then the squeegee lowers and pushes the screen onto the substrate forcing the paste through openings in the screen during its horizontal motion (Figure 3.34B).[71] During the last step, the screen snaps back, the thick-film paste which adheres between the screening frame and the substrate shears, and the printed pattern is formed on the substrate (Figure 3.34C). The resolution of the process depends on the openings in the screen and the nature of the pastes. With a 325-mesh screen (i.e., 325 wires per inch or 40-μm holes) and a typical paste, a lateral resolution of 100 μm can be obtained. For difficult-to-print pastes, a shadow mask may complement the process, such as a thin metal foil with openings. However, the resolution of this method is inferior (>500 μm). After printing, the wet films are allowed to settle for 15 minutes to flatten the surface while drying. This removes the solvents from the paste. Subsequent firing burns off the organic binder, metallic particles are reduced or oxidized, and glass particles are sintered. Typical temperatures range from 500 to 1000°C. After firing, the thickness of the film ranges from 10 to 50 μm.

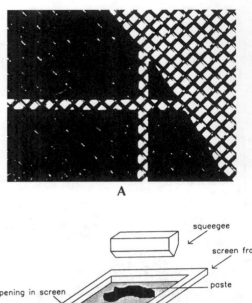

**A**

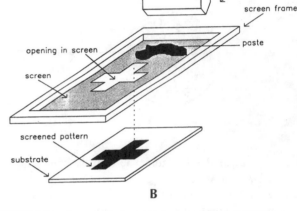

**B**

**FIGURE 3.33** Screen printing. (A) Screen with a 0.002-in. line opening oriented at 45° to the mesh weave. (B) Schematic representation of the screen-printing process. (From Lambrechts, M. and W. Sansen, *Biosensors: Microelectrical Devices*, IOP, 1992. With permission.)

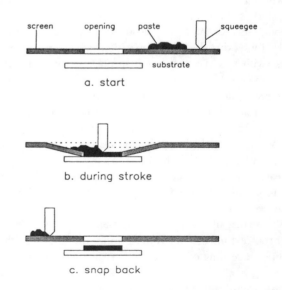

**FIGURE 3.34** The three different steps during the silk-screening process. (From Lambrechts, M. and W. Sansen, *Biosensors: Microelectrical Devices*, IOP, 1992. With permission.)

# Types of Inks

## Traditional Inks

Inks are formulated to exhibit *pseudoplastic behavior*,[72–74] to prevent flowing through the screen until the squeegee applies sufficient pressure. Almost all materials compatible with the high firing temperature and the other ink constituents can be used to screen print. Different pastes — conductive (e.g., Au, Pt, Ag/Pd, etc.), resistive (e.g., $RuO_2$, $IrO_2$), overglaze and dielectric pastes (e.g., $Al_2O_3$, $ZrO_2$), are commercially available.

The conductive pastes are based on metal particles, such as Ag, Pd, Au, Pt, or a mixture of these combined with glass. Glass is necessary for the adhesion of the metal conductor to the ceramic ($Al_2O_3$) substrate. According to the bonding mechanism, one distinguishes between the following conductive pastes:

- Glass bonded or fritted pastes are inks where adhesion of the metal is achieved with the addition of a glass mixture (30%) (a typical glass composition is 65% PbO, 25% $SiO_2$, and 10% $Bi_2O_3$). A fritted ink contains powdered glass which binds the ink to the substrate material when fired at a temperature of 850°C. Thus a fritted ink will generally consist of the glass frit, a material defining the desired ink property and an organic vehicle which renders the ink printable. Vehicles typically consist of solvents mixed with slightly more viscous materials, such as resins, in a ratio designed to give the optimum overall viscosity to the ink.

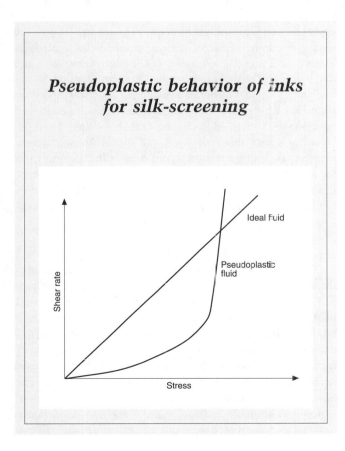

*Pseudoplastic behavior of inks for silk-screening*

- Oxides or fritless-bonded inks adhere the metal via the addition of copper oxide (3%).
- Mixed bonded pastes are inks for which adhesion is achieved by use of both glass and copper oxide.

Resistive pastes are based on $RuO_2$ or $Bi_2Ru_2O_7$ mixed with glass (65% PbO, 25% $SiO_2$, 10% $Bi_2O_3$). The resistivity is determined by the mixing ratio. Overglaze and dielectric pastes are based on glass mixtures. According to composition, different melting temperatures can be achieved.

Thick-film technology with the above type of traditional inks has application in the construction of a wide variety of hybrid sensors, such as sensors for radiant signals, pressure sensors, strain gauges, displacement sensors, humidity sensors, thermocouples, capacitive thick-film temperature sensors, and pH sensors (see Middlehoek et al.[74] and references therein). Also, with Si-based sensors (for example, pressure and accelerometers) die-mounted on a ceramic substrate, thick-film resistors are used for calibration by trimming resistors on the ceramic substrate.

## New Inks for Sensors

More recently, inks specifically developed for sensor applications are becoming available or are under development. For example, $SnO_2$ pastes incorporating Pt, Pd, and Sb dopants have been developed for the construction of high temperature (>300°C) semiconductor gas sensors for reducing gases (so-called Taguchi sensors).[23] Thick metal phthalocyanine films have been deposited on alumina to form the active material in low temperature (<180°C) gas sensors.[75] For biosensor applications, thick-film technology based on pastes which can be deposited at room temperature is crucial. Special grades of polymer based pastes (e.g., for carbon, Ag and Ag/AgCl electrodes) are becoming commercially available for this purpose.[76] Polymer thick films with a thickness anywhere from 5 to 50 μm can be screen-printed on cheap polymer substrates. The first commercial planar electrochemical glucose sensor (the ExacTech by MediSense) resulted from a screen-printed sensor based on such polymer-based inks (see also Chapter 10).

In the research phase of new chemical sensors, pastes must be developed from their pure components. Some examples follow. Pace et al.[77] screen-printed a PVC/ionophore layer for a pH sensor. Belford et al.[78] investigated pH-sensitive glass mixtures and proceeded to screen-print them on a multilayer metal conductor to make planar pH sensors. In Pace et al.,[79] a thick-film, multi-layered oxygen sensor with screen-printed chemical membranes of PVA and silicone rubber is detailed (see also Karagounis et al.[80]) A thick film glucose sensor is presented by Lewandowski et al.[81] and by Lambrechts et al.[71] The latter authors developed an enzyme-based thick film glucose sensor with $RuO_2$ electrodes. All the above thick-film sensors were fabricated on $Al_2O_3$ substrates. Weetall et al.[82] present an extremely low cost, silk-screened immunosensor on cardboard. Cha et al.[33] compare the performance of thick film Au and Pt electrodes with conventional bulk electrodes.

When faced with adapting a biosensor membrane to an IC process vs. adaptation to a thick film process, it now seems that, once the specialty inks are available, a thick film approach is easier, less expensive, and more accessible. For chemical sensors where small size is not as important, we expect more research to result in a switch from the overly ambitious IC approach to the more realistic thick film approaches as sketched above. Below we compare the pros and cons of the two technologies.

## Comparison of Thin- vs. Thick-Film Deposition

A comparison of thick-film vs. thin-film deposition is presented in Table 3.11. Resolution and minimum feature size for thin films is obviously superior. Also, the porosity, roughness, and purity of deposited metals is less reproducible with thick films. Finally, the geometric accuracy is poorer with thick films. On the other hand, the thick-film method displays versatility which is often key in chemical sensor manufacture. Silk-screening forms an excellent alternative when size does not matter but cost in relatively small production volumes does. For biomedical applications the size limitations, clear from Table 3.11, make thick-film sensors more appropriate for *in vitro* applications. For *in vivo* sensors where size is more crucial, IC-based technologies might be more appropriate.

Thin-film technology does not necessarily involve IC integration of the electronic functions. A comparison table of the economic and technical aspects of implementing thin-film technology, IC fabrication, thick-film, and classic construction for sensors is presented in Table 3.12.[71] It should be mentioned that the sensor fabrication cost consists of 60 to 80% of packaging, an aspect not addressed in Table 3.12. In comparison with CMOS-compatible sensors, the packaging of thick-film

sensors usually is more straightforward. Since packaging expenses overshadow all other costs, this is a decisive criterion. In the case of chemical sensors where the sides of the conductive Si substrate might shunt the sensing function through contact with the electrolyte, encapsulation is especially difficult compared to the packaging of an insulating plastic or ceramic substrate (see also Chapter 8).

## Deposition Methods of Organic Layers

Materials rarely dealt with in the IC industry are encountered when manufacturing chemical sensors. Organic membranes for the manufacture of ion selective electrodes (ISEs), room temperature gas sensors, enzyme sensors, immunosensors, etc. especially present difficulties. Most chemical membranes are based on classical polymers, such as PVC (polyvinylchloride), PVA (polyvinylalcohol), PHEMA (polyhydroxyethylmethacrylate), or silicone rubber, and may incorporate biological materials such as enzymes, antigens, and antibodies. Because of their importance, we briefly list the different techniques available to coat these materials on a flat substrate. For the organic film deposition techniques reviewed earlier, we provide a short summary.

### Spin Coating and Photolithography
#### Spin Coating

Spin coating technology has been optimized for deposition of thin layers of photoresist, about 1 to 2 μm thick, on round and nearly ideally flat surface Si wafers. Resists are applied by dropping the resist solution, a polymer, a sensitizer (for two-component resists), and a solvent on the wafer and rotating the wafer on a spinning wheel at high speed so that centrifugal forces push the excess solution over the edge of the wafer, and the residue on the wafer remains due to surface tension. In this way films down to 0.1 μm can be made. An empirical expression (Equation 1.1) relating film thickness to solution viscosity and rotation speed was given in Chapter 1 on Lithography. However, biosensor substrates rarely are round or flat; many chemical membranes require a thickness considerably thicker than 1 μm for proper functioning. For example, a typical ISE membrane measures 50 μm. Consequently, spin coating technology does not necessarily fit in with thick chemical membranes on a variety of substrates.

#### Photolithography Process

Spinning, UV exposure, and development of photosensitive materials are well-known, low-cost, mass-production procedures. One has to take into account, though, that photosensitized membrane materials needed in biosensors may not be available commercially. To cope with this problem, one has to prepare the photosensitive material from high-purity materials. Water-soluble polymers such as PVA can be used while adding a photosensitizer. After spin coating, the polymer film is photochemically cross linked by UV light. In the exposed regions, an insoluble hydrogel materializes. The development or removal of unexposed regions is carried out in (warm) water.

**TABLE 3.11** A Comparison of Thin- vs. Thick-Film Technology

| Property | Si/Thin Film | Hydbrid/Thick Film |
|---|---|---|
| In-plane resolution | 0.25 μm and better | 12 μm |
| Minimum feature size | 0.75 μm and better | 90 μm |
| Temperature range | <125°C | ≥125°C |
| Sensor size | Smaller | Small |
| Geometric accuracy | Very high | Poor |
| Deposition methods | Evaporation, sputtering, CVD | Screen printing, stencil printing |
| Patterning methods | Etch-through photomask, lift-off, stencil | Screen photomask, etched stencil, machined stencil |
| Reliability nonencapsulated device | Low | High |
| Electronic compatibility | Good | Moderate |
| Versatility | Low | Very good |
| Roughness, purity, and porosity of deposited materials | Superior | Moderate |
| Energy consumption | Low | Moderate |
| Handling | Difficult | Easy |
| Appropriate capital costs per unit | Very large, but very low in large volumes | Low; very low in moderate volumes |

TABLE 3.12 A comparison Between Different Sensor Technologies: Economic and Technical Aspects

| | Classic Construction | Thick Film Technology | Thin Film Technology | IC Technology |
|---|---|---|---|---|
| Technology substrate | Wires and tubes | Screen printing $A_2O_3$, plastic | Evaporation-sputtering $Al_2O_3$, glass, quartz | IC techniques silicon, GaAs |
| Initial investment | Very low | Moderate | High | High |
| Production line cost | >10 k$ | >100 k$ | >400 k$ | >800 k$ |
| Production | Mannual production | Mass production | Mass production | Mass production |
| Units per year | 1–1000 | 1000–1,000,000 | 10,000–10,000,000 | 100,000– |
| Prototype | Cheap | Cheap | Moderate | Expensive |
| Sensor price | Expensive sensor | Low cost per sensor | Low cost per sensor | Low cost per sensor |
| Use | Multiple use, *in vitro - in vivo* | Disposable, *in vitro* | Disposable, *in vivo* | Disposable, *in vivo* |
| Markets | Research, aerospace | Automotive, industrial | Industrial, medical | Medical, consumer |
| Dimension | Large | Moderate | Small | Extreme miniaturization |
| Solidity | Fragile | Robust | Robust | Robust |
| Reproducibility | Low | Moderate | High | High |
| Maximum temperature | 800°C | 800°C | 1000°C | 150°C (Si) |
| Interfacing | External discrete devices | Smart sensors, surface mount | Smart sensors, surface mount | Smart sensors, CMOS, bipolar |

(From Lambrechts, M. and W. Sansen, *Biosensors: Microelectrical Devices*, IOP, Philadelphia, 1992. With permission.

As a well-known system, we can cite PVA in combination with $(NH_4)_2Cr_2O_7$ as a photosensitizer.[71]

Lift-off techniques for lithographically patterning materials was introduced in Chapter 1. A problem with lift-off for thick organic membranes is the thickness needed for the photoresist layer. As discussed earlier, a typical resist layer measures about 1 μm thick. The thickness needed for chemical membranes can reach 50 μm, demanding thicker resist layers for patterning. Also, lift-off can be used only with materials resistant to the solvent necessary for the resist removal. Although attractive for batch fabrication, this technology can be applied only in specific cases.

### Glow Discharge (Plasma) Polymerization

During glow discharge, polymerization of polymer films ensues from plasmas containing organic vapors. Figure 3.35 illustrates the set-up we used in our own work on plasma polymerization of doped polymers. The apparatus consisted of a quartz tube reaction chamber; a gas-handling system to introduce the carrier gas, monomer, and dopant into the system as well as to remove the unreacted material; and a power supply (operating at 27 MHz) to provide the RF energy necessary to create and maintain a plasma within the system. Despite the complex chemistry of this process, good conformal coatings often result. Guckel applied this technique to obtain deposition of PMMA in layers >100 μm.[84] During our research at SRI International we synthesized electro-active plasma polymers using $I_2$ or $N_2O$ as dopants from the following monomers: thiophene, furan, aniline, benzoaldehyde, benzene, indole, diphenylacetylene, and 1-methylpyrrole. In general, plasma-polymerized materials offer the following advantages:

- The plasma-polymerized films are uniform, pinhole-free, chemically resistant, and mechanically strong.

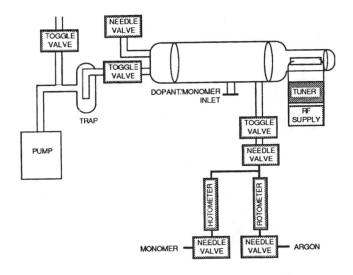

FIGURE 3.35 Schematic of plasma deposition system.

- A thin to thick film (200 Å to >100 μm) can be formed in a flawless manner at ambient temperature onto any substrate.
- The organic film deposited by the plasma process adheres strongly to the substrate.
- The choice of monomers is unlimited; almost any organic compound convertible into vapor can be polymerized.
- The plasma process (one-step process from a vapor source) is compatible with conventional CMOS technology.
- Some functional groups can be introduced onto the surface of the organic film by subsequent glow discharge treatment in a reactive gas atmosphere.

• Highly irregular surfaces can be coated and patterned by depositing a light-sensitive polymer by plasma polymerization (e.g., polymerized PMMA).

For more details on plasma polymerization, see for example Yasuda in Reference 15.

### Plastic Spraying and Dip Coating

Plastic spray-coating techniques may involve liquids, gases, or solids. Spray coating of a liquid involves pressurization by compressed air or, in an airless method, pushing liquid mechanically through tiny orifices. Vapors are carried in an inert dry vapor carrier. In the case of a solid, a powdered plastic resin is melted and blown through a flame-shrouded nozzle. In electrostatic spraying a negatively charged plastic powder is spray gunned onto grounded conductive parts.[11]

Thermoplastic and thermosetting coatings are formed by dipping a heated part into a container of resin particles set in motion by a stream of low pressure air (fluidized-solid bed). Dip coating of a substrate in a dissolved polymer typifies the simplest method to apply an organic layer to a substrate. It is especially suited for wire type ion selective electrodes (ISEs) where the membrane forms a droplet at the end of a wire. A substrate, e.g., a chloridized silver wire, is dipped into a solution containing the polymer and a solvent. After evaporation of the solvent, a thin membrane forms on the surface of the sensor. To obtain pinhole-free membranes, the dipping is repeated several times interspersed with drying periods. Even though the eventual goal is typically the production of a planar sensor structure, an Ag wire may be applied in the research phase to quickly evaluate a new membrane composition. Most biosensors today are made by dip coating.

### Casting

Casting is based on the application of a given amount of dissolved material on the surface of a mounted sensor and letting the solvent evaporate. A rim structure is fashioned around the substrate, providing a 'flat beaker' for the solution. This method provides a more uniform and a more reproducible membrane than dip coating. Membranes in planar ion selective electrodes are often made this way.

### Langmuir-Blodgett Film Approach

Use of monolayer electron-beam resists may lead to extremely good lithography resolution (see Chapter 1). The sensor industry manufactures monolayers of organics for use in immunosensors or to provide anchor points for subsequent membranes. The Langmuir-Blodgett deposition propelled this method into acceptance. The technique, invented by Irving Langmuir and Katharine Blodgett, allows the controlled deposition of monomolecular layers. This ultra-thin film deposition technique is limited to materials that consist of amphilic long chain molecules with a hydrophobic molecule at one end and a hydropholic molecule at the other end.

During the Langmuir-Blodgett process, a monolayer of film-forming molecules (stearic acid is a model molecule) on an aqueous surface is compressed into a compact floating film and transferred to a solid substrate by passing a substrate through the water surface at a constant speed and film surface tension (Figure 3.36). Thus, layered films can be built up in thickness, up to 100 layers by consecutive dippings in the Langmuir trough. Phthalocyanine films sensitive to oxidizing gases such as $NO_2$ and biological materials sensitive to odors resulted via this method. More recent studies show that self-assembling monolayers can be formed as well. Most of the difficulties with Langmuir-Blodgett films stem from the need to make the material pinhole-free and to overcome the problem of their lack of mechanical and thermal stability.

### Ink-Jet Printing and Other Drop-Dispensing Systems

Ink-jet drop delivery is based on the same principle as commercial ink-jet printing. On a somewhat larger scale, epoxy-delivery systems used in the IC industry similarly deliver drops in a serial fashion on specific spots on a substrate. A typical commercial drop-dispensing system (e.g., the Ivek Digispense 2000) delivers 0.20 to 0.50 μl in a drop and has a cycle time per dispense of one second. A vision system verifies substrate position and accurate dispense location to within ±25 μm of a specified location. In the case of an ink-jet printing head, the ink-jet nozzle, connected to a reservoir filled with the chemical membrane solution, is placed above a computer-controlled XY stage. Depending on the ink expulsion method, even temperature-sensitive enzyme formulations can be delivered. The substrate to be coated is placed on the XY stage, and under computer control liquid drops (50 μm in diameter) are expelled through the nozzle onto a well-defined place on the wafer (see Figure 3.37). Different nozzles print different membranes in parallel. In the authors' experience, this method, although promising, must still overcome a lot of problems such as excessive splashing, clogging of the nozzle, and poor uniformity of the deposit. The clogging problem can be alleviated somewhat by using a solvent-saturated deposition environment. Although these drop delivery systems are serial, they can be very fast as evidenced by epoxy delivery stations in an IC manufacturing line.

### Silk-Screening or Screen-Printing

Earlier we discussed the current trend to use silk-screening for disposable chemical and biological sensors. Although some inks suitable for depositing thick film electrodes on plastics are now available, the inks for most chemical and biological organic membranes still need further research before becoming commercially available.

Table 3.13 summarizes the comparison of the above-reviewed membrane deposition and patterning techniques.[71] At first glance, photolithography seems very promising, but the chemistry is complex and very few results have been published to date. Since a lot more development work is needed, most developers have opted for screen-printing, i.e., the safest and least expensive approach today.

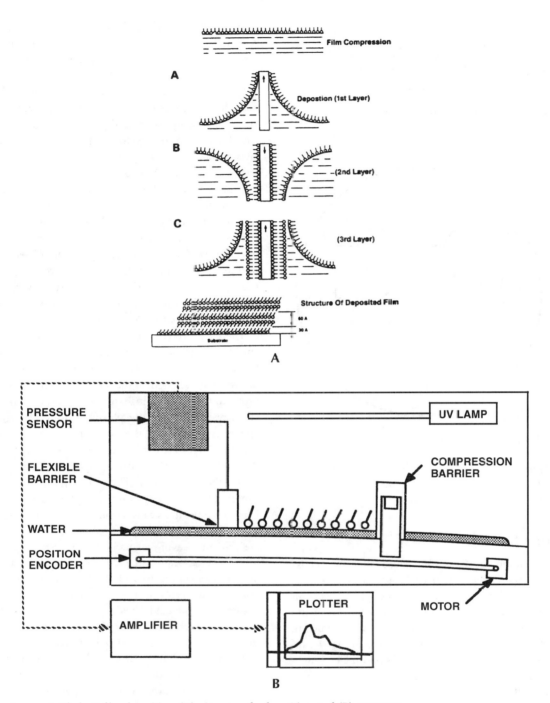

**FIGURE 3.36** Langmuir-Blodgett film deposition: (A) sequence of a deposition, and (B) apparatus.

# Planarization

## Introduction

Microlithography process latitude can become severely limited by nonsmooth topography. Resist films crossing over steps have their local thickness altered; thinning occurs over high features and pile-up in low-lying regions. During exposure the thin resist may get overexposed and the thick regions underexposed. Moreover, resist pile-up regions may exhibit standing wave effects, leading to reduction of resolution. Reflective notching on buried topographical features could decrease resolution even

further. As seen before, practical depth of focus (DOF) should encompass device topography, resist thickness, wafer flatness, focus, and tilt errors in order for the projected image to remain sharp over the whole wafer (see Chapter 1, Lithography). Shallow DOF of high numerical aperture (NA) optical exposure tools, standing waves, and reflective notching can be solved by using very thin resist layers on planar substrates or by planarizing resist layers on nonplanar substrates. Some chip makers stack metal lines four layers high, and five and six level devices are being considered. These 'high-rise' chips will need more and more rigorous interlevel planarization. The extremely high features in micromachining pose yet more challenging

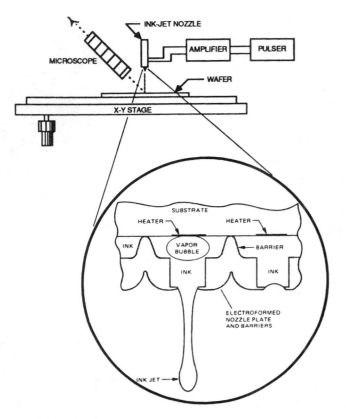

**FIGURE 3.37**   Ink-jet deposition system.

planarization problems. Since planarization strategies often rely on deposition of a sacrificial layer (e.g., resist), we briefly revisit this important topic here.

## Planarization Strategies

Nonsmooth topography in ICs is often caused by CVD-deposited dielectric layers. Conventional CVD deposition methods experience difficulty depositing oxides conformably. More film tends to build on flat surfaces than on vertical surfaces, showing a pronounced tendency for narrow spaces to be covered before being completely filled in, creating voids. Smoothing should contain two components, i.e., filling in gaps and planarizing a film's top surface. Planarization methods provide degrees of smoothing rather than true absolute flatness.

In the planarizing etch back process illustrated in Figure 3.38, hilly contours left behind by a conventional CVD oxide deposition technique are planarized by spinning on glasses (SOG) or resists sacrificial layers, after which both sacrificial layer and oxide are etched back. As its name implies, a spin-on glass dielectric is spun onto a wafer as a liquid and then cured at elevated temperatures. The etch-back process is adjusted, usually by adding oxygen to the etch gases to cause the sacrificial layer and the underlying film, principally oxide or other dielectrics such as PECVD silicon nitride to etch at equal rates. In the etch-back planarization, photoresists are more popular than SOG because they provide a more planar result.[85]

SOG often is used in the dielectric sandwich approach for narrow gap filling. Voids in the narrow spaces between metal lines are averted by employing many thin dielectric layers piled on top of each other until the spaces are completely filled. In this sandwich, layers of CVD oxide alternate with coats of a SOG dielectric. The top surface of the film may then be planarized by etching back with photoresist.

There are two types of SOGs: silicates (inorganic) and siloxanes (organic). The film properties of both resemble those of low-temperature CVD glasses. In both cases films tend to crack due to tensile strength as the films cure. Silicate glasses are more prone to this, but doping with phosphorous makes them more crack resistant. Even the phosphosilicate glasses can only be applied in layers of 1000 Å, so often two or three coats must be applied. A siloxane can be applied easier in a thicker film, e.g., > 3000 Å. Recently, an organosiloxane SOG able to deposit films thicker than 1 μm was developed by Hitachi Chemical Company America, Ltd. Narrow gap-filling to planarize a surface is becoming more difficult as gaps become more narrow. At gap spacings below 0.3 μm, current SOGs fail to fill the gap completely and are prone to leaving a small circular or teardrop-shaped void. Such voids appear with increasing frequency as the gap size decreases and as the gap aspect ratio increases. The effect is even more pronounced with re-entrant angles.[86]

Replacing the oxides with a type of dielectric that fills and planarizes much the same as a liquid film may help to eliminate the severe topologies CVD oxides create. Polyimides fit that dielectric by forming crack-free films which are virtually free of pinholes, absorb stress, and generally exhibit a much lower dielectric constant than oxides. Currently, polyimides that will

**TABLE 3.13**   Deposition and Patterning Techniques for Planar Chemical Membranes

|  | Typical Use | Thickness Range (μm) | Cost | Uniformity | Reproducibility | Patterning |
|---|---|---|---|---|---|---|
| Dip coating | Wire ISEs | 0.1–50 | Low | Poor | Poor | No |
| Casting | Planar ISEs | 0.1–>100 | Low | Moderate | Moderate | No |
| Photolithography | Planar sensors (PVA, PHEMA) | 1–10 | Moderate | Good | Good | Yes |
| Lift-off | Immunosensors | 0.1–3 | Moderate | Moderate | Good | Yes |
| Plasma etching | PVC, Teflon | 1–10 | High | Good | Good | Yes |
| Ink-jet printing | Universal | 1–5 | Moderate | Poor | Moderate | Yes |
| Screen-printing | Universal | 5–50 | Low | Moderate | Moderate | Yes |

(From Lambrechts, M. and W. Sansen, *Biosensors: Microelectrical Devices*, IOP, Philadelphia, 1992. With permission.)

planarize a large pitch geometry to a 90% level with a single coat are available. Reluctance to expand the use of polyimide at this point is linked to its poor barrier properties towards moisture and ions. Also, the tendency to trust inorganic CVD oxides more than organic polyimides prevails.[85]

An ideal solution would be if the CVD process itself filled the spaces between metal lines and planarized the top surface without breaking the vacuum within a single processing chamber. One example of this type of processing is ECRCVD (Figure 3.39). ECRCVD uses electron cyclotron resonance (ECR) to generate a high-energy plasma that can deposit CVD $SiO_2$ at high rates while temperatures and pressures remain low. As

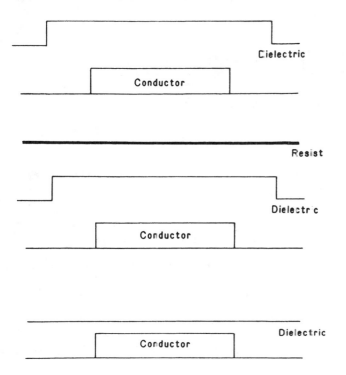

**FIGURE 3.38** Planarizing etch-back process. Photoresists and spin-on glasses are the common sacrificial materials. The removal rates of dielectric and sacrificial layer are adjusted to 1:1 by adding oxygen to the etch gases.

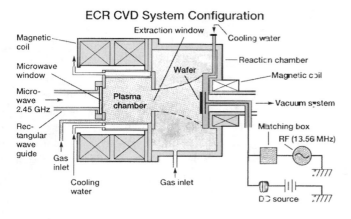

**FIGURE 3.39** In ECRCVD, the plasma can deposit a film while simultaneously sputtering.

deposition proceeds, the wafer is RF-biased so that argon atoms from the plasma can simultaneously sputter etch the substrate. The sputtering keeps submicron spaces open until they are filled. By carefully balancing the two processes, one can deposit a film and sputter it back in such a way that high aspect ratio spaces are completely filled while the oxide surface planarizes. If planarization needs improvement, an ECRCVD can be combined with an etch back. The set-up is ideal for a cluster tool approach.

Chemical-mechanical polishing (CMP) typifies another emerging planarization technique where a wafer with an uneven surface is polished in an abrasive slurry on a polishing pad. Oxides, polysilicon, and metal topography can be planarized this way. CMP is gaining popularity with micromachinists as well. Sniegowski[87] is using CMP to enhance the manufacturability of polysilicon surface micromachinery. CMP planarization alleviates processing problems associated with the fabrication of multi-level structures, eliminates design constraints linked with nonplanar topography, and provides an avenue for integrating different process technologies (see also Chapter 5).

A thin film of Au, Al, or Cu can effectively planarize by briefly melting them with an optical laser. Planarization occurs rapidly due to the high surface tension and low viscosity of clean liquid metals.[88]

## Plasma Spraying

Except for silk screening, most of the above-described additive techniques pertain to the IC industry. Microstructures can also be crafted economically with non-IC equipment and materials. Because of the need to incorporate more and different materials in a thickness ranging anywhere from monolayers to a few hundred microns, micromachinists are broadening their horizon beyond IC deposition techniques. The potential use of plasma spraying exemplifies this trend. With plasma spraying almost any material can be coated on any type of substrate. Applications include corrosion- and temperature-protective coatings, superconductive materials, and abrasion resistance coatings.[16] Today, turbine blades and other components of aircraft engines are plasma coated with corrosion- and temperature-resistant coatings.

The CVD and PVD techniques discussed above rely mostly on atomistic deposition, i.e., atoms or molecules were individually depositing onto a surface to form a coating. There are also techniques in which particles, a few microns to 100 μm in diameter, are transported from source to substrate. Figure 3.40 shows a schematic diagram of a typical particulate deposition technology, i.e., plasma spraying.

In *plasma spraying*, a high intensity arc is operated between a stick-type cathode and a nozzle-shaped, water-cooled anode as illustrated in Figure 3.41. For atmospheric spraying, one typically works at power levels from 10 to 100 kW. Plasma gas, pneumatically fed along the cathode, is heated by the arc to plasma temperatures, leaving the anode nozzle as a plasma jet or plasma flame. Argon and mixtures of argon with other noble

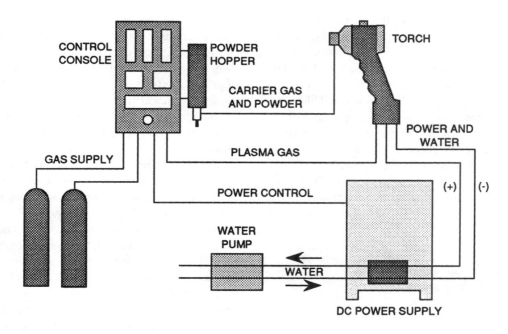

FIGURE 3.40   Set-up for plasma spray.

(He) or molecular gases (H$_2$, N$_2$, O$_2$, etc.) are frequently used for plasma spraying. Fine powder suspended in a carrier gas is injected into the plasma jet where the particles are accelerated and heated. The plasma jet may reach temperatures of 20,000 K and velocities up to 1000 ms$^{-1}$. The temperature of the particle surface is lower than the plasma temperature, and the dwelling time in the plasma gas is very short. The lower surface temperature and short duration prevent the spray particles from being vaporized in the gas plasma. The particles in the plasma assume a negative charge owing to the different thermal velocities of electrons and ions. As the molten particles splatter with high velocities onto a substrate, they spread, freeze, and form a more or less dense coating, typically forming a good bond with the substrate. The resulting coating is a layered structure (lamellae). As shown in Figure 3.41, the particle goes through different regions of temperature and flow velocity. Ideally, the particles should arrive at the substrate at high velocities in a completely molten state to form the densest coating with little porosity. To produce porous films in our own work,[89] we positioned the substrate somewhere between regions 4 and 5. To produce dense films, we positioned the substrate in region 4. With this technique a minimum thickness is about 25 μm and very thick coats up to millimeters thick became possible. When attempting to deposit gas-sensitive layers on thermally isolated, thin Si membranes to make a power efficient gas sensor, we found that the kinetic energy of the plasma was sufficient to break the thin suspended Si membranes. The high temperature and kinetic energy preclude the potential for integrating Si with high-temperature plasma spraying. In Chapter 7 we demonstrate how plasma spraying of yttria stabilized zirconia may be used in the batch fabrication of all solid state oxygen sensors.

## Selection Criteria for Deposition Method

Selection criteria for the additive processes reviewed in this chapter depend on a variety of considerations, such as:

1. Limitations imposed by the substrate or the mask material: T$_{max}$ (maximum temperature), surface morphology, substrate structure and geometry, etc.
2. Apparatus requirement and availability.
3. Limitations imposed by the material to be deposited: chemistry, purity, thickness, T$_{max}$, morphology, crystal structure, etc.
4. Rate of deposition to obtain the desired film quality.
5. Adhesion of deposit to the substrate; necessity of adhesion layer or buffer layer.
6. Total running time, including set-up time and post-coating processes.
7. Cost.
8. Ease of automation.
9. Safety and ecological considerations.

The decisive factor determining a deposition process is intertwined with the choice of an optimum micromachining process. Table 3.14 compares some of the additive processes in microsensors and micromachining. This table and the above criteria complement the questions in the check-off list presented in Table 8.1. The check-off table introduced in Chapter 8 is meant as a guide toward a more intelligent choice of an optimum machining substrate for the micromachining task at hand.

## Plasma spraying station

# Doping

## Two Main Doping Techniques

Elements having one valence electron, more or less, than Si (4) are used as substitutional donors (n-type) or acceptors (p-type), respectively. Boron (−1), phosphorous (+1), arsenic (+1), and antimony (+1) represent the more commonly used dopant elements. Upon incorporation into the crystal lattice, the dopant either gives up (donor) or receives (acceptor) an electron from the crystal. Fabrication of circuit elements and micromachines requires a method for selective n- or p-type doping of the silicon substrate. The two means of doping Si are diffusion and ion implantation, as compared in Table 3.15.

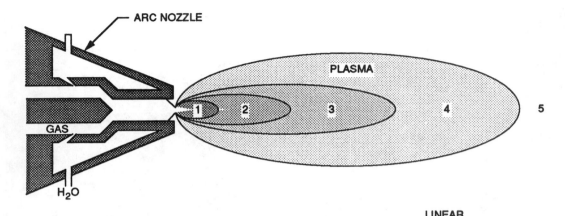

| REGION | DISTANCE FROM NOZZLE | TEMPERATURE RANGE (K) | LINEAR FLOW VELOCITY (m/s) |
|--------|---------------------|----------------------|---------------------------|
| 1 | < 1 cm | $1.5 \times 10^4 - 1 \times 10^4$ | ~ 400 |
| 2 | 1 < d < 5 cm | $1 \times 10^4 - 5 \times 10^3$ | ~ 400 — 200 |
| 3 | 5 < d < 10 cm | $5 \times 10^3 - 2 \times 10^3$ | ~ 200 — 100 |
| 4 | 10 < d < 20 | $2 \times 10^3 - 1 \times 10^3$ | < 100 |
| 5 | 20 < d | < 500 | |

A

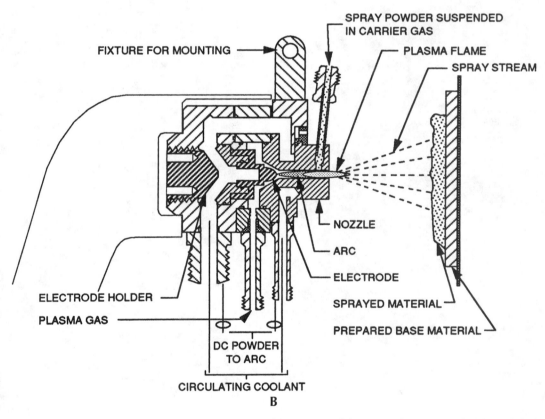

B

**FIGURE 3.41**   Plasma spray nozzle. (A) Typical ranges of temperature and flow velocity with distance from the nozzle. (B) Torch for powder spraying.

## Incorporation by Diffusion

In the past, dopants were typically diffused thermally into the substrate in a furnace at temperatures between 950 and 1280°C. To a first approximation, Fick's first and second laws describe the diffusion of dopants in silicon:[25]

1. The dopant flux is proportional to the concentration gradient:

$$J = -D \frac{dN}{dx}$$

3.56

**TABLE 3.14** Comparison of Additive Processes Important in Microsensors and Micromachining

|  | Evaporation | Sputtering | CVD | Electrodeposition | Thermal Spraying |
|---|---|---|---|---|---|
| Mechanism of producing deposition species | Thermal energy | Momentum transfer | Chemical reaction | Deposition from solution | From flames or plasmas |
| Deposition rate | Very high, up to 750,000 Å/min | Low, except for pure metals (e.g., Cu — 10,000 Å/min) | Moderate (200–2500 Å/min) | Low to high | Very high |
| Depositing species | Atoms and ions | Atoms and ions | Atoms | Ions | Droplets |
| Coverage of complex-shaped objects | Poor line-of-sight coverage | Good, but nonuniform thickness distribution | Good | Good | No |
| Coverage into a small blind hole | Poor | Poor | Limited | Limited | Very limited |
| Metal deposition | Yes | Yes | Yes | Yes, limited | Yes |
| Alloy deposition | Yes | Yes | Yes | Limited | Yes |
| Refractory compound deposition | Yes | Yes | Yes | Limited | Yes |
| Energy of depositing species | Low (0.1–0.5 eV) | Can be high (1–100 eV) | Can be high for PECVD | Can be high | Can be high |
| Bombardment of substrate/deposit by inert ions | Not normally | Yes or no, depending on geometry | Possible | No | Yes |
| Substrate heating by external means | Yes, normally | Not generally | Yes | No | Not normally |
| Cost | Low | High | High | Low | Very high |

2. The flux gradient is proportional to the time ratio of change:

$$\frac{dN}{dt} = -\frac{dJ}{dx} \rightarrow D\frac{d^2N}{dx^2} \qquad 3.57$$

In gaseous doping of a silicon wafer, the dopant concentration at the Si surface is constant during the doping process, resulting in the following relation:

$$N(x,t) = N_0 \, \mathrm{erfc}\left(\frac{x^2}{4Dt}\right)^{\frac{1}{2}} \qquad 3.58$$

where erfc is the complementary error function and $\sqrt{Dt}$ is called the diffusion length, $N_0$ is the surface concentration of the dopant (in $cm^{-3}$). Typical dopant sources used are $POCl_3$ (liquid), $BN_3$ (solid), and $PH_3$ (gas) or $B_2H_6$ (gas).

If the dopant concentration at the silicon surface is limited, the dopant concentration profile is Gaussian instead of an erfc, because one boundary condition has changed with the surface dopant concentration falling as the material supply is depleted. The expression then is

$$N(x,t) = \frac{Q}{(\pi Dt)^{\frac{1}{2}}} \exp\left(-\frac{x^2}{4Dt}\right) \qquad 3.59$$

with $Q$ = surface dose ($cm^{-2}$). Typical dopant sources in this case are doped oxides and implantation. The D in the above equations is the dopant diffusivity:

$$D = D_0(T_0)\exp\left(-\frac{E_a}{kT}\right) \qquad 3.60$$

with $E_a$ the activation energy with a value that depends upon the transport mechanism (typically 0.5 to 1.5 eV). The diffusion coefficient, $D_0$, is only a constant for a given reaction and concentration. Dopant atoms take on several different ionization states and diffuse by as many different mechanisms. The effective diffusion coefficient is actually the sum of the diffusivities of each species, which are in turn a function of temperature and dopant concentration. Hence, actual dopant concentration profiles deviate from the simple first order functions given above.

The dopant diffusivities of phosphorous, arsenic, and boron in silicon at 1100°C typically are $2 \times 10^{-13}$, $1 \times 10^{-13}$, and $1 \times 10^{-13}$ $cm^2$/sec, respectively. Below the intrinsic carrier concentration (~$5 \times 10^{18}$ $cm^{-3}$ for Si at 1000°C), D is independent of the concentration. Above the intrinsic concentration, the diffusion increases with carrier concentration and generally follows a power law due to the diffusion mechanism becoming point defect assisted (vacancies and interstitials). It is common practice to make use of calibration charts to determine the junction depth $x_J$ based upon:

$$x_j = k\sqrt{D_s t} \qquad 3.61$$

**TABLE 3.15**   Characteristics of Two Means of Semiconductor Doping

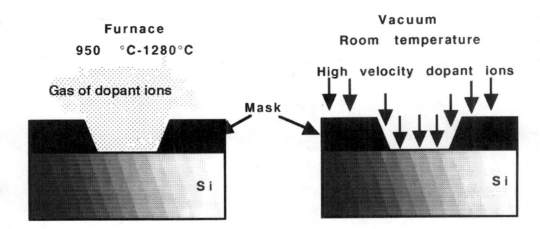

| | | |
|---|---|---|
| Dopant uniformity and reproducibility | ±5% on wafer, ±15% overall | ±1% overall |
| Contamination danger | High | Low |
| Delineation | Refractory insulators and refractory metals, polysilicon | Refractory and nonrefractory materials, metals |
| Environment | Furnace | Vacuum |
| Temperature | High | Low |

where k represents a coefficient that depends upon the diffusivity regime and can take a value from 1.6 to 0.87; $D_s$ is the surface diffusivity and is a function of the dopant material and the temperature; and t is the diffusion time.[25] It is important to remember that the concentration profile of the dopant in the semiconductor cannot change abruptly in this case, as this would imply a zero or infinite flux, so diffusion also takes place gradually around the edges of the mask.

## Ion Implantation

The principle method of doping today centers on high energy ion implantation. Implantation offers the advantage of being able to place any ion at any depth in the sample, independent of the thermodynamics of diffusion and problems with solid solubility and precipitation. Ion beams produce crystal damage that can reduce electrical conductivity, but most of this damage can be eliminated by annealing at 700 to 1000°C.

A beam of energetic ions 'implants' dopants into the substrate. Depth and dopant concentration are controlled by the acceleration energy and the beam current. The stopping mechanism of the ions involves nuclear collisions at low energy and electronic interactions at high energy.

The jargon associated with ion implantation includes:[25]

- Projected range $R_p$, i.e., average distance traveled by ions parallel to the beam
- Projected straggle, $\Delta R_p$, i.e., fluctuation in the projected range

- Lateral straggle, $\Delta R//$, i.e., fluctuation in the final rest position, perpendicular to the beam
- Peak concentration, $N_p$, i.e., concentration of implanted ions at $R_p$

The concentration profile, to a first order approximation, is Gaussian, i.e.,

$$N(x) = N_p \exp\left[ -\frac{\left(x - R_p\right)^2}{2\left(\Delta R_p\right)^2} \right]$$

$$N_p = \frac{Q}{\sqrt{2\pi}\Delta R_p}$$

**3.61**

The range is determined by the acceleration energy, the ion mass, and the stopping power of the material. Orientation of the substrate surface away from perpendicular to the beam prevents channeling which can occur along crystal planes, ensuring reproducibility of $R_p$. Ion-channeling leads to an exponential tail in the concentration vs. depth profile. This tail is due to the crystal lattice and is not observed in an amorphous material. The dose is determined by the charge per ion, zq; the implanted area, A; and the charge per unit time (current) arriving at the substrate. In other words:

$$\int i \, dt = Q \quad \text{and} \quad \frac{Q}{[zqA]} = \text{Dose (atoms/cm}^2) \qquad \textbf{3.62}$$

The technique is now commonly used with penetration depths in silicon of As, P, and B typically being 0.5, 1, and 2 μm at 1000 kEV.

Thermal annealing at temperatures above 900°C is required to remove damage to the silicon lattice and to activate the implanted impurities. For deep diffusions (>1 μm), implantation is used to create a dose of dopants, and thermal diffusion (limited source) is used to drive in the dopant.

Ion beams can implant enough material to actually form new materials, e.g., oxides and nitrides, some of which show improved wear and strength characteristics.

# Examples of Pattern Transfer with Additive Techniques

The two examples of additive processes we choose are typical for sensor rather than for IC manufacture: spray pyrolysis to make planar gas sensors and electro-deposition of Ni micromechanical structures.

## Example 1. Spray Pyrolysis

The use of a ceramic cylindrical tube in a classical Taguchi sensor (Figure 3.42) maximizes the utilization of heater power, so that most of the power is used to heat the gas-sensitive tin oxide. The tin oxide paste is applied by dip coating the outside ceramic body. The tin oxide covers thick-film resistance measuring pads and is sintered at high temperature. The resistance of tin oxide at high temperature gives a measure for the amount of reducing gases in the contacting atmosphere. Sintering stabilizes the intergranular contacts (necks) where the sensitivity of the gas sensor resides.[23] The design shown in Figure 3.42 is not suited for mass production as it involves excessive hard labor. The problems with the thick-film structure mainly reside in the areas of reproducibility: compressing and sintering a powder, the deposition of the catalyst, and the use of binders and other ceramics (e.g., for filtering). Application of IC techniques could improve the state of the art dramatically if one could only make a thin semiconductor oxide film (e.g., tin oxide) with the same sensitivity as the traditional thick sintered film. Micromachined heater elements as shown in Figure 3.43* improve the reproducibility and the absolute power budget needed to bring the sensor to the required temperature. The thermal efficiency and low power budget are obtained by thermally isolating the heater element on a thin suspended membrane. Whereas low power budget sensors have proven quite feasible, it has been more difficult to make a compatible thin film as sensitive as the ceramic-type, thick-film devices. When making thin films with PVD methods, the films have a lower surface area and buried intergranular contacts (see Figure 3.44). The buried intergranular contacts cannot be reached by the contacting gas as easily as in the powder case, making not only for a lower sensitivity, but also slower responding gas sensors. The slower response of the thin-film sensors is due to the fact that oxygen diffusion in

the grain boundaries between compacted grains involves long time constants. When the grain boundaries are blocked on purpose, the sensor reacts faster but displays even less sensitivity.[23]

Using silk screening of thick films on a thin membrane would seem to provide the solution — a low energy heater combined with a fast and sensitive thick tin oxide film. Unfortunately, during our attempts to deposit thick films this way, usually the thin Si membrane broke due to the pushing action of the squeegee.[90] Spray pyrolysis could provide the answer here as its deposition does not involve mechanical pressure on the silicon membrane. Fast responding, planar Taguchi $SnO_2$ gas sensors were made by spraying, with oxygen as the carrier gas, an organic solution of $(CH_3COO)_2SnCl_2$ in ethylacetate onto quartz, glass, and $Al_2O_3$ ceramic substrates, heated to 300°C.[56] This result carries significance; other micromachining attempts based on thin PVD $SnO_2$ film fail to reach the sensitivity attained with the classical thick films and sintered $SnO_2$ films. Spray pyrolysis, with its very small grain size deposits, holds the promise of reaching the objective of providing films with freely accessible polycrystalline grains. As a planar technology it is also

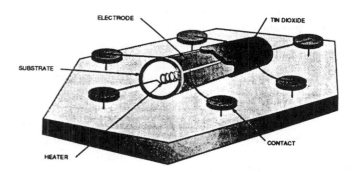

FIGURE 3.42   Classical Taguchi sensor gas sensor.

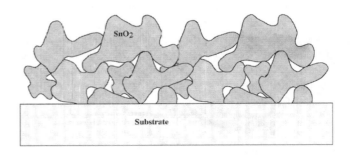

A

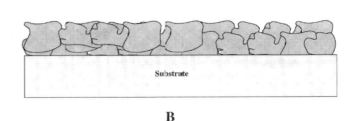

B

FIGURE 3.44   Grain structure in a sintered powder, $SnO_2$ film (A) and a PVD, $SnO_2$ film (B).

---

*   Figure 3.43 appears as a color plate after page 14.

suited for large substrates and considered a viable alternative to the dip-coating process.

## Example 2. Nickel Micromechanical Structures

Nickel is an attractive material for micromechanical structures due to its mechanical, thermal, magnetic, and optical properties, such as hardness, melting point, magnetic permeability, and reflectance. Parameswaran et al.[91] fabricated Ni micromechanical structures on a silicon substrate employing bulk and surface micromachining. They found the patterned films resistant to anisotropic silicon etchants, making the structures suitable for subsequent maskless silicon micromachining. The composition of the electroless plating solution consisted of 33.58 g/L $NiSO_4$. 6 $H_2O$, 10 g/L $NaH_2PO_4$, $H_2O$, 65 g/L $C_6H_6O_7(NH_4)_2$, and the pH was 8.5 ± 0.5 (adjusted by titration with a 10% $NH_4OH$ solution). The temperature was maintained at 90 ± 3°C. A deposition rate of 12 μm/hr was reached this way.

The building of a bulk micromachined Ni structure is illustrated in Figure 3.45, which represents an Si wafer with an AZ 1213 positive resist coat treated as follows: 30 min softbake at 100°C; patterned, developed, and hardbaked for 50 min at 120°C. Electroless for 20 minutes resulted in plating on the exposed Si only (Figure 3.45A). The thickness of the Ni measured 4 μm. The photoresist was then stripped and the wafers annealed at 120°C for 1 hour in a nitrogen ambient. This annealing step was introduced to release the residual stress in the polycrystalline Ni films. It has been shown that the pH of the plating solution has a strong effect on the magnitude of this residual stress. In a next step after the anneal, the exposed silicon was anisotropically etched using ethylene diamine pyrocathechol (EDP) in water for 60 min without any further masking step. The resulting anisotropic cavity depth was measured to be about 75 μm deep. Figure 3.45B displays an SEM photograph of two Ni cantilevers.

A surface micromachined Ni structure can be made as shown in Figure 3.46. A high viscosity, 1.2-μm thick photoresist (AZ 4620) layer was patterned. The remaining photoresist was activated using a catalyst (SHIPLEY CATAPREP 404) to make the resist surface amenable to catalytic Ni deposition. Then 1 μm of Ni was deposited on the exposed Si as well as on the activated resist. After the stress-releasing anneal, the film was patterned. In the end, photoresist was stripped, leaving free-standing Ni microstructures on the silicon surface.

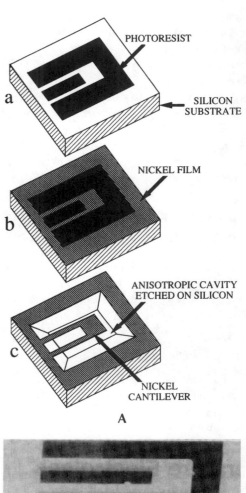

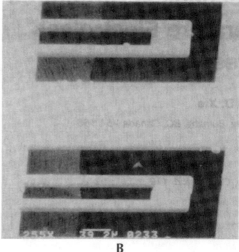

**FIGURE 3.45** Illustration of bulk Si micromachining process incorporating nickel micromechanical structures. (A) Process steps, and (B) SEM photograph of a Ni cantilever fabricated using bulk micromachining.

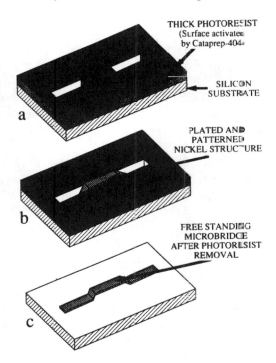

THICK PHOTORESIST
(Surface activated
by Cataprep-404)

SILICON
SUBSTRATE

a

PLATED AND
PATTERNED
NICKEL STRUCTURE

b

FREE STANDING
MICROBRIDGE
AFTER PHOTORESIST
REMOVAL

c

**FIGURE 3.46** Illustration of surface micromachining process incorporating Ni micromechanical structures. (From Parameswaran, M., D. Xie, and P. G. Glavina, *J. Electrochem. Soc.*, 140, L11-L112, 1993. With permission.)

# References

1. Ko, W. H., J. T. Suminto, and G. J. Yeh, "Bonding Techniques for Microsensors", in *Micromachining and Micropackaging of Transducers*, Fung, C. D., Cheung, P. W., Ko, W. H. and Fleming, D. G., Eds., Elsevier, Amsterdam, 1985, p. 41-61.

2. Sze, S. M., Ed. *VLSI Technology*, McGraw-Hill, New York, 1988.

3. Bockris, J. O. M. and A. K. N. Reddy, *Modern Electrochemistry*, Plenum Press, New York, 1977.

4. Mallory, G. O. and J. B. Hadju, Eds. Electroless Plating, Fundamentals and Aplications, American Electroplaters and Surface Finishers Society (AESF), Orlando, FL, 1990.

5. Menz, W. and P. Bley, *Mikrosystemtechnik fur Ingenieure*, VCH Publishers, Weinheim, Germany, 1993.

6. Jaeger, R. C., *Introduction to Microelectronic Fabrication*, Addison-Wesley, Reading, MA, 1988.

7. Cotell, C. M., "Pulsed Laser Deposition and Processing of Biocompatible Hydroxylapatite Thin Films", *Appl. Surf. Sci.*, 69, 140-148, 1993.

8. Pearce, C. W., "Crystal Growth and Wafer Preparation", in *VLSI Technology*, Sze, S. M., Ed., McGraw-Hill, New York, 1988.

9. Hess, D. W. and K. F. Jensen, Eds. *Microelectronics Processing, Chemical Engineering Aspects*, American Chemical Society, Washington, DC., 1989.

10. Parker, E. H. C., Ed. *The Technology and Physics of Molecular Beam Epitaxy*, Plenum Press, New York, 1985.

11. Licari, J. J., *Plastic Coatings for Electronics*, McGraw-Hill, New York, 1970.

12. Harper, C. A., Ed. *Handbook of Thick Film Hybrid Microelectronics*, McGraw-Hill, New York, 1982.

13. Colclaser, R. A., *Microelectronics: Processing and Device Design*, John Wiley & Sons, New York, 1980.

14. Mooney, J. B. and S. B. Radding, "Spray Pyrolysis Processing", *Ann. Rev. Mater. Sci.*, 12, 81-101, 1982.

15. Vossen, J. L. and W. Kern, Eds. *Thin Film Processes*, Academic Press, Orlando, FL, 1978.

16. Pfender, E., "Fundamental Studies Associated with the Plasma Spray Process", *Surf. Coat. Technol.*, 34, 1-14, 1988.

17. Lischner, D. J., H. Basseches, and F. A. D'Altroy, "Observations of the Temperature Gradient Zone Melting Process for Isolating Small Devices", *J. Electrochem. Soc.*, 132, 2991-2996, 1985.

18. Brown, A. R., "Theory of Transport Processes in Semiconductor Crystal Growth from the Melt", in *Microelectronics Processing-Chemical Engineering Aspects*, Hess, D. W. and Jensen, K. F., Eds., American Chemical Society, Washington, DC, 1989.

19. Wong, A., "Silicon Micromachining", 1990, Viewgraphs, Presented in Chicago, IL.

20. Deal, B. E. and A. S. Grove, "General Relationship for the Thermal Oxidation of Silicon", *J. Appl. Phys.*, 36, 3770-3778, 1965.

21. Katz, L. E., "Oxidation", in *VLSI Technology*, Sze, S. M., Ed., McGraw-Hill, New York, 1988, p. 98-140.

22. Fair, R. B., "Diffusion and Oxidation of Silicon", in *Microelectronics Processing: Chemical Engineering Aspects*, Hess, D. W. and Jensen, K., F., Eds., American Chemical Society, Washington, DC, USA, 1989, p. 265-323.

23. Madou, M. J. and S. R. Morrison, *Chemical Sensing with Solid State Devices*, Academic Press, New York, 1989.

24. Moreau, W. M., *Semiconductor Lithography*, Plenum Press, New York, 1988.

25. Brodie, I. and J. J. Muray, *The Physics of Microfabrication*, Plenum Press, New York, 1982.

26. Kendall, D. L., "Vertical Etching of Silicon at Very High Aspect Ratios", *Ann. Rev. Mater. Sci.*, 9, 373-403, 1979.

27.  Seidel, H., "The Mechanism of Anisotropic Electro-chemical Silicon Etching in Alkaline Solutions", Technical Digest of the 1990 Solid State Sensor and Actuator Workshop., Hilton Head Island, SC, 1990, p. 86-91.

28.  Schmidt, P. F. and W. Michel, "Anodic Formation of Oxide Films on Silicon", *J. Electrochem. Soc.*, 104, 230-236, 1957.

29.  Madou, M. J., S. R. Morrison, and V. P. Bondarenko, "Introduction of Impurities in Anodically Grown Silica", *J. Electrochem. Soc.*, 135, 229-235, 1988.

30.  Madou, M. J., W. P. Gomes, F. Fransen, and F. Cardon, "Anodic Oxidation of p-type Silicon in Methanol as Compared to Glycol", *J. Electrochem. Soc.*, 129, 2749-2752, 1982.

31.  Murarka, S. P., "Metallization", in *VLSI Technology*, Sze, S. M., Ed., McGraw-Hill, New York, 1988, p. 375-421.

32.  ERL, "SAMPLE Version 1.6a User's Guide", Electronics Research laboratory (ERL), UC Berkeley, CA, USA, Berkeley, CA, 1985.

33.  Fichtner, W., "Process Simulation", in *VLSI Technology*, Sze, S. M., Ed., McGraw-Hill, New York, 1988, p. 422-465.

34.  Studt, T., "Better Living Through Sputtering", *Res. & Dev.*, May, 54, 1992.

35.  Comello, V., "Tough Coatings Are a Cinch With New PVD Method", *Res. & Dev.*, January, 59-60, 1992.

36.  Maissel, L. I. and R. Glang, Eds. *Handbook of Thin Film Technology*, McGraw-Hill, New York, 1970.

37.  Kinoshita, K. and M. J. Madou, "Electrochemical Measurement on Pt, Ir, and Ti Oxides as pH Probes", *J. Electrochem. Soc.*, 131, 1098-1094, 1984.

38.  Editorial, Say Yes to I.S.T, Brochure, . 1994, Innovative Sputtering Technology (IST): Zulte B-9870, Belgium.

39.  Comello, V., "Silicon MBE Research Races Toward Production", *Res. & Dev.*, April, 87-88, 1992.

40.  Dutta, B., X. D. Wu, A. Inam, and T. Venkatesan, "Pulsed Laser Deposition: A Viable Process for Superconducting Thin Films?", *Solid State Technol.*, February, 106-110, 1989.

41.  Inam, A., "Pulsed Laser Takes the Heat Off HTS Materials Deposition", *Res. & Dev.*, February, 90-92, 1991.

42.  Cotell, C. M. and K. S. Grabowski, "Novel Materials Applications of Pulsed Laser Deposition", MRS Bull., XVII, 44-53, 1992.

43.  Cotell, C. M., D. B. Chrisey, K. S. Grabowski, J. A. Sprague, and C. R. Gossett, "Pulsed Laser Deposition of Hydroxylapatite Thin Films on Ti-6Al-4V", *J. Appl. Biomater.*, 3, 87-93, 1992.

44.  Jensen, K., F., "Chemical Vapor Deposition", in *Microelectronics Processing: Chemical Engineering Aspects*, Hess, D. W. and Jensen, K. F., Eds., American Chemical Society, Washington, DC, USA, 1989, p. 199-263.

45.  Granger, R. A., *Fluid Mechanics*, Dover Publications, Inc., New York, 1995.

46.  Wolf, S. and R. N. Tauber, *Silicon Processing for the VLSI Era*, Lattice Press, Sunset Beach, 1987.

47.  Lee, H. H., *Fundamentals of Microelectronics Processing*, McGraw-Hill, New York, 1990.

48.  Coronell, D. G. and K. F. Jensen, "Simulation of Rarified Gas Transport and Profile Evolution in Nonplanar Substrate Chemical Vapor Deposition", *J. Electrochem. Soc.*, 141, 2545-2551, 1994.

49.  Adams, A. C., "Dielectric and Polysilicon Film Deposition", in *VLSI Technology*, Sze, S. M., Ed., McGraw-Hill, 1988, p. 233-71.

50.  Ehrlich, D. J., R. M. J. Osgood, and J. Deutsch, "Photodeposition of Metal Films with Ultraviolet Laser Light", *J. Vac. Sci. Technol.*, 21, 23-32, 1982.

51.  Wallenberger, F. T. and P. C. Nordine, "Inorganic Fibers and Microstructures by Laser Assisted Chemical Vapor Deposition", *Mater. Technol.*, 8, 198-202, 1993.

52.  Compton, R., D., "PECVD: A Versatile Technology", *Semicond. Int.*, July, 60-65, 1992.

53.  Wu, T. H. T. and R. S. Rosler, "Stress in PSG and Nitride Films as Related to Film Properties and Annealing", *Solid State Technol.*, May, 65-71, 1992.

54.  Hey, H. P. W., B. G. Sluijk, and D. G. Hemmes, "Ion Bombardement Factor in Plasma CVD", *Solid State Technol.*, April, 139-144, 1990.

55.  Burggraaf, P., "The Status of MOCVD Technology", *Semicond. Int.*, 16, 80-83, 1993.

56.  Tomar, M. S. and F. J. Garcia, "Spray Pyrolysis in Solar Cells and Gas Sensors", *Prog. Cryst. Growth Charact.*, 4, 221-248, 1988.

57.  Peters, L., "SOI Takes Over Where Silicon Leaves Off", *Semicond. Int.*, March, 48-51, 1993.

58.  Anderson, T. J., "Liquid-Phase Epitaxy and Phase Diagrams of Compound Semiconductors", in *Microelectronics Processing-Chemical Engineering Aspects*, Hess, D. W. and Jensen, K. F., Eds., American Chemical Society, Washington, D.C, 1989.

59.  Borland, J., R. Wise, Y. Oka, M. Gangani, S. Fong, and Y. Matsumoto, "Silicon Epitaxial Growth for Advanced Device Structures", *Solid State Technol.*, January, 111-119, 1988.

60.  Rehrig, D., L., "In Search of Precise Epi Thickness Measurements", *Semicond. Int.*, 13, 90-95, 1990.

61.  Singer, P., "CVC Builds Cluster Tool Through Partnering", *Semicond. Int.*, June, 46, 1992.

62.  Glanz, J., "Is 1992 the Year of the Mini In Cleanroom Technology?", *Res. & Dev.*, May, 97-101, 1992.

63.  Romankiw, L. T., "Pattern Generation in Metal Films Using Wet Chemical Techniques", *Etching for Pattern Definition*, Washington, DC, 1976, p. 137-139.

64.  Ohno, I., "Methods for Determination of Electroless Deposition Rate", Proceedings of the Symposium on Electroless Deposition of Metals and Alloys, Honolulu, HW, 1988, p. 129-141.

65. Fleischmann, M., S. Pons, D. R. Rolison, and P. P. Schmidt, "Ultramicroelectrodes", The Utah Conference on Ultramicroelectrodes, Utah, 1987.

66. Madou, M. J. and S. R. Morrison, "High-field Operation of Submicrometer Devices at Atmospheric Pressure", 6th International Conference on Solid-State Sensors and Actuators (Transducers '91), San Francisco, CA, 1991, p. 145-149.

67. van der Putten, A. M. T. and J. W. G. Bakker, "Geometrical Effects in the Electroless Metallization of Fine Metal Patterns", *J. Electrochem. Soc.*, 140, 2221-2228, 1993.

68. Masuko, N., T. Osaka, and Y. Ito, Eds. Electrochemical Technology: Innovation and New Developments, Co-published by Kodansha (Japan), Ltd. and Gordon and Breach Science Publishers S.A. (The Netherlands), Tokyo, Japan and Amsterdam, The Netherlands, 1996.

69. Dukovic, J. O., "Feature-Scale Simulation of Resist Patterned Electrodeposition", *IBM Journal of Research and Development*, 37, 125-140, 1993.

70. Leyendecker, K., W. Bacher, W. Stark, and A. Thommes, "New Microelectrodes for the Investigation of the Electroforming of LIGA Microstructures", *Electrochim. Acta*, 39, 1139-1143, 1994.

71. Lambrechts, M. and W. Sansen, *Biosensors Microelectrical Devices*, Institute of Physics Publishing, Philadelphia, PA, 1992.

72. Riemer, D. E., "Analytical Engineering Model of the Screen Printing Process: Part I", *Solid State Technol.*, August, 107-111, 1988.

73. Sze, S. M., Ed. *VLSI Technology*, John Wiley & Sons, New York, 1983.

74. Middlehoek, S., D. J. W. Noorlag, and G. K. Steenvoorden, "Silicon and Hybrid Micro-Electronic Sensors", *Electrocomponent Science and Technology*, 10, 217-229, 1983.

75. White, N. M. and A. W. J. Cranny, "Design and Fabrication of Thick Film Sensors", *Hybrid Circuits*, 12, 32-35, 1987.

76. Acheson, "Product Data Sheets on Screen-printable Ag, Ag/AgCl and C Pastes, Brochure", Acheson Coloiden BV, Scheemda, Netherlands, 1991.

77. Pace, S. J. and M. A. Jensen, "Thick-Film Multi-Layer pH Sensor for Biomedical Applications", 2nd International Meeting on Chemical Sensors, Bordeaux, France, 1986, p. 557-561.

78. Belford, R. E., A. E. Owen, and R. G. Kelly, "Thick-Film Hybrid pH Sensors", *Sensors Actuators*, 11, 387-398, 1987.

79. Pace, S. J., P. P. Zarzycki, R. T. McKeever, and L. Pelosi, "A Thick-Film Multi-Layered Oxygen Sensor", 3rd International Conference on Solid-State Sensors and Actuators (Transducers '85), Philadelphia, PA, 1985, p. 406-409.

80. Karagounis, V., L. Lun, and C. Liu, "A Thick-Film Multiple Component Cathode Three-Electrode Oxygen Sensor", *IEEE Trans. Biomed. Eng.*, BME-33, 108-112, 1986.

81. Lewandowski, J. J., P. S. Malchesky, M. Zborowski, and Y. Nose, "Assessment of Microelectronic Technology for Fabrication of Electrocatalytic Glucose Sensor", Proceedings of the Ninth Annual Conference of the IEEE Engineering in Medicine and Biology Society, Boston, MA, 1987, p. 784-785.

82. Weetall, H. H. and T. Hotaling, "A Simple, Inexpensive, Disposable Electrochemical Sensor for Clinical and Immuno-Assay", *Biosensors*, 3, 57-63, 1987.

83. Cha, C. S., M. J. Shao, and C. C. Liu, "Electrochemical Behavior of Microfabricated Thick-Film Electrodes", *Sensors and Actuators B*, B2, 277-281, 1990.

84. Guckel, H., J. Uglow, M. Lin, D. Denton, J. Tobin, K. Euch, and M. Juda, "Plasma Polymerization of Methyl Methacrylate : A Photoresist for 3D Applications", Technical Digest of the 1988 Solid State Sensor and Actuator Workshop., Hilton Head Island, SC, 1988, p. 9-12.

85. Comello, V., "Planarizing Leading Edge Devices", *Semicond. Int.*, November, 60-66, 1990.

86. Wiesner, J. R., "Gap Filling of Multilevel Metal Interconnects with 0.25-μm Geometries", *Solid State Technol.*, October, 63-64, 1993.

87. Sniegowski, J. S., "Chemical-Mechanical Polishing: Enhancing the Manufacturability of MEMS", Micromachining and Microfabrication Process Technology II, Austin, Texas, USA, 1996, p. 104-115.

88. Ong, E., H. Chu, and S. Chen, "Metal Planarization with an Excimer Laser", *Solid State Technol.*, August, 63-68, 1991.

89. Oh, S. and M. Madou, "Planar-Type, Gas Diffusion-Controlled Oxygen Sensor Fabricated by the Plasma Spray Method", *Sensors and Actuators B*, B14, 581-582, 1992.

90. Madou, M., "Solid-State Gas Sensors: World Markets and New Approaches to Gas Sensing", NIST, Report No. NIST Special Publication 865, Gaithersburg, MD, 1994.

91. Parameswaran, M., D. Xie, and P. G. Glavina, "Fabrication of Nickel Micromechanical Structures Using a Simple Low-Temperature Electroless Plating Process", *J. Electrochem. Soc.*, 140, L111-L112, 1993.

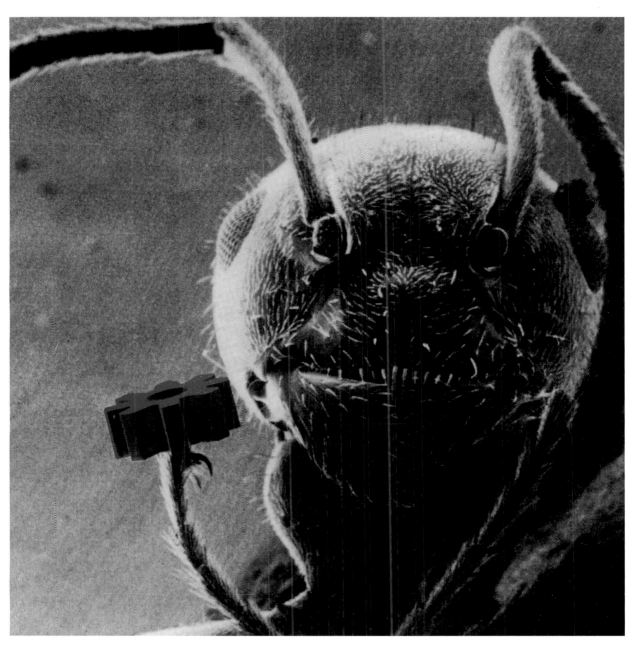

**Figure 1.28** An eye catching example of a LIGA 'product'. The ant holds a Ni gear in its claw. (From Forschungszentrum Karlsruhe GmbH Technik und Umwelt, Projekt Mikrosystemtechnik (PMT). With permission.)

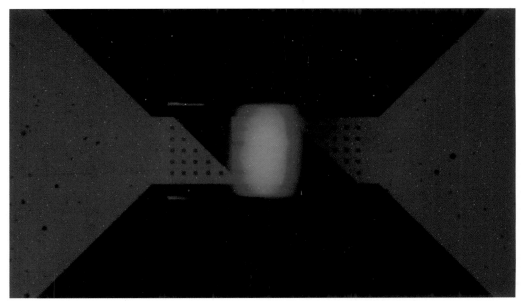

**Figure 3.43** Glowing micromachined heater element in Si.

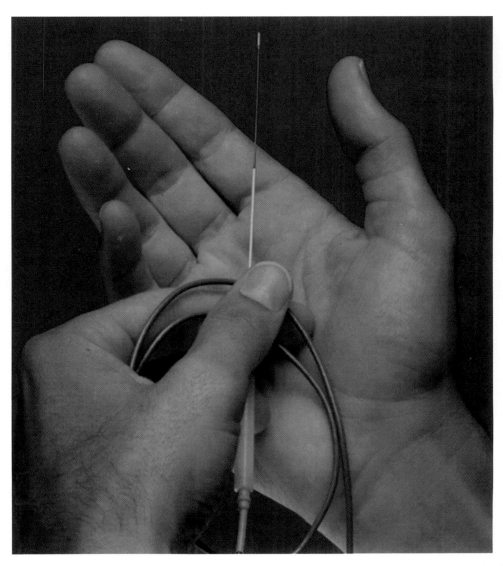

**Figure 4.66** An *in vivo* pH, $CO_2$ and $O_2$ sensor based on a linear array of electrochemical cells.

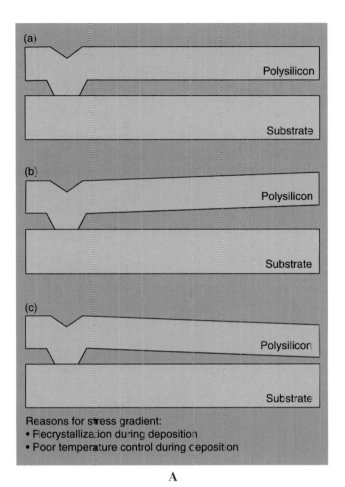

**Figure 5.11** Micro-cantilever deflection for measuring stress non-uniformity. (A) Micro-cantilever deflection for measuring stress non-uniformity. (a) No gradient; (b) higher tensile stress near the surface; and (c) lower tensile stress near the surface. (B) Topographical contour map of polysilicon cantilever array. (From Core, T.A., W.K. Tsang, and S.J. Sherman, *Solid State Technol.*, 36, 39-47, 1993. With permission.)

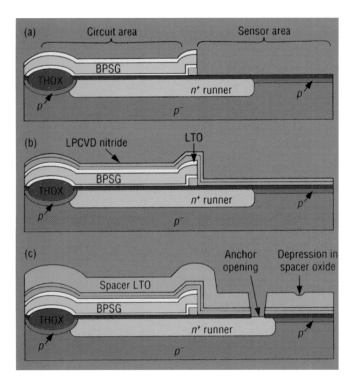

**Figure 5.36** Preparation of IC chip for poly-Si. (a) Sensor area post-BPSG planatization and moat mask. (b) Blanket deposition of thin oxide and thin nitride layer. (c) Bumps and anchors made in LTO spacer layer. (From Core, T.A., W.K. Tsang, and S.J. Sherman, *Solid State Technol.*, 36, 39-47, 1993. With permission.)

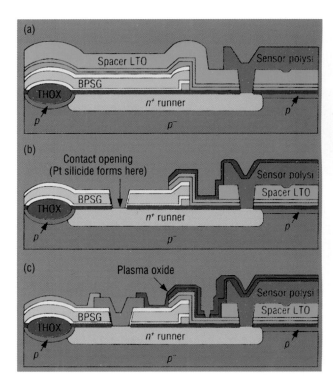

**Figure 5.37** Poly-Si deposition and IC metallization. (a) Cross-sectional view after polysilicon depostion, implant, anneal, and patterning. (b) Sensor area after removal of dielectrics from circuit area, contact mask, and Pt silicide. (c) Metallization scheme and plasma oxide passivation and patterning. (From Core, T.A., W.K. Tsang, and S.J. Sherman, *Solid State Technol.*, 36, 39-47, 1993. With permission.)

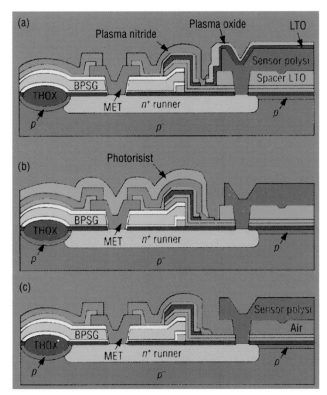

**Figure 5.38** Pre-release preparation and release. (a) Post-plasma nitride passivation and patterning. (b) Photoresist protection of the IC. (c) Freestanding, released poly-Si beam. (From Core, T.A., W.K. Tsang, and S.J. Sherman, *Solid State Technol.*, 36, 39-47, 1993. With permission.)

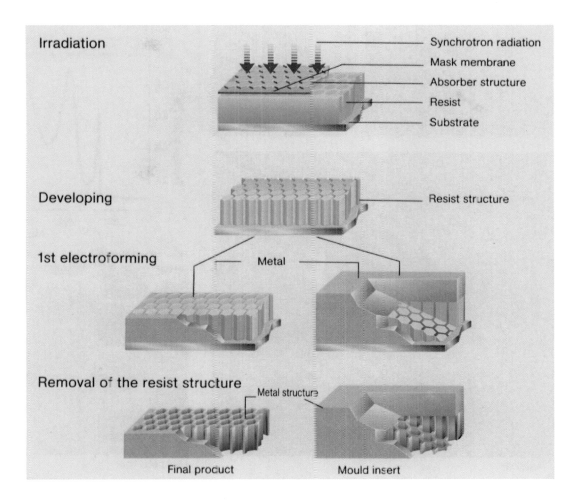

**Figure 6.1A**  Basic LIGA process steps X-ray deep-etch lithography and 1st electroforming. (From Lehr, H. and Schmidt, M., The LIGA Technique, commercial brochure, IMM Institut für Mikrotechnik GmbH, Mainz -Hechtsheim, 1995. With permission).

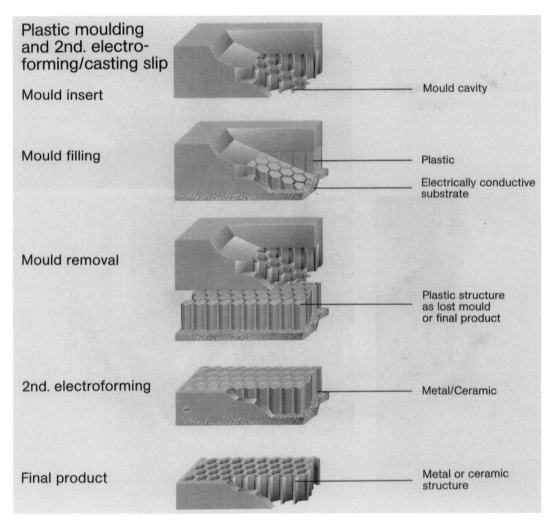

**Figure 6.1B**  Plastic molding and 2nd electroforming/slip casting.(From Lehr, H. and Schmidt, M., The LIGA Technique, commercial brochure, IMM Institut für Mikrotechnik GmbH, Mainz - Hechtsheim, 1995. With permission).

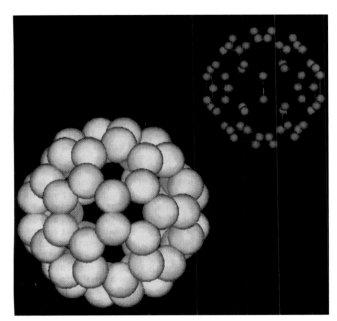

*A Buckminsterfullerene(icosohedral $C_{60}$)- the most symmetrical molecule in nature. Each of the 60 carbon atoms that are located at the vertices of the truncated icosahedron are equidistant from the center-of-mass of the "buckyball".*

A

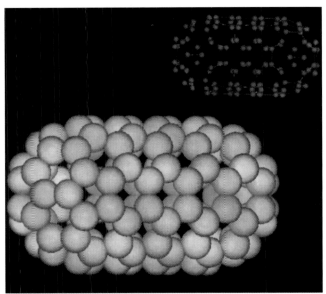

*An Isomer of $C_{120}$, which can also be considered to be a short carbon nanotube: extension along the long axis of symmetry would yield a carbon nanotube similar to those actually produced and studied by TEM.*

B

**Figure 7.31** (A) Buckyball. (B) Buckytube.

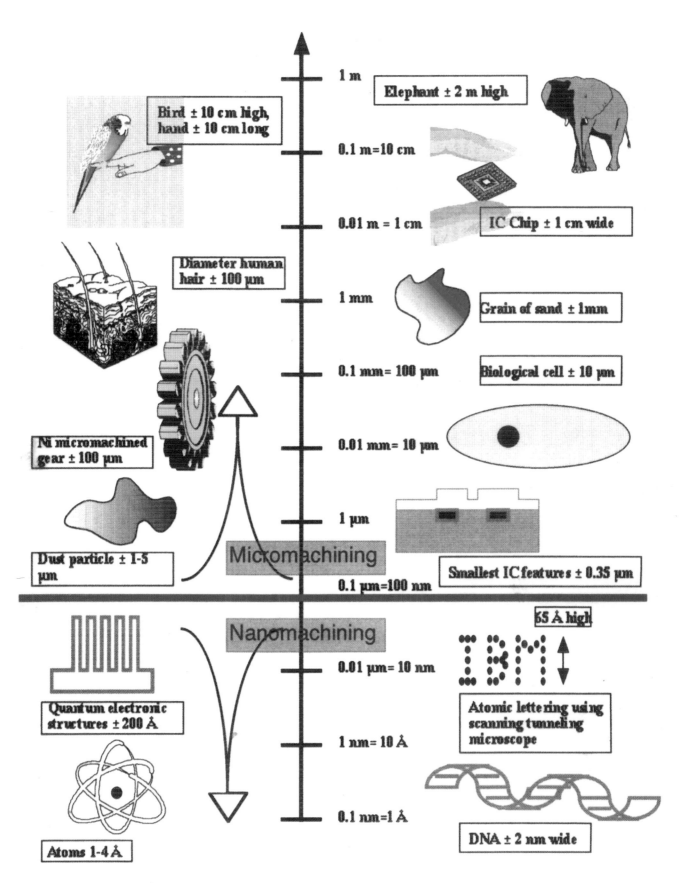

**Figure 9.1** Sizes of 'things' familiar and less familiar.

**Figure 10.2A** Si accelerometer based on hybrid technology; bulk machined 50 g accelerometer bonded into a TO-5 header. Signal conditioning electronics are on a separate chip. (Photomicrograph from Lucas Novasensor.)

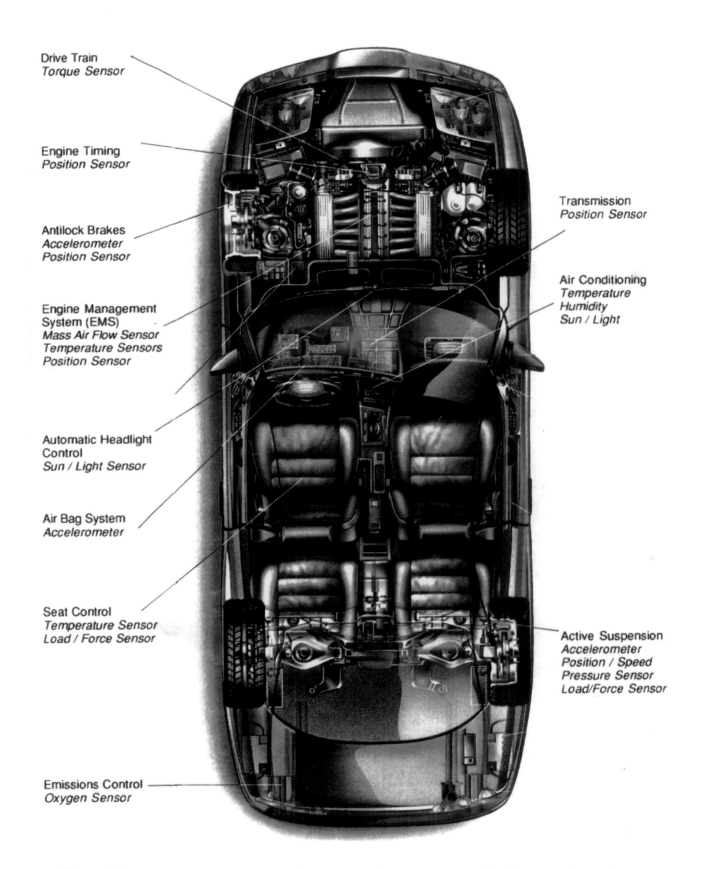

**Figure 10.9** Automotive sensors. Some of the opportunities for micromachining in automotive sensing.

Top: These tiny airpressure sensors are made from etched silicon which is onded to glass. They are used in General Motors' computerized fuel economy and pollution control unit.

Left: The sensors are small enough to fit inside an engine inlet manifold. With the aid of micromechanics, automobiles in the near future will contain computer chips connected to sensors for many different functions.

**Figure 10.10**    General Motors MAP sensor. Notice how packaging dwarfs the sensor. (Courtesy of General Motors.)

**Figure 10.13** i-STAT's handheld clinical analyzer and disposable cartridge (center) (courtesy of Imants Lauks, i-STAT), Biologix's pH sensor cube and reader (top left) (courtesy of Michael Grandon, Biologix, Inc.), and Porton Diagnostic's potassium sensor and reader (bottom right). Development was discontinued.

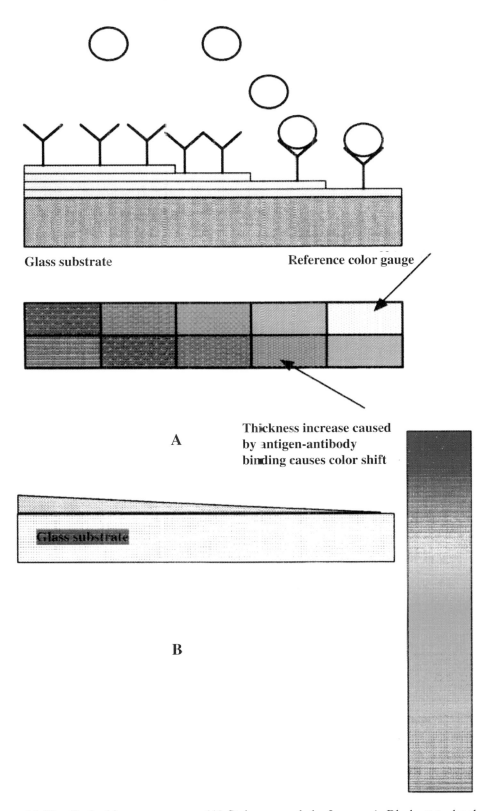

**Figure 10.22** Optical immunosensors. (A) Staircase made by Langmuir-Blodgett technology or etching (additive technique); (B) wedge made by pulling sample out of an etchant (subtractive technique).

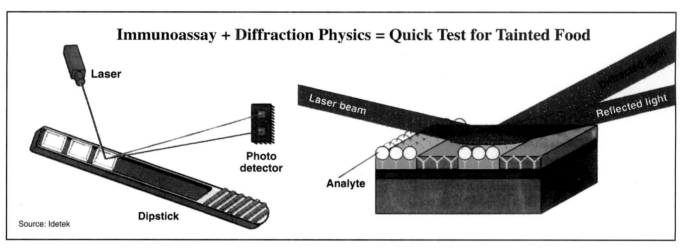

**Immunoassay + Diffraction Physics = Quick Test for Tainted Food**

Laser

Laser beam

Diffracted light

Reflected light

Photo detector

Analyte

Dipstick

Source: Idetek

***Up to six 4 × 5 mm BioChips*** *can be mounted on each disposable plastic holder, or dipstick*

***Antibodies on chip are illuminated*** *through a mask. Analytes bind only to active rows, forming a diffraction grating.*

**Figure 10.26** Ideteck, Inc., immunosensor based on grating. (Courtesy of Mark Platshon, Ideteck, Inc., Sunnyvale, CA.)

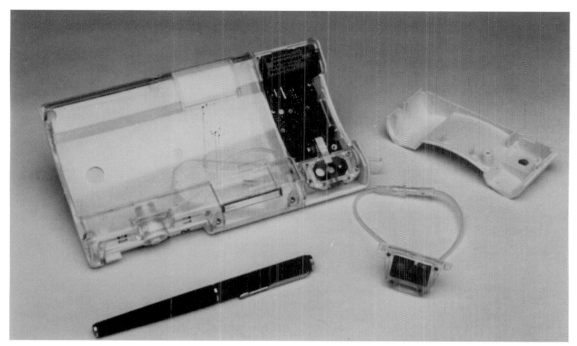

**Figure 10.30A**  Micropump technology. Debiotech's micropump is a Si micromachined piezo-pump, it is part of a beltworn drug delivery system. The Oncojet is disposable and may deliver a drug at a rate of 10 to 50 ml/day. The glass-Si sandwich pump is about 2 by 1 cm in size. (Courtesy Ary Saaman, Debiotech, Switzerland).

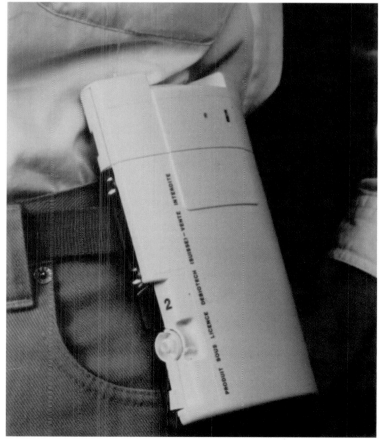

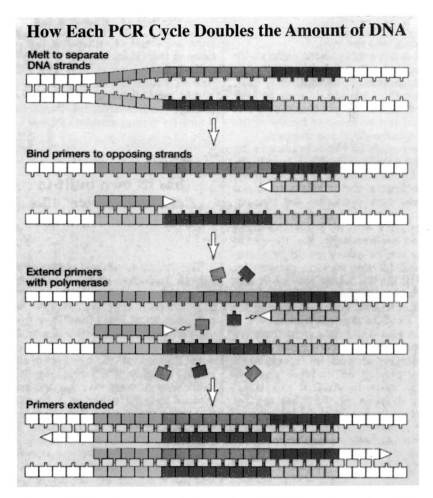

**Figure 10.35A** Polymerase chain reaction (PCR). Illustration on how PCR cycle doubles DNA.

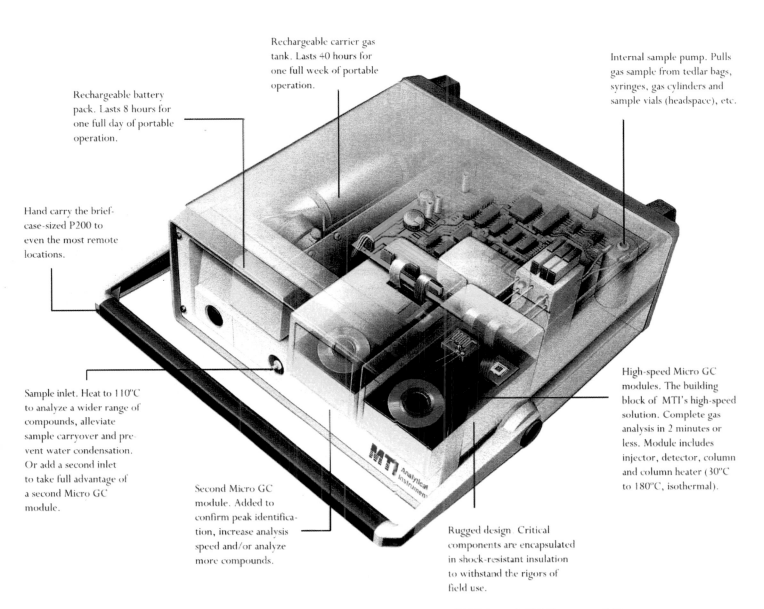

Rechargeable carrier gas tank. Lasts 40 hours for one full week of portable operation.

Internal sample pump. Pulls gas sample from tedlar bags, syringes, gas cylinders and sample vials (headspace), etc.

Rechargeable battery pack. Lasts 8 hours for one full day of portable operation.

Hand carry the brief-case-sized P200 to even the most remote locations.

Sample inlet. Heat to 110°C to analyze a wider range of compounds, alleviate sample carryover and prevent water condensation. Or add a second inlet to take full advantage of a second Micro GC module.

Second Micro GC module. Added to confirm peak identification, increase analysis speed and/or analyze more compounds.

High-speed Micro GC modules. The building block of MTI's high-speed solution. Complete gas analysis in 2 minutes or less. Module includes injector, detector, column and column heater (30°C to 180°C, isothermal).

Rugged design. Critical components are encapsulated in shock-resistant insulation to withstand the rigors of field use.

**Figure 10.38A**    MTI's Gas Chromatrograph. (Courtesy Bill Hidgon, MIT, Fremont, CA).

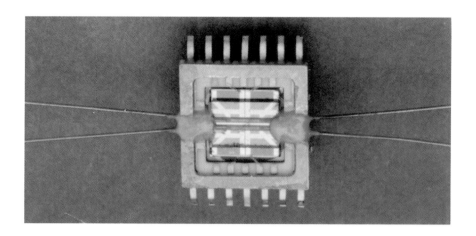

**Figure 10.38B**  Micromachined thermal conductivity detector (top) and micromachined injector (bottom).

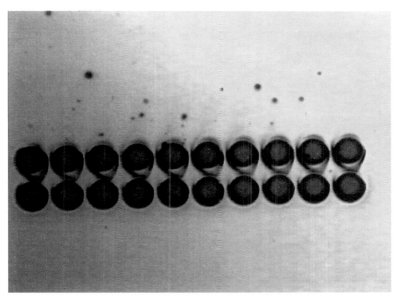

**Figure 10.45**  Holes in PMMA photoresist for the fabrication of quadrupole array using LIGA (1 mm high rods and 80 μm diameter). (Courtesy of Dr. Keith Jackson, LBNL, Berkeley, CA.)

**Figure 10.46A** An example of IR absorption spectrometer instrument, the so-called Argus, made with traditional machining methods at NASA Ames. (Courtesy of Mr. Steve Wegener, NASA Ames, Mountain View, CA.)

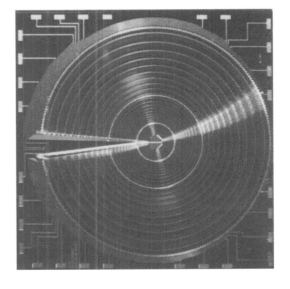

**Figure 10.50** Photo of the retina sensor. (Courtesy of Dr. Lou Hermans, IMEC, Belgium.)

**Figure 10.51B** Photograph of SAW hygrometer. The SAW device (long, retangular bar) is mounted next to a platinum temperature sensor (square, white package), on top of a minature two-stage thermo-electric cooler. The device size is approximately 1 cm by 1 cm. (Courtesy of Dr. Michael Hoemk, JPL, Pasadena, CA.)

# Wet Bulk Micromachining

## Introduction

In wet bulk micromachining, features are sculpted in the bulk of materials such as silicon, quartz, SiC, GaAs, InP, Ge, and glass by orientation-dependent (anisotropic) and/or by orientation-independent (isotropic) wet etchants. The technology employs pools as tools,[1] instead of the plasmas studied in Chapter 2. A vast majority of wet bulk micromachining work is based on single crystal silicon. There has been some work on quartz, some on crystalline Ge and GaAs, and a minor amount on GaP, InP, and SiC. Micromachining has grown into a large discipline, comprising several tool sets for fashioning microstructures from a variety of materials. These tools are used to fabricate microstructures either in parallel or serial processes. Table 7.7 summarizes these tools. It is important to evaluate all the presented micromanufacturing methods before deciding on one specific machining method optimal for the application at hand — in other words, to zero-base the technological approach.[2] The principle commercial Si micromachining tools used today are the well-established wet bulk micromachining and the more recently introduced surface micromachining (Chapter 5). A typical structure fashioned in a bulk micromachining process is shown in Figure 4.1. This type of piezoresistive membrane structure, a likely base for a pressure sensor or an accelerometer, demonstrated that batch fabrication of miniature components does not need to be limited to integrated circuits (ICs). Despite all the emerging new micromachining options, Si wet bulk micromachining, being the best characterized micromachining tool, remains most popular in industry. The emphasis in this chapter is on the wet etching process itself. Other machining steps typically used in conjunction with wet bulk micromachining, such as additive processes and bonding processes, are covered in Chapters 3 and 8, respectively.

Wet bulk micromachining had its genesis in the Si IC industry, but further development will require the adaptation of many different processes and materials. To emphasize the need for micromachinists to look beyond Si as the ultimate substrate and/or building material, we have presented many examples of non-Si micromachinery throughout the book. The need to incorporate new materials and processes is especially urgent for progress in chemical sensors and micro-instrumentation which rely on non-IC materials and often are relatively large. The merging of bulk micromachining with other new fabrication tools such as surface micromachining and electroplating and the adaptation of new materials such as Ni and polyimides has

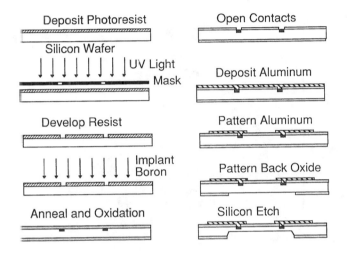

**FIGURE 4.1**  A wet bulk micromachining process is used to craft a membrane with piezoresistive elements. Silicon micromachining selectively thins the silicon wafer from a starting thickness of about 400 µm. A diaphragm having a typical thickness of 20 µm or less with precise lateral dimensions and vertical thickness control results.

fostered a powerful new, nontraditional precision engineering method. A truly multidisciplinary engineering education will be required to design miniature systems with the most appropriate building philosophy.

After a short historical note on wet bulk micromachining we begin this chapter with an introduction to the crystallography of single crystal Si and a listing of its properties, clarifying why Si is such an important sensor material. Some empirical data on wet etching are reviewed and different models for anisotropic and isotropic etching behavior follow. Then, etch stop techniques, which catapulted micromachining into an industrial manufacturing technique, are discussed. Subsequently, a discussion of problems associated with bulk micromachining such as IC incompatibility, extensive real-estate usage, and issues involving corner compensation is presented. Finally, examples of applications of wet bulk micromachining in mechanical and chemical sensors are given.

## Historical Note

The earliest use of wet etching of a substrate, using a mask (wax) and etchants (acid-base), appears to be in the fifteenth century for *decorating armor*.[1] Engraving hand tools were not hard

## *Decorating armor*

(From Harris, T. W., *Chemical Milling*, Clarendon Press, Oxford, 1976. With permission.)

enough to work the armor and more powerful acid-base processes took over. By the early seventeenth century, etching to decorate arms and armor was a completely established process. Some pieces stemming from that period have been found where the chemical milling was accurate to within 0.5 mm. The masking in this traditional chemical milling was accomplished by cutting the maskant with a scribing tool and peeling the maskant off where etching was wanted. Harris[1] describes in detail all of the improvements that, by the mid 1960s, made this type of chemical milling a valuable and reliable method of manufacturing. It is especially popular in the aerospace world. The method enables many parts to be produced more easily and cheaply than by other means and in many cases provides a means to design and produce parts and configurations not previously possible. Through the introduction of photosensitive masks by Niépce in 1822 (see Chapter 1), chemical milling in combination with lithography became a reality and a new level of tolerances came within reach. The more recent major applications of lithography-based chemical milling are the manufacture of printed circuit boards, started during the Second World War, and, by 1961, the fabrication of Si-integrated circuitry. Photochemical machining is also used for such precision parts as color television shadow masks, integrated circuit lead frames, light chopper and encoder discs, and decorative objects such

as costume jewelry.[3] The geometry of a 'cut' produced when etching silicon integrated circuits is similar to the chemical-milling cut of the aerospace industry, but the many orders of magnitude difference in size and depth of the cut account for a major difference in achievable accuracy. Accordingly, the tolerances for fashioning integrated circuitry are many orders of magnitude smaller than in the chemical milling industry.

In this book we are concerned with lithography and chemical machining used in the IC industry and in microfabrication. A major difference between these two fields is in the aspect ratio (height-to-width ratio) of the features crafted. In the IC industry one deals with mostly very small, flat structures with aspect ratios of 1 to 2. In the microfabrication field, structures typically are somewhat larger and aspect ratios might be as high as 400.

Isotropic etching has been used in silicon semiconductor processing since its beginning in the early 1950s. Representative work from that period is the impressive series of papers by Robbins and Schwartz[4-7] on chemical isotropic etching, and Uhlir's paper on electrochemical isotropic etching.[8] The usual chemical isotropic etchant used for silicon was HF in combination with $HNO_3$ with or without acetic acid or water as diluent.[4-7] The early work on isotropic etching in an electrochemical cell (i.e., 'electropolishing') was carried out mostly in nonaqueous solutions, avoiding black or red deposits that formed on the silicon surface in aqueous solutions.[9] Turner showed that if a critical current density is exceeded, silicon can be electropolished in aqueous HF solutions without the formation of any deposits.[10]

In the mid-1960s, the Bell Telephone Laboratories started the work on anisotropic Si etching in mixtures of, at first, KOH, water, and alcohol and later in KOH and water. This need for high aspect ratio cuts in silicon arose for making dielectrically isolated structures in integrated circuits such as for beam leads. Chemical and electrochemical anisotropic etching methods were pursued.[11-21] In the mid-1970s, a new surge of activity in anisotropic etching was associated with the work on V-groove and U-groove transistors.[22-24]

The first use of Si as a micromechanical element can be traced back to a discovery and an idea from the mid-1950s and early 1960s, respectively. The discovery was the large piezoresistance in Si and Ge by Smith in 1954.[25] The idea stems from Pfann et al. in 1961,[26] who proposed a diffusion technique for the fabrication of Si piezoresistive sensors for stress, strain, and pressure. As early as 1962, Tufte et al.,[27] at Honeywell, followed up on this suggestion. By using a combination of a wet isotropic etch, dry etching, and oxidation processes, Tufte et al. made the first thin Si piezoresistive diaphragms, of the type shown in Figure 4.1, for pressure sensors.[27] Sensym/National Semiconductor (sold to Hawker Siddley in 1988) became the first to make stand-alone Si sensor products (1972). By 1974, National Semiconductor described a broad line of Si pressure transducers, in the first complete silicon pressure transducer catalog.[28] Other early commercial suppliers of micromachined pressure sensor products were Foxboro/ICT, Endevco, Kulite, and Honeywell's Microswitch. Other micromachined structures began to be explored by the mid-to late-1970s: Texas Instruments produced a thermal print head (1977);[29] Hewlett Packard made thermally isolated diode detectors (1980);[30] fiberoptic alignment structures were made at Western Electric,[31] and IBM produced ink jet nozzle arrays (1977).[32] Many *Silicon Valley microsensor companies* played and continue to play a pivotal role in the development of the market for Si sensor products.

European and Japanese companies followed the U.S. lead more than a decade later; for example, Druck Ltd. in the U.K. started exploiting Greenwood's micromachined pressure sensor in the mid-1980s.[33]

Petersen's 1982 paper, extolling the excellent mechanical properties of single crystalline silicon, helped galvanize academia to get involved in Si micromachining in a major way.[34]

Before that time most efforts had played out in industry and practical needs were driving the technology (market pull). The new generation of micromachined devices often constitutes gadgetry only, and the field is perceived by many as a technology looking for applications (technology push). It has been estimated that today there are more than 10,000 scientists worldwide involved in Si sensor research and development.[35] In order to justify the continued investment, it has become an absolute priority to understand the intended applications better and to be able to select a more specific micromachining tool set intelligently and to identify large market applications (see Chapter 10).

## *Silicon Valley microsensor companies*

Silicon Valley Micromachining

- 1972 Foxboro ICT
- 1972 Sensym/National Semiconductor (sold to Hawker Siddley in 1988)
- 1975 Endevco
- 1975 IBM Micromachining
- 1976 Cognition (sold to Rosemount in 1978)
- 1980 Lawrence Livermore Lab
- 1981 Microsensor Technology (sold to Tylan in 1986)
- 1982 Transensory Devices (sold to ICSensors in 1987)
- 1982 ICSensors (sold to EG&G in1994)
- 1984 SRI International (group left to form TSDC in 1989)
- 1985 NovaSensor (sold to Lucas in 1990)
- 1986 Captor (sold to Dresser in 1991)
- 1988 Redwood Microstructures
- 1988 TiNi Alloys
- 1988 Nanostructures
- 1989 Teknekron Sensor Development Corporation-TSDC (SRI-group dissolved in 1993)
- 1990 Microflow
- [a] 1991 Sentir
- 1992 Silicon Microstructures
- 1992 Rohm Micromachining
- 1993 Microfabrication Applications
- 1993 Silicon Micromachines
- 1993 Fluid IC (dissolved in 1995)
- 1993 Next Sensors
- 1994 Berkeley Microsystems Incorporated (BMI)
- 1994 Piedmont Microactuators
- 1995 MicroScape
- 1996 Caliper

# Silicon Crystallography

## Miller Indices

The periodic arrangement of atoms in a crystal is called the lattice. The unit cell in a lattice is a segment representative of the entire lattice. For each unit cell, basis vectors ($a_1$, $a_2$, and $a_3$) can be defined such that if that unit cell is translated by integral multiples of these vectors, one arrives at a new unit cell identical to the original. A simple cubic-crystal unit cell for which $a_1 = a_2 = a_3$ and the axes angles are $\alpha = \beta = \gamma = 90°$ is shown in Figure 4.2. In this figure, the dimension 'a' is known as the lattice constant. To identify a plane or a direction, a set of integers h, k, and l called the Miller indices are used. To determine the Miller indices of a plane, one takes the intercept of that plane with the axes and expresses these intercepts as multiples of the basis vectors $a_1$, $a_2$, $a_3$. The reciprocal of these three integers is taken, and, to obtain whole numbers, the three reciprocals are multiplied by the smallest common denominator. The resulting set of numbers is written down as (hkl). By taking the reciprocal of the intercepts, infinities ($\infty$) are avoided in the plane identification. A direction in a lattice is expressed as a vector with components as multiples of the basis vectors. The rules for determining the Miller indices of an orientation are: translate the orientation to the origin of the unit cube and take the normalized coordinates of its other vertex. For example, the body diagonal in a cubic lattice as shown in Figure 4.2 is 1a, 1a, and 1a or a diagonal along the [111] direction. Directions [100], [010], and [001] are all crystallographically equivalent and are jointly referred to as the family, form, or group of <100> directions. A form, group, or family of faces which bear like relationships to the crystallographic axes — for example, the planes (001), (100), (010), (00$\bar{1}$), ($\bar{1}$00), and (0$\bar{1}$0) — are all equivalent and they are marked as {100} planes. For illustration, in Figure 4.3, some of the planes of the {100} family of planes are shown.

## Crystal Structure of Silicon

Crystalline silicon forms a covalently bonded structure, the diamond-cubic structure, which has the same atomic arrangement as carbon in diamond form and belongs to the more general zinc-blend classification.[36] Silicon, with its four covalent bonds, coordinates itself tetrahedrally, and these tetrahedrons make up the diamond-cubic structure. This structure can also be represented as two interpenetrating face-centered cubic lattices, one displaced (1/4,1/4,1/4)a with respect to the other, as shown in Figure 4.4. The structure is face-centered cubic (fcc), but with two atoms in the unit cell. For such a cubic lattice, direction [hkl] is perpendicular to a plane with the three integers (hkl), simplifying further discussions about the crystal orientation, i.e., the Miller indices of a plane perpendicular to the [100] direction are (100). The lattice parameter 'a' for silicon is 5.4309 Å and silicon's diamond-cubic lattice is surprisingly wide open, with a packing density of 34%, compared to 74% for a regular face-centered cubic lattice. The {111} planes present the highest packing density and the atoms are oriented such that three

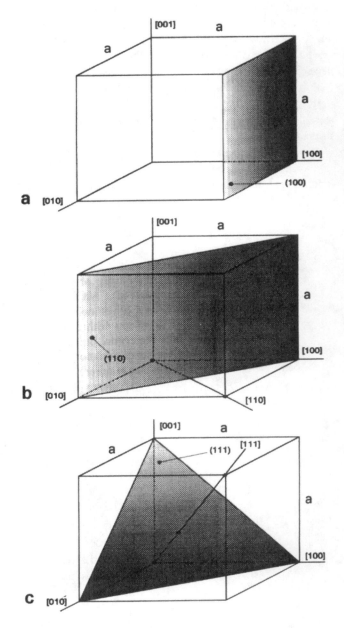

**FIGURE 4.2** Miller indices in a cubic lattice: planes and axes. Shaded planes are: a (100), b (110), c (111).

bonds are below the plane. In addition to the diamond-cubic structure, silicon is known to have several stable high-pressure crystalline phases[37] and a stress-induced metastable phase with a wurtzite-like structure, referred to as diamond-hexagonal silicon. The latter has been observed after ion-implantation[38] and hot indentation.[39]

When ordering silicon wafers, the crystal orientation must be specified. The most common orientations used in the IC industry are the <100> and <111> orientation; in micromachining, <110> wafers are used quite often as well. The <110> wafers break or cleave much more cleanly than other orientations. In fact, it is the only major plane that can be cleaved with exactly perpendicular edges. The <111> wafers are used less, as they cannot be etched anisotropically except when using laser-assisted etching.[40] On a <100> wafer, the <110> direction is often made evident by a flat segment, also called an orientation flat. The precision on the flat is about 3°. The position of the flat on (110)-oriented wafers varies from manufacturer to manufacturer, but often parallels a (111) direction. Flat areas help orientation determination, placement of slices in cassettes and fabrication equipment (large primary flat), and help identify orientation and conductivity type (smaller secondary flat) (see also Chapter 3 under Si growth). Primary and secondary flats on <111> and <100> silicon wafers are indicated in Figure 4.5.

## Geometric Relationships Between Some Important Planes in the Silicon Lattice

To better appreciate the different three-dimensional shapes resulting from anisotropically etched single crystal Si (SCS) and to better understand the section further below on corner compensation, some of the more important geometric relationships

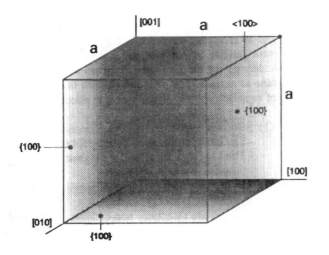

**FIGURE 4.3** Miller indices for some of the planes of the {100} family of planes.

between different planes within the Si lattice need further clarification. We will consider only silicon wafers with a (100) or a (110) as the surface planes. We will also accept, for now, that in anisotropic alkaline etchants the {111} planes, which have the highest atom-packing density, are nonetching compared to the other planes. As the {111} planes are essentially not attacked by the etchant, the sidewalls of an etched pit in SCS will ultimately be bounded by this type of plane, given that the etch time is long enough for features bounded by other planes to be etched away. The types of planes introduced initially depend on the geometry and the orientation of the mask features.

### [100]-Oriented Silicon

In Figure 4.6, the unity cell of a silicon lattice is shown together with the correct orientation of a [100]-type wafer relative to this cell.[41] It can be seen from this figure that intersections of the nonetching {111} planes with the {100} planes (e.g., the wafer surface) are mutually perpendicular and lying along the <110> orientations. Provided a mask opening (say, a rectangle or a square) is accurately aligned with the primary orientation flat, i.e., the [110] direction, only {111} planes will be introduced as sidewalls from the very beginning of the etch. Since the nonetching character of the {111} planes renders an exceptional degree of predictability to the recess features, this is the mask arrangement most often utilized in commercial applications. During etching, truncated pyramids (square mask) or truncated V-grooves (rectangular mask) deepen but do not widen (Figure 4.7). The edges in these structures are <110> directions, the ribs are <211> directions, the sidewalls are {111} planes, and the bottom is a (100) plane parallel with the wafer surface. After prolonged etching, the {111} family of planes is exposed down to their common intersection and the (100) bottom plane disappears, creating a pyramidal pit (square mask) or a V-groove (rectangular mask) (Figure 4.7). As shown in Figure 4.7, no underetching of the etch mask is observed, due to the perfect alignment of the concave oxide mask opening with

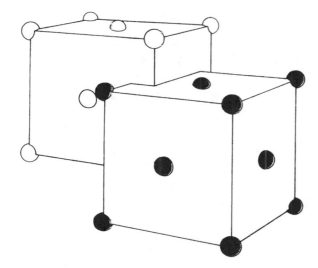
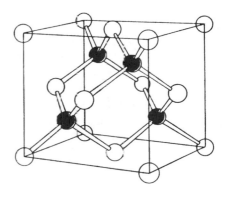

**FIGURE 4.4** The diamond-type lattice can be constructed from two interpenetrating face-centered cubic unit cells. Si forms four covalent bonds, making tetrahedrons.

the <110> direction. Misalignment still results in pyramidal pits, but the mask will be undercut. For a mask opening with arbitrary geometry and orientation (for example, a circle) and for sufficiently long etch times, the anisotropically etched recess in a {100} wafer is pyramidal with a base perfectly circumscribing the circular mask opening.[41] Convex corners (>180°) in a mask opening will always be completely undercut by the etchant after sufficiently long etch times. This can be disadvantageous (for example, when attempting to create a mesa rather than a pit) or it can be advantageous for undercutting suspended cantilevers or bridges. In the section of this chapter on corner

compensation, the issue of undercutting will be addressed in detail. The slope of the sidewalls in a cross-section perpendicular to the wafer surface and to the wafer flat is determined by the angle α as in Figure 4.6 depicting the off-normal angle of the intersection of a (111) sidewall and a (110) cross-secting plane, and can be calculated from:

$$\text{or } R_{111} = ($$

4.1

with $L = a \times \dfrac{\sqrt{2}}{2}$  or $\alpha = \arctan \dfrac{\sqrt{2}}{2} = 35.26°$, or $54.74°$ for the complementary angle. The tolerance on this slope is determined by the alignment accuracy of the wafer surface with respect to the (100) plane. Wafer manufacturers typically specify this misalignment to 1° (0.5° in the best cases).

The width of the rectangular or square cavity bottom plane, $W_0$, in Figure 4.8, aligned with the <110> directions, is completely defined by the etch depth, z, the mask opening, $W_m$, and the above-calculated sidewall slope:

$$W_0 = W_m - 2\cot(54.74°)z$$

or

$$W_0 = W_m - \sqrt{2}\,z$$

4.2

The larger the opening in the mask, the deeper the point at which the {111} sidewalls of the pit intersect. The etch stop at the {111} sidewalls' intersection occurs when the depth is about 0.7 times the mask opening. If the oxide opening is wide enough, $W_m > 849$ μm (for a typical 6-inch wafer with thickness $t_{si} = z = 600$ μm), the {111} planes do not intersect within the wafer.

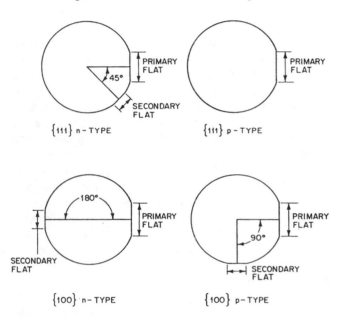

FIGURE 4.5   Primary and secondary flats on silicon wafers.

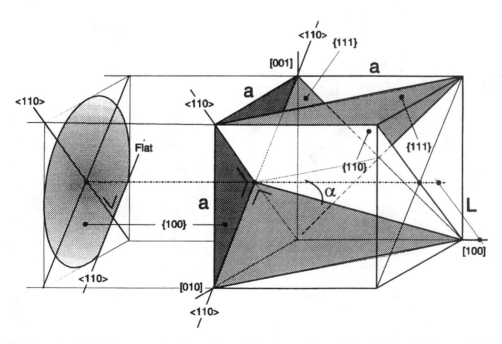

FIGURE 4.6   (100) silicon wafer with reference to the unity cube and its relevant planes. (From Peeters, E., *Process Development for 3D Silicon Microstructures with Application to Mechanical Sensor Design*, KUL, Belgium, 1994. With permission.)

The etched pit in this particular case extends all the way through the wafer, creating a *small orifice or via.* If a high density of such vias through the Si is required, the wafer must be made very thin.

Corners in an anisotropically etched recess are defined by the intersection of crystallographic planes, and the resulting corner radius is essentially zero. This implies that the size of a silicon

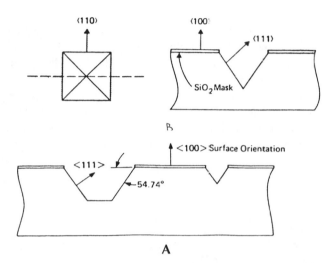

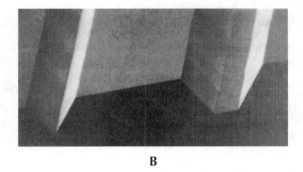

**A**

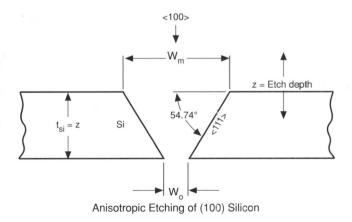

**B**

**FIGURE 4.7** Anisotropically etched features in a (100) wafer with (A) Square mask (schematic) and (B) Rectangular mask (scanning electron microscope micrograph of resulting actual V- and U-grooves).

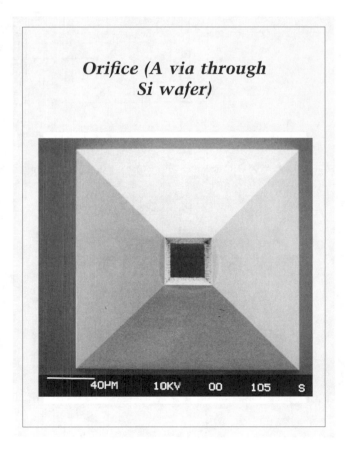

diaphragm is very well defined, but it also introduces a considerable stress concentration factor. The influence of the zero corner radius on the yield load of diaphragms can be studied with finite element analysis (FEA).

One way to obtain vertical sidewalls instead of 54.7° sidewalls using a [100]-oriented Si wafer is illustrated in Figure 4.9. It can be seen in this figure that there are {100} planes perpendicular to the wafer surface and that their intersections with the wafer surface are <100> directions. These <100> directions enclose a 45° angle with the wafer flat (i.e., the <110> directions). By aligning the mask opening with these <100> orientations, {100} facets are initially introduced as sidewalls. The {110} planes etch faster than the {100} planes and are not introduced. As the bottom and sidewall planes are all from the same {100} group, lateral underetch equals the vertical etch rate and rectangular channels, bounded by slower etching {100} planes, result (Figure 4.10). Since the top of the etched channels is exposed to the etchant longer than the bottom, one might have expected the channels in Figure 4.10 to be wider at the top than at the bottom. With some minor corrections in Peeters' derivation,[41] we can use his explanation for why the sidewalls stay vertical. Assume the width of the mask opening to be $W_m$. At a given depth, z, into the wafer, the underlying Si is no longer masked by $W_m$, but rather by the intersection of the previously formed {100} facets with the bottom surface. The width of this new mask is larger than the lithography mask $W_m$ by the amount the latter is being undercut. Let's call the new mask width $W_2$ the effective mask width at a depth z. The relation

**Orifice (A via through Si wafer)**

**FIGURE 4.8** Relation of bottom cavity plane width with mask opening width.

between $W_m$ and $W_z$ is given by the lateral etch rate of a {100} facet and the time that facet was exposed to the etchant at depth z, i.e.:

$$W_z = W_m + 2R_{xy}\Delta t_z \qquad 4.3$$

where $R_{xy}$ is the lateral underetch rate (i.e., etch rate in the x-y plane) and $\Delta t_z$ the etch time at depth z. The underetching, $U_z$, of the effective mask opening $W_z$ is given by:

$$U_z = TR_{xy} - R_{xy}\Delta t_z \qquad 4.4$$

where T is the total etch time so far. The width of the etched pit, $W_{tot}$, at depth z is further given by the sum of $W_m$ and twice the underetching for that depth:

$$W_{tot} = W_z + 2U_z = W_m + 2TR_{xy} \qquad 4.5$$

Or, since T can also be written as the measured total etch depth z divided by the vertical etch rate $R_z$, Equation 4.5 can be rewritten as

$$W_{tot} = W_m + 2z\left(R_{xy}/R_z\right) \text{ or since } R_{xy} = R_z,$$

$$W_{tot} = W_m + 2z \qquad 4.6$$

The width of the etched recess is therefore equal to the photolithographic mask width plus twice the etch depth — independent of that etch depth, in other words; the walls remain vertical independent of the depth z.

For sufficiently long etch times, {111} facets take over eventually from the vertical {100} facets. These inward sloping {111} facets are first introduced at the corners of a rectangular mask and grow larger at the expense of the vertical sidewalls until

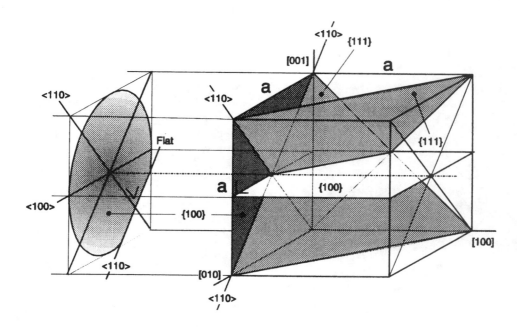

FIGURE 4.9    (100) silicon wafer with <100> mask-aligned features introduces vertical sidewalls. (From Peeters, E., *Process Development for 3D Silicon Microstructures with Application to Mechanical Sensor Design*, KUL, Belgium, 1994. With permission.)

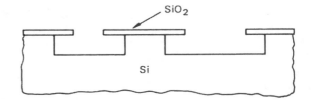

FIGURE 4.10    Vertical sidewalls in a (100) wafer.

the latter ultimately disappear altogether. Alignment of mask features with the <100> directions in order to obtain vertical sidewalls in {100} wafers, therefore, is not very useful for the fabrication of diaphragms. However, it can be very effective for anticipating the undercutting of convex corners on {100} wafers. This useful aspect will be revisited when discussing corner compensation.

### [110]-Oriented Silicon

In Figure 4.11, we show a unit cell of Si properly aligned with the surface of a (110) Si wafer. This drawing will enable us to predict the shape of an anisotropically etched recess on the basis of elementary geometric crystallography. Whereas the intersections of the {111} planes with the (100) wafer surface are mutually perpendicular, here they enclose an angle γ in the (110) plane. Moreover, the intersections are not parallel (<110>) or perpendicular (<100>) to the main wafer flat (assumed to be <110> in this case), but rather enclose angles δ or δ + γ. It follows that a mask opening that will not be undercut (i.e., oriented such that resulting feature sidewalls are exclusively made up by {111} planes) cannot be a rectangle aligned with the flat, but must be a parallelogram skewed by γ – 90° and δ degrees off-axis. The angles γ and δ are calculated as follows[41] (see Figure 4.11 and Figure 4.12):

$$\tan\beta = \frac{\frac{1}{2}a\frac{\sqrt{2}}{2}}{\frac{a}{2}} = \frac{\sqrt{2}}{2} \qquad 4.7$$

$$\gamma = 180° - 2\beta = 180° - 2\arctan\left(\frac{\sqrt{2}}{2}\right) = 109.47°$$

$$\delta = 90° + \beta = 90° + \arctan\left(\frac{\sqrt{2}}{2}\right) = 125.26° \qquad 4.8$$

$$\phi = 270° - \delta = 144.74°$$

From Figure 4.11, it can also be seen that the {111} planes are oriented perpendicular to the (110) wafer surface. This makes it possible to etch pits with vertical sidewalls (Figure 4.12). The bottom of the pit shown here is bounded by {110} and/or {100} planes, depending on the etch time. At short etch times, one mainly sees a flat {110} bottom. As the {110} planes are etching slightly faster than the {100} planes, the flat {110} bottom is getting smaller and smaller and a V-shaped bottom bounded by {100} planes eventually results. The angle ε as shown in Figure 4.12 equals 45°, being the angle enclosed by the intersections of a {100} and a {110} bottom plane. The general rule does apply that an arbitrary window opening is circumscribed by a parallelogram with the given orientation and skewness for sufficiently long etch times. Another difference between (100)- and (110)-oriented Si wafers is that on the (110) wafers it is possible to etch under microbridges crossing at a 90° angle a shallow V-groove (formed by {111} planes). In order to undercut a bridge on a (100) plane, the bridge cannot be perpendicular to the V-groove; it must be oriented slightly off normal.[42]

### Selection of [100]- or [110]-Oriented Silicon

In Table 4.1, we compare the main characteristics of etched features in [100]- and [110]-oriented wafers. This guide

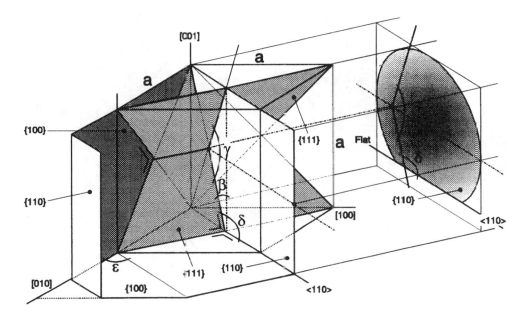

FIGURE 4.11  (110) silicon wafer with reference to the unity cube and its relevant planes. The wafer flat is in a <110> direction. (From Peeters, E., *Process Development for 3D Silicon Microstructures with Application to Mechanical Sensor Design*, KUL, Belgium, 1994. With permission.)

can help decide which orientation to use for a specific micro-fabrication application at hand.

From this table it is obvious that for membrane-based sensors, [100] wafers are preferred. The understanding of the geometric considerations with [110] wafers is important, though, if one wants to fully appreciate all the possible single crystal silicon (SCS) micromachined shapes, and it is especially helpful to understand corner compensation schemes (see below). Moreover, all processes for providing dielectric isolation require that the silicon be separated into discrete regions. To achieve a high component density with anisotropic etches on (100) wafers, the silicon must be made very thin because of the aspect ratio limitations due to the sloping walls (see above). With vertical sidewall etching in a (100) wafer, the etch mask is undercut in all directions to a distance approximately equal to the depth of the etching. Vertical etching in (110) surfaces relaxes the etching requirement dramatically and

enables more densely packed structures such as beam leads or image sensors. Kendall describes and predicts a wide variety of applications for (110) wafers such as fabrication of trench capacitors, vertical multi-junction solar cells, diffraction gratings, infrared interference filters, large area cathodes, and filters for bacteria.[43,44]

## Silicon as a Substrate and Structural Material

### Silicon as Substrate

For many mechanical sensor applications, single crystal Si, based on its intrinsic mechanical stability and the feasibility of integrating sensing and electronics on the same substrate, often presents an excellent substrate choice. For chemical sensors, on

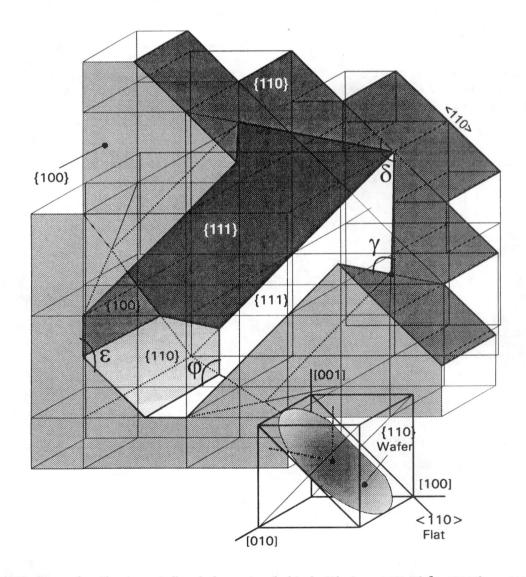

**FIGURE 4.12** (110) silicon wafer with anisotropically etched recess inscribed in the Si lattice. $\gamma = 109.47°$; $\delta = 125.26°$; $\varphi = 144.74°$; and $\varepsilon = 45°$. (From Peeters, E., *Process Development for 3D Silicon Microstructures with Application to Mechanical Sensor Design*, KUL, Belgium, 1994. With permission.)

**TABLE 4.1**  Selection of Wafer Type

| [100] Orientation | [110] Orientation |
|---|---|
| Inward sloping walls (54.74°) | Vertical {111} walls |
| The sloping walls cause a lot of lost real estate | Narrow trenches with high aspect ratio are possible |
| Flat bottom parallel to surface is ideal for membrane fabrication | Multifaceted cavity bottom ({110} and {100} planes) makes for a poor diaphragm |
| Bridges perpendicular to a V-groove bound by (111) planes cannot be underetched | Bridges perpendicular to a V-groove bound by (111) planes can be undercut |
| Shape and orientation of diaphragms convenient and simple to design | Shape and orientation of diaphragms awkward and more difficult to design |
| Diaphragm size, bounded by nonetching {111} planes, is relatively easy to control | Diaphragm size is difficult to control; the <100> edges are not defined by nonetching planes |

**TABLE 4.2**  Performance Comparison of Substrate Materials

| Substrate | Cost | Metallization | Machinability |
|---|---|---|---|
| Ceramic | Medium | Fair | Poor |
| Plastic | Low | Poor | Fair |
| Silicon | High | Good | Very good |
| Glass | Low | Good | Poor |

the other hand, Si, with few exceptions,* is merely the substrate and the choice is not always that straightforward.

In Table 4.2, we show a performance comparison of substrate materials in terms of cost, metallization ease, and machinability. Both ceramic and glass substrates are difficult to machine, and plastic substrates are not readily amenable to metallization. Silicon has the highest material cost per unit area, but this cost often can be offset by the small feature sizes possible in a silicon implementation. Si with or without passivating layers, due to its extreme flatness, relative low cost, and well established coating procedures, often is the preferred substrate especially for thin films. A lot of thin-film deposition equipment is built to accommodate Si wafers and, as other substrates are harder to accommodate, this lends Si a convenience advantage. There is also a greater flexibility in design and manufacturing with silicon technology compared to other substrates. In addition, although much more expensive, the initial capital equipment investment is not product specific. Once a first product is on line, a next generation or new products will require changes in masks and process steps but not in the equipment itself.

Disadvantages of using Si usually are most pronounced with increasing device size and low production volumes and when electronics do not need or cannot be incorporated on the same Si substrate. The latter could be either for cost reasons (e.g., in the case of disposables such as glucose sensors) or for technological reasons (e.g., the devices will be immersed in conductive liquids or they must operate at temperatures above 150°C).

An overwhelming determining factor for substrate choice is the final package of the device. A chemical sensor on an insulating substrate almost always is easier to package than a piece of Si with conductive edges in need of insulation.

Sensor packaging is so important in sensors that as a rule sensor design should start from the package rather than from the sensor. In this context, an easier to package substrate has a

huge advantage. The latter is the most important reason why recent chemical sensor development in industry has retrenched from a move towards integration on silicon in the 1970s and early 1980s to a hybrid thick film on ceramic approach in the late 1980s and early 1990s. In academic circles in the U.S. chemical sensor integration with electronics continued until the late 1980s; in Europe and Japan, such efforts are still going on.[45]

In Chapter 8 we further refine decision criteria about what substrate material to use for a given micromachining application (Table 8.1) and we compare Si with other important sensor substrate materials (Table 8.3).

## Silicon as a Structural Element in Mechanical Sensors

### Introduction

In mechanical sensors the active structural elements convert a mechanical external input signal (force, pressure, acceleration, etc.) into an electrical signal output (voltage, current, or frequency). The transfer functions in mechanical devices describing this conversion are mechanical, electro-mechanical, and electrical.

In the mechanical conversion, a given external load is concentrated and maximized in the active member of the sensor. Structurally active members are typically high aspect ratio elements such as suspended beams or membranes. The electromechanical conversion is the transformation of the mechanical quantity into an electrical quantity such as capacitance, resistance, charge, etc. Often the electrical signal needs further electrical conversion into an output voltage, frequency, or current. For electrical conversion into an output voltage, a Wheatstone bridge may be used as in the case of a piezoresistive sensor, and a charge amplifier may be used in the case of a piezoelectric sensor. To optimize all three transfer functions, detailed electrical and mechanical modeling is required. One of the most important inputs required for the mechanical models are the experimentally determined independent elasticity constants or moduli. In what follows, we describe what makes Si such an important structural element in mechanical sensors and present its elasticity constants.

### Important Characteristics of Mechanical Structural Elements: Stress-Strain Curve and Elasticity Constants

Yield, tensile strength, hardness, and creep of a material all relate to the elasticity curve, i.e., the stress-strain diagram of the material as shown in Figure 4.13. For small strain values, Hooke's law applies, i.e., stress (force per unit area, $N/m^2$) and

---

*    A notable exception is the ion-sensitive field effect transistor (ISFET) where the Si space charge is modulated by the presence of chemicals for which the ISFET chemical coating is sensitive (see Chapter 10).

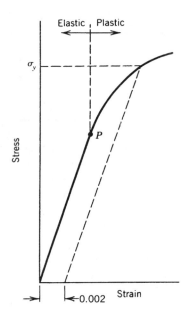

**FIGURE 4.13** Typical stress-strain behavior for a material showing elastic and plastic deformations, the proportional limit P, and the yield strength $\sigma_y$, as determined using the 0.002 strain offset method (see text).

strain (displacement per unit length, dimensionless) are proportional and the stress-strain curve is linear, with a slope corresponding to the elastic modulus E (Young's modulus-N/m²). This regime as in Figure 4.13 is marked as the elastic deformation regime. For isotropic media such as amorphous and polycrystalline materials, the applied axial force per unit area or tensile stress, $\sigma_a$, and the axial or tensile strain, $\varepsilon_a$, are thus related as:

$$\sigma_a = E\varepsilon_a \qquad 4.9$$

with $\varepsilon_a$ given by the dimensionless ratio of $L_2 - L_1/L_1$, i.e., the ratio of the wire's elongation to its original length. The elastic modulus may be thought of as stiffness or a material's resistance to elastic deformation. The greater the modulus, the stiffer the material. A tensile stress usually also leads to a lateral strain or contraction (Poisson effect), $\varepsilon_1$, given by the dimensionless ratio of $D_2 - D_1/D_1$ ($\Delta D/D_1$), where $D_1$ is the original wire diameter and $\Delta D$ is the change in diameter under axial stress (see Figure 4.14). The Poisson ratio is the ratio of lateral over axial strain:

$$\nu = -\frac{\varepsilon_1}{\varepsilon_a} \qquad 4.10$$

The minus sign indicates a contraction of the material. For most materials, $\nu$ is a constant within the elastic range. Normally, some slight volume change does accompany the deformation, and, consequently, $\nu$ is smaller than 0.5. The magnitude of the Young's modulus ranges from $4.1 \times 10^4$ MPa (the N/m² unit is called the Pascal, Pa) for magnesium to $40.7 \times 10^4$ MPa for tungsten and 144 GPa for Invar. With increasing temperature, the elastic modulus diminishes. The Poisson ratios for aluminum and cast steel are 0.34 and 0.28, respectively. The value of $\nu$

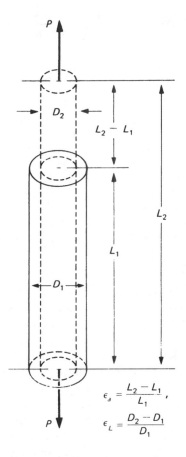

$$\varepsilon_a = \frac{L_2 - L_1}{L_1},$$

$$\varepsilon_L = \frac{D_2 - D_1}{D_1}$$

**FIGURE 4.14** Metal wire under axial or normal stress; normal stress creates both elongation and lateral contraction.

fluctuates for different materials over a relatively narrow range. Generally, it is on the order of 0.25 to 0.35. In extreme cases, values as low as 0.1 (certain types of concrete) and as high as 0.5 (rubber) occur. A value of 0.5 is the largest value possible. It is attained by materials during plastic flow and indicates a constant volume. For an elastic isotropic medium subjected to a triaxial state of stress, the resulting strain component in the x direction, $\varepsilon_x$, is given by the summation of elongation and contraction:

$$\varepsilon_x = \frac{1}{E}\left[\sigma_x - \nu\left(\sigma_y + \sigma_z\right)\right] \qquad 4.11$$

and so on for the y and z directions (3 equations in total).

For an analysis of mechanical structures we must consider not only compressional and tensile strains but also shear strains. Whereas normal stresses create elongation plus lateral contraction with accompanying volume changes, shear stresses (e.g., by twisting a body) create shape changes without volume changes, i.e., shear strains. The one-dimensional shear strain, $\gamma$, is produced by the shear stress, $\tau$ (N/m²). For small strains, Hooke's law may be applied again:

$$\gamma = \frac{\tau}{G} \qquad 4.12$$

where G is called the elastic shear modulus or the modulus of rigidity. For any three-dimensional state of shear stress, three equations of this type will hold. Isotropic bodies are characterized by two independent elastic constants only, since the shear modulus G, it can be shown,[46] relates the Young's modulus and the Poisson ratio as:

$$G = \frac{E}{2(1+\nu)} \qquad 4.13$$

Crystal materials, whose elastic properties are anisotropic, require more than two elastic constants, the number increasing with decreasing symmetry. Cubic crystals (bcc, fcc), for example, require three elastic constants, hexagonal crystals require five, and materials without symmetry require 21.[5,46] The relation between stresses and strains is more complex in this case and depends greatly on the spatial orientation of these quantities with respect to the crystallographic axes. Hooke's law in the most generic form is expressed in two formulas:

$$\sigma_{ij} = E_{ijkl} \cdot \varepsilon_{kl} \quad \text{and} \quad \varepsilon_{ij} = S_{ijkl} \cdot \sigma_{kl} \qquad 4.14$$

where $\sigma_{ij}$ and $\sigma_{kl}$ are stress tensors of rank 2 expressed in N/m²; $\varepsilon_{ij}$ and $\varepsilon_{kl}$ are strain tensors of rank 2 and are dimensionless; $E_{ijkl}$ is a stiffness coefficient tensor of rank 4 expressed in N/m²; and $S_{ijkl}$ is a compliance coefficient tensor of rank 4 expressed in m²/N. The first expression is analogous to Equation 4.9 and the second expression is the inverse, giving the strains in terms of stresses. The tensor representations in Equation 4.14 can also be represented as two matrices:

$$\sigma_m = \sum_{n=1}^{6} E_{mn} \varepsilon_n \quad \text{and} \quad \varepsilon_m = \sum_{n=1}^{6} S_{mn} \sigma_n \qquad 4.15$$

Components of tensors $E_{ijkl}$ and $S_{ijkl}$ are substituted by elements of the matrices $E_{mn}$ and $S_{mn}$, respectively. To convert the ij indices to m and the kl indices to n, the following scheme applies:

11 → 1, 22 → 2, 33 → 3, 23 and 32 → 4, 13 and 31 → 5, 12 and 21 → 6, $E_{ijkl} → E_{mn}$ and $S_{ijkl} → S_{mn}$ when m and n = 1,2,3; $2S_{ijkl} → S_{mn}$ when m or n = 4,5,6; $4S_{ijkl} → S_{mn}$ when m and n = 4,5,6; $\sigma_j →$ $\sigma_m$ when m = 1,2,3; and $\varepsilon_{ij} → \varepsilon_m$ when m = 4,5,6

With these reduced indices there are thus six equations of the type:

$$\sigma_x = E_{11}\varepsilon_x + E_{12}\varepsilon_y + E_{13}\varepsilon_z$$
$$+ E_{14}\gamma_{yz} + E_{15}\gamma_{zx} + E_{16}\gamma_{xy} \qquad 4.16$$

and hence 36 moduli of elasticity or $E_{mn}$ stiffness constants. There are also six equations of the type:

$$\varepsilon_x = S_{11}\sigma_x + S_{12}\sigma_y + S_{13}\sigma_z$$
$$+ S_{14}\tau_{yz} + S_{15}\tau_{zx} + S_{16}\tau_{xy} \qquad 4.17$$

defining 36 $S_{mn}$ constants which are called the compliance constants (see also Equation 9.29). It can be shown that the matrices $E_{mn}$ and $S_{mn}$, each composed of 36 coefficients, are symmetrical; hence, a material without symmetry elements has 21 independent constants or moduli. Due to symmetry of crystals, several more of these may vanish until, for our isotropic medium they number two only (E and $\nu$). The stiffness coefficient and compliance coefficient matrices for cubic-lattice crystals with the vector of stress oriented along the [100] axis are given as:

$$E_{mn} = \begin{bmatrix} E_{11} & E_{12} & E_{12} & 0 & 0 & 0 \\ E_{12} & E_{11} & E_{12} & 0 & 0 & 0 \\ E_{12} & E_{12} & E_{11} & 0 & 0 & 0 \\ 0 & 0 & 0 & E_{44} & 0 & 0 \\ 0 & 0 & 0 & 0 & E_{44} & 0 \\ 0 & 0 & 0 & 0 & 0 & E_{44} \end{bmatrix}$$

$$S_{mn} = \begin{bmatrix} S_{11} & S_{12} & S_{12} & 0 & 0 & 0 \\ S_{12} & S_{11} & S_{12} & 0 & 0 & 0 \\ S_{12} & S_{12} & S_{11} & 0 & 0 & 0 \\ 0 & 0 & 0 & S_{44} & 0 & 0 \\ 0 & 0 & 0 & 0 & S_{44} & 0 \\ 0 & 0 & 0 & 0 & 0 & S_{44} \end{bmatrix} \qquad 4.18$$

In cubic crystals, the three remaining independent elastic moduli are usually chosen as $E_{11}$, $E_{12}$, and $E_{44}$. The $S_{mn}$ values can be calculated simply from these $E_{mn}$ values. Expressed in terms of the compliance constants, one can show that $1/S_{11} =$ E = Young's modulus, $-S_{12}/S_{11} = \nu$ = Poisson's ratio, and $1/S_{44}$ = G = shear modulus. In the case of an isotropic material, such as a metal wire, there is an additional relationship:

$$E_{44} = \frac{E_{11} - E_{12}}{2} \qquad 4.19$$

reducing the number of independent stiffnesses constants to two. The anisotropy coefficient $\alpha$ is defined as:

$$\alpha = \frac{2E_{44}}{E_{11} - E_{12}} \qquad 4.20$$

making $\alpha = 1$ for an isotropic crystal. For an anisotropic crystal, the degree of anisotropy is given by the deviation of $\alpha$ from 1. Single crystal silicon has moderately anisotropic elastic properties,[47,48] with $\alpha$ = 1.57. Brantly[47] gives the non-zero stiffness components, referred to the [100] crystal orientation as: $E_{11} = E_{22} = E_{33} = 166 \times 10^9$ N/m², $E_{12} = E_{13} = E_{23} = 64 \times 10^9$ N/m², and $E_{44} = E_{55} = E_{66} = 80 \times 10^9$ N/m²:

$$\begin{matrix} \sigma_x \\ \sigma_y \\ \sigma_z \\ \tau_{xy} \\ \tau_{xz} \\ \tau_{yz} \end{matrix} = \begin{vmatrix} 166(E_{11}) & 64(E_{12}) & 64(E_{12}) & 0 & 0 & 0 \\ 64(E_{12}) & 166(E_{11}) & 64(E_{12}) & 0 & 0 & 0 \\ 64(E_{12}) & 64(E_{12}) & 166(E_{11}) & 0 & 0 & 0 \\ 0 & 0 & 0 & 80(E_{44}) & 0 & 0 \\ 0 & 0 & 0 & 0 & 80(E_{44}) & 0 \\ 0 & 0 & 0 & 0 & 0 & 80(E_{44}) \end{vmatrix} \times \begin{matrix} \varepsilon_x \\ \varepsilon_y \\ \varepsilon_z \\ \gamma_{xy} \\ \gamma_{xz} \\ \gamma_{yz} \end{matrix} \qquad 4.21$$

with σ normal stress, τ shear stress, ε normal strain, and γ shear strain. The values for $E_{mn}$, in Equation 4.21, compare with a Young's modulus of 207 GPa for a low carbon steel. Variations on the values of the elastic constants on the order of 30%, depending on crystal orientation, must be considered; doping level (see below) and dislocation density have minor effects as well. From the stiffness coefficients the compliance coefficients of Si can be calculated as $S_{11} = 7.68 \times 10^{-12}$ m²/N, $S_{12} = -2.14 \times 10^{-12}$ m²/N, and $S_{44} = 12.6 \times 10^{-12}$ m²/N.[49] A graphical representation of elastic constants for different crystallographic directions in Si and Ge is given in Worthman et al.[50] and is reproduced in Figure 4.15.

Figure 4.15A to D displays E and ν for Ge and Si in planes (100) and (110) as functions of direction. Calculations show that E, G, and ν are constant for any direction in the (111) plane. In other words, a plate lying in this plane can be considered as having isotropic elastic properties. A review of independent determinations of the Si stiffness coefficients, with their respective temperature coefficients, is given in Metzger et al.[51] Some of the values from that review are reproduced in Table 4.3. Values for Young's modulus and the shear modulus of Si can also be found in Greenwood[52] and are reproduced in Table 4.4 for the three technically important crystal orientations.

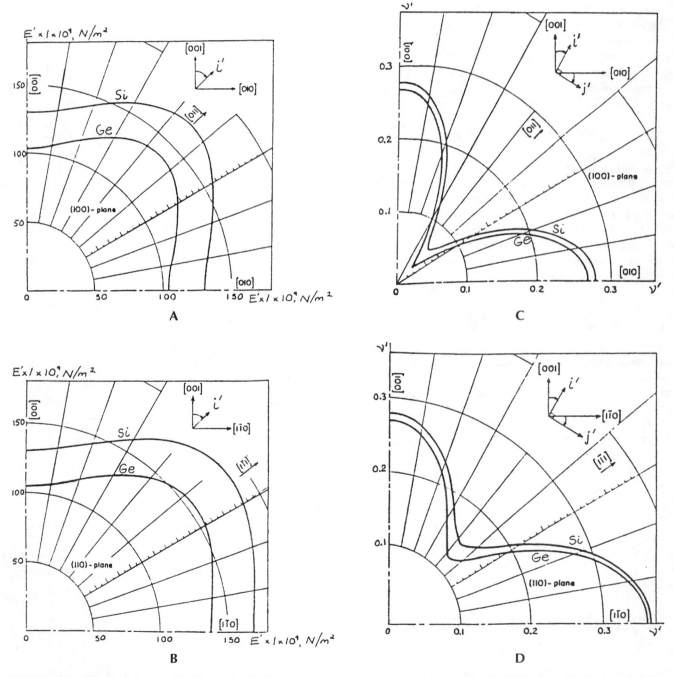

**FIGURE 4.15** Elasticity constants for Si and Ge. (A) Young's modulus as a function of direction in the (100) plane. (B) Young's modulus as a function of direction in the (110) plane. (C) Poisson ratio as a function of direction in the (100) plane. (D) Poisson's ratio as a function of direction in the (110) plane. (From Worthman, J. J. and R. A. Evans, *J. Appl. Phys.*, 36, 153–156, 1965. With permission.)

## Residual Stress

Most properties, such as the Young's modulus, for lightly and highly doped silicon are identical. In Chapter 5 we will see that the Young's modulus for polycrystalline silicon is about 161 GPa. Residual stress and associated stress gradients in highly boron doped single crystal silicon present some controversy. Highly boron-doped membranes, which are usually reported to be tensile, also have been reported compressive.[53,54] From a simple atom-radius argument, one expects that a large number of substitutional boron atoms would create a net shrinkage of the lattice compared to pure silicon and that the residual stress would be tensile with a stress gradient corresponding to the doping gradient. That is, an etched cantilever would be expected to bend up out of the plane of the silicon wafer. Maseeh et al.[54] believe that the appearance of compressive behavior in heavily boron-doped single crystal layers results from the use of an oxide etch mask. They suggest that plastic deformation of the p⁺ silicon beneath the compressively stressed oxide can explain the observed behavior. Ding et al.,[55] who also found compressive behavior for nitride-covered p⁺ Si thin membranes, believe that the average stress in p⁺ silicon is indeed tensile, but great care is required to establish this fact because the combination of heavy boron doping and a high-temperature drive-in under oxidizing conditions can create an apparent reversal of both the net stress (to compressive) and of the stress gradient (opposite to the doping gradient). A proposed explanation is that at the oxide-silicon interface, a thin compressively stressed layer is formed during the drive-in which is not removed in buffered HF. It can be removed by reoxidation and etching in HF, or by etching in KOH.

## Yield, Tensile Strength, Hardness, and Creep

As a material is deformed beyond its elasticity limit, yielding or plastic deformation — permanent, non-recoverable deformation — occurs. The point of yielding in Figure 4.13 is the point of initial departure from linearity of the stress-strain curve and is sometimes called the proportional limit indicated by a letter "P". The Young's modulus of mild steel is ±30,000,000 psi,* and its proportional limit (highest stress in the elastic range) is approximately 30,000 psi. Thus, the maximum elastic strain in mild steel is about 0.001 under a condition of uniaxial stress. This gives an idea as to the magnitude of the strains we are dealing with. A convention has been established wherein a straight line is constructed parallel to the elastic portion of the stress-strain curve at some specified strain offset, usually 0.002. The stress corresponding to the intersection of this line and the stress-strain curve as it bends over in the plastic region is defined as the yield strength, $\sigma_y$ (see Figure 4.13). The

TABLE 4.3   Stiffness Coefficients and Temperature Coefficient of Stiffness for Si[51]

| Stiffness Coefficients (GPa = $10^9$ N/m²) | Temperature Coefficient of Stiffness ($\delta E/\delta T/E$) |
|---|---|
| $E_{11} = 164.8 \pm 0.16$ | $-122 \times 10^{-6}$ |
| $E_{12} = 63.5 \pm 0.3$ | $-162 \times 10^{-6}$ |
| $E_{44} = 79.0 \pm 0.06$ | $-97 \times 10^{-6}$ |

TABLE 4.4   Derived Values for Young's Modulus and Shear Modulus for Si[52]

| Miller Index for Orientation | Young's Modulus (E) (GPa) | Shear Modulus (G) (GPa) |
|---|---|---|
| [100] | 129.5 | 79.0 |
| [110] | 168.0 | 61.7 |
| [111] | 186.5 | 57.5 |

magnitude of the yield strength of a material is a measure of its resistance to plastic deformation. Yield strengths may range from 35 MPa (5000 psi) for a soft and weak aluminum to over 1400 MPa (200,000 psi) for high strength steels. The tensile strength is the stress at the maximum of the stress-strain curve (Figure 4.16). This corresponds to the maximum stress that can be sustained by a structure in tension; if the stress is applied and maintained, fracture will result. Both tensile strength and hardness are indicators of a metal's resistance to plastic deformation. Consequently, they are roughly proportional.[56] Material deformation occurring at elevated temperatures and static material stresses is termed creep. It is defined as a time-dependent and permanent deformation of materials when subjected to a constant load or stress.

Silicon exhibits no plastic deformation or creep below 800°C; therefore, Si sensors are inherently very insensitive to fatigue failure when subjected to high cyclic loads. Silicon sensors have actually been cycled in excess of 100 million cycles with no observed failures. This ability to survive a very large number of duty cycles is due to the fact that there is no energy absorbing or heat generating mechanism due to intergranular slip or movement of dislocations in silicon at room temperature. However, single crystal Si, as a brittle material, will yield catastrophically, when stress beyond the yield limit is applied, rather than deform plastically as metals do (see Figure 4.16). At room temperature, high modulus materials, such as Si, SiO₂, and Si₃N₄ often exhibit linear-elastic behavior at lower strain and transition abruptly to brittle-fracture behavior at higher strain. Plastic deformation in metals is based on stress-induced dislocation generation in the grain boundaries and a subsequent dislocation migration that results in a macroscopic deformation from intergrain shifts in the material. No grain boundaries exist in single crystal silicon (SCS), and plastic deformation can only occur through migration of the defects originally present in the lattice or of those that are generated at the surface. As the number of these defects is very low in SCS, the material can be considered

---

*   In the sensor area it is still mandatory to be versatile in the different unit systems especially with regards to expressing units for quantities such as pressure and stress. In this book we are using mostly Pascal, Pa (= N/m²) but in industry it is still customary to use psi when dealing with metal properties, torr when dealing with vacuum systems, and dyne/cm² when dealing with surface tension.

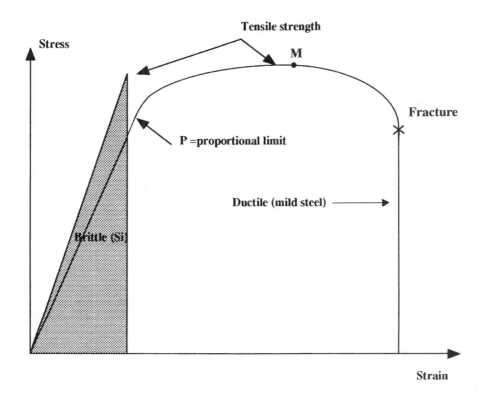

**FIGURE 4.16** Stress-strain curve. Tensile strength of a metal is the stress at the maximum of this curve. Abrupt brittle fracture of a high modulus material with no plastic deformation region like Si is indicated as well.

a perfect elastic material at normal temperatures. Perfect elasticity implies proportionality between stress and strain (i.e., load and flexure) and the absence of irreversibilities or mechanical hysteresis. The absence of plastic behavior also accounts for the extremely low mechanical losses in SCS, which enable the fabrication of resonating structures that exhibit exceptionally high Q-factors. Values of up to $10^8$ in vacuum have been reported. At elevated temperatures, and with metals and polymers at ordinary temperatures, complex behavior in the stress-strain curve can occur. Considerable plasticity can be induced in SCS at elevated temperatures (>800°C), when the mobility of defects in the lattice is substantially increased. Huff and Schmidt[53] actually report a pressure switch exhibiting hysteresis based on buckling of plastically deformed silicon membranes. To eliminate plastic deformation of Si wafers, it is important that during high temperature steps the presence of films that could stress or even warp the wafer in an asymmetric way, typically oxides or nitrides, be avoided.

### Piezoresistivity in Silicon

Piezoresistance is the fractional change in bulk resistivity induced by small mechanical stresses applied to a material. Most materials exhibit piezoresistivity, but the effect is particularly important in some semiconductors (more than an order of magnitude higher than that of metals). Monocrystalline silicon has a high piezoresistivity and, combined with its excellent mechanical and electronic properties, makes a superb material for the conversion of mechanical deformation into an electrical signal. Actually, the history of silicon-based sensors started with

the discovery of the piezoresistance effect in Si and Ge more than four decades ago.[25] The two main classes of piezoresistive sensors are membrane-type structures (typically pressure and flow sensors) and cantilever beams (typically acceleration sensors) with in-diffused resistors (boron, arsenic, or phosphorus) strategically placed in zones of maximum stress.

For a three-dimensional anisotropic crystal, the electrical field vector (E) is related to the current vector (i) by a 3-by-3 resistivity tensor.[49] Experimentally the nine coefficients are always found to reduce to six, and the symmetric tensor is given by:

$$\begin{bmatrix} E_1 \\ E_2 \\ E_3 \end{bmatrix} = \begin{bmatrix} \rho_1 & \rho_6 & \rho_5 \\ \rho_6 & \rho_2 & \rho_4 \\ \rho_5 & \rho_4 & \rho_3 \end{bmatrix} \cdot \begin{bmatrix} i_1 \\ i_2 \\ i_3 \end{bmatrix} \qquad 4.22$$

For the cubic Si lattice, with the axes aligned with the <100> axes, $\rho_1$, $\rho_2$, and $\rho_3$ define the dependence of the electric field on the current along the same direction (one of the <100> directions); $\rho_4$, $\rho_5$, and $\rho_6$ are cross-resistivities, relating the electric field to the current along a perpendicular direction.

The six resistivity components in Equation 4.22 depend on the normal ($\sigma$) and shear ($\tau$) stresses in the material as defined in the preceding section. Smith[25] was the first to measure the resistivity coefficients $\pi_{11}$, $\pi_{12}$, and $\pi_{44}$ for Si at room temperature. Table 4.5 lists Smith's results.[25] The piezoresistance coefficients are largest for $\pi_{11}$ in n-type silicon and $\pi_{44}$ in p-type silicon, about $-102.10^{-11}$ and $138.10^{-11}$ Pa$^{-1}$, respectively.

**TABLE 4.5**  Resistivity and Piezoresistance at Room Temperature[25,49]

|  | $\rho$ ($\Omega$ cm) | $\pi_{11}$[a] | $\pi_{12}$[a] | $\pi_{44}$[a] |
|---|---|---|---|---|
| p-Si | 7.8 | +6.6 | −1.1 | +138.1 |
| n-Si | 11.7 | −102.2 | +53.4 | −13.6 |

[a] Expressed in $10^{-12}$ cm² dyne⁻¹ or $10^{-11}$ Pa¹.

Resistance change can now be calculated as a function of the membrane or cantilever beam stress. The contribution to resistance changes from stresses that are longitudinal ($\sigma_l$) and transverse ($\sigma_t$) with respect to the current flow is given by:

$$\frac{\Delta R}{R} = \sigma_l \pi_l + \sigma_t \pi_t \qquad 4.23$$

where

$\sigma_l$ = Longitudinal stress component, i.e., stress component parallel to the direction of the current.

$\sigma_t$ = Transversal stress component, i.e., the stress component perpendicular to the direction of the current.

$\pi_l$ = Longitudinal piezoresistance coefficient.

$\pi_t$ = Transversal piezoresistance coefficient.

The piezoresistance coefficients $\pi_l$ and $\pi_t$ for (100) silicon as a function of crystal orientation are reproduced from Kanda in Figure 4.17A (for p-type) and B (for n-type).[57] By maximizing the expression for the stress-induced resistance change in Equation 4.23, one optimizes the achievable sensitivity in a piezoresistive silicon sensor.

The orientation of a membrane or beam is determined by its anisotropic fabrication. The surface of the silicon wafer is usually a (100) plane; the edges of the etched structures are intersections of (100) and (111) planes and are thus <110> directions. p-Type piezoresistors are most commonly used because the orientation of maximum piezoresistivity (<110>) happens to coincide with the edge orientation of a conventionally etched diaphragm and because the longitudinal coefficient is roughly equal in magnitude but opposite in sign as compared to the transverse coefficient (Figure 4.17A).[41] With the values in Table 4.5, $\pi_l$ and $\pi_t$ now can be calculated numerically for any orientation. The longitudinal piezoresistive coefficient in the <110> direction is $\pi_l = 1/2(\pi_{11} + \pi_{12} + \pi_{44})$. The corresponding transverse coefficient is $\pi_t = -1/2(\pi_{11} + \pi_{12} - \pi_{44})$. From Table 4.5 we know that, for p-type resistors, $\pi_{44}$ is more important than the other two coefficients and Equation 4.23 is approximated by:

$$\frac{\Delta R}{R} = \frac{\pi_{44}}{2}\left(\sigma_l - \sigma_t\right) \qquad 4.24$$

For n-type resistors, $\pi_{44}$ can be neglected, and we obtain:

$$\frac{\Delta R}{R} = \frac{\pi_{11} + \pi_{12}}{2}\left(\sigma_l + \sigma_t\right) \qquad 4.25$$

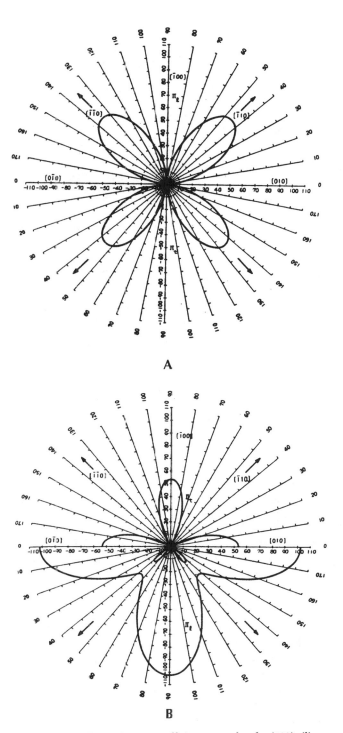

**FIGURE 4.17**  Piezoresistance coefficients $\pi_l$ and $\pi_t$ for (100) silicon. (A) For p-type in the (001) plane ($10^{-12}$ cm²/dyne). (B) For n-type in the (001) plane ($10^{-12}$ cm²/dyne). (From Kanda, Y., *IEEE Trans. Electr. Dev.*, ED-29, 64–70, 1982. With permission.)

Equations 4.24 and 4.25 are valid only for uniform stress fields or if the resistor dimensions are small compared with the membrane or beam size. When stresses vary over the resistors they have to be integrated, which is most conveniently done by computer simulation programs.

To convert the piezoresistive effect into a measurable electrical signal, a Wheatstone bridge is often used. A balanced Wheatstone bridge configuration is constructed as in Figure 4.18A by

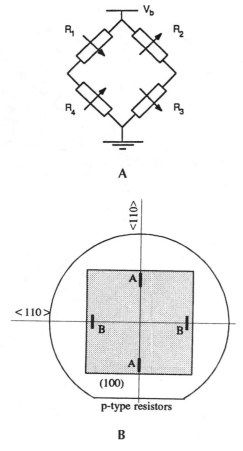

FIGURE 4.18  Measuring on a membrane with piezoresistors. (A) Wheatstone-bridge configuration of four in-diffused piezoresistors. The arrows indicate resistance changes when the membrane is bent downward. (B) Maximizing the piezoresistive effect with p-type resistors. The A resistors are stressed longitudinally and the B resistors are stressed transversally. (From Peeters, E., Ph.D. Thesis, KUL, Belgium, 1994. With permission.)

locating four p-piezoresistors midway along the edges of a square diaphragm as in Figure 4.18B (location of maximum stress). Two resistors are oriented so that they sense stress in the direction of their current axes and two are placed to sense stress perpendicular to their current flow. Two longitudinally stressed resistors (A) are balanced against two transversally stressed resistors (B); two of them increase in value and the other two decrease in value upon application of a stress. In this case, from Equation 4.24,

$$\frac{\Delta R}{R} \approx 70 \cdot 10^{-11}\left(\sigma_1 - \sigma_t\right)$$  4.26

with σ in Pa. For a realistic stress pattern where $\sigma_1$ = 10 MPa and $\sigma_t$ = 50 MPa, Equation 4.26 gives us a $\Delta R/R \approx$ 2.8%.[41]

By varying the diameter and thickness of the silicon diaphragms, piezoresistive sensors in the range of 0 to 200 MPa have been made. The bridge voltages are usually between 5 and 10 volts, and the sensitivity may vary from 10 mV/kPa for low pressure to 0.001 mV/kPa for high pressure sensors.

Peeters[41] shows how a more sensitive device could be based on n-type resistors when all the n-resistors oriented along the <100> direction are subjected to an uniaxial stress pattern in the longitudinal axis, as shown in Figure 4.19. The overall maximum piezoresistivity coefficient ($\pi_1$ in the <100> direction) is substantially higher for n-silicon than it is for p-type silicon in any direction (maximum $\pi_t$ and $\pi_1$ in the <100> direction, Figure 4.17B). Exploitation of these high piezoresistivity coefficients is less obvious, though, since the resistor orientation for maximum sensitivity (<100>) is rotated over 45° with respect to the <110> edges of an anisotropically etched diaphragm. Also evident from Figure 4.17B is that a transversally stressed resistor cannot be balanced against a longitudinally stressed resistor. Peeters has circumvented these two objections by an uniaxial, longitudinal stress pattern in the rectangular diaphragm represented in Figure 4.19. With a (100) substrate and a <100> orientation (45° to wafer flat) we obtain:

$$\frac{\Delta R}{R} \approx 53 \cdot 10^{-11} \cdot \sigma_t - 102 \cdot 10^{-11} \cdot \sigma_1$$  4.27

with σ in Pa. Based on Equation 4.27, with $\sigma_t$ = 10 MPa and $\sigma_1$ = 50 MPa, $\Delta R/R \approx$ −4.6%. In the proposed stress pattern it is important to minimize the transverse stress by making the device truly uniaxal as the longitudinal and transverse stress components have opposite effects and can even cancel out. In practice, a pressure sensor with an estimated 65% gain in pressure sensitivity over the more traditional configurations could be made in the case of an 80% uniaxiality.[41]

The piezoresistive effect is often described in terms of the gauge factor, G, defined as:

$$G = \frac{1}{\varepsilon} \frac{\Delta R}{R}$$  4.28

which is the relative resistance change divided by the applied strain. The gauge factor of a metal strain gauge is typically

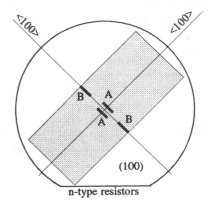

FIGURE 4.19  Higher pressure sensitivity by strategic placement of in-diffused piezoresistors proposed by Peeters.[41] The n-resistors are stressed longitudinally with the A resistors under tensile stress, and the B resistors under compressive stress. (From Peeters, E., Ph.D. Thesis, KUL, Belgium, 1994. With permission.)

around 2, for single crystal Si it is 90, and for poly-crystalline Si it is about 30 (see also Chapter 5).

## Thermal Properties of Silicon

In Figure 4.20, the expansion coefficient of Si, W, SiO₂, Ni-Co-Fe alloy and Pyrex® is plotted vs. absolute temperature. Single crystal silicon has a high thermal conductivity (comparable with metals such as steel and aluminum) and a low thermal expansion coefficient. Its thermal expansion coefficient is closely matched to Pyrex® glass but exhibits considerable temperature dependence. A good match in thermal expansion coefficient between the device wafer (e.g. Si) and the support substrate (e.g. Pyrex) is required. A poor match introduces stress, which degrades the device performance. This makes it difficult to fabricate composite structures of Pyrex and Si that are stress-free over a wide range of temperatures. Drift in silicon sensors often stems from packaging. In this respect, several types of stress-relief, subassemblies for stress-free mounting of the active silicon parts play a major role; using silicon as the support for silicon sensors is highly desirable. The latter aspect will be addressed in more detail when discussing anodic bonding of Pyrex® glass to Si and fusion bonding of Si to Si in Chapter 8.

Although the Si band-gap is relatively narrow, by employing silicon on insulator (SOI) wafers (see Chapter 5) high temperature sensors can be fashioned. For the latter application, relatively highly doped Si, which is relatively linear in its temperature coefficient of resistance and sensitivity over a wide range, typically is employed.

When fabricating thermally isolated structures on Si, such as the miniature heater element in Figure 3.43, the large thermal conductivity of Si poses a considerable problem as the major heat leak occurs through the Si material. For thermally isolated structures, machining in glass or quartz with their lower thermal conductivity represents an important alternative.

# Wet Isotropic and Anisotropic Etching

## Wet Isotropic and Anisotropic: Empirical Observations

### Introduction

Wet etching of Si is used mainly for cleaning, shaping, polishing, and characterizing structural and compositional features.[8] Wet chemical etching provides a higher degree of selectivity than dry etching techniques. Wet etching often is also faster; compare a few microns to several tens of microns per minute for isotropic etchants and about 1 μm/min for anisotropic etchants vs. 0.1 μm/min in typical dry etching. More recently though, with ECR dry etching, rates of up to 6 μm/min were achieved (see Chapter 2). Modification of wet etchant and/or temperature can alter the selectivity to silicon dopant concentration and type and, especially when using alkaline etchants, to crystallographic orientation. Etching proceeds by reactant transport to the surface (1), surface reaction (2), and reaction product transport away from the surface (3). If (1) or (3) is rate determining, etching is diffusion limited and may be increased by

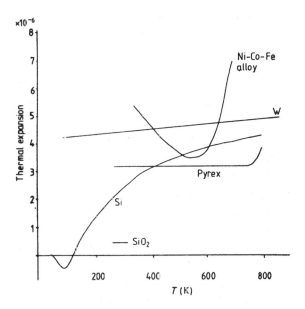

**FIGURE 4.20** Thermal expansion coefficient vs. absolute temperature. (From Greenwood, J.C., *J. Phys. E: Sci. Instrum.*, 21 1114–1128, 1988. With permission.)

stirring. If (2) is the rate-determining step, etching is reaction rate limited and depends strongly on temperature, etching material, and solution composition. Diffusion-limited processes have lower activation energies (of the order of a few Kcal/mol) than reaction-rate controlled processes and therefore are relatively insensitive to temperature variations. In general, one prefers reaction rate limitation as it is easier to reproduce a temperature setting than a stirring rate. The etching apparatus needs to have both a good temperature controller and a reliable stirring facility.[58,59]

Isotropic etchants, also polishing etchants, etch in all crystallographic directions at the same rate; they usually are acidic, such as HF/HNO₃/CH₃COOH (HNA), and lead to rounded isotropic features in single crystalline Si. They are used at room temperature or slightly above (<50°C). Historically they were the first Si etchants introduced.[4-10,60,61] Later it was discovered that some alkaline chemicals will etch anisotropically, i.e., they etch away crystalline silicon at different rates depending on the orientation of the exposed crystal plane. Typically the pH stays above 12, while more elevated temperatures are used for these slower type etchants (>50°C). The latter type of etchants surged in importance in the late 1960s for the fabrication of dielectrically isolated structures in silicon.[11-24,62] Isotropic etchants typically show diffusion limitation, while anisotropic etchants are reaction rate limited.

Preferential or selective etching (also structural etchants) usually are isotropic etchants that show some anisotropy.[63] These etchants are used to produce a difference in etch rate between different materials or between compositional or structural variations of the same material on the same crystal plane. These type of etches often are the fastest and simplest techniques to delineate electrical junctions and to evaluate the structural perfection of a single crystal in terms of slip, lineage, and stacking faults. The artifacts introduced by the defects etch into small pits of characteristic shape. Most of the etchants used for this purpose are acids with some oxidizing additives.[64-68]

### Isotropic Etching

#### *Usage of Isotropic Etchants*

When etching silicon with aggressive acidic etchants, rounded isotropic patterns form. The method is widely used for:

1. Removal of work-damaged surfaces
2. Rounding of sharp anisotropically etched corners (to avoid stress concentration)
3. Removing of roughness after dry or anisotropic etching
4. Creating structures or planar surfaces in single-crystal slices (thinning)
5. Patterning single-crystal, polycrystalline, or amorphous films
6. Delineation of electrical junctions and defect evaluation (with preferential isotropic etchants)

For isotropic etching of silicon, the most commonly used etchants are mixtures of nitric acid ($HNO_3$) and hydrofluoric acids (HF). Water can be used as a diluent, but acetic acid ($CH_3COOH$) is preferred because it prevents the dissociation of the nitric acid better and so preserves the oxidizing power of $HNO_3$ which depends on the undissociated nitric acid species for a wide range of dilution.[5] The etchant is called the HNA system; we will return to this etch system below.

#### *Simplified Reaction Scheme*

In acidic media, the Si etching process involves hole injection into the Si valence band by an oxidant, an electrical field, or photons. Nitric acid in the HNA system acts as an oxidant; other oxidants such as $H_2O_2$ and $Br_2$ also work.[69] The holes attack the covalently bonded Si, oxidizing the material. Then follows a reaction of the oxidized Si fragments with $OH^-$ and subsequent dissolution of the silicon oxidation products in HF. Consider the following reactions that describe these processes.

The holes are, in the absence of photons and an applied field, produced by $HNO_3$, together with water and trace impurities of $HNO_2$:

$$HNO_3 + H_2O + HNO_2 \rightarrow$$
$$2HNO_2 + 2OH^- + 2h^+$$

**Reaction 4.1**

The holes in Reaction 4.1 are generated in an autocatalytic process; $HNO_2$ generated in the above reaction re-enters into the further reaction with $HNO_3$ to produce more holes. With a reaction of this type, one expects an induction period before the oxidation reaction takes off, until a steady-state concentration of $HNO_2$ has been reached. This has been observed at low $HNO_3$ concentrations.[69] After hole injection, $OH^-$ groups attach to the oxidized Si species to form $SiO_2$, liberating hydrogen in the process:

$$Si^{4+} + 4OH^- \rightarrow SiO_2 + H_2$$

**Reaction 4.2**

Hydrofluoric acid (HF) dissolves the $SiO_2$ by forming the water-soluble $H_2SiF_6$. The overall reaction of HNA with Si looks like:

$$Si + HNO_3 + 6HF \rightarrow H_2SiF_6 + HNO_2$$
$$+ H_2O + H_2 \text{ (bubbles)}$$

**Reaction 4.3**

The simplification in the above reaction scheme is that only holes are assumed. In the actual Si acidic corrosion reaction, both holes and electrons are involved. The question of hole and/or electron participation in Si corrosion will be considered after the introduction of the model for the Si/electrolyte interfacial energetics. We will learn from that model that the rate-determining step in acidic etching involves hole injection in the valence band, whereas in alkaline anisotropic etching it involves electron injection in the conduction band by surface states. The reactivity of a hole injected in the valence band is significantly greater than that of an electron injected in the conduction band. The observation of isotropy in acidic etchants and anisotropy in alkaline etchants centers on this difference in reactivity.

#### *Iso-Etch Curves*

By the early 1960s, the isotropic HNA silicon etch was well characterized. Schwartz and Robbins published a series of four very detailed papers on the topic between 1959 and 1976.[4-7] Most of the material presented below is based on their work.

HNA etching results, represented in the form of iso-etch curves, for various weight percentages of the constituents are shown in Figure 4.21. For this work, normally available concentrated acids of 49.2 wt% HF and 69.5 wt% $HNO_3$ are used. Water as diluent is indicated by dash-line curves and acetic acid by solid-line curves. Also, as in Wong's representation,[70] we have recalculated the curves from Schwartz et al to express the etch rate in µm/min and divided the authors' numbers by 2 as we are considering one-sided etching only. The highest etch rate is observed around a weight ratio HF-$HNO_3$ of 2:1 and is nearly 100 times faster than anisotropic etch rates. Adding a diluent slows down the etching. From these curves, the following characteristics of the HNA system can be summarized:

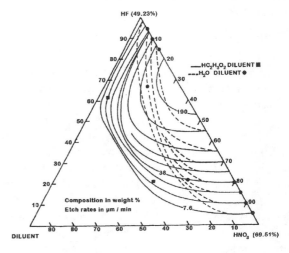

**FIGURE 4.21**   Iso-etch curves. From Robbins et al.,[5] recalculated for one-sided Si etching and expressed in µm/min.

1. At high HF and low $HNO_3$ concentrations, the iso-etch curves describe lines of constant $HNO_3$ concentrations (parallel to the HF-diluent axis); consequently, the $HNO_3$ concentration controls the etch rate. Etching at those concentrations tends to be difficult to initiate and exhibits an uncertain induction period (see above). In addition, it results in relatively unstable silicon surfaces proceeding to slowly grow a layer of $SiO_2$ over a period of time. The etch is limited by the rate of oxidation, so that it tends to be orientation dependent and affected by dopant concentration, defects, and catalysts (sodium nitrate often is used). In this regime the temperature influence is more pronounced, and activation energies for the etching reaction of 10 to 20 Kcal/mol have been measured.

2. At low HF and high $HNO_3$ concentrations, iso-etch curves are lines parallel to the nitric-diluent axis, i.e., they are at constant HF composition. In this case, the etch rate is controlled by the ability of HF to remove the $SiO_2$ as it is formed. Etches in this regime are isotropic and truly polishing, producing a bright surface with anisotropies of 1% or less (favoring the <110> direction) when used on <100> wafers.[71] An activation energy of 4 Kcal/mol is indicative of the diffusion limited character of the process; consequently, in this regime, temperature changes are less important.

3. In the region of maximal etch rate both reagents play an important role. The addition of acetic acid, as opposed to the addition of water, does not reduce the oxidizing power of the nitric acid until a fairly large amount of diluent has been added. Therefore, the rate contours remain parallel with lines of constant nitric acid over a considerable range of added diluent.

4. In the region around the HF vertex the surface reaction rate-controlled etch leads to rough, pitted Si surfaces and sharply peaked corners and edges. In moving towards the $HNO_3$ vertex, the diffusion-controlled reaction results in the development of rounded corners and edges and the rate of attack on (111) planes and (110) planes becomes identical in the polishing regime (anisotropy less than 1%; see point 2).

In Figure 4.22 we summarize how the topology of the Si surfaces depends strongly on the composition of the etch solution. Around the maximum etch rates the surfaces appear quite flat with rounded edges, and very slow etching solutions lead to rough surfaces.[7]

### *Arrhenius Plot for Isotropic Etching*

The effect of temperature on the reaction rate in the HNA system was studied in detail by Schwartz and Robbins.[6] An Arrhenius plot for etching Si in 45% $HNO_3$, 20% HF, and 35% $HC_2H_3O_2$, culled from their work, is shown in Figure 4.23. Increasing the temperature increases the reaction rate. The graph shows two straight-line segments, indicating a higher activation energy below 30°C and a lower one above this temperature. In

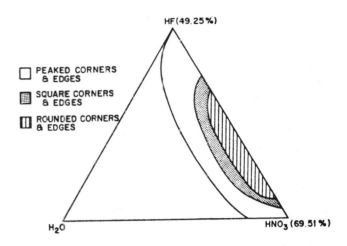

FIGURE 4.22 Topology of etched Si surfaces. (From Schwartz, B. and H. Robbins, *J. Electrochem. Soc.*, 123, 1903–1909, 1976. With permission.)

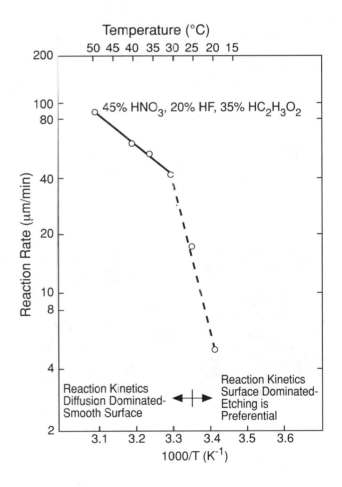

FIGURE 4.23 Etching Arrhenius plot. Temperature dependence of the etch rate of Si in HF:$HNO_3$:$CH_3$:COOH (1:4:3). (From Schwartz, B. and H. Robbins, *J. Electrochem. Soc.*, 108, 365–372, 1961. With permission.)

the low temperature range, etching is preferential and the activation energy is associated with the oxidation reaction. At higher temperatures the etching leads to smooth surfaces and the activation energy is lower and associated with diffusion limited dissolution of the oxide.[6]

With isotropic etchants the etchant moves downward and outwards from an opening in the mask, undercuts the mask, and enlarges the etched pit while deepening it (Figure 4.24). The resulting isotropically etched features show more symmetry and rounding when agitation accompanies the etching (the process is diffusion limited). This agitation effect is illustrated in Figure 4.24. With agitation the etched feature approaches an ideal round cup; without agitation the etched feature resembles a rounded box.[34] The flatness of the bottom of the rounded box generally is poor, since the flatness is defined by agitation.

### Masking for Isotropic Silicon Etchants

Acidic etchants are very fast; for example, an etch rate for Si of up to 50 μ min$^{-1}$ can be obtained with 66% $HNO_3$ and 34% HF (volumes of reagents in the normal concentrated form).[60,63] Isotropic etchants are so aggressive that the activation barriers associated with etching the different Si planes are not differentiated; all planes etch equally fast, making masking a real challenge.

Although $SiO_2$ has an appreciable etch rate of 300 to 800 Å/min in the $HF:HNO_3$ system, one likes to use thick layers of

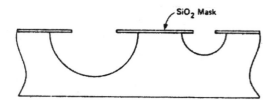

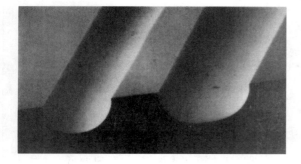

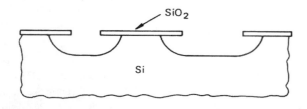

**FIGURE 4.24** Isotropic etching of Si with (A) and without (B) etchant solution agitation.

$SiO_2$ as a mask anyway, especially for shallow etching, as the oxide is so easy to form and pattern. A mask of nonetching Au or $Si_3N_4$ is needed for deeper etching. Photoresists do not stand up to strong oxidizing agents such as $HNO_3$, and neither does Al.

Silicon itself is soluble to a small extent in pure HF solutions; for a 48% HF, at 25°C, a rate of 0.3 Å/min was observed for n-type, 2-ohm cm (111)-Si. It was established that Si dissolution in HF is not due to oxidation by dissolved oxygen. Diluted HF etches Si at a higher rate because the reaction in aqueous solutions proceeds by oxidation of Si by OH$^-$ groups.[72] A typically buffered HF (BHF) solution has been reported to etch Si at radiochemically measured rates of 0.23 to 0.45 Å/min, depending on doping type and dopant concentration.[73]

By reducing the dopant concentration (n or p) to below $10^{17}$ atoms/cm$^3$ the etch rate of Si in HNA is reduced by ~150.[74] The doping dependence of the etch rate provides yet another means of patterning a Si surface (see next section). A summary of masks that can be used in acidic etching is presented in Table 4.6.

### Dopant Dependence of Silicon Isotropic Etchants

The isotropic etching process is fundamentally a charge-transfer mechanism. This explains the etch rate dependence on dopant type and concentration. Typical etch rates with an HNA system (1:3:8) for n- or p-type dopant concentrations above $10^{18}$ cm$^3$ are 1 to 3 μm/min. As presented in the preceding section, a reduction of the etch rate by 150 times is obtained in n- or p-type regions with a dopant concentration of $10^{17}$ cm$^{-3}$ or smaller.[74] This presumably is due to the lower mobile carrier concentration available to contribute to the charge transfer mechanisms. In any event, heavily doped silicon substrates with high conductivity can be etched more readily than lightly doped materials. Dopant-dependent isotropic etching can also be exploited in an electrochemical set-up as described in the next section. Although doping does change the chemical etch rate, attempts to exploit these differences for industrial production have failed so far.[75] This situation is different in electrochemical isotropic etching (see next section).

### Electrochemical Isotropic Silicon Etch-Etch Stop

Sometimes, a high temperature or extremely aggressive chemical etching process can be replaced by an electrochemical procedure utilizing a much milder solution, thus allowing a simple photoresist mask to be employed.[63] In electrochemical acidic etching, with or without illumination of the corroding Si electrode, an electrical power supply is employed to drive the chemical reaction by supplying holes to the silicon surface (W-EL see Figure 4.25). A voltage is applied across the silicon wafer and a counter electrode (C-EL usually platinum) arranged in the same etching solution. Oxidation is promoted by a positive bias applied to the silicon causing an accumulation of holes in the silicon at the silicon/electrolyte interface. Under this condition, oxidation at the surface proceeds rapidly while the oxide is readily dissolved by the HF solution. No oxidant such as $HNO_3$ is needed to supply the holes; excess electron-hole pairs are created by the electrical field at the surface and/or by optical excitation, thereby increasing the etch rate. This technique proved successful in removing heavily doped layers, leaving

**TABLE 4.6** Masking Materials for Acidic Etchants[a]

| | Etchants | | |
| | Piranha (4:1, $H_2O_2$:$H_2SO_4$) | Buffered HF ($\leq$1 $NH_4F$:conc. HF) | HNA |
| --- | --- | --- | --- |
| Masking | | | |
| Thermal $SiO_2$ | | 0.1 µm/min | 300–800 Å/min; limited etch time, thick layers often are used due to ease of patterning |
| CVD (450°C) $SiO_2$ | | 0.48 µm/min | 0.44 µm/min |
| Corning 7740 glass | | 0.063 µ/min | 1.9 µ/min |
| Photoresist | Attacks most organic films | Good for short while | Resists do not stand up to strong oxidizing agents like $HNO_3$ and are not used |
| Undoped Si, polysilicon | Forms 30 Å of $SiO_2$ | 0.23 to 0.45 Å/min | Si 0.7 to 40 µm/min at room temperature; at a dopant concentration <$10^{17}$ cm$^{-3}$ (n or p) |
| Black wax | | | Usable at room temperature |
| Au/Cr | Good | Good | Good |
| LPCVD $Si_3N_4$ | | 1 Å/min | Etch rate is 10–100 Å/min; preferred masking material |

[a] The many variables involved necessarily means that the given numbers are approximate only.

behind the more lightly doped membranes in all possible dopant configurations: p on p$^+$, p on n$^+$, n on p$^+$, and n on n$^+$.[76,77] This electrochemical etch-stop technique is demonstrated in the inset of Figure 4.25.[78] A 5% HF solution is used, the electrolyte cell is kept in the dark at room temperature, and the distance between the Si anode and the Pt cathode in the electrochemical cell is 1 to 5 cm. Instead of using HF, one can substitute $NH_4F$ (5 wt%) for the electrochemical etching as described by Shengliang.[79] Shengliang reports a selectivity of n-silicon to n$^+$-silicon (0.001 Ωcm) of 300 with the latter etchant. In the inset in Figure 4.25 the current density vs. applied voltage across the anode and cathode during dissolution is plotted. The current density is related to the dissolution rate of silicon. It can be seen that p-type and heavily doped n-type materials can be dissolved at relatively low voltages, whereas n-type silicon with a lower doping level does not dissolve at the same low voltages. Experiments in this same set-up with homogeneously doped silicon wafers show that n-type silicon of about $3.10^{18}$ cm$^{-3}$ (<0.01 Ωcm) completely dissolves in these etching conditions, whereas n-type silicon of donor concentrations lower than $2.10^{16}$ cm$^{-3}$ (>0.3 Ωcm) barely dissolves. For p-type silicon, dissolution is initiated when the acceptor concentration is higher than $5.10^{15}$ cm$^{-3}$ (<3 Ωcm) and the dissolution rate further increases with increasing acceptor concentration. Under specific circumstances, namely high HF concentrations and low etching currents, porous Si may form.[80]

The acidic electrochemical technique has not been used much in micromachining and is primarily used to polish surfaces. Since the etching rate increases with current density, high spots on the surface are more rapidly etched and very smooth surfaces result. The method of isotropic electrochemical etching has some major advantages which could make it a more important micromachining tool in the future. The etched surfaces are very smooth (say with an average roughness, R$_a$ of 7 nm), the process is room temperature and IC compatible, simpler resists schemes can be used as the process is much milder than etching in HNA, and etching can be controlled simply by switching a voltage on or off. We will pick up the discussion of anodic polishing, photo etching, and formation of porous silicon in HF solutions after gathering more insight in various etching models.

### Preferential Etching

A variety of additives to the HNA system, mainly oxidants, can be included to modify the etch rate, surface finish, or isotropy, rendering the etching baths preferential. It is clear that the effect of these additives will only show up in the reaction-controlled regime. Only additives that change the viscosity of the solution could modify the etch rate in the diffusion-limited regime, thereby changing the diffusion coefficient of the reactants.[69,81] We will not review the effect of these additives any further, refer to Table 4.7 and the cited literature for further study.[64-68]

### Problems with Isotropic Etchants

There are several problems associated with isotropic etching of Si. First, it is difficult to mask with high precision using a desirable and simple mask such as $SiO_2$ (etch rate is 2 to 3% of the silicon etch rate). Second, the etch rate is very agitation sensitive in addition to being temperature sensitive. This makes it difficult to control lateral as well as vertical geometries. Electrochemical isotropic etching (see above) and the development of anisotropic etchants in the late 1960s (see below) overcame many of these problems.

A comprehensive review of isotropic etchants solutions can be found in Kern et al.[63] These authors also give a review of different techniques practiced in chemical etching such as immersion etching, spray etching, electrolytic etching, gas-phase etching, and molten salt etching (fusion techniques). In Table 4.7 some isotropic and preferential etchants and their specific applications are listed.

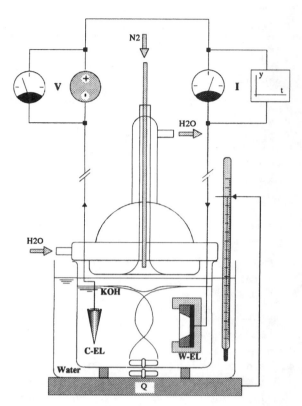

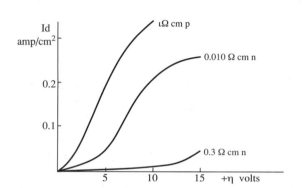

**FIGURE 4.25**   Electrochemical etching apparatus. W-EL: working electrode (Si), C-EL: counter electrode (e.g. Pt), Q = heat supplied. Inset: Current-voltage ($I_d/\eta$) curves in electrochemical etching of Si of various doping. Etch rate dependence on dopant concentration and dopant type for HF-anodic etching of silicon. (From van Dijk, H.J.A. and de Jonge, J., *J. Electrochem. Soc.*, 117, 553–554, 1970. With permission.)

## Anisotropic Etching

### Introduction

Anisotropic etchants shape, also 'machine', desired structures in crystalline materials. When carried out properly, anisotropic etching results in geometric shapes bounded by perfectly defined crystallographic planes. Anisotropic wet etching techniques, dating back to the 1960s at the Bell Laboratories, were developed mainly by trial and error. Going over some experimental data, before embarking upon the models, seems fitting. Moreover, we must keep in mind that for higher index planes most models fail.

Figure 4.1 shows a cross-section of a typical shape formed using anisotropic etching. The thinned membranes with diffused resistors could be used for a piezoresistive pressure sensor or an accelerometer. In the usual application, the wafer is selectively thinned from a starting thickness of 300 to 500 μm to form a diaphragm having a final thickness of 10 to 20 μm with precisely controlled lateral dimensions and a thickness control of the order of 1 μm or better. A typical procedure involves the steps summarized in Table 4.8.[42]

The development of anisotropic etchants solved the lateral dimension control lacking in isotropic etchants. Lateral mask geometries on planar photoengraved substrates can be controlled with an accuracy and reproducibility of 0.5 μm or better, and the anisotropic nature of the etchant allows this accuracy to be translated into control of the vertical etch profile. Different etch stop techniques, needed to control the membrane

thickness, are available. The invention of these etch stop techniques truly made items as shown in Figure 4.1 manufacturable.

While anisotropic etchants solve the lateral control problem associated with deep etching, they are not without problems. They are slower, even in the fast etching <100> direction, with etch rates of 1 μm/min or less. That means that etching through a wafer is a time-consuming process: to etch through a 300-μm thick wafer one needs 5 hours. They also must be run hot to achieve these etch rates (85 to 115°C), precluding many simple masking options. Like the isotropic etchants, their etch rates are temperature sensitive; however, they are not particularly agitation sensitive, considered to be a major advantage.

### Anisotropic Etchants

A wide variety of etchants have been used for anisotropic etching of silicon, including alkaline aqueous solutions of KOH, NaOH, LiOH, CsOH, $NH_4OH$, and quaternary ammonium hydroxides, with the possible addition of alcohol. Alkaline organics such as ethylenediamine, choline (trimethyl-2-hydroxyethyl ammonium hydroxide) or hydrazine with additives such as pyrocathechol and pyrazine are employed as well. Etching of silicon occurs without the application of an external voltage and is dopant insensitive over several orders of magnitude, but in a curious contradiction to its suggested chemical nature, it has been shown to be bias dependent.[82,83] This contradiction will be explained with the help of the chemical models presented below.

**TABLE 4.7** Isotropic and Preferential Defect Etchants and Their Specific Applications

| Etchant | Application | Remark |
|---|---|---|
| HF: 8 vol%, HNO$_3$: 75 vol%, and CH$_3$COOH: 17 vol% | n- and p-type Si, all planes, general etching | Planar etch; e.g., 5 μm/min at 25°C |
| 1 part 49% HF, 1 part of (1.5 $M$ CrO$_3$) (by volume) | Delineation of defects on (111), (100), and (110) Si without agitation | Yang etch[a] |
| 5 vol parts nitric acid (65%), 3 vol parts HF (48%), 3 vol parts acetic acid (96%), 0.06 parts bromine | Polishing etchant used to remove damage introduced during lapping | So-called CP4 etchant; Heidenreich, U.S. Patent 2619414 |
| HF | SiO$_2$ | Si etch rate for 48% HF at 25°C is 0.3 Å/min with n-type 2 Ω cm (111) Si |
| 1HF, 3HNO$_3$, 10CH$_3$COOH (by volume) | Delineates defects in (111) Si; etches p$^+$ or n$^+$ and stops at p$^-$ or n$^-$ | Dash etch; p- and n-Si at 1300 Å/min in the [100] direction and 46 Å/min in the [111] direction at 25°C[b] |
| 1HF, 1(5 $M$ CrO$_3$) (by volume) | Delineates defects in (111); needs agitation; does not reveal etch pits well on (100) well | Sirtl etch[c] |
| 2HF, 1(0.15 $M$ K$_2$Cr$_2$O$_7$) (by volume) | Yields circular (100) Si dislocation etch pits; agitation reduces etch time | Secco etch[d] |
| 60 ml HF, 30 ml HNO$_3$, 30 ml (5 $M$ CrO$_3$), 2 g Cu(NO$_3$)$_2$, 60 ml CH$_3$COOH, 60 ml H$_2$O | Delineates defects in (100) and (111) Si; requires agitation | Jenkins etch[e] |
| 2HF, 1 (1 $M$ CrO$_3$) (by volume) | Delineates defects in (100) Si without agitation; works well on resistivities 0.6–15.0 Ωcm n- and p-types) | Schimmel etch[f] |
| 2HF, 1 (1 $M$ CrO$_3$), 1.5 (H$_2$O) (by volume) | Works well on heavily doped (100) silicon | Modified Schimmel[f] |
| HF/KMnO$_4$/CH$_3$COOH | Epitaxial Si | |
| H$_3$PO$_4$ | Si$_3$N$_4$ | 160–180°C |
| KOH + alcohols | Polysilicon | 85°C |
| H$_3$PO$_4$/HNO$_3$/HC$_2$H$_3$O$_2$ | Al | 40–50°C |
| HNO$_3$/BHF/water | Si and polysilicon | 0.1 μm min$^{-1}$ for single crystal Si |

[a] Yang, K.H., *J. Electrochem. Soc.*, 131, 1140 (1984).
[b] Dash, W.C., *J. Appl. Phys.*, 27, 1193 (1956).
[c] Sirtl, E. and Adler, A., *Z. Metallkd.*, 52, 529 (1961).
[d] Secco d'Aragona, F., *J. Electrochem. Soc.*, 119, 948 (1972).
[e] Jenkins, M.W., *J. Electrochem. Soc.*, 124, 757 (1977).
[f] Schimmel, D.G., *J. Electrochem. Soc.*, 126, 479 (1979).

Alcohols such as propanol and isopropanol butanol typically slow the attack on Si.[84,85] The role of pyrocatechol[86] is to speed up the etch rate through complexation of the reaction products. Additives such as pyrazine and quinone have been described as catalysts by some;[87] but this is contested by other authors.[88] The etch rate in anisotropic etching is reaction rate controlled and thus temperature dependent. The etch rate for all planes increases with temperature and the surface roughness decreases with increasing temperature, so etching at the higher temperatures gives the best results. In practice, etch temperatures of 30 to 85°C are used to avoid solvent evaporation and temperature gradients in the solution.

### Arrhenius Plots for Anisotropic Etching

A typical set of Arrhenius plots for <100>, <110>, and <111> silicon etching in an anisotropic etchant (EDP, or ethylene-diamine/pyrocatechol) is shown in Figure 4.26.[90] It is seen that the temperature dependence of the etch rate is quite large and is less dependent on orientation. The slope differs for the different planes, i.e., (111) > (100) > (110). Lower activation energies in Arrhenius plots correspond to higher etch rates. The anisotropy ratio (AR) derived from this figure is

$$AR = (hkl)_1 \text{ etch rate}/(hkl)_2 \text{ etch rate} \qquad 4.29$$

The AR is approximately 1 for isotropic etchants and can be as high as 400/200/1 for (110)/(100)/(111) in 50 wt% KOH/H$_2$O at 85°C. Generally, the activation energies of the etch rates of EDP are smaller than those of KOH. The (111) planes always etch slowest but the sequence for (100) and (110) can be reversed (e.g., 50/200/8 in 55 vol% ethylenediamine ED/H$_2$O; also at 85°C). The (110) Si plane etches 8 times slower and the (111) 8 times faster in KOH/H$_2$O than in ED/H$_2$O, while the (100) etches at the same rate.[89] Working with alcohols and other organic additives often changes the relative etching rate of the different Si planes. Along this line, Seidel et al.[88,90] found that the decrease in etch rate by adding isopropyl alcohol to a KOH solution was 20% for <100>, but almost 90% for <110>. As a result of the much stronger decrease of the etch rate on a (110) surface, the etch ratio of (100):(110) is reversed.

**TABLE 4.8** Summary of the Process Steps Required for Anisotropic Etching of a Membrane[42]

| Process | Duration | Process Temperature (°C) |
|---|---|---|
| Oxidation | Variable (hours) | 900–1200 |
| Spinning at 5000 rpm | 20–30 sec | Room temperature |
| Prebake | 10 min | 90 |
| Exposure | 20 sec | Room temperature |
| Develop | 1 min | Room temperature |
| Post-bake | 20 min | 120 |
| Stripping of oxide (BHF:1:7) | ±10 min | Room temperature |
| Stripping resist (acetone) | 10–30 sec | Room temperature |
| RCA1 ($NH_3$(25%) + $H_2O$ + $H_2O_2$:1:5:1) | 10 min | Boiling |
| RCA2 (HCl + $H_2O$ + $H_2O_2$:1:6:1) | 10 min | Boiling |
| HF-dip (2% HF) | 10 sec | Room temperature |
| Anisotropic etch | From minutes up to one day | 70–100 |

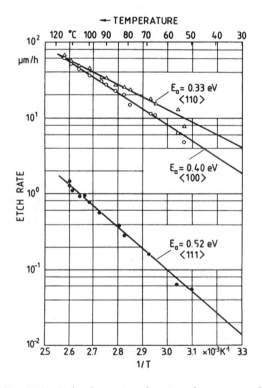

**FIGURE 4.26** Vertical etch rates as a function of temperature for different crystal orientations: (100), (110), and (111). Etch solution is EDP (133 ml $H_2O$, 160 g pyrocatechol, 6 g pyrazine, and 1 l ED). (From Seidel, H. et al., *J. Electrochem. Soc.*, 137, 3612–3626, 1990. With permission.)

## Important Anisotropic Etchant Systems

In choosing an etchant, a variety of issues must be considered:

- Ease of handling
- Toxicity
- Etch rate
- Desired topology of the etched bottom surface
- IC-compatibility
- Etch stop
- Etch selectivity over other materials
- Mask material and thickness of the mask

The principal characteristics of four different anisotropic etchants are listed in Table 4.9. The most commonly used are KOH[11-24,88-94,96-98] and ethylene-diamine/pyrocatechol + water (EDP);[86,87,99] hydrazine-water rarely is used.[100,101] More recently, quaternary ammonium hydroxide solutions such as tetraethyl ammonium hydroxide (TEAH) have become more popular.[102,103] Each has its advantages and problems. NaOH is not used much anymore.[95]

Hydrazine-water is explosive at high hydrazine concentrations (rocket fuel) and is a suspected carcinogen. Its use should be avoided for safety reasons. A 50% hydrazine/water solution is stable, though, and, according to Mehregany,[101] excellent surface quality and sharply defined corners are obtained in Si. Also on the positive side, the etchant has a very low $SiO_2$ etch rate and will not attack most metal masks except for Al, Cu, and Zn. According to Wise, on the other hand, Al does not etch in hydrazine either, but the etch produces rough Si surfaces.[104]

Ethylenediamine in EDP reportedly causes allergic respiratory sensitization, and pyrocatechol is described as a toxic corrosive. The material is also optically dense, making end-point detection harder, and it ages quickly; if the etchant reacts with oxygen, the liquid turns to a red-brown color, and it loses its good properties. If cooled down after etching, one gets precipitation of silicates in the solution. Sometimes one even gets precipitation during etching, spoiling the results. When preparing the solution, the last ingredient added should be the water, since water addition causes the oxygen sensitivity. All of the above make the etchant quite difficult to handle. But, a variety of masking materials can be used in conjunction with this etchant and it is less toxic than hydrazine. No sodium or potassium contamination occurs with this compound and the etch rate of $SiO_2$ is slow. The ratio of etch rates of Si and $SiO_2$ using EDP can be as large as 5000:1 (about 2 Å/min of $SiO_2$ compared to 1 μm/min of Si) which is much larger than the ratio in KOH a ratio of as high as 400:1 has been reported.[105] Importantly, the etch rate slows down at a lower boron concentration than with KOH. A typical fastest-to-slowest hierarchy of Si etch rates with EDP at 85°C according to Barth[106] is (110) > (411) > (311) > (511) > (211) > (100) > (331) > (221) > (111).

The simple KOH water system is the most popular etchant. A KOH etch, in near saturated solutions (1:1 in water by weight) at 80°C, produces a uniform and bright surface. Nonuniformity of etch rate gets considerably worse above 80°C. Plenty of bubbles are seen emerging from the Si wafer while etching in KOH. The etching selectivity between Si and $SiO_2$ is not very good in KOH, as it etches $SiO_2$ too fast. KOH is also incompatible with the IC fabrication process and can cause blindness when it gets in contact with the eyes. The etch rate for low index planes is maximal at around 4 $M$ (see Figure 4.27A[41] and Lambrechts et al.[107]). The

**TABLE 4.9** Principal Characteristics of Four Different Anisotropic Etchants[a]

| Etchant/Diluent/Additives/ Temperature | Etch Stop | Etch Rate (100) ($\mu$m/min) | Etch Rate Ratio (100)/(111) | Remarks | Mask (Etch Rate) |
|---|---|---|---|---|---|
| KOH/water, isopropyl alcohol additive, 85°C | B > $10^{20}$ cm$^{-3}$ reduces etch rate by 20 | 1.4 | 400 and 600 for (110)/(111) | IC incompatible, avoid eye contact, etches oxide fast, lots of H$_2$ bubbles | Photoresist (shallow etch at room temperature); Si$_3$N$_4$ (not attacked); SiO$_2$ (28 Å/min) |
| Ethylene diamine pyrocatechol (water), pyrazine additive, 115°C | $\geq 5 \times 10^{19}$ cm$^{-3}$ reduces the etch rate by 50 | 1.25 | 35 | Toxic, ages fast, O$_2$ must be excluded, few H$_2$ bubbles, silicates may precipitate | SiO$_2$ (2–5 Å/min); Si$_3$N$_4$ (1 Å/min); Ta, Au, Cr, Ag, Cu |
| Tetramethyl ammonium hydroxide (TMAH) (water), 90°C | >$4 \times 10^{20}$ cm$^{-3}$ reduces etch rate by 40 | 1 | From 12.5 to 50 | IC compatible, easy to handle, smooth surface finish, few studies | SiO$_2$ etch rate is 4 orders of magnitude lower than (100) Si LPCVD Si$_3$N$_4$ |
| N$_2$H$_4$/(water), isopropyl alcohol, 115°C | >$1.5 \times 10^{20}$ cm$^{-3}$ practically stops the etch | 3.0 | 10 | Toxic and explosive, okay at 50% water | SiO$_2$ (<2 Å/min) and most metallic films; does not attack Al according to some authors[104] |

[a] Given the many possible variables, the data in the table are only typical examples.

surface roughness continuously decreases with increasing concentration as can be gleaned from Figure 4.27B. Since the difference in etch rates for different KOH concentrations is small, a highly concentrated KOH (e.g., 7 *M*) is preferred to obtain a smooth surface on low index planes.

Except at very high concentrations of KOH, the etched (100) plane becomes rougher the longer one etches. This is thought to be due to the development of hydrogen bubbles, which hinder the transport of fresh solution to the silicon surface.[108] Average roughness, $R_a$, is influenced strongly by fluid agitation. Stirring can reduce the $R_a$ values over an order of magnitude, probably caused by the more efficient removal of hydrogen bubbles from the etching surface when stirring.[109] The silicon etch rate as a function of KOH concentration is shown in Figure 4.27C.[90]

Herr[110] found that the high-index crystal planes exhibit the highest etch rates for 6 *M* KOH and that for lower concentrations the etch bottoms disintegrate into microfacets. In 6 *M* KOH, the etch-rate order is (311) > (144) > (411) > (133) > (211) > (122). These authors could not correlate the particular etch rate sequence with the measured activation energies. This is in contrast to lower activation energies corresponding to higher etching rates for low index planes as shown in Figure 4.26. Their results obtained on large open area structures differ significantly from previous ones obtained by underetching special mask patterns. The vertical etching rates obtained here are substantially higher than the underetching rates described elsewhere, and the etch rate sequence for different planes is also significantly different. These results suggest that crevice effects may plan an important role in anisotropic etching.

Besides KOH,[111] other hydroxides have been used, including NaOH,[82,95] CsOH,[112] and NH$_4$OH.[113] A major disadvantage of KOH is the presence of alkali ions, which are detrimental to the fabrication of sensitive electronic parts. Work is under way to

find anisotropic etchants that are more compatible with CMOS processing and that are neither toxic nor harmful. Two examples are ammonium hydroxide-water (AHW) mixtures[113] and tetramethyl ammonium hydroxide-water (TMAHW) mixtures (Tabata et al.[103] and Schnakenberg et al.[113]). TMAHW solutions do not decompose at temperatures below 130°C, a very important feature from the viewpoint of production. They are nontoxic and can be handled easily. TMAHW solutions also exhibit excellent selectivity to silicon oxide and silicon nitride. At a solution temperature of 90°C and 22 wt% TMAH, a maximum (100) silicon etch rate of 1.0 $\mu$m/min is observed, 1.4 $\mu$m/min for (110) planes (this is higher than those observed with EDP, AHW, hydrazine water, and tetraethyl ammonium hydroxide (TEA), but slower than those observed for KOH) and an anisotropy ratio, AR(100)/(111), of between 12.5 and 50.[114] From the viewpoint of fabricating various silicon sensors and actuators, a concentration above 22 wt% is preferable, since lower concentrations result in larger roughness on the etched surface. However, higher concentrations give a lower etch rate and lower etch ratio (100)/(111). Tabata[115] also studied the etching characteristics of pH-controlled TMAHW. To obtain a low aluminum etching rate of 0.01 $\mu$m/min, pH values below 12 for 22 wt% TMAHW were required. At those pH values the Si(100) etching rate is 0.7 $\mu$m/min.

### Surface Roughness and Notching

Anisotropic etchants frequently leave too rough a surface behind, and a slight isotropic etch is used to 'touch-up'. A distinction must be made between **macroscopic** and **microscopic** **roughness**. Macroscopic roughness, also referred to as notching or pillowing, results when centers of exposed areas etch with a seemingly lower average speed compared with the borders of the areas, so that the corners between sidewalls and (100)

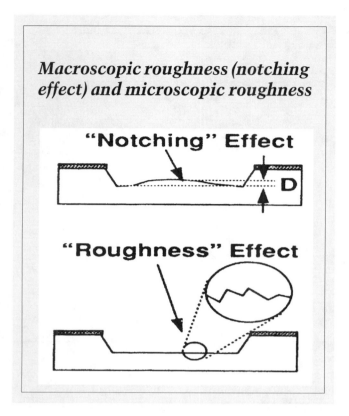

ground planes are accentuated. Membranes or double-sided clamped beams (microbridges) therefore tend to be thinner close to the clamped edges than in the center of the structure. This difference can be as large as 1 to 2 μm, which is quite considerable if one is etching 10 to 20 μm thick structures. Notching increases linearly with etch depth but decreases with higher concentrations of KOH. The microscopic smoothness of originally mirror-like polished wafers can also be degraded into microscopic roughness. It is this type of short-range roughness we referred to in discussing Figure 4.27B above. For more background on metrology techniques to measure surface roughness, see Appendix A.

*Masking for Anisotropic Etchants*

Etching through a whole wafer (400 to 600 μm) takes several hours (a typical wet anisotropic etch rate being 1.1 μm/min), definitely not a fast process. When using KOH, $SiO_2$ cannot be used as a masking material for features requiring that long an exposure to the etchants. The $SiO_2$ etch rate as a function of KOH concentration at 60°C is shown in Figure 4.28. There is a distinct maximum at 35 wt% KOH of nearly 80 nm/hr. The shape of this curve will be explained further below on the basis of Seidel et al.'s model. Experiments have shown that even a 1.5-μm thick oxide is not sufficient for the complete etching of a 380-μm thick wafer (6-hours) because of pinholes in the oxide.[107] The etch rate of thermally grown $SiO_2$ in KOH-$H_2O$ somewhat varies and apparently depends not only on the quality of the oxide, but also on the etching container and the age of the etching solution, as well as other factors.[44] The Si/$SiO_2$ selectivity ratio at 80°C in 7 *M* KOH is 30 ± 5. This ratio increases with decreasing temperature; reducing the temperature from 80 to 60°C increases the selectivity ratio from 30 to 95 in 7 *M*

KOH.[43] Thermal oxides are under strong compressive stress due to the fact that in the oxide layer one silicon atom takes nearly twice as much space as in single crystalline Si (see also Chapter 3). This might have severe consequences; for example, if the oxide mask is stripped on one side of the wafer, the wafer will bend. Atmospheric pressure chemical vapor deposited (APCVD) $SiO_2$ tends to exhibit pinholes and etches much faster than thermal oxide. Annealing of APCVD oxide removes the pinholes but the etch rate in KOH remains greater by a factor of 2 to 3 than that of thermal oxide. Low pressure chemical vapor deposited (LPCVD) oxide is a mask material of comparable quality as thermal oxide. The etch rate of $SiO_2$ in EDP is smaller by two orders of magnitude than in KOH.

For prolonged KOH etching, a high density silicon nitride mask has to be deposited. A low pressure chemical vapor deposited (LPCVD) nitride generally serves better for this purpose than a less dense plasma deposited nitride.[116] With an etch rate of less than 0.1 nm/min, a 400-Å layer of LPCVD nitride suffices to mask against KOH etchant. The etch selectivity Si/$Si_3N_4$ was found to be better than $10^4$ in 7 *M* KOH at 80°C. The nitride also acts as a good ion-diffusion barrier, protecting sensitive electronic parts. Nitride can easily be patterned with photoresist and etched in a $CF_4/O_2$-based plasma or, in a more severe process, in $H_3PO_4$ at 180°C (10 nm/min).[117] Nitride films are typically under a tensile stress of about $1 \times 10^9$ Pa. If in the overall processing of the devices, nitride deposition does not pose a problem, KOH emerges as the preferential anisotropic wet etchant. For dopant dependent etching, EDP is the better etchant and generally better suited for deep etching since its oxide etch rate is negligible (<5 Å/min).

Oxide and nitride are masking for anisotropic etchants to varying degrees with both mask types being used. When these layers are used to terminate an etch in the [100] direction, a low etch rate of the mask layer allows overetching of silicon, to compensate for wafer thickness variations. A KOH solution etches $SiO_2$ at a relatively fast rate of 1.4 to 3 nm/min so that $Si_3N_4$ or Au/Cr must be used as a mask against KOH for deep and long etching.

*Backside Protection*

In many cases it is necessary to protect the backside of a wafer from an isotropic or anisotropic etchant. The backside is either mechanically or chemically protected. In the mechanical method the wafer is held in a holder, often made from Teflon. The wafer is fixed between Teflon-coated O-rings which are carefully aligned in order to avoid mechanical stress in the wafer. In the chemical method, waxes or other organic coatings are spun onto the back side of the wafer. Two wafers may be glued back to back for faster processing.

*Etch Rate and Etch Stops*

The Si etch rate, R, as a function of KOH concentration at 72°C, was shown in Figure 4.27C. The etch rate has a maximum at about 20% KOH. The best fit for this experimentally determined etch rate, for most KOH concentrations, is[88,90]

$$R = k[H_2O]^4 [KOH]^{\frac{1}{4}} \qquad \textbf{4.30}$$

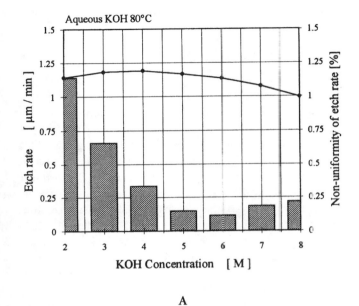

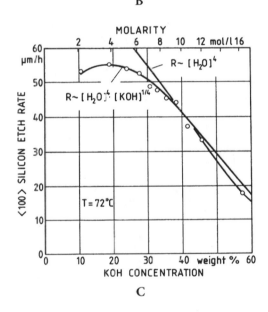

FIGURE 4.27 Anisotropic etching of silicon. (A) Silicon (100) etch rate (line) and nonuniformity of etch rate (column) in KOH at 80°C as a function of KOH concentration. The etch rate for all low index planes is maximal at around 4 *M*. (B) Silicon (100) surface roughness (Ra) in aqueous KOH at 80°C as a function of concentration for a 1-hour etch time (thin line) and for an etch depth of 60 µm (thick line). (C) Silicon (100) etch rate as a function of KOH concentration at a temperature of 72°C. (See also Appendix B.) (A and B from Peeters, E., Ph.D. thesis, 1996, KUL, Belgium. With permission. C from Seidel et al., *J. Electrochem. Soc.*, 137, 3612–3626, 1990. With permission.)

Any model of anisotropic etching will have to explain this peculiar dependency on the water and KOH concentration, as well as the fact that all anisotropic etchant systems of Table 4.9 exhibit drastically reduced etch rates for high boron concentrations in silicon ($\geq 5 \times 10^{19}$ cm$^{-3}$ solid solubility limit). Other impurities (P, Ge) also reduce the etch rate, but at much higher concentrations (see Figure 4.29[90]). Boron typically is incorporated using ion implantation (thin layers) or liquid/solid source deposition (thick layers > 1 µm). These doped layers are used as very effective etch stop layers (see below). Hydrazine or EDP, which display a smaller (100) to (111) etch rate ratio (~ 35) than KOH, exhibit a stronger boron concentration dependency. The etch rate in KOH is reduced by a factor of 5 to 100 for a boron concentration larger than $10^{20}$ cm$^{-3}$. When etching in EDP, the factor climbs to 250.[118] With TMAHW solutions, the Si etch rate decreases to 0.01 µm/min for boron concentrations of about $4 \times 10^{20}$ cm$^{-3}$.[119] The mechanism eludes us, but Seidel et al.'s model (see below) gives the most plausible explanation

for now. Some of the different mechanisms to explain etch stop effects that have been suggested follow:

1. Several observations suggest that doping leads to a more readily oxidized Si surface. Highly boron- or phosphorus-doped silicon in aqueous KOH spontaneously can form a thin passivating oxide layer.[120,121] The boron-oxides and -hydroxides initially generated on the silicon surface are not soluble in KOH or EDP etchants.[34] The substitutional boron creates local tensile stress in the silicon, increasing the bond strength so that a passivating oxide might be more readily formed at higher boron concentrations. Boron-doped silicon has a high defect density (slip planes), encouraging oxide growth.

2. Electrons produced during oxidation of silicon are needed in a subsequent reduction step (hydrogen evolution in Reaction 4.2). When the hole density passes $10^{19}$ cm$^{-3}$ these electrons combine with holes instead, thus

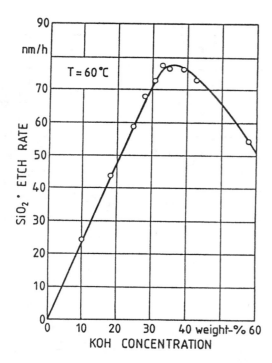

FIGURE 4.28    The $SiO_2$ etch rate in nm/hr as a function of KOH concentration at 60°C. (From Seidel, H. et al., *J. Electrochem. Soc.,* 137, 3612–3626, 1990. With permission.)

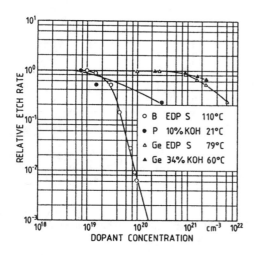

FIGURE 4.29    Relative etch rate for (100) Si in EDP and KOH solutions as a function of concentration of boron, phosphorus, and germanium. (From Seidel, H. et al., *J. Electrochem. Soc.,* 137, 3626–3632, 1990. With permission.)

stopping the reduction process.[121] Seidel et al.'s model follows this explanation (see below).

3. Silicon doped with boron is under tension as the smaller boron atoms enter the lattice substitutionally. The large local tensile stress at high boron concentration makes it energetically more favorable for the excess boron (above 5 $\times 10^{19}$ cm$^{-3}$) to enter interstitial sites. The strong B-Si bonds bind the lattice rigidly. With high enough doping the high binding energy can stop etching.[34] This hypothesis is similar to item 1, except that no oxide formation is invoked.

In what follows we review the results of some typical anisotropic etching experiments.

## Anisotropically Etched Structures

### Examples

In Figure 4.30 we compare a wet isotropic etch (a) with examples of anisotropic etches (b to e). In the anisotropic etching examples, a square (b and c) and a rectangular pattern (d) are defined in an oxide mask with sides aligned along the <110> directions on a <100>-oriented silicon surface. The square openings are precisely aligned (within one or two degrees) with the <110> directions on the (100) wafer surface to obtain pits that conform exactly to the oxide mask rather than undercutting it. Most (100) silicon wafers have a main flat parallel to a <110> direction in the crystal, allowing for an easy alignment of the mask (see Figure 4.5). Etching with the square pattern results in a pit with well-defined {111} sidewalls (at angles of 54.74° to the surface) and a (100) bottom.

The dimensions of the hole at the bottom of the pit, as we saw above, are given by Equation 4.2. The larger the square opening in the mask, the deeper the point where the {111}

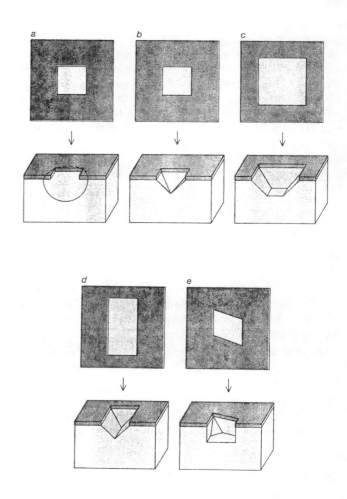

FIGURE 4.30    Isotropic and anisotropic etched features in <100> and <110> wafers. (a) isotropic etch; (b) to (e) anisotropic etch. (a) to (d): <100> oriented wafers and (e): <110> oriented wafer.

sidewalls of the pit intersect. If the oxide opening is wide enough, i.e., $W_m > \sqrt{2}\, z$ (with $z = 600\ \mu m$ for a typical 6-in. wafer, this means $W_m > 849\ \mu m$), the {111} planes do not intersect within the wafer (see also earlier in this chapter). The etched pit in this particular case extends all the way through the wafer, creating a small square opening on the bottom surface. As shown in Figure 4.30 (b to d), no under-etching of the etch mask is observed due to the perfect alignment of the concave oxide mask opening with the <110> direction. In Figure 4.30a, an undercutting isotropic etch (acidic) is shown. Misalignment in the case of an anisotropic etch still results in pyramidical pits, but the mask will also be severely undercut. A rectangular pattern aligned along the <110> directions on a <100> wafer leads to *long V-shaped grooves* (see Figure 4.30d) or an open slit, depending on the width of the opening in the oxide mask.

Using a properly aligned mask on a <110> wafer, holes with four vertical walls ({111} planes) result (see Figure 4.30e and Figure 4.31 A and C). Figure 4.31B shows that a slight mask misorientation leads to all skewed sidewalls. A U-groove based on a rectangular mask with the long sides along the <111> directions and anisotropically etched in (110) silicon has a complex shape delineated by six {111} planes, four vertical and two slanted (see Figure 4.31D). Before emergence of {111} planes, the U-groove is defined by four vertical (111) planes and a horizontal (110) bottom. Self-stopping occurs when the tilted end planes intersect at the bottom of the groove. It is easy to etch long, narrow U-grooves very deeply into a <110> silicon wafer. However, it is impossible to etch a short, narrow U-groove deeply into a slice of silicon,[44] because the narrow dimension of the groove is quickly limited by slow-etching {111}

planes that subtend an angle of 35° to the surface and cause etch termination. At a groove of length $L = 1\ mm$ on the top surface, etching will stop when it reaches a depth of 0.289 mm, i.e., $D_{max} = L/2\sqrt{3}$. For very long grooves, the tilted end planes are too far apart to intersect in practical cases, making the end effects negligible compared to the remaining U-shaped part of the groove.

A laser can be used to melt or 'spoil' the shallow (111) surfaces, making it possible to etch deep vertical-walled holes through a (110) wafer as shown in Figure 4.32A.[122-124] The technique is illustrated in Figure 4.32B. The absorbed energy of a Nd:YAG laser beam causes a local melting or evaporation zone enabling etchants to etch the shallow (111) planes in the line-of-sight of the laser. Etching proceeds till 'unspoiled' (111) planes are encountered. Some interesting resulting possibilities, including partially closed microchannels, are shown in Figure 4.32C.[122,125] Note that with this method it is possible to use <111> wafers for micromachining. The light of the Nd:YAG laser is very well suited for this micromachining technique due to the 1.17-eV photon energy, just exceeding the band gap energy of Si. Details on this laser machining process can be found, for example, in Alavi et al.[40,125]

Especially when machining surface structures by undercutting, the orientation of the wafer is of extreme importance. Consider, for example, the formation of a bridge in Figure 4.33A.[123] When using a (100) surface, a suspension bridge cannot form across the etched V-groove; two independent truncated V-grooves flanking a mesa structure result instead. To form a suspended bridge it must be oriented away from the <110> direction. This in contrast with a (110) wafer where a

## Long V-shaped grooves in a (100) Si wafer

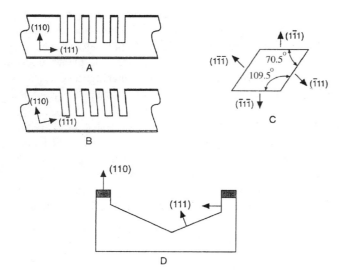

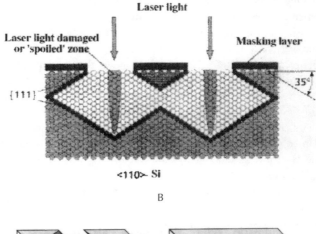

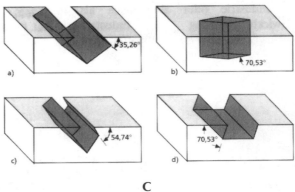

**FIGURE 4.31** Anisotropic etching of <110> wafers. (A) Closely spaced grooves on correctly oriented (110) surface. (B) Closely spaced grooves on misoriented <110> wafer. (C) Orientations of the {111} planes looking down on a (110) wafer. (D) Shallow slanted (111) planes eventually form the bottom of the etched cavity.

microbridge crossing a V-groove with a 90° angle will be undercut. Convex corners will be undercut by etchant, allowing formation of cantilevers as shown in Figure 4.33B. The diving board shown forms by undercutting starting at the convex corners.

To create vertical (100) faces, as shown in Figure 4.10, in general only KOH works (not EDP or TMAHW) and it has to happen in high-selectivity conditions (low temperature, low concentration: 25 wt% KOH, 60°C). Interestingly, high concentration KOH (45 wt%) at higher temperatures (80°C) produces a smooth sidewall, controllable and repeatable at an angle of 80°. EDP produces 45° angled planes and TMAHW usually makes a 30° angle.[120]

*Alignment Patterns*

When alignment of a pattern is critical, pre-etch alignment targets become useful to delineate the planes of interest since the wafer flats often are aligned to ±1° only. In order to find the proper alignment for the mask, a test pattern of closely spaced lines can be etched (see Figure 4.34). The groove with the best vertical walls determines the proper final mask orientation. Along this line, Ciarlo[127] made a set of lines of 3 mm long and 8 μm wide, fanning out like spokes in a wagon wheel at angles 0.1° apart. This target was printed near the perimeter of the wafer and then etched 100 μm into the surface. Again, by evaluating the undercut in this target the correct crystal direction could be determined. Alignment with better than 0.05° accuracy was accomplished this way. Similarly, to obtain detailed experimental data on crystal orientation dependence of etch rates, Seidel et al.[90] used a wagon-wheel or star-shaped mask (e.g., made from CVD-$Si_3N_4$; $SiH_4$ and $NH_3$ at 900°C), consisting of radially divergent segments with an angular

**FIGURE 4.32** Laser/KOH machining. (A) Holes through a (110) Si wafer created by laser spoiling and subsequent KOH etching. The two sets of (111) planes making an angle of 70° are the vertical walls of the hole. The (111) planes making a 35° angle with the surface tend to limit the depth of the hole but laser spoiling enables one to etch all the way through the wafer (see also B (c)). (B) After laser spoiling, the line-of-sight (111) planes are spoiled and etching proceeds until unspoiled (111) planes are reached. (C) Some of the possible features rendered by laser spoiling. (From Schumacher, A. et al., *Technische Rundschau*, 86, 20–23, 1994. With permission.)

separation of one degree. Yet finer 0.1° patterns were made around the principal crystal directions. The etch pattern emerging on a <100> oriented wafer covered with such a mask is shown in Figure 4.35A. The blossom-like figure is due to the total underetching of the passivation layer in the vicinity of the

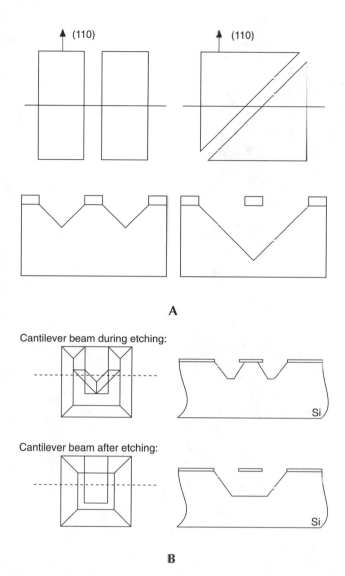

**FIGURE 4.33** How to make (A) a suspension bridge from a (100) Si wafer and (B) a diving board from a (110) Si wafer.[123]

center of the wagon wheel, leaving an area of bare exposed Si. The radial extension of the bare Si area depends on the crystal orientation of the individual segments, leading to a different amount of total underetching. The observation of these blossom-like patterns was used for qualitative guidance of etching rates only. In order to establish quantitative numbers for the lateral etch rates, the width, w, of the overhanging passivation layer was measured with an optical line-width measurement system (see Figure 4.35B). Laser beam reflection was used to identify the crystal planes and ellipsometry was used to monitor the etching rate of the mask itself. Lateral etch rates determined in this way on <100>- and <110>-oriented wafers at 95°C in EDP (470 ml water, 1 1 ED, 176 g pyrocatechol) and at 78°C in a 50% KOH solution are shown in Figure 4.36. Etch-rates shown are normal to the actual crystal surface and are conveniently described in a polar plot in which the distance from the origin to the polar plot surface (or curve in two dimensions) indicates the etch rate for that particular direction. Note the deep minima at the {111} planes. It can also be seen that in KOH the peak

etch rates are more pronounced. A further difference is that with EDP the minimum at {111} planes is steeper than with KOH. For both EDP and KOH, the etch rate depends linearly on misalignment. All the above observations have important consequences for the interpretation of anisotropy of an etch (see below). The difference between KOH and EDP etching behavior around the {111} minima has the direct practical consequence that it is more important for etching in EDP to align the crystallographic direction more precisely than in KOH.[42]

When determining etch rates without using underetching masks but by using vertical etching of beveled silicon samples, results are quite different than when working with masked silicon.[110] The etch rates on open areas of beveled structures are much larger than in underetching experiments with masked silicon, and different crystal planes develop. Herr et al. conclude that crevice effects may play an important role in anisotropic etching. Elwenspoek et al.'s model,[128,129] analyzed below, is the only model which predicts such a crevice effect. He explains why, when etchants are in a small restricted crevice area and are not refreshed fast enough, etching rates slow down and increase anisotropy.

## Chemical Etching Models

### Introduction

Lots of conflicting data exist in the literature on the anisotropic etch rates of the different Si planes, especially for the higher index planes. This is not too surprising, given the multiple parameters influencing individual results: temperature, stirring, size of etching feature (i.e., crevice effect), KOH concentration, addition of alcohols and other organics, surface defects, complexing agents, surfactants, pH, cation influence, etc. More rigorous experimentation and standardization will be needed, as well as better etching models, to better understand the influence of all these parameters on etch rates.

Several chemical models explaining the anisotropy in etching rates for the different Si orientations have been proposed. Presently we will list all of the proposed models and compare the two most recent and most detailed models (the one by Seidel et al.[88,90] and the one by Elwenspoek et al.[128,129]). Different Si crystal properties have been correlated with the anisotropy in silicon etching.

1. It has been observed that the {111} Si planes present the highest density of atoms per $cm^2$ to the etchant and that the atoms are oriented such that *three bonds are below the plane*. It is possible that these bonds become chemically shielded by surface bonded (OH) or oxygen, thereby slowing the etch rate.

2. It also has been suggested that etch rate correlates with available bond density, the surfaces with the highest bond density etching faster.[85] The available bond densities in Si and other diamond structures follow the sequence 1:0.71:0.58 for the {100}:{110}:{111} surfaces. However, Kendall[44] commented that bond density alone is an unlikely explanation because of the magnitude of etching

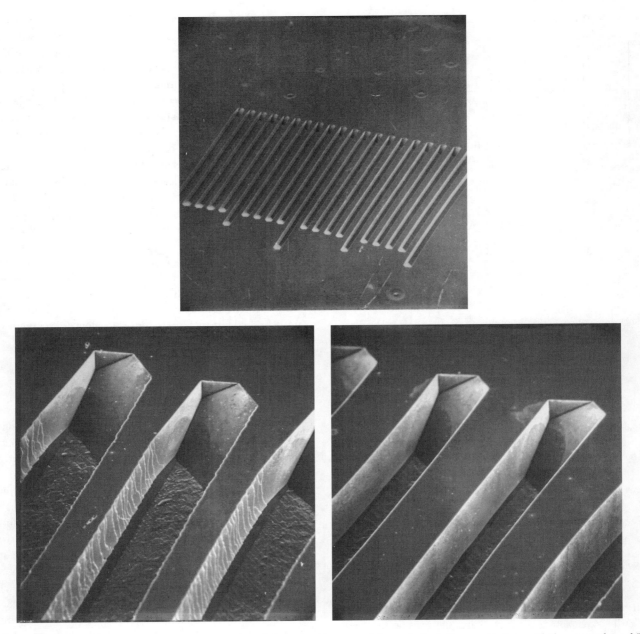

**FIGURE 4.34**  Test pattern of U-grooves in a <110> wafer to help in the alignment of the mask; final alignment is done with the groove that exhibits the most perfect long perpendicular walls.

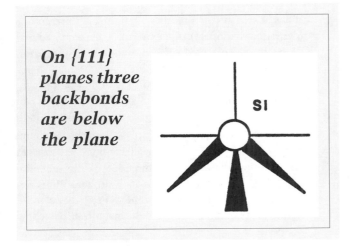

*On {111} planes three backbonds are below the plane*

**SI**

anisotropy (e.g., a factor of 400), compared to the bond density variations of at most a factor of two.

3. Kendall[44] explains the slow etching of {111} planes on the basis of their faster oxidation during etching; this does not happen on the other faces, due to greater distance of the atoms on planes other than (111). Since they oxidize faster, these planes may be better protected against etching. The oxidation rate in particular follows the sequence {111} > {110} > {100}, and the etch rate often follows the reverse sequence (see also Chapter 3 on Si oxidation). In the most used KOH-H$_2$O, however, the sequence is {110} > {100} > {111}.

4. In yet another model, it is assumed that the anisotropy is due to differences in activation energies and backbond geometries on different Si surfaces.[130]

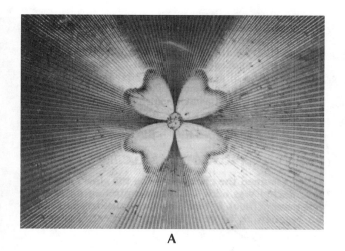

A

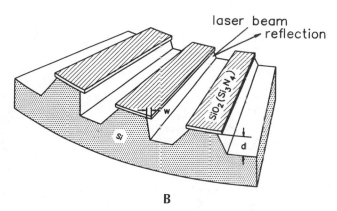

B

FIGURE 4.35  (A) Etch pattern emerging on a wagon wheel-masked, <100>-oriented Si wafer after etching in an EDP solution. (B) Schematic cross-section of a silicon test chip covered with a wagon wheel-shaped masking pattern after etching. The measurement of 'w' is used to construct polar diagrams of lateral underetch rates as shown in Figure 4.36. (From Seidel et al., *J. Electrochem. Soc.*, 177, 3612–3626, 1990. With permission.)

5. Seidel et al.'s model[88,90] supports the previous explanation. They detail a process to explain anisotropy based on the difference in energy levels of backbond-associated surface states for different crystal orientations.

6. Finally, Elwenspoek et al.[128,129] propose that it is the degree of atomic smoothness of the various surfaces that is responsible for the anisotropy of the etch rates. Basically, this group argues that the kinetics of smooth faces (the (111) plane is atomically flat) is controlled by a nucleation barrier that is absent on rough surfaces. The latter, therefore, would etch faster by orders of magnitude.

The reason why acidic media lead to isotropic etching and alkaline media to anisotropic etching was, until recently, not addressed in any of the models surveyed. In the following we will give our own model as well as Elwenspoek et al.'s to explain isotropic vs. anisotropic etching behavior.

It is our hope that the reading of this section will inspire more detailed electrochemistry work on Si electrodes. The refining of an etching model will be of invaluable help in writing more predictive Si etching software code.

## Seidel et al.'s Model

Seidel et al's model is based on the fluctuating energy level model of the silicon/electrolyte interface and assumes the injection of electrons in the conduction band of Si during the etching process. Consider the situation of a piece of Si immersed in a solution without applied bias at open circuit. After immersion of the silicon crystal into the alkaline electrolyte a negative excess charge builds up on the surface due to the higher original Fermi level of the $H_2O/OH^-$ redox couple as compared to the Fermi level of the solid, i.e., the work function difference is equalized. This leads to a downward bending of the energy bands on the solid surface for both p- and n-type silicon (Figure 4.37A and B). The downward bending is more pronounced for p-type than for n-type due to the initially larger difference of the Fermi levels between the solid and the electrolyte.

Next, hydroxyl ions cause the Si surface to oxidize, consuming water and liberating hydrogen in the process. The detailed steps, based on suggestions by Palik,[120] are:

$$Si + 2OH^- \rightarrow Si(OH)_2^{2+} + 2e^-$$

$$Si(OH)_2^{2+} + 2OH^- \rightarrow Si(OH)_4 + 2e^- \qquad \text{Reaction 4.4}$$

$$Si(OH)_4 + 4e^- + 4H_2O \rightarrow Si(OH)_6^{2-} + 2H_2$$

Silicate species were observed by Raman spectroscopy.[131]

The overall silicon oxidation reaction consumes four electrons first injected into the conduction band, where they stay near the surface due to the downward bending of the energy bands (see Figure 4.37). Evidence for injection of four electrons rather than a mixed hole and electron mechanism was first presented by Raley et al.[132] The authors could explain the measured etch-rate dependence on hole concentration only by assuming that the proton or water reduction reaction is rate determining and that a four-electron injection mechanism with the conduction band is involved. These injected electrons are highly 'reducing' and react with water to form hydroxide ions and hydrogen:

$$4H_2O + 4e^- \rightarrow 4H_2O^- \qquad \text{Reaction 4.5}$$

$$4H_2O^- \rightarrow 4OH^- + 4H^+ + 4e^- \rightarrow 4OH^- + 2H_2 \quad \text{Reaction 4.6}$$

It is thought that the hydroxide ions in Reaction 4.6, generated directly at the silicon surface, react in the oxidation step. The hydroxide ions from the bulk of the solution may not play a major role, as they will be repelled by the negatively charged Si surface, whereas the hydroxide ions formed *in situ* do not need to overcome this repelling force. This would explain why the etch rates for an EDP solution with an $OH^-$ concentration of 0.034 mol/l are nearly as large as those for KOH solutions with a hundred-fold higher $OH^-$ concentration of 5 to 10 mol/l.[90] The hydrogen formed in Reaction 4.6 can inhibit the reaction and surfactants may be added to displace the hydrogen (IBM, U.S. Patent 4,113,551, 1978). Additional support for the

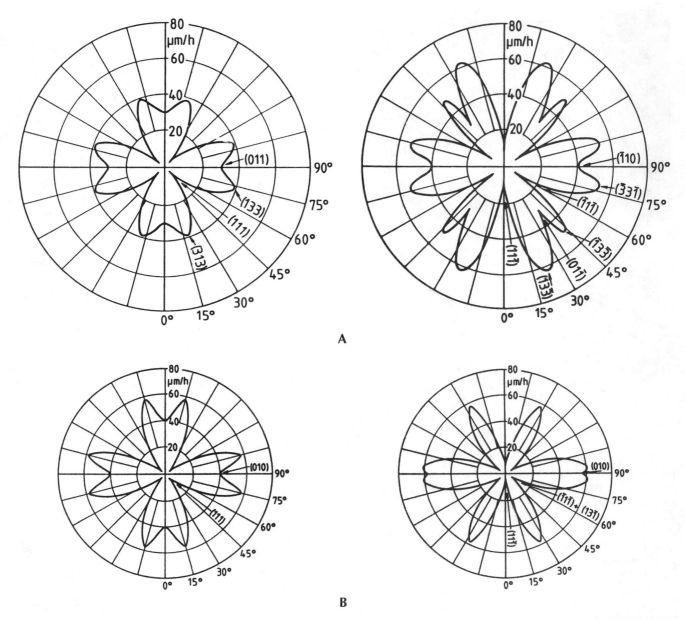

**FIGURE 4.36**  Lateral underetch rates as a function of orientation for (A) EDP (470 ml water, 1 l ED, 176 g pyrocatechol) at 95°C. (B) KOH (50% solution) at 78°C. Left, <100>- and right, <110>-oriented Si wafers. (From Seidel, H. et al., *J. Electrochem. Soc.*, 137, 3612–3626, 1990. With permission.)

involvement of four water molecules (Reaction 4.5) comes from the experimentally observed correlation between the fourth power of the water concentration and the silicon etch rate for highly concentrated KOH solutions (Equation 4.30 and Figure 4.27C). The weak dependence of the etching curve on the KOH concentration (~1/4 power) supports the assumption that the hydroxide ions involved in the oxidation reactions are mostly generated from water. A strong influence of water on the silicon etch rate was also observed for EDP solutions. In molar water concentrations of up to 60%, a large increase of the etch rate occurs.[86] The driving force for the overall Reaction 4.4 is given by the larger Si-O binding energy of 193 kcal/mol as compared to a Si-Si binding energy of only 78 kcal/mol. The role of cations, $K^+$, $Na^+$, $Li^+$, and even complicated cations such as $NH_2(CH_2)_2NH_3^+$ can probably be neglected.[90]

The four electrons in Reaction 4.5 are injected into the conduction band in two steps. In the case of {100} planes there are two *dangling bonds* per surface atom for the first two of the four hydroxide ions to react with, injecting two electrons into the conduction band in the process. As a consequence of the strong electronegativity of the oxygen atoms, the two bonded hydroxide groups on the silicon atom reduce the strength of the two *silicon backbonds*. With two new hydroxide ions approaching, two more electrons (now stemming from the Si-Si backbonds) are injected into the conduction band and the silicon-hydroxide complex reacts with the two additional hydroxide ions. Seidel et al.[90] claim that the step of activating the second two electrons from the backbonds into the conduction band is the rate-limiting step, with an associated thermal activation energy of about 0.6 eV for {100} planes. The electrons in the

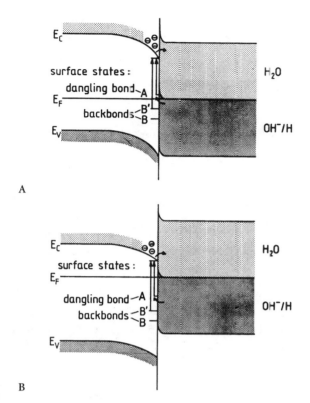

A

B

**FIGURE 4.37** Band model of the silicon/electrolyte interface for moderately doped Si (electrolyte at pH > 12): (A) p-type Si and (B) n-type Si. We assume no applied bias and no illumination. The energy scale functions in respect to the saturated calomel electrode (SCE), an often-used reference in electrochemistry. Notice that p-type Si exhibits more band bending as its Fermi level is lower in the band gap. For simplicity, we show only one energetic position for surface states associated with dangling bonds and backbonds; in reality, there will be new surface states arising during reactions as the individual dangling bonds and backbonds are taking on different energies as new Si-OH bonds are introduced.

backbonds are associated with surface states within the bandgap (see Figure 4.37). The energy level of these surface states is assumed to be varying for different surface orientations, being lowest for {111} planes. The thermal activation of the backbonds corresponds to an excitation of the electrons out of these surface states into the conduction band. Since the energy for the backbond surface state level is the lowest within the bandgap for {111} planes, these planes will be hardest to etch. The {111} planes have only one dangling bond for a first hydroxide ion to react with. The second rate-limiting step involves breaking three lower energy backbonds. The lower energy of the backbond surface states for {111} Si atoms can be understood from the simple argument that their energy level is raised less by the electronegativity of a single binding hydroxide ion, compared to two in the case of the silicon atoms in {100} planes. The high etch rate generally observed on {110} surfaces is similarly explained by a high energy level of the backbond-associated surface states for these planes. Elwenspoek et al.[128,129] do not accept this 'two vs. three backbonds' argument. They point out that the silicon atoms in the {110} planes also have three backbonds, and activation energy in these crystallographic directions should be comparable to that of {111} planes in

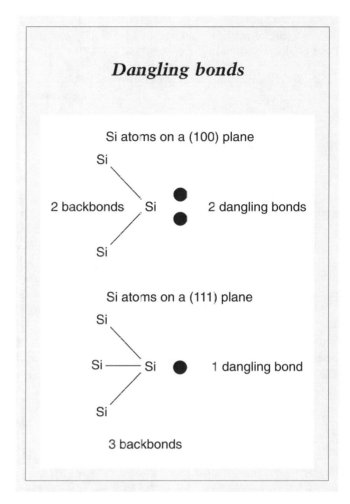

*Dangling bonds*

Si atoms on a (100) plane

Si
2 backbonds    Si    2 dangling bonds
Si

Si atoms on a (111) plane

Si
Si ——— Si    1 dangling bond
Si

3 backbonds

contrast to experimental evidence. Seidel et al. would probably counter-argue here that the backbonds and the energy levels of the associated surface states is not necessarily the same for {111} and {110} planes, as that energy will also be influenced by the effect of the orientation of these bonds. Another argument in favor of the high etching rates of {110} planes is the easier penetrability of {110} surfaces for water molecules along channels in that plane.

The final step in the anisotropic etching is the removal of the reaction product $Si(OH)_4$ by diffusion. If the production of $Si(OH)_4$ is too fast, for solutions with a high water concentration, the $Si(OH)_4$ leads to the formation of a $SiO_2$-like complex before $Si(OH)_4$ can diffuse away. This might be observed experimentally as a white residue on the wafer surface.[133] The high pH values in anisotropic etching are required to obtain adequate solubility of the $Si(OH)_4$ reaction product and to remove the native oxide from the silicon surface. From silicate chemistry it is known that for pH values above 12 the $Si(OH)_4$ complex will undergo the following reaction by the detachment of two protons:

$$Si(OH)_4 \rightarrow SiO_2(OH)_2^{2-} + 2H^+ \qquad \text{Reaction 4.7}$$

$$2H^+ + 2OH^- \rightarrow 2H_2O \qquad \text{Reaction 4.8}$$

Pyrocathechol in an ethylenediamine etchant acts as complexing agent for reaction products such as $Si(OH)_4$, converting these products into more complex anions:

$$Si(OH)_4 + 2OH^- + 3C_6H_4(OH)_2 \rightarrow$$
$$Si(C_6H_4O_2)_3^{2-} + 6H_2O \qquad \text{Reaction 4.9}$$

There is evidence by Abu-Zeid et al.[134] of diffusion control contribution to the etch rate in EDP, probably because the hydroxide ion must diffuse through the layer of complex silicon reaction products (see Reaction 4.9). The same authors also found that the etch rate depends on the effective Si area being exposed and on its geometry (crevice effect). That is why the silicon wafer is placed in a holder and the solution is vigorously agitated in order to minimize the diffusion layer thickness. For KOH solutions, no effect of stirring on etching rate was noticed. Stirring here is mainly used to decrease the surface roughness, probably through removal of hydrogen bubbles.

The influence of alcohol on the KOH etching rate in Equation 4.30 mainly is due to a change in the relative water concentration and its concomitant pH change; it does not participate in the reaction (this was confirmed by Raman studies by Palik et al.[131]). The reversal of etch rates for {110} and {100} planes through alcohol addition to KOH/water etchants can be understood by assuming that the alcohol covers the silicon surface,[131] thus canceling the 'channeling advantage' of the {110} planes. In the case of EDP, alcohol has no effect, as the water concentration can be freely adjusted without significantly influencing the pH value due to the incomplete dissociation of EDP.

For the etching of $SiO_2$ shown in Figure 4.28, Seidel et al. propose the following reaction:

$$SiO_2 + 2OH^- \rightarrow SiO_2(OH)_2^{2-} \qquad \text{Reaction 4.10}$$

At KOH concentrations up to 35% a linear correlation occurs between etch rate and KOH concentration. The $SiO_2$ etch rate in KOH solutions exceeds those in EDP by close to three orders of magnitude. For higher concentrations, the etch rate decreases with the square of the water concentration, indicating that water plays a role in this reaction. Seidel et al. speculate that at high pH values the silicon electrode is highly negatively charged (the point of zero charge of $SiO_2$ is 2.8), repelling the hydroxide ions while water takes over as reaction partner. An additional reason for the decrease is that the hydroxide concentration does not continue to increase with increasing KOH concentration for very concentrated solutions. The decrease of the $Si/SiO_2$ etch rate ratio with increasing temperature and pH value of the solution follows out of the larger activation energy of the $SiO_2$ etch rate (0.85 eV) and its linear correlation with the hydroxide concentration, whereas the silicon etch rate mainly depends on the water concentration.

The effect of water concentration and pH value on the etching process in the Seidel et al. model is summarized in Table 4.10.[135] Etch rates for Si in $\mu m/hr$ and for thermally grown $SiO_2$ in $nm/hr$ for various KOH concentrations and etch temperatures are given in Appendix C.[90] For aqueous KOH solutions within a concentration range from 10 to 60% the following empirical

**TABLE 4.10** Effect of Water Concentration and pH Value on the Characteristics of Silicon Etching[135]

|  | $-H_2O+$ | $-pH+$ |
|---|---|---|
| $SiO_2$ etch rate | No effect | $- \Leftrightarrow +$ |
| Si etch rate | $- \Leftrightarrow +$ | Little effect |
| Solubility | No effect | $- \Leftrightarrow +$ |
| $Si/SiO_2$ ratio | $- \Leftrightarrow +$ | $+ \Leftrightarrow -$ |
| Diffusion effects | $- \Leftrightarrow +$ | $+ \Leftrightarrow -$ |
| Residue formation | $- \Leftrightarrow +$ | $+ \Leftrightarrow -$ |
| $p^+$ etch stop | $- \Leftrightarrow +$ | $+ \Leftrightarrow -$ |
| p, n etch stop | $- \Leftrightarrow +$ | $+ \Leftrightarrow -$ |

formula for the calculation of the silicon etch rate R proved to be in close agreement with the experimental data:

$$R = k_0 [H_2O]^4 [KOH]^{\frac{1}{4}} e^{-\frac{E_a}{kT}} \qquad 4.31$$

The values for the fitting parameters were $E_a = 0.595$ eV and $k_0 = 2480$ $\mu m/hr$ $(mol/l)^{-4.25}$; for a (100) wafer $E_a = 0.6$ and $k_0 = 4500$ $\mu m/hr$ $(mol/l)^{-4.25}$. For the $SiO_2$ etch, an activation energy of 0.85 eV was used.

In the section on etch stop techniques, we will see that the Seidel et al. model also nicely explains why all alkaline etchants exhibit a strong reduction in etch rate at high boron dopant concentration of the silicon; at high doping levels the conduction band electrons for the rate-determining reduction step are not confined to the surface anymore and the reaction basically stops.

The key points of the Seidel et al. model can be summarized as follows (see also Table 4.10):[135]

1. The rate-limiting step is the water reduction.

2. Hydroxide ions required for oxidation of the silicon are generated through reduction of water at the silicon surface. The hydroxide ions in the bulk do not contribute to the etching, since they are repelled from the negatively charged surface. This implies that the silicon etch rate will depend on the molar concentration of water and that cations will have little effect on the silicon etch rate.

3. The dissolution of silicon dioxide is assumed to be purely chemical with hydroxide ions. The $SiO_2$ etch rate depends on the pH of the bulk electrolyte.

4. For boron concentrations in excess of $3 \times 10^{19}$ $cm^{-3}$, the silicon becomes degenerate, and the electrons are no longer confined to the surface. This prevents the formation of the hydroxide ions at the surface and thus causes the etching to stop.

5. Anodic biases will prevent the confinement of electrons near the surface as well and lead to etch stop as in the case of a $p^+$ material.

Points 4 and 5 will become more clear when we discuss the workings of etch-stop techniques. This model applies well for lower index planes (i.e., {nnn} with n < 2) where high etch rates always correspond to low activation energies. But, for higher index planes (i.e., {n11} and {1nn} with n = 2,3,4), Herr et al.[110]

found no correlation between activation energies and etch rates. For higher index planes, we must rely mainly on empirical data.

The Si etching reactions suggested by Seidel et al. are only the latest; in earlier proposed schemes, according to Ghandhi[136] and Kern,[60] the silicon oxidation reaction steps suggested were injection of holes into the Si (raising the oxidation state of Si), hydroxylation of the oxidized Si species, complexation of the silicon reaction products and dissolution of the reaction products in the etchant solution. In this reaction scheme, etching solutions must provide a source of holes as well as hydroxide ions and they must contain a complexing agent with soluble reacted Si species in the etchant solution, e.g., pyrocathechol forming the soluble $Si(C_6H_4O_2)_3^{2-}$ species. This older model still seems to be guiding the current thinking of many micromachinists, although Seidel et al.'s energy level based model of the silicon/electrolyte interface proves more satisfying.

### Elwenspoek et al. Model

Elwenspoek et al.[128,129] introduced an alternative model for anisotropic etching of Si, a model built on theories derived from crystal growth. According to these authors, the Seidel et al. model does not clearly explain the fast etching of {110} planes. Those planes, having three backbonds like the {111} planes, should etch equally slowly. The activation energy of the anisotropic etch rate depends on the etching system used; for example, etching in KOH is faster than in EDP, even when the pH of the solution is the same. Seidel et al. attribute this dependence to diffusion that plays a greater role in EDP than in KOH solutions. But Elwenspoek et al. point out that, at least for slow etching, the etch rate should not be diffusion controlled but governed by surface reactions. With surface reactions, diffusion should have a minor effect, analogous to growth at low pressure in an LPCVD reactor (see Chapter 3). Another comment focuses on the lack of understanding why certain etchants etch isotropically and others etch anisotropically.

Elwenspoek et al. note the parallels in the process of etching and growing of crystals; slowly growing crystal planes also etch slowly! A key to understanding both processes, growing or dissolution (etching), pertains to the concept of the energy associated with the creation of a critical nucleus on a single crystalline smooth surface, i.e., the free energy associated with the creation of an island (growth) or a cavity (etching). Etching or growing of a material starts at active kink sites on steps. Kink sites are atoms with as many bonds to the crystal as to the liquid. Kinetics depend critically on the number of such kink sites. This aspect remained neglected in the discussion of etch rates of single crystals up to now.

The free energy change, $\Delta G$, involved in creating an island or digging a cavity (of circular shape in an isotropic material) of radius r on or in an atomically smooth surface, is given by:

$$\Delta G = -N\Delta\mu + 2\pi r\gamma \qquad 4.32$$

where N is the number of atoms forming the island or the number of atoms removed from the cavity, $\Delta\mu$ is the chemical potential difference between silicon atoms in the solid state and in the solution, and $\gamma$ is the step free energy. The step free energy

in Equation 4.32 will be different at different crystallographic surfaces. This can easily be understood from the following example. A perfectly flat {111} surface in the Si diamond lattice has no kink positions (three backbonds, one dangling bond per atom), while on the {001} face every atom has two backbonds and two dangling bonds, i.e., every position is a kink position. Consequently, creating an adatom-cavity pair on {111} surfaces costs energy: three bonds must be broken and only one is reformed. In the case of {001} faces, the picture is quite different. Creating an adatom-cavity pair now costs no energy because one has to break two bonds in order to remove an atom from the {001} face, but one gets them back by placing the atom back on the surface. The binding energy $\Delta E$ of an atom in a crystal slice with orientation (hkl) divided by kT (Boltzmann constant times absolute temperature) is known as the $\alpha$ factor of Jackson of that crystal face,[137] or:

$$\alpha = \frac{\Delta E}{kT} \qquad 4.33$$

At sufficiently low temperature, where entropy effects can be ignored, $kT\alpha$ is proportional to the step free energy $\gamma$ and the number of adatom-cavity pairs is proportional to exp $(-\alpha)$. This number is very small on the {111} silicon faces at low temperature, but 1 on the {001} silicon faces at any temperature. The consequence for {111} and {001} planes is that, at sufficiently low temperatures, the first are atomically smooth and the latter are atomically rough. N in Equation 4.32 can be further written out as:

$$N = \pi r^2 h\rho \qquad 4.34$$

where h is the height of the step, r is the diameter of the hole or island, and $\rho$ is the density (atoms per cm³) of the solid material. The result is

$$\Delta G = -\pi r^2 h\rho\Delta\mu + 2\pi r\gamma \qquad 4.35$$

where $\Delta\mu$ is counted positive and $\gamma$ is positive in any case. In Figure 4.38 we show a plot of $\Delta G$ vs. r. Equation 4.35 exhibits a maximum at:

$$r^* = \frac{\gamma}{h\rho\Delta\mu} \qquad 4.36$$

At $r^*$ the free energy is

$$\Delta G^* = \Delta G\left(r^*\right) = \frac{\pi\gamma^2}{h\rho\Delta\mu} \qquad 4.37$$

Consequently, an island or an etch cavity of critical size exist on a smooth face. If by chance a cavity is dug into a crystal plane smaller than $r^*$, it will be filled rather than allowed to grow and an island that is too small will dissolve rather than continue to grow, since that is the easy way to decrease the free energy. With $r = r^*$, islands or cavities do not have any course of action, but

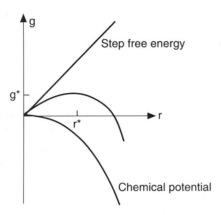

**FIGURE 4.38**  A plot of $\Delta G$ vs. r based on Equation 4.35 exhibits a maximum.

in case of $r > r^*$ the islands or cavities can grow until the whole layer is filled or removed. In light of the above nucleation barrier theory, to remove atoms directly from flat crystal faces such as the {111} Si faces seems very difficult, since the created cavities increase the free energy of the system and filling of adjacent atoms is more probable than removal; in other words, a nucleation barrier has to be overcome. The growth and etch rates, R, of flat faces are proportional to:

$$R \sim \exp\left(-\frac{\Delta G^*}{kT}\right) \qquad \textbf{4.38}$$

Since $\Delta G^*$ is proportional to $\gamma^2$, the activation energy is different for different crystallographic faces and both the etch rate and the activation energy are anisotropic. If $\Delta G^*/kT$ is large, the etch rate will be very small, as is the case for large step free energies and for small undersaturation (i.e., the 'chemical drive' or $\Delta\mu$ is small) (see Equation 4.37). Both $\Delta\mu$ and $\gamma$ depend on the temperature and type of etchant and these parameters might provide clues to understanding the variation of etch rate, degree of anisotropy, temperature dependence, etc., giving this model more bandwidth than the Seidel et al.'s model. According to Elwenspoek et al., the chemical reaction energy barrier and the transport in the liquid are isotropic and the most prominent anisotropy effect is due to the step free energy (absent on rough surfaces) rather than the surface free energy. The surface free energy and the step free energies are related, though: when comparing flat faces, those having a large surface free energy have a small step free energy, and vice versa. The most important difference in these two parameters is that the step-free energy is zero for a rough surface, whereas the surface energy remains finite.

Flat faces grow and etch with a rate proportional to $\Delta G^*$ which predicts that faces with a large free energy to form a step will grow and etch much slower than faces with smaller free energy. Elementary analysis indicates that the only smooth face of the diamond lattice is the (111) plane. There may be other flat faces, but with lower activation energies, due to reconstruction and/or adsorption, prominent candidates in this category are {100} and

{110} planes. On the other hand, a rough crystal face grows and etches with a rate directly proportional to $\Delta\mu$. The temperature at which $\gamma$ vanishes and a face transitions from smooth to rough is called the roughening transition temperature $T_R$.[138,139] Above $T_R$, the crystal is rough on the microscopic scale. Because the step free energy is equal to zero, new Si units may be added or removed freely to the surface without changing the number of steps. Rough crystal faces grow and dissolve with a rate proportional to $\Delta\mu$ and therefore proceed faster than flat surfaces. Imperfect crystals, e.g., surfaces with screw dislocations, etch even faster with R proportional to $\Delta\mu^2$.

For the state of a surface slightly above or below the roughening temperature, $T_R$, thermal equilibrium conditions apply. Etching, in most practical cases, is far from equilibrium and kinetic roughening might occur. Kinetic roughening[139] occurs if the super- or undersaturation of the solution is so large that the thermally created islands or cavities are the size of the critical nucleus. One can show that if the super- or undersaturation is larger than $\Delta\mu_c$, given by:

$$\Delta\mu_c = \frac{\pi f_o \gamma^2}{kT} \qquad \textbf{4.39}$$

($f_0$ being the area one atom occupies in a given crystal plane), the growth and etch mechanism changes from a nucleation barrier-controlled mechanism to a direct growth/etch mechanism. The growth rate and etch rate again become proportional to the chemical potential difference. It can thus be expected that if the undersaturation becomes high enough, even the {111} faces could etch isotropically, as they indeed do in acidic etchants. If the undersaturation becomes so large that $\Delta G^* \ll kT$, the nucleation barrier breaks down. Each single-atom cavity acts as a nucleus made in vast numbers by thermal fluctuations. The face in question etches with a rate comparable to the etch rate of a rough surface. This situation is called kinetic roughening. If all faces are kinetically rough, the etch rate becomes isotropic.

Isotropic etching requires conditions of kinetic roughening, because the etch rate is no longer dominated by a nucleation barrier but by transport processes in solution and the chemical reaction. To test this aspect of the model, Elwenspoek et al. show that there is a transition from isotropic to anisotropic etching if the undersaturation becomes too small. This can occur if one etches with an acidic etchant very long or if one etches through very small holes in a mask (crevice effect). In both cases, anisotropic behavior becomes evident as aging or limited transport of the solution causes the undersaturation to become very small. No proof is available to indicate that acidic etchants are much more undersaturated than the alkaline etchants. Still, the above explains some phenomena that the Seidel et al.'s model fails to address. Another nice confirmation of the Elwenspoek et al. model is in the effect of misalignment on etch rate. A misalignment of the mask close to smooth faces implies steps; there is no need for nucleation in order to etch. Since the density of steps is proportional to the angle of misalignment, the etch rate should be proportional to the misalignment angle, provided the distance between steps is not too large. Nucleation of new cavities becomes very probable. This has indeed been observed for the etch rate close to the <112> directions.[90]

Where the Elwenspoek et al.'s model becomes a bit murky is in the classification of which surfaces are smooth and which ones are rough. Elementary analysis classifies only the {111} planes as smooth at low temperatures. At this stage the model does not explain anything more than other models; every model has an explanation for the slower {111} etch rate. But these authors invoke the possibility of surface reconstruction and/or adsorption of surface species which, by decreasing the surface-free energy, could make faces such as {001} and {110} flat as well but with lower activation energies. They also take heart in the fact that CVD experiments often end up showing flat {110}, {100}, {331}, and, strongest of all {111} planes. Especially where the influence of the etchant is concerned a lot more convincing thermodynamic data to estimate $\Delta\mu$ and $\gamma$ are needed.

### Isotropic vs. Anisotropic Etching of Silicon

In contrast to alkaline etching, with an acidic etchant such as HF, holes are needed for etching Si. An n-type Si electrode immersed in HF in the dark will not etch due to lack of holes. The same electrode in an alkaline medium etches readily. A p-type electrode in an HF solution, where holes are available under the proper bias, will etch even in the dark. For HF etchants one might assume that the Ghandi and Kern model,[60,147] relying on the injection of holes, applies. In terms of the band model, this must mean that the silicon/electrolyte interface in acidic solutions exhibits quite different energetics from the alkaline case. It is not directly obvious why the energetics of the silicon/electrolyte interface would be pH dependent. To the contrary, since the flatband potential of most oxide semiconductors as well as most oxide-covered semiconductors (such as Si) and the Fermi level of the $H_2O/OH^-$ redox couple in an aqueous solution are both expected to change by 59 mV for each pH unit change,[140] one would expect the energetics of the interface to be pH independent. Since the electronegativity in a Si-F bond is higher than for a Si-OH bond, one might even expect the backbond surface states to be raised higher in an HF medium, making an electron injection mechanism even more likely than in alkaline media. To clarify this contradiction we will analyze the band model of a Si electrode in an acidic medium in more detail. The band model shown in Figure 4.39 was constructed on the basis of a set of impedance measurements on an n-type Si electrode in a set of aqueous solutions at different pH values. From the impedance measurements Mott-Schottky plots were constructed to determine the pH dependency of the flat-band potential. From that, the position of the conduction band and valence band edges ($E_{cs}$ and $E_{vs}$, respectively) of the Si electrode in an acidic medium at a fixed pH of 2.2 (the point of zero charge) was calculated at 0.74 eV vs. SCE (saturated calomel electrode as reference) for $E_{cs}$, and –0.36 eV vs. SCE for $E_{vs}$.[141,142]

We have assumed in Figure 4.39 that the bands are bent upwards at open circuit (see below for justification), so that

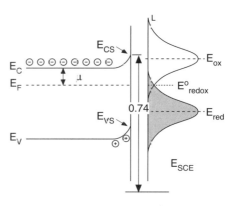

**FIGURE 4.39** Band diagram for n-type Si in pH = 2.2 (no bias or illumination). Reference is the saturated calomel electrode (SCE). In Figure 4.37 no energy values were given; here we provide actual positions of the conduction band edge, $E_{cs}$ = 0.74 eV vs. SCE, and the valence band edge, $E_{vs}$ = 0.74 eV –1.1 eV = –0.36 eV vs. SCE (1.1 eV is the band gap of Si). These values were determined by means of Mott-Schottky plots.[141] The separation between the Fermi energy and the bottom of the conduction band is indicated by $\mu$.

Holes in the valence band are driven to the interface where they can react with Si atoms or with competing reducing agents from the electrolyte. Since we want to etch Si, we are only interested in the reactions where Si itself is consumed. Reactions of holes with reducing agents are of great importance in photoelectrochemical solar cells.[142] For n-type Si, holes can be (1) injected by oxidants from the solution (e.g., by adding nitric acid to the HF solution), (2) supplied at the electrolyte/semiconductor interface by shining light on a properly biased n-Si wafer, or (3) created by impact ionization, i.e., Zener breakdown, of a sufficiently high reverse biased n-Si electrode.[143] With a p-Si wafer, a small forward bias supplies all the holes needed for the oxidation of the lattice even without light, as the conduction happens via a hole mechanism. An important finding, explaining the different reaction paths in acidic and alkaline media comes from plotting the flat band potential as a function of pH. It was found that the band diagram of Si shifts with less than 59 mV per pH unit. Actually the shift is only about 30 mV per pH unit.[142] As shown in Figure 4.40, with increasing pH, the energy levels of the solution rise faster than the energy levels in the semiconductor. As a consequence it is more likely that electron injection takes place in alkaline media as the filled levels associated with the $OH^-$ are closer to the conduction band, whereas in acidic media the filled levels of the redox system overlap better with the valence band, favoring a hole reaction. A lower position of the redox couple with respect to the conduction band edge, $E_{cs}$, in acidic media explains the upward bending of the bands as drawn in Figure 4.39. With isotropic etching in acidic media, the reaction starts with a hole in the valence band, equivalent to a broken Si-Si bond. In this case, the relative position of backbond related surface states in different crystal orientations are of no consequence and all planes etch at the same rate. A study of the interfacial energetics helps to understand why isotropic etching occurs in acidic media and anisotropic etching in alkaline media.

A few words of caution on our explanation for the reactivity difference between acidic and alkaline media are in order. Little is known about the width of the bell-shaped curves describing the redox levels in solution.[140] Not knowing the surface concentrations of the reactive redox species involved in the etching reactions further hinders a better understanding of the surface energetics as the bell-shaped curves for oxidant and reductant will only be the same height (as they were drawn in Figures 4.39 and 4.40) if the concentration of oxidant and reductant are the same. Clearly, the above picture is oversimplified; several authors have found that the dissolution of Si in HF might involve both the conduction and valence band, a claim confirmed by photocurrent multiplication experiments.[144,145] These photomultiplication experiments showed that one or two holes generated by light in the Si valence band were sufficient to dislodge one Si[47] unit, meaning that the rest of the charges were injected into the conduction band. Our contention here is only that the low pH dependence of the flat-band potential of a silicon electrode makes a conduction band mechanism more favorable with alkaline type etchants and a hole mechanism more favorable in acidic media.

Continued attempts at modeling the etch rates of all Si planes are under way. For example, Hesketh et al.[146] attempted to model the etch rates of the different planes developing on silicon spheres in etching experiments with KOH and CsOH by calculating the surface free energy. The number of surface bonds per centimeter square on a Si plane is indicative of the surface free energy, which can be estimated by counting the bond density and multiplying by the bond energy. Using the unit cell dimension "a" of Si of 5.431 Å, and a silicon-to-silicon bond energy

of 42.2 kcal/mol, the surface free energy, $\Delta G$, can be related to, $N_B$, the bond density, by the following expression:

$$\Delta G = \frac{N_B}{2} \times 2.94 \times 10^{-19} \, \frac{J}{m^2} \qquad \textbf{4.40}$$

Although Hesketh et al. could not explain the etching differences observed between CsOH and KOH (these authors identified a cation effect on the etch rate!), a plot of the calculated surface free energy vs. orientation yielded minima for all low index planes such as {100}, {110}, and {111}, as well as for the high index {522} planes. Fewer bonds per unit area on the low index planes produce a lower surface energy and lower etching rate. When Hesketh et al. added the in-plane bond density to the surface bond density, producing a total bond density, a correlation with the hierarchy of etch rates in CsOH and KOH was found, i.e., {311}, {522} > {100} > {111}. The surfaces with the higher bond density etched faster, suggesting that the etch rate might be a function of the number of electrons available at the surface. Hesketh et al. imply that their result falls in line with the Seidel et al.'s model, although it is unclear how the total bond density relates to surface state energies of backbonds. Moreover, Hesketh's model does not take into account the angles of the bonds, and in Elwenspoek's view, the surface free energy actually does not determine the anisotropy.

More research could focus on the modeling of Si etch rates. The semiconductor electrochemistry of corroding Si electrodes will be a major tool in further developments. Interested readers may consult Sundaram et al.[147] in an article on Si etching in

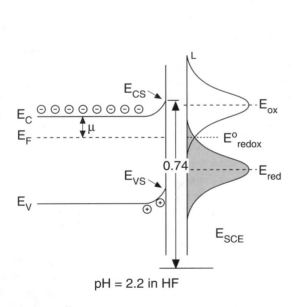

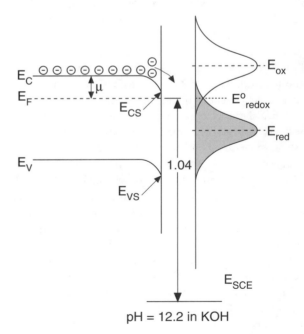

**FIGURE 4.40**  Band model comparison of the Si/electrolyte interface at low and high pH. Increasing the pH by 10 units shifts the redox-levels up by 600 mV, whereas the Si bands only move up by 300 mV. This leads to a different band-bending and a different reaction mechanism, i.e., electron injection in the alkaline media (anisotropic) and hole injection in acidic media (isotropic).

hydrazine and Palik et al.[120] on the etch-stop mechanism in heavily doped silicon; both explain in some detail the silicon/electrolyte energetics. A more generic treatise on Semiconductor Electrochemistry can be found in Reference 148.

## Etching With Bias and/or Illumination of the Semiconductor

The isotropic and anisotropic etchants discussed so far require neither a bias nor illumination of the semiconductor. The etching in such cases proceeds at open circuit and the semiconductor is shielded from light. In a cyclic voltammogram, as shown in Figure 4.41, the operational potential is the rest potential, $V_r$, where anodic and cathodic currents are equal in magnitude and opposite in sign, resulting in the absence of flow of current in an external circuit. This does not mean that macroscopic changes do not occur at the electrode surface, since the anodic

and cathodic currents may be part of different chemical reactions. Consider isotropic etching of Si in an HF/HNO$_3$ etchant at open circuit where the local anodic reaction is associated with corrosion of the semiconductor:

$$Anode : Si + 2H_2O + nh^+ \rightarrow SiO_2 + 4H^+ + (4-n)e^-$$

$$\text{Reaction 4.11}$$

$$SiO_2 + 6HF \rightarrow H_2SiF_6 + 2H_2O \qquad \text{Reaction 4.12}$$

the involvement of holes 'h$^+$' in the acidic reaction was discussed in the preceding section), while the local cathodic reaction could be associated with reduction of HNO$_3$:

$$Cathode : HNO_3 + 3H^+ \rightarrow NO + H_2O + 3H^+ \qquad \text{Reaction 4.13}$$

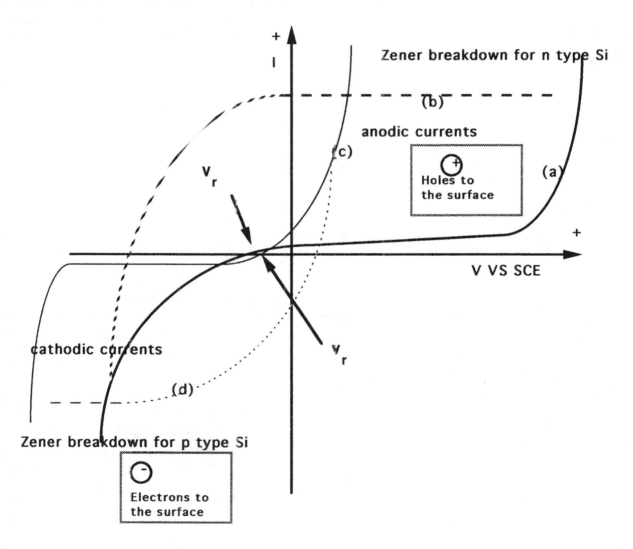

**FIGURE 4.41**  Basic cyclic voltammograms (I vs. V) for n- and p-type Si in an HF solution, in the absence of a hole injecting oxidant: (a) n-type Si without illumination; (b) n-type Si under illumination; (c) p-type Si without illumination; (d) p-type Si under illumination. If the reactions on the dark Si electrode determining the rest potential $V_r$ for n- and p-type are the same, then $V_r$ is expected to be the same as well. For the clarity of the figure we have chosen the $V_r$s different here; in practice, $V_r$ for n- and p-type Si in HF are found to be identical.

We will now explore Si etching while illuminating and/or applying a bias to the silicon sample. To simplify the situation, we will first consider the case of an oxidant-free solution so that all the holes must come from within the semiconductor. Anodic dissolution of n-Si in an HF containing solution requires a supply of holes to the surface. For an n-type wafer under reverse bias, very few holes will show at the surface unless the high reverse (anodic) bias is sufficient to induce impact ionization or Zener breakdown (see Figure 4.41a). Alternatively, the interface can be illuminated, creating holes in the space charge region which the field pushes towards the semiconductor/electrolyte interface (Figure 4.41b). In the forward direction, electrons from the Si conduction band (majority carriers) reduce oxidizing species in the solution (e.g., reduction of protons to hydrogen). A p-type Si sample exhibits high anodic currents even without illumination at small anodic (forward) bias (Figure 4.41c). Here the current is carried by holes. A p-type electrode illuminated under reverse bias gives rise to a cathodic photocurrent (Figure 4.41d). At relatively low light intensities, the photocurrent plateaus for both n- and p-types (Figure 4.41b and 4.41d, respectively)

depend linearly on light intensity. The photocurrent is cathodic for p-type Si (species are reduced by photoproduced electrons at the surface, e.g., hydrogen formation) and anodic for n-type Si (species are oxidized by photoproduced holes at the surface; either the lattice itself is consumed or reducing compounds in solution are). In Figure 4.42 we show the cyclic voltammograms of n-type and p-type Si in the presence of a hole-injecting oxidant. The most obvious effect is on the dark p-type Si electrode. The injection of holes in the valence band increases the cathodic dark current dramatically. The current level measured in this manner for varying oxidant concentration or different oxidants could be used to estimate the efficiency of different isotropic Si etchants; a pointer to the fact that semiconductor electrochemistry has been underutilized as a tool to study Si etching. When n-type Si is consumed under illumination, we experience photocorrosion (see Figure 4.41b). This photocorrosion phenomenon has been a major barrier to the long-term viability of photoelectrochemical cells.[141] In what follows, photocorrosion is put to use for electropolishing and formation of microporous and macroporous layers.[143]

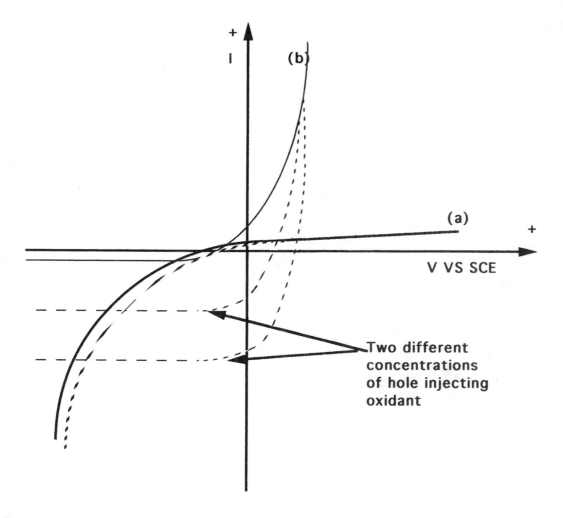

**FIGURE 4.42**  Basic cyclic voltammograms for n- and p-type Si in an HF solution and in the presence of a hole injecting oxidant, e.g., $HNO_3$: (a) n-type Si in the dark; (b) p-type Si in the dark. An increase of the cathodic dark current on the p-type electrodes is most obvious. The current level is proportional to the oxidant concentration.

# Electropolishing and Microporous Silicon

## Electropolishing

Photoelectrochemical etching (PEC-etching) involves photocorrosion in an electrolyte in which the semiconductor is generally chemically stable in the dark, i.e., no hole-injecting oxidants are present, as in the case of an HF solution (see Figure 4.41). For carrying out the experiments a set-up as shown in Figure 4.25 may be used with a provision to illuminate the semiconductor electrode. At high light intensities the anodic curves for n- and p-type Si are the same, except for a potential shift of a few hundred mV (see Figure 4.43). Because of this equivalence, several of the etching processes described apply for both forward biased p-type and n-type Si under illumination. The anodic curves in Figure 4.43 present two peaks, characterized by $i_{CRIT}$ and $i_{MAX}$. At the first peak, $i_{CRIT}$, partial dissolution of Si in reactions such as:

$$Si + 2F^- + 2h^+ \rightarrow SiF_2 \qquad \text{Reaction 4.14}$$

$$SiF_2 + 2HF \rightarrow SiF_4 + H_2 \qquad \text{Reaction 4.15}$$

leads to the formation of porous Si while hydrogen formation occurs simultaneously. The porous Si typically forms when using low current densities in a highly concentrated HF solution — in other words, by limiting the oxidation of silicon due to a hole and $OH^-$ deficiency. Above $i_{CRIT}$, the transition from the charge-supply-limited to the mass-transport-limited case, the porous film delaminates and bright electropolishing occurs at potentials positive of $i_{MAX}$. With dissolution of chemical reactants in the electrolyte rate limiting, HF is depleted at the electrode surface and a charge of holes builds up at the interface. Hills on the surface dissolve faster than depressions because the current density is higher on high spots. As a result, the surface becomes smoother, i.e., electropolishing takes place.[149] Electropolishing in this regime can be used to smooth silicon surfaces or to thin epitaxially grown silicon layers. The peak and oscillations in Figure 4.43 are explained as follows: at current densities exceeding $i_{MAX}$ an oxide grows first on top of the silicon, leading to a decrease of the anodic photocurrent (explaining the $i_{MAX}$ peak), until a steady state is reached in which dissolution of the oxide by HF through formation of a fluoride complex in solution ($SiF_6^{2-}$) equals the oxide growth rate. The oscillations observed in the anodic curve in Figure 4.43 can be explained by a nonlinear correlation between formation and dissolution of the oxide.[143]

## Porous Silicon

### Introduction

The formation of porous Si was first discovered by Uhlir in 1956.[8] His discovery is leading to all types of interesting new devices from quantum structures, permeable membranes, photoluminescent and electroluminescent devices to a basis for making thick $SiO_2$ and $Si_3N_4$ films.[80,150] Two types of pores exist: micropores and macropores. Their sizes can differ by three orders of magnitude and the underlying formation mechanism

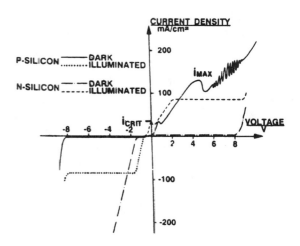

**FIGURE 4.43** Cyclic voltammograms identifying porous Si formation regime and electropolishing regime. (From Levy-Clement, C. et al., *Electrochima Acta*, 37, 877–888, 1992. With permission.)

is quite different. Some important features of porous Si, as detailed later, are

- Pore sizes in a diameter range from 20 Å to 10 µm.
- Pores that follow crystallographic orientation.[151]
- Very high aspect ratio (~250) pores in Si maintained over several millimeters' distance.[149]
- Porous Si is highly reactive, oxidizes and etches at very high rate.
- Porosity varies with the current density.

These important attributes contribute to the essential role porous Si plays in both micromachining[152] and the fashioning of quantum structures.[153]

### Microporous Silicon

Whether one is in the regime of electropolishing or porous silicon formation depends on both the anodic current density and the HF concentration. The surface morphology produced by the Si dissolution process critically depends on whether mass transport or hole supply is the rate-limiting step. Porous silicon formation is favored for high HF concentrations and low currents (weak light intensities for n-type Si), where the charge supply is limiting, while etching is favored for low HF concentration and high currents (strong light intensity for n-type Si), where mass transport is limiting. For current densities below $i_{MAX}$ (Figure 4.43) holes are depleted at the surface and HF accumulates at the electrode/electrolyte interface. As a result, a dense network of fine holes forms.[154,155] The formation of a porous silicon layer (PSL) in this regime has been explained as a self-adjusting electrochemical reaction due to hole depletion by a quantum confinement in the microporous structure.[156] The structure of the pores in PSL can best be observed by transmission electron microscopy (TEM) and its thickness can be monitored with an IR microscope. The structure of the porous layer primarily depends on the doping level of the wafer and on the illumination during etching. The pore sizes decrease when etching occurs under illumination.[143] The pores formed in p-type Si show much smaller diameters than n-type ones for the same formation conditions. It is believed that the porous silicon layer (PSL) consists of silicon hydrides and oxides. The pore diameter typically ranges from 40 to 200 Å.[157–159] This very reactive porous material etches or oxidizes rapidly. Heat treatment in an oxidizing atmosphere (1100°C in oxygen for 30 minutes is sufficient to make a 4-μm thick film) leads to oxidized porous silicon (OPS). The oxidation occurs throughout the whole porous volume and several micrometers-thick $SiO_2$ layers can be obtained in times that correspond to the growth of a few hundred nanometers on regular Si surfaces. Porous silicon is low density and remains single crystalline, providing a suitable substrate for epitaxial Si film growth. These properties have been used to obtain dielectric isolation in ICs and to make silicon-on-insulator wafers.[155]

Porous Si can also be formed chemically through Reactions 4.11 and 4.13. In this case the difference between chemical polishing and porous Si formation conditions is more subtle. In the chemical polishing case, all reacting surface sites switch constantly from being local anode (Reaction 4.11) to being local cathode (Reaction 4.13), resulting in nonpreferential etching. If surface sites do not switch fast between being local anode and cathode, charges have time to migrate over the surface. In this case, the original local cathode site remains a cathode for a longer time, and the corresponding local anode site, somewhere else on the surface, also remains an anode in order to keep the overall reaction neutral. A preferential etching results at the localized anode sites, making the surface rough and causing a porous silicon to form.[160] Any inhomogeneity, e.g., some oxide or a kink site at the surface might increase such preferential etching.[160] Unlike PSLs fabricated by electrochemical means, the chemically etched porous Si film thickness is self limiting.

Besides its use for dielectric isolation and the fabrication of SOI wafers, porous silicon has been introduced in a wide variety of other applications: Luggin capillaries for electrochemical reference electrodes, high surface area gas sensors, humidity sensors, sacrificial layers in silicon micromachining, etc. Recently PSL was shown to exhibit photoluminescent and electroluminescent behavior; light-emitting porous silicon (LEPOS) was demonstrated. Visible light emission from regular Si is very weak due to its indirect bandgap. Pumping porous Si with a green light laser (argon) caused it to emit a red glow. If a LEPOS device could be integrated monolithically with other structures on silicon, a big step in micro-optics, photon data transmission, and processing would be achieved. To explain the blue shift of the absorption edge of LEPOS of about 0.5 eV compared to bulk silicon[156] and room temperature photoluminescence,[153] Searson et al.[161] have proposed an energy-level diagram for porous silicon where the valence band is lowered with respect to bulk silicon to give a band-gap of about 1.8 eV. Not only may PSL formation, as seen above, be explained invoking quantum structures, but its remarkable optical properties may also be explained this way. Canham believes that the thin Si filaments may act as quantum wires. Significant quantum effects require structural sizes below 5 nm (see Chapter 7) and the porous Si definitely can have structures

of that size. By treatment of the porous Si with $NH_3$ at high temperatures, it is possible to make thick $Si_3N_4$ films. Even at 13 μm, these films show little evidence of stress in contrast to stoichiometric LPCVD nitride films.[150]

Porous Si might represent a simple way of making the quantum structures of the future. Pore size of PSL can be influenced by both light intensity and current density. The quantum aspect adds significantly to the sudden big interest taken in porous Si.

### Macroporous Silicon

In addition to micropores, well-defined macropores can also be made in Si by photo and/or bias etching in HF solutions. Macropores have sizes as large as 10 μm, visible with a scanning electron microscope (SEM) rather than TEM. The two types of pores often coexist, with micropores covering the walls of macropores. Sizes can differ by three orders of magnitude. This is not a matter of a broad fractal type distribution of pores, but the formation mechanism is quite different.

Electrochemical etching of macropores or macroholes has been reported for n-type silicon in 2.5 to 5% HF under high voltage (>10 V), low current density (10 mAcm⁻²), under illumination, and in the dark.[143] In the latter case, Zener breakdown in silicon (electric field strength in excess of $3 \times 10^5$ V/cm) causes the hole formation. The macropores are formed only with lightly doped n-silicon at much higher anodic potentials than those used for micropore formation.[162] By using a pore initiation pattern the macropores actually can be localized at any desired location. This dramatic effect is illustrated by comparing Figures 4.44A and B.[149,157] Pores orthogonal to the surface with depths up to a whole wafer thickness can be made and aspect ratios as large as 250 become possible. The formation mechanism in this case cannot be explained on the basis of depletion of holes due to quantum confinement in the fine porous structures, given that these macropores exhibit sizes well beyond 5 nm. As with microporous Si, the surface morphology produced by the dissolution process depends critically on whether mass transport or charge supply is the rate-limiting step. For pore formation one must work again in a charge depletion mode. Macropore formation, as micropore formation, is a self-adjusting electrochemical mechanism. In this case, the limitation is due to the depletion of the holes in the pore walls in n-Si wafers causing them to passivate. Holes keep on being collected by the pore tip where they promote dissolution. No passivating layer is involved to protect the pore wall. The only decisive differences between pore tips and pore walls are their geometry and their location. Holes generated by light or Zener breakdown are collected at pore tips. Every depression or pit in the surface initiates pore growth because the electrical field at a curved pore bottom is much larger than that of a flat surface due to the effect of the radius of curvature. The latter leads to higher current and enhances local etching.[162] Zener breakdown and illumination of n-Si lead to different types of pore geometry.[149,163,164] Branched pores with sharp tips form if holes are generated by breakdown (See Figure 4.45A).[149] Unbranched pores with larger tip radii result from holes created by illumination (see Figure 4.45B).[164] The latter difference can be understood as follows: the electric field strength is a function of bias, doping density, and geometry.

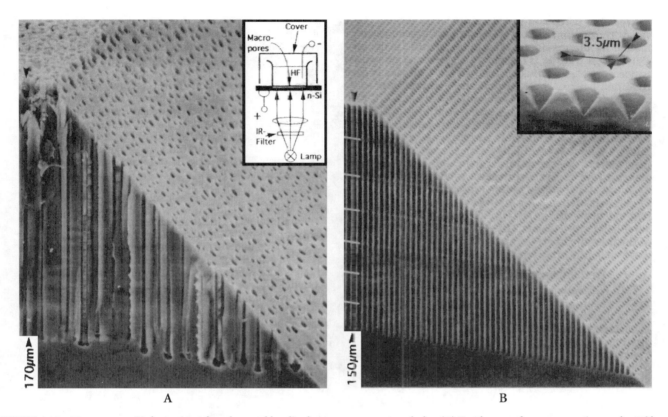

**FIGURE 4.44** Macroporous Si; formation of random and localized macropores or macroholes. (A) Random: surface, cross-section, and a 45° bevel of an n-type sample ($10^{15}/cm^3$ phosphorus-doped) showing a random pattern of macropores. Pore initiation was enhanced by applying 10-V bias in the first minute of anodization followed by 149 min at 3 V. The current density was kept constant at 10 mA/cm² by adjusting the backside illumination. A 6% aqueous solution of HF was used as an electrolyte. The set-up used for anodization is sketched in the upper right corner. (B) Localized: surface, cross-section, and a 45° bevel of an n-type sample ($10^{15}/cm$ phosphorus-doped) showing a predetermined pattern of macropores (3 V, 350 min, 2.5% HF). Pore growth was induced by a regular pattern of pits produced by standard lithography and subsequent alkaline etching (inset upper right). To measure the depth dependence of the growth rate, the current density was kept periodically at 5 mA/cm² for 45 min and then reduced to 3.3 mA/cm² for 5 min. This reduction resulted in a periodic decrease of the pore diameter, as marked by white labels in the figure. (From Lehmann, V., *J. Electrochem. Soc.*, 140, 2836–2843, 1993. With permission.)

High doping level density or sharp pore tips will lower the required bias for breakdown, so macropores will tend to follow pores with the sharpest tips. Since every tip causes a new breakdown and hole generation, the position of the original pore tip becomes independent of the other pores, branching of the pores is possible, and fir-tree type pores can be observed. With illumination the pore radius may be larger as the breakdown field strength is not necessary to generate charge carriers, so the pores remain unbranched.

The Si anisotropy common with alkaline etchants surprisingly shows up here with an isotropic etchant such as HF. For example, with breakdown-supplied holes, <100>-directed macroholes with <110> branches form (see also Figure 4.45A),[149] leading to a complex network of caverns beneath the silicon surface. Pyramidal pore tips[149] also were observed when the current density was limited by the bias ($i < i_{CRIT}$). Isotropic pore pits form when the current is larger than the critical current density, i.e., isotropy in HF etching can be changed into anisotropy when the supply of holes is limited. We refer here to the Elwenspoek et al. model (see above) which predicts that in confined spaces etching will tend to be more anisotropic, even when using a normally isotropic etchant such as HF. It was also determined that macrohole formation depends on the

wavelength of the light used. No hole formation occurs below about 800 nm. Depending on the wavelength, the shape of the hole can be manipulated as well.[164] For wavelengths above 867 nm, the depth profile of the holes changed from conical to cylindrical. The latter was interpreted in terms of the influence of the local minority carrier generation rate. Carriers generated deep in the bulk would promote the hole growth at the tips, whereas near-surface generation would lead to lateral growth.

Cahill et al.[165] reported in 1993 the creation of 1- to 5-μm size pores with pore spacings (center to center) from 200 to 1000 μm. Until this finding, the pores typically formed were spaced in the range of 4 to 30 μm center-to-center, while being 0.6 to 10 μm in diameter. Making highly isolated pores presents quite another challenge. In the previous work, the relative close spacing of the pores allowed the authors to conclude that the regions between the pores were almost totally depleted and that practically all carriers were collected by the pore tips. In such a case, neither the pore side walls nor the wafer surface etched as all holes were swept to the pore tips. Since the surface was not attacked by pore-forming holes, the quality of the pore initiation mask lost its relevance. In Cahill et al.'s case, on the other hand, a long-lived mask (>20 hours) needed to be developed to help prevent pore formation everywhere except at initiation pits. The

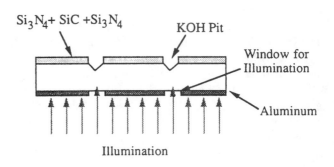

**FIGURE 4.45**  Comparison of macropores made with breakdown holes (A) and macropores made with light created holes (B). (A) An oxide replica of pores etched under weak backside illumination visualizes the branching of pores produced by generation of charge carriers due to electrical breakdown (5 V, 3% HF, room temperature, $10^{15}/cm^3$ phosphorus-doped). (From Lehmann, V., *J. Electrochem. Soc.*, 140, 2836–2844, 1993. With permission.) (B) Single pore associated with KOH pit. (From Lehmann, V. and Foll, H., *J. Electrochem. Soc.*, 137, 653–659, 1990. With permission.)

mask eventually used is shown in Figure 4.46; it shows a SiC layer sandwiched between two layers of silicon nitride. The silicon nitride directly atop the silicon served to insulate the silicon carbide from the underlying substrate. As the silicon carbide proves very resistant to HF, loss of thickness did not show during the procedure. The top nitride served to protect the carbide during anisotropic pit formation. By lowering the bias to less than two volts with respect to a saturated calomel electrode (SCE) side-branching was avoided.

It seems very likely that this macropore formation phenomenon could extend to all types of n-type semiconductors. Some evidence with InP and GaAs supports this statement.[149]

**FIGURE 4.46**  Long-lived macropore initiation mask.

# Etch-Stop Techniques

## Introduction

In many cases it is desirable to stop etching in silicon when a certain cavity depth or a certain membrane thickness is reached. Nonuniformity of etched devices due to nonuniformity of the silicon wafer thickness can be quite high. Taper of double-polished wafers, for example, can be as high as 40 μm![107] Even with the best wafer quality the wafer taper is still around 2 μm. The taper and variation in etch depth lead to intolerable thickness variations for many applications. Etch-rate control typically requires monitoring and stabilization of:

- Etchant composition
- Etchant aging
  - Stabilization with $N_2$ sparging (especially with EDP and hydrazine)
  - Taking account of the total amount of material etched (loading effects)
- Etchant temperature
- Diffusion effects (constant stirring is required, especially for EDP)
  - Stirring also leads to a smoother surface through bubble removal
  - Trenching (also pillowing) and roughness decrease with increased stirring rate
- Light may affect the etch rate (especially with n-type Si)
- Surface preparation of the sample can have a big effect on etch rate (the native oxide retards etch start; a dip in dilute HF is recommended)

With good temperature, etchant concentration and stirring control the variation in etch depth typically is 1% (see Figure 4.47).[107] A good pretreatment of the surface to be etched is a standard RCA clean combined with a 5% HF dip to remove the native oxide immediately prior to etching in EOH.

In the early days of micromachining, one of the following techniques was used to etch a Si structure anisotropically to a predetermined thickness. In the simplest mode the etch time was monitored (Table 4.9 lists some etch rates for different etchants) or a bit more complex; the infrared transmittance was followed. For thin membranes the etch stop cannot be determined by a constant etch time method with sufficient precision. The spread in etch rates becomes critical if one etches membranes down to thicknesses of less than 20 μm; it is almost impossible to etch structures down to less than 10 μm with a timing technique. In the V-groove technique, V-grooves with precise openings (see Equation 4.2) were used such that the V-groove stopped etching at the exact moment a desired membrane thickness was reached (see Figure 4.48).[166] One can also design wider mask openings on the wafer's edge so that the

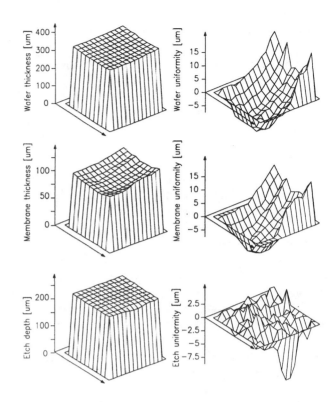

FIGURE 4.47  A map of the wafer thickness, the membrane thickness, and the etch depth. (From Lambrechts, S.W. and W. Sansen, *Biosensors: Microelectrochemical Devices*, Institute of Physics Publishing, 1992. With permission.)

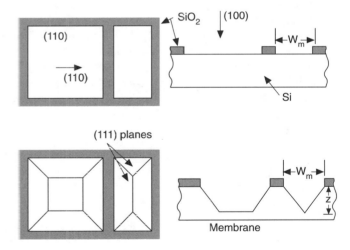

FIGURE 4.48  V-groove technique to monitor the thickness of a membrane. At the precise moment the V-groove is developed, the membrane has reached the desired thickness.

wafer is etched through at those sites at the moment the membrane has reached the appropriate thickness. Although Nunn and Angell[167] claimed that an accuracy of about 1 μm could be obtained using the V-groove method, none of the mentioned techniques are found to be production worthy. Nowadays the above methods are almost completely replaced by etch-stop techniques based on a change in etch rate dependant on doping level or chemical nature of a stopping layer. High resolution silicon micromachining relies on the availability of effective etch-stop layers. It is actually the existence of impurity-based etch stops in silicon that has allowed micromachining to become a high-yield production process.

## Boron Etch Stop

The most widely used etch-stop technique is based on the fact that anisotropic etchants, especially EDP, do not attack boron-doped (p+) Si layers heavily. This effect was first noticed by Greenwood in 1969.[168] He assumed that the presence of a p-n junction was responsible.

Bogh in 1971[118] found that at an impurity concentration of about $7 \times 10^{19}/cm^3$ resulted in the etch rate of Si in EDP dropping sharply (see also Table 4.9), but without any requirement for a p-n junction. For KOH-based solutions, Price[85] found a significant reduction in etch rate for boron concentrations above $5 \times 10^{18} cm^{-3}$. The model by Seidel et al., discussed above, provides an elegant explanation for the etch stop at high boron concentrations. At moderate dopant concentration, we saw that the electrons injected into the conduction band stay localized near the semiconductor surface due to the downward bending of the bands (Figure 4.37). The electrons there have a small probability of recombining with holes deeper into the crystal even for p-type Si. This situation changes when the doping level in the silicon increases further. At a high dopant concentration, silicon degenerates and starts to behave like a metal. For a degenerate p-type semiconductor, the space charge thickness shrinks and the Fermi level drops into the valence band as indicated in Figure 4.49. The injected electrons shoot (tunnel) right through the thin surface charge layer into deeper regions of the crystal where they recombine with holes from the valence band. Consequently, these electrons are not available for the subsequent reaction with water molecules (Reaction 4.5), the reduction of which is necessary for providing new hydroxide ions in close

proximity to the negatively charged silicon surface. These hydroxide ions are required for the dissolution of the silicon as $Si(OH)_4$. The remaining etch rate observed within the etch stop region is then determined by the number of electrons still available in the conduction band at the silicon surface. This number is assumed to be inversely proportional to the number of holes and thus the boron concentration. Experiments show that the decrease in etch rate is nearly independent of the crystallographic orientation and the etch rate is proportional to the inverse fourth power of the boron concentration in all alkaline etchants.

From the above it follows that a simple boron diffusion or implantation, introduced from the front of the wafer, can be used to create beams and diaphragms by etching from the back. The boron etch-stop technique is illustrated in Figure 4.50 for the fabrication of a micromembrane nozzle.[169] The $SiO_2$ mesa in Figure 4.50b leads to the desired boron p+ profile. The anisotropic etch from the back clears the lightly doped p-Si (Figure 4.50d). Layers of p+ silicon having a thickness of 1 to 20 μm can be formed with this process. The boron etch stop constitutes a very good etch stop; it is not very critical when the operator takes the wafer out of the etchant. One important practical note is that the boron etch stop may become badly degraded in EDP solutions that were allowed to react with atmospheric oxygen. Since boron atoms are smaller than silicon, a highly doped, freely suspended membrane or diaphragm will be stretched; the boron-doped silicon is typically in tensile stress and the microstructures are flat and do not buckle. While doping with boron decreases the lattice constant, doping with germanium increases the lattice constant. A membrane doped with B and Ge still etches much slower than undoped silicon, and the stress in the layer is reduced. A stress-free, dislocation-free and slow etching layer (±10 nm/min) is obtained at doping levels of $10^{20} cm^{-3}$ boron and $10^{21} cm^{-3}$ germanium.[88,170] One disadvantage with this etch-stop technique is that the extremely high boron concentrations are not compatible with standard CMOS or bipolar techniques, so they can only be used for microstructures without integrated electronics. Another limitation of this

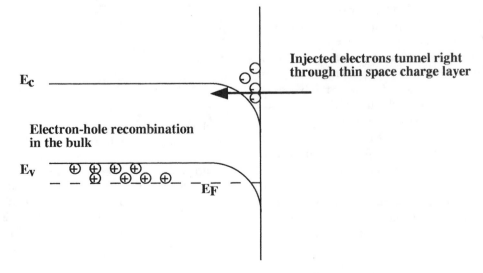

**FIGURE 4.49**   Si/electrolyte interface energetics at high doping level explaining etch-stop behavior.

process is the fixed number and angles of (111) planes one can accommodate. The etch stop is less effective in KOH compared to EDP. Besides boron, other impurities have been tried for use in an etch stop in anisotropic etchants. Doping Si with germanium has hardly any influence on the etch rate of either the KOH or EDP solutions. At a doping level as high as $5 \times 10^{21}$ cm$^{-3}$ the etch rate is barely reduced by a factor of two.[88]

By burying the highly doped boron layer under an epitaxial layer of lighter doped Si the problem of incompatibility with active circuitry can be avoided. A ±1% thickness uniformity is possible with modern epilayer deposition equipment (see for example Semiconductor International, July 1993, pp. 80–83). A widely used method of automatically measuring the epi thickness is with infrared (IR) instruments, especially Fourier transform infrared (FTIR)[171] (see also Epitaxy in Chapter 3 and Chapter 5).

## Electrochemical Etch-Stop Technique

For the fabrication of piezoresistive pressure sensors, the doping concentration of the piezo resistor must be kept smaller than $1 \times 10^{19}$ cm$^{-3}$ because the piezoresistive coefficients drop considerably above this value and reverse breakdown becomes an issue. Moreover, high boron levels compromise the quality of the crystal by introducing slip planes and tensile stress and prevent the incorporation of integrated electronics. As a result, a boron stop often cannot be used to produce well-controlled thin membranes unless, as suggested above, the highly doped boron layer is buried underneath a lighter doped Si epi layer. Alternatively, a second etch-stop method, an electrochemical technique, can be used. In this case, a lightly doped p-n junction

is used as an etch stop by applying a bias between the wafer and a counter electrode in the etchant. This technique was first proposed by Waggener in 1970.[17] Other early work on electrochemical etch-stops with anisotropic etchants such as KOH and EDP was performed by Palik et al.,[121] Jackson et al.,[172] Faust, J. W. and E. D. Palik,[173] and Kim, S. C. and K. Wise.[174] In electrochemical anisotropic etching a p-n junction is made, for example, by the epitaxial growth of an n-type layer (phosphorous-doped, $10^{15}$ cm$^{-3}$) on a p-type substrate (boron-doped, 30 Ωcm). This p-n junction forms a large diode over the whole wafer. The wafer is usually mounted on an inert substrate, such as a sapphire plate, with an acid-resistant wax and is partly or wholly immersed in the solution. An ohmic contact to the n-type epilayer is connected to one pole of a voltage source and the other pole of the voltage source is connected via a current

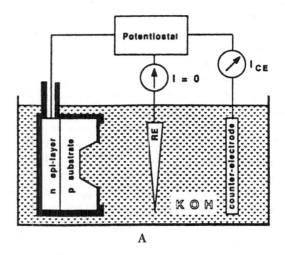

A

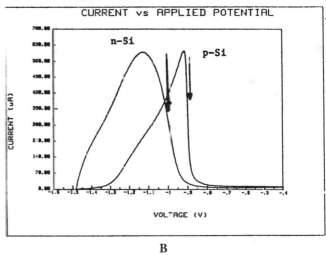

B

**FIGURE 4.51** Electrochemical etch stop. (A) Electrochemical etching set-up with potentiostatic control (three-electrode system). Potentiostatic control, mainly used in research studies, enables better control of the potential as it is referenced now to a reference electrode such as a saturated calomel electrode (SCE). In industrial settings, electrochemical etching is often carried out in a simpler two-electrode system, i.e., a Pt counter electrode and Si working electrode.[178] (B) Cyclic voltammograms of n- and p-type silicon in an alkaline solution at 60°C. Flade potentials are indicated with an arrow.

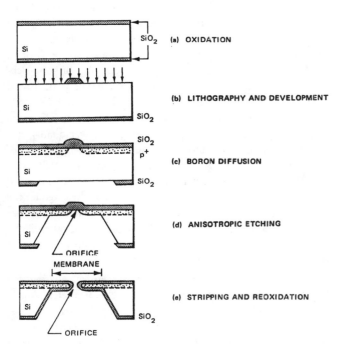

**FIGURE 4.50** Illustration of the boron etch stop in the fabrication of a membrane nozzle. (From Brodie, I. and J. J. Muray, *The Physics of Microfabrication*, Plenum Press, New York, 1982. With permission.)

meter to a counterelectrode in the etching solution (see Figure 4.51A). In this arrangement the p-type substrate can be selectively etched away and the etching stops at the p-n junction, leaving a membrane with a thickness solely defined by the thickness of the epilayer. The incorporation of a third electrode (a reference electrode) in the three-terminal method depicted in Figure 4.51A allows for a more precise determination of the silicon potential with respect to the solution than a two-terminal set-up as we illustrated in Figure 4.25.

At the Flade potential in Figure 4.51B, the oxide growth rate equals the oxide etch rate; a further increase of the potential results in a steep fall of the current due to complete passivation of the silicon surface. At potentials positive of the Flade potential, all etching stops. At potentials below the Flade potential, the current increases as the potential becomes more positive. This can be explained by the formation of an oxide which etches faster than it forms, i.e., the silicon is etched away. Whereas electrochemists like to talk about the Flade potential, physicists like to discuss matters in terms of the flat-band potential. The flat-band potential is that applied potential at which there are no more fields within the semiconductor, i.e., the energy band diagram is flat throughout the semiconductor. The passivating $SiO_2$ layer is assumed to start growing as soon as the negative surface charge on the silicon electrode is cancelled by the externally applied positive bias, a bias corresponding to the flat band potential. At these potentials, the formation of $Si(OH)_x$ complexes does not lead to further dissolution of silicon, because two neighboring Si-OH HO-Si groups will react by splitting of water, leading to the formation of Si-O-Si bonds. As can be learned from Figure 4.51B, the value of the Flade potential depends on the dopant type. Consequently, if a wafer with both n and p regions exposed to the electrolyte is held at a certain potential in the passive range for n-type and the active range for p-type, the p-regions are etched away, whereas the n-regions are retained. In the case of the diode shown in Figure 4.51A, where at the start of the experiment only p-type Si is exposed to the electrolyte, one starts by applying a positive bias to the n-type epilayer ($V_n$). This reverse biases the diode, and only a reverse bias current can flow. The potential of the p-type layer in this regime is negative of the flatband potential and active dissolution takes place. At the moment that the p-n junction is reached, a large anodic current can flow and the applied positive potential passivates the n-type epilayer. Etching continues on the areas where the wafer is thicker until the membrane is reached there, too. The thickness of the silicon membrane is thereby solely defined by the thickness of the epilayer; neither the etch uniformity nor the wafer taper will influence the result. A uniformity of better than 1% can be obtained on a 10-μm thick membrane. The current vs. time curve can be used to monitor the etching process; at first, the current is relatively low, i.e., limited by the reverse bias current of the diode, then as the p-n junction is reached a larger anodic current can flow until all the p-type material is consumed and the current falls again to give a plateau. The plateau indicates we have reached the current associated with a passivated n-type Si electrode. The etch procedure can be stopped at the moment that the current plateau has been reached, and since the etch stop is thus basically one of anodic passivation one sometimes terms it an anodic oxidation etch stop. Registering an I vs. V curve as in Figure 4.51B will

establish an upper limit on the applied voltage, $V_n$,[175] and such curves are used for *in situ* monitoring and controlling of the etch stop. A crude endpoint monitoring can be accomplished by the visual observation of cessation of hydrogen bubble formation accompanying Si etching. Palik et al. presented a detailed characterization of Si membranes made by electrochemical etch stop.[176]

Hirata et al.,[177] using the same anodic oxidation etch stop in a hydrazine-water solution at 90 ± 5°C and a simple two-terminal electrochemical cell, obtained a pressure sensitivity variation of less than 20% from wafer to wafer (pressure sensitivity is inversely proportional to the square of membrane thickness). A great advantage of this etch-stop technique is that it works with Si with low doping levels of the order of $10^{16}$ cm$^{-3}$. Due to the low doping levels, it is possible to fabricate structures with a very low, or controllable, intrinsic stress. Moreover, active electronics and piezo membranes can be built into the Si without problems. A disadvantage is that the back of the wafer with the aluminum contact has to be sealed hermetically from the etchant solution which requires complex fixturing and manual wafer handling. The fabrication of a suitable etch holder is no trivial matter. The holder must (1) protect the epi-contact from the etchant, (2) provide a low-resistance ohmic contact to the epi, and (3) must not introduce stress into wafers during etching.[41,42] Stress introduced by etch holders easily leads to diaphragm or wafer fracture and etchant seepage through to the epi-side.

Using a four-electrode electrochemical cell, controlling the potentials of both the epitaxial layer and the silicon wafer, as shown in Figure 4.52, can further improve the thickness control of the resulting membrane by directly controlling the p/n bias voltage. The potential required to passivate n-type Si can be measured using the three-electrode system in Figure 4.51A, but this system does not take into account the diode leakage. If the reverse leakage is too large, the potential of the n-Si, $V_n$, will approach the potential of the p-Si, $V_p$. If there is a large amount of reverse diode leakage, the p-region may passivate prior to reaching the n-region and etching will cease. In the four-electrode configuration, the reverse leakage current is measured separately via a p-region contact, and the counter-electrode current may be monitored for end-point detection. The four-electrode approach allows etch stopping on lower quality epis (larger leakage current) and should also enable etch stopping of p-epi on n-substrates. Kloeck et al.[178] demonstrated that using such an electrochemical etch-stop technique and with current monitoring, the sensitivity of pressure sensors fabricated on the same wafer could be controlled to within 4% standard deviation. These authors used a 40% KOH solution at 60 ± 0.5°C. Without the etch stop, the sensitivity from sensor to sensor on one wafer varied by a factor of two.

The etching solution used in electrochemical etching can either be isotropic or anisotropic. Electrochemical etch stop in isotropic media was discussed above. In this case one uses $HF/H_2O$ mixtures to etch the highly doped regions of p$^+$p, n$^+$n, n$^+$p, and p$^+$n systems.[76,77,163,180,181] The rate-determining step in etching with isotropic etchants does not involve reducing water with electrons from the conduction band as it does in anisotropic etchants and the etch stop mechanism, as we learned earlier is obviously different. In isotropic media, the etch stop is simply a

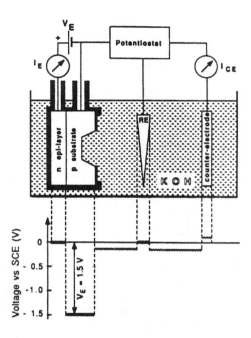

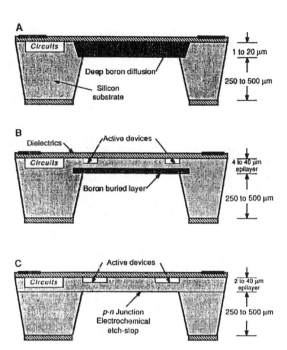

FIGURE 4.52 Four-electrode electrochemical etch-stop configuration. Voltage distribution with respect to the SCE reference electrode (RE) for the four-electrode case. The fourth electrode enables an external potential to be applied between the epitaxial layer and the substrate, thus maintaining the substrate at etching potentials. (From Kloeck et al., *IEEE Trans. Electron. Dev.*, 36, 663–669, 1989. With permission.)

FIGURE 4.53 Typical cross-sections of bulk micromachined wafers with various methods for etch-stop formation shown. (A) Diffused boron etch stop. (B) Boron etch stop as a buried layer. (C) Electrochemical etch stop. (From Wise, K.D., in *Silicon Micromachining and Its Application to High Performance Integrated Sensors*, Fung, C.D., Ed., Elsevier, New York, 1985. With permission.)

consequence of the fact that higher conductivity leads to higher corrosion currents and the etch slows down on lower conductivity layer. A major advantage of the KOH electrochemical etch is that it retains all of the anisotropic characteristics of KOH without needing a heavily doped p+ layer to stop the etch.[179]

In Figure 4.53, we review the etch-stop techniques discussed so far: diffused boron etch stop, buried boron etch stop, and electrochemical etch stop.[104] A comparison of the boron etch stop and the electrochemical etch stop reveals that the IC compatibility and the absence of built-in stresses, both due to the low dopant concentration, are the main assets of the electrochemical etch stop.

## Photo-Assisted Electrochemical Etch Stop (for n-Type Silicon)

A variation on the electrochemical diode etch-stop technique is the photo-assisted electrochemical etch-stop method illustrated in Figure 4.54.[182] An n-type silicon region on a wafer may be selectively etched in an HF solution by illuminating and applying a reverse bias across a p-n junction, driving the p-layer cathodic and the n-layer anodic. Etch rates up to 10 µm/min for the n-type material and a high resolution etch stop render this an attractive potential micromachining process. Advantages also include the use of lightly doped n-Si, bias- and illumination intensity-controlled etch rates, *in situ* process monitoring using the cell current, and the ability to spatially control etching with optical masking or laser writing. Using this method, Mlcak et al.[182] prepared stress-free cantilever beam test samples. They diffused boron into a $10^{15}$ cm$^{-3}$ (100) n-Si substrate through a patterned oxide mask, leaving

exposed a small n-type region which defines two p-type cantilever beams (see Figure 4.54).

The boron diffusion resulted in a junction 3.3 µm underneath the surface. An ohmic contact on the backside of the wafer was used to apply a variable voltage across the p-n junction, and both p and n areas were exposed to the HF electrolyte. The exposed n-region was etched to a depth of 150 µm by shining light on the whole sample. The resulting n-type Si surface was at first found to be rough, as porous Si up to 5 µm in height forms readily in HF solutions. The Si surface could be made smoother by etching at higher bias (4.3 V vs. SCE) and higher light intensity (2 W/cm$^2$) to a finish with features of the order of 0.4 µm in height. Smoothing could also be accomplished by a 5-second dip in HNO$_3$:HF:H$_3$COOH or a 30-second dip in 25 wt%, 25°C, KOH. Yet another way of removing unwanted porous Si is a 1000°C wet oxidation to make oxidized porous Si (OPS) followed by an HF etch (see next section).[183]

## Photo-Induced Preferential Anodization, PIPA (for p-Type Silicon)

Electrochemical etching requires the application of a metal electrode to apply the bias. The application of such a metal electrode often induces contamination and constitutes at least one extra process step, and extra fixturing is needed. With photo-induced preferential anodization (PIPA) it is not necessary to deposit metal electrodes. Here one relies on the illumination of a p-n junction to bias the p-type Si anodically, and the p-Si converts automatically into porous Si while the n-type Si acts as a cathode for the reaction. The principle of photo-biasing for etching

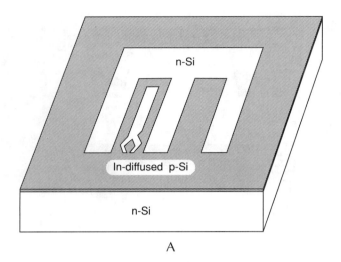

A

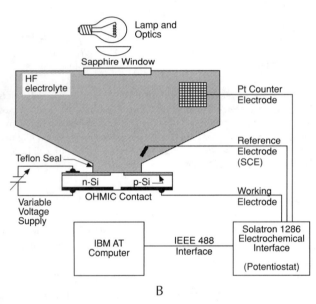

B

**FIGURE 4.54**  Photoelectrochemical etching. (A) Schematic of the photoelectrochemical etching experiment apparatus. (B) Schematic of the spatial geometry of the diffused p-Si layer into n-Si used to form cantilever beam structures. (From Mlcak, R., Photo-Assisted Electrochemical Machining of Micromechanical Structures, presented at Micro Electromechanical Systems, Ft. Lauderdale, FL, 1993. With permission.

purposes was known and patented by Shockley as far back as 1963.[184] In U.S. patent 3,096,262, he writes "… light can be used in place of electrical connections" … for biasing of the sample …. This means a small isolated area of p-type material on an n-type body may be preferentially biased for removal of material beyond the junction by etching." The method was reinvented by Yoshida et al.[183] in 1992 and in 1993 by Peeters et al.[185,186] The latter group called the method 'PHET' for photovoltaic electrochemical etch-stop technique, and the former group coined the PIPA acronym.

In PIPA, etch rates of up to 5 μm/min result in the formation of porous layers readily removed with Si etching solutions. An important advantage of the technique is that very small and isolated p-type islands can be anodized at the same time. The method also lends itself to fabricate three-dimensional structures using p-type Si as sacrificial layers.[183] A disadvantage of the technology is that one cannot control the process very well, as the

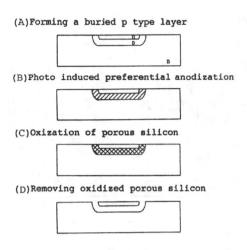

**FIGURE 4.55**  Photo-assisted electrochemical etch stop (for n-type Si). Fabrication process for a microbridge and SEM picture of Si structure before and after PIPA. (From Yoshida et al., Photo-Induced Preferential Anodization for Fabrication of Monocrystalline Micromechanical Structures, presented at Micro Electromechanical Systems, Travemunde, 1992. With permission.)

current cannot be measured for endpoint detection. Application of PIPA to form a microbridge is shown in Figure 4.55. First, a buried p-Si layer doped to $10^{18}$ $cm^{-3}$ and an n-type layer doped to $10^{15}$ $cm^{-3}$ are formed on an n-type substrate using epitaxy (Figure 4.55a). Then p-type is preferentially anodized in 10% HF solution under 30 mW/$cm^2$ light intensity for 180 minutes (Figure 4.55b), forming porous Si. The porous Si is then oxidized in wet oxygen at 1000°C (Figure 4.55c). Finally, the sacrificial layer of oxidized porous silicon is etched and removed with an HF solution (Figure 4.55d). The resulting surfaces of the n-type silicon are very smooth. It is interesting to consider making complicated three-dimensional structures by going immediately to the electropolishing regime instead. Yoshida et al.[183] believe that porous silicon as a sacrificial intermediate is more suitable for fabricating complicated structures. The authors are probably referring to the fact that electropolishing is much more aggressive and could not be expected to lead to the same retention of the shape of the buried, sacrificial p-layers. Peeters et al.[185] carry out their photovoltaic etching in KOH, thus skipping the porous Si stage of Yoshida et al. They, like Yosida et al., stress the fact that in one single etch step this technique can make a variety of complex shapes that would be impossible with electrochemical etching techniques. These authors found it necessary to coat the n-Si part of the wafer with Pt to get enough photovoltaic drive for the anodic dissolution; this metallization step makes the process more akin to the photoelectrochemical etching and some of the advantage of the photo-biasing process is being lost.

## Etch Stop at Thin Films-Silicon on Insulator

Yet another distinct way (the fourth) to stop etching is by employing a change in composition of material. An example is an etch stopped at a $Si_3N_4$ diaphragm. Silicon nitride is very strong, hard, and chemically inert and the stress in the film can be controlled by changing the Si/N ratio in the LPCVD deposition process. The stress turns from tensile in stoichiometric films to compressive in silicon-rich films (for details see Chapter 5). A great number of materials are not attacked by anisotropic etchants. Hence, a thin film of such a material can be used as an etch stop.

Another example is the $SiO_2$ layer in a silicon-on-insulator structure (SOI). A buried layer of $SiO_2$, sandwiched between two layers of crystalline silicon, forms an excellent etch stop because of the good selectivity of many etchants of Si over $SiO_2$. The oxide does not exhibit the good mechanical properties of silicon nitride and is consequently used rarely as a mechanical member in a microdevice. As with the photo-induced preferential anodization (PiPA), no metal contacts are needed with an SOI etch stop, greatly simplifying the process over an electrochemical etch stop technique.

We have classified SOI micromachining under surface micromachining in Chapter 5. More details about this very promising micromachining alternative are presented there.

# Problems with Wet Bulk Micromachining

## Introduction

Despite the introduction of better controllable etch-stop techniques, bulk micromachining remains a difficult industrial process to control. It is also not an applicable submicron technology, because wet chemistry is not able to etch reliably on that cale. For dubmicron structure definitin dry etching is required (dry etching is also more environmentally safe). We will look now into some of the other problems associated with bulk micromachining, such as the extensive real estate con-

sumption and difficulties in etching at convex corners, and detail the solutions that are being worked on to avoid, control, or alleviate those problems.

## Extensive Real Estate Consumption

### Introduction

Bulk micromachining involves extensive real estate consumption. This quickly becomes a problem in making arrays of devices. Consider the diagram in Figure 4.56, illustrating two membranes created by etching through a <100> wafer from the backside until an etch stop, say a $Si_3N_4$ membrane, is reached. In creating two of these small membranes a large amount of Si real estate is wasted and the resulting device becomes quite fragile.

### Real Estate Gain by Etching from the Front

One solution to limiting the amount of Si to be removed is to use thinner wafers, but this solution becomes impractical below 200 μm as such wafers break too often during handling. A more elegant solution is to etch from the front rather than from the back. Anisotropic etchants will undercut a masking material an amount dependent on the orientation of the wafer with respect to the mask. Such an etchant will etch any <100> silicon until a pyramidal pit is formed, as shown in Figure 4.57. These pits have sidewalls with a characteristic 54.7° angle with respect to the surface of a <100> silicon wafer, since the delineated planes are {111} planes. This etch property makes it possible to form cantilever structures by etching from the front side, as the cantilevers will be undercut and eventually will be suspended over a pyramidal pit in the silicon. Once this pyramidal

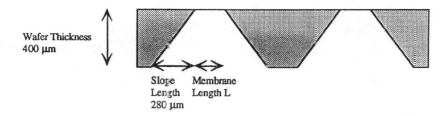

**FIGURE 4.56** Two membranes formed in a <100>-oriented silicon wafer.

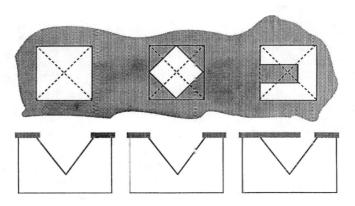

**FIGURE 4.57** Three anisotropically etched pits etched from the front in a <100>-oriented silicon wafer.

pit is completed, the etch rate of the {111} planes exposed is extremely slow and practically stops. Process sequences, which depend on achieving this type of a final structure, are therefore very uniform across a wafer and very controllable. The upper drawings in Figure 4.57 represent patterned holes in a masking material: a square, a diamond, and a square with a protruding tab. The drawings immediately below represent the etched pit in the silicon produced by the anisotropic etchant. Note in the first drawing that the square mask produces a four-sided pyramidal pit. In the second drawing, a similar shape oriented at 45° produces an etched pit which is oriented parallel to the pit etched in the first drawing. In the second drawing the corners of the diamond are undercut by the etchant as it produces the final etch pit. The third drawing illustrates that any protruding member is eventually undercut by the anisotropic etchant, leaving a cantilever structure suspended over the etch pit.

### Real Estate Gain by Using Silicon Fusion-Bonded Wafers

Using silicon fusion-bonded (SFB) wafers rather than conventional wafers also makes it possible to fabricate much smaller microsensors. The process is clarified in Figure 4.58 for the fabrication of a gauge pressure sensor.[187] The bottom, handle wafer has a standard thickness of 525 μm and is anisotropically etched with a square cavity pattern. Next, the etched handle wafer is fusion bonded to a top sensing wafer (the SFB process itself is detailed in Chapter 5 dealing with surface micromachining and in Chapter 8 on packaging). The sensor wafer consists of a p-type substrate with an n-type epilayer corresponding to the required thickness of the pressure-transducing membrane. The sensing wafer is thinned all the way to the epilayer by

electrochemical etching and resistors are ion implanted. The handle wafer is ground and polished to the desired thickness. For gauge measurement, the anisotropically etched cavity is truncated by the polishing operation, exposing the backside of the diaphragm. For an absolute pressure sensor the cavity is left enclosed. With the same diaphragm dimensions and the same overall thickness of the chip, an SFB device is almost 50% smaller than a conventional machined device (see Figure 4.59).

## Corner Compensation

### Underetching

Underetching of a mask which contains no convex corners, i.e., corners turning outside in, in principle stems from mask misalignment and/or from a finite etching of the {111} planes. Peeters measured the widening of {111}-walled V-grooves in a (100) Si wafer after etching in 7 *M* KOH at 80 ± 1° over 24 hours as 9 ± 0.5 μm.[41] The sidewall slopes of the V-groove are a well-defined 54.74°, and the actual etch rate $R_{111}$ is related to the rate of V-groove widening $R_v$ through:

$$R_{111} = \frac{1}{2}\sin(54.74°)R_v$$

$$\text{or } R_{111} = 0.408 \cdot R_v \qquad \qquad 4.41$$

with $R_{111}$ the etch rate in nm/min and $R_v$ the groove widening, also in nm/min. The V-groove widening experiment then results in a $R_{111}$ of 2.55 ± 0.15 nm/min. In practice, this etch rate implies

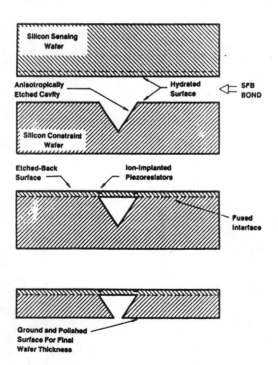

FIGURE 4.58  Fabrication process of an SFB-bonded gauge pressure sensor. (From Bryzek, J. et al., *Silicon Sensors and Microstructures*, Novasensor, Fremont, CA, 1990) With permission.)

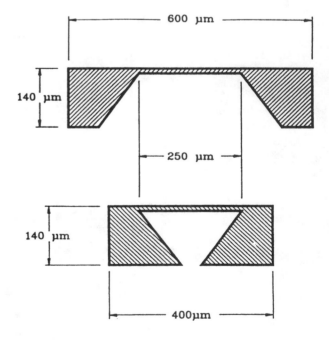

FIGURE 4.59  Comparison of conventional and SFB processes. The SFB process results in a chip which is at least 50% smaller than the conventional chip. (From Bryzek, J. et al., *Silicon Sensors and Microstructures*, Novasensor, Fremont, CA, 1990. With permission.)

a mask underetching of only 0.9 μm for an etch depth of 360 μm. For a 1-mm long V-groove and a 1° misalignment angle, a total underetching of 18 μm is theoretically expected, with 95% due to misalignment and only 5% due to etching of the {111} sidewalls.[41] The total underetching will almost always be determined by misalignment, rather than by etching of {111} walls.

Mask underetching with masks that do include convex corners is usually much larger than the underetching just described, as the etchant tends to circumscribe the mask opening with {111} walled cavities. This is usually called undercutting rather than underetching. It is advisable to avoid mask layouts with convex corners. Often mesa-type structures are essential, though, and in that case there are two possible ways to reduce the undercutting. One is by chemical additives, reducing the undercut at the expense of a reduced anisotropy ratio, and the other is by a special mask compensating the undercut at the expense of more lost real estate.

### Undercutting

When etching rectangular convex corners, deformation of the edges occurs due to undercutting. This is an unwanted effect, especially in the fabrication of, say, acceleration sensors, where total symmetry and perfect 90° convex corners on the proof mass are mandatory for good device prediction and specification. The undercutting is a function of etch time and thus directly related to the desired etch depth. An undercut ratio is defined as the ratio of undercut to etch depth ($\delta/H$).

Saturating KOH solutions with isopropanol (IPA) reduces the convex corner undercutting; unfortunately, this happens at the cost of the anisotropy of the etchant. This additive also often causes the formation of pyramidical or cone-shaped hillocks.[41 109] Peeters claims that these hillocks are due to carbonate contamination of the etchant and he advises etching under inert atmosphere also and stock-piling all etchant ingredient under an inert nitrogen atmosphere.[41]

Undercutting can also be reduced or even prevented by so-called corner compensation structures which are added to the corners in the mask layout. Depending on the etching solution, different corner compensation schemes are used. Commonly used are square corner compensation (EDP or KOH) and rotated rectangle corner compensation methods (KOH). In Figure 4.60, these two compensation methods are illustrated. In the square corner compensation case, the square of $SiO_2$ in the mask, outlining the square proof mass feature for an accelerometer, is

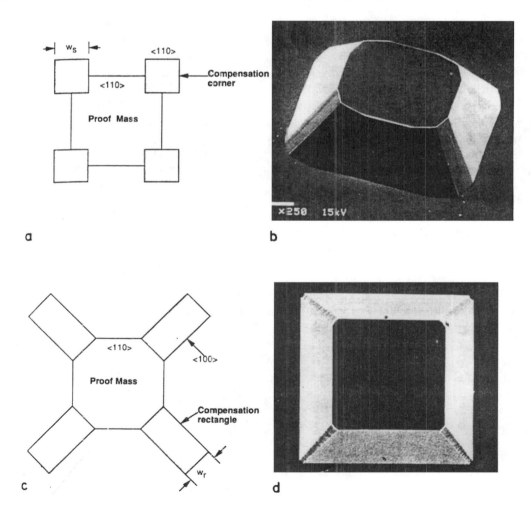

a
b
c
d

**FIGURE 4.60** Formation of a proof mass by silicon bulk micromachining. (a,b) Square corner compensation method, using EDP as the etchant. (c,d) Rotated rectangle corner compensation method, using KOH as the etchant.

enhanced by adding an extra $SiO_2$ square to each corner (Figure 4.60a). Both the proof mass and the compensation squares are aligned with their sides parallel to the <110> direction. In this way, two concave corners are created at the convex corner to be protected. Thus, direct undercutting is prevented. The three 'sacrificial' convex corners at the protective square are undercut laterally by the fast etching planes during the etch process. The dimension of the compensation square, $w_s$, for a 500-μm thick wafer (4-in.) is about 500 μm. The resulting mesa structure after EDP or KOH etching is shown in Figure 4.60b. In the rotated rectangle corner compensation method shown in Figure 4.60c, a properly scaled rectangle ($w_r$ should be twice the thickness of the wafer) is added to each of the mask corners. The four sides of the mesa square are still aligned along the <110> direction, but the compensation rectangles are rotated (45°) with their longer sides along the <100> directions. Using KOH as an etchant reveals the mesa shown in Figure 4.60d. A proof mass is frequently dislodged by simultaneously etching from the front and the back. Corner compensation requires significant amount of space around the corners, making the design less compact, and the method is often only applicable for simple geometries.

Different groups around the world have been using different corner compensation schemes and they all claim to have optimized spatial requirements. For an introduction to corner compensation, refer to Puers et al.[188] Sandmaier et al.[189] for KOH etching use <110>-oriented beams, <110>-oriented squares, and <010>-oriented bands for corner compensation. They found that spatial requirements for compensation structures could be reduced dramatically by combining several of these compensation structures. The mask layouts for some of the different compensation schemes used by Sandmaier et al. are shown in Figure 6.61. To understand the choice and dimensioning of these compensation structures, as well as those in Figure 6.60, we will first look at the planes emerging at convex corners during KOH etching. Mayer et al.[190] found that the undercutting of convex corners in pure KOH etch is determined exclusively by {411} planes. The {411} planes of the convex under-etching corner, as shown in Figure 4.62, are not entirely laid free, though; rugged surfaces, where only fractions of the main planes can be detected, overlap the {411} planes under a diagonal line shown as AB in this figure. The ratio of {411} to {100} etching does not depend on temperature between 60 and 100°C. The value does decline with increasing KOH concentration from about 1.6 at 15% KOH to 1.3 at above 40% where the curve flattens out.[190] Ideally one avoids rugged surfaces and searches for well-defined planes bounding the convex corner. In Figure 4.63 it is shown how a <110> beam is added to the convex corner to be etched. The fast etching {411} planes, starting at the two convex corners, are laterally underetching a <110>-oriented beam (broken lines in Figure 4.63). It is clear that the longer this <110>-oriented beam is, the longer the convex corner is protected from undercutting. It is essential that by the end of the etch the beam has disappeared to maintain a minimum of rugged surface at the convex edge. On the other hand, as is obvious from Figure 4.63, a complete disappearance of the beam leads to a beveling at the face of the convex corner. The dimensioning of the

compensating <110> beam works then as follows: the length of the compensating beam is calculated primarily from the required etch depth (H) and the etch rate ratio R {411}/R{100} (≈δ/H), at the concentration of the KOH solution used:

$$L = L_1 - L_2 = 2H\frac{R\{411\}}{R\{100\}} - \frac{B_{\langle110\rangle}}{2\tan(30.9°)} \qquad 4.42$$

with H the etch depth; $B_{\langle110\rangle}$ the width of the <110>-oriented beam; tan (30.9°) the geometry factor.[190] The factor 2 is the first term of this equation results as the etch rate of the {411} plane is determined normal to the plane and has to be converted to the <110> direction. The second term in Equation 4.42 takes into account that the <110> beam needs to disappear completely by the time the convex corner is reached. The resulting beveling in Figure 4.63 can be reduced by further altering the compensation structures. This is done by decelerating the etch front, which largely determines the corner undercutting. One way to accomplish this is by creating more concave shapes right before the convex corner is reached. In Figure 4.61A, splitting of the compensation beam creates such concave corners, and by arranging two such double beams a more symmetrical final structure is achieved. By using these split beams, the bevelling at the corner is reduced by a factor of 1.4 to 2 and leads to bevel angles under 45°.

Corner compensation with <110>-oriented squares (as shown in Figure 4.60A) features considerably higher spatial requirements than the <110>-oriented beams. Since these squares are again undercut by {411} planes that are linked to the rugged surfaces described above, the squares do not easily lead to sharp {111}-defined corners. Dimensioning of the compensation square is done by using Equation 4.42 again, where $L_1$ is half the side length of the square, and for $B_{\langle110\rangle}$ the side length is used. All fast-etching planes have to reach the convex corner at the same time. As before, the spatial requirements of this compensation structure can be reduced if it is combined with <110>-oriented beams. Such a combination is shown in Figure 4.61B. The three convex corners of the compensation square are protected from undercutting by the added <110> beams. During the first etch step, the <110>-oriented side beams are undercut by the etchant. Only after the added beams have been etched does the square itself compensate the convex corner etching. The dimensioning of this combination structure is carried out in two steps. First the <110>-oriented square is selected with a size that is permitted by the geometry of the device to be etched. From these dimensions, the etch depth corresponding to this size is calculated from Equation 4.42. For the remaining etch depth the <110>-oriented beams are dimensioned like any other <110>-oriented beam. If the side beam on corner b is selected about 30% longer than the other two side beams, the quality of the convex corner can be further improved. In this case, the corner is formed by the etch fronts starting at the corners a and b (Figure 4.61B).

A drawback to all the above proposed compensation schemes is the impossibility, due to rugged surfaces always accompanying {411} planes, of obtaining a clean corner in both the top

**A**

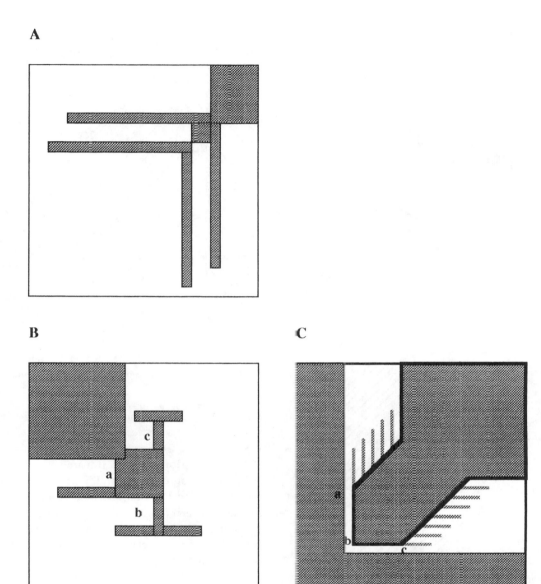

**B**

**C**

**FIGURE 4.61**  Mask layout for various convex corner compensation structures. (From Sandmaier, H. et al., Corner Compensation Techniques in Anisotropic Etching of (100) Silicon Using Aqueous KOH, presented at Transducers '91, San Francisco, CA, 1991. With permission.)

and the bottom of a convex edge. Buser et al.[191] introduced a compensation scenario where a convex corner was formed by two {111} planes which were well defined all the way from the mask to the etch bottom. No rugged, undefined planes show in this case. The mask layout to create such an ideal convex corner has bands that are added to convex corners in the <100> direction (see Figure 4.61C and Figure 4.60 c and d). These bands will be underetched by vertical {100} planes from both sides. With suitable dimensioning of such a band, a vertically oriented membrane results, thinning, and eventually freeing the convex edge shortly before the final intended etch depth is reached. In contrast to compensation structures undercut by {411} planes, posing problems with undefined rugged surfaces (see above), this compensation structure is mainly undercut by {100} planes. Over the temperature range of 50 to 100°C and KOH concentrations ranging from 25 to 50 wt%, no undefined surfaces could be detected in the case of structures undercut by {100}

planes.[189] The width of these <010>-oriented compensation beams, which determines the minimum dimension of the structures to etch, has to be twice the etching depth. These beams can either connect two opposite corners and protect both from undercutting simultaneously, or they can be added to the individual convex corners (open beam). With an open beam approach (see Figure 4.64) one has to be certain that the {100} planes reach the corner faster than the {411} planes. For that purpose the beams have to be wide enough to avoid complete underetching by {411} planes moving in from the front side, before they are completely underetched by {100} planes moving in from the side. For instance, in a 33% KOH etchant a ratio between beam length and width of at least 1.6 is required. To make these compensation structures smaller while at the same time maintaining {100} undercutting to define the final convex corner, Sandmaier et al. remarked that the shaping {100} planes do not need to be present at the beginning of the etching

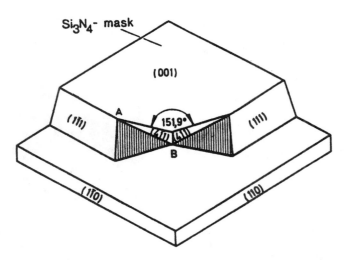

**FIGURE 4.62** Planes occurring at convex corners during KOH etching. (From Mayer, G.K. et al., *J. Electrochem. Soc.*, 137, 3947–3951, 1990. With permission.)

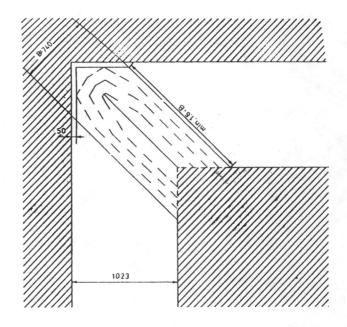

**FIGURE 4.64** Beam structure open on one side. The beam is oriented in the <010> direction. Dimensions in microns. B is the width of the beam.

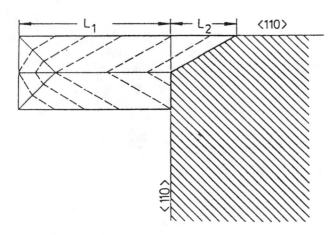

**FIGURE 4.63** Dimensioning of the corner compensation structure with a <110>-oriented beam. (From Sandmaier, H. et al., Corner Compensation Techniques in Anisotropic Etching of (100) Silicon Using Aqueous KOH, presented at Transducers '91, San Francisco, CA, 1991. With permission.)

process. These authors implement delaying techniques by adding fan-like <110>-oriented side beams to a main <100>-oriented beam (see Figure 4.61C). As described above, these narrow beams are underetched by {411} planes and the rugged surfaces they entail until reaching the <100>-oriented beam. Then the {411} planes are decelerated in the concave corners between the side beams by the vertical {100} planes with slower etching characteristics. The length of the <110>-oriented side beams is calculated from:

$$L_{\langle 110 \rangle} = \left( H - \frac{B_{\langle 010 \rangle}}{2} \right) \frac{R\{411\}}{R\{100\}} \qquad \textbf{4.43}$$

with H being the etching depth at the deepest position of the device.

The width of the side beams does not influence the calculation of their required length. In order to avoid the rugged

surfaces at the convex corner, the width of the side beams as well as the spaces between them should be kept as small as possible. For an etching depth of 500 μm, a beam width of 20 μm and a space width of 2 μm are optimal.[189]

Depending on the etchant, different planes are responsible for undercutting. From the above we learned that in pure KOH solutions undercutting, according to Mayer et al.[190] and Sandmaier et al.,[189] mainly proceeds through {411} planes or {100} planes. That the {411} planes are the fastest undercutting planes was confirmed by Seidel;[192] at the wafer surface, the sectional line of a (411) and a (111) plane point in the <410> direction, forming an angle of 30.96° with the <110> direction, and it was in this direction that he found a maximum in the etch rate. In KOH and EDP etchants Bean[105] identified the fast undercutting planes as {331} planes. Puers et al.,[188] for alkali/alcohol/water, identified the fast underetching planes as {331} planes, as well. Mayer et al.,[190] working with pure KOH, could not confirm the occurrence of such planes. Lee indicated that in hydrazine-water the fastest underetching planes are {211} planes.[91] Abu-Zeid[193] reported that the main beveling planes are {212} planes in ethylene-diamine-water solution (no added pyrocatechol). Wu et al.[194] found the main beveling planes at undercut corners to be {212} planes whether using KOH, hydrazine, or EPW solutions are used. In view of our earlier remarks on the sensitivity of etching rates of higher index rates on a wide variety of parameters (temperature, concentration, etching size, stirring, cation effect, alcohol addition, complexing agent, etc.) these contradictory results are not too surprising. Along the same line, Wu et al.[195] and Peurs et al.[188] have suggested triangles to compensate for underetching, but Mayer et al.[190] found them to lead to rugged surfaces at the convex corner. Combining a chemical etchant with more limited undercutting (IPA in KOH) with Sandmaier's reduced compensation structure schemes could further decrease the required size of the compensation features while retaining an acceptable anisotropy.

Corner compensation for <110>-oriented Si was explored by Ciarlo.[127] Ciarlo comments that both corner compensation and corner rounding can be minimized by etching from both surfaces so as to minimize the etch time required to achieve the desired features. This requires accurate front-to-back alignment and double-sided polished wafers.

Employing corner compensation offers access to completely new applications such as rectangular solids, orbiting V-grooves, truncated pyramids with low cross-sections on the wafer surface, bellow structures for decoupling mechanical stresses between micromechanical devices and their packaging, etc.[189]

# Wet Bulk Micromachining Examples

## Example 1. Dissolved Wafer Process

Figure 4.65 illustrates the dissolved wafer process Cho is optimizing for the commercial production of low-cost inertial instruments.[196] This process, also in use by Draper Laboratory[197,198] for the same application, involves a sandwich of a silicon sensor anodically bonded to a glass substrate. The preparation of the silicon part requires only two masks and three processing steps. A recess is KOH-etched into a p-type (100) silicon wafer (step 1 with mask 1), followed by a high-temperature boron diffusion (step 2 with no mask). In step one RIE may be used as well. Cho claims that by maintaining a high-temperature uniformity in the KOH etching bath (±0.1°C), the accuracy and absolute variation of the etch across the wafer,

wafer-to-wafer, and lot-to-lot, can be maintained to <0.1 μm using pre-mixed, 45 wt% KOH. Cho is also using low-defect oxidation techniques (e.g., nitrogen annealing and dry oxidation) to form defect-free silicon surfaces. In the boron diffusion the key is optimizing the oxygen content. In general, the optimal flow of oxygen is on the order of 3 to 5% of the nitrogen flow, in which case the doping uniformity is on the order of ±0.2 μm. Varying the KOH etch depth and the shallow boron diffusion time, a wide variety of operating ranges and sensitivities for sensors can be obtained. Next, the silicon is patterned for a reactive ion etching (RIE) etch (step 3 with mask 2). Aspect ratios above 10 are accessible. Using some of the newest dry etching techniques depths in excess of 500 μm at rates above 4 μm/min (with a $SF_6$ chemistry) are now possible.[199] The glass substrate (#7740 Corning glass) preparation involves etching a recess, depositing and in a one-mask step patterning a multi-metal system of Ti/Pt/Au. The electrostatic bonding of glass to silicon takes place at 335°C with a potential of 1000 V applied between the two parts (electrostatic or anodic bonding is explained in detail in Chapter 8). Commercial bonders have alignment accuracies on the order of <1 μm. The lightly doped silicon is dissolved in an EDP solution at 95°C. The keys to uniform EDP etching are temperature uniformity and suppression of etchant depletion through chemical aging or restricted flow (e.g., through bubbles). These effects can be minimized by techniques that optimize temperature control and reduce bubbling (e.g., proper wafer spacing, lower temperature, large bath). The structures are finally rinsed in DI water and a hot methanol bath.

Draper Laboratory, although obtaining excellent device results with the dissolved wafer process, is now exploring an SOI

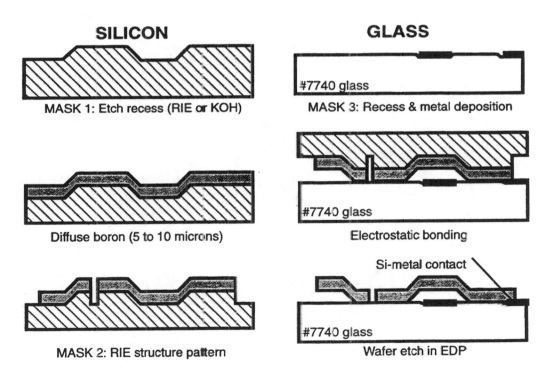

FIGURE 4.65   Dissolved wafer process. (From Greiff, P., SOI-Based Micromechanical Process, presented at Micromachining and Microfabrication Process Technology, Austin, TX, 1995. With permission.)

process as an alternative. The latter yields an all-silicon device while preserving many of the dissolved wafer process advantages (see also Chapter 5 under SOI Surface Micromachining).

## Example 2. An Electrochemical Sensor Array Measuring pH, $CO_2$, and $O_2$ in a Dual Lumen Catheter

This sensor array developed by the author is shown in Figure 4.66,* packaged and ready for *in vivo* monitoring of blood pH, $CO_2$, and $O_2$. The linear electrochemical array fits inside a 20-gauge catheter (750 μm in diameter) without taking up so much space as to distort the pressure signal monitored by a pressure sensor outside the catheter. A classical (macro) reference electrode, making contact with the blood through the saline drip, was used for the pH signal, while the $CO_2$ and $O_2$

---

* Figure 4.66 appears as a color plate after page 144.

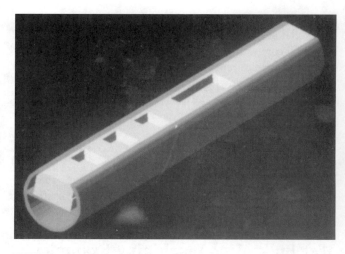

FIGURE 4.67 CAD of the electrochemical sensor array showing two pieces of Si (each 250 μm thick) on top of each other, mounted in a dual lumen catheter. The bottom part of the catheter is left open so pressure can be monitored and blood samples can be taken.

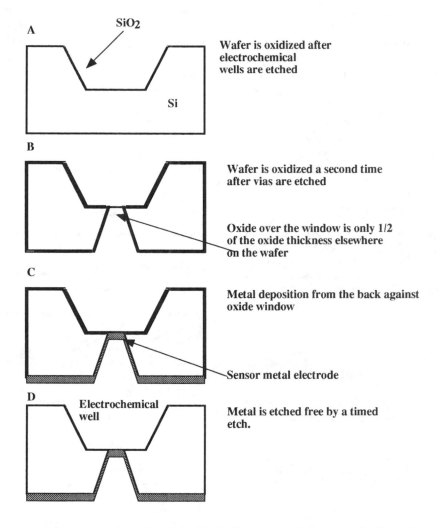

A    SiO₂    **Wafer is oxidized after electrochemical wells are etched**    Si

B    **Wafer is oxidized a second time after vias are etched**

**Oxide over the window is only 1/2 of the oxide thickness elsewhere on the wafer**

C    **Metal deposition from the back against oxide window**

**Sensor metal electrode**

D    **Electrochemical well**    **Metal is etched free by a timed etch.**

FIGURE 4.68 Fabrication sequence for a generic electrochemical cell in Si. Depth of the electrochemical well, number of electrodes, and electrode materials can be varied.

had their own internal reference electrodes. The high impedance of the small electrochemical probes makes a close integration of the electronics mandatory; otherwise, the high impedance connector leads, in a typical hospital setting, act as antennas for the surrounding electronic noise. As can be seen from the computer-aided design (CAD) picture in Figure 4.67, the thickness of the sensor comes from two silicon pieces, the top piece containing electrochemical cells and the bottom piece containing the active electronics. Each wafer is 250 μm thick. The individual electrochemical cells are etched anisotropically into the top silicon wafer. The bottom piece is fabricated in a custom IC housing using standard IC processes. The process sequence to build a generic electrochemical cell in a 250-μm thick Si wafer with one or more electrodes at the bottom of each well is illustrated in Figure 4.68. Electrochemical wells are etched from the front of the wafer and, after an oxidation step, access cavities for the metal electrodes are also etched from the back. The etching of the vias stops at the oxide-covered bottoms of the electrochemical wells (Figure 4.68A). Next, the wafers are oxidized a second time with the oxide thickness doubling everywhere except in the suspended window areas where no more Si can feed further growth (Figure 4.68B). The desired electrode metal is subsequently deposited from the back of the wafer into the access cavities and against the oxide window (Figure 4.68C). Finally, a timed oxide etch removes the sacrificial oxide window from above the underlying metal, while preserving the thicker oxide layer in the other areas on the chip (Figure 4.68D).[200–202] An

SEM micrograph demonstrating step D in Figure 4.68 is shown in Figure 4.69. The electrodes in the electrochemical cells of the top wafer are further connected to the bottom wafer electronics by solder balls in the access vias of the top silicon wafer (see Figure 4.70). Separating the chemistry from the electronics in this way provides extra protection for the electronics from the electrolyte as well as from the electronics, and chemical sensor manufacture can proceed independently. Depending on the type of sensor element, one or more electrodes are fabricated at the bottom of the electrochemical cells. For example, shown in Figure 4.71 is an almost completed (Severinghaus) $CO_2$ sensor with an Ag/AgCl electrode as the reference electrode (left in the figure) and an IrOx pH-sensitive electrode, both at the bottom of one electrochemical well. The metal electrodes are electrically isolated from each other by the $SiO_2$ passivation layer over the surface of the silicon wafer. To complete the $CO_2$ sensor, we silk-screen a hydrogel containing an electrolytic medium into the silicon sensor cavity and dip-coat the sensor into a silicone-polycarbonate rubber solution to form the gas-permeable membrane. For hydrogel inside the micromachined well, we use poly(2-hydroxyethyl)methacrylate (PHEMA) or polyvinylalcohol (PVA).

The concept of putting the sensor chemistries and the electronics on opposite sides of a substrate is a very important design feature we decided upon several years ago in view of the overwhelming problems encountered in building chemical sensors based on ISFETs or EGFETs.[140]

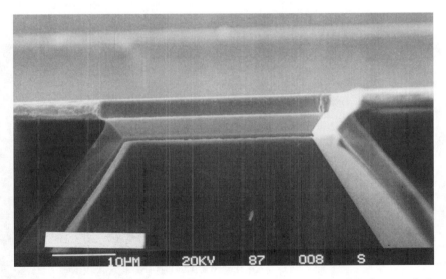

FIGURE 4.69    SEM micrograph illustrating process step D from Figure 4.68. A 30 × 30 μm Pt electrode is shown at the bottom of an electrochemical well. This Pt electrode is further contacted to the electronics from the back.

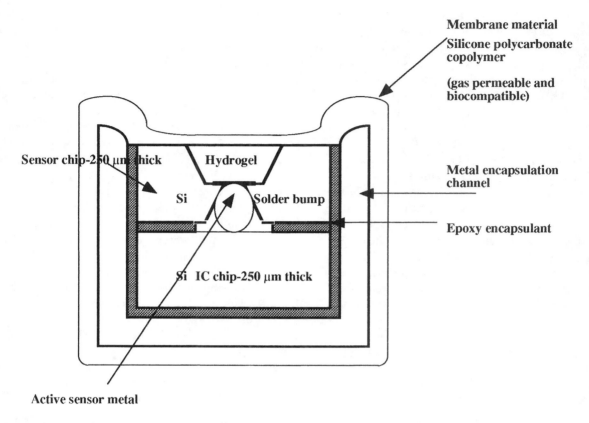

**FIGURE 4.70**   Schematic of the bonding scheme between sensor wafer and IC. The schematic is a cut-through of the catheter.

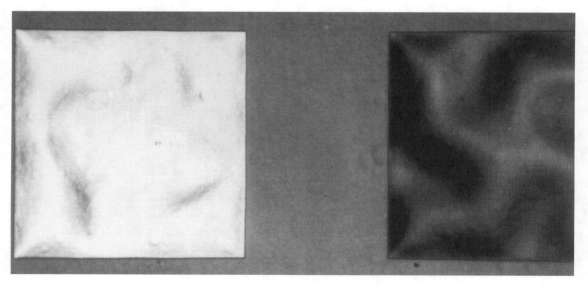

**FIGURE 4.71**   SEM micrograph of an Ag/AgCl (left) and an IrO$_x$ (right) electrode at the bottom of an anisotropically etched well in Si. This electrochemical cell forms the basis for a Severinghaus CO$_2$ sensor.

# References

1. Harris, T. W., *Chemical Milling*, Clarendon Press, Oxford, 1976.

2. Private Communication, Block, B., *Zero-Base the Technological Approach*, Private Communication.

3. Allen, D., M., *The Principles and Practice of Photochemical Machining and Photoetching*, Adam Hilger, Bristol and Boston, 1986.

4. Robbins, H. R. and B. Schwartz, "Chemical Etching of Silicon-I. The System, HF, HNO$_3$ and H$_2$O", *J. Electrochem. Soc.*, 106, 505–508, 1959.

5. Robbins, H. and B. Schwartz, "Chemical Etching of Silicon-II. The System HF, HNO$_3$, H$_2$O, and HC$_2$C$_3$O$_2$", *J. Electrochem. Soc.*, 107, 108–111, 1960.

6. Schwartz, B. and H. Robbins, R., "Chemical Etching of Silicon–III. A Temperature Study in the Acid System", *J. Electrochem. Soc.*, 108, 365–372, 1961.

7. Schwartz, B. and H. Robbins, "Chemical Etching of Silicon-IV. Etching Technology", *J. Electrochem. Soc.*, 123, 1903–1909, 1976.

8. Uhlir, A., "Electrolytic Shaping of Germanium and Silicon", *Bell Syst. Tech. J.*, 35, 333–347, 1956.

9. Hallas, C. E., "Electropolishing Silicon", *Solid State Technol.*, 14, 30–32, 1971.

10. Turner, D. R., "Electropolishing Silicon in Hydrofluoric Acid Solutions", *J. Electrochem. Soc.*, 105, 402–408, 1958.

11. Stoller, A. I. and N. E. Wolff, "Isolation Techniques for Integrated Circuits", Proc. Second International Sym. on Microelectronics, Munich, Germany, 1966.

12. Stoller, A. I., "The Etching of Deep Vertical–Walled Patterns in Silicon", *RCA Rev.*, 31, 271–275, 1970.

13. Forster, J. H. and J. B. Singleton, "Beam–Lead Sealed Junction Integrated Circuits", *Bell Laboratories Record*, 44, 313–317, 1966.

14. Kenney, D. M., *Methods of Isolating Chips of a Wafer of Semiconductor Material*, in U.S. Patent #3,332,137, 1967.

15. Lepselter, M. P., "Beam Lead Technology", *Bell. Sys. Tech. J.*, 45, 233–254, 1966.

16. Lepselter, M. P., *Integrated Circuit Device and Method*, in U.S. Patent # 3,335,338, 1967.

17. Waggener, H. A., "Electrochemically Controlled Thinning of Silicon", *Bell. Sys. Tech. J.*, 49, 473–475, 1970.

18. Kragness, R. C. and H. A. Waggener, *Precision Etching of Semiconductors*, in U.S. Patent #3,765,969, 1973.

19. Waggener, H. A., R. C. Kragness, and A. L. Tyler, "Anisotropic Etching for Forming Isolation Slots in Silicon Beam Leaded Integrated Circuits", IEEE International Electron Devices Meeting. Technical Digest. IEDM '67, Washington, D.C., 1967, p. 68.

20. Waggener, H. A., R. C. Krageness, and A. L. Tyler, "Two–way Etch", *Electronics*, 40, 274, 1967.

21. Bean, K. E. and W. R. Runyan, "Dielectric Isolation: Comprehensive, Current and Future", *J. Electrochem. Soc.*, 124, 5C–12C, 1977.

22. Rodgers, T. J., W. R. Hiltpold, B. Frederick, J. J. Barnes, F. B. Jenné, and J. D. Trotter, "VMOS Memory Technology", *IEEE J. Solid-State Circuits*, Vol. SC–12, 515–523, 1977.

23. Rodgers, T. J., W. R. Hiltpold, J. W. Zimmer, G. Marr, and J. D. Trotter, "VMOS ROM", *IEEE J. Solid-State Circuits*, SC–11, 614–622, 1976.

24. Ammar, E. S. and T. J. Rodgers, "UMOS Transistors on (110) Silicon", *IEEE Trans. Electron Devices*, ED–27, 907–914, 1980.

25. Smith, C. S., "Piezoresistance Effect in Germanium and Silicon", *Phys. Rev.*, 94, 42–49, 1954.

26. Pfann, W. G., "Improvement of Semiconducting Devices by Elastic Strain", *Solid State Electron.*, 3, 261–267, 1961.

27. Tufte, O. N., P. W. Chapman, and D. Long, " Silicon Diffused–element Piezoresistive Diaphragms", *J. Appl. Phys.*, Vol.33, 3322, 1962.

28. Editorial, "Transducers, Pressure and Temperature, Catalog.", National Semiconductor, Sunnyvale, CA, 1974.

29. Editorial, *Thermal Character Print Head*, Texas Instruments, T.I. Austin, USA, 1977.

30. O'Neill, P., "A Monolithic Thermal Converter", *Hewlett-Packard J.*, 31, 12–13, 1980.

31. Boivin, L. P., "Thin Film Laser–to Fiber Coupler", *Appl. Opt.*, 13, 391–395, 1974.

32. Bassous, E., H. H. Taub, and L. Kuhn, "Ink Jet Printing Nozzle Arrays Etched in Silicon", *Appl. Phys. Lett.*, 31, 135–137, 1977.

33. Greenwood, J. C., "Etched Silicon Vibrating Sensor", *J. Phys. E, Sci. Instrum.*, 17, 650–652, 1984.

34. Petersen, K. E., "Silicon as a Mechanical Material", *Proceedings of the IEEE*, 70, 420–457, 1982.

35. Middlehoek, S. and U. Dauderstadt, "Haben Mikrosensoren aus Silizium eine Zukunft?", *Technische Rundschau*, July, 102–105, 1994.

36. Kittel, C., *Introduction to Solid State Physics*, John Wiley & Sons, 1976.

37. Hu, J. Z., L. D. Merkle, C. S. Menoni, and I. L. Spain, "Crystal Data for High–Presure Phases of Silicon", *Phys. Rev. B*, 34, 4679–4684, 1986.

38. Tan, T. Y., H. Foll, and S. M. Hu, "On the Diamond-Cubic to Hexagonal Phase Transformation in Silicon", *Phil. Mag. A*, 44, 127–140, 1981.

39. Eremenko, V. G. and V. I. Nikitenko, "Electron Microscope Investigation of the Microplastic Deformation Mechanisms by Indentation", *Phys. Stat. Sol. A*, 14, 317–330, 1972.

40. Alavi, M., S. Buttgenbach, A. Schumacher, and H. J. Wagner, "Fabrication of Microchannels by Laser Machining and Anisotropic Etching of Silicon", *Sensors and Actuators A*, A32, 299–302, 1992.

41. Peeters, E., "Process Development for 3D Silicon Microstructures, with Application to Mechanical Sensor Design", Ph.D. Thesis, Catholic University of Louvain, Belgium, 1994.

42. Elwenspoek, M., H. Gardeniers, M. de Boer, and A. Prak, "Micromechanics", University of Twente, Report No. 122830, Twente, Netherlands, 1994.

43. Kendall, D. L., "On Etching Very Narrow Grooves in Silicon", *Appl. Phys. Lett.*, 26, 195–198, 1975.

44. Kendall, D. L., "Vertical Etching of Silicon at Very High Aspect Ratios", *Ann. Rev. Mater. Sci.*, 9, 373–403, 1979.

45. Madou, M. J., "Compatibility and Incompatibility of Chemical Sensors and Analytical Equipment with Micromachining", Technical Digest of the 1994 Solid State Sensor and Actuator Workshop., Hilton Head Island, SC, 1994, p. 164–171.

46. Chou, P. C. and N. J. Pagano, *Elasticity: Tensor, Dyadic, and Engineering Approaches*, Dover Publications, Inc., New York, 1967.

47. Brantley, W. A., "Calculated Elastic Constants for Stress Problems Associated with Semiconductor Devices", *J. Appl. Phys.*, 44, 534–535, 1973.

48. Nikanorov, S. P., Y. A. Burenkov, and A. V. Stepanov, "Elastic Properties of Silicon", *Sov. Phys.-Solid State*, 13, 2516–2518, 1972.

49. Khazan, A. D., *Transducers and Their Elements*, PTR Prentice Hall, Englewood Cliffs, NJ, 1994.

50. Worthman, J. J. and R. A. Evans, "Young's Modulus, Shear Modulus and Poisson's Ratio in Silicon and Germanium", *J. Appl. Phys.*, 36, 153–156, 1965.

51. Metzger, H. and F. R. Kessler, "Der Debye-Sears Effect zur Bestimmung der Elastischen Konstanten von Silicium", *Z. Naturf.*, A25, 904–906, 1970.

52. Greenwood, J. C., "Silicon in Mechanical Sensors", *J. Phys. E, Sci. Instrum.*, 21, 1114–1128, 1988.

53. Huff, M. A. and M. A. Schmidt, "Fabrication, Packaging and Testing of a Wafer Bonded Microvalve", Technical Digest of the 1992 Solid State Sensor and Actuator Workshop., Hilton Head Island, SC, 1992, p. 194–197.

54. Maseeh, F. and S. D. Senturia, "Plastic Deformation of Highly Doped Silicon", *Sensors and Actuators A*, A23, 861–865, 1990.

55. Ding, X. and W. Ko, "Buckling Behavior of Boron-Doped P+ Silicon Diaphragms", 6th International Conference on Solid-State Sensors and Actuators (Transducers '91), San Francisco, CA, 1991, p. 201–204.

56. Callister, D. W., *Materials Science and Engineering*, John Wiley & Sons, New York, 1985.

57. Kanda, Y., "A Graphical Representation of the Piezoresistance Coefficients in Silicon", *IEEE Trans. Electron Devices*, ED-29, 64–70, 1982.

58. Kaminsky, G., "Micromachining of Silicon Mechanical Structures", *J. Vac. Sci. Technol.*, B3, 1015–1024, 1985.

59. Stoller, A. I., R. F. Speers, and S. Opresko, "A New Technique for Etch Thinning Silicon Wafers", *RCA Rev.*, 265–270, 1970.

60. Kern, W., "Chemical Etching of Silicon, Germanium, Gallium Arsenide, and Gallium Phosphide", *RCA Rev.*, 39, 278–308, 1978.

61. Klein, D., L. and D. J. D'Stefan, "Controlled Etching of Silicon in the HF-HNO$_3$ System", *J. Electrochem. Soc.*, 109, 37–42, 1962.

62. Schnable, G. L. and P. F. Schmidt, "Applications of Electrochemistry to Fabrication of Semiconductor Devices", *J. Electrochem. Soc.*, 123, 310C–315C, 1976.

63. Kern, W. and C. A. Deckert, "Chemical Etching", in *Thin Film Processes*, Vossen, J. L. and Kern, W., Eds., Academic Press, Orlando, 1978.

64. Yang, K. H., "An Etch for Delineation of Defects in Silicon", *J. Electrochem. Soc.*, 131, 1140–1145, 1984.

65. Chu, T. L. and J. R. Gavaler, "Dissolution of Silicon and Junction Delineation in Silicon by the CrO$_3$-HF-H$_2$O System", *Electrochim. Acta*, 10, 1141–1148, 1965.

66. Archer, V. D., "Methods for Defect Evaluation of Thin <100> Oriented Silicon in Epitaxial Layers Using a Wet Chemical Etch", *J. Electrochem. Soc.*, 129, 2074–2076, 1982.

67. Schimmel, D. G. and M. J. Elkind, "An Examination of the Chemical Staining of Silicon", *J. Electrochem. Soc.*, 125, 152–155, 1973.

68. Secco d'Aragona, F., "Dislocation Etch for (100) Planes in Silicon", *J. Electrochem. Soc.*, 119, 948–951, 1972.

69. Tuck, B., "Review-The Chemical Polishing of Semiconductors", *J. Mater. Sci.*, 10, 321–339, 1975.

70. Wong, A., "Silicon Micromachining", 1990, Viewgraphs, Presented in Chicago,IL.

71. Wise, K. D., M. G. Robinson, and W. J. Hillegas, "Solid State Processes to Produce Hemispherical Components for Inertial Fusion Targets", *J. Vac. Sci. Technol.*, 18, 1179–1182, 1981.

72. Hu, S. M. and D. R. Kerr, "Observation of Etching of n-type Silicon in Aqueous HF Solutions", *J. Electrochem. Soc.*, 114, 414, 1967.

73. Hoffmeister, W., "Determination of the Etch Rate of Silicon in Buffered HF Using a $^{31}$Si Tracer Method", *Int. J. Appl. Radiation and Isotopes*, 2, 139, 1969.

74. Muraoka, H., T. Ohashi, and T. Sumitomo, "Controlled Preferential Etching Technology", Semiconductor Silicon 1973, Chicago, IL, 1973, p. 327–338.

75. Seidel, H., "Nasschemische Tiefenatztechnik", in *Mikromechanik*, Heuberger, A., Ed., Springer, Heidelberg, 1989.

76. Theunissen, M. J., J. A. Apples, and W. H. C. G. Verkuylen, "Applications of Preferential Electrochemical Etching of Silicon to Semiconductor Device Technology", *J. Electrochem. Soc.*, 117, 959–965, 1970.

77. Meek, R. L., "Electrochemically Thinned N/N+ Epitaxial Silicon-Method and Applications", *J. Electrochem. Soc.*, 118, 1240–1246, 1971.

78. van Dijk, H. J. A. and J. de Jonge, "Preparation of Thin Silicon Crystals by Electrochemical Thinning of Epitaxially Grown Structures", *J. Electrochem. Soc.*, 117, 553–554, 1970.

79. Shengliang, Z., Z. Zongmin, and L. Enke, "The NH₄F Electrochemical Etching Method of Silicon Diaphragm for Miniature Solid-State Pressure Transducer", 4th International Conference on Solid-State Sensors and Actuators (Transducers '87), Tokyo, Japan, 1987, p. 130–133.

80. Bomchil, G., R. Herino, and K. Barla, "Formation and Oxidation of Porous Silicon for Silicon on Insulator Technologies", in *Energy Beam-Solid Interactions and Transient Thermal Processes*, Nguyen, V. T. and Cullis, A., Eds., Les Editions de Physique, Les Ulis, 1986, p. 463.

81. Bogenschutz, A. F., K.-H. Locherer, W. Mussinger, and W. Krusemark, "Chemical Etching of Semiconductors in HNO₃-HF-CH₃COOH", *J. Electrochem. Soc.*, 114, 970–973, 1967.

82. Allongue, P., V. Costa-Kieling, and H. Gerischer, "Etching of Silicon in NaOH Solutions, Part I and II", *J. Electrochem. Soc.*, 140, 1009–1018 (Part I) and 1018–1026 (Part II), 1993.

83. Palik, E. D., O. J. Glembocki, and J. I. Heard, "Study of Bias-Dependent Etching of Si in Aqueous KOH", *J. Electrochem. Soc.*, 134, 404–409, 1987.

84. Linde, H. and L. Austin, "Wet Silicon Etching with Aqueous Amine Gallates", *J. Electrochem. Soc.*, 139, 1170–1174, 1992.

85. Price, J. B., "Anisotropic Etching of Silicon with KOH–H₂O-Isopropyl Alcohol", Semiconductor Silicon 1973, Chicago, IL, 1973, p. 339–353.

86. Finne, R. M. and D. L. Klein, "A Water-Amine-Complexing Agent System for Etching Silicon", *J. Electrochem. Soc.*, 114, 965–970, 1967.

87. Reisman, A., M. Berkenbilt, S. A. Chan, F. B. Kaufman, and D. C. Green, "The Controlled Etching of Silicon in Catalyzed Ethylene-Diamine-Pyrocathechol-Water Solutions", *J. Electrochem. Soc.*, 126, 1406–1414, 1979.

88. Seidel, H., L. Csepregi, A. Heuberger, and H. Baumgartel, "Anisotropic Etching of Crystalline Silicon in Alkaline Solutions-Part II. Influence of Dopants", *J. Electrochem. Soc.*, 137, 3626–3632, 1990.

89. Kendall, D. L. and G. R. de Guel, "Orientation of the Third Kind: The Coming of Age of (110) Silicon", in *Micromachining and Micropackaging of Transducers*, Fung, C. D., Ed., Elsevier, New York, 1985, p. 107–124.

90. Seidel, H., L. Csepregi, A. Heuberger, and H. Baumgartel, "Anisotropic Etching of Crystalline Silicon in Alkaline Solutions–Part I. Orientation Dependence and Behavior of Passivation Layers", *J. Electrochem. Soc.*, 137, 3612–3626, 1990.

91. Lee, D. B., "Anisotropic Etching of Silicon", *J. Appl. Phys.*, 40, 4569–4574, 1969.

92. Noworolski, J. M., E. Klaassen, J. Logan, K. Petersen, and N. Maluf, "Fabrication of SOI Wafers With Buried Cavities Using Silicon Fusion Bonding and Electrochemical Etchback", 8th International Conference on Solid-State Sensors and Actuators (Transducers '95), Stockholm, Sweden, June, 1995, p. 71–4.

93. Waggener, H. A. and J. V. Dalton, "Control of Silicon Etch Rates in Hot Alkaline Solutions by Externally Applied Potentials", *J. Electrochem. Soc.*, 119, 236C, 1972.

94. Weirauch, D. F., "Correlation of the Anisotropic Etching of Single-Crystal Silicon Spheres and Wafers", *J. Appl. Phys.*, 46, 1478–1483, 1975.

95. Pugacz-Muraszkiewicz, I. J. and B. R. Hammond, "Application of Silicates to the Detection of Flaws in Glassy Passivation Films Deposited on Silicon Substrates", *J. Vac. Sci. Technol.*, 14, 49–53, 1977.

96. Clemens, D. P., "Anisotropic Etching of Silicon on Sapphire", 1973, p. 407.

97. Bean, K. E., R. L. Yeakley, and T. K. Powell, "Orientation Dependent Etching and Deposition of Silicon", *J. Electrochem. Soc.*, 121, 87C, 1974.

98. Declercq, M. J., J. P. DeMoor, and J. P. Lambert, "A Comparative Study of Three Anisotropic Etchants for Silicon", *Electrochem. Soc. Abstracts*, 75–2, 446, 1975.

99. Wu, X. P., Q. H. Wu, and W. H. Ko, " A Study on Deep Etching of Silicon Using Ethylene-Diamine-Pyrocathechol-Water", *Sensors and Actuators*, 9, 333–343, 1986.

100. Declercq, M. J., L. Gerzberg, and J. D. Meindl, "Optimization of the Hydrazine-Water Solution for Anisostropic Etching of Silicon in Integrated Circuit Technology", *J. Electrochem. Soc.*, 122, 545–552, 1975.

101. Mehregany, M. and S. D. Senturia, "Anisotropic Etching of Silicon in Hydrazine", *Sensors Actuators*, 13, 375–390, 1988.

102. Asano, M., T. Cho, and H. Muraoko, "Applications of Choline in Semiconductor Technology", *Electrochem. Soc. Extend. Abstr.*, 354, 76–2, 911–913, 1976.

103. Tabata, O., R. Asahi, and S. Sugiyama, "Anisotropic Etching with Quarternary Ammonium hydroxide Solutions", 9th Sensor Symposium. Technical digest, Tokyo, Japan, 1990, p. 15–18.

104. Wise, K. D., "Silicon Micromachining and Its Applications to High Performance Integrated Sensors", in *Micromachining and Micropackaging of Transducers*, Fung, C. D., Cheung, P. W., Ko, W. H. and Fleming, D. G., Eds., Elsevier, New York, 1985, p. 3–18.

105. Bean, K., "Anisotropic Etching of Silicon", *IEEE Trans. Electron Devices*, ED–25, 1185–1193, 1978.

106. Barth, P., "Si in Biomedical Applications", *Micro-electronics—Photonics, Materials, Sensors and Technology*, 1984.

107. Lambrechts, M. and W. Sansen, *Biosensors: Micro-electrochemical Devices*, Institute of Physics Publishing, Philadelphia, PA, 1992.

108. Ternez, L., Ph.D. Thesis, Uppsala University, 1988.

109. Gravesen, P., "Silicon Sensors", (Status report for the industrial engineering thesis), DTH Lyngby, Denmark, 1986.

110. Herr, E. and H. Baltes, "KOH Etch Rates of High–Index Planes from Mechanically Prepared Silicon Crystals", 6th International Conference on Solid-State Sensors and Actuators (Transducers '91), San Francisco, CA, 1991, p. 807–810.

111. Clark, L. D. and D. J. Edell, "KOH:$H_2$O Etching of (110) Si, (111) Si, $SiO_2$, and Ta: an Experimental Study", Proceedings of the IEEE Micro Robots and Tele-operators Workshop, Hyannis, MA, 1987, p. 5/1–6.

112. Clark, J. D. L., J. L. Lund, and D. J. Edell, "Cesium Hydroxide [CsOH]: A Useful Etchant for Micro-machining Silicon", Technical Digest of the 1988 Solid State Sensor and Actuator Workshop., Hilton Head Island, SC, 1988, p. 5–8.

113. Schnakenberg, U., W. Benecke, and B. Lochel, "$NH_4OH$-Based Etchants for Silicon Micromachining", *Sensors and Actuators A*, A23, 1031–1035, 1990.

114. Tabata, O., R. Asahi, H. Funabashi, K. Shimaoka, and S. Sugiyama, "Anisotropic Etching of Silicon in TMAH Solutions", *Sensors and Actuators A*, A34, 51–57, 1992.

115. Tabata, O., "pH-Controlled TMAH Etchants for Silicon Micromachining", 8th International Conference on Solid-State Sensors and Actuators (Transducers '95), Stockholm, Sweden, June, 1995, p. 83–86.

116. Puers, R., "Mechanical Silicon Sensors at K.U. Leuven", Proceedings. Themadag: SENSOREN, Rotterdam, Netherlands, 1991, p. 1–8.

117. Buttgenbach, S., *Mikromechanik*, Teubner Studienbucher, Stuttgart, 1991.

118. Bogh, A., "Ethylene Diamine-Pyrocatechol-Water Mixture Shows Etching Anomaly in Boron-Doped Silicon", *J. Electrochem. Soc.*, 118, 401–402, 1971.

119. Steinsland, E., M. Nese, A. Hanneborg, R. W. Bernstein, H. Sandmo, and G. Kittilsland, "Boron Etch-Stop in TMAH Solutions", 8th International Conference on Solid-State Sensors and Actuators (Transducers '95), Stockholm, Sweden, June, 1995, p. 190–193.

120. Palik, E. D., V. M. Bermudez, and O. J. Glembocki, "Ellipsometric Study of the Etch-Stop Mechanism in Heavily Doped Silicon", *J. Electrochem. Soc.*, 132, 135–141, 1985.

121. Palik, E. D., J. W. Faust, H. F. Gray, and R. F. Green, "Study of the Etch-Stop Mechanism in Silicon", *J. Electrochem. Soc.*, 129, 2051–2059, 1982.

122. Schumacher, A., H.-J. Wagner, and M. Alavi, "Mit Laser und Kalilauge", Technische Rundschau, 86, 20–23, 1994.

123. Barth, P. W., P. J. Shlichta, and J. B. Angel, "Deep Narrow Vertical-Walled Shafts in <110> Silicon", 3rd International Conference on Solid-State Sensors and Actuators, Philadelphia, PA, 1985, p. 371–373.

124. Seidel, H. and L. Csepregi, "Advanced Methods for the Micromachining of Silicon", 7th Sensor Symposium. Technical digest, Tokyo, Japan, 1988, p. 1–6.

125. Alavi, M., S. Buttgenbach, A. Schumacher, and H. J. Wagner, "Laser Machining of Silicon for Fabrication of New Microstructures", 6th International Conference on Solid-State Sensors and Actuators (Transducers '91), San Francisco, CA, 1991, p. 512–515.

126. Private Communication, Slater, T., *Vertical (100) Etching, Personal Communication*, October, 1995.

127. Ciarlo, D. R., "Corner Compensation Structures for (110) Oriented Silicon", Proceedings of the IEEE Micro Robots and Teleoperators Workshop, Hyannis, MA, 1987, p. 6/1–4.

128. Elwenspoek, M., "On the Mechanism of Anisotropic Etching of Silicon", *J. Electrochem. Soc.*, 140, 2075–2080, 1993.

129. Elwenspoek, M., U. Lindberg, H. Kok, and L. Smith, "Wet Chemical Etching Mechanism of Silicon", IEEE International Workshop on Micro Electro Mechanical Systems, MEMS '94, Oiso, Japan, 1994, p. 223–8.

130. Glembocki, O. J., R. E. Stahlbush, and M. Tomkiewicz, "Bias-Dependent Etching of Silicon in Aqueous KOH", *J. Electrochem. Soc.*, 132, 145–151, 1985.

131. Palik, E. D., H. F. Gray, and P. B. Klein, "A Raman Study of Etching Silicon in Aqueous KOH", *J. Electrochem. Soc.*, 130, 956–959, 1983.

132. Raley, N. F., F. Sugiyama, and T. Van Duzer, "(100) Silicon Etch-Rate Dependence on Boron Concentration in Ethylenediamine-Pyrocatechol-Water Solutions", *J. Electrochem. Soc.*, 131, 161–171, 1984.

133. Wu, X. P., Q. H. Wu, and W. H. Ko, "A Study on Deep Etching of Silicon Using EPW", 3rd International Conference on Solid-State Sensors and Actuators (Transducers '85), Philadelphia, PA, 1985, p. 291–4.

134. Abu-Zeid, M. M., D. L. Kendall, G. R. de Guel, and R. Galeazzi, "Abstract 275", JECS, Toronto, Canada, 1985, p. 400.

135. Seidel, H., "The Mechanism of Anisotropic Electrochemical Silicon Etching in Alkaline Solutions", Technical Digest of the 1990 Solid State Sensor and Actuator Workshop., Hilton Head Island, SC, 1990, p. 86–91.

136. Ghandi, S., K., *The Theory and Practice of Micro-electronics*, John Wiley & Sons, New York, 1968.

137. Jackson, K. A., "A Review of the Fundamental Aspects of Crystal Growth", Crystal Growth, Boston, MA, 1966, p. 17–24.

138. Elwenspoek, M. and J. P. van der Weerden, "Kinetic Roughening and Step Free Energy in the Solid-on-Solid Model and on Naphtalene Crystals", *J. Phys. A: Math. Gen.*, 20, 669–678, 1987.

139. Bennema, P., "Spiral Growth and Surface Roughening: Developments since Burton, Cabrera, and Frank", *J. Cryst. Growth*, 69, 182–197, 1984.

140. Madou, M. J. and S. R. Morrison, *Chemical Sensing with Solid State Devices*, Academic Press, New York, 1989.

141. Madou, M. J., K. W. Frese, and S. R. Morrison, "Photoelectrochemical Corrosion of Semiconductors for Solar Cells", *SPIE*, 248, 88–95, 1980.

142. Madou, M. J., B. H. Loo, K. W. Frese, and S. R. Morrison, "Bulk and Surface Characterization of the Silicon Electrode", *Surf. Sci.*, 108, 135–152, 1981.

143. Levy-Clement, C., A. Lagoubi, R. Tenne, and M. Neumann-Spallart, "Photoelectrochemical Etching of Silicon", *Electrochim. Acta*, 37, 877–888, 1992.

144. Matsumura, M. and S. R. Morrison, "Photoanodic Properties of an n-Type Silicon Electrode in Aqueous Solution Containing Fluorides", *J. Electroanal. Chem.*, 144, 113–120, 1983.

145. Lewerenz, H. J., J. Stumper, and L. M. Peter, "Deconvolution of charge injection steps in quantum yield multiplication on silicon", *Phys. Rev. Lett.*, 61, 1989–92, 1988.

146. Hesketh, P. J., C. Ju, S. Gowda, E. Zanoria, and S. Danyluk, "Surface Free Energy Model of Silicon Anisotropic Etching", *J. Electrochem. Soc.*, 140, 1080–1085, 1993.

147. Sundaram, K. B. and H.-W. Chang, "Electrochemical Etching of Silicon by Hydrazine", *J. Electrochem. Soc.*, 140, 1592–1597, 1993.

148. Morrison, S. R., *Electrochemistry on Semiconductors and Oxidized Metal Electrodes*, Plenum Press, New York, 1980.

149. Lehmann, V., "The Physics of Macropore Formation in Low Doped n-Type Silicon", *J. Electrochem. Soc.*, 140, 2836–2843, 1993.

150. Smith, R. L. and S. D. Collins, "Thick Films of Silicon Nitride", *Sensors and Actuators A*, A23, 830–834, 1990.

151. Chuang, S. F. and R. L. Smith, "Preferred Crystallographic Directions of Pore Propagation in Porous Silicon Layers", Technical Digest of the 1988 Solid State Sensor and Actuator Workshop, Hilton Head Island, SC, 1988, p. 151–153.

152. Barret, S., F. Gaspard, R. Herino, M. Ligeon, F. Muller, and I. Ronga, "Porous Silicon as a Material in Microsensor Technology", *Sensors and Actuators A*, A33, 19–24, 1992.

153. Canham, L., T., "Silicon Quantum Wire Array Fabrication by Electrochemical and Chemical Dissolution of Wafers", *Appl. Phys. Lett.*, 57, 1046–1050, 1990.

154. Unagami, T., "Formation Mechanism of Porous Silicon Layer by Anodization in HF Solution", *J. Electrochem. Soc.*, 127, 476–483, 1980.

155. Watanabe, Y., Y. Arita, T. Yokoyama, and Y. Igarashi, "Formation and Properties of Porous Silicon and Its Applications", *J. Electrochem. Soc.*, 122, 1351–1355, 1975.

156. Lehmann, V. and U. Gosele, "Porous Silicon Formation: A Quantum Wire Effect", *Appl. Phys. Lett.*, 58, 856–858, 1991.

157. Lehmann, V., "Porous Silicon-A New Material for MEMS", The Ninth Annual International Workshop on Micro Electro Mechanical Systems, MEMS '96, San Diego, CA, USA, 1996, p. 1–6.

158. Yamana, M., N. Kashiwazaki, A. Kinoshita, T. Nakano, M. Yamamoto, and W. C. Walton, "Porous Silicon Oxide Layer Formation by the Electrochemical Treatment of a Porous Silicon Layer", *J. Electrochem. Soc.*, 137, 2925–7, 1990.

159. Arita, Y. and Y. Sunohara, "Formation and Properties of Porous Silicon Film", *J. Electrochem. Soc.*, 124, 285–295, 1977.

160. Jung, K. H., S. Shih, and D. L. Kwong, "Developments in Luminescent Porous Si", *J. Electrochem. Soc.*, 140, 3046–3064, 1993.

161. Searson, P. C., S. M. Prokes, and O. J. Glembocki, "Luminescence at the Porous Silicon/Electrolyte Interface", *J. Electrochem. Soc.*, 140, 3327–3331, 1993.

162. Zhang, X. G., "Mechanism of Pore Formation on n-type Silicon", *J. Electrochem. Soc.*, 138, 3750–3756, 1991.

163. Theunissen, M. J. J., "Etch Channel Formation During Anodic Dissolution of n-type Silicon in Aqueous Hydrofluoric Acid", *J. Electrochem. Soc.*, 119, 351–360, 1972.

164. Lehmann, V. and H. Foll, "Formation Mechanism and Properties of Electrochemically Etched Trenches in n-Type Silicon", *J. Electrochem. Soc.*, 137, 653–659, 1990.

165. Cahill, S. S., W. Chu, and K. Ikeda, "High Aspect Ratio Isolated Structures in Single Crystal Silicon", 7th International Conference on Solid-State Sensors and Actuators (Transducers '93), Yokohama, Japan, 1993, p. 250–253.

166. Samaun, S., K. D. Wise, and J. B. Angell, "An IC Piezoresistive Pressure Sensor for Biomedical Instrumentation", *IEEE Trans. Biomed. Engr.*, 20, 101–109, 1973.

167. Nunn, T. and J. Angell, "An IC Absolute Pressure Transducer with Built–in Reference Chamber", Workshop on Indwelling Pressure Transducers and Systems, Cleveland, OH, 1975, p. 133–136.

168. Greenwood, J., C., "Ethylene Diamine-Cathechol-Water Mixture Shows Preferential Etching of p-n Junction", *J. Electrochem. Soc.*, 116, 1325–1326, 1969.

169. Brodie, I. and J. J. Muray, *The Physics of Micro-fabrication*, Plenum Press, New York, 1982.

170. Heuberger, A., *Mikromechanik*, Springer Verlag, Heidelberg, 1989.

171. Rehrig, D., L., "In Search of Precise Epi Thickness Measurements", *Semicond. Int.*, 13, 90–95, 1990.

172. Jackson, T. N., M. A. Tischler, and K. Wise, D., "An Electrochemical P-N Junction Etch-Stop for The Formation of Silicon Microstructures", *IEEE Electron Device Lett.*, EDL-2, 44–45, 1981.

173. Faust, J. W. and E. D. Palik, "Study of The Orientation Dependent Etching and Initial Anodization of Si in Aqueous KOH", *J. Electrochem. Soc.*, 130, 1413–1420, 1983.

174. Kim, S. C. and K. Wise, "Temperature Sensitivity in Silicon Piezoresistive Pressure Transducers", *IEEE Trans. Electron Devices*, ED–30, 802–810, 1983.

175. McNeil, V. M., S. S. Wang, K.-Y. Ng, and M. A. Schmidt, "An Investigation of the Electrochemical Etching of (100) Silicon in CsOH and KOH", Technical Digest of the 1990 Solid State Sensor and Actuator Workshop., Hilton Head Island, SC, 1990, p. 92–97.

176. Palik, E. D., O. J. Glembocki, and R. E. Stahlbush, "Fabrication and Characterization of Si Membranes", *J. Electrochem. Soc.*, 135, 3126–3134, 1988.

177. Hirata, M., K. Suzuki, and H. Tanigawa, "Silicon Diaphragm Pressure Sensors Fabricated by Anodic Oxidation Etch-Stop", *Sensors Actuators*, 13, 63–70, 1988.

178. Kloeck, B., S. D. Collins, N. F. de Rooij, and R. L. Smith, "Study of Electrochemical Etch-Stop for High-Precision Thickness Control of Silicon Membranes", *IEEE Trans. Electron Devices*, 36, 663–669, 1989.

179. Saro, P. M. and A. W. van Herwaarden, "Silicon Cantilever Beams Fabricated by Electrochemically Controlled Etching for Sensor Applications", *J. Electrochem. Soc.*, 133, 1722–1729, 1986.

180. Wen, C. P. and K. P. Weller, "Preferential Electro-chemical Etching of p+ Silicon in an Aqueous HF–H$_2$SO$_4$ Electrolyte", *J. Electrochem. Soc.*, 119, 547–548, 1972.

181. van Dijk, H. J. A., "Method of Manufacturing a Semiconductor Device and Semiconductor Device Manufactured by Said Method", in U.S. Patent 3,640,807, 1972.

182. Mlcak, R., H. L. Tuller, P. Greiff, and J. Sohn, "Photo Assisted Electromechanical Machining of Micro-mechanical Structures", Proceedings. IEEE Micro Electro Mechanical Systems, (MEMS '93), Fort Lauderdale, FL, 1993, p. 225–229.

183. Yoshida, T., T. Kudo, and K. Ikeda, "Photo-Induced Preferential Anodization for Fabrication of Mono-crystalline Micromechanical Structures", Proceedings. IEEE Micro Electro Mechanical Systems, (MEMS '92), Travemunde, Germany, 1992, p. 56–61.

184. Shockley, W., *Method of making Thin Slices of Semiconductive Material*, in U.S. Patent 3,096,262, 1963.

185. Peeters, E., D. Lapadatu, W. Sansen, and B. Puers, "PHET, an Electrodeless Photovoltaic Electrochemical Etch-Stop Technique", 7th International Conference on Solid-State Sensors and Actuators (Transducers '93), Yokohama, Japan, 1993, p. 254–257.

186. Peeters, E., D. Lapadatu, W. Sansen, and B. Puers, "Developments in Etch-Stop Techniques", 4th European Workshop on Micromechanics (MME '93), Neuchatel, Switzerland, 1993.

187. Bryzek, J., K. Petersen, J. R. Mallon, L. Christel, and F. Pourahmadi, *Silicon Sensors and Microstructures*, Novasensor, Fremont, CA, 1990.

188. Puers, B. and W. Sansen, "Compensation Structures for Convex Corner Micromachining in Silicon", *Sensors and Actuators A*, A23, 1036–1041, 1990.

189. Sandmaier, H., H. Offereins, L., K. Kuhl, and W. Lang, "Corner Compensation Techniques in Anisotropic Etching of (100)-Silicon Using Aqueous KOH", 6th International Conference on Solid-State Sensors and Actuators (Transducers '91), San Francisco, CA, 1991, p. 456–459.

190. Mayer, G. K., H. L. Offereins, H. Sandmeier, and K. Kuhl, "Fabrication of Non-Underetched Convex Corners in Ansisotropic Etching of (100)-Silicon in Aqueous KOH with Respect to Novel Micromechanic Elements", *J. Electrochem. Soc.*, 137, 3947–3951, 1990.

191. Buser, R. A. and N. F. de Rooij, "Monolithishes Kraftsensorfeld", *VDI-Berichte*, Nr. 677, 1988.

192. Seidel, H., "Doctoral Thesis", Ph.D. Thesis, FU Berlin, Germany, 1986.

193. Abu-Zeid, M. M., "Corner Undercutting in Aniso-tropically Etched Isolation Contours", *J. Electrochem. Soc.*, 131, 2138–2142, 1984.

194. Wu, X.-P. and W. H. Ko, "Compensating Corner Undercutting in Anisotropic Etching of (100) Silicon", *Sensors Actuators*, 18, 207–215, 1989.

195. Wu, X. and W. A. Ko, "A Study on Compensating Corner Undercutting in Anisotropic Etching of (100) Silicon", 4th International Conference on Solid-State Sensors and Actuators (Transducers '87), Tokyo, Japan, 1987, p. 126–129.

196. Cho, S. T., "A Batch Dissolved Wafer Process for Low Cost Sensor Applications", Micromachining and Microfabrication Process Technology, (Proceedings of the SPIE), Austin, TX, 1995, p. 10–17.

197. Greiff, P., "SOI-based Micromechanical Process", Micromachining and Microfabrication Process Technology, (Proceedings of the SPIE), Austin, TX, 1995, p. 74–81.

198. Weinberg, M., J. Bernstein, J. Borenstein, J. Campbell, J. Cousens, B. Cunningham, R. Fields, P. Greiff, B. Hugh, L. Niles, and J. Sohn, "Micromachining Inertial Instruments", Micromachining and Micro-fabrication Process Technology II, Austin, Texas, USA, 1996, p. 26–36.

199. Craven, D., K. Yu, and T. Pandhumsoporn, "Etching Technology for "Through-The-Wafer" Silicon Etching", Micromachining and Microfabrication Process Technology, (Proceedings of the SPIE), Austin, TX, 1995, p. 259–263.

200. Joseph, J., M. Madou, T. Otagawa, P. Hesketh, and A. Saaman, "Catheter-Based Micromachined Electrochemical Sensors", Catheter-Based Sensing and Imaging Technology, (Proceedings of the SPIE), Los Angeles, CA, USA, 1989, p. 18–22.

201.  Madou, M. J. and  T. Otagawa, *Microelectrochemical Sensor and Sensor Array*, in U.S. Patent 4,874,500, 1989.

202.  Holland, C. E., E. R. Westerberg, M. J. Madou, and T. Otagawa, *Etching Method for Producing an Electrochemical Cell in a Crystalline Substrate*, in U.S. Patent 4,764,864, 1988.

# Surface Micromachining

5

## Introduction

Bulk micromachining means that three-dimensional features are etched into the bulk of crystalline and noncrystalline materials. In contrast, surface micromachined features are built up, layer by layer, on the surface of a substrate (e.g., a single crystal silicon wafer). Dry etching defines the surface features in the x,y plane and wet etching releases them from the plane by undercutting. In surface micromachining, shapes in the x,y plane are unrestricted by the crystallography of the substrate. For illustration, in Figure 5.1 we compare an absolute pressure sensor based on poly-Si and made by surface micromachining with one made by bulk micromachining in single crystal Si.

The nature of the deposition processes involved determines the very flat surface micromachined features (Hal Jerman from EG&G's ICSensors called them 2.5 D features[1]). Specifically, low-pressure chemical vapor deposited (LPCVD) polycrystalline silicon (poly-Si) films generally are only a few microns high (low z). In contrast with wet bulk micromachining only the wafer thickness limits the feature height. A low z may mean a drawback for some sensors. For example, it will be difficult to fashion a large inertial mass for an accelerometer from those thin poly-Si plates. Not only do many parameters in the LPCVD polysilicon process need to be controlled very precisely, subsequent high temperature annealing (say, at temperatures of about 580°C) is needed to transform the as-deposited amorphous silicon into polysilicon — the main structural material in surface micromachining. Even with the best possible process control, polysilicon has some material disadvantages over single crystal Si. For example, it generates a somewhat smaller yield strength (values between 2 and 10 times smaller have been reported)[2,3] and has a lower piezoresistivity.[4] Moreover, since a grain diameter may constitute a significant fraction of the thickness of a mechanical member, the effective Young's modulus may exhibit significant variability from sensor to sensor.[5] An important positive attribute of poly-Si is that its material properties, although somewhat inferior to single crystal, are far superior compared to those of metal films, and, most of all, since they are isotropic, design is rendered dramatically simpler than with single crystal material. Dimensional uncertainties are of greater concern than material issues. Although absolute dimensional tolerances obtained with lithography techniques may be submicron, relative tolerances are poor, perhaps 1% on the length of a 100-μm long feature. The situation becomes critical with yet smaller feature sizes (see Figure 7.1). Although the relatively coarse dimensional control in the microdomain is not specific to surface micromachining, there is no crystallography to rely on for improved dimensional control as in the case of wet bulk micromachining. Moreover, since the mechanical members in surface micromachining tend to be smaller, more post-fabrication adjustment of the features is required to achieve reproducible characteristics. Finally, the wet process for releasing structural elements from a substrate tends to cause sticking of suspended structures to the substrate, so-called stiction, introducing another disadvantage associated with surface micromachining.

Some of the mentioned problems associated with surface micromachining have recently been resolved by process modifications and/or alternative designs, and the technique has rapidly gained commercial interest, mainly because it is the most IC-compatible micromachining process developed to date. Moreover, in the last 5 to 10 years, processes such as silicon on insulator (SOI),[6] hinged poly-Si,[7] Keller's molded milli-scale polysilicon,[8] thick poly-Si (10 μm and beyond),[9] as well as LIGA* and LIGA-like processes, have further enriched the surface micromachining arsenal. Some preliminary remarks on each of these surface micromachining extensions follow.

Silicon crystalline features, anywhere between fractions of a micron to 100 μm high, can readily be obtained by surface micromachining of the epi-silicon or fusion-bonded silicon layer of SOI wafers.[10] Structural elements made from these single crystalline Si layers result in more reproducible and reliable sensors. SOI machining combines the best features of surface micromachining (i.e., IC compatibility and freedom in x,y shapes) with the best features of bulk micromachining (superior single crystal Si properties). Moreover, SOI surface micromachining frequently involves fewer process steps and offers better control over the thickness of crucial building blocks. Given the poor reproducibility of mechanical properties and generally poor electronic characteristics of polysilicon films, SOI machining may surpass the poly-Si technology for fabricating high performance devices.

The fabrication of poly-Si planar structures for subsequent vertical assembly by mechanical rotation around micromachined hinges dramatically increases the plethora of designs feasible with poly-Si.[7] Today, erecting these poly-Si structures with the probes of an electrical probe station or, occasionally, assembly by chance in the HF etch or DI water rinse,[11] represent

---

\* LIGA is the acronym for the German Lithographie, Galvanoformung, Abformung (see Chapter 6).

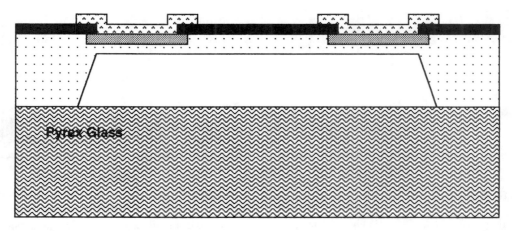

**Surface Micromachined Absolute Pressure Sensor**

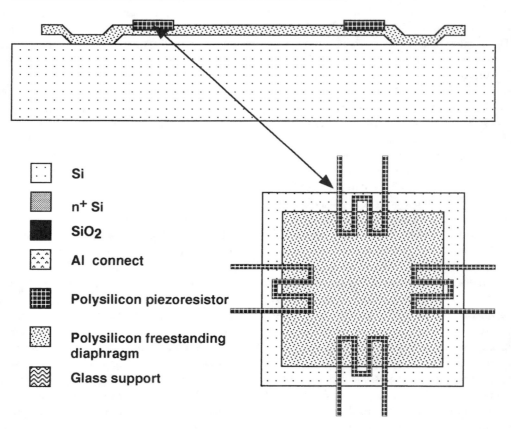

| | Si |
|---|---|
| | n⁺ Si |
| | SiO₂ |
| | Al connect |
| | Polysilicon piezoresistor |
| | Polysilicon freestanding diaphragm |
| | Glass support |

**FIGURE 5.1** Comparison of bulk micromachined and surface micromachined absolute pressure sensors equipped with piezoresistive elements. (Top) Bulk micromachining in single crystal Si. (Bottom) Surface micromachining with poly-Si.

too complicated or unreliable postrelease assembly methods for commercial acceptance.

At the University of California at Berkeley, Keller introduced a combination of surface micromachining and LIGA-like molding processes[8] in the so-called HEXSIL* process, a technology enabling the fabrication of tall three-dimensional microstructures without postrelease assembly. Using CVD processes one can generally only deposit thin films (~2 to 5 μm) on flat surfaces. If however, these surfaces are the opposing faces of deep narrow trenches, the growing films will merge to form solid

beams. Releasing of such polysilicon structures and the incorporation of electroplating steps expand the surface micromachining bandwidth in terms of choice of materials and accessible feature heights. In this fashion, high aspect ratio structures normally associated with LIGA can now also be made of CVD polysilicon.

Applying classical LPCVD to obtain poly-Si deposition is a slow process. For example, a layer of 10 μm typically requires a deposition time of 10 hours. Consequently, most micromachined structures are based on layer thicknesses in the 2- to 5-μm range. Based on dichlorosilane (SiH₂Cl₂) chemistry, Lange et al.[9] developed a CVD process in a vertical epitaxy batch

---

\*      HEXSIL is the acronym for HEXagonal honeycomb polySILicon.

reactor with deposition rates as high as 0.55 μm/min at 1000°C. The process yields acceptable deposition times for thicknesses in the 10-μm range. The highly columnar poly-Si films were deposited on sacrificial $SiO_2$ layers and exhibit low internal tensile stress making them suitable for surface micromachining.

Finally, thick layers of polyimide and other new UV resists also receive a lot of attention as important new extensions of surface micromachining. Due to their transparency to exposing UV light, they can be transformed into tall surface structures with LIGA-like high aspect ratios. They may be both electroplated and molded.

In this chapter, we first review thin film material properties in general, focusing on significant differences with bulk properties of the same material, followed by a review of the main surface micromachining processes. Next, we clarify the recent extensions of the surface micromachining technique listed above. Because of the complexity of the many parameters influencing thin film properties, we then present a set of case studies on the most commonly used thin film materials.

In the surface micromachining examples at the end of this chapter, we first look at a lateral resonator. Resonators have found an important industrial application in accelerometers introduced by Analog Devices. The other example concerns a micromotor: an electrostatic linear motor with a slider stepping between two poly-Si rails. Micromotors may lack practical use as of yet, but, just as the ion-sensitive field effect transistor (ISFET) galvanized the chemical sensor community in trying out new chemical sensing approaches, micromotors energized the micromachining research community to fervently explore miniaturization of a wide variety of mechanical sensors and actuators. Micromotors also brought about the christening of the micromachining field into 'Microelectromechanical Systems' or MEMS.

# Historical Note

The first example of a surface micromachine for an electromechanical application consisted of an underetched metal cantilever beam for a resonant gate transistor made by Nathanson in 1967.[12] By 1970, a first suggestion for a magnetically actuated metallic micromotor emerged.[13] Because of fatigue problems metals rarely are counted on as mechanical components. The surface micromachining method as we know it today was first demonstrated by Howe and Muller in the early 1980s and relied on polysilicon as the structural material.[14] These pioneers and Guckel,[15] an early contributor to the field, produced free-standing LPCVD polysilicon structures by removing the oxide layers on which the polysilicon features were formed. Howe's first device consisted of a resonator designed to measure the change of mass upon adsorption of chemicals from the surrounding air. A gas sensor does not necessarily represent a good application of a surface micromachined electrostatic structure since humidity and dust foul the thin air gap of such an unencapsulated microstructure in a minimal amount of time. Later mechanical structures, especially hermetically sealed mechanical devices, provided proof that the IC revolution could be extended to

electromechanical systems.[5] In these structures, the height (z-direction) typically is limited to less than 10 μm, ergo the name surface micromachining.

The first survey of possible applications of poly-Si surface micromachining was presented by Gabriel et al. in 1989.[16] Microscale movable mechanical pin joints, springs, gears, cranks, sliders, sealed cavities, and many other mechanical and optical components have been demonstrated in the laboratory.[17,18] In 1991, Analog Devices announced the first commercial product based on surface micromachining, namely the ADXL-50, a 50-g accelerometer for activating air-bag deployment.[19] A new wave of major commercial applications for surface micromachining could be based on Texas Instruments' Digital Micromirror Device™. This surface micromachined movable mirror is a digital light switch that precisely controls a light source for projection display and hard copy applications.[20] The commercial acceptance of this application will likely determine the staying power of the surface micromachining option.

# Mechanical Properties of Thin Films

## Introduction

Thin films in surface micromachines must satisfy a large set of rigorous chemical, structural, mechanical, and electrical requirements. Excellent adhesion, low residual stress, low pinhole density, good mechanical strength, and chemical resistance all may be required simultaneously. For many microelectronic thin films, the material properties depend strongly on the details of the deposition process and the growth conditions. In addition, some properties may depend on post-deposition thermal processing, referred to as annealing. Furthermore, the details of thin-film nucleation and/or growth may depend on the specific substrate or on the specific surface orientation of the substrate. Although the properties of a bulk material might be well characterized, its thin-film form may have properties substantially different from those of the bulk. For example, thin films generally display smaller grain size than bulk materials. An overwhelming reason for the many differences stem from the properties of thin films, which exhibit a higher surface-to-volume ratio than large chunks of material and are strongly influenced by the surface properties.

For more details on deposition techniques in general, refer to Chapter 3; now we focus on the physical characteristics of the resulting thin deposits. Some terminology characterizing thin films and their deposition is introduced in Table 5.1.

Since thin films were not intended for load-bearing applications, their mechanical properties have largely been ignored. The last 10 years saw the development of a strong appreciation for understanding the mechanical properties as essential for improving the reliability and life-time in thin films, even in nonstructural applications.[21] Surface micromachining contributes heavily to this understanding. One of the most influential long-term contributions of surface micromachining might well lie in the elucidation of stress mechanism in thin films.

TABLE 5.1　Thin Film Terms Used in Characterizing Deposition

| Term | Remark |
| --- | --- |
| Physisorbed film | Bond energy < 10 kcal/mole |
| Chemisorbed film | Bond energy > 20 kcal/mole |
| Nucleation | Adatoms forming stable clusters |
| Condensation | Initial formation of nuclei |
| Island formation | Nuclei grow in three dimensions, especially along the substrate surface |
| Coalescence | Nuclei contact each other and larger, rounded shapes form |
| Secondary nucleation | Areas between islands are filled in by secondary nucleation, resulting in a continuous film |
| Grain size of thin film | Generally smaller than for bulk materials and function of deposition and annealing conditions (higher T, larger grains) |
| Surface roughness | Lower at high temperatures except when crystallization starts; at low temperature the roughness is higher for thicker films; also oblique deposition and contamination increase roughness |
| Epitaxial and amorphous films | Very low surface roughness |
| Density | More porous deposits are less dense; density reveals a lot about the film structure |
| Crystallographic structure | Adatom mobility: amorphous, polycrystalline, single crystal or fiber texture, or preferred orientation |

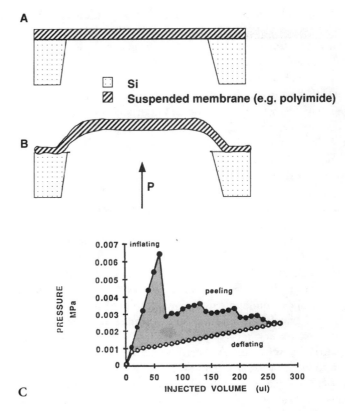

**FIGURE 5.2** Micromachined structure to evaluate adhesion. (A) Suspended membrane. (B) Partially detached membrane-outward peel. (C) Pressure-volume curve during inflate-deflate cycle. (From Senturia, S., Can We Design Microrobotic Devices Without Knowing the Mechanical Properties of Materials?, presented at Micro Robots and Tekoperators Workshop, Hyannis, MA, 1987. With permission.)

## Adhesion

One cannot stress enough the importance of adhesion of various films to one another and to the substrate in overall IC performance and reliability. As mechanical pulling forces might be involved, adhesion is even more crucial in micromachining. If films lift from the substrate under a repetitive, applied mechanical force, the device will fail. Classical adhesion tests include the scotch tape test, abrasion method, scratching, deceleration (ultrasonic and ultracentrifuge techniques), bending, pulling, etc.[22] Micromachined structures, because of their sensitivity to thin-film properties, enable some innovative new ways of *in situ* adhesion measurement. Figure 5.2 illustrates how a suspended membrane may be used for adhesion measurements. Figure 5.2A shows the membrane suspended but still adherent to the substrate. Figure 5.2B shows the membrane after it has been peeled from its substrate by an applied load (gas pressure). Figure 5.2C illustrates the accompanying P(ressure)-V(olume) cycle, in which the membrane is inflated, peeled, and then deflated. The shaded portion of Figure 5.2C illustrates the P-V work creating the new surface, which equals the average work of adhesion for the film-substrate interface times the area peeled during the test.[23]

Cleanliness of a substrate is a *conditio sine qua non* for good film adhesion. Roughness, providing more bonding surface area and mechanical interlocking, further improve it. Adhesion also improves with increasing adsorption energy of the deposit and/or increasing number of nucleation sites in the early growth stage of the film. Sticking energies between film and substrate, range from less than 10 kcal/mole in physisorption to more than 20 kcal/mole for chemisorption. The weakest form of adhesion involves Van der Waals forces only (see also Table 5.1).

It is highly advantageous to include a layer of oxide-forming elements between a metal and an oxide substrate. These adhesion layers, such as Cr, Ti, Al, etc., provide good anchors for subsequent metallization. Intermediate film formation allowing a continuous transition from one lattice to the other results in the best adhesion. Adhesion also improves when formation of intermetallic metal alloys takes place.

## Stress in Thin Films

### Stress in Thin Films — Qualitative Description

Film cracking, delamination, and void formation may all be linked to film stress. Nearly all films foster a state of residual stress, due to mismatch in the thermal expansion coefficient, nonuniform plastic deformation, lattice mismatch, substitutional or interstitial impurities, and growth processes. Figure 5.3 lists stress-causing factors categorized as either intrinsic or extrinsic.[24] The intrinsic stresses (also growth stresses) develop during the film nucleation. Extrinsic stresses are imposed by unintended external factors such as temperature gradients or sensor package-induced stresses. Thermal stresses, the most

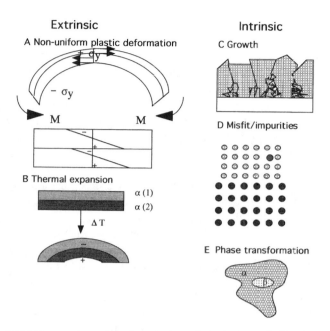

**FIGURE 5.3** Examples of intrinsic and extrinsic residual stresses. (A) Nonuniform plastic deformation results in residual stresses upon unloading. M = bonding moment. (B) Thermal expansion mismatch between two materials bonded together. $\alpha$ (1) and $\alpha$ (2) are thermal expansion coefficients. (C) Growth stresses evolve during film deposition. (D) Misfit stresses due to mismatches in lattice parameters in an epitaxial film and stresses from substitutional or interstitial impurities. (E) Volume changes accompanying phase transformations cause residual stresses.[21] (After Krulevitch, P.A., Micromechanical Investigations of Silicon and Ni-Ti-Cu Thin Films, Ph.D. thesis, University of California, Berkeley, 1994.)

common type of extrinsic stresses, are well understood and often easy to calculate (see below). They arise either in a structure with inhomogeneous thermal expansion coefficients subjected to a uniform temperature change or in a homogeneous material exposed to a thermal gradient.[24] Intrinsic stresses in thin films often are larger than thermal stresses. They usually are a consequence of the nonequilibrium nature of the thin-film deposition process. For example, in chemical vapor deposition, depositing atoms (adatoms) may at first occupy positions other than the lowest energy configuration. With too high a deposition rate and/or too low adatom surface mobility, these first adatoms may become pinned by newly arriving adatoms, resulting in the development of intrinsic stress. Other types of intrinsic stresses illustrated in Figure 5.3 include: transformation stresses occurring when part of a material undergoes a volume change during a phase transformation, misfit stresses arising in epitaxial films due to lattice mismatch between film and substrate, and impurities either interstitial or substitutional which cause intrinsic residual stresses due to the local expansion or contraction associated with point defects. Intrinsic stress in a thin film does not suffice to result in delamination unless the film is quite thick. For example, to overcome a low adsorption energy of 0.2 eV, a relatively high stress of about $5 \times 10^9$ dyn cm$^{-2}$ ($10^7$ dyn/cm$^2$ = 1 MPa) is required.[22] High stress can result in buckling or cracking of films.

The stress developing in a film during the initial phases of a deposition may be compressive (i.e., the film tends to expand parallel to the surface), causing buckling and blistering or delamination in extreme cases (especially with thick films). Alternatively, thin films may be in tensile stress (i.e., the film tends to contract), which may lead to cracking if forces high enough to exceed the fracture limit of the film material are present. Subsequent rearrangement of the atoms, either during the remainder of the deposition or with additional thermal processing, can lead to further densification or expansion, decreasing remaining tensile or compressive stresses, respectively.

The mechanical response of thin-film structures is affected by the residual stress, even if the structures do not fail. For example, if the residual stress varies in the direction of film growth, the resulting built-in bending moment will warp released structures, such as cantilever beams. The presence of residual stress also alters the resonant frequency of thin-film, resonant microstructures (see Equation 5.16 below).[25] Also, residual stress can lead to degradation of electrical characteristics and yield loss through defect generation, such as {311} defects.[24] Another observation found the resistivity of stressed metallic films to be higher than that of their annealed counterparts. In a few cases residual stress has been used advantageously, such as in self-adjusting microstructures,[26] and for altering the shape-set configuration in shape memory alloy films.[24]

In general, the stresses in films, by whatever means produced, are in the range of $10^8$ to $5 \times 10^{10}$ dyn/cm$^{-2}$ and can be either tensile or compressive. For normal deposition temperatures (50 to a few hundred degrees C), the stress in metal films typically ranges from $10^8$ to $10^{10}$ dyn cm$^{-2}$ and is tensile, the refractory metals at the upper end, the soft ones (Cu, Ag, Au, Al) at the lower end. At low substrate temperatures, metal films tend to exhibit tensile stress. This often decreases in a linear fashion with increasing substrate temperature, finally going through zero or even becoming compressive. The changeover to compressive stress occurs at lower temperatures for lower melting point metals. The mobility of the adatoms is key to understanding the ranking for refractory and soft metals. A metal such as aluminum has a low melting temperature and a corresponding high diffusion rate even at room temperature, thus usually it is fairly stress free. By comparison, tungsten has a relatively high melting point and a low diffusion rate and tends to accumulate more stress when sputter-deposited. With dielectric films, stresses often are compressive and have slightly lower values than commonly noted in metals.

Tensile films result, for example, when a process byproduct is present during deposition and later driven off as a gas. If the deposited atoms are not sufficiently mobile to fill in the holes left by these departing byproducts, the film will contract and go in tension. Nitrides deposited by plasma CVD usually are compressive due to the presence of hydrogen atoms in the lattice. By annealing, driving the hydrogen out, the films can turn highly tensile. Annealing also has a dramatic effect on most oxides. Oxides often are porous enough to absorb or give off a large amount of water. Full of water, they are compressive; devoid of water they are tensile. Thermal $SiO_2$ is compressive, though, even when dry. If atoms are jammed in place (such as with sputtering) the film tends to act compressively. The stress in a thin film also varies with depth. The RF power of a plasma-enhanced CVD (PECVD) deposition influences stress, e.g., a thin film may start out tensile, decrease as the power increases, and finally become compressive with further RF power increase. CVD equipment manufacturers concentrate on building stress-control capabilities into new equipment by controlling plasma frequency (see also Chapter 3).

### Stress in Thin Films on Thick Substrates — Quantitative Analysis

The total stress in a thin film typically is given by:

$$\sigma_{tot} = \sigma_{th} + \sigma_{int} + \sigma_{ext} \qquad 5.1$$

i.e., the sum of any intentional external applied stress ($\sigma_{ext}$), the thermal stress ($\sigma_{th}$, an unintended external stress), and different intrinsic components ($\sigma_{int}$). With constant stress through the film thickness, the stress components retain the form of

$$\sigma_x = \sigma_x(x,y)$$
$$\sigma_y = \sigma_y(x,y)$$
$$\tau_{xy} = \tau_{xy}(x,y) \qquad 5.2$$
$$\tau_{xz} = \tau_{yz} = \sigma_z = 0$$

That is, the three nonvanishing stress components are functions of x and y alone. No stress occurs in the direction normal to the substrate (z). With x,y as principal axes, the shear stress $\tau_{xy}$ also vanishes[27] and Equation 5.2 reduces to the following strain-stress relationships:

$$\varepsilon_x = \frac{\sigma_x}{E} - \frac{\nu\sigma_y}{E}$$
$$\varepsilon_y = \frac{\sigma_y}{E} - \frac{\nu\sigma_x}{E} \qquad 5.3$$
$$\sigma_z = 0$$

In the isotropic case $\varepsilon = \varepsilon_x = \varepsilon_y$ so that $\sigma_x = \sigma_y = \sigma$, or:

$$\sigma = \left(\frac{E}{1-\nu}\right)\varepsilon \qquad 5.4$$

where the Young's modulus of the film and the Poissons's ratio of the film act independently of orientation. The quantity E/l – ν often is called the biaxial modulus. Uniaxial testing of thin films is difficult, prompting the use of the biaxial modulus, rather than Young's uniaxial modulus. Plane stress, as described here, presents a good approximation several thicknesses away from the edge of the film (say, 3 thicknesses from the edge).

### *Thermal Stress*

Thermal stresses develop in thin films when high temperature deposition or annealing are involved, and usually are unavoidable due to mismatch of thermal expansion coefficients

between film and substrate. The problem of a thin film under residual thermal stress can be modeled by considering a thought experiment involving a stress-free film at high temperature on a thick substrate. Imagine detaching the film from the high-temperature substrate and cooling the system to room temperature. Usually, the substrate dimensions undergo minor shrinkage in the plane while the film's dimensions may reduce significantly. In order to reapply the film to the substrate with complete coverage, the film needs stretching with a biaxial tensile load to a uniform radial strain ε, followed by perfect bondage to the rigid substrate and load removal. The film stress is assumed to be the same in the stretched and free-standing film as in the film bonded to the substrate, i.e., no relaxation occurs in the bonding process. To calculate the thermal residual stress from Equation 5.4 the elastic moduli of the film must be known, as well as the volume change associated with the residual stress, i.e., the thermal strain, $\varepsilon_{th}$, resulting from the difference in the coefficients of thermal expansion between the film and the substrate.

Let us now consider whether, qualitatively, the above assumptions apply to the measurement of thin films on Si wafers. Such films typically measure 1 μm thick and are deposited on 4-in. wafers nominally 550 μm thick. In this case, the substrate measures nearly three orders of magnitude thicker than the film, and, because the bending stiffness is proportional to the thickness cubed, the substrate essentially is rigid relative to the film. The earlier assumptions clearly apply.

Figure 5.4 portrays a quantitative example where a polyimide film, strain-free at the deposition temperature ($T_d$) of 400°C, is cooled to room temperature $T_r$ (25°C) on an Si substrate with a different coefficient of thermal expansion. The resulting strain is given by:

$$\varepsilon_{th} = \int(\alpha_f(T) - \alpha_s(T))\,dT \qquad 5.5$$

where $\alpha_f$ and $\alpha_s$ represent the coefficients of thermal expansion for the polyimide film and the Si substrate, respectively. The thermal strain can be of either sign, based on the relative values of $\alpha_f$ and $\alpha_s$: positive is tensile, negative is compressive. Polyimide features a thermal expansion ($\alpha_f = 70 \times 10^{-6}°C^{-1}$) larger than the thermal expansion coefficient of Si ($\alpha_s = 2.6 \times 10^{-6}°C^{-1}$); hence, a tensile stress is expected. With $SiO_2$ grown or deposited on silicon at elevated temperatures, a compressive component ($\alpha_f (0.35 \times 10^{-6}°C^{-1}) < \alpha_s (2.6 \times 10^{-6}°C^{-1})$) is expected. Assuming that the coefficients of thermal expansion are temperature independent, Equation 5.5 simplifies to:

$$\varepsilon_{th} = (\alpha_f - \alpha_s)(T_d - T_r) \qquad 5.6$$

The calculated thermal strain $\varepsilon_{th}$ for polyimide on Si then measures $25 \times 10^{-3}$ at room temperature. The biaxial modulus (E/1 – ν), with E = 3 GPa and ν = 0.4, equals 5 GPa, and the residual stress σ, from Equation 5.4, is 125 MPa and tensile.

### *Intrinsic Stress*

The intrinsic stress, $\sigma_i$, reflects the internal structure of a material and is less clearly understood than the thermal stress which it often dominates.[28] Several phenomena may contribute

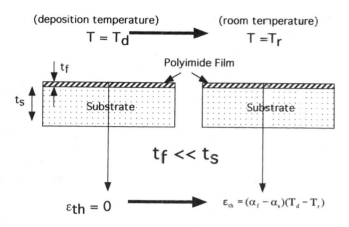

(deposition temperature)        (room temperature)
$$T = T_d \longrightarrow T = T_r$$

$$t_f \ll t_s$$

$$\varepsilon_{th} = 0 \longrightarrow \varepsilon_{th} = (\alpha_f - \alpha_s)(T_d - T_r)$$

$\alpha_f > \alpha_s$ : tension (polyimide on Si)

$\alpha_f < \alpha_s$ : compression (SiO$_2$ on Si)

**FIGURE 5.4** Thermal stress. Tension and compression are determined by the relative size of thermal expansion coefficients of film and substrate. Suppose a strain-free film at deposition temperature, $T_d$, is cooled to room temperature, $T_r$, on a substrate with a different coefficient of thermal expansion.

to $\sigma_i$, making its analysis very complex. Intrinsic stress depends on thickness, deposition rate (locking in defects), deposition temperature, ambient pressure, method of film preparation, type of substrate used (lattice mismatch), incorporation of impurities during growth, etc. Some semiquantitative descriptions of various intrinsic stress causing factors follow:

- Doping ($\sigma_{int} > 0$ or $\sigma_{int} < 0$): When doping Si, the atomic or ionic radius of the dopant and the substitutional site determine the positive or negative intrinsic stress ($\sigma_{int} > 0$ or tensile, and $\sigma_{int} < 0$ or compressive). With boron-doped poly-Si, an atom small compared to Si, the film is expected to be tensile ($\sigma_{int} > 0$); with phosphorous doping, an atom large compared to Si, the film is expected to be compressive ($\sigma_{int} < 0$).

- Atomic peening ($\sigma_{int} < 0$): Ion bombardment by sputtered atoms and working gas densifies thin films, rendering them more compressive. Magnetron sputtered films at low working pressure (<1 Pa) and low temperature often exhibit compressive stress. This topic was discussed at length in Chapter 3.

- Microvoids ($\sigma_{int} > 0$): Microvoids may arise when byproducts during deposition escape as gases and the lateral diffusion of atoms evolves too slowly to fill all the gaps, resulting in a tensile film.

- Gas entrapment ($\sigma_{int} < 0$): As an example we can cite the hydrogen trapped in Si$_3$N$_4$. Annealing removes the hydrogen, and a nitride film, compressive at first, may become tensile if the hydrogen content is sufficiently low.

- Shrinkage of polymers during cure ($\sigma_{int} > 0$): The shrinkage of polymers during curing may lead to severe tensile stress, as becomes clear in the case of polyimides. Special problems are associated with measuring the mechanical properties of polymers as they exhibit a time-dependent

mechanical response (viscoelasticity), a potentially significant factor in the design of mechanical structures where polymers are subjected to sustained loads.[29]

- Grain boundaries ($\sigma_{int}$ = ?): Based on intuition one expects that the interatomic spacing in grain boundaries differs depending on the amount of strain, thus contributing to the intrinsic stress. But the origin of, for example, the compressive stress in polysilicon and how it relates to the grain structure and interatomic spacing are not yet completely clear (see also below on coarse and fine-grated Si).

For further reading on thin-film stress, refer to Hoffman.[30,31]

## Stress-Measuring Techniques

### Introduction

A stressed thin film will bend a thin substrate by a measurable degree. A tensile stress will bend and render the surface concave; a compressive stress renders the surface convex. The most common methods for measuring the stress in a thin film are based on this substrate bending principle. The deformation of a thin substrate due to stress is measured either by observing the displacement of the center of a circular disk or by using a thin cantilevered beam as a substrate and calculating the radius of curvature of the beam and hence the stress from the deflection of the free end. More sophisticated local stress measurements use analytical tools such as X-ray,[32] acoustics, Raman spectroscopy,[33] infrared spectroscopy,[34] and electron-diffraction techniques. Local stress does not necessarily mean the same as the stress measured by substrate bending techniques, since stress is defined microscopically, while deformations are induced mostly macroscopically. The relation between macroscopic forces and displacements and internal differential deformation, therefore, must be modeled carefully. Local stress measurements may also be made using *in situ* surface micromachined structures such as strain gauges made directly out of the film of interest itself.[35] The deflections of thin suspended and pressurized micromachined membranes may be measured by mechanical probe,[36] laser,[37] or microscope.[38] Intrinsic stress influences the frequency response of microstructures (see Equation 5.16) which can be measured by laser,[39] spectrum analyzer,[40] or stroboscope. Whereas residual stress can be determined from wafer curvature and microstructure deflection data, material structure of the film can be studied by X-ray diffraction and transmission electron microscopy (TEM). Krulevitch, among others, attempted to link the material structure of poly-Si and its residual strain.[24] Following, we will review stress-measuring techniques, starting with the more traditional ones and subsequently clarifying the problems and opportunities in stress measuring with surface micromachined devices.

### Disk Method

For all practical purposes only stresses in the x and y directions are of interest in determining overall thin-film stress, as a film under high stress can only expand or contract by bending the substrate and deforming it in a vertical direction. Vertical deformations will not induce stresses in a substrate because it freely moves in that direction. The latter condition enables us to obtain quite accurate stress values by measuring changes in bow or radius of curvature of a substrate. The residual stresses in thin films are large and sensitive optical or capacitive gauges may measure the associated substrate deflections.

The disk method, most commonly used, is based on a measurement of the deflection in the center of the disk substrate (say, a Si wafer) before and after processing. Since any change in wafer shape is directly attributable to the stress in the deposited film, it is relatively straightforward to calculate stress by measuring these changes. Stress in films using this method is found through the Stoney equation,[30] relating film stress to substrate curvature:

$$\sigma = \frac{1}{R}\frac{E}{6(1-v)}\frac{T^2}{t} \qquad 5.7$$

where R represents the measured radius of curvature of the bent substrate, $E/(1 - v)$ the biaxial modulus of the substrate, T the thickness of the substrate, and t the thickness of the applied film.[41] The underlying assumptions include:

- The disc substrate is thin and has transversely isotropic elastic properties with respect to the film normal.
- The applied film thickness is much less than the substrate thickness.
- The film thickness is uniform.
- Temperature of the disk substrate/film system is uniform.
- Disc substrate/film system is mechanically free.
- Disc substrate without film has no bow.
- Stress is equi-biaxial and homogeneous over the entire substrate.
- Film stress is constant through the film thickness.

For most films on Si we assume that $t \le T$; for example, $t/T$ measures $\sim 10^{-3}$ for thin films on Si. The legitimacy of the uniform thickness, homogeneous, and equi-biaxial stress assumptions depend on the deposition process. Chemical vapor deposition (CVD) is such a widely used process as it produces relatively uniform films; however, sputter-deposited films can vary considerably over the substrate. In regard to the assumption of stress uniformity with film thickness, residual stress can vary considerably through the thickness of the film. Equation 5.7 only gives an average film stress in such cases. In cases where thin films are deposited onto anisotropic single crystal substrates, the assumption of a substrate with transversely isotropic elastic properties with respect to the film normal is not always satisfied. Using single crystal silicon substrates possessing moderately anisotropic properties (Equation 4.20) (<100> or <111> oriented wafers) satisfies the transverse isotropy argument. Any curvature inherent in the substrate must be measured before film deposition and algebraically added to the final measured radius of curvature. To give an idea of the degree of curvature, 1 μm of thermal oxide may cause a 30-μm

warp of a 4-in. silicon wafer, corresponding to a radius of curvature of 41.7 m.

Presently, five companies offer practical disk method-based instruments to measure stress on wafers: ADE Corp. (Newton, MA), GCA/Tropel (Fairport, NY), Ionic Systems (Salinas, CA), Scientific Measurements Systems, Inc. (San Jose, CA), and Tencor Instruments (Mountain View, CA).[41] Figure 5.5A illustrates the sample output from Tencor's optical stress analysis system. Figure 5.5B represents the measuring principle of Ionic Systems' optical stress analyzer. None of the above techniques satisfies the need for measuring stress in low modulus materials such as polyimide. For the latter applications the suspended membrane approach (see below) is more suited.

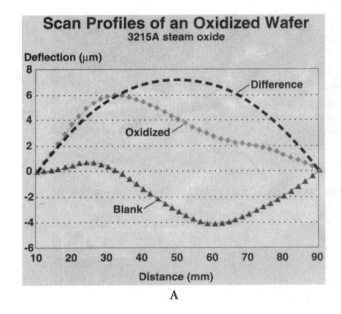

A

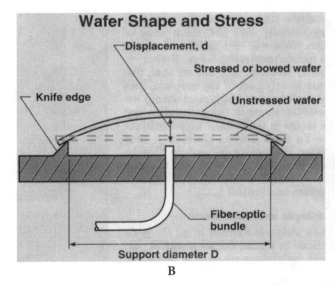

B

FIGURE 5.5   Curvature measurement for stress analysis. (A) Sample output from Tencor's FLX stress analysis instrument, showing how stress is derived from changes in wafer curvature. (B) The reflected light technique, used by Ionic Systems to measure wafer curvature. (From Singer, P., *Semicond. Int.*, 15, 54–58, 1992. With permission.)

*Uniaxial Measurements of Mechanical Properties of Thin Films*

Many problems associated with handling thin films in stress test equipment may be bypassed by applying micromachining techniques. One simple example of problems encountered with thin films is the measurement of uniaxial tension to establish the Young's modulus. This method, effective for macroscopic samples, proves problematic for small samples. The test formula is illustrated in Figure 5.6A. The gauge length L in this figure represents the region we allow to elongate and the area A (= W × H) is the cross section of the specimen. A stress F/A is applied and measured with a load cell; the strain δL/L is measured with an LVDT or another displacement transducer (a typical instrument used is the Instron 1123). The Young's modulus is then deduced from:

$$E = \left(\frac{F}{A}\right)\left(\frac{L}{\delta L}\right) \qquad 5.8$$

The obvious problem, for small or large samples, is how to grip onto the sample without changing A. Under elongation, A will indeed contract by $(w + H)\delta vL$. In general, making a dog-bone shaped structure (Instron specimen) solves that problem as shown in Figure 5.6B. Still, the grips introduced in an Instron sample can produce end effects and uncertainties in determining L. Making Instron specimens in thin films is even more of a challenge since the thin film needs to be removed from the surface, possibly changing the stress state, while the removal itself may modify the film.

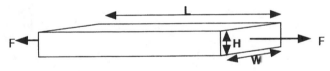

Stress : F/A (area A= W*H)

Strain : δL/L

L ="gauge length" (region we allow to elongate)

A, under elongation,  will contract by (W+H) δ v L unless a dog-bone-structure is used

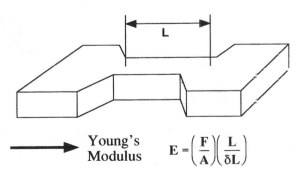

Young's Modulus    $E = \left(\frac{F}{A}\right)\left(\frac{L}{\delta L}\right)$

**FIGURE 5.6** Measuring Young's modulus. (A) With a bar-shaped structure. (B) With a dog-bone, Instron specimen. (After Reference 44.)

As in the case of adhesion, some new techniques for testing stress in thin films, based on micromachining, are being explored. These microtechniques prove more advantageous than the whole wafer disc technique in that they are able to make local measurements.

The fabrication of micro-Instron specimens of thin polyimide samples is illustrated in Figure 5.7A. Polyimide is deposited on a p$^+$ Si membrane in multiple coats. Each coat is prebaked at 130°C for 15 min. After reaching the desired thickness the film is cured at 400°C in nitrogen for 1 hr (A). The polyimide is then covered with a 3000-Å layer of evaporated aluminum (B). The aluminum layer is patterned by wet etching (in phosphoric-acetic-nitric solution referred to as PAN etch) to the Instron specimen shape (C). Dry etching transfers the pattern to the polyimide (D). After removing the Al mask by wet etching, the p$^+$ support is removed by a wet isotropic etch (HNA) or a SF$_6$ plasma etch (E), and finally the side silicon is removed along four pre-etched scribe lines, releasing the residual stress (F). The remaining silicon acts as supports for the grips of the Instron.[42] The resulting structure can be manipulated as any macrosample without the need for removal of the film from its substrate. This technique enables the gathering of stress/strain data for a variety of commercially available polyimides.[43] A typical measurement result for Dupont's polyimide 2525, illustrated in Figure 5.7B, gives a break stress and strain of 77 MPa ($\sigma_b$) and 2.7% ($\varepsilon_b$), respectively, and 3350 MPa for the Young's modulus.

Figure 5.8A illustrates a micromachined test structure able to establish the strain and the ultimate stress of a thin film.[23] A suspended rectangular polymer membrane is patterned into an asymmetric structure before removing the thin supporting Si. Once released, the wide suspended strip (width $w_1$) pulls on the thinner necks (total width $w_2$), resulting in a deflection δ from its original mask position toward the right to its final position after release. The residual tensile stress in the film drives the deformation δ as shown in Figure 5.8A. By varying the geometry, it is possible to create structures exhibiting small strain in the thinner sections as opposed to others that exceed the ultimate strain of the film. For structures where the strain is small enough to be modeled with linear elastic behavior, the deflection δ can be related to the strain as follows:

$$\varepsilon = \frac{\sigma}{E} = \frac{\delta\left(\dfrac{W_1}{L_1} + \dfrac{W_2}{L_2}\right)}{W_1 - W_2} \qquad 5.9$$

where the geometries are defined as illustrated in Figure 5.8A. Figure 5.8B displays a photograph of two released structures: one with thicker necks, the other with necks so thin that they fractured upon release of the film. Based on the residual tensile strain of the film and the geometry of the structures that failed, the ultimate strain of the particular polyimide used was determined to be 4.5%.

Using similar micromachined tensile test structures Biebl et al.[2] measured the fracture strength of undoped and doped

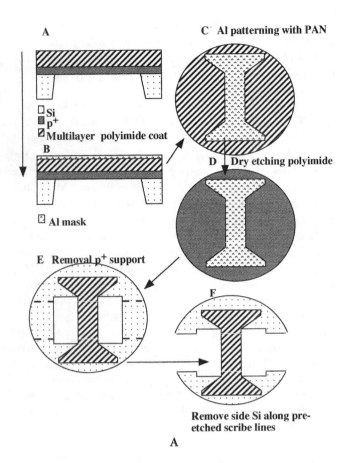

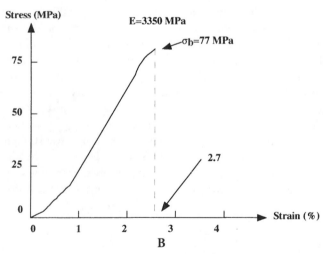

**FIGURE 5.7** Uniaxial stress measurement. (A) Fabrication process of a dog bone sample for measurement of uniaxial strain. (B) Stress vs. strain for Du Pont's 2525 polyimide. (Courtesy of Dr. F. Maseeh-Tehrani.)

polysilicon and found 2.84 ± 0.09 GPa for undoped material and 2.11 ± 0.10 GPa in the case of phosphorous doping, 2.77 ± 0.08 GPa for boron doping, and 2.70 ± 0.09 GPa for arsenic doping. No statistically significant differences were observed between samples released using concentrated HF or buffered HF. However, a 17% decrease of the fracture stress was observed for a 100% increase in etching time. These data contrast with Greek et al.'s[3] *in situ* tensile strength test result of 768 MPa for an undoped poly-Si film. A mean tensile strength almost ten

times less than that of single crystal Si (6 GPa).[45] We normally expect polycrystalline films to be stronger than single crystal films (see below under Strength of Thin Films). Greek et al.[3] explain this discrepancy for poly-Si by pointing out that their polysilicon films have a very rough surface compared to single crystal material, containing many locations of stress concentration where a fracture crack can be initiated.

### Biaxial Measurements of Mechanical Properties of Thin Films: Suspended Membrane Methods

We noted earlier that none of the disk stress-measuring techniques is suitable for measuring stress in low modulus tensile materials such as polyimides. Suspended membranes are very convenient for this purpose. The same micromachined test structure used for adhesion testing, sometimes called the blister test, as shown in Figure 5.2, can measure the tensile stress in low modulus materials. This type of test structure ensues from shaping a silicon diaphragm by conventional anisotropic etching, followed by applying the coating, and, finally, removing the supporting silicon from the back with an $SF_6$ plasma.[23] By pressurizing one side of the membrane and measuring the deflection, one can extract both the residual stress and the biaxial modulus of the membrane. Pressure to the suspended film can be applied by a gas or by a point-load applicator.[21] The load-deflection curve at moderate deflections (strains less than 5%) answers to:

$$p = C_1 \frac{\sigma t d}{a^2} + C_2 \left( \frac{E}{1-\nu} \right) \frac{t d^3}{a^4} \qquad 5.10$$

where p represents the pressure differential across the film; d, the center deflection; a, the initial radius; t, the membrane thickness; and $\sigma$, the initial film stress. In the simplified Cabrera model for circular membranes the constants $C_1$ and $C_2$ equal 4 respectively 8/3. For more rigorous solutions for both circular and rectangular membranes and references to other proposed models, refer to Maseeh-Tehrani.[42] The relation in Equation 5.10 can simultaneously determine $\sigma$ and the biaxial modulus $E/1 - \nu$; plotting $pa^2/dt$ vs. $(d/a)^2$ should yield a straight line. The residual stress can be extracted from the intercept and the biaxial modulus from the slope of the least-squares-best-fit line.[23] A typical result obtained via such measurements is represented in Figure 5.9. For the same Dupont polyimide 2525, measuring a Young's modulus of 3350 MPa in the uniaxial test (Figure 5.7B), the measurements give 5540 MPa for the biaxial modulus and 32 MPa for the residual stress. The residual stress-to-biaxial modulus ratio, also referred to as the residual biaxial strain, thus reaches 0.6%. The latter quantity must be compared to the ultimate strain when evaluating potential reliability problems associated with cracking of films. By loading the membranes to the elastic limit point, yield stress and strain can be determined as with the uniaxial test.

### Poisson Ratio for Thin Films

The Poisson ratio for thin films presents us with more difficulties to measure than the Young's modulus as thin films tend to bend out of plane in response to in-plane shear. Maseeh

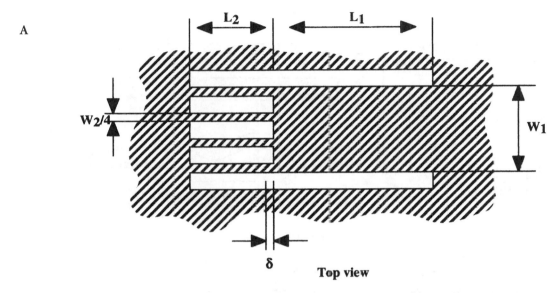

A

Top view

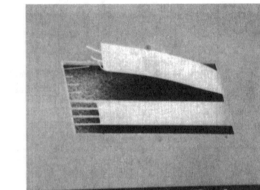

B

**FIGURE 5.8** Ultimate strain. (A) Test structures for stress-to-modulus (strain) and ultimate stress measurements. (B) Two released structures, one of which has exceeded the ultimate strain of the film, resulting in fracture of the necks. (From Senturia, S., Can We Design Microbiotic Devices without Knowing the Mechanical Properties of the Materials?, presented at Micro Robots and Teleoperators Workshop, Hyannis, MA, 1987. With permission.)

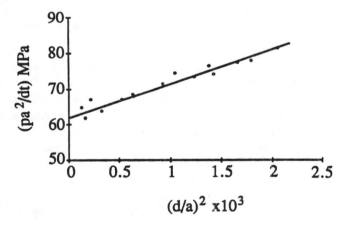

**FIGURE 5.9** Load deflection data of a polyimide membrane (Du Pont 2525). From the intercept, a residual stress of 32 MPa was calculated and from the slope, a biaxial modulus of 5540 MPa. (From Senturia, S., Can We Design Microbiotic Devices without Knowing the Mechanical Properties of the Materials?, presented at Micro Robots and Teleoperators Workshop, Hyannis, MA, 1987. With permission.)

and Senturia[43] combine uniaxial and biaxial measurements to calculate the in-plane Poisson ratio of polyimides. For example, for the Dupont polyimide 2525, they determined 3350 MPa for E and 5540 MPa for the biaxial modulus (E/1 − ν) leading to 0.41 ± 0.1 for the Poisson's ratio (ν). The errors on both the biaxial and uniaxial measurements need to be reduced in order to develop more confidence in the extracted value of the Poisson's ratio. At present, the precision on the Poisson's ratio is limited to about 20%.

*Other Surface Micromachined Structures to Gauge Intrinsic Stress*

Various other surface micromachined structures have been used to measure mechanical properties of thin films. We will give a short review here, but the interested reader might want to consult the original references for more details.

**Clamped-Clamped Beams** Several groups have used rows of clamped-clamped beams (bridges) with incrementally increasing lengths to determine the critical buckling load and hence deduce the residual compressive stress in polysilicon films (Figure 5.10A).[46,47] The residual strain, ε = σ/E, is obtained from the critical length, $L_c$, at which buckling occurs (Euler's formula for elastic instability of struts):

$$\varepsilon = \frac{4\pi^2}{A} \frac{I}{L_c^2} \qquad 5.11$$

where A is the beam cross-sectional area and I the moment of inertia. As an example, with a maximum beam length of 500 μm

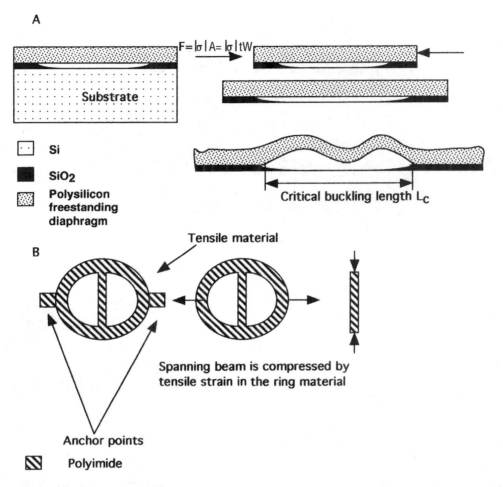

**FIGURE 5.10**  Some micromachined structures used for stress measurements. (A) Clamped-clamped beams: measuring the critical buckling length of clamped-clamped beams enables measurement of residual stress. (B) Crossbar rings: tensile stress can be measured by buckling induced in the crossbar of a ring structure. (C) A schematic of a strain gauge capable of measuring tensile or compressive stress. (D) SEM microphotograph of two strain gauges. (C and D from Lin, L., Selective Encapsulations of MEMS: Micro Channels, Needles, Resonators and Electromechanical Filters, Ph.D. thesis, University of California, Berkeley, 1993. With permission.)

and a film thickness of 1.0 µm, the buckling beam method can detect compressive stress as small as 0.5 MPa. This simple Euler approach does not take into considerations additional effects such as internal moments resulting from gradients in residual stress.

**Ring Crossbar Structures**  Tensile strain can be measured by a series of rings (Figure 5.10B) constrained to the substrate at two points on a diameter and spanned orthogonally by a clamped-clamped beam. After removal of the sacrificial layer, tensile strain in the ring places the spanning beam in compression; the critical buckling length of the beam can be related to the average strain.[49]

**Vernier Gauges**  Both clamped-clamped beams and ring structures need to be implemented in entire arrays of structures. They do not allow easy integration with active microstructures due to space constraints. As opposed to proof structures one might use vernier gauges to measure the displacement of structures induced by residual strain.[48] The idea was first explored by Kim,[50] whose device consisted of two cantilever beams fixed at two opposite points. The end movement of the

beams caused by the residual strain was measured by a vernier gauge. This method only requires one structure, but the best resolution for strain measurement reported is only 0.02% for 500 µm beams. Moreover, the vernier gauge device may indicate an erroneous strain when an out-of-plane strain gradient occurs.[48] Other types of direct strain measurement devices are the T- and H-shaped structures from Allen et al.[38] and Mehregany et al.[35] Optical measurement of the movement at the top of the T- or H-shape structures only becomes possible with very long beams (greater than 2.5 mm). They occupy large areas and their complexity requires finite element methods to analyze their output. The same is true for the strain magnification structure by Goosen et al.[51] This structure measures strain by interconnecting two opposed beams such that the residual strain in the beams causes a third beam to rotate as a gauge needle. The rotation of the gauge needle quantifies the residual strain. A schematic of a micromachined strain gauge capable of measuring tensile or compressive residual stress, as shown in Figure 5.10C, was developed by Lin at University of California, Berkeley.[48] Figure 5.10D represents a scanning electron microscope (SEM) photograph of Lin's strain gauge. This gauge by far outranks the various *in situ* gauges explored. The strain

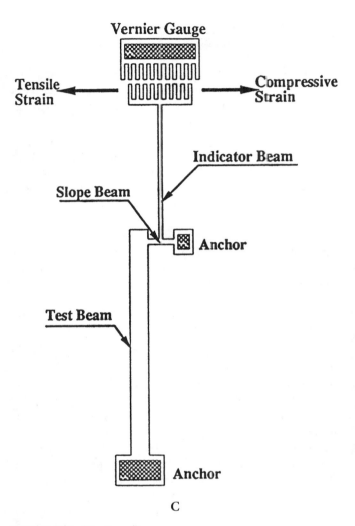

C

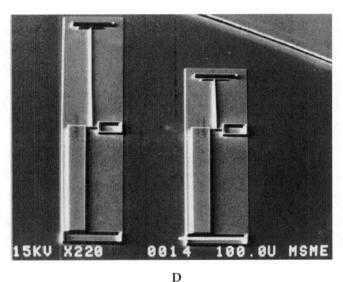

D

FIGURE 5.10 (Continued)

gauge uses only one structure, can be fabricated *in situ* with active devices, determines tensile or compressive strain under optical microscopes, has a fine resolution of 0.001%, and resists to the out-of-plane strain gradient. When the device is released in the sacrificial etch step, the test beam (length $L_t$) expands or contracts, depending on the sign of the residual stress in the film, causing the compliant slope beam (length $L_s$) to deflect into an 's' shape. The indicator beam (length $L_i$) attached to the deforming beam at its point of inflection, rotates through an angle θ, and the deflection δ is read on the vernier scale. The residual strain is calculated as:[48]

$$\varepsilon_f = \frac{2L_s\delta}{3L_iL_tC} \qquad 5.12$$

where C is a correction factor due to the presence of the indicator beam.[48] This equation was derived from simple beam theory relations and assumes that no out-of-plane motion occurs. The accuracy of the strain gauge is greatly improved because its output is independent of both the thickness of the deposited film and the cross-section of the microstructure. Krulevitch used these devices to measure residual stress in *in situ* phosphorous-doped poly-Si films,[24] while Lin tested LPCVD silicon-rich silicon nitride films with it.[48]

An improved micromachined indicator structure, inspired by Lin's work, was built by Ericson et al.[52] By reading an integrated nonius scale in an SEM or an optical microscope, internal stress was measured with a resolution better than 0.5 MPa.[52,53] Both thick (10 μm) and thin (2 μm) poly-Si films were characterized this way.

**Lateral Resonators** Biebl et al.[54] extracted the Young's modulus of *in situ* phosphorus-doped polysilicon by measuring the mechanical response of poly-Si linear lateral comb-drive resonators (see Figure 5.18). The results reveal a value of 130 ± 5 GPa for the Young's modulus of highly phosphorus-doped films deposited at 610°C with a phosphine-to-silane mole ratio of $1.0 \times 10^{-2}$ and annealing at 1050°C. For a deposition at 560°C with a phosphine to silane ratio of $1.6 \times 10^{-3}$, a Young's modulus of 147 ± 6 GPa was extracted.

### Stress Nonuniformity Measurement by Cantilever Beams and Cantilever Spirals

The uniformity of stress through the depth of a film introduces an extremely important property to control. Variations in the magnitude and direction of the stress in the vertical direction can cause cantilevered structures to curl toward or away from the substrate. Stress gradients present in the polysilicon film must thus be controlled to ensure predictable behavior of designed structures when released from the substrate. To determine the thickness variation in residual stress, noncontact surface profilometer measurements on an array of simple cantilever beams[55,56] or cantilever spirals can be used.[57]

**Cantilever Beams**   The deflections resulting from stress variation through the thickness of simple cantilever beams after their release from the substrate is shown in Figure 5.11A.* The bending moment causing deflection of a cantilever beam follows out of pre-release residual stress and is given by (see Figure 5.11C):

$$M = \int_{-t/2}^{t/2} zb\sigma(z)\,dZ \qquad\qquad 5.13$$

where $\sigma(z)$ represents the residual stress in the film as a function of thickness and b stands for cantilever width. Assuming a linear strain gradient $\Gamma$ (physical dimensions 1/length) such that $\sigma(z) = E\Gamma z$, Equation 5.13 converts to:

$$\Gamma = \frac{M(12)}{Ebt^3} = \frac{M}{EI} \qquad\qquad 5.14$$

where the moment of inertia, I, for a rectangular cross-section is given by $I = bt^3/12$. The measured deflection z, i.e., the vertical deflection of the cantilever's endpoint, from beam theory for a cantilever with an applied end moment, is given as:

$$z = \frac{ML^2}{2EI} = \frac{\Gamma L^2}{2} \qquad\qquad 5.15$$

Figure 5.11B represents a topographical contour map of an array of polysilicon cantilevers. The cantilevers vary in length from 25 to 300 $\mu$m by 25-$\mu$m increments. Notice that the tip of the longest cantilever resides at a lower height (approximately 0.9 $\mu$m closer to the substrate) than the anchored support, indicating a downward bending moment.[55] The gradients can be reduced or eliminated with a high-temperature anneal. With integrated electronics on the same chip, long high-temperature processing must be avoided. Therefore, stress gradients can limit the length of cantilevered structures used in surface-micromachined designs.

**Cantilever Spirals**   Residual stress gradients can also be measured by Fan's cantilever spiral as shown in Figure 5.12A.[57] Spirals anchored at the inside spring upwards, rotate, and contract with positive strain gradient (tending to curl a cantilever upwards), while spirals anchored at the outside deflect in a similar manner in response to a negative gradient. Theoretically, positive and negative gradients produce spirals with mirror symmetry.[57] The strain gradient can be determined from spiral structures by measuring the amount of lateral contraction, the change in height, or the amount of rotation. Krulevitch presented the computer code for the spiral simulation in his doctoral thesis.[24]

Figure 5.12B shows a simulated spiral with a bending moment of $\Gamma = \pm 3.0$ mm$^{-1}$ after release.

Krulevitch compared all the above surface micromachined structures for stress and stress gradient measurements on poly-Si films. His comments are summarized in Table 5.2.[24] Krulevitch found that the fixed-fixed beam structures for determining compressive stress from the buckling criterion produced remarkably self-consistent and repeatable results. Wafer curvature stress profiling proved reliable for determining average stress and the true stress gradient as compared with micromachined spirals. Measurements of curled cantilevers could not be used much as the strain gradients mainly were negative for poly-Si, leading to cantilevers contacting the substrate. The strain gauge dial structures were useful over a rather limited strain-gradient range. With too large a strain gradient curling of the long beams overshadows expansion effects and makes the vernier indicator unreadable.

## Strength of Thin Films

Due to the high activation energy for dislocation motion in silicon (2.2 eV), hardly any plastic flow occurs in single crystalline silicon for temperatures lower than 673°C. Grain boundaries in poly-Si block dislocation motion; hence, polysilicon films can be treated as an ideal brittle material at room temperature.[54] High yield strengths often are obtained in thin films with values up to 200 times as large as those found in the corresponding bulk material.[22] In this light, the earlier quoted fracture stresses of poly-Si, between two and ten times smaller than that of bulk single crystal, are surprising. Greek et al.[3] explain this deviation by pointing at the high surface roughness of poly-Si films compared to single crystal Si. They believe that a reduction in surface roughness would improve the tensile fracture strength considerably.

Indentation (hardness) testing is very common for bulk materials where the direct relationship between bulk hardness and yield strength is well known. It can be measured by pressing a hard, specially shaped point into the surface and observing indentation. This type of measurement is of little use for measuring thin films below $5 \times 10^4$ Å. Consequently, very little is known about the hardness of thin films. Recently, specialized instruments have been constructed (e.g., the Nanoindenter) in which load and displacement data are collected while the indentation is being introduced in a thin film. This eliminates the errors associated with later measurement of indentation size and provides continuous monitoring of load/displacement data similar to a standard tensile test. Load resolution may be 0.25 $\mu$N, displacement resolution 0.2 to 0.4 nm, and x-y sample position accuracy 0.5 $\mu$m. Empirical relations have correlated hardness with Young's modulus and with uniaxial strength of thin films. Hardness calculations must include both plastic and long-distance elastic deformation. If the indentation is deeper than 10% of the film, corrections for elastic hardness contribution of the substrate must also be included.[21] Mechanical properties such as hardness and modulus of elasticity can be determined on the micro- to picoscales using AFM.[58] Bushan provides an excellent introduction to this field in *Handbook of Micro/Nanotribology.*[59]

---

*   Figure 5.11 appears as a color plate after page 144.

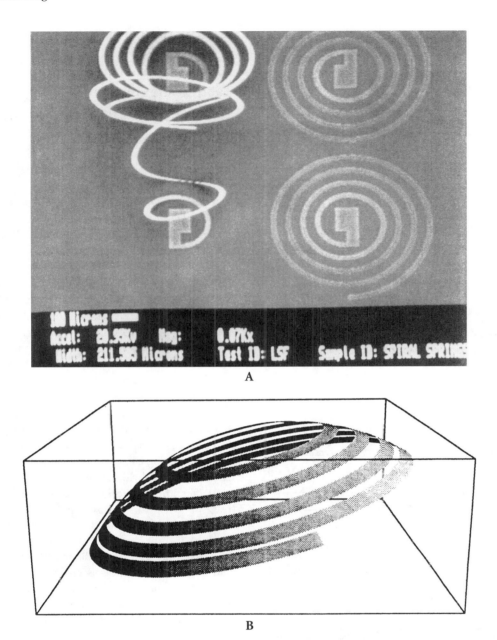

**FIGURE 5.12** Cantilever spirals for stress gradient measurement. (A) SEM of micrographs of spirals from an as-deposited poly-Si. (Courtesy of Dr. L.S. Fan.) (B) Simulation of a thin-film micromachined spiral with $\Gamma = 3.0$ mm$^{-1}$. (From Krulevitch, P.A., Micromechanical Investigations of Silicon and Ni-Ti-Cu Thin Films, Ph.D. thesis, University of California, Berkeley, 1994. With permission.)

## Surface Micromachining Processes

### Basic Process Sequence

A surface micromachining process sequence for the creation of a simple free-standing poly-Si bridge is illustrated in Figure 5.13.[60,61] A sacrificial layer, also called a spacer layer or base, is deposited on a silicon substrate coated with a dielectric layer as the buffer/isolation layer (Figure 5.13A). Phosphosilicate glass (PSG) deposited by LPCVD stands out as the best material for the sacrificial layer because it etches even more rapidly in HF than SiO$_2$. In order to obtain a uniform etch rate, the PSG film must be densified by heating the wafer to 950–1100°C in a furnace or a rapid thermal annealer (RTA).[62] With a first mask the base is patterned as shown in Figure 5.13B. Windows are opened up in the sacrificial layer and a microstructural thin film, whether consisting of polysilicon, metal, alloy, or a dielectric material, is conformably deposited over the patterned sacrificial layer (Figure 5.13C). Furnace annealing, in the case of polysilicon at 1050°C in nitrogen for one hour, reduces stress stemming from thermal expansion coefficient mismatch and nucleation and growth of the film. Rapid thermal annealing has been found effective for reducing stress in polysilicon as well.[62] With a second mask, the microstructure layer is patterned, usually by dry etching in a CF$_4$ + O$_2$ or a CF$_3$Cl + Cl$_2$ plasma (Figure 5.13D).[63] Finally, selective wet etching of the sacrificial layer, say in 49% HF, leaves a free-standing micromechanical structure (Figure 5.13E). The surface micromachining technique is applicable to combinations of thin films and lateral dimensions where the sacrificial layer can be etched without

**TABLE 5.2**  Summary of Various Techniques for Measuring Residual Film Stress

| Measurement Technique | Measurable Stress State | Remarks |
| --- | --- | --- |
| Wafer curvature | Stress gradient, average stress | Average stress over entire wafer; provides true stress gradient; approx. 5 MPa resolution |
| Vernier strain gauges | Average stress | Local stress; small dynamic range; resolution = 2 MPa |
| Spiral cantilevers | Stress gradient | Local stress, provides equivalent linear gradient |
| Curling beam cantilevers | Large positive stress gradient | Local stress, provides equivalent linear gradient |
| Fixed-fixed beams | Average compressive stress | Local stress measurement |

*Source:*  Krulevitch, P.A., *Micromechanical Investigations of Silicon and Ni-Ti-Cu Thin Films*, Ph.D. thesis, University of California, Berkeley, 1994. With permission.

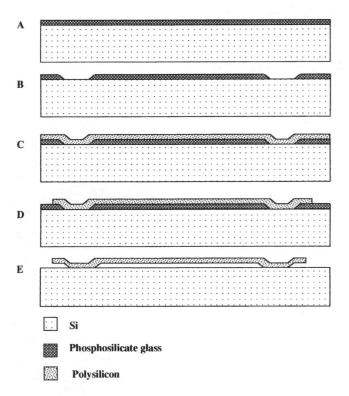

**FIGURE 5.13**  Basic surface micromachining process sequence. (A) Spacer layer deposition (the thin dielectric insulator layer is not shown). (B) Base patterning with mask 1. (C) Microstructure layer deposition. (D) Pattern microstructure with mask 2. (E) Selective etching of spacer layer.

significant etching or attack of the microstructure, the dielectric, or the substrate.

## Fabrication Step Details

### Pattern Transfer to SiO$_2$ Buffer/Isolation Layer

A blanket n$^+$ diffusion of the Si substrate, defining a ground plane, often outlines the very first step in surface micromachining, followed by a passivation step of the substrate, for example, with 0.15-μm thick LPCVD nitride on a 0.5-μm thermal oxide. Suppose the buffer/isolation passivation layer as illustrated in Figure 5.14 itself needs to be patterned, perhaps to make a metal contact pad onto the Si substrate. Then, the appropriate fabrication step is a pattern transfer to the thin isolation film as shown in Figure 5.14, illustrating a wet pattern transfer to a 1-μm thick thermal SiO$_2$ film with a 1-μm resist layer. Typically, an isotropic etch such as buffered HF e.g., BHF (5:1), which is 5 parts NH$_4$F and 1 part Conc. HF, is used (unbuffered HF attacks the photoresist). This solution etches SiO$_2$ at a rate of 100 nm/min, and the creation of the opening to the underlying substrate takes about 10 min. The etch progress may be monitored optically (color change) or by observing the hydrophobic/hydrophilic behavior[64] of the etched layer. With a resist opening, L$_m$, the undercut typically measures the same thickness as the oxide thickness, t$_{SiO_2}$. In other words, the contact pad will have a size of L$_m$ + 2t$_{SiO_2}$. The undercut worsens with loss of photoresist adhesion during etching. Using an adhesion promoter such as HMDS (hexamethyldisilazane) prove useful in such case. A new bake after 5 min of etching is a good procedure to maintain the resist integrity. After the isotropic etch, the resist is stripped in a piranha etch bath. This strong oxidizer grows about 3 nm of oxide back in the cleared window. To remove the oxide resulting from the piranha, a dip in diluted BHF suffices. After cleaning and drying, the substrate is ready for contact metal and base material deposition. Applying a dry etch (say, a CF$_4$-H$_2$ plasma) to open up the window in the oxide would eliminate undercutting of the resist, but requires a longer set-up time. With an LPCVD low stress nitride on top of a thermal SiO$_2$, an often-used combination for etching the buffer/isolation layer is a dry etch (say, a SF$_6$ plasma) followed by a 5:1 BHF.[65]

### Base Layer (also Spacer or Sacrificial Layer) Deposition and Etching

A thin LPCVD phosphosilicate glass (PSG) layer (say, 2 μm thick) is a preferred base, spacer, or sacrificial layer material. Adding phosphorous to SiO$_2$ to produce PSG enhances the etch rate in HF.[66,67] Other advantages to using doped SiO$_2$ include its utility as a solid state diffusion dopant source to make subsequent polysilicon layers electrically conductive and helping to control window taper (see below). As deposited phosphosilicate displays a nonuniform etch rate in HF and must be densified, typically carried out in a furnace at 950°C for 30 min to 1 hr in a wet oxygen ambient. The etch rate in BHF can be used as the measure of the densification quality. The base window etching

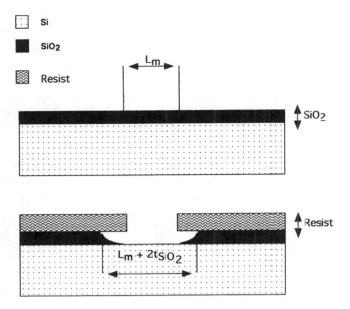

FIGURE 5.14 Wet etch pattern transfer to a thin thermal SiO$_2$ film for the fabrication of a contact pad to the Si substrate.

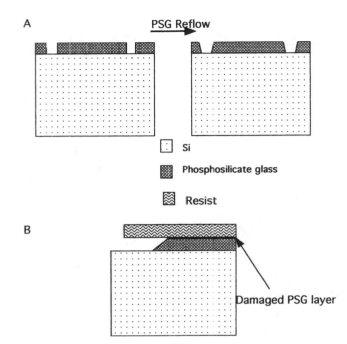

FIGURE 5.15 Edge taper of spacer layer. (A) Taper by reflow of PSG. (B) Taper by ion implantation of PSG.

stops at the buffer isolation layer, often a Si$_3$N$_4$/SiO$_2$ layer which also forms the permanent passivation of the device. Windows in the base layer are used to make anchors onto the buffer/isolation layer for mechanical structures.

The edges of the etched windows in the base may need to be tapered to minimize coverage problems with subsequent structural layers, especially if these layers are deposited with a line-of-sight deposition technique. An edge taper is introduced through an optimization of the plasma etch conditions (see Chapter 2), through the introduction of a gradient of the etch rate, or by reflow of the etched spacer. In Figure 5.15A we depict the reflow process of a PSG spacer after patterning in a dry etch. Viscous flow at higher temperature smoothes the edge taper. The ability of PSG to undergo viscous deformation at a given temperature primarily is a function of the phosphorous content in the glass; reflow profiles get progressively smoother the higher the phosphorous concentration — reflecting the corresponding enhancement in viscous flow.[68] In Figure 5.15B, ion implantation of PSG has created a rapidly etching, damaged PSG layer.[69-71] The steady state taper is a function of the etch rate in PSG etchants (e.g., BHF).

### Deposition of Structural Material

For the best step coverage of a structural material over the base window, chemical vapor deposition is preferred. If a physical deposition method must be used, sputtering is preferred over line-of-sight deposition techniques which lead to the poorest step coverage (see Chapter 3). In the latter case, edge taper could be introduced advantageously.

The most widely used structural material in surface micromachining is polysilicon (poly-Si or simply poly). Polysilicon is deposited by low pressure (25 to 150 Pa) chemical vapor deposition in a furnace (a poly chamber) at about 600°C. The undoped material is usually deposited from pure silane, which thermally decomposes according to the reaction:

$$SiH_4 \rightarrow Si + 2H_2 \qquad \text{Reaction 5.1}$$

Typical process conditions may consist of a temperature of 605°C, a pressure of 550 mTorr (73 Pa), and a silane flow rate of 125 sccm. Under those conditions a normal deposition rate is 100 Å/min. To deposit a 1-μm film will take about 90 min. Sometimes the silane is diluted by 70 to 80% nitrogen. The silicon is deposited at temperatures ranging from 570 to 580°C for fine-grained poly-Si to 620 to 650°C for coarse-grained poly-Si. The characteristics of these two types of poly-Si materials will be compared in the case studies presented below. Furnace annealing of the poly-Si film at 1050°C in nitrogen for one hour is employed commonly to reduce stress stemming from thermal expansion coefficient mismatch and nucleation and growth of the polysilicon film.

To make parts of the microstructure conductive, dopants can be introduced in the poly-Si film by adding dopant gases to the silane gas stream, by drive-in from a solid dopant source, or by ion implantation. When doping from the gas phase, the dopant can be readily controlled in the range of 10$^{19}$ to 10$^{21}$/cm$^3$. Polysilicon deposition rates, in the case of gas phase doping (*in situ* doping), may be significantly impacted. For example, decreases in poly-Si deposition rate by as much as a factor of 25 have been reported in phosphine and arsine doping. The effect is associated with the poisoning of reaction sites by phosphine and arsine.[63] The lower deposition rate of *in situ* phosphine doping can be mitigated by reducing the ratio of phosphine to silane flow by one third.[72] With the latter flow regime, deposition at 585°C (for a 2-μm thick film), followed by 900°C rapid thermal annealing for 7 min, results in a polysilicon with low residual stress, negligible stress gradient, and low resistivity.[72,73] *In situ* boron doping, in contrast to arsine and phosphine doping, accelerates the polysilicon deposition rate through an enhancement of silane adsorption induced by the boron presence.[74] Film thickness uniformity for doped films typically is less than 1%

and sheet resistance uniformity less than 2%. Alternatively, poly-Si may be doped from PSG films sandwiching the undoped poly-Si film. By annealing such a sandwich at 1050°C in $N_2$ for one hour, the polysilicon is symmetrically doped by diffusion of dopant from the top and the bottom layers of PSG. Symmetric doping results in a polysilicon film with a moderate compressive stress. The resulting uniform grain texture avoids gradients in the residual stress which would cause bending moments warping microstructures upon release. Finally, ion implantation of undoped polysilicon, followed by high temperature dopant drive-in, also leads to conductive polysilicon. This polysilicon has a moderate tensile stress, with a strain gradient that causes cantilevers to deflect toward the substrate.[55] The poly-Si is now ready for patterning by RIE in, say, a $CF_4$-$O_2$ plasma.

Although the mechanical properties still are not well understood, microstructures based on poly-Si as a mechanical member have been commercialized.[19,55] Other structural materials used in surface micromachining include single crystal Si (epi-Si or etched back, fusion bonded Si), $SiO_2$, silicon nitride, silicon oxynitride, polyimide, diamond, SiC, GaAs, tungsten, $\alpha$-Si:H, Ni, W, Al, etc. A few words about the merit of some of these materials as structural components follow.

Silicon nitride and silicon oxide also can be deposited by CVD methods but usually exhibit too much residual stress which hampers their use as mechanical components. However, CVD of mixed silicon oxynitride can produce substantially stress-free components.

Amorphous Si ($\alpha$-Si) can be stress annealed at temperatures as low as 400°C.[75] This low-temperature anneal makes the material compatible with almost any active electronic component. Unfortunately, very little is known about the mechanical properties of amorphous Si.

Hydrogenated amorphous Si ($\alpha$-Si:H), with its interesting electronic properties, is even less understood in terms of its mechanical properties. If the mechanical properties of hydrogenated amorphous Si were found as good as those of poly-Si, the material might make a better choice than poly-Si as a MEMS material given its better electronic characteristics.

Tungsten CVD deposition is IC compatible. The material has some unique mechanical properties (see Table 8.4). Moreover, the material can be applied selectively. Selective CVD tungsten has the unique property that tungsten will only nucleate on silicon or metal surfaces but does not deposit on dielectrics such as oxides and nitrides.[76]

Metals and polyimides, because they are easily deformed, usually do not qualify as mechanical members but have been used, for example, in plastically deformable hinges.[77,78] Aluminum constitutes the mirror material in Texas Instruments' flexure-beam micromirror devices (FBMDs). The metal is used both for the L-shaped flexure hinges and the mirror itself.[79] Polycrystalline diamond films, deposited by CVD, are potential high-temperature, harsh environment MEMS candidates.[80] Problems include oxidation above 500°C for nonpassivated films, difficulty of making reliable ohmic contact to the material, and the reproducibility and surface roughness of the films.[81] Poly-SiC films have been deposited by an APCVD process on four-inch, polysilicon-coated, silicon wafers. A surface micromachining process using the underlying polysilicon film as the sacrificial layer was developed. Poly-SiC is projected for use as structural material for high temperatures and harsh environments, and to reduce friction and wear between moving components.[82] Surface micromachining of thin single crystalline Si layers in SOI and with polyimides is discussed separately below.

## Selective Etching of Spacer Layer
### Selective Etching

To create movable micromachines the microstructures must be freed from the spacer layers. The challenge in freeing microstructures by undercutting is evident from Figure 5.16. After patterning the poly-Si by RIE in, say, a $SF_6$ plasma, it is immersed in an HF solution to remove the underlying sacrificial layer, releasing the structure from the substrate. Commonly, a layer of sacrificial phosphosilicate glass, between 1 and 2000 $\mu$m long and 0.1 to 5 $\mu$m thick, is etched in concentrated, dilute or buffered HF. The spacer etch rate, $R_s$, should be faster than the attack on the microstructural element, $R_m$, and that of the insulator layer, $R_i$. For this type of complete undercutting, only wet etchants can be used. Etching narrow gaps and undercutting wide areas with BHF can take hours. To shorten the etch time, extra apertures in the microstructures sometimes are provided for additional access to the spacer layer. The etch rate of PSG, the most common spacer material, increases monotonically with dopant concentration, and thicker sacrificial layers etch faster than thinner layers.[83]

The selectivity ratios for spacer layer, microstructure, and buffer layer are not infinite,[84] and in some instances even silicon substrate attack by BHF was observed under polysilicon/spacer regions.[85,86] Heavily phosphorous-doped polysilicon is especially prone to attack by BHF. Silicon nitride deposited by LPCVD etches much more slowly in HF than oxide films, making it a more desirable isolation film. When depositing this film with a silicon-rich composition, the etch rate is even slower (15 nm/min).[65] Eaton et al.[87] compared oxide and nitride etching in a 1:1 $HF:H_2O$ and in a 1:1 HF:HCl solution and concluded that the HCl-based etch yielded both faster oxide etch rates (617 nm/min vs. 330 nm/min) and slower nitride etch rates (2 nm/min vs. 3.6 nm/min), providing a much greater selectivity of the oxide to silicon nitride (310 vs. 91!). The same authors also studied the optimum composition of a sacrificial oxide for the fastest possible etching in their most selective 1:1 HF:HCl etch. The faster sacrificial layer etch limited the damage to nitride structural elements. Their results are summarized in Table 5.3. A densified CVD $SiO_2$ was used as a control, and a 5%/5% borophosphosilicate glass (BPSG) was found to etch the fastest.

Watanabe et al.[88] using low pressure vapor HF, found high etch ratios of PSG and BPSG to thermal oxides of over 2000, with the BPSG etching slightly faster than the PSG. We will see further that low pressure vapor HF also leads to less stiction of structural elements to the substrate. In Table 5.4 we present etch rate and etch ratios for $R_i$ and $R_s$ in BHF (7:1) for a few selected materials.

Detailed studies on the etching mechanism of oxide spacer layers were undertaken by Monk et al.[83,89] They found that the

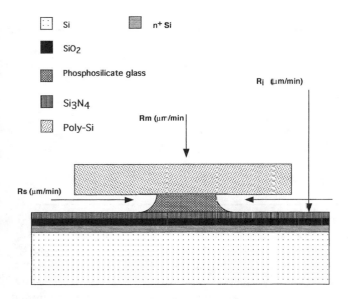

Si

n⁺ Si

SiO₂

Phosphosilicate glass

Si₃N₄

Poly-Si

**FIGURE 5.16**  Selective etching of spacer layer.

**TABLE 5.3**  Etch Rate in 1:1 HF:HCl of a Variety of Sacrificial Oxides

| Thin Oxide | Lateral Etch Rate ($Å$/min) |
|---|---|
| CVD SiO₂ (densified at 1050°C for 30 min) | 6170 |
| Ion-implanted and densified CVD SiO₂ (P, $8 \times 10^{15}$/cm², 50 keV) | 8330 |
| Phosphosilicate (PSG) | 11,330 |
| 5%/5% borophosphosilicate (BPSG) | 41,670 |

*Source:* Adapted from Eaton, W.P. and Smith, J.H., A CMOS-compatible, Surface Micromachined Pressure Sensor for Aqueous Ultrasonic Application, presented at SPIE Smart Structure and Materials, 1995.

**TABLE 5.4**  Etching of Spacer Layer and Buffer Layer in BHF (7:1)

| Property | Material | | |
|---|---|---|---|
| | LPCVD Si₃N₄ | LPCVD SiO₂ | LPCVD 7% PSG |
| Etch rate | 7–12 $Å$/min ($R_i$) | 700 $Å$/min ($R_s$) | ~10,000 $Å$/min ($R_s$) |
| Selectivity ratio | 1 | 60–100 | ~800–1200 |

etching reaction shifts from kinetic controlled to diffusion controlled as the etch channel becomes longer. This affects mainly large-area structures, as diffusion limitations were observed only after approximately 200 μm of channel etching or 15 min in concentrated HF. Eaton and Smith[90] developed a release etch model which is an extension of the work done by Monk et al.[83,89] and Liu et al.[91]

Etching is followed by rinsing and drying. Extended rinsing causes a native oxide to form on the surface of the polysilicon structure. Such a passivation layer often is desirable and can be formed more easily by a short dip in 30% H₂O₂.

*Etchant-Spacer-Microstructure Combinations*

A wide variety of etchant, spacer, and structural material combinations have been used; a limited listing is presented in

Table 5.5. One interesting case concerns poly-Si as the sacrificial layer. This was used, for example, in the fabrication of a vibration sensor at Nissan Motor Co.[92] In this case poly-Si is etched in KOH from underneath a nitride/polysilicon/nitride sandwich cantilever. Also, a solution of HNO₃ and BHF can be used to etch poly-Si, but it proves difficult to control. Using aqueous solutions of NR₄OH, where R is an alkyl group, provides a better etching solution for poly-Si, with greater selectivity with respect to silicon dioxide and phosphosilicate glass. The relatively slow etch rate enables better process control[93] and the etchant does not contain alkali ions, making it more CMOS compatible. With tetramethylammonium hydroxide (TMAH) the etch rate of CVD poly-Si, deposited at 600°C from SiH₄, follows the rates of the (100) face of single crystal Si and is dopant dependent. The selectivity of Si/SiO₂ and Si/PSG, at temperatures below 45°C are measured to be about 1000. Hence, a layer of 500-$Å$ PSG can be used as the etch mask for 10,000 $Å$ of poly-Si.

**Stiction**

*Stiction During Release*

The use of sacrificial layers enables the creation of very intricate movable polysilicon surface structures. An important limitation of such polysilicon shapes is that large-area structures tend to deflect through stress gradients or surface tension induced by trapped liquids and attach to the substrate/isolation layer during the final rinsing and drying step, a stiction phenomenon that may be related to hydrogen bonding or residual contamination. Recently, great strides were made towards a better understanding and prevention of stiction.

The sacrificial layer removal with a buffered oxide etch followed by a long, thorough rinse in deionized water and drying under an infrared lamp typically represent the last steps in the surface micromachining sequence. As the wafer dries, the surface tension of the rinse water pulls the delicate microstructure to the substrate where a combination of forces, probably van der Waals forces and hydrogen bonding, keeps it firmly attached (see Figure 5.17).[55] Once the structure is attached to the substrate by stiction, the mechanical force needed to dislodge it usually is large enough to damage the micromechanical structure.[84,94,95] Basically, the same phenomena are thought to be involved in room temperature wafer bonding (Chapter 8). We will not further dwell upon the mechanics of the stiction process here, but the reader should refer to the theoretical and experimental analysis of the mechanical stability and adhesion of microstructures under capillary forces by Mastrangelo et al.[96,97]

Creating stand-off bumps on the underside of a poly-Si plate[65,98] or adding meniscus-shaping microstructures to the perimeter of the microstructure are mechanical means to help reduce sticking.[99] Fedder et al.[100] used another mechanical approach to avoid stiction by temporarily stiffening the microstructures with polysilicon links. These very stiff structures are not affected by liquid surface tension forces and the links are severed afterwards with a high current pulse once the potentially destructive processing is complete. Yet another mechanical approach to avoid stiction involves the use of sacrificial supporting polymer columns. A portion of the sacrificial layer is substituted by polymer spacer material, spun-on after partial

**TABLE 5.5**   Etchants-Spacer and Microstructural Layer

| Etchant | Buffer/Isolation | Spacer | Microstructure | Ref. |
|---|---|---|---|---|
| Buffered HF (5:1, NH$_4$F:conc. HF) | LPCVD Si$_3$N$_4$/thermal SiO$_2$ | PSG | Poly-Si | 60, 61, 102 |
| RIE using CHF$_3$ BHF (6:1) | LPCVD Si$_3$N$_4$ | LPCVD SiO$_2$ | CVD Tungsten | 76 |
| KOH | LPCVD Si$_3$N$_4$/thermal SiO$_2$ | Poly-Si | Si$_3$N$_4$ | 103, 104 |
| Ferric chloride | Thermal SiO$_2$ | Cu | Polyimide | 105 |
| HF | LPCVD Si$_3$N$_4$/thermal SiO$_2$ | PSG | Polyimide | 77 |
| Phosphoric/acetic acid/nitric acid (PAN, or 5:8:1:1 Water:phosphoric:acetic:nitric) | Thermal SiO$_2$ | Al | PECVD Si$_3$N$_4$ Nickel | 75, 106 |
| Ammonium iodide/iodine alcohol | Thermal SiO$_2$ | Au | Ti | 107 |
| Ethylene-diamine/pyrocathecol (EDP) | Thermal SiO$_2$ | Poly-Si | SiO$_2$ | — |

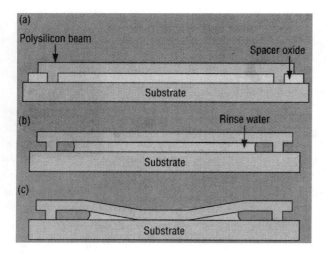

**FIGURE 5.17**   Stiction phenomenon in surface micromachining and the effect of surface tension on micromechanical structures. (a) Unreleased beam. (b) Released beam before drying. (c) Released beam pulled to the substrate by capillary forces as the wafer dries [conc].

etch of the oxide glass. After completion of the oxide etch, the polymer spacer prevents stiction during evaporative drying. Finally, an isotropic oxygen plasma etches the polymer to release the structure.[101]

Ideally, to ensure high yields, one should avoid contact between structural elements and the substrate during processing. In a liquid environment, however, this may become impossible due to the large surface tension effects. Consequently, most solutions to the stiction problem involve reducing the surface tension of the final rinse solution by physico-chemical means. Lober et al.,[84] for example, tried HF vapor and Guckel et al.[108,109] used freeze-drying of water/methanol mixtures. Freezing and sublimating the rinse fluid in a low-pressure environment gives improved results by circumventing the liquid phase. Takeshima et al.[110] used t-butyl alcohol freeze-drying. Since the freezing point of this alcohol lies at 25.6°C, it is possible to perform freeze-drying without special cooling equipment. More recently, attempts at supercritical drying resulted in high microstructure yields.[111] With this technique, the rinse fluid is displaced with a liquid that can be driven into a supercritical phase under high pressure. This supercritical phase does not exhibit surface tension. Typically, CO$_2$ under about 75 atm is used (see also Chapter 1 under Cleaning).

Kozlowski et al.[112] substituted HF in successive exchange steps by the monomer divinylbenzene to fabricate very thin (500 nm) micromachined polysilicon bridges and cantilevers. The monomer was polymerized under UV light at room temperature and was removed in an oxygen plasma. Analog Devices applied a proprietary technique involving only standard IC process technology in the fabrication of a micro-accelerometer to eliminate yield losses due to stiction.[55]

### Stiction After Release, i.e., In-Use Stiction

Stiction remains a fundamental reliability issue due to contact with adjacent surfaces after release. Stiction free passivation that can survive the packaging temperature cycle is not known at present.[5] Attempted solutions are summarized below.

Adhesive energy may be minimized in a variety of ways, for example, by forming bumps on surfaces (see above) or roughening of opposite surface plates.[113] Also, self-assembled monolayer coatings have been shown to reduce surface adhesion and to be effective at friction reduction in bearings at the same time.[114] Making the silicon surface very hydrophobic or coating it with diamond-like carbon are other potential solutions for preventing postrelease stiction of polysilicon microstructures. Man et al.[115] eliminate post-release adhesion in microstructures by using a thin conformal fluorocarbon film. The film eliminates the adhesion of polysilicon beams up to 230 μm long even after direct immersion in water. The film withstands temperatures as high as 400°C and wear tests show that the film remains effective after 10$^8$ contact cycles. Along the same line, ammonium fluoride-treated Si surfaces are thought to be superior to the HF-treated surfaces due to a more complete hydrogen termination, leading to a cleaner hydrophobic surface.[116] Gogoi et al.[117] introduced electromagnetic pulses for postprocessing release of stuck microstructures.

## Control of Film Stress

After reviewing typical surface micromachining process sequences we are ready to investigate some of the mechanical properties of the fabricated mechanical members. Consider a lateral resonator as shown in Figure 5.18A. Electrostatic force is applied by a drive comb to a suspended shuttle. Its motion is detected capacitively by a sense comb. For many applications a

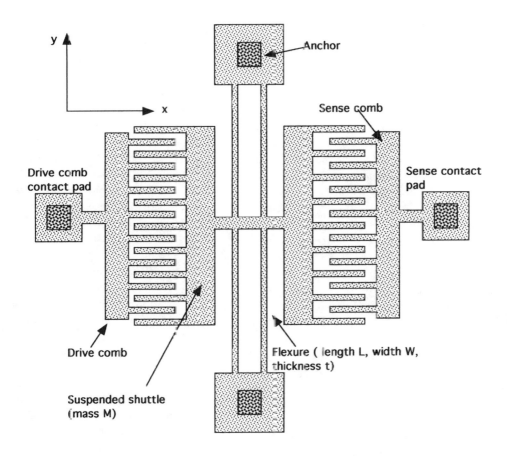

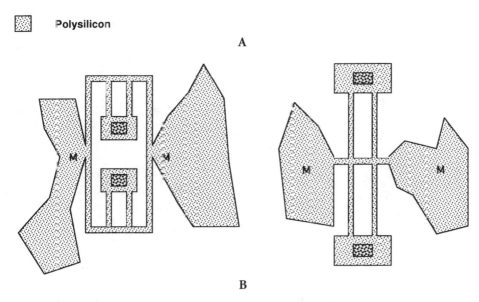

**FIGURE 5.18** Layout of a lateral resonator with straight flexures. (A) Folded flexures (left) to release stress are compared with straight flexures (right). (B) M is shuttle mass.

tight control over the resonant frequency, $f_0$, is required. A simple analytical approximation for $f_0$ of this type of resonator can be deduced from Rayleigh's method:[118]

$$f_o \approx \frac{1}{2\pi} \sqrt{\frac{4EtW^3}{ML^3} + \frac{24\sigma_r tW}{5ML}} \qquad 5.16$$

where E represents the Young's modulus of polysilicon; L, W, and t are the length, width, and thickness of the flexures; and M stands for the mass of the suspended shuttle (of the order of $10^{-9}$ kg, or less). For typical values (L = 150 μm and W = t = 2 μm) and a small tensile residual stress, the resonant frequency $f_0$ is between 10 and 100 kHz.[5] For typical values of L/W the stress term in Equation 5.16 dominates the bending term. Any residual stress, $\sigma_r$, obviously will affect the resonant frequency. Consequently, stress and stress gradients represent critical stages for microstructural design. One of the many challenges of any surface-micromachining process is to control the intrinsic stresses in the deposited films. Several techniques can be used to control film stress. Some we detailed before, but we list them again for completeness:

- Large-grained poly-Si films, deposited around 625°C, have a columnar structure and are always compressive. Compressive stress can cause buckling in constrained structures. Annealing at high-temperatures, between 900 and 1150°C, in nitrogen significantly reduces the compressive stress in as-deposited poly-Si[47,119] and can eliminate stress gradients. No significant structural changes occur when annealing a columnar poly-Si film. The annealing process is not without danger in cases where active electronics are integrated on the same chip. Rapid thermal annealing might provide a solution (see IC Compatibility, below).

- Undoped poly-Si films are in an amorphous state when deposited at 580°C or lower. The stress and the structure of this low-temperature material depend on temperature and partial pressure of the silane. A low-temperature anneal leads to a fine-grained poly-Si with low tensile stress and very smooth surface texture.[108] Tensile rather than compressive films are a necessity if lateral dimensions of clamped structures are not to be restricted by compressive buckling. Conducting regions are formed in this case by ion implantation. Fine-grained and large-grained poly-Si are compared in more detail further below.

- Phosphorous,[120-122] boron,[121,123,124] arsenic,[121] and carbon[125] doping have all been shown to affect the state of residual stress in poly-Si films. In the case of single crystal Si, to compensate for strain induced by dopants, one can implant with atoms with the opposite atomic radius vs. silicon. Similar approaches would most likely be effective for poly-Si as well.

- Tang et al.[65,126] developed a technique that sandwiches a poly-Si structural layer between a top and bottom layer of PSG and lets the high temperature anneal drive in the phosphorous symmetrically, producing low stress poly-Si with a negligible stress gradient.

- Another stress reduction method is to vary the materials composition, something readily done in CVD processes. An example of this method is the Si enrichment of $Si_3N_4$ which reduces the tensile stress.[46,127]

- During plasma-assisted film deposition processes, one can influence stress dramatically. In a physical deposition process such as sputtering, stress control involves varying gas pressure and substrate bias. In plasma-enhanced chemical vapor deposition (PECVD) the RF power, through increased ion-bombardment, influences stress. In this way, the stress in a thin film starts out tensile, decreases as the power increases, and finally becomes compressive with further RF power increase. PECVD equipment manufacturers also are working to build stress control capabilities into new equipment by controlling plasma frequency (see also Chapter 3). In CVD, stress control involves all types of temperature treatment programs.

- A clever mechanical design might facilitate structural stress relief.[128] By folding the flexures in the lateral resonator in Figure 5.18B, and by the overall structural symmetry, the relaxation of residual polysilicon stress is possible without structural distortion. By folding the flexures, the resonant frequency, $f_0$, becomes independent of $\sigma_r$ (see Equation 5.16). The springs in the resonator structure provide freedom of travel along the direction of the comb-finger motions (x) while restraining the structure from moving sideways (y), thus preventing the comb fingers from shorting out the drive electrodes. In this design, the spring constant along the y direction must be higher than along the x direction, i.e., $k_y \gg k_x$. The suspension should allow for the relief of the built-in stress of the polysilicon film and the axial stress induced by large vibrational amplitudes. The folded-beam suspensions meet both criteria. They enable large deflections in the x direction (perpendicular to the length of the beams) while providing stiffness in the y direction (along the length of the beams). Furthermore, the only anchor points (see Figure 5.18B) for the whole structure reside near the center, thus allowing the parallel beams to expand or contract in the y direction, relieving most of the built-in and induced stress.[65] Tang also modeled and built spiral and serpentine springs supporting torsional resonant plates. An advantage of the torsional resonant structures is that they are anchored only at the center, enabling radial relaxation of the built-in stress in the polysilicon film.[65] For some applications, the design approach with folded flexures is an attractive way to eliminate residual stress. However, a penalty for using flexures is increased susceptibility to out-of-plane warpage from residual stress gradients through the thickness of the polysilicon microstructure.[5]

- Corrugated structural members, invented by Jerman for bulk micromachined sensors,[129] also reduce stress effectively. In the case of a single crystal Si membrane, stress may be reduced by a factor of 1000 to 10,000.[130] One of the applications of such corrugated structures is the decoupling of a mechanical sensor of its encapsulation, by reducing the influence of temperature changes and packaging stress.[130,131] Thermal stress alone can be reduced by a factor of 120.[132] Besides stress release, corrugated structures enable much larger deflections than do similar planar structures. This type of structural stress release was studied in some detail for single crystal Si[133] (see also Chapter 4), polyimide,[134] and LPCVD silicon nitride membranes;[135] but the quantitative influence of corrugated poly-Si structures still requires investigation. Figure 5.19 illustrates a fabrication sequence for a polyimide corrugated structure. The sacrificial Al in step 4 may be etched away by a mixture of phosphoric acidic: acetic acid: nitric acid (PAN, see Table 5.5).

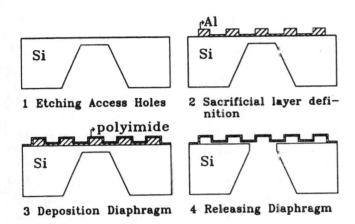

1 Etching Access Holes

2 Sacrificial layer defi-
nition

3 Deposition Diaphragm

4 Releasing Diaphragm

FIGURE 5.19  Schematic view of the fabrication process of a polyimide corrugated diaphragm. (From van Mullem, C.J. et al., Large Deflection Performance of Surface Micromachined Corrugated Diaphragms, presented at Transducers '91, San Francisco, CA, 1991. With permission.)

## Dimensional Uncertainties

The often-expressed concerns about run-to-run variability in material properties of polysilicon or other surface micromachined materials are somewhat misplaced, Howe points out.[72] He contrasts the relatively large dimensional uncertainties inherent to any lithography technique with poly-Si quality factors of up to 100,000 and long-term (>3 years) resonator frequency variation of less than 0.02 Hz. We follow his calculations here to prove the relative importance of the dimensional uncertainties. The shuttle mass M of a resonator as shown in Figure 5.18 is proportional to the thickness (t) of the polysilicon film, and neglecting the residual stress term, Equation 5.16 reduces to:

$$f_o \propto \left(\frac{W}{L}\right)^{\frac{3}{2}} \qquad 5.17$$

In case the residual stress term dominates in Equation 5.16, the resonant frequency is expressed as:

$$f_o \propto \left(\frac{W}{L}\right)^{\frac{1}{2}} \qquad 5.18$$

The width-to-length ratio is affected by systematic and random variations in the masking and etching of the microstructural polysilicon. For 2-$\mu$m thick structural polysilicon, patterned by a wafer stepper and etched with a reactive-ion etcher, a reasonable estimate for the variation in linear dimension of etched features, $\Delta$ is about 0.2 $\mu$m (10% relative tolerance).

From Equation 5.17 the variation $\Delta$ in lateral dimensions will result in an uncertainty $\delta f_0$ in the lateral frequency of:

$$\frac{\delta f_o}{f_o} \approx \frac{3}{2}\left(\frac{\Delta}{W}\right) \qquad 5.19$$

for a case where the residual stress can be ignored. With a nominal flexure width of W = 2 $\mu$m, the resulting uncertainty

in resonant frequency is 15%. For the stress-dominated case, Equation 5.18 indicates that the uncertainty is

$$\frac{\delta f_o}{f_o} \approx \frac{1}{2}\left(\frac{\Delta}{W}\right) \qquad 5.20$$

The same 2-$\mu$m wide flexure would then lead to a 5% uncertainty in resonant frequency.

Interestingly, the stress-free case exhibits the most significant variation in the resonant frequency. In either case, resonant frequencies must be set by some postfabrication frequency trimming or other adjustment.

In Chapter 7 we draw further attention to the increasing loss of relative manufacturing tolerance with decreasing structures size (see Figure 7.1).

## Sealing Processes in Surface Micromachining

Sealing cavities to hermetically enclose sensor structures is a significant attribute of surface micromachining. Sealing cavities often embodies an integral part of the overall fabrication process and presents a desirable chip level, batch packaging technique. The resulting surface packages (microshells) are much smaller than typical bulk micromachined ones (see Chapter 8 on packaging).

## IC Compatibility

Putting detection and signal conditioning circuits right next to the sensing element enhances the performance of the sensing system, especially when dealing with high impedance sensors. A key benefit of surface micromachining, besides small device size and single-sided wafer processing, is its compatibility with CMOS processing. IC compatibility implies simplicity and economy of manufacturing. In the examples at the end of this chapter we will discuss how Analog Devices used a mature 4-$\mu$m BICMOS process to integrate electronics with a surface micromachined accelerometer.

To develop an appreciation of integration issues involved in combining a CMOS line with surface micromachining, we highlight Yun's[52] comparison of CMOS circuitry and surface micromachining processes in Table 5.6A. Surface micromachining processes are similar to IC processes in several aspects. Both processes use similar materials, lithography, and etching techniques. CMOS processes involve at least 10 lithography steps where lateral small feature size plays an important role. Some processing steps, such as gate and contact patterning, are critical to the functionality and performance of the CMOS circuits. Furthermore, each processing step is strongly correlated with other steps. Change in any one of the processing steps will lead to modifications in a number of other steps in the process. In contrast, surface micromachining is relatively simple. It usually consists of two to six masks, and the feature sizes are much larger. The critical processing steps, such as structural poly-Si, often are self-aligned which eliminates lithographic alignment. The CMOS process is mature, quite generic, and fine-tuned,

**TABLE 5.6** Surface Micromachining and CMOS

A. Comparison of CMOS and Surface Micromachining

| | CMOS | Surface Micromachining |
|---|---|---|
| Common features | Silicon-based processes; same materials, same etching principles | |
| Process flow | Standard | Application specific |
| Vertical dimension | ~1 μm | ~1–5 μm |
| Lateral dimension | <1 μm | 2–10 μm |
| Complexity (# masks) | >10 | 2–6 |

B. Critical Process Temperatures for Microstructures

| | Temperature (°C) | Material |
|---|---|---|
| LPCVD deposition | 450 | Low temperature oxide (LTO)/PSG |
| LPCVD deposition | 610 | Low stress poly-Si |
| LPCVD deposition | 650 | Doped poly-Si |
| LPCVD deposition | 800 | Nitride |
| Annealing | 950 | PSG densification |
| Annealing | 1050 | Poly-Si stress annealing |

(After Yun, W., A Surface Micromachined Accelerometer with Integrated CMOS Detection Circuitry, Ph.D. Thesis, U.C. Berkeley, 1992.)

while surface micromachining strongly depends on the application and still needs maturing.

Table 5.6B presents the critical temperatures associated with the LPCVD deposition of a variety of frequently used materials in surface micromachining. Polysilicon is used for structural layers and thermal $SiO_2$, LPCVD $SiO_2$, and PSG are used as sacrificial layers; silicon nitride is used for passivation. The highest temperature process in Table 5.6B is 1050°C and is associated with the annealing step to release stress in the polysilicon layers. Doped polysilicon films deposited by LPCVD under conventional IC conditions usually are in a state of compression that can cause mechanically constrained structures such as bridges and diaphragms to buckle. The annealing step above about 1000°C promotes crystallite growth and reduces the strain. If one wants to build polysilicon microstructures after the CMOS active electronics have been implemented (a so-called post-CMOS procedure), one has to avoid temperatures above 950°C, as junction migration will take place at those temperatures. This is especially true with devices incorporating shallow junctions where migration might be a problem at temperatures as low as 800°C. The degradation of the aluminum metallization presents yet a bigger problem. Aluminum typically is used as the interconnect material in the conventional CMOS process. At temperatures of 400 to 450°C, the aluminum metallization will start suffering. Anneal temperatures (densification of the PSG and stress anneal of the poly-Si) only account for some of the concerns; in general, several compatibility issues must be considered: (1) deposition and anneal temperatures, (2) passivation during micromachining etching steps, and (3) surface topography. Yun[62] compared three possible approaches to build integrated microdynamic systems: pre-, mixed, and post-

CMOS microstructural processes as shown in Figure 5.20. He concluded that building up the microstructures after implementation of the active electronics offers the best results.

In a post-CMOS process, the electronic circuitry is passivated to protect it from the subsequent micromachining processes. The standard IC processing may be performed at a regular IC foundry, while the surface micromachining occurs as an add-on in a specialized sensor fabrication facility. LPCVD silicon nitride (deposited at 800°C, see Table 5.6B) is stable in HF solutions and is the preferred passivation layer for the IC during the long release etching step. PECVD nitride can be deposited at around 320°C, but it displays relatively poor step coverage, while pin holes in the film allow HF to diffuse through and react with the oxide underneath. LPCVD nitride is conformably deposited. While it shows fewer pin holes, circuitry needs to be able to survive the 800°C deposition temperature.

Aluminum metallization must be replaced by another interconnect scheme in order to raise the post-CMOS temperature ceiling higher than 450°C. Tungsten, which is refractory, shows low resistivity and has a thermal expansion coefficient matching that of Si, is an obvious choice. One problem with tungsten metallization is that tungsten reacts with silicon at about 600°C to form $WSi_2$, implying the need for a diffusion barrier. The process sequence for the tungsten metallization developed at Berkeley is shown in Figure 5.21. A diffusion barrier consisting of $TiSi_2$ and TiN is used. The TiN film forms during a 30-sec sintering step to 600°C in $N_2$. Rapid thermal annealing with its reduced time at high temperatures (10 sec to 2 min) and high ramp rates (~150°C/sec) allows very precise process control as well as a dramatic reduction of thermal budgets, reducing duress for the active on-chip electronics. Titanium silicide is formed at the interface of titanium and silicon while titanium nitride forms simultaneously at the exposed surface of the titanium film. The $TiSi_2$/TiN forms a good diffusion barrier against the formation of $WSi_2$ and at the same time provides an adhesion and contact layer for the W metallization.

To avoid the junction migration in a post-CMOS process, rapid thermal annealing is used for both the PSG densification and polysilicon stress anneal: 950°C for 30 sec for the PSG densification and 1000°C for 60 sec for the stress anneal of the poly-Si. Alternatively, one could consider the use of fine-grained polysilicon which can yield a controlled tensile strain with low-temperature annealing.[136]

Despite some advantages the post-CMOS process with tungsten metallization is not the preferred implementation. Hillock formation in the W lines during annealing and high contact resistance remain problems.[72] Moreover, the finely tuned CMOS fabrication sequence may also be affected by the heavily doped structural and sacrificial layers.

The mixed CMOS/micromachining approach implements a processing sequence which puts the processes in a sequence to minimize performance degradation for both electronic and mechanical components. According to Yun this requires significant modifications to the CMOS fabrication sequence. Nevertheless, Analog Devices relied on such an interleaved process sequence to build the first commercially available integrated

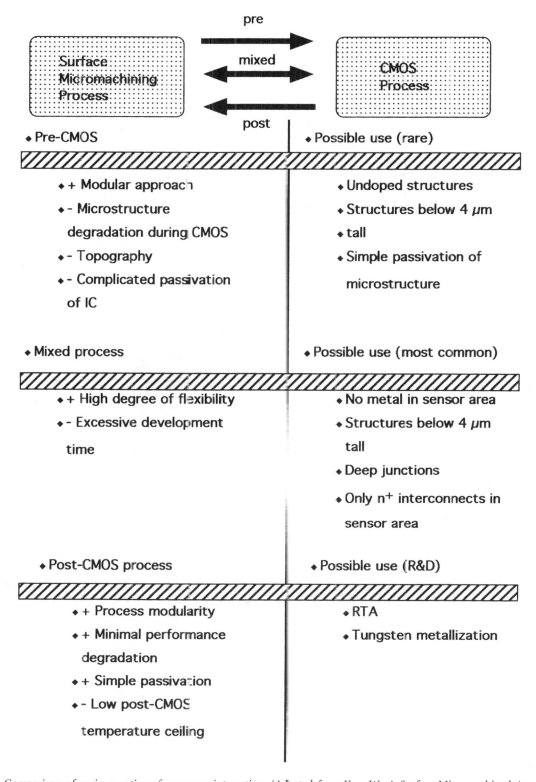

**FIGURE 5.20** Comparison of various options for process integration. (Adapted from Yun, W., A Surface Micromachined Accelerometer with Integrated CMOS Detection Circuitry, Ph.D. thesis, University of California, Berkeley, 1992. With permission.)

micro-accelerometer (see Example 1). The modifications required on a standard BICMOS line were minimal to facilitate integration of the IC and to surface micromachine the thickness of deposited microstructural films the line was limited to 1 to 4 μm. Relatively deep junctions permitted thermal processing for the sensor poly-Si anneal and interconnections to the sensor

were made only via n+ underpasses. No metallization is present in the sensor area. This industrial solution remains truer to the traditional IC process experience than the post-CMOS procedure. Howe recently detailed another example of such a mixed process.[72] In Howe's scenario, the micromachining sequence is inserted after the completion of the electronic structures, but

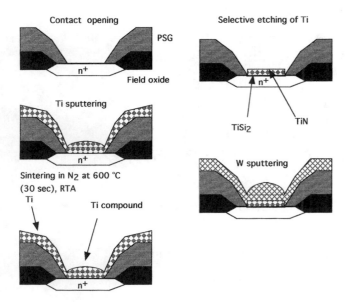

◆ Self-aligned process

◆ No peeling of W up to 1050 °C
  in Ar

◆ Thin layer of nitride forms
  during RTA in $N_2$

**FIGURE 5.21** Tungsten metallization process in a modified CMOS process. (Adapted from Yun, W., A Surface Micromachined Accelerometer with Integrated CMOS Detection Circuitry, Ph.D. thesis, University of California, Berkeley, 1992. With permission.)

prior to contact etching or aluminum metallization. By limiting the polysilicon annealing to 7 min at 900°C, only minor dopant redistribution is expected. Contact and metallization lithography and etching become more complex now due to the severe topography of the poly-Si microstructural elements.

The pre-CMOS approach is to fabricate microstructural elements before any CMOS process steps. At first glance this seems like an attractive approach as no major modifications would be needed for process integration. However, due to the vertical dimensions of microstructures, step coverage is a problem for the interconnection between the sensor and the circuitry (the latest approach introduced by Howe faces the same dilemma). Passivation of the microstructure during the CMOS process can also become problematic. Furthermore, the fine-tuned CMOS fabrication sequence, such as gate oxidation, can be affected by the heavily doped structural layers. Consequently, this approach is only used for some special applications.[62]

A unique pre-CMOS process was developed at Sandia National Labs.[137] In this approach, micromechanical devices are fabricated in a trench etched in a Si epilayer. After the mechanical components are complete, the trench is filled with oxide, planarized using chemical-mechanical polishing (CMP) (see Chapter 3), and sealed with a nitride membrane. The flat wafer with the embedded micromechanical devices is then processed by means of conventional CMOS processing. Additional steps are added at the end of the CMOS process in order to expose and release the embedded micromechanical devices.

1. Deposit CVD silicon nitride on silicon wafer.

2. Pattern and plasma-etch silicon nitride.

3. Deposit CVD polysilicon.

4. Deposit CVD silicon nitride.

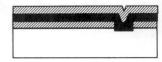

5. Pattern and plasma-etch nitride and polysilicon.

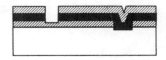

**FIGURE 5.22** Masking and process sequence preparing a wafer for laterally grown porous polysilicon. (From Field, L., *Low-Stress Nitrides for Use in Electronic Devices* (University of California, Berkeley, 1987) pp. 42, 43. With permission.)

The SPIE "Smart Structures and Materials 1996" meeting in San Diego, CA, had two complete sessions dedicated to the crucial issue of integrating electronics with polysilicon surface micromachining.[138]

## Poly-Si Surface Micromachining Modifications

### Porous Poly-Si

In Chapter 4 we discussed the transformation of single crystal Si into a porous material with porosity and pore sizes determined by the current density, type and concentration of the dopant, and the hydrofluoric acid concentration. A transition from pore formation to electropolishing is reached by raising the current density and/or by lowering the hydrofluoric acid concentration.[139,140] Porous silicon can also be formed under similar conditions from LPCVD poly-Si.[141] In this case pores roughly follow the grain boundaries of the polysilicon. Figure 5.22 illustrates the masking and process sequence to prepare a wafer to make thin layers of porous Si between two insulating layers of low stress silicon nitride.[142] The wafer, after the process steps outlined in Figure 5.22, is put in a Teflon test fixture, protecting the back from HF attack. An electrical contact is established on the back of the wafer and a potential is applied with respect to a Pt-wire counterelectrode immersed in the same HF solution. Electrolytes consisting of 5 to 49% HF (wt) and current densities from 0.1 to 50 A cm$^{-2}$ are used. The advance of a pore-etching front, growing parallel to the wafer surface, may be monitored using a line-width measurement tool. The highest observed rate of porous-silicon formation is 15 μm min$^{-1}$ (in 25%, wt HF). In the electro-polishing regime, at the highest currents, the etch rate is diffusion limited but the reaction is controlled by surface reaction kinetics in the porous Si growth regime.

By changing the conditions from pore formation to electropolishing and back to porous Si, an enclosed chamber may be formed with porous poly-Si walls (plugs) and 'floor and ceiling' silicon nitride layers. Sealing of the cavities by clogging the microporous poly-Si was attempted by room temperature oxidation in air and in a $H_2O_2$ solution. Leakage through the porous plug persisted after those room temperature oxidation treatments, but with a Ag deposition from a 400 m$M$ AgNO$_3$ solution and subsequent atmospheric tarnishing (48 h, Ag$_2$S) the chambers appeared to be sealed, as determined from the lack of penetration by methanol. This technology might open up possibilities for filling cavities with liquids and gases under low-temperature conditions. The chamber provided with a porous plug might also make a suitable on-chip electrochemical reference electrolyte reservoir.

Hydrofluoric acid can penetrate thin layers of poly-Si either at foreign particle inclusion sites, or at other critical film defects such as grain boundaries (see above). This way the HF can etch underlying oxide layers, creating, for example, circular regions of free-standing poly-Si, so-called 'blisters'. The poly-Si permeability associated with blistering of poly-Si films has been applied successfully by Judy et al.[143,144] to produce thin-shelled hollow beam electrostatic resonators from thin poly-Si films deposited onto PSG. The possible advantage of using these hollowed structures is to obtain a yet higher resonator quality factor Q. The devices were made in such a way that the 0.3-μm thick undoped poly-Si completely encased a PSG core. After annealing, the structures were placed in HF which penetrated the poly-Si shell and dissolved away the PSG, eliminating the need for etch windows. It was not possible to discern the actual pathways through the poly-Si using TEM.

Lebouitz et al.[145] apply permeable polysilicon etch-access windows to increase the speed of creating microshells for packaging surface micromachined components. After etching the PSG through the many permeable Si windows, the shell is sealed with 0.8 μm of low stress LPCVD nitride.

## Hinged Polysilicon

One way to achieve high vertical structures with surface micromachining is building large flat structures horizontally and then rotating them on a hinge to an upright position. Pister et al.[146–150] developed the poly-Si hinges shown in Figure 5.23A; on these hinges, long structural poly-Si features (1 mm and beyond) can be rotated out of the plane of a substrate. To make the hinged structures, a 2-μm thick PSG layer (PSG-1) is deposited on the Si substrate as the sacrificial material, followed by the deposition of the first polysilicon layer (2-μm thick poly-1). This structural layer of polysilicon is patterned by photolithography and dry etching to form the desired structural elements, including hinge pins to rotate them. Following the deposition and patterning of poly-1, another layer of sacrificial material (PSG-2) of 0.5-μm thickness is deposited. Contacts are made through both PSG layers to the Si substrate, and a second layer of polysilicon is deposited and patterned (poly-2), forming a staple to hold the first polysilicon layer hinge to the surface. The first and second layers of poly are separated everywhere by PSG-2 in order for the first polysilicon layer to freely rotate off the wafer surface when the PSG is removed in a sacrificial etch. After the sacrificial etch, the structures are rotated in their respective positions. This is accomplished in an electrical probe station by skillfully manipulating the movable parts with the probe needles. Once the components are in position, high friction in the hinges tends to keep them in the same position. To obtain more precise and stable control of position, additional hinges and supports are incorporated. To provide electrical contact to the vertical poly-Si structures one can rely on the mechanical contact in the hinges, or poly-Si beams (cables) can be attached from the vertical structure to the substrate.

Pister's research team made a wide variety of hinged microstructures, including hot wire anemometers, a box dynamometer to measure forces exerted by embryonic tissue, a parallel plate gripper,[146] a micro-windmill,[147] and a micro-optical bench for free-space integrated optics,[148] and a standard CMOS single piezoresistive sensor to quantify rat single heart cell contractile forces.[50] One example from this group's efforts is illustrated in Figure 5.23B, showing an SEM photograph of an edge-emitting laser diode shining light onto a collimating micro-Fresnel lens.[148] The micro-Fresnel lens in the SEM photo is surface micromachined in the plane and erected on a polysilicon hinge. The lens has a diameter of 280 μm. Alignment plates at the front and the back sides of the laser are used for height adjustment of the laser spot so that the emitting spot falls exactly onto the optical axis of the micro-Fresnel lens. After assembly, the laser is electrically contacted by silver epoxy. Although this hardly outlines standard IC manufacturing practices, excellent collimating ability for the Fresnel lenses has been achieved. The eventual goal of this work is a micro-optical bench (MOB) in which microlenses, mirrors, gratings, and other optical components are pre-aligned in the mask layout stage using computer-aided design. Additional fine adjustment would be achieved by on-chip micro-actuators and micropositioners such as rotational and translational stages.

Today, erecting these poly-Si structures with the probes of an electrical probe station or, occasionally, assembly by chance in the HF etch or deionized water rinse represent too complicated or too unreliable postrelease assembly methods for commercial acceptance.

Friction in poly-Si joints, as made by Pister, is high because friction is proportional to the surface area (s$^2$) and becomes dominant over inertial forces (s$^3$) in the microdomain (see Chapter 9). Such joints are not suitable for microrobotic applications. Although attempts have been made to incorporate poly-Si hinges in such applications,[149] plastically deformable hinges make more sense for microrobot machinery involving rotation of rigid components. Noting that the external skeleton of insects incorporates hard cuticles connected by elastic hinges, Suzuki et al.[77] fabricated rigid poly-Si plates (E = 140 GPa) connected by elastic polyimide hinges (E = 3 GPa) as shown in Figure 5.24 (see also section below on Polimide Surface Structures). Holes in the poly-Si plates shorten the PSG etch time compared to plates without holes. The plates without holes remain attached to the substrate while the ones with holes are completely freed. Using electrostatic actuators, a structure as shown in Figure 5.24 can be made to flap like the wing of a butterfly. By applying an AC voltage of 10 kHz, resonant vibration of such a flapping wing was observed.[77]

More recently, Hoffman et al.[78] demonstrated aluminum plastically deformable hinges on oxide movable thin plates. Oxide plates and Al hinges were etched free from a Si substrate by using $XeF_2$, a vapor phase etchant exhibiting excellent selectivity of Si over Al and oxide. According to the authors, this process, due to its excellent CMOS compatibility, might open the way to designing and fabricating sophisticated integrated CMOS-based sensors with rapid turnaround time (see also Chapter 2).

## Thick Polysilicon

Applying classical LPCVD to obtain poly-Si deposition is a slow process. For example, a layer of 10 μm typically requires a deposition time of 10 hr. Consequently, most micromachined structures are based on layer thicknesses in the 2- to 5-μm range. Basing their process on dichlorosilane ($SiH_2Cl_2$) chemistry, Lange et al.[9] developed a CVD process in a vertical epitaxy batch reactor with deposition rates as high as 0.55 μm/min at 1000°C. The process yields acceptable deposition times for thicknesses in the 10-μm range (20 min). The highly columnar poly-Si films are deposited on sacrificial $SiO_2$ layers and exhibit low internal tensile stress making them suitable for surface micromachining. The surface roughness comprises about 3% of the thickness, which might preclude some applications.

Kahn et al.[151] made mechanical property test structures from thick undoped and *in situ* B-doped polysilicon films. The elastic modulus of the B-doped polysilicon films was determined as 150 ± 30 GPa. The residual stress of as-deposited undoped thick polysilicon was determined as 200 ± 10 MPa.

## Milli-Scale Molded Polysilicon Structures

The assembly of tall three-dimensional features in the described hinged polysilicon approach is complicated by the manual assembly of the fabricated microparts resembling building a miniature boat in a bottle. Keller at University of California, Berkeley, came up with an elegant alternative for building tall, high aspect ratio microstructures in a process that does not require postrelease assembly steps.[8] The technique involves deep dry etching of trenches in a Si substrate, deposition of sacrificial and structural materials in those trenches, and demolding of the deposited structural materials by etching away the sacrificial materials. CVD processes can typically only deposit thin films (~ 1 to 2 μm) on flat surfaces. If, however, these surfaces are the opposing faces of deep narrow trenches, the growing films will merge to form solid beams. In this fashion, high aspect ratio structures that would normally be associated with LIGA now also can be made of CVD polysilicon. The procedure is illustrated in Figure 5.25.[8,152] The first step is to etch deep trenches into a silicon wafer. The depth of the trenches equals the height of the desired beams and is limited to about 100 μm with aspect

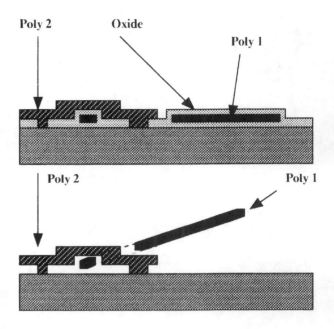

**Poly 2 staple holding poly 1 plate**

A

**FIGURE 5.23** Microfabricated hinges. (A) Cross-section, side view and top view of a single-hinged plate before and after the sacrificial etch. (B) Schematic (top) and SEM micrograph of the self-aligned hybrid integration of an edge-emitting laser with a micro-Fresnel lens. (From Lin, L.Y. et al., Micromachined Integrated Optics for Free-Space Interconnections, presented at MEMS '96, Amsterdam, 1995. With permission.)

ratios of about 10 (say a 10-μm diameter hole with a depth of 100 μm). For trench etching, Keller uses a $Cl_2$ plasma etch with the following approximate etching conditions: flow rates of 200 sccm for He and 180 sccm for $Cl_2$, a working pressure of 425 mTorr, a power setting of 400 W, and an electrode gap of 0.8 cm. The etch rate for Si in this mode equals 1 μm/min. Thermal oxide and CVD oxide act as masks with 1 μm of oxide needed for each 20 μm of etch depth. Before the $Cl_2$ etch a short 7-sec $SF_6$ pre-etch removes any remaining native oxide in the mask

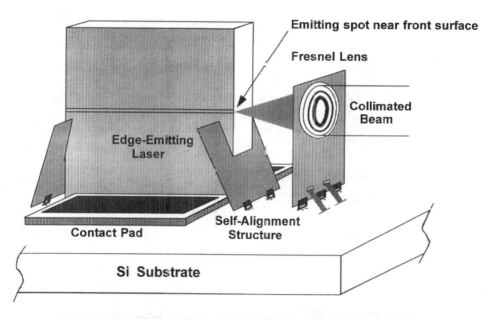

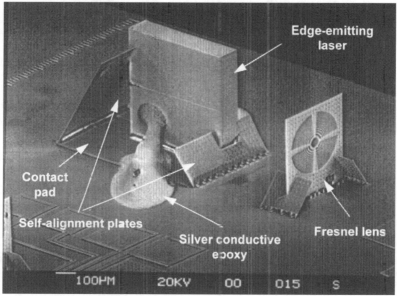

B

FIGURE 5.23  (Continued)

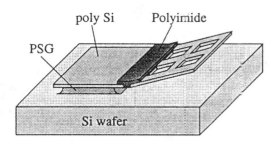

FIGURE 5.24  Flexible polyimide hinge and poly-Si plate (butterfly wing). (From Suzuki, K. et al., *J. Microelectromech. Sys.*, 3, 4–9, 1994. With permission.)

openings. During the chlorine etch a white sidewall passivating layer must be controlled to maintain perfect vertical sidewalls. After every 30 min of plasma etching, the wafers are submerged in a silicon isotropic etch long enough to remove the residue.[153] Beyond 100 μm, severe undercutting occurs and the trench cross-section becomes sufficiently ellipsoidal to prevent molded parts from being pulled out. Advances in dry cryogenic etching are continually improving attainable etch depths, trench profiles, and minimum trench diameter. We can expect continuous improvements in the tolerances of this novel technique. After plasma etching, an additional 1 μm of silicon is removed by an isotropic wet etch to obtain a smoother trench wall surface.

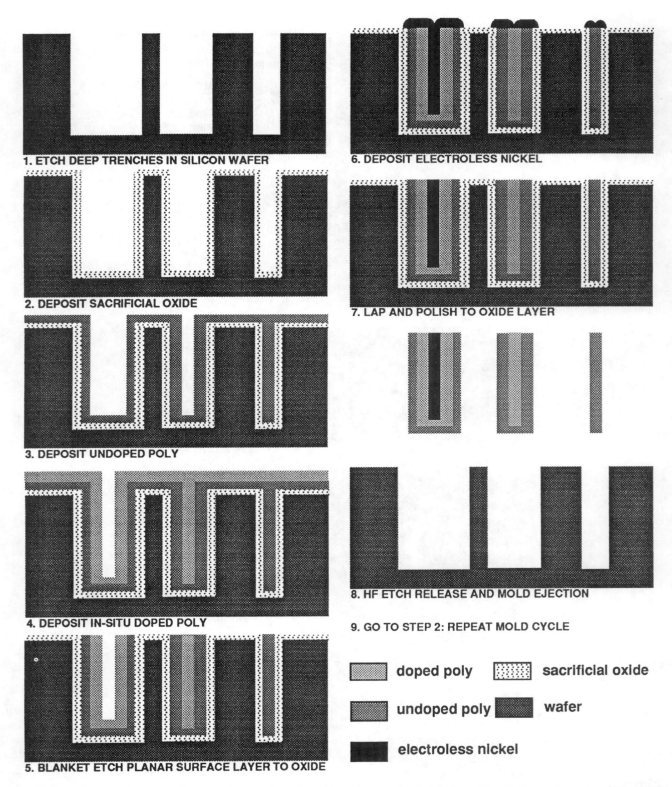

1. ETCH DEEP TRENCHES IN SILICON WAFER

2. DEPOSIT SACRIFICIAL OXIDE

3. DEPOSIT UNDOPED POLY

4. DEPOSIT IN-SITU DOPED POLY

5. BLANKET ETCH PLANAR SURFACE LAYER TO OXIDE

6. DEPOSIT ELECTROLESS NICKEL

7. LAP AND POLISH TO OXIDE LAYER

8. HF ETCH RELEASE AND MOLD EJECTION

9. GO TO STEP 2: REPEAT MOLD CYCLE

doped poly          sacrificial oxide

undoped poly        wafer

electroless nickel

**FIGURE 5.25**   Schematic illustration of HEXSIL process. The mold wafer may be part of an infinite loop. (From Reference 8, courtesy of Mr. C. Keller.)

Alternatively, to smooth sidewalls and bottom of the trenches a thermal wet oxide is grown and etched away. The sacrificial oxide in step 2 is made by CVD phosphosilicate glass (PSG at 450°C, 140 Å/min), CVD low-temperature oxide (LTO at 450°C), or CVD polysilicon (580°C, 65 Å/min). The latter is completely converted to $SiO_2$ by wet thermal oxidation at 1100°C. The PSG needs an additional reflowing and densifying anneal at 1000°C in nitrogen for 1 hr. This results in an etching rate of the sacrificial layer of ~20 μm/min in 49% HF. The mold shown in Figure 5.25 displays three different trench widths and can be used to build integrated micromachines incorporating doped and undoped poly-Si parts as well as metal parts. The remaining volume of the narrowest trench after oxide deposition is filled completely with the first deposition of undoped polysilicon (poly 1) in step 3. The undoped poly will constitute the insulating regions in the micromachine. Undoped CVD polysilicon was formed in this case at 580°C, with a 100 SCCM silane flow rate, and a 300 mTorr reactor pressure, resulting in a deposition rate of 0.39 μm/hr. The deposited film under these conditions is amorphous or very fine-grained. Since the narrowest trenches are completely filled in by the first deposition, they cannot accept material from later depositions. The trenches of intermediate width are lined with the first material and then completely filled in by the second deposition. In the case illustrated the second deposition (step 4) consists of *in situ* doped poly-Si and forms the resistive region in the micromachine under construction. To prevent diffusion of P from the doped poly deposited on top of the narrow undoped beams, a blanket etch in step 5 is used to remove the doped surface layer prior to the anneal of the doped poly. The third deposition, in step 6 of the example case, consists of electroless nickel plating on poly-Si surfaces but not on oxides surfaces and results in the conducting parts of the micromachine. By depositing structural layers in order of increasing conductivity, as done here, regions of different conductivity can be separated by regions of narrow trenches containing only nonconducting mate-

rial. Lapping and polishing in step 7 with a 1-μm diamond abrasive in oil planarizes the top surface, readying it for HF etch release and mold ejection in step 8. Annealing of the polysilicon is required to relieve the stress before removing the parts from the wafer so they remain straight and flat. In step 8 the sacrificial oxide is dissolved in 49% HF. A surfactant such as Triton X100 is added to the etch solution to facilitate part ejection by reducing surface adhesion between the part and the mold. The parts are removed from the wafer, and the wafer may be returned to step 2 for another mold cycle. An example micromachine, resulting from the described process, is the thermally actuated tweezers shown in Figure 5.26. These HEXSIL tweezers measure 4 mm long, 2 mm wide, and 80 μm tall. The thermal expansion beam to actuate the tweezers consists of the *in situ* doped poly-Si; the insulating parts are made from the undoped poly-Si material. Ni-filled poly-Si beams are used for the current supply leads. It is possible to combine the HEXSIL process with classical poly-Si micromachining, as illustrated in Figure 5.27, where HEXSIL forms a stiffening rib for a membrane filter fashioned by surface micromachining of a surface poly-Si layer. The surface poly-Si is deposited after HEXSIL. A critical need in HEXSIL technology is controlled mold ejection. Keller et al.[153] have experimented with HEXSIL-produced bimorphs, making the structure spring up after release.

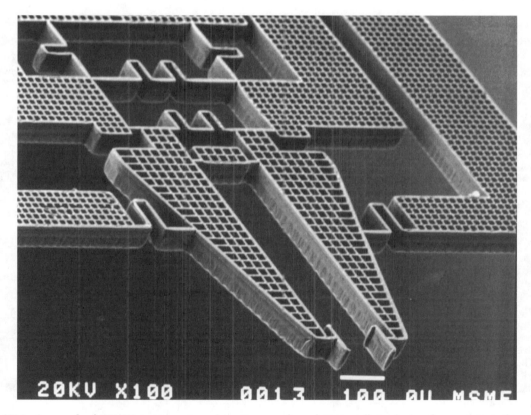

**FIGURE 5.26** SEM micrograph of HEXSIL tweezers: 4 mm long, 2 mm wide, and 80 μm tall. Lead wires for current supply are made from Ni-filled poly-Si beams; *in situ* phosphorus-doped polysilicon provides the resistor part for actuation. The width of the beam is 8 μm: 2 μm poly-Si, 4 μm Ni, and 2 μm poly-Si. (Courtesy of Mr. C. Keller.)

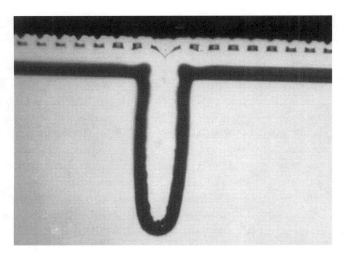

**FIGURE 5.27** SEM micrograph of surface micromachined membrane filter with a stiffening rib (50 μm high). Original magnification 1000×. (Courtesy of Mr. C. Keller.)

# Non-Poly-Si Surface Micromachining Modifications

## Silicon on Insulator Surface Micromachining

### Introduction

A major sensor use of silicon on insulator wafers is established in the production of high-temperature sensors. Compared to p-n junction isolation, which is limited to about 125°C, much higher temperature devices are possible based on the dielectric insulation of SOI. Recently, a wide variety of SOI surface micromachined structures have been explored, including pressure sensors, accelerometers, torsional micromirrors, light sources, optical choppers, etc.[6,10]

Three major techniques currently are applied to produce *SOI* wafers (see also under Epitaxy in Chapter 3): *SIMOX* (Separated by IMplanted OXygen), the Si fusion bonded (SFB) wafer

---

## *How SOI wafers are made*

(From Dunn, P.N., *Solid State Technol.*, October, 32–35 (1993). With permission.)

### How SOI wafers are made

Many different silicon-on-insulator materials have been developed over the years, but two are currently being used for IC production: SIMOX (Separated by IMplanted OXygen) and bonded wafers.

In the SIMOX process, a standard silicon wafer is implanted with oxygen ions, and then annealed at high temperatures; the oxygen and silicon combine to form a silicon oxide layer beneath the wafer surface. To minimize wafer damage, the oxygen is sometimes implanted in two or more passes, each followed by an anneal. The oxide layer's thickness and depth are controlled by varying the energy and dose of the implant and the anneal temperature. In some cases, a CVD process is used to deposit additional silicon on the top layer.

The bonded wafer process starts with an oxide layer of the desired thickness (typically 0.25 to 2 microns) being grown on a stan-

dard silicon wafer. That wafer is then bonded at high temperatures to another wafer, with the oxide sandwiched between. One of the wafers is then ground to a thickness of a few microns using a mechanical tool.

Because advanced devices require an even thinner layer, more silicon must be removed. In Hughes Danbury's AcuThin process, the wafer is etched with a confined plasma, between 3 and 30 mm wide, which is stepped across the wafer surface. A film thickness map is made

for each wafer, and used to compute the dwell time for the plasma etcher at each stop. The process can be repeated for additional precision; Hughes Danbury offers silicon thicknesses of as little as 1000 to 3000 angstroms, with total thickness variation of 200 angstroms. IBM has also developed an etch-back process for bonded wafers. ∎

**SIMOX wafer production**

Oxygen
⬇ ⬇ ⬇ ⬇ ⬇

Silicon

Silicon

SiO₂

Silicon

*Process flow*

Implant oxygen

↓

Anneal to form SiO₂

↓

Grow CVD EPI to required Si thickness (optional)

↓

Fabricate wafer

**Bonded wafer production**

Bulk Si wafer

Oxide layer

Wafer bonding

Mechanical grinding

Final thinning by etch

technique, and zone-melt recrystallized (ZMR) polysilicon. With SIMOX, standard Si wafers are implanted with oxygen ions and then annealed at high temperatures (1300°C). The oxygen and silicon combine to form a silicon oxide layer beneath the silicon surface. The oxide layer's thickness and depth are controlled by varying the energy and dose of the implant and the anneal temperature. In some cases, a CVD process deposits additional epitaxial silicon on the top silicon layer. Attempts have also been made to implant nitrogen in Si to create abrupt etch stops. At high enough energies the implanted nitrogen is buried 1/2 to 1 μm deep. At a high enough dose, the etching in that region stops. It is not necessary to implant the stoichiometric amount of nitrogen concentration; a dose lower by a factor of 2 to 3 suffices. After implantation, it is necessary to anneal the wafer because the implantation destroys the crystal structure at the surface of the wafer.

The bonded wafer process starts with an oxide layer grown (typically about 1 μm) on a standard Si wafer. That wafer is then bonded to another wafer, with the oxide sandwiched between. For the bonding no mechanical pressure or other forces are applied. The sandwich is annealed at 1100°C for 2 hr in a nitrogen ambient leading to a strong binding between the two wafers. One of the wafers is then ground to a thickness of a few microns using mechanical and CMP.

A third process for making SOI structures is to recrystallize polysilicon (e.g., with a laser, an electron-beam, or a narrow strip heater) deposited on an oxidized silicon wafer. This process is called zone melting recrystallization (ZMR). This technique is used primarily for local recrystallization and has not yet been explored much in micromachining applications.

The crystalline perfection of conventional silicon wafers in SFB and ZMR is completely maintained in the SOI layer as the wafers do not suffer from implant-induced defects. By using plasma etching, wafers with a top Si layer thickness of as little as 1000 to 3000 Å with total thickness variations of less than 200 Å can be made.[154] SOI layers of 2 μm thick are more standard.[155]

Kanda[156] reviews different types of SOI wafers in terms of their micromachining and IC applications. Working with SOI wafers offers several advantages over bulk Si wafers: fewer process steps are needed for feature isolation, parasitic capacitance is reduced, and power consumption is lowered. In the IC industry these wafers are used for high-speed CMOS ICs, smart power ICs, three-dimensional ICs, and radiation-hardened devices.[157] In micromachining, SOI wafers are employed to produce an etch stop in such mechanical devices as pressure and acceleration sensors and in high-temperature sensors, and ISFETs. Etched-back fusion-bonded Si wafers and SIMOX are already employed extensively to build micromachines. The two types of SOI wafers are commercially available. A tremendous amount of effort is spent in the IC industry on controlling the SOI thin Si layer thickness which will benefit any narrow tolerance micromachine. The silicon fusion bonded (SFB) method offers the more versatile MEMS approach due to the associated potential for thicker single crystal layers and the option of incorporating buried cavities, facilitating micromachine packaging. Sensors manufactured by means of SFB now are commercially available.[158] The SIMOX approach is less labor intensive and holds

better membrane thickness control. An important expansion of the SOI technique is selective epitaxy. The latter enables a wide range of new mechanical structures (see also Chapter 3) and enables novel etch-stop methods,[159] as well as electrical and/or thermal separation and independent optimization of active sensor and readout electronics.[160] In all cases SOI machining involves dry anisotropic etching to etch a pattern into the Si layer on top of the insulator. These structures then become free by etching the sacrificial buried SiO$_2$ insulator layer, which displays a thickness with very high reproducibility (400 ± 5 nm) and uniformity (<±5 nm), especially in the case of SIMOX. Etched free cantilevers and membranes consist of single crystalline silicon with thicknesses ranging from microns and submicrons (SIMOX) up to hundreds of microns (SFB). Below we review three implementations of SOI techniques that may be crucial for future MEMS development.

### Silicon Fusion Bonded Micromachining

Silicon fusion bonding enables the formation of thick single crystal layers with cavities built in. An example is shown in Figure 5.28.[10] The device pictured involves two 4-in. <100> wafers: a handle wafer and a wafer used for the SOI surface. The p-type (3 to 7 Ωcm) handle wafer is thermally oxidized at 1100°C to obtain a 1-μm thick oxide. Thermal oxidation enables thicker oxides than the ones formed in SIMOX by ion implantation and avoids the potential implantation damage in the working material. To make a buried cavity, the oxide is patterned and etched. To produce yet deeper cavities, the Si handle wafer may be etched as well (as in Figure 5.28). In the case shown, the top wafer consists of the same p-type substrate material as the handle wafer with a 2- to 30-μm thick n-type epitaxial layer. The epitaxial layer determines the thickness of the final mechanical material. The epitaxial layer is fusion bonded to the cavity side of the handle wafer (2 hr at 1100°C). The top wafer is then partially thinned by grinding and polishing (Figure 5.28A). An

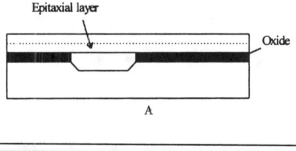

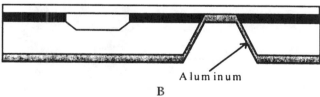

FIGURE 5.28 (A) A wafer sandwich after grind-and-polish step. (B) A wafer after electrochemical etch-back in KOH, buried oxide removal, and aluminum deposition. (From Noworolski, J.M. et al., Fabrication of SOI Wafers with Buried Cavities Using Silicon Fusion Bending and Electrochemical Etchback, presented at Transducers '95, Stockholm, Sweden, 1995. With permission.)

insulator is deposited and patterned on the back side of the handle wafer to etch access holes to the insulator. After the insulator at the bottom of the etch hole is removed by a buffered oxide etch (BOE), aluminum is sputtered and sintered to make contact to the n-type epilayer for the electrochemical etch back of the remaining p-type material (Figure 5.28B). The final single crystal silicon thickness is uniform to within ±0.05 μm (std. dev.) and does not require a costly, high accuracy polish step.

Draper Laboratory is using SOI processes in the development of inertial sensors, gyros, and accelerometers as an alternative to their current devices fabricated by the dissolved wafer process (see Example 1, Chapter 4). The main advantage is that the former consists of an all Si process rather than a Si/Pyrex sandwich.[161]

## SIMOX Surface Micromachining

Both capacitive and piezoresistive pressure sensors were microfabricated from SIMOX wafers.[6] Figure 5.29 illustrates the process sequence by Diem et al.[6] for fabricating an absolute capacitive pressure sensor. The 0.2-μm silicon surface layer of the SIMOX wafer is thickened with doped epi-Si to 4 μm. An access hole is RIE etched in the Si layer, and vacuum cavity and electrode gap are obtained by etching the $SiO_2$ buried layer. Since the buried thick oxide layer exhibits a very high reproducibility and homogeneity over the whole wafer (0.4 μm ± 5 nm), the resulting vacuum cavity and electrode gap after etching also are very well controlled. The small gap results in relatively high capacitance values between the free membrane and bulk

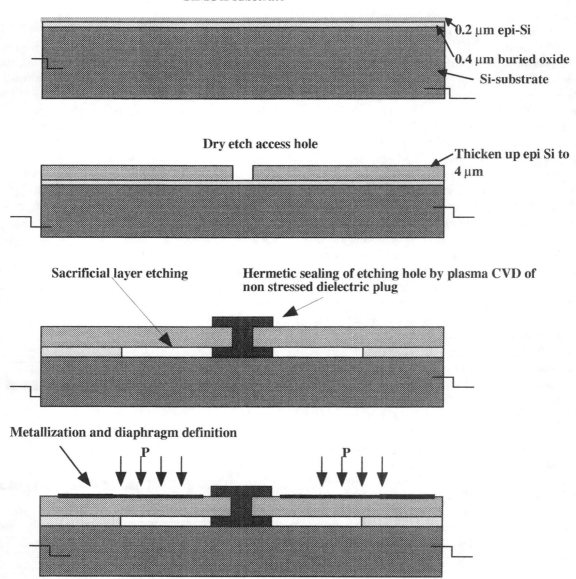

**FIGURE 5.29** Process sequence of a SIMOX absolute capacitive pressure sensor by Diem et al. (From Diem, B. et al., SOI (SIMOX) as a Substrate for Surface Micromachining of Single Crystalline Silicon Sensors and Actuators, presented at Seventh Int. Conf. on Solid State Sensors and Actuators, Yokohama, Japan, 1993. With permission.)

substrate (20 pf/mm²). Diaphragm diameter, controlled by the SiO₂ etching is up to several hundreds micrometers (±2 μm). The etching hole is hermetically sealed under vacuum by plasma CVD deposition of nonstressed dielectric layer plugs.

With the above scheme Diem et al. realized an absolute pressure sensor with a size of less than 1.5 mm². The temperature dependence of a capacitive sensor is mainly due to the temperature coefficient of the offset capacitance. Therefore, a temperature compensation is needed for high accuracy sensors. A drastic reduction of the temperature dependence is obtained by a differential measurement, especially if the reference capacitor resembles the sensing capacitor. A reference capacitor was designed with the membrane blocked by several plugs for pressure insensitivity. The localization and the number of plugs were modeled by finite element analysis (FEA) (ANSYS software was used) to get a deformation lower than 1% of the active sensor's deformation. Even without temperature calibration the high output of the differential signal resulted in an overall output error better than ±2% over the whole temperature range (−40°C to +125°C) compared to 10% for nondifferential measurements. The temperature coefficient of the sensitivity is about 100 ppm/°C which agrees with the theoretical variation of the Young modulus of silicon. A piezoresistive sensor could be achieved by implanted strain gauges in the membrane. Although SIMOX wafers are more expensive than regular wafers, they come with several process steps embedded and they make packaging easier.

### Selective Epitaxy Surface Micromachining

In the discussion on epitaxy in Chapter 3 we drew attention to the potential of selective epitaxy for creating novel microstructures. The example in Figure 3.21 illustrates the selective deposition of epi-Si on a Si substrate through a SiO₂ window. The same figure also demonstrates the simultaneous deposition of poly-Si on SiO₂ and crystalline epi-Si on Si, creating the basis for a structure featuring an epi-Si anchor with poly-Si side arms.

Neudeck et al.[162,163] at Purdue and Gennissen et al.[159,160] at Twente proved that selective epitaxy can also be applied for automatic etch stop on buried oxide islands. Figure 5.30A demonstrates how epitaxial lateral overgrowth (ELO) can bury oxide islands. After removal of the native oxides from the seed windows, epi is grown for 20 min at 950°C and at 60 torr using a Si₂H₂Cl₂-HCl-H₂ gas system. The epi growth front moves parallel to the wafer surface while growing in the lateral direction, leaving a smooth planar surface. During epi growth the HCl prevents poly nucleation on the nonsilicon areas. The epi quality is strongly dependent on the orientation of the seed holes in the oxide. Seed holes oriented in the <100> direction lead to the best epi material and surface quality. Selective epi's other big problem for fabrication remains sidewall defects.[154] The buried oxide islands stop the KOH etch of the substrate, enabling formation of beams and membranes as shown in Figure 5.30B. This technique might form the basis of many high performance microstructures. The Purdue and Twente groups also work on confined selective epitaxial growth (CSEG), a process pioneered by Neudeck et al.[163] In this process a micromachined cavity is formed above a silicon substrate with a seed contact window to

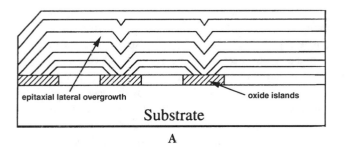

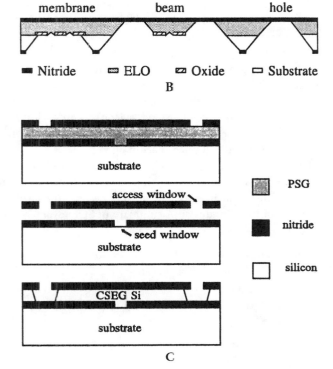

**FIGURE 5.30** Micromachining with epi-Si. (A) Lateral overgrowth process of epi-Si (ELO, epitaxial lateral overgrowth) out of <100>-oriented holes in an oxide mask. (B) KOH etch stop on buried oxide islands or front side nitride. (C) Principle of confined selective epitaxial growth. (From papers presented at Transducers '95, Stockholm, Sweden, 1995. With permission. )

the silicon substrate and access windows for epi-Si (Figure 5.30C).[160] Low-stress, silicon-rich nitride layers act as structural layers to confine epitaxial growth; PSG is used as sacrificial material. This confined selective epitaxial growth technique allows electrical and/or thermal isolation separation, as well as independent optimization of active sensor and readout electronic areas.

### SOI vs. Poly-Si Surface Micromachining

The power of poly-Si surface micromachining mainly lays in its CMOS compatibility. When deposited on an insulator, both poly-Si and single crystal layers enable higher operating temperatures (>200°C) than bulk micromachined sensors featuring p-n junction isolation only (130°C max).[165] An additional benefit for SOI-based micromachining is IC compatibility combined with single crystal Si performance excellence. The maximum gauge factor (see Equation 4.28) of a poly-Si piezoresistor is about 30, roughly 15 times larger than that of a metal strain gauge but only one third of that of an indiffused resistor in single crystal Si.[166] Higher piezoresistivity and fracture stress would seem to favor SOI for sensor manufacture. But there is an important counter argument: the piezoresistivity and fracture stress in poly-Si are isotropic, a major design simplification. Moreover, by laser recrystallization the gauge factor of poly-Si might increase to above 50,[167] and by appropriate boron doping the temperature coefficient of resistance (TCR) can actually reach 0 vs. a TCR of, say, $1.7 \times 10^{-3}$ $^\circ K^{-1}$ for single crystal p-type Si. Neither technical nor cost issues will be the deciding factor in determining which technology will become dominant in the next few years. Micromachining is very much a hostage to trends in the IC industry: promising technologies such as GaAs and micromachining do not necessarily take off in no small part because of the invested capital in some limited sets of standard silicon technologies. On this basis SOI surface micromachining is the favored candidate: SOI extends silicon's technological relevance and experiences increasing investment from the IC industry, benefiting SOI micromachining.[156]

Based on the above we believe that SOI micromachining not only introduces an improved method of making many simple micromachines, but it also will probably become the favored approach of the IC industry. A summary of SOI advantages is listed below:

- IC industry use in all type of applications such as MOS, bipolar digital, bipolar linear, power devices, BICMOS, CCDs, heterojunction bipolar,[168] etc.
- Batch packaging through embedded cavities
- CMOS compatibility
- Substrate industrially available at lower and lower cost (about $200 today)
- Excellent mechanical properties of the single crystalline surface layer
- Freedom of shapes in the x-y dimensions and continually improving dry etching techniques, resulting in larger aspect ratios and higher features
- Freedom of choice of a very well-controlled range of thicknesses of epi surface layers
- $SiO_2$ buried layer as sacrificial and insulating layer and excellent etch stop
- Dramatic reduction of process steps as the SOI wafer comes with several 'embedded' process steps
- High temperature operation

# Resists as Structural Elements and Molds in Surface Micromachining

## Introduction

Polyimide and deep UV photoresists were covered already in Chapter 1 on lithography. We now reiterate some of the material covered there in the context of surface micromachining. Novel deep UV photoresists enable the molding of a wide variety of high aspect ratio microstructures in a wide variety of moldable materials or they are used directly as structural elements. LIGA (covered in Chapter 6), employing X-rays to pattern resists, is really just an extension of the same principles

## Polyimide Surface Structures

Polyimide surface structures, due to their transparency to exposing UV light, can be made very high and exhibit LIGA-like high aspect ratios. By using multiple coats of spun-on polyimide, thick suspended plates are possible. Moreover, composite polyimide plates can be made, depositing and patterning a metal film between polyimide coats. Polyimide surface microstructures are typically released from the substrate by selectively etching an aluminum sacrificial layer (see Figure 5.19), although Cu and PSG (e.g., in the butterfly wing in Figure 5.24) have been used as well (see Table 5.5).

An early result in this field was obtained at SRI International, where polyimide pillars (spacers) about 100 μm in height were used to separate a Si wafer, equipped with a field emitter array, from a display glass plate in a flat panel display.[169] The flat panel display and an SEM picture of the pillars are shown in Figure 5.31. The Probimide 348 FC formulation of Ciba-Geigy was used. This viscous precursor formulation (48% by weight of a polyamic ester, a surfactant for wetting, and a sensitizer) with a 3500-cs viscosity was applied to the Si substrate and formed into a film of a 125-μm thickness by spinning. A 30- to 40-min prebake at about 100°C removed the organic solvents from the precursor. The mask with the pillar pattern was then aligned to the wafer coated with the precursor and subjected to about 20 min of UV radiation. After driving off moisture by another baking operation the coating, still warm, was spray developed (QZ 3301 from Ciba-Geigy), revealing the desired spacer matrix. By baking the polyimide at 100°C in a high vacuum ($10^{-9}$ Torr) the pillars shrunk to about 100 μm, while the polyimide became more dense and exhibited greater structural integrity.

More recently Frazier et al.[170] obtained a height-to-width aspect ratio of about 7 with polyimide structures. Ultraviolet was used to produce structures with heights in the range of 30 to 50 μm. At greater heights, the verticality of the sidewalls was relatively poor. Spun-on thickness in excess of 60 μm in a single coat was obtained for both Ciba-Geigy and Du Pont commercial UV-exposable, negative-tone polyimides. Using a G-line mask aligner, an exposure energy of 350 mJ/cm$^2$ was sufficient to develop a pattern with the Ciba-Geigy QZ 3301 developer. Allen and his team combined polyimide insert molds with electrodeposition to make a wide variety of metal structures.[171] This polyimide application will be contrasted with LIGA in Chapter 6.

## UV Depth Lithography

Besides polyimides, research on novolak-type resists also is leading to higher three-dimensional features. Lochel et al.[172–174] use novolak, positive tone resists of high viscosity (e.g., AZ 4000 series, Hoechst). They deposit in a multiple-coating process

A

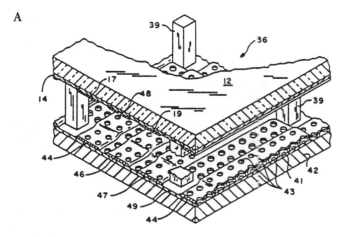

B

**FIGURE 5.31** Polyimide structural elements. (A) Micromachined flat panel display. Number 39 represents one of the spacer pillars in the matrix of polyimide pillars (100 μm high). The spacer array separates the emitter plate from the front display plate. (B) SEM of the spacer matrix. (Courtesy, Dr. I. Brodie.) The height of the pillars is similar to what can be accomplished with LIGA. Since only a simple UV exposure was used, this polyimide is referred to as poor man's LIGA process, or pseudo-LIGA. (From Brodie, I. et al., U.S. Patent 4,923,421, 1990.)

layers up to 200 μm thick in a specially designed spin coater incorporating a co-rotating cover. The subsequent UV lithography yields patterns with aspect ratios up to 10, steep edges (more than 88°), and a minimum feature size down to 3 μm. By combining this resist technology with sacrificial layers and electroplating, a wide variety of three-dimensional microstructures resulted.

Along the same line, researchers at IBM have started experimenting with Epon SU-8 (Shell Chemical), an epoxy-based, onium-sensitized, UV transparent negative photoresist used to produce high aspect ratio (>10:1) features as well as straight sidewalled images in thick film (>200 μm) using standard lithography.[175,176] SU-8 imaged films were used as stencils to plate permalloy for magnetic motors.[175]

Patterns generated with these thick resist technologies should now be compared with LIGA-generated patterns, not only in terms of aspect ratio, where LIGA presumably still produces

better results, but also in terms of sidewall roughness and sidewall run-out. Such a comparison will determine which surface machining technique to employ for the job at hand.

## Comparison of Bulk Micromachining with Surface Micromachining

Surface and bulk micromachining have many processes in common. Both techniques rely heavily on photolithography; oxidation; diffusion and ion implantation; LPCVD and PECVD for oxide, nitride, and oxynitride; plasma etching; use of polysilicon; metallizations with sputtered, evaporated, and plated Al, Au, Ti, Pt, Cr, and Ni. Where the techniques differ is in the use of anisotropic etchants, anodic and fusion bonding, (100) vs. (110) starting material, p+ etch stops, double-sided processing and electrochemical etching in bulk micromachining, and the use of dry etching in patterning and isotropic etchants in release steps for surface micromachines. Combinations of substrate and surface micromachining also frequently appear. The use of polysilicon avoids many challenging processing difficulties associated with bulk micromachining and offers new degrees of freedom for the design of integrated sensors and actuators. The technology combined with sacrificial layers also allows the nearly indispensable further advantage of *in situ* assembly of the tiny mechanical structures, because the structures are preassembled as a consequence of the fabrication sequence. Another advantage focuses on thermal and electrical isolation of polysilicon elements. Polycrystalline piezoresistors can be deposited and patterned on membranes of other materials, e.g., on a $SiO_2$ dielectric. This configuration is particularly useful for high-temperature applications. The p-n junctions act as the only electrical insulation in the single crystal sensors, resulting in high leakage currents at high temperatures, whereas current leakage for the poly-Si/$SiO_2$ structure virtually does not exist. The limits of surface micromachining are quite striking. CVD silicon usually caps at layers no thicker than 1 to 2 μm because of residual stress in the films and the slow deposition process (thick poly-Si needs further investigation). A combination of a large variety of layers may produce complicated structures, but each layer is still limited in thickness. Also, the wet chemistry needed to remove the interleaved layers may require many hours of etching (except when using the porous Si option discussed above), and even then stiction often results.

The structures made from polycrystal silicon exhibit inferior electronic and slightly inferior mechanical properties compared to single crystal silicon. For example, poly-Si has a lower piezoresistive coefficient (resulting in a gauge factor of 30 vs. 90 for single crystal Si) and it has a somewhat lower mechanical fracture strength. Poly-Si also warps due to the difference of thermal expansion coefficient between polysilicon and single crystal silicon. Its mechanical properties strongly depend on processing procedures and parameters.

Table 5.7 extends a comparison of surface micromachining with polysilicon and wet bulk micromachining. The status depicted reflects the mid-1990s and only includes poly-Si surface micromachining. As discussed, SOI micromachining, thick poly-Si, hinged poly-Si, polyimide, and millimeter-molded poly-Si structures have dramatically expanded the application bandwidth of surface micromachining. In the chapter on LIGA we will see how X-ray lithography can further expand the z direction for new surface micromachined devices with unprecedented aspect ratios and extremely low surface roughness. In Table 5.8 we compare physical properties of single crystal Si with those of poly-Si.

Summarizing: although polysilicon can be an excellent mechanical material, it remains a poor electronic material. Reproducible mechanical characteristics are difficult and complex to consistently realize. Fortunately, SOI surface micromachining and other newly emerging surface micromachining techniques can alleviate many of the problems.[177]

**TABLE 5.7**  A Comparison of Bulk Micromachining with Surface Micromachining

| Bulk Micromachining | Surface Micromachining |
|---|---|
| Large features with substantial mass and thickness | Small features with low thickness and mass |
| Utilizes both sides of the wafer | Multiple deposition and etching required to build up structures |
| Vertical dimensions: one or more wafer thicknesses | Vertical dimensions are limited to the thickness of the deposited layers (~2 μm), leading to compliant suspended structures with the tendency to stick to the support |
| Generally involves laminating Si wafer to Si or glass | Surface micromachined device has its built-in support and is more cost effective |
| Piezoresistive or capacitive sensing | Capacitive and resonant sensing mechanisms |
| Wafers may be fragile near the end of the production | Cleanliness critical near end of process |
| Sawing, packaging, testing are difficult | Sawing, packaging, testing are difficult |
| Some mature products and producers | No mature products or producers |
| Not very compatible with IC technology | Natural but complicated integration with circuitry; integration is often required due to the tiny capacitive signals |

*Source:*  Adapted from Jerman, H., *Bulk Silicon Micromachining*, (Banf, Canada, 1994).

# Materials Case Studies

## Introduction

Thin-film properties prove not only difficult to measure but also to reproduce, given the many influencing parameters. Dielectric and polysilicon films can be deposited by evaporation, sputtering, and molecular beam techniques. In VLSI and surface micromachining, none of these techniques are as widely used as CVD techniques. The major problems associated with the former methods are defects caused by excessive wafer handling, low throughput, poor step coverage, and nonuniform depositions. From the comparison of CVD techniques in Table 5.9 we can conclude that LPCVD, at medium temperatures, prevails above all others. VLSI devices and integrated surface micromachines require low processing temperatures to prevent movement of shallow junctions. uniform step coverage, few process-induced defects (mainly from particles generated during wafer handling and loading), and high wafer throughput to reduce cost. These requirements are best met by hot-wall, low-pressure depositions (see also Chapter 3).[179] While depositing a material with LPCVD the following process parameters can be varied: deposition temperature, gas pressure, flow rate, and deposition time.

**TABLE 5.8** Comparison of Material Properties of Si Single Crystal with Crystalline Polysilicon

| Material Property | Single Crystal Si | Poly-Si |
|---|---|---|
| Thermal conductivity (W/cm°K) | 1.57 | 0.34 |
| Thermal expansion (10⁻⁶/°K) | 2.33 | 2–2.8 |
| Specific heat (cal/g°K) | 0.169 | 0.169 |
| Piezoresistive coefficients | n-Si ($\pi_1 = -102.2$); p-Si ($\pi_{44} = +138.1$); e.g., gauge factor of 90 | Gauge factor of 30 (>50 with laser recrystallization) |
| Density (cm³) | 2.32 | 2.32 |
| Fracture strength (GPa) | 6 | 0.8 to 2.84 (undoped poly-Si) |
| Dielectric constant | 11.9 | Sharp maxima of 4.2 and 3.4 eV at 295 and 365 nm, respectively |
| Residual stress | None | Varies |
| Temperature resistivity coefficient (TCR) (°K⁻¹) | 0.0017 (p-type) | 0.0012 nonlinear, + or – through selective doping, increases with decreasing doping level, can be made 0! |
| Poisson ratio | 0.262 max for (111) | 0.23 |
| Young's modulus (10¹¹N/m²) | 1.90 (111) | 1.61 |
| Resistivity at room temperature (ohm.cm) | Depends on doping | 7.5 10⁻⁴ (always higher than for single crystal silicone) |

*Source:* Based on References 48, 63, 178. (See also Table 8.5.)

Table 5.10 cites some approximate mechanical properties of microelectronic materials. The numbers for thin film materials must be approached as approximations; the various parameters affecting mechanical properties of thin films will become clear in the case studies below.

## Polysilicon Deposition and Material Structure

### Introduction

The IC industry applies polysilicon in applications ranging from simple resistors, gates for MOS transistors, thin-film transistors (TFT) (with amorphous hydrogenated silicon: $\alpha$-Si:H), DRAM cell plates, and trench fills, as well as in emitters in bipolar transistors and conductors for interconnects. For the last application, highly doped polysilicon is especially suited; it is easy to establish ohmic contact, it is light insensitive, corrosion resistant, and its rough surface promotes adhesion of subsequent layers. Doping elements such as arsenic, phosphorous or boron reduce the resistivity of the polysilicon. Polysilicon also has emerged as the central structural/mechanical material in surface micromachining, and a closer look at the influence of deposition methodology on its materials characteristics is warranted.

### Undoped Poly-Si

The properties of low-pressure chemical vapor deposited (LPCVD) undoped polysilicon films are determined by the nucleation and growth of the silicon grains. LPCVD Si films, grown slightly below the crystallization temperature (about 600°C for LPCVD), initially form an amorphous solid that subsequently may crystallize during the deposition process.[24,109] The CVD method results in amorphous films when the deposition temperatures are well below the melting temperature of Si (1410°C). The subsequent transition from amorphous to crystalline depends on atomic surface mobility and deposition rate. At low temperatures, surface mobility is low, and nucleation and growth are limited. Newly deposited atoms become trapped in random positions and, once buried, require a substantial amount of time to crystallize as solid state diffusion is significantly lower than surface mobility. That is why, for low temperature deposition, amorphous layers only start to crystallize after sufficient time at temperature in the reactor. Working at temperatures between 580 and 591.5°C, Guckel et al.[109] produced mostly amorphous films. But Krulevitch, working at only slightly higher temperatures (605°C) and probably leaving the films longer in the LPCVD set-up, produced crystallized films. Upon crossing the transition temperature between amorphous and crystalline growth (see Figure 3.18), crystalline growth immediately initiates at the substrate due to the increased surface mobility which allows adatoms to find low energy, crystalline positions from the start of the deposition process. The deposition temperature at which the transition from amorphous to a crystalline structure occurs depends on many parameters, such as deposition rate, partial pressure of hydrogen, total pressure, presence of dopants, and presence of impurities (O, N, or C).[63] In the crystalline regime, numerous nucleation sites form, resulting in a transition zone of a multitude of small grains at the film/substrate interface to columnar crystallites on top, as shown in the schematic of a 620–650°C columnar film in Figure 5.32. In this figure, a transition zone of small, randomly oriented grains is sketched near the SiO₂ layer. The rate of crystallization is faster here than the deposition rate. Columnar grains ranging between 0.03 and 0.3 μm in diameter form on top of the small grains.[63] The columnar coarse grain structure arises from a process of growth competition among the small grains, during which those grains preferentially oriented for fast vertical growth survive at the expense of misoriented, slowly growing grains.[181,182] The lower the deposition temperature, the smaller the initial grain size will be. At 700°C, films also are columnar; however, the grains are cylindrical extending through the thickness of the entire film and there is no transition zone near the SiO₂ interface.[24]

Stress in poly-Si films was found to vary significantly with deposition temperature and silane pressure. Guckel et al.[109] found that their mainly amorphous films, deposited at temperatures below 600°C, proved highly compressive with strain levels as high as –0.67%. At temperatures barely above 600°C, Krulevitch reports tensile films, whereas for yet higher temperatures

**TABLE 5.9**  Comparison of Different Deposition Techniques

|  | Atmospheric Pressure CVD (APCVD) | Low Temperature LPCVD | Medium Temperature LPCVD | Plasma-Enhanced CVD (PECVD) |
|---|---|---|---|---|
| Temp (°C) | 300–500 | 300–500 | 500–900 | 100–350 |
| Materials | $SiO_2$, P-glass | $SiO_2$, P-glass, BP-glass | Poly-Si, $SiO_2$, P-glass, BP-glass, $Si_3N_4$, SiON | SiN, $SiO_2$, $SiO_2$, SiON |
| Uses | Passivation, insulation, spacer | Passivation, insulation spacer | Passivation, gate metal, structural element, spacer | Passivation, insulation, structural elements |
| Throughput | High | High | High | Low |
| Step coverage | Poor | Poor | Conformal | Poor |
| Particles | Many | Few | Few | Many |
| Film properties | Good | Good | Excellent | Poor |

*Note:*  P-glass = phosphorus-doped glass; BP-glass = borophosphosilicate glass.

*Source:*  Adapted from Adams, A.C., in *VLSI Technology,* Sze, S.M., Ed. (McGraw-Hill, New York, 1988) pp. 233–271.

**TABLE 5.10**  Approximate Mechanical Properties of Microelectronic Materials

|  | E(GPa) | $\nu$ | $\alpha(1/°C)$ | $\sigma_0$ |
|---|---|---|---|---|
| **Substrates** |  |  |  |  |
| Silicon | 190 | 0.23 | $2.6 \times 10^{-6}$ | — |
| Alumina | ~415 | — | $8.7 \times 10^{-6}$ | — |
| Silica | 73 | 0.17 | $0.4 \times 10^{-6}$ | — |
| **Films** |  |  |  |  |
| Polysilicon | 160 | 0.23 | $2.8 \times 10^{-6}$ | Varies |
| Thermal $SiO_2$ | 70 | 0.20 | $0.35 \times 10^{-6}$ | Compressive, e.g., 350 MPa |
| PECVD $SiO_2$ | — | — | $2.3 \times 10^{-6}$ | — |
| LPCVD $Si_3N_4$ | 270 | 0.27 | $1.6 \times 10^{-6}$ | Tensile |
| Aluminum | 70 | 0.35 | $25 \times 10^{-6}$ (high!) | Varies |
| Tungsten (W) | 410 (stiff!) | 0.28 | $4.3 \times 10^{-6}$ | Varies |
| Polyimide | 3.2 | 0.42 | $20–70 \times 10^{-6}$ (very high) | Tensile |

*Note:*  E = Young's modulus, $\nu$ = Poisson ratio, $\alpha$ = coefficient of thermal expansion, $\sigma_0$ = residual stress.

*Source:*  Based on lecture notes from Senturia, S.D. and R.T. Howe.

(≥620°C) the stress again turns compressive. While films deposited at temperatures greater than 630°C all turned out compressive, the magnitude of the compression decreased with increasing temperature. The stress gradient in the poly-Si films explains why compressive undoped and unannealed poly-Si beams tend to curl upwards (positive stress gradient) when released from the substrate.[183]

Using high resolution transmission electron microscopy, Guckel et al.[109] and Krulevitch[24] found a strong correlation between the material's microstructure and the exhibited stress. Guckel et al. found that in their mainly amorphous films, deposited at temperatures below 600°C, a region near the substrate interface crystallized during growth with grains between 100 and 4000 Å. Krulevitch found that tensile, low-temperature films (605°C) have Si grains dispersed throughout the film thickness. Krulevitch suggests that the compressive stress in the higher temperature compressive films (≥620°C) relates to the competitive growth mechanism of the columnar grains. The

same author concluded that thermal sources of stress are insignificant.

Importantly, Guckel et al. discovered that annealing in nitrogen or under vacuum converts the low-temperature films with compressive built-in strain (–0.007) to a tensile strain with controllable strain levels between 0 and + 0.003 (see Figure 5.33). During anneal no grain size increase was noticed (100 to 4000 Å) but a slight increase in surface roughness was measured. This type of poly-Si is referred to as fine-grain poly-Si (also Wisconsin poly-Si). Guckel et al. explain this strain field reversal as follows: as the amorphous region of the film crystallizes, it attempts to contract, but due to the substrate constrained newly crystallized region, a tensile stress results. Higher temperature films, during an anneal, also become less strained, but the strain remains compressive (see lower curve in Figure 5.33). Moreover, in this case grain size does increase and the surface turns considerably rougher. The latter is called coarse-grain poly-Si. Fine-grain poly-Si with its tensile strain is preferable; however, it

**Columnar polysilicon**

**Random small polysilicon grains**

**Oxide**

**Si substrate**

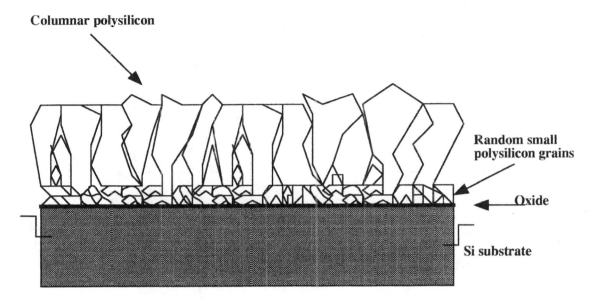

**FIGURE 5.32** Schematic of compressive poly-Si formed at 620–650°C. The columnar coarse-grain structure arises from a process of grain growth competition among the small grains, during which those grains preferentially oriented for fast vertical growth survive at the expense of misoriented, slowly growing grain. (After Krulevitch, P.A., *Micromechanical Investigations of Silicon and Ni-Ti-Cu Thin Films*, Ph.D. thesis, University of California, Berkeley, 1994.)

cannot be doped to as low a resistivity as coarse-grain polysilicon. Hence, fine-grain polysilicon should be considered as a structural material rather than an electronic material.

Summarizing stress in poly-Si depends on the material's microstructure, with tension arising from the amorphous to crystalline transformation during deposition and compression from the competitive grain growth mechanism.

Polysilicon deposited at 600 to 650°C has a {110}-preferred orientation. At higher temperatures, the {100} orientation dominates. Dopants, impurities, and temperature influence this preferred orientation.[63] Drosd and Washburn[184] introduced a model explaining the experimental observation that regrowth of amorphized Si is faster for {100} surfaces, followed by {110} and {111} which are 2.3 and 20 times slower, respectively. Interestingly, the latter also pinpoints the order of fastest to slowest etching of the crystallographic planes in alkaline etchants. As discussed in Chapter 4, Elwenspoek et al.[185] used this observation of symmetry between etching and growing of Si planes as an important insight to develop a new theory explaining anisotropy in etching. In Table 5.11 we compare the discussed coarse- and fine-grain poly-Si forms.

The above picture might further be complicated by the controlling nature of the substrate. For example, depositing amorphous Si ($\alpha$ – Si) at even lower temperatures of 480°C from disilane ($Si_2H_6$) and crystallizing it by subsequent annealing at 600°C demonstrated a large dependency of crystallite size on the underlying $SiO_2$ surface condition. Treating the surface with $HF:H_2O$ or $NH_4OH:H_2O_2:H_2O$ leads to poly-Si films with a large grain size, two or three times as large as without $SiO_2$ treatment, believed to be the consequence of nucleation rate suppression.[186]

Abe, T. and M.L. Reed made low strain polysilicon thin film by DC-magnetics sputtering and post-annealing. The films showed very small regional stress and very smooth texture. The deposition rate was 193Å/min and the substrate was neither cooled nor heated. The average roughness was found to be comparable to the surface roughness of polished, bare silicon substrates.[187]

**Doped Poly-Si**

To produce micromachines, doped poly-Si is used far more frequently than undoped poly-Si. Dopants decrease the resistivity to produce conductors and control stress. Polysilicon can be doped by diffusion, implantation, or the addition of dopant gases during deposition (*in situ* doping). We only detail material properties of *in situ* doped poly-Si here.

Doping poly-Si films *in situ* reduces the number of processing steps required for producing doped micro-devices and also provides the potential for uniform doping through the film thickness. *In situ* doping of poly-Si with phosphorous is accomplished by maintaining a constant $PH_3$ to $SiH_4$ gas flow ratio of about 1 vol % in a hot-wall LPCVD set-up. At this ratio the phosphorous content in the film appears above the saturation limit and the excess dopant segregates at the grain boundaries.[63] *In situ* phosphorus-doped poly-Si undergoes the same amorphous to crystalline growth transformation observed in the undoped film, with the material's microstructure depending on deposition temperature as well as deposition pressure. The temperature of transformation is lower for the doped films than for the undoped poly-Si and occurs between 580 and 620°C.[188,189] Phosphorus doping thus enhances crystallization in amorphous silicon,[190] and, due to passivation of the poly-Si surface by the phosphine gas, reduces the poly-Si deposition rate.[188] Decreases in deposition rate by as much as a factor of 25 have been reported.[191] Slower deposition rates allow more time for adatoms to find crystalline sites, resulting in crystalline growth at lower temperatures. From Table 5.11 we read that the grain size of phosphorus-doped poly-Si tends to be larger (240 to 400 Å) than for the undoped material and that {311} planes show up as a texture facet in the doped material.

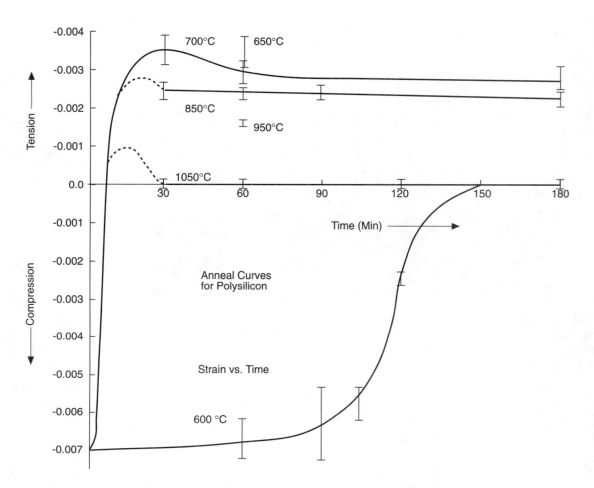

**FIGURE 5.33** Anneal curves for poly-Si. Strain versus anneal time. Upper curves: low-temperature film. Lower curve: high-temperature film. (From Guckel, H. et al., *IEEE Trans. Electron. Dev.*, 35, 800–801, (1988). With permission.)

**TABLE 5.11** Comparison of Coarse-Grain and Fine-Grain Poly-Si

|  | Coarse-Grain Poly-Si | Fine-Grain Poly-Si |
|---|---|---|
| Temperature of deposition (°C) | 620–650 | 570–591.5 |
| Surface roughness | Rough, >50Å | Smooth, <15Å |
| Grain size | Undoped: 160–320 Å as deposited; *in situ* P-doped: 240–400 Å | Very small grains |
| As deposited strain | –0.007 (compressive) | –0.007 (compressive) |
| Effect of high-temperature anneal | Grains size increases, residual strain decreases but remains compressive, reduced bending moment | Grain size increases to 100 Å;[49] others have found 700–900 Å; large variation in strain (see Figure 5.33) from compressive to tensile |
| Dry and wet etch rate | Higher for doped material; depends on dopant concentration | Higher for doped material; depends on dopant concentration |
| Texture | <110> as deposited, <311> *in situ* P-doped | No texture as deposited; depends on dopant concentration, <111> after 900–1000°C anneal[192] |

At lower deposition temperatures and higher pressures, the microstructure again consists of amorphous and crystalline regions, while at higher temperatures and lower pressures columnar films result and as deposited films exhibit compressive residual stress. The columnar films have a stress gradient that increases towards the film surface, as opposed to the gradient found in undoped columnar poly-Si. This gradient in stress most likely is due to nonuniform distribution of phosphorus throughout the film. Annealing at 950°C for 1 hour results in the same stress and stress gradient for initially columnar and initially amorphous/crystalline films (i.e., $\sigma_f = -45$ MPa and $\Gamma = +0.2$ mm$^{-1}$, respectively).[24]

As with undoped poly-Si, phosphorus-doped poly-Si films with smooth surfaces (fine-grain) can be obtained by depositing *in situ* doped films in the amorphous state and then annealing.[192,193] Phosphorus doped poly-Si oxidizes faster than undoped poly-Si. The rate of oxidation is determined by the dopant concentration at the poly-Si surface.[63]

Two drawbacks of *in situ* phosphorus doping are slower deposition rates,[194] and reduced film thickness uniformity;[191] although uniformity can be improved by modifying the reactor geometry.[188] As discussed before, the lower deposition rate of *in situ* phosphine doping can be mitigated by reducing the flow of phosphine/silane ratio by one third.[72] In contrast to *in situ* phosphine and arsine doping, which both decrease the deposition rate, diborane doping of poly-Si to make it $p^+$ accelerates the deposition rate.[63]

The addition of oxygen to poly-Si increases the film's resistivity and the resulting coating, semi-insulating poly-Si (SIPOS) acts as a passivating coating for high voltage devices in the IC industry. The latter material does not seem to have emerged in surface micromachining yet.

The values for the fracture stress of boron-, arsenic-, and phosphorus-doped polysilicon are 2.77 ± 0.08 Gpa, 2.70 ± 0.09 Gpa, and 2.11 ± 0.1 Gpa, respectively, compared with 2.84 ± 0.09 Gpa for undoped polysilicon. The lower value for phosphorus-doped material has been attributed to high surface roughness and with the large number of defects associated with extensive grain growth in highly phosphorus-doped films.[2]

The quest for low temperature poly-Si deposition makes the 320°C PECVD deposition method especially interesting. PECVD films, deposited in a 50 kHz parallel-plate diode reactor, can be doped *in situ* and crystallized by rapid thermal annealing (RTA: 1100°C, 100 seconds). It was shown that small-grained PECVD films annealed by RTA have good electrical properties and gauge factors between 20 and 30 i.e., similar to those reported for other alternative types of polycrystalline silicon.[196]

## Amorphous and Hydrogenated Amorphous Silicon

Amorphous silicon behaves quite different from either fine- or coarse-grain poly-Si. The amorphous material produces a high breakdown strength (7 to 9 MV/cm) oxide with low leakage currents (vs. a low breakdown voltage and large leakage currents for polycrystalline Si oxides). Amorphous polysilicon also attains a broad maximum in its dielectric function without the characteristic sharp structures near 295 and 365 nm (4.2 and 3.4 eV) of crystalline poly-Si. Approximate refractive index values at a wavelength of 600 nm are 4.1 for crystalline polysilicon and 4.5 for amorphous material.[63] As deposited, the material is under compression, but an anneal at temperatures as low as 400°C reduces the stress significantly, even leading to tensile behavior.[75]

Hydrogenated amorphous silicon ($\alpha$-Si:H) enables the fabrication of active semiconductor devices on foreign substrates at temperatures between 200 and 300°C. The technology, first applied primarily to the manufacturing of photovoltaic panels, now is quickly expanding into the field of large-area microelectronics such as active matrix liquid crystal displays (AMLCD). It is somehow surprising that micromachinists have not taken more advantage of this material either for powering surface micromachines or to implement electronics cheaply on non-Si substrates.

Spear and LeComber[197] showed that, in contrast to $\alpha$-Si, $\alpha$-Si:H could be doped both n- and p-type. Singly bonded hydrogen, incorporated at the Si dangling bonds, reduces the electronic defect density from ~$10^{19}$/cm³ to ~$10^{16}$/cm³ (typical H concentrations are 5 to 10 atomic percent — several orders of magnitude higher than needed to passivate all the Si dangling

**TABLE 5.12** Typical Opto-Electronic Parameters Obtained for PECVD $\alpha$-Si:H

| | Symbol | Parameter |
|---|---|---|
| Undoped | | |
| Hydrogen content | | ~10% |
| Dark conductivity at 300 K | $\sigma_D$ | ~$10^{-10}$ (Ω-cm)$^{-1}$ |
| Activation energy | $E_\sigma$ | 0.8–0.9 eV |
| Pre-exponent conductivity factor | $\sigma_0$ | >$10^3$ (Ω-cm)$^{-1}$ |
| Optical bank gap at 300 K | $E_g$ | 1.7–1.8 eV |
| Temperature variation of band gap | $E_g(T)$ | $2–4 \times 10^{-4}$ eV/K |
| Density of states at the minimum | $g_{min}$ | >$10^{15}$–$10^{17}$ cm³/eV |
| Density of states at the conduction band edge | | ~$10^{15}$/cm³ |
| ESR spin density | $N_s$ | ~$10^{21}$/cm³–eV |
| Infrared spectra | | 2000/640 cm$^{-1}$ |
| Photoluminescence peak at 77K | | ~1.25 eV |
| Extended state mobility | | |
|   Electrons | $\mu_n$ or $\mu_e$ | >10 cm²/V–s |
|   Holes | $\mu_p$ or $\mu_h$ | ~1 cm²/V–s |
| Drift mobility | | |
|   Electrons | $\mu_n$ or $\mu_e$ | ~1 cm²/V–s |
|   Holes | $\mu_p$ or $\mu_h$ | ~$10^{-2}$ cm²/V–s |
| Conduction band tail slope | | 25 meV |
| Valence band tail slope | | 40 meV |
| Hole diffusion length | | ~1 μm |
| Doped amorphous | | |
|   n-Type[a] | $\sigma_D$ | $10^{-2}$ (Ω–cm)$^{-1}$ |
| | $E_g$ | ~0.2 eV |
|   p-Type[b] | $\sigma_D$ | $10^{-3}$ (Ω–cm)$^{-1}$ |
| | $E_g$ | ~0.3 eV |
| Doped microcrystalline | | |
|   n-Type[c] | $\sigma_D$ | ≥1 (Ω–cm)$^{-1}$ |
| | $E_g$ | ≤0.05 eV |
|   p-Type[d] | $\sigma_D$ | ≥1 (Ω–cm)$^{-1}$ |
| | $E_g$ | ≤0.05 eV |

[a] ~1% pH$_3$ added to gas phase.

[b] ~1% B$_2$H$_6$ added to gas phase.

[c] ~1% PH$_3$ added to dilute SiH$_4$/H$_2$, or 500 vppm PH$_3$ added to SiF$_4$/H$_2$ (8:1) gas mixtures. Relatively high powers are involved.

[d] ~1% B$_2$H$_6$ added to dilute SiH$_4$/H$_2$.

*Source:* Crowley, J.L., *Solid State Technol.*, February, 94–98 (1992). With permission.

bonds). The lower defect density results in a Fermi level that is free to move, unlike in ordinary amorphous Si, where it is pinned. Other interesting electronic properties are associated with $\alpha$-Si:H — exposure of $\alpha$-Si:H to light increases photoconductivity by four to six orders of magnitude, and its relatively high electron mobility (~1 cm²/V sec$^{-1}$) enables fabrication of useful thin-film transistors. Only recently did Lee et al.[198] note that hydrogenated amorphous silicon solar cells are an attractive means to realize an on-board power supply for integrated micromechanical systems. They point out that the absorption coefficient of $\alpha$-Si:H is more than an order of magnitude larger than that of single crystal Si near the maximum solar photon energy region of 500 nm. Accordingly, the optimum thickness of the active layer in an α-Si:H solar cell can measure 1 μm, much smaller than that of single crystal Si solar cells. By interconnecting 100 individual solar cells in series, the measured open circuit potential reaches as high as 150 V under AM 1.5 conditions, a voltage high enough to drive on-board electrostatic actuators.

Hydrogenated amorphous silicon is manufactured by plasma-enhanced chemical vapor deposition from silane. Usually, planar RF-driven diode sources using $SiH_4$ or $SiH_4/H_2$ mixtures are used. Typical pressures of 75 mTorr and temperatures between 200 and 300°C allow silane decomposition with Si deposition as the dominant reaction. Decomposition occurs by electron impact ionization, producing many different neutral and ionic species.[199] Deposition rates for usable device quality α-Si:H generally do not exceed ~2 to 5 Å/sec, due to the effects of temperature, pressure, and discharge power. Table 5.12 gives state-of-the-art parameters for α-Si:H prepared by PECVD. Although its semiconducting properties are inferior to single crystal Si, the material is finding more and more applications. Some examples are TFT switches for picture elements in AML-CDs,[200] page-wide TFT-addressed document scanners, high-voltage TFTs capable of switching up to 500 V, etc.[201] An excellent source for further information on amorphous silicon is the book *Plasma Deposition of Amorphous Silicon-Based Materials* (Reference 202).

## Silicon Nitride

Silicon nitride is a commonly used material in microcircuit and microsensor fabrication due to its many superior chemical, electrical, optical, and mechanical properties. The material provides an extremely good barrier to the diffusion of water and to ions, particularly of $Na^+$. It also oxidizes slowly (about 30 times less than silicon) and has highly selective etch rates over $SiO_2$ and Si in many etchants. Some applications of silicon nitride are optical waveguides (nitride/oxide), encapsulant (diffusion barrier to water and ions), insulator (high dielectric strength), mechanical protection layer, etch mask, oxidation barrier, and ion implant mask (density is 1.4 times that of $SiO_2$). Silicon nitride is also hard and can be used as, for example, a bearing material in micromotors.[203]

Silicon nitride can be deposited by a wide variety of CVD techniques: APCVD, LPCVD, and PECVD.

Nitride often is deposited from $SiH_4$ or other Si containing gases and $NH_3$ in a reaction such as:

$$3SiCl_2H_2 + 4NH_3 \rightarrow Si_3N_4 + 6HCl + 6H_2 \qquad \textbf{Reaction 5.2}$$

In this CVD process, the stoichiometry of the resulting nitride can be moved toward a silicon-rich composition by providing excess silane or dichlorosilane compared to ammonia.

### PECVD Nitride

Plasma-deposited silicon nitride, also plasma nitride or SiN, is used as the encapsulating material for the final passivation of devices. The plasma-deposited nitride provides excellent scratch protection, serves as a moisture barrier, and prevents sodium diffusion. Because of the low deposition temperature, 300 to 350°C, the nitride can be deposited over the final device metallization. Plasma-deposited nitride and oxide both act as insulators between metallization levels, particularly useful when the bottom metal level is on aluminum or gold. The silicon nitride that results from PECVD in the gas mixture of Reaction 5.2 has two shortcomings: high hydrogen content (in the range of 20 to 30 atomic percent) and high stress. The high compressive stress (up to $5 \times 10^9$ dyn/cm²) can cause wafer warping and voiding and cracking of underlying aluminum lines.[206] The hydrogen in the nitride also leads to degraded MOSFET lifetimes. To avoid hydrogen incorporation, one may employ low or no hydrogen-containing source gases such as nitrogen instead of ammonia as the nitrogen source. Also, a reduced flow of $SiH_4$ results in less Si-H in the film. The hydrogen content and the amount of stress in the film are closely linked. Compressive stress, for example, changes towards tensile stress upon annealing to 490°C in proportion to the Si-H bond concentration. By adding $N_2O$ to the nitride deposition chemistry an oxynitride forms with lower stress characteristics; however, oxynitrides are somewhat less effective as moisture and ion barriers than nitrides.

We discussed the effect of RF frequency on nitride stress, hydrogen content, density, and the wet etch rate in detail in Chapter 3. We can conclude that low frequency (high energy bombardment) results in films with low compressive stress, lower etch rates, and higher density. The higher the ion bombardment during PECVD, the higher the stress is. Stress also is affected by moisture exposure and temperature cycling (for a good review, see Wu et al.[207]). We compare properties of silicon nitride formed by LPCVD and PECVD in Tables 5.13 and 5.14. Table 5.14 highlights typical process parameters.

### LPCVD Nitride

In the IC industry, stoichiometric silicon nitride ($Si_3N_4$) is LPCVD deposited at 700 to 900°C and functions as an oxidation mask and as a gate dielectric in combination with thermally grown $SiO_2$. In micromachining, LPCVD nitride serves as an important mechanical membrane material and isolation/buffer layer. By increasing the Si content in silicon nitride the tensile film stress reduces (even to compressive), the film turns more transparent, and the HF etch rate lowers. Such films result by increasing the dichlorosilane:ammonia ratio.[208]

Figure 5.34 illustrates the effect of gas flow ratio and deposition temperature on stress and the corresponding refractive index and HF etch rate.[208]

Silicon-rich or low-stress nitride emerges as an important micromechanical material. Low residual stress means that relatively thick films can be deposited and patterned without fracture. Low etch rate in HF means that films of silicon-rich nitride survive release etches better than stoichiometric silicon nitride. The etch characteristics of LPCVD SiN are summarized in Table 5.15. The properties of a LPCVD $Si_xN_y$ film, deposited by reaction of $SiCl_2H_2$ and $NH_3$ (5:1 by volume) at 850°C were already summarized in Table 5.13.

**TABLE 5.13** Properties of Silicon Nitride[a]

| Deposition | LPCVD | PECVD |
|---|---|---|
| Temperature (°C) | 700–800 | 250–350 |
| Density (g/cm³) | 2.9–3.2 | 2.4–2.8 |
| Pinholes | No | Yes |
| Throughput | High | Low |
| Step coverage | Conformable | Poor |
| Particles | Few | Many |
| Film quality | Excellent | Poor |
| Dielectric constant | 6–7 | 6–9 |
| Resistivity (Ωcm) | $10^{16}$ | $10^6$–$10^{15}$ |
| Refractive index | 2.01 | 1.8–2.5 |
| Atom % H | 4–8 | 20–25 |
| Energy gap | 5 | 4–5 |
| Dielectric strength ($10^6$ V/cm) | 10 | 5 |
| Etch rate in conc. HF | 200 Å/min | |
| Etch rate in BHF | 5–10 Å/min | |
| Residual stress ($10^9$ dyn/cm²) | 1T | 2 C–5T |
| Poisson ratio | 0.27 | |
| Young's modulus | 270 GPa | |
| TCE | $1.6 \times 10^{-6}$/°C | |

*Note:* C = compressive; T = tensile.

[a] See References 63, 204, 205.

**TABLE 5.14** Silicon Nitride PECVD Process Conditions

| Flow (sccm) | $SiH_4$ | 190–270 |
|---|---|---|
| | $NH_3$ | 1900 |
| | $N_2$ | 1000 |
| Temperature (°C) | T.C. | 350 or 400 |
| | Wafer | ~330 or 380 |
| Pressure (torr) | | 2.9 |
| rf power (watt) | 1100 | |
| Deposition rate (Å/min) | | 1200–1700 |
| Refractive index | | 2.0 |

*Source:* Wu, T.H.T. and Rosler, R.S., *Solid State Technol.,* May, 65–71 (1992). With permission.

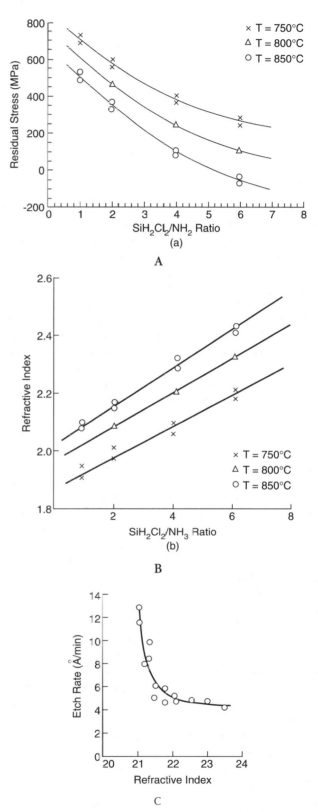

**FIGURE 5.34** Silicon nitride LPCVD deposition parameters. (A) Effect of gas-flow ratio and deposition temperature on stress in nitride films. (B) The corresponding index of refraction. (C) The corresponding HF etch rate. (From Sakimoto, M. et al., *J. Vac. Sci. Technol.,* 21, 1017–1021 (1982). With permission.)

**TABLE 5.15**   Etching Behavior of LPCVD $Si_3N_4$

| Etchant | Temperature (°C) | Etch Rate (Å/min) | Selectivity of $Si_3N_4$:$SiO_2$:Si |
|---|---|---|---|
| $H_3PO_4$ | 180 | 100 | 10:1:0.3 |
| $CF_4$–4% $O_2$ plasma | — | 250 | 3:2.5:17 |
| BHF | 25 | 5 to 10 | 1:200:±0 |
| HF (40%) | 25 | 200 | 1:>100:0.1 |

## CVD Silicon Dioxides

Silicon oxides, like other dielectrics, function as insulation between conducting layers, for diffusion and ion implementation masks, for diffusion from doped oxides, for capping doped oxides to prevent the loss of dopants, for gettering impurities, and for passivation to protect devices from impurities, moisture, and scratches. In micromachining, silicon oxides serve the same purposes but also act as sacrificial material. Phosphorus-doped glass (PSG), also called P-glass or phosphosilicate glass, and borophosphosilicate glasses (BPSG) soften and flow at lower temperatures enabling the smoothing of topography. They etch much faster than $SiO_2$, which benefits their application as sacrificial material. The deposition of thermal $SiO_2$ was covered in Chapter 3. Now we consider CVD techniques to deposit doped and undoped $SiO_2$.

### CVD Undoped $SiO_2$

Several CVD methods will deposit $SiO_2$. CVD silicon dioxide can be deposited on the wafer out of the vapor phase from the reaction of silane ($SiH_4$) with oxygen at relative low temperatures (300 to 500°C) and low pressure. The silane/oxygen gas mixture has been applied at atmospheric pressure (AP) and plasma-enhanced conditions. The main advantage of silane/oxygen is the low deposition temperature; the main disadvantage is the poor step coverage.

In general, tetraethylorthosilicate (TEOS) brings about a better starting chemistry than the traditional silane-based CVD technologies. TEOS-based depositions lead to superior film quality in terms of step coverage and reflow properties. With an LPCVD reactor, the deposition temperature for TEOS is as high as 650 to 750°C, precluding its use over aluminum lines. In contrast, silicon dioxide is deposited by PECVD at 300 to 400°C from TEOS. CVD oxides in general feature porosity and low density. Low frequency (high energy) ion bombardment results in more compressive films, higher density, and lower etch rates as well as better moisture resistance.[206] Use of TEOS oxide, to replace $SiH_4$-based oxide in spin-on-glass (SOG) and photoresist planarization schemes, now has become commonplace for devices with small features.

Yet another promising silicon dioxide deposition technology is the subatmospheric pressure CVD (SACVD). Both undoped and borophosphosilicate glass have been deposited in this fashion. The process involves an ozone and TEOS reaction at a pressure of 600 Torr and at a temperature below 400°C. While offering the same good step coverage over submicron gaps, the films from the thermal reaction of ozone and TEOS exhibit relatively neutral stress and have a higher film density compared to low pressure processes. This increased density gives the oxide greater moisture resistance, lower wet etch rate, and smaller thermal shrinkage. Compared to a 60 Torr process, the film density has increased from 2.09 to 2.15 g/cm³; the wet etch rate decreased by more than 40%; and the thickness shrinkage changed from 12 to 4% after a 30-min anneal in dry $N_2$ at 1000°C.[209]

### CVD Phosphosilicate Glass Films

Adding a few percent of phosphine to the gas stream during deposition to obtain a lower melting point for the oxide ($PH_3$) results in a phosphosilicate glass. Numerous applications of this material exist, such as:

1. Interlevel dielectric to insulate metallization levels
2. Gettering and flow capabilities
3. Passivation overcoat to provide mechanical protection for the chip from its environment
4. Solid diffusion source to dope silicon with phosphorus
5. Fast etching sacrificial material in surface micromachining

Both wet and dry etching rates of PSG are faster than the undoped material and depend on the dopant concentration. Profiles over steps get progressively smoother with higher phosphorous concentrations, reflecting the corresponding enhancement in viscous flow.[68] Increased ion bombardment (energy or density) in a PECVD PSG results in more stable phosphosilicate glass film with compressive as-deposited stress. Table 5.16 summarizes typical PSG process conditions.[207]

The addition of boron to P-glass further lowers its softening temperature. Flow occurs at temperatures between 850 and 950°C, even with phosphorous concentration as low as 4 wt%. A BPSG doped at 4% boron and 6% phosphorus is normal. BPSG deposition conditions are shown in Table 5.17.[210]

**TABLE 5.16**   Typical PSG Process Conditions

| Flow (sccm) | $SiH_4$ | 150–230 |
|---|---|---|
| | $N_2O$ | 4500 |
| | $N_2$ | 1500 |
| | $PH_3$ | 150[a] |
| Temperature (°C) | T.C. | 400 |
| | Wafer | ~380–390 |
| Pressure (Torr) | | 2.2 |
| RF power (watt) | | 1200 |
| Deposition rate (Å/min) | | 4000–5000 |
| Refractive index | | 1.46 |

[a]   10% $PH_3$ in $N_2$.

[b]   5% $PH_3$ in $N_2$.

*Source:*   Wu, T.H.T. and Rosler, R.S., *Solid State Technol.*, May, 65–71 (1992). With permission.

**TABLE 5.17** Typical BPSG Process Conditions

| Parameter | Dimension | NSG | BPSG |
|---|---|---|---|
| Deposition temp. | °C | 400 | 400 |
| TEOS flow | g/min | 0.33 | 0.66 |
| $O_2$ flow | SLM | 7.5 | 7.5 |
| $O_3/O_2$ | volume % | 1 | 4.5 |
| Carrier $N_2$ flow | SLM | 18.0 | 18.0 |
| B conc. | atomic % | | 4 |
| P conc. | atomic % | | 6 |
| Exhaust | $mmH_2O$ | 2.0 | 2.0 |
| Growth rate | Å/min | 1200 | 1300 |
| Thickness | Å | 1000 | 5300 |

*Note:* NSG = non-doped silicate glass; BPSG = borophos-phosilicate glass.

From Bonifield, T., K. Hewes, B. Merritt, R. Robinson, S. Fisher, and D. Maisch, *Semicond. Int.*, July, 200–204, 1993. With permission.

# Polysilicon Surface Micromachining Examples

## Example 1. Analog Devices Accelerometer

Accelerometers based on lateral resonators represent the main application of surface micromachining today. To facilitate integration of their 50-g surface micromachined accelerometer (ADXL-50) with on-board electronics, Analog Devices opted for a mature 4-μm BICMOS process.[55] Figure 5.35 presents a photograph of the finished capacitive sensor (center) with on-chip excitation, self-test, and signal conditioning circuitry. The polysilicon-sensing element only occupies 5% of the total die area. The whole chip measures 500 × 625 μm and operates as an automotive airbag deployment sensor. The measurement accuracy is 5% over the ±50-g range. Deceleration in the axis of sensitivity exerts a force on the central mass that, in turn, displaces the interleaved capacitor plates, causing a fractional change in capacitance. In operation, the device has a force-balance electronic control loop to prevent the mass from actual movement. Straight flexures were used for the layout of the lateral resonator shuttle mass (see Figure 5.18B).

In the sensor design, n+ underpasses connect the sensor area to the electronic circuitry, replacing the usual heat sensitive aluminum connect lines. Most of the sensor processing is inserted into the BICMOS process right after the borophospho-silicate glass planarization. After planarization, a designated sensor region or moat is cleared in the center of the die (Figure 5.36a).* A thin oxide is then deposited to passivate the n+ underpass connects, followed by a thin, low-pressure,

---
* Figure 5.36 appears as a color plate after page 144.

**FIGURE 5.35** Analog Devices' ADXL-50 accelerometer with a surface micromachined capacitive sensor (center), on-chip excitation, self-test, and signal-conditioning circuitry. (From Core, T.A. et al., *Solid State Technol.*, October, 39–47 (1993). With permission.)

vapor-deposited nitride to act as an etch stop (buffer layer) for the final poly-Si release etch (Figure 5.36b). The spacer or sacrificial oxide used is a 1.6-μm densified low-temperature oxide (LTO) deposited over the whole die (Figure 5.36c). In a first timed etch, small depressions that will form bumps or dimples on the underside of the polysilicon sensor are created in the LTO layer. These will limit stiction in case the sensor comes in contact with the substrate. A subsequent etch cuts anchors into the spacer layer to provide regions of electrical and mechanical contact (Figure 5.36c). The 2-μm thick sensor poly-Si is then deposited, implanted, annealed, and patterned (Figure 5.37a).* The relatively deep junctions of the BICMOS process permit the polysilicon thermal anneal as well as brief dielectric densifications without resulting in degradation of the electronic functions. Next is the IC metallization which starts with the removal of the sacrificial spacer oxide from the circuit area along with the LPCVD nitride and LTO layer. A low-temperature oxide is deposited on the poly-Si sensor part and contact openings appear in the IC part of the die where platinum is deposited to form a platinum silicide (Figure 5.37b). The trimmable thin-film material, TiW barrier metal, and Al/Cu interconnect metal are sputtered on and patterned in the IC area. The circuit area is then passivated in two separate deposition steps. First, plasma oxide is deposited and patterned (Figure 5.37c), followed by a plasma nitride (Figure 5.38a) to form a seal with the earlier deposited LPCVD nitride. The nitride acts as an HF barrier in the subsequent long etch release. The plasma oxide left on the sensor acts as an etch stop for the removal of the plasma nitride (Figure 5.38a).** Subsequently, the sensor area is prepared for the final release etch. The undensified dielectrics are removed from the sensor and the final protective resist mask is applied. The photoresist protects the circuit area from the long-term

---

\*      Figure 5.37 appears as a color plate after page 144.
\*\*    Figure 5.38 appears as a color plate after page 144.

buffered oxide etch (Figure 5.38b). The final device cross-section is shown in Figure 5.38c.

## Example 2. Polysilicon Stepping Slider

The principle of operation of the stepping slider is illustrated in Figure 5.39.[211,212] Figure 5.39A shows a cross-sectional view of the polysilicon slider plate (length = 50 μm, width = 30 μm, height = 1.0 μm) and bushing (height = 1 μm) on an insulator film ($Si_3N_4$) as it sets one step. Figure 5.39B displays a schematic of the complete slider. At the rise of an applied voltage pulse on the slider rail, the polysilicon plate is pulled down. Since one end of the plate, supported by the bushing, cannot move, the other part is pulled down to come into contact with the surface of the insulator. The warp of the plate causes the bushing to shift. At the fall of the pulse, the distortion is released and the plate snaps back to its original shape. If the size of the polysilicon step is $\Delta x$, the velocity of the polysilicon microstructure is given by:

$$v = \Delta x \cdot f \qquad\qquad 5.21$$

where f is the frequency of the pulse. The size of the step itself is given by:

$$\Delta x = \frac{h^2}{2(I - I')} \qquad\qquad 5.22$$

where I′ stands for the length of the plate touching the insulator film.

Figure 5.40 illustrates the sequence of a stepping slider fabrication process. A layer of 0.3-μm silicon nitride is deposited at 900°C by the reaction of $SiH_4$ and $NH_3$ in $N_2$. Next, *in situ* phosphorus-doped polysilicon is deposited and patterned in a $SF_6$ plasma to form the sliding rail for the power supply of the

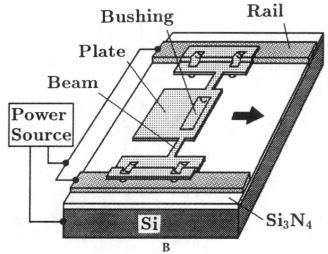

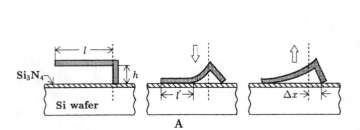

**FIGURE 5.39**   Principle of operation of stepping slider. (A) Cross-sectional view of the polysilicon plate and bushing on an insulator ($Si_3N_4$) on Si as the plate moves one step. (B) Schematic diagram of stepping slider.

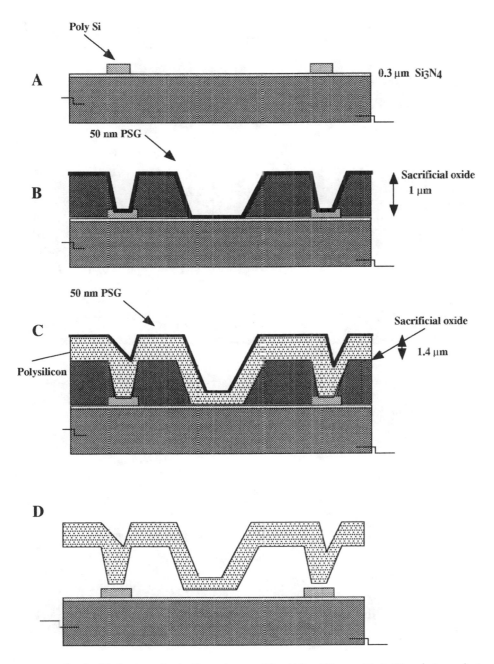

**FIGURE 5.40** Process sequence for the fabrication of polysilicon stepper slider. (After Akiyama, T., *J. Microelectromech. Sys.*, 2, 106–110 (1993). With permission.)

slider (A) (mask 1). Thick sacrificial $SiO_2$ is deposited at 510°C by CVD of $SiH_4$ and $O_2$ in $N_2$. The thickness of the oxide (1 μm) determines the future bushing height. The oxide is patterned with a second mask exposing the silicon nitride in the area of the bushing. Fifty nm of CVD PSG is then deposited over the whole wafer at 610°C from the thermal reaction of $SiH_4$, $PH_3$, and $O_2$ in $N_2$ (B). In step C, 1.4-μm polysilicon is deposited at 610°C from $SiH_4$, coated with PSG, and patterned with mask 3 to shape the slider plate. The silicon wafer is then heated in $N_2$ at 1050°C for 60 min to release the residual strain in the polysilicon and at the same time to activate and diffuse phosphorous from the PSG into the polysilicon slider. The last step (D) in the fabrication process is the release of the polysilicon slider, by

dipping the wafer into 50% HF to fully dissolve the sacrificial oxide. Finally, the wafer is rinsed in deionized water and IPA (isopropyl-alcohol) and dried in $N_2$. In the fabrication process, to avoid washing away the slider in the rinse, the slider is attached by polysilicon springs to the end of the rail. The springs are cut by the step motion of the slider and subsequently the sliders can move along the rail. The velocity at the peak voltage of 150 V is 30 μm/sec with a 1200-Hz pulse. Figure 5.41 shows an SEM photo of the resulting slider (top). Components of the slider are named on the scanned image (bottom).

**For surface micromachining on the internet, see Appdendix B.**

A

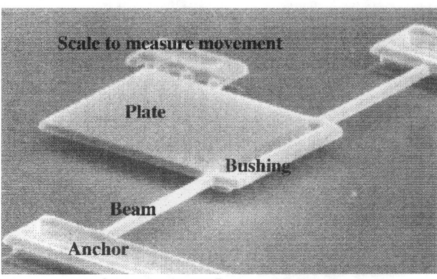

B

**FIGURE 5.41** SEM micrograph of the stepping slider (top) and scanned image (bottom). (From Akiyama, T., *J. Microelectromech. Sys.*, 2, 106–110 (1993). With permission.)

# References

1. Jerman, H., "Bulk Silicon Micromachining", Hardcopies of viewgraphs, 1994, Banf, Canada.

2. Biebl, M. and H. von Philipsborn, "Fracture Strength of Doped and Undoped Panical Filters", Ph.D. Thesis, UC Berkeley, 1993.

3. Greek, S., F. Ericson, S. Johansson, and J.-Å. Schweitz, "In Situ Tensile Strength Measurement of Thick-Film and Thin-Film Micromachined Structures", 8th International Conference on Solid-State Sensors and Actuators (Transducers '95), Stockholm, Sweden, June, 1995, p. 56–9.

4. Le Berre, M., P. Kleinmann, B. Semmache, D. Barbier, and P. Pinard, "Electrical and Piezoresian, V. K. and P. J. McWhorter, Eds. Smart Electronics and MEMS, Proceedings of the Smart Structures and Materials 1996 Meeting, Vol. 2722, SPIE, San Diego, CA, USA, 1996.

5. Howe, R. T., "Recent Advances in Surface Micromachining", 13th Sensor Symposium. Technical Digest, Tokyo, Japan, 1995, p. 1–8.

6. Diem, B., M. T. Delaye, F. Michel, S. Renard, and G. Delapierre, "SOI(SIMOX) as a Substrate for Surface Micromachining of Single Crystalline Silicon Sensors and Actuators", 7th International Conference on Solid-State Sensors and Actuators (Transducers '93), Yokohama, Japan, 1993, p. 233–6.

7. Pister, K. S. J., "Hinged Polysilicon Structures with Integrated CMOS TFT's", Technical Digest of the 1992 Solid State Sensor and Actuator Workshop., Hilton Head Island, SC, 1992, p. 136–9.

8. Keller, C. and M. Ferrari, "Milli-Scale Polysilicon Structures", Technical Digest of the 1994 Solid State Sensor and Actuator Workshop., Hilton Head Island, SC, 1994, p. 132–137.

9. Lange, P., M. Kirsten, W. Riethmuller, B. Wenk, G. Zwicker, J. R. Morante, F. Ericson, and J. Å. Schweitz, "Thick Polycrystalline Silicon for Surface Micro-mechanical Applications: Deposition, Structuring and Mechanical Characterization", 8th International Conference on Solid-State Sensors and Actuators (Transducers '95), Stockholm, Sweden, June, 1995, p. 202–5.

10. Noworolski, J. M., E. Klaassen, J. Logan, K. Petersen, and N. Maluf, "Fabrication of SOI Wafers With Buried Cavities Using Silicon Fusion Bonding and Electrochemical Etchback", 8th International Conference on Solid-State Sensors and Actuators (Transducers '95), Stockholm, Sweden, June, 1995, p. 71–4.

11. Chu, P. B., P. R. Nelson, M. L. Tachiki, and K. S. J. Pister, "Dynamics of Polysilicon Parallel-Plate Electrostatic Actuators", 8th International Conference on Solid-State Sensors and Actuators (Transducers '95), Stockholm, Sweden, June, 1995, p. 356–359.

12. Nathanson, H. C., W. E. Newell, R. A. Wickstrom, and J. R. Davis, "The Resonant Gate Transistor", *IEEE Trans. Electron Devices*, ED–14, 117–133, 1967.

13. Dutta, B., "Integrated Micromotor Concepts", Int. Conf. on Microelectronic Circuits and Systems Theory, Sydney, Australia, 1970, p 36–37.

14. Howe, R. T. and R. S. Muller, "Polycrystalline Silicon Micromechanical Beams", Spring Meeting of The Electrochemical Society, Montreal, Canada, 1982, p. 184–5.

15. Guckel, H. and D. W. Burns, "A Technology for Integrated Transducers", International Conference on Solid-State Sensors and Actuators, Philadelphia, PA, 1985, p. 90–2.

16. Gabriel, K., J. Jarvis, and W. Trimmer, "Small Machines, Large Opportunities: A Report on the Emerging Field of Microdynamics", National Science Foundation, 1989.

17. Muller, R. S., "From IC's to Microstructures: Materials and Technologies", Proceedings of the IEEE Micro Robots and Teleoperators Workshop, Hyannis, MA, 1987, p. 2/1–5.

18. Fan, L.-S., Y.-C. Tai, and R. S. Muller, "Pin Joints, Gears, Springs, Cranks, and Other Novel Micromechanical Structures", 4th International Conference on Solid-State Sensors and Actuators (Transducers '87), Tokyo, Japan, 1987, p. 849–52.

19. Editor, "Analog Devices Combines Micromachining with BICMOS", *Semicond. Int.*, 14, 17, 1991.

20. Hornbeck, L. J., "Projection Displays and MEMS: Timely Convergence for a Bright Future", Micro-machining and Microfabrication Process Technology, (Proceedings of the SPIE), Austin, TX, 1995, p. 2.

21. Vinci, R. P. and J. C. Braveman, "Mechanical Testing of Thin Films", 6th International Conference on Solid-State Sensors and Actuators (Transducers '91), San Francisco, CA, 1991, p. 943–8.

22. Campbell, D. S., "Mechanical Properties of Thin Films", in *Handbook of Thin Film Technology*, Maissel, L. I. and R., G., Eds., McGraw-Hill, New York, 1970.

23. Senturia, S., "Can we Design Microrobotic Devices Without Knowing the Mechanical Properties of Materials ?", Proceedings of the IEEE Micro Robots and Teleoperators Workshop, Hyannis, MA, 1987, p. 3/1–5.

24. Krulevitch, P. A., "Micromechanical Investigations of Silicon and Ni-Ti-Cu Thin Films", Ph.D. Thesis, UC Berkeley, 1994.

25. Pratt, R. I., G. C. Johnson, R. T. Howe, and D. J. J. Nikkel, "Characterization of Thin Films Using Micro-mechanical Structures", *Mat. Res. Soc. Symp. Proc.*, 276, 197–202, 1992.

26. Judy, M. W., Y. H. Cho, R. T. Howe, and A. P. Pisano, "Self-Adjusting Microstructures (SAMS)", Proceedings. IEEE Micro Electro Mechanical Systems, (MEMS '91), Nara, Japan, 1991, p. 51–6.

27. Chou, P. C. and N. J. Pagano, *Elasticity: Tensor, Dyadic, and Engineering Approaches*, Dover Publications, Inc., New York, 1967.

28. Guckel, H., D. W. Burns, H. A. C. Tilmans, C. C. G. Visser, D. W. DeRoo, T. R. Christenson, P. J. Klomberg, J. J. Sniegowski, and D. H. Jones, "Processing Conditions for Polysilicon Films with Tensile Strain for Large Aspect Ratio Microstructures", Technical Digest of the 1988 Solid State Sensor and Actuator Workshop., Hilton Head Island, SC, 1988, p. 51–6.

29. Maseeh, F. and S. D. Senturia, "Viscoelasticity and Creep Recovery of Polyimide Thin Films", Technical Digest of the 1990 Solid State Sensor and Actuator Workshop., Hilton Head Island, SC, 1990, p. 55–60.

30. Hoffman, R. W., "Mechanical Properties of Non-Metallic Thin Films", in *Physics of Nonmetallic Thin Films, (NATO Advanced Study Institutes Series : Series B, Physics)*, Dupuy, C. H. S. and Cachard, A., Eds., Plenum Press, 1976, p. 273–353.

31. Hoffman, R. W., "Stresses in Thin Films : The Relevance of Grain Boundaries and Impurities", *Thin Solid Films*, 34, 185–90, 1975.

32. Wong, S. M., "Residual Stress Measurements on Chromium Films by X-ray Diffraction Using the sin2 Y Method", *Thin Solid Films*, 53, 65–71, 1978.

33. Nishioka, T., Y. Shinoda, and Y. Ohmachi, "Raman Microprobe Analysis of Stress in Ge and GaAs/Ge on Silicon Dioxide-Coated Silicon Substrates", *J. Appl. Phys.*, 57, 276–81, 1985.

34. Marco, S., J. Samitier, O. Ruiz, J. R. Morante, J. Esteve-Tinto, and J. Bausells, "Stress Measurements of SiO₂-Polycrystalline Silicon Structures for Micro-mechanical Devices by Means of Infrared Spectroscopy Technique", 6th International Conference on Solid-State Sensors and Actuators (Transducers '91), San Francisco, CA, 1991, p. 209–12.

35. Mehregany, M., R. T. Howe, and S. D. Senturia, "Novel Microstructures for the In Situ Measurement of Mechanical Properties of Thin Films", *J. Appl. Phys.*, 62, 3579–84, 1987.

36. Jaccodine, R. J. and W. A. Schlegel, "Measurements of Strains at Si-SiO2 Interface", *J. Appl. Phys.*, 37, 2429–34, 1966.

37. Bromley, E. I., J. N. Randall, D. C. Flanders, and R. W. Mountain, "A Technique for the Determination of Stress in Thin Films", *J. Vac. Sci. Technol.*, B1, 1364–6, 1983.

38. Allen, M. G., M. Mehregany, R. T. Howe, and S. D. Senturia, "Microfabricated Structures for the In Situ Measurement of Residual Stress, Young's Modulus and Ultimate Strain of Thin Films", *Appl. Phys. Lett.*, 51, 241–3, 1987.

39. Zhang, L. M., D. Uttamchandani, and B. Culshaw, "Measurement of the Mechanical Properties of Silicon Microresonators", *Sensors and Actuators A*, A29, 79–84, 1991.

40. Pratt, R. I., G. C. Johnson, R. T. Howe, and J. C. Chang, "Micromechanical Structures for Thin Film Characterization", 6th International Conference on Solid-State Sensors and Actuators (Transducers '91), San Francisco, CA, 1991, p. 205–8.

41. Singer, P., "Film Stress and How to Measure It", *Semicond. Int.*, 15, 54–8, 1992.

42. Maseeh-Tehrani, F., "Characterization of Mechanical Properties of Microelectronic Thin Films", Ph.D. Thesis, Massachusetts Institute of Technology, 1990.

43. Maseeh, F. and S. D. Senturia, "Elastic Properties of Thin Polyimide Films", in *Polyimides: Materials, Chemistry and Characterization*, Feger, C., Khojasteh, M. M. and McGrath, J. E., Eds., Elsevier Science Publishers B.V., Amsterdam, 1989, p. 575–584.

44. Senturia, S. D. and R. T. Howe, "Mechanical Properties and CAD", Lecture Notes, MIT, Boston, MA, 1990.

45. Ericson, F. and J. Å. Schweitz, "Micromechanical Fracture Strength of Silicon", *J. Appl. Phys.*, 68, 5840–4, 1990.

46. Sekimoto, M., H. Yoshihara, and T. Ohkubo, "Silicon Nitride Single-Layer X-Ray Mask", *J. Vac. Sci. Technol.*, 21, 1017–21, 1982.

47. Guckel, H., T. Randazzo, and D. W. Burns, "A Simple Technique for the Determination of Mechanical Strain in Thin Films with Applications to Polysilicon", *J. Appl. Phys.*, 57, 1671–5, 1985.

48. Lin, L., "Selective Encapsulations of MEMS: Micro Channels, Needles, Resonators and Electromechanical Filters", Ph.D. Thesis, UC Berkeley, 1993.

49. Guckel, H., D. W. Burns, C. C. G. Visser, H. A. C. Tilmans, and D. Deroo, "Fine-grained Polysilicon Films with Built–in Tensile Strain", *IEEE Trans. Electron Devices*, 35, 800–1, 1988.

50. Kim, C. J., "Silicon Electromechanical Microgrippers: Design, Fabrication, and Testing", Ph.D. Thesis, University of California, Berkeley, 1991.

51. Goosen, J. F. L., B. P. van Drieenhuizen, P. J. French, and R. F. Wolfenbuttel, "Stress Measurement Structures for Micromachined Sensors", 7th International Conference on Solid-State Sensors and Actuators (Transducers '93), Yokohama, Japan, 1993, p. 783–6.

52. Ericson, F., S. Greek, J. Soderkvist, and J. Å. Schweitz, "High Sensitive Internal Film Stress Measurement by an Improved Micromachined Indicator Structure", 8th International Conference on Solid-State Sensors and Actuators (Transducers '95), Stockholm, Sweden, June, 1995, p. 84–7.

53. Benitez, M. A., J. Esteve, M. S. Benrakkad, J. R. Morante, J. Samitier, and J. Å. Schweitz, "Stress Profile Characterization and Test Structures Analysis of Single and Double Ion Implanted LPCVD Polycrystalline Silicon", 8th International Conference on Solid-State Sensors and Actuators (Transducers '95), Stockholm, Sweden, June, 1995, p. 88–91.

54. Biebl, M., G. Brandl, and R. T. Howe, "Young's Modulus of In Situ Phosphorous-Doped Polysilicon", 8th International Conference on Solid-State Sensors and Actuators (Transducers '95), Stockholm, Sweden, June, 1995, p. 80–3.

55. Core, T. A., W. K. Tsang, and S. J. Sherman, "Fabrication Technology for an Integrated Surface-Micromachined Sensor", *Solid State Technol.*, 36, 39–47, 1993.

56. Chu, W. H., M. Mehregany, X. Ning, and P. Pirouz, "Measurement of Residual Stress-Induced Bending Momemt of p+ Silicon Films", *Mat. Res. Soc. Symp.*, 239, 169, 1992.

57. Fan, L. S., R. S. Muller, W. Yun, J. Huang, and R. T. Howe, "Spriral Microstructures for the Measurement of Average Strain Gradients in Thin Films", Proceedings. IEEE Micro Electro Mechanical Systems, (MEMS '90), Napa Valley, CA, 1990, p. 177–81.

58. Bushan, B., "Nanotribology and Nanomechanics of MEMS Devices", Proceedings IEEE, The Ninth Annual International Workshop on Micro Electro Mechanical Systems, San Diego, CA, USA, 1996, p. 91–98.

59. Bushan, B., Ed. *Handbook of Micro/Nanotribology*, CRC, Boca Raton, FL, USA, 1995.

60. Howe, R. T. and R. S. Muller, "Polycrystalline Silicon Micromechanical Beams", *J. Electrochem. Soc.*, 130, 1420–1423, 1983.

61. Howe, R., T., "Polycrystalline Silicon Microstructures", in *Micromachining and Micropackaging of Transducers*, Fung, C. D., Cheung, P. W., Ko, W. H. and Fleming, D. G., Eds., Elsevier, New York, 1985, p. 169–87.

62. Yun, W., "A Surface Micromachined Accelerometer with Integrated CMOS Detection Circuitry", Ph.D. Thesis, U.C.Berkeley, 1992.

63. Adams, A. C., "Dielectric and Polysilicon Film Deposition", in *VLSI Technology*, Sze, S. M., Ed., McGraw-Hill, 1988, p. 233–71.

64. Hermansson, K., U. Lindberg, B. Hok, and G. Palmskog, "Wetting Properties of Silicon Surfaces", 6th International Conference on Solid-State Sensors and Actuators (Transducers '91), San Francisco, CA, 1991, p. 193–6.

65. Tang, W. C.-K., "Electrostatic Comb Drive for Resonant Sensor and Actuator Applications", Ph.D. Thesis, University of California at Berkeley, 1990.

66. Monk, D. J., D. S. Scane, and R. T. Howe, "LPCVD Silicon Dioxide Sacrificial Layer Etching for Surface Micromachining", Smart Materials Fabrication and Materials for Micro-Electro-Mechanical Systems, San Francisco, CA, 1992, p. 303–310.

67. Tenney, A. S. and M. Ghezzo, "Etch Rates of Doped Oxides in Solutions of Buffered HF", *J. Electrochem. Soc.*, 120, 1091–1095, 1973.

68. Levy, R. A. and K. Nassau, "Viscous Behavior of Phosphosilicate and Borophosphosilicate Galsses in VLSI Processing", *Solid State Technol.*, October, 123–30, 1986.

69. North, J. C., T. E. McGahan, D. W. Rice, and A. C. Adams, "Tapered Windows in Phosphorous-Doped Silicon Dioxide by Ion Implantation", *IEEE Trans. Electron Devices*, ED-25, 809–12, 1978.

70. Goetzlich, J. and H. Ryssel, "Tapered Windows in Silicon Dioxide, Silicon Nitride, and Polysilicon Layers by Ion Implantation", *J. Electrochem. Soc.*, 128, 617–9, 1981.

71. White, L. K., "Bilayer Taper Etching of Field Oxides and Passivation Layers", *J. Electrochem. Soc.*, 127, 2687–93, 1980.

72. Howe, R. T., "Polysilicon Integrated Microsystems: Technologies and Applications", 8th International Conference on Solid-State Sensors and Actuators (Transducers '95), Stockholm, Sweden, June, 1995, p. 43–6.

73. Biebl, M., G. T. Mulhern, and R. T. Howe, "In Situ Phosphorous-Doped Polysilicon for Integrated MEMS", 8th International Conference on Solid-State Sensors and Actuators (Transducers '95), Stockholm, Sweden, June, 1995, p. 198–201.

74. Fresquet, G., C. Azzaro, and J.-P. Couderc, "Analysis and Modeling of In Situ Boron-Doped Polysilicon Deposition by LP CVD", *J. Electrochem. Soc.*, 142, 538–47, 1995.

75. Chang, S., W. Eaton, J Fulmer, C. Gonzalez, B. Underwood, J. Wong, and R. L. Smith, "Micro-mechanical Structures in Amorphous Silicon", 6th International Conference on Solid-State Sensors and Actuators (Transducers '91), San Francisco, CA, 1991, p. 751–4.

76. Chen, L.-Y. and N. C. MacDonald, "A Selective CVD Tungsten Process for Micromotors", 6th International Conference on Solid-State Sensors and Actuators (Transducers '91), San Francisco, CA, 1991, p. 739–42.

77. Suzuki, K., I. Shimoyama, and H. Miura, "Insect–Model Based Microrobot with Elastic Hinges", *J. Microelectromech. Syst.*, 3, 4–9, 1994.

78. Hoffman, E., B. Warneke, E. Kruglick, J. Weigold, and K. S. J. Pister, "3D Structures with Piezoresistive Sensors in Standard CMOS", Proceedings. IEEE Micro Electro Mechanical Systems, (MEMS '95), Amsterdam, Netherlands, 1995, p. 288–93.

79. Lin, T.-H., "Flexure-Beam Micromirror Devices and Potential Expansion for Smart Micromachining", Proceedings SPIE, Smart Electronics and MEMS, San Diego, CA, USA, 1996, p. 20–29.

80. Herb, J. A., M. G. Peters, S. C. Terry, and J. H. Jerman, "PECVD Diamond Films for Use in Silicon Microstructures", *Sensors and Actuators A*, A23, 982–7, 1990.

81. Obermeier, E., "High Temperature Microsensors Based on Polycrystalline Diamond Thin Films", 8th International Conference on Solid-State Sensors and Actuators (Transducers '95), Stockholm, Sweden, June, 1995, p. 178–81.

82. Fleischman, A. J., S. Roy, C. A. Zorman, M. Mehregany, and L. Matus, G., "Polycrystaline Silicon Carbide for Surface Micromachining", The Ninth Annual International Workshop on Micro Electro Mechanical Systems, San Diego, CA, USA, 1996, p. 234–238.

83. Monk, D. J., D. S. Soane, and R. T. Howe, "Hydrofluoric Acid Etching of Silicon Dioxide Sacrificial Layers. Part I. Experimental Observations.", *J. Electrochem. Soc.*, 141, 264–269, 1994.

84. Lober, T. A. and R. T. Howe, "Surface Micromachining for Electrostatic Microactuator Fabrication", Technical Digest of the 1988 Solid State Sensor and Actuator Workshop., Hilton Head Island, SC, 1988, p. 59–62.

85. Fan, L. S., Y. C. Tai, and R. S. Mulller, "Integrated Movable Micromechanical Structures for Sensors and Actuators", *IEEE Trans. Electron Devices*, 35, 724–30, 1988.

86. Mehregany, M., K. J. Gabriel, and W. S. N. Trimmer, "Integrated Fabrication of Polysilicon Mechanisms", *IEEE Trans. Electron Devices*, 35, 719–23, 1988.

87. Eaton, W. P. and J. H. Smith, "A CMOS-compatible, Surface-Micromachined Pressure Sensor for Aqueous Ultrasonic Application", Smart Structures and Materials 1995. Smart Electronics, (Proceedings of the SPIE), San Diego, CA, 1995, p. 258–65.

88. Watanabe, H., S. Ohnishi, I. Honma, H. Kitajima, H. Ono, R. J. Wilhelm, and A. J. L. Sophie, "Selective Etching of Phosphosilicate Glass with Low Pressure Vapor HF", *J. Electrochem. Soc.*, 142, 237–43, 1995.

89. Monk, D. J., D. S. Soane, and R. T. Howe, "Hydrofluoric Acid Etching of Silicon Dioxide Sacrificial Layers. Part II. Modeling", *J. Electrochem. Soc.*, 141, 270–274, 1994.

90. Eaton, W. P. and J. H. Smith, "Release-Etch Modeling for Complex Surface Micromachined Structures", SPIE Proceedings, Micromachining and Microfabrication Process Technology II, Austin, Texas, USA, 1996, p. 80–93.

91. Liu, J., Y.-C. Tai, J. Lee, K.-C. Pong, Y. Zohar, and C.-H. Ho, "In Situ Monitoring and Universal Modeling of Sacrificial PSG Etching Using Hydrofluoric Acid", Proceedings of Micro Electro Mechanical Systems, IEEE, Fort Lauderdale, FL, 1993, p. 71–76.

92. Nakamura, M., S. Hoshino, and H. Muro, "Monolithic Sensor Device for Detecting Mechanical Vibration", Densi Tokyo, 24 (IEE Tokyo Section), 87–8, 1985.

93. Bassous, E. and C.-Y. Liu, *Polycrystalline Silicon Etching with Tetramethylammonium Hydroxide*, in US Patent 4,113,551, 1978.

94. Guckel, H., D. W. Burns, C. K. Nesler, and C. R. Rutigliano, "Fine Grained Polysilicon and its Application to Planar Pressure Transducers", 4th International Conference on Solid-State Sensors and Actuators (Transducers '87), Tokyo, Japan, 1987, p. 277–282.

95. Alley, R. L., G. J. Cuan, R. T. Howe, and K. Komvopoulos, "The Effect of Release Etch Processing on Surface Microstructure Stiction", Technical Digest of the 1988 Solid State Sensor and Actuator Workshop., Hilton Head Island, SC, 1988, p. 202–7.

96. Mastrangelo, C. H. and C. H. Hsu, "Mechanical Stability and Adhesion of Microstructures Under Capillary Forces-Part I: Basic Theory", *J. Micro-electromech. Syst.*, 2, 33–43, 1993.

97. Mastrangelo, C. H. and C. H. Hsu, "Mechanical Stability and Adhesion of Microstructures Under Capillary Forces-Part II: Experiments", *J. Micro-electromech. Syst.*, 2, 44–55, 1993.

98. Fan, L. S., "Integrated Micromachinery: Moving Structures on Silicon Chips", Ph.D. Thesis, University of California at Berkeley, 1989.

99. Abe, T., W. C. Messner, and M. L. Reed, "Effective Methods to Prevent Stiction During Post–Release-Etch Processing", Proceedings. IEEE Micro Electro Mechanical Systems, (MEMS '95), Amsterdam, Netherlands, 1995, p. 94–9.

100. Fedder, G. K., J. C. Chang, and R. T. Howe, "Thermal Assembly of Polysilicon Microactuators with Narrow–Gap Electrostatic Comb Drive", Technical Digest of the 1992 Solid State Sensor and Actuator Workshop., Hilton Head Island, SC, 1992, p. 63–8.

101. Mastrangelo, C. H. and G. S. Saloka, "A Dry-Release Method Based on Polymer Columns for Microstructure Fabrication", Proceedings. IEEE Micro Electro Mechanical Systems, (MEMS '93), Fort Lauderdale, FL, 1993, p. 77–81.

102. Guckel, H. and D. W. Burns, "Planar Processed Polysilicon Sealed Cavities for Pressure Transducer Arrays", IEEE International Electron Devices Meeting. Technical Digest, IEDM '8 '84, San Francisco, CA, 1984, p. 223–5.

103. Sugiyama, S., T. Suzuki, K. Kawahata, K. Shimaoka, M. Takigawa, and I. Igarashi, "Micro-Diaphragm Pressure Sensor", IEEE International Electron Devices Meeting. Technical Digest, IEDM '86, Los Angeles, CA, 1986, p. 184–7.

104. Sugiyama, S., K. Kawakata, M. Abe, H. Funabashi, and I. Igarashi, "High-Resolution Silicon Pressure Imager with CMOS Processing Circuits", 4th International Conference on Solid-State Sensors and Actuators (Transducers '87), Tokyo, Japan, 1987, p. 444–447.

105. Kim, Y. W. and M. G. Allen, "Surface Micromachined Platforms Using Electroplated Sacrificial Layers", 6th International Conference on Solid-State Sensors and Actuators (Transducers '91), San Francisco, CA, 1991, p. 651–4.

106. Scheeper, P. R., W. Olthuis, and P. Bergveld, "Fabrication of a Subminiature Silicon Condenser Microphone Using the Sacrificial Layer Technique", 6th International Conference on Solid-State Sensors and Actuators (Transducers '91), San Francisco, CA, 1991, p. 408–11.

107. Yamada, K. and T. Kuriyama, "A New Modal Mode Controlling Method for A Surface Format Surrounding Mass Accelerometer", 6th International Conference on Solid-State Sensors and Actuators (Transducers '91), San Francisco, CA, 1991, p. 655–8.

108. Guckel, H., J. J. Sniegowski, T. R. Christenson, S. Mohney, and T. F. Kelly, "Fabrication of Micromechanical Devices from Polysilicon Films with Smooth Surfaces", *Sensors Actuators*, 20, 117–21, 1989.

109. Guckel, H., J. J. Sniegowski, T. R. Christenson, and F. Raissi, "The Application of Fine-Grained, Tensile Polysilicon to Mechanically Resonant Transducers", *Sensors and Actuators A*, A21, 346–351, 1990.

110. Takeshima, N., K. J. Gabriel, M. Ozaki, J. Takahashi, H. Horiguchi, and H. Fujita, "Electrostatic Parallelogram Actuators", 6th International Conference on Solid-State Sensors and Actuators (Transducers '91), San Francisco, CA, 1991, p. 63–6.

111. Mulhern, G. T., D. S. Soane, and R. T. Howe, "Supercritical Carbon Dioxide Drying of Micro-structures", 7th International Conference on Solid-State Sensors and Actuators (Transducers '93), Yokohama, Japan, 1993, p. 296–9.

112. Kozlowski, F., N. Lindmair, T. Scheiter, C. Hierold, and W. Lang, "A Novel Method to Avoid Sticking of Surface Micromachined Structures", 8th International Conference on Solid-State Sensors and Actuators (Transducers '95), Stockholm, Sweden, June, 1995, p. 220–3.

113. Alley, R. L., P. Mai, K. Komvopoulos, and R. T. Howe, "Surface Roughness Modifications of Interfacial Contacts in Polysilicon Microstructures", 7th International Conference on Solid-State Sensors and Actuators (Transducers '93), Yokohama, Japan, 1993, p. 288–91.

114. Alley, R. L., R. T. Howe, and K. Komvopoulos, "The Effect of Release-Etch Processing on Surface Micro-structure Stiction", Technical Digest of the 1992 Solid State Sensor and Actuator Workshop., Hilton Head Island, SC, 1992, p. 202–7.

115. Man, P. F., B. P. Gogoi, and C. H. Mastrangelo, "Elimination of Post–Release Adhesion in Micro-structures Using Thin Conformal Fluorocarbon Films", The Ninth Annual International Workshop on Micro Electro Mechanical Systems, MEMS'96, San Diego, CA, USA, 1996, p. 55–60.

116. Houston, M. R., R. Maboudian, and R. Howe, "Ammonium Fluoride Anti-Stiction Treatment for Polysilicon Microstructures", 8th International Conference on Solid-State Sensors and Actuators (Transducers '95), Stockholm, Sweden, June, 1995, p. 210–3.

117. Gogoi, B. P. and C. H. Mastrangelo, "Post-Processing Release of Microstructures by Electromagnetic Pulses", 8th International Conference on Solid-State Sensors and Actuators (Transducers '95), Stockholm, Sweden, June, 1995, p. 214–7.

118. Howe, R. T., "Resonant Microsensors", 4th International Conference on Solid-State Sensors and Actuators (Transducers '87), Tokyo, Japan, 1987, p. 843–8.

119. Howe, R. T. and R. S. Muller, "Stress in Polycrystalline and Amorphous Silicon Thin Films", *J. Appl. Phys.*, 54, 4674–5, 1983.

120. Murarka, S. P. and T. F. J. Retajczyk, "Effect of Phosphorous Doping on Stress in Silicon and Polycrystalline Silicon", *J. Appl. Phys.*, 54, 2069–2072, 1983.

121. Orpana, M. and A. O. Korhonen, "Control of Residual Stress in Polysilicon Thin Films by Heavy Doping in Surface Micromachining", 6th International Conference on Solid-State Sensors and Actuators (Transducers '91), San Francisco, CA, 1991, p. 957–60.

122. Lin, L., R. T. Howe, and A. P. Pisano, "A Novel In Situ Micro Strain Gauge", Proceedings. IEEE Micro Electro Mechanical Systems, (MEMS '93), Fort Lauderdale, FL, 1993, p. 201–206.

123. Choi, M. S. and E. W. Hearn, "Stress Effects in Boron-Implanted Polysilicon Films", *J. Electrochem. Soc.*, 131, 2443–6, 1984.

124. Ding, X. and W. Ko, "Buckling Behavior of Boron-Doped P+ Silicon Diaphragms", 6th International Conference on Solid-State Sensors and Actuators (Transducers '91), San Francisco, CA, 1991, p. 201–204.

125. Hendriks, M., R. Delhez, and S. Radelaar, "Carbon Doped Polycrystalline Silicon Layers", in *Studies in Inorganic Chemistry*, Elsevier, Amsterdam, 1983, p. 193.

126. Tang, W. C., T.-C. H. Nguyen, and R. T. Howe, "Laterally Driven Polysilicon Resonant Microstructures", Proceedings. IEEE Micro Electro Mechanical Systems, (MEMS '89), Salt Lake City, UT, 1989, p. 53–9.

127. Guckel, H., D. K. Showers, D. W. Burns, C. R. Rutigliano, and C. G. Nesler, "Deposition Techniques and Properties of Strain Compensated LP CVD Silicon Nitride", Technical Digest of the 1986 Solid State Sensor and Actuator Workshop, Hilton Head Island, SC, 1986.

128. Tang, W. C., T. H. Nguyen, and R. T. Howe, "Laterally Driven Polysilicon Resonant Microstructures", *Sensors Actuators*, 20, 25–32, 1989.

129. Jerman, J. H., "The Fabrication and Use of Micro-machined Corrugated Silicon Diaphragms", *Sensors and Actuators A*, A23, 988–92, 1990.

130. Spiering, V. L., S. Bouwstra, R. M. E. J. Spiering, and M. Elwenspoek, "On-Chip Decoupling Zone for Package-Stress Reduction", 6th International Conference on Solid-State Sensors and Actuators (Transducers '91), San Francisco, CA, 1991, p. 982–5.

131. Offereins, H. L., H. Sandmaier, B. Folkmer, U. Steger, and W. Lang, "Stress Free Assembly Technique for a Silicon Based Pressure Sensor", 6th International Conference on Solid-State Sensors and Actuators (Transducers '91), San Francisco, CA, 1991, p. 986–9.

132. Spiering, V. L., S. Bouwstra, J. Burger, and M. Elwenspoek, "Membranes Fabricated with a Deep Single Corrugation for Pckage Stress Reduction and Residual Stress Relief", 4th European Workshop on Micromechanics (MME '93), Neuchatel, Switzerland, 1993, p. 223–7.

133. Zhang, Y. and K. D. Wise, "Performance of Non-Planar Silicon Diaphragms under Large Deflections", *J. Microelectromech. Syst.*, 3, 59–68, 1994.

134. van Mullem, C. J., K. J. Gabriel, and H. Fujita, "Large Deflection Performance of Surface Micromachined Corrugated Diaphragms", 6th International Conference on Solid-State Sensors and Actuators (Transducers '91), San Francisco, CA, 1991, p. 1014–7.

135. Scheeper, P. R., W. Olthuis, and P. Bergveld, "The Design, Fabrication, and Testing of Corrugated Silicon Nitride Diaphragms", *J. Microelectromech. Syst.*, 3, 36–42, 1994.

136. Guckel, H., D. W. Burns, H. A. C. Tilmans, D. W. DeRoo, and C. R. Rutigliano, "Mechanical Properties of Fine Grained Polysilicon: the Repeatability Issue", Technical Digest of the 1988 Solid State Sensor and Actuator Workshop., Hilton Head Island, SC, 1988, p. 96–9.

137. Smith, J., S. Montague, J. Sniegowski, and P. McWhorter, "Embedded Micromechanical Devices for Monolithic Integration of MEMs with CMOS", IEEE International Electron Devices Meeting. Technical Digest, IEDM '95, Washington, DC, 1995, p. 609–12.

138. Varadan, V. K. and P. J. McWhorter, Eds. *Smart Electronics and MEMS, Proceedings of the Smart Structures and Materials 1996 Meeting*, Vol. 2722, SPIE, San Diego, CA, USA, 1996.

139. Memming, R. and G. Schwandt, "Anodic Dissolution of Silicon in Hydrofluoric Acid Solutions", *Surf. Sci.*, 4, 109–24, 1966.

140. Zhang, X. G., S. D. Collins, and R. L. Smith, "Porous Silicon Formation and Electropolishing of Silicon by Anodic Polarization in HF Solution", *J. Electrochem. Soc.*, 136, 1561–5, 1989.

141. Anderson, R. C., R. S. Muller, and C. W. Tobias, "Porous Polycrystalline Silicon: A New Material for MEMS", *J. Microelectromech. Syst.*, 3, 10–18, 1994.

142. Field, L., "Low-Stress Nitrides for Use in Electronic Devices", EECS/ERL Res. Summary, 42–43, Univ. of California at Berkeley, 1987.

143. Judy, M. W. and R. T. Howe, "Hollow Beam Polysilicon Lateral Resonators", Proceedings. IEEE Micro Electro Mechanical Systems, (MEMS '93), Fort Lauderdale, FL, 1993, p. 265–271.

144. Judy, M. W. and R. T. Howe, "Highly Compliant Lateral Suspensions Using Sidwall Beams", 7th International Conference on Solid-State Sensors and Actuators (Transducers '93), Yokohama, Japan, 1993, p. 54–57.

145. Lebouitz, K. S., R. T. Howe, and A. P. Pisano, "Permeable Polysilicon Etch-Access Windows for Microshell Fabrication", 8th International Conference on Solid-State Sensors and Actuators (Transducers '95), Stockholm, Sweden, June, 1995, p. 224–227.

146. Burgett, S. R., K. S. Pister, and R. S. Fearing, "Three Dimensional Structures Made with Microfabricated Hinges", ASME 1992, Micromechanical Sensors, Actuators, and Systems, Anaheim, CA, 1992, p. 1–11.

147. Ross, M. and K. Pister, "Micro-Windmill for Optical Scanning and Flow Measurement", Eurosensors VIII, (Sens. Actuators A), Toulouse, France, 1994, p. 576–9.

148. Lin, L. Y., S. S. Lee, M. C. Wu, and K. S. J. Pister, "Micromachined Integrated Optics for Free-Space Interconnections", Proceedings. IEEE Micro Electro Mechanical Systems, (MEMS '95), Amsterdam, Netherlands, 1995, p. 77–82.

149. Yeh, R., E. J. Kruglick, and K. S. J. Pister, "Towards an Articulated Silicon Microrobot", ASME 1994, Micromechanical Sensors, Actuators, and Systems, Chicago, IL, 1994, p. 747–754.

150. Lin, G., K. S. J. Pister, and K. P. Roos, "Standard CMOS Piezoresistive Sensor to Quantify Heart Cell Contractile Forces", The Ninth Annual International Workshop on Micro Electro Mechanical Systems, MEMS '96, San Diego, CA, USA, 1996, p. 150–155.

151. Kahn, H., S. Stemmer, K. Nandakumar, A. H. Heuer, R. L. Mullen, R. Ballarini, and M. A. Huff, "Mechanical Properties of Thick, Surface Micromachined Polysilicon Films", The Ninth Annual International Workshop on Micro Electro Mechanical Systems, San Diego, CA, USA, 1996, p. 343–348.

152. Keller, C. G. and R. T. Howe, "Nickel-Filled Hexsil Thermally Actuated Tweezers", 8th International Conference on Solid-State Sensors and Actuators (Transducers '95), Stockholm, Sweden, June, 1995, p. 376–379.

153. Keller, C. G. and R. T. Howe, "Hexsil Bimorphs for Vertical Actuation", 8th International Conference on Solid-State Sensors and Actuators (Transducers '95), Stockholm, Sweden, June, 1995, p. 99–102.

154. Dunn, P. N., "SOI:Ready to Meet CMOS Challenge", *Solid State Technol.*, October, 32–5, 1993.

155. Abe, T. and J. H. Matlock, "Wafer Bonding Technique for Silicon-on-Insulator Technology", *Solid State Technol.*, November, 39–40, 1990.

156. Kanda, Y., "What Kind of SOI Wafers are Suitable for What Type of Micromachining Purposes?", 6th International Conference on Solid-State Sensors and Actuators (Transducers '91), San Francisco, CA, 1991, p. 452–5.

157. Kuhn, G. L. and C. J. Rhee, "Thin Silicon Film on Insulating Substrate", *J. Electrochem. Soc.*, 120, 1563–6, 1973.

158. Pourahmadi, F., L. Christel, and K. Petersen, "Silicon Accelerometer With New Thermal Self–Test Mechanism", Technical Digest of the 1992 Solid State Sensor and Actuator Workshop., Hilton Head Island, SC, 1992, p. 122–5.

159. Gennissen, P. T. J., M. Bartke, P. J. French, P. M. Sarro, and R. F. Wolffenbuttel, "Automatic Etch Stop On Buried Oxide Using Epitaxial Lateral Overgrowth", 8th International Conference on Solid-State Sensors and Actuators (Transducers '95), Stockholm, Sweden, June, 1995, p. 75–8.

160. Bartek, M., P. T. J. Gennissen, P. J. French, and R. F. Wolffenbuttel, "Confined Selective Epitaxial Growth: Potential for Smart Silicon Sensor Fabrication", 8th International Conference on Solid-State Sensors and Actuators (Transducers '95), Stockholm, Sweden, June, 1995, p. 91–94.

161. Greiff, P., "SOI–based Micromechanical Process", Micromachining and Microfabrication Process Technology, (Proceedings of the SPIE), Austin, TX, 1995, p. 74–81.

162. Neudeck, G. W. and et. al., "Three Dimensional Devices Fabricated by Silicon Epitaxial Lateral Overgroth", *J. Electron. Mater.*, 19, 1111–1117, 1990.

163. Schubert, P., J and G. Neudeck, W., "Confined Lateral Selective Epitaxial Growth of Silicon for Device Fabrication", *IEEE Electron Device Lett.*, 11, 181–183, 1990.

164. Bashir, R. and et. al., "Characterization and Modeling of Sidewall Defects in Selective Epitaxial Growth of Silicon", *Journal of Vacuum Science &*.

165. Luder, E., "Polcrystalline Silicon-Based Sensors", *Sensors Actuators*, 10, 9–23, 1986.

166. Obermeier, E. and P. Kopystynski, "Polysilion as a Material for Microsensor Applications", *Sensors and Actuators A*, A30, 149–155, 1992.

167. Voronin, V. A., A. A. Druzhinin, I. I. Marjamora, V. G. Kostur, and J. M. Pankov, "Laser–Recrystallized Polysilicon Layers in Sensors", *Sensors and Actuators A*, A30, 143–147, 1992.

168. Burggraaf, P., "Epi's Leading Edge", *Semicond. Int.*, June, 67–71, 1991.

169. Brodie, I., H. R. Gurnick, C. E. Holland, and H. A. Moessner, *Method for Providing Polyimide Spacers in a Field Emission Panel Display*, in US Patent 4,923,421, 1990.

170. Frazier, A. B. and M. G. Allen, "Metallic Microstructures Fabricated Using Photosensitive Polyimide Electro-plating Molds", *J. Microelectromech. Syst.*, 2, 87–94, 1993.

171. Ahn, C. H., Y. J. Kim, and M. G. Allen, "A Planar Variable Reluctance Magnetic Micromotor With a Fully Integrated Stator and Wrapped Coils", Proceedings. IEEE Micro Electro Mechanical Systems, (MEMS '93), Fort Lauderdale, FL, 1993, p. 1–6.

172. Lochel, B., A. Maciossek, H. J. Quenzer, and B. Wagner, "UV Depth Lithography and Galvanoforming for Micromachining", Electrochemical Microfabrication II, Miami Beach, FL, 1994, p. 100–111.

173. Lochel, B., A. Maciossek, H. J. Quenzer, and B. Wagner, "Ultraviolet Depth Lithography and Galvanoforming for Micromachining", *J. Electrochem. Soc.*, 143, 237–244, 1996.

174. Loechel, B., M. Rothe, S. Fehlberg, G. Gruetzner, and G. Bleidiessel, "Influence of Resist Baking on the Pattern Quality of Thick Photoresists", SPIE–Micromachining and Microfabrication Process Technology II, Austin, Texas, USA, 1996, p. 174–181.

175. Acosta, R. E., C. Ahn, I. V. Babich, E. I. Cooper, J. M. Cotte, W. J. Horkans, C. Jahnes, S. Krongelb, K. T. Kwietniak, N. C. Labianca, E. J. M. O'Sullivan, A. T. Pomerene, D. L. Rath, L. T. Romankiw, and J. A. Tornello, "Integrated Variable Reluctance Magnetic Mini–Motor", *J. Electrochem. Soc.*, 95–2, 494–495, 1995.

176. LaBianca, N. C., J. D. Gelorme, E. Cooper, E. O'Sullivan, and J. Shaw, "High Aspect Ratio Optical Resist Chemistry for MEMS Applications", JECS 188th Meeting, Chicago, IL, 1995, p. 500–501.

177. Petersen, K., D. Gee, F. Pourahmadi, R. Craddock, J. Brown, and L. Christel, "Surface Micromachined Structures Fabricated with Silicon Fusion Bonding", 6th International Conference on Solid-State Sensors and Actuators (Transducers '91), San Francisco, CA, 1991, p. 397–399.

178. Heuberger, A., *Mikromechanik*, Springer Verlag, Heidelberg, 1989.

179. Iscoff, R., "Hotwall LP CVD Reactors: Considering the Choices", *Semicond. Int.*, June, 60–64, 1991.

180. Kermani, A., K. E. Johnsgard, and W. Fred, "Single Wafer RT CVD of Polysilicon", *Solid State Technol.*, May, 71–73, 1991.

181. van der Drift, A., "Evolutionary Selection, A Principle Governing Growth Orientation in Vapour–Deposited Layers", *Philips Res. Repts.*, 22, 267–288, 1967.

182. Matson, E. A. and S. A. Polysakov, "On the Evolutionary Selection Principle in Relation to the Growth of Polycrystalline Silicon Films", *Phys. Sta. Sol. (a)*, 41, K93–K95, 1977.

183. Lober, T. A., J. Huang, M. A. Schmidt, and S. D. Senturia, "Characterization of the Mechanisms Producing Bending Moments in Polysilicon Micro-Cantilever Beams by Interferometric Deflection Measurements", Technical Digest of the 1988 Solid State Sensor and Actuator Workshop., Hilton Head Island, SC, 1988, p. 92–95.

184. Drosd, R. and J. Washburn, "Some Observation on The Amorphous to Crystalline Transfomation in Silicon", *J. Appl. Phys.*, 53, 397–403, 1982.

185. Elwenspoek, M., U. Lindberg, H. Kok, and L. Smith, "Wet Chemical Etching Mechanism of Silicon", IEEE International Workshop on Micro Electro Mechanical Systems, MEMS '94, Oiso, Japan, 1994, p. 223–8.

186. Shimizu, T. and S. Ishihara, "Effect of $SiO_2$ Surface Treatment on the Solid-Phase Crystallization of Amorphous Silicon Films", *J. Electrochem. Soc.*, 142, 298–302, 1995.

187. Abe, T. and M. L. Reed, "Low Strain Sputtered Polysilicon for Micromechanical Structures", The Ninth Annual International Workshop on Micro Electro Mechanical Systems, San Diego, CA, USA, 1996, p. 258–262.

188. Mulder, J. G. M., P. Eppenga, M. Hendriks, and J. E. Tong, "An Industrial LP CVD Process for In Situ Phosphorus-Doped Polysilicon", *J. Electrochem. Soc.*, 137, 273–279, 1990.

189. Kinsbron, E., M. Sternheim, and R. Knoell, "Crystallization of Amorphous Silicon Films During Low Pressure Chemical Vapor Deposition", *Appl. Phys. Lett.*, 42, 835–837, 1983.

190. Lietoila, A., A. Wakita, T. W. Sigmon, and J. F. Gibbons, "Epitaxial Regrowth of Intrinsic, $^{31}P$–Doped and Compensated ($^{31F+11}B$-Doped) Amorphous Si", *J. Appl. Phys.*, 53, 4399–4405, 1982.

191. Meyerson, B. S. and W. Olbricht, "Phosphorous-Doped Polycrystalline Silicon Via LPCVD I. Process Characterization.", *J. Electrochem. Soc.*, 131, 2361–2365, 1984.

192. Harbeke, G., L. Krausbauer, E. F. Steigmeier, and A. E. Widmer, "LP CVD Polycrystalline Silicon: Growth and Physical Properties of In-Situ Phosphorous Doped and Undoped Films", *RCA Rev.*, 44, 287–313, 1983.

193. Hendriks, M. and C. Mavero, "Phosphorous Doped Polysilicon for Double Poly Structures. Part I. Morphology and Microstructure", *J. Electrochem. Soc.*, 138, 1466–1470, 1991.

194. Kurokawa, H., "P–Doped Polysilicon Film Growth Technology", *J. Electrochem. Soc.*, 129, 2620–4, 1982.

195. Learn, A. J. and D. W. Foster, "Deposition and Electrical Properties of In Situ Phosphorous-Doped Silicon Films Formed by Low–Pressure Chemical Vapor Deposition", *J. Appl. Phys.*, 61, 1898–1904, 1987.

196. Compton, R., D., "PECVD: A Versatile Technology", *Semicond. Int.*, July, 60–65, 1992.

197. Spear, W. E. and P. G. Le Comber, "Substitutional Doping of Amorphous Silicon", *Solid State Commun.*, 17, 1193–1196, 1975.

198. Lee, J. B., Z. Chen, M. G. Allen, A. Rohatgi, and R. Arya, "A Miniaturized High-Voltage Solar Cell Array as an Electrostatic MEMS Power Supply", *J. Microelectromech. Syst.*, 4, 102–108, 1995.

199. Crowley, J. L., "Plasma Enhanced CVD for Flat Panel Displays", *Solid State Technol.*, February, 94–98, 1992.

200. Holbrook, D. S. and J. D. McKibben, "Microlithography for Large Area Flat Panel Display Substrates", *Solid State Technol.*, May, 166–172, 1992.

201. Bohm, M., "Advances in Amorphous Silicon Based Thin Film Microelectronics", *Solid State Technol.*, September 1988, 125–131, 1988.

202. Bruno, G., P. Capezzuto, and A. Madan, Eds. *Plasma Deposition of Amorphous Silicon-Based Materials*, Academic Press, Boston, 1995.

203. Pool, R., "Microscopic Motor Is a First Step", *Res. News*, October, 379–380, 1988.

204. Sinha, A. K. and T. E. Smith, "Thermal Stresses and Cracking Resistance of Dielectric Films", *J. Appl. Phys.*, 49, 2423–2426, 1978.

205. Retajczyk, T. F. J. and A. K. Sinha, "Elastic Stiffness and Thermal Expansion Coefficients of Various Refractory Silicides and Silicon Nitride Films", *Thin Solid Films*, 70, 241–247, 1980.

206. Rosler, R. S., "The Evolution of Commercial Plasma Enhanced CVD Systems", *Solid State Technol.*, June, 67–71, 1991.

207. Wu, T. H. T. and R. S. Rosler, "Stress in PSG and Nitride Films as Related to Film Properties and Annealing", *Solid State Technol.*, May, 65–71, 1992.

208. Sakimoto, M., H. Yoshihara, and T. Ohkubo, "Silicon Nitride Single-Layer X-Ray Mask", *J. Vac. Sci. Technol.*, 21, 1017–1021, 1982.

209. Lee, J. G., S. H. Choi, T. C. Ahn, C. G. Hong, P. Lee, K. Law, M. Galiano, P. Keswick, and B. Shin, "SA CVD : A New Approach for 16 Mb Dielectrics", *Semicond. Int.*, May, 115–120, 1992.

210. Bonifield, T., K. Hewes, B. Merritt, R. Robinson, S. Fisher, and D. Maisch, " Extended Run Evaluation of TEOS/Ozone BPSG Deposition", *Semicond. Int.*, July, 200–204, 1993.

211. Akiyama, T., "Controlled Stepwise Motion in Polysilicon Microstructures", *J. Microelectromech. Syst.*, 2, 106–110, 1993.

212. Akiyama, T. and K. Shono, "A New Step Motion of Polysilicon Microstructures", Proceedings. IEEE Micro Electro Mechanical Systems, (MEMS '93), Fort Lauderdale, FL, 1993, p. 272–277.

# 6

# LIGA

## Introduction

LIGA is the German acronym for X-ray lithography (X-ray lithographie), electrodeposition (galvanoformung), and molding (abformtechnik). The process involves a thick layer of X-ray resist — from microns to centimeters — and high-energy X-ray radiation exposure and development to arrive at a three-dimensional resist structure. Subsequent electrodeposition fills the resist mold with a metal and, after resist removal, a free standing metal structure results (see Figure 6.1).[1]* The metal shape may be a final product or serve as a mold insert for precision plastic injection molding. Injection-molded plastic parts may in turn be final products or lost molds. The plastic mold retains the same shape, size, and form as the original resist structure but is produced quickly and inexpensively as part of an infinite loop. The plastic lost mold may generate metal parts in a second electroforming process or ceramic parts in a slip casting process.

Micromachining techniques are reshaping manufacturing approaches for a wide variety of small parts; frequently, IC-based, batch microfabrication methods are being considered for their fabrication together with more traditional, serial machining methods. LIGA, as a 'handshake-technology' between IC and classical manufacturing technologies, has the potential of speeding up this process. The method borrows lithography from the IC industry and electroplating and molding from classical manufacturing. The capacity of LIGA for creating a wide variety of shapes from different materials makes it akin to classical machining with the added benefit of unprecedented aspect ratios and absolute tolerances. The LIGA bandwidth of possible sizes in three dimensions makes it potentially useful, not only for microstructure manufacture itself (micron and submicron dimensions), but also for the manufacture of microstructure packages (millimeter and centimeter dimensions) as well as for connectors from those packages (electrical, e.g., through-vias or physical, e.g., gas in- and outlets) to the 'macroworld'.

Given the cost of the LIGA equipment, various pseudo-LIGA processes are under development. They involve replication of molds created by alternate means such as deep, cryogenic dry etching and novel ultraviolet thick photoresists.

Some aspects of X-ray lithography for use in LIGA were explored in Chapter 1. After an historical introduction we will analyze all of the process steps depicted in Figure 6.1 in detail,

starting with a description of the different applications and technical characteristics of synchrotron radiation, followed by an introduction to the crucial issues involved in making X-ray masks optimized for LIGA. Alternative process sequences, popular in LIGA micromanufacturing today, will be reviewed as well. After an introduction to the technology we will review current and projected LIGA applications and discuss key technological barriers to be overcome and consider competing 'LIGA-like' processes.

## LIGA Processes — Introduction to the Technology

### History

LIGA combines the sacrificial wax molding method, known since the time of the Egyptians, with X-ray lithography and electrodeposition. Combining electrodeposition and X-ray lithography was first carried out by Romankiw and coworkers at IBM in 1975.[2] These authors made high aspect ratio metal structures by plating gold in X-ray-defined resist patterns of up to 20 μm thick. They had, in other words, already invented 'LIG'; i.e., LIGA without the abformtech (molding).[2] This IBM work formed an extension to through mask plating, also pioneered by Romankiw in 1969, geared toward the fabrication of thin-film magnetic recording heads.[3] The addition of molding to the lithography and plating process was realized by Ehrfeld et al.[4] at KfK in 1982. By adding molding, Ehrfeld and co-workers made the technology potentially low cost. These pioneers recognized the broader implications of LIGA as a new means of low-cost manufacturing of a wide variety of microparts previously impossible to batch fabricate.[4] In Germany, LIGA originally developed almost completely outside of the semiconductor industry. It was Guckel in the U.S. who repositioned the field in light of existing semiconductor capabilities and brought it closer to standard manufacturing processes.

The development of the LIGA process initiated by the Karlsruhe Nuclear Research Center (Kernforschungszentrum Karlsruhe, or KfK) was intended for the mass production of micron-sized nozzles for uranium-235 enrichment (see Figure 6.2).[4] The German group used synchrotron radiation from a 2.5-GeV storage ring for the exposure of the poly(methylmethacrylate) (PMMA) resist.

---

\* Figure 6.1 appears as a color plate after page 144.

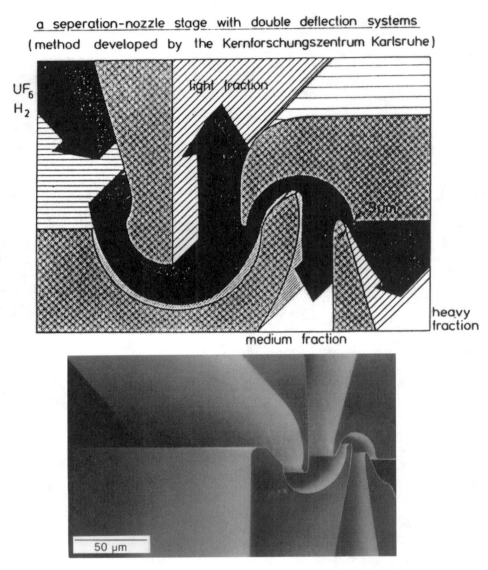

**FIGURE 6.2** Scanning electron micrograph of a separation nozzle structure produced by electroforming with nickel using a micromolded PMMA template. This nozzle represents the first actual product ever made by LIGA. (From Hagmann, P. et al., Fabrication of Microstructures with Extreme Structural Heights by Reaction Injection Molding, presented at First Meeting of the European Polymer Federation, European Symp. on Polymeric Materials, Lyon, France, 1987. With permission.)

Today LIGA is researched in many laboratories around the world. The technology has reinvigorated attempts at developing alternative micromolds for the large scale production of precise micromachines. Mold inserts, depending on the dimensions of the microparts, the accuracy requirements, and the fabrication costs are realized by e-beam writing, deep UV resists, excimer laser ablation, electrodischarge machining, laser cutting, and X-ray lithography as involved in the LIGA technique.

## Synchrotron Orbital Radiation — General Applications

Lithography based on synchrotron radiation, also called synchrotron orbital radiation (SOR), is primarily being developed with the aim of adopting the technology as an industrial tool for the eventual large-scale manufacture of microelectronic circuits with characteristic dimensions in the submicron range.[5,6]

Synchrotron radiation sources outshine all other sources of X-rays. They emit a much higher flux of usable collimated X-rays, thereby allowing shorter exposure times and larger throughputs. Pros and cons of X-ray radiation for lithography in IC manufacture, identified in Chapter 1, are summarized in Table 6.1.

Despite the many promising features of X-ray lithography, the technique lacks mainstream acceptance in the IC industry. Continued improvements in optical lithography outpace the industrial use of X-ray lithography for IC applications. Its use for prototype development on a small scale will no doubt continue its course though. In 1991, experts projected that X-ray lithography could be in use by 1995 for 64 megabit DRAM manufacture, with critical dimensions (CDs) around 0.3 to 0.4 μm. With more certainty, they projected that the transition to X-rays would occur with the 0.2- to 0.3-μm CDs of 256 megabit DRAMs, expected to be in production by 1998.[6] The

**TABLE 6.1**  Pros and Cons of SOR X-Ray Lithography for IC Manufacture

| Pros | Cons |
|---|---|
| Lithography process insensitive to resist thickness, exposure time, development time (large depth of focus) | Resist not very sensitive (not too important because of the intense light source) |
| Absence of backscattering results in insensitivity to substrate type, reflectivity and topography, pattern geometry and proximity, dust and contamination | Masks very difficult and expensive to make |
| High resolution < 0.2 μm | Very high start-up investment |
| High throughput | Not proven as a manufacturing system yet |
| | Radiation effects on $SiO_2$ |

first projected date has passed without materialization of the industrial use of X-rays, many are doubtful that the second prediction will be met.[7]

In addition to the primary intended application — the production of the newest generation of DRAM chips — X-rays may also be used in the fabrication of microsensors and other microstructures, especially since the LIGA method enables the fabrication of high aspect ratio microstructures from a wide variety of materials. In LIGA, synchrotron radiation is used in the lithography step only. Other micromachining applications for SOR exist. Urisu and his colleagues, for example, explored the use of synchrotron radiation for radiation-excited chemical-vapor deposition and etching.[3] Micromachinists are well aware of the need to piggy-back X-ray lithography research and development efforts for the fabrication of micromachines onto major IC projects. The use of X-ray lithography for fabricating microdevices, other than integrated circuits, does not appear to present a large business opportunity yet. Not having a major IC product line associated with X-ray lithography makes it extra hard to justify the use of X-ray lithography for micromachining, especially since other, less expensive micromachining technologies have not yet opened up the type of mass markets one is used to in the IC world. The fact that the X-rays used in LIGA are shorter wavelength than in the IC application (2 to 10 Å vs. 20 to 50 Å) also puts micromachinists at a disadvantage. For example, the soft X-rays in the IC industry may eventually be generated from a much less expensive source, such as a transition radiation source.[9] Also, nontraditional IC materials are frequently employed in LIGA. Fabricating X-ray masks pose more difficulties than masks for IC applications. Rotation and slanting of the X-ray masks may be required to craft nonvertical walls. All these factors make exploring LIGA in the current economic climate a major challenge. However, given sufficient research and development money, large markets are almost certain to emerge over the next 10 years. These markets may be found in the manufacture of devices with stringent requirements imposed on resolution, aspect ratio, structural height, and parallelism of structural walls. Optical applications seem particularly important product targets. The commercial exploitation of LIGA is

being aggressively pursued by two German companies: Microparts* (formed in 1990) and IMM** (formed in 1991).

So far, the research community has primarily benefited from the availability of SOR photon sources. With its continuously tunable radiation across a very wide photon range, also highly polarized and directed into a narrow beam, SOR provides a powerful probe of atomic and molecular resonances. Other types of photon sources prove unsatisfactory for these applications in terms of intensity or energy spread. As can be concluded from Table 6.2, applications of SOR beyond lithography range from structural and chemical analysis to microscopy, angiography, and even to the preparation of new materials.

**TABLE 6.2**  SOR Applications

| Application Area | Instruments/Technologies Needed |
|---|---|
| Structural analysis | |
|   Atoms | Photoelectron spectrometers |
|   Molecules | Absorption spectrometers |
|   Very large molecules | Fluorescent spectrometers |
|   Proteins | Diffraction cameras |
|   Cells | Scanning electron microscope (to view topographical radiographs) |
|   Polycrystals | Time-resolved X-ray diffractometers |
| Chemical analysis | |
|   Trace | Photoelectron spectrometers |
|   Surface | (Secondary ion) mass spectrometer |
|   Bulk | Absorption/fluorescence spectrometers |
| | Vacuum systems |
| Microscopy | |
|   Photoelectron | Photoemission microscopes |
|   X-ray | X-ray microscopes SEM (for viewing) |
| | Vacuum systems |
| Micro/nanofabrication | |
|   X-ray lithography | Steppers, mask making |
|   Photochemical deposition of thin films | Vacuum systems |
|   Etching | LIGA process |
| Medical diagnostics | |
|   Radiography | X-ray cameras and equipment |
|   Angiography and tomography | Computer-aided display |
| Photochemical reactions | |
|   Preparation of novel materials | Vacuum systems |
| | Gas-handling equipment |

*Source:*  After Muray, J.J. and Brodie, I., Study on Synchroton Orbital Radiation (SOR) Technologies and Applications, SRI International 2019, 1991.

---

\*  Dortmund, Griesbachstrasse 10, D-7500 Karlsruhe 21; tel: 0721/84990; fax: 0721/857865.

\*\*  Mainz-Hechtsheim, Carl-Zeiss-Str. 18-20, D-55129; tel: 06131/990-0; fax: 06131/990-205.

## Synchrotron Radiation — Technical Aspects

Some important concepts associated with synchrotron radiation, such as the bending radius of the synchrotron magnet, magnetic field strength, beam current, critical wavelength, and total radiated power require introduction. Figure 6.3 presents a schematic of an X-ray exposure station. The cone of radiation in this figure is the electromagnetic radiation emitted by electrons due to the radial acceleration that keeps them in orbit in an electron synchrotron or storage ring. For high energy particle studies, this radiation, emitted tangential to the circular electron path, limits the maximum energy the electrons can attain. This so-called 'Bremsstrahlung' is a nuisance for studies of the composition of the atomic nucleus in which high energy particles are smashed into the nucleus. To minimize this problem, physicists desire ever bigger synchrotrons. For X-ray lithography applications, electrical engineers want to maximize the X-ray emission and build small synchrotrons instead (the radius of curvature for a compact, superconducting synchrotron, for example, is 2 m). The angular opening of the radiation cone in Figure 6.3 is determined by the electron energy, E, and is given by:

$$\theta \approx \frac{mc^2}{E} = \frac{0.5}{E(GeV)} \ (mrad) \qquad 6.1$$

The X-ray light bundle with the cone opening, $\theta$, describes a horizontal line on an intersecting substrate as the X-ray bundle is tangent to the circular electron path. In the vertical direction, the intensity of the beam exhibits a Gaussian distribution and the vertical exposed height on the intersecting substrate can be calculated knowing $\theta$ and R, the distance from the radiation point, P, to the substrate. With E = 1 GeV, $\theta$ = 1 mrad and R = 10 m, the exposed area in the vertical direction measures about 0.5 cm. To expose a substrate homogeneously over a wider vertical range, the sample must be moved vertically through the irradiation band with a precision scanner (e.g., done at a speed of 10 mm/sec over a 100 mm scanning distance). Usually the substrate is stepped up and down repeatedly until the desired X-ray dose is obtained.

The electron energy E in Equation 6.1 is given by:

$$E(SeV) = 0.29979 \ B \ (Tesla) \ \rho \ (meters) \qquad 6.2$$

with B the magnetic field and $\rho$ the radius of the circular path of the electrons in the synchrotron.

The total radiated power can be calculated from the energy loss of the electrons per turn and is given by:

$$P(kW) = \frac{88.47E^4 i}{\rho} \qquad 6.3$$

with i the beam current.

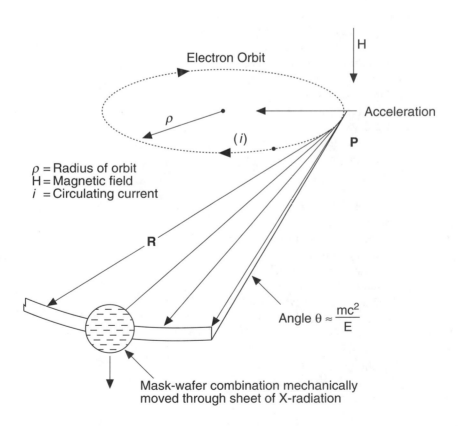

FIGURE 6.3 Schematic of an X-ray exposure station with a synchrotron radiation source. The X-ray radiation cone (opening, $\theta$) is tangential to the electron's path, describing a line on an intersecting substrate.

The spectral emission of the synchrotron electrons is a broad spectrum without characteristic peaks or line enhancements, and its distribution extends from the microwave region through the infrared, visible, ultraviolet, and into the X-ray region. The critical wavelength, $\lambda_c$, is defined so that the total radiated power at lower wavelengths equals the radiated power at higher wavelengths, and is given by:

$$\lambda_c(\text{Å}) = \frac{5.59\rho(m)}{E^3(\text{GeV})} \qquad 6.4$$

Equation 6.3 shows that the total radiated power increases with the fourth power of the electron energy. From Equation 6.4 we appreciate that the spectrum shifts towards shorter wavelengths with the third power of the electron energy.

The dose variation absorbed in the top vs. the bottom of an X-ray resist should be kept small so that the top layer does not deteriorate before the bottom layers are sufficiently exposed. Since the depth of penetration increases with decreasing wavelength, synchrotron radiation of very short wavelength is needed to pattern thick resist layers. In order to obtain good aspect ratios in LIGA structures, the critical wavelength ideally should be 2 Å. Bley et al.[12] at KfK, designed a new synchrotron optimized for LIGA. They proposed a magnetic flux density, B, of 1.6285 T; a nominal energy, E, of 2.3923 GeV; and a bending radius, $\rho$, of 4.9 m. With those parameters, Equation 6.4 results in the desired $\lambda_c$ of 2 Å and an opening angle of radiation, based on Equation 6.1, of 0.2 mrad (in practice this angle will be closer to 0.3 mrad due to electron beam emittance).

The X-rays from the ring to the sample site are held in a high vacuum. The sample itself is either kept in air or in a He atmosphere. The inert atmosphere prevents corrosion of the exposure chamber, mask, and sample by reactive oxygen species, and removal of heat is much faster than in air (the heat conductivity of He is high compared to air). In He, the X-ray intensity loss is also 500 times less than in air. A beryllium window separates the high vacuum from the inert atmosphere. For wavelengths shorter than 1 nm, Be is very transparent — i.e., an excellent X-ray window.

## Access to Technology

Unfortunately, the construction cost for a typical synchrotron today totals over 30 million dollars, restricting the access to LIGA dramatically. Obviously, one would like to find less expensive alternatives for generating intense X-rays. In Japan, companies such as Ishikawajima-Harima Heavy Industries (IHI) are building compact synchrotron X-ray sources (e.g., a 800-MeV synchrotron will be about 30 feet per side).

By the end of 1993, eight nonprivately owned synchrotrons were in use in the U.S. The first privately owned synchrotron was put into service in 1991 at IBM's Advanced Semiconductor Technology Center (ASTC) in East Fishkill, NY. Table 6.3 lists the eight U.S. synchrotron facilities.

Some of the facilities listed in Table 6.3 can perform micromachining work. For example, MCNC (Research Triangle Park,

**TABLE 6.3** Access to Synchrotron Radiation Available at the Following Facilities in the U.S.

| Facility | Institute | Contact (1997) |
|---|---|---|
| Advanced Photon Source | Argonne National Laboratory | Susan Barr, Derek Mancini |
| Cornell High Energy Synchrotron Source | Cornell University | Proposal Administrator |
| National Synchrotron Light Source | Brookhaven National Laboratory | Susan White-DePace |
| Stanford Synchrotron Radiation Laboratory | Stanford University | Katherine Cantwell |
| SURF | National Institute of Standards and Technology | Robert Madden |
| Synchrotron Radiation Center | University of Wisconsin-Madison | Pamela Layton |
| Center for Advanced Microdevices (CAMD) | Louisiana State University | Volker Saile |
| Advanced Light Source (ALS) | Lawrence Berkeley Laboratory | Alfred Schlachter |

NC), in collaboration with the University of Wisconsin-Madison, announced its first multi-user LIGA process sponsored by ARPA in September 1993. CAMD at Louisiana State University has three beam lines on their synchrotron exclusively dedicated to micromachining work; ALS at Berkeley has one beam line available for micromachining.

Forschungszentrum Karlsruhe GmbH* also offers a multi-user LIGA service (LEMA, or LIGA-experiment for multiple applications).

## X-Ray Masks

### Introduction

A short introduction to X-ray masks was presented in Chapter 1 on lithography. X-ray masks should withstand many exposures without distortion, be alignable with respect to the sample, and be rugged. A possible X-ray mask architecture and its assembly with a substrate in an X-ray scanner are shown in Figure 6.4. The mask shown here has three major components: an absorber, a membrane or mask blank, and a frame. The absorber contains the information to be imaged onto the resist. It is made up of a material with a high atomic number (Z), often Au, patterned onto a membrane material with a low Z. The high-Z material absorbs X-rays, whereas the low-Z material transmits X-rays. The frame lends robustness to the membrane/absorber assembly so that the whole can be handled confidently.

The requirements for masks in LIGA differ substantially from those for the IC industry. A comparison is presented in Table 6.4.[13] The main difference lies in the absorber thickness. In order to achieve a high contrast (>200), very thick absorbers (>10 µm vs. 1 µm) and highly transparent mask blanks (transparency > 80%) must be used because of the low resist sensitivity and the great

* Forschungszentrum Karlsruhe GmbH; Postfach 3640; D-76021 Karlsruhe.

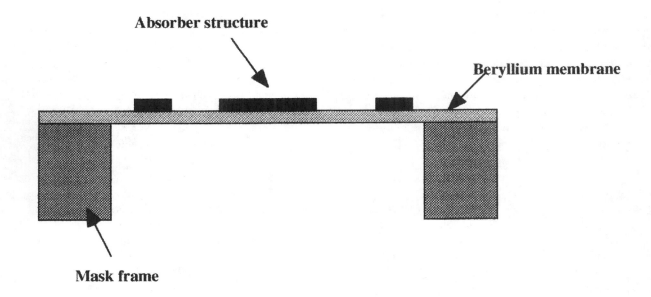

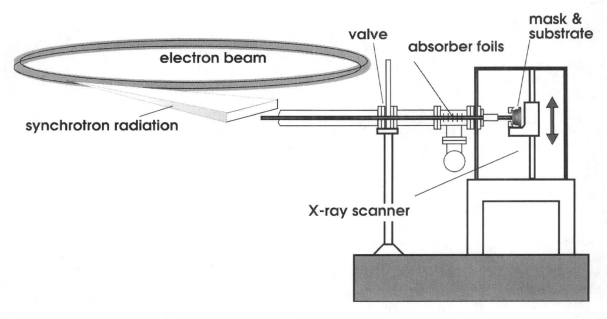

**FIGURE 6.4** Schematic of a typical X-ray mask and mask and substrate assembly in an X-ray scanner. (The latter from IMM, 1995 Brochure. With permission.)

**TABLE 6.4**  Comparison of Masks in LIGA and the IC Industry[13]

| Attribute | Semiconductor Lithography | LIGA Process |
|---|---|---|
| Transparency | ≥50% | ≥80% |
| Absorber thickness | ±1 µm | 10 µm or higher |
| Field size | 50 × 50 mm² | 100 × 100 mm² |
| Radiation resistance | = 1 | = 100 |
| Surface roughness | <0.1 µm | <0.5 µm |
| Waviness | < ±1 µm | < ±1 µm |
| Dimensional stability | <0.05 µm | <0.1–0.3 µm |
| Residual membrane stress | ~ 10⁸ Pa | ~ 10⁸ Pa |

depth of resist. Another difference focuses on the radiation stability of membrane and absorber. For conventional optical lithography, the supporting substrate is a relatively thick, near optically flat piece of glass or quartz highly transparent to optical wavelengths. It provides a highly stable (>10⁶ µm) basis for the thin (0.1 µm) chrome absorber pattern. In contrast, the X-ray mask consists of a very thin membrane (2 to 4 µm) of low-Z material carrying a high-Z thick absorber pattern.[14] A single exposure in LIGA results in an exposure dose a hundred times higher than in the IC case.

We will look into these different mask aspects separately before detailing a process with the potential of obviating the need for a separate X-ray mask by using a so-called transfer mask.

## X-Ray Membrane (Mask Blank)

The low-Z membrane material in an X-ray mask must have a transparency for rays with a critical wavelength, $\lambda_c$, from 0.2 to 0.6 nm of at least 80% and should not induce scattering of those rays. To avoid pattern distortion, the residual stress, $\sigma_r$, in the membrane should be less than $10^8$ dyn/cm². Mechanical stress in the absorber pattern can cause in-plane distortion of the supporting thin membrane, necessitating a high Young's modulus for the membrane material. Humidity or high deposited doses of X-ray might also distort the membrane. During one typical lithography step the masks may be exposed to 1 MJ/cm² of X-rays. Since most membranes must be very thin for optimum transparency, a compromise must be found between transparency, strength, and form stability. Important X-ray membrane materials are listed in Table 6.5. The higher radiation dose in LIGA prevents the use of BN as well as compound mask blanks which incorporate a polyimide layer. Those mask blanks are perfectly appropriate for classical IC lithography work but will not do for LIGA work. Mask blanks of metals such as titanium (Ti) and beryllium (Be) were specifically developed for LIGA applications because of their radiation hardness.[15,15] In comparing titanium and beryllium membranes, beryllium can have a much greater membrane thickness, d, and still be adequately transparent. For example, a membrane transparency of 80%, essential for adequate exposure of a 500-μm thick PMMA resist layer can be achieved with a thin 2-μm titanium film, whereas with beryllium a thick 300-μm membrane must be used. The beryllium membrane can thus be used at a thickness permitting easier processing and handling. In addition, beryllium has a greater Young's modulus E than titanium (330 vs. 140 kN/mm²) and, since it is the product of E × d which determines the amount of mask distortion, distortions due to absorber stress should be much smaller for beryllium blanks.[15,16] Beryllium comes forward as an excellent membrane material for LIGA because of its high transparency and excellent damage resistance. Such a mask should be good for up to 10,000 exposures and may cost $20 to 30 K ($10 to 15 K in quantity). Stoichiometric silicon nitride (Si₃N₄) used in X-ray mask membranes may contain numerous oxygen impurities absorbing X-rays and thus producing heat. This heat often suffices to prevent the use of nitride as a good LIGA mask. Single crystal silicon masks have been made (1 cm square and 0.4 μm thick and 10 cm square and 2.5 μm thick) by electrochemical etching techniques. For Si and Si₃N₄, the Young's modulus is quite low compared to CVD-grown diamond and SiC films, which are as much as three times higher. These higher stiffness materials are more desirable because the internal stresses of the absorbers, which can distort mask patterns, are less of an issue.

## Absorber

The requirements on the absorber are high attenuation (>10 db), stability under radiation over extended periods of time, negligible distortion (stress < $10^8$ dyn/cm²), ease of patterning, repairable and low defect density. A listing of typical absorber materials is presented in Table 6.6. Gold is used most commonly. Several groups are looking at the viability of tungsten and other materials. In the IC industry an absorber thickness of 0.5 μm might suffice, whereas LIGA deals with thicker layers of resist requiring a thicker absorber to maintain the same resolution.

Figure 6.5 illustrates how X-rays, with a characteristic wavelength of 0.55 nm, are absorbed along their trajectory through a Kapton preabsorber filter, an X-ray mask, and resist.[17] The low energy portion of the synchrotron radiation is absorbed mainly in the top portion of the resist layer, since absorption increases with increasing wavelength. The Kapton preabsorber filters out much of the low energy radiation in order to prevent overexposure of the top surface of the resist. The X-ray dose at which the resist gets damaged, $D_{dm}$, and the dose required for development of the resist, $D_{dv}$, as well as the 'threshold-dose' at which the resist starts dissolving in a developer, $D_{th}$, are all indicated in Figure 6.5. In the areas under the absorber pattern of the X-ray mask, the absorbed dose must stay below the 'threshold-dose', $D_{th}$; otherwise, the structures partly dissolve, resulting in poor feature definition. From Figure 6.5 we can deduce that the height of the gold absorbers must exceed 6 μm to reduce the

**TABLE 6.5** Comparison of Membrane Materials for X-Ray Masks

| Material | X-Ray Transparency | Nontoxicity | Dimensional Stability | Remark |
|---|---|---|---|---|
| Si | 0 | −+ | 0 | Single crystal Si, well developed, rad hard, stacking faults cause scattering, material is brittle |
| SiNₓ | 0 | −+ | 0 | Amorphous, well developed, rad hard if free of oxygen, resistant to breakage |
| SiC | + | −+ | ++ | Poly and amorphous, rad hard, some resistance to breakage |
| Diamond | + | −+ | ++ | Poly, research only, highest stiffness |
| BN | + | −+ | 0 | Not rad hard, i.e., not applicable for LIGA |
| Be | ++ | − | ++ | Research, especially suited for LIGA; even at 100 μm the transparency is good, 30 μm typical; difficult to electroplate; toxic material |
| Ti | − | ++ | 0 | Research, used for LIGA, not very transparent, films must not be more than 2 to 3 μm thick |

**TABLE 6.6** Comparison of Absorber Materials for X-Ray Masks

| Material | Remark |
|----------|--------|
| Gold | Not the best stability (grain growth), low stress, electroplating only, defects repairable |
| Tungsten | Refractory and stable; special care is needed for stress control; dry etchable; repairable |
| Tantalum | Refractory and stable; special care is needed for stress control; dry etchable; repairable |
| Alloys | Easier stress control, greater thickness required to obtain 10 db |

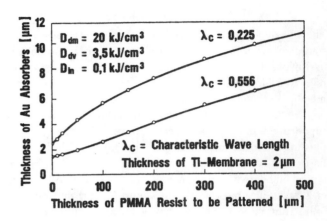

**FIGURE 6.6** Necessary thickness of the gold absorbers of an X-ray mask. (From Bley, P. et al., Application of the LIGA Process in Fabrication of Three-Dimensional Mechanical Microstructures, presented at 1991 Int. Micro Process Conf. With permission.)

is that larger resist areas can be exposed since one does not depend on a fragile membrane/absorber combination.[20]

## Thick Absorber Fabrication

To make a mask with gold absorber structures of a height above 10 µm, one must first succeed in structuring a resist of that thickness. The height of the resist should in fact be a bit higher (say, 20%) than the absorber itself in order to accommodate the electrodeposited material fully in between the resist features. Currently, no means to structure a resist of that height with sufficient accuracy and perfect verticality of the walls exist, unless X-rays are used. Different procedures for producing X-ray masks with thicker absorber layers using a two-stage lithography process have been developed.

The KfK solution calls for first making an intermediate mask with photo or e-beam lithography. This intermediate mask starts with a 3-µm thick resist layer, in which case the needed accuracy and steepness of the walls of printed features can be obtained through several means. After gold plating in between the resist features and stripping of the resist, this intermediate mask is used to write a pattern with X-rays in a 20-µm thick resist. After electrodepositing and resist stripping, the latter will become the actual X-ray mask, i.e., the master mask.

Since hardly any accuracy is lost in the copying of the intermediate mask with X-rays to obtain the master mask, it is the intermediate mask quality that determines the ultimate quality of the LIGA-produced microstructures. The structuring of the resist in the intermediate mask is handled with optical techniques when the requirements on the LIGA structures are somewhat relaxed. The minimal lateral dimensions for optical lithography in a 3-µm thick resist typically measure about 2.5 µm. Under optimum conditions, a wall angle of 88° is achievable. With electron-beam lithography, a minimum lateral dimension of 1 µm is feasible. The most accurate pattern transfer is achieved through reactive ion etching of a tri-level resist system. In this approach, a 3- to 4-µm thick polyimide resist is first coated onto the titanium or beryllium membrane, followed

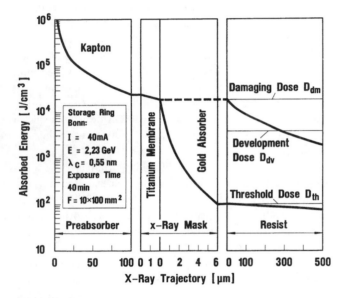

**FIGURE 6.5** Absorbed energy along the X-ray trajectory including a 500-µm thick PMMA specimen, X-ray mask, and a Kapton preabsorber. (From Bley, P. et al., Application of the LIGA Process in Fabrication of Three-Dimensional Mechanical Microstructures, presented at 1991 Int. Micro Process Conf. With permission.)

absorbed radiation dose of the resist under the gold pattern to below the threshold dose, $D_{th}$. In Figure 6.6, the necessary thickness of the gold absorber patterns of an X-ray mask is plotted as a function of the thickness of the resist to be patterned; the Au must be thicker for thicker resist layers and for shorter characteristic wavelengths, $\lambda_c$, of the X-ray radiation. In order to pattern a 500-µm high structure with a $\lambda_c$ of 0.225 nm, the gold absorber must be more than 11 µm high.

Exposure of yet more extreme photoresist thicknesses is possible if proper X-ray photon energies are used. At 3000 eV, the absorption length in PMMA roughly measures 100 µm, which enables the above-mentioned 500-µm exposure depth.[18] Using 20,000-eV photons results in absorption lengths of 1 cm. PMMA structures up to 10 cm (!) thick have been exposed this way.[19] A high energy mask used by Guckel for these high energy exposures has an Au absorber 50 µm thick and a blank membrane of 400-µm thick Si. Guckel obtained an absorption contrast of 400 when exposing a 1000-µm thick PMMA sheet with this mask. An advantage of being able to use these thick Si blank membranes by a coat of 10 to 15 nm titanium deposited with magnetron sputtering. The thin layer of titanium is an excellent etch mask for the polyimide; in an optimized oxygen plasma, the titanium etches 300 times slower than the polyimide. To structure the thin titanium layer itself, a 0.1-μm thick optical resist is used. Since this top resist layer is so thin, excellent lateral tolerances result. The thin Ti layer is patterned with optical photolithography and etched in an argon plasma. After etching the thin titanium layer, exposing the polyimide locally, an oxygen plasma helps to structure the polyimide down to the titanium or beryllium membrane. Lateral dimensions of 0.3 μm can be obtained in this fashion. Patterning the top resist layer with an electron-beam increases the accuracy of the three-level resist method even further. Electrodeposition of gold on the titanium or beryllium membrane and stripping of the resist finishes the process of making the intermediate LIGA mask. To make a master mask, this intermediate mask is printed by X-ray radiation onto a PMMA resist coated master mask. The PMMA thickness corresponds to a bit more than the desired absorber thickness. Since the resist layer thickness is in the 10- to 20-μm range, a synchrotron X-ray wavelength of 10 Å is adequate for the making of the master mask. A further improvement in LIGA mask making is to fabricate intermediate and master mask on the same substrate, greatly reducing the risk for deviations in dimensions caused, for example, by temperature variations during printing.[21] The ultimate achievement would be to create a one-step process to make the master mask. Along this line of work, Hein et al.[16] have started an investigation in the direct patterning of 10-μm high resist layers with a 100-kV electron-beam.

A completely different approach is suggested by Friedrich from the Institute of Manufacturing in Louisiana. He suggested the fabrication of an X-ray mask by traditional machining methods, such as micromilling, micro-EDM (electrodischarge machining), and lasers.[22] The advantages of this proposal are rapid turnaround (less than 1 day per mask), low cost, and flexibility, as no intermediate steps interfere. Disadvantages are

less dimensional edge acuity and nonsharp interior corners, as well as less absolute tolerance.

## Alignment of the X-Ray Mask to the Substrate

The mask and resist-coated substrate must be properly registered to each other before they are put in an X-ray scanner. Alignment of an X-ray mask to the substrate is problematic since no visible light can pass through most X-ray membranes. To solve this problem, Schomburg et al.[15] etched windows in their Ti X-ray membrane. Diamond membranes have a potential advantage here, as they are optically transparent and enable easy alignment for multiple irradiations without a need for etched holes.

Figure 6.7 illustrates an alternative, innovative X-ray alignment system involving capacitive pickup between conductive metal fingers on the mask and ridges on a small substrate area; Si, in this case (U.S. Patent 4,607,213 [1986] and 4,654,581 [1987]). When using multiple groups of ridges and fingers, two axis lateral and rotational alignment become possible.

Another alternative may involve liquid nitrogen-cooled Si (Li) X-ray diodes as alignment detectors, eliminating the need for observation with visible light.[23]

## Transfer Mask for High Aspect Ratio Micro-Lithography

Recently, Vladimirsky et al.[24] developed a procedure to eliminate the need for an X-ray mask membrane. Unlike conventional masks, the so-called 'X-ray transfer mask' does not treat a mask as an independent unit. The technique is based on forming an absorber pattern directly on the resist surface. An example process sequence is shown in Figure 6.8. In this process, a transfer mask plating base is first prepared on the PMMA substrate plate by evaporating 70 Å of chromium (as adhesion layer) followed by 500 Å of gold using an electron beam evaporator. A 3-μm thick layer of standard novolak-based AZ-type resist S1400-37 (Shipley Co.) is then applied over the plating

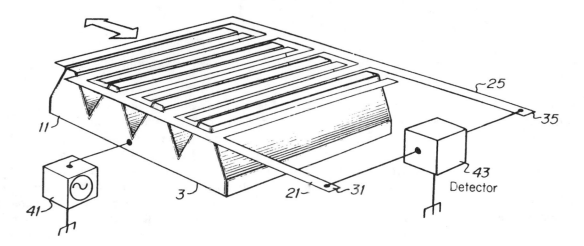

**FIGURE 6.7** Mask alignment system in X-ray lithography. Conductive fingers on the mask and ridges on the Si are used for alignment. (From U.S. Patents 4,654,581 [1987] and 4,607,213 [1986].)

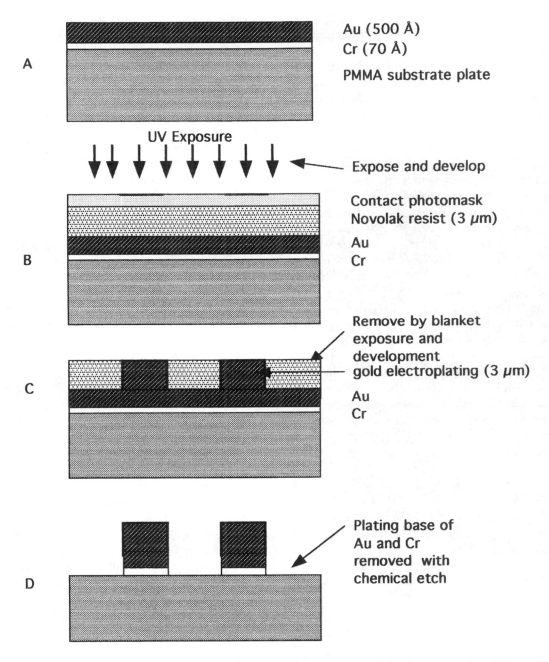

**FIGURE 6.8**  Sample transfer mask formation. (After Vladimirsky, Y. et al., Transfer Mask for High Aspect Ratio Micro-Lithography, presented at Microlithography '95, Santa Clara, CA, 1995.)

base and exposed in contact mode through an optical mask using an ultraviolet exposure station. After development, the transfer mask is further completed by 3-μm gold electroplating on the exposed plating base. The remaining resist is removed by a blanket exposure and subsequent development, and the 500 Å of Au plating base is dissolved by a dip of 20 to 30 sec in a solution of KI (5%) and I (1.25%) in water; the Cr adhesion layer is removed by a standard chromium etch (from KTI). Fabrication of the transfer mask can thus be performed using standard lithography equipment available at almost any lithography shop. Depending on the resolution required, the X-ray transfer mask can be fabricated using known photon, electron-beam, or X-ray lithography techniques. New lithography and post lithography techniques, such as *in situ* development,

etching, and deposition can now be realized by using a transfer mask. The presence of the membrane in a conventional mask practically precludes use of *in situ* development and is not compatible with etching and deposition due to physical obstruction and material resistance considerations. The patterning of the PMMA resist with a transfer mask can be accomplished in multiple steps of exposure and development. An example of a cylindrical resonator made this way is shown in Figure 6.9. Each exposure/development step involved an exposure dose of about 8 to 12 J/cm². Subsequent 5-min development steps removed ~30 μm of PMMA. With seven steps, a self-supporting 1.5-mm thick PMMA resist was patterned to a depth of more than 200 μm. The resist pattern shown in Figure 6.9 is 230 μm thick and exhibits a 2-μm gap between the inner cylinder and the

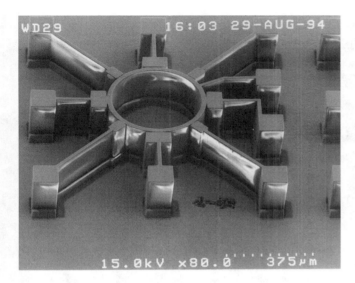

**FIGURE 6.9** SEM micrograph of a cylindrical PMMA resonator made by the Transfer Mask method and multiple exposure/development steps. (From Vladimirsky, Y. et al., Transfer Mask for High Aspect Ratio Micro-Lithography, presented at Microlithography '95, Santa Clara, CA, 1995. With permission.) (Courtesy of Dr. V. Saile.)

pickup electrodes (aspect ratio is 100:1). The pattern was produced using soft (=10 Å) X-rays and a 3-μm thick Au absorber only.

Vladimirsky et al.[24] suggest that the forming of the transfer mask directly on the sample surface creates several additional new opportunities, besides the already mentioned *in situ* development, etching, and deposition, these include exposure of samples with curved surfaces and dynamic deformation of a sample surface during the exposure (hemispherical structures would become possible this way). Elegant and cost-saving innovations like these could mainstream LIGA faster than currently believed.

## LIGA Process Steps and Materials

### Choice of Primary Substrate

In the LIGA process, the primary substrate, or base plate, must be a conductor or an insulator coated with a conductive top layer. A conductor is required for subsequent electrodeposition. Some examples of primary substrates that have been used successfully are austenite steel plate, Si wafers with a thin Ti or Ag/Cr top layer,[25] and copper plated with gold, titanium, or nickel.[21] Other metal substrates as well as metal-plated ceramic, plastic, and glass plates have been employed.[26] It is important that the plating base provide good adhesion for the resist. For that purpose, prior to applying the X-ray resist on copper or steel, the surface sometimes is mechanically roughened by microgrit blasting with corundum. Microgrit blasting may lead to an average roughness, $R_a$, of 0.5 μm, resulting in better physical anchoring of the microstructures to the substrate.[27] In the case of a polished metal base, chemical preconditioning may be used to improve adhesion of the resist microstructures. During chemical preconditioning, a titanium layer, sputter-deposited onto the polished metal base plate (e.g., a Cu plate), is oxidized

for a few minutes in a solution of 0.5 $M$ NaOH and 0.2 $M$ H$_2$O$_2$ at 65°C. The oxide produced typically measures 30 nm thick and exhibits a microrough surface instrumental to better securing resist to the base plate. The Ti adhesion layer may further be covered with a thin nickel seed layer (~150 Å) for electroless or eletroplating of nickel in between the resist microstructures. When using a highly polished Si surface, adhesion promoters need to be added to the resist resin (see also Chapter 1 and Resist Adhesion further below). A substrate of special interest is a processed silicon wafer with integrated circuits. Integrating the LIGA process with IC circuitry on the same wafer might greatly improve sensor applications.

Thick resist plates can act as plastic substrates themselves. For example, using 20,000-eV rather than the more typical 3000-eV radiation, Guckel et al.[18,19] exposed plates of PMMA up to 10 cm thick.

### Resist Requirements

An X-ray resist ideally should have high sensitivity to X-rays, high resolution, resistance to dry and wet etching, thermal stability of greater than 140°C and a matrix or resin absorption of less than 0.35 μm$^{-1}$ at the wavelength of interest (see Chapter 1).[28] These requirements equal those for IC production with X-ray lithography.[29] To produce high aspect ratio microstructures with very tight lateral tolerances demands an additional set of requirements. The unexposed resist must be absolutely insoluble during development. This means that a high contrast ($\gamma$) is required. The resist must also exhibit very good adhesion to the substrate and be compatible with the electroforming process. The latter imposes a resist glass transition temperature ($T_g$) greater than the temperature of the electrolyte bath used to electrodeposit metals between the resist features remaining after development (say, at 60°C). To avoid mechanical damage to the microstructures induced by stress curing development, the resist layers should exhibit low internal stresses.[30] If the resist structure is the end product of the fabrication process, further specifications depend on the application itself, e.g., optical transparency and refractive index for optical components or large mechanical yield strength for load-bearing applications. For example, PMMA exhibits good optical properties in the visible and near infrared range and lends itself to the making of all types of optical components.[31]

Due to excellent contrast and good process stability, known from electron-beam lithography, poly(methylmethacrylate) (PMMA) is the preferred resist for deep-etch synchrotron radiation lithography. Two major concerns with PMMA as a LIGA resist are a rather low lithographic sensitivity of about 2J/cm$^2$ at a wavelength $\lambda_c$ of 8.34 Å and a susceptibility to stress cracking. For example, even at shorter wavelengths, $\lambda_c$ = 5 Å, over 90 min of irradiation are required to structure a 500-μm thick resist layer with an average ring storage current of 40 mA and a power consumption of 2 MW at the 2.3-GeV ELSA synchrotron (Bonn, Germany).[32] The internal stress arising from the combination of a polymer and a metallic substrate can cause cracking in the microstructures during development, a phenomenon PMMA is especially prone to, as illustrated in the scanning electron microscope in Figure 6.10.

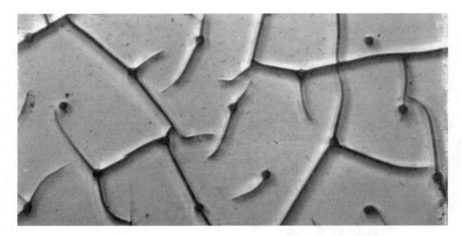

**FIGURE 6.10**  Cracking of PMMA resist. Method to test stress in thick resist layers. The onset of cracks in a pattern of holes with varying size (say 1 to 4 μm) in a resist is shifted towards smaller hole diameter the lower the stress in the film. The SEM picture displays extensive cracking incurred during development of the image in a 5-μm thick PMMA layer on an Au covered Si wafer. The 5-μm thick PMMA layer resulted from five separate spin coats. Annealing pushed the onset of cracking toward smaller holes until the right cycle was reached and no more cracks were visible.[34]

To make throughput for deep-etch lithography more acceptable to industry, several avenues to more sensitive X-ray resists have been pursued. For example, copolymers of PMMA were investigated: methylmethacrylate combined with methacrylates with longer ester side chains show sensitivity increases of up to 32% (with tertiary butylmethacrylate). Unfortunately, a deterioration in structure quality was observed.[33] Among the other possible approaches for making PMMA more X-ray sensitive we can count on the incorporation of X-ray absorbing high-atomic-number atoms or the use of chemically amplified photoresists (see Chapter 1). X-ray resists explored more recently for LIGA applications are poly(lactides), e.g., poly(lactide-co-glycolide) (PLG); polymethacrylimide (PMI); polyoxymethylene (POM); and polyalkensulfone (PAS). Poly(lactide-co-glycolide) is a new positive resist developed by BASF AG, more sensitive to X-rays by a factor of 2 to 3 compared with PMMA. Its processing is less critical but it is not commercially available yet. From the comparison of different resists for deep X-ray lithography in Table 6.7, PLG emerges as the most promising LIGA resist. POM, a promising mechanical material, may also be suited for medical applications given its biocompatability. All of the resists shown in Table 6.7 exhibit significantly enhanced sensitivity compared to PMMA, and most exhibit a reduced stress corrosion.[32] Negative X-ray resists have inherently higher sensitivities compared to positive X-ray resists, although their resolution is limited by swelling. Poly(glycidylmethacrylate-co-ethyl acrylate) (PGMA), a negative electron-beam resist (not shown in Table 6.7), has also been used in X-ray lithography. In general, resist materials sensitive to electron-beam exposure also display sensitivity to X-rays and function in the same fashion; materials positive in tone for electron-beam radiation typically are also positive in tone for X-ray radiation. A strong correlation exists between the resist sensitivities observed with these two radiation sources, suggesting that the reaction mechanisms might be similar for both types of irradiation (see Chapter 1). IMM, in Germany, started developing

**TABLE 6.7**  Properties of Resists for Deep X-Ray Lithography

|                       | PMMA | POM | PAS | PMI | PLG |
|-----------------------|------|-----|-----|-----|-----|
| Sensitivity           | −    | +   | ++  | 0   | 0   |
| Resolution            | ++   | 0   | − − | +   | ++  |
| Sidewall smoothness   | ++   | − − | − − | +   | ++  |
| Stress corrosion      | −    | ++  | +   | − − | ++  |
| Adhesion on substrate | +    | +   | +   | −   | +   |

*Note:*  PMMA = poly(methylmethacrylate), POM = polyoxymethylene, PAS = polyalkensulfone, PMI = polymethacrylimide, PLG = poly(lactide-co-glycolide). ++ = excellent; + = good; 0 = reasonable; − = bad, − − = very bad.

*Source:*  After Ehrfeld, W., *LIGA at 1MM*, Baniff, Canada, 1994.

a negative X-ray resist 20 times more sensitive than PMMA, but the exact chemistry has not been disclosed yet.[1]

### Resist Application

#### Multiple Spin Coats

Different methods to apply ultra-thick layers of PMMA have been studied. In the case of multilayer spin coating, high interfacial stresses between the layers can lead to extensive crack propagation upon developing the exposed resist. For example, in Figure 6.10 we present an SEM picture of a 5-μm thick PMMA layer, deposited in five sequential spin coatings. Development resulted in the cracked riverbed mud appearance with the most intensive cracking propagating from the smallest resist features. The test pattern used to expose the resist consisted of arrays of holes ranging in size from 1 to 4 μm. We found that annealing the PMMA films shifted the cracking towards holes with smaller diameter compared to the unannealed film shown in Figure 6.10.[34] The described experiment might comprise a generic means of studying the stress in thick LIGA resist layers. It was found at CAMD that multiple spin coating can be used for up to 15-μm thick resist layers and that with applying the appropriate annealing and developer (see below) no cracking results.[36] Further on we will learn that a prerequisite for low stress and small lateral tolerances in PMMA films is a high mean molecular weight. The spin-coated resist films in Figure 6.10 might not have a high enough molecular weight to lead to good enough selectivity between radiated and nonradiated PMMA during a long development process.

### Commercial PMMA Sheets

High molecular weight PMMA is commercially available as prefabricated plate (e.g., GS 233; Rohm GmbH, Darmstadt, Germany) and several groups have employed free-standing or bonded PMMA resist sheets for producing LIGA structures.[20,27] After overcoming the initial problems encountered when attempting to glue PMMA foils to a metallic base plate with adhesives, this has become the preferred method.[27] Guckel used commercially available thick PMMA sheets (thickness > 3mm), XY-sized and solvent-bonded them to a substrate, and, after milling the sheet to the desired thickness, exposed the resist without cracking problems.[20]

### Casting of PMMA

PMMA also can be purchased in the form of a casting resin, e.g., Plexit 60 (PMMA without added cross-linker) and Plexit 74 (PMMA with cross-linker added) from Rohm GmbH in Darmstadt, Germany. In a typical procedure, PMMA is *in situ* polymerized from a solution of 35 wt% PMMA of a mean molecular weight of anywhere from 100,000 g/mol up to $10^6$ g/mol in methylmethacrylate (MMA). Polymerization at room temperature takes place with benzoyl peroxide (BPO) catalyst as the hardener (radical builder) and dimethylaniline (DMA) as the starter or initiator.[27,33] The oxygen content in the resin, inhibiting polymerization, and gas bubbles, inducing mechanical defects, are reduced by degassing while mixing the components in a vacuum chamber at room temperature and at a pressure of 100 mbar for 2 to 3 min.

In a practical application, resin is then dispensed on a base plate provided with shims to define pattern and thickness and subsequently covered with a glass plate to avoid oxygen absorption. The principle of polymerization on a metal substrate is schematically represented in Figure 6.11. Due to the hardener, polymerization starts within a few minutes after mixing of the components and comes to an end within 5 min. The glass cover plate is coated with an adhesion preventing layer (e.g., Lusin L39; Firma Lange u. Seidel, Nurnberg). After polymerization the anti-adhesion material is removed by diamond milling and a highly polished surface results. *In situ* polymerization and commercial-cast PMMA sheets top the list of thick resist options in LIGA today.

### Plasma Polymerized PMMA

Guckel et al. explored plasma-polymerized PMMA to make conformal coatings more than 100 µm thick.[37] As discussed in Chapter 3, plasma polymerization offers several advantages. Surprisingly, little work focuses on plasma polymerized X-ray resists.

### Resist Adhesion

Adhesion promotion by mechanically or chemically modifying the primary substrate was introduced above under Choice

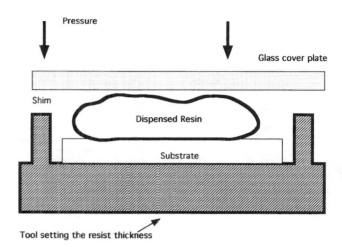

**FIGURE 6.11** Principle of *in situ* polymerization of a thick resist layer on a metal substrate.

of Primary Substrate. Smooth surfaces such as Si wafers with an average roughness, $R_a$, smaller than 20 nm pose additional adhesion challenges often solved by modifying the resist itself. To promote adhesion of resist to polished untreated surfaces, such as a metal-coated Si wafers, coupling agents must be used to chemically attach the resist to the substrate. An example of such a coupling agent is methacryloxypropyl trimethoxy silane (MEMO). With 1 wt% of MEMO added to the casting resin, excellent adhesion results. The adherence is brought about by a siloxane bond between the silane and the hydrolyzed oxide layer of the metal. As illustrated in Figure 6.12, the integration of this coupling agent in the polymer matrix is achieved via the double bond of the methacryl group of MEMO.[33]

Hydroxyethyl methacrylate (HEMA) can improve PMMA adhesion to smooth surfaces but higher concentrations are needed to obtain the same adhesion improvement. Silanization of polished surfaces prior to PMMA casting, instead of adding adhesion promoters to the resin, did not seem to improve the PMMA adhesion. In the case of PMMA sheets, as mentioned before, one option is solvent bonding of the layers to a substrate. In another approach Galhotra et al.[38] simply mechanically clamped the exposed and developed self-supporting PMMA sheet onto a 1.0 mm-thick Ni sheet for subsequent Ni plating. Rogers et al.[39] have shown that cyanoacrylate can be used to bond PMMA resist sheets to a Ni substrate and that it can be lithographically patterned using the same process sequence used to pattern PMMA. For a 300 µm-thick PMMA sheet on a sputtered Ni coating on a silicon wafer a 10 µm-thick cyanoacrylate bonding layer was used. Such a thick cyanoacrylate layer caused some problems for subsequent uniform electrodeposition of metal. The dissolution rate of the cyanoacrylate is faster than the PMMA resist, resulting in metal posts with a wide profile at the base.

### Stress-Induced Cracks in PMMA

The internal stress arising from the combination of a polymer on a metallic substrate can cause cracking in the microstructures during development. To reduce the number of

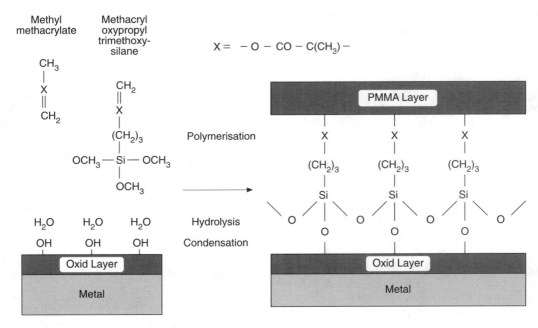

**FIGURE 6.12**   Schematic presentation of the adherence mechanism of methacryloxypropoyl trimethoxy silane (MEMO). (From Mohr, J. et al., *J. Vac. Sci. Technol.*, B6, 2264–2267 (1988). With permission.)

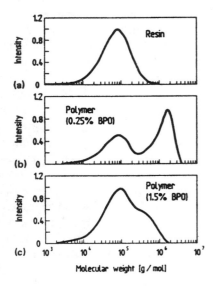

**FIGURE 6.13**   Molecular weight distribution of (a) the casting resin, (b) a resist layer polymerized at low hardener content, and (c) a resist layer polymerized at high hardener content determined by gel permeation chromatography. (From Mohr, J. et al., *J. Vac. Sci. Technol.*, B6, 2264–2267 (1988). With permission.)

stress-induced cracks (see Figure 6.10), both the PMMA resist and the development process must be optimized. Detailed measurements of the heat of reaction, the thermomechanical properties, the residual monomer content, and the molecular weight distribution during polymerization and soft baking have shown the necessity to produce resist layers with a high molecular weight and with only a very small residual monomer content.[27,33] Figure 6.13 compares the molecular weight distribution determined by gel permeation chromatography of a polymerized PMMA resist (two hardener concentrations were used) with the molecular weight distribution of the casting resin. The

casting resin is unimodal, whereas the polymerized resist layer typically shows a bimodal distribution with peak molecular weights centered around 90,000 g/mol and 300,000 g/mol. The first low-molecular-weight peak belongs to the PMMA oligomer dissolved in the casting resin, and the second molecular weight peak results from the polymerization of the monomer. The molecular weight distribution is constant across the total resist thickness, except for the boundary layer at the base plate, where the average molecular weight can be significantly higher (~450,000 g/mol).[30]

The amount of the high molecular weight portion in the polymerized resist depends on the concentration of the hardener. A low hardener content leads to a high molecular weight dominance and vice versa (see Figure 6.13).[33] Since high molecular weight is required for low stress, a hardener concentration of less than 1% BPO (benzoyl peroxide) must be used. Ideally, for a low stress resist, the residual monomer content should be less than 0.5%. The residual monomer content decreases with increasing hardener content, and >1% BPO is needed to reduce the residual monomer content below 0.5%. The problem resulting from these opposite needs can be overcome by the addition of 1% of a cross-linking dimethacrylate (triethylene glycol dimethacrylate, TEDMA) to the resin. In such cross-linked PMMA, a smaller amount of BPO suffices to suppress the residual monomer content; crack-free PMMA can be obtained with 0.8% of BPO.[27]

For solvent removal and to further minimize the defects caused by stress, the polymerized resin is cured at 110°C for one hour (soft bake). The measurement of the reaction enthalpy shows that post-polymerization reactions occur at room temperature and during heating to the glass transition temperature.[33] The rate of heating up to that temperature is 20°C/hr; after curing, the samples are cooled down from 110°C to room temperature at a very low rate of 5 to 10°C/hr.[27,30] The soft-bake temperature is slightly below the glass transition temperature measured to be 115°C.

Another important factor reducing stress in thick PMMA resist layers is the optimization of the developer. Stress-induced cracking can be minimized with solvent mixtures whose dissolution parameters lie near the boundary of the PMMA solubility range, i.e., a nonaggressive solvent is preferred. This is discussed in more detail below, under Development. Small amounts of additives such as described above for reducing stress or to promote adhesion do not influence the mechanical stability of the microstructures or the sensitivity of the resist.

### Optimum Wavelength and Deposited Dose

#### Optimum Wavelength

For a given polymer, the lateral dimension variation in a LIGA microstructure could, in principle, result from the combined influence of several mechanisms, such as Fresnel diffraction, the range of high energy photoelectrons generated by the X-rays, the finite divergence of synchrotron radiation, and the time evolution of the resist profiles during the development process. The theoretical manufacturing precision obtainable by deep X-ray lithography was investigated by means of computer simulation of both the irradiation step and the development step by Becker et al.[40] and by Munchmeyer,[41] and further tested experimentally and confirmed by Mohr et al.[27] The theoretical results demonstrate that the effect of Fresnel diffraction (edge diffraction), which increases as the wavelength increases, and the effect of secondary electrons in PMMA, which increases as the wavelength decreases, lead to minimal structural deviations when the characteristic wavelength ranges between 0.2 and 0.3 nm (assuming an ideal development process and no X-ray divergence). To fully utilize the accuracy potential of a 0.2- to 0.3-nm wavelength, the local divergence

of the synchrotron radiation at the sample site should be less than 0.1 mrad. Under these conditions, the variation in critical lateral dimensions likely to occur between the ends of a 500-μm high structure due to diffraction and secondary electrons are estimated to be 0.2 μm. The estimated Fresnel diffraction and secondary electron scattering effects are shown as a function of characteristic wavelength in Figure 6.14.

Using cross-linked PMMA, or linear PMMA with a unimodal and extremely high molecular-weight distribution (peak molecular weight greater than 1,000,000 g/mol), the experimentally determined lateral tolerances on a test structure as shown in Figure 6.15 are 0.055 μm per 100 μm resist thickness, in good

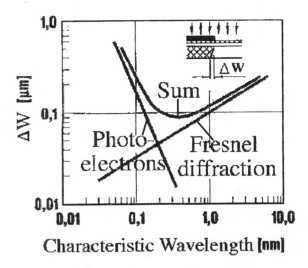

**FIGURE 6.14** Fresnel diffraction and photoelectron generation as a function of characteristic wavelength, $\lambda_c$, and the resulting lateral dimension variation ($\Delta W$). (After Menz, W.P., *Mikrosystemtechnik für Ingenieure* (VCH Publishers, Germany, 1993.)

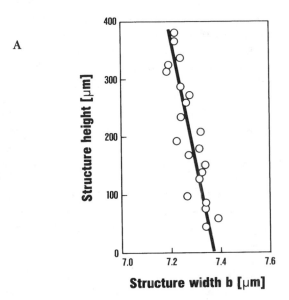

**FIGURE 6.15** Structural tolerances. (A) SEM micrograph of a test structure to determine conical shape. (B) Structural dimensions as a function of structure height. The tolerances of the dimensions are within 0.2 μm over the total structure height of 400 μm.[33] (Courtesy of the Karlsruhe Nuclear Research Center.)

agreement with the 0.2 μm over 500 μm expected on a theoretical basis.[27] These results are obtained only when a resist/developer system with a ratio of the dissolution rates in the exposed and unexposed areas of approximately 1000 is used.

The use of resist layers, not cross-linked and displaying a relatively low bimodal molecular-weight distribution, as well as the application of excessively strong solvents such as used to develop thin PMMA resist layers in the IC industry, lead to more pronounced conical shapes in the test structures of Figure 6.15. An illustration of the effect of molecular weight distribution on lateral geometric tolerances is that linear PMMA with a peak molecular weight below 300,000 g/mol shows structure tolerances of up to 0.15 μm/100 μm.[27] To obtain the best tolerances necessitates a PMMA with a very high molecular weight, also a prerequisite for low stress in the developed resist. Finally, if the synchrotron beam is not parallel to the absorber wall but at an angle greater than 50 mrad, greater coning angles may also result.[27]

*Deposited Dose*

As shown in Figure 6.16, depicting the average molecular weight of PMMA as a function of radiation dose, the X-ray irradiation of PMMA reduces the average molecular weight.[42] For one-component positive resists, this lowering of the average molecular weight ($M_n^*$; see Equation 1.7) causes the solubility of the resist in the developer to increase dramatically. The average molecular weight making dissolution possible is a sensitive function of the type of developer used and the development temperature. Interestingly, it can be observed from Figure 6.16 that above a certain dose (15 to 20 kJ/cm³) the average molecular weight does not decrease any further. This can be understood by recalling the G(s) and G(x) values of a resist introduced in Chapter 1. These G values express the number of main chain scissions (in case of a positive resist), and cross-links (in case of a negative resist), respectively, per 100 eV of absorbed radiation energy. The average molecular weight decreases with radiation dose as long as G(s) is larger than G(x). When the G-values are equal the average molecular weight does not change any further. Our example in Figure 6.16 shows this to occur at a radiation dose between 15 and 20 kJ/cm³.

The molecular weight distribution, measured after resist exposure, is unimodal with peak molecular weights ranging from 3000 g/mol to 18,000 g/mol, dependent on the dose deposited during irradiation (see Figure 6.16). The peak molecular weight increases nearly linearly with increasing resist depth, i.e., decrease of the absorbed dose.[30] At the bottom of the resist layer, the absorbed dose must be higher than the development dose, $D_{dv}$, while at the top of the resist the absorbed dose must be lower than the damaging dose, $D_{dm}$. In Figure 6.5, where the absorption of X-rays along the path from source to sample was illustrated, the exposure time and the preabsorber were chosen so that the bottom of a 500-μm thick PMMA layer received the necessary development dose $D_{dv}$, while the dose at the top of the layer stayed well below $D_{dm}$. Exposure of PMMA with longer wavelengths results in correspondingly longer exposure times and can lead to an overexposure of the top surface, where the lower energy radiation is mainly absorbed.

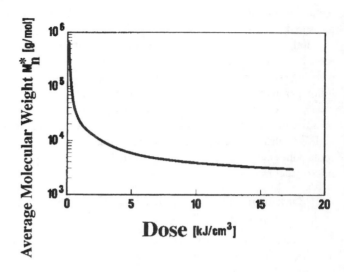

**FIGURE 6.16**   Average molecular weight $M_n^*$ vs. X-ray radiation dose. (From Menz, W.P., *Mikrosystemtechnik fur Ingenieure* (VCH Publishers, Germany, 1993). With permission.)

Menz and Bley[42] describe the influence of the radiation dose on the quality of the resulting LIGA structures in a slightly different manner. Following their approach, Figure 6.17A illustrates a typical bimodal molecular weight distribution of PMMA before radiation, exhibiting an average molecular weight of 600,000. The gray region in this figure indicates the molecular weight region where PMMA readily dissolves, i.e., below the 20,000 g/mol level for the temperature and developer used. Since the fraction of PMMA with a 20,000 molecular weight is very small in nonirradiated PMMA, the developer hardly attacks the resist at all. After irradiation with a dose $D_{dv}$ of 4 kJ/cm³, the average molecular weight becomes low enough to dissolve almost all of the resist (Figure 6.17B). With a dose $D_{dm}$ of 20 kJ/cm³, all of the PMMA dissolves swiftly (Figure 6.17C). At a dose above $D_{dm}$, the microstructures are destroyed by the formation of bubbles. It follows that to dissolve PMMA completely and to make defect-free microstructures, the radiation dose for the specific type of PMMA used must lay between 4 and 20 kJ/cm³. These two numbers also lock in a maximum value of 5 for the ratio of the radiation dose at the top and bottom of a PMMA structure. To make this ratio as small as possible, the soft portion of the synchrotron radiation spectrum is usually filtered out by a preabsorber, e.g., a 100-μm thick polyimide foil (Kapton) in order to reduce differences in dose deposition in the resist.

X-ray exposure equipment developed by KfK has a vibration-free bedding; the exposure chamber is under thermostatic control (±0.2°C) and includes a precision scanner for the periodic movement of the sample through the irradiation plane. The polyimide window isolates the vacuum of the accelerator from the helium atmosphere (200 mbar) which serves as coolant for substrate and mask in the irradiation chamber.[21] IMM in collaboration with Jenoptik GmbH developed an X-ray scanner for deep lithography, enabling irradiation of up to 1000-μm thick resist. A mask-to-resist registration within ±0.3 μm is claimed.[43]

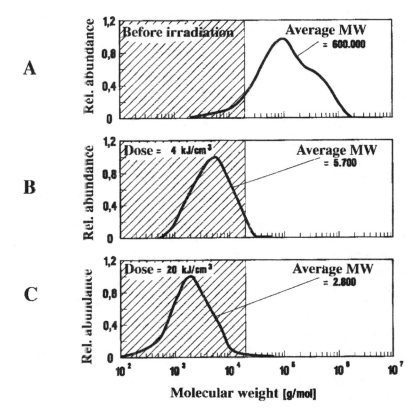

**FIGURE 6.17** Molecular weight distribution of PMMA before (A) and after irradiation with 4 (B) to 20 kJ/cm² (C). The shaded areas indicate the domain in which PMMA is minimally 50% dissolved (at 38°C in the LIGA developer described in the text). (From Menz, W.P., *Mikrosystemtechnik fur Ingenieure* (VCH Publishers, Germany, 1993). With permission.)

### Development

X-ray radiation changes the polymer in the unmasked areas, and chemicals etch away the regions that have been exposed during development. To fully utilize the accuracy potential of synchrotron radiation lithography, it is essential that the resist/developer system has a ratio of dissolution rate in the exposed and unexposed areas of approximately 1000 (see above). The developer empirically arrived at by KfK consists of a mixture of 20 vol% tetrahydro-1,4-oxazine (an azine), 5 vol% 2-aminoethanol-1 (a primary amine), 60 vol% 2-(2-butoxyethoxy)ethanol (a glycolic ether), and 15 vol% water.[27,30,44] This developer causes an infinitely small dissolution of unexposed, high-molecular-weight, cross-linked PMMA and achieves a sufficient dissolution rate in the exposed area. It also exhibits much less stress-induced cracking than developers conventionally used for thin PMMA resists. During development the developing agent flows toward the resist surface being developed, circulates, and is filtered continuously and the temperature remains controlled at 35°C. To stop development, less concentrated developer solutions are applied to prevent the precipitation of already dissolved resist.

Systematic investigation of different organic solvents and mixtures of the above developer systems showed that solvents with a solubility parameter at the periphery of the solubility range of PMMA dissolve exposed PMMA slowly but selectively and without stress-induced cracking or swelling of unexposed areas. Solvents with a solubility parameter close to those of MMA show a much higher dissolution rate, but cause serious problems related to cracking and swelling. As we saw before, the application of excessively strong solvents also led to more pronounced conical shapes in the test structures shown in Figure 6.15. An improved developer found in the above systematic investigation is a mixture of tetrahydro-1,4-oxazine and 2-aminoethanol-1. Its sensitivity is 30% higher, but the process latitude is much narrower compared to the developer described above.[30]

At KfK, a dedicated machine was built for the development process, enabling the continuous and homogeneous transport of developing and rinsing agents into deep structure elements and the removal of the dissolved resist from these structures. Several substrates are arranged vertically on a rotor, with each structure surface facing to the outside. Three independent medium circuits are available for immersion and spraying processes.[21]

After development the microstructures are rinsed with deionized water and dried in a vacuum. Alternatively, drying is done by spinning and blasting with dry nitrogen. At this stage the devices can be the final product, e.g., as micro-optical components, or they are used for subsequent metal deposition.

### Electrodeposition/Electroless Deposition

#### Electrodeposition

**Overview** In the electroforming of microdevices with LIGA (see Figure 6.1), a conductive substrate, carrying the resist structures, serves as the cathode. The metal layer growing on the substrate fills the gaps in the resist configuration, thus forming a complementary metal structure. The use of a solvent-containing development agent ensures a substrate surface completely free of grease and ready for plating. Most of the plating involves Ni. We will discuss Ni electroplating in detail and list some of the other materials that have been or could be plated.

The fabrication of metallic relief structures is a well-known art in the electroforming industry (see also Chapters 3 and 7). The technology is used, for example, to make the fabrication tools for records and videodiscs where structural details in the submicron range are transferred. Electrodeposition is a powerful tool for creating microstructures, not only in LIGA where resist molds are made by X-ray lithography, but also in combination with molds made by laser ablation, dry etching, Keller's polysilicon molding technique (see Chapter 5), bulk micromachining, etc. Because of the extreme aspect ratio, several orders of magnitude larger than in the crafting of CDs, electroplating in LIGA poses new challenges. The near-ultimate in packing density, the extreme height-to-width aspect ratio, and fidelity of reproduction of electrodeposition were first demonstrated in Romankiw's precursor work to LIGA in which gold was plated through polymeric masks generated by X-ray lithography. This author, using X-ray exposure, developed holes in PMMA resist features of 1-μm width and depths of up to 20 μm and electroplated gold in such features up to thicknesses of 8 μm (see also Chapter 7).[2] By 1975, the same author and his team had used the electroplating approach to create microwires $300 \times 300$ atoms in cross-section.[45]

**Nickel Electrodeposition**   The electrodeposition reaction of nickel on the cathode, i.e.,

$$Ni^{2+} + 2e^- \rightarrow Ni \qquad \textbf{Reaction 6.1}$$

competes with the hydrogen evolution reaction on the same electrode, i.e.,

$$2H^+ + 2e^- \rightarrow H_2 \qquad \textbf{Reaction 6.2}$$

The theoretical amount of Ni deposited from Reaction 6.1 can be calculated from Faraday's law (see Chapter 3, Equation 3.29). Since some of the current goes into hydrogen evolution, the actual amount of Ni deposited usually is less than calculated from that equation. The electrodeposition yield, $\eta$, is then defined as the actual amount of deposited material, $m_{act}$, over the total calculated from Faraday's law, $m_{max}$, and can be calculated from[45]

$$\eta[\%] = \frac{m_{act}}{m_{max}} \times 100 = \frac{m_{act}}{\frac{1}{z} iAt \frac{1}{F} M} \times 100 \qquad 6.5$$

where

|   |   |   |
|---|---|---|
| A | = | electrode surface |
| t | = | electrodeposition time |
| F | = | Faraday constant (96487 A/sec/mol) |
| i | = | current density |
| z | = | electrons involved in Reaction 6.1 |
| M | = | atomic weight of Ni |

The amount of hydrogen evolving and competing with Ni deposition depends on the pH, the temperature, and the current density. Since one of the most important causes of defects in

the metallic LIGA microstructures is the appearance of hydrogen bubbles, these three parameters need very precise control. Pollutants cause hydrogen bubbles to cling to the PMMA structures, resulting in pores in the nickel deposit, so the bath must be kept clean, e.g., by circulating through a membrane filter with 0.3-μm pore openings.[46] Besides typical impurities, such as airborne dust or dissolved anode material, the main impurities are nickel hydroxide formed at increased pH-values in the cathode vicinity and organic decomposition products from the wetting agent. The latter two can be avoided to some degree by monitoring and controlling the pH and by adsorption of the organic decomposition products on activated carbon.

Another cause of defects in electroplating is an incomplete wetting of the microchannels in the resist structure. The contact angle between PMMA and the plating electrolyte at 50°C lies between 70 and 80°. A wetting agent is thus indispensable for wetting the surface of the plastic structures in order for the electrolyte to penetrate into the microchannels. With a wetting agent, the contact angle between PMMA and the plating electrolyte can be reduced from 80 to 5°.[46] In the electroforming of microdevices, a much higher concentration of wetting agent is necessary than in conventional electroplating. A dramatic illustration of this wetting effect can be seen in Figure 6.18A and B. In the set-up illustrated in Figure 6.18A, where only 2.5 ml/l of a wetting agent is added to the sulfamate nickel deposition solution, nickel posts with a diameter of 50 μm often fail to form. In the same experiment no posts with a diameter of 5 and 10 μm form at all. Increasing the wetting agent concentration to 10 ml/l results in the perfectly formed nickel posts with a diameter of 5 μm as shown in Figure 6.18B.

A microelectrode of the same size or less than the diffusion layer thickness, as shown in Figure 6.18B, could be expected to plate faster than a larger electrode because of the extra increment of current due to the nonlinear diffusion contribution (Equation 3.53). On the other hand, as derived from Equation 3.55 describing the current to a metal electrode recessed in a resist layer, we learned that the nonlinear diffusion contribution increases with decreasing radius r but decreases with increasing resist thickness, L. High aspect ratio features consequently will plate slower than low aspect ratio features. Moreover, the consumption of hydrogen ions in the high aspect ratio features causes the pH to locally increase. As no intense agitation is possible in these crevices, an isolating layer of nickel hydroxide might form, preventing further metal deposition. This all contributes to making the deposition rate, important for an economical production, much smaller than the rates expected from the linear diffusion model of current density in large, low aspect ratio structures.

A nickel sulfamate bath composition, optimized for Ni electrodeposition, is given in Table 6.8. In addition to nickel sulfamate and boric acid as a buffer, a small quantity of an anion-active wetting agent is added. Sulfur-depolarized nickel pellets are used as anode materials and are held in a titanium basket.[47] Table 6.8 also lists operational parameters. Nickel sulfamate baths produce low internal stress deposits without the need for additional agents avoiding a cause for more defects. The bath is operated at 50 to 62°C and at a pH value between

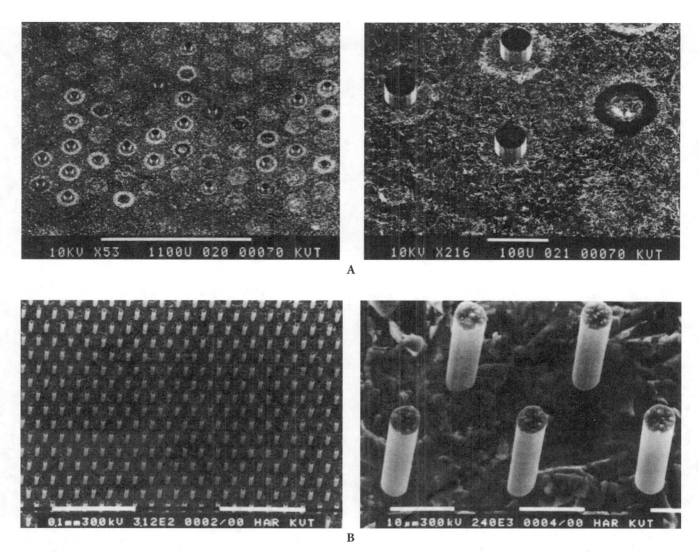

**FIGURE 6.18** Electrodeposition of nickel posts of varying diameter. (A) nickel posts with diameter of 50 µm. Only 2.5 ml/l wetting agent added to the nickel sulfamate solution. Many 50-µm posts are missing and posts with a diameter of 10 or 5 µm do not even form. (B) Nickel posts with diameter of 5 µm. 10 ml/l wetting agent was added; all posts developed perfectly. (Courtesy of the Karlsruhe Nuclear Research Center.)

**TABLE 6.8** Composition and Operating Conditions of Nickel Sulfamate Bath

| Parameter | Value |
|---|---|
| Nickel metal (as sulfamate) | 76–90 g/l |
| Boric acid | 40 g/l |
| Wetting agent | 10 ml/l |
| Current density | 1–10 A/dm$^2$ |
| Temperature | 50–62°C |
| pH | 3.5–4.0 |
| Anodes | Sulfur depolarized |

3.5 and 4.0. Metal deposition is carried out at current densities up to 10 A/dm$^2$. Growth rates vary from 12 (at 1 A/dm$^2$) to 120 µm/hr (at 10 A/dm$^2$).

**Physical Properties of Electrodeposited Nickel** The hardness of the Ni deposits can be adjusted from 200 to 350 Vickers by varying the operating conditions. The hardness decreases with increasing current density. To reach a high hardness of 350 Vickers, the electroforming must proceed at a reduced current of 1 A/dm$^2$. Also, for low compressive stress of 20 N/mm$^2$ or less, a reduced current density must be used. Internal stress in the Ni deposits is not only influenced by current density but also by the layer thickness, pH, temperature, and solution agitation. In the case of pulse plating (see below), pulse frequency has a distinct influence as well.[46] From Figure 6.19 we can derive that for thin Ni deposits the stress is high and decreases very fast as a function of thickness. For thick Ni deposits (>30 µm), the stress as a function of thickness reaches a plateau. At a current density of 10 A/dm$^2$ these thick Ni films are under compressive stress; at 1 A/dm$^2$ they are under tensile stress; at 5 A/dm$^2$ the internal stress reduces to practically zero. Stirring of the plating solution reduces stress dramatically, indicating that mass transport to the cathode is an important factor in determining the ultimate internal stress. Consequently, since high aspect ratio features do not experience the same agitation of a stirred solution as bigger features do, this results

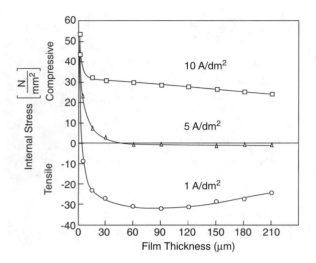

**FIGURE 6.19** Influence of the nickel layer thickness on the internal stress. The electrolyte used is described in Table 6.5 (pH = 4; bath temperature = 52°C).[46] (Courtesy of the Karlsruhe Nuclear Research Center.)

in higher stress concentration in the smallest features of the electroplated structure. Since the stress is most severe in the thinnest Ni films, Harsch et al.[46] undertook a separate study of internal stresses in 5-μm thick Ni films. For three plating temperatures investigated (42°, 52°, and 62°) 5-μm thick films were found to exhibit minimal or no stress at a current density of 2 A/dm². At 62°C, the 5-μm thick films show no stress at 2 A/dm² and remain at a low compressive stress value of about 20 N/mm² for the whole current range (1 to 10 A/dm²). The internal stress at 2 A/dm² is minimal at a pH value between 3.5 and 4.5 of a sulfamate electrolyte (Table 6.8). Ni concentration (between 76 and 100 g/l) and wetting agent concentration do not seem to influence the internal stress. The higher the frequency of the pulse in pulse plating (see below), the smaller the internal stress.

The long-term mechanical stability of Ni LIGA structures was investigated by electromagnetic activation of Ni cantilever beams by Mohr et al.[48] The number of stress cycles, N, necessary to destroy a mechanical structure depends on the stress amplitude, S, and is determined from fatigue curves or S-N curves (applied stress on the x-axis and number of cycles necessary to cause breakage on the y-axis). Experimental results show that for Ni cantilevers produced by LIGA the long-term stability reaches the range of comparable literature data for bulk annealed and hardened nickel specimens. Usually, stress leads to crack initiation, which often starts at the surface of the structure. Since microstructures have a higher surface-to-volume ratio, one might have expected the S-N curves to differ from macroscopic structures, but so far this has not been observed. To the contrary, it seems that the smaller structures are more stable.

**Pulse Plating**  For the fabrication of microdevices with high deposition rates exceeding 120 μm/hr, a reduction of the internal stress can occur only by raising the temperature of the bath or by using alternative electrodeposition methods. Raising of the bath temperature is not an attractive option, but using

alternative electrochemical deposition techniques deserves further exploration. For example, using pulsed (≥500 Hz) galvanic deposition instead of a DC method can be used to influence several important properties of the Ni deposit. Properties such as grain size, purity, and porosity can be manipulated this way without the addition of organic additives.[46] In pulse plating, the current pulse is characterized by three parameters: pulse current density, $i_p$; pulse duration, $t_d$; and pulse pause, $t_p$. These three independent variables determine the average current density, $i_a$, which is the important parameter influencing the deposit quality and is given by:

$$i_a = \frac{t_d}{t_d + t_p} * i_p \qquad\qquad 6.6$$

Pulse plating leads to smaller metal grain size and smaller porosity due to a higher deposition potential. Because each pulse pause allows some time for Ni²⁺ replenishment at the cathode (Ni²⁺ enrichment) and for diffusion away of undesirable reaction products which might otherwise get entrapped in the Ni deposit, a cleaner Ni deposit results. The higher frequency of the pulse leads to the smaller internal stress in the resulting metal deposit.[46]

Pulse plating represents only one of many emerging electrochemical plating techniques considered for microfabrication. For more background on techniques such as laser-enhanced plating, jet plating, laser-enhanced jet plating, and ultrasonically enhanced plating, refer to the review of Romankiw et al.[49] and references therein, as well as Chapters 3 and 7.

**Primary Metal Microstructure and Metal Mold Inserts**  The backside of electrodeposited microdevices attaches to the primary substrate but can be removed from the substrate if necessary. In the latter case, the substrate may be treated chemically or electrochemically to induce poor adhesion. Ideally, excellent adhesion exists between substrate and resist and poor adhesion exists between the electroplated structure and plating base. Achieving these two contradictory demands is one of the main challenges in LIGA.

Slight differences in metal layer thickness cannot be avoided in the electroforming process. Finish-grinding of the metal samples with diamond paste is used to even out microroughness and slight variations in structural height. Finally, the remaining resist is stripped out from between the metal features (see also under PMMA Stripping) and a primary metal microstructure results.

If the metal part needs to function as a mold insert, it can be left on the primary plating base or metal can be plated several millimeters beyond the front faces of the resist structures to produce a monolithic micromold (see Figure 6.20). In the latter case, to avoid damage to the mold insert when separating it from the plating base, an intermediate layer sometimes is deposited on the base plate, ensuring adhesion of the resist structures while facilitating the separation of the electroformed mold insert from its plating base. In addition, it helps to prevent burrs from forming at the front face of the mold insert as a result of underplating of the resist structures. Underplating can occur

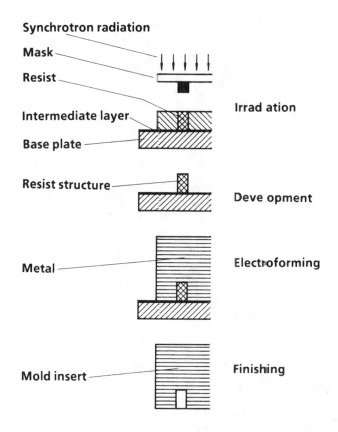

**Synchrotron radiation**
**Mask**
**Resist**

**Irradiation**

**Intermediate layer**
**Base plate**

**Resist structure**

**Development**

**Metal**

**Electroforming**

**Mold insert**

**Finishing**

**FIGURE 6.20** Fabrication of a LIGA mold insert. (From Hagmann, P. et al., Fabrication of Microstructures with Extreme Structural Heights by Reaction Injection Molding, presented at First Meeting of the European Polymer Federation, European Symp. on Polymeric Materials, Lyon, France, 1987. With permission.)

because the resist does not adhere well to the substrate, allowing electrolyte solution to penetrate between the two, or the plating solution might attack the substrate/resist interface. Finally, microcracks at the interface might contribute to underplating. Burrs are easily eliminated then by dissolving the thin auxiliary metal layer with a selective etchant, removing the mold without the need for a mechanical load. In view of the observed underplating problems it is surprising that Galhotra et al [38] did obtain good plating results when simply mechanically clamping the exposed and developed PMMA sheet to an Ni plating box.

**Plating Automation** An automated galvanoforming facility used by KfK is shown in Figure 6.21.[50] This set-up includes provisions for on-line measurement of each electrolyte constituent in flow-through cells and concentration corrections when tolerance limits are exceeded. A computer-controlled transport system moves the individual plating racks, holding the microdevice substrates, through the process stages whereby the substrates are degreased, rinsed, pickled, electroplated, and dried and are then returned to a magazine, which can accommodate up to seven racks. The facility is designed as cleanroom equipment, since contamination of the microstructures must be prevented.[47] An instrument as described here can be bought, for example, from Reinhard Kissler (the µGLAV 750).[50] Although this type of automation is a must for the

eventual commercialization of the LIGA technique, it should be recognized that at this explanatory stage over-automatization may be counterproductive.

*Plating Issues*

Two major sources of difficulty associated with plating of tall structures in photoresist molds are chemical and mechanical incompatibility. The chemical incompatibility means that the photoresist mask may be attacked by the plating solution, while the mechanical incompatibility is film stress in the plated layer which could cause the plated structure to lose adhesion to the substrate.

If the plating solution attacks the photoresist even slightly, considerable damage to the photoresist layer may have occurred by the time a 200-µm thick structure has been created. The limiting thickness of a plated structure due to a chemical interaction is therefore dependent on both the photoresist chemistry and the plating bath itself. In general, the plating bath must have a pH in the range of 3 to around 9.5 (acidic to mildly alkaline), a fairly wide range that can accommodate many of the commercially available photoresists. A surprising number of plating baths do not fall into that range though. Baths, either very strongly acidic or more than mildly alkaline, tend to attack and destroy photoresist.

The plated structure must have extremely low stress to avoid cracking or peeling during the plating process. In addition, if the structure contains narrow features, the stress must be tensile as any compressive stress would result in buckling of the structure. The primary concern with regard to stresses is the incorporation of 'brightening agents' into the plating solution. These can be selenium, arsenic, thallium, and ammonium ions, as well as others. The brightening materials exhibit different atomic configurations and sizes compared to most materials which we would like to plate such as Cu, Au, Ni, FeNi, Pt, Ag, NiCo, etc. This represents the source of a good deal of the stress intrinsic to the plating process. Ag and FeNi are notorious for their high stresses. The Ni-sulfamate bath, presently used extensively in LIGA, is used without brighteners, leading to Ni deposits with extremely low stress. Some other systems low in stress are the electroless plating baths (see section below). These auto-catalytic baths, such as Cu and Ni, coat indiscriminately, but have very low stress. They should be compatible with the LIGA process for certain applications, such as mold insert formation where significant overplating is required.

A third difficulty in finding plating baths compatible with LIGA is that most plating baths are not intended for use in the semiconductor industry. Information including normal deposition parameters, uniformity, stress, compatibility with various semiconductor processes, and particulate contamination normally supplied by a vendor, is not available. Since the majority of the semiconductor industry does not incorporate these kinds of "thin-film" processes, it may be difficult to determine whether a plating bath will be suitable simply from conversations with the vendor. In many instances, trial and error decide.

*Electroless Metal Deposition*

Electroless plating may be elected above electrodeposition due to the simplicity of the process since no special plating base

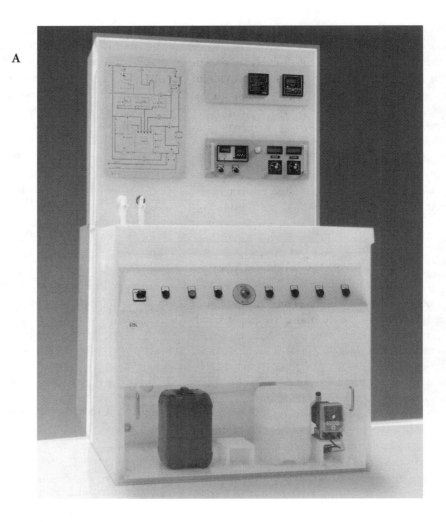

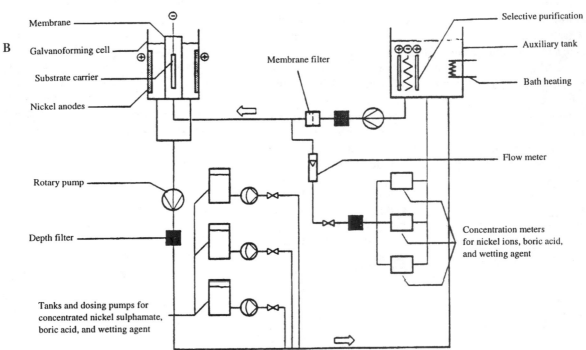

**FIGURE 6.21**   (A) The µGALV 750 comprises the galvanoforming cell with nickel anodes and substrate carrier and an auxiliary tank incorporating heating and purification auxiliary equipment. Three concentration meters measure nickel ions, boric acid, and the concentration of the wetting agent. The concentrations are adjusted automatically by adding via metering pumps.[50] (With permission from Reinhard Kissler, GmbH, 1994.) (B) Schematic drawing of a galvanoforming unit. (From Maner, A. et al., *Plating and Surf. Finishing*, 60–65 (1988). With permission.)

is needed. This represents a major simplification for combining LIGA structures with active electronics where a plating base could short out the active electronics. Also, electroless Ni exhibits less stress than electrodeposited Ni, a fact of considerable importance in most mechanical structures. The major concern is the temperature of the electroless plating processes, which is often considerably higher than for electrochemical processes. Electroless deposition was reviewed in Chapter 3. Few studies of electroless plating in LIGA molds have been reported.[46] We want to draw attention, though, to electroless plating applications with large market potential, i.e., the making of very high density read-write magnetic heads and vias and interconnects for three-dimensional ICs. It is interesting to note the similarity in traditional via plating efforts and making of magnetic read-write heads and the more recent LIGA work. Combining innovative plating techniques such as the ones pioneered by Romankiw[51] and van der Putten[52] with high aspect ratio LIGA molds will create many new opportunities for magnetic heads and vias and interconnects with unprecedented packing densities. Possibly, the progress made in LIGA microfabrication for these three-dimensional IC and magnetic head applications may pay off the most.

## Summary of Electrochemical and Electroless Processes and Materials

Figure 6.22 features a list of choices for material deposition employing electrochemical and electroless techniques.[35] The first metal LIGA structures consisted of nickel, copper, or gold electrodeposited from suitable electrolytes.[47] Nickel-cobalt and nickel-iron alloys were also experimented with. A nickel-cobalt electrolyte for deposition of the corresponding alloys has been developed especially for the generation of microstructures with increased hardness (400 Vickers at 30% cobalt) and elastic limit.[53] The nickel-iron alloys permit tuning of magnetic and thermal properties of the crafted structures.[53,54] From Figure 6.22 it is obvious that many more materials could be combined with LIGA molds.

Beyond the deposition of a wide variety of materials in LIGA molds, additional shaping and manipulation of the electroless or electrodeposited metals during or after deposition will bring a new degree of freedom to micromachinists. Of special interest in this respect is the 'bevel plating' from van der Putten.[52] With 'bevel plating' van der Putten introduced anisotropy in the electroless Ni plating process resulting in beveled Ni microstructures (see Figure 7.10). Combining van der Putten's bevel plating with LIGA one can envision LIGA-produced contact leads with sharp contact points. In our own efforts we are attempting to make sharp Ni needles for use in scanning tunnelling microscopy work. A process to sharpen electrodeposited Ni posts is being developed by applying anodic potentials between an array of posts and a gate electrode, as shown in Figure 1.8.[34,55] The anodic potentials force the posts to corrode in the electrolyte. Because of the concentric field distribution between gate and posts, the corrosion works it way concentrically inwards, leaving sharp tips. We believe that the only limitation in tip sharpness will be associated with the Ni crystal grain size which might be made very small by controlling the Ni deposition current density.

## PMMA Stripping

After finish grinding an electroplated LIGA work piece, a primary metal shape results from removing the photoresist by ashing in an oxygen plasma or stripping in a solvent. In the case

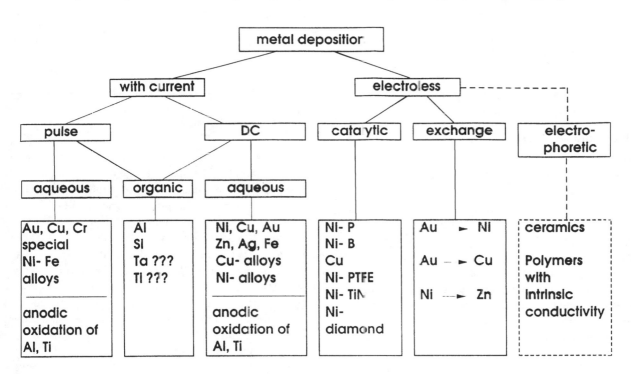

**FIGURE 6.22** Processes and materials for electrochemical deposition of LIGA structures. (From Ehrfeld, W., *LIGA at IMM*, Banff, Canada, 1994. With permission.)

of cross-linked PMMA, the resist is exposed again to synchrotron radiation, guaranteeing sufficient solubility before being stripped.

## Comparing Micromolds

LIGA PMMA features, as small as 0.1 μm, are replicated in the metal with almost no defects. When comparing mold inserts made by spark erosive cutting (see Chapter 7) and by X-ray lithography combined with electroforming, the latter proves superior.[56] The electroformed structures have a superior surface quality with a surface roughness, $R_{max}$, of less than 0.02 μm.[10] A significant application of LIGA might thus be the fabrication of those metal molds that cannot be accomplished with other techniques because of the tight wall roughness tolerances, small size, and high aspect ratios. Discussions with classical electroforming industries might prove useful in determining the application shortcomings of their technology.

Micromolds may also be formed by such techniques as precision electro discharge machines (EDM), excimer layer ablation and e-beam writing. In Table 6.9 LIGA molds are compared with laser machined and precision engineered molds.[57]

**TABLE 6.9**　Comparison of Micromolds

| Parameters | LIGA | LASER | EDM |
|---|---|---|---|
| Aspect ratio | 10–50 | <10 | Up to 100 (related to dimensions) |
| Surface roughness | <50 nm | 100 nm | 0.3–1 μm |
| Accuracy | <1 μm | A few microns | Some microns |
| Mask needed? | Yes | No | No |
| Maximum height | Some microns | A few 100 μm | Microns to millimeters |

From Weber, L., W. Ehrfeld, H. Freimuth, M. Lacher, H. Lehr, and P. Pech, *SPIE, Macromachining and Microfabrication Process Technology II*, Austin, Texas, 1996, p. 156–167. With permission.

From the table it is obvious that LIGA micromolds excel both in very low surface roughness and excellent accuracy.

## Molding and Demolding

### *Overview*

The previously described LIGA steps to produce primary PMMA structures or complementary primary metal mold inserts require intensive, slow, and very costly labor. In order for LIGA to become an economically viable micromachining alternative, one needs to succeed in replicating microstructures successively, in a so-called infinite loop, without having to remake the primary structures. In the plastic molding process illustrated in Figure 6.1, a metal structure produced by electroforming serves as a mold insert used over and over without reverting back to X-ray exposure. For mass-produced plastic devices, one of the following molding techniques is suitable: reaction injection molding (RIM), thermoplastic injection molding, and impression molding. As of the time of this writing, most of the plastic molding in LIGA has been executed in Germany with only some early efforts at CAMD in the U.S.[38,39]

During reaction injection molding, the polymer components are mixed shortly before injection into the mold and the polymerization takes place in the mold itself (e.g., polyurethane and PMMA). Thermoplastic injection molding means heating the polymer above its glass transition temperature and introducing it in a more or less viscous form into the mold, where it hardens again by cooling — e.g., PVC (polyvinylchloride), PMMA, and ABS (polyacrylnitrilbutadienstyrol). The pros and cons of these technologies are summarized in Table 6.10. Compression molding means heating the mold material above its glass transition temperature and patterning it in vacuum by impression of the molding tool.

**TABLE 6.10**　A Comparison of Various Molding Techniques

| | Injection Molding | Reaction Molding |
|---|---|---|
| Gas bubbles | Less of a problem | Reduced by evacuating the mold insert and the mold material; contributes to a high cost |
| Surface quality of mold insert | Less of an issue | Because of low viscosity demands on the surface quality of the mold, insert is very high |
| Chemicals | Easy to handle the molten plastic | Very reactive and explosive reagents |
| Reagent temperature (RT), form temperature (FT), and injection pressure (IP) of mold material | RT = 200°C, FT = 25°C, IP = 100 bar; temperature, time, and pressure need to be controlled very accurately | RT = 40°C FT = 70°C IP = 10–100 bar |
| Slow steps, cycle time | Injection time is long | Medium, with optimized release agent, 11.5 min cycle time |
| Viscosity of mold material | $10^2$–$10^5$ Pa/sec | 0.1–1 Pa/sec |

*Source:*　Based on Menz, W. and P. Bley, *Mikrosystemtechnik for Ingenieure* (VCH Publishers, Germany, 1993).

The high aspect ratio, electrodeposited metal mold inserts generate new problems compared to the molding and demolding processes for the production of records and videodiscs. Even high aspect ratio metal structures can be molded and demolded with polymers quite easily, as long as a polymer with a small adhesive power and rubber-elastic properties, such as silicone rubber, is used. However, rubber-like plastics have low shape stability and would not be adequate. Shape-preserving polymers on the other hand, after hardening, require a mold with extremely smooth inner surfaces to prevent form-locking between the mold and the hardened polymer. A mold release agent may be required for demolding. Using external mold release agents may prove difficult because of the small dimensions of the microstructures to be molded, as typically they are sprayed onto the mold. Consequently, internal mold release compounds need to be mixed with the polymer, without significantly changing the polymer characteristics. Given the early stage of development of molding and demolding plastic LIGA shapes in this country, we have analyzed the German plastic replication work in some detail.

### Reaction Injection Molding (RIM)

A laboratory-size reaction injection molding set-up, as shown in Figure 6.23, was used by the KfK group for their LIGA work.[56] The set-up consists of a container for mixing the various reactants, a vacuum chamber for evacuation of the mold cavity, the molding tool, and a hydraulic clamping unit to open and close the vacuum chamber and the molding tool. After the vacuum chamber has been closed and evacuated, the tool is closed by the clamping unit. The mixed reagents are degassed in the materials container and, under a gas overpressure of up to 3 MPa, pushed through the opened inlet valve into the tool holding the evacuated insert mold. To compensate for shrinkage due to polymerization, an overpressure of up to 30 MPa can be applied during hardening of the casting resin. If the holding overpressure is too low, sunken spots appear in the plastic due to shrinkage of the polymer. For PMMA, the volume shrinkage is about 14%. If the mold material is not degassed and the mold cavity is not evacuated, bubbles develop in the molded piece, resulting in defects and possible partial filling of the mold. To harden the mold material and anneal material stress, the reaction injection molding machine can be operated at temperatures up to 150°C.

A variety of resins were tried to fabricate RIM products, including epoxy resins, silicone resins, and acrylic resins. The most promising results are obtained with resins on a methylmethacrylate base to which an internal mold release agent is added in order to reduce adhesion of the molded piece to the walls of the metal mold. The mold insert in the evacuated tool is covered by means of an electrically conductive perforated plate, i.e., the gate plate (Figure 6.24A). For filling the mold, injection holes are positioned above large, free spaces in the metal structure and the low-viscosity reactants fill the smaller sections laterally. After hardening of the molded polymer, a form-locking connection between the produced part and the gate plate is established at the injection holes, permitting the demolding of the part from the insert (Figure 6.24B). No damage to the mold insert is observed after up to 100 mold-demold cycles, even at the level of a scanning electron microscope picture. The secondary plastic structures formed are exact replicas of the original PMMA structures obtained after X-ray irradiation and development. If the final product is plastic, the molding process can be simplified considerably by omitting the gate plate. The mold insert is then cast over with the molding polymer in a vacuum vessel, building a stable gate block over the mold insert after hardening. The polymeric gate block is used to demold the part from the insert.

The secondary plastic templates usually are not the final product but may be filled with a metal by electrodeposition as with the primary molds. In this case, the secondary plastic templates must be provided with an electrode or plating base. The gate plate can be used directly as the electrode for the deposition of metal in the secondary plastic templates (see Figure 6.24C). It is important that the mold insert is pressed tight against the gate plate so that no uniform isolating plastic film forms over the whole interface between the gate plate and the mold insert; otherwise, electrodeposition might be impossible. Safe sealing between mold insert and gate plate is achieved by use of soft-annealed aluminum gate plates into which the insert is pressed when closing the tool. It was found that the condition

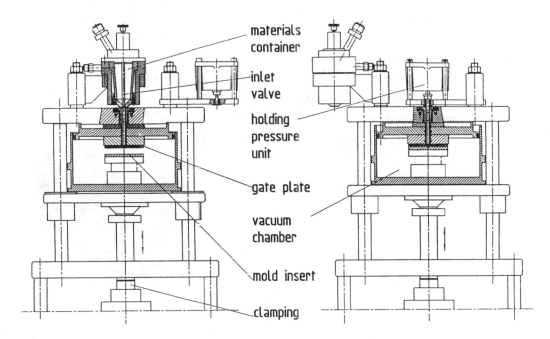

**FIGURE 6.23** Schematic presentation of a vacuum molding set-up. With minor changes, this set-up can be used for reaction injection molding, thermoplastic injection molding, and compression molding. (From Hagman, P.W., *J. Polymer Process. Soc.,* IV, 188–195 (1988). With permission.)

to be met for perfect deposition of metal does not require that the gate plate be entirely free of plastic.[56] It suffices that a number of electrical conducting points emerge from the plastic film. Transverse growth of the metal layer produced in

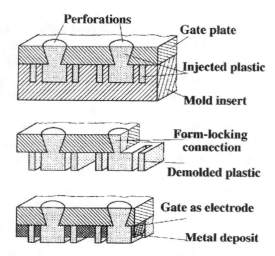

**FIGURE 6.24** Molding process with a perforated gate plate. (A) Filling of the insert mold with plastic material. (B) Demolding. (C) Electrodeposition with the gate plate as plating base. (After Menz, W. and P. Bley, *Mikrosystemtechnik fur Ingenieure* (VCH Publishers, Germany, 1993.)

electrodeposition allows a fault-free, continuous layer to be produced. The secondary metal microstructures are replicas of the primary metal structures and a comparison of, for example, a master separation nozzle mold insert with a secondary separation nozzle structure did not reveal any differences in quality. After the metal forming, the top surface must be polished and the plastic structures must be removed. Depending on the intended use of the secondary metal shape, one can leave it connected to the gate plate or the gate plate can be selectively dissolved or mechanically removed.

Gate plates can be applied only if the microstructures are interconnected by relatively large openings. The injection holes in the gate plate are produced by mechanical means with openings of about 1 mm in diameter and must be aligned with large openings in the mold insert. Many desirable microstructures cannot be built this way. For example, in Figure 6.25 we show a Ni honeycomb structure with a cell diameter of 80 μm, a wall thickness of 8 μm, and a height of 70 μm.[58] The holes in a gate plate are too large to inject plastic into this structure and the interconnections in the honeycomb are too small to provide a good transverse movement of the polymerizing resin. Alternative approaches enabling the plating of any shape are discussed next.

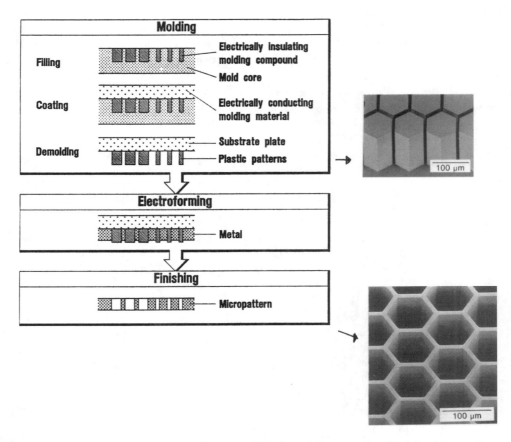

**FIGURE 6.25** Ni honeycomb structure with cell diameter of 80 μm, wall thickness of 8 μm, and wall height of 70 μm. A structure like this could not be produced using a gate plate; a conductive plastic is used instead. The Ni plating occurs on the conductive plastic carrier plate, which is fused with the insulating plastic of the microstructure. (From Harmening, M. et al., *Makromol. Chem. Macromol. Symp.*, 50, 277–284 (1991). With permission.)

**Electrically Conductive Plastic Plates as Carriers and Cathodes** It is possible to use electrically conducting plastic as a starting layer for electrodeposition.[58] In this process, illustrated in Figure 6.25, the mold insert is filled initially with the electrically insulating, e.g., unfilled, casting resin. The surplus molding material is scraped off the front of the mold insert followed by covering of the mold insert with the electrically conducting carrier plate. While the casting resin cures, the carrier plate is forced against the front of the mold insert. In the curing process, the electrically insulating material in the honeycomb structures fuses with the carrier plate. After demolding one obtains insulating plastic structures firmly connected with the electrically conducting carrier plate. The conductive plastic is used for the electrodeposition between the insulating plastic features and, after removal of the plastic with a solvent, a plate with metal microstructures results. To achieve the desired conductivity in the plastic carrier plates one adds metal powders (e.g., copper, silver, or gold) or carbon powders of an average grain diameter of less than 3 µm to the cast resin. From the metallic powder additives, copper oxidizes too quickly, leading to a high resistivity composite material. Silver-filled resins, on the other hand, process easily and set into homogeneously conducting composites, i.e., free from bubbles and blisters. Carbon-filled resins only lead to homogeneous composite materials to a certain extent. At an Ag filler (Ag, Demetron type 6321-8000) content of 75 wt%, the PMMA attained a value of composite resistivity of $2.4 \times 10^{-4}$ $\Omega$cm. The PMMA composite must also contain an internal mold release agent to make the release of the metal shapes easy (see Demolding). The addition of a demolding agent decreases the glass transition temperature ($T_g$) and the temperature for the beginning of the softening process ($T_s$) with respect to the values for pure PMMA, but both properties are practically independent of filler concentration. The thermal and mechanical properties of the composites are dependent on the filler content. With increasing filler content, the composite becomes stiffer, as indicated by an increasing Young's modulus and a drop of the maximum loss factor, tan $\delta$, at the glass transition temperature. The coefficient of thermal expansion decreases with increase in filler content. To ensure high dimensional stability, both a high Young's modulus and a low thermal expansion coefficient are beneficial. For uniform electrodeposition the composite material must also have a low, homogeneous resistivity. Metal deposition starts at the filler particles exposed at the surface. Those particles are connected to the bulk of the electrode by a network of conducting particles. Fault-free electroforming of patterns in the micrometer range requires a very small distance between the electrical points of contact, the starting points of electrodeposition, on the surface of the composite material. At a filler content of 75 wt%, the mean distance between the starting points of electrodeposition attained is 1.5 µm.

The micro-pattern illustrated in Figure 6.25 was made by molding electrically insulating micro-patterns of PMMA on the surface of substrate plates of PMMA filled with 75 wt% of silver. The pattern consists of a total of 74,400 single patterns on an area of 500 mm² and serves as an infrared high-pass filter.[58]

*Injection Molding*

Thermoplastic injection molding is the technique used to make compact discs involving features about 0.1 µm high and minimum lateral dimensions of 0.6 µm, i.e., an aspect ratio of 0.16. In LIGA, the aspect ratio easily can be 1000 times larger. Injection molding can work here as well, as long as the mold insert walls are kept at a uniform temperature above the glass transition temperature ($T_g$) of the plastic material, so that the injected molten plastic mass does not harden prematurely. The latter requires a rather slow filling of the mold insert with the plastic mass. To be able to fill features with aspect ratios higher than 10, the temperature of the tool holding the insertion mold should be higher than what is typically used in more conventional PMMA injection molding applications. One should stay well below the PMMA melting temperature of 240°C (say, 70°C less), as the microstructure might show temperature-induced defects if one works at a temperature closer to the melting temperature.[59] For injection molding, an apparatus very similar to the one shown in Figure 6.23 can be used.

The mold insert must again have extremely smooth walls to be able to demold the structures. Walls with the required smoothness cannot be produced with a classical technique such as spark erosion. Because PMMA absorbs water in air (up to 0.3%), leading to poorer wall quality, the PMMA polymer flakes need to be dried for 4 to 6 hours at 70° before use in injection molding.

*Compression Molding (also Relief Printing or Hot Embossing)*

Compression molding or relief printing is the method of choice for incorporating LIGA structures on a processed Si wafer. Hot embossing takes place in a machine frame similar to that of a press. In this method, microstructures are generated by molding of a thermoplastic polymer (mold material) using a molding process as illustrated in Figure 6.26. In a first step, the mold material is applied by polymerization onto the processed wafer with an electrically isolating layer and overlayed with a metal plating base (Figure 6.26A). At the glass transition temperature, the plastic is in a viscoelastic condition suitable for the impression molding step. Above this temperature (about 160°C for PMMA) the mold material is patterned in vacuum by impression with the molding tool. In order to avoid damaging the electronic circuits by contact with the molding tool, the mold material is not completely displaced during molding. A thin, electrically insulating residual layer is left between the molding tool and the plating base on the processed wafer (Figure 6.26B). The residual layer is removed by reactive ion etching (RIE) in an oxygen plasma etch (Figure 6.26C), freeing the plating base for subsequent electrodeposition. The oxygen RIE process is as anisotropic as possible, so that the sidewalls of the structures do not deteriorate and the amount of material removed from the top of the plastic microstructures is small compared to the total height. After electrodeposition on the plating base the metal microstructures are laid bare by dissolution of the mold material (Figure 6.26D). Finally, using argon sputtering or chemical etching, the plating base in between the

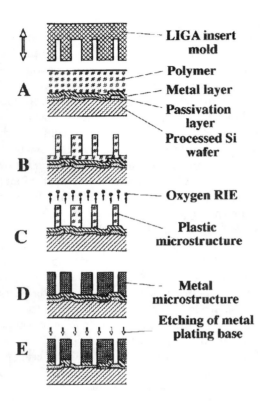

**FIGURE 6.26** Fabrication of microstructures on a processed Si wafer. (A) Patterned isolation layer, conductive plating base, and *in situ* polymerization of PMMA. (B) Impression molding using a LIGA primary insert mold. (C) Removal of remaining plastic from the plating base by a highly directional oxygen etch. (D) Electrodeposition of metal shape. (E) Removal of plating base from between the metal structures by argon sputtering. (After Menz, W. and P. Bley, *Mikrosystemtechnik fur Ingenieure*, VCH, Weinheim, Germany, 1993.)

metal microstructures is removed so that the metal parts do not shortcircuit (Figure 6.26E). In the case of chemical etching, a very fast etch should be employed to avoid etching of the plating base from underneath the electrodeposited structures.[25,59]

At KfK, hot embossing is mainly used for small series production, whereas injection molding is applied to mass production. Both techniques are applied to amorphous (PMMA, PC, PSU) and semicrystalline thermoplastics (POM, PA, PVDF, PFA). Reaction injection molding is used for molding of thermoplastics (PMMA, PA), duroplastics, (PMMA, expoxide), or elastomers (silicones).[60]

*Demolding*

The demolding step is carried out by means of a clamping unit at preset temperatures and rates (see Figure 6.23). Demolding is facilitated by an internal mold release agent such as PAT 665 (Wurtz GmbH, Germany), normally employed with polyester resins.[56] The optimum yield with this agent and Plexit M60 PMMA occurs at 3 to 6 wt%. The yield drops very fast below 3 wt% as adhesion between the molded piece and the mold insert becomes stronger. These adhesive forces are estimated by qualitatively determining the forces necessary to remove the plastic structures from the mold insert. An upper limit is 10 wt% where the MMA does not polymerize anymore. Above 5 wt%,

the Young's modulus and, hence, the mechanical stability decrease, and above 6 wt%, pores start forming in the microstructures. The internal mold release agent also has a marked influence on the optimum demolding temperature. The demolding yield decreases quickly above 60°C for a 4 wt% PAT internal mold release agent. With 6 wt%, one can only obtain good yields at 20°C.[10,56] The molding process initially led to a production cycle time of 120 min. For a commercial process, much faster cycle times are needed, for instance by optimizing the mold release agent. With that in mind, the KfK group started to work with a special salt of an organic acid leading to a 100% yield at a release agent content of 0.2 wt% only and a temperature of 40°C (at 0.05 wt% a 95% yield is still achieved). At 80°C, a 100% demolding yield was obtained and, significantly, a cycle time of 11.5 min was reached. During these experiments the mold was filled at 80°C and heated to 110°C within 7.5 min. As the curing occurs at 110°C, the material needs to be cooled down to 80°C for demolding. Moreover, the 0.2 wt% of the 'magic release agent' did not impact the Young's modulus and the glass transition temperature of Plexit M60.[56]

### LIGA Extensions: Movable, Stepped, and Slanted Structures

Modifying the LIGA process enables movable microstructures and microstructures with different level heights, i.e., stepped and inclined sidewalls. These features often are essential for mechanical and optical devices.[61]

*Movable Structures*

As in surface micromachining (see Chapter 5), sacrificial layers make it possible to fabricate partially attached and freed metal structures in the primary mold process.[62] The ability to implement these features leads to assembled micromechanisms with submicron dimension accuracies, opening up many additional applications for LIGA, especially in the field of sensors and actuators. The sacrificial layer may be polyimide, silicon dioxide, polysilicon, or some other metal.[63] The sacrificial layer is patterned with photolithography and wet etching before polymerizing the resist layer over it. At KfK, a several micron thick titanium layer often acts as the sacrificial layer because it provides good adhesion of the polymer and it can be etched selectively against several other metals used in the process. If for exposure the X-ray mask is adjusted to the sacrificial layer, some parts of the microstructures will lie above the openings in the sacrificial layer, whereas other parts will be built up on it. These latter parts will be able to move after removal of the sacrificial layer.[64] The fabrication of a movable LIGA structure is illustrated in Figure 6.27A.

*Stepped Microstructures*

In principle, stepped LIGA structures can be accomplished by means of X-ray lithography with a single mask with stepped absorber structures. In this manner, variable dose depositions can be achieved at the same resist heights. The variable dose results in different molecular weights and, hence, in a different developing behavior. This technique unfortunately leads to rounded features and poor step-height control.

Labels for figure:
- LIGA insert mold
- Polymer
- Metal layer
- Passivation layer
- Processed Si wafer
- Oxygen RIE
- Plastic microstructure
- Metal microstructure
- Etching of metal plating base

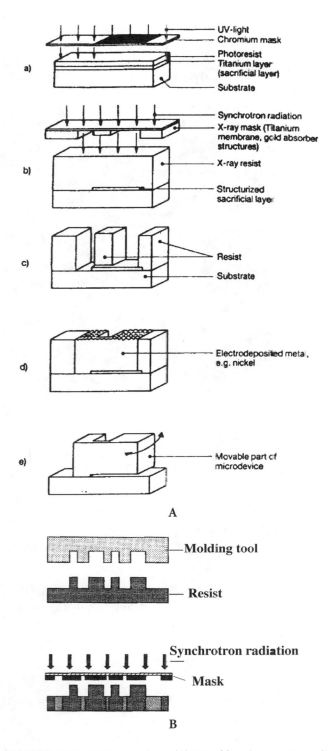

a)
- UV-light
- Chromium mask
- Photoresist
- Titanium layer (sacrificial layer)
- Substrate

b)
- Synchrotron radiation
- X-ray mask (Titanium membrane, gold absorber structures)
- X-ray resist
- Structurized sacrificial layer

c)
- Resist
- Substrate

d)
- Electrodeposited metal, e.g. nickel

e)
- Movable part of microdevice

**A**

—**Molding tool**

—**Resist**

**Synchrotron radiation**

—**Mask**

**B**

**FIGURE 6.27** LIGA extensions. (A) Movable microstructures. (B) Stepped microstructures. (From Mohr et al. *Microsyst. Technol. 90*, Springer, Berlin, 1990. With permission.)

Alternatively, one can first relief print a PMMA layer, e.g., by using a Ni mold insert made from a first X-ray mask. Subsequently, the relief structure may be exposed to synchrotron radiation to further pattern the polymer layer through a precisely adjusted second X-ray mask. To carry out this process, a two-layer resist system needs to be developed consisting of a top PMMA layer which fulfills the requirement of the relief printing process, and a bottom layer which fulfills the requirements of

the X-ray lithography.[65] The bottom resist layer promotes high molecular weight and adhesion, while the top PMMA layer is of lower molecular weight and contains an internal mold-release agent. This process sequence combining plastic impression molding with X-ray lithography is illustrated in Figure 6.27B. The two-step resist then facilitates the fabrication of a mold insert by electroforming, which can be used for the molding of two-step plastic structures. Extremely large structural heights can be obtained from the additive nature of the individual microstructure levels. Alternatively, to make stepped structures one can also employ multiple masks and irradiation steps. The latter method only requires one blend of PMMA resist.

### Slanted Microstructures

Changing the angle at which synchrotron radiation is incident upon the resist, usually 90°, enables the fabrication of microstructures with inclined sidewalls.[66] Slanted microstructures may be produced by a single oblique irradiation or by a swivel irradiation. One application of microstructures incorporating inclined sidewalls is the vertical coupling of light into waveguide structures using a 45° prism.[67] Such optical devices must have a wall roughness of less than 50 nm. Due to the sharp decrease of the dose in the resist underneath the edge of the inclined absorber and because of the sharp decrease of the dissolution of the resist as a function of the molecular weight in the developer, usually no deviation of the inclination of the sidewall over the total height of the microstructure is found.

### Alternative Materials in LIGA

Alternative X-ray resist materials and a variety of other metals for electroplating besides Ni were discussed above. Following we will discuss some alternative molding materials. Besides PMMA and POM, used in a commercially available form, semicrystalline polyvinylidenefluoride (PVDF), a piezoelectric material, has been used to make polymeric microstructures.[65] The optimum molding temperature of the PVDF was found to be 180°C. The PVDF structures could be molded without using mold release agents. Fluorinated polymers such as PVDF will also enable higher temperature applications than PMMA. Polycarbonate (PC), well known for molding compact discs, is under development and should perform well.[26]

Other materials that have the ability to flow or can be sintered, such as glasses and ceramics, can be incorporated in the LIGA process, as well. In the case of ceramics, for example, the plastic microstructures are used as disposable or lost molds. The molds get filled with a slurry in a slip-casting process. Before sintering at high temperatures, the plastic mold and the organic slurry components are removed completely in a burnout process at lower temperatures. First results have been obtained for zirconium oxide and aluminum oxide. The process is illustrated in Figure 6.1. An important application for ceramic microstructures is the fabrication of arrays of piezoceramic columns embedded in a plastic matrix. Since the performance of these actuators is linked to the height and width ratio of the individual ceramic columns, as well as the distance between the columns in the array, LIGA, with its tremendous capacity for tall, dense, and high aspect ratio features is an ideal fabrication tool.[68,69]

A very interesting method to make LIGA ceramic or glass structures in the future may be based on sol-gel technology. This technique involves relatively low temperatures and may enable LIGA products such as glass capillaries for gas chromatography (GC) or possibly even high-TC superconductor actuators. Sol-gel techniques should work well with LIGA-type molds if they are filled under a vacuum to eliminate trapped gas bubbles. In the sol-gel technique a solution is spun on a substrate which is then given an initial firing at around 200°C, driving off the solvent in the film. Subsequently, the substrate is given a high-temperature firing at 800 to 900°C to drive out the remaining solvents and crystallize the film. A major issue deals with the large shrinkage of the sol-gel films during the initial firing. For example, the maximum thickness of high-TC superconductor films currently achievable measures approximately 2 μm due to the high stress in the film caused during shrinkage. The latter is related to the ceramic yield or the amount of ceramic in the sol-gel compared to the amount of solvent. The sol-gel technique, in general, suits the production of thicker films well, as long as high ceramic yield sol-gels are used. As LIGA-style films tend to be thick, a reduction in the amount of solvents to allow a high ceramic yield after firing would be in order to make sol-gel compatible with LIGA. This would require some development. Also lead zirconate titanate (PZT) devices could be made with sol-gel technology and LIGA, most likely by applying the metal plated structure rather than the direct PMMA template for the creation of the device, as PZT must be processed in several elevated temperature steps. The PZT sol-gel contains no particles that would prevent flow into small channels in a plated mold. The sol-gel contains high molecular-weight polymer chains that hold the constituent metal salts, which are further processed into the final ceramic film. Thus, the sol-gel could be formed into the mold in a process much like reaction-injection molding, with the mold being filled with the chemistry under vacuum.

# LIGA Applications

## Introduction

A plethora of potential applications of LIGA have been investigated. Examples include a capacitive accelerometer, a *copper coil* for an inductive position sensor, a *resonant mesh infrared band pass filter* with a cross-shaped dipole pattern of 18.5-μm slit length and 3-μm slit width and a distance between two crosses of 1.5 μm, electrical and optical interconnections, micron-sized nozzles, microfiltration membranes, spinneret plates for synthetic fiber production, microvalves, nozzles for fuel injection, ink-jet nozzles, microchannels, and mixing elements. The economic and technical justification for applying LIGA for several of those applications requires further exploration. Nevertheless, these studies illustrate the potential of the method.

The following list of LIGA microstructures includes the more significant research devices realized as of this writing. The majority remain experimental and their performance cannot accurately be assessed yet, or they represent simple test structures designed to prove the feasibility of the process. Geometrical

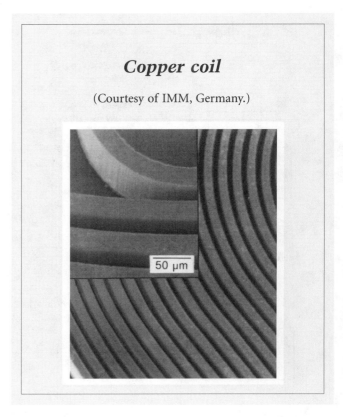

*Copper coil*

(Courtesy of IMM, Germany.)

specifications of LIGA structures depend on the design, the fabrication process, and the material. Typical values of LIGA features incorporated in the listed applications are:[26]

- Structural height (typical): 20–500 μm
- Minimum dimension: 1–2 μm
- Smallest x, y surface detail: 0.25–0.5 μm
- Surface roughness: 0.03–0.05 μm (max. peak to valley)
- Maximum x, y dimension: 20 × 60 mm

The description of LIGA examples includes a discussion on the appropriateness of LIGA as a fabrication tool and complements the microfabrication applications discussed in Chapter 10.

## Microfluidic Elements

### Introduction

In fluid and gas dynamics a number of devices must have micrometer dimensions or have their efficiency and performance improved by miniaturization.[70] The impact of miniaturization in fluidics is discussed in detail in Chapter 9. Fluidic devices such as nozzles, valves, mixing chambers, and pumps may at first glance appear good LIGA applications because both high aspect ratio and accuracy often drive the manufacturing choice. Moreover, the flexibility in the choice of materials and the wide dynamic range in size gives LIGA a significant advantage over bulk Si micromachining approaches. Surface micromachining in poly-Si does not suit microfluidics; the height of achievable structures is too low and the tendency of components to stick to each other would only be exacerbated when in contact

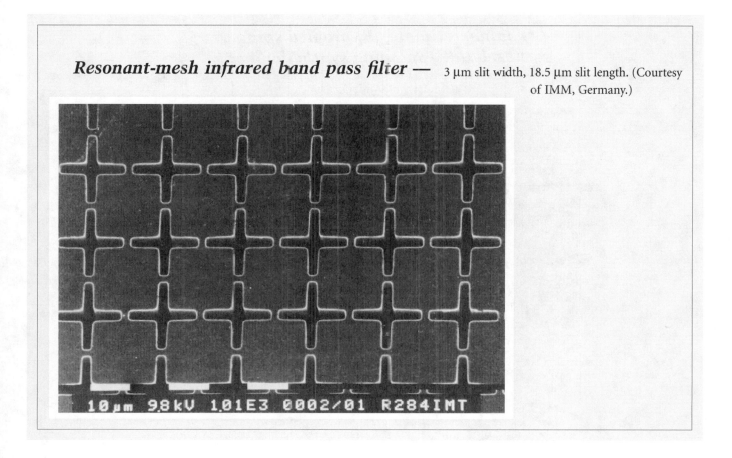

***Resonant-mesh infrared band pass filter*** — 3 μm slit width, 18.5 μm slit length. (Courtesy of IMM, Germany.)

with fluids. Fluidic devices (for example, fluidic logic devices) mostly made from photosensitive glass, reached their peak of popularity in the mid-1960s but lost out to electronic devices by the early 1970s (see also Chapter 7). A new manufacturing technique will not revive fluidic logic; an approach proven non-competitive. In the case of other fluidic devices, such as pneumatic sensors and analytical equipment incorporating fluidic elements, photosensitive glass again is an inexpensive alternative. LIGA might yet play an important role in those cases where tolerances are very tight, or extreme aspect ratios or very low wall roughness dominate. Examples include gas separation nozzles, spinneret nozzles, and very sensitive flow meters.

### Double Deflecting Gas Separation Nozzles for Uranium Enrichment

In separation nozzles, the optimum operating pressure is inversely proportional to the characteristic dimension of the flow; smaller nozzles make for an increased separation efficiency. The manufacture of tiny separation nozzles for separation of uranium isotopes at KfK[4,21] actually initiated the development work of LIGA (see Figure 6.2 for a picture of these micronozzles). In 1982, these nozzles were made by electroplating Ni in the primary resist structure; by 1985 they resulted from a combination of electroforming and plastic molding. The minimum slit width in the curved nozzle measures 3 μm, and the slit length is about 300 μm. Since a higher gas pressure results in a corresponding reduction of the compressor size and the piping system for a given throughput of a separation nozzle

plant, the dimension reduction of the individual nozzles has a direct economic advantage.[70]

The feasibility of the LIGA separation nozzles was demonstrated, but the parts were never implemented as government funding for the project stopped (the KfK work was carried out in Germany for the Brazilian government and the U.S. finally objected to the sale of the uranium separation system to Brazil). The accuracy and aspect ratio required for the reliable gas separation of uranium isotopes still make the gas separation nozzles one of the best demonstrations for the correct application of LIGA. No other technique can yet produce a mold insert with the required aspect ratios and tolerances (see also Spinneret Nozzles).

### Spinneret Nozzles

Profiled capillaries (nozzles) in a ***spinneret plate*** for ***spinning synthetic fibers*** from a molten or dissolved polymer normally are produced by spark erosion (see Chapter 7). This process establishes a practical lower limit of 20 to 50 μm for the minimum characteristic dimension. Smaller, more precise nozzles can be produced by LIGA.[71]

Spinneret nozzles lend themselves to LIGA from the technical point of view. Compared to fabrication by spark erosion, the minimum characteristic dimensions with LIGA can be reduced by an order of magnitude, and a high capillary length with excellent surface finish can easily be obtained. Moreover, the LIGA process makes all the nozzles in parallel while the spark erosion operates as a serial technique. The market might focus

## (A) Spinneret plate; (B) profiled spinneret nozzles; (C) spinning synthetic fiber.

(Courtesy of IMM, Germany.)

**A**

**B**

**C**

on niche applications such as specialty, multilumen catheters. Also, medical use of atomizers for dispensing drugs is projected to increase dramatically in the coming years, this might cause a run for a wide variety of precise, inexpensive micronozzles.

### LIGA Turbine

A small, released nickel *turbine rotor*, rotated by blowing gas through channels embedded in the surrounding structure, was produced using the LIGA technique. The turbine exemplifies one of the first movable LIGA structures. The axle, rotor, and channels of the turbine are fabricated of plated nickel.[72] As the shaft and the gear are produced in one mask step, no alignment procedure of the finished components is necessary. Some data on the performance of the turbine were presented by Mohr et al.[64,73] The maximum speed measured 2500 rotations per second, and a total of $85 \times 10^6$ rotations on one turbine have been recorded. To make this turbine into a volumetric flow sensor requires a very tight packaging scheme. The packaging must minimize gas flow around the turbine and channel it all through the turbine. Another requirement is for very hard materials to enhance life-time. An integrated optical fiber measures rotation speed in the microturbine flow sensor. Flow rates from 10 to 50 sccm min$^{-1}$ have been detected this way.[74]

More materials-related development work is needed to further mature this application. For example, the bonding of a top layer to enclose LIGA structures is an area that still needs a good solution. In principle, flow sensors embody a good LIGA opportunity since they require both accuracy and a high aspect ratio.

## Turbine rotor with integrated optical fiber

(Courtesy KfK, Karlsruhe, Germany.)

Improved fabrication accuracy may make for more uniformity of flow sensors from and between large batches. The high aspect ratio combined with high manufacturing accuracy will ensure more sensitive flow sensors for special applications.

### Microvalve

Microvalves represent one of the most eagerly pursued microdevices today, either as separate components or for integration in micromachined fluidic systems (pumps). No good solutions have been presented; valves are either too large or too expensive or they do not seal tightly enough.

Systems of *four LIGA microvalves* assembled from thin titanium membranes, an electroplated valve body in nickel or gold, and glass covers have been assembled.[75] The microvalves find a common support in the center of the four-valve system and can be actuated individually by an external pneumatic system. To close the flow channel, the 2.7-$\mu$m thin titanium membrane is pressed pneumatically onto a valve seat dividing each valve chamber into two parts. If the membrane is not actuated, it is deflected off the valve seat by the pumped medium. Besides the lithography and plating of the valve body, no other LIGA steps are involved and the valve mechanism is completely traditional.[76] One wonders why the manufacture of the current microvalves did not follow the traditional thermoplastic molding method as in the case of the injection molded polysulfone pump discussed below.

The LIGA manufacturing technique introduced here does not improve upon the technological shortcomings of these types of valves, i.e., the high cost and poor valve sealing. Several commercially available non-LIGA micromachined valves are compared in Chapter 10.

### Micropump

For analysis and dosage of small amounts of liquids and gases, pumps capable of handling minute amounts of well-defined fluid quantities are needed. Many attempts to make such micropumps have been made but no clear solution yet stands out. Current micropumps are too expensive and, as we discussed above, do not have adequate valving systems. The following short history of LIGA pumps clearly illustrates that LIGA does not present the answer to every micromachining problem at hand.

The *first LIGA pump* for both water and air was a micromembrane pump with a case made of electroplated gold and incorporating a complex 1-mm diameter valve consisting of a stiff titanium membrane working in tandem with a flexible polyimide membrane.[77] The pump was manufactured by KfK using X-ray lithography to fabricate the gold pump case, photolithography for defining both the titanium and the polyimide membrane layers, and gluing on of a top glass plate. The LIGA pump was tested with an external pneumatic drive.[77,78] The complexity of the first device led to the fabrication of a second *micro-pump* not involving any LIGA steps at all by the same KfK group. In this case, the device was made by combining traditional thermoplastic molding with membrane techniques.[79] The pump case consisted of two pieces made by injection molding of

## LIGA microvalves. Four valves connected to common inlet

(From Schomburg et al., *J. Micromech. Microeng.*, 4, 186–191, 1994. With permission.)

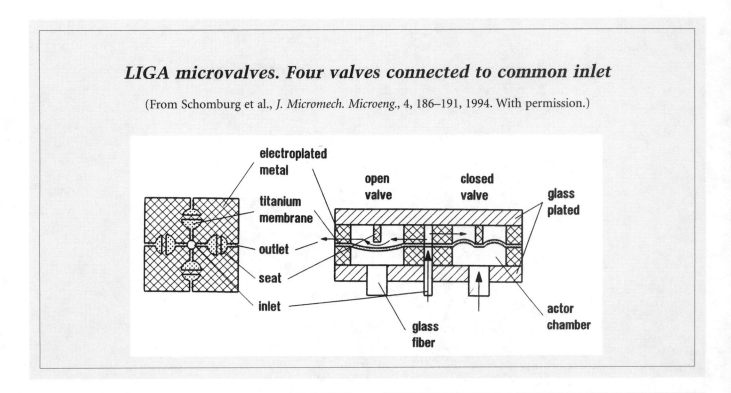

## First LIGA pump

(From Rapp et al., *Sensors and Actuators A*, A40, 57–61, 1994. With permission.)

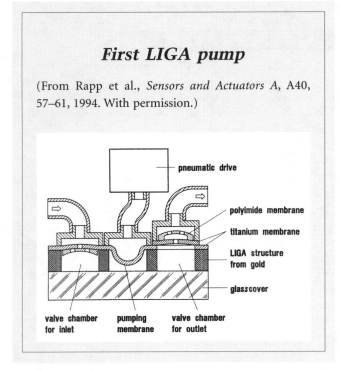

## Alternative non-LIGA pump

(From Bustgens et al., Proceedings MEM '94, Oiso, Japan, 1994. With permission.)

polysulfone and had lateral dimensions of 7 by 10 mm. The depth and diameter of the pump chamber were 100 μm and 4 mm, respectively, and the chamber was covered by a 2.5-μm thick polyimide membrane. The inlet and outlet valves were formed by orifices in the membrane positioned over a valve seat which is a part of the case. The membrane was driven in this case by an integrated thermopneumatic actuator.

The above short history of the LIGA pump demonstrates the fallacy of applying a new micromachining method to every micromachining issue without establishing a clear advantage beforehand. A new and expensive manufacturing technology should only be applied to a commercial need if it enables some new approaches (e.g., LIGA might provide much smoother valve seat surfaces to make a stronger actuator, etc.) or if it ultimately affords an old approach to be implemented more efficiently and at lower cost. In the case of the LIGA microvalves and micropumps, neither is the case and we are still struggling today to find a better, low-cost tightly closing microvalve. The micropump closest to market introduction is a bulk Si micromachined pump incorporating a piezoelectrically activated membrane developed by Debiotech (see Chapter 10).

## Fluidic flip-flop

Bistable wall attachment amplifier. (From Vollmer et al., *Sensors and Actuators A*, A43, 330–334, 1994. With permission.)

### Fluidic Amplifiers

A LIGA *fluidic flip-flop* also called a wall-attachment amplifier, is based on the Coanda effect. A jet emanating from a supply nozzle is attached to one of the two attachment walls. The jet can be switched from one position to the other by the temporary application of a small control pressure at the control port next to the attached jet. The jet reaching one of the two output ports builds up a pressure which can be used, for example, to drive a microfluidic actuator. Using LIGA, bistable wall attachment amplifiers have been manufactured with heights up to 500 μm.[80,81]

The LIGA work from the early and mid-1990s as applied to fluidic amplifiers lacks originality. More sophisticated fluidic components providing functions such as flip-flops, logic elements such as OR-NOR gates and AND gates, proportional amplifiers, and fluid resistors already existed in the 1960s. The fluidic components made in those days still provide an excellent demonstration for the capabilities of photofabrication with photosensitive glass (see Chapter 7 and Humphrey and Tarumoto[82]). Humphrey and colleagues made detailed technical and economic comparisons of the different manufacturing options for fluidic elements before embarking upon photofabrication as the manufacturing choice, a practice every micromachinist should adapt. In the 1980s, Si micromachining was applied to make the same fluidic logic devices, as well as fluidic valves and analytical equipment (for a review of the more recent work, see Elwenspoek et al.[83]). Despite these attempts, neither LIGA nor any of the Si microfluidics produced today surpass the degree of maturity achieved with the photofabrication method. The 'fluidic' logic elements could achieve many of the

functions of an electronic circuit and were more robust, simpler, and better able to operate in hostile environments. They reacted much slower (milliseconds instead of nanoseconds) and were larger than their electronic counterparts. The two latter factors added up to the eventual demise of microfluidics in the 1960s.

In view of the above, the use of LIGA micromachining for fluidic logic is hardly justifiable. If fluidic logic would, for some unforeseen reason, become competitive again with an electronic approach, the earlier photofabrication machining approach would still be the better approach for most applications since cost, not accuracy, is the major concern.

As demonstrated in the fabrication of some of the fluidic devices discussed so far, zero-basing the validity of a new approach to the micromanufacture of a microdevice at hand should be the first step in any micromachine design. This requires a thorough knowledge of the intended application, suppression of the urge to apply a new manufacturing tool to anything small, and a literature search stretching beyond the last 5 years. The role of LIGA in analytical instrumentation, especially equipment involving optics, may still represent an opportunity as tolerances on wall roughness and accuracy may be very stringent. Even in these latter applications, advantages over established techniques are not always self evident.

### Membrane Filters

A key component in all of today's filtration membranes is a microporous structure with uniform and well-defined pore sizes and a high overall porosity. Most current microfiltration membranes are characterized by a relatively broad distribution with respect to the size and position of the pores. When using membranes made by the nuclear track technique, isoporous membranes result. However, these type of filters suffer from limited porosity. The process to make LIGA membrane filters, on the other hand, allows the production of isoporous structures with a very high porosity.[21] Also, much smaller capillaries can be realized than when applying spark erosion. Since the membrane thickness may be some hundred micrometers, a high mechanical stability can be realized in combination with an extremely high transparency.[84]

The porous membrane application makes a lot of sense for LIGA. This application would be of great interest, especially if we could make large sheets of the porous material in a continuous process. In principle, large primary molds could be manufactured by rolling long sheets of PMMA, in a continuous process, through the X-ray exposure zone. Table 6.11 presents a summary of the expected benefits of LIGA microfilters.[85]

A *transverse flow microfilter* (cross-flow) with porous membrane builtup by X-ray lithography is described in a 1989 patent by Ehrfeld et al. (U.S. Patent 4,797,211).

## Electronic Microconnectors and Packages

The decrease in critical dimensions in ICs results in a corresponding demand for higher density electrical connections with subminiaturized dimensions. *Electrical multipin microconnectors* can be equipped with precise guiding structures which allow multiple nondestructive insertions and separations.[86]

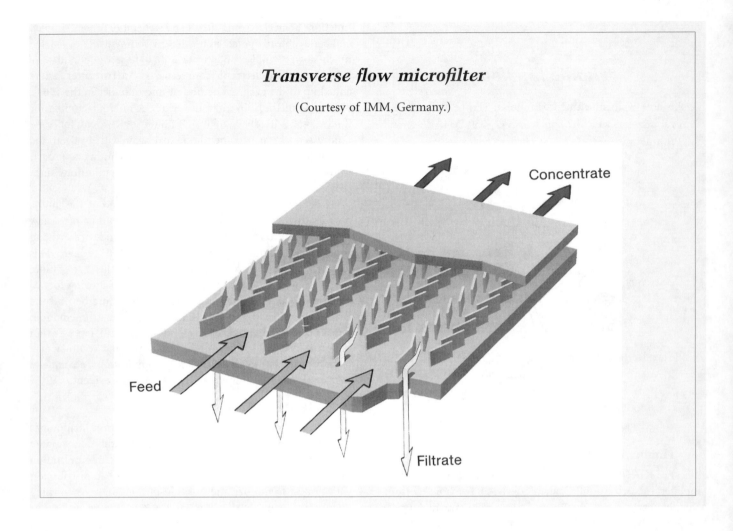

### Transverse flow microfilter

(Courtesy of IMM, Germany.)

**Table 6.11**   LIGA Microfilter Characteristics

| | |
|---|---|
| Uniform pore size | Sharp separation boundary |
| Extremely high porosity | High filtration flow |
| Designable pore shape | Optimized filtration characteristics |
| Constant channel cross-section | Ideal surface filter |
| Integrated supporting structures | High strength |
| Wide range of materials | Suitable for corrosive and high temperature media |

*Source:*   From *The LIGA Technique,* courtesy of MicroParts.

Applications of these arrays include detachable interconnectors of high quality ICs or miniaturized multipin sensor heads.[86] The typical mating force of such a LIGA multipin electrical connector is 2 cN/pin and 15 cN/pin for the latching. The typical current capacity is 15 mA/pin and values up to 4 A before destruction have been observed.

Given that the required tolerances for microconnectors will become increasingly stringent, the electronic interconnect application does seem promising. Pseudo-LIGA techniques (for example, based on UV sensitive polyimides) may prove more cost effective solutions for this application in the short run. But, further down the line, we may envision microconnectors being fabricated as part of the parallel batch IC process. LIGA with its wide dynamic range can cover the submicron dimensions of an IC as well as the millimeter dimensions of the connectors. Once the insertion point of X-ray lithography in the IC industry has become a fact, this 'integrated connector strategy' may become very important.

Most of the costs for producing an integrated chip relate to the package and introduction of the circuit into the package. Part of the cost differences follow from the vast difference between integrated circuit technology and packaging technology. In the future, we foresee packaging becoming more and more part of the IC process. The IC package, for example, could be replaced by a LIGA sealing wall around the edge of the chip and a seal of a metal diaphragm over the top. It may also be possible to plate completely over the top of the chip rather than sealing with a diaphragm. The leads to the chip would remain outside of the LIGA wall and the entire assembly could then be wire bonded to an inexpensive plastic package and still have the hermetic seal associated with more expensive packages. Additionally, this approach could be used for packaging of multi-chip modules. Other advantages would be built-in EMI (electromagnetic interference) protection and easy assembly of multi-chip modules, for instance for stacking die vertically.

### Electrical multipin microconnectors

(From Ehrfeld, W., Proceedings, Micro System Technologies '90, Berlin, 521–528, 1990. With permission.)

50 µm     50 µm

## Micro-Optical Components

### Introduction

Before LIGA, techniques suitable to micromanufacture micro-optical components such as prisms, lenses, beam splitters, and optical couplings in a freely selectable configuration, material and structural size were hard to come by.[31] LIGA enables the integrated fabrication of most of these components in a wide variety of materials and sizes. Micro-optics may develop into a significant LIGA application, especially for cases where cost can be overlooked. When very smooth vertical walls and high aspect ratio of the parts play a major role, LIGA may be the only available technique.

PMMA is transparent in the visible and near IR regions, making it a natural optical material. The simplest LIGA optical structure based on PMMA is a *multimode waveguide* of rectangular cross-section consisting of a three-layer resist structure.[67] To make such a planar waveguide, an epoxy phenolic resin substrate, insensitive to X-rays, is first coated with a cladding resist layer. A cladding material well-suited to X-ray patterning is a copolymer of PMMA (78%) and tetrafluoropropyl methacrylate (TFPMA) (22%) with a refractive index of 1.476. With this composition and a PMMA core layer (refractive index of 1.49), a numerical aperture of 0.2 is obtained. By varying the composition of the copolymer each refractive index between 1.49 and 1.425 can be obtained precisely, allowing the numerical aperture of the polymer waveguide to be fitted to the light-guiding fibers used to inject the light into the component. The planar cladding layer is milled down to a thickness corresponding to

### Multimode waveguide

(From Gottert et al., Integrierte Optik und Mikrooptik mit Polymeren, Mainz, Germany, 1992. With permission.)

PMMA
P(MM + TFPMA)

the thickness of a connecting optical fiber cladding. A foil of PMMA core material is then welded onto the first cladding layer and the core is milled down to the thickness of the same fibre core. Welding is performed slightly above the glass transition temperature and under pressure. The procedure causes only a small amount of interdiffusion of the two layers and a well-defined refractive index profile results; the variation of the refractive index between two successive layers takes place within an interface layer of less than 10 μm thickness. A cover foil consisting of the same copolymer is finally welded and milled to cladding thickness on the PMMA core foil. Bulk PMMA specimens show a mean attenuation of 0.2 dB/cm in the 600- to 1300-nm wavelength range. At absorption peaks of 900 and 1180 nm, this value rises to 0.4 db/cm and 3 dB/cm, respectively. To manufacture low-loss PMMA material for the near-infrared region, deuterated PMMA may be used as a core material. This causes the absorption peak at about 900 nm, caused by C-H resonances, to be shifted to longer wavelengths. Attenuation of the deuterated material in the spectral range from 600 to 1300 nm is only about 0.1 dB/cm. For the described three-layer resist structure, the attenuation is somewhat larger than for the bulk PMMA specimens due to the relatively large roughness of the side walls of the thus fabricated waveguides and by defects in the three-resist layer.[87]

Although LIGA is producing projected features rather than three-dimensional true shapes (such as, say, in a contact lens), this technique has brought micromachinists much closer to the possible realization of a totally integrated micro-optical bench. The lithographic alignment of the separate optical components; the high, extremely smooth, vertical walls; and the excellent optical properties of one of the key LIGA resists, PMMA, all combine to propel LIGA as an ideal technique for micro-optical

applications within reach of commercialization.[88] Micro-optical benches were attempted via polysilicon surface micromachining[89] and bulk micromachining techniques.[90] In the former method, the very thin optical components had to be lifted from the surface and secured upright mechanically in a postfabrication assembly step (see Chapter 5). In the latter method, the proposed structures consisted of complex multi components. We suggest that given the required tolerances neither of these alternative optical benches will ever be manufactured outside a research laboratory.

### Fiberoptic Connectors

LIGA elements for *coupling monomode fibers* with integrated optical chips are under development. These prealignment arrays may utilize fixed nickel guiding structures in combination with leaf springs to ensure a precise alignment of the optical fibers relative to the optical chip with an accuracy in the submicron range. The thermal expansion coefficient of the substrate material is matched to the optical chip material — e.g., a glass substrate is used for coupling fibers to glass chips — and the use of spring elements simplifies the handling of the fibers.[91] Alternatives involve the use of optical adhesives and the use of silicon V-groove arrays (see Chapter 4). Both technologies are more labor intensive and suffer from thermal expansion problems.

The use of microptical LIGA components is especially attractive for the coupling of multimode fibers; here the capability of exact lithographic positioning of mechanical mounting supports is used advantageously to position the multimode fibers

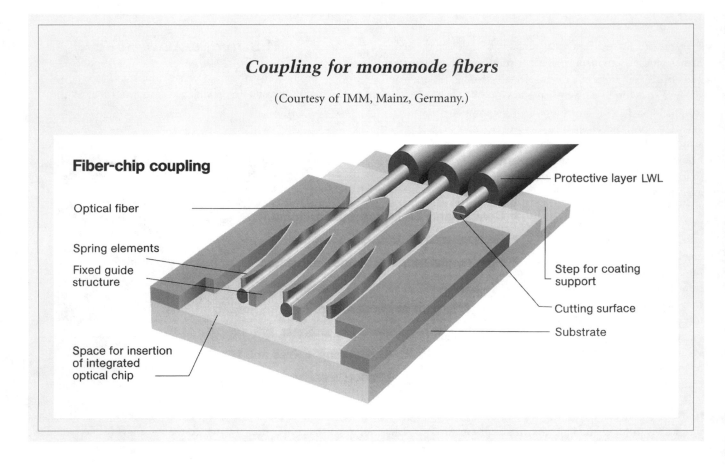

## *Coupling for monomode fibers*

(Courtesy of IMM, Mainz, Germany.)

**Fiber-chip coupling**

Optical fiber

Spring elements

Fixed guide structure

Space for insertion of integrated optical chip

Protective layer LWL

Step for coating support

Cutting surface

Substrate

very precisely with respect to the micro-optical components without need for any additional adjusting operations.[67,92]

A *pull-push LIGA* connector for single-mode fiber ribbons, incorporating a set of *precision microsprings*, coupling up to 12 fibers spaced at 250 μm, is close to commercial implementation and is one of the LIGA products with mass market appeal. To obtain good coupling efficiency, the fibers are positioned horizontally with a precision of 1 μm. LIGA is an excellent method to provide that precision. These connectors are under development at IMM.[93] In line with the micromanufacturing philosophy presented in this book — to optimize the use of each micromachining technique for its optimum cost/performance application — those parts of the mold insert which do not require such high accuracy are fabricated using other methods of precision machining such as EDM (electro discharge machining) (see Chapter 7).

### Fresnel Zone Plates

Self-supporting *Fresnel zone plates* have been made by LIGA.[94] One such lens was made of nickel with a distance between the rings of 74 μm on the inner side, decreasing to 10 μm on the external side. The width of the radial web was also 10 μm. A *Fresnel zone plate array in PMMA* with a 10-mm focal length and a zone plate diameter of 10 mm was made as well.[95] In a similar way, self-supporting gratings or infrared polarizers might be fabricated.

### Planar Grating Spectrograph-Integrated Optical Bench

Using the above described planar polymer waveguide in combination with a self-focusing reflection grating, a *spectrometer* 18 × 6.4 mm in size was developed.[96] The polychromatic light, injected into the waveguide by an optical fiber is dispersed at a curved grating coated with a sputtered gold layer. The spectrally divided portions of the light are focused on ten optical fibers, or, alternatively, on an array of diodes positioned at the focal line. The intensity at the focal line is about 25% of the incoupled intensity, and the spectral resolution is 0.2 nm/μm.

As the fiber grooves are aligned exactly to the grating, no adjustment is necessary to make the component work. The grating is designed to work in a spectral range between 720 and 900 nm and is blazed to diffract in the second order. The grating pitch (groove spacing) is 3.4 μm, and the step of the groove is 0.5 μm (more recently 0.25 μm was realized), which is well within the accuracy of deep-etch X-ray lithography.[87] Although the spectral resolution of this microspectrometer is relatively low, this type of grating can be used for optical analysis in the whole visible and near-IR-wavelength region (up to 1100 nm), e.g., for photometric analysis of gas mixtures, optical film thickness measurement, and as a spectral detector in liquid chromatography.[97] Another set of applications of the described microspectrometer is in the area of combining and separation of optical wavelengths. In this technology, signals of several different wavelengths are multiplexed on a single optical fiber. Multiplexer (MUX) and demultiplexer (DMUX) devices are required to combine and separate multichannel signals. Among the many proposed systems, the self-focusing concave reflection

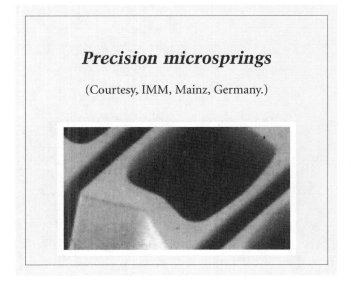

*Precision microsprings*

(Courtesy, IMM, Mainz, Germany.)

*Pull-push LIGA connector*

(Courtesy, IMM, Mainz, Germany.)

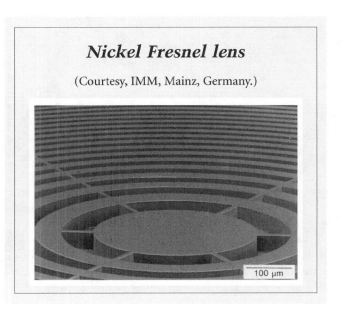

*Nickel Fresnel lens*

(Courtesy, IMM, Mainz, Germany.)

100 μm

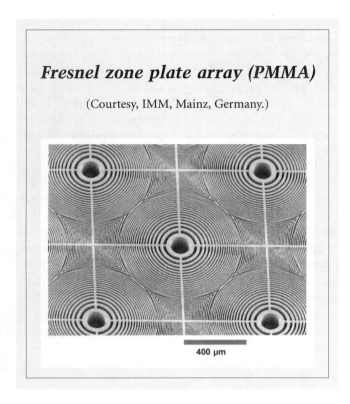

# Fresnel zone plate array (PMMA)

(Courtesy, IMM, Mainz, Germany.)

400 μm

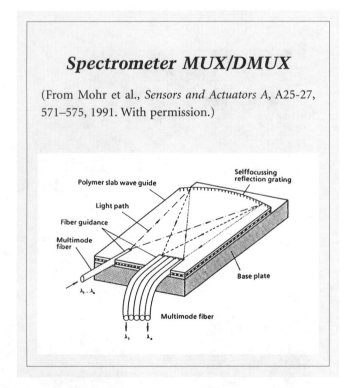

# Spectrometer MUX/DMUX

(From Mohr et al., *Sensors and Actuators A*, A25-27, 571–575, 1991. With permission.)

grating is the only one which can handle a large channel number (>5) with good performance and without the need of additional reflecting and focusing elements.[96]

For operation at wavelengths greater than 1100 nm, air must be used as a transmission medium instead of a PMMA-based waveguide. At NASA Ames, we have embarked upon the development of a gas analysis spectrometer including the capability

of $CO_2$ detection at 4.2 μm. Instead of piping polychromatic light into the spectrometer on a multimode fiber, the intent here is to use a microfabricated heated filament integrated into the spectrometer. The microheater filament has already been realized by a combination of surface and bulk micromachining on an SOI substrate. Monochromatic light, at 4.2 μm, will be isolated from the broad-band filament radiation by reflecting the light off a self-focusing grating and then bouncing the monochromatic light from several LIGA micromirrors in an absorption cell-type arrangement. A 10-cm light path is required for a 1-ppm $CO_2$ sensitivity, and we found that such a path length is easily designed by arranging the mirrors on a $1 \times 1$-cm footprint. Crucial technical problems yet to be overcome are, first, the considerable self-diffraction of a small light beam at 4.2 μm in air spreads the light spot quickly beyond the individual mirror size, and, second, the light emanating from the microfilament, fabricated in the substrate plane, must be brought from the vertical into a horizontal plane so it can be reflected off the vertical self-focusing grating and micromirrors. Finally, the integrity of the micro-optical components (e.g., reflectivity) must be preserved when exposed to a gaseous uncontrolled environment (see also Chapter 10).[98]

### Other Passive Optical Components

Other passive optical components that have been made using LIGA are Y-couplers,[99] cylindrical and prismatic lenses, and beam splitters. The compact design of splitters and couplers with the light paths between two fiber ends in the range of the fiber diameter makes attenuation losses very small. For example, the excess loss of a *bidirectional fiber coupling element (1 × 2 beam splitter)* with 100/140 μm multimode fibers is less than 1 dB.

## LIGA Accelerometer

The sacrificial layer technique was used to fabricate a LIGA *acceleration sensor* with a capacitive read-out. The width of the gap between stationary electrodes and a seismic mass (Ni in this case) changes when they are accelerated. The resulting change in capacity is what is measured. The advantage of a LIGA accelerometer over Si-technology is the material variety and therefore the choice of density of the seismic mass, as well as the choice of the Young's modulus of the bending beams. This flexibility could enable a family of accelerometers with a much wider g-range than bulk Si micromachined devices and definitely a wider range than poly-Si surface micromachined accelerometers which, due to the thin inertial mass, are limited in this respect. Also, the extreme aspect ratios attainable with LIGA bring about a better signal. The thickness of the sensor has no influence on the sensitivity, but it does affect the value of the absolute capacitance, which increases with greater thickness. A large sensor thickness also improves the stiffness of the cantilever which is advantageous to the cross sensitivity. The width of the capacitor gap influences the sensitivity: the smaller the gap, the more sensitive the accelerometer. LIGA enables very narrow gaps and thick inertial plates. For example, the zero-point capacity given by a gap between the seismic mass and

*Fiber coupling
(1 × 2 beam splitter)*

(Courtesy, KfK, Karlsruhe, Germany.)

the electrodes of only 3 to 4 μm at a structural height of 100 μm is 0.7 pF. A sensor with this zero-point capacity was shown to have a sensitivity of 15% capacity change per g and a resonance frequency of 570 Hz. The temperature coefficient was found to be about 0.5% K$^{-1}$. By optimized sensor design, the zero-point capacity and the temperature behavior can further be improved. Extremely temperature resistant accelerometers were achieved by designing the sensor partly with a positive and partly with a negative temperature coefficient. In this way, a LIGA accelerometer for 1 g with a measured temperature coefficient of offset (TCO) of $1.02 \times 10^{-4}$ g/K in the temperature range −10 to +100°C was realized.[100] For automotive applications a resonance frequency of more than 500 Hz is required, and for 1 g of acceleration the change in capacity should be larger than 20%.[62] In case the accelerometer is also electrostatically actuated, other benefits of using LIGA result as explained further below under electrostatic LIGA actuators.

It has been shown that LIGA may be compatible with substrates containing active electronics, and in principle integrated LIGA accelerometers could represent a good business opportunity. On the other hand, there is more accessible bulk and surface micromachining technology available, which is now feeding most of the industrial needs.

## Interlocking Gears

A series of interlocking gears was fabricated in one process sequence without assembly. The gears are free to rotate and could be used in watches or in audio, video, and measuring technologies. The precision of the LIGA process allows the realization of gears with minimized backlash and friction.[26] At this point, it is not clear if the accuracy demands in those different application fields really do require LIGA or if LIGA-like processes could fulfill those needs.

## Electrostatic Actuators

For the fabrication of actuator elements, where free lateral patterning is required, one usually resorts to surface micromachining. With polysilicon surface micromachining, the thickness of actuator elements is usually limited to heights of a few micrometers, which is particularly disadvantageous in the case of microactuators. This can be understood from our discussions on scaling laws in Chapter 9. In practical realizations, linear electrostatic microactuators often have two *comb-like structures*. By applying a voltage the movable part of the comb moves closer to the fixed part. For a rectangular design, the force and the microactuator displacement is proportional to the ratio of structure height, T, vs. separation gap, d, and the number of fingers, N (see also Chapter 9):

$$F = -\frac{\varepsilon_0 T V^2}{2d} N \qquad 6.7$$

An efficient comb drive can be thus achieved by designing many high comb fingers with narrow gaps. In surface micromachining, the thickness, T, is limited to about 2 to 3 μm, whereas with LIGA several hundreds of microns are possible. Clearly, LIGA has the potential to produce better electrostatic actuators. Mohr et al.[101] further showed that by making the comb fingers trapezoidal the maximum displacement of the interdigitated fingers can be further increased and smaller capacitor gaps are possible. For a more rigorous mathematical derivation of Equation 6.7, we refer to the finite-element simulation of comb drives by Tang.[102] Muller and others have used LIGA electrostatic comb drives micromachined on top of a silicon wafer to fabricate micro-optical switches.[103]

For better electrostatic actuators, LIGA will have to compete with surface micromachined structures on SOI wafers where the thickness of the epilayer can be much thicker than that of a poly-Si layer and where dry etching can also make good vertical walls and narrow gaps. But perhaps where ceramic actuators are involved, LIGA has a unique advantage.

## LIGA Inductors

LIGA coils could find applications in inductive position or proximity sensing and all types of magnetic actuators. It is possible to achieve a high Q-factor as well as a high driving current as a result of the high aspect ratio and the high surface quality of the conductors. A very important application is to expand the work, initiated by Romankiw at IBM, to make finer and finer, batch-fabricated, thin-film magnetic recording heads (see Figure 7.8).[104] Other applications of LIGA inductors are in RF oscillators or in inductive power transmission for rotating systems.[26] Rogge et al.[105] successfully demonstrated a fully batch fabricated magnetic micro-actuator using a two-layer LIGA process. The contact forces at 1A for a permalloy relay made this way were well beyond those of miniaturized relays. Based on cost, only the most demanding applications such as the read-write heads might justify the use of LIGA, for most other applications LIGA-like methods will probably suffice.

## *Acceleration sensor with capacitive read-out*

(Courtesy, IMM, Mainz, Germany.)

200 µm

50 µm

**Acceleration sensor**

Seismic mass and central electrode

Counter electrodes

Contact pads

Silicon wafer with integrated electronic circuit

Support block

Leaf spring

**Micro-optical switch with electrostatic comb-like structures**

**Large aspect ratio magnetic motor**

(Courtesy IMM, Mainz, Germany.)

## Electromagnetic Micromotor

We have shown already how LIGA makes better electrostatic actuators than other Si micromachining techniques; the same is true for electromagnetic actuators. In Chapter 9 on scaling laws we will see that most Si micromachined motors today are producing negligible amounts of torque. In practical situations, what is needed are actuators in the millimeter range delivering torques of $10^{-6}$ to $10^{-7}$ Nm. Such motors can be fabricated with classical precision engineering or a combination of LIGA and precision engineering.

The performance with respect to torque and speed of a traditional miniature magnetic motor with a 1-mm diameter to 2-mm long permanent magnetic rotor was demonstrated, in practice, to be incomparably better than the surface micromachined motors discussed in Chapter 5.[106] The small torque of the surface micromachined electrostatic motors is, to a large extent, due to the fact that they are so flat. The magnetic device made with conventional three-dimensional metal-working techniques has an expected maximum shaft torque of $10^{-6}$ to $10^{-7}$ Nm. The torque depends on the Maxwell shearing stress on the rotor surface integrated over the area of the latter; the longer magnetic rotor easily outstrips the flat electrostatic one with the same rotor diameter.[106] In both electric and magnetic microactuators, force production is proportional to changes in stored energy. The amount of force that may be generated per unit substrate area is proportional to the height of the actuator. Large aspect ratio structures are therefore preferred.

With LIGA techniques *large aspect ratio magnetic motors* can be built.[35] Essentially, a soft magnetic rotor follows a rotating magnetic field produced by the currents in the stator coils. The motor manufacture is an example of how micromachining and precision engineering can be complementary techniques for producing individual parts, which have to be assembled afterward. Only the components with the smallest features are

produced by means of microfabrication techniques, whereas the other parts are produced by traditional precision mechanical methods.

A magnetic LIGA drive was also demonstrated by the Wisconsin team[107] and consisted of a plated nickel rotor which was free to rotate about a fixed shaft. Large plated poles (stators) came close to the rotor. The rotor was turned by spinning a magnet beneath the substrate. Further work by the same group involved *enveloping coils*, to convert current to magnetic field *in situ* rather than externally.[108] It is necessary to plate two different materials — a high-permeability material such as nickel and a good conductor such as gold — to accomplish this.

The motor can also be used as electrostatic motor. By applying voltages to the stator arms, the rotor (which is grounded through the post about which it rotates) is electrostatically attracted and, by poling the voltages, can be made to turn. Design rules and tests of electrostatic LIGA micromotors were presented by Wallrabe et al.[109] Minimum driving voltages needed were measured to be about 60 V; optimized design torques of the order of some μNm are expected. Cost of electromagnetic LIGA micromotors will often urge the investigation

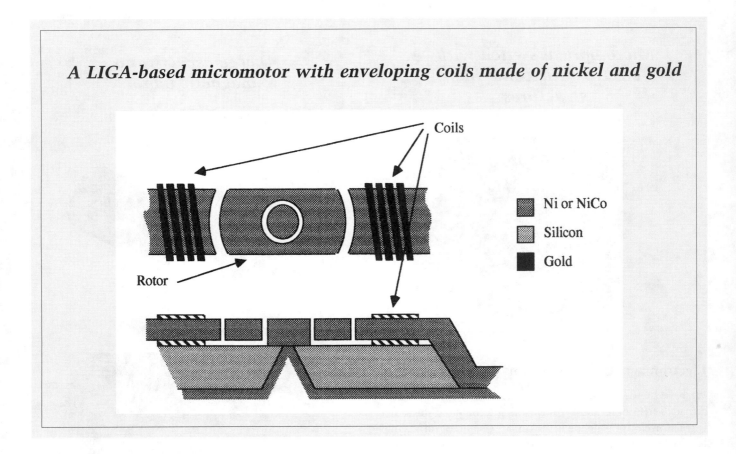

**A LIGA-based micromotor with enveloping coils made of nickel and gold**

Coils

Rotor

Ni or NiCo

Silicon

Gold

of LIGA-like technologies or hybrid approaches (LIGA combined with traditional machining) as more accessible and adequate machining alternatives.

## Other Potential LIGA Applications

Besides the interconnect and packaging applications already mentioned, LIGA might have other applications in the microelectronics industry. Some of the circuit elements currently very expensive to fabricate on an integrated circuit could be replaced with LIGA circuit elements. The expense associated with large capacitors on an integrated circuit, for example, is due to the large amount of space they use. Shrinking capacitors to use less real estate is an area of great interest for research. A LIGA capacitor can be manufactured with tall, thin vertical plates to achieve a high capacitance without using a large amount of real estate.

An additional area of application would be the manufacture of supercapacitors, small devices with very large values of capacitance. These are used as discrete devices for a large range of applications, including the generation of extremely large current pulses (by trickle-charging the capacitor, then driving the capacitor plates together with explosive devices) for use in military equipment.

There are a number of methods for using LIGA as a means for creating much more precise and inexpensive optical components. Some possible applications, including a spectrograph for liquid and gas analysis, were mentioned already and some additional opportunities follow.

Presently, one of the largest areas for sensor research is in optical sensors. In general, an optical sensor is measuring a change in an optical property with the use of optical fibers. This change can be seen by measuring an amplitude change in light passing through or reflected off of some media or by a phase change resulting in interference patterns.

Different methods for connecting fiberoptic waveguides to sensor platforms, which can hold various chemistries or fluids or be inserted into a human body, still have not been satisfactorily developed. LIGA has a great deal of potential in this area, due to the ability to create both a fiberoptic connector and waveguide in one step. The waveguides can be of arbitrary shape, making it easy to include small cups to hold chemistries or to wrap extremely long waveguides into a small area. The number of ways which LIGA could be used for optical sensing is tremendous.

An additional item of note is that one of the limiting factors for the wide application of fiber optic sensors is the cost associated with connecting the sensor to a fiber with the required accuracy. This would not be a problem with LIGA-based sensor platforms. The optical applications of LIGA might have potential independent of developments in the IC industry as higher prices are often acceptable for precise optical components.

The use of LIGA as a method for creating on-chip waveguides would have a tremendous impact on the area of optical communication between chips. Presently, there is great interest in producing light-emitting diodes or semiconductor lasers on a silicon substrate by growing gallium arsenide (or other III-V semiconductors) directly on a silicon substrate and using this to create the light source. This source would then be used as a means of communication with the outside world, rather than electrical connections.

Coupling these light sources will require on-chip waveguides and a means for connecting the chips to optical fibers. LIGA is well suited to creating optical waveguides, as discussed above, and is an obvious and potentially inexpensive solution.

Micromanipulation is a field which is on the verge of explosive growth. The combination of techniques developed for use with integrated circuit lithography and probing completed microchips is also being applied to other areas of microelectronics — for instance, automated testing of wire bonds or failure analysis of integrated circuits.

Traditional micromanipulators are now quite large in size and movement compared to the devices they are to manipulate or probe. Not only can LIGA-based micromanipulator machines be built of extremely tiny LIGA parts, the actual microprobes can be manufactured by LIGA.

The ability to plate refractory materials such as rhenium, osmium, and ruthenium would have use in the creation of incredibly precise two- and four-point probes for much higher accuracy in testing applications for integrated circuits. The relative precision with which a two- or four-point probe can test a substrate is related to how closely the probe points can be spaced. For two-point probes, the smallest spacing available is about 10 μm, and this can only be achieved by hand assembly and grinding operations by a skilled technician. With LIGA, the probe points could easily be spaced as close as 0.5 μm, perhaps as close as 0.2 μm. The utility of this for probing silicon wafers during the manufacturing process would be very high.

An additional area for application would be the creation of LIGA probe cards for testing of integrated circuits. These probe cards are difficult to manufacture and expensive, and periodic realignment of the probe points is necessary due to wear and tear. A cheap, disposable LIGA probe card would be of great use for probing integrated circuit test patterns, which are increasingly complicated and decreasing remarkably in size. Many major integrated circuit manufacturers are relocating electrical test patterns into the scribe lines of silicon wafers to save on real estate. This necessitates a major size reduction in the contact pads that the probe card must contact. In the area of extremely precise probe cards, LIGA would have a definite advantage.

# Technological Barriers and Competing Technologies

## Insertion Point of X-Ray Lithography

The original work demonstrating the LIGA process was performed by a group of scientists led by Dr. E. W. Becker at Karlsruhe. Subsequently, a group led by Prof. Guckel, replicated most of the aspects of this process and demonstrated some new technologies as well. This achievement was accomplished in a rather short amount of time, approximately 18 months. This

was due to the extensive literature available on the topic (the Karlsruhe group has been very prolific in publishing on the various aspects of LIGA processing), but it also shows that LIGA is a technology which, in principle, is ready to be taken up by industry. Wider commercial acceptance of micromachining will be linked to cost/performance advantages and availability of the technique. The latter may well be dependent on the insertion point of X-ray lithography in the production of the next generation of industrial ICs. As pointed out earlier in this chapter and in Chapter 1, the X-ray lithography insertion point keeps on being postponed, possibly keeping LIGA a tool for the research community only.

## Competing Technologies

Deep dry etching (often cryogenic) and thick UV-sensitive resists such as polyimides, AZ-4000, and SU-8 (see Chapters 1 and 5) are contenders for several of the applications for which LIGA is being promoted. With respect to dry etching, higher and higher aspect ratios are being achieved; especially when using highly anisotropic etching conditions and cryogenic cooling, remarkable results are being achieved (see Chapter 2). With respect to photosensitive polyimides, 100-μm high features can be accomplished readily, and minimum feature sizes of 2 μm are possible (see Chapter 1). Both competing techniques are more accessible than LIGA and will continue to improve, taking more opportunities away from LIGA. Like LIGA, both alternative techniques can be coupled with plating, but neither technique can yet achieve the extreme low surface roughness and vertical walls LIGA is capable of. The lesson is that more comparative studies of the different micromachining technologies should be undertaken before one is being decided upon.

Other less obvious competing technologies for LIGA-like applications are ion-beam milling, laser ablation methods, and even ultra-precision machining (see Chapter 7). The latter three are serial processes and rather slow, but if we are considering making a mold then these technologies might be competitive.

For LIGA on the internet, see Appendix B.

## References

1. IMM, "The LIGA Technique", Commercial Brochure, IMM, 1995.
2. Spiller, E., R. Feder, J. Topalian, E. Castellani, L. Romankiw, and M. Heritage, "X-Ray Lithography for Bubble Devices", *Solid State Technol.*, April, 62–68, 1976.
3. Romankiw, L. T., I. M. Croll, and M. Hatzakis, "Batch-Fabricated Thin-Film Magnetic Recording Heads", *IEEE Trans. Magn.*, MAG-6, 597–601, 1970.
4. Becker, E. W., W. Ehrfeld, D. Munchmeyer, H. Betz, A. Heuberger, S. Pongratz, W. Glashauser, H. J. Michel, and V. R. Siemens, "Production of Separation Nozzle Systems for Uranium Enrichment by a Combination of X-Ray Lithography and Galvanoplastics", *Naturwissenschaften*, 69, 520–523, 1982.

5.  Waldo, W. G. and A. W. Yanof, "0.25 Micron Imaging by SOR X-Ray Lithography", *Solid State Technol.*, 34, 29–31, 1991.

6.  Hill, R., "Symposium on X-ray Lithography in Japan", 1991, National Academy of Sciences.

7.  Stix, G., "Toward Point One", Sci. Am., 272, 90–95, 1995.

8.  Urisu, T. and H. Kyuragi, "Synchrotron Radiation-Excited Chemical-Vapor Deposition and Etching", *J. Vac. Sci.Technol.*, B5, 1436–1440, 1987.

9.  Goedtkindt, P., J. M. Salome, X. Artru, P. Dhez, N. Maene, F. Poortmans, and L. Wartski, "X-Ray Lithography with a Transition Radiation Source", *Microelectron. Eng.*, 13, 327–330, 1991.

10. Hagmann, P., W. Ehrfeld, and H. Vollmer, "Fabrication of Microstructures with Extreme Structural Heights by Reaction Injection molding", First Meeting of the European Polymer Federation, European Symp. on Polymeric Materials, Lyon, France, 1987, p. 241–251.

11. Muray, J. J. and I. Brodie, "Study on Synchrotron Orbital Radiation (SOR) Technologies ans Appli-cations", SRI International, Menlo Park, CA, Report No. 2019, 1991.

12. Bley, P., D. Einfeld, W. Menz, and H. Schweickert, "A Dedicated Synchrotron Light Source for Micro-mechanics", EPAC92. Third European Particle Accelerator Conference, Berlin, Germany, 1992, p. 1690–2.

13. Ehrfeld, W., W. Glashauer, D. Munchmeyer, and W. Schelb, "Mask Making for Synchrotron Radiation Lithography", *Microelectron. Eng.*, 5, 463–470, 1986.

14. Lawes, R. A., "Sub-Micron Lithography Techniques", *Appl. Surf. Sci.*, 36, 485–499, 1989.

15. Schomburg, W., K., H. J. Baving, and P. Bley, "Ti- and Be-X-Ray Masks with Ailgnement Windows for the LIGA Process", *Microelectron. Eng.*, 13, 323–326, 1991.

16. Hein, H., P. Bley, J. Gottert, and U. Klein, "Elektro-nen-strahllithographie zur Herstellung von Rontgen-masken fur das LIGA-Verfahren", *Feinw.tech. Messtech*, 100, 387–389, 1992.

17. Bley, P., W. Menz, W. Bacher, K. Feit, M. Harmening, H. Hein, J. Mohr, W. K. Schomburg, and W. Stark, "Application of the LIGA Process in Fabrication of Three-Dimensional Mechanical Microstructures", 4th International Symposium on MicroProcess Conference, Kanazawa, Japan, 1991, p. 384–389.

18. Guckel, H., T. R. Christenson, T. Earles, J. Klein, J. D. Zook, T. Ohnstein, and M. Karnowski, "Laterally Driven Electromagnetic Actuators", Technical Digest of the 1994 Solid State Sensor and Actuator Workshop, Hilton Head Island, SC, 1994, p. 49–52.

19. Siddons, D. P. and E. D. Johnson, "Precision Machining Using Hard X-Rays", *Synchrotron Radiation News*, 7, 16–18, 1994.

20. Guckel, H., "Deep Lithography", (Notes from Handouts), 1994, Banff, Canada.

21. Becker, E. W., W. Ehrfeld, P. Hagmann, A. Maner, and D. Munchmeyer, "Fabrication of Microstructures with High Aspect Ratios and Great Structural Heights by Synchrotron Radiation Lithography, Galvanoforming, and Plastic Molding (LIGA process)", *Microelectron. Eng.*, 4, 35–56, 1986.

22. Friedrich, C., "Complementary Micromachining Processes", (Notes from Handouts), 1994, Banff, Canada.

23. Henck, R., "Detecteurs au Silicium Pour Electrons et Rayons X, Principes de Fonctionnement, Fabrication et Performance", *J. Microsc. Spectrosc. Electron.*, 9, 131–133, 1984.

24. Vladimirsky, Y., O. Vladimirsky, V. Saile, K. Morris, and J. Klopf, M., "Transfer Mask for High Aspect Ratio Micro-Lithography", Microlithography '95. Proceedings of the Spie - the International Society for Optical Engineering, Santa Clara, CA, 1995, p. 391–396.

25. Michel, A., R. Ruprecht, and W. Bacher, "Abformung von Mikrostrukturen auf Prozessierten Wafern", KfK, Report No. 5171,1993, Karlsruhe, Germany, 1993.

26. Rogner, A., J. Eichner, D. Munchmeyer, R.-P. Peters, and J. Mohr, "The LIGA Technique-What Are the New Opportunities?", J*ournal of Micromechanics and Microengineering*, 2, 133–140, 1992.

27. Mohr, J., W. Ehrfeld, and D. Munchmeyer, "Analyse der Defectursachen und der Genauigkeit der Structuru-bertragung bei der Rontgentiefenlithographie mit Synchrotronstrahlung", KfK, Report No. 4414, Karlsruhe, Germany, 1988.

28. Moreau, W. M., *Semiconductor Lithography*, Plenum Press, New York, 1988.

29. Lingnau, J., R. Dammel, and J. Theis, "Recent Trends in X-Ray Resists: Part 1*", *Solid State Technol.*, 32, 105–112, 1989.

30. Mohr, J., W. Ehrfeld, D. Munchmeyer, and A. Stutz, "Resist Technology for Deep-Etch Synchrotron Radiation Lithography", *Makromol. Chem. Macromol. Symp.*, 24, 231–251, 1989.

31. Gottert, J., J. Mohr, and C. Muller, "Mikrooptische Komponenten aus PMMA, Hergestellt Durch Roent-gentiefenlithographie Werkstoffe der Mikrotechnik-Bais fur neue Producte", *VDI Berichte*, 249–263, 1991.

32. Wollersheim, O., H. Zumaque, J. Hormes, J. Langen, P. Hoessel, L. Haussling, and G. Hoffman, "Radiation Chemistry of Poly(lactides) as New Polymer Resists for the LIGA Process", *J. Micromech. Microeng.*, 4, 84–93, 1994.

33. Mohr, J., W. Ehrfeld, and D. Munchmeyer, "Requirements on Resist Layers in Deep-Etch Synchrotron Radiation Lithography", *J. Vac. Sci. Technol.*, B6, 2264–2267, 1988.

34. Madou, M. J. and M. Murphy, "A Method for PMMA Stress Evaluation", *Unpublished Results*, 1995.

35. Ehrfeld, W., "LIGA at IMM", (Notes from Handouts), 1994, Banff, Canada.

36. Private Communication, Vladimirsky, O., *Spin Coating PMMA*, LSU, May 1995.

37. Guckel, H., J. Uglow, M. Lin, D. Denton, J. Tobin, K. Euch, and M. Juda, "Plasma Polymerization of Methyl Methacrylate : A Photoresist for 3D Applications", Technical Digest of the 1988 Solid State Sensor and Actuator Workshop., Hilton Head Island, SC, 1988, p. 9–12.

38. Galhotra, V., C. Marques, Y. Desta, K. Kelly, M. Despa, A. Pendse, and J. Collier, "Fabrication of LIGA Mold Inserts Using a Modified Procedure", SPIE, Micro-machining and Microfabrication Process Technology II, Austin, Texas, USA, 1996, p. 168–173.

39. Rogers, J., C. Marques, and K. Kelly, "Cyanoacrylate Bonding of Thick Resists for LIGA", SPIE Micro-lithography and Metrology in Micromachining II, Austin, Texas, USA, 1996, p. 177–182.

40. Becker, E. W., W. Ehrfeld, and D. Munchmeyer, "Untersuchungen zur Abbildungsgenauigkeit der Rontgentiefenlitographie mit Synchrotonstrahlung", KfK, Report No. 3732, Karlsruhe, Germany, 1984.

41. Munchmeyer, D., "PhD. Thesis", Ph.D. Thesis, University of Karlsruhe, Germany, 1984.

42. Menz, W. and P. Bley, *Mikrosystemtechnik fur Ingenieure*, VCH Publishers, Weinheim, Germany, 1993.

43. Editorial, "X-Ray Scanner for Deep Lithography", (Commercial brochure), 1994.

44. Ghica, V. and W. Glashauser, *Verfahren fur die Spannungsrissfreie Entwicklung von Bestrahlthen Polymethylmethacrylate-Schichten*, in Deutsche Offenlegungsschrift, Germany, Patent No. 3039110, 1982.

45. Romankiw, L. T., ""Think Small," One Day It May Be Worth A Billion", *Interface*, Summer, 17–57, 1993.

46. Harsch, S., W. Ehrfeld, and A. Maner, "Untersuchungen zur Herstellung von Mikrostructuren grosser Strukturhohe durch Galvanoformung in Nickel-sulfamatelektrolyten", KfK, Report No. 4455, Karlsruhe, Germany, 1988.

47. Maner, A., S. Harsch, and W. Ehrfeld, "Mass Production of Microdevices with Extreme Aspect Ratios by Electroforming", *Plating and Surface Finishing*, 60–65, 1988.

48. Mohr, J. and M. Strohmann, "Examination of Long-term Stability of Metallic LIGA Microstructures by Electromagnetic Activation", *Journal of Micromechanics and Microengineering*, 2, 193–195, 1992.

49. Romankiw, L. T. and T. A. Palumbo, "Electrodeposition in the Electronic Industry", Proceedings of the Symposium on Electrodeposition Technology, Theory and Practice, San Diego, CA, 1987, p. 13–41.

50. GmbH, R. K., "µGALV 750", Sales Brochure, RK Kissler, Daimlerstrasse 8, Speyer 67346, 1994.

51. Romankiw, L. T., "Electrochemical Technology in Electronics Today and Its Future: A Review", *Oberflache-Surface*, 25, 238–247, 1984.

52. van der Putten, A. M. T. and J. W. G. de Bakker, "Anisotropic Deposition of Electroless Nickel-Bevel Plating", *J. Electrochem. Soc.*, 140, 2229–2235, 1993.

53. Harsch, S., D. Munchmeyer, and H. Reinecke, "A New Process for Electroforming Movable Microdevices", Proceedings of the 78th AESF Annual Technical Conference (SUR/FIN '91), Toronto, Canada, June, 1991.

54. Thomes, A., W. Stark, H. Goller, and H. Liebscher, "Erste Ergebnisse zur Galvanoformung von LIGA Mikrostrukturen aus Eisen-Nickel Legierungen", Symp. Mikroelektrochemie, Friedrichsroda, 1992.

55. Akkaraju, S., Y. M. Desta, B. Q. Li, and M. C. Murphy, "A LIGA-Based Family of Tips for Scanning Probe Applications", SPIE, Microlithography and Metrology in Micromachining II, Austin, Texas, USA, 1996, p. 191–198.

56. Hagmann, P. and W. Ehrfeld, "Fabrication of Microstructures of Extreme Structural Heights by Reaction Injection Molding", *The Journal of Polymer Processing Society*, IV, 188–195, 1988.

57. Weber, L., W. Ehrfeld, H. Freimuth, M. Lacher, H. Lehr, and B. Pech, "Micro-Molding-A Powerful Tool for the Large Scale Production of Precise Microstructures", SPIE, Micromachining and Microfabrication Process Technology II, Austin, Texas, USA, 1996, p. 156–167.

58. Harmening, M., W. Bacher, and W. Menz, "Molding Plateable Micropatterns of Electrically Insulating and Electrically Conducting Poly(Methyl Methacrylate)s By the LIGA Technique", *Makromol. Chem. Macromol. Symp.*, 50, 277–284, 1991.

59. Eicher, J., R. P. Peters, and A. Rogner, VDI-Verlag, VDI-Verlag, Dusseldorf, Report No. VDI-Bericht 960, 1992.

60. Ruprecht, R., Bacher,W.,Hausselt,J.,H. and Piotter,V., "Injection Molding of LIGA and LIGA-Similar Microstructures Using Filled and Unfilled Thermoplastics", Micromachining and Microfabrication Process Technology, (Proceedings of the SPIE), Austin, Texas, 1995, p. 146–157.

61. Mohr, J., W. Bacher, P. Bley, M. Strohmann, and U. Wallrabe, "The LIGA Process-A Tool for the Fabrication of Microstructures Used in Mechatronics", Proceedings. 1st Japanese-French Congress of Mechatronics, Besancon, France, October, 1992, p. 1–7.

62. Burbaum, C., J. Mohr, P. Bley, and W. Ehrfeld, "Fabrication of Capacitive Acceleration Sensors by the LIGA Technique", *Sensors and Actuators A*, A25, 559–563, 1991.

63. Guckel, H., K. J. Skrobis, T. R. Christenson, J. Klein, S. Han, B. Choi, and E. G. Lovell, "Fabrication of Assembled Micromechanical Components via Deep X-ray Lithography", Proceedings. IEEE Micro Electro Mechanical Systems, (MEMS '91), Nara, Japan, 1991, p. 74–79.

64. Mohr, J., C. Burbaum, P. Bley, W. Menz, and U. Wallrabe, "Movable Microstructures Manufactured by the LIGA Process as Basic Elements for Micro-systems", in *Microsystem Technologies 90*, Reichl, H., Ed., Springer, Berlin, 1990, p. 529–537.

65.  Harmening, M., W. Bacher, P. Bley, A. El-Kholi, H. Kalb, B. Kowanz, W. Menz, A. Michel, and J. Mohr, "Molding of Three-Dimensional Microstructures by the LIGA Process", Proceedings. IEEE Micro Electro Mechanical Systems, (MEMS '92), Travemunde, Germany, 1992, p. 202–207.

66.  Bley, P., J. Gottert, M. Harmening, M. Himmelhaus, W. Menz, J. Mohr, C. Muller, and U. Wallrabe, "The LIGA Process for the Fabrication of Micromechanical and Microoptical Components", in *Microsystem Technologies '91*, Krahn, R. and Reichl, H., Eds., VDE-Verlag, Berlin, 1991, p. 302–314.

67.  Gottert, J., J. Mohr, and C. Muller, "Examples and Potential Applications of LIGA Components in Microoptics", Integrierte Optik und Mikrooptik mit Polymeren, Mainz, Germany, 1992.

68.  Preu, G., A. Wolff, D. Cramer, and U. Bast, "Microstructuring of Piezoelectric Ceramic", Proceedings of the Second European Ceramic Society Conference (2nd ECerS '91), Augsburg, Germany, 1991, p. 2005–2009.

69.  Lubitz, K., "Mikrostrukturierung von Piezokeramik", *VDI-Tagungsbericht*, 796, 1989.

70.  Ehrfeld, W., P. Bley, F. Gotz, P. Hagmann, A. Maner, J. Mohr, H. O. Moser, D. Munchmeyer, W. Schelb, D. Schmidt, and E. Becker, W., "Fabrication of Microstructures Using the LIGA Process", Proceedings of the IEEE Micro Robots and Teleoperators Workshop, Hyannis, MA, 1987, p. 1–11.

71.  Maner, A., S. Harsch, and W. Ehrfeld, "Mass Production of Microstructures with Extreme Aspect Ratios by Electroforming", Proceedings of the 74th AESF Annual Technical Conference (SUR/FIN '87), Chicago, IL, July, 1987, p. 60–65.

72.  Menz, W., W. Bacher, M. Harmening, and A. Michel, "The LIGA Technique-a Novel Concept for Micro-structures and the Combination with Si-Technologies by Injection Molding", Proceedings. IEEE Micro Electro Mechanical Systems, (MEMS '91), Nara, Japan, 1991, p. 69–73.

73.  Mohr, J., P. Bley, C. Burbaum, W. Menz, and U. Wallrabe, "Fabrication of Microsensor and Microactuator Elements by the LIGA-Process", 6th International Conference on Solid-State Sensors and Actuators (Transducers '91), San Francisco, CA, 1991, p. 607–609.

74.  Himmelhaus, M., P. Bley, J. Mohr, and U. Wallrabe, "Integrated Measuring System for the Detection of the Revolution of LIGA Microturbines in View of a Volumetric Flow Sensors", *J. Micromech. Microeng.*, 2, 196–198, 1992.

75.  Schomburg, W. K., J. Fahrenberg, D. Maas, and R. Rapp, "Active Valves and Pumps for Microfluidics", *J. Micromech. Microeng.*, 3, 216–218, 1993.

76.  Schomburg, W. K., J. Vollmer, B. Bustgens, J. Fahrenberg, H. Hein, and W. Menz, "Microfluidic Components in LIGA Technique", *J. Micromech. Microeng.*, 4, 186–191, 1994.

77.  Rapp, R., W. K. Schomburg, D. Maas, J. Schulz, and W. Stark, "LIGA Micropump for Gases and Liquids", *Sensors and Actuators A*, A40, 57–61, 1994.

78.  Rapp, R., P. Bley, W. Menz, and W. K. Schomburg, "Konzeption und Entwicklung einer Mikromembranpumpe in LIGA-Technik", KfK, Report No. 5251, Karlsruhe, Germany, 1993.

79.  Bustgens, B., W. Bacher, W. Menz, and W. Schomburg, K., "Micropump Manufactured by Thermoplastic Molding", Proceedings. IEEE Micro Electro Mechanical Systems, (MEMS '94), Oiso, Japan, January, 1994, p. 18–21.

80.  Vollmer, J., H. Hein, W. Menz, and F. Walter, "Bistable Fluidic Elements in LIGA Technique for Flow Control in Fluidic Microactuators", *Sensors and Actuators A*, A43, 330–334, 1994.

81.  Vollmer, J., H. Hein, W. Menz, and F. Walter, "Bistable Fluidic Elements in LIGA-Technique as Micro-actuators", *Sensors and Actuators A*, A43, 330–4, 1994.

82.  Humphrey, E. F. and D. H. Tarumoto, Eds. *Fluidics*, Fluidic Amplifier Associates, Inc., Boston, MA, 1965.

83.  Elwenspoek, M., T. S. J. Lammerink, R. Miyake, and J. H. J. Fluitman, "Micro Liquid Handling Systems", Proceedings of the Symposium, Onder the Loep Genomen, Koninghshof Veldhoven, Netherlands, 1994.

84.  Ehrfeld, W., R. Einhaus, D. Munchmeyer, and H. Strathmann, "Microfabrication of Membranes with Extreme Porosity and Uniform Pore Size", *J. Membr. Sci.*, 36, 67–77, 1988.

85.  MicroParts, "The LIGA Technique", Commercial Brochure.

86.  Ehrfeld, W., "The LIGA Process for Microsystems", Proceedings. Micro System Technologies '90, Berlin, Germany, 1990, p. 521–528.

87.  Mohr, J., B. Anderer, and W. Ehrfeld, "Fabrication of a Planar Grating Spectrograph by Deep-etch Lithography with Synchrotron Radiation", *Sensors and Actuators A*, A25–27, 571–575, 1991.

88.  Menz, W., "LIGA and Related Technologies for Industrial Application", 8th International Conference on Solid-State Sensors and Actuators (Transducers '95), Stockholm, Sweden, June, 1995, p. 552–555.

89.  Lin, L. Y., S. S. Lee, M. C. Wu, and K. S. J. Pister, "Micromachined Integrated Optics for Free-Space Inter-connections", Proceedings. IEEE Micro Electro Mechanical Systems, (MEMS '95), Amsterdam, Netherlands, 1995, p. 77–82.

90. Wolffenbuttel, R. F. and T. A. Kwa, "Integrated Monochromator Fabricated in Silicon Using Micromachining Techniques", 6th International Conference on Solid-State Sensors and Actuators (Transducers '91), San Francisco, CA, 1991, p. 832–835.

91. Rogner, A., W. Ehrfeld, D. Munchmeyer, P. Bley, C. Burbaum, and J. Mohr, "LIGA-Based Flexible Microstructures for Fiber-Chip Coupling", *J. Micromech. Microeng.*, 1, 167–170, 1991.

92. Gottert, J., J. Mohr, and C. Muller, "Coupling Elements for Multimode Fibers by the LIGA Process", Proceedings. Micro System Technologies '92, Berlin, Germany, 1992, p. 297–307.

93. Editorial, "Fibre Ribbon Ferrule Insert Made by LIGA", (Commercial brochure), 1994.

94. Munchmeyer, D. and W. Ehrfeld, "Accuracy Limits and Potential Applications of the LIGA Technique in Integrated Optics", 4th International Symposium on Optical and Optoelectronic Applied Sciences and Engineering, (Proceedings of the SPIE), The Hague, Netherlands, 1987, p. 72–9.

95. Editorial, "Micro-Optics at IMM", (Commercial brochure), 1994.

96. Anderer, B., W. Ehrfeld, and D. Munchmeyer, "Development of a 10–Channel Wavelength Division Multiplexer/Demultiplexer Fabricated by an X-Ray Micromachining Process", *SPIE*, 1014, 17–24, 1988.

97. Staerk, H., A. Wiessner, C. Muller, and J. Mohr, "Design Considerations and Performance of a Spectro-Streak Apparatus Applying a Planar LIGA Microspectrometer for Time-Resolved Ultrafast Fluorescence Spectrometry", *Rev. Sci. Instrum.*, 67, 2490–2495, 1996.

98. Desta, Y. M., M. Murphy, M. Madou, and J. Hines, "Integrated Optical Bench for a CO2 Gas Sensor", Microlithography and Metrology in Micromachining, (Proceedings of the SPIE), Austin, TX, 1995, p. 172–177.

99. Rogner, A., "Micromoulding of Passive Network Components", Proceedings. Plastic Optical Fibres and Applications Conference, Paris, France, 1992, p. 102–4.

100. Strohrmann, M., P. Bley, O. Fromhein, and J. Mohr, "Acceleration Sensor with Integrated Compensation of Temperature Effects Fabricated by the LIGA Process", *Sensors and Actuators A*, A41–42, 426–429, 1994.

101. Mohr, J., Bley P., Strohrmann, M., and Wallrabe, U., "Microactuators Fabricated by the LIGA process", *J. Micromech. Microeng.*, 2, 234–241, 1992.

102. Tang, W. C.-K., "Electrostatic Comb Drive for Resonant Sensor and Actuator Applications", Ph.D. Thesis, University of California at Berkeley, 1990.

103. Muller, A., J. Gottert, and J. Mohr, "LIGA Micro-structures on Top of Micromachined Silicon Wafers Used to Fabricate a Micro-Optical Switch", *J. Micromech. Microeng.*, 3, 158–160, 1993.

104. Romankiw, L. T., "Thin Film Inductive Heads; From One to Thirty One Turns", Proceedings of the Symposium on Magnetic Materials, Processes, and Devices, Hollywood, FL, 1989, p. 39–53.

105. Rogge, B., J. Schulz, J. Mohr, A. Thommes, and W. Menz, "Fully Batch Fabricated Magnetic Micro-actuator", 8th International Conference on Solid-State Sensors and Actuators (Transducers '95), Stockholm, Sweden, June, 1995, p. 552–555.

106. Goemans, P. A. F. M., E. M. H. Kamerbeeek, and P. L. A. J. Klijn, "Measurement of the Pull-out Torque of Synchronous Micromotors with PM Rotor", 6th International Conference on Electrical Machines and Drives, Oxford, 1993, p. 4–8.

107. Guckel, H., K. J. Skrobis, T. R. Christenson, J. Klein, S. Han, B. Choi, E. G. Lovell, and T. W. Chapma, "Fabrication and Testing of the Planar Magnetic Micromotor", *J. Micromech. Microeng.*, 4, 40–45, 1991.

108. Guckel, H., T. R. Christenson, and K. Skrobis, "Metal Micromechanisms via Deep X-Ray Lithography, Electroplating and Assembly", *J. Micromech. Microeng.*, 2, 225–228, 1992.

109. Wallrabe, U., P. Bley, B. Krevet, W. Menz, and J. Mohr, "Design Rules and Test of Electrostatic Micromotors Made by the LIGA Process", *J. Micromech. Microeng.*, 4, 40–45, 1994.

# Comparison of Micromachining Techniques

## Introduction and Terminology

In manufacturing, forming and removing are the two primary processes. Primary forming processes create an original shape from a molten mass, gaseous state, or from solid particles. During such processes, cohesion is created among particles;[1] examples include plastic molding, metal deposition by evaporation or sputtering, and electroforming. Removing processes remove material, destroying cohesion among particles; examples are wet chemical etching, electrodischarge machining (EDM), and traditional mechanical turning and drilling. In this book, we are interested in those forming and removing processes enabling the manufacture of precision micromachines.

Precision machining has been defined as machining in which the relative accuracy is $10^{-4}$ or less of a feature/part size.[2] For comparison, a relative accuracy of $10^{-3}$ in the construction of a house is considered excellent. It is important to realize that, while silicon micromachining can achieve excellent absolute tolerances, relative tolerances are rather poor compared to those achieved by more traditional techniques. The decrease in manufacturing accuracy with decreasing size is rarely mentioned in discussions of Si micromachines; this probably is related to the fact that most Si micromachinists are electrical engineers rather than mechanical engineers. The contrast is striking; in traditional machining, relative tolerances of $10^{-6}$ (ppm) are becoming standard, whereas in the IC industry a $10^{-2}$ relative tolerance is considered good. The definition of precision machining, with relative tolerances of $10^{-4}$, does actually exclude micromachining. To capture micromachining with the same terminology, it must be expanded to cover all machining methods where the relative accuracies are at least $10^{-4}$ or where absolute size in one or more dimensions is in the micrometer range.

The term 'precision machining', when used by mechanical engineers, has typically been reserved for removal processes only, whereas 'micromachining', as used to describe IC-based fabrication technology, covers both removal and forming processes. Precision machining and Si micromachining are complementary; both are striving to improve relative tolerances, with Si based technology better at obtaining smaller features and traditional methods better at obtaining tighter relative tolerances. In this book, precision machining encompasses removal and forming processes, with micromachining as one of its newest components.

The terms 'nontraditional' or 'nonconventional machining' have been used by mechanical engineers for at least the last 30 years[3] to describe nonmechanical machining methods such as thermal, electrochemical, and chemical machining of high strength and corrosion- and wear-resistant materials. It often involves small, precise, and complex structures and may incorporate computer-integrated machining methods.[4] The name nontraditional in this sense is becoming obsolete though, as the so-called nonconventional machining methods have found a wide range of commercial applications even in more readily machined materials.[5] In what follows we will refer to all non-IC manufacturing techniques as traditional machining techniques. In this nomenclature, IC-based micromachining techniques are the nontraditional machining methods of today.

In Figure 7.1 the application field of precision machining in terms of absolute size and absolute and relative tolerances is illustrated. In traditional mechanical precision machining, typically used to machine the biggest objects, one is mostly concerned with objects of an absolute size of 10 cm or less, although a parabolic astronomic mirror with a diameter of several meters must obviously be considered a precision machined object as well.[6] Absolute tolerances of the dimensions dealt with in precision engineering are 10 μm or below, depending on the technique. The application domains of normal machining, precision machining, and ultraprecision machining are depicted in Figure 7.2. This illustration combines the Taniguchi curves,[7] dramatically illustrating the improvement in traditional manufacturing accuracies achieved in the second half of this century, and Moore's law, demonstrating the logarithmic progress in density of transistors on a chip over the last few decades of this century. Traditional machining and IC technology-based machining methods enabling those accuracies are listed on the left-hand side of this figure (see also Evans[8]). From this figure we can expect ultra-precision diamond machining to reach a manufacturing accuracy of better than 1 nm by the year 2000. To obtain those accuracies one cannot, of course, work with materials such as wood or brick, but one must employ form-stable

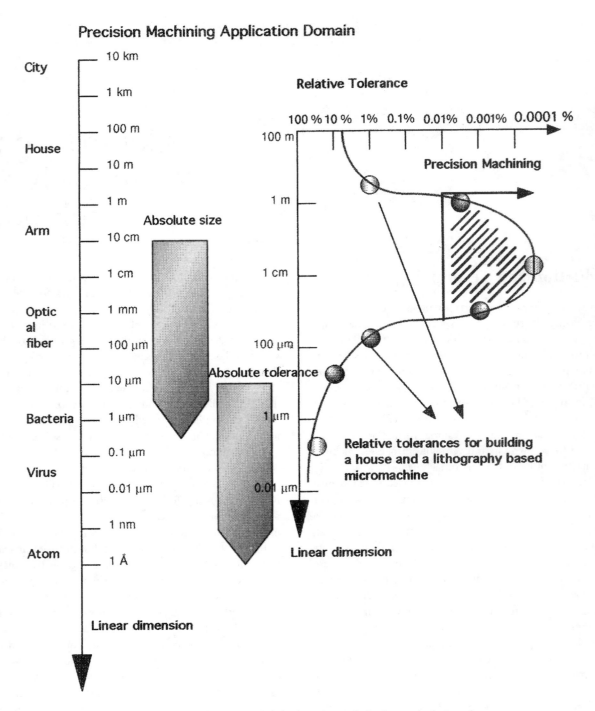

**FIGURE 7.1**   Application field for precision-machining in terms of absolute sizes and absolute and relative tolerances.

and workable materials such as aluminum, stainless steel, ceramics, or glasses; in other words, material and machining accuracy are intertwined. From Figure 7.2 we also observe that the slope of both the Taniguchi and Moore plot decreases as the year 2000 approaches. Progress in mechanical machining is to a large part based on continued improvement in positioning stages, computer control, and metrology techniques. Photolithography equipment is based on the same ruling engine design philosophy from mechanical technology.[2] Reasons for the slowdown in the Moore curve are a bit more complicated though. In lithography, subsequent layers need to be aligned, an

operation not needed when diamond milling a high precision lens. The alignment operation may be performed by laser-interferometer-controlled stages to an absolute positioning accuracy of <0.1 μm, thereby allowing the registration of successive layers of a semiconductor device to a similar precision. Alignment errors (A) depict only one part of the total device ground-rule tolerances. Tolerance (T) or the ability to overlay (align) spatially one masking layer with the previously etched or deposited pattern is given by the standard deviations on the mask (M), etch (E), and alignment (A) errors (see Chapter 1).[9] Typical mask errors range from 0.1 to 0.3 μm, etch errors from 0.2 to

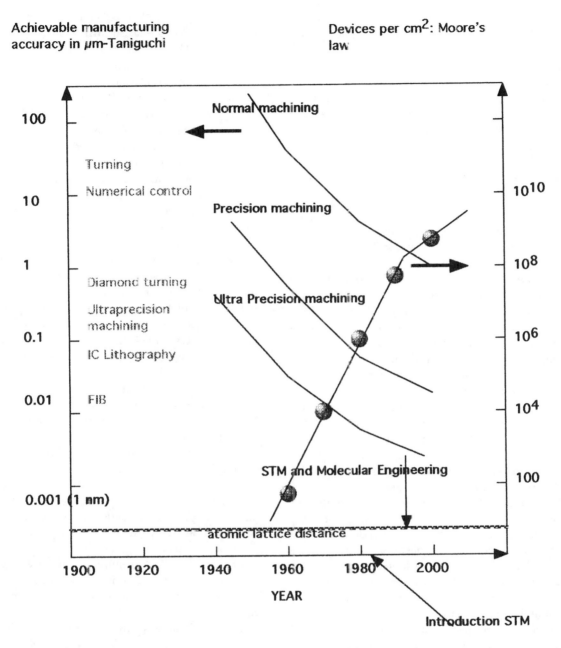

**Achievable manufacturing accuracy in μm-Taniguchi**

**Devices per cm²: Moore's law**

Normal machining

Turning

Numerical control

Precision machining

Diamond turning

Ultraprecision machining

Ultra Precision machining

IC Lithography

FIB

STM and Molecular Engineering

atomic lattice distance

YEAR

Introduction STM

**FIGURE 7.2** Definition for normal, precision, and ultraprecision machining. Left side ordinate: increase of manufacturing accuracy over time according to Taniguchi.[7] Right side ordinate: increase in transistor density over time according to Moore's law.

0.4 μm, and alignment errors from 0.2 to 0.5 μm. For the next generation of submicrometer ground rules, a fourfold reduction in alignment and other errors will be necessary. Since interferometric schemes enable sensitivities to below 10 nm, it appears feasible that aligners compatible with linewidths of 0.1 μm or even 50 nm can be developed.[10] Unfortunately, there is a second reason for the slowdown in electronic device development; below 0.1 μm quantum effects start playing a role. To further improve manufacturing accuracy in electronics one has to look beyond current manufacturing methodologies and consider methods such as molecular engineering and nanofabrication. Micromachining plays a pivotal role in those new machining approaches. For example, the scanning tunnelling microscope

(STM), invented in 1982, creates machining opportunities with accuracies on the atomic level, well beyond the predictions of the Taniguchi and Moore curves in Figure 7.2. The crucial component in an STM is a micromachined cantilever with an integrated sharp tip.

For mechanical structures, operating in the microdomain imposes new design philosophies; one must adopt more error-insensitive designs with simpler mechanisms, smaller numbers of parts, and, wherever possible, flexible members rather than rigid ones. Nature, an excellent teacher in nanomachining, builds living organisms this way.[11]

After a historical note we start this chapter by reviewing manufacturing tools. They are grouped according to the energy source involved in removal or addition of materials at the workpiece. The classes we distinguish are chemical (including photochemical), electrochemical, electrothermal, and mechanical. Subsequently, we introduce an overview of available microfabrication tools as a guide for deciding upon the ideal tool for the micromachining job at hand. Finally, we discuss some of the more obvious shortfalls of micromachining technology and speculate how molecular engineering and nanofabrication might be able to go guide us beyond these limitations.

## Historical Note

In mechanical machining, mechanical energy is the *modus operandi* behind every operation, from grinding and drilling to broaching (removing very thin slices) and milling. Toolmaking for mechanical machining began perhaps a million years ago when humans learned to walk erect and had hands free to grasp objects of wood and stone. Stronger tools that lasted longer resulted when metals such as copper and iron replaced wood and stone. Humans have been cutting metal for thousands of years to make the myriad tools, machines, and other devices our civilization demands. Hipparchus in the 2nd century B.C. and Ptolemy in 150 A.D. used graduated instruments for astronomy and sailing. Precision machining with a hand-operated lathe emerged from the demands for more precision in building instruments for astronomy and time measurement in the mid-1600s.[6] Ruling engines for the manufacture of diffraction gratings have driven the state of the art of precision machine tool design since the 1700s, and nearly a century ago, Rowland and John Hopkins were building ruling engines with part-per-million accuracy.[2] The founders of Browne and Sharpe, the inventors of the first diffraction grating, were clockmakers; the first lathes and many other machine tools are also rooted in watch and clock making.[12] Tools driven by mechanical power brought on the machine age in the late 1700s.[13] Diamond as a cutting tool was apparently first used in 1779 for making threads in hardened steel, and by 1920 the Lord's Prayer was engraved with a diamond tool into an area of $100 \times 40$ µm.[12] Cutting with very small diamond tools (say, 25-µm diameter) was developed in Japan in the 1980s. The liquid metal ion source used in focused ion-beam milling (FIBM) was demonstrated in the period 1978 to 1980. The latter technique, although very slow and expensive, represents the latest and most accurate means of mechanical precision machining.

In modern times, precision engineering was initially pushed by nuclear programs and eventually by semiconductor manufacture. Today, the mechanical technology upon which many photolithographic systems are based comes from the ruling engine design philosophy.[2] In the last 25 years great progress was made in building machines that can be operated and controlled automatically. Highly precise instruments such as servomotors, feedback devices, and computers were implemented by 1977, and many types of machine tools are now equipped for computer numerical control, commonly called CNC. In flexible manufacturing systems (FMS), CNC workstations are linked by automatic workpiece transfer and handling, with flexible routing and automatic workpiece loading and unloading.[4] Despite the

tremendous progress reflected in the Taniguchi curves (see Figure 7.2), it is somewhat surprising that it still takes a 2-ton machine to fabricate microparts with cutting forces in the milli- to micro-Newton range. This presents a clear indication that a 'from the ground up' machine tool redesign for the manufacture of micromachines is in order.[12] Along this line, in Japan the concept of desktop flexible manufacturing systems for the building of micromachines was proposed in the early 1990s.[14]

Operations more sophisticated than the just described mechanical processes can be performed, often more effectively, by using other forms of energy — such as chemical (including photochemical), electrochemical, and electrothermal. These methods, together with mechanical machining, are illustrated with some examples in Figure 7.3.

Among electrothermal methods, electrodischarge machining (EDM), using electrical sparks to cut, is one of the oldest. In an 'uncontrolled' fashion, EDM has been known ever since the first thunderstorm. The first widespread 'controlled' application for shaping some of the toughest metals was developed during World War II. By the 1960s, in the U.S., EDM machining was used to fabricate devices in the 75-µm size range, and even more precise EDM, employing rotating spindles, was pioneered in Japan in the late 1980s (5-µm size range). Electron-beam machining (EBM) and laser-beam machining (LBM), two other examples of thermal techniques, came about in the 1960s and 1980s, respectively. In EBM, high velocity electrons are used instead of sparks, and in LBM intense photon pulses are employed.

Electrochemical grinding (ECG), a combination of electrochemical etching and mechanical grinding, was an early forerunner of several electrochemical methods. Electrochemical machining (ECM), in which a shaped cathode cuts into the anode workpiece, is currently the best known example, with wide application in the aircraft, automotive, and business machine industries.

Chemical and photochemical etching, as we saw in Chapters 1 and 4, eventually led to micromachining. About 1400 A.D., armor was etched with acids through a linseed oil maskant, marking the first use of chemical milling in conjunction with some type of a mask. In 1822, Niépce developed the photoprocess and in 1827, 5 years later, Lemaître did the subsequent chemical milling of a photoplate he received from Niépce, constituting the first deliberate use of photolithography in combination with chemical milling. The IC revolution in the 1950s employed these same photolithography techniques and brought along the micromachining discipline grounded in the late 1960s with the fabrication of the first Si strain gauges.

In this work we stress how traditional precision engineering and more modern micromachining techniques are complementary and that in making a manufacturing decision it is not so much a question of which manufacturing technique is better or newer (invariably, since the 1960s, silicon has been the answer, often independent of the question), but which one is appropriate to use for the problem at hand.

The IC philosophy of batch manufacturing is being applied today to a wide variety of non-IC products, and traditional precision manufacturing methods are beginning to be combined

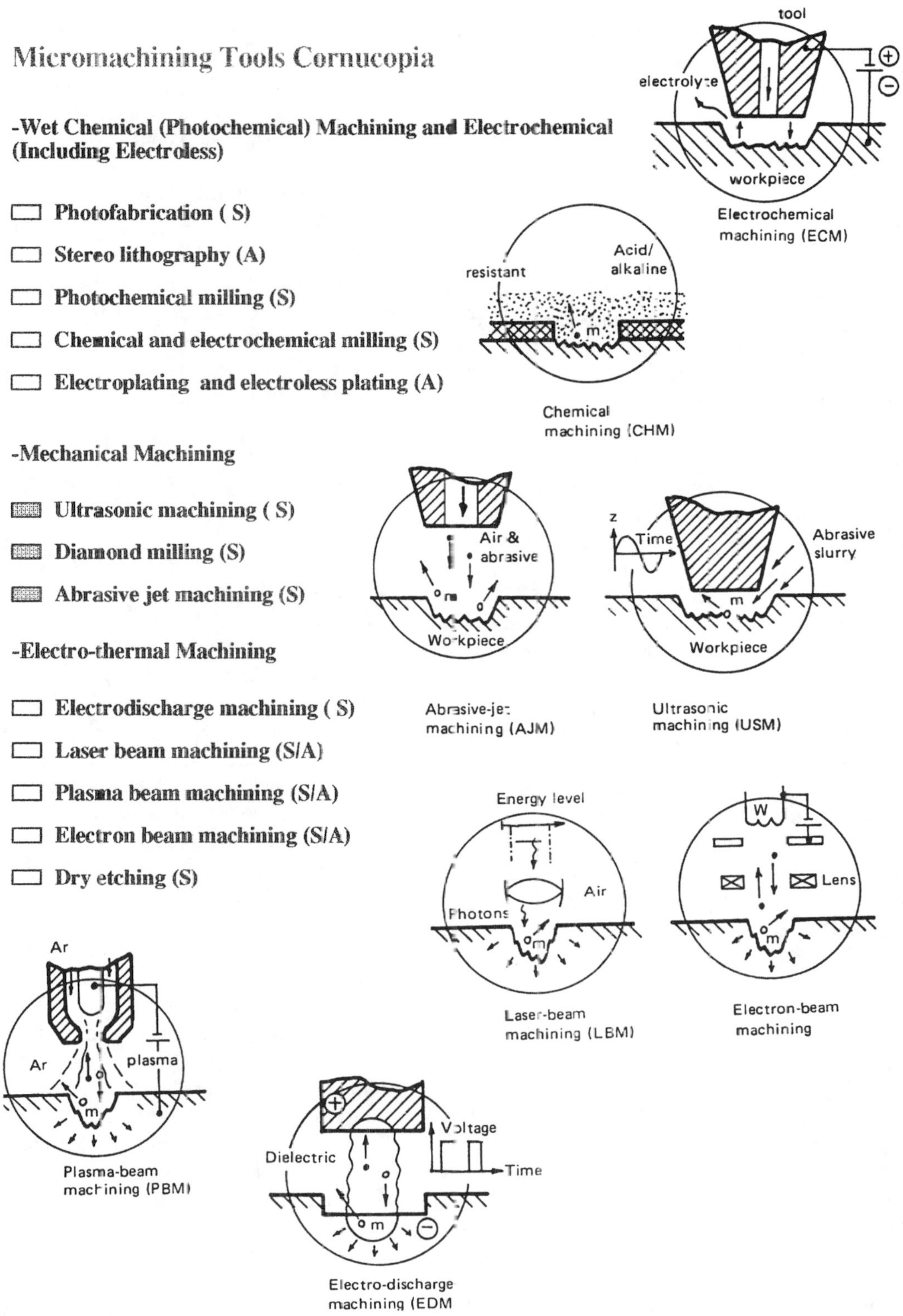

# Micromachining Tools Cornucopia

**-Wet Chemical (Photochemical) Machining and Electrochemical (Including Electroless)**

☐ Photofabrication ( S)

☐ Stereo lithography (A)

☐ Photochemical milling (S)

☐ Chemical and electrochemical milling (S)

☐ Electroplating and electroless plating (A)

**-Mechanical Machining**

▦ Ultrasonic machining ( S)

▦ Diamond milling (S)

▦ Abrasive jet machining (S)

**-Electro-thermal Machining**

☐ Electrodischarge machining ( S)

☐ Laser beam machining (S/A)

☐ Plasma beam machining (S/A)

☐ Electron beam machining (S/A)

☐ Dry etching (S)

**FIGURE 7.3** Examples of machining methods (S = subtractive; A = additive). Classification based on the applied energy appearance at the workpiece. (After Snoeys, R., Non-Conventional Machining Techniques, The State of the Art, presented at Advances in Non-Traditional Machining, Anaheim, CA, 1986.)

with those IC techniques. It is to be expected that the merging of traditional precision engineering and IC manufacturing methodology will lead to interesting new hybrid manufacturing processes.

# Machining Tools

## Introduction

Not all manufacturing techniques can be covered in one book chapter; we have chosen to cover in most detail those methods offering the best potential for building improved microstructures. For some manufacturing methods, a definition and some special attributes are all we are able to accommodate here. The micromachining techniques compared in this chapter include all of the nontraditional batch techniques covered in previous chapters: bulk micromachining, surface micromachining, and LIGA, as well as the more traditional chemical, electrothermal, and mechanical techniques. We do expand mostly on electrochemical processes and chemical machining of metals, as well as photofabrication with photosensitive materials such as glasses, ceramics, and plastics, because these methods all have been underutilized for micromachining in comparison with Si-based processes. Photofabrication, a batch technology, with potential for inexpensive, high aspect ratio microstructures, has too often been ignored as an important micromanufacturing option. Serial machining techniques such as electrothermal methods, including electrodischarge machining, electron- and laser-beam machining, and mechanical machining techniques such as ultrasonic drilling and focused ion-beam milling, are expanded upon because we believe that in micromachining, as opposed to IC fabrication, serial processes can often be utilized cost effectively. For a more in-depth study on traditional manufacturing techniques, refer to the literature. We recommend Harris,[15] Shaw,[16] Devries,[17] Slocum,[18] and Evans.[8]

## Chemical Processes

### Introduction

In chemical removal processes, reactions destroy the atomic bonds of the material to be removed. In chemical forming, chemical reactions at the surface form a new compound. Chemical forming reactions were studied in Chapter 3 and etching of Si (including photochemical etching) was studied in Chapter 2 (dry etching) and Chapter 4 (wet etching). Below we present additional insights into chemical and photochemical milling (removal) of metals and photofabrication of glasses, ceramics, and plastics. The industrial application of chemical and photochemical etching is mainly in the shaping of thin metals foils, whereas in photofabrication photosensitive glasses, plastics, and ceramics are patterned into intricate three-dimensional shapes. The latter method offers often simple, inexpensive, high aspect ratio three-dimensional microstructures for small to medium batch sizes without the need to resort to expensive Si micromachining or LIGA techniques.

The best known examples of additive chemical processes fall in the category of plastic molding. These additive technologies are often still the least expensive manufacturing techniques available. Their combination with improved mold making, such as in LIGA or LIGA-like techniques, is bound to lead to many new exciting microfabricated products.

### Chemical and Photochemical Milling

Chemical machining involves material removal through etching of preferentially exposed (masked) surfaces, producing typically stress-free parts in nearly any type of material. There are two major categories of chemical machining applicable: chemical milling and photoetching. The major difference between the two techniques is in the masking, the size of the substrate which can be accommodated, and in the depth and width of the etched cut.

In chemical milling, a traditional machining method illustrated in Figure 7.4,[15] the mask is usually a tape or chemically resistant paint on large area substrates that are up to two centimeters thick. The etch factor, or eat-back ratio, defined in Figure 7.4A, is equal to A/R, with R the etch-depth and A the overhang of the maskant. A scribing tool, of the surgical-knife type, is used to cut through the maskant along the line defined by the edge of a template (Figure 7.4B).

In photoetching, a nontraditional machining method illustrated in Figure 7.5, photomasks and parts with much smaller areas (maximum $60 \times 60$ cm) and thicknesses (<0.5 mm) than employed in chemical etching are used. To ensure high accuracy, oversized drawings are reduced photographically to actual size. The negative is then used to contact-print the component image onto a metal sheet covered with a photosensitive coating. Careful masking of the part is crucial in achieving precise parts in both chemical and photochemical cutting. The approach used for photochemical milling is very similar to the chemical milling processes, except that, through the absence of scribing errors, the accuracy stemming from the photosensitive maskant leads to much better tolerances on small intricate parts. Lateral tolerances of $\pm 0.25$ to $\pm 0.5$ mm are usual for chemical milling, an accuracy mainly determined by the manual scribing process. For the IC industry and micromachining, this lateral tolerance would be a very poor performance. Photochemical milling was in fact designed specifically to alleviate the lateral-accuracy problem of chemical milling, and this it does extraordinarily well with lateral accuracies, for example, for printed circuit production, of $\pm 0.013$ mm. Another reason why the tolerances in photochemical etches are so much better is related to the attempted etch depths. The geometry of the cut produced, when photo-etching integrated circuits, is similar to the chemical-milling cut of, say, the aerospace industry, but one must remember the many orders in magnitude difference in the depth of the cut.[15] As the dimensions of the components decrease (i.e., from precision products, through printed circuits, to integrated circuits), the necessity for chemical cleanliness of the material surfaces increases and leads to the need for clean-room manufacture in the case of IC fabrication. The chemical and photochemical etching of Si under clean-room conditions was explored in great detail in Chapters 2 and 4.

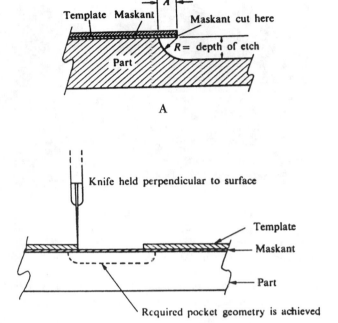

FIGURE 7.4 Chemical milling, showing the effect upon the workpiece. (A) Undercut caused by chemical milling. (B) Scribing of the maskant. (From Harris, T.W., *Chemical Milling*, Clarendon Press, Oxford, 1976. With permission.)

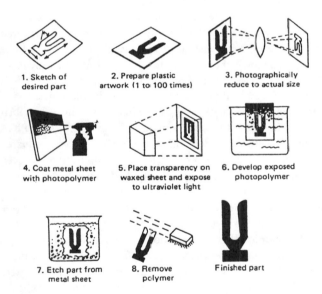

FIGURE 7.5 Photoetching process.

Photoetching of metals is less generic than chemical milling of metals but can be used, for example, with the following metals (in order of increasing difficulty): copper, nickel, carbon steel, stainless steel, aluminum, titanium, and molybdenum.[19,20] Etching rates — with sodium hydroxide for aluminum, hydrochloric and nitric acids for steel, and iron chloride with nitric acid for copper — vary from 1.3 to $7.6 \times 10^{-3}$ cm/min.

Both chemical and photochemical etching techniques are isotropic in nature, and aspect ratios are consequently low. To create a high aspect ratio (>1) microfluidic device from etched channels in stainless steel sheets would necessitate laminating several of these sheets; therefore, this type of technology is better suited for etching shallow features in single substrate devices.

Photoetching of metals requires moderate investments and competence in two fields: etching of metals and lithography. Chemical milling of metals on the other hand is more widely applicable, simpler, and requires little operator skill. Practical cutting depths are generally limited to under 2 cm, because the etchant process is relatively slow and undercutting occurs behind the edges of the protective mask. Large shallow areas are especially suited for chemical machining. Aluminum airplane wing parts have been chemically machined in 50-foot-long tanks. For smaller parts and finer structures and much higher lateral accuracies, photochemical milling is the best bet. Major applications are lead frames, printed circuit boards, and shadow masks for color televisions. Applying photochemical etching to large surface areas is a very expensive proposition.[21] A good further review with many relevant references on wet chemical etching as it relates to the IC and electronics industry was presented by Romankiw.[22]

### Photofabrication

#### Introduction

The industrial application of photofabrication, creating sharp images in ceramics, plastics, and glasses, preceded the Si microfabrication era. Given the current interest in high aspect ratio, three-dimensional structures, this technique is experiencing a rebirth.

Photofabrication is a process where ultraviolet exposure through a mask results in a strong modification of the solubility in an optically clear layer. This permits the direct photochemical production of quite thick three-dimensional structures without the need for a masking step during development (e.g., acid etching). Examples of useful materials are Foturan™ glass from Glaswerk-Schott,[23] Corning's Photosensitive Fotoform™ glasses and Fotoceram™ ceramics,[24] and Du Pont's light-sensitized nylon 'Dycril' and 'Templex' photopolymers.[19] For most photoplastics, aspect ratios do not exceed 3:1, but with the photosensitive glasses and ceramics it is possible to make a 0.13-mm diameter hole through a 2.7-mm thick plate, with an aspect ratio of ~21; this is no mean achievement.

Note: Stereo lithography, commented on in Chapter 1, is an additive technique in which plastic shapes are generated from a molten plastic mass; the technology is sometimes referred to as microphotoforming and should not be confused with subtractive photofabrication discussed here.

#### How It Works

Electrochromic glasses temporarily change color when exposed to strong light. Photosensitive glasses, on the other hand, develop an invisible, permanent 'latent image' after ultraviolet (UV) illumination through a mask. Subsequent heat treatment can develop the latent image into a permanent structural and/or color change with accompanying differences in solubility. Corning Glass Works (now Corning, Inc.), in collaboration with the Harry Diamond Laboratories, was an early pioneer in this field.[24] Typical photosensitive glass compositions developed at that time were $SiO_2$, 81.5; $Li_2O$, 12.0; $K_2O$, 3.5; $Al_2O_3$, 3.0; $CeO_2$, 0.03; and Ag, 0.02 (all in weight percent). Upon exposure to UV light photo-electrons released by cerium, the photosensitizing ingredient, are trapped at specific sites in the silicate-glass network to form a latent image. By heating the exposed glass to around 500°C, the electrons are released from the traps and silver ions are reduced to silver atoms which aggregate into minute colloidal islands. The minute silver clumps serve as nuclei on which crystals such as lithium meta-silicate, sodium bromide, or sodium fluoride precipitate. If suf-ficient numbers of fairly large crystals form (~4 μm), an opalescent image results while the masked regions remain glassy and transparent. In HF, opal regions etch 15 to 30 times faster than the unexposed glass. No masking is required during the etching process. In this way, by exposing a photosensitive glass plate to UV and etching in HF after heat treatment, a wide variety of microstructures can be made. Further heat treatment of the finished glass pieces converts the glass to a higher strength, partially crystalline material, i.e., Fotoceram™.

Through-holes in photosensitive glass were experimentally found to have a sidewall taper of 2 to 3° only; this means that substantially parallel holes may be cut, having a very high depth-to-diameter ratio. The sloping effect can be reduced even further if the HF etching occurs from both sides of the glass simulta-neously. Etching from both sides produces a profile with two truncated cones and gives a minimum overall increase of hole diameter. Although etching of holes with diameters appreciably less than one tenth of the thickness of the glass is not recom-mended, aspect ratios of 20 are quite possible. In contrast, with photoetching in metal, as we saw above, holes should not be made with a diameter less than the thickness of the sheet. If sidewall taper in photosensitive glass is desired, it can be increased to 20 to 30°. Maximum sheet thickness is 0.6 cm, but several Fotoceram™ pieces can be thermally bonded, forming excellent seals. One disadvantage of the bonding process is that it causes the photoglasses to become dark which makes the approach less attractive for certain optical applications.

Fluidic components providing functions such as flip-flops, logic elements such as OR-NOR gates and AND gates, propor-tional amplifiers, and fluid resistors are examples of the capa-bilities of the photofabrication technology. In Figure 7.6 a fluidic-control device is shown;[15] the photo shows the individual layers of an AND device after through-etching. Fusion of stacks of photosensitive glass plates provides a permanently sealed and rugged unit, complete with the necessary input and output vents. Through-etching provides the connection between the different planes. The geometrical accuracy of these fluidic devices is maintained within ±0.025 mm, and the tolerance that may be expected generally on this type of etched glass product is ±0.05 mm on etched edges up to 12.7 mm in length. With hole spacings up to 25 mm apart, accuracy to within 0.025 mm center-to-center is possible.

None of the Si microfluidics produced today have reached the degree of complexity achieved with the photofabrication method of three decades ago. The 'fluidic' elements made in the 1960s could achieve many of the functions of an electronic

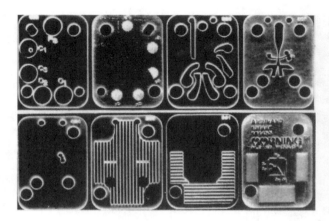

**FIGURE 7.6**  Fluidic-control device employing chemically milled photosensitive glass. The individual layers of an AND device after through-etching with HF. (From Harris, T.W., *Chemical Milling* (Clar-endon Press, Oxford, 1976). With permission.)

circuit, and they were more robust, simpler, and more suited to operate in a hostile environment. Of course, they were much slower (milliseconds instead of nanoseconds) and larger than their electronic counterparts. The latter spelled the eventual demise of microfluidics in the late 1960s. In the 1980s, several Si micromachined fluidic devices were developed for analytical applications, without much reference to this earlier sophisti-cated work in photosensitive glasses. For a review of the more recent work in microfluidics for analytical applications, we refer to Elwenspoek et al.[25] and Gravesen et al.[26]

The relatively large size of fluidic elements, the need for a robust environment, the relative low numbers of required devices, and the fact that electronics usually do not need to be integrated render the use of Si for this application hardly justi-fiable. The photofabrication machining approach, which is inexpensive and simple, coupled with the chemical inertness of glass, remains the better approach. Fluidic oscillators,[27,28] fabri-cated using the yet more expensive and hardly accessible LIGA technique, are even more difficult to justify when compared with photoformed devices. The latter misapplication has been recognized lately, and photoformed glass (Foturan™ from Schott Glaswerke) was reintroduced as the best solution for this application by one of the inventors of the LIGA technology (see also Figure 7.29).[29]

Photofabrication offers additional opportunities beyond high aspect ratio and inexpensive manufacture. A second, more severe UV exposure followed by another heat treatment frees additional silver from the glass matrix, which tends to precipi-tate needle-like onto the tips of the sodium bromide pyramids precipitated earlier. These needles are large enough to absorb visible light. The absorption of a narrow band of light by the silver needles imparts brilliant color to the glass. By varying the UV exposure, the needle size and thus the color of the workpiece can be modulated.

Besides making photosensitive glass ornaments, photosensi-tive glass has been used to make microlenses (±100 μm in diameter), fluidic elements (see Figure 7.6), spacers in photo-multiplier tubes, color filters, cell sheets in gas-discharge dis-plays, and charge plates and nozzles in ink-jet printers.

In a very intriguing development it was found that the different properties of opal and glass areas on photosensitive glass enable the direct formation of electrical circuitry on glass. Immersing a sample with an opal pattern in a molten salt bath and heating it in a hydrogen atmosphere forms a conducting silver film on the opal regions. The film that forms on the glassy regions remains nonconducting.[24]

DuPont, in the early 1960s, developed Templex™, a photosensitive plastic sold in sheet form, either bonded to a substrate such as carbon steel, aluminum, or polyester film (Cronar) or freestanding. The plastic is an alcohol-soluble nylon which can be made photosensitive by adding a cross-linking agent such as methylene bis-acrylamide and an activator such as benzophenone. Upon exposure to UV, the benzophenone forms free radicals, which react with both the other compounds and hence insolubilizes the nylon; in other words, the material acts as a negative tone resist. The unexposed areas are then washed away with a dilute aqueous sodium hydroxide spray, forming channels and cavities. By using the benzophenone in a low concentration, say about 5% by weight, a layer with relatively low absorption coefficient results and thick layers can be insolubilized. Stencils with 1-mm high vertical sidewalls with a taper of 2° can be produced this way, and 25-μm deep channels of a 25-μm width do not show any undercutting. The aspect ratio cannot exceed 3:1, though, and the operating temperature range is limited to −100 and +200°F. To seal cavities or channels (say, for the fabrication of fluidic devices), the first Templex™ plastic sheet can be laminated with a Cronar film.[19]

Recent work in depth lithography with UV and deep dry etching, especially cryogenic dry etching, is considerably expanding the horizons for polymer microfabrication. For example, positive resists in the AZ 4000 series which have a high transparency and high viscosity can be deposited in a multi-spin coating process to thicknesses of up to 80 μm. Steep resist profiles and aspect ratios of 7 are obtained by edge bead removal and vacuum printing.[30] Also, with polyimides[31] and epoxies,[32] thick resist layers, between 100 and 200 μm, and improved aspect ratios of about 10 have been demonstrated. Moreover, by using a Ti-mask and selective anisotropic etching, an aspect ratio of more than 20 was achieved in a 2-μm pattern using magnetically controlled RIE etching of polyimide resist layers.[33] For more details on UV depth lithography, see Chapters 1 and 5.

Photofabrication is an attractive alternative for microfabrication of all types of high aspect ratio, passive elements, especially for intricate shapes which are >1 cm² in size and are only to be produced in modest production volumes. The current interest in micromachined fluidic elements for analytical purposes is reviving interest in photosensitive glasses, as it surely makes more sense to use photofabrication of insulators than to fabricate fluidic elements from Si, since the latter, after etching, must be passivated, a passivation which may undergo electrical breakdown, especially in an aqueous environment.

### Plastic Molding

By far the cheapest method for producing microstructures is by plastic molding. Plastic molding, as we saw in Chapter 6 on LIGA, includes reaction injection molding, thermoplastic injection molding, and hot embossing. One only has to look around the home or office to recognize the pervasive nature of quite intricate, very low-cost plastic parts surrounding us. The method has the capability to make truly three-dimensional parts with a wide range of feature sizes.

A typical application example of traditional plastic molding in miniaturization is in medical diagnostics such as in Biotrack's cholesterol sensor.[34] Here, a series of miniature cavities and capillaries is fabricated from ultrasonically welded injection molded ABS plastic. The sensors enable the drawing of a biological sample, mixing the sample with reagents, defining a flow path, and reading the result. To succeed in this application area, nontraditional micromachining methods (say, using a LIGA mold instead of a precision machined mold) must demonstrate a significant cost/performance advantage to become commercially viable. So far, all commercial medical diagnostics remain exclusively based on traditional manufacturing technologies.

An important challenge for micromachinists, in the coming 5 years, is to adapt plastic molding processes to the smaller and more precise molds that are becoming available through micromachining. Micromachining can indeed be successfully combined with plastic molding as compact discs (CD) demonstrate vividly. The master from which CDs are made is produced from a glass plate covered with a thin layer of photosensitive polymer. The polymer is exposed with a laser and then developed to create micrometer-sized pits. Metal is electroformed on this master to make the mold from which the familiar polycarbonate plastic CDs are made. This illustrates not only the resolution achievable through plastic molding but also how a micromachined prototype can be used to mass-produce plastic parts.

## Electrochemical Machining Processes: Electrochemical Microfabrication

### Introduction

The term 'electrochemical machining' has often been used to describe electrochemical removing processes only.[35] In this book, both electrodeposition and electrochemical removal processes are covered by the same term, which we also refer to as electrochemical microfabrication. Despite the performance, speed, and cost advantages, electrochemical machining is used remarkably little in the IC field when other techniques, especially dry processes, are available. This seems to be connected in part with the overall trend towards dry processing and the disproportionately large number of electrical engineers vs. chemical and electrochemical engineers in the IC and micromachining field.[36] Despite this bias, electrodeposition has been used extensively in the electronic industry in many stages of the manufacturing process, from the device stage, chip carriers, and PC boards, to corrosion protection and electromagnetic shielding of the electronic enclosures (see Table 7.1). The processes include electrodeposition and electroless deposition of copper, nickel, tin, tin-lead alloys, and precious metals such as gold, gold alloys, palladium, and palladium alloys, as well as NiFe, CoP, NiCoP, and other magnetic alloys.[37]

**TABLE 7.1**   Electrochemical Applications in the IC Industry and Micromachining[37,38]

*Photocircuit boards*
- Single and multilayer epoxy boards
- Flexible circuit boards
- Electrophoretically glazed steel boards

*Contacts and connectors*
- Beam leads
- Contacts (pins and sockets)
- Reed switches, etc.

*Cabinets and enclosures*
- Corrosion protective surfaces
- Electromagnetic shielding
- Decorative purposes (anodization, electrophoretic painting)

*Auxiliary equipment used in device fabrication*
- Paste-screening masks
- Metal evaporation masks
- X-ray lithography masks
- Diamond saws and cutting tools

*Active elements*
- IC chips
- Magnetic recording heads
- Wire memories
- Magnetic bubble chips
- Recording surfaces
- Displays
- Wear-resistant surfaces

*Chip carriers and packages*
- Chip in tape packages
- Surface mount boards
- Dual in-line packages
- Pin-grid array packages
- Multilayer ceramic packages
- Hybrid packages

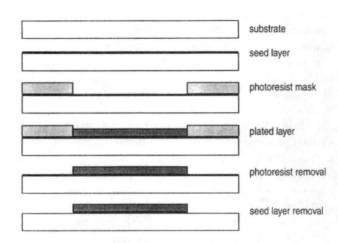

photolithography and plating process

**FIGURE 7.7**   Schematic representation of individual steps of preparing the dielectric surface for through-mask plating.

The fundamentals of electroless and electrodeposition were already reviewed in Chapter 3 and touched upon again in Chapter 6. In the current chapter, we emphasize new principles and the myriad non-Si applications in the precision microengineering field.

Electrochemical removal (machining) is induced by an electrical field in an electrolyte, and it destroys the atomic bonds of the material.[35] The fundamentals of electrochemical etching of semiconductors were reviewed in Chapter 4; here we will review non-Si machining with wide industrial use for metals and other conductive materials.

## Forming Processes

### Electrodeposition

As evident from Table 7.1, electrodeposition plays a key role in electronic device fabrication; because of its more recent utilization in LIGA and LIGA-like processes, where electroforming is a crucial step in the overall process, this technology might play an even bigger role in the future of micromachines.

Electrodeposition through polymer masks, as illustrated in Figure 7.7, represents one of the most powerful techniques available for formation of very high density patterns and circuits with extremely large height-to-width ratios. Most other methods, such as chemical isotropic etching, sputter etching, ion milling, reactive ion etching, cannot be used to produce metal patterns with better than 1:1 height-to-width aspect ratio. To understand this contrasting behavior, we must look into the fundamental nature of the two different processes. In the non-electrochemical technologies, a polymeric or inorganic mask is used to cover the existing metal while either a chemical solution or an active gas phase is removing the metal from the open areas. Hence, the limitation in the height-to-width aspect ratio, the maximum circuit density, and the fidelity of reproduction of the features are caused by shadowing, which is the mechanism by which the patterning takes place. In contrast, with electrochemical techniques, one obtains the exact replication of the recesses in the mask.[38]

A comparison of conventional subtractive etching with electrochemical additive processes was presented as early as 1973 by Romankiw et al.[39] For the fabrication of highly miniaturized magnetic bubble memory devices, this IBM group showed clearly that the additive plating process is superior over dry etching-based techniques.

In a typical electrodeposition through a polymer mask, metal conductors must be deposited on a dielectric as shown in Figure 7.7. The dielectric is usually made conductive by sputtering a thin adhesion metal layer (Ti, Cr, etc.) and a conductive seed layer (Au, Pt, Cu, Ni, NiFe, etc.). The thickness of the thin refractory metal adhesion layer may be as small as 50 to 100 Å, while the thickness of the conducting layer can range from 150 to 300 Å.[39] The key requirement for the seed layer is that it is electrically continuous and offers a low sheet resistance. After forming a pattern in a spin-coated polymer by UV exposure, electron-beam, or X-ray radiation and developing away the exposed resist, contact is made to the seed layer, and electrodeposition or electroless plating is carried out. Electrodeposition reproduces with the greatest fidelity the finest features of the mask. This is not surprising when one considers that in plating, the metal ions discharge from solution present inside the mold, and in doing so the metal displaces the solution from the mold atom by atom, conforming to the smallest features that exist in the mold.[37]

The plating-through mask technology is successfully used in volume production of beam leads and bumps on IC wafers, fabrication of thin film magnetic heads, fabrication of PC boards, bubble memory devices, X-ray lithography mask gratings, and diffraction gratings. Some of these applications are expanded upon next.

**Printed Circuit Boards** Today's printed circuit boards may consist of as many as 20 to 30 layers of metal and dielectric. Printed circuit board fabrication is carried out using either an all-additive electroless copper process or a combination of electroless and electroplating processes. Due to the continuing effort to make active devices smaller and faster and to minimize the length of connecting wire and to make them narrower and thicker with smaller spaces in between, more layers per board and smaller diameter holes with a larger ratio of hole length to diameter must be made (10:1 length to diameter hole is routine and 20:1 is feasible). Without the electrochemical technology, such a degree of integration would not have been possible.[37]

**Contacts and Connectors** Whereas photocircuit boards use electroless or electroplated Cu, for contacts and connectors, electroplated Au and Au alloys are used or, alternatively, low cost precious metal substitutes such as Pd, PdNi, PdAg, NiP, NiAs, and NiB with a thin gold overcoat are employed. Separable, low-force, low-voltage contact applications require a contact finish whose resistance is stable for the projected contact life and which has sufficient wear resistance to withstand the projected number of insertions. The surface of one of the two mating parts is usually pure soft gold, while the other is a hard gold alloy. In less critical applications, tin and tin-lead alloys are used as contact materials. Connecting chips with chip carriers or packages also often involves Au contacts. Gold does not adhere directly to silicon dioxide or silicon nitride, and Ti is typically used as an adhesion layer. Since Au and Ti form intermetallics, a Pt barrier is usually deposited between Ti and Au. The near-ultimate in connector packing density, the highest height-to-width aspect ratio, and fidelity of reproduction of electrodeposition were best demonstrated in Romankiw's precursor work to LIGA in which gold was plated through polymeric masks generated by X-ray lithography.[40,41]

**Magnetic Media** It is possible to achieve many desirable magnetic properties by electroless or electroplating of magnetic media, but the merits of sputtered films seem to outweigh an electrochemical approach. Indeed, sputtered films can be codeposited with Cr, which affords additional wear and particularly corrosion resistance. And, since it is popular to overcoat magnetic media by sputtered carbon, the low cost of plating is not so important since the disc will have to see a sputtering station to be coated with carbon anyway.

**Thin-Film Magnetic Heads** It was also plating-through mask technology that Romankiw, in another pioneering IBM effort, used to make thin film heads as shown in Figure 7.8.[42] The development of batch-produced, thin-film heads is an excellent example of how micromachining has been gainfully practiced in industry: with a definite, practical, well-specified application in mind and using the machining tools best suited for the problem at hand.

Read-write heads, which were initially horseshoe magnets hand wound with insulated copper wire, are of utmost importance in magnetic storage. The fabrication of traditional ferrite heads reached its limit more than a decade ago and its further extension was hard to imagine. The IBM objective was to build the next generation read-write head, using, as much as possible, batch fabrication and lithography techniques. To develop a multiturn head as shown in Figure 7.8A, it was necessary to develop a technology which would handle dimensions between those of printed circuit boards and semiconductor devices. Plating of copper conductors through thick resist masks (Figure 7.8B) and of thin films of nickel-iron alloys (Permalloy) (Figure 7.8C) made these thin-film heads possible. The Cu coils in Figure 7.8B carry a lot of current and have to be at least 2 to 3 $\mu$m thick and nearly square in cross-section. The more turns in a given length, the higher the writing field at the pole tip gap and the higher the read back signal. The head in Figure 7.8C has 8 turns; the present day 3380-K IBM head has 32 turns (Figure 7.8A). The Permalloy plating in the thin film head is particularly challenging. Due to the very high sensitivity of composition to agitation of the plating solution, a plating cell had to be developed which assured uniform agitation over the entire part. This was achieved by a specially shaped paddle which moves at a predetermined frequency with a precisely defined separation from the surface being plated. With the right choice of conditions, it was eventually possible to achieve both thickness and composition uniformity better than ±5%.[43] As a dielectric, hard-baked AZ photoresist was used, greatly simplifying the fabrication process. As a substrate, Si was used in the early days; today $Al_2O_3 \cdot TiC$ is employed.

Without these new micromachined thin-film heads, the large increases in reading and writing the bit densities of magnetic media in the last few years (200 million bits of information on a square inch in 1993) would probably not have been possible. The thin-film magnetic head production estimated to be 250 million pieces in 1996 at revenue of 3 billion dollars. This feat makes it probably the most successful micromachined device to date.

*Maskless Electrochemical Metal Deposition*

Masking is costly and requires extra steps, but there are several electrochemical techniques being proposed which would obviate the need for masks. One of the simpler approaches utilizes impinging microjets of electrolyte. This increases the mass transport greatly, and plating selectivity is achieved by virtue of the fact that the jet serves as the current path. The smallest spot size of the area being plated is usually limited to twice the diameter of the jet, which can be operated without too frequent plugging. This size may be anywhere from 1 mm

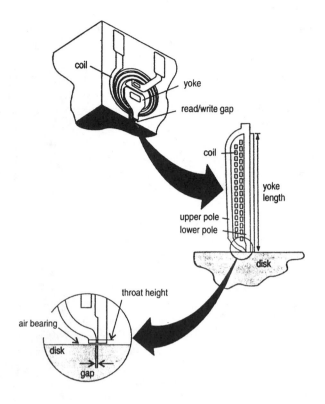

*Read and write thin film head*

**A**

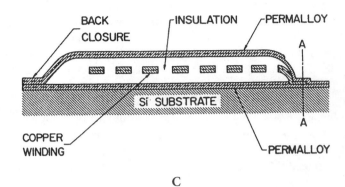

**C**

**Photo resist mask**
**on copper seed layer**

**Electroplated copper coil**
**after resist removal**

**B**

**FIGURE 7.8**   Thin film head. (A) Schematic of a multiturn thin-film head with inset of pole tip structure on the air-bearing surface. (B) Resist mold and electroplated copper coil after resist removal. (C) Schematic cross-section of the head. (From Romankiw, L.T. et al., *IEEE Trans. Magn.*, MAG-6, 597–601 (1970). With permission.)

to as little as ~0.01 mm in diameter. The lower limit depends strongly on the degree of filtration of the solution. A newer technique utilizes laser-enhanced plating and laser-enhanced jet plating. In the laser-enhanced plating, the selectivity and greatly increased plating rates are achieved through the use of properly chosen lasers whose energy is not absorbed by the

plating solution but is absorbed by the solid. The local heating of the substrate (up to 150°C) results in highly increased deposition. Using this technique, plating enhancement of up to 1000 times has been obtained with gold. Plating spots and lines as small as 2 μm wide have been obtained. Laser enhancement also works with electroless deposition, and etching is of course feasible as well. The most recent maskless, selective, high-speed plating (etching) technique is a combination of the jet and laser techniques. The jet is used here as a light pipe for the laser and at the same time as a means for the high rate of supply of ions. Current densities of up to 15 A/cm$^2$ and copper plating rates of 50 μm/sec (!) have been obtained. Etching rates, for example for stainless steel, of 10 μm/sec have been demonstrated. For more background on laser-enhanced plating and etching, jet plating, laser-enhanced jet plating, and ultrasonically enhanced plating, refer to the review by Romankiw et al.[37] and references therein. Several of these techniques offer tremendous opportunities for alternative micromachining methods to be developed. Maciossek[44] relies on overplating of a patterned plating base to create all types of novel slanted and curved metallic microstructures. The angle of the deposited wedge shapes is adjustable in a range from 0° to 45°. Besides linearly increasing structure profiles a sinusoidal or parabolic surface can also be generated. The angles of the deposited wedges depend on the distance between and the width of a set of parallel metallic strips in the plating base. At the start of the plating process only a central metal strip is biased? and during the isotropic growing process the outlaying metal strips get contacted one at a time by the electrodepositing metal. When the contact has been made, electrodeposition will occur also in this area while further deposition in the original contact area (the central strip) results in a higher deposit on that first electrode. This innovative 3D fabrication process enlarges the MEMs arsenal of tools considerably.

### Electroless Plating

Here we add one interesting recent development beyond the treatise on electroless plating presented in Chapter 3: electroless plating of beveled structures. Usually electroless plating is an isotropic process with no means to control the profile of plated features; however, van der Putten et al.[45,46] were able to induce anisotropy in the plating process by taking advantage of the increased poisoning effect a stabilizer has at the edges of any given pattern (see Figure 7.10A). Since the edges of a given pattern experience additional contribution from nonlinear

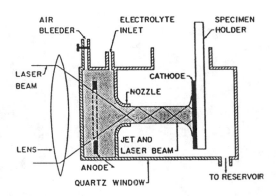

**FIGURE 7.9** Schematic representation of laser-enhanced jet plating showing the co-incident laser beam and solution jet with the jet acting as the waveguide. (From Romankiw, L.T. and Palumbo, T.A., Electrodeposition in the Electronic Industry, presented at Electrodeposition Technology, Theory and Practice, 1987. With permission.)

diffusion, mass transport to these edges is enhanced compared to material supply to the bulk of a substrate (see Chapter 3). As a result, the surface concentration of the adsorbed stabilizer ($Pb^{2+}$) is higher at these edges. By choosing the proper $Pb^{2+}$ concentration, the edges can be selectively poisoned: bevels can be grown with an angle, $\alpha$, which is a function of the stabilizer concentration in the solution (see Figure 7.10B). Importantly, van der Putten's bevel-plating technique eliminates the commonly observed lateral overgrowth of resist patterns, and the technique might have broader applications in the area of micromachining. The technique is less controllable than Maciossek's maskless electrodeposition[44] discussed above but does not require a patterned plating base.

### Removal Processes

#### Electrochemical Grinding

In electrochemical grinding (ECG), an electrically conductive abrasive wheel, the cathode, and an anode workpiece are connected to a low-voltage DC power source (Figure 7.11A).[47] An electrolyte is flushed onto the abrasive wheel, and the abrasive particles in the wheel's surface remove electrochemically dissolved material, always keeping the surface fresh for further electrochemical attack. The abrasive action accounts for as much as 10% of the metal removal; the remainder is electrochemical. All hard conductive materials are candidates for ECG (e.g., sharpening of tungsten-carbide tool bits). The method is also good for fragile parts that would be damaged by conventional mechanical grinding. The method produces parts free of residual stresses and heat damage and with surface finishes ranging from 0.1 to 0.5 µm. The surfaces are also free of scratches and burrs.

#### Electrochemical Machining

Electrochemical machining (ECM), which historically followed electrochemical grinding, employs a cathode shaped to provide a form in the anode workpiece (Figure 7.11B). A rapidly flowing electrolyte stream separates the cutting tool from the workpiece. Metal removal is accomplished by passing a DC current of up to 100 A/cm² through the salt solution cell.

The electrolyte serves two important functions in ECM. First, the electrolyte provides the medium for the anodic dissolution of the workpiece to take place; second, it removes the heat that is generated as a result of the high current flow. The cathode is advanced into the workpiece anode at a rate matching the dissolution rate, which is between 0.5 and 10 mm/min when applying current densities of 10 to 100 A/cm². The supply voltage commonly used in ECM ranges from 5 to 20 V, the lower values being used for finish machining and the higher voltages for rough machining. The rate of material removal is the same for hard or soft materials, and surface finishes are between 0.3 and 1 µm. Odd-shaped and complex cavities are formed using ECM tools. Although ECM may be used to machine any electrically conductive material, high-strength and very hard materials are most appropriate for this process. To date ECM has found its widest acceptance in the aircraft, automotive, and business machine industries. In the aircraft industry, ECM is used extensively for shaping turbine blades. In the automotive industry, the technique is used to make connecting rods, pistons, and fuel-injection nozzles. Overall, the technique is best suited for mass production of high-precision, small, complex shapes in hard-to-machine materials (accuracy better than 10 µm).[4] The fact that metal removal in ECM is not achieved by mechanical shearing (as in mechanical machining) or by melting and vaporization of the metal (as in EDM, see below) means that as with chemical machining no thermal damage occurs and no residual stresses are produced on the worked surface. For this reason, ECM is often selected for highly stressed or fatigue-sensitive applications in the aerospace industries. Large systems may cost up to $400,000 (about the same as multi-axis milling machines).

Sometimes chemical and electrochemical machining are in competition. Chemical machining might be preferred when very large workpieces have to be machined because of the large currents involved in ECM and also for ultrasmall parts, where the electric connection causes problems for ECM.

Another electrochemical process is electrochemical discharge grinding (EDG) (Figure 7.11C). It forms a combination of two other techniques, i.e., electrochemical machining and electrical discharge machining (EDM), an electrothermal process (see below). A rotating graphite electrode is used to shape electrically conductive metals. Unlike electromechanical grinding, there is no contact between the graphite wheel and the workpiece. Metal removal is caused by a sparking action which is guided by the breakdown of the anodic film at random sites on the workpiece. Spark intensity and frequency determine metal removal rate and surface finish. The process is much faster than electrical discharge grinding, but what it gains in speed it loses in precision. Normal tolerances run about 0.025 mm.[48]

The electrochemical discharge technique has been adapted to insulating materials as well.[49] In this case, the electrochemical discharge phenomenon generated in the vicinity of a pin electrode causes the workpiece and the chemicals in the solution to reach high temperatures, inducing thermochemically machining of the insulating material. Esashi et al.[50] used electrochemical discharge drilling to make electrical feedthroughs in Pyrex glass anodically bonded to silicon in absolute pressure sensors and

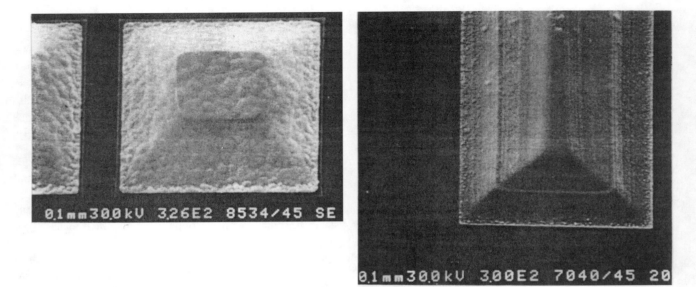

A

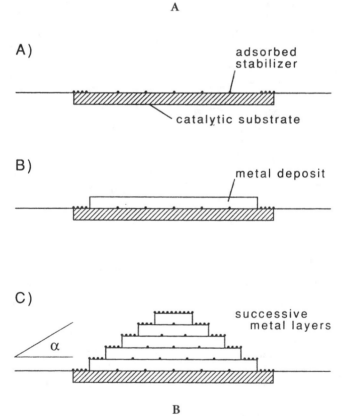

B

**FIGURE 7.10** Electroless micromachining: bevel plating. (A) Examples of bevel-plated Ni structures. (B) Method to make bevel-plated devices. (From van der Putten, A.M.T. and de Bakker, J.W.G., *J. Electrochem. Soc.*, 140, 2229–2235 (1993). With permission.)

in the fabrication of small holes in Pyrex glass to connect small stainless steel pipes (diameter 630 μm) to an integrated chemical analyzing system. A narrow through-hole in 300-μm thick Pyrex glass is bored in a few seconds.

### Wet vs. Dry Micromachining

There is an overwhelming push in the IC industry and in micromachining towards dry processing and deep dry-etching developments are particularly very swift. Electrochemical machining, on the other hand, is often still considered a 'black art' and 'dirty', with lack of cleanliness and particles making it an unacceptable technique. In reality, besides the above mentioned major technological advantages, electrochemistry offers several intrinsic advantages such as low capital equipment, high deposition rate, and relatively simple operation.

There are even cases where only wet chemical and electrochemical processes are feasible; for example, line-of-sight deposition techniques cannot readily be used on curved or irregular

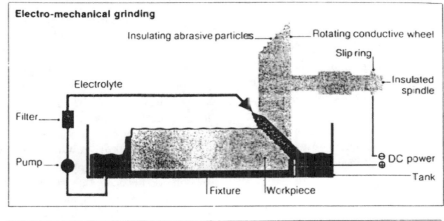

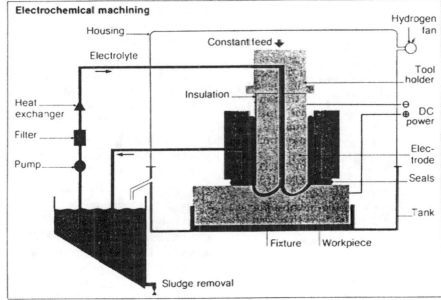

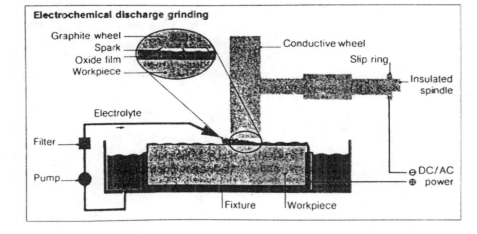

**FIGURE 7.11** Electrochemical removal processes. (A) Electromechanical grinding. (B) Electrochemical machining. (C) Electrochemical discharge grinding. (From Rain, C., *High Technol.*, November/December, 55–61 (1981). With permission.)

surfaces or to plate into via holes or blind vias. Chemical vapor deposition (CVD) often leaves voids when metallizing (say, tungsten) in a high aspect ratio structure, whereas with electrodeposition such voids can be avoided completely. Multilayer PC boards with through-holes connecting the individual metal layers would not have been possible without electroless or electrochemical technology. As long as a solution can wet and fill a cavity to be plated, a metal pattern can be generated precisely, replicating the mold down to angstrom dimensions.

The need for higher aspect ratio structures ensures that the role of electrochemistry will become even more important in years to come. The technology provides higher resolution, better pattern definition, and much higher aspect ratio patterns than chemical or even plasma ion etching can achieve. In Table 7.2 we define several situations in which an aqueous chemical or electrochemical technique may be the only means by which a desired property or quality of a deposit can be achieved.

Electrochemical microfabrication techniques such as plating through masks, electrochemical etching, laser-enhanced and jet plating, and jet etching make the electrochemical technology particularly useful for replicating high aspect ratio, micron, sub-micron, and nanometer mask patterns and constitute a whole new set of micromachining tools hardly explored today.

# Electrothermal Processes

## Introduction

In electrothermal removing and forming processes, thermal energy, provided by a heat source, melts and/or vaporizes the volume of the material to be removed or deposited. Among thermal removal methods, electrodischarge machining (EDM) is the oldest and most widely used. Electron-beam and laser machining are the newer thermal techniques becoming industrially accepted. Plasma-arc cutting using a plasma arc torch is mostly used for cutting relatively thick materials in the range of 3 to 75 mm, but the plasma deposition technique is also important for the deposition of coatings of intermediate thickness, say from 25 $\mu$m to 1 mm. In thermal removal processes, one always is left with a small heat-damaged layer, sometimes called a recast layer. In electron-beam, laser, and arc machining deposition forming as well as removal methods are available.

**TABLE 7.2**  Situations for which an Aqueous Electrochemical or Chemical Technique is the Only Means To Obtain the Desired Result[37,38]

- Uniform plating on irregular surfaces
- Leveling of rough surfaces
- Plating of via holes and blind vias
- Formation of alloys of metastable phases
- Compositionally modulated structures
- Amorphous metal films
- Maskless, high-speed, selective deposition of metals and alloys
- Most faithful reproduction of features of polymeric masks
- Smallest dimension features with highest height-to-width aspect ratio (highest packing density of conductors)
- Metal deposits with incorporated particles (diamond, WC, TiC, $Al_2O_3$, teflon, oil, etc.)
- Wear-resistant surfaces
- Corrosion-resistant surfaces
- Uniform thickness nonporous anodic films
- Anodic films with porous structure
- Very uniform, thick, pore-free, polymer films on irregular surfaces
- Very uniform thickness, glazed, and devitrified glass surfaces
- Shaping of almost any metal, semiconductor or conductive cer electromachining without introducing stress
- Electropolishing of metal surfaces to mirror bright finishes

## Electrodischarge Machining

Electrical discharge machining, more commonly known as EDM or spark machining, has traditionally been used to machine unusual designs in hard, brittle metals, but it has had a limited application in micromachining.[51] In EDM, metal is removed by high-frequency electrical sparks generated by pulsating a high voltage between the cathode tool, shaped in the form of the desired cavity, and a conductive work piece anode. The workpiece and the tool are submerged in a dielectric fluid as shown in Figure 7.12A. With a gap between tool and workpiece of 25 $\mu$m and a voltage of (for example) 80 V, intense sparking occurs across the gap, melting and vaporizing metal from both pieces (ideally more from the workpiece than from the tool). Resolidified small hollow spheres are washed away by the recirculating dielectric oil. A servo-mechanism adjusts the workpiece-tool gap and DC current pulses produce sparks at a rate of up to 500,000 per second. Each spark generates a localized high temperature on the order of 12,000°C in its immediate vicinity (the power concentration is $10^5$ W/mm$^2$ to $10^7$ W/mm$^2$). The tool wear ratio, a measure of workpiece material removal to tool material removal, varies from about 3:1 for metallic electrodes to about 70:1 for the better carbon electrodes. The tool wear can actually be reduced to 1% or less with an adequate selection of tool and workpiece materials and appropriate generator settings.[52] Volumetric metal removal rates, ranging from 0.001 to 0.1 cm$^3$/hr, are determined by thermal characteristics such as conductivity and melting temperature of tool and workpiece rather than hardness and are small compared to more conventional machining methods. Metal removal rates increase as spark intensity (current) increases, but so does the surface roughness. Increasing the spark frequency, while keeping the other parameters constant, results in a decrease in surface roughness, because the energy available is shared between more sparks, and smaller-sized surface craters are produced in the workpiece. The frequency range of EDM machines is from 180 Hz, for roughing cuts, to several hundred kilohertz, for fine finishing. With carefully controlled conditions it is possible to achieve tolerances as fine as 0.005 mm, but 0.02 mm is more typical. The accuracy of the process is closely related to the spark-gap width: the smaller the gap, the higher the accuracy, but a smaller gap results in a lower working voltage and a lower removal rate. A typical surface finish is in the 1- to 3-$\mu$m range. A layer of melted and resolidified material usually from 0.0025 to 0.05 mm thick, known as recast, is left on the surface produced by EDM. This layer tends to be very hard and brittle and may have to be removed if high levels of fatigue resistance are required. In Table 7.3 some key EDM features are summarized.

With EDM, holes as small as 0.3 mm in materials 20 mm or more in thickness can be readily achieved. With efficient flushing, holes with an aspect ratio as high as 100:1 have been produced. The process has been used successfully to produce, for example, very small diameter holes in hardened fuel-injector nozzles. EDM is traditionally used to drill holes for punch and dies. A recent variant, electrodischarge wire cutting, in contrast with classical EDM, has already made an important impact in microfabrication.

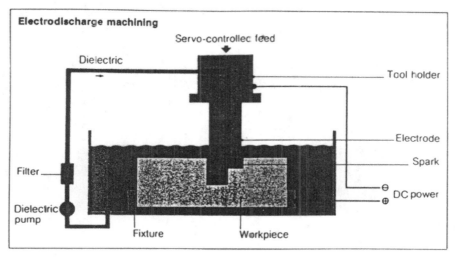

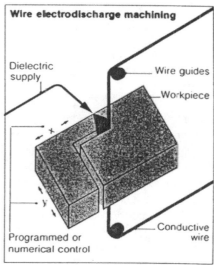

**FIGURE 7.12** Discharge machining. (A) Electrodischarge machining. (B) Wire electrodischarge machining. (From Rain, C., *High Technol.*, November/December, 55–61, 1981. With permission.)

**TABLE 7.3** Electrodischarge Machining Characteristics

| Feature | Characteristics |
|---|---|
| Typical Use | Hard, brittle metals, toolmaking |
| Tool | Carbon, zinc, brass, copper, silver-tungsten, and copper-tungsten |
| Dielectric medium | Petroleum oils, silicones, triethylene, glycol-water mixtures |
| Aspect ratio of holes | As high as 100:1 |
| Surface finish | 1 to 3 μm, but even 0.25 μm has been reported |
| Gap size/voltage | 25 μm/80 V |
| Removal rate | 0.001 to 0.1 cm³/hr |
| Workpiece | Conductor |

## Electrodischarge Wire Cutting (EDWC)

One of the most interesting adaptions of EDM is the development of electrodischarge wire cutting (EDWC),[48] which can be used for cutting complex two- and three-dimensional shapes from electrically conducting materials. In wire EDM, the cutting electrode is a continuously moving thin copper or brass wire, 0.15 to 0.3 mm in diameter or, for high-precision cutting, stronger molybdenum-steel wire (0.15 to 0.03 mm). The wire is pulled from a supply reel past two fixed sapphire or diamond guides, one located above and one below the workpiece, as shown in Figure 7.12B, and collected on a take-up reel. This continuously delivers fresh wire to the work area. Deionized water is used as dielectric and as a flushing fluid. Flushing is essential for achieving high accuracy and preventing shortcircuiting between the wire and part. The main advantage of EDWC is that it can mill through metals at higher rates (four or five times faster) than EDM. Moreover, no expensive, shaped EDM electrodes are required. Linear cutting rates are still relatively slow, especially for thick substrates: from 38 to 115 mm/hr in 25-mm thick steel.

Numerically controlled wire EDM has revolutionized die making, particularly for plastic molders. Wire EDM is now common in tool-and-die shops. Shape accuracy in EDWC in a working environment with temperature variations of about 3°C is about 4 μm. If temperature control is within ±–1°C, the obtainable accuracy is closer to 1 μm.[53]

The EDWC technique can also be used to make circular rods, by rotating the workpiece against the cutting wire as shown in Figure 7.13A, resulting in diameters down to 5 μm and a surface roughness of 0.1 μm. The rods can have other than circular cross-sections such as triangular, rectangular, slit-type, and polygonal. The way to make a triangular shape is illustrated in Figure 7.13B.[54] Kuo et al.[55] used EDWC to make micropipes with an inner diameter of 23 μm, outer diameter of 186 μm, and a length of 3 mm with a variety of micronozzle exit shapes and sizes at the end of these pipes; slit-openings of $100 \times 10$ μm, triangular openings with a base of 50 μm and circular openings of 6-μm radius were demonstrated. The manufacturing process of those micropipes is illustrated in Figure 7.14.[55] EDWC is first used to machine a stainless steel rod (a), then electroforming is used to deposit a Ni coat (b), the Ni deposit is subsequently

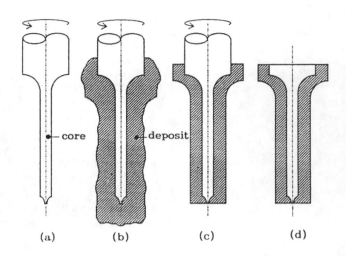

FIGURE 7.14 Micronozzle fabrication process. (From Kuo, C.-L. et al., High-Precision Micronozzle Fabrication, presented at Micro Electro Mechanical Systems, Travemunde, Germany, 1992. With permission.)

sized by EDWC (c), and finally the stainless steel core is removed (d). In order to separate the core from the deposit after plating, it is necessary to reduce the deposit tension stress as much as possible. A sulfamate nickel-plating solution with a plating temperature of 50°C and a pH of 4 was selected for this experiment because it can be operated with high current density and results in low residual stress. A schematic side view of a typical micropipe with a circular exit is shown in Figure 7.15. None of the lithography techniques discussed in the preceding chapters can produce this type of intricate pipes as easily as EDWC. It is of great value for a micromachinist to recognize at a glance which shapes are better produced by a lithography-based method and which are better produced by a truly three-dimensional lathe-based method.

## Electron-Beam Machining

### Electron-Beam Removal Process

Electron-beam removal of materials is another fast-growing thermal technique. Instead of electrical sparks, this method uses a stream of focused high-velocity electrons from an electron gun to melt and vaporize the workpiece material.[56] Originally, in the early 1960s, the method was used for welding, but today it is used for many micromachining tasks. An electromagnetic coil or lens focuses the diverging electron beam to a power density of about $10^8$ W/cm², high enough to vaporize the workpiece material. A typical cross-sectional diameter of the beam is 0.01 mm at the point of impact on the workpiece. A bias electrode switches the electron beam energy from the cathode (electron gun) on and off for pulsed drilling operations. The bias electrode also controls beam intensity. With additional coils, the beam can be deflected in any direction. With mechanical means, holes of a diameter of 0.8 mm are difficult to drill, and below 0.1 mm nearly impossible. In contrast, EBM works well for deep holes of diameter below 0.1 mm and is generally effective with a diameter-to-depth ratio of 1:10 in hard-to-machine materials. With multiple pulses, diameter-to-depth

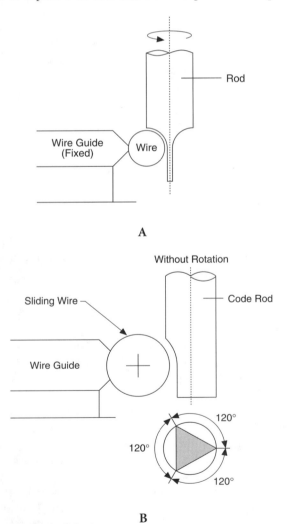

FIGURE 7.13 Electrodischarge wire cutting. (A) For a round rod profile. (B) For a triangular rod profile. (From Kuo, C.-L. et al., High-Precision Micronozzle Fabrication, presented at Micro Electro Mechanical Systems, Travemunde, Germany, 1992. With permission.)[54,55]

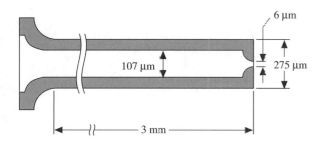

**FIGURE 7.15** Schematic side view of a circular pipe with 6-μm exit hole. (From Kuo, C.-L. et al., High-Precision Micronozzle Fabrication, presented at Micro Electro Mechanical Systems, Travemunde, Germany, 1992. With permission.)

ratios can reach 1:100. For example, hole diameters of 0.1 mm can be drilled in materials up to 10 mm thick.

Once the vacuum is established ($10^{-4}$ mmHg), EBM can be extremely fast. The material removal rate is about 0.1 mg/sec. Drilling rates of 2000 holes per second are routine with EBM. Typical tolerances are about 10% of the hole's diameter (say, 0.005 mm on a 0.05-mm hole). Today, scanning electron-beam lithography remains the only demonstrated method for micro-fabrication in the sub-100 nm regime. The most suitable application is for workpieces requiring large numbers of simple small holes to be drilled or for drilling holes in materials that are hard and difficult to machine with other processes. The equipment is very expensive, and the need to operate in vacuum adds considerably to machining time.

### Electron-Beam Forming Processes

Electron-beam induced metal deposition is a rather slow but flexible and versatile process for the fabrication of micro-structures with high lateral accuracies. For example, using a precursor gas at a pressure of $2.10^{-2}$ mbar, W/C needles 0.2 μm in diameter have been deposited. The diameter of the needles is bigger than the electron-beam diameter, which is less than 100 nm, because of electron scattering. To increase the growth rate, the local gas flow is increased by directing the gas through a small nozzle onto the surface. The total gas flow needs to be kept below a maximum value to maintain the base pressure below $10^{-5}$ mbar. For the deposition of other metals, numerous gases have been reported in the literature as precursors (mainly for ion-beam-induced deposition); $Al(CH_3)_3$ for Al, $C_7H_7O_2F_6Au$ for Au, and (methylcyclopentadientyl)trimethyl platinum for Pt.[57] A schematic of an electron-beam fabrication set-up is shown in Figure 7.16, together with an SEM photograph of W/C needles produced in this type of set-up.[57] The deposition time of the needles shown was 1 to 2 min. Using the same type of technology Matsui demonstrated tips with a diameter of 15 nm for use with a tunneling microscope.[58]

Electron beam deposition obviously can achieve submicron dimensions, but the removal of material is more complicated if the same tool is to be used. An etching gas can be used, but this process is likely to be selective.[59] For processes such as controlled submicrometer repair of masks, focused ion-beam-induced

deposition appears to be the preferred technique (see below), although both lasers[60,61] and electron beams[62] have been used.

### Laser Machining

#### Introduction

Lasers have been used in materials processing for more than two decades. Applications in micromachining include:

- Heat treatment
- Welding
- Ablation
- Deposition
- Etching
- Lithography
- Photopolymerization (stereo lithography)
- Microelectroforming
- Focused-beam milling of plastics, glasses, ceramics, metals

From this list it is clear that lasers can be used for removal as well as for additive processes. Interestingly, the laser can often serve as its own process monitor. This multi-use, multi-role, and *in situ* capability is not offered with any of the other advanced materials processing techniques such as molecular beam epitaxy (MBE), chemical vapor deposition (CVD), focused ion-beam etching (FIB), etc.[63]

In laser beam cutting, coherent light replaces electrons as the cutting tool. No vacuum is required but the removal rate is much slower than in electron-beam machining. Short pulses vaporize the substrate for cutting (subtractive process). Long pulses or continuing light melt the metal for welding (additive process). In subtractive processes, the laser energy may also be used to assist the dry and wet chemical etching and building of complex three-dimensional micromachines. The process is called laser-induced chemical etching in which, for example, chlorine is used as the etchant and is photodissociated to react with the substrate.[64,65] Both solid-state and gas lasers are commercially available for drilling via holes, repair of embedded interconnects, and for thick/thin film resistor and capacitor trimming.

Laser-assisted chemical vapor deposition (LCVD) is an additive technique and has been employed for growing three-dimensional microstructures and thin films in elemental or compound form, and at least 17 elements of the periodic chart have been deposited (W, Ni, Ti, etc.).[66]

Lasers also are used in stereo lithography to make polymeric microstructures; this additive technology was reviewed in Chapter 1 and will not be repeated here. In general, laser machining is touted for post-assembly processing, where site-specific action is necessary. Table 7.4 summarizes common laser parameters which can be controlled and their induced effect on processing.[67] For micromachining purposes, the spot size (i.e., the minimum diameter of the focused laser beam, $d_{min}$) corresponding to the system's resolution, the average laser beam intensity,

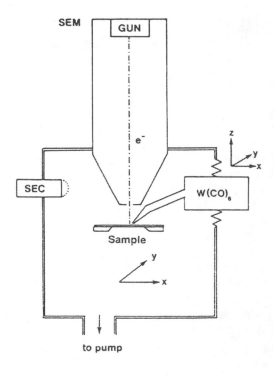

**FIGURE 7.16**  Electron-beam machining set-up and example of an electron-beam-formed W/C microstructure. (From Brunger, W.H. and Kohl-mann, K.T., E-Beam-Induced Fabrication of Microstructures, presented at Micro Electro Mechanical Systems, Travemunde, Germany, 1992. With permission.)

and the depth of focus are three of the most important listed laser parameters.

The minimum diameter of the focused laser beam, $d_{min}$ (mm), is normally determined by the diffraction limits of the imaging system, so the theoretical resolution is given by:

$$d_{min} = \frac{\lambda f}{\pi d_o} \qquad 7.1$$

where $\lambda$ (mm) is the wavelength of the radiation, $d_o$ (mm) is the diameter of the beam at the focusing lens, and f is the focal length of the lens. Thus, the principal way of increasing the resolution is by reducing the wavelength.

The average laser beam intensity I (W/mm²) at focus is given by:

$$I = \frac{P}{\pi W_o^2} \qquad 7.2$$

where P = laser power in watts. The depth of focus (DOF), also defocus tolerance, is given by:

$$DOF = \pm \frac{0.32\pi W_o^2}{\lambda} \qquad 7.3$$

where C is constant.

From Equations 7.1 and 7.3 it follows that there is an inverse relationship between the need for high resolution and a practical depth of field. Material processing with a very short depth of focus requires a very flat surface. If the surface has a corrugated topology, a servo-loop connected with an interferometric autor-anging device must be used.[63]

Flexible laser machining set-ups as shown in Figure 7.17 can be used for the machining of three-dimensional parts by utiliz-ing highly localized laser microchemical reactions and an accu-rate computer-controlled x,y,x,θ translation table. This type of set-up can accommodate either additive[68] or subtractive

**TABLE 7.4**  Laser Parameter and Related Processing Parameters

| Laser Parameter | Influence on Material Processing |
|---|---|
| Power (average) | Temperature (steady state), process throughput |
| Wavelength (μm) | Optical absorption, reflection, transmission, resolution, and photochemical effects |
| Spectral linewidth (nm) | Temporal coherence, chromatic aberration |
| Beam size (mm) | Focal spot size, depth of focus, intensity |
| Lasing modes | Intensity distribution, spatial uniformity, speckle, spatial coherence, modulation transfer function |
| Peak power (W) | Peak temperature, damage/induced stress, nonlinear effects |
| Pulsewidth (sec) | Interaction time, transient processes |
| Stability (%) | Process latitude |
| Efficiency (%) | Cost |
| Reliability | Cost |

*Source:* Liu, Y.S., in *Laser Microfabrication — Thin Film Processes and Lithography*, Tsao, J.Y. and Ehrlich, D.J., Eds. (Academic Press, New York, 1989) p. 3. With permission.)

processes but it is more convenient to build separate dedicated equipment, as illustrated in Figure 7.17A and B.

## Types of Lasers Used for Laser Machining

For laser micromachining, the carbon dioxide ($CO_2$) laser and neodymium yttrium aluminum garnet (Nd:YAG) are favored over other traditional laser types such as the He:Ne or dye laser because of their higher power output. The $CO_2$ lasers use a mixture of helium (83%), nitrogen (16%), and carbon dioxide (6%) as the lasing material. The $CO_2$ laser produces a collimated coherent beam in the infrared region (10.6 μm) characteristic of the active material ($CO_2$). The wavelength is strongly absorbed by glasses or polymers and so either mirrors or ZnSe lenses with excellent infrared transparency are used to handle the beam. The Nd:YAG laser produces a collimated coherent beam in the near-infrared region of wavelength 1.06 μm and can be run pulsed or continuously. In a pulsed mode, peak power pulses of up to 20 kW are now available. The shorter wavelength allows the use of optical glasses to control the beam path.

In 1975, the excited dimer (excimer) laser was invented in which a diatomic molecule such as $N_2$ or $H_2$ is used as the lasing material. Since then rare gas halide lasers have become more common but are still referred to as excimer lasers. Excimer lasers cover a range of wavelengths in the ultraviolet range from 157 nm ($F_2$) to 353 nm (XeF) with 193 nm (ArF), 248 nm (KrF), and 308 nm (XeU) being particularly useful intermediate wavelengths. Peak powers typically range from 3 MW ($F_2$) to 50 MW (KrF or XeU) in pulses lasting a few tens of nanoseconds.

The excimer laser has several distinct advantages in the dry etching of polymers over the $CO_2$ and Nd:YAG lasers. For instance, the wavelengths are more compatible with the chemical bond energies in organic compounds and tend to produce less thermal damage. Excimer lasers, as we learned in Chapter 1, can also be used as more efficient ultraviolet sources for photolithographic exposure of resists for VLSI circuits. For any laser, high-rate pulsing or continuous operation is required for cutting applications.

### Laser-Beam Removal Processes (Drilling, Scribing, and Trenching)

Conventional lithographic techniques for making three-dimensional shapes are limited to unidirectional extensions of two-dimensional patterns. By using laser-induced etching, truly three-dimensional structures can be made without the need of masks. Three-dimensional structures are first constructed using a commercial solid-modeling software package. Then these structures are digitized into a stack of planar software masks, each comprising an array of pixels. In a laser machining set-up, such as shown in Figure 7.17A, depth contouring is accomplished by dynamically refocusing the laser beam. Numerous different profiles (masks) can be stored in the CNC positioning system and recalled as required. The power levels of the highly collimated, monochromatic, and coherent light beam at the workpiece, which can be held in open air, may reach $10^6$ W/mm², enough to melt any metal. A very simple example is a hole, laser drilled through glass or Si, which can produce high aspect ratio, through-the-wafer vias.[69,73] For illustration we show the electrical contact topology for a solid state capacitive pressure sensor in Figure 7.18.[69] This compact pressure sensor incorporates electrically conducting vias drilled through the glass cap by a $CO_2$ laser. In contrast, the through-the-wafer via connection fabricated by anisotropic etching has a large opening angle of (54.7°) associated with the slow etching crystallographic planes of the Si substrate. Typically the closest spacing for vias that are wet etched in a 300-μm thick wafer is approximately 425 μm, thus limiting this method to a low density of via holes.

When holes are laser drilled in a substrate, the expelled material solidifies into dust; therefore, a moving protective tape is used in front of the laser lens to keep it clean. Also, debris will collect around the perimeter of the openings in, say, a silicon substrate, when drilling, scribing, or trenching is done in air or an inert atmosphere. However, if the operation is performed in chlorine gas, the silicon vapor and droplets leaving the hole combine with the chlorine to form $SiCl_4$ gas and are carried away from the work area. Silicon, when heated to its melting point, reacts nearly at the gas-transport-limited rate with chlorine, enabling removal rates exceeding $10^5$ μm³/sec.[71] The resulting hole is cleaner and more sharply defined, and debris does not splatter on the surface of the substrate; thus an excellent spatial resolution results (1 μm x,y resolution at a removal rate of $2.10^5$ μm³/sec).[71,72] The latter removal rate is several orders of magnitude faster than with electrodischarge machining methods.

At the Massachusetts Institute of Technology Lincoln Laboratories, optical scanning of the laser beam at a rate of 7500 μm/sec is employed to remove Si layers in a $Cl_2$ etchant. After a plane of 1 μm is removed, the focusing objective is lowered 1 μm, and a new pattern is etched. With this technique

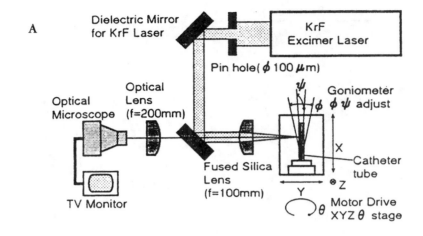

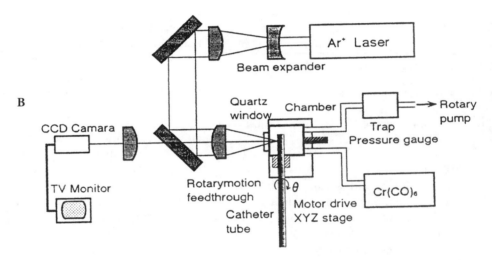

**FIGURE 7.17**  Laser micro-etching/deposition experimental set-up. (A) Etching/ablation of a catheter tube. (B) Laser CVD deposition system. (B is from Maeda, S. et al., KrF Excimer Laser-Induced Selective Non-Planar Metallization, presented at Micro Electro Mechanical Systems, Oiso, Japan, 1994. With permission.)

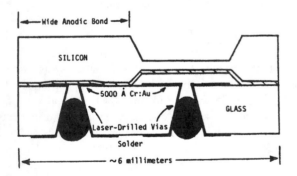

**FIGURE 7.18**  Electrical contact in a silicon capacitive pressure sensor using laser-drilled vias. (From Bowman, L., in *Micromachining and Micropackaging of Transducers,* Fung, C. et al., Eds. (Elsevier, Amsterdam, 1985), pp. 79–84. With permission.)

it is very easy to make multilevel-type structures. Some example devices made this way are shown in Figure 7.19. The Lincoln group also demonstrated that structures as shown can be replicated by using simple hot embossing molding techniques.[71]

In addition to $Cl_2$, $Br_2$, HCl, $XeF_2$, KOH, etc. can also be used as the etchant in the laser photochemical etching of silicon. Table 7.5 summarizes several kinds of substrates etched by laser processes with different etchants and laser wavelengths.[63,66]

Laser cutting machines are now used quite extensively for cutting complex profiles in sheet and plate materials. Hard materials such as diamond, tungsten, and titanium can be cut without problems of tool wear since there is no contact between tool and workpiece. For many applications, laser micromachining offers the advantage of rapid prototyping (as compared to, say, FIB machining or lithography techniques). The laser has other distinct advantages; aspect ratios for hole diameter-to-depth on the order of 1:50 are possible and holes can be drilled in hard-to-reach areas and at difficult angles (e.g., 10° to the surface). A laser-based scribe can make narrower trenches than the current mechanical diamond stylus. Nd:YAG and excimer lasers have been used to drill holes in the 30 to 50 μm diameter range in a variety of materials, and the smallest feature for excimer lasers is about 0.8 μm in thin metal foils. When hundreds of holes are to be drilled, all of the holes being at the same angle,

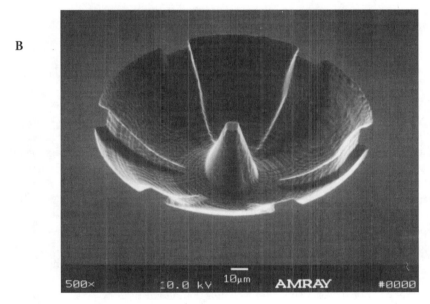

**FIGURE 7.19** Some examples of laser-machined shapes.[73] The devices are built using chlorine assisted laser etching (100 Torr $Cl_2$). The beam spot used is 1 μm and the objective is lowered in 1-μm increments after each plane is scanned at 7500 μm/sec. Etch rate 100,000 μm³/sec. (A) SEM micrograph of a microfluidic device layer cut into silicon. The white bar is a 100 μm marker. (B) SEM micrograph of a cone inside a dish cut into silicon. The dish has an upper radius of 80 μm and a lower radius of 35 μm with a radius curvature of 82 μm and spacer layers extending 8 μm inward. The clipped cone has an upper radius of 4 μm and lower radius of 16 μm. The laser was swept in a circular scan in each plane. The white bar is a 10 μm marker. (Courtesy of Dr. D. Ehrlich, Revise, Inc.)

makes the job better handled by the electron beam or the electric discharge machine. But if holes have to be drilled at many angles, the laser may be the most cost-effective method. lasers are most cost-effective for drilling holes between 1.5 and 0.01 mm. At the upper bound the cost of the large laser oscillator needed to produce the energy for vaporization becomes prohibitive. And, because of the large amounts of material being vaporized, the hole produced suffers in terms of quality and controllable tolerance. With the very small holes, the high-power focusing lenses required to withstand the energies necessary for drilling holes smaller than 0.01 mm are limiting factors. The requirement for even smaller features necessitates the use of EBM (see above) or FIB (see below). Like other thermal removing techniques, laser machining does leave a small heat-affected zone on the workpiece.

### Forming Processes with Laser Machining

The set-up shown in Figure 7.17B is used to laser deposit materials. With laser-assisted CVD (LCVD), part of the energy needed for deposition is provided by photons (see also Chapter 3 on pattern transfer with additive techniques). This method fills the need for a low-temperature deposition process. With a laser source, it is possible to write a pattern on a surface directly by scanning the microsized light beam over the substrate in the

**TABLE 7.5**  Examples of Etchants and Substrates Etched by the Laser Photochemical Process[63,66]

| Substrate | Etchant(s) | Etch Rate | Laser/Intensity |
|---|---|---|---|
| Semiconductors | | | |
| Si | $Cl_2$, HCl | 7 µm/sec | $Ar^+$/>5 MW/cm² |
| Si | KOH | 15 µm/sec | $Ar^+$/~10⁷ W/cm² |
| Si | NaOH | 2 µm/min | $CO_2$/NA |
| Ge | $Br_2$ | 36 µm/sec | $Ar^+$/0.1 kW/cm² |
| GaAs | $HNO_3$ | 2 µm/sec | $Ar^+$/60 MW/cm² |
| GaAs | $CH_3Br$ (750 Torr) | 60 Å/sec | $Ar^+$ (257 nm)/1 KW/cm²) |
| GaP | KOH | 600 Å/sec | $Ar^+$ (351 nm/3.5 KW/cm²) |
| InP | $CH_3Br$ (750 Torr) | 9.4 Å/sec | $Ar^+$ (257 nm)/100 W/cm²) |
| Insulators | | | |
| $SiO_2$ | $SiH_4$ | 40 Å/sec | KrF/0.3 J/cm² |
| $SiO_2$ | $NF_3$:$H_2$ | 0.12 nm/sec | ArF/7.7 mJ/cm² |
| Polyimide | KOH | 0.3 µm/sec | $Ar^+$/0.03 MW/cm² |
| Polyimide | Air | 0.15 µm/pulse | KrF/03 J/cm² |
| PMMA | Air | 0.3 µm/pulse | ArF/0.25 J/cm² |
| Diamond | $Cl_2$, $O_2$, $NO_2$ | 1400 Å/pulse | ArF/30 J/cm² |
| Metals | | | |
| Ti | $NF_3$ | 0.29 Å/pulse | 0.115 J/cm² |
| W | $Cl_2$ (0.1 Torr) | 240 Å/min | 0.12 J/cm² |
| Al | $Cl_2$ (0.1 Torr) | 1.4 µm/sec | 1.0 J/cm² |
| Ag | $Cl_2$ (0.1 Torr) | 500 Å/min | $N_2$ (337 nm)/0.12 J/cm² |

presence of the suitable reactive gases. Pulsed laser deposition (PLD) is a technique which rivals MBE for epitaxial growth quality of a variety of films such as diamond and all types of semiconductors. Evaporation stabilities of better than 1% over 5 hours have been shown. But lasers and particle beams, in general, add another important capability to materials processing: the ability to process selected materials at specific sites and at low bulk temperatures. Of the two techniques, the laser is the more versatile as it can be operated in air, is not affected by surface charging, and can more easily be incorporated in a manufacturing assembly line.[63] The species to be deposited is usually chemically encapsulated in a precursor gas with hexacarbonyls or alkyl 'backbones'. Typical laser driven deposition reactions are:

- Tungsten deposited by:

$$WF_6 + 3H_2 \rightarrow W + 6HF \qquad \textbf{Reaction 7.1}$$

- Nickel deposited by:

$$Ni(CO)_4 \rightarrow Ni + 4CO \qquad \textbf{Reaction 7.2}$$

- Silicon deposited by:

$$SiH_4 \rightarrow Si + 2H_2$$

$$Si_2H_6 \rightarrow 2Si + 3H_2 \qquad \textbf{Reaction 7.3}$$

$$Si_3H_8 \rightarrow 3Si + 4H_2$$

Doped polysilicon lines with conductivities as low as $10^{-2}$ to $10^{-3}$ ohm-cm may be obtained by adding $PH_3$ to the silane. Typical deposition rates are in the order of µm/sec. By optimizing the process, however, very high deposition rates have been obtained in LCVD; for example, in the case of silicon, deposition rates of 500 µm/sec have been reported by using 100% $SiH_4$ at one atmosphere, and the writing speed of tungsten on silicon substrates can exceed several centimeters per second.[74]

Not only 'flat' lines but more complex three-dimensional structures such as fibers and springs can be grown by LCVD. This is accomplished by adjusting the focal point of the laser continuously by moving the substrate using a three-dimensional linear micropositioning system (Figure 7.20).[74] Microscale Si and boron rods and helical structures, freestanding tungsten coils, and a tungsten helix on a cylindrical silicon substrate (for a microsolenoid) have been demonstrated this way (see Figure 7.20).[74] For deposition of the tungsten helix on the Si rod a rotating goniometer was used.

Table 7.6 lists examples of localized, electroless, and maskless photochemical laser depositions. Besides two- and three-dimensional writing patterns, there are two other site-specific actions that are possible by the laser irradiation technique: (1) the driving of dopants into semiconductor materials and (2) the inducing of oxidation at a surface. Examples of the latter two laser machining applications are listed in Table 7.6 as well.

### Conclusions

Subtractive and additive laser machining present several application opportunities in micromachining due to their versatility, site-specific operation, and rapid prototyping capability.

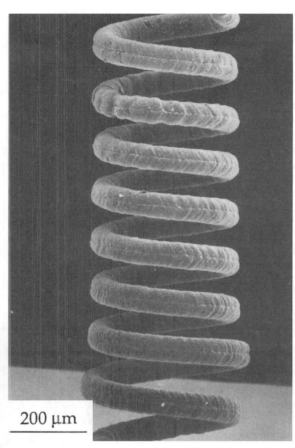

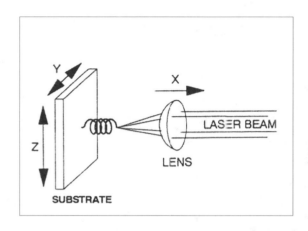

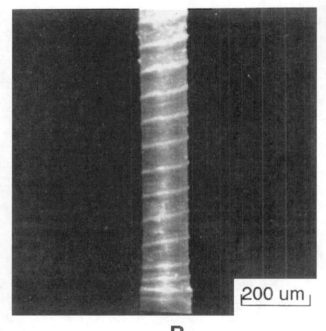

**FIGURE 7.20** Examples of LCVD microfabricated shapes made by adjusting the focal point of a laser (Tungsten helix) (A) and rotating goniometer (Si microsolenoid with tungsten helix) (B). (From Boman, M. et al., Helical Microstructures Grown by Laser-Assisted Chemical Vapour Deposition, presented at Micro Electro Mechanical Systems, Travemünde, Germany, 1992. With permission.)

There are some obvious drawbacks with these direct write techniques; they are serial processes and the deposition or removal rates limit the speed of the micromanufacture. However, for rapid prototyping, mold fabrication, and site-specific manufacturing, laser machining has a very bright future and, as we remarked before, some microsystems might carry a bigger price tag than an IC, making a more expensive manufacturing technology acceptable.

A good further introduction to laser machining can be found in *Laser Machining: Theory and Practice*, by George Chryssolouris[1] and in an excellent chapter by Henry Helvajian:

Laser Material Processing: A Multifunctional *In Situ* Processing Tool for Microinstrument Development (see Reference 63).

### Plasma-Beam Machining

*Forming Processes*

The basic configuration of a plasma-arc torch for layer deposition was discussed in Chapter 3 and the system set-up is shown in Figures 3.40 and 3.41. Here, we present an example of how to use this technology to batch produce solid-state oxygen sensors based on yttria-stabilized $ZrO_2$ (YSZ) solid electrolyte films.[75] Plasma spraying is a particulate method geared

**TABLE 7.6**  Examples of Localized Photochemical Deposition, Doping, and Oxidation

**Material Deposition**

| Deposit | Precursor | Laser |
|---|---|---|
| Titanium | $TiCl_4$ | $Ar^+$ |
| Titanium/Aluminum | $TiCl_4$, $Al(CH_3)_3$ | $Ar^+$ |
| Chromium | $Cr(CO)_6$ | $Ar^+$ |

**Laser-Induced Doping**

| Dopant/Substrate or Film | Precursor | Laser |
|---|---|---|
| Boron in silicon | Evaporated boron | Ruby |
| Phosphorus in polysilicon | Phosphorus-doped glass | $N_2$-dye |
| Arsenic in silicon | $AsH_3$ | XeCl |

**Laser Oxidation of Substrates**

| Substrate | Oxide Species | Laser |
|---|---|---|
| Titanium | $TiO_2$ | Nd:YAG |
| Zirconium | $ZrO_2$ | $CO_2$ |

toward fast deposition of thicker films (>30 μm) and the method might enable the batch fabrication of solid state oxygen sensors at a fraction of the current cost ($2 vs. $12 and up).

Traditional automotive solid-state oxygen sensors for combustion control (so-called lambda probes) are nose-shaped three-dimensional structures fabricated by sintering a molded green tape zirconia body (see Figure 7.21A). The resulting dense ceramic YSZ solid electrolyte and the silk-screened oxygen-sensing Pt electrode contacting it are protected by a plasma-sprayed, porous, gas-diffusion barrier, typically a spinel structure oxide. Newer designs for oxygen sensors, so-called wide range air-to-fuel ratio sensors, incorporate two oxygen pumping cells and one potentiometric gauge. They are planar and fabricated by laminating and co-firing at high temperatures several layers of ceramic green tape, some of which are metal coated and have openings through them (see Figure 7.21B).[76] As both metal and ceramic layers constitute the sensor, the materials should be well prepared and the cofiring process must be tightly controlled to avoid metal diffusion or reaction with the ceramic. Because of the process complexities this sensor is too expensive. In Figure 7.21C an alternative planar oxygen sensor fabricated using plasma-spray deposition[75,77] is compared with the traditional oxygen sensor. In this planar oxygen sensor, the metal electrodes are deposited by sputtering and the plasma-sprayed YSZ film acts both as a gas-diffusion layer and as an oxygen-conducting electrolyte. The ionic conductivity of the plasma-deposited YSZ films did not reach the same level as YSZ electrolytes sintered at high temperatures as the films are not as dense. However, the relatively thin-film geometry of the present sensor allowed for using the plasma-sprayed YSZ film as an oxygen-pumping cell for the wide range, air-to-fuel ratio sensor. The major challenge in the manufacture is the control of the porosity gradient in the gas diffusion barrier and electrolyte material which was accomplished by controlling the size of the

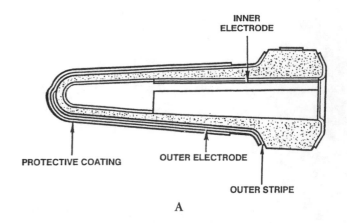

A

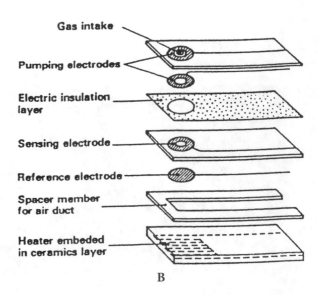

B

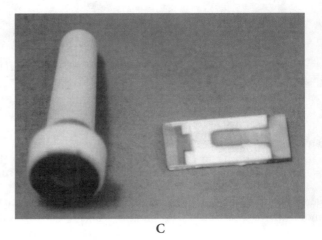

C

**FIGURE 7.21**  Solid state oxygen sensors. (A) Traditional oxygen probe only protective, gas diffusion barrier is applied by plasma spray. (B) Wide-range air-to-fuel ratio sensor. (C) Planar oxygen probe, sensor, and gas diffusion layer made by plasma spray and compared with a classical oxygen probe. (Part B from Suzuki, S. et al., *SAE Paper 860408*, 1986. With permission.)

spraying powder and spraying conditions.[77] The plasma-sprayed films adhered very well to the substrates and had exceptionally high integrity. The plasma method produces almost fully activated YSZ films onto a substrate carrying the thin-film, sputter-deposited Pt electrodes and there is no need for additional sintering. This simple manufacture of the solid-state oxygen sensors can be performed in large batches by using simple shadow masks and laser cutting the separate sensor elements.

We believe that this approach — a combination of thin- and thick-film methods — opens up the potential for planarization of many types of gas sensor devices. In the manufacture of chemical sensors, thick-film technology on hybrid substrates is more prevalent than in the IC industry, and plasma deposition of all types of chemical sensor materials is, in the author's opinion, fertile ground for research. Plasma deposition of sensor materials such as $ZrO_2$, $SnO_2$, ZnO, etc. on large inert carrier substrates could provide wafers coated with chemical sensor material very quickly and inexpensively.

### Removal Processes

Plasma arc cutting is mainly used for cutting thick sections of electrically conductive materials. A high-temperature plasma stream (20,000 to 33,000°C) interacts with the workpiece, causing rapid melting. A typical plasma torch is constructed in such a way that the plasma is confined in a narrow column about 1 mm in diameter. The electrically conductive workpiece is positively charged and the electrode is negatively charged. Relatively large cutting speeds can be obtained: e.g., 380 mm/min for a stainless steel plate 75 mm thick at an arc current of 800 A. Tolerances of ±0.8 mm can be achieved in materials of thicknesses less than 25 mm, and tolerances of ±3 mm are obtained for greater thicknesses. The heat-affected zone for plasma arc cutting varies between 0.7 and 5 mm in thickness. The technique is of marginal use in micromachining, except perhaps to cut the plasma arc-coated sensor substrates discussed in the previous section.

## Mechanical Processes

### Introduction

In mechanical removing (machining), stresses induced by a tool overcome the strength of the material. How well a part made from a given material holds its shape with time and stress is referred to as the dimensional stability of the part and the material. In order to maximize dimensional stability, the machine design engineer tries to minimize the ratios of applied and residual stress to yield strength of the material. A good rule of thumb is to keep the static stress below 10 to 20% of yield. Thermal errors are often the dominant type of error in a precision machine, and thermal characteristics such as thermal expansion coefficient and thermal conductivity deserve special attention.[18]

### Precision Mechanical Machining vs. Si Micromachining

For over a hundred years, precise macromachines have been designed and built with mechanical removal techniques

achieving part-per-million accuracies; recently macromechanical machines have entered the realm of 100 parts-per-billion accuracies.[2]

In comparing Si micromachines with mechanically machined macrostructures Slocum makes some important observations.[2] He notes that, while micromachines are impressive for their small size, their relative accuracy is two orders of magnitude worse than is typically achieved in macromachines, which moreover are much more complex (see also Figure 7.1). Typical micromachines today, Slocum points out, are comparable with the macromachines of the early 1700s with respect to complexity and accuracy. The surface roughness of micromachines looks large compared to the specular finishes of bearing surfaces. In reality, the absolute roughness is about the same but in micromachines surface forces such as friction are relatively more important. Thermal errors in a micromachine are generally more relaxed; the small scale and fast thermal equilibrium of the smaller structures are their saving graces. Another key issue Slocum brings up is that of position measurement systems; verification of fabricated geometry and tolerances is much more difficult for micromachines since measurement of a displacement of one part in ten thousand will typically get one down to the nanometer level.

Higher accuracy machining is needed to provide computer memory disks and optical mirrors and lenses with accuracies to a fraction of the wavelength of light. A better understanding of the limits of traditional and nontraditional machining methods would help the mechanical engineer make the best choice in machining tools. At the same time, as micromachine mechanical complexity increases, it seems appropriate for the designers of micromachines, who are typically electrical engineers, to study macromachine design.

### Ultra-High Precision Mechanical Machining

Ultra-high precision machines with sharp single-crystal diamond tools have made submicrometer precision machining possible. The first computer numerical controlled (CNC) machines became available in 1977, and the smallest movement the machines could reproducibly make was 0.5 μm. By 1993, 0.05 μm became possible, and today there is equipment available featuring a 0.01-μm resolution.[78] And, as we saw in Figure 7.2, Taniguchi predicts a machining accuracy for ultra-precision machining of about 1 nm by the year 2000. The resolution of the machine is, of course, a determining factor for the manufacturing accuracy of the workpiece. A typical single-point diamond mounted on a lathe is shown in Figure 7.22. For the production of computer hard discs, mirrors for X-ray applications, photocopier drums, commercial optics such as polygon mirrors for laser-beam printers, consumer electronics such as mold inserts for the production of compact disc reader heads, and camcorder viewfinders, in addition to high-definition television (HDTV) projection lenses and VCR scanning heads, ultraprecision manufacturing is now the commercially preferred technique.

The single hard point on the lathe in Figure 7.22 may be made from natural diamond, which can be applied to most

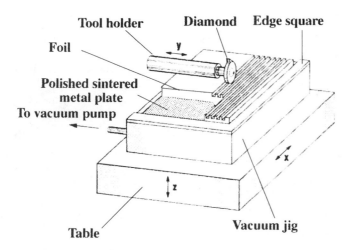

**FIGURE 7.22**  Diamond tool on a lathe for precision mechanical machining.

nonferrous metals, polymers, and crystals, or, for precision cutting of hardened ferrous metals and high temperature alloys, it may be made from Borazon. In Figure 7.23A we show an SEM photograph of a single-crystal diamond machine tool. The radius of the edge on this tool is estimated at less than 0.05 μm; with such a tool submicron grooves equivalent to these produced by silicon micromachining can be produced (Figure 7.23B).[79] Using such a diamond tip and numerical control (NC) rice tip-sized cars were machined at Nippondenso[80] (one car was 4.5 mm long and another was 7.5 mm long) (Figure 7.23C). The shell of the car was made by the sacrificial mold technique. A piece of Al was NC machined and plated with electroless Ni (30 μm thick); after cutting off the lower part of the body with electrodischarge machining, the Al was removed by KOH etching. Finally, the Ni car shell was Au coated to protect it from oxidation. The stainless chassis and wheels were also made by NC machining, and the core shaft and coil of the motor were made by NC machining and electrodischarge machining. The zirconia wheel shaft was machined down to a 250-μm diameter. The electromagnetic step motor is driven by an external magnetic field, and the car's permanent barium ferrite rotor runs at a maximum speed of 100 mm/sec developing a torque of about $10^{-6}$ Nm (at 3V and 20 mA). To assemble the various microparts into a complete car, a mechanical micro-manipulator, ordinarily used for handling biological cells, was employed.[80]

The single-crystal diamond tool refinements are not the only reason for the high precision achieved nowadays; submicron precision is also being achieved through high-stiffness machine beds, air bearings (air bearings with a rotational precision of 0.01 μm are available now), and measurement systems such as laser interferometry. Furthermore, highly precise instruments such as servomotors, feedback devices, and computers were implemented, and many types of machine tools are now equipped for computer numerical control, further improving precision and reproducibility of the manufactured parts. The latest trend is toward flexible manufacturing. In response to the need of automation and the demands created by frequent design

changes over a broad variety of products, the flexible manufacturing system (FMS) was developed. FMS is a combination of several technologies such as computers, CNC workstations, robots, transport bands, computer-aided design (CAD), and automatic storage.

The technique was developed to produce many varieties of a certain product in smaller quantities rather than many devices of one type. CNC workstations are linked by automatic workpiece transfer and handling, with flexible routing and automatic workpiece loading and unloading.[4]

Besides high accuracy cutting, there is also significant progress to report in die punching. Aoki et al.[81] have pushed the art of die forming with a press for three-dimensional, metallic medical microcomponents to new limits. Medical forceps to be inserted into an endoscope used to remove diseased tissue were fabricated as an example of the new capabilities. The diameter of the forceps is 0.6 to 1 mm and the length is 10 mm. Assembly holes in the forceps are 0.3 mm and the teeth in the cutting tool are 0.2 mm. These authors found that the final manufacturing time, including assembly, is about 3 minutes per piece.

Despite all of this progress, the fact that it often takes a 2-ton machine tool to fabricate microparts where cutting forces are in the milli- to micro-Newton range is a clear indication that a complete machine tool redesign is required for the fabrication of micromachines.[12] Along that line, in Japan the concept for desktop flexible manufacturing systems (DFMS) for building micromachines was proposed in the early 1990s.[14] The manufacturing units would be tabletop and include universal chuck modules to which workpieces would be continuously clamped through most of the manufacturing process. The miniature die press from Aoki and Takahashi is one of the first examples of progress in that direction.[81]

### Ultrasonic Machining

In ultrasonic machining (USM), also called ultrasonic impact grinding, high-frequency vibrations delivered to a tool embedded in an abrasive slurry create accurate cavities of virtually any shape, i.e., 'negatives' of the tool. Almost any hard material, including aluminum oxides, silicon, silicon carbide, silicon nitride, glass, quartz, sapphire, ferrite, fiber optics, etc. can be ultrasonically machined. The tool does not exert any pressure on the workpiece (drilling without drills)[82] and is often made from brass, cold-rolled steel or stainless steel and wears only slightly. The tool, typically vibrating with an amplitude of 0.025 mm at a frequency of 20 to 100 kHz, is gradually fed into the workpiece to form a cavity corresponding to the tool shape. The high frequency power supply for the magnetostrictive or piezoelectric transducer stack that drives the tool is typically rated between 0.1 and 40 kW. Since this method is nonthermal, nonelectrical, and nonchemical, it produces virtually stress-free shapes even in hard and brittle workpieces. Ultrasonic drilling is most effective for hard and brittle materials; soft materials absorb too much sound energy and make the process less efficient. In Figure 7.24 key components of a typical ultrasonic machining (USM) set-up are shown.[83] In Figure 7.25 SEM photomicrographs are shown of 640-μm holes drilled in alumina

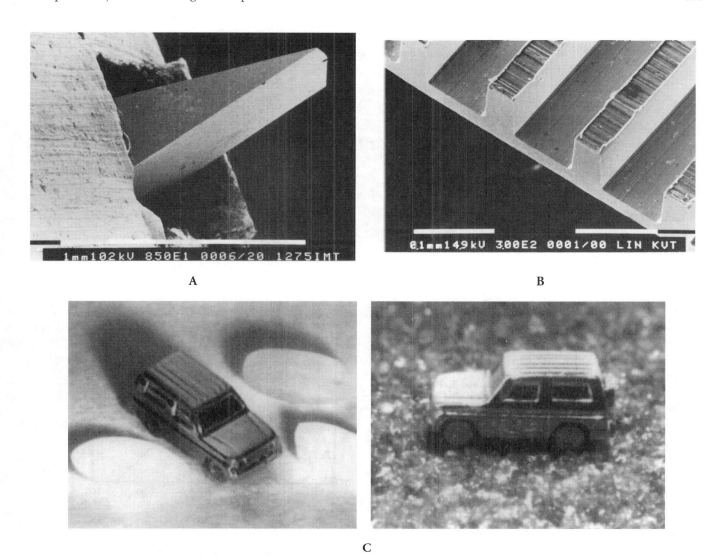

**FIGURE 7.23** Diamond ultra-high precision tools. (A) Single-crystal diamond tool. Angle of tool is 20°. Maximum depth of cut is 500 μm. (B) Microgroove produced by diamond machining a 100 μm thick Al foil. Width is 85 μm, depth is 70 μm. (C) Microcar and rice tips (left) and microcar on sandpaper (grain size = 200 μm) (right). The car is 7 mm long. 2.3 mm wide, and 3 mm high. (A and B is from Bier, W., G. Linder, D. Seidel, and K. Schubert, *KfK Nachrichten*, 1991, 165–173. With permission. C is from Teshiguhara, A., M. Watanabe, N. Kawahara, Y. Ohtsuka, and T. Hattori, *J. Microelectromech. Syst.*, 4, 76–80, 1995. With permission.)

by two different techniques: (A) ultrasonic machining and (B) laser-beam machining. The edges of the hole are much better defined with the ultrasonic technique.[83] In Figure 7.26 posts and holes drilled in a polycrystalline Si slab are shown.[83]

The key to ultrasonic machining, besides the tool itself, is the abrasive: a slurry of water or oil and small abrasive particles (large = 0.5 mm average particle diameter or small = 0.008 mm average particle diameter) of boron carbide, silicon carbide, or aluminum oxide. Boron carbide is the hardest abrasive and lasts the longest. The abrasives are suspended in water or oil at a concentration of 20 to 50 wt% and recirculated constantly to the workpiece to supply fresh abrasive at the cutting site and remove abraded particles. The particle size and the amplitude of the vibrations used are typically made about the same. Size of the abrasive particles determines the roughness or surface finish and the speed of the cut. Material removal rates are quite low, usually less than 50 mm³/min. The mechanical properties and fracture behavior of the workpiece materials also play a

large role in both roughness and cutting speed. For a given grit size of the abrasive, the resulting surface roughness depends on the ratio of the hardness (H) to the modulus of elasticity (E). As this ratio increases, the surface roughness increases. For example, under identical ultrasonic cutting conditions, carbide with an H/E ratio of 1.26 has a resulting surface roughness of 0.5 μm, and glass with an H/E of 8.1 ends up with a surface finish of 1.55 μm.[83]

Higher H/E ratios also lead to higher removal rates: ≈ mm³/min for carbide and 11 mm³/min for glass. Thus, a low fracture toughness will result in a higher material removal rate and an increase in workpiece roughness values. For a given material, a larger particle size results in a rougher surface. With a very fine particle size of 0.008 mm and a low H/E ratio (e.g., alumina with H/E 2.36), a 0.5-μm surface roughness can be achieved.[20] Holes as small as 0.076 mm have been drilled, but in production applications 0.25 mm is more the norm. The upper limit on cavity size is approximately 75 mm. With a 250-μm

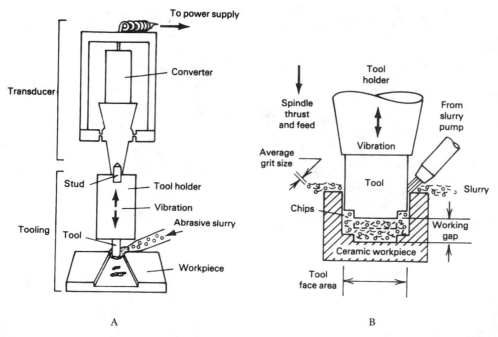

A                                          B

**FIGURE 7.24**  Schematic showing key components of a typical ultrasonic machining installation. (A) Transducer assembly coupled to tooling assembly. (B) Close-up view of tooling assembly used to machine ceramics. (From Moreland, M.A., in *Ultrasonic Machining, Engineered Materials Handbook,* Vol. 4, *Ceramics and Glasses* (ASM Int., 1992). With permission.)

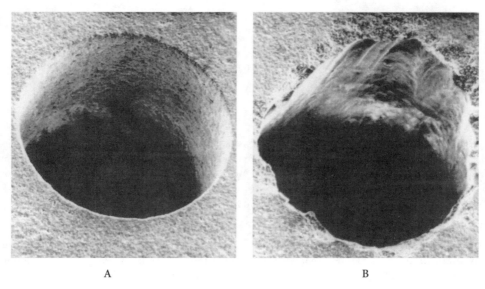

A                                          B

**FIGURE 7.25**  Scanning electron microscopy photomicrographs of 640-μm holes drilled into alumina with two different techniques. (A) Ultrasonic machining. (B) Laser-beam machining. (From Moreland, M.A., in *Ultrasonic Machining, Engineered Materials Handbook,* Vol. 4, *Ceramics and Glasses* (ASM Int., 1992). With permission.)

diameter hole, the aspect ratio is about 2.5; with increasing hole size the aspect ratio increases. It is safe to assume that a ±10-μm tolerance can be achieved with the finest abrasive powders.

Rotary ultrasonic machining (RUM) enhances material removal rates, finish capabilities, and overall drilling efficiency. A 1/4-in. hole can be drilled 5 in. deep into glass in about 2 min with a rotating ultrasonic tool. To make such high aspect ratio holes and deep cavities it is often required to remove the workpiece periodically to ensure uninterrupted abrasive slurry flow to the cutting zone. Conventional USM is not capable of drilling

microholes much smaller than 100 μm for lack of corresponding co-axial microtools. Using electrodischarge wire cutting (EDWC) to make co-axial microtools (similar to the pipe structures shown in Figure 7.14), Sun et al.[84] made holes as small as 15 μm in diameter. Surface roughness was typically about 0.2 μm and ultrasonic drilling speed ranged from 2.0 μm/min to 6.0 μm/min. Computer-controlled standard USM equipment is available with a position resolution on all axes of 0.25 μm.

Summarizing with ultrasonic machining operating costs are reasonably low and required operator skill level is modest.

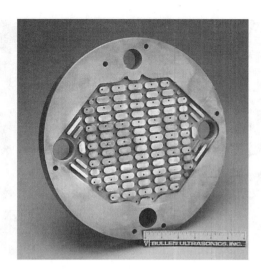

FIGURE 7.26    Channels and holes ultrasonically machined in a poly-crystalline silicon wafer. (From Moreland, M.A., in *Ultrasonic Machining, Engineered Materials Handbook*, Vol. 4, *Ceramics and Glasses* (ASM Int., 1992). With permission.)

Machines cost up to $20,000 and production rates of about 2500 parts per machine per day are typical. If the machined part is a complex element (say, a fluidic element) of a size > 1 cm² and the best material to be used is an inert, hard ceramic, this machining method might well be the most appropriate.

## Water Jet, Abrasive Water Jet, and Abrasive Jet Machining

In water jet machining, material is removed through mechanical impact, i.e., erosion effects of a high-velocity (600 to 900 m/sec with pressures up to 400 MPa), small-diameter (in the range of 0.07 to 0.5 mm) jet of water.[85] The technique is mainly used to cut soft, nonmetallic materials such as Plexiglas, cardboard, foam rubber, wood, etc. With a water jet nozzle of 0.13 mm and a pressure of 379 MPa, a 3-mm thick Plexiglas plate can be cut at speeds of 0.9 m/min. Tolerances are a function of the material type and thickness, and they are usually within ±0.1 to 0.2 mm.[86]

By adding abrasives to the water jet, a much wider range of materials, including metals and ceramics, can be machined. The method is principally used for edge finishing, deburring, and polishing.

In abrasive jet machining, also called microabrasive blasting, the workpiece material (hard, brittle materials such as glass, silicon, tungsten, and ceramics) is removed by mechanical impact of a high-velocity (300 m/sec at 0.7 MPa) stream of abrasive-laden inert gas.[87] Jet diameters are usually 0.12 to 1.25 mm. The material removal rate is low — usually about 0.015 cm³/min. Tolerances of ±0.12 mm are readily maintained, together with surface finishes ranging from 0.25 to 1.25 μm. Typical applications are the etching of shallow intricate holes in electronic components such as resistor paths in insulators and patterns in semiconductors.[5] This often involves components that cannot tolerate normal batch-processing and the operator directs the blasting stream at the part.

## Focused Ion-Beam Milling

Some authors classify focused ion-beam (FIB) milling as a thermomechanical technique. It is more widely accepted now as a pure mechanical machining technique in which the drill bit is replaced with a stream of energetic ions. This technique uses a very bright liquid metal ion (LMI) source (e.g., gallium) rather than argon and focuses the ion beam down to a submicron diameter. The liquid metal source employs a capillary feed (or needle) system and an acceleration voltage of 5 to 50 keV. Beam spots of 50 nm are possible (see Chapter 1). FIB has been used for maskless implantation, metal patterning, IC repair, etc. (see Table 1.6 in Chapter 1). More recently, it has been shown that FIB also provides for a production technique of micron-sized objects in an arbitrary material without the requirements of anisotropic etching and lithography. Vasile et al.,[88,89] for example, describe a 360° rotation mounting device used analogously to a lathe, with a 20-keV Ga ion beam as the cutting tool for in-vacuum micromachining. The ion beam they use is 0.3 μm in diameter, and they crafted tungsten needles, hooks, tuning forks, and specialized scanning probe tips. Some of these tungsten microstructures are shown in Figure 7.27. The first microstructure (Figure 7.27A and B) is a parasol-shaped scanning probe which can be used to probe underneath resist lips and other concave features. The disc structure is sitting on a 1.5-μm square shank. The disc is 4 μm in diameter and about 0.7 μm thick. The shank segment just prior to the disc is deliberately thinned to 0.4 μm at the thinnest point. The microshovel shown in Figure 7.27C is another illustration of the truly three-dimensional aspect of this new manufacturing technique. The resolution limits to the fabrication process in these various tungsten structures mainly come from the grain structure of the material. Start-to-finish fabrication time for the objects shown was ~2 hr. It should be noted that this expensive, serial technique might make sense for the fabrication of an atomic force microscope tool set but to make a microshovel as shown in Figure 7.27C is merely a fancy demonstration of the technique.

A variation on FIB machining is fast atom-beam machining (FAB). In FAB, ionized atoms, accelerated to about 100 km/sec are neutralized and used as the mechanical cutting tool. The etching rate of GaAs with chlorine atoms at 60°C is 0.15 μm/min and vertical walls are obtained. The workpiece is mounted on a lathe as in FIB, and by rotating it with respect to a Ni mask the different faces of the GaAs workpiece are etched and three-dimensional machines are created. Hatamura et al.[90] made a micro-Japanese temple (Gojunoto) of 280 by 100 by 100 μm this way. A hole etched in a polyimide sheet enabled the erection of the temple with the help of the rotational robot holding the temple.

## Other Ion Machining Techniques

In ion-beam sputtering a broad surface area is bombarded with energetic ions from inert gases such as argon, and atoms from the target surface are removed by momentum transfer. The technique was discussed in detail in Chapter 2. In reactive ion-beam etching (RIBE), also reviewed in Chapter 2, a flux of reactive particles, such as $CF_4^+$ and $F^+$, is directed at the workpiece in a narrow beam. In the latter process,

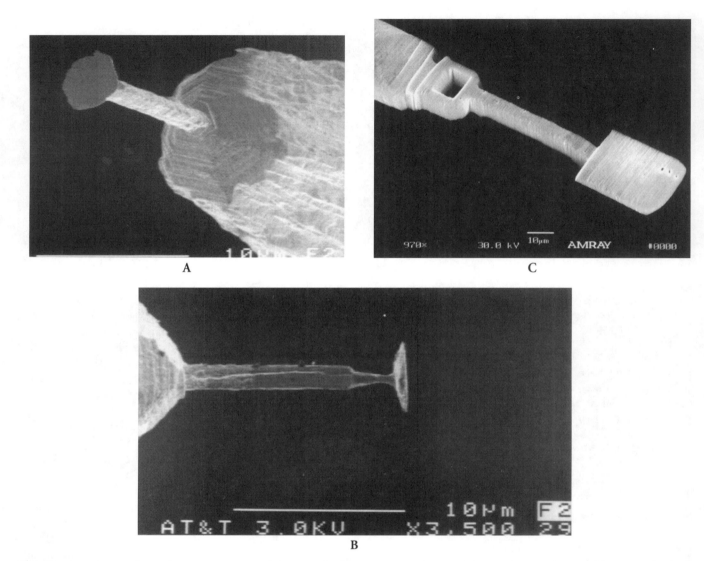

**FIGURE 7.27**   FIB manufactured tungsten structures. (A) A specialized scanning tip probe parasol. (B) A side view of a parasol structure. The shank is 1.5 μm square and thinned down locally to 0.4 μm. The disk is 4 μm in diameter. (C) Microshovel. (From Vasile, M.J. et al., *J. Vac. Sci. Technol.*, B12, 2388–2393, 1994. With permission.) (Courtesy of Dr. M. J. Vasile.)

momentum transfer and chemical reactions are involved in the etching process and the process cannot be solely characterized as mechanical. In reactive ion etching (RIE), the etching is almost exclusively by chemical reaction with the reactive species formed in the plasma and must be seen as a dry chemical etching technique rather than a mechanical technique. The latter techniques are almost exclusively used in semiconductor settings. It is possible with these ion machining techniques, in combination with a mask, to etch almost any material in a very controlled fashion (dimensions < 100 nm).

# Comparison of Micromachining Techniques

## Design from the Package Inwards

The choice of processes for manufacturing a three-dimensional microdevice can best be made after studying the detailed requirements of the application. Since the package serves as the interface between the microstructure and the macroworld and since it is the major contributor to cost and size, one should start the design with the package and work toward the best machining process for the micromachine inside. The package is often made with a traditional machining method such as turning and grinding or possibly with ultrasonic or wire electro-discharge machining.

For the micromachine inside the package there may be many machining options, as both batch IC and traditional serial production techniques must be considered. Several new, batch Si micromachining techniques derived from the IC industry are gaining acceptance, but serial, precision machining techniques with specialized tools (e.g., ultrasonic machining or laser-beam machining) are still often the only commercially available method to make intricate small three-dimensional microcomponents.

## Batch vs. Serial

Most nontraditional micromachining carried out today, like IC fabrication, relies on batch fabrication, i.e., repetitive features are simultaneously photographically defined on a silicon wafer, and many wafers are then processed further to fabricate desired structures. Other micromachining techniques, such as laser drilling or chemical-jet etching are serial techniques. Batch fabrication techniques lend themselves to economies of scale which are unavailable with serial techniques. Although micromachining is often compared to microelectronics, it lacks the high degree of generality characteristic of microelectronics. ICs can be grouped in a limited number of classes, within which design and production follow well-defined and common steps. As a consequence the price performance ratio allows industry to make profits on complex ICs and it makes ICs suited for mass production. Sensors and actuators, on the other hand, are very specific. They have to be in contact with their surroundings, and each environment imposes its own constraints. This makes the packaging more important, further illustrating our earlier claim that the design should start from the packaging constraints and work its way to the miniature sensor inside rather than the other way around. For micromachines, the economics dictating profitable IC manufacture luckily do not necessarily apply; a microinstrument — say, a micro-gas chromatograph — may cost considerably more than a typical IC. Moreover, sensors and actuators often use quite esoteric non-IC type materials, are produced in relatively low volumes, and are application specific.

## Comparison Table

Micromachining is emerging as a set of new manufacturing tools to solve specific industrial problems rather than as a monolithic new industry with generic solutions for every manufacturing problem. In Table 7.7 some of the more popular machining techniques to craft micromachines are compared. From this table, Si micromachining emerges as only one of the many options for precision machining. The renewed interest in some of the non-Si based machining methods often stems from two major deficiencies of IC-based machining techniques: i.e., the difficulty to create truly three-dimensional microstructures and the problem of interfacing microstructures with the macroworld. A machining method with a range that covers macroscopic to microscopic scale devices is needed, as often micromachines cannot easily be accessed. Laser beam lithography, electrodischarge machining, ultraprecision machining, LIGA, and pseudo-LIGA techniques are steps in that direction, forming so-called 'hand-shake' technologies, bridging micro- to macroworld. These hand-shake technologies should be of importance beyond micromachines — for example, in packaging of ICs where one is faced with the same dilemma of packaging, mounting, and wiring LSIs and VLSIs. Several of these new technologies are maskless and enable very fast prototyping.

Silicon-based micromachining techniques, often resulting in stunningly beautiful SEM micrographs, are the most visual and 'sexy' micromachining tools used today (the MTV, or music television, of micromachining), but in the trend toward miniaturization, almost a law of nature ever since the oil crises and the IC revolution, many other precision machining techniques on many different materials have found much larger commercial applications. Wet bulk Si micromachining, the most traditional form of Si micromachining and dominated by electrical engineers, is gaining acceptance in traditional precision machining circles, dominated by mechanical engineers, and vice versa; Si micromachinists are starting to look at precision engineering techniques as a way to extend their tool kit. The most important characteristic for both disciplines is the ever increasing quest for improved accuracies, higher aspect ratios, and reduced cost, and practitioners are starting to recognize the complementary nature of their respective fields.

Whereas mechanical sensor manufacture is moving towards more integration of sensing function with electronics, embodied in increased reliance on surface micromachining, chemical and biosensors are moving away from integration toward hybrid technology. This trend in chemical and biological sensors is caused by the need for modularity and the tremendous compatibility problems involved when attempting to integrate chemical sensor materials with ICs. In the case of instrumentation, there is a trend toward using laser machining and some early exploration of LIGA use and very recently a rediscovery of photoformed glass.

Often micromachinists are not aware of the capabilities of traditional machining and use Si micromachining for parts that could have been made better with conventional technology. By applying the right tool to the machining job at hand we hope that micromachining will lead to many more successful commercial applications than we have seen so far. A better insight of the characteristics of the various machining options, compared in Table 7.7 should be most useful to that end.

## From Perception to Realization

Often Si micromachining seems to be applied to prove to the world that group 'X' also has a cleanroom or can make a yet smaller micromotor. In order for micromachining to lose the stigma of a technology looking for an application which is used in research backrooms and excess cleanrooms only, it will be important to describe most new results in academia on micromachinery as fine-tuning micromachining skills rather than to proclaim each new result as the latest breakthrough in sensor technology or analytical chemistry.

Micromachining in industry must be seen as a set of tools to solve practical problems in subminiaturized three-dimensional structures. Once a thorough understanding of the micromachining application is reached, it will be easier to decide upon the correct manufacturing technique. The micromachining tool used in the early prototyping phase does not need to be the same as the eventual manufacturing technique decided upon. In many cases a method enabling faster prototyping is preferred.

Micromachining is thus very engineering oriented and application specific, and therefore it can best be carried out in

**TABLE 7.7**  Precision Engineering Machining Tools

| Machining Method | Material/Application | Typical Min./Max. Size Feature | IC Compatible | Tolerance | Aspect Ratio (Depth/Width) | Important Reference on Technique | Initial Investment Cost/Access |
|---|---|---|---|---|---|---|---|
| **Group: Traditional Techniques** (Not Involving Photolithography-Defined Masks) | | | | | | | |
| Chemical milling (S), (Ba) | Almost all metals | From submillimeters to a few meters (x,y); max thickness (z) ± 1 cm | Yes | Lateral tolerance 0.25 to 0.5 mm | ±1 | Harris[15] | Low/good |
| Electrochemical machining (S/A), (Ba) | Hard and soft metals, turbine blades, pistons, fuel-injection nozzles | Minimum size devices larger than in chemical milling because of the contacting need | Fair | Lateral tolerance < 10 μm | (S) 100 | (A) Romankiw,[37] (S) Phillips[47] | ±$400,000/good |
| Electrodischarge machining (EDM) (S), (Se) | Hard, brittle, conductive materials used for tools and dies | Minimum holes of 0.3 mm in 20-mm thick plate | No | Lateral tolerance, 5–20 μm | 100 | Kalpajian[51] | High/good: equipment with numerical control is common |
| Electrodischarge wire cutting (EDWC) (S), (Se) | Hard, brittle materials; many punch-and-die applications | Minimum rods 20 μm in diameter and 3 mm long | No | Lateral tolerance, 1 μm | >100 | Saito[53] | High/good: does not require special electrodes |
| Electron-beam machining (EBM) (S/A), (Se) | Hard-to-machine materials | (S) most suited for large numbers of simple holes (<0.1 mm) | Fair | (S) ~ 10% of feature size (5 μm on a 50-μm hole) | (S) 10 is typical but 100 is possible | Taniguchi[56] | Very high/fair ±100,000 when using modified SEM |
| Continuous deposition (A), (C); e.g., doctor's blade technology | With all materials available in inks, e.g., glucose sensors | Most suited for inexpensive disposables, from 100 μm to a few millimeters | No | 15 μm | — | Harper[91] | Low/good |
| Focused ion-beam on a lathe (S/A), (Sc) | Very pure IC materials | From submicrons to millimeters | Yes | (S) 50–100 nm | — | Vasile[88] | High/poor |
| Hybrid thick film (A), (Ba) | Wide variety of materials available in inks | Minimum feature size 90 μm | Fair | 12 μm | — | Harper[91] | ±$30,000/good |
| Laser-beam machining (LBM) (S/A), (Se) | Complex profiles in hard materials | (S) Holes from 10 μm to 1.5 mm at all angles | Fair | 1 μm | (S) 50 | Helvajian[63] | ±$50,000 but up to $400,000 for a five-axis system/good |
| Plasma-beam machining (PBM) (S/A), (Se/Ba) | Very high temperature materials | (A) only used for thick films > 25 μm; (S) for very thick films > 2.5 mm | No | (A) 20 μm for a 25-μm thick film; (S) typical ±3 mm but 0.8 mm is possible | — | Pfender[92] | ±$600,000/fair |
| Stereo lithography (A), (Se) | Polymeric photosensitive materials | Max. 10 × 10 × 10 mm (x,y,z) | Yes | Minimum solidification unit 5,5,3 μm (x,y,z) | — | Ikuta[93] | Low/good |
| Ultra-precision mechanical machining (S), (Se) | Form-stable materials | From submillimeters (e.g., 0.2-mm hole) to meters | No | 1 nm by the year 2000 | — | Boothroyd[4] | $400 k/good |
| Ultrasonic machining (S), (Se) | Hard and brittle materials | Holes from 50 μm to 75 mm | No | Lateral tolerance 10 μm | 2.5 μm for a 250-μm hole | Bellows[82] | $20 k/good |

**Group: Nontraditional (Involving Photolithography-Defined Masks)**

| Machining Method[a] | Material Application | Typical Min./Max. Size Feature | IC Comparable | Tolerances | Important Reference on Technique | Aspect Ratio (Depth/Width) | Shape and Height/Depth | Initial Investment Cost/Access |
|---|---|---|---|---|---|---|---|---|
| Photofabrication (S) | Plastic, glass (ceramic), e.g., fluidic elements | Max. x,y = 40 × 40 cm and max z = 0.6 cm | Yes | Lateral tolerance 20 μm | Trotter[24] | ~3 for photoplastics; ~20 for photoglass | x,y is free; z = up to 6 mm | Medium/poor to medium |
| Photochemical milling (S) | Printed circuit boards, lead frames, shadow masks | Max. 60 × 60 cm, max. thickness < 0.5 mm | Yes | 13 μm (printed circuits) | Allen[21] | ±1 | x,y is free; z = up to 0.5 mm | Medium/good |
| Wet etching of anisotropic materials (S) | Crystal Si, GaAs, quartz, SiC, InP | Max. wafer size, min. feature a few microns | Fair | 1 μm | Kern[94] | 100 | x,y,z shape locked in by crystallography, z height of the wafer | Low/good |
| Dry etching (S) | Most solids | Max. wafer size, min. feature submicron | Good | 0.1 μm | Manos[93] | 10 | x,y shape free; z = up to 200 μm | High/good |
| Polysilicon surface micromachining (S/A) | Poly-Si, Al, Ti, etc. | Max. wafer size, min. feature submicron | Good | 0.5 μm | Howe[96] | — | x,y free; z = 0.1 to 10 μm, but preferably 1–2 μm | High/fair |
| SOI (S) | Crystalline Si | Max. wafer size, min. feature submicron | Good | 0.1 μm | Diem et al.[97] | — | x,y free; z height depending on the epilayer, e.g., 100 μm | High/poor |
| LIGA (S/A) | Ni, PMMA, Au, ceramic, etc. | 10 × 10 cm or more, 0.2 μm | Fair | 0.3 μm | Ehrfeld[98] | >100 | x,y free; z up to several cm | High (>M$35)/ poor |
| UV transparent resists (S/A) | Polyimide, SU-8, AZ-4000 | Max. wafer size | Good | 0.5 μm | Ahn et al.[31] | 10 | x,y free, z up to 100 μm | Medium/fair |
| Molded polysilicon HEXSIL (Keller) (S/A) | Poly-Si, Ni, etc. | | Good | 0.5 μm | Keller[99] | 10 | x,y free; z up to 100 μm | High/poor |
| Erect polysilicon (Pister) (S/A) | Poly Si | | Good | 0.5 μm | Pister[97] | 10 | x,y free; z up to mm | High/poor |

*Note:* S = subtractive, A = additive, Ba = batch, Se = serial, C = continuous.

a All batches.

industry or in very applied research and development groups. Indeed, all recent examples of triumphs of precision engineering were realized in industry, and the Si micromachining community often failed to appreciate accomplishments of the more traditional machining methods. The HP Kittyhawk Personal Storage Module (PSM), a 1.3-in. disk drive and the Schwarzschild CD pickup, with a distance from bottom of pickup to disc of 6.5 mm, shown in Figure 7.28, are triumphs of precision engineering, but neither involve Si. Romankiw's thin-film head, shown in Figure 7.8, is another example and probably constitutes the micromachine with the highest sales today (3 billion was predicted for 1996). The manufacture of the thin-film head is principally based on electroplating and does not involve any Si either.

Because there are no good engineering practices established on deciding upon an ideal manufacturing technique for a given microapplication, manufacturing techniques often seem to be rediscovered. The latter is illustrated by the evolution of manufacturing methods which have been attempted for the construction of microfluidic elements as illustrated in Figure 7.29. Microfluidic elements in the 1960s were principally made from photoglass. After attempting to craft the same devices by dry etching in Si in the 1980s, LIGA was employed in the early 1990s but by the mid-1990s photoglass was rediscovered as the best machining option for this application.

From the academic point of view, micromachining is of extreme importance as it provides an excellent opportunity to create totally new devices often operating in a regime where continuum theory breaks down, with the potential of discovering important new phenomena and developing new instruments of the same fundamental significance as the scanning tunneling microscope. Unfortunately, funding pressures are forcing scientists to sell research as development and development as manufacturing. By putting their work in the context of a practical solution too early, academics might have lost some of industry's trust in micromachining. A breakthrough in deep, cryogenic etching or yet thicker vertical photoresist walls should stand on itself and does not necessarily need to be illustrated with the fabrication of wheels for micromotors! A better practice in the future might be to compare these new academic results (say, on photoresists) with other, more traditional approaches so that the work becomes of generic value to a potential user for his specific application.

# Micromachining: What's Next? Nanomachinery by Molecular Engineering and Nanofabrication

## Introduction

Moore's law projecting the progress in achievable transistor density on a chip and Taniguchi's curves predicting accuracy improvements in mechanical machines are both showing a slowdown as we approach the year 2000 (see Figure 7.2). Around 2000 the accuracy of precision machining will result in precision

A

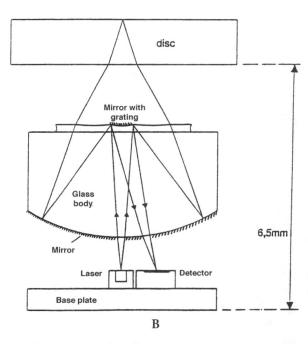

B

**FIGURE 7.28** Triumphs of mechatronics: (A) HP's Kittyhawk. (B) Schwarzschild CD pickup.

devices with a manufacturing accuracy of 1 nm, and lithographically defined ICs will approach critical dimensions of 0.1 μm. For electronic devices at those dimensions, bulk matter reveals itself as a crowd of individual atoms, and integrated circuits barely function as the continuum theory of electrical conduction breaks down. Further miniaturization of nanostructures will be based on methods originating in molecular engineering and nanofabrication of quantum structures.

Nanostructures are assemblies of bonded atoms that have dimensions in the range of 1 to 100 nm.[101] They can be built up from their atomic constituents through additive molecular techniques (bottom up) or by building down from larger structures

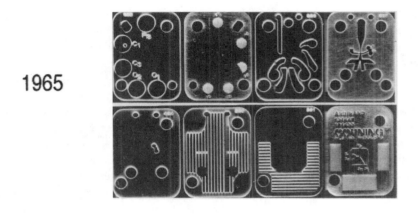

1965

Fluid Amplifier Associates
Photoformed Glass

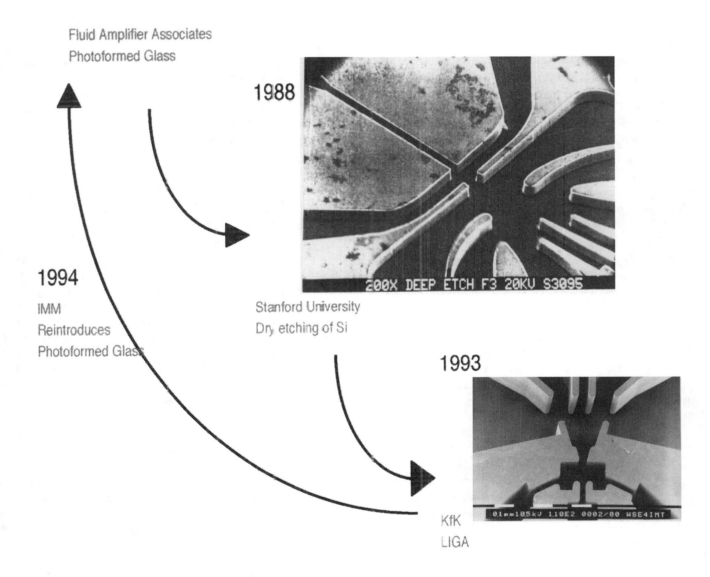

1988

1994

IMM
Reintroduces
Photoformed Glass

Stanford University
Dry etching of Si

1993

KfK
LIGA

FIGURE 7.29   The rediscovery of photoformed glass for the manufacture of microfluidics.

by subtractive techniques in nanofabrication (top down). We will look first at molecular engineering and then into nanofabrication as it applies to making quantum structures.

Nanostructures, in molecular engineering, are assembled from their molecular or atomic constituents and enable a plethora of interesting new machining opportunities. Molecular engineering actually holds the promise of versatility of design offered by nature itself. The latter is a consequence of the building blocks being similar in size and/or nature to those employed by nature. Molecular self-assembly/replication and biomemetic systems are all contributing to the appeal of this type of nanotechnology.[102]

Nanofabrication is principally based on subtractive processes, akin to traditional micromachining techniques, and this top-down approach constitutes an alternative path to nanotechnology. Whereas using molecular engineering to make nanoscale devices is still in the research phase, nanofabrication has already led to commercially available quantum devices.

## Molecular Engineering

### Nature as a Guide

In dealing with miniaturization, nature is a good teacher in the design of efficient microsystems. Nature makes the best of scaling laws and exploits the quantum-size effects of its components. In biological systems the energy efficiency is approximately proportional to 2/3 power of the linear dimension. This follows because metabolism is proportional to the second power of the linear dimension (surface of the organism, $1^2$), and energy uptake (feeding) is proportional to the body volume ($1^3$), so the smaller organism has the higher efficiency (see also scaling laws in Chapter 9). Microscopic sizes are consequently the normal working size for the biological world — a biological cell of about 5 to 10 μm being one of the larger components. Smaller components of biological cells exhibit fascinating feats of nanoengineering. For example, molecular motors 25 nm wide are powering the flagellum of many bacteria. These motors appear to be powered by a flow of protons across a pH gradient, spin at 18,000 revolutions per minute, and can push an average-sized cell at 30,000 nm/sec or 15 body-lengths per second — and it is reversible, too![103] In comparison, the surface micromachined motors of Chapter 5 are hopelessly impractical. Even more notable is the protein manufacture by DNA in the cell's ribosomes as illustrated in Figure 7.30. Biomolecules can process electrical and/or optical signals but what proteins are really good at is recognizing and reacting to three-dimensional shapes as exemplified by transport RNA (tRNA) molecules herding only the proper amino acids towards the ribosomes to build a new peptide chain (see Figure 7.30) or, for example, by ferritin, a liver protein that forms 8-nm wide cages with strong affinity for iron oxide, keeping this toxic compound safely caged.

Every living body is thus a unique molecular machine in which function at the nanoscale dictates the cellular and whole organ behavior on the macroscale.[104] In micromachining and traditional machining one builds down from big chunks of material toward smaller and smaller structures, whereas in molecular engineering one builds from the atomic and molecular level up towards bigger and bigger molecular entities. Si micromachining involves Si wafer slabs as thick as 500 μm, insulating layers up to a micron thick, Al and Au metal layers between a few 100 Å and thousands of angstroms thick, and in general a very limited choice of other materials compared to nature's arsenal. However, by using relatively large building blocks micromachining is fast, while one might expect nature, using very small building blocks, i.e., atoms with a diameter of 0.0003 μm, to be very slow. To offset the time it takes to work with those very small, basic building blocks, nature, in growing an organism, relies on an additive process featuring massive parallelism. In molecular engineering one mimics those natural processes in the hope of getting the same diversity in shapes, materials, functions, potential activities, and memory size offered by nature itself.

There is obviously something very attractive about the small size of biological cells and the smaller building blocks mentioned above. Biosystems have been tested and selected by eons of evolution and at a time where we find that our subtractive machining tools are limiting the fabrication of yet smaller devices, exploration of some of nature's tools might indeed be rewarding.

### Some Molecular Engineering Approaches

*Self-Assembly, Self-Organization, and Self-Replication*

Most biological events and many chemical changes are presided over by noncovalent bonding interactions in and between and beyond molecules. Chemistry beyond the molecule has given rise to supramolecular chemistry — a new field of chemistry.[105,106] The fundamental concepts underpinning supramolecular science are processes in which atoms, molecules, aggregates of molecules, and components arrange themselves into ordered, functioning entities without human intervention.[107] Through supramolecular chemistry with weak ionic and hydrogen bonds and van der Waals interactions (0.1 to 5 kcal/mol), structures with dimensions of 1 to 100 nm are possible. For chemical synthesis the largest of these dimensions is often too large and for microfabrication it is often too small. The different fundamental processes at work in supramolecular chemistry all can be illustrated by examples from nature.

Self-assembly is the lock-and-key assembly of complex molecules such as in the case of neurotransmitters assembling with molecular recognition sites on proteins that elicit chemical responses, e.g., acetylcholine recognition at nerve-muscle junctions resulting in muscle contractions.[101]

Self-organization leads to large supramolecular architectures such as molecular crystals, liquid crystals, colloids, micelles, spherical bilayer structures of cell walls, phase-separated polymers, and self-assembled monolayers. No chemist needs to specify the individual atom or molecule positions in these systems.[108] Colloid chemistry, for example, is able to precipitate inorganic solids such as silver halide crystals with astonishing

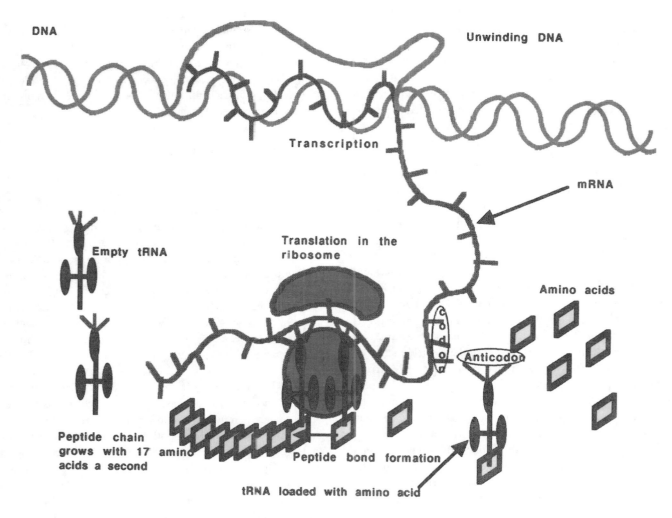

**DNA**

**Unwinding DNA**

**Transcription**

**mRNA**

**Empty tRNA**

**Translation in the ribosome**

**Amino acids**

**Anticodon**

**Peptide chain grows with 17 amino acids a second**

**Peptide bond formation**

**tRNA loaded with amino acid**

**FIGURE 7.30** Transport RNA (tRNA) herding proteins to the ribosome protein factory and STM herding atoms on a surface.

regularity in size and properties, comparable to what can be accomplished with microlithography, and it has become one of several routes to building quantum structures (see below).

In self-replication, molecules such as DNA encode the genetic information for cells and the resulting organisms, thus allowing the information to be passed on from cell to cell and from generation to generation.[109]

Some examples of nanoconstructs follow. Buckminster-fullerene and its analogs ($C_{60}$ and $C_{70}$) are nanostructures made by self-organization. These carbon buckyballs are named for their resemblance to the geodesic domes of architect Buckminster Fuller (Figure 7.31A).* Ever since fullerenes were discovered by Smalley of Rice University in 1985, expectations have been high that these hollow, pure carbon nanostructures could be used as drug delivery agents, environmental tracers, superconductors, etc. The most tantalizing applications are those involving buckyballs filled with a specific atom or a molecule, analogous to the ferretin liver protein caging an iron oxide molecule. Until recently, due to the extreme stability of these molecules, it has been very difficult to open and close buckyballs. But, by rendering one of the 30 double bonds making up

a $C_{60}$ cage more reactive, Wudl at University of California, Santa Barabara, has succeeded in opening and closing the cage, contributing to a widening class of container molecules tailored to fit a variety of target molecules.[110]

Self-organization has also produced tiny graphite tubes (nanotubes) that are among the smallest electrical wires ever made.[107] These tubes are called buckytubes, because they are structurally similar to the carbon buckyballs. Buckytubes consist of several nested, concentric cylinders with nanometer-scale diameters (say 1 to 6 nm). Like fullerenes, they may be formed in macroscopic quantities in the so-called Huffman-Kratschmer carbon arc.[111] In Figure 7.31B we show an example of a very short buckytube. Nanotubes are far from the only self-organized nanoproducts formed in a carbon arc fullerene generator. At the extremely high temperatures in the arc plasma, a wide range of carbon materials are formed. This may include amorphous carbon, fullerenes, nanotubes, endohedrals, nanocrystallites, and graphite crystals. When using metals and metal compounds as anodes in the plasma arc, encapsulation of metals (transition metals such as iron, cobalt, and nickel) and metal compounds (rare earth carbides) was observed.[112] These coated nanoparticles are of tremendous potential importance to study encapsulated materials of small dimensions in a closed-off

---

\*    Figure 7.31 appears as a color plate after page 144.

environment, for example for protection against oxidation, nanowires, magnetic data storage, magnetic toner xerography, magnetic inks, ferrofluids, etc.

For a good introductory reading on fullerenes we suggest *The Chemistry of the Fullerenes*, by Hirsch.[113] For a good set of recent papers on nanotubes, refer to: Fullerenes. Recent Advances in the Chemistry and Physics of Fullerenes and Related Materials, 1994, JECS, San Francisco Proceedings.

### *Mechanosynthesis*

As an example of an additive molecular technique, Drexel[114] discusses mechanosynthesis. In this additive technique two reactive molecules are held in contact with each other in a controlled orientation for a specific amount of time with a controlled amount of force.

One concrete way of realizing mechanosynthesis is by using STM. Rorher and Binnig submitted their first patent disclosure on this technology in mid-1979 and by 1982 they produced images of Si surfaces, showing individual atoms. Some time later, in 1987 at AT&T Bell Laboratories, it was demonstrated that atoms and molecules may be manipulated in a variety of ways by using the interactions present in the tunnel junction of such an STM.[115] This approach enables a variety of different atomic manipulation processes, classified as parallel and perpendicular processes. In parallel processes, adsorbate atoms or molecules are moved in parallel to the surface, and in perpendicular motion atoms or molecules are transferred to the tip and redeposited elsewhere on the surface. By using such processes in 1990, the IBM logo was written with the tip of an STM by herding supercooled xenon atoms into the right place on a Ni substrate. Less than a year later, at Hitachi Central Research Laboratory in Tokyo, "Peace '91" was carved into a sulfur medium — at room temperature! Soon thereafter, with further improvements, the time it took for such atomic writing was reduced from hours to seconds.[116]

The potential is mind boggling: in theory, a 1-cm$^2$ surface containing 1 quadrillion atoms could store in this way all recorded human knowledge.[116] There is one major caveat: massive parallelism is the only way to manufacture even microscale three-dimensional objects in a reasonable amount of time using such mechanosynthesis; for example, for a $10 \times 10$ μm-sized Ag pad, depositing one Ag atom a second, it would take about 9 months to finish the job! In Drexel's approach, armies of assemblers are instructed to build such micro three-dimensional objects.[117] Of course, this is exactly how nature has solved the problem: with DNA in separate cells as the assemblers, all working in parallel.[118] In Figure 7.30 it was shown how a peptide chain is fashioned amino acid by amino acid in the cell's ribosomes. The writing with STM, by herding atoms into the right place on a substrate, mimics in a very modest fashion the herding by tRNA of the proper amino acids towards the ribosomes to build a new peptide chain. In a further development on this theme, proximal probe tips have been fitted with specific antibodies which will bind exclusively with their three-dimensional complementary protein or antigen, and the scanning probe thus has become a selective transport vehicle for only one type of

molecule and the probe could now be used to move the piece around and assemble it with other pieces.

A good set of introductory articles on the topic of molecular electronics was presented in the February/March 1994 issue of *IEEE, Engineering in Medicine and Biology*, and the journal *Nanobiology* (Carfax Publishing) is a good source for monitoring developments in this field.

## Nanofabrication of Quantum Structures

Nanofabrication, the heir to microfabrication, uses the newest microfabrication tools such as molecular beam epitaxy and high resolution electron-beam lithography to write features in the range below 100 Å. By confining electrons to multi-atom dimensions, quantum effect devices can make faster, less power-consuming electronics. The technology used to fashion these quantum devices is often referred to as nanofabrication and should not be confused with molecular engineering. If nanoelectronics below 0.1 μm succeeds, ICs with $10^{12}$ devices on a chip will become possible.[119] Both molecular engineering and nanomachining are considered nanotechnology and for future developments we speculate on the merging of micromachining and nanomachining with molecular engineering.

To go beyond 0.1 μm in minimum feature size in an electronic device and to still ensure low-leakage transistors, the operating temperature can be lowered or the transistor operating voltage can be lowered, both rather unattractive propositions. One also might more realistically consider quantum devices. The basic principle behind quantum wells, wires, and dots is the same: confine electrons in a restricted region of semiconductor by hemming it in with another semiconductor material that has a higher bandgap, a measure of the amount of energy that has to be pumped into the material to get electrons flowing from region to region. Like water, the electrons will flow to the lowest potential energy quantum well. The lowest energy well might be formed by a region of a 100- to 200-Å thick slice of GaAs, built up by vapor deposition on a base of a higher bandgap material such as aluminum gallium arsenide. A second layer of aluminum gallium arsenide on top of the thin slice of GaAs closes off the top. Confined in the slice, the electrons have so little headroom that their energy states are forced to cluster around specific peaks. In these stacked layers the electrons show fundamentally new electrical and optical properties. Quantum wires confine the electrons on four sides rather than two, squeezing them into a linear channel and thus achieving an even sharper clustering of energy states. A quantum structure should emit and absorb at unusual frequencies, corresponding to the spacing of the quantized energy levels, and it should allow currents to pass only at specific energies. The ultimate would be to make a quantum dot, i.e., electrons caged at a single point rather than a line or a plane. Quantum dots should achieve virtually perfect quantum confinement, forcing all states into the same set of quantized energy states.

Quantum devices require accurate fabrication equipment that can control all dimensions to within tens of nanometers. Quantum structures are structured using sophisticated techniques such as molecular beam epitaxy, resulting in dimensions that are measured in hundreds or tens of atoms. Volume production will be a major challenge but one kind of a quantum structure, the quantum well, has already found its way into the transistors in satellite microwave receivers and the lasers in some communications systems and in run-of-the mill compact disc players.

# References

1. Chryssolouris, G., *Laser Machining*, Springer Verlag, New York, 1991.

2. Slocum, A. H., "Precision Machine Design: Macromachine Design Philosophy and its Applicability to the Design of Micromachines", Proceedings. IEEE Micro Electro Mechanical Systems, (MEMS '92), Travemunde, Germany, 1992, p. 37–42.

3. van Osenbruggen, C., "High-Precision Spark Machining", *Philips Technical Review*, 30, 195–208, 1969.

4. Boothroyd, G. and W. A. Knight, *Fundamentals of Machining and Machine Tools*, Marcel Dekker, Inc., New York, 1989.

5. Snoeys, R., "Non-Conventional Machining Techniques, The State of the Art", Advances in Non-Traditional Machining, Anaheim, California, 1986, p. 1–20.

6. Zuurveen, F., " Precisie-Technologie: Kennis en Kunde, Inzicht en Uitzicht", in *Precisie-Technologie-Jaarboek 1994*, NVFT, Eindhoven, 1994, p. 99–118.

7. Taniguchi, N., "Current Status in, and Future Trends of, Ultraprecision Machining and Ultrafine Materials Processing", *Annals of the CIRP*, 32, S573–582, 1983.

8. Evans, C., *Precision Engineering: An Evolutionary View*, Cranfield Press, Cranfield, Bedford, England, 1989.

9. Moreau, W. M., *Semiconductor Lithography*, Plenum Press, New York, 1988.

10. Smith, H. I. and M. L. Schattenburg, "Why Bother with X-Ray Lithography ?", SPIE, 1671, 282–298, 1992.

11. Hayashi, T., "Micromechanism and Their Characteristics", IEEE International Workshop on Micro Electro Mechanical Systems, MEMS '94, Oiso, Japan, 1994, p. 39–44.

12. Friedrich, C., "Complementary Micromachining Processes", (Notes from Handouts), 1994, Banff Canada.

13. Compton, *Compton's Interactive Encyclopedia (Interactive Multimedia). Computer Data and Program*, Compton's New Media, Carlsbad, 1994.

14. Higuchi, T. and Y. Yamagata, "Micro Machining by Machine Tools", Proceedings. IEEE Micro Electro Mechanical Systems, (MEMS '93), Fort Lauderdale, FL, 1993.

15. Harris, T. W., *Chemical Milling*, Clarendon Press, Oxford, 1976.

16. Shaw, M. C., *Metal Cutting Principles*, Clarendon Press, Oxford, 1984.

17. DeVries, W. R., *Analysis of Material Removal Processes*, Springer-Verlag, New York, 1992.

18. Slocum, A. H., *Precision Machine Design*, Prentice Hall, Englewood Cliffs, New Jersey, 1992.

19. Stevens, G. W. W., *Microphotography; Photography and Photofabrication at Extreme Resolution*, John Wiley & Sons, New York, 1968.

20. Humphrey, E. F. and D. H. Tarumoto, Eds. *Fluidics*, Fluidic Amplifier Associates, Inc., Boston, MA, 1965.

21. Allen, D., M., *The Principles and Practice of Photo-chemical Machining and Photoetching*, Adam Hilger, Bristol and Boston, 1986.

22. Romankiw, L. T., "Pattern Generation in Metal Films Using Wet Chemical Techniques", Etching for Pattern Definition, Washington, DC, 1976, p. 137–139.

23. Glaswerk-Schott, "Duran-Laborglass-Katalog, Nr.50020/1991", 1991.

24. Trotter, D. M., "Photochromic and Photosensitive Glass", Sci. Am., 124, 124–129, 1991.

25. Elwenspoek, M., T. S. J. Lammerink, R. Miyake, and J. H. J. Fluitman, "Micro Liquid Handling Systems", Proceedings of the Symposium, Onder the Loep Genomen, Koninghshof Veldhoven, Netherlands, 1994.

26. Gravesen, P., J. Branebjerg, and O. S. Jensen, "Microfluidics: A Review", *J. Micromech. Microeng.*, 3, 168–182, 1993.

27. Vollmer, J., H. Hein, W. Menz, and F. Walter, "Bistable Fluidic Elements in LIGA Technique for Flow Control in Fluidic Microactuators", *Sensors and Actuators A*, A43, 330–334, 1994.

28. Schomburg, W. K., J. Vollmer, B. Bustgens, J. Fahrenberg, H. Hein, and W. Menz, "Microfluidic Components in LIGA Technique", *J. Micromech. Microeng.*, 4, 186–191, 1994.

29. Ehrfeld, W., "LIGA at IMM", (Notes from Handouts), 1994, Banff, Canada.

30. Engelmann, G., O. Ehrmann, J. Simon, and H. Reichl, "Fabrication of High Depth-to-Width Aspect Ratio Microstructures", Proceedings. IEEE Micro Electro Mechanical Systems, (MEMS '92), Travemunde, Germany, 1992, p. 93–98.

31. Ahn, C. H., Y. J. Kim, and M. G. Allen, "A Planar Variable Reluctance Magnetic Micromotor With a Fully Integrated Stator and Wrapped Coils", Proceedings. IEEE Micro Electro Mechanical Systems, (MEMS '93), Fort Lauderdale, FL, 1993, p. 1–6.

32. LaBianca, N. C., J. D. Gelorme, E. Cooper, E. O'Sullivan, and J. Shaw, "High Aspect Ratio Optical Resist Chemistry for MEMS Applications", JECS 188th Meeting, Chicago, IL, 1995, p. 500–501.

33. Shimokawa, F., A. Furuya, and S. Matsui, "Fast and Extremely Selective Polyimide Etching with a Magnetically Controlled Reactive Ion Etching System", Proceedings. IEEE Micro Electro Mechanical Systems, (MEMS '91), Nara, Japan, 1991, p. 192–196.

34. Hillman, R. S., M. E. Cobb, J. D. Allen, I. Gibbons, V. E. Ostoich, and L. Winfrey, J., *Capillary Flow Device*, Patent No. US Patent 4,963,498, 1990.

35. McGeough, J. A., *Principles of Electrochemical Machining*, John Wiley & Sons, New York, 1974.

36. Romankiw, L. T., "Electrochemical Technology in Electronic Industry and its Future", in *New Materials & New Processes: View from the Top*, JEC Press, Cleveland, 1985, p. 39–51.

37. Romankiw, L. T. and T. A. Palumbo, "Electrodeposition in the Electronic Industry", Proceedings of the Symposium on Electrodeposition Technology, Theory and Practice, San Diego, CA, 1987, p. 13–41.

38. Romankiw, L. T., "Electrochemical Technology in Electronics Today and Its Future: A Review.", *Oberflache-Surface*, 25, 238–247, 1984.

39. Romankiw, L. T., S. Krongelb, E. E. Castellani, J. Powers, A. Pfeiffer, and B. Stoeber, "Additive Electroplating Technique for Fabrication of Magnetic Devices", International Conference on Magnetics-ICM-73, 1973, p. 104–111.

40. Romankiw, L. T., ""Think Small," One Day It May Be Worth A Billion", *Interface*, Summer, 17–57, 1993.

41. Spiller, E., R. Feder, J. Topalian, E. Castellani, L. Romankiw, and M. Heritage, "X-Ray Lithography for Bubble Devices", *Solid State Technol.*, April, 62–68, 1976.

42. Romankiw, L. T., I. M. Croll, and M. Hatzakis, "Batch-Fabricated Thin-Film Magnetic Recording Heads", *IEEE Trans. Magn.*, MAG-6, 597–601, 1970.

43. Romankiw, L. T., "Thin Film Inductive Heads; From One to Thirty One Turns", Proceedings of the Symposium on Magnetic Materials, Processes, and Devices, Hollywood, FL, 1989, p. 39–53.

44. Maciossek, A., "Electrodeposition of 3D Microstructures without Molds", SPIE, Micromachining and Microfabrication Process Technology II, Austin, Texas, USA, 1996, p. 275–279.

45. van der Putten, A. M. T. and J. W. G. Bakker, "Geometrical Effects in the Electroless Metallization of Fine Metal Patterns", *J. Electrochem. Soc.*, 140, 2221–2228, 1993.

46. van der Putten, A. M. T. and J. W. G. de Bakker, "Anisotropic Deposition of Electroless Nickel-Bevel Plating", *J. Electrochem. Soc.*, 140, 2229–2235, 1993.

47. Phillips, R. E., "Electrochemical Grinding-What is it? How Does it Work?", *SME Technical Paper*, MR-85, 383, 1985.

48. Rain, C., "Nontraditional Methods Advance Machining Industry", *High Technology*, November/December, 55–61, 1981.

49. Kamada, H., H. Daiku, and H. Maehata, "A Study on the Machining of Ceramics Using the Electrochemical Discharge Machining Process", *Hitachi Zosen Technical Report*, 1986, 16–22.

50. Esashi, M., Y. Matsumoto, and S. Shoji, "Absolute Pressure Sensors by Air-Tight Electrical Feedthrough Structure", *Sensors and Actuators A*, A23, 1048–1052, 1990.

51. Kalpajian, S., *Manufacturing Processes for Engineering Materials*, Addison-Wesley, Reading, MA, 1984.

52. Dauw, D. F., R. Snoeys, and F. Staelens, "A Real Time Tool Wear Monitoring Function to be Used for EDM Adaptive Control", Advances in Non-Traditional Machining, Anaheim, CA, 1986, p. 21.

53. Saito, N., "Recent Electrical Discharge Machining (EDM) Techniques in Japan", *Bull. Jpn. Soc. Precis. Eng.*, 18, 110–16, 1984.

54. Kuo, C.-L., T. Masuzawa, and M. Fujino, "High-Precision Micronozzle Fabrication", Proceedings. IEEE Micro Electro Mechanical Systems, (MEMS '92), Travemunde, Germany, 1992, p. 116–121.

55. Kuo, C.-L., T. Masuzawa, and M. Fujino, "A Micropipe Fabrication Process", Proceedings. IEEE Micro Electro Mechanical Systems, (MEMS '91), Nara, Japan, 1991, p. 80–85.

56. Taniguchi, N., "Research on, and Develoment of Energy Beam Processing of Materials in Japan", *Bull. Jpn. Soc. Precis. Eng.*, 18, 117–25, 1984.

57. Brunger, W. H. and K. T. Kohlmann, "E-Beam Induced Fabrication of Microstructures", Proceedings. IEEE Micro Electro Mechanical Systems, (MEMS '92), Travemunde, Germany, 1992, p. 168–170.

58. Matsui, S., "Electron Beam Induced Selective Etching and Deposition Technology", *Superlattices Microstruct.*, 7, 295–301, 1990.

59. Matsui, S. a. M., K., "Direct writing onto Si by electron beam stimulated etching", *Appl. Phys. Lett.*, 51, 1498–9, 1987.

60. Bauerle, D., *Chemical Processing with Lasers*, Springer, Berlin, Germany, 1986.

61. Ehrlich, D. J. and J. Y. Tsao, *J. Vac. Sci. Technol.*, B4, 299, 1986.

62. Kunz, R. R. and T. M. Mayer, "Catalytic growth rate enhancement of electron beam deposited iron films", *Appl. Phys. Lett.*, 50, 962–4, 1987.

63. Helvajian, H., "Laser Material Processing: A Multifunctional In-Situ Processing Tool for Microinstrument Development", in *Microengineering Technology for Space Systems*, Helvajian, H., Ed., The Aerospace Corporation, El Segundo, California, 1995.

64. Ehrlich, D. J., D. J. Silversmith, R. W. Mountain, and J. Tsao, "Fabrication of Through-Wafer Via Conductors in Si by Laser Photochemical Processing", *IEEE Trans. Compon. Hybrids Manuf. Technol.*, CHMT-5, 520–521, 1982.

65. von Gutfeld, R. J. and R. T. Hodgson, "Laser Enhanced Etching in KOH", *Appl. Phys. Lett.*, 40, 352–354, 1982.

66. Eden, J. G., "Photochemical Processing of Semi-conductors: New Applications for Visible and Ultraviolet Lasers", *IEEE Circuits Devices Mag.*, 2, 18–24, 1986.

67. Liu, Y. S., "Sources, Optics and Laser Microfabrication Systems for Direct Write and Projection Lithography", in *Laser Microfabrication—Thin Film Processes and Lithography*, Tsao, J. Y. and Ehrlich, D. J., Eds., Academic Press, New York, 1989, p. 3.

68. Maeda, S., K. Minami, and M. Esashi, "KrF Excimer Laser Induced Selective Non-Planar Metallization", IEEE International Workshop on Micro Electro Mechanical Systems, MEMS '94, Oiso, Japan, 1994, p. 75–80.

69. Bowman, L., J. M. Schmitt, and J. D. Meindl, "Electrical Contacts to Implantable Integrated Sensors by $CO_2$ Laser Drilled Vias Through Glass", in *Micromachining and Micropackaging of Transducers*, Fung, C. D., Cheung, P. W., Ko, W. H. and Fleming, D. G., Eds., Elsevier, Amsterdam, Netherlands, 1985, p. 79–84.

70. Anthony, T. R. and H. R. Cline, *J. Appl. Phys.*, 47, 25556, 1976.

71. Bloomstein, T. M. and D. J. Ehrlich, "Laser-Chemical Three-Dimensional Writing for Microelectromechanics and Application to Standard-Cell Microfluidics", *J. Vac. Sci. Technol. B*, 10, 2671–2674, 1992.

72. Shlichta, P. J., "Laser Micromachining in a Reactive Atmosphere", *NASA Tech. Briefs*, 12, 84, 1988.

73. Bloomstein, T. M. and D. J. Ehrlich, "Laser Deposition and Etching of Three-Dimensional Microstructures", 6th International Conference on Solid-State Sensors and Actuators (Transducers '91), San Francisco, CA, 1991, p. 507–511.

74. Boman, M., H. Westberg, S. Johansson, and J.-A. Schweitz, "Helical Microstructures Grown by Laser Assisted Chemical Vapour Deposition", Proceedings. IEEE Micro Electro Mechanical Systems, (MEMS '92), Travemunde, Germany, 1992, p. 162–176.

75. Oh, S. and M. Madou, "Planar-Type, Gas Diffusion-Controlled Oxygen Sensor Fabricated by the Plasma Spray Method", *Sensors and Actuators B*, B14, 581–582, 1992.

76. Suzuki, S., T. Sasayama, M. Miki, M. Ohsuga, S. Tanake, S. Ueno, and N. Ichikawa, "Air-Fuel Ratio Sensor for Rich, Stoichiometric and Lean Ranges", *SAE paper 860408*, 1986.

77. Oh, S., "A Planar-Type Sensor for Detection of Oxidizing and Reducing Gases", *Sensors and Actuators B*, B20, 33–41, 1994.

78. Szepesi, D., "Sensoren en Actuatoren in Ultraprecisie Draaibanken", in *Sensoren en Actuatoren in de Werktuigbouw/Machinebouw*, Centrum voor Micro-Electronica, The Hague, 1993, p. 99–107.

79. Bier, W., G. Linder, D. Seidel, and K. Schubert, "Mechanische Mikrotechnik", *KfK Nachrichten*, 1991, 165–173.

80. Teshigahara, A., M. Watanabe, N. Kawahara. Y. Ohtsuka, and T. Hattori, "Performance of a 7–mm Micro-fabricated Car", *J. Microelectromech. Syst.*, 4, 76–80, 1995.

81. Aoki, I. and T. Takahashi, "Development of Metal-Forming Machine for Fabricating Micromechanical Component", SPIE, Micromachining and Micro-fabrication Process Technology II, Austin, Texas. USA, 1996.

82. Bellows, G. and J. B. Kohls, "Drilling Without Drills", *Am. Mach. Special Report*, 743, 187, 1982.

83. Moreland, M. A., "Ultrasonic Machining", in *Engineered Materials Handbook*, Schneider, S. J., Ed., ASM International, Metals Park, OH, 1992, p. 359–362.

84. Sun, X.-Q., T. Masuzawa, and M. Fujino, "Micro Ultrasonic Machining and Self-Aligned Multilayer Machining/Assembly Technologies for 3D Micro-machines", The Ninth Annual International Workshop on Micro Electro Mechanical Systems, MEMS 96, San Diego, CA, USA, 1996, p. 312–317.

85. Koenig, W. and C. Wulf, "Wasserstrahlschneiden", *Industrie-Anzeiger*, 106, 1984.

86. Benedict, G., *Nontraditional Manufacturing Processes*, Marcel Dekker, Inc., New York, 1987.

87. Dombrowski, T. J., "The How and Why of Abrasive Jet Machining", *Modern Machine Shop*, 76, 1983.

88. Vasile, M. J., C. Biddick, and S. A. Schwalm, "Microfabrication by Ion Milling: the Lathe Technique", Micromechanical Systems, , 1993.

89. Vasile, M. J., C. Biddick, and S. Schwalm, A., "Microfabrication by Ion Milling: The Lathe Technique", *J. Vac. Sci. Technol.*, B12, 2388–2393, 1994.

90. Hatamura, Y., M. Nakao, K. Sato, K. Koyano, K. Ichiki, H. Sangu, M. Hatakeyama, T. Kobata, and K. Nagai, "Construction of 3–D Micro Structure by Multi-Face FAB, Co-Focus Rotational Robot and Various Mechanical Tools", IEEE International Workshop on Micro Electro Mechanical Systems, MEMS '94, Oiso, Japan, 1994, p. 297–302.

91. Harper, C. A., Ed. *Handbook of Thick Film Hybrid Microelectronics*, McGraw-Hill, New York, 1982.

92. Pfender, E., "Fundamental Studies Associated with the Plasma Spray Process", *Surf. Coat. Technol.*, 34, 1–14, 1988.

93. Ikuta, K., K. Hirowatari, and T. Ogata, "Three Dimensional Integrated Fluid Systems (MIFS) Fabricated by Stereo Lithography", IEEE International Workshop on Micro Electro Mechanical Systems, MEMS '94, Oiso, Japan, 1994, p. 1–6.

94. Kern, W., "Chemical Etching of Silicon, Germanium, Gallium Arsenide, and Gallium Phosphide", *RCA Rev.*, 39, 278–308, 1978.

95. Manos, D. M. and D. L. Flamm, Eds. *Plasma Etching An Introduction*, Academic Press, Boston, 1989.

96. Howe, R. T., "Recent Advances in Surface Micromachining", 13th Sensor Symposium. Technical Digest, Tokyo, Japan, 1995, p. 1–8.

97. Diem, B., M. T. Delaye, F. Michel, S. Renard, and G. Delapierre, "SOI(SIMOX) as a Substrate for Surface Micromachining of Single Crystalline Silicon Sensors and Actuators", 7th International Conference on Solid-State Sensors and Actuators (Transducers '93), Yokohama, Japan, 1993, p. 233–6.

98. Ehrfeld, W., "The LIGA Process for Microsystems", Proceedings. Micro System Technologies '90, Berlin, Germany, 1990, p. 521–528.

99.  Keller, C. and M. Ferrari, "Milli-Scale Polysilicon Structures", Technical Digest of the 1994 Solid State Sensor and Actuator Workshop., Hilton Head Island, SC, 1994, p. 132–137.

100. Pister, K. S. J., "Hinged Polysilicon Structures with Integrated CMOS TFT's", Technical Digest of the 1992 Solid State Sensor and Actuator Workshop., Hilton Head Island, SC, 1992, p. 136–9.

101. Whitesides, G. M., J. P. Mathias, and C. T. Seto, "Molecular Self-Asembly and Nanochemistry: A Chemical Strategy for the Synthesis of Nanostructures", *Science*, 254, 1312–1318, 1991.

102. Preece, J. A. and J. F. Stoddart, "Concept Transfer from Biology to Materials", *Nanobiology*, 3, 149–166, 1994.

103. Kudo, S., Y. Magariyama, and S. Aizawa, "Abrupt Changes in Flagellar Rotation Observed by Laser Dark-Field Microscopy", *Nature*, 346, 677–679, 1990.

104. Robinson, D. W., "Medicine, Molecules and Microengineering", *Nanobiology*, 3, 147–148, 1994.

105. Lehn, J. M., "Supramolecular Chemistry-Scope and Perspectives (Nobel Lecture)", *Angew. Chem., Int. Ed. Engl.*, 27, 89–112, 1988.

106. Lehn, J.-M., "Supramolecular Chemistry", *Nature*, 260, 1762–1763, 1993.

107. Whitesides, G. M., "Self-Assembling Materials", *Sci. Am.*, 146–149, 1995.

108. Ringsdorf, H., B. Schlarb, and J. Venzmer, "Molecular Architecture and Function of Polymeric Oriented Systems: Models for the Study of Organization, Surface Recognition, and Dynamics", *Angewandte Chemie, International Edition in English*, 27, 113–158, 1988.

109. Achilles, T. and G. Von Kiedrowski, "A Self-Replicating System from Three Starting Materials", *Angewandte Chemie, International Edition in English*, 32, 1198–1201, 1993.

110. Nemecek, S., "A Tight Fit", *Sci. Am.*, 34–36, 1995.

111. Saito, Y., "Synthesis and Characterization of Carbon Nanocapsules Encaging Metal and Carbide Crystallites", Fullerenes, San Francisco, CA, 1994, p. 1419–1432.

112. Seraphin, S., "Single-Walled Tubes and Encapsulation of Nanocrystals into Carbon Clusters", Fullerenes, San Francisco, CA, 1994, p. 1433–1447.

113. Hirsch, A., *The Chemistry of the Fullerenes*, George Thieme Verlag, Stuttgart, Germany, 1994.

114. Drexel, K. E., *Nanosystems: Molecular Machinery, Manufacturing, and Computation*, John Wiley & Sons, New York, 1992.

115. Stroscio, J. A. and D. M. Eigler, "Atomic and Molecular Manipulation with the Scanning Tunneling Microscope", *Science*, 254, 1319–1326, 1991.

116. Crawford, R., "Japan Starts a Big Push Toward the Small Scale", *Science*, 254, 1304–1305, 1991.

117. Drexel, K. E., *Engines of Creation: The Coming Era of Nanotechnology*, Double Day, New York, 1986.

118. Drexel, K. E. and C. Peterson, *Unbounding the Future*, Quill William Morrow, New York, 1991.

119. Bate, T. R., "Nanoelectronics", Solid State Technol., November, 101–108, 1989.

# Micromachine Development and Packaging

## Introduction

This chapter approaches micromachine development and sensor packaging. For a generic micromachine development strategy we introduce a checklist to decide upon a preferred material and machining approach. For packaging methods we will consider those micromachine packaging issues that differ the most from ICs. With micromachines one faces the paradox of needing hermetic isolation from the environment that must be sensed. This has proven to be a formidable challenge for mechanical sensors, such as for pressure and speed and to a lesser extent for accelerometers, but it has been the Achilles heel of chemical and biological sensors. We distinguish between wafer level (batch) packaging steps and serial packaging processes. We will learn that micromachining itself is playing an important role in transforming costly serial packaging steps into less costly, wafer level processes. Many of the new-found ways of hermetically sealing and bonding different layers and making contacts between them increasingly interest IC manufacturers. An important opportunity for micromachinists is to transfer the developed three-dimensional machining technologies to the newest generations of three-dimensional ICs. We also will review software helpful in the design phase, i.e., software for computer-aided design (CAD) of microstructures.

## Micromachine Development

### Introduction

When approaching the development of a microdevice, we suggest using the terminology 'miniaturization science' rather than 'micromachining', a term too loaded with the notion that Si technology represents the only solution. Applied miniaturization starts with a thorough understanding of the sensor or microsystem specification list, interviews with potential users, understanding of the application environment, and, most importantly, a firm market appreciation.

## Checklist

Some 10 years ago, Senturia posed a rhetorical question about building micromachines: "Can we design microrobotic devices without knowing the mechanical properties of the materials involved?"[1] In subsequent years, characterizing mechanical properties of MEMS materials became popular. However, an equally important question — "Would you build a microstructure without knowing all the available construction methods?" — remained largely unaddressed. Choices of materials and machining processes are usually intertwined and both must be confronted early on in the design phase. Table 7.7 gives a good idea about the many available machining options. In Chapter 4 we compared Si as a substrate choice with ceramic, glass, and plastics in terms of traditional mechanical machinability, metallization ease, and cost (Table 4.2). Quite a few more criteria need consideration before deciding on an ideal substrate and machining method for a given sensor application. Following we introduce a more complete checklist to help in the decision process of substrate and machining tool.

In Chapter 10, we learn how different markets dictate different machining approaches. For example, chemical sensors, serving a fragmented market, are better constructed in a modular hybrid thick film approach, but automotive accelerometers, serving a huge and uniform market, are better built in an integrated polysilicon design. The choice of materials for miniaturization of sensors, actuators, and power sources will depend also on the particular sensing principle employed. Available transduction choices are summarized in a $6 \times 6$ matrix introduced in Chapter 10, Table 10.1. For our checklist we will start from the assumption that the market and specifications list are well understood and that the detailed specifications list and market understanding agree best with one particular sensing approach. In an example exercise illustrated in Figure 10.8, the first round of brainstorming for a new disposable glucose sensor design resulted in the suggestion of an electrochemical approach. Now we are ready to face the first design iteration by looking at material and machining options for the proposed electrochemical glucose sensor.

**TABLE 8.1**    Checklist To Determine Substrate for a New Micromachining Application

1. How will the package of the sensor or system most likely look and how does it interface with the real world? The package and the interface with the environment determine size and cost of the total product and the nature of the microdevice inside, as well as the answers to most of the following questions.

2. Is there a need for integration of electronic functions in the micromachine? Such need (e.g., because the sensor has a very high impedance) necessitates a semiconductor substrate.

3. How many of the microdevices will be made (production volume or number of units) and what is the unit complexity (number of devices in a unit)? The number might suggest a serial (small number), hybrid (large number), or batch (very large number) approach. High complexity might point to a batch approach.

4. Cost per device? The cost may suggest a serial (high cost, > \$40.00), or Si batch (low cost, < \$2.00) approach. For very low cost (glucose sensor strips, < \$.20), a continuous fabrication process becomes mandatory.

5. Does the substrate merely function as a support? If so, glass, ceramic, or even plastic and cardboard all become options. If the substrate has a mechanical function, Si is an excellent candidate. If the substrate must have good optical properties, materials such as GaAs and poly(methylmethacrylate) (PMMA) are candidates.

6. Is there a need for modularity? Modularity is important with chemical sensor arrays, where integration often is counterproductive because of the incompatibility issues faced when depositing different chemical sensor coatings on an integrated chip.

7. What are the expected relative tolerances on the lateral dimensions and what is the required depth-to-width aspect ratios of features built into the substrate? Very small relative tolerances on lateral dimensions cannot be achieved yet; 1% lateral tolerance on a 100-$\mu$m line, in optical lithography, is considered good, but $10^{-3}$ on 1 cm in diamond milling is very poor (see Figure 7.1). Large aspect ratios (say, > 20) might necessitate wet anisotropic etching of a single crystalline material, room temperature and cryogenic dry etching, or LIGA and pseudo LIGA processes.

8. What environment (air, water, or other) will the device be exposed to? Sensors exposed to aqueous environments such as blood pose more packaging issues and make integration of electronics difficult.

9. Which substrate makes the packaging requirements less stringent? For a sensor in aqueous solutions, for example, a ceramic substrate requires no protection of the sides of the ceramic strip; a Si sensor, on the other hand, is difficult to insulate and package, since a conductive medium might short out the chemical sensor signal via the conductive Si sidewalls.

10. Is the desired microstructure a truly three-dimensional part (e.g., a contact lens) or is it a projected form? A truly three-dimensional part suggests traditional precision engineering, e.g., diamond turning or a molding process; projected forms allow for a lithography process.

11. Thermal requirements?
    - Maximum temperature of exposure (temperatures > 150°C may make integrated Si functions impossible). If integration and high temperature are needed, specially processed silicon, SOI, or GaAs become candidates. In Figure 8.1 we show the maximum temperature of use vs. band-gap for various semiconductor substrates.
    - Is thermal conductivity needed?
    - Is a thermal match with other materials important?

12. Flatness requirements (often in connection with the optical properties of the substrate)?
    - Average roughness, $R_a$?
    - One or both sides polished?

13. Optical requirements?
    - Transparency in certain wavelength regions?
    - Index of refraction?
    - Reflectivity?

14. Electrical and magnetic requirements?
    - Conductor vs. insulator?
    - Dielectric constant?
    - Magnetic properties?

15. Process compatibility?
    - Is the substrate part of the process?
    - Chemical compatibility?
    - Ease of metallization?
    - Machinability?

16. Strain-dependent properties?
    - Piezoresistivity?
    - Piezoelectricity?
    - Fracture behavior?
    - Young's modulus?

The checklist in Table 8.1 starts with a question about the design of the interface of the micromachine with the real world — in other words, with its package. Once the package design is established, one can look for the best manufacturing technique for the micromachined part inside. This strategy dictates itself because the packaging cost contributes largely to the overall cost of any micromachined product and often overwhelms size specifications as well. The reason for expensive packaging can be ascribed to the many serial labor processes involved. Also, each device may require individual attention. The size of the package is dictated by the need to interface to a bigger structure, by limitations of traditional machining methods, or by the need for manual handling of the structure. Micromachining techniques such as fusion bonding, anodic bonding, and other integrated encapsulation techniques, commonly used in bulk and surface micromachining, might help transform more and more of the serial packaging steps into parallel batch steps even for ICs which will prove advantageous in keeping costs of the total structure down. In the author's opinion, micromachined packaging solutions represent one of the most unmined potentials for new micromachining applications.

Next in the checklist one needs to establish whether to integrate active electronics on the substrate and confirm the expected temperature range of operation (if too high, the latter might void the integration option). A further consideration concerns number and cost of the devices and which substrate will make packaging easier. When faced with a new micromachining problem, one should consider all available precision manufacturing options, i.e., zero-base the approach. Zero-basing the technological approach to a micromachining problem should include the previous questions, as well as the questions summarized in the checklist in Table 8.1.

## Scaling Characteristics of Micromechanisms

Superimposed on answers to the questions listed in Table 8.1 is a required understanding of the scaling characteristics of micromechanisms. As these effects will be explored in detail in Chapter 9, some intuitive guidance, summarized in Table 8.2, should suffice for now.[2,3]

For the use of the checklist in Table 8.1 and the scaling laws in Table 8.2, we again refer to Figure 10.8, which exemplifies the process for a glucose sensor design. The confrontation of the electrochemical approach with the manufacturing options of Table 7.7 leads, after several iterations, to the suggestion of a thick-film manufacturing process with a continuous printing process on a plastic substrate. In this iterative process the checklist in Table 8.1 functions as a control gate.

## Comparison of Important Micromachining Materials

### General Comments

A rough comparison of important substrate materials is presented in Table 8.3, including materials that are merely 'passive' substrates (e.g., plastic and cardboard support structures)

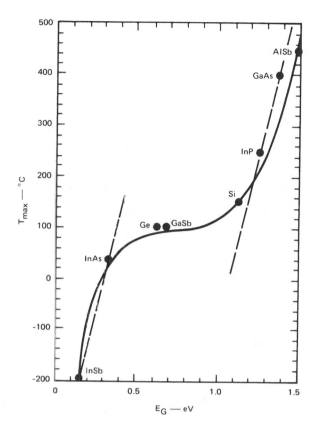

**FIGURE 8.1**  Maximum operational temperature for a number of semiconductors as a function of their bandgap. In case of Si, higher temperatures can be tolerated through special processing or by using SOI.

and other that are 'active' and may constitute an active member of the micromachine (e.g., GaAs and quartz).[4] In the next section, some of the most important 'active' candidate micromachining materials are detailed.

If the substrate to be machined must have good mechanical properties, single crystal Si is the favored material in micromachining. By comparing the yield strength, density, specific strength (i.e., ratio of yield strength to density), Knoop hardness, Young's modulus, thermal conductivity, and thermal expansion coefficient for some important technological materials (Table 8.4), one realizes how outstanding Si performs as a mechanical sensor substrate. Although 95% of bulk micromachining relies on single crystal Si, the need for all types of micromachines with a wide variety of functions is slowly changing this picture. This is particularly true for chemical sensors, optical devices, and micro-instrumentation. But even for mechanical devices, Si is far from the only choice, with quartz, for example, being an excellent contender. For very fast devices and high temperature and optical applications, GaAs is a good candidate.

### Silicon

Mechanical stability is crucial for mechanical sensing applications. Any sensing device must be free of drift to avoid recalibration at regular intervals. Part of the drift in mechanical sensors may be associated with movement of crystal dislocations

TABLE 8.2   Some Important Effects of Scaling on Micromachine Design

1.    Size effect on forces: weak forces are preferred in the micro-domain. Electrostatic forces (with a quadratic dependency on dimension $(1^2)$ under certain conditions (see Chapter 9) outperform magnetic torque (with a cubic dependency on dimension $(1^3)$) since they become relatively stronger in small devices.

2.    Increasing strength of materials in the microdomain. Single crystal whiskers may be more than 1000 times stronger than the bulk material.

3.    The dominance of surface effects in the microworld: for example, due to an increase in S/V (surface-to-volume ratio), better heat dissipation results (thermal isolation is difficult and cooling is excellent) and frictional forces increase.

4.    Decrease of manufacturing accuracy of micromachines (see Figure 7.1).

5.    The need for error-insensitive design (fewer parts, flexible materials) and the use of nature as a guide in design philosophy for very small devices.

6.    Mechanical, thermal, electromagnetic response time constants and many other time constants shrink with advanced miniaturization, generally resulting in shorter response times. Microsystems with their small inertial mass are faster, leading to higher speed or frequency, but higher frequency of a cyclic process generally leads to larger losses and consequently to lower efficiency.

7.    Power consumption is often dramatically reduced, especially if good thermal isolation is possible for elements with small heat capacity.

8.    Energy sources scale very poorly $(1^3)$, making on-board energy sources (e.g., batteries and fuel cells) less attractive for powering micromachines than radiating in energy (e.g., solar or laser light with micromachined photovoltaic converters or microwaves if extremely small receivers and converters were available).

9.    Most actuators, which rely on power for inducing a movement also scale poorly and it is often better to consider a 'macro-actuator' which can produce a micromotion than a micro-machined actuator with too small of a stroke.

in the loaded mechanical part. In ductile materials, such as metals, dislocations move readily. In contrast, in brittle materials such as semiconductors, dislocations hardly move. Mechanical engineers often avoid using brittle materials and opt for ductile materials, even though these plastically deform, meaning that they are subject to mechanical hysteresis. Single crystal Si can be made virtually without defects and under an applied load no dislocation lines can move, so that at room temperature Si can only be elastically deformed. The last property coupled with an extremely high yield strength — comparable to steel — makes Si a material superior to any metal in most applications. As a consequence, Si has been quite successful as a structural element in mechanical sensors, particularly over the last 15 years. Pressure and acceleration sensors, based on simple piezoresistive elements embedded in a Si movable mechanical member, have turned into major commercial applications. Besides the desirable mechanical and known electrical properties, this success also can be attributed to the available fabrication technology of integrated circuits.

The significant properties that have made Si a successful material not only for electronic applications, but also for

mechanical applications are reviewed in Table 8.5. From this table we reiterate some reasons behind the success of Si as a mechanical sensor element:

- Silicon surpasses stainless steel in yield strength and also displays a density lower than that of aluminum. In fact, silicon's specific strength, defined as the ratio of yield strength to density, is significantly higher than for most common engineering materials.

- The hardness of Si is slightly better than that of stainless steel; it approaches that of quartz and is higher than most common glasses.

- The Young's modulus of Si has a value approaching that of stainless steel and is well above that of quartz. From Tables 8.4 and 8.5 we also note that $Si_3N_4$, a coating often used on silicon, has a hardness topped only by a material such as diamond. The combination of Si with Si nitride coatings, therefore, can be used for highly wear-resistant components as required in micromechanisms such as micromotors (see Chapter 5).

### Quartz

Quartz in some respects presents us with an even more ideal sensor material than Si because of its nearly temperature-independent thermal expansion coefficient. A variety of quartz micromachined piezoelectric devices are on the market such as electronic filters, resonators, and wristwatch tuning forks. Natural quartz has been replaced by quartz grown in large single crystals. Wafers up to 3 in. in diameter are commercially available as starting material.

Machining methods for quartz include mechanical machining such as diamond saw cutting, grinding, lapping, polishing, and, more recently, ultrasonic machining and wet and dry chemical etching. Lapping and polishing can be used for fashioning quartz plates as thin as 100 µm. These techniques do not lend themselves to making small, complex shapes, though. For the latter applications ultrasonic machining has been applied to make, for example, high precision bulk wave oscillators and complex shaped piezoelectric resonator sensors, in the millimeter size range, but the technique is not suitable for mass production. Wet chemical etching of quartz using $HF/NH_4F$ on Z-cut quartz can be used for submillimeter fashioning of complex shapes and is currently used for mass production of wristwatch tuning fork resonators.[5] Etching rates strongly depend on the crystallographic orientation with the Z-cut exhibiting the highest etching rates: 1 µm/min at 50°C, with typical etch ratios between 50:1 to 500:1 with respect to X-cut and Y-cut plates.[6,7] Conventional photolithography is used to pattern shapes on quartz with chromium/gold for masking layers. High repeatability and automatic batch processing yield very low cost devices. Similar technologies are used for the fabrication of force sensors[8] and accelerometers. The tolerances of these devices for mechanical sensing applications are, as expected, better than for Si devices. Quartz also allows more possible geometrical shapes than Si and there is no need for deposition of an insulator between conductor and substrates. Dufour et al.[9] presented a balanced comparison between micromachined pressure sensors

**TABLE 8.3** Comparison of Substrate Properties

| | | Cost | Fracture | Metalization | Machinability (Common Methods) | Dielectric Constant | Piezo Properties | Young's Modulus E(GPa) | Thermal Conductivity (W/mK) |
|---|---|---|---|---|---|---|---|---|---|
| Single Crystals | Si | $ | b,s | Good | Very good | 11.8 | Piezoresistive | 165 | 150 |
| | Quartz | $ | b,s | Good | Poor | 4.4 | Piezoelectric | 87 | 7 |
| | GaAs | $$$ | b,f | Good | Poor | 13.1 | Piezoelectric | 119 | 50 |
| | Sapphire | $$$ | b,s | Good | Poor | 9.4 | No | 490 | 40 |
| Amorphous | Fused silica | $–$$ | b,f | Good | Poor | 3.9 | No | 72 | 1.4 |
| | Plastic | c | t,s | Poor | Fair | — | No | — | — |
| | Paper/cardboard | c | t,s | Poor | Fair | — | No | — | — |
| | Glass | c-$ | b,f | Good | Poor | 4.6 | No | 64 | 1.1 |
| Polycrystalline | Alumina | cc-$$ | b,s | Fair | Poor | 9.4 | No | 400 | ~30 |
| | Aluminum | $ | t,s | Good | Very good | — | No | 77 | ~240 |

*Note:* b = brittle, t = tough, s = strong, f = fragile; $ = dollars, $$ = more dollars, $$$ = most dollars, c = cents, cc = more cents.

*Source:* Based on Senturia, S., *Chapter 9: Materials Properties* (Massachusetts Institute of Technology, Boston, MA, 1994).

**TABLE 8.4** Mechanical Properties of Single Crystal Silicon (SCS), Among Other Technological Materials

| | Yield Strength (GPa) | Specific Strength ($10^3$ m$^2$ sec$^{-}$) | Knoop Hardness (kg/mm$^2$) | Young's Modulus E(GPa) | Density ($10^3$ kg/m$^3$) | Thermal Conductivity (W/cm/°C) | Thermal Expansion ($10^{-6}$/°C) |
|---|---|---|---|---|---|---|---|
| Diamond (SC) | 53 | 15,000 | 7000 | 1035 | 3.5 | 20 | 1.0 |
| **SCS** | **2.8–6.8** | **3040** | **850–1100** | **190 (111)** | **2.32** | **1.57** | **2.35** |
| GaAs(SC) | 2.0 | | | 75 | 5.3 | .81 | 6.0 |
| Si$_3$N$_4$ | 14 | 4510 | 3486 | 385 | 3.1 | .19 | 0.8 |
| SiO$_2$ (fibers) | 8.4 | | 820 | 73 | 2.5 | .014 | 0.4–0.55 |
| SiC (6H-SiC) | 21 | 6560 | 2480 | 448 | 3.2 | 3.50 | 3.3 |
| Iron | 12.6 | | 400 | 196 | 7.8 | .803 | 12 |
| Tungsten | 4 | 210 | 485 | 410 | 19.3 | 1.78 | 4.5 |
| Aluminum | 0.17 | 75 | 130 | 70 | 2.7 | 2.36 | 25 |
| Al$_2$O$_3$ | 15.4 | | 2100 | 530 | 4.0 | 0.5 | 5.4–8.7 |
| Stainless steel | 0.5–1.5 | | 660 | 206–235 | 7.9–8.2 | 0.329 | 17.3 |
| Quartz values ‖ Z ⊥ Z | 0.5–0.7 | | 850 | 97 76 | 2.65 | 0.014 | 7.1 13.2 (increases with T) |
| Polysilicon | 1.8 (annealed) | | | 161 | | | 2.8 |

*Note:* SC = single crystal; SCS = single crystal silicon; T = temperature; ‖ Z = parallel with Z-axis; ⊥ Z = perpendicular to Z-axis.

using quartz or Si vibrating beams and concluded that both materials are strong candidates for mass-produced micromachines. Companies such as Systron Donner,[10] that have been selling traditional precision quartz products such as high-temperature, high shock-resistant accelerometers, are appreciating the potential competition from a Si technology and are developing Si technology in parallel. Some quartz properties are summarized in Table 8.6.

Hjort et al.[11] developed a most interesting, novel anisotropic dry/wet etching scheme for single crystalline quartz. By bombarding quartz with fast heavy ions (e.g., $^{197}$Au at 11.6 MeV/a and $^{129}$X$_e$ at 11.4 MeV/a specific energy (a is the atomic mass (unit)) deep amorphous nanocylinders latent tracks (±10 mm wide) are formed. The bombarded wafers are then Au-Cr masked and etched in a 20 MKOH at 143°C. By masking parts

of the wafer prior to etching microstructures can be generated along the ion track direction. The masked zone remains latent and the tracks can be eliminated in an annealing step. When etching the ion tracks in the mask windows the latent pores are opened and they widen until they overlap creating the desired micro features. By irradiating the sample of its normal slanted microstructures can be generated. The aspect ratio of the nanocylinders reaches a whopping $10^4$ and after etching aspect ratios larger than 100 result. Features as high as 80 µm were demonstrated in this fascinating new micromachining approach.

### GaAs

Single crystal GaAs lends itself as a material for micromechanics because of its attractive opto-electronic and thermal

**TABLE 8.5**   Single Crystal Silicon Material Characteristics

| Silicon Parameter | Value and Comment |
|---|---|
| Atomic weight | 28.1 |
| Atoms/cm³ | $5 \times 10^{22}$ |
| Lattice constant (Å) | 5.43 |
| Intrinsic resistivity (Ωcm) | $2.3 \times 10^5$ vs. $10^8$ for GaAs |
| Intrinsic carrier concentration (cm⁻³) | $1.45 \times 10^{10}$ |
| Breakdown field | $3 \times 10^5$ |
| Minority carrier lifetime (sec) | $2.5 \times 10^{-3}$ |
| Yield strength ($10^9$ N/m²) | 7 (steel is 2.1), IC grade Si is stronger than steel. |
| Knoop hardness (kg/mm²) | 850 (stainless steel is 820). Silicon is harder than steel and can readily be coated with Si nitride, providing high abrasion resistance. |
| Young's modulus (E(GPa)) | 190 [111] direction. The elastic modulus is similar to that of steel (steel is 2). |
| Specific heat (J/gk) | 0.7 |
| Thermal diffusivity (cm²/cm) | 0.9 |
| Dielectric constant | 11.9 vs. 13.1 for GaAs |
| Density (g/cm³) | 2.3. Silicon has a lower density than aluminum (2.7), thus it has a high stiffness-to-weight ratio. |
| Thermal conductivity at 300 K (W/cm/°C) | 1.57. Silicon has a high thermal conductivity comparable to metals such as carbon steel (0.97) and Al (2.36). |
| Linear coefficient of thermal expansion ($10^{-6}$/°C) | 2.33. The low expansion coefficient of Si is closer to quartz than to metal, making it insensitive to thermal shock. |
| Dislocation density | <100 cm². IC grade Si contains virtually no imperfections, thus it is relatively insensitive to cycling and fatigue failure. |
| Melting point | 1350°C. Silicon is a high melting material, making it suitable for high temperature applications. |
| Band gap at 300 K | 1.12 eV. Silicon has a high bandgap, making it useful electrically at high temperatures. Indirect bandgap in the near infrared. It is opaque to ultraviolet and transparent to infrared. |
| Chemical resistance | High. Silicon is resistant to most acids, except combinations of HF/HNO₃ and certain bases. |
| Oxide growth | Silicon grows a dense, strong, chemically resistant, passivating layer of SiO₂. This oxide is an excellent thermal insulator with a low expansion coefficient. |
| Silicon nitride | A typical coating for Si with a hardness and wear resistance topped only by diamond. |

properties. A comparison of Si and GaAs for micromachining applications is presented in Table 8.7, from which we can conclude that GaAs is a better material for thermal isolation and for higher temperature operation (see also Figure 8.1). The material is less attractive for mechanical devices with a yield load smaller by a factor of two compared to silicon.[12] The many heterostructures possible with GaAs make a wide variety of optical components such as lasers and optical waveguides feasible. The piezoelectric effect enables piezoelectric transducers. Based on the high electron mobility the material is also ideal for the measurement of a magnetic field through the Hall effect. 'Macro' pressure, temperature and vibration sensors, making use of the influence of external pressure and temperature on the bandgap, have been built. Besides wet[12] and dry[13] bulk micromachining of GaAs, some surface micromachining work has been reported.[14] Surface micromachined sensor and actuator structures of metal on GaAs substrate have been demonstrated in a scheme compatible with normal IC processing of GaAs. It was also shown that surface micromachined structures in epitaxial GaAs using $Al_xGa_{1-x}As(x \geq 0.5)$ as sacrificial layers are possible. By using several steps of MOCVD epitaxial layer regrowth, structures of polysilicon-like complexity were built.[15]

Even though GaAs enables the production of faster devices in the IC industry, its use is prohibitive due to its high cost. In the micromachining world possible applications are optical shutters or choppers, actuators in monolithic microwave integrated circuits (MMIC), sensors using piezo- or opto-electrical properties of GaAs, or applications favoring integration of micromechanical devices with electronic circuitry for fast signal processing, high operating temperature, or high radiation tolerance. The latter applications so far have mainly lured research and development money from military agencies.

As the micromachining industry largely evolves with the IC industry, we cannot expect GaAs micromachines to succeed unless GaAs use penetrates the IC industry. In this context Karam et al.[17] are investigating gallium arsenide micromachining techniques using HEMT (High Electron Mobility Transistor) and the MESFET (Metal Schottky Field Effect Transistor) foundry processes. Perhaps work on automated processes to deposit GaAs on Si could benefit both IC and micromachining industries. For example, GaAs/Si wafers will be larger and stronger, have a better thermal conductivity, and be lighter than GaAs wafers.[18] Along this line, Yeh et al. trapped GaAs laser diodes in micromachined wells in a Si substrate (see Figure 8.21).[19] For a listing of dry and wet GaAs etchants see Reference 17.

**TABLE 8.6** Some Properties of Quartz

| Property | Value ‖ Z | Value ⊥ Z | Temperature Dependency |
|---|---|---|---|
| Thermal conductivity (cal/cm/sec/°C) | $29 \times 10^{-3}$ | $16 \times 10^{-3}$ | Decreases with T |
| Dielectric constant | 4.6 | 4.5 | Decreases with T |
| Thermal expansion coefficient (1/°C) | $7.1 \times 10^{-6}$ | $13.2 \times 10^{-6}$ | Increases with T |
| Electrical resistivity (Ωcm) | $0.1 \times 10^{15}$ (ionic) | $20 \times 10^{15}$ (electronic) | Decreases with T |
| Density (kg/m³) | $2.66 \times 10^{3}$ | $2.66 \times 10^{3}$ | |
| Curie temperature (°C) | 573 | 573 | |
| Lattice constant | 4.904 | 4.904 | Increases with T |
| Band gap energy (eV) | ≈ 8 | ≈ 8 | Increases with T |
| Melting point (°C) | 1710 | 1710 | |
| Fracture strength (GPa) | 1.7 | 1.7 | Decreases with T |
| Stiffness constants | | | |
| $C_{11}$ (GPa) | 86.8 | 86.8 | Decreases with T |
| $C_{12}$ | 7.04 | 7.04 | Decreases with T |
| $C_{13}$ | 58.2 | 58.2 | Decreases with T |
| $C_{14}$ | 11.91 | 11.91 | Decreases with T |
| $C_{15}$ | −18.04 | −18.04 | Decreases with T |
| $C_{16}$ | 105.75 | 105.75 | Decreases with T |
| Hardness (GPa) | 12 | 12 | |

*Note:* ‖ Z = parallel with Z; ⊥ Z = perpendicular to Z; T = temperature.

### Other Active Substrate Materials for Micromachining

Various other micromachining materials are discussed elsewhere in this book. Poly-Si and other thin-film microelectronic materials are discussed in Chapters 3 and 5. Typical actuator materials such as ZnO, NiTi, PVDF, etc. and materials for integrated power generation such as Li and α-H:Si are covered in Chapter 9.

From the research point of view a few additional interesting micromachining materials deserve our attention. SiC, for example, is an attractive candidate as a semiconductor material for high-temperature and high-power applications because of its large bandgap, good carrier mobility, and excellent thermal and chemical stability. The fabrication technology of SiC electronic devices moreover is based on the Si microelectronics industry. Tong et al.[20] evaluated both epitaxial SiC and sputtered amorphous films as a new micromechanical material. One property which makes SiC films particularly attractive for micromachining is that these film can easily be patterned by dry etching using Al masks. Patterned SiC films can further be used as passivation layers in the micromachining of the underlying Si substrate (SiC can withstand both KOH and HF etching!).[21] Madou et al.[22] used single crystal SiC to fabricate high temperature pH sensors. For further reading on SiC as a MEMS material, refer to the set of papers in the Transducers '95 proceedings.[23,24]

# Design Software

## Introduction

After brainstorming about sensor specifications and sensor transduction principles and making preliminary designs on paper and the white board, it is a good idea to initiate a computer-aided design of the overall microsystem in order to better grasp how all the components, including the package, fit together. In Figure 10.4B we show an example of such a three-dimensional rendering for the case of a NiTi-based valve. Next come the design, simulation, and verification of the individual microparts.

Especially for microparts involving 'regular' IC manufacture some very evolved CAD software packages are now available (e.g., the well-known SUPREM 3[25] and SAMPLE[26–28]). A good introduction to the available CAD programs for IC processes comes from Fichtner.[29] In general, these CAD software packages are structured as sketched in Figure 8.2 with the design aids used to create the design, simulation to develop the technology, and verification to check the design. The final verification, of course, always happens in the lab. The goal, though, is to avoid wasteful and slow experiments by carrying out less costly computer work in order to get the fabrication 'right' the first time.[30] These IC development software packages are being expanded with mechanical modeling programs and materials data bases. There is indeed a growing need for the ability to perform mechanical analysis of microelectronic devices, both in assuring structural reliability against failure of thin-film layers and in evaluating the effects of various external loads, including temperature and humidity effects. In addition, with the development of increasingly high aspect ratio micromechanical devices, including pumps, valves, and micromotors, and with the increasing performance demands being placed on these devices, notably in precision and accuracy, there is a critical need for CAD tools which will permit rational design of these devices and evaluate the effects of parameters such as temperature, strain, acceleration, etc. In order to construct the required data base for these new CAD tools, one must understand all the parameters influencing the microstructure. At this point in time a good understanding of the processes and materials properties required to ensure reproducibility for commercial MEMS products is emerging slowly. Several CAD systems, which might facilitate the wider acceptance of MEMS, are under development. Most of them include a materials database which can be updated by the user.

## CAD for MEMS

### CAD from the IC Industry Applicable to MEMS Designers

Computer-aided design tools are essential to the evolution of high aspect ratio micromachining from laboratory status

**TABLE 8.7**  A Comparison of Silicon with GaAs for Micromachining Purposes

| Property | GaAs | Si |
|---|---|---|
| Qualitative | | |
| Speed | + | − |
| Heterostructures | Many | ? |
| Opto-electronics | + | − |
| Piezo effect | Yes | No |
| Thermal conductivity | Relatively low | Relatively high |
| Integration density | + | ++ |
| Cost | High | Low |
| Bonding to other substrates | Difficult | Relatively easy |
| Fracture | Brittle, fragile | Brittle, strong |
| Etching behavior | Isotropic/anisotropic | Isotropic/anisotropic |
| Operation temperature | High | Low |
| Quantitative | | |
| Bandgap (eV) | 1.424 | 1.12 |
| Bandgap transition | Direct | Indirect |
| Melting point (°C) | 1238 | 1415 |
| Maximum operating temperature | 460 | 300 |
| Physical stability | Fair: sublimation of arsenic is a problem | Very good |
| Coefficient of thermal expansion ($°C^{-1} \times 10^{-6}$) | 6.86 | 2.6 |
| Thermal conductivity (W/cm/K) | 0.5 | 1.5 |
| Breakdown voltage ($\times 10^6$ V/cm) | 0.4 | 0.3 |
| Electron saturation speed ($10^7$ cm/sec) | 2 | 1 |
| Young's modulus (GPa) | 75 | 190 |
| Dielectric constant | 13.1 | 11.9 |
| Piezoelectric coefficient | $d_{14} = 2.6$ pN/C | − |
| Hole mobility ($cm^2$/V/sec) | 400 | 600 |
| Electron mobility ($cm^2$/V/sec) | 8500 | 1500 |
| Hardness (GPa) | 7 | 10 |
| Fracture strength (GPa) | 2.7 | 6 |
| Debey temperature (K) | 370 | 463 |
| Lattice constant (Å) | 5.65 | 5.43 |
| Density (g/cm³) | 5.3 | 2.3 |
| Specific heat (J/°C) | 0.35 | 0.7 |

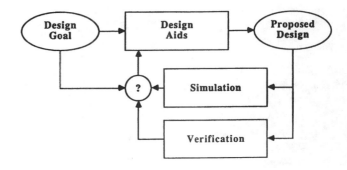

**FIGURE 8.2**  Design, simulation, and verification. (After Senturia, S.D. and R.T. Howe, Lecture Notes, MCT, Boston, MA, 1990.)

For bulk micromachining and surface micromachining, significant progress is being made toward software design platforms as expansions of IC design software. The new CAD software will benefit both IC and micromachining production. For very high aspect micromachining methods such as LIGA, software design packages are in the early development stage.[30]

One typically separates the design process in a conceptual design-and-simulation phase and a phase of final design of masks and processes. The ideal suite of CAD tools required for each activity is summarized in Table 8.8.[4]

Several of the required tools listed in Table 8.8 are already commercially available for use in the semiconductor industry and can be transposed onto Si bulk and poly-Si surface micromachining. For mask layout a variety of tools are available from vendors such as TMA and Cadence; in the public domain one can turn to KIC from the University of California, Berkeley.[31] Mechanical drafting tools with interfaces to maskmaking, such as AutoCAD, have full geometric flexibility. Other popular mechanical CAD tools for three-dimensional, solid model construction, meshing, application of loads and boundary conditions, and visualization of results include Pro/Engineer, PATRAN, and I-DEAS. Suppliers of CAD systems to the semiconductor industry also offer one- and two-dimensional process simulators. These simulators cover standard semiconductor processes such as oxidation, diffusion, thin-film deposition, and plasma etching. An example is SAMPLE for simulation of projection lithography, deposition, and etching. Simulation tools for basic physical properties include ABAQUS for structural and thermal finite-element simulation and FASTCAP for electrostatic boundary-element simulation. For some of the simplest micromachined structures, modest modification of existing IC software packages enables designing of the micromachine as in regular IC processing. Marshall et al.,[32] for example, have modified the standard 'Magic' CAD package to permit the design of simple micromachined structures using CMOS processes. This modification makes standard foundry work through the MOSIS service possible.[33] Tools for design-rule checking from the commercially available semiconductor design-rule checkers should be easy to adapt for use with MEMS. In practice, the MEMS structures are usually simple enough that design-rule checking has not been a high priority.[4]

to a bona fide and accessible manufacturing process. Without CAD tools, fabrication remains in the domain of experts, and evolution of the design process relies on empirical approaches. Development of a consistent CAD framework will increase access to the limited high aspect ratio micromachining facilities as more design engineers will try to gain from using microscale devices.

**TABLE 8.8**   Ideal Tool Suites for MEMS

| Conceptual Design and Simulation | Final Design of Masks and Process |
|---|---|
| Rapid construction and visualization of three-dimensional solid models | Process simulation or process database, including:<br>    Lithographic and etch process biases (the difference between as-drawn and as-fabricated dimensions)<br>    Process tolerances on thicknesses, lateral dimensions, doping, and resistivity levels |
| A database of materials properties | Design optimization and sensitivity analysis:<br>    Variation of device sizing to optimize performance<br>    Analysis of effects of process tolerances |
| Simulation tools for basic physical phenomena, for example:<br>    Thermal analysis: heat flow<br>    Mechanical and structural analysis: deformation<br>    Electrostatic analysis: capacitance and charge density<br>    Magnetostatic analysis: inductance and flux density<br>    Fluid analysis: pressures and flow | Mask layout |
| Coupled-force simulators, for example:<br>    Thermally induced deformation<br>    Electrostatic and magnetostatic actuators<br>    Interaction of fluids with deformable structures | Design verification, including:<br>    Construction of a three-dimensional solid model of the design, using the actual masks and process sequence<br>    Checking the design for violation of any design rules imposed by the process<br>    Simulation of the expected performance of the design, including the construction of macromodels of performance useable in circuit simulators to assess overall system performance |
| Formulation and use of macromodels, for example:<br>    Lumped mechanical equivalents for complex structures<br>    Equivalent electric circuit of a resonant sensor<br>    Feedback representation for coupled-force problems | |

*Source:* After Senturia, S.D. and Howe, R.T., *Chapter 8 Mechanical Properties and CAD* (Massachusetts Institute of Technology, Boston, MA, 1990).

## Examples of MEMS CAD Software Under Development

Some examples of MEMS CAD programs under development are IBM's Oyster,[34] MIT's MEMCAD,[35] the University of Michigan's CAEMEMS,[36] and ETH's (Zurich) SESES[37] system. These programs mainly address bulk Si micromachining and poly-Si surface micromachining applications.

Oyster facilitates the construction of a three-dimensional polyhedral-based solid mask set and gives a rudimentary process description. MEMCAD is directed at conceptual design and simulation, as well as design verification. CAEMEMS (acronym for computer-aided engineering for micro electromechanical systems) is geared toward design optimization and sensitivity analysis. SESES addresses conceptual design and simulation and design verification. We will limit our detailed description of the programs to CAEMEMS and MEMCAD.

The CAEMEMS program provides a database of material properties and process model parameters, a process modeler that takes process information as input and produces a solid model of the device to be fabricated, a solid modeler that allows for visualization and design verification, and a device modeler that performs finite element simulations of components and systems. The database can be updated by the user as new information becomes available. A block diagram of the CAEMEMS software package is shown in Figure 8.3.[36] The architecture of

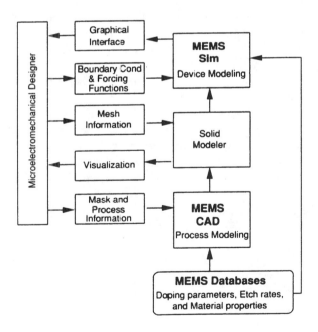

**FIGURE 8.3**   Block diagram of the CAEMEMS system. (Crary, S. and Y. Zhang, Proceedings IEEE Micro Electro Mechanical Systems, (MEMS '90), Napa Valley, CA, 1990, p. 113–114. With permission.)

MEMCAD illustrated in Figure 8.4 integrates various simulators, databases, and a solid state modeler with a user interface.[4] As outlined by the dashed blocks in Figure 8.4, the CAD system consists of three sections: the microelectronic CAD section, the mechanical CAD section, and the material property simulator. The interactions among these three sections and the flow of information is denoted by the direction of the arrows. In MEM-CAD Version 1.0, the primary interface for mechanical modeling was through PATRAN. MEMCAD Version 2.0 provides this function from I-DEAS. In the microelectronic section the mask is created using KIC. SUPREM 3 and SAMPLE are integrated to provide depth and cross-sectional modeling capabilities. Inspired by MEMCAD, IntelliSense Corp. sells IntelliCAD,[38] including the MEMaterial database. The material database contains electrical, mechanical, optical, and physical properties of semiconductor thin films collected from the literature. As in the case of CAE-MEMS, the database can be updated by the user. The architecture of IntelliCAD, with a central graphical user interface (GUI), is shown in the inset in Figure 8.4. Figure 8.5 shows a typical output of MEMaterial. The selected three-dimensional plot shows the density variation of a plasma-enhanced chemical vapor deposited (PECVD) Si nitride film as a function of deposition temperature and pressure.

### CAD for High Aspect Ratio and Truly Three-Dimensional Micromachines

The micromachining field is changing so fast that most of the approaches described above may become obsolete soon except for those machining methods that are most similar to IC manufacturing (e.g., poly-Si surface micromachining). Better simulators for anistropic etching of Si, for example, are badly needed (see Chapter 4) and are now under development at several locations. Two- and three-dimensional simulation programs for anisotropic etching profiles have been developed by Sequin at University of California, Berkeley,[39] Koide et al. at Hitachi,[40] and Hubbard et al. at CalTech.[41] These programs enable the prediction of a change in cross-sectional shape of a feature in a Si wafer with arbitrary crystallographic orientation and with a mask including concave and convex edges. IntelliSense is close to introducing its own anisotropic etching simulation program which will be embedded in the general MEMCAD architecture.

Current developments in CAD in the U.S. and Europe, which have evolved from the two-dimensional IC circuit work, lack the tools for *a priori* design of micromechanical devices in 2 1/2 or 3 dimensions essential for high aspect ratio microsystems. Given the freedom to fabricate three-dimensional systems, designers think in terms of the whole system — not a series of two-dimensional reticulations of the system (i.e., masks). The mask-to-model approach might still be suitable for surface micromachined devices but becomes less useful as the number of processes, materials, and mechanical degrees of freedom increases, as in the case of high aspect ratio

micromachining (especially for a process such as LIGA).[42] CAD programs addressing generic miniaturization problems rather than just micromachining for Si and poly-Si only would engage many of the traditional industrial manufacturers to become interested in miniaturization of their products. The architecture should have both lithographic and nonlithographic machining options and adhere to accepted format standards for data communication and integrate available design, modeling, and simulation software from both traditional and nontraditional (i.e., IC manufacturing) industries wherever possible. A glaring example for the need of a 2 1/2 or 3-dimensional design tool is in micromolding of LIGA shapes in which nontraditional (lithography) and traditional (plastic molding) technologies are combined. Hill et al.[43] are exploring the use of the computer software I-DEAS Master Series™ Thermoplastic Molding to model the micromolding aspect of such LIGA parts and found so far that it accurately described the filling characteristics of the microparts. What is needed now is to extend this package to include the lithography steps.

A most detailed account of simulation and design issues of microsystems and microstructures is by Adey et al.[44] and for continuously updated information on MEMS modeling visit the MEMS Interchange on the WWW. An example of a well-documented new MEMS software package on the web is MISTIC from the University of Michigan (for more related URL's, see Appendix B).

# Packaging

## Introduction

In electronic device fabrication, processes can be divided into three major groups: IC fabrication with additive and subtractive processes as described in Chapters 1 to 6; packaging involving processes such as bonding, wafer scribing, lead attachment, and encapsulation in a protective body; and testing including package leak tests and electrical integrity. The last two process groups incorporate the most costly steps. In the case of a micromachine, which often interfaces with a hostile environment and where testing might involve chemical and/or mechanical parameters, packaging and testing become even more difficult and expensive. More than 70% of the sensor cost may be determined by its package, and the physical size of a micromachined sensor often is dwarfed by the size of its package (see for example GM's MAP sensor in Figure 10.10). Most conventional packaging approaches are space inefficient, with volumetric efficiencies, even in the case of ICs, of less than 1%.[45]

Packaging in IC technologies provides at least four functions: signal redistribution, mechanical support, power distribution, and thermal management. The signal redistribution is easy to understand: the electrical contacts of an IC are too closely spaced to accommodate the interconnection capacity of a traditional printed wiring board (PWB). Packaging redistributes contacts over a larger, more manageable surface — it fans out

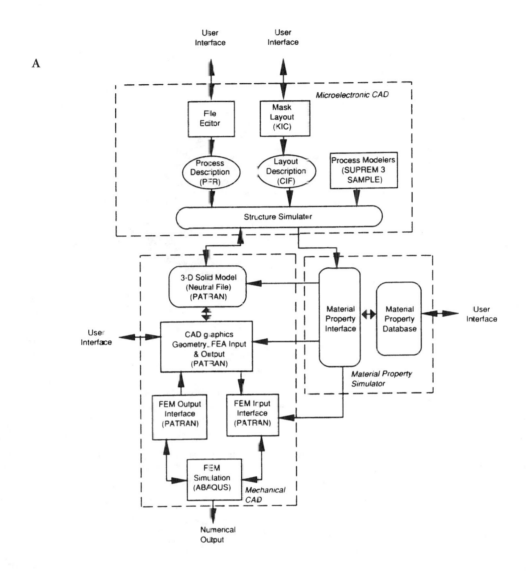

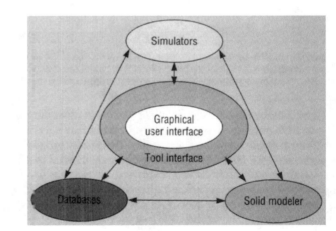

**FIGURE 8.4** (A) Outline of the Massachusetts Institute of Technology MEMCAD.[4] (B) Inset: Schematic of IntelliCAD architecture. (Courtesy of Dr. F. Maseeh, IntelliSense Corp.)

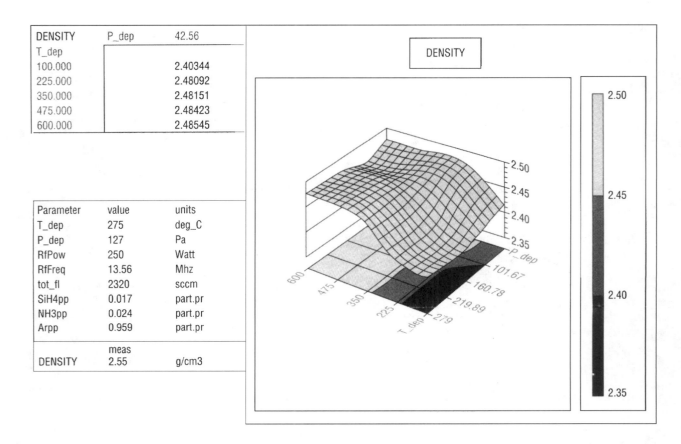

DENSITY      P_dep      42.56

| T_dep | | |
|---|---|---|
| 100.000 | | 2.40344 |
| 225.000 | | 2.48092 |
| 350.000 | | 2.48151 |
| 475.000 | | 2.48423 |
| 600.000 | | 2.48545 |

| Parameter | value | units |
|---|---|---|
| T_dep | 275 | deg_C |
| P_dep | 127 | Pa |
| RfPow | 250 | Watt |
| RfFreq | 13.56 | Mhz |
| tot_fl | 2320 | sccm |
| SiH4pp | 0.017 | part.pr |
| NH3pp | 0.024 | part.pr |
| Arpp | 0.959 | part.pr |
| | meas | |
| DENSITY | 2.55 | g/cm3 |

**FIGURE 8.5**  The MEMaterial design window from IntelliSense Corp. The three-dimensional graph shows the variation of Si nitride density deposited by PECVD. The table on the bottom, left-hand side of the graph shows the process parameters, their numerical values, and units. The resulting density is shown at the bottom of the table. The user can alter any of the parameters. (Courtesy Dr. F. Maseeh, IntelliSense Corp.)

the electrical path. The mechanical support provides rigidity, stress release, and protection from the environment (e.g., against electromagnetic interference). Power distribution is similar to signal distribution except that power delivery systems are more robust than signal delivery systems and the thermal management function is there to support adequate thermal transport to sustain operation for the product lifetime.[45]

The various levels of packaging and their interconnections in the IC industry and in micromachining are summarized in Table 8.9[45] and illustrated in Figure 8.6.[46,47] The lowest level in the IC electronics (L0 in the hierarchy) is single IC features interconnected on a single die with IC metallization lines into a level 1 IC or discrete component. Single-chip packages and multi-chip modules are at level 2. Levels 3 and 4 are the PWB (L3) and chassis or box (L4), respectively. Level 5 is the system itself. The hierarchy for a large electronic system is illustrated in Figure 8.6A and for a micromachine such as a pressure sensor in Figure 8.6B and C.

Packages for micromachines are considerably more complex as they serve to protect from the environment, while, somewhat in contradiction, enabling interaction with that environment in order to measure or affect the desired physical or chemical parameters. With the advent of integrated sensors, the protection problem became further complicated. Silicon circuitry is sensitive to (to name just a few) temperature, moisture, magnetic field, electromagnetic interference, and light. The package must

**TABLE 8.9**  Connection Between Packaging Elements in Various Levels in the Hierarchy for ICs and Micromachines

| Level | Element | Interconnected by |
|---|---|---|
| Level 0 | Transistor within IC or resonator in a micromachine | IC metallization |
| Level 1 | ICs, discrete components such as a Si/glass pressure sensor sandwich | Package leadframes single-chip or multichip module interconnections system |
| Level 2 | Single- and multi-chip packages (a pressure sensor in a TO header) | Printed wiring boards |
| Level 3 | Printed wiring boards | Connectors/backplanes (busses) |
| Level 4 | Chassis or box | Connectors/cable harnesses |
| Level 5 | System itself (a computer or a gas alarm) | |

*Source:*  After Lyke J.C., Packaging technologies for space-based microsystems and their elements, in *Microengineering Technologies for Space Systems,* Helvajian, H., Ed. (The Aerospace Corp., El Segundo, CA, 1995, pp. 131–180.)

then protect the on-board circuitry while simultaneously exposing the sensor to the effect it measures. For example, in an *in vivo* integrated sensor, a true hermetic seal is necessary to protect the circuitry from the effects of blood. Sometimes the

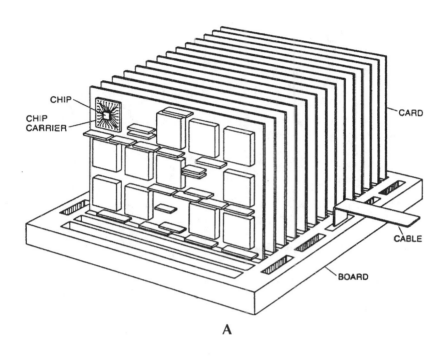

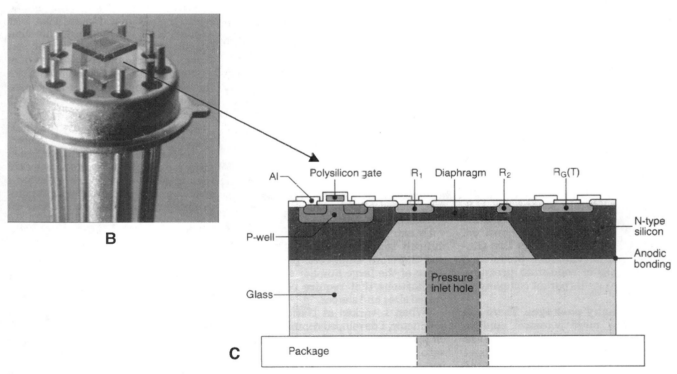

**FIGURE 8.6** The packaging hierarchy in the IC world and in micromachining. (A) IC: a large electronic system. Features on chip: L0, chip; L1, chip carrier; L2, PC board (card); L3, chassis (board); L4, and cable to system (L5). (From Jensen, J.R., *Microelectronics Processing*, ACS, Washington, DC, 441–504, 1989. With permission.) (B and C) Micromachining piezoresistive pressure sensor from NEC. Diaphragm on die; L0, Si/glass piece; L1, sensor mounted in TO header; L2.

circuit itself can be used to reduce packaging concerns, for example, to transmit data about effects which cannot be screened out, as in the removal of a temperature effect from a pressure signal. But more frequently integration in micromachines causes more problems than it solves. A package must also provide communication links, remove heat, and provide a means for handling and testing. The material used for the package must be one that will afford physical protection against normal process handling during and after assembly, testing, and prescribed mechanical shock. Chemical protection is also required during assembly and in practical usage. For example, during the process of installing dies on substrates, the package undergoes a series of cleanings and, under normal use, the package may be exposed to oxygen, moisture, oil, gasoline, and salt water. The package must also be capable of providing an interior environment compatible with device performance and reliability, for example, a high Q resonator might need a good vacuum.

The sensor community is still searching for the perfect way to protect sensors from their environment while at the same time probing that environment. The packaging problem is the least severe for a physical sensor such as an accelerometer which can be sealed and protected from all chemical environment effects while probing the inertial effects it measures and is most severe for chemical and biological sensors, which must be exposed directly to an unfriendly world. Generalizing, chemical and biological sensors have been an order of magnitude more difficult to commercialize than physical sensors, in no small part due to packaging problems. In the latter case, problems are especially severe with electronic components and sensors integrated on the same side of a single Si substrate. Physical sensors usually have a physical barrier layer or fluid, e.g., a pressure transmission fluid interposed between the sensing Si elements and the environment outside, making the encapsulation problem less severe.

In what follows we are reviewing dicing of MEMS wafers and wafer level packaging methods employed in sensors, specifically bonding and interconnections of micromachines. Subsequently, higher level of packaging issues are briefly addressed. For more information on advanced packaging schemes in ICs we refer to the excellent reviews by Lyke[45] and Jensen.[46]

# Dicing

One of the final steps in fabrication of a three-dimensional microstructure is usually sawing the finished Si wafer into individual 'dice'. After the wafer is probed it is mounted onto sticky-tape and put onto the dicing saw. A typical sawblade consists of a thick-shaped hub with a thin rim impregnated with diamond grit. Rotating at a speed of several thousands revolutions per minute, the blade encounters the Si wafer at a feed rate on the order of a centimeter per second, sawing partially or completely through the wafer.

Except when cutting a silicon-glass bonded wafer combination, standard IC cutting practices regarding surfactants, cleanliness, and blade width/depth ratio apply to MEMS wafers. Typical kerf widths are 50 to 200 μm, while typical roughness along the kerf edge is 10 to 50 μm. The restrictions of mechanical sawing include this roughness, which precludes the fabrication of smooth structures along the outside edge of the die. In addition, the vibration inherent in the sawing process makes it difficult to securely hold down dice during sawing if these dice are less than 0.5 mm on a side. For smaller dice, micromachining is necessary for edge definition and separation of individual devices. In the latter case, anisotropically etched V-grooves extending almost through the depth of the wafer are separating the individual dice which might be broken apart by applying a small mechanical force. Alternatively, the dice are etched completely free at the moment critical components, e.g., a pressure-sensitive membrane, have reached the specified thickness (see V-groove thickness control in Chapter 4).

# Wafer Level Packaging

## Introduction

Hardly any microdevice has been fabricated without making use of some type of wafer bonding or cavity sealing. Wafer bonding and cavity sealing can serve as a 'batch'-compatible zero and first level packaging technique by encapsulating a die feature (L0) or a whole die at a time (L1). In bulk micromachining (Chapter 4) and Si fusion-bonded (SFB) surface micromachining (Chapter 5), cavities are fabricated by bonding, respectively, a glass plate (anodic bonding) or Si wafer (fusion bonding) over etched cavities in a bottom Si wafer. The Si and glass layers bonded typically are rather thick and do not lend themselves well to die feature level packaging. In poly-Si and selective epitaxy surface micromachining (see Chapter 5), on the other hand, sealing cavities often is a more integral part of the overall fabrication process and both die and die features might be packaged this way. These and other lithography-defined packages, such as those involving ultraviolet patternable polymers, lend themselves to inexpensive, batch solutions and represent a new technology area where micromachining could provide solutions even for the IC industry.

### Sealing Processes of Die Features and Whole Dice: L0 and L1 Level Packaging

In Figure 8.7A we illustrate how a polysilicon vacuum shell encapsulates a polysilicon resonator element. The micromachined surface package (microshell) illustrated is much smaller than typical bulk micromachined packages. Microshells can be made by defining thin gaps (~100 nm) between the substrate and the perimeter of the structural elements by etching away a sacrificial layer sandwiched between the two and then sealing the resulting gaps. In so-called reactive sealing, demonstrated in the schematic fabrication sequence in Figure 8.7B, thermal oxidation of the polysilicon and Si substrate at 1000°C seals the narrow openings left after removal of the spacer phosphosilicate glass (PSG).[48-50] Reactive sealing is possible even in vacuum due to the reaction of oxygen, trapped inside the cavity. Alternatively, sealant films, such as oxides and nitrides, can be deposited over small etchant holes,[49,51] as illustrated in the fabrication sequence in Figure 8.8A. In the latter case the cavity is sealed by low-pressure chemical vapor deposited (LPCVD) Si nitride deposition. The excellent coverage of this method ensures that the nitride closes the etch channel quickly before too much deposition in the chamber itself can take place. Typical deposition conditions during the sealing step are 850°C and 250 mTorr, so the residual pressure in the microshell, at room temperature (assuming ideal gases), should be about 67 mTorr. Some researchers have reported residual pressures of 200 to 300 mTorr.[52] Eaton et al.[53] established that the resulting cavity pressure is stable for a given membrane, but the residual pressure is variable and nonrepeatable across a substrate and even less from substrate to substrate.

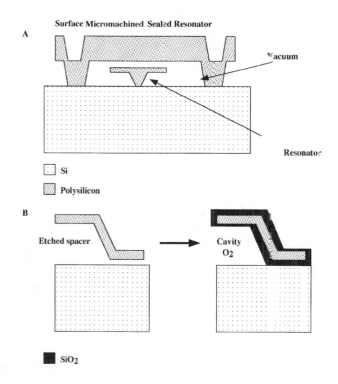

**FIGURE 8.7** Sealed cavities in surface micromachining. (A) Typical sealed cavity with a resonator structure inside. (B) Reactive sealing.

The fabrication time of packaging shells makes for a long process because of the location of narrow etch holes at the perimeter of the shell. The University of California, Berkeley, micromachining team recently introduced an interesting means to speed up the fabrication of microshells.[54] The process is outlined in Figure 8.9. Permeable polysilicon windows (see Chapter 5) are used as an etch access for removing the underlying sacrificial PSG. Using concentrated HF, shells 3 μm high and as wide as 1 mm have been cleared of PSG in less than 120 sec. Subsequent low-pressure hermetic sealing using low-stress Si nitride leads to deposition of less than 100 Å inside the package.

Sealed cavities as shown here lend themselves well to the pressure sensing application. Depending on the atmosphere to which the chip is exposed during the sealing process, gauge, vacuum, or absolute pressure sensors can be created.[55,56] The process outline in Figure 8.8A actually illustrates the case of an absolute pressure sensor with a $Si_3N_4$ membrane.[57] The sensor consists of a circular Si nitride diaphragm which forms the top of a sealed vacuum cavity, providing the pressure reference. Polysilicon strain gauges are fabricated on top of the diaphragm and the measured resistance changes are, to first order, directly proportional to the applied pressure. In Figure 8.8B we show a schematic of the sensor and in Figure 8.8C we feature an SEM microphotograph of a finished device.

Of reactive sealing and sealant films, the reactive sealing process is by far the most elegant and highest performance process. In industry (e.g., at SSI Technologies, Honeywell, and Foxboro) progress is being made in optimizing the reactive sealing process for pressure sensors. Recently, the first commercial absolute

poly-Si pressure sensor, incorporating a reactively sealed vacuum shell, was introduced for automotive applications.[57] Another application of the above type of sealed surface shells is the vacuum packaging of lateral surface resonators. Most resonator applications share a need for resonance quality factors from 100 to 10,000. However, the operation of comb-drive microstructures in ambient atmosphere results in low quality factors of less than 100 due to air damping above and below the moving microstructure.[58] Vacuum encapsulation is thus essential for high Q applications. Recent work has shown that lateral comb-drive microresonators,[59] besides their use in commercial accelerometers, have potential applications in areas such as mechanical filters for signal processing,[60] noncontact electrostatic voltage sensing,[61] and rotation-rate sensing.[62]

Deposition of a set of epitaxial Si layers with varying doping has also been shown to afford the formation of a hermetically sealed surface cavity.[63] The cavity in this example of selective epitaxy surface micromachining is formed by selective etching of p+ epitaxial Si over more heavily doped Si p++ layers. The process is illustrated in Figure 8.10. The fabrication starts with a HCl dry etch at 1050°C in an epi-reactor through a hole in an oxide mask (steps a and b). There follows a selective epitaxial growth of Si in this sequence: doped p+, p++, and p+, p++ as indicated in steps c, d, e, and f. These steps are all carried out in the same epi-reactor, simply by changing the concentration of the $B_2H_6$ dopant. The next step consists of stripping of the oxide in HF (g), followed by selective electrochemical etching in hydrazine (h). In this electrochemical etching step, the n-substrate is passivated against the etching by the imposed potential, and the p++ structures by the boron etch-stop mechanism. At this juncture one ends up with a microbridge covered by a cap, both made out of single crystalline, heavily boron-doped Si (p++). The cap is finally sealed by growth of an n-type epitaxial layer (i). The residual hydrogen in the cavity after sealing is diffused through the epitaxial layer by high-temperature annealing in a nitrogen ambient, resulting in a residual pressure inside the cavity of 1 mTorr (j).[63,64] In order to make a pressure sensor that incorporates such an encapsulated resonator in a suspended membrane, one has to fabricate a membrane by etching from the backside of the wafer, a rather trivial step compared to the process just described. In the particular device shown here, the resonance of the encapsulated beam is activated by a Lorenz force and detected by measuring the resulting inductance.

University of California, Berkeley, developed another vacuum encapsulation method suitable for L0 and L1 level packaging involving the wafer-to-wafer transfer of micromachined caps as demonstrated in Figure 8.11A.[65] Reactive sealing requires a temperature of 1000°C, and thick SiN sealing requires 850°C. Moreover, in the latter technique some sealant gas deposits on the encapsulated microdevices. In the alternative University of California, Berkeley, packaging method, tethered cap structures are sealed down to the substrate employing a low temperature Au-Si eutectic bond (the gold-silicon eutectic at 363°C is safely below that for aluminum-silicon). The encapsulation caps are made in a HEXSIL process as demonstrated in Figure 8.11B (see Chapter 5 for details on the HEXSIL process). The caps have the general shape of a top hat and are suspended by

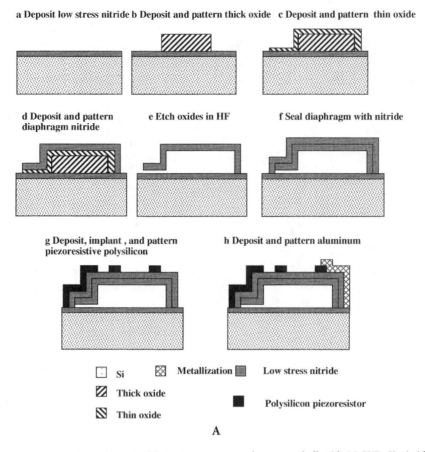

**a Deposit low stress nitride  b Deposit and pattern thick oxide   c Deposit and pattern  thin oxide**

**d Deposit and pattern diaphragm nitride**     **e Etch oxides in HF**     **f Seal diaphragm with nitride**

**g Deposit, implant , and pattern piezoresistive polysilicon**     **h Deposit and pattern aluminum**

☐ Si     ⊠ Metallization     ▨ Low stress nitride

▨ Thick oxide

◼ Polysilicon piezoresistor

▨ Thin oxide

**A**

**FIGURE 8.8**    Absolute pressure sensor:[53] (A) Schematic fabrication sequence of vacuum shell with LPCVD Si nitride sealing. The vacuum shell constitutes the reference chamber of an absolute pressure sensor. (B) Schematic of absolute pressure sensor with Si nitride membrane. (Courtesy J.H. Smith, Sandia National Laboratories.)

breakaway polysilicon flexures in the handle wafer. The brim of the hat is coated with 0.7 μm of gold. The HEXSIL wafer with the embedded caps is positioned (Δx <10 μm) over the Si wafer with the active devices to be encapsulated. *In situ* annealing in vacuum by infrared heating forms the eutectic bond between the Si of the bottom wafer and the gold of the cap. The transfer occurs at a chamber pressure of $10^{-5}$ Torr, with 10 psi of mechanical pressure applied to the wafer sandwich. When the handle wafer is withdrawn the polysilicon flexures break, leaving the caps sealed to the substrate wafer. In some limited test runs in a laboratory setting, a transfer yield of 100% was obtained for arrays of 30 caps. The HEXSIL mold wafer is reusable. An example of a transferred encapsulation cap is shown in the SEM micrograph in Figure 8.11C.

### Sealing Processes of Dice: L1 Level Packaging

*Field-Assisted Thermal Bonding*

Field-assisted thermal bonding also known as anodic bonding, electrostatic bonding, or the Mallory process is commonly used for joining glass to silicon. The main utility of the process stems from the relatively low process temperature. Since the glass and Si remain rigid during anodic bonding, it is possible to attach glass to Si surfaces, preserving etched features in either the glass or the silicon. This method is mostly applicable to wafer scale die bonding (L1).

A bond can be established between a sodium-rich glass, say Corning #7740 (Pyrex), and virtually any metal.[66] Besides Pyrex, Corning #7070, soda lime #0080, potash soda lead #0120, and aluminosilicate #1720 are suitable as well.[67] In the case of Si, Pyrex is most commonly used. Bonding can be accomplished on a hot plate in atmosphere or vacuum at temperatures between 180 and 500°C; typical voltages, depending on the thickness of the glass and the temperature, range from 200 to 1000 V. The operating temperatures are near the glass-softening point but well below its melting point, as well as below the sintering temperature of standard AlSi metallization. At the most elevated temperatures the wafers are bonded in 5 to 10 min dependent on voltage and bonded area.[67] Compared to Si fusion bonding(see below), anodic bonding has the advantage of being a lower temperature process with a lower residual stress and less stringent requirements for the surface quality of the wafers. Figure 8.12 represents a schematic of an anodic bonding set-up. Generally, one places a glass plate on top of the Si wafer and makes a pin-point contact to the uppermost surface of the glass piece which is held at a constant negative bias with respect to the electrically grounded silicon. The bonding is easy to follow. Looking through the glass, the bonded region moves from the contact cathode pin-point outward and can be detected visually through the glass, by the disappearance of the interference fringes. When the whole area displays a dark gray color,

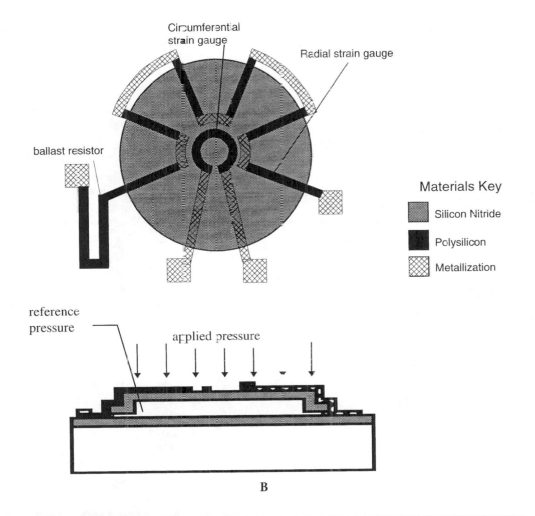

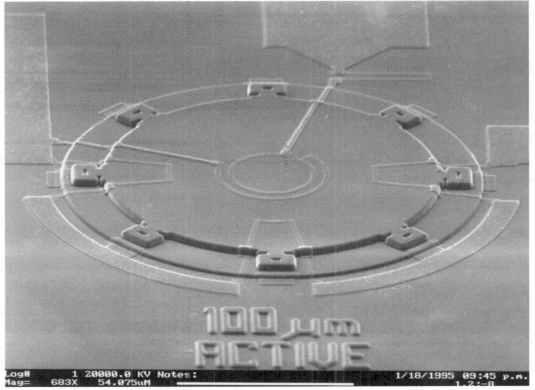

FIGURE 8.8 (Continued)

the bonding is completed. A constant current, instead of constant voltage, could also be used but is avoided since dielectric breakdown may occur after the bonding is completed and the interface becomes an insulator (see bonding mechanism below). The contacting surfaces need to be flat (surface roughness $R_a <$ 1 μm) and dust free for a good bond to form. The native or thermal oxide layer on the Si must be thinner than 200 nm. The thermal expansion coefficients of the bonded materials must match in the range of bonding. In Figure 8.13 we show the thermal expansion coefficient of Si and Pyrex as function of temperature (see also Figure 4.20).[68] Above 450°C the thermal properties of the materials begin to deviate seriously; therefore, the process should be limited to 450°C. One also would expect that Si would be under compression for seal temperatures below

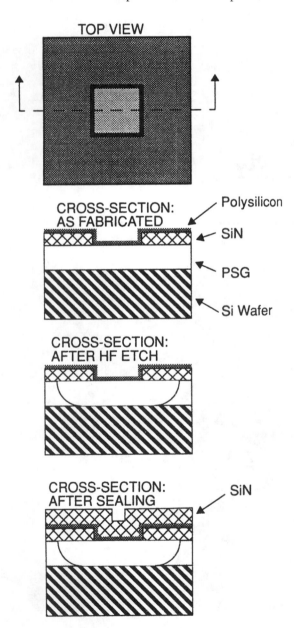

**FIGURE 8.9** Schematic of etch access window design, operation, and sealing. (From Lebovitz et al., 8th Int'l Conf. on Solid-State Sensors and Actuators (Transducers '95), Stockholm, Sweden, 224–227, 1995. With permission.)

280°C and under tension for temperatures in excess of 280°C.[68] Wafer curvature measurements indicate, however, that the transition from concave 7440 glass/Si sandwiches (Si under compression) to convex sandwiches (Si under tension) lies around a seal temperature of 315°C.[68,69] This indicates that other non-negligible, stress-inducing effects add an additional compressive component. As we learned before, tensile stress is preferred over compressive stress, and a considerable safety margin toward higher bonding temperatures must be respected to avoid buckled Si membranes and bridges.

The anodic bonding mechanism is not yet completely understood. Electrochemical, electrostatic, and thermal mechanisms and combinations thereof have been suggested to explain bond formation, but the dominant mechanism has not been clearly defined. It is suggested that at elevated temperatures the glass becomes a conductive solid electrolyte and the bonding results through the migration of sodium (Pyrex contains approximately 3 to 5% sodium) toward the cathode. As it moves, it leaves a space charge (bound negative charges) in the region of the glass-silicon interface. Most of the applied voltage drop occurs across this space charge region and the high electrical field between the glass and Si results in an electrostatic force which pulls the glass and Si into intimate contact. The elevated temperatures result in covalent bonds forming between the surface atoms of the glass and the silicon. A good quantitative discussion on the many important effects in anodic bonding is by Anthony.[70]

Field-assisted bonding has also been applied to bond GaAs to glass. Corning #0211 is used at 360°C, and a bias of 800 V is applied for 30 min to complete the bonding process. It is well known that GaAs forms very poorly adhering oxides, leading to poor anodic bonding prebake of the glass at 400°C for 15 h in a reducing atmosphere ($H_2$ and $N_2$) is reported to lead to better bonding.[71] Von Arx et al. bonded glass capsules to a smooth poly-Si surface to form a hermetically sealed cavity large enough to contain hybrid circuitry of a biocompatible implant.[72]

The high electrical field and the migration of sodium are making anodic bonding of glass plates to Si a rather difficult technology. The mismatch in thermal expansion coefficient between the glass and the Si causes both thermally induced and built-in mechanical stress. In addition, the viscous behavior of the glass results in degraded long-term stability of the components. As a result of these problems, several modifications of the basic technology have been introduced (see below).

### Field-Assisted Thermal Bonding Modifications

The pin-point method for anodic bonding requires a very high bias voltage and a long period of time to bond areas far removed from the cathode point since the electrical field in the glass substrate diminishes fast as the distance from the pin-point cathode increases. At NEC, a Ti mesh bias electrode is deposited over the whole glass wafer to accomplish faster bonding. Because of the mesh assistance, the whole wafer may be Si bonded at 400°C and 600 V in less than 5 min compared to over an hour at the same temperature and voltage without the mesh.[73]

One modification of anodic bonding by Sander[74] involves deposition of intermediate layers of Si dioxide and aluminum to screen the underlying Si from harm from the high electrical

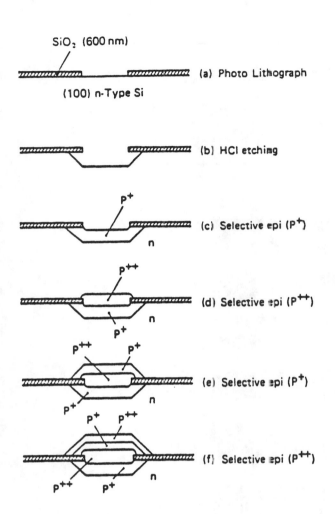

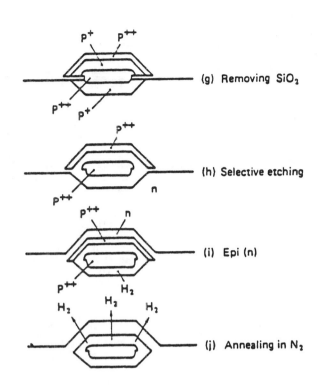

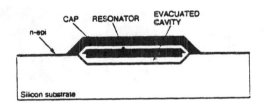

**FIGURE 8.10**　Epilayer-based fabrication process of a vacuum-encapsulated resonant beam. Process from (b) to (f) is carried out in one batch epitaxial process. (From Ikeda et al., *Sensors and Actuators A*, A21, 146–150, 1990. With permission.)

fields. First Si dioxide is thermally grown on the Si surface. Then a layer of aluminum is deposited on the oxide surface. Finally, a piece of glass is bonded to the aluminum. This technique produces a good hermetic seal, but the soft aluminum may be expected to creep after bonding, producing drift in the sensor output. In addition, the aluminum is not corrosion resistant for *in vivo* applications, so that the bond area may corrode rapidly. A similar modification of glass-to-silicon bonding also uses a sandwich structure, but a layer of polycrystalline Si (poly-Si) is deposited on the oxide surface instead of aluminum, and a piece of glass is bonded to the poly-Si (see also Von Arx et al.[72]).

It is also possible to anodically bond two Si wafers with a thin intermediate sputtered[77,78] or evaporated[79] borosilicate glass layer (4 to 7 μm thick). Using a thin intermediate film makes the devices akin to a monolithic structure. Residual stresses and thermal expansion coefficient mismatches create only minor effects on device performance. The surfaces of the Si to be bonded should be polished and one of the two wafers coated with the thin glass layer. This intermediate layer may be sputtered from a Pyrex target, resulting in a rather slow process. Hanneborg, for example, reports a growth rate of 100 nm/hr at a pressure of 5 mTorr argon and sputtering power of 1.6 W/cm².[78] The as-deposited films are highly stressed and must be annealed before bonding at the annealing point of 565°C in a wet oxygen atmosphere.[67] With a 4-μm thick sputtered glass layer, 50 V and a temperature of 450 to 550°C may suffice to

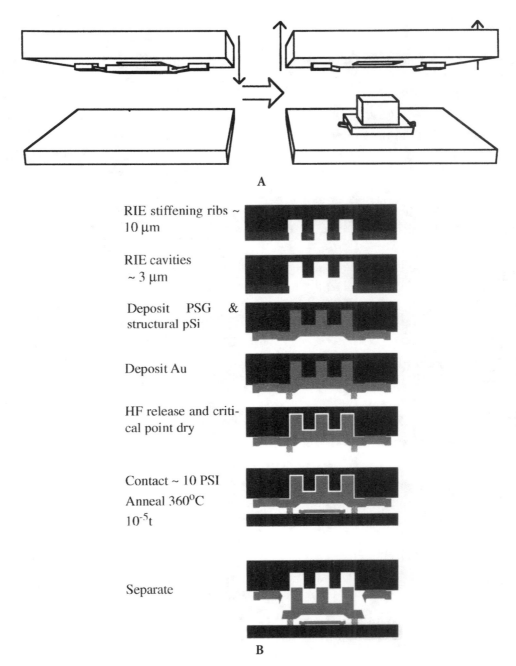

RIE stiffening ribs ~
10 μm

RIE cavities
~ 3 μm

Deposit    PSG    &
structural pSi

Deposit Au

HF release and criti-
cal point dry

Contact ~ 10 PSI
Anneal 360°C
$10^{-5}$t

Separate

**FIGURE 8.11**    Wafer-to-wafer transfer of encapsulation structures. (From Hok, B., Dubon, C., and Ovren, C., *Appl. Phys. Lett.*, 43, 267–269 (1983). With permission.) (A) Principle of the wafer-to-wafer transfer method of encapsulation caps. (B) Fabrication of the tethered caps in a HEXSIL process. (C) SEM micrograph of a transferred cap.[65] (Courtesy of Mr. M. Cohen, UCB, USA.)

create a hermetic bond. With sputtered layers below 4 μm thick, a thermally grown $SiO_2$ must be included between the glass and the Si to avoid electrical breakdown during anodic bonding. The bond strength with Pyrex as an intermediate layer, measured by pulling the wafers apart, ranges from 2 to 3 MPa. As the Pyrex thin films etch very slowly in KOH, they can be used as a mask material for anisotropic etching.[80] Esashi et al.[81] report room temperature anodic bonding using a 0.5- to 4-μm thick film sputtered from a Corning glass #7570 target. This glass has a softening point of 440°C compared to 821°C for Pyrex, and bonding is completed within 10 min with 30 to 60 V applied. A pull test reveals a resulting bond strength in excess of 1.5 MPa. Application of pressure up to 160 kPa on the wafer sandwich apparently supports the bonding process — at a pressure of 160 kPa, the minimum voltage to achieve bonding drops by a factor of two.

The slow deposition rate of glass sputtering presents a major disadvantage for this technique, but evaporated glass layers can be deposited with much higher rates up to 4 μm/min and could provide a solution; unfortunately, in this case pinholes tend to reduce the breakdown voltage.[79]

Field-assisted bonding between two Si wafers both provided with a thermally grown oxide film is also reported to be successful. Both Si wafers must be covered with 1 μm of oxide.

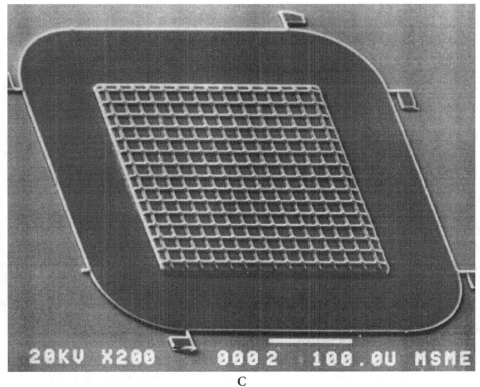

C

**FIGURE 8.11** (Continued)

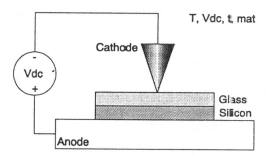

**FIGURE 8.12** Principle sketch of anodic glass-to-Si bonding. Control parameters are temperature (300–400°C), bias voltage (700–1200 V), time (~2′), and materials (glasses, Si, SiO$_2$).

Bonding of a bare Si wafer to a second wafer with oxide failed because of oxide breakdown under very small applied bias. Temperatures range from 850 to 950°C and a voltage of 30 V must be applied for 45 min at the chosen bonding temperature.[82]

### Thermal Silicon Fusion Bonding

The ability to bond two Si wafers directly without intermediate layers simplifies the fabrication of many devices (see, for example, Figure 5.28 and Figure 4.58). This direct bonding of Si to silicon, Si fusion bonding or SFB, is based on a chemical reaction between OH-groups present at the surface of the native or grown oxides covering the wafers. The method is of great interest for the fabrication of Si on insulator (SOI) structures. The processes involved in making these bonds are simpler than other currently employed bonding techniques, the yields are higher, the costs are lower, and the mechanical sensors built according to this principle exhibit an improved performance.

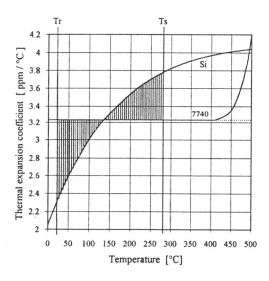

**FIGURE 8.13** Thermal expansion coefficients of Si and Corning 7740 Pyrex. Tr = room temperature; Ts = seal temperature. The temperature Ts is a variable. (From Peeters, E., Process Development for 3D Silicon Microstructures with Application to Mechanical Sensor Design, Ph.D. thesis, Catholic University of Louvain, Belgium, 1994. With permission.)

Finally, as demonstrated in Figure 4.58, the size of an SFB-type device can be almost 50% smaller compared to a conventional anodically bonded chip. Wafer bonding may be achieved by placing the surfaces of two wafers in close contact and inserting them in an oxidizing ambient at temperatures greater than 800°C. It may be noted that the term 'silicon fusion bonding' is somewhat misleading; the fusion point of Si at atmospheric pressure is 1410°C, well below the relevant process temperatures.[80] The quality of the bond depends critically on temperature and the roughness of the surface. Flatness requirements are much more stringent than for anodic bondings, with a microroughness less than 4 nm vs. 1 µm in anodic bonding. Because of the high temperature involved active electronics cannot be incorporated before the bonding takes place (the temperature limit for bonding IC-processed standard Si substrates is about 420 to 450°C; see IC Compatibility in Chapter 5). Higher temperature (above 1000°C) is usually required to get voidless and high strength bonding.[83] Bond strength up to 20 MPa has been reported. The application of a small pressure during the bonding process further increases the final bond strength.[84] The bonding can be done successfully with one oxidized Si wafer to another bare Si wafer, two oxidized wafers, or two bare Si wafers, and even between one wafer with a thin layer of nitride (100 to 200 nm) and one bare wafer or two wafers with a thin layer of nitride. The same fusion-bonding technique can be applied for bonding quartz wafers, GaAs to silicon, and Si to glass. Provided that the surfaces are mirror smooth and can be hydrated, the bonding process proceeds in an identical fashion as for Si-Si bonding (see Schmidt[85] and references therein).

According to most references, before fusion bonding the oxidized Si surfaces must undergo hydration. Hydration is usually accomplished by soaking the wafers in a $H_2O_2$-$H_2SO_4$ mixture, diluted $H_2SO_4$, or boiling nitric acid. After this treatment, a hydrophilic top layer consisting of O–H bonds is formed on the oxide surface. An additional treatment in an oxygen plasma greatly enhances the number of OH groups at the surface.[86] Then the wafers are rinsed in deionized water and dried. Contacting the mirrored surfaces at room temperature in clean air forms self-bonding throughout the wafer surface with considerable bonding forces.[87] In a transmission infrared microscope, a so-called bonding wave can be seen to propagate over the whole wafer in a matter of seconds. The bonded pair of wafers can be handled without danger of the sandwich falling apart during transportation. A subsequent high temperature anneal increases the bond strength by more than an order of magnitude. The self-bonding is the same phenomenon described under stiction of surface micromachined features (see Chapter 5).

At present the fusion bonding mechanism is not completely clear. However, the polymerization of silanol bonds is believed to be the main bonding reaction:

$$Si-OH + OH-Si \rightarrow H_2O + Si-O-Si \qquad \text{Reaction 8.1}$$

Figure 8.14 shows the suggested bonding mechanism including the transformation from silanol bonds to siloxane bonds. Silanol groups give rise to hydrogen bonding, which takes place spontaneously even at room temperature and without applying pressure as long as the two wafers are extremely clean and smooth.[87] Measurements of bond strength as a function of anneal temperature indicate three distinct regions according to Schmidt.[85] For anneal temperatures below 300°C, the bond strength remains the same as the spontaneous bond strength measured before the anneal. For self-bonded wafers with hydrogen bonding between the silanol groups of opposite surfaces, Tong et al.[88] measured an activation energy of 0.07 eV for

temperatures <110°C, fairly close to the theoretical value of around 0.05 eV for hydrogen bonding. At about 300°C, according to Schmidt, the –OH groups form water molecules, and the voids that are sometimes observed at this temperature are believed to be due to water vapor formed in the process.[89] Stengl et al. report that the formation of voids can be avoided by first contacting the wafers at a temperature of 50°C.[90] The voids tend to disappear also at temperatures above 300°C, while the bond strength increases and then levels out. It is assumed that in this regime Si-O-Si bonds (siloxane) start forming between the surfaces (Reaction 8.1). Tong et al.[88] found a pronounced increase of interface energy of room-temperature, self-bonded hydrophilic Si/Si, Si/$SiO_2$, and $SiO_2$/$SiO_2$ wafers after storage in air at room temperature and up to 150°C for periods between 10 and 400 hr. The interfacial energy increase is ascribed to the generation of additional –OH groups due to a reaction with water and the strained oxide and/or Si below 100°C, as well as the formation of stronger siloxane bonds which this group claims already forms from temperatures below 150°C. They find that the siloxane groups have a much higher bond energy with an activation energy of 1.8 to 2.1 eV.[88] After prolonged storage, interface bubbles were observed which seem to contain hydrogen and hydrocarbons. At the highest temperatures (>800°C) Schmidt finds the bond strength starts increasing again, and at 1000°C the bond is reaching the fracture strength of single crystalline silicon: 10 to 20 MPa. In this third region it has been suggested that surfaces can deform more easily (oxide flow), bringing them into better contact.[85] At these temperatures, oxygen also diffuses into the Si lattice.

It should be noted that spontaneous bonding of Si wafers has also been claimed with hydrophobic surfaces. It was suggested that such low temperature bonding is due to van der Waals forces, whereas hydrogen bridging is involved for hydrophilic surfaces. The bonding energy obtained with hydrophobic wafers was as low as 26 mJ/m²; but after annealing at 600°C, the bonding energy reached a value of 2.5 J/m², only attainable at 900°C with hydrophilic wafers.[91] This is in contradiction with most other research which concludes that the wafers must be hydrophilic for SFB.

Some years back it was reported that $Si_3N_4$ was a nonbondable surface.[92] Based on Reaction 8.1, one would assume that the low-temperature direct bonding of $Si_3N_4$ could take place only if oxidation of the nitride introduced silanol groups. Indeed, an oxidized surface of nitride was reported to bond at high temperatures consistent with Si fusion bonding.[93] Unexpectedly, Bower et al.[94] demonstrated extremely strong bonds between Si wafers coated with a smooth, clean layer of LPCVD $Si_3N_4$ at temperatures ranging between 90 and 300°C. A 500-Å thick nitride was first deposited in the LPCVD reactor and then, without breaking the vacuum, activated in an ammonia stream for several minutes. This group speculates that a chemical reaction such as:

$$\equiv Si-\left(N_xH_y\right)+\left(N_xH_y\right)-Si\equiv \rightarrow$$
$$\equiv Si-\left(2N_x\right)-Si\equiv + yH_2 \qquad \text{Reaction 8.2}$$

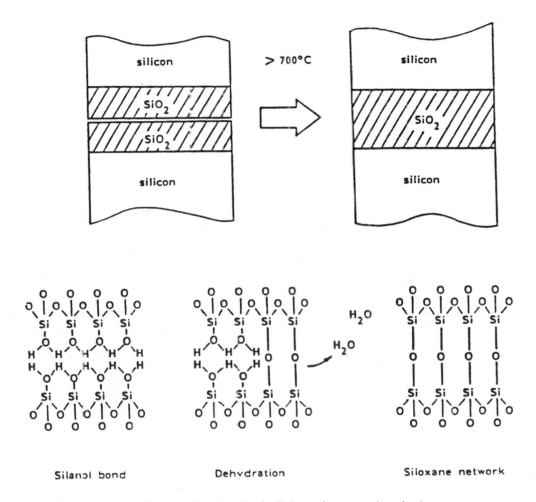

**FIGURE 8.14** Schematic illustration of the silicon-to-silicon bonding by fusion and a proposed mechanism.

is responsible for bonding where the released hydrogen diffuses from the bonding interface. Fracture occurs at about 2 MPa. In a further extension of this work, it was found that a $NH_3$ plasma treatment of Si, $SiO_2$, and TiN for 5 min or longer also provided direct bondable surfaces. Using $Si_3N_4$ as an intermediate layer, Bower et al. bonded silicon to silicon, Si to $SiO_2$, and silicon to GaAs.[95] This low activation energy bonding allows bonding of opto-electronic components such as GaAs with completed electronic circuitry on Si with Al metalization in place.

Suga et al.[96] are taking the use of self-bonding yet one step further by introducing it as a generic approach to assembly and interconnection of MEMS components. Surfaces to be bonded in 'surface activated bonding' or 'SAB' are rendered atomically clean by energetic particle bombardment (argon fast atom beam irradiation of 1.5 keV for several minutes) in ultrahigh vacuum ($<10^{-7}$) and the surfaces are then bonded by contacting them at room temperature. A micro-assembly manipulator, a multiaxial stage with 12 degrees of freedom of motion, affords the contacting of a variety of microparts in the vacuum chamber, and their positions are monitored *in situ* by an SEM. Two pieces of single crystalline Al bonded by SAB at room temperature exhibited a bond strength above 100 MPa.

## Thermal- and Photolithography-Based Bonding with Intermediate Layers: L0 and L1 Level Packaging

### Introduction

When incorporating an intermediate layer between two substrates many thermal bonding techniques are feasible. If the intermediate layers can be patterned by lithography, both L0 and L1 level packaging become possible. Anodic bonding occurs uniformly but the high fields are detrimental to field effect devices. Some intermediate layer materials with a low melting point (e.g., PSG glasses) need no electric field for bonding, but the thickness and uniformity are hard to control. An ideal method would require neither an electrical field nor high temperatures and be very uniform. With films that become soft at low temperatures, all features on the wafer may be covered smoothly as long as they are not a significant fraction of the intermediate film thickness itself.

### Intermediate Thin-Layer Thermal Glass Bonding

LPCVD phosphosilicate glass (PSG) has been used for Si to Si wafer bonding. Fusion of two Si wafers coated with 1- to 2-$\mu$m thick PSG layers is fast. An excellent bond results provided the wafers are clean and reasonably flat. Unfortunately, the bonding process occurs at a high temperature of 1100°C for 30 min.[97]

Several suppliers of glass materials offer low-temperature sealing glasses. Corning Glass Works, for example, has introduced a series of glass frits (#75xx) with sealing temperatures ranging from 415 to 650°C.[98] These glasses can be applied by spraying, screen-printing, extrusion, or sedimentation. After the glass is deposited, it needs to be preglazed to remove the organic residues produced by vehicle and binder decomposition (see Inks in Chapter 3). The substrate-glass-substrate sandwich is then heated to the sealing temperature while a slight pressure is applied (>1 psi). Devitrifying glasses and low-melting-point glasses are offered. The devitrifying glasses are thermosetting and they crystallize during the sealing procedure, thereby changing their mechanical properties. The low melting point vitreous glasses are thermoplastic and they melt and flow during sealing but do not change material properties after sealing is completed. An important problem today remains the uncertainty regarding the mechanical and chemical behavior of these glasses.[74]

Ko et al.[67] used RF sputtering from a target he made from a Corning 7593 glass frit to obtain a thinner (8000 Å) and more uniform intermediate glass film for bonding. The sputtered glass did not need glazing. Annealing at 650°C in an oxygen atmosphere resulted in an excellent bond strength of two Si wafers coated with these films.

Legtenberg et al.[99] used a thin film of atmospheric pressure chemical vapor deposited (APCVD) boron oxide with a softening temperature of 450°C. However, boron oxide is hygroscopic and does not present a viable solution. Field et al.[100] used boron-doped Si dioxide which also becomes soft at 450°C. The doping is performed in a solid source drive-in furnace. The bond seal is hermetic and a crack propagates right through the material rather than along the bonded surface. A problem with this method is the sensitivity of the film to phosphorous contamination.

Spin-on-glass (SOG) was reported by Yamada et al.[102] and Quenzer et al.[103] In Yamada et al.'s effort SOG ($Si(OH)_x$ (with $2 < x < 4$)) was coated on the wafer surface to be bonded. After baking the 50-nm thick film at a temperature of 250°C for 10 min, the wafers are contacted and pressurized in vacuum at 250°C for 1 hr. The bonding is already strong after this step, but wafers are further annealed for 1 hr at 1150°C in air to sinter the SOG layer and to improve the breakdown voltage of the layer. Quenzer et al. use sodium silicate as an intermediate layer. The wafers are made hydrophilic in an RCA1 cleaning step, and the glass is spun on from a dilute sodium silicate solution. The film thickness is between 3 and 100 nm, and annealing at 200°C produces good bonds between bare Si wafers, Si wafers with native oxide, thermal oxide, and Si nitride. No need for further annealing is mentioned in this work.

To bond the unpolished backside of a Si die to another Si part, an aqua-gel based on a hydrophilic pyrogenic silica powder (grain diameter smaller than 7 nm) and polyvinyl alcohol (PVA) as a binder may be used. In this scheme, the adhesion depends on hydrogen bonding between the –OH groups of the PVA, the $SiO_2$ grains, and the Si surfaces. No special surface treatment is required, and the bonding can be achieved in 15 min at room temperature. Bond strength is in excess of 1 MPa. This new technique works for bonding Si to silicon, Si to $SiO_2$, $SiO_2$ to $SiO_2$, and glass to Si or $SiO_2$. Surprisingly, this technique was also shown to work for bonding III–V materials. This type of bonding is proposed for the fashioning of multichip-on-silicon packaging (L2 level).[103]

### Eutectic Bonding

Silicon microstructures can be sealed together by eutectic bonding. The Au-Si eutectic bonding takes place at a temperature of 363°C, which is well below the critical temperature for Al metallized components, and Au is a quite commonly used thin-film material. These are good reasons to select the Au/Si eutectic, but other material combinations are possible. It is possible to bond bare Si against Au covered Si, or Au covered Si against gold-covered silicon. Another alternative is to use Au/Si preform with a composition close to the eutectic concentration. Eutectic bonding in packaging was demonstrated in Figure 8.11. Tiensuu et al. have demonstrated a mean fracture strength of the Au-Si bond of 148 MPa.[104] This compares well with fusion bonding with a typical bond strength of 5 to 15 MPa.

There are also some considerable disadvantages associated with Au/Si eutectic bonding. It is difficult to obtain complete bonding over large areas, and even native oxides prevent the bonding to take place and eutectic preform (e.g., with Au/Sn) bonding is reported to introduce substantial mounting stress in piezoresistive sensors, causing long-term drift due to relaxation of the built-in stress.[67]

### Bonding with Organic Photopatternable Layers

A most interesting new option for packaging is through lithographic patterning of thick resist layers. Prominent new candidates for packaging micromachines are all of the thick ultraviolet photoresists such as polyimides, AZ-4000, and SU-8 reviewed in Chapter 1, as well as LIGA resists such as PMMA and PLG discussed in Chapter 5. Using polymers, very low bonding temperatures are possible. The bond strength can be quite high, no metal ions are present, and the elastic properties can reduce stress. One can envision that with thick enough polymer layers, higher level packaging will become possible (up to level 2). Disadvantages include: impossibility of hermetic seals (see Figure 8.19), high vapor pressure, and poor mechanical properties.

An example of photopatterned bonding is provided by the flexible polysiloxane interconnection between two substrates demonstrated by Arquint et al.[105] UV-sensitive crosslinking of polysiloxane layers with a thickness of several hundreds micrometers is used to pattern the polymer onto the first substrate, and a condensation reaction is used to form a chemical bond to the second surface. The procedure is illustrated in Figure 8.15.[105]

Good bonding results with photopatternable resists have also been obtained by Den Besten et al.,[106] who used negative photoresist at a bonding temperature of 130°C. In an important potential application UV-curable encapsulant resins have been employed for wafer-level packaging of ion-sensitive field effect transistors (ISFETs).[107]

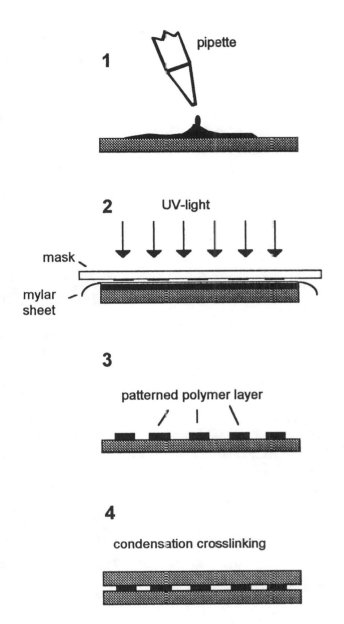

**FIGURE 8.15** Procedure for silicon/polymer/silicon bonding: (1) the monomer solution is deposited either by spin coating or casting. (2) Exposure to ultraviolet light for photopolymerization. 3) The polymeric pattern is developed in xylene. Then, (4) the second wafer is pressed onto the polymeric pattern and left for humidity-induced polymerization in ambient air (10 hours). (From Arquint et al., Flexible Polysiloxane Interconnection between Two Substrates for Microsystem Assembly, 8th Intl. Conf. Solid-State Sensors and Actuators, Stockholm, Sweden, 263–264, 1995. With permission.)

## Alignment During Bonding

Obtaining good alignment between the device wafer and support substrate (glass or another Si wafer) during bonding poses some extra challenges because a transferring action from aligner to annealing furnace is generally required after alignment. One technique that has been used for bonding alignment is to generate holes in both the device wafer and the substrate, followed by putting them in a specially designed fixture to perform the

bonding. The accuracy achievable with this option is only around 50 μm, though. Higher alignment accuracy (~2.5 μm) can be achieved by using a bonding machine equipped with an *in situ* optical alignment set-up. Such aligned wafer bonding has been worked on extensively by Bower et al.[108] They used an infrared aligner modified to hold two imprinted wafers face to face while projecting an infrared image of the surfaces to a viewing screen (see Figure 8.16).[108] An array of V-grooves etched into the surface of the Si wafers was then precisely aligned and the wafers were brought in contact for initial bonding. The (111) planes that define the walls of the V-groove create a shadow image without the necessity of metal features commonly used for two-sided backside alignment. The initial bonding is nothing more than the normal soft contact provided by the aligner itself. Hydrogen bonds are formed between the two surfaces in intimate contact. These relatively weak bonds of about 0.05 eV are sufficient to hold the two pieces firmly together without relative displacement during the transport from the aligner to the annealing furnace. Subsequent high temperature annealing (950°C for 30 min) was used to strengthen and complete the chemical bonding. The aligned-wafer technique can also be used when dealing with dissimilar substrates.

Shoaf et al.[109] developed an alignment technique to assist in precise Au-Si eutectic bonding of Si structures. In this technique a (100) Si wafer is anisotropically etched to create V-grooves around the periphery of the structure to be bonded. Gold is deposited onto one of the wafers prior to dicing into individual dice. Optical fibers are placed into the orthogonal V-grooves, as demonstrated in Figure 8.17, and used as precision location keys for assembly prior to bonding. The entire structure is placed on a hot chuck at 400°C for bonding, and the fibers are removed after bonding. Results have shown a maximum misalignment of 5 μm for a 1 × 1-cm die. This technique allows a sensor die to be precisely bonded to an electronics die without the aid of a microscope or micropositioners.

## Imaging and Bond Strength and Package Hermeticity Tests

Methods for imaging a bonded pair of Si wafers include infrared transmission (voids larger than 20 to 30 μm can be seen this way), ultrasonic (qualitative information about the bond quality), and X-ray topography. The most common mechanical techniques to characterize the bond strength are illustrated in Figure 8.18.[85]

Both the burst test (Figure 8.18A) and the tensile/shear test (Figure 8.18B) yield important engineering insights for sensor construction but do not yield information about the detailed nature of the bond because of the complicated loading of the interface.[85] In the Maszara method[110] a thin blade is inserted between the bonded wafers and a crack is introduced (Figure 8.18C). The length of the crack, measured by infrared imaging, gives a value for the surface energy inferred through a knowledge of the sample and blade thickness and the elastic properties of the wafer. This method has the advantage of creating a well-defined loading on the bonded interface.

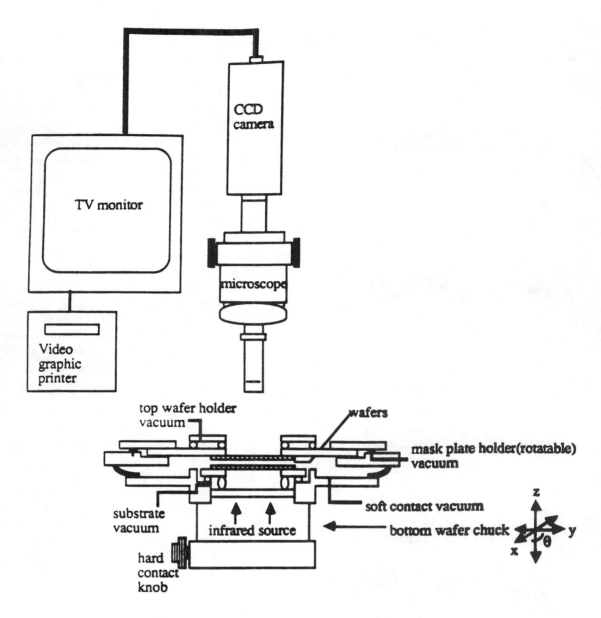

**FIGURE 8.16** Alignment fixture for bonding. The schematic diagram shows the infrared aligner system used for wafer bonding. Wafer holders are designed with a large opening at the center to allow maximum infrared energy to pass through to the camera. (Courtesy Dr. R. Bower, UC Davis, USA.)

Unfortunately, the surface energy is a fourth power of the crack length and uncertainties in that length make for large errors.

The hermeticity of sealed cavities is important for physical protection purposes and in some cases for the performance of the sensor inside. For example, the quality factor, Q, of a resonator, the vacuum reference of an absolute pressure sensor, and the cavity of a pneumatic (i.e., a Golay cell) infrared sensor are all critically dependent on a good hermetic package. Testing of the hermeticity of electronic devices usually is carried out by helium leak detection. While this technique, with a minimum detectable leak rate of $5 \times 10^{-11}$ to $5 \times 10^{-10}$ Torr 1/sec, is appropriate for testing of relatively large packages, it is unacceptable for nearly every Si sensor application. Nese et al. have introduced Fourier Transform infrared spectroscopy (FTIR) for measuring the gas concentration inside a sealed Si cavity using $N_2O$ as a tracer gas.[111] Although the sensitivity for leakage is about

the same as with a helium leak detector, the method is accumulative so that by prolonging the exposure times the sensitivity can be increased, which is not possible with the dynamic helium leakage testing method.

Although anodic bonding is carried out in a vacuum chamber, the pressure inside a Si cavity differs from the chamber pressure because of gas generation, probably oxygen, during bonding. To control cavity pressure for critical damping of packaged micromechanical devices, Minami et al.[112] use nonevaporable getters (e.g., a Ni/Cr ribbon covered with a mixture of porous Ti and Zr-V-Fe alloy which absorbs gases after activation at 400°C) built into the microdevice. These authors monitored the effectiveness of their method to control cavity pressure by measuring deflection of a thin membrane covering the cavity.

Figure 8.19 shows the relative capabilities of several materials to exclude moisture from the encapsulated components over

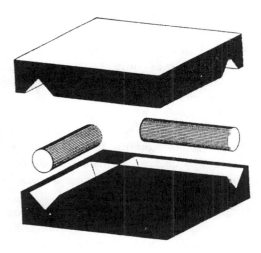

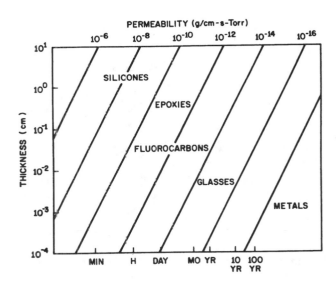

FIGURE 8.17 Schematic representation of the anisotropically etched V-groove/optical fiber alignment technique. Optical fibers are placed into the V-grooves of the bottom Si die. A second die with etched V-grooves is then placed on top of the first die. The fibers act as precision locating keys and align the two Si die. (From Shoaf, S.E. and Feinerman, A.D., Aligned Au-Si Eutectic Bonding of Silicon Structures, *J. Vac. Sci. Technol.*, A12, 19–22, 1994. With permission.)

FIGURE 8.19 The calculated time for moisture to permeate various sealant materials (to 50% of the exterior humidity) in one defined geometry. Organics are orders of magnitude more permeable than materials typically used for hermetic seals. (From Striny, K.M., Assembly Techniques and Packaging of VLSI Devices, *VLSI Technology*, Sze, S.M., Ed., McGraw-Hill, New York, 1988. With permission.)

a truly hermetic seal for a particular package is often prohibitive. In that case, in addition to polymer lid sealing, surface die coats, for example with silicones, are applied. From Figure 8.19[113] we can conclude that silicones do not act as a moisture barrier; the exact mechanism by which they protect the die when applied as a surface coat is not yet well understood.[114]

## Corrugated Structures for Decoupling Micromachines from Package Induced Stress

Corrugated structural members, invented by Jerman,[115] decouple the sensor from its encapsulation, reducing the influence of temperature changes and packaging stress. Corrugated members have been implemented in single crystal silicon, poly-Si, and polyimide (see Figure 5.19). The technique was described in Chapter 5 as one of the methods to control stress in micromachined structures. The stress-decoupling, corrugated bellows are either surrounding the sensor structure itself[116,117] or the sensor chip is mounted via an intermediate with a bellow structure to the housing.[118] The latter implementation is sketched in Figure 8.20.

## Connections between Layers: Multichip Packages

Micromachined chips may be packed laterally as done in multichip modules (MCMs) or they may be stacked on top of each other using micromachining. In the latter case, individual dice, blocks of dice, and entire wafers may be stacked with electrical interconnections running vertically from plane to plane.[119] There are several options to make the vertical interconnects. At first sight, chemical etching is the simplest way to form holes (vias) through Si. The author applied this method for the fabrication

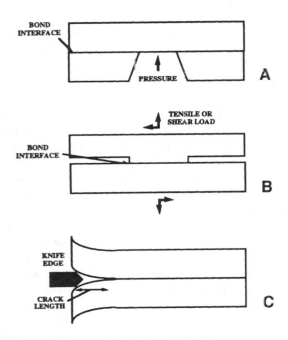

FIGURE 8.18 Three methods for bond strength measurements. (A) Burst test. (B) Tensile and shear test. (C) Maszera test. (From Schmidt, M.A., Silicon Wafer Bonding for Micromechanical Devices, Technical Digest 1994 Solid State Sensor and Actuator Workshop, South Carolina, 1994. With permission.)

long periods of time. Organic materials are not good candidates for hermetic packages. For almost all high-reliability applications, the hermetic seal is made with glass or metal. For measuring moisture penetration in a package, temperature-accelerated soak tests may be performed. Moisture penetration can be followed, for example, with an integrated on-chip dew-point sensor.[72] Striny[113] points out that the cost or difficulty of obtaining

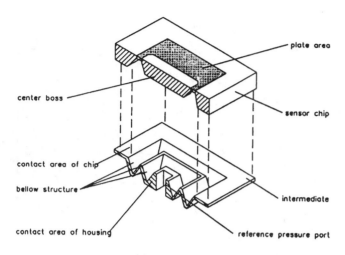

center boss

contact area of chip

bellow structure

contact area of housing

plate area

sensor chip

intermediate

reference pressure port

**FIGURE 8.20**  Stress release bellows. A corrugated intermediate Si piece decouples the sensor from its housing. (From Offereins et al., Stress Free Assembly Technique for a Silicon Based Pressure Sensor, 6th Int'l. Conference on Solid-State Sensors and Actuators, San Francisco, CA, 1991. With permission.)

of an electrochemical sensor array bonded to a bottom Si chip carrying the electronics (see Figures 4.66 to 4.71). Unfortunately, the aspect ratio, length vs. diameter, is low, less than 1. So, integration is limited by space between the holes. With dry etching of vias, problems so far included a lack of sufficiently fast etch rates and sufficient selectivity over the mask. Conventional reactive ion etching (RIE), does not match these requirements, but recent advances in high density plasma sources allow for etch rates up to 4 $\mu$m/min, a selectivity of Si vs. $SiO_2$ of over 150 (70:1 for resists) and aspect ratios of 30:11.[120,121] Other alternatives include laser and ultrasonic drilling and Al thermomigration. With laser drilling an aspect ratio of 1:50 can be reached, and relative high drilling speeds (e.g., 10 holes per sec) are possible at a power density of $10^{11}$ W/cm$^2$. An example application in the production of a capacitive pressure sensor is shown in Figure 7.18. Ultrasonic drilling makes for cleaner vias than does laser machining which, unless there is a reactive gas involved, leaves debris on the via rims. But the technique is limited to 'large' diameter holes (100 $\mu$m and up). Laser-machined holes are compared with ultrasonically drilled holes in Figure 7.25.

After etching the hole, a metal must be deposited for electrical connection. Most methods described to deposit the metal in the formed vias combine sputtering and electroless or electroplating. When the deposition process is limited by the diffusion of species in the hole chemical vapor deposition may be used (see also Chapter 7).

Localized, very deep (throughout the thickness of a wafer) doping of Si is possible by temperature gradient zone melting (TGZM).[122] In this case, via formation and metal deposition are one and the same process. An example is the fabrication of Al interconnects through a Si wafer. At sufficiently high temperatures, aluminum will form an alloy with Si. If the Si substrate is subjected to a temperature gradient, the molten alloy zone will migrate to the hotter side of the wafer. In practice, Al thermomigration in Si starts with electron-beam deposition of

a thick layer of Al (>5 $\mu$m) and photolithographically defining the aluminum into the desired pattern on one side of the Si wafer. The wafer is then radiatively heated from the other side to temperatures considerably higher (1000 to 1200°C) than the eutectic temperature to form the molten zone. The one-sided heating imposes a thermal gradient of up to 50°C/cm across the molten zone, resulting in a highly aluminum-doped ($2 \times 10^{19}$ cm$^{-3}$) zone of single crystal Si. Due to the speed (the zone moves through the silicon, on the order of 10 $\mu$m/min) wafers can be doped through their thickness in minutes. With TGZM, the concomitant side diffusion is a few microns compared to mils for solid-state diffusion.

In principle, this technique offers very exciting opportunities for novel three-dimensional structures, as connections can be made front to back on a Si wafer. The lines through the silicon, being active junctions, are light sensitive and the Al on the exit side can turn rough so that a new polishing step warrants itself. The process, being so extreme, also needs to precede the implementation of any other structure on the wafer.

## Self-Assembly

In micromachining one often faces the challenge of manually assembling components from different technologies. The magnet in Figure 9.12, for example, is glued manually onto its Si support. An attractive option to integrate dissimilar process technologies, such as the macromachined permanent magnet and the micromachined Si wafer substrate in Figure 9.12, or in general CMOS with MEMS and photonics devices involve self-assembly. To achieve self-assembly there must be bonding forces present, the bonding must occur selectively, and the assembling parts must be moving randomly so that they can come together by chance. The essence of the approach is to fabricate large numbers of microstructures, and then 'place' them at predetermined sites on a target substrate using a batch process. The sites, which may be defined by low-resolution lithography, are designed to have an affinity for the microstructures, e.g., via electrostatic or magnetic forces. Placement proceeds spontaneously and in parallel since the potential energy wells represent lower energy states for the wandering microstructures. Yando, for example, employs an array of magnetic sites to assemble magnetically coated semiconductor dice in a square array. The dice are vibrated at a gradually attenuated amplitude to 'place' the array elements.[123] Yeh et al.[19,124] trap semiconductor LEDs, suspended in a liquid, in micromachined wells on a wafer by solvent-surface forces. As illustrated in Figure 8.21, carrier fluid containing the GaAs dice is dispensed over the host Si wafer with etched holes to assemble the GaAs blocks.[19] Because of the trapezoidal design, the blocks fit preferentially into the holes in the design orientation. Random mechanical vibration of the microstructures enables large numbers of microstructures to be positioned into precise registration with the sites. Greater than 90% of the holes etched in Si were correctly filled by the blocks before the carrier fluid evaporated.

Cohn et al.[125] demonstrated an interesting variation on Yeh et al.'s work by using alignment capabilities over electrostatic traps. The process is illustrated in Figure 8.22. A critical problem

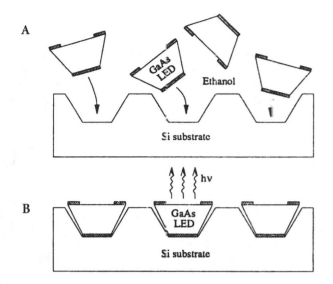

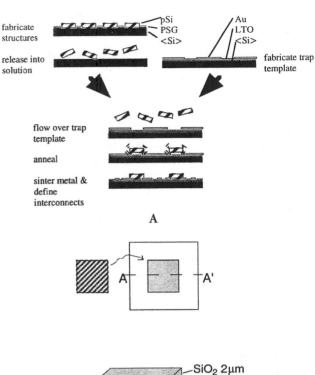

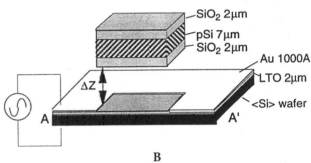

**FIGURE 8.21** Fluid self-assembly of GaAs LEDs on Si micromachined substrate. (A) Solution containing the GaAs blocks is dispensed over the patterned silicon substrate. (B) Silicon substrate with GaAs LEDs integrated by fluidic self-assembly. (From Yeh, H.-J. and J.S. Smith, Fluidic Self-Assembly of Microstructures and its Application to the Integration of GaAs on Si, IEEE Intl. Workshop on Microelectro Mechanical Systems, Oiso, Japan, 1994. With permission.)

in attempting to vibrate microstructures into specific sites from a liquid is that when structures adhere to the substrate or to each other, their progress toward optimal, precise registration is halted. This is caused by the same stiction phenomena we described for surface micromachined structures in Chapter 5. Cohn et al. are addressing this problem by levitating the microstructures a short distance above the target electrostatic trap site before letting them settle (see Figure 8.22). As a result random sticking is prevented and critical positioning and orientation of the microstructure with respect to the trap become possible with a relatively low applied field (10 V/µm). The levitation over the trap site is accomplished by creating a short-range repulsive force between the microstructure and the target site. Microstructures with a relatively high permitivity ($\varepsilon = 10$) are suspended in a low-permitivity solvent (hexane), resulting in a net attraction to the charged electrodes comprising the sites. However, a small amount of a more polar solvent (acetone) is added which segregates into the high-field region between the two attracted parts. This prevents contact between the site and the suspended microstructure. Critical positioning and orientation become possible, in the manner of a compass needle resting on a low-friction bearing.[125] Once the sites are populated, the polar solvent is titrated out, gradually lowering the trapped devices to the target sites and pinning them onto the surface. A subsequent sintering step could effect permanent attachment say by eutectic bonding. Hosokawa et al.[126] have demonstrated two-dimensional self assembly of thin film parts consisting of a polyimide layer on a polysilicon layer (assembly units are 400 µm in size) floating on water. The units are selectively bonded to each other using the characteristics of surface tension (a dominant force in the microscale, see Chapter 9). The floating parts that are at equal height are attracted to each other. Sharp

**FIGURE 8.22** Self-assembly using electrostatic levitation. (A) Batch assembly of microstructures into binding sites on a substrate. Structures may include MEMS, CMOS, opto-electronics, magnets, etc. (B) Detail of an electrostatic binding site. The section view shows position of a levitated microstructure. The levitation height $\Delta z$ is in the range of 0 to 100 µm. The microstructure, a $SiO_2$-pSi-$SiO_2$ sandwich, has an average dielectric constant $\varepsilon$ of 10, between that for hexane and acetone. The aperture in the Au film may range from ~2 to 100 µm diameter, depending on the size of the microstructure to be trapped. (Courtesy of Mr. M. Cohen, UC Berkeley, CA.)

corners induce yet larger selective attractive forces and parts located at different heights are mutually repulsed. To disturb the self-assembly system, magnetic or fluid forces are used. The floating units tend to align, with their sharpest features pointing toward each other. Successful assembly of microstructures will require many novel approaches like the ones described above.

## Higher Levels of Packaging: L2 to L5

### Introduction

Micromachining deviates most dramatically from ICs in the nonstandard nature of packages from level 2 up. In the case of some sensors, standard TO-5 and TO-8 headers might be used (see Figure 8.6B for example). In most cases the package is dictated by the specific application. A TO-8 packaged sensor and a packaged entrance slit for a precision spectrophotometer are presented as examples of these two extremes.

### Sensor Die Attach and Wire Bond in a TO-8 Header

After dicing a sensor die (for example, a Si/glass piezoresistive pressure sensor as illustrated in Figure 8.6B) the die attach step follows, for example, to a TO-8 header. A cross-section of a TO-8 header is shown in Figure 8.23. Gold-plated pins are hermetically sealed to the header base with glass eyelets which provide a matched hermetic feed through with both high electrical and mechanical integrity. To protect the mounted sensor a metal can is hermetically resistance-welded to the header. In the header shown in Figure 8.23, the pressure is interfaced to the sensor via a drawtube integrated in the header cap. Alternatively, in the pressure sensor illustrated in Figure 8.6B the pressure tube is interfaced through the header base. Since the thermal expansion coefficient of the package material to which the die is attached typically differs substantially from that of Si and glass, the die attachment must compensate for this. The three methods used the most for die attach are eutectic (Au-Si alloy), epoxy, and silicone rubber, in order of highest to lowest stress.[45,46] The next step in the process consists of the electrical connection to the Si sensor. Gold or aluminum wire is bonded by thermosonic, thermocompression, or wedge-wedge ultrasonic wire bonding. Only in the case of very fragile structures do the ultrasonics damage the device. The fine gold wires in the TO-8 package illustrated in Figure 8.6B are barely visible.

### Die Protection

The last assembly step involves the protection of the silicon die and electrical leads. Various methods are available including:[45,46]

- Vapor-deposited organics (e.g., parylene)
- Silicone gel coating over the die (see above)
- A plastic or ceramic cap for particle and handling protection
- A welded-on nickel cap with pressure port (see Figure 8.23)

### Nonstandard Packages

Pressure sensors and accelerometers aside, few micromachines fit in standard packages. Examples of nonstandard packages we discussed in this book are the catheter-based electrochemical sensor shown in Figures 4.66 to 4.71, where a dual lumen catheter forms the basis of the packaging scheme, the thin-film magnetic head illustrated in Figure 7.8, and GM's MAP pressure sensor displayed in Figure 10.10. The very fragmented nature of micromachining applications and the nonstandardness of the package necessitate a design approach starting from the package.

An excellent additional example of nontraditional packaging is that of a variable entrance slit for a precision spectrophotometer. The slit system pictured in Figure 8.24 also represents a good example of combining Si micromachining with conventional mechanical parts.[127] The slit system is composed of two aluminum units (Figure 8.24A). The cover unit houses the entrance window and protects the slit displacement mechanism mounted on the front face of the body unit. The slit system itself is shown in more detail in Figure 8.24B. It involves two superimposed Si plates whose frames are assembled together on reference pins. The upper plate has a fixed aperture in the center defining the slit height H (see schematic in Figure 8.24C). The lower plate defines the slit width and comprises a set of five slits and their guiding structure for positioning the slits under the aperture hole. The elastic beam structure has the very important function of allowing millimeter-long linear translations while restricting lateral excursions to a fraction of a micrometer. Translations of the slits are operated by means of transmission gears connected to a standard stepping motor embedded in the aluminum body. The slit plates are made by bulk Si micromachining, resulting in a slit width accuracy of ±1 μm. The reproducibility of the slit center position is better than 0.01 μm. The rest of the system is composed of conventional parts fabricated with the usual tools of the watch and fine mechanics industry. The last example illustrates a typical challenge every micromachinist faces, i.e., how to adapt a Si microstructure to a completely unconventional package. For the mechanical aperture application, the adaptation is relatively simple. In the case of the biosensor shown in Figure 4.66 to 4.71, it is an extremely challenging proposition.

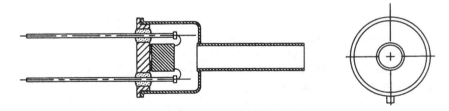

**FIGURE 8.23** Crossection of a pressure sensor die-mounted in a TO-8 header. The metal cap holds an integrated drawtube to interface the external pressure to the sensor inside.

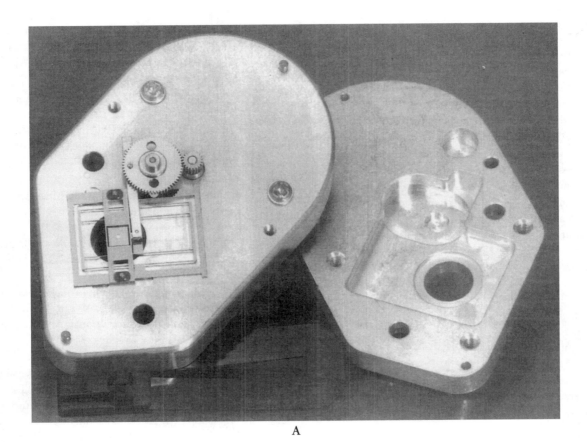

A

B

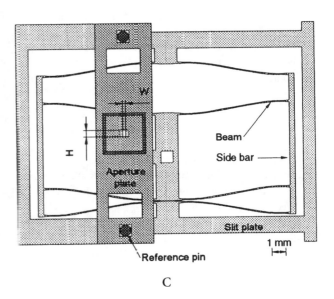

C

**FIGURE 8.24** Variable entrance slit system for precision spectrophotometers. (A) Overall view of the variable slit system with package, showing slit displacement mechanism on the aluminum body (left) and cover (right). (B) Close-up view of the variable slit system showing the Si plate assembly together with the driving gears and connecting bar. (C) Dual Si plate assembly showing the aperture plate (defining the slit height, H) over the slit plate (defining the slit width, W). (From Vuilleumier, R., Variable Entrance Slit System for Precision Spectrophotometers, Proceedings, IEEE Micro Electro Mechanical Systems, Amsterdam, Netherlands, 1995. With permission.)

# References

1.  Senturia, S., "Can We Design Microrobotic Devices Without Knowing the Mechanical Properties of Materials?", Proceedings of the IEEE Micro Robots and Teleoperators Workshop, Hyannis, MA, 1987, p. 3/1–5.

2.  Hayashi, T., "Micromechanism and Their Characteristics", IEEE International Workshop on Micro Electro Mechanical Systems, MEMS '94, Oiso, Japan, 1994, p. 39–44.

3.  Goemans, P. A. F. M., "Microsystems and Energy: The Role of Energy", Microsystem technology: exploring opportunities, 1994.

4.  Senturia, S. D. and R. T. Howe, "Mechanical Properties and CAD", Lecture Notes, MIT, Boston, MA, 1990.

5.  Studer, B. and W. Zingg, "Technology and Characteristics of Chemically Milled Miniature Quartz Crystals", 4th European Frequency and Time Forum, Neuchatel, Switzerland, 1990, p. 653–658.

6.  Tellier, C. R. and F. Jouffroy, "Orientation Effects in Chemical Etching of Quartz Plates", *J. Mater. Sci.*, 18, 3621–3632, 1983.

7.  Ueda, T., F. Kohsaka, T. Lino, and D. Yamazoki, "Theory to Predict Etching Shapes in Quartz and Applications to Design Devices", *Trans. Soc. Inst. Control. Eng.*, 23, 1–6, 1987.

8.  Chuang, S. S., "Force Sensor Using Double Ended Tuning Fork Quartz Crystals", Proceedings of the 37th Annual Frequency Control Symposium 1983, Philadelphia, PA, 1983, p. 248–254.

9.  Dufour, M., M. T. Delaye, F. Michel, J. S. Danel, B. Diem, and G. Delapierre, "A Comparison Between Micromachined Pressure Sensors Using Quartz or Silicon Vibrating Beams", 6th International Conference on Solid-State Sensors and Actuators (Transducers '91), San Francisco, CA, 1991, p. 668–671.

10.  Editorial, "Quartz Flexure Accelerometer", Systron Donner, BEI, 2700 Systron Drive, Concord California 94518, 1994.

11.  Hjort, K., G. Thornell, R. Spohr, and J.-Å. Schweitz, "Heavy Ion Induced Etch in Single Crystalline Quartz", The Ninth Annual International Workshop on Micro Electro Mechanical Systems, San Diego, CA, USA, 1996, p. 267–271.

12.  Ericson, F., S. Johansson, and J.-Å. Schweitz, "Hardness and Fracture Toughness of Semiconducting Materials Studied by Indentation and Erosion Techniques", *J. Mater. Sci. Eng.*, A105/106, 131–41, 1988.

13.  Takebe, T., T. Yamamoto, M. Fujii, and K. Kobayashi, "Fundamental Selective Etching Characteristics of $HF+H_2O_2+H_2O$ Mixtures for GaAs", *J. Electrochem. Soc.*, 140, 1169–1180, 1993.

14.  Zhang, Z. L. and N. C. MacDonald, "Fabrication of Submicron High-Aspect-Ratio GaAs Actuators", *J. Microelectromech. Syst.*, 2, 66–73, 1993.

15.  Hjort, K., J. A. Schweitz, and B. Hok, "Bulk and Surface Micromachining of GaAs Structures", Proceedings. IEEE Micro Electro Mechanical Systems, (MEMS '90), Napa Valley, California, 1990, p. 73–76.

16.  Hjort, K., J. A. Schweitz, S. Andersson, O. Kordina, and E. Janzen, "Epitaxial Regrowth in Surface Micromachining of GaAs", Proceedings. IEEE Micro Electro Mechanical Systems, (MEMS '92), Travemunde, Germany, 1992, p. 83–86.

17.  Karam, J. M., B. Courtois, M. Holjo, J. L. Leclercq, and P. Viktorotovitch, "Collective Fabrication of Gallium Arsenide Based Microsystems", SPIE, Micromachining and Microfabrication Process Technology II, Austin, Texas, USA, 1996, p. 315–324.

18.  Morkoc, H., H. Unlu, H. Zabel, and N. Otsuka, "Gallium Arsenide on Silicon: A Review", *Solid State Technol.*, March, 71–76, 1988.

19.  Yeh, H.-J., J. Smith, J. S., "Fluidic Self-Assembly of Microstructures and its Application to the Integration of GaAs on Si", IEEE International Workshop on Micro Electro Mechanical Systems, MEMS '94, Oiso, Japan, 1994, p. 279–284.

20.  Tong, L., M. Mehregany, and L. G. Matus, "Silicon Carbide as a New Micromechanics Material", Technical Digest of the 1992 Solid State Sensor and Actuator Workshop., Hilton Head Island, SC, 1992, p. 198–201.

21.  Krotz, G., W. Legner, C. Wagner, H. Moller, H. Sonntag, and G. Muller, "Silicon Carbide as A Mechanical Material", 8th International Conference on Solid-State Sensors and Actuators (Transducers '95), Stockholm, Sweden, June, 1995, p. 186–189.

22.  Madou, M. J. and A. M. Agarwal, "SiC as a High-Temperature Semiconductor pH Sensor", 2nd International Meeting on Chemical Sensors, Bordeaux, France, 1986.

23.  Yamaguchi, Y., H. Nagasawa, T. Shoki, and N. Annaka, "Properties of Heteroepitaxial 3C-SiC Films Grown by LPCVD", 8th International Conference on Solid-State Sensors and Actuators (Transducers '95), Stockholm, Sweden, June, 1995, p. 190–193.

24.  Wagner, C., G. Krotz, H. Sonntag, H. Moller, G. Muller, and S. Kalbitzer, "Applications of Crystalline-Amorphous Phase Transitions in Silicon Carbide", 8th International Conference on Solid-State Sensors and Actuators (Transducers '95), Stockholm, Sweden, June, 1995, p. 194–197.

25.  Ho, C. P., J. D. Plummer, S. E. Hansen, and R. W. Dutton, "VLSI process modeling SUPREM 3", *IEEE Trans. Electron Devices*, ED-30, 1438–1452, 1983.

26.  ERL, "SAMPLE Version 1.6a User's Guide", Electronics Research laboratory (ERL), UC Berkeley, CA, USA, Berkeley, CA, 1985.

27.  Oldham, W. G., S. N. Nandgaonkar, A. R. Neureuther, and M. O'Toole, "A General Simulator for VLSI Lithography and Etching Processes: Part 1 – Application to Projection Lithography", *IEEE Trans. Electron Devices*, ED-26, 717–722, 1979.

28. Oldham, W. G., A. R. Neureuther, J. L. Reynolds, S. N. Nandgaonkar, and C. Sung, "A General Simulator for VLSI Lithography and Etching Processes: Part 2–Application to Deposition and Etching", *IEEE Trans. Electron Devices*, ED-27, 1455–1559, 1980.

29. Fichtner, W., "Process Simulation", in *VLSI Technology*, Sze, S. M., Ed., McGraw-Hill, New York, 1988, p. 422–465.

30. Maseeh, F., "A CAD Architecture for Microelectromechanical Systems", Proceedings. IEEE Micro Electro Mechanical Systems, (MEMS '90), Napa Valley, CA, 1990, p. 44–49.

31. Billingsley, G. C., "Program Reference for KIC", UC Berkeley, Report No. UCB/ERL M83/62, 1983.

32. Marshall, J., C., Parameswaran,M., Zaghloul, M. E. and Gaitan, M., "High-Level CAD Melds Micromachine Devices with Foundries", *IEEE Circuits Devices Mag.*, 8, 10–17, 1992.

33. Tomovich, C., Ed., "MOSIS User Manual", University of Southern California, 1988.

34. Koppelman, G. M., "OYSTER, a 3D Structural Simulator for Micro Electromechanical Design", Proceedings. IEEE Micro Electro Mechanical Systems, (MEMS '89), Salt Lake City, UT, 1989, p. 88–93.

35. Gilbert, J. R., G. K. Ananthasuresh, and S. D. Senturia, "3D Modeling of Contact Problems and Hysteresis in Couple Electro-Mechanics", The Ninth Annual International Workshop on Micro Electro Mechanical Systems, MEMS '96, San Diego, CA, USA, 1996, p. 127–132.

36. Crary, S. and Y. Zhang, "CAEMEMS: An Integrated Computer Aided Engineering Workbench for Micro-Electr-Mechanical Systems", Proceedings. IEEE Micro Electro Mechanical Systems, (MEMS '90), Napa Valley, CA, 1990, p. 113–114.

37. Korvink, J. G., J. Funk, M. Roos, G. Wachutka, and H. Baltes, "SESES: A Comprehensive MEMS Modelling System", IEEE International Workshop on Micro Electro Mechanical Systems, MEMS '94, Oiso, Japan, 1994, p. 22–27.

38. Maseeh, F., "A Novel Multidimensional Semiconductor Material Analysis Tool", *Solid State Techno.*, 37, 83–4, 1994.

39. Sequin, C. H., "Computer Simulation of Anisotropic Crystal Etching", *Sensors and Actuators A*, A34, 225–241, 1992.

40. Koide, A., Sato, K. and Tanaka, S., "Simulation of Two-Dimensional Etch Profile of Silicon During Orientation-Dependent Anisotropic Etching", Proceedings. IEEE Micro Electro Mechanical Systems, (MEMS '91), Nara, Japan, 1991, p. 216–220.

41. Hubbard, T. J. and E. K. Antonsson, "Emergent Faces in Crystal Etching", *J. Microelectromech. Syst.*, 3, 19–28, 1994.

42. Private Communication, Murphy, M., *CAD for High Aspect Ratio and truly 3-D Micromachines*, Personal communication, 1995.

43. Hill, S. D. J., Kamper, K. P., Dasbach, U., Dopper, J., Ehrfeld, W., Kaupert, M., "An Investigation of Computer Modelling for Micro-Injection Moulding", in *Simulation and design of Microsystems and Microstructures*, Adey, R. A., Lahrmann, A., and Lessmollmann, C., Ed., Computational Mechanics Publications, Boston, 1995, p. 276–283.

44. Adey, R. A., A. Lahrmann, and C. Lessmollmann, Eds. *Simulation and Design of Microsystems and Microstructures*, Computational Mechanics Publications, Southampton, 1995.

45. Lyke, J. C., "Packaging Technologies for Space-Based Microsystems and Their Elements", in *Microengineering Technologies for Space Systems*, Helvajian, H., Ed., The Aerospace Corporation, El Segundo, CA, 1995, p. 131–180.

46. Jensen, J. R., "Interconnection and Packaging of High-Performance Integrated Circuits", in *Microelectronics Processing*, Hess, D. W. and Jensen, K. F., Eds., American Chemical Society, Washington, DC, 1989, p. 441–504.

47. Allen, R., "Sensors in Silicon", *High Technology*, September, 43–81, 1984.

48. Guckel, H. and D. W. Burns, "Planar Processed Polysilicon Sealed Cavities for Pressure Transducer Arrays", IEEE International Electron Devices Meeting. Technical Digest, IEDM '84, San Francisco, CA, 1984, p. 223–5.

49. Guckel, H. and D. W. Burns, "Fabrication Techniques for Integrated Sensor Microstructures", IEEE International '86, Los Angeles, CA, 1986, p. 176–179.

50. Guckel, H. and D. W. Burns, "A Technology for Integrated Transducers", International Conference on Solid-State Sensors and Actuators, Philadelphia, PA, 1985, p. 90–2.

51. Guckel, H., D. W. Burns, C. K. Nesler, and C. R. Rutigliano, "Fine Grained Polysilicon and its Application to Planar Pressure Transducers", 4th International Conference on Solid-State Sensors and Actuators (Transducers '87), Tokyo, Japan, 1987, p. 277–282.

52. Lin, L., K. M. McNair, R. T. Howe, and A. P. Pisano, "Vacuum-Encapsulated Lateral Microresonators", 7th International Conference on Solid-State Sensors and Actuators (Transducers '93), Yokohama, Japan, 1993, p. 270–273.

53. Eaton, W. P. and J. H. Smith, "A CMOS-compatible, Surface-Micromachined Pressure Sensor for Aqueous Ultrasonic Application", Smart Structures and Materials 1995. Smart Electronics, (Proceedings of the SPIE), San Diego, CA, 1995, p. 258–65.

54. Lebouitz, K. S., R. T. Howe, and A. P. Pisano, "Permeable Polysilicon Etch-Access Windows for Microshell Fabrication", 8th International Conference on Solid-State Sensors and Actuators (Transducers '95), Stockholm, Sweden, June, 1995, p. 224–227.

55. Guckel, H., D. W. Burns, and C. R. Rutigliano, "Design and Construction Techniques for Planar Polysilicon Pressure Transducers with Piezoresistive Read-out", Technical Digest of the 1986 Solid State Sensor and Actuator Workshop., Hilton Head Island, SC, 1986.

56. Erskine, J. C., "Polycrystalline Silicon-on-Metal Strain Gauge Transducers", *IEEE Trans. Electron Devices*, 30, 796–801, 1983.

57. SSI Technologies, I., "Solid-State Integrated Pressure Sensor", 1995, SSI Technologies, Inc., Janesville, WI.

58. Cho, Y.-H., B. M. Kwak, A. P. Pisano, and R. T. Howe, "Viscous Energy Dissipation in Laterally Oscillating Planar Microstructures", Proceedings. IEEE Micro Electro Mechanical Systems, (MEMS '93), Fort Lauderdale, FL, 1993, p. 93–98.

59. Tang, W. C., T. H. Nguyen, and R. T. Howe, "Laterally Driven Polysilicon Resonant Microstructures", *Sensors Actuators*, 20, 25–32, 1989.

60. Lin, L., T. C. Nguyen, R. T. Howe, and A. P. Pisano, "Micro Electromechanical Filters for Signal Processing", Proceedings. IEEE Micro Electro Mechanical Systems, (MEMS '92), Travemunde, Germany, 1992, p. 226–231.

61. Hsu, C. H. and R. S. Muller, "Micromechanical Electrostatic Voltmeter", 6th International Conference on Solid-State Sensors and Actuators (Transducers '91), San Francisco, CA, 1991, p. 659–662.

62. Bernstein, J., S. Cho, A. T. King, A. Kourepenis, P. Maciel, and M. Weinberg, "A Micromachined Comb-Drive Tuning Fork Rate Gyroscope", Electro Mechanical Systems Workshop (MEMS '93), Fort Lauderdale, FL, 1993, p. 143–148.

63. Ikeda, K., H. Kuwayama, T. Kobayashi, T. Watanabe, T. Nishikawa, T. Yoshida, and K. Harada, "Three-Dimensional Micromachining of Silicon Pressure Sensor Integrating Resonant Strain Gauge on Diaphragm", *Sensors and Actuators A*, A23, 1007–1010, 1990.

64. Ikeda, K., H. Kuwayama, T. Kobayashi, T. Watanabe, T. Nishikawa, T. Yoshida, and K. Harada, "Silicon Pressure Sensor Integrates Resonant Strain Gauge on Diaphragm", *Sensors and Actuators A*, A21, 146–150, 1990.

65. Cohn, M. B., Y. Liang, R. T. Howe, and A. P. Pisano, "Wafer-to-Wafer Transfer of Microstructures for Vacuum Packaging", Technical Digest of the 1996 Solid State Sensor and Actuator Workshop., Hilton Head Island, SC, 1996, p. 32–35.

66. Wallis, G. and D. I. Pomerantz, "Field assisted glass-metal sealing", *J. Appl. Phys.*, 40, 3946–9, 1969.

67. Ko, W. H., J. T. Suminto, and G. J. Yeh, "Bonding Techniques for Microsensors", in *Micromachining and Micropackaging of Transducers*, Fung, C. D., Cheung, P. W., Ko, W. H., and Fleming, D. G., Eds., Elsevier, Amsterdam, 1985, p. 41–61.

68. Peeters, E., "Process Development for 3D Silicon Microstructures, with Application to Mechanical Sensor Design", Ph.D. Thesis, Catholic University of Louvain, Belgium, 1994.

69. Puers, B., Cozma, A., Van De Weyer, E., "Technology Establishment for Silicon-Glass Electrostatic Bonding", Intermediate Project Report, KU Leuven, 1993.

70. Anthony, T., R., "Anodic Bonding of Imperfect Surfaces", *J. Appl. Phys.*, 54, 2419–28, 1983.

71. Hok, B., C. Dubon, and C. Ovren, "Anodic Bonding of Galium Arsenide to Glass", *Appl. Phys. Lett.*, 43, 267–269, 1983.

72. Von Arx, J., B. Ziaie, M. Dokmeci, and K. Najafi, "Hermeticity Testing of Glass-Silicon Packages with On-Chip Feedthroughs", 8th International Conference on Solid-State Sensors and Actuators (Transducers '95), Stockholm, Sweden, June, 1995, p. 244–247.

73. Ito, N., K. Yamad, H. Okada, M. Nishimura, and T. Kuriyama, "A Rapid and Selective Anodic Bonding Method", 8th International Conference on Solid-State Sensors and Actuators (Transducers '95), Stockholm, Sweden, June, 1995, p. 277–280.

74. Sander, C. S., "A Bipolar-Compatible Monolithic Capacitive Pressure Sensor", Technical Report No.G558–10, Stanford University Integrated Circuits Laboratory, '85–98, Stanford, CA 94305, 1980.

75. Barth, P. W., B. E. Burns, and K. F. Lee, "Polycrystalline Silicon Intermediate Layers for Hermetic Seal Technology", Laboratory report, EE 412, Stanford University Department of Electrical Engineering (Spring)., Stanford University, 1980.

76. Yamada, K., *Semiconductor Absolute Pressure Transducer Assembly and Method*, in U.S. Patent No. 4,291,293 (Sept. 22), 1981.

77. Brooks, A. D. and R. P. Donovan, "Low Temperature Electrostatic Si-to-Si Seals Using Sputtered Borosilicate Glass", *J. Electrochem. Soc.*, 119, 545–546, 1972.

78. Hanneborg, A., M. Nese, H. Jacobsen, and R. Holm, "Silicon-to-Thin-Film Anodic Bonding", *J. Micromech. Microeng.*, 2, 117–121, 1992.

79. Krause, P., M. Sporys, E. Obermeier, K. Lange, and S. Grigull, "Silicon to Silicon Anodic Bonding Using Evaporated Glass", 8th International Conference on Solid-State Sensors and Actuators (Transducers '95), Stockholm, Sweden, June, 1995, p. 228–231.

80. Elwenspoek, M., H. Gardeniers, M. de Boer, and A. Prak, "Micromechanics", University of Twente, Report No. 122830, Twente, Netherlands, 1994.

81. Esashi, M., A. Nakano, S. Shoji, and H. Hebiguchi, "Low-Temperature Silicon-to-Silicon Bonding with Intermediate Low Melting Point Glass", *Sensors and Actuators A*, A23, 931–934, 1990.

82. Anthony, T. R., "Dielectric Isolation of Silicon by Anodic Bonding", *J. Appl. Phys.*, 58, 1240–1247, 1985.

83. Ohashi, H., J. Ohura, T. Tsukakoshi, and M. Shimbo, "Improved Dielectrically Isolated Device Integration by Silicon-Wafer Direct Bonding Technique", IEEE International Electron Devices Meeting. Technical Digest, IEDM '86, Los Angeles, CA, 1986, p. 211–213.

84. Kissinger, G. and W. Kissinger, "Void-Free Silicon-Wafer-Bond Strenghtening in the 200–400°C Range", *Sensors and Actuators A*, A36, 149–156, 1993.

85. Schmidt, M. A., "Silicon Wafer Bonding for Micromechanical Devices", Technical Digest of the 1994 Solid State Sensor and Actuator Workshop., Hilton Head Island, SC, 1994, p. 127–130.

86. Sun, L. G., J. Zhan, Q. Y. Tong, S. J. Xie, Y. M. Caim, and S. J. Lu, "Cool Plasma Activated Surface in Silicon Wafer Direct Bonding Technology", *J. Physique Colloq. C*, 49, 79–82, 1988.

87. Shimbo, M., K. Furukawa, and K. Tanzawa, "Silicon-to-Silicon Direct Bonding Method", *J. Appl. Phys.*, 60, 2987–2989, 1986.

88. Tong, Q.-Y., G. Cha, R. Gafiteanu, and U. Gosele, "Low Temperature Wafer Direct Bonding", *Journal of Micromechanical Systems*, 3, 29–35, 1994.

89. Barth, P. W., "Silicon Fusion Bonding for Fabrication of Sensors, Actuators and Microstructures", *Sensors and Actuators A*, A23, 919–926, 1990.

90. Stengl, R., T. Tan, and U. Gosele, "A Model for the Silicon Wafer Bonding Process", *Jap. J. Appl. Phys. Part I*, 28, 1735–1741, 1989.

91. Backlund, Y., K. Ljungberg, and A. Soderbarg, "A Suggested Mechanism for Silicon Direct Bonding from Studying Hydrophilic and Hydrophobic Surfaces", *J. Micromech. Microeng.*, 2, 158–160, 1992.

92. Lasky, J. B., "Wafer Bonding for Silicon-on-Insulator Technologies", *Appl. Phys. Lett.*, 48, 78–80, 1986.

93. Ismail, M. S., R. W. Bower, J. L. Veteran, and O. J. Marsh, "Silicon nitride direct bonding", *Electron. Lett.*, 26, 1045–1046, 1990.

94. Bower, R. W., M. S. Ismail, and B. E. Roberds, "Low Temperature Si3N4 Direct Bonding", *Appl. Phys. Lett.*, 62, 3485–3497, 1993.

95. Ismail, M. S., R. W. Bower, B. E. Roberds, and S. N. Farrens, "One Step Direct Bonding Process of Low Temperature Si3N4 and TiN Technology", 7th International Conference on Solid-State Sensors and Actuators (Transducers '93), Yokohama, Japan, 1993 p. 188–193.

96. Suga, T. A. H. N., "A Novel Approach to Assembly and Interconnection for Micro Electro Mechanical Systems", Proceedings. IEEE Micro Electro Mechanical Systems, (MEMS '95), Amsterdam, Netherlands, 1995, p. 413–418.

97. Anacker, W., E. Bassous, F. F. Fang, R. E. Mundie, and H. N. Yu, "Fabrication of Multiprobe Miniature Electrical Connector", *IBM Tech. Bull.*, 19, 372–374, 1976.

98. Editorial, "Sealing Glass", Corning Technical Publication, Corning Glass Works, 1981.

99. Legtenberg, R., S. Bouwstra, and M. Elwenspoek, "Low-Temperature Glass Bonding for Sensor Applications Using Boron Oxide Thin Films", *J. Micromech. Microeng.*, 1, 157–160, 1991.

100. Field, L. A. and R. Muller, "Fusing Silicon Wafers with Low Melting Temperature Glass", *Sensors and Actuators A*, A23, 935–938, 1990.

101. Yamada, A., T. Kawasaki, and M. Kawashima, "SOI Wafer Bonding with Spin-on Glass as Adhesive", *Electronic Lett.*, 23, 39–40, 1987.

102. Quenzer, H. J. and W. Benecke, "Low-Temperature Silicon Wafer Bonding", *Sensors and Actuators A*, A32, 340–344, 1992.

103. Guerin, L., M. A. Schaer, R. Sachot, and M. Dutoit, "Proposal for New MultiChip-on-Silicon Packaging Scheme", 8th International Conference on Solid-State Sensors and Actuators (Transducers '95), Stockholm, Sweden, June, 1995, p. 252–255.

104. Tiensuu, A.-L., J.-Å. Schweitz, and S. Johansson, "In Situ Investigation of Precise High Strength Micro Assembly Using Au-Si Eutectic Bonding", 8th International Conference on Solid-State Sensors and Actuators (Transducers '95), Stockholm, Sweden, June, 1995, p. 236–239.

105. Arquint, P., P. D. van der Wal, B. H. van der Schoot, and N. F. de Rooij, "Flexible Polysiloxane Interconnection Between Two Substrates for Microsystem Assembly", 8th International Conference on Solid-State Sensors and Actuators (Transducers '95), Stockholm, Sweden, June, 1995, p. 263–264.

106. den Besten, C., R. E. G. van Hal, J. Munoz, and P. Bergveld, "Polymer Bonding of Micromachined Silicon Structures", Proceedings. IEEE Micro Electro Mechanical Systems, (MEMS '92), Travemunde, Germany, 1992, p. 104–109.

107. Munoz, J., A. Bratov, R. Mas, N. Abramova, C. Dominguez, and J. Bartroli, "Packaging of ISFETs at the Wafer Level by Photopatternable Encapsulat Resins", 8th International Conference on Solid-State Sensors and Actuators (Transducers '95), Stockholm, Sweden, June, 1995, p. 248–251.

108. Bower, R. W., M. S. Ismail, and S. N. Farrens, "Aligned Wafer Bonding: A Key to Three Dimensional Microstructures", *J. Electron. Mater.*, 20, 383–387, 1991.

109. Shoaf, S. E. and Feinerman, A. D., "Aligned Au-Si Eutectic Bonding of Silicon Structures", J. Vac. Sci. Technol., A12, 19–22, 1994.

110. Maszara, W. P., G. Goetz, A. Caviglia, and J. B. McKitterick, "Bonding of Silicon Wafers for Silicon-on-Insulator", *J. Appl. Phys.*, 64, 4943–4950, 1988.

111. Nese, M., R. W. Bernstein, I.-R. Johansen, and R. Spooren, "New Method for Testing Hermeticity of Silicon Sensor Structures", 8th International Conference on Solid-State Sensors and Actuators (Transducers '95), Stockholm, Sweden, June, 1995, p. 260–262.

112. Minami, K., T. Moriuchi, and M. Esashi, "Cavity Pressure Control for Critical Damping of Packaged Micro Mechanical Devices", 8th International Conference on Solid-State Sensors and Actuators (Transducers '95), Stockholm, Sweden, June, 1995, p. 240–243.

113. Striny, K. M., "Assembly Techniques and Packaging of VLSI Devices", in *VLSI Technology*, Sze, S. M., Ed., MacGraw-Hill Book Company, New York, 1988, p. 566–611.

114. Traeger, R. K., "Hermeticity of Polymeric Lid Sealants", 26th Electronic Components Conference, San Francisco, CA, 1976, p. 361–7.

115. Jerman, J. H., "The Fabrication and Use of Micromachined Corrugated Silicon Diaphragms", *Sensors and Actuators A*, A23, 988–92, 1990.

116. Spiering, V. L., S. Bouwstra, J. Burger, and M. Elwenspoek, "Membranes Fabricated with a Deep Single Corrugation for Package Stress Reduction and Residual Stress Relief", 4th European Workshop on Micromechanics (MME '93), Neuchatel, Switzerland, 1993, p. 223–7.

117. Spiering, V. L., S. Bouwstra, R. M. E. J. Spiering, and M. Elwenspoek, "On-Chip Decoupling Zone for Package-Stress Reduction", 6th International Conference on Solid-State Sensors and Actuators (Transducers '91), San Francisco, CA, 1991, p. 982–985.

118. Offereins, H. L., H. Sandmaier, B. Folkmer, U. Steger, and W. Lang, "Stress Free Assembly Technique for a Silicon Based Pressure Sensor", 6th International Conference on Solid-State Sensors and Actuators (Transducers '91), San Francisco, CA, 1991, p. 986–9.

119. Linder, S., H. Baltes, F. Gnaedinger, and E. Doering, "Fabrication Technology for Wafer Through-Hole Interconnections and Three-Dimensional Stacks of Chips and Wafers", Proceedings. IEEE Micro Electro Mechanical Systems, (MEMS '94), Oiso, Japan, January, 1994, p. 349–354.

120. Bhardwaj, J. K. and H. Ashraf, "Advanced Silicon Etching Using High Density Plasma", Micromachining and Microfabrication Process Technology, (Proceedings of the SPIE), Austin, TX, 1995, p. 224–233.

121. Craven, D., K. Yu, and T. Pandhumsoporn, "Etching Technology for "Through-The-Wafer" Silicon Etching.", Micromachining and Microfabrication Process Technology, (Proceedings of the SPIE), Austin, TX, 1995, p. 259–263.

122. Lischner, D. J., H. Basseches, and F. A. D'Altroy, "Observations of the Temperature Gradient Zone Melting Process for Isolating Small Devices", *J. Electrochem. Soc.*, 132, 2991–2996, 1985.

123. Yando, S., *Method and Apparatus for Fabricating an Array of Discrete Elements*, in U.S. Patent #3,439,416, 1969.

124. Yeh, H. J. and J. S. Smith, "Integration of GaAs Vertical-Cavity Surface-Emitting Laser On Si By Substrate Removal", *Appl. Phys. Lett.*, 64, 1466–1468, 1994.

125. Cohn, M. B., R. T. Howe, and A. P. Pisano, "Self-Assembly of Microsystems Using Non-Contact Electrostatic Traps", Proceedings of the ASME International Congress and Exposition, Symposium on Micromechanical Systems, (IC '95), San Francisco, CA, 1995, p. 893–900.

126. Hosokawa, K., I. Shimoyama, and H. Miura, "Two-Dimensional Micro-Self-Assembly Using the Surface Tension of Water", The Ninth Annual International Workshop on Micro Electro Mechanical Systems, MEMs'96, San Diego, CA, USA, 1996, p. 67–72.

127. Vuilleumier, R. A. K. K., "Variable Entrance Slit System for Precision Spectrophotometers", Proceedings. IEEE Micro Electro Mechanical Systems, (MEMS '95), Amsterdam, Netherlands, 1995, p. 181–185.

# 9

# Scaling Laws, Actuators, and Power in Miniaturization

## Introduction

The electronics revolution in the 1960s drove miniaturization of radio, television, hard disk drives, camcorders, personal digital assistants, etc. to the point where miniaturization now appears natural. Newer is the miniaturization of non-IC hardware such as physical and biological sensors and analytical instruments. The latter progress is derived mainly from developments in mature, traditional technologies such as precision engineering but also to some minor extent from the application of nontraditional technologies borrowed from microelectronics.[1] The realization of the encroaching limits of earthly resources and the continuing deterioration of the environment in the last quarter of this century have added urgency to the miniaturization trend. We hope to contribute to the science of miniaturization by clarifying the consequences of miniaturizing a given system and by aiding in making the right choice of micromachining method to build the microsystem.

In this chapter we are addressing scaling laws and the size regimes where macrotheories start requiring corrections with the aim of better understanding the physical consequences of downscaling, electrostatic, electromagnetic, fluidic, optical, and chemical devices. We aspire to elucidate the unexpected behavior of micromachines and to better understand why, in some cases, it makes sense to miniaturize a device for reasons beyond economics, volume, and weight considerations.

Throughout the chapter we compare actuator mechanisms to illustrate scaling of different forces. An actuator is a device that converts energy from one form into another. We compare different actuators on the basis of their scaling and other properties of main interest which, besides size and cost, are displacement (linear, angular), force or torque, response time, and power consumption. The results must be viewed as guide posts only; other considerations such as the size of the absolute forces involved, the potential for integration with electronics, materials choice, material defect structure and purity, etc. will dominate the design rules for a particular microstructure. Finally, we describe power sources for micromachines. When fabricating actuators or integrated power sources, the results show that the advantages of micromachining are not always self-evident.

## Microintuition

Before introducing the mathematics behind miniaturization and the breakdown of macroscale theory, we present some general insights on scaling laws based mainly on observations from nature. We want to develop an intuition about how systems are likely to behave when they are downsized, i.e., 'microintuition' as Trimmer calls it.[2]

In Figure 9.1* some well- and not so well-known structures and their respective sizes are shown. Humans are accustomed to think in distances and sizes. Even microsizes have become quite intuitive, more so than, say, microtimes. To accommodate this human lack of comprehending short amounts of times, Isaac Asimov introduced a new method of measuring microtime based on the speed of light. He introduced the 'light-meter' as the time required for light (in a vacuum) to cover a distance of a meter. An illustrative case is the half-life of radioactive particles: saying that the half-life of particle (1) is of the order of a hundredth of a millionth of a second and that of particle (2) is of a thousandth of a trillionth of a second leaves very little impression on the mind. On the other hand, if we say that the half-life of one type of particle is of the order of a 'light-meter' and the other is of a 'tenth of a light-micrometer', visualization becomes easier.[3] Linear extrapolation of lengths comes easy to us. But we are quickly at a loss when considering the implications that shrinking of length has on area and volume ratios and on the relative strength of external forces working on microstructures. In dealing with microdevices our 'macrointuition' on their operation is misleading. If a system is reduced isomorphically in size (i.e., scaled down with all dimensions of the system decreasing uniformly), the changes in length, area, and volume ratios alter the relative influence of various physical effects which determine the overall operation in unexpected ways. As objects shrink, the ratio of surface area to volume increases, rendering surface forces more important. For example, if we build two bridges geometrically similar, the larger will be the weaker of the two and will be so in the ratio of their linear dimensions (l). The strength of the iron girders in the

---

* Figure 9.1 appears as a color plate after page 144.

405

bridges varies with the square of the linear dimension, ($l^2$) but the weight of the whole structure varies with the cube of its linear dimension ($l^3$). The larger the structure, the more severe the strain becomes. By reducing the size of a device, the structural stiffness generally increases relative to inertially imposed loads. Another striking example is that of surface tension. The mass of a liquid in a capillary tube, and hence the weight, scales as $l^3$ and decreases more rapidly than the surface tension, which scales as $l$ as the system becomes smaller. That is why it is more difficult to empty liquids from a capillary compared to spilling coffee from a cup.[2]

In his marvelous book *On Growth and Form*, Thompson tells of the profound effects scaling has in nature.[4] Land animals fighting the resistance of gravity can grow only so big without becoming too clumsy and inefficient. The small birds and beasts are quick and agile. The diameter of a tall homogeneous body such as a tree must grow as the power 3/2 of its height, which accounts for Goethe's 'Es ist dafür gesorgt, dass die Bäume nicht in den Himmel wachsen' as Bonner muses in the introductory chapter to Thompson's book. For mammals, problems arise with being too large, but there are also significant problems associated with being too small. The pygmy shrew, for example, must eat continuously or freeze to death. The heat loss from a living creature is proportional to the animal's surface area ($l^2$), and the rate of compensating heat generation through eating is proportional to its volume ($l^3$). As animals get smaller, a greater percentage of their energy intake is required to balance the heat loss. A warm-blooded animal much smaller than a mouse becomes improbable; it could neither obtain nor digest the food required to maintain its constant temperature. Insects circumvent this problem by being cold-blooded. Scaling laws also impose a lower limit to size, of about 5 to 10 μm, even for cold-blooded animals. Smaller organisms cannot retain their vital fluids long enough to survive.[5] Water-based life increases its range of sizes both above and below that of terrestrial animals by evading gravity and drying out. For creatures of the sea, the same physical barrier of gravity to indefinite growth does not exist. The resistance swimmers must overcome is not gravity but 'skin-friction' which only increases as the square of the linear dimensions. In this case, larger size leads to a distinct advantage in that the larger the creature grows, the greater its swimming speed is. This can be understood as follows. The available energy, E, for swimming speed (v) depends on the mass of the creature's muscles ($l^3$), while its motion through water is opposed by the skin-friction resistance, R, ($l^2$). This leads to $E \sim Rv^2$ or $v \sim \sqrt{l}$; in other words, the bigger fish or the bigger ship moves faster but only in the ratio of the square root of the increasing length. A detailed analysis of the natural laws between speed and body length in water, on a surface, and in air recently was presented by Hayashi.[6] Hayashi also analyzes the size effect on forces, strength increase of materials, surface phenomena, and decrease of manufacturing accuracy.

One can easily recognize how insects are taking advantage of several of the phenomena described here. For example, insects can jump farther in proportion to their size than can man and some even walk on water. The effort involved in jumping is proportional to the mass (m) and to the height (h) to which that weight is raised; $E \sim mh$. The biological force in a muscle available for this work is proportional to the mass of the muscle or to the mass of the animal. It follows that h is, or tends to be, a constant. In other words, animals tend to jump to the same actual height independent of their size. The walking on water is based on the surface tension at the water surface ($l^1$) which easily supports an insect's weight ($l^3$). To circumvent problems with excessive heat loss, they are cold-blooded. Having adapted so well to so many niches in nature, it is no wonder insects are so abundant.

As we are comparing actuators in this chapter, a few more words are in order about mammalian muscle, an example of a chemomechanical actuator and nature's ubiquitous actuator for larger organisms. The maximum static muscle force generated per unit cross-sectional area (i.e., tension, stress) is about 0.350 MPa, a constant number for all vertebrate muscle fibers. In vertebrate muscle the maximum force generated can only be held for short periods of time because of muscle fatigue. The maximum sustainable force usually is only about 30% of the peak value. For this reason, the maximum static sustainable stress generated by muscle is about 0.100 MPa. The maximum power per unit mass (in W/kg) is an important figure for robotic and prosthetic actuators. For human muscle it typically measures about 50 W/kg but can be as high as 200 W/kg for some muscles for a brief period of time. To illustrate the excellent cycle lifetime of a muscle we need only look at cardiac muscle. The heart beats over $3 \times 10^9$ times in the lifetime of an average person; an excellent lifetime compared to any artificial actuator. One of the most difficult properties of muscle to mimic with an artificial actuator is the extreme changes in stiffness which occur between the resting muscle and maximally activated muscle. Stiffness can increase as much as 5 times from rest to a 100% contraction.[7]

Without relying on complicated math we can already appreciate that, as the scale of structures decreases, so does the importance of phenomena that vary with the largest power of the linear dimension l: gravity ($l^3$), inertia ($l^3$), magnetism ($l^2$, $l^3$, or $l^4$, depending on the exact configuration), flow ($l^4$), and thermal emission ($l^2$ to $l^4$). Phenomena that are more weakly dependent on size dominate in small dimensions: electrostatics ($l^2$), friction ($l^2$), surface tension ($l$), diffusion ($l^{1/2}$), and van der Waal's forces ($l^{1/4}$) (see Figure 9.2).

# Scaling in Electrostatics and Electromagnetics

## Electrostatics

To appreciate scaling issues in electrostatic devices, we will follow Trimmer's analysis of isometric scaling of a simple parallel plate capacitor.[2,8,9] In this analysis, l represents a linear dimension. If the system becomes one tenth the size, $l = 1/10$, all the dimensions decrease by a tenth. The dimensions of the capacitor

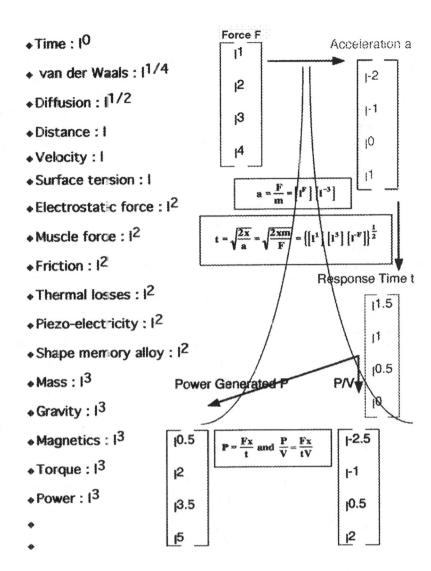

**FIGURE 9.2**   Taking advantage of scaling laws.

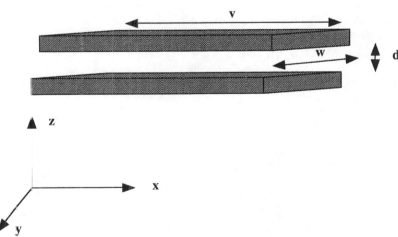

**FIGURE 9.3**   A parallel plate capacitor of plate size, w × v, and separation, d, stores a maximum potential energy, $E_{e,m}$.

in Figure 9.3 are w, v, and d and scale with l. The maximum electrostatic potential energy, $E_{e,m}$, stored in this capacitor is[*]

$$E_{e,m} = \frac{\varepsilon_0 w v V_b^2}{2d} \qquad 9.1$$

where $V_b$ represents the electrical breakdown voltage and $\varepsilon_0 w v/d$ the capacitance C. The dielectric constant, $\varepsilon_0$ in Asec/Vm, remains unchanged with scaling so we assign it a $l^0$ dimension. We also assume, for now, that $V_b$ scales linearly with d.

---

[*]   More correctly Equation 9.1 should include the product $\varepsilon_0 \times \varepsilon_r$ with $\varepsilon_r$ the relative permittivity of the dielectric which is 1 for vacuum and very close to 1 for air.

Intuitively, we expect a smaller gap d to result in a lower break-down voltage, $V_b$. However, as we learned in Chapter 2, counter to our macrointuition, at very small capacitor plate separation the linear relationship between $V_b$ and d breaks down and the breakdown voltage of the capacitor actually starts increasing with decreasing plate separation (see discussion below on continuum breakdown and the discussion of the Paschen curve in Figure 2.7).

Writing out all the dimensions in Equation 9.1 in terms of l we obtain:

$$E_{e,m} = \frac{(l^0)(l^1)(l^1)(l^1)^2}{(l^1)} \qquad 9.2$$

and the maximum energy stored in the capacitor scales as ($l^3$). It follows that if l decreases by a factor of ten, the stored maximum energy in the capacitor decreases by a factor of 1000.

When moving one plate of the capacitor with respect to the other in any direction, the electrostatic force involved is the negative spatial derivative in that direction of the stored energy $E_e$.[8] A translational movement of one plate in a microactuator in the direction, x, leads to a force, $F_x$, as:

$$F_x = -\frac{\partial E_e}{\partial x} = -\frac{V^2}{2}\frac{\partial C}{\partial X} = \frac{(l^3)}{(l^1)} = (l^2) \qquad 9.3$$

where V is the applied voltage. The electrostatic force for a constant field is thus found to scale as ($l^2$). This is an advantage because the mass and, hence, inertial forces scale as ($l^3$). The electrostatic force gains over inertial forces as the size of the system is decreased. A decrease in size by a factor of 10 leads to a decrease of the inertial forces by a factor of 1000 whereas the electrostatic force decreases by a factor of only 100.

If the two plates of the parallel capacitor in a microactuator are displaced perpendicular to each other (z-direction) the force, based on Equation 9.3, is given as:

$$F_x = \frac{\varepsilon_0 wvV^2}{2d^2} \qquad 9.4$$

Vertically driven resonant microstructures operate in this mode (see Figure 9.4A). Vibration in the z-direction is excited electrostatically and motion is detected electrostatically, as well, by sensing the change in capacitance.[10]

A large displacement in a micromachined electrostatic actuator element can be achieved only if the actuator moves parallel to the capacitor plate (see Figure 9.4B).[10] For a parallel movement of the plates (x- and y-directions) Equation 9.3 results in:

$$F_x = -\frac{\varepsilon_0 wV^2}{2d} \quad \text{and} \quad F_y = -\frac{\varepsilon_0 vV^2}{2d} \qquad 9.5$$

In this case, a force parallel to the plates tends to realign the plates. This is exploited in many laterally driven linear and rotary electrostatic motors (see Figure 9.5A[8] and B). The force remains constant during the movement as long as the fringing fields can be neglected (see Figure 9.4B). Besides longer strokes, driving and sensing of planar microstructures parallel to the substrate have two other major advantages over a vertical movement of the plates: the forces change linearly with distance (see Equation 9.5) and dissipative squeeze damping is avoided.[11-13] The latter is related to the magnitude of the viscous losses when moving structures displace fluids (usually air) from small separator gaps. Losses are larger for vertical displacement of fluids, leading to squeeze-film damping and a resultant low quality factor Q for resonating elements. Couette flow in the gap between the structure and the substrate for lateral motion of the actuator is much less dissipative than squeeze-film damping and higher Q resonators are obtained.

Lateral electrostatic actuation has become important in linear resonators with comb-like structures as shown in Figure 5.18. By applying a voltage a movable shuttle moves towards the fixed part. For a rectangular design of the resonator beams, the force and the microactuator displacement is proportional to the ratio of structure height, T, vs. separation gap, d, and the number of fingers, N (see also Chapter 6, Equation 6.7):

$$F = \frac{\varepsilon_0 TV^2}{2d}N \qquad 9.6$$

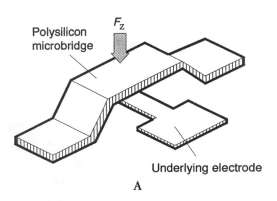

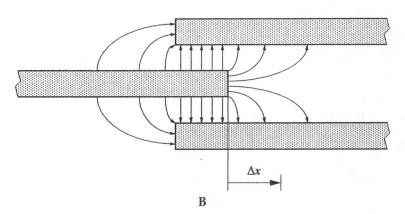

A

B

**FIGURE 9.4**  Electrostatic actuation. (A) Vertically driven polysilicon microbridge. (B) Laterally driven electrostatic actuator. (From Tang, W.C., Ph.D. thesis, UC Berkeley, 1990. With permission.)

An efficient comb drive can thus be achieved by designing many high comb fingers with narrow gaps. In surface micromachining, the thickness T is typically limited to about 2 to 3 μm, whereas with LIGA or pseudo-LIGA methods several hundreds of microns are possible. Clearly the latter machining methods have the potential to produce stronger electrostatic actuators. Mohr et al.[14] further showed that by making the comb fingers trapezoidal in shape the maximum displacement of the intercigitated fingers can be further increased as smaller capacitor gaps are possible (see Chapter 6 on LIGA, Electrostatic Actuators). For a more rigorous mathematical derivation of Equation 9.6 we refer to the finite-element simulation of comb drives by Tang.[10]

The electrostatic energy density, $E_e'$, between the capacitor plates is obtained from Equation 9.1 by dividing the total potential energy, $E_e$, by the volume, $V_0$, of the capacitor, $V_0 = w \cdot v \cdot d$:

$$E_e' = \frac{\varepsilon_0 E^2}{2} \qquad 9.7$$

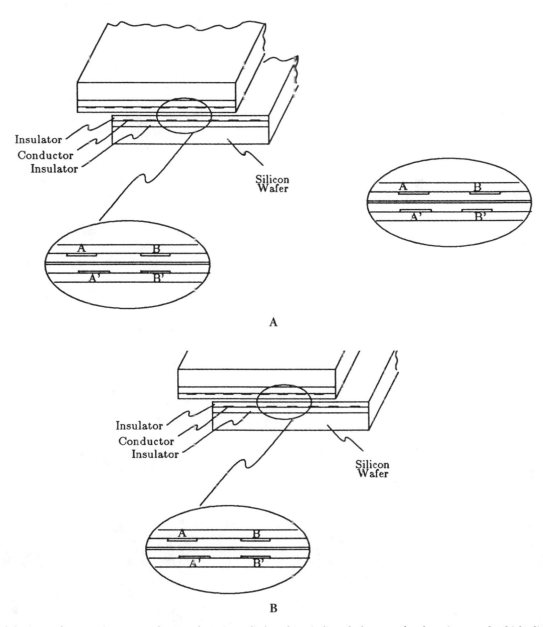

**FIGURE 9.5** (A) Linear electrostatic motor: when a voltage is applied to the misaligned plates A-A′, a force is exerted, which aligns plates A-A′. Now, plates B-B′ are misaligned and in a position to be activated to cause a motion. (From Trimmer, W.S.N. and Gabriel, K.J., *Sensors Actuators*, 1, 189–206, 1987. With permission.) (B) Rotational electrostatic motor: SEM micrograph of a polysilicon 12:8 salient pole micromotor. The rotor sits atop a 0.5-μm thick layer of polysilicon that acts as an electrostatic shield. Rotor, hub, and stators are formed from 1.5-μm thick polysilicon. A 2.0-μm thick polysilicon disk is attached to the rotor. In turn, the hub overlaps this disk to pin the rotor onto the substrate. (Courtesy of Mr. D. Koester, MCNC-MEMS Technology Applications Center.)

where E (= V/d) represents the electrical field magnitude, limited to approximately $3 \times 10^6$ V m$^{-1}$ by the electrical breakdown of air (E$_b$). Thus, the stored energy for an air capacitor is about 40 Jm$^{-3}$.[15] Electrostatics is a surface force, and the surface force density, F′, is given by:

$$F' = \frac{\partial F}{\partial A} = \frac{\varepsilon_0 E^2}{2} \qquad 9.8$$

where A stands for the surface area. The surface force density equals the energy density in the field. Hence, one wants to use the maximum field possible for the largest possible force.

From Equation 9.3 translational motor action is associated with a change in the capacitance: an increase in capacitance for a motor, a decrease in capacitance for a generator. For a rotational force, F$_r$, in analogy to Equation 9.3 one has:

$$F_r = -\frac{\partial U_e}{\partial \theta} \qquad 9.9$$

where $\partial \theta$ is the differential angular displacement in radians. From Equations 9.3 and 9.9, F$_r$ can again be expressed in terms of the change in capacitance:

$$F_r = -\frac{V^2}{2}\frac{\partial C}{\partial \theta} = -\frac{Q^2}{2C^2}\frac{\partial C}{\partial \theta} \qquad 9.10$$

The second expression on the right is obtained by introducing the relation V = Q/C, where Q is the total charge on the capacitor plates. Detailed calculations of the capacitance of electrostatic motors can be found in Mahadevan[15,16] and in Kumar et al.[17] Torque of a rotational motor is given by:

$$\mathbf{T} = \mathbf{r} \times \mathbf{F_r} \qquad 9.11$$

With r the radius and F$_r$ perpendicular, Equation 9.11 simply becomes T = rF$_r$. The power, P, generated by the motor can easily be calculated from the torque as:

$$P = T\omega = T2\pi f \qquad 9.12$$

where $\omega$ represents the angular frequency and f the frequency at which the motor is rotating. From Equation 9.12 one would expect that, as electrostatic motors decrease, the inherent increase in frequency could help to offset the decrease in torque. However, static and dynamic frictional forces (surface forces!) are a major barrier to overcome. Despite these problems, surface micromachined electrostatic motors with rotational speeds of 15,000 rpm and continuous operation for over a week have proven to be possible at voltages below 300 V.[19] Friction reduction methods include deposition of a silicon nitride sliding surface; electrostatic, magnetic, or other types of levitation; and replacing sliding contacts with rolling contacts in *wobble motors*.[20]

To drive a load several hundred microns in thickness and several millimeters in diameter, the required torque of a motor is of the order of 10$^{-5}$ Nm. However, the torque generated with most surface-micromachined electrostatic motors (say, 100 to 150 μm in diameter and with air-gaps of about 2 μm (see Figure 9.5B)), is only in the nano-Newton-meter or pico-Newton-meter range.[20] Calculation for an outer rotor surface-micromachined, wobble motor,* with copper electroplated structural elements, indicates that to achieve a torque of 10$^{-4}$ to

---

\* A detailed description of an electrostatic eccentric drive micro-motor (wobble motor) can be found in Price et al.[21]

## Four wobble motor designs

(a) Conventional inner rotor design requiring stator insulation; (b) conventional one with bearing; (c) new outer rotor design requiring stator insulation; (d) new outer rotor design with spacer. Types (b) and (d) do not require any insulation.

(a)          (b)          (c)          (d)

▪ stators          □ rotor

$10^{-5}$ Nm, the stator radius must be a few millimeters and the thickness needs to be of the order of 10 μm.[22] As frequently observed, actuators do not scale as advantageously as sensors. The conversion effect is a volume effect; small-size devices have a limited range of force available for actuation. In view of this recognition, perhaps a too extraordinary amount of research went into surface-micromachined electrostatic motors. Following we will compare electrostatic motors with the more complicated and larger micromachined magnetic motors which can perform actual work.

Besides different types of monolithic electrostatic micromotors[19,23] electrostatic actuators have been used for hybrid mounted micromotors,[24] microvalves,[25] mechanical resonators,[10] and switches.[26] An electrostatic microswitch developed by this author is illustrated in Figure 9.6.

## Trimmer's Vertical Bracket Notation to Represent Scaling Laws

Trimmer introduced an elegant method to express different scaling laws by using a vertical bracket notation. For different possible forces he writes:[2]

$$F = \begin{bmatrix} 1^1 \\ 1^2 \\ 1^3 \\ 1^4 \end{bmatrix} \qquad 9.13$$

The top element in this notation refers to the case where the force scales as ($1^1$); the next one down refers to a case where the force scales as ($1^2$), etc. We can appreciate the usage of this representation with the following example: as systems become smaller, the scaling of the force also determines the acceleration, a; transit time, t; and the amount of power per unit volume ($P/V_0$) generated and dissipated. The mass of a system, m, scales as ($1^3$). For a generalized case with a force F scaling as ($1^F$), we obtain:

$$a = \frac{F}{m} = \left[ 1^F \right]\left[ 1^{-3} \right]$$

$$t = \sqrt{\frac{2xm}{F}} = \left( \left[ 1^1 \right]\left[ 1^3 \right]\left[ 1^{-F} \right] \right)^{\frac{1}{2}} \qquad 9.14$$

$$\frac{P}{Vo} = \frac{Fx}{tVo}$$

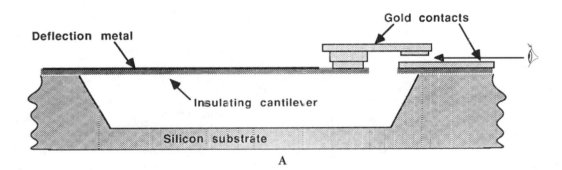

**FIGURE 9.6** Electrostatic microswitch: (A) schematic; (B) SEM micrograph looking straight into the opened contact.

where x = distance. From Equation 9.13 it follows that:

$$F = \begin{bmatrix} 1^1 \\ 1^2 \\ 1^3 \\ 1^4 \end{bmatrix} \Rightarrow a = \begin{bmatrix} 1^{-2} \\ 1^{-1} \\ 1^0 \\ 1^1 \end{bmatrix} \Rightarrow t = \begin{bmatrix} 1^{1.5} \\ 1^1 \\ 1^{0.5} \\ 1^0 \end{bmatrix}$$

$$\frac{P}{Vo} \Rightarrow \begin{bmatrix} 1^{-2.5} \\ 1^{-1} \\ 1^{0.5} \\ 1^2 \end{bmatrix}$$

9.15

For the case of electrostatic discussed above, with $1^F = 1^2$, one obtains $a = 1^{-1}$, $t = 1^1$, and the power density is $P/Vo = 1^{-1}$ (an increase of a factor of 10). For force laws with a power higher than $1^2$, the power generated per volume degrades as the scale decreases. Even in the worst case with $F = 1^4$, the time required to perform a task remains constant when the system is scaled down. This is an observation we know intuitively: small things tend to be quick.[2] It should be kept in mind that beneficial effects can easily be overshadowed by loss mechanisms, which scale in the same way or become even more important in the micro-domain.

## Breakdown of Continuum Theory for Electrostatics

We will now consider a first case where macroscopic laws err in the microscopic regime. At small distances between two conductors in air the electrical field is not isotropic as assumed above. This can be gleaned from the so-called Paschen curve in Figure 2.7. This figure represents the breakdown voltage between two conductors, plotted vs. the product of the distance between them and the surrounding gas pressure, P. The general shape of the Paschen curve can be understood easier by assuming that the pressure, P, remains constant at 1 atm, thus reducing the x-axis to a simple distance axis. We notice that on the right size of the Paschen curve, at large electrode distances, the field is constant and the earlier deduced scaling laws pertain. But for smaller electrode gaps (below 5 μm in air) the curve sharply reverses, bending upwards and leading to higher electrical breakdown fields for thinner air gaps. Electrical breakdown in these small gaps does not occur at the predicted voltages.[27] From Bart et al.[15] and Busch-Vishniac[28] and references therein we quote observed electrical fields of $10^8$ V m$^{-1}$ (1.5 μm air-gap), $1.7\ 10^8$ V m$^{-1}$ (2 μm air-gap) to $3.2\ 10^7$ V m$^{-1}$ (12.5 μm air-gap). The upper limit is the electrical field measured for small gaps in vacuum, i.e., $3.0 \times 10^8$ Vm$^{-1}$.[29] Surface roughness results in a lower average breakdown voltage, but even then these fields are significantly higher than the $3 \times 10^6$ V m$^{-1}$ quoted before for macro breakdown. Assuming a field strength of $3 \times 10^8$ V m$^{-1}$, Equation 9.7 predicts an energy density of about $4 \times 10^5$ J m$^{-3}$, compared to 40 J m$^{-3}$ without the Paschen effect. This unexpected effect is caused by the fact that there are not enough ionizing collisions to induce avalanche over that short of a distance. A lot higher fields can be achieved before the critical breakdown voltage is reached. The Paschen curve also illustrates how miniaturization can be considered the equivalent to reducing the gas pressure. When the gap between two conductors approaches the mean free path of the molecules, λ, statistically fewer molecules are present to be ionized between the closely spaced conductors. The same situation arises if we lower the pressure in a macroscopic set-up.

Besides the previously mentioned advantage of micromachined electrostatic systems, the Paschen effect suggests yet another advantage for a micro-electrostatic system. The field in the nonisotropic region scales more like $1^{-1/2}$, whereas the force scales like $1^1$. With a factor of 10 size reduction, the inertial force still decreases by a factor of 1000, but the electrostatic force decreases only by a factor of 10. Acceleration and transit times are higher than for the isotropic system. Superficially this seems to put magnetic motors, based on volume forces, at a disadvantage compared to electrostatic micromotors. We shall see, though, that this is not the end of the story. For example, magnetic field energy can be made two orders of magnitude greater than the best one can achieve with electrostatic fields and small air-gaps in air or vacuum.[28]

Working on the left side of the Paschen curve renders high field operation possible for a wide variety of electrostatic microdevices without incurring catastrophic sparking. Examples include the electrostatic motors illustrated in Figure 9.5, the electrostatic switch in Figure 9.6, and the author's novel ionization sources presented in Figure 10.49.

## Piezoelectricity

### Introduction

Pierre Curie and his brother Paul-Jacques discovered in 1880–1881 that external forces applied to certain crystals generate a charge on the surface of the crystal found to be roughly proportional to the applied mechanical stress (force per unit area). These so-called piezoelectric materials exhibit the converse effects as well: an applied voltage generates a deformation of the crystal (the Curie brothers discovered this inverse effect one year later). Piezoelectric actuators follow mammalian muscle as the most ubiquitous actuator principle in nature. For example, human skin and bones possess piezoelectric properties. Piezoelectricity must not be confused with ferroelectricity, which is the property of a spontaneous or induced electric dipole moment. All ferroelectric materials are piezoelectric, but the contrary is not always true. Piezoelectricity relates to the crystalline ionic structure. Ferroelectricity instead relates to electron spin.

Ferroelectric materials exhibit strong electrostriction. Electrostriction, a property of all dielectrics but only pronounced for ferroelectrics, is similar to the piezoelectric effect in that it involves an increase in length parallel to an applied electric field. In electrostrictive materials, in contrast with piezoelectric materials, the direction of this small change in geometry does not reverse if the direction of the electrical field is reversed.

### Mechanism

A simplified model of piezoelectricity entails the notion of anions (−) and cations (+) moving in opposite directions under the influence of an electric field or a mechanical force. The forces generated by this motion cause lattice deformation for noncentrosymmetric crystals due to the presence of both high and low stiffness ionic bonds. The effect for quartz is illustrated in Figure 9.7. If the cell shown here is deformed along the x- or y-axis, the O-ion is displaced, and positive or negative charges are formed. As a result, all piezoelectric materials are necessarily anisotropic; in cases of central symmetry, an applied force does not yield an electric polarization. By applying mechanical deformations to piezoelectric crystals, electric dipoles are generated and a potential difference develops when changing those mechanical deformations. It is important to remember that the potential and the associated currents in piezoelectric materials are a function of the continuously changing mechanical deformation. Therefore, typical and practical uses are in situations involving dynamic strains of an oscillatory nature.[30] Piezoelectric properties are present in 20 of the 32 different crystallographic point groups, although only a few of them are used. They are found also in amorphous ferroelectric materials. Of the 20 crystallographic classes, only 10 display ferroelectric properties.

A more detailed understanding of piezoelectricity is based on an understanding of the piezoelectric equations describing the

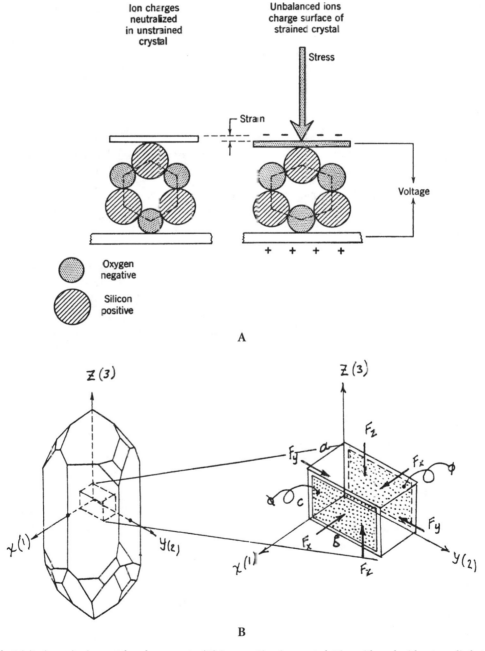

**FIGURE 9.7**  Piezoelectricity in an ionic crystal such as quartz. (A) Ion position in quartz lattice with and without applied stress. (B) Element cut from quartz crystal under stress.

coupling between electric and mechanical strain in a piezoelectric material. When a stress, $\sigma$(F/A), is applied to a slab of material cut from a piezoelectric material such as quartz, the resulting one-dimensional strain, $\varepsilon$, in the elastic range, can be written as (see Chapter 4, Equation 4.14):

$$\varepsilon = S\sigma \qquad\qquad 9.16$$

which represents Hooke's law, where S stands for the compliance (1/S = E = Young's modulus). With a potential difference applied across the faces of the piezoelectric, an electric field, E (V/m), is established and we obtain:

$$D = \varepsilon_r E = \varepsilon_0 E(1+\chi) = \varepsilon_0 E + P \qquad 9.17$$

where D is the electric displacement (or electric flux density), $\varepsilon_r$ the relative permittivity of the material (dimensionless), E the electric field, $\varepsilon_0$ the permittivity of vacuum in A sec/Vm, $\chi$ the electrical susceptibility, and P the polarization, i.e., the electric dipole moment per unit volume of material. Equation 9.16 contains only mechanical quantities, and Equation 9.17 contains only electrical parameters. Piezoelectric materials possess, in addition, a special interlocking behavior in which electrical charges are produced by straining the material and internal forces are produced by subjecting the material to an electric field.[31] For a one-dimensional piezoelectric material with electric field and stress in the same direction, according to the principle of energy conservation from thermodynamics we can describe this situation as:[32]

$$D = d\sigma + \varepsilon(d)^\sigma E$$
$$D = e\varepsilon + \varepsilon(d)^\varepsilon E \qquad\qquad 9.18$$

where d represents a piezoelectric constant expressed in coulombs divided by Newtons (C/N) and $\varepsilon(d)^\sigma$ is the permittivity at constant stress (we added a '(d)' to distinguish the dielectric constant from the strain $\varepsilon$). In the second expression, e stands for a piezoelectric constant in C/m and $\varepsilon(d)^\varepsilon$ is the dielectric constant at a constant strain. Solving for E we can rewrite Equation 9.18 as:

$$E = -g\sigma + \frac{D}{\varepsilon(d)^\sigma}$$
$$E = -h\varepsilon + \frac{D}{\varepsilon(d)^\varepsilon} \qquad\qquad 9.19$$

where $g = d/\varepsilon(d)^\sigma$ is a piezoelectric constant in V/M/N and $h = e/\varepsilon(d)^\sigma$, a piezoelectric constant in V/m. The above set of equations describe the direct piezoelectric effect. The equations for the inverse effect are written as:

$$\varepsilon = dE + S^E \sigma$$
$$\varepsilon = gD + S^D \sigma \qquad\qquad 9.20$$

with d and g piezoelectric constants given in M/V and m²/C, respectively; $S^E$ is the compliance at constant field; and $S^D$ is the compliance at constant electric flux density. Solving Equation 9.20 for $\sigma$ results in:

$$\sigma = -eE + E(Y)^E \varepsilon$$
$$\sigma = -hD + E(Y)^D \varepsilon \qquad\qquad 9.21$$

where $e = d/S^E$ and $h = g/S^D$ are piezoelectric constants given, respectively, in N/V/m and N/C, and $E(Y)^E$ is the Young's modulus (we added a '(Y)' here to distinguish the Young's modulus from the electric field E) under constant electrical field, and $E(Y)^D$ is the Young's modulus under constant electric flux density. Equations 9.18 to 9.21 are known as the piezoelectric constitutive relations. They are summarized in Table 9.1, together with the definition of the four piezoelectric constants d, e, g, and h.

If no external field is imposed, according to Equations 9.17 and 9.18, a stress $\sigma$ will lead to the following polarization P:

$$P = d\sigma \qquad\qquad 9.22$$

and, according to Equation 9.20, if there is no external stress applied, a field, E, will lead to the following strain, $\varepsilon$:

$$\varepsilon = dE \qquad\qquad 9.23$$

so that the same constant is used for the direct and reverse effects.

A parameter often used when comparing different piezoelectric materials is the electromechanical coupling coefficient, k. It can be interpreted as the square root of the ratio of the electrical energy converted to mechanical energy per unit volume to the total electrical energy input per unit volume. It is thus a measure of the interchange of electrical and mechanical energy. It can be shown that k equals the geometric mean of the piezoelectric voltage coefficient (g) and the piezoelectric stress coefficient (e) and is indicative of the ability of a material to both detect and generate mechanical vibrations.[33]

$$k = \sqrt{ge} = \sqrt{\frac{d^2}{\varepsilon^\sigma S^E}} \qquad\qquad 9.24$$

In order to activate the maximum piezoelectric strain in a given crystal, the orientation dependence of the piezoelectric effect should be carefully considered. For the general case of a piezoelectric crystal, rather than a one-dimensional piezoelectric, the piezoelectric constitutive relations must be generalized. The generalized elastic response of a piezoelectric crystal to an electric field (see Equation 9.23) may then be expressed as:

$$\varepsilon_k = \sum_{i=1}^{3} d_{ik} E_i \qquad\qquad 9.25$$

**TABLE 9.1** Piezoelectric Constitutive Equations and Definitions of Piezoelectric Parameters

| Piezoelectric Equations | Definitions of Constants | M.K.S. Units |
|---|---|---|
| $D = d\sigma + \varepsilon(d)^\sigma E$ | $d = \dfrac{\text{charge density developed}}{\text{applied stress}}$ | C/N |
| $D = e\varepsilon + \varepsilon(d)^\varepsilon E$ | $e = \dfrac{\text{charge density developed}}{\text{applied strain}}$ | C/m$^2$ |
| $E = -g\sigma + \dfrac{D}{\varepsilon(d)^\sigma}$ | $g = \dfrac{\text{field developed}}{\text{applied stress}}$ | V/m/N |
| $E = -h\varepsilon + \dfrac{D}{\varepsilon(d)^\varepsilon}$ | $h = \dfrac{\text{field developed}}{\text{applied strain}}$ | V/m |
| $\varepsilon = dE + S^E\sigma$ | $d = \dfrac{\text{strain developed}}{\text{applied field}}$ | M/V |
| $\varepsilon = gD + S^D\sigma$ | $g = \dfrac{\text{strain developed}}{\text{applied charge density}}$ | m$^2$/C |
| $\sigma = -eE + E(Y)^E\varepsilon$ | $e = \dfrac{\text{stress developed}}{\text{applied field}}$ | N/V/m |
| $\sigma = -hD + E(Y)^D\varepsilon$ | $h = \dfrac{\text{stress developed}}{\text{applied charge density}}$ | N/C |

and the polarization in the absence of a field as:

$$P_i = \sum_{k=1}^{6} d_{ik}\sigma_k \qquad 9.26$$

In these equations, i = 1, 2, and 3 make up the indices of the components of polarization and k = 1, 2, ... 6 the indices of the components of mechanical stress and strain. In the piezoelectric constants, the first subscript refers to the direction of the field; the second subscript refers to the direction of the stress. The subscripts 1, 2, and 3 indicate the x-, y-, and z-axes, respectively. For example in $d_{ik} = d_{33}$, i = 3 indicates that the polarizing electrodes are perpendicular to the 3 axis (i.e., z-axis), and the piezoelectric-induced strain or applied stress is in the 3-direction (along the z-axis). For the coupling coefficient, k (Equation 9.24), the same notation is used. The above expressions are equivalent to writing (strain tensor) = (d tensor) × (electric field vector) and (polarization vector) = (d tensor) × (stress tensor), respectively. Taking advantage of this tensor notation, the constitutive Equations 9.18 to 9.21 may be generalized as:

$$\mathbf{D}_i = \mathbf{d}_{ik}\sigma_k + \varepsilon(\mathbf{d})_{ik}\sigma E_i \qquad 9.27$$

and

$$\varepsilon_k = \mathbf{d}_{ik}E_i + S_{ik}^E\sigma_k \qquad 9.28$$

and similar generalized equations for the other piezoelectric relations.

From Equation 4.17 in Chapter 4 we recall that the stress tensor has six components: $\sigma_x$, $\sigma_y$, $\sigma_z$ for compression or tension and $\tau_{xy}$, $\tau_{xz}$, and $\tau_{yz}$ for the shear components. The piezoelectric behavior of a crystal can be completely described if the set of data related to the piezoelectric constants ($d_{ik}$), elastic compliances ($S_{ik}$), and permittivities ($\varepsilon_{ik}$) are given. This set can be introduced by a 9 × 9 matrix in which columns are associated with mechanical stresses and field strengths, and rows refer to strains and polarizations:[34]

$$
\begin{array}{ccccccccc}
\sigma_x & \sigma_y & \sigma_z & \tau_{xy} & \tau_{xz} & \tau_{yz} & E_x & E_y & E_z
\end{array}
$$

$$
\begin{array}{c}
\varepsilon_x \\ \varepsilon_y \\ \varepsilon_z \\ \gamma_{xy} \\ \gamma_{xz} \\ \gamma_{yz}
\end{array}
\begin{bmatrix}
S_{11} & S_{12} & S_{13} & S_{14} & S_{15} & S_{15} \\
S_{21} & S_{22} & S_{23} & S_{24} & S_{25} & S_{26} \\
S_{31} & S_{32} & S_{33} & S_{34} & S_{35} & S_{36} \\
S_{41} & S_{42} & S_{43} & S_{44} & S_{45} & S_{46} \\
S_{51} & S_{52} & S_{53} & S_{54} & S_{55} & S_{56} \\
S_{61} & S_{62} & S_{63} & S_{64} & S_{65} & S_{66}
\end{bmatrix}
\begin{bmatrix}
d_{11} & d_{21} & d_{31} \\
d_{12} & d_{22} & d_{32} \\
d_{13} & d_{23} & d_{33} \\
d_{14} & d_{24} & d_{34} \\
d_{15} & d_{25} & d_{35} \\
d_{16} & d_{26} & d_{36}
\end{bmatrix}
\qquad 9.29
$$

$$
\begin{array}{c}
P_x \\ P_y \\ P_z
\end{array}
\begin{bmatrix}
d_{11} & d_{12} & d_{13} & d_{14} & d_{15} & d_{15} \\
d_{21} & d_{22} & d_{23} & d_{24} & d_{25} & d_{26} \\
d_{31} & d_{32} & d_{33} & d_{34} & d_{35} & d_{36}
\end{bmatrix}
\begin{bmatrix}
\varepsilon_{11} & \varepsilon_{12} & \varepsilon_{13} \\
\varepsilon_{21} & \varepsilon_{22} & \varepsilon_{23} \\
\varepsilon_{31} & \varepsilon_{32} & \varepsilon_{33}
\end{bmatrix}
$$

This matrix is symmetrical ($S_{ik} = S_{ki}$, $d_{ik} = d_{ki}$, and $\varepsilon_{ik} = \varepsilon_{ki}$), composed of 45 terms including 21 compliances, 6 permittivities, and 18 piezoelectric constants. In the top left we recognize Equation 4.17. A simplified notation with one subscript usually is used for dielectric constants: $\varepsilon_{11} = \varepsilon_1$, $\varepsilon_{22} = \varepsilon_2$, and $\varepsilon_{33} = \varepsilon_3$. For the defined orientation of the crystallographic axes, $\varepsilon_{ik} = 0$ for $i \neq k$.

For crystalline quartz, the 18 remaining piezoelectric constants further reduce to 2:

$$d_{ik} = \begin{bmatrix} d_{11} & d_{12} & 0 & d_{14} & 0 & 0 \\ 0 & 0 & 0 & 0 & d_{25} & d_{26} \\ 0 & 0 & 0 & 0 & 0 & 0 \end{bmatrix} \qquad 9.30$$

with $d_{11} = -d_{12} = -d_{26}/2 = 2.31$ pC/N and $d_{14} = -d_{25} = 0.73$ pC/N. The meaning of $d_{14}$ is as follows: a torsion stress of 1 N/m² around the axis 1 (x) of quartz (direction 4 or xz) induces a charge density of 0.73 pC/N in two metal plates on the material in the direction 1 (x).

### Piezoelectric Actuation

The small strains (usually less than 0.1%) and high stresses (e.g., 35 MPa) generated by piezoelectric devices have spawned a diverse range of actuator applications ranging from motors (see Figure 9.8) to pumps (see Figure 10.30B). Bulk ceramic piezoelectric devices on the basis of lead zirconate titanates (PZT) have been widely used for decades but thin-film applications are new arrivals. Thin-film piezoelectric elements have been used in surface acoustic wave (SAW) sensors, micromotors, and tip-positioning stages for scanning tunneling microscopes (STM) systems. High mechanical forces are achieved for small amounts of power. But high voltages are needed to achieve appreciable deformations; therefore, stacked structures are made to obtain larger displacements even though they are difficult to implement in a microstructure. Zinc oxide and ferroelectrics of the lead zirconate titanate (PZT) family have been prepared in thin films by sputtering and sol-gel deposition techniques. Typical change in length per electric field unit (1 perpendicular to the field E) equals $5 \times 10^{-10}$ C/N for ZnO and 100 to 200 $10^{-10}$ C/N for PZT-type ceramics.

Piezoelectric actuators, as Flynn et al.[35] point out, offer significant advantages over both electrostatic and electromagnetic actuators. With motors, for example, the greater the energy density that can be stored in the gap between rotor and stator, the greater the potential for converting to torques or useful work. With electrostatic motors, the maximum energy density storable in the air gap is determined by $E_b$, the maximum electric field before breakdown (approximately $3 \times 10^8$ V/m for 1-μm gaps). From Equation 9.7:

$$E'_e = \frac{\varepsilon_{air} E_b^2}{2} \qquad 9.31$$

where $\varepsilon_{air}$ represents the permittivity of air (equal to that of free space). This results in an energy density of $4 \times 10^5$ Jm⁻³. With

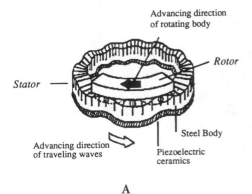

A

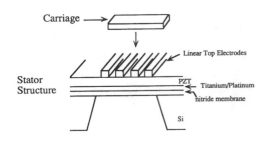

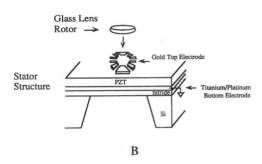

B

**FIGURE 9.8** Piezoelectric motors. (A) Schematic of commercially available Panasonic ultrasonic motor: an electrically induced wave of mechanical deformation travels through a piezoelectric medium in the stator, moving the rotor body along through friction. (B) Thin-film PZT motors: (Top) Linear motor — the PZT stator is patterned onto a silicon membrane. The stator can deflect more because the nitride membrane is thin; titanium and platinum form the ground electrode; and the linear gold stripes are the top electrodes. A carriage is deposited on the stator by hand. (Bottom) Rotary motor — identical to the linear motor except that the top electrodes are patterned in a circle. (From Flynn, A.M. et al., *J. Microelectromech. Sys.*, 1(1), 44–52 (1992). With permission.)

a piezoelectric motor made from ferroelectric material such as PZT, the energy density again is determined by the maximum electric field before breakdown. Thin-film PZT can withstand high electric fields, about the same as an air-gap ($E_b\pm = 3 \times 10^8$ V/m), but the dielectric constant is three orders of magnitude larger ($\varepsilon_{PZT} = 1300 \, \varepsilon_{air}$) than air:

$$E'_e = \frac{\varepsilon_{PZT} E_b^2}{2} \qquad 9.32$$

In principle, the energy density also should be three orders of magnitude larger. The magnetic energy density, Em', stored in an air gap is given by:

$$U'_m = \frac{B^2}{2\mu_0} \qquad 9.33$$

where B is the magnitude of the magnetic flux density in the gap and $\mu_0$ the magnetic permeability of free space. With magnetic actuators the gap energy density by pushing B into saturation (about 1.5 T) leads to $\approx 50,000$ $Jm^{-3}$. The energy density stored for the three actuation means reviewed and that of shape-memory alloy (SMA) actuation (see below) are compared in Table 9.2.

Obviously, piezoelectric ultrasonic motors have a distinct energy advantage. Other advantages, summarized from Flynn et al.[35] include:

- Low voltages: no air-gap is needed; mechanical forces are generated by applying a voltage directly across the piezoelectric film. With a 0.3-$\mu$m thin film only a few volts are required, as opposed to hundreds of volts needed in air-gap electrostatic motors.

- Geardown: motors can be fabricated without the need of a gearbox. Electrostatic wobble motors are also able to produce an inherent gear reduction but do not have the high dielectric advantage.

- No levitation: with electrostatic motors levitation and flatness are very important to obtain good sliding motion of the rotor around the bearing. The piezoelectric motor depends on friction so that no levitation is required and it can be freely sized.

- Axial coupling: electrostatic motors require axial symmetry around the bearing. Since height is difficult to obtain with most nontraditional micromachining techniques (except for LIGA and pseudo-LIGA methods) limited area is available for energy transduction. With the piezoelectric traveling-wave motors, linear or rotary motors can be built. As the stator is flat, with the rotor sitting on top of it, planar technologies are very capable of creating extra area to couple power out.

- Rotor material: the rotor can be of any material. Very importantly the conductivity of fluids being pumped with such motors does not affect the device, whereas an electrostatic motor would be shunted by a conductive liquid.

- Holding torque: because they are based on friction, piezoelectric ultrasonic motors can maintain holding torque even in the absence of applied power.

A schematic of a commercially available piezoelectric motor from Panasonic which uses an ultrasonic traveling wave is shown in Figure 9.8A. Two bulk ceramic PZT layers, both segmented with alternating poled regions, are placed on top of each other. For a given polarity of the applied bias one segment contracts and the neighboring segment expands. With the two ceramic pieces put on top of each other with their segmented areas out of phase, any point on the stator then moves with the rotor being pulled along the stator surface through frictional coupling.

Thin-film PZT linear and rotary motors by Flynn et al.[35] and Udayakumar et al.[36] are illustrated in Figure 9.8B. The stators in both cases are microfabricated using lithography techniques. Carriages and rotors may be made out of any type of material (in the rotational case a glass lens was used for a rotor). The work demonstrates that for a 5-V excitation $1.6 \times 10^{-12}$ $N/m/V^2$ normalized torque could be achieved as opposed to $1.4 \times 10^{-15}$ $N/m/V^2$ for electrostatic motors operating at 100 V.

The piezoelectric effect scales down with the bulk of the material; therefore, miniaturization opportunities are limited and hybrid type microactuators, where larger pieces of piezoelectric material are glued onto a substrate, are sometimes more reasonable. This has been done for making micropumps (Figure 10.30B)[37,38] and microvalves.[39]

### Piezoelectric Materials

The most extensively used natural piezoelectric materials are crystals — quartz and tourmaline. In synthetic piezoelectrics, ceramics formed by many tightly compacted monocrystals (about 1 $\mu$m in size) are most popular. These ceramics, such as lead zirconate titanate (PZT), barium titanate, and lead niobate, are ferroelectrics. To align the dipoles in the monocrystals in the same direction (i.e., to polarize or pole them), they are subjected to a strong electric field during their fabrication process. Above a certain temperature, known as the Curie point, the dipole directions in ferroelectric materials have random orientations. To align the dipoles, fields of 10 kV/cm are common at temperatures slightly above the Curie temperature. The ceramic is then cooled while maintaining the field. When the field is removed, the crystallites cannot reorder in random form because of the mechanical stresses accumulated, resulting in a

**TABLE 9.2**  Comparing Energy Densities for Magnetic, Shape-Memory Alloy, Electrostatic, and Piezoelectric Actuation

| Principle | Maximum Energy Density | Equation | Special Drive Condition |
|---|---|---|---|
| Magnetic | $9.5 \times 10^5$ $Jm^{-3}$ | $1/2 \ B^2/\mu_0$ | 1.5 T |
| Shape-memory alloy | $10.4 \times 10^6$ $Jm^{-3}$ (from stress-strain isotherms) | — | 1.4 W $mm^{-3}$ |
| Electrostatic | $4 \times 10^5$ $Jm^{-3}$ | $1/2 \ \varepsilon_{air}E_b^2$ | $3 \times 10^8$ $Vm^{-1}$, 1 $\mu$m gap in air |
| Piezoelectric | $5.2 \times 10^7$ $Jm^{-3}$ | $1/2 \ \varepsilon_{PZT}E_b^2$ | $3 \times 10^8$ $Vm^{-1}$, 1 $\mu$m-thick PZT film |

permanent electric polarization. PZT, as we saw above, can be grown in thin-film form by sputtering and by the sol-gel process. One obtains a piezoelectric material with a relative dielectric constant of more than 1000.[35] Abe et al.[40] argue that the sol-gel technique may not be feasible for micro electromechanical systems (MEMS) applications, given that 1-$\mu$m thick film would require repeated spin-coatings and pyrolysis steps. They prepared PZT films by sputtering from a composite target and found substrate heating during deposition to be crucial to transform the films into the perovskite phase during subsequent annealing. A problem with these materials relates to their temperature sensitivity and aging (loss of piezo properties) when approaching the Curie temperature.

Polymers such as polyvinylidene fluoride (PVF$_2$ or PVDF), lacking central symmetry, also display piezoelectric properties. Like traditional piezo materials, PVDF converts mechanical energy to electrical response and electrical signals to mechanical motion. This piezo polymer is also pyroelectric, capable of converting changes in temperature to electrical output (see below). Compared to quartz and ceramics, piezofilm is more pliant and lighter in weight. In addition, it is rugged, inert, and low cost. This engineering material could provide a sensing solution for vibration sensors, acceleration and shock sensors, passive infrared sensors, solid state switches, and acoustic and ultrasonic sensors. It should be noted that the peak reversible stress and strain developed by PVDF approximates 3 MPa and 0.1%, respectively. The latter values are rather small and as for piezoelectric ceramics and magnetostrictive materials (see below) mechanical amplifiers are required to produce the larger displacements needed for many applications. A 200-fold mechanical amplification is required to make its peak strain similar to that of mammalian muscle.[7]

Poled, cast, or spun-on films of PVDF can be employed to integrate piezoelectrics inexpensively into silicon micromachines. Gluing of commercial PVDF film with a urethane adhesive also works. Kolesar et al.,[41] for example, realized a two-dimensional electrically multiplexed tactile sensor this way. A 40-$\mu$m thick PVDF film was glued onto an 8 × 8 array of taxel electrodes (400 × 400 $\mu$m each) electrically coupled to a set of 64 MOSFET amplifiers located around the periphery of this array. The response of the tactile sensor was found to be linear for loads between 0.008 and 1.35 N. Crude imaging of the applied loads also was demonstrated. Table 9.3 summarizes some of the physical properties making PVDF such an interesting sensor material.

Some typical piezoelectric materials are compared in Table 9.4. To further improve the mechanical properties of piezoelectric sensors, piezoelectric composite materials sometimes are

**TABLE 9.3** Typical Properties of Piezofilm-Uniaxially Oriented Film

| | |
|---|---|
| Film thickness | (10–100 $\mu$m) |
| Piezoelectric strain constants (pC/N) | |
| $d_{31}$ | 28 |
| $d_{33}$ | −35 |
| Young's modulus (10$^9$ N/m$^2$) | |
| $E_{33}$ | 5.4 |
| Poisson ratio | |
| $v_{21}$ | 0.25 |
| $v_{31}$ | 0.57 |
| $v_{32}$ | 0.45 |
| Pyroelectric coefficient | |
| (10$^{-5}$ C/m$^2$/°K) | −30 |
| Mass density (10$^3$ kg/m$^3$) | 1.78 |
| Temperature range (°C) | −40 to 145 |
| Breakdown voltage (V/$\mu$m) | 80 |
| Maximum operating voltage (V/$\mu$m) | 30 |
| Yield strength (10$^6$ N/m$^2$) | 45–55 |
| Permittivity (10$^{-12}$ F/m) @ 10 kHz | 106 to 113 |
| Capacitance (pF/cm$^2$) @ 10 kHz | 380 |
| Electromechanical coupling factor (10 Hz) | |
| $k_{31}$ | 13% |
| $k_{32}$ | 1.7% |
| Piezoelectric stress constant (10$^{-3}$ m/mC/m$^2$) | |
| $g_{31}$ | 216 |
| $g_{33}$ | −339 |
| Volume resistivity ($\Omega$m) | >10$^{13}$ |

Based on References 42 and 43.

**TABLE 9.4** Comparison of Some Important Piezoelectric Materials

| Material | Piezoelectric Constant (d(pC/N)) | | | Dielectric Constant | Curie Temp. (°C) | Maximum k (Coupling Coefficient) |
|---|---|---|---|---|---|---|
| Quartz | 2.31 ($d_{11}$) | 0.73 ($d_{14}$) | — | $\varepsilon_1 = 4.52$ $\varepsilon_3 = 4.63$ | 550 | 0.1 |
| PZT (depending on composition) | 80–593 ($d_{33}$) | −94 to −274 ($d_{31}$) | 494–784 ($d_{15}$) | $\varepsilon_3 = 425$–1900 | 193–490 | 0.69–0.75 |
| PVDF (Kynar) | 23 ($d_{31}$) | 4 ($d_{32}$) | −35 ($d_{33}$) | 4 | >MP (150) | 0.2 |
| ZnO | −12 ($d_{15}$) | 12 ($d_{33}$) | −4.7 ($d_{31}$) | $\varepsilon_3 = 8.2$ | — | — |
| Sol-gel PZT | 220 ($d_{33}$) | −88.7 ($d_{31}$) | — | $\varepsilon = 1300$ | — | 0.49 |

*Note:* Data compiled from References 32, 34, and 35.
MP = melting point.

used. These are heterogeneous systems consisting of two or more different phases of which at least one shows piezoelectric properties. Hirata et al.,[44] for example, used LIGA to make arrays of PZT rods with diameter below 20 μm and an aspect ratio of over 5 (see also Chapter 6 and slurry casting of piezoceramics by Preu et al.[45] and Lubitz[46]). The PZT rods are embedded in a polymer matrix to make a PZT/polymer composite. A PZT slurry, average particle size 0.4 μm, is injected into a LIGA resist mold. After solidification, the resist is removed by plasma etching and the array of ceramic posts is cast in an epoxy resin. Resolution of acoustic imaging transducers, for example for medical applications, should gain resolution as the PZT rods further undergo miniaturization as a result of minimized cross-talk in the array.

For thin-film deposition of ZnO, plasma-magnetron sputtering appears to give the best piezoelectric and pyroelectric characteristics. Using planar magnetron sputtering, highly oriented ZnO films have been deposited on $SiO_2$, polycrystalline silicon, and bare silicon substrates. Muller[47] found the best thin-film crystallinity at a sputtering power of 200 W with a 10 mTorr ambient gas mixture consisting of an equal mix of oxygen and argon. The distance between the substrate and target measures about 4 cm with the substrate temperature maintained at 230°C during deposition.[47] Tjhen et al.[48] characterized the thin-film properties of sputtered ZnO and sol-gel deposited PZT and found the electrical properties of both to be sensitive to substrate material, stress, and surface condition. The compatibility of ZnO with silicon processing is demonstrated in a commercially available acoustic sensor from Honeywell Corp. as shown in Figure 9.9.[49] The microphone chip illustrated incorporates ZnO deposited on a silicon substrate including signal-conditioning circuitry. A unique annular electrode design permits the parasitic signal due to temperature variations in the zinc oxide film to be canceled while doubling the output due to pressure (see insert in Figure 9.9).

Besides PZT and ZnO, AlN is another thin-film piezoelectric material popular with the sensor industry. The considerations in designing sensors with these films include:

1. Value of electromechanical coupling
2. Good adhesion to the substrate
3. Resistance to environmental effects (e.g., humidity, temperature)
4. VLSI process compatibility
5. Cost effectiveness

For a more detailed review of ferroelectric polymers, refer to *Ferroelectric Polymer-Chemistry, Physics, and Applications.*[43] For a survey on inorganic ferroelectrics, refer to *Ferroelectric Transducers and Sensors.*[33] For a mathematical treatise of piezoelectric

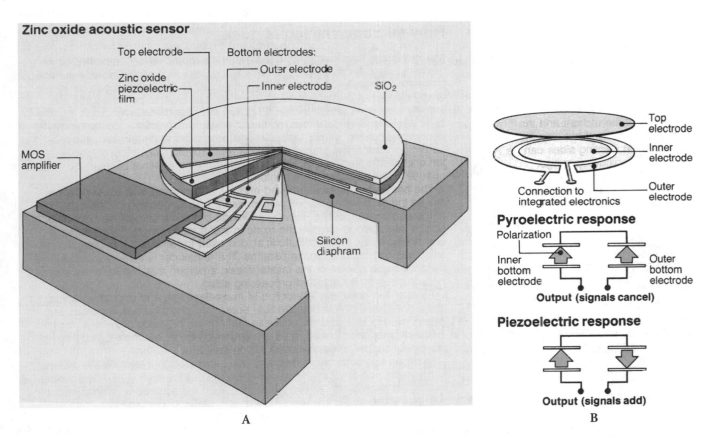

**FIGURE 9.9** Zinc oxide acoustic sensor. Two parallel plate electrodes in Honeywell's acoustic sensor act as capacitors. Voltages due to temperature variations (pyroelectric effect) on zinc oxide film cancel, while those due to pressure add, doubling the output. (From Allen, R., *Sensors in Silicon, High Technology*, 43–81, 1984. With permission.)

devices we refer to the 1969 *Design of Resonant Piezoelectric Devices*[50] and the more recent *The Theory of Piezoelectric Shells and Plates.*[51]

## Pyroelectricity

Pyroelectricity, first recorded in the scientific literature in 1824 and derived from the Greek for 'heat electricity', is the development of electric polarization in classes of noncentrosymmetric crystals subjected to a temperature change.[30] The phenomenon is present in crystallized dielectrics with one or more polar axes of symmetry. While all pyroelectric materials are piezoelectric, the opposite is not true, e.g., quartz is piezoelectric but not pyroelectric. Both zinc oxide with its wurtzite crystal structure and PZT with a ferroelectric perovskite structure feature an intrinsic dipole moment and show pyroelectricity even in thin film form. The strongest pyroelectric effect for a thin film is measured with the sol-gel processed tetragonal $PbTiO_3$.[48,52] The pyroelectric constant of these materials is defined as the differential change in polarization with temperature in the case of uniform heating, constant stress, and low electric field in the crystal:

$$P_\sigma = \left(\frac{\partial P}{\partial T}\right)_{\sigma, E \approx 0} \qquad 9.34$$

$P_\sigma$ is expressed in $C/cm^2K$. Thin films of ZnO, PZT, and $PbTiO_3$ all show pyroelectric coefficients similar to the bulk material — $0.95$–$1.05 \times 10^{-9}$ $C/cm^2K$ for ZnO, $50$–$70 \times 10^{-9}$ $C/cm^2K$ for PZT, and $95$–$125 \times 10^{-9}$ $C/cm^2K$ for $PbTiO_3$.[48,52] Using $LiTaO_3$ Zemel et al.[53] produced sensitive, wide-range anemometers. In all thermal anemometers the goal is to measure heat loss; usually that is accomplished by measuring the resistance of a temperature-sensitive resistor, but with a pyroelectric sensor heat loss generates a current directly. Gas flows of 0.1 ml/min to over 20 l/min were detected. It should be remembered that only changing temperatures can be detected this way. For a crystal maintained at a constant temperature, the charges will decay rapidly.

## Electrostriction

Electrostrictive materials, basically all dielectrics, develop mechanical deformations when subjected to an external electrical field, i.e., they show an increase in length parallel to the field. The phenomenon is attributed to the rotation of small electrical domains in the material upon exposure to a field. In the absence of the field the domains are oriented randomly. The alignment of the electrical domains parallel to the electrical field results in the development of a deformation field in the electrostrictive material. Usually the effect is small but a few materials, such as certain titanates and zirconates, show a large effect in which case they are ferroelectric and exhibit properties analogous to ferromagnetic materials. In both groups of materials, spontaneous polarization occurs within small regions or domains. The electric dipoles in ferroelectric materials or the elementary magnets

in ferromagnetic materials are in parallel alignment within each domain. Two example electrostrictive materials are barium titanate and lead-magnesium-niobate.[30,31]

## Electrets

A permanently charged dielectric, an electret, forms an elegant base to improve upon electrostatic actuation by setting up a permanent electric field, thereby reducing or eliminating the need for a large applied bias. There are many different methods to form electrets from high impedance dielectrics such as the corona, the liquid contact, and the electron-beam methods.[54,55] An electret-based micromachine may incorporate a capacitor charged due to the presence of an electret between both capacitor plates, while one of the plates is sensitive to an external force (e.g., pressure). Micromachined electret microphones were built this way and exhibit an open-circuit sensitivity of about 2.5 mV/μbar at 1 kHz.[56] By charging up $SiO_2$ with a 300-V corona, electrets in these subminiature microphones exhibit a charge decay time constant amounting to more than 100 years of expected operation. Similarly, an electret-based pressure sensor with a permanently charged polymer foil (commercially available Teflon-FEP) as the electret was manufactured. The foil was charged by electron-beam exposure. A maximum sensitivity of 10 mA/A/100 mmHg (i.e., the measured relative change of drain current) about ten times higher than the sensitivity of piezoresistive pressure sensors with comparable dimensions (1 mm by 2 mm by 0.3 mm), was determined.

Wolffenbuttel et al.[57] point out that an insulating electret-implanted rotor circumvents two major limitations of electrostatic motors, i.e., the relatively large voltages and the friction between the rotor and the stator during rotation. Replacing the polysilicon rotor in a motor as shown in Figure 9.5B allows propulsion at smaller drive voltages as well as electrostatic levitation and lateral alignment of the rotor, making it virtually contactless. Temperature and humidity sensitivity constitute some of the biggest drawbacks of employing electrets in actuators.

## Electrorheological Fluids

Electrorheological (ER) fluids are a class of colloidal dispersions which exhibit large reversible changes in their rheological behavior when subjected to external electrical fields. These changes in rheological behavior typically are manifested by a pseudo-phase change from the liquid phase to solid state or a dramatic increase in flow resistance (the viscosity might increase by several orders of magnitude). The voltages required to activate the phase-change in ER fluids typically are of the order of 1 to 4 kV/mm. The total power required is rather low and the response of ER fluids to an electrical stimulus typically is less than a millisecond. The electrical field induces dipoles in the particulate phase, forging an interaction between the individual particles which form columnar structures as presented in Figure 9.10.[30] Typical products proposed with these actuators include clutches, hydraulic valves, and vibration isolation systems such as engine mounts and shock absorbers. Given the

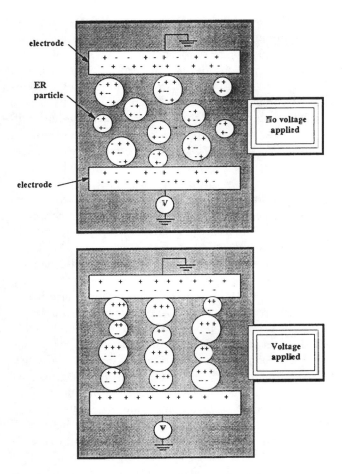

**FIGURE 9.10** The electrorheological (ER) phenomenon. (From Gandhi, M.V. and Thompson, B.S., *Smart Materials and Structures*, Chapman & Hall, London, 1992. With permission.)

fact that it is easy to obtain large fields in the microdomain, research on adapting electrorheological fluids to MEMS should follow soon.

## Dielectric Induction Effects

In an induction motor, localized charges are induced in the rotor by the electrostatic field generated by the stator electrodes. Since a finite relaxation time is needed for this excess of charge to diffuse out, a driving torque is generated. The induced charges can be attractive or repulsive with respect to the inducing field, depending on the dielectric constants of the particle and the medium in which the particle is embedded. The principle can be used for dielectric linear and rotational induction motors as well as for sorting, moving, and separating small objects such as biological cells.[58-60] An induction motor is driven by a traveling wave of voltage and uses charge relaxation rather than the physically salient poles in a variable-capacitance motor as shown in Figure 9.5B.[15] Consequently, the rotor can be a smooth uniform conductor and may even be a fluid.[61] In the case of the salient rotor blades, fabrication involves some undesirable planarization steps. The conductivity of the induction motor materials, on the other hand, strongly affects performance, so fabrication difficulties associated with variable capacitance

motors are traded for those associated with conductivity control in induction motors.[62]

In electrohydrodynamic (EHD) pumping, fluid forces are generated by the interaction of electric fields and charges in the fluid. In contrast to forces generated by mechanical pumping using an impeller or bellows, EHD pumping requires no moving parts.[63] This type of pumping is compared with other fluid propulsion methods under Fluidic Propulsion Mechanisms in the next section on Scaling in Fluidics.

## Scaling in Magnetics

### Introduction

Electromagnetic (EM) forces dominate in the development of actuators in the macroworld. Electrostatic macromotors, on the other hand, such as the 1889 capacitive motor made by Zipernowsky,[64] are rarities rather than commodities. Due to the three-dimensional nature of magnets and solenoids, EM systems proved difficult to micromachine using planar IC processes. Electrostatic motors are simpler and more compatible with IC fabrication, and electrostatics are scaling more favorably into the microdomain. These factors explain the current preponderance of electrostatic actuators in micromachining. We shall learn though from the following detailed comparison of electrostatic vs. magnetic actuation that both methods present advantages and disadvantages and, as all actuators do, present severe problems when high power output is demanded.

### Scaling in Magnetics

A good starting point for an understanding of the scaling in magnetics is Ampère's circuital law used to calculate the magnetic induction field, B, analogous to the way we use Gauss' law to calculate the electrical field, E. The law states that the line integral of B · dl over a closed curve C is equal to $\mu_0$, the magnetic permeability of free space, times the current through the loop, C:

$$\oint_C B \cdot dl = \mu_o \int_A J \cdot dA = \mu_o I \qquad 9.35$$

This equation describes the creation of a magnetic field by a current. In this expression A defines the area of the surface bounded by C, and J is the current density. In the last term on the right, I represents the net current that crosses A. An example calculation illustrating the concept is shown in Figure 9.11.[2] As the system becomes smaller, the area dA decreases more rapidly than dl, and B decreases unless we increase the current density J. When keeping the current density constant, as the size of the system is decreased, the force between two electromagnetic wires or coils scales as $(l^4)$. In other words, a size reduction of 10 means a magnetic force reduction of 10,000. The scaling results of the interaction between a coil and a permanent magnet is better — $(l^3)$.[2] To improve the situation we can increase the current density. The resulting heat, as we will see below, is more effectively removed from microstructures, thus avoiding overheating and improving the scaling. Different assumptions

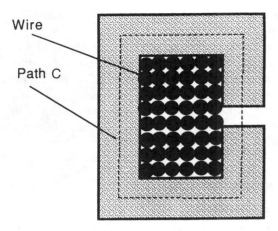

**FIGURE 9.11** The line integral of Bdl over a closed curve C is equal to $\mu_0$, the magnetic permeability of free space times the current through this loop C. As the system becomes smaller, the area decreases more rapidly than the length of the dashed line. (From Trimmer, T., Integrated Micro-Motion — *Micromachining, Control, and Applications* (Elsevier Science, Amsterdam, 1990). With permission.)

about current density and heat transfer modes lead to different scaling laws. Constant current density, as just mentioned, leads to $(l^4)$ $((l^3)$ with a permanent magnet); constant heat flow per unit area of the windings gives $(l^3)$ $((l^{2.5})$ with a permanent magnet); a constant temperature difference between windings and environment yields $(l^2)$ (also $(l^2)$ for the permanent magnet case) (see Appendix in Trimmer's article[9]). For the magnetic case where the force scales as $(l^2)$, the power that must be dissipated per unit volume scales as $(l^{-1})$ (see Equation 9.15), or, when the scale is decreased by a factor of 10, 10 times as much power must be dissipated. The use of superconductors could eliminate this problem.

## Comparing Electrostatic and Magnetic Actuation

The debate about the relative merits of electrostatic vs. magnetic actuation, especially for driving micromotors, continues.[21,28] Fujita et al., for example, has argued that electrostatic actuation is preferred,[24,65,66] pointing to the following attributes of surface micromachined electrostatic actuators:

1. Thin insulating layers such as $SiO_2$ or $Si_3N_4$ exhibit breakdown strengths as high as 2 MV/cm. The power density in this field is $7 \times 10^5$ J/m³; this value is equal to the power density of a 1.3-T magnetic field. The contracting pressure induced by this field is 1.3 MPa. A voltage of about 100 is sufficient to generate the strong fields mentioned.

2. The electrostatic force is a surface force exhibiting a favorable scaling law. The actuation is simple as it only involves a pair of electrodes separated by an insulator.

3. The electrostatic actuator is driven by voltage and voltage switching is far easier and faster than current switching (as in electromagnetic actuators). Energy loss through Joule heating is also lower.

4. Weight and power consumption are low.

The following comparison between electrostatic and magnetic microactuators demonstrates that besides scaling many other factors need to be considered when deciding upon a certain type of actuation principle. Whereas the magnetic power, in some cases, might scale disadvantageously into the microscale, the absolute forces achievable are very large. In conventional motors, using iron, the magnetic induction is limited to 1.5 T because of saturation, producing an energy density of about $9 \times 10^5$ J m⁻³ (see Table 9.2), more than twice the achievable electrostatic energy in a 1-μm gap. It is important to remember that, in the case of electrostatic motors, there is little room for further improvement since the electrostatic energy density, assumed above, is close to what is achievable in vacuum. In contrast, there is plenty of room for further improvement in magnetic energy density. Thin ferromagnetic films have yielded 2-T fields, and substantially higher fields are generated, at much larger expense, using superconductors.[28] For example, using small-bore 10- to 15-T superconducting magnets, the achieved energy density of 4 to $9 \times 10^7$ J m⁻³ is roughly two orders of magnitude larger than the highest possible electric field density.

The 10- to 15-T superconducting magnets fall outside the realm of micromachining. But Busch-Vishniac[28] argues that from a micro-actuation perspective, defined as the generation of micron-scale motion, it does not matter that the actuator itself is physically large. In any event, this same author goes on to show that, theoretically at least, even when microfabricated, a magnetic actuator's force can be the same as that of a similarly sized electrostatic actuator (say about $2.0 \times 10^{-5}$ N for a 100-μm motor). She also maintains that such an actuator can be built simply and as cost competitively as the IC compatible electrostatic micromotors. The work by Wagner et al.[67] discussed below seems to confirm her points.

The efficiency, the ratio of power used in performing a desired task to the total amount of power consumed for a current driven magnetic motor, will be worse than for a voltage-driven electrostatic device (see above). But, Busch-Vishniac argues,[28] since friction in small gaps rather than electrical losses are the major losses in micro-actuators, mechanical losses render nearly all types of micro-actuators equally inefficient. Moreover, since the gap in magnetics may be larger, friction is actually easier to avoid in magnetic actuators. One could add that a wider separation gap also will make the actuator less sensitive to dust and humidity.

The above analysis puts magnetic actuation at an advantage over electrostatic actuation and suggests that micromachined actuators are not always possible or useful. Wise summarizes the status in micro-actuators by concluding that electrostatics is useful in dry environments and over limited distances and that magnetics is still difficult to collapse into integrated structures. Overall, we do not yet have good micromachined actuators.[64]

While there is validity, especially in the case of micromotors, to the above arguments favoring magnetic actuators, the criticism of micromachined electrostatic actuators may be too harsh where linear micromachined electrostatic devices such as resonators are concerned. In the latter application, power output is of little concern and these devices have found a major application in the Analog Devices accelerometer described in

Chapter 5 (see Figure 5.35). Similarly, for most optical applications the amount of power required to deflect a laser beam with a micromachined mirror, for example, is minimal.

Micromagnetics for many years have found major use in Hall chips and thin-film heads for magnetic discs (see Figure 7.8). Besides micromotors, new efforts concentrate on making denser thin-film magnetic heads;[69] high-performance magnetographic printing heads;[70] electronic components;[71] different types of vertical, torsional, and multiaxial actuators;[67] and contactless magnetic transmission of force to ferrofluids.[72] An advantage of magnetic actuators is the long range of the force. With dimensions above 1 mm and for larger forces, actuators based on permanent magnets become a good choice. To conclude this comparison of electromagnetic vs. electrostatic actuators we summarize the key points in Table 9.5.

## Merit of Micromachined Motors

On November 28, 1960, William McLellan collected $1000.00 from Richard Feynman for having made the first operating electric motor only 1/64 in.[21] in size. The handmade motor was a more-or-less standard two-phase permanent-magnet motor. The MEMS challenge is geared towards making micromotors in batch fashion. So far MEMS activities in micromachined motors has generated a lot of new insights in phenomena such as stiction, friction, and wear of various materials but has not yet produced a viable, batch-produced micromotor which can compete with a proven product such as the Panasonic piezoelectric device pictured in Figure 9.8A.

When motors with torques larger than $10^{-5}$ Nm are needed, theory and recent results prove that actuators in the millimeter range, fabricated with classical precision engineering or a combination of LIGA or pseudo-LIGA and precision engineering

need to be applied (see Chapter 6, LIGA Applications Electromagnetic Micromotor). Besides the problem of miniscule amounts of power produced, a problem with heat dissipation in a micromotor below a certain size requires solving. Busch-Vishniac calculated that a generic problem of heat dissipation exists if microfabricated micro-actuators continue to shrink in size, unless the amount of power used in the system is reduced. Accordingly, a motor dissipating 100 mW should not have a rotor smaller than 20 µm as the temperature might rise above 250°C resulting in thermal breakdown of the metals in the microstructure. The classical models for heat convection were used to reach this conclusion. A more accurate model would result in a reduction of the heat lost by convection as the effective viscosity in the gap is greater than the classical model predicts (see section on heat transfer). In other words, the results generated with the classical model should be viewed as a conservative approximation; real microdevices might run hotter yet.

## Levitation

Friction and wear on MEMS primarily relate to the surface contact between solids, in particular to the surface contact in bearing surfaces supporting the load of micromachinery. They scale as $l^2$ and become increasingly important in the microdomain. The friction in early electrostatic micromotors was actually found to be comparable to the friction of brake materials on cast iron.[73] Levitation can eliminate wear and friction, two factors affecting reliability and control. Levitation might even enable sensors to measure very weak forces resulting from various physical effects. Levitation can be obtained by magnetic forces,[74] electrostatic forces[75] and fluidic forces.[76] We will only consider magnetic levitation here.

### Levitation with Permanent Magnets

Levitation with magnetic devices (maglev) can be achieved by various methods using a permanent magnet, an electromagnet (including superconducting magnets), and a diamagnetic body. Working with permanent magnets leads to a better scaling behavior than working with electromagnets. Wagner et al.[72] levitated a small rare earth permanent magnet ($1.5 \times 1.5 \times 10$ mm³) out of the plane of a silicon wafer equipped with a planar coil as shown in Figure 9.12A. The permanent magnet is glued onto a silicon micromachined plate suspended by a thin silicon spring made from suspending beams parallel to the substrate edge. The monolithically integrated 17-turn planar coil is used to generate a magnetic field force which forces the magnet on its spring to move vertically (z-direction) (see also Benecke).[77] Using planar IC technologies to manufacture the coil ensures an optimum heat flow within the device (see below). If $\mu_z$, the magnetization of the permanent magnet, is independent of the magnetic induction field, B, the vertical force acting on the permanent magnet is given by:

$$F_z = \mu_z \int \frac{dB_z}{dz} dV \qquad 9.36$$

**TABLE 9.5** Comparison of Electrostatic vs. Magnetic Actuators

|  | Electrostatic | Magnetic |
|---|---|---|
| Field energy density | $4 \times 10^5$ Jm⁻³ (max) | $4 \times 10^7$ Jm⁻³ (10 T) |
| Scaling | ($l^2$) | ($l^3$) (constant current) |
| Gap contamination sensitivity | Very sensitive to humidity and attracts dust | Fairly insensitive |
| IC compatibility | Good | Not very good |
| Range | Short range | Long range |
| Power efficiency | Very good in the absence of mechanical friction | More power-consuming even in the absence of mechanical friction |
| Implementation of levitation schemes | Possible | Easier implemented because larger gaps are possible |
| Miniaturization | Excellent | Difficult |
| Complexity | Low | High |
| Control | Voltage switching control is faster, easier to make, and more efficient | Current switching control is less efficient, more complex |

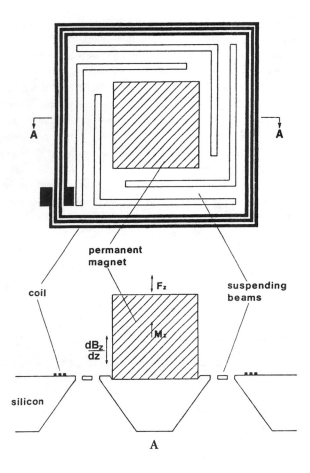

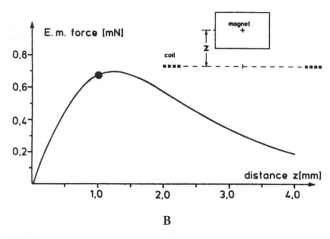

**FIGURE 9.12** Magnetic actuator. (A) Schematic of a vertical magnetic micro-actuator with integrated planar coil and hybrid mounted permanent magnet. (B) Electromagnetic force on a magnet. The dot indicates the magnet position in the fabricated device. (From Benecke, W., *Silicon-Microactuators: Activation Mechanisms and Scaling Problems*, Intl. Conference on Solid-State Sensors and Actuators, San Francisco, CA, 46–50, 1991. With permission.)

where $B_z$ represents the vertical component of the magnetic field produced by the planar coil. The magnetic force is proportional to the volume of the magnet. Thin magnetic layers will, in general, be insufficient to generate high forces. With commercially available permanent magnets other problems included the

fact that NdFeB permanent magnets, like other rare earth based magnets, incorporate on their side surfaces (nonpole surfaces) a magnetically reversed layer of about 20 μm. This surface demagnetization limits the miniaturization of these permanent magnets to values of a few 100 μm (they are available down to a size of about 0.3 mm). The deflection of the described actuator is given by Hooke's law as:

$$\Delta z = \frac{F_z}{k} \qquad\qquad 9.37$$

with k symbolizing the spring constant. The magnitude of the magnetic field will depend on the current densities achieved in the windings generating the magnetic field, and this current is limited by heating of the solenoid metal (Au in this case). Sheet resistance of thin-film conductors, electromigration phenomena, and thermal and geometric constraints all conspire to reduce the achievable magnetic energy density. But, Wagner et al.[67,72] assume constant current and very efficient power dissipation and derive a magnetic force showing quadratic scaling and a deflection showing linear scaling. To obtain constant current this group has the number of coil turns scaling linearly with the cross-section of the coil wire. The quadratic force scaling, also found for electrostatics, lends itself well to the microrange. In practice, the magnetic force in the example case first rapidly increases and then slowly decreases with z exhibiting a maximum value of 0.68 mN at a distance of 1.25 mm from the coil plane (see Figure 9.12B).[77] Increasing the volume of the permanent magnet thus works only up to a point as the magnetic field slowly decreases with z. For a driving current range between –300 and +300 mA, an elevation of the permanent magnet of 143 μm was achieved, resulting in a mean slope of 24 μm/100 mA. This example indicates that magnetic actuation, at least for linear actuators, can be designed to scale as well as electrostatics and, when allowing for some final manual assembly, these actuators are easily implemented. Electromagnetic microactuators with glued magnets of millimeter dimensions have been used for different types of vertical, torsional, and multiaxial actuation. A variety of motors including linear sliding, linear rolling, and rotational permanent magnets, as well as contactless magnetic transmission of force to a ferrofluid, were also demonstrated.[72]

### Levitation with Superconductors

The above-described permanent magnets are ferromagnets with a magnetic susceptibility $\chi_m$ positive and very large ($\chi_m$ usually is only constant in a small range of magnetization). Self-levitation can also be based on diamagnetic materials for which $\chi_m$ is small and negative. In diamagnetism, as a magnet approaches a diamagnetic material, magnetic dipole moments are induced in the diamagnetic material that oppose the applied field. These dipole moments lead to magnetic forces that tend to push away the magnet. Some ordinary diamagnetic materials such as carbon and bismuth, with a magnetic susceptibility, $\chi_m$, of about $-1 \times 10^{-6}$ (compared to $-0.20 \times 10^{-6}$ for Au or $-0.11 \times 10^{-6}$ for Cu), can levitate magnets at room temperature. Although only a small force per unit mass, this kind of levitation does not require power, operates at room temperature, and can be used with a variety of materials. It is well known that superconductors can levitate magnets because of their diamagnetism (Meissner effect). Most high-purity metals when cooled down to temperatures nearing 0K, exhibit a gradually decreasing electrical resistivity, approaching some small, yet finite value characteristic to the particular metal. For a few superconducting materials the resistivity abruptly plunges from a finite value to one that is virtually zero at very low temperatures and remains there upon further cooling. The loss of resistance to electrical current flow occurs below a critical temperature, $T_c$. Below that temperature superconductors offer an energy transfer medium with virtually no power losses. Until 1986 the critical temperature $T_c$ was 23 K for the best of the low-temperature superconductors, requiring liquid He for maintaining the superconductivity state. Bednorz and Muller, in 1986 at the IBM Zurich Research Laboratory, discovered higher temperature superconductivity in a lanthanium-barium-copper-oxide. Their invention triggered a race to superconducting materials with a transition temperature above 77 K, the boiling point of liquid nitrogen. Liquid nitrogen is obtained much more easily and cheaply than liquid helium. Given that power sources and actuators, in general, exhibit poor scaling in the microdomain, superconductors could play a major role in powering and actuating micromechanisms more efficiently. Kim et al. levitated a permanent-magnet mover by applying the Meissner effect.[73] A major micromachining challenge for these new micromachines is to find ways around using liquid nitrogen cooling. Perhaps thermoelectric coolers could be applied if we would only learn how to make those more energy efficient.

## Magnetostriction

In magnetostrictive or piezomagnetic materials, magnetization in an external field causes a dimensional change of the material. The relative change in length — $\Delta l/l = \lambda$, typically in the parts-per-million range (e.g., $40 \times 10^{-6}$) — is called the magnetostriction, which can be positive or negative in a direction parallel to the magnetic field and is independent of the direction of that field, indicating a square-law type of relationship such as:

$$\frac{\Delta l}{l} = \varepsilon = cB^2 \qquad\qquad 9.38$$

where $\varepsilon$ defines the static strain produced by a flux density B, and c ($m^4/Wb^2$) is a material constant. For small AC-driven field, B, Equation 9.38 reduces to the basic magnetostrictive strain equation:

$$\varepsilon = \beta B \qquad\qquad 9.39$$

where $\beta$ defines a magnetostrictive strain constant with the dimensions of $m^2/Wb$. As in the case of piezoelectricity the inverse effect exists as well. A changing flux density results when modulating the stress on a piezomagnet. The magnetostriction phenomenon is attributed to the rotations of small magnetic

domains in the material, randomly orientated in the absence of a magnetic field. The orienting of these small domains by the imposition of a magnetic field results in the development of a strain field. As the intensity of the field increases more and more domains line up with the field until saturation. This actuation mechanism is important when large forces must be obtained over small distances. Materials with high magnetostriction coefficients were developed in the 1960s for underwater radar with as a typical material: Terfenol-D ($Tb_xDy_{1-x}Fe_y$ with x between 0.27 and 0.3 and y between 1.9 and 1.98).[30] For automotive torque sensing in the 1990s ferromagnetic materials such as $Co_{75}Si_{15}B_{10}$[78] are being investigated. An uniaxial crystal growth technique was developed for the latter material, resulting in a bigger magnetostrictive effect with a smaller magnetic field. Terfenol-D is a commercially available magnetostrictive material incorporating the rare-earth element dysprosium. This material offers strains up to 0.002, an order of magnitude superior to other piezoceramic materials. The effect is often referred to as the giant magnetostrictive effect.

Sputter-deposited magnetostrictive films present an interesting opportunity for actuation in micromachines where contactless, high-frequency operation is desired. Terbium-iron alloys are typical magnetostrictive materials that can be deposited by sputtering. Quandt et al.,[79] using an RF-sputtered TbDyFe film (10 µm thick) on a silicon cantilever 2 cm long and 50 µm thick could deflect the silicon cantilever by more than 200 µm in an external field of only 30 mT. The cantilever was operated at a frequency of 500 Hz and no degradation could be observed after more than $10^7$ operations. The same group is exploring the use of these films for valves and pumps. Honda et al.[80] discovered that amorphous Tb-Fe films exhibit positive magnetostriction while amorphous Sm-Fe thin films exhibit negative magnetostriction and built magnetostrictive bimorph cantilevers and traveling machines as illustrated in Figure 9.13A and B. The actuation behavior of the magnetostrictive bimorph cantilever is as follows: a 1-cm long polyimide beam (thicknesses of 7.5 µm, 50 µm, and 125 µm were experimented with) is sandwiched between an upper Tb-Fe film and a bottom Sm-Fe film, each 1 µm thick. When a magnetic field is applied along the cantilever length direction, the Tb-Fe film expands and the Sm-Fe film contracts, bending the polyimide beam downwards. With a magnetic field along the width direction, the Tb-Fe film contracts and the Sm-Fe film expands, deflecting the beam upwards. The traveling machine consists of a 7.5-mm thick polyimide film equipped with the magnetostrictive bimorph actuator layers and with two legs inclined so that it travels in one direction. With an alternating magnetic field of 100 Oe at 50 Hz applied along the machine length direction, it vibrates and travels at an average speed of 0.5 mm/sec in the arrow direction.

The power requirements for giant magnetostriction are greater than for piezoelectric materials but the actuation offers a larger displacement and the ratio of mass per unit stress is greater than with a PZT actuator.[81] Moreover, no contacts or heaters must be incorporated in the microdevice. Radiating in energy is an attractive alternative for micromachine since power sources scale so disadvantageously in the microdomain.

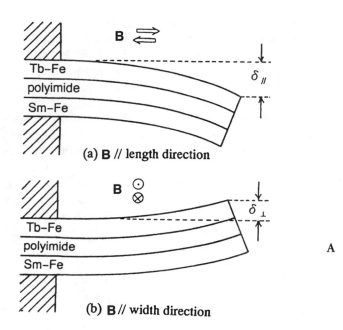

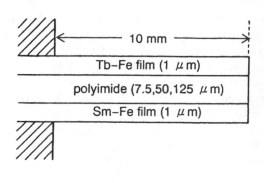

A

B

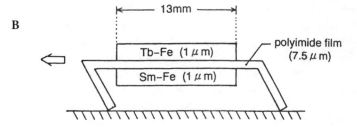

**FIGURE 9.13**  Magnetostrictive actuators. (A) Actuation behavior of a magnetostrictive bimorph cantilever. (B) Schematic view of traveling machine. (From Honda et al., *Fabrication of Actuators Using Magnetostrictive Thin Films*, MEMS, Osio, Japan, 51–56, 1994. With permission.)

A major application of magnetostriction is in noncontact torque sensing. An integrated silicon micromachined sensor head for torque and force measurements was first proposed in 1994 by Rombach et al.[78] It is based on a magnetic yoke with an exciting and receiving coil that detects the change of permeability of an amorphous $Co_{75}Si_{15}B_{10}$ ribbon fixed to the shaft surface of an automotive vehicle.

## Scaling In Fluidics

### Introduction

Fluid mechanics (fluidics) is the study of motion of fluids, liquid, or gas and the forces associated with that motion.[12] Micromachined fluid components and systems pioneered in the 1960s for fluidic logic[82] reemerged in the 1970s at Stanford University in the form of a miniature gas chromatograph[83] and at IBM with ink-jet printer nozzles.[84] In the mid-1990s microfluidics became fashionable for crafting miniaturized analytical equipment. Fluidic microsystems are being explored as a better approach to advance chemical analysis techniques compared to chemical microsensors. Examples of this recent wave of activities include Ceiba Geigy's work on micro-electrophoresis,[85] Redwood Microstructures' thermopneumatic 'Fluistor' valve[86] and Lawrence Livermore National Laboratories' work on micro-PCR

devices.[87] Microfluidic systems comprising nozzles, pumps, channels, reservoirs, mixers, oscillators, diodes, amplifiers, valves, etc. can be used for a variety of applications including drug dispensation, ink-jet printing, and general transport of liquids, gases, and liquid/gas mixtures. The advantages of these devices include lower fabrication costs, enhancement of analytical performance, lower power budget, and lower consumption of chemicals than traditional pumping and spraying systems. Ultimately, these developments will revolutionize applications where precise control of fluid flow is a necessity. As this area is quite new it brings a lot of excitement to the micromachining world.

Our review will focus on macroscale and microscale behavior of fluids. Fundamental work on the physical behavior of fluids in micromachined vessels started only recently (1988–1990).[88-91] For gases, breakdown of macroscale physics is readily recognized and sometimes taken advantage of. For liquids, clear evidence for deviation from macroscale laws has yet to be presented.

### Macroscale Laws for Fluid Flow

A fluid is a material that cannot sustain a shearing stress in the absence of motion.[12] The two most important parameters characterizing a fluid are density and viscosity. Density, $\rho$, of a fluid is the mass per unit volume; the viscosity, $\eta$, is the fluid property that causes the shear stresses when the fluid is moving; without viscosity in a fluid there would be no fluid resistance. Viscosity is measured in pascal × seconds (Pa × sec) (in the c.g.s. system it was measured in poise (1 poise = $10^{-1} Nsm^{-2}$)) and is given as the ratio of shear stress, $\tau_s$, to shear rate, $\Gamma$:

$$\eta = \frac{\tau_s}{\Gamma_s} \qquad 9.40$$

Equation 9.40 might be understood from inspection of Figure 9.14, where we illustrate two large plates of area, A, separated by a fluid layer of uniform thickness, h. When one plate is moved in a straight line relative to the other plate at a constant speed, U, the force required to obtain that constant speed is given by F. The shear stress equals F/A (force/area) and the shear rate is given as U/h (1/time). For Newtonian fluids the shearing stress and the velocity are thus linearly related.[92] The derivation of the fundamental laws governing fluid motion, called the Navier-Stokes equations, is beyond the scope of this treatment. We will only introduce the most important expressions derived from the Navier-Stokes equations. For more detailed study, refer to Allen[93] and Denn.[12]

Experimental determination of fluid viscosity is possible from the Hagen-Poiseuille law for slow flow of a fluid through a capillary:

$$Q = \frac{\pi r^4 \Delta p}{8 \eta l} \qquad 9.41$$

where Q defines the volumetric flow rate through a capillary of radius r, $\Delta p$ the pressure drop over length l, and $\eta$ the fluid viscosity. Either the pressure difference over the capillary is measured for a known flow rate or the flow rate may be measured for a fixed pressure drop. The average velocity, U, often is used for convenience instead of flow rate, Q; the two are related through $U = Q/\pi r^2$.

The pressure gradient in macroscopic laminar flow in the capillary is given as:

$$\frac{\Delta p}{\Delta x} = -\frac{8 \eta U}{r^2} \qquad 9.42$$

Since flow occurs from a higher pressure to a lower pressure, the pressure change over l is negative in sign. U is directly proportional to the pressure gradient. The pressure change over a length l of the pipe, based on Equation 9.42 is

$$\Delta p = -\frac{8 \eta U l}{r^2} \qquad 9.43$$

clearly predicting the pressure drop over narrow capillaries to be very high. The fact that the volumetric flow rate reduces as the fourth power of the radius can be understood from the fact that a gas or a liquid is at rest at the capillary wall and that the fluid velocity, which reduces as the square of the channel radius (from Equation 9.42), is maximal at the center of the tube. In a wide tube, larger velocities can thus be achieved. This, coupled with the trivial factor of the larger diameter, results in a Q proportional to $r^4$. Consequently, distributed micropumping with each pump responsible for only a fraction of the total fluid movement is expected to be important in microsystems. The required high pressures are a good reason to look at alternative propulsion techniques with surface forces rather than a volume force like pressure. Piezoelectric, osmotic, electrowetting, and electro-hydrodynamic pumping (see below) are all surface forces. As they scale more favorably they might bring some advantages in moving fluids in narrow capillaries.

Two other important expressions need to be introduced before we can analyze fluidics in the microdomain, i.e., the Reynolds number and the friction factor. Both are dimensionless or reduced parameters, introduced in many engineering disciplines to reduce the number of variables to work with. The Reynolds number characterizes fluid flow as laminar or turbulent and includes fluid properties, $\rho$ and $\eta$; the characteristic length, L; and average velocity, U:

$$R_e = \frac{\rho U L}{\eta} \qquad 9.44$$

The characteristic length, as the term implies, is the length which best represents the body under consideration. In a capillary, for example, the diameter, L, is far more important in determining the nature of the flow than the length, l. The Reynolds number may be viewed as the ratio of shear stress due to turbulence (i.e., inertial forces or forces set up by acceleration or deceleration of the fluid) to shear stress due to viscosity. The friction factor is given by:

$$f = \frac{\Delta p}{2 \rho U^2} \frac{L}{l} \qquad 9.45$$

with l the length of the pipe. The friction factor can be regarded as the ratio of the net imposed external force to the inertial force. The relation between the two dimensionless groups defined above is

$$\frac{\Delta p}{2 \rho U^2} = \frac{l}{L} \times \text{function only of } \frac{\rho U L}{\eta} \qquad 9.46$$

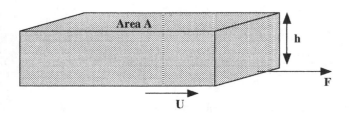

**FIGURE 9.14** Schematic of shearing experiment.

Equation 9.46 can be rewritten as:

$$f = F(R_e) \qquad\qquad 9.47$$

The friction factor for smooth pipe flow of all incompressible Newtonian fluids is thus a unique function of the Reynolds number. This is a very important insight. Only two groups of variables need to be studied experimentally to obtain a relation that is universally valid for a wide class of fluids, geometries, and flow parameters.[12] Figure 9.15A represents $R_e$ vs. friction factor, f, for different pipe sizes, relative pipe roughness, densities, and a broad range of viscosities.[12] Data from all pipe sizes and the entire viscosity range overlap. The data cover more than four decades in $R_e$. At Reynolds numbers below 2100 the viscous shear effects are so large to balance the driving effect of the pressure, a laminar, creeping, or Stokes flow condition result. For this type of flow, acceleration is negligible. This condition is either met by a slow flow, i.e., U is small, L is small, or $\eta$ is large. At large enough Reynolds numbers, above 4000, a laminar flow becomes unstable and eventually turns turbulent.[93] The region between 2100 and 4000 is a transition region. This tran-

sition depends on the size and frequency of disturbances inherent to the type of flow, the roughness of the boundaries, the temperature, boundary flexibility, etc. For macroscopic laminar flow in long tubes it is found that the friction factor correlates to the Reynolds number as:

$$f = \frac{C}{R_e} \qquad\qquad 9.48$$

where C is a constant, also called the Poiseuille number, that depends on the cross-sectional shape but typically has a value of 16 for tube flow. It is also in this regime that Equation 9.41, the Hagen-Poiseuille equation, is valid. For turbulent flow the friction factor-Reynolds number dependence is weaker. An empirical equation for turbulent flow is $f = 0.079\, R_e^{-1/4}$.[12] The onset of turbulent flow, as is obvious from Figure 9.15A, is related to the relative roughness of the pipewall. As illustrated in Figure 9.15B, the relative roughness is represented by the dimensionless group k/L. The larger that ratio, the earlier turbulence starts. For micromachined tubes one might expect

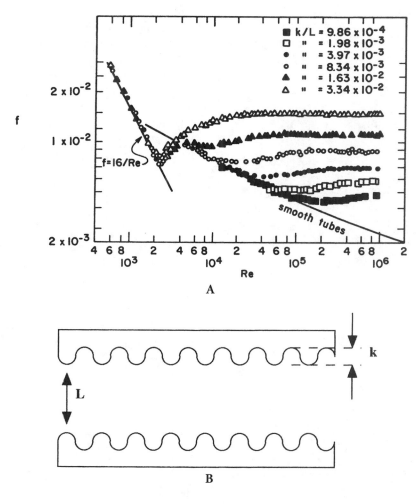

A

B

**FIGURE 9.15** (A) The Reynolds number, $R_e$, is plotted vs. friction factor, f, for different pipe sizes, densities, and a broad range of viscosities. Data from all pipe sizes and the entire viscosity range overlap. Different relative roughness factors k/L (caused by sand-roughened pipes) cause different onsets for turbulence. (From Denn, M.M., *Process Fluid Mechanics*, Prentice-Hall, Inc., Englewood Cliffs, NJ, 1980. With permission.) (B) Schematic of a pipe wall of uniform, regular roughness.

a large k/L and turbulence to kick in quickly. But some simple calculations show that the Reynolds numbers are so low that turbulence is not likely to be observed in microsystems even with a large k/L.

Applying Equation 9.44 to calculate the Reynolds number for the flow of air ($\eta = 1 \times 10^{-6}$ m²/sec) and water ($\eta = 15 \times 10^{-6}$ m²/sec) in a 50-μm diameter capillary, assuming $U = 500 \times 10^{-6}$ m/sec (10 times the diameter of the pipe per second), we obtain 0.0016 for air and 0.025 for water, both considerably less than 1, indicating that, if fluids behave as a continuum, as is assumed in classical fluids, microsystems will operate in a viscous-dominated, Stokes regime. A tiny rotor blade in an air-filled, 10-μm wide capillary will only be able to rotate in milliseconds rather than microseconds. The consideration of moving small things in water is even more fearsome as the water is like syrup to the submerged micromachine. Fluid systems in the microdomain are heavily damped and exhibit a slow response time.

## Macroscale Laws in Heat Transfer

Consider heat transfer, Q, through a material between two parallel plates at temperatures $T_1$ and $T_2$ (Figure 9.16A). Heat is transported from one plate to the other and assuming a linear temperature gradient Q is given as:

$$Q = \frac{\kappa}{d}\left(T_1 - T_2\right) \qquad 9.49$$

where $\kappa$ is the bulk heat conductivity and d the plate separation. The heat flow into a material of volume, $V_0$, at a uniform temperature, $T_0$, through boundary surface, A (Equation 9.49), is equated to the rate of increase of the internal energy of the material in volume, $V_0$ (see Figure 9.16B), or:

$$Q = -\lambda A\left(T_0 - T_\infty\right) = \rho c_p V_0 \frac{dT}{dt} \qquad 9.50$$

where $\rho$ defines the density of the material, $c_p$ the specific heat under constant pressure, $\lambda$ the heat transfer coefficient at the surface, and $T_\infty$ the environment temperature. Solving the above differential equation for the case of a solid at a uniform temperature, $T_0$, then suddenly immersed at time $t = 0$ in a well-stirred environment at $T_\infty$ gives us the temperature of the solid as a function of time:

$$\frac{T - T_\infty}{T_0 - T_\infty} = e^{-\left(\frac{\lambda A}{\rho c_p V_0}\right)t} \qquad 9.51$$

The term on the left represents the dimensionless temperature, and it can be verified that the temperature in the solid decays exponentially with time and that the shape of the curve is determined by the time constant, $\tau_c$, given by:

$$\tau_c = \frac{\rho c_p V_0}{\lambda A} \qquad 9.52$$

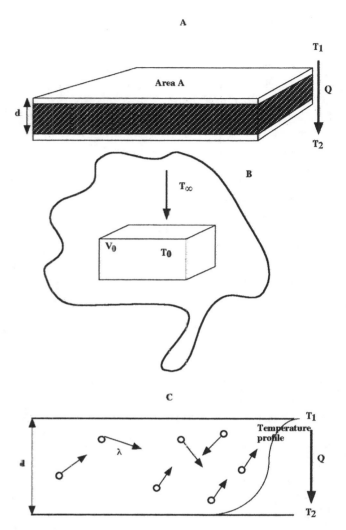

**FIGURE 9.16** Heat conduction. (A) Heat conduction in a solid between two plates at different temperatures. (B) Heat conduction from a solid at temperature $T_0$ to the environment at $T_\infty$. (C) Heat conduction in a gas between two parallel plates.

To develop a feeling for the above expression we compare how fast a Ni rod of 1/2-in. diameter and a 50-μm radius come to equilibrium with their surroundings. For a cylinder, with $V_0 = \pi r^2 l$ and $A = 2\pi r l$, the above equation transforms to:

$$\tau_c = \frac{\rho c_p r}{2\lambda} \qquad 9.53$$

assuming $\lambda = 10$ W/m²/K, $\rho = 8900$ kg/m³, and $c_p = 0.44$ J/kg/K gives a $\tau_c = 1.24$ sec for the bigger cylinder. If we assume equilibration with the surrounding temperature, $T_\infty$, in 4 $\tau_c$'s, equilibration is reached in 5 sec. For the 50-μm Ni rod equilibrium is reached after 0.039 sec.

Efficient heat sinking in the microworld is of particular interest. A practical example involves a microchannel heat sink for a semiconductor laser diode array. The heat sink consists of water flowing through an array of parallel channels underneath the diode array. The silicon micromachined channels measure 300 μm high and 50 μm wide and flow rates of up to 8.6 cm²/sec are achieved.[94] This microchannel cooling can effectively dissipate up to 1 kW/cm², approximately 40 times greater than that for conventional heat sinks. Another example involves microrefrigeration. For cooling to very low temperatures, microminiature refrigeration through the Joule-Thomson effect may be employed. The cooling originates from the expansion of high-pressure gas in narrow capillaries and may cool devices from ambient to 80 K in minutes (see Figure 10.42 for a commercial product based on microrefrigeration). The scaling analysis of this type of microrefrigerator was presented by Little.[95,96]

When a solid body is placed in a surrounding of a different temperature, the body heats up or cools down and internal temperature gradients that cause thermal stresses are set up. The dimensionless number which characterizes the magnitude of the established thermal gradient is the Biot number (Bi), the ratio of the surface heat-transfer coefficient, $\lambda$, to the unit conductance of the solid, $\kappa$, over a characteristic dimension, L:

$$Bi = \frac{\lambda L}{\kappa} \qquad 9.54$$

As L becomes smaller, the Biot number goes down. For $Bi \ll 1$, internal temperature gradients become small and the body can be treated as having a uniform temperature. Consequently, with small devices there is less worry about thermal stresses induced by thermal gradients, and systems with larger internal heat generation capacity by volume can be built.[97] A representative application of the described effect is the micromachined planar Taguchi gas sensor shown in Figure 3.43. Not only is the power required to operate at 350°C reduced by more than a factor of 4 (225 mW instead of 1 W), it also can be heated very quickly to high temperatures without breaking.[98]

With small things we can heat and cool down many times a second (~ 20 Hz in the example of the 50-μm Ni rod) while thermal stress is minimized. This finds its application in fast thermal detectors such as thermocouples, high-performance heat sinks for high-density microelectronic devices, and fast actuators such as thermopneumatic valves, in μ-PCR chambers, fast gas sensors, etc.

## Breakdown of Continuum Theory in Fluidics

### Breakdown of Heat Transfer Continuum

Some simple equations and rules of thumb providing guidance in the evaluation of when macroscale theory breaks down in heat transfer follow. Based on Equation 9.49, in the case of a gas, the effectiveness of heat transport from one plate to the other is controlled by the mean free path, $\lambda$, of the gas molecules, i.e., the average distance before the next collision (see Figure 9.16C). If the characteristic dimension of the system, d in this case, is much larger than the mean free path, $\lambda$, continuum equations apply. Within the continuum approximation, $\kappa$ of a gas is given from kinetic theory by:

$$\kappa = \frac{1}{3}\rho U \lambda c_v$$
$$U = \sqrt{\frac{8kT}{\pi m}} \qquad 9.55$$
$$\lambda \propto \frac{1}{\rho}$$

where $c_v$ represents the constant volume specific heat capacity, $\lambda$ the mean free path, U the mean velocity of the molecules, $\rho$ the gas density, and m the molecular weight of the gas molecule. From the last expression we can conclude that an increase in density decreases the mean free path so that the product, $\rho\lambda$, remains constant. Consequently, $\kappa$ is not a function of density. We also confirm that $\kappa$ will increase with increasing temperature. What happens, though, when $\lambda$ approaches d? As the density $\rho$ decreases, no corresponding increase in $\lambda$ occurs since $\lambda$ is limited by d. In other words, boundary conditions start limiting the average value of $\lambda$ within the gas and the thermal conductivity goes down. A qualitative corrective factor to the heat flux in Equation 9.49, considering the effect on both plates, is sometimes used for small gaps:

$$Q = \frac{\kappa \Delta T}{d + 2g} \qquad 9.56$$

where for conduction in air, oxygen, nitrogen, carbon dioxide, methane, and helium, experimental g-values satisfy: $2.4\lambda < g < 2.9\lambda$. The macroscopic prediction for the heat flux exceeds the microscale prediction by less than 5% when $d > 12\lambda$. For hydrogen, however, the experimental value for g is $11.7\lambda$, and the condition $d > 47\lambda$ must be satisfied.[90]

Kinetic theory also determines the boundary for microscale conduction in solids and liquids. For thermal conduction in solids an equation similar to Equation 9.55 holds, in the case of metals, with $\lambda$ the electron mean free path, $C_v$ the electron specific heat, and U the Fermi velocity. In dielectrics, fluids, and semiconductors, $\lambda$ is represented by the phonon mean free path, $C_v$ the phonon specific heat, and U the average speed of sound. Experimental g-values for solids are not yet available. But calculations indicate that macroscale exceeds microscale results by less than 5% for $d > 7\lambda$. Consequently, the boundary between the microscale and macroscale regimes equals $d = 7\lambda$ for the conduction across a layer, where d is the layer thickness and $\lambda$ the dominant carrier of heat.[90]

Maps for heat transfer regimes in microstructures, showing immediately whether or not a certain heat transport in a given device can be analyzed with macroscale theory, have been developed.[90]

### Breakdown of Flow Continuum

The macro expression for the pressure drop through a tube is given in Equation 9.42 with $\Delta p/\Delta x \sim -1/r^2$. However, when $\lambda$ of a gas approaches the separation, d, of the container plates in a set-up as shown in Figure 9.16C, conditions easily met in a rarefied gas, the laws of high-pressure behavior begin to err. As the density is lowered, the gas seems to lose its grip, so to speak, upon solid surfaces; in viscous flow it begins to slip over the surface, and in the conduction of heat a discontinuity of temperature develops at the boundary of the gas (see above). All of the same phenomena can be seen at high pressure but with fluidic vessels of microdimensions. The no-slip condition does not apply anymore and the friction factor actually starts to decrease with diameter reduction. For tube flow with C = 16, the corrected Stokes-Navier Equation 9.48 is written as:

$$f = \left(\frac{16}{R_e}\right)\frac{1}{1 + 8\dfrac{\lambda}{D}} \qquad 9.57$$

For $\lambda$ of a size approaching D, f is approximately 1/9 the value expected from analysis, assuming that the velocity at the wall equals 0. Experiments of flow resistance in 0.5-$\mu$m channels by Pfahler et al.[99] show indeed a higher flow rate. This phenomenon is important for gas damping of oscillating microstructures, as they are almost always operating in the size regime where $\lambda$ approaches D.[88]

Liu et al.[100] measured the pressure distribution of gaseous flow in microchannels by lining the microchannels with a series of pressure sensors. This group could confirm the Navier-Stokes equations with slip boundary conditions (first used by Arkilic et al.).[101]

Deviations from the Navier-Stokes equations in liquid are much more difficult to demonstrate. We assume usually that the viscosity is independent from the dimensions of the flow channel. Some experiments by Harley et al.,[89] however, may indicate that in microchannels, liquid flow takes on values as much as an order of magnitude greater than expected from a constant viscosity. Pfahler et al.[99] similarly measured a flow significantly higher than expected from macrotheory when using isopropanol and silicone oil. Gravesen et al.[88] in 1993 reviewed some of the literature describing deviations from Navier-Stokes equations in liquid flow in narrow channels but could not find enough consistency in the data to conclude that there is a breakdown in the macro theory.

Given the dimensions of a phonon at least 10 times smaller than $\lambda$ of a gas at atmospheric pressure, we might not observe deviations from the Navier-Stokes equations in liquids until the dimensions of the containing vessels are in the 100-Å range.

## Fluid Propulsion Mechanisms

### Introduction

Equation 9.43 clearly predicts the pressure drop over narrow capillaries to be high, which offers a good reason to look at alternative fluid propulsion methods. Piezoelectric, electro-osmotic, electrowetting, and electrohydrodynamic pumping all scale more favorably in the microdomain and lend themselves

better to distributed pumping. Fluid propulsion is also described in Chapter 10 on generic challenges in miniaturization of analytical equipment.

### Piezo-Pumps

In a tube vibrating with an acoustic energy, an axially directed, acoustic-streaming force is generated along the inner surface.[132] This driving force can be generated along the entire length of a narrow tube to create a distributed pump that moves fluids without an externally applied pressure. The pump surface area and the driving force per unit volume of enclosed fluid increase as the tube diameter decreases. The axially-directed, steady-state force acting on the fluid is proportional to the square of the acoustic-wave amplitude and diminishes exponentially with distance from the wall.[102] This force is sketched in Figure 9.17A. Because the force reaches a maximum near the walls of the tube, blunt flow is established in the fluid, as shown in Figure 9.17B, establishing an attractive technology to choose for miniaturization. Unique properties of acoustic streaming pumps include the ability to pump without directly contacting the fluid so that a wide variety of liquids and gases may be pumped and a simple compact structure with zero internal dead-volume. Tens of individual acoustic pumps could be operated in coordination to make an active network of interconnecting channels. One implementation of such a pump is based on an acoustic-wave delay line, such as the flexural-plate-wave (FPW) delay line. This device consists of a thin (~1-$\mu$m thick) membrane supported on all sides by a silicon chip. The membrane is coated with a piezoelectric film (say ZnO) and aluminum interdigital transducers (IDTs). An RF voltage is applied to one IDT, producing mechanical stress in the piezoelectric layer which generates flexural acoustic waves in the membrane. Magnified, these waves look like the ripples in a flag waving in the breeze, as sketched in the inset of Figure 9.18. This wave motion induces acoustic streaming in the fluid next to the membrane. Wave propagation and fluid velocity both operate from left to right as shown in Figure 9.18. The structure

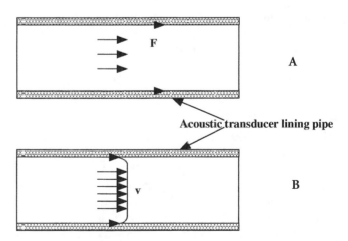

**FIGURE 9.17** Acoustic streaming in a fluid inside a pipe. (A) Streaming force generated by an axially directed, traveling acoustic wave that grazes the inside walls of the tube. (B) The resulting steady-state velocity distribution (blunt flow).

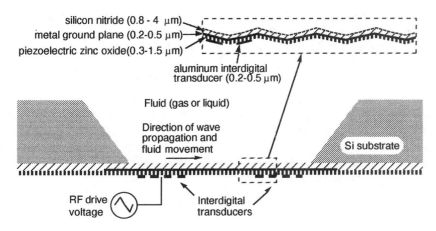

**FIGURE 9.18**   Micromachined FPW pump. The RF drive voltage applied to the piezoelectric interdigital transducers induces the flexural wave motion shown in the inset. Fluid motion is induced near the surface by acoustic stress. (Courtesy of Dr. D. White, UC Berkeley, CA.)

generates large acoustic-wave amplitudes and is particularly effective for producing streaming; it has been proven to move air at speeds of 30 mm/sec and water at 0.3 mm/sec with only 5 V of RF drive.[103,104] Acoustic streaming has also been established with the related surface acoustic wave (SAW) delay line.[106] Miyazaki et al.,[106] using a thin metal pipe that has been flattened and bonded to a piezoelectric plate, produced a head pressure of about 10 mmH$_2$O and a flow rate of 0.02 cm$^3$/sec at 40 V, peak to peak, applied to the piezoelectric ceramic plate.

### Electro-Osmosis Pumping

Electro-osmotic flow results from the effect of an electrical field on both charged particles in a fluid and the fluid itself when the fluid is placed in a narrow capillary. The force on the particles in the fluid leads to electrophoresis while the force on the fluid in a narrow column leads to electro-osmosis. The electrophoretic and electro-osmotic velocities are both given by:

$$u = \mu \frac{V}{l} \qquad 9.58$$

where u is the velocity of ions or fluid column, V the applied voltage, $\mu$ is the vector sum of the electroosmotic ($\mu_{eo}$) and electrophoretic mobilities ($\mu_{ef}$) and is given by:

$$\mu = \frac{\varepsilon \zeta}{4\pi\eta} \qquad 9.59$$

where $\varepsilon$ defines the dielectric constant, $\zeta$ the electrokinetic potential at the capillary wall, and $\eta$ the viscosity of the fluid. In absolute terms the electro-osmotic mobility generally is larger than the electrophoretic mobility. A thin layer of liquid along the capillary wall never moves; liquid movement begins at a distance, $x_0$, away from the wall. The electrokinetic potential, $\zeta$ — from which all electrokinetic effects can be derived — represents the value of the electrostatic potential, $\psi$, at $x_0$. Its value at any point is determined from the profile of the electrostatic potential as a function of distance from the capillary wall into the moving fluid. The amount and type of ionic species in the solution can greatly influence these potentials. For example, the larger the

concentration of added indifferent electrolyte in the moving solution is, the smaller $\zeta$ is (see Figure 9.19). Surface impurities also might affect the electrokinetic potential and this all conspires to make electro-osmosis a rather difficult to control pumping mechanism.

In free-flow, traditional capillary electrophoresis work by Jorgenson et al.,[107] electro-osmotic flow of 1.7 mm/sec was reported, certainly fast enough for many purposes. The technology for generating this flow rate does not involve moving parts and is easily implemented in micromachined channels. One needs only one electrode in reservoirs at each end of the flow channel. Harrison et al.[118] achieved electro-osmotic pumping with flow rates up to 1 cm/sec in 20-μm capillaries micromachined in glass. Also, injection, mixing, and reaction of fluids in a manifold of micromachined flow channels without the use of valves was demonstrated. The key aspect for tight valving of liquids at intersecting capillaries in such a manifold is the suppression of convective and diffusion

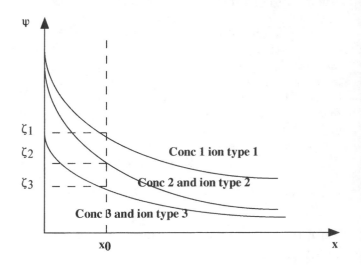

**FIGURE 9.19**   The electrokinetic potential, $\zeta$, is the electrostatic potential, $\psi$, at a distance, $x_0$, from the surface, i.e., the distance where the fluid movement begins. The position of the curves and thus the electrokinetic potential are influenced by the concentration and type of ions in the solution. Ions and impurities at the surface influence these curves as well.

effects. Harrison et al.[108] demonstrated that these effects can be controlled by the appropriate application of voltages to the intersecting channels simultaneously.

The main disadvantage of electro-osmotic pumping is the sensitivity of the electrokinetic potential to the nature of the electrolyte. This brings with it the insidious problem of pressure build-up at intersections where liquid columns of different ionic strength meet. From a practical point of view the connection between the outside world and the capillaries still needs addressing by micromachinists. Gluing glass reservoirs for holding the platinum bias electrodes to the glass plate substrate with RTV silicone is not an attractive option.[109]

### Electro-Hydrodynamic Pumping

In electro-hydrodynamic (EHD) pumping, fluid forces are generated by the interaction of electric fields and charges in the fluid. In contrast to forces generated by mechanical pumping using an impeller or bellows, EHD pumping requires no moving parts.[63] The interaction of electrical fields on induced electrical charge in the fluid yields a force which then transfers momentum to the fluid.[110] A practical requirement for pumping a fluid in a closed conduit is the induction of free electric charge in the volume of the fluid to be pumped. Reasonable electric fields can only be built up with acceptable voltage levels within microstructures; therefore, this principle becomes more and more effective with decreasing dimensions. A requirement for the continued existence of free charge is the presence of a spatial gradient in the conductivity or permittivity. Free charge injected into a region without gradient will relax in a time characterized by the relaxation time of the charge. One way to accomplish a conductivity gradient in the bulk of a slightly conductive fluid is by imposing a temperature gradient. A big limitation of EHD is its reduced effectiveness with conductive fluids, such as in biological environments.

### Electro-Wetting

Electrical control of interfacial tension between a liquid and a solid provides yet another means of direct fluid pumping with no moving mechanical parts. To increase the surface area of any liquid requires effort and this work per unit surface area is the liquid's surface tension, $\gamma$. When liquid comes in contact with both a solid and a gas, as shown in Figure 9.20A, a 'contact angle' of between 0 and 180° is set up. The force balance between the three interfacial tensions in Figure 9.20A represents the Young's equation:

$$\gamma_{SL} + \gamma_{LG} \cos\theta = \gamma_{SG} \qquad 9.60$$

where $\theta$ is the contact angle. In the example shown, $\cos\theta$ is positive, indicating that the solid-liquid interface is of lower energy than the solid-gas interface, in which case liquid in a capillary will rise in order to seek a lower interfacial energy (see Figure 9.15B). The liquid will rise until it is lifting a column of liquid balancing the energy reduction gained by raising. The energy reduction by rising over a distance, l, in the capillary is given by $(\gamma_{SG} - \gamma_{SL})$ $2\pi r l$, where r is the capillary radius and the work performed in raising the liquid column is given by $\Delta p \pi r^2 l$, with $\Delta p$ the constant

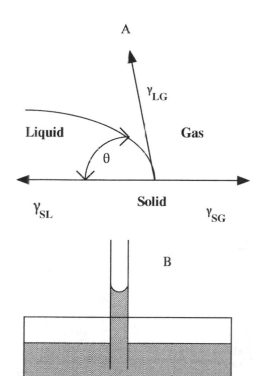

**FIGURE 9.20** Electrowetting: (A) contact angle for liquid/solid/gas interface; (B) liquid rising in a capillary. (After Colgate et al.[111])

pressure difference between the top and bottom of the column. By equating these two energies we can deduce the pressure drop across a liquid-gas interface in a small capillary:

$$\Delta p = \frac{2(\gamma_{SG} - \gamma_{SL})}{r} = \frac{2\gamma_L \cos\theta}{r} \qquad 9.61$$

where it has been assumed, as is generally the case, that $\gamma_L \approx \lambda_{LG}$. The pressure produced by electrowetting, calculated from Equation 9.61 for typical values gas-liquid surface tension and a capillary radius of 10 μm, is in the range of 0.01 MPa, roughly the same as achieved with piezoelectric pumps which are much larger in physical size. Colgate et al.[111] report that the pressure difference, $\Delta p_b$, needed to move a gas bubble in a straight channel is empirically given by:

$$\Delta p_b = \frac{2\gamma_f}{r} \qquad 9.62$$

where $\gamma_f$ is a frictional surface tension. The pressure differences in the last two given equations both increase with decreasing channel dimension.

Altering the apparent surface tension at the solid-liquid interface becomes possible by applying a potential over the interface. This phenomenon is known as electrowetting and is described by the Lippman equation:

$$\gamma_{SL} = \gamma_{SL}^{max} - \frac{\varepsilon_r \varepsilon_0}{2\delta}\left(V - V_{pzc}\right)^2 \qquad 9.63$$

where $V_{pzc}$ represents the potential of zero charge for the solid electrolyte interface, and $\gamma_{SL}^{max}$ is the surface tension corresponding to $V = V_{pzc}$. A voltage applied across the interface will alter the surface tension and cause the liquid to move within the capillaries.

Surface tension becomes important only when dimensions drop well below 1 mm. As in the case of electro-osmotic pumping, subtle uncontrollable changes at the solid electrolyte interface make this actuation principle difficult to control and electrolyte specific.

### Other Pumping Mechanisms

Other valveless pumping mechanisms being explored include the centrifugal force and phase-changes of fluids. In the former case, a disc with a network of chambers and channels is spun at different rates, propelling the liquids from the center towards outlying fluidic channels and reservoirs. This type of pumping is implemented in Abaxis' Piccolo™ blood analyzer.[112] In the Piccolo™ a disposable plastic rotor disc incorporates dry reagent beads and aqueous diluent to carry and dilute the blood sample. Spinning the disc brings the diluted blood sample in contact with the reagents sites, which are read by multiple wavelength photometry. In the case of the phase-change pump, periodic phase changes of a fluid are used for pumping. The pressure source in this case is the switching from liquid to vapor phase by heating. The pressure losses depend on the different kinematic viscosities dictated by the phase condition of the fluid and the fluidic channel widths and act as the valving mechanism. Considering the speed of heat exchange and the value of the Reynolds number in narrow capillaries, this mechanism scales well into the microdomain.[113]

## Separation Performance in Flow Channels (Capillaries)

Miniaturization in chromatography and electrophoresis flow channels has been a study topic for many years, not only to separate smaller amounts of chemical compounds but also because theory predicts that a reduction in dimensions of the separation column should result in an enhancement of analytical performance.[91,114] Analytical separation of compounds is effected by forcing the sample compounds, suspended in a carrier medium or mobile phase, through a selectively absorbing medium or stationary phase immobilized in the flow channels. To force the sample and carrier phase through the column either a pressure or electrical gradient, respectively defining chromatography techniques and electrophoresis techniques, can be employed. The following theoretical analysis will shed light on scaling laws applicable to these separation systems. We are closely following the derivations as presented by Manz et al.[91]

The volumetric flow rate, Q in $m^3sec^{-1}$, of an incompressible fluid in a capillary with inner diameter, d (m), and length, l (m), resulting from an applied pressure, $\Delta p$ ($N/m^2$), derived from Equation 9.41 is given by:

$$Q = \frac{\pi}{128\eta}\left(\frac{\Delta p}{l}\right)d^4 \qquad 9.64$$

Conversely, the pressure drop per unit length, $\Delta p/l$ ($N/m^3$), sustaining a constant linear flow rate (m/sec) equals (see Equation 9.42):

$$\frac{\Delta p}{l} = -32\eta\frac{U}{d^2} \qquad 9.65$$

where $\eta$ ($N/sec/m^2$) is the viscosity, and U (m/sec) the average linear flow rate. These equations form the starting point for any analysis of pressure-driven chromatographic systems. Using diameter d instead of radius r yields a Poiseuille number of 32 instead of 8 (compare Equations 9.64 and 9.65 with respectively 9.41 and 9.42).

We already mentioned that an electrical field can induce flow in a liquid. An electrical field works on both charged particles in a fluid and the fluid itself when the fluid is placed in a narrow capillary. The force on the particles in the fluid leads to electrophoresis; the force on the fluid in a narrow column leads to electro-osmosis, i.e., both move under the influence of a field. Mathematically, it does not make a difference whether the fluid passes a charged wall (i.e., electro-osmosis) or a charged particle moves through a fluid (i.e., electrophoresis). Therefore, the expression for the electrophoretic and electroosmotic is given by Equation 9.58:

$$u = \mu\frac{V}{l} \qquad 9.58$$

where u is the velocity of ions and/or fluid column, V the applied voltage, l the length of the capillary, and $\mu$ ($m^2/V/sec$) the sum of the electro-osmotic and electrophoretic mobilities.

The mobility $\mu$ is given by Equation 9.59:

$$\mu = \frac{\varepsilon\zeta}{4\pi\eta} \qquad 9.59$$

where $\varepsilon$ stands for the dielectric constant, $\zeta$ the electrokinetic potential at the capillary wall (in the expression for electro-osmosis) or the charged particle (in the expression for electrophoresis), and $\eta$ the viscosity of the fluid. The latter two equations form the starting point for any analysis involving field driven separation.

Different compounds are retained for different amounts of time, called retention times, onto the immobilized medium through which they are forced. Consequently, different compounds in an applied sample separate into bands. The narrower these bands are the better the separation performance of the set-up. The band broadening of a compound is influenced by longitudinal and radial diffusion, as well as diffusion in the stationary phase. The bands associated with a specific compound have a Gaussian concentration distribution. The standard deviation of this distribution curve, $\sigma_x$, is used to describe the bandwidth length. Golay[115] calculated the height equivalent of a theoretical plate, H, as:

$$H = \frac{2D_m}{u} + \frac{1+6k'+11k'^2}{96(1+k')^2}\left(\frac{d^2}{D_m}\right)u + \frac{2k'}{3(1+k')^2}\left(\frac{d_f^2}{D_s}\right)u \qquad 9.66$$

In this expression, $D_m$ and $D_s$ (m²/sec) correspond to the diffusion coefficients of, respectively, the sample molecules in the mobile and stationary phase, u (m/sec) is the linear flow rate of the mobile phase; $k'$ the capacity factor of the specific compound; d the diameter of the capillary; and $d_f$ the thickness of the stationary phase layer. The height equivalent to a theoretical plate, H, and the bandwidth, $\sigma_x$, are related by:

$$H = \frac{\sigma_x^2}{l} \qquad 9.67$$

In capillary electrophoresis, Equation 9.66 can be simplified to:

$$H = \frac{2D_m}{u} \qquad 9.68$$

This simpler equation results as a consequence of the large longitudinal diffusion terms. By comparing this electrophoresis expression with the chromatography equation the importance of the thickness of the stationary phase in the case of chromatography emerges. The thickness uniformity is one of the more difficult parameters to control in micromachined columns. As a result, the current commercial micro gas chromatographs have miniaturized detectors and injectors but no micromachined columns (see Figure 10.30). Glass capillary columns with inner diameter as small as 100 μm continue to outperform micromachined columns due to the difficulty in maintaining coating uniformity in micromachined structures (see also Chapter 10).

Individual solute bands generally assume symmetric concentration profiles that can be described in terms of the Gaussian distribution curve and the standard deviation, $\sigma$, of this curve. The above equations only describe the band broadening in the column itself, but not the contributions from injection, detection, and connection elements. The total band broadening is calculated from a summing of the squares of the standard deviations, i.e., the variances:

$$\sigma_{total}^2 = \Sigma \sigma_i^2 \qquad 9.69$$

The bandwidth parameter, $\sigma$, for a flow channel can be expressed in terms of length in mm ($\sigma_x$ see above), time in ms ($\sigma_t$), and volume in pL ($\sigma_v$). These are all important parameters to establish when designing a separation system. For example, to contribute less than 10% to the total band broadening, injection and detection volume must be less than $\sigma_v/2$ of the column, and the detector response time must be faster than $\sigma_t/2$. In practice, the column is the least of the problems. The difficulty lies in designing systems such that the dead volume is kept minimal and in the case of gas chromatography the capillary can be heated and cooled very quickly and uniformly with a minimal of power budget, i.e., the heat capacity of the system must be minimized. By fixing the number of theoretical plates, N; the retention time, t; and the heating power per length (for capillary electrophoresis), the resulting flow channel dimensions and operating conditions for different separation techniques can be compared. In Table 9.6 we reproduce calculations by Manz et al.[91,114] comparing capillary electrophoresis (CE),

**TABLE 9.6** Calculated Parameter Sets for a Given Separation Performance[a]

| Parameter | | Electro-osmotic Chromatography (EC) | | | Liquid Chromatography (LC) | | Supercritical Fluid Chromatography (SFC) | |
|---|---|---|---|---|---|---|---|---|
| No. theoretical plates | N | 100k | 1M | 10M | 100k | 1M | 100k | 1M |
| Analysis time | 1 ($k' = 5$) (min) | 1 | 1 | 1 | 1 | 1 | 1 | 1 |
| Heating power | P/L (W/m) | 1.1 | 1.1 | 1.1 | | | | |
| Capillary i.d. | d (μm) | 24 | 7.6 | 2.4 | 2.8 | 0.9 | 6.9 | 2.2 |
| Capillary length | L (cm) | 6.5 | 21 | 65 | 8.1 | 26 | 20 | 64 |
| Pressure drop | p (atm) | | | | 26 | 2600 | 1.4 | 140 |
| Voltage | U (kV) | 5.3 | 53 | 580 | | | | |
| Peak capacity | n | 180 | 570 | >2000 | 220 | 700 | 220 | 700 |
| Signal bandwidth | σ (mm) | 0.21 | 0.21 | 0.21 | 0.56 | 0.56 | 1.4 | 1.4 |
| | σ (msec) | 42 | 13 | 4.2 | 70 | 22 | 70 | 22 |
| | σ (pl) | 94 | 9.4 | 0.94 | 3.3 | 0.33 | 52 | 5.2 |
| Detection volume | V (pl) | 47 | 4.7 | 0.47 | 0.8 | 0.08 | 1.2 | 12 |
| Response time | t (msec) | 21 | 6.5 | 2.1 | 16 | 5 | 16 | 5 |
| Injection pulse | p × t (sec × atm) | | | | 1.5 | 49 | 0.075 | 2.4 |
| | U × t (sec × kV) | 0.41 | 1.3 | 4.1 | | | | |
| Stop time | t (sec) | 3.3 | 3.3 | 3.3 | 5.1 | 5.1 | 5.1 | 5.1 |

[a] Obtained with capillary electro-osmotic (EC), liquid (LC), and supercritical fluid chromatography (SFC). Assumed constants are diffusion coefficients $1.6 \times 10^{-9}$ m²/sec (LC, EC) and $10^{-8}$ m²/sec (SFC); viscosities of the mobile phase $10^{-3}$ Nsec/m² (LC, EC) and $5 \times 10^{-5}$ Nsec/m² (SFC); electrical conductivity of the mobile phase 0.3 Siemens/m (EC); electrical permittivity × zeta potential, $5.6 \times 10^{-11}$ N/V (EC); heating power 1.1 W/m (EC).

*Source:* From References 91 and 114.

liquid chromatography (LC), and supercritical fluid chromatography (SFC). Listed are flow channel lengths, diameter, operating voltage or pressure, minimum sample bandwidths (expressed in $\sigma_x$, $\sigma_t$, and $\sigma_v$), peak capacity, length/diameter ratio of eluting peaks ($\sigma_x/d$), and the required detector and injector specifications.

Still following Manz et al.'s analysis,[114] we introduce now a set of dimensionless or reduced parameters to make comparisons of separation systems easier. First we define some of these reduced parameters in terms of those parameters that can be assumed constant over the entire range of interest, including inner capillary diameter (d), mobile phase viscosity ($\eta_\mu$), average diffusion coefficient ($D_m$) of the sample in the mobile phase, and a Poiseuille number, C, of 32 for a circular cross-section. The diameter, d, is also the characteristic length here. With those constants, other quantities can be grouped into dimensionless forms, e.g., volume, V; column length, L; linear flow rate, u; retention time, t; and pressure drop, $\Delta p$. A reduced volume parameter, for example, is obtained by division through $d^3$. Thus, one obtains:

$$w = \frac{V}{d^3} \quad \text{and} \quad s_v = \frac{\sigma_v}{d^3} \qquad 9.70$$

Similarly, time-related parameters such as the migration time, $t_0$; the retention, t, for a compound with a capacity factor $k'$; and the time bandwidth, $\sigma_t$, can be reduced to their dimensionless terms, known as Fourier numbers, in the following fashion:

$$\tau_o = \frac{t_o D_m}{d^2}, \quad \tau = \frac{t D_m}{d^2}(k'+1) \quad \text{and} \quad s_t = \frac{\sigma_t D_m}{d^2} \qquad 9.71$$

The first expression describes the ratio of the time required for a molecule to migrate in a field-driven flow through the capillary from end to end to the time required to diffuse from wall to wall. The second term defines the analogous expression for chromatography. Terms with a length dimension are reduced by dividing by d:

$$\lambda = \frac{l}{d}, \quad h = \frac{H}{d} \quad \text{and} \quad s_x = \frac{\sigma_x}{d} \qquad 9.72$$

where $\lambda$ is the so-called reduced length. The linear flow rate of a mobile phase is reduced to the so-called Péclet number, $\nu$, by:

$$\nu = \frac{ud}{D_m} \qquad 9.73$$

which represents the average linear flow rate divided by the absolute value of the average rate of diffusion orthogonal to the direction of flow, $D_m/d$. The applied pressure can be reduced to the so-called Bodenstein number as:

$$\Pi = \frac{\Delta p d^2}{\eta_m D_m \Phi} = \frac{ul}{D_m} \qquad 9.74$$

The Bodenstein number is thus the average linear flow rate divided by the absolute value of the average rate of longitudinal diffusion, $D_m/L$. The applied voltage can be reduced to the electrical Bodenstein number, $\psi$:

$$\Psi = \frac{V \varepsilon \xi}{\eta_m D_m} \qquad 9.75$$

Both Bodenstein numbers, describing the reduced pressure and voltage, can be rewritten as the so-called flow equations:

$$\Psi = \lambda \nu \qquad 9.76$$

in electro-osmotic and electrophoretic flow and

$$\Pi = \lambda \nu \qquad 9.77$$

for pressure-driven chromatography. Also, the Fourier numbers can be further simplified as

$$\tau_o = \frac{\lambda}{\nu} \quad \text{and} \quad \tau = \frac{\lambda}{\nu}(k'+1) \qquad 9.78$$

The equations for reduced plate height, h, and the Péclet number, $\nu$, lead to the following simplified Golay equation for band broadening in pressure-driven flow, neglecting the third term in Equation 9.66:

$$h = \frac{2}{\nu} + \frac{1 + 6k' - 11k'^2}{96(1+k')^2} \nu \qquad 9.79$$

and for a field-driven flow:

$$h = \frac{2}{\nu} \qquad 9.80$$

The reduced bandwidth is then given by:

$$s_x^2 = \lambda h \qquad 9.81$$

and the plate number by:

$$N = \frac{\lambda}{h} \qquad 9.82$$

With these reduced parameters the examples of Table 9.6 are compared again in Table 9.7. We conclude that the values for the reduced variables for LC and SFC are identical, regardless of differences in capillary diameter, lengths, diffusion coefficients, and viscosities. This can now be conveniently used to deduce the influence of changing the capillary diameter on retention time, pressure, and signal bandwidth for a given number of theoretical plates and a single set of reduced parameters, as illustrated in Table 9.8.

TABLE 9.7    Reduced Parameter Set for the Example Separation Systems in Table 9.6

| Parameter | CE | Capillary LC | Capillary SFC |
|---|---|---|---|
| Number of theoretical plates, N | 100,00 | 100,000 | 100,000 |
| Analysis time, t (k' = 5) (min) | 1 | 1 | 1 |
| Capillary inner diameter, $d_c$ (μm) | 24 | 2.8 | 6.9 |
| Capillary length, L (cm) | 6.5 | 8.1 | 20 |
| Reduced length, λ | 2700 | 29,000 | 29,000 |
| Péclet number (reduced flow rate), ν | 100 | 14 | 14 |
| Fourier number (reduced retention time), τ | 28 | 2100 | 2100 |
| Bodenstein number (reduced pressure drop), Π | — | 400,000 | 400,000 |
| Electric Bodenstein number (reduced voltage drop), Ψ | 260,000 | — | — |

*Note:*    CE = capillary electrophoresis; LC = liquid chromatography; SFC = supercritical flow chromatography.

*Source:*    From References 91 and 114.

TABLE 9.8    Calculated Parameter Set for an Open Tubular Column LC System[a]

| | Diameter d (μm) | | | | | Reduced |
| | 1 | 2 | 5 | 10 | 20 | Parameter |
|---|---|---|---|---|---|---|
| Length, L (m) | 0.45 | 0.9 | 2.3 | 4.5 | 9 | λ = 450,000 |
| Time, t (min) | 0.12 | 0.5 | 3 | 12 | 50 | τ = 11,800 |
| Pressure, Δp (atm) | 8700 | 2200 | 350 | 87 | 22 | Π = 17,000,000 |
| Peak, $\sigma_z$ (μm) | 450 | 890 | 2200 | 4500 | 8900 | $s_x$ = 447 |
| Peak, $\sigma_t$ (msec) | 7.4 | 30 | 190 | 740 | 3000 | $s_t$ = 12 |
| Peak, $\sigma_v$ (pl) | 0.35 | 2.8 | 44 | 350 | 2800 | $s_v$ = 354 |

[a]    One million theoretical plates at zero retention (Péclet number ν = 38).

[b]    Assume diffusion coefficient is $D_m = 10^{-3} m^2$ sec.

*Source:*    From References 91 and 114.

The scaling for diffusion-controlled separations, in which the time scale is proportional to $c^2$, is summarized in Table 9.9. Reducing the characteristic length d (i.e., the tube diameter) by a factor of 10 makes for a 100-times faster analysis. The pressure required is 100 times higher and, importantly, the voltage requirements remain unchanged in electrophoresis/electro-osmosis systems. Miniaturization in the latter case leads to a higher rate of separation while maintaining separation efficiency. The latter is again confirmed in the following derivations.

Separation efficiency in capillary electrophoresis, in terms of theoretical plates per second, can be estimated from Equation 9.80 and 9.82:

$$N = \frac{\Psi}{2} \propto V \qquad\qquad 9.83$$

where ψ is the reduced voltage so that N is proportional to V, the applied voltage; the higher the applied voltage, the higher the number of theoretical plates. The voltage cannot be increased too much as heat evolution quickly becomes the limiting factor. In standard capillary electrophoresis the maximum

TABLE 9.9    Proportionality Factors for Some Mechanical Parameters in Relation to the Characteristic Length, d, in a Diffusion-Controlled System

| | Diffusion-Controlled System |
|---|---|
| Space, d | d |
| Time, t | $d^2$ |
| Linear flow rate, u | l/d |
| Volume flow rate, F | d |
| Pressure drop (laminar flow) Δp | $l/d^2$ |
| Voltage (electro-osmotic flow), U | Constant |
| Electric field, UL | l/d |
| Reynolds number, $R_e$ | Constant |
| Péclet number, reduced flow rate, ν | Constant |
| Fourier number, reduced elution time, τ | Constant |
| Bodenstein number, reduced pressure, Π | Constant |
| Reduced voltage, ψ | Constant |

*Source:*    From References 91 and 114.

allowable heat generation is about 1 W/m. Due to the higher surface-to-volume ratio in micromachined channels, heat dissipation is faster which might permit higher electric fields than in standard capillary electrophoresis. For now, we will admit 1 W/m as an upper limit. Keeping the power per unit length constant means:

$$\frac{P}{l} = \frac{VI}{l} = \text{const} \qquad 9.84$$

where I is the current through the capillary. The resulting upper limit for the voltage is then determined by the geometry of the capillary in terms of the characteristic length ($\lambda = l/d$):

$$V_{max} \propto \lambda \qquad 9.85$$

The plate number N can thus reach values up to $N_{max}$ where:

$$N_{max} \propto \lambda \qquad 9.86$$

The minimum migration time is given by:

$$t_0 \propto \frac{l^2}{V} \qquad 9.87$$

and

$$t_{min} \propto ld \qquad 9.88$$

and the maximum number of plates obtainable per second is then given by:

$$\frac{N_{max}}{t_{min}} \propto \frac{1}{l^2} \qquad 9.89$$

very clearly demonstrating the benefits of reducing the inner diameter of the capillary for rapid separation in capillary electrophoresis.

The reduced height, h, of a chromatographic system according to Equation 9.79 shows a minimum, $h_{min}$, at an optimum reduced flow rate, $\nu_{opt}$ (Péclet number). To reduce losses in pressure drop and to minimize analysis time one must operate close to this optimum. Analogous to Equation 9.82, we can write the maximum number of theoretical plates as:

$$N_{max} = \frac{\lambda}{h_{min}} \qquad 9.90$$

where $h_{min}$ is a constant for an optimum Péclet number, $\nu_{opt}$, at a fixed capacity factor k′. This makes Equation 9.90 for pressure-driven chromatography equivalent to Equation 9.86 for electrophoresis. Equations 9.88 and 9.89 are also equally valid for capillary electrophoresis and capillary chromatography.

As we glean from Table 9.6 miniaturizing the detectors in a miniaturized analysis system is quite a challenge, with sizes in the order of picoliters. Different detection schemes exhibit different signal scaling factors. An amperometric technique in which an analytical current is the signal output exhibits an ($l^2$) scaling (i.e., proportional to the area of the electrode); an optical absorption techniques scales as ($l^3$), and a potentiometric technique is size independent.

To conclude this section on scaling in separation techniques we want to approach the resolution in the electrophoretic separation of two analytes (1 and 2). As we deduced from Equation 9.58, the mobility of an analyte in an electrophoresis channel is given by the vector sum of the electrophoretic ($\mu_{ef}$) and electroosmotic mobility ($\mu_{eo}$). The resolution of two analytes is empirically given by:

$$R_s = 0.177\left(\mu_{ef,1} - \mu_{ef,2}\right)\left[\frac{E}{D\left(\mu_{ef,av} - \mu_{eo}\right)}\right] \qquad 9.91$$

where $\mu_{ef,1}$ and $\mu_{ef,2}$ are the electrophoretic mobilities of the two analytes, D the average of their diffusion coefficient, $\mu_{ef,av}$ the average of their electrophoretic mobilities, and $\mu_{eo}$ the electroosmotic mobility. From this equation we conclude that the greatest resolution will be obtained when the electro-osmotic flow roughly equals in magnitude but is opposite in sign (direction) to the analyte's mobilities. This will yield higher resolution, but at the expense of longer analysis times (u in Equation 9.58 is smaller).

## Scaling in Optics

As an illustration of scaling in optics in Chapter 10 we will explore the miniaturization of an infrared absorption instrument for detection of $CH_4$ (absorption at 3.3 μm) and $NO_2$ (absorption at 4.5 μm). Summarizing, we will see that a penalty in sensitivity is paid for miniaturization of absorption based optical equipment. The optical absorption path can be put on a small footprint, but the reflectivity of the mirrors limits the maximum number of bounces to about 100, thereby limiting the ultimate total optical path-length and thus sensitivity of the instrument.

## Scaling in Chemistry and Electrochemistry

According to the equation for the random walk problem the diffusion length, x, of a molecule in solution is given by:

$$x = \sqrt{2D\tau} \qquad 9.92$$

where D is the diffusion coefficient and τ the time required for the molecule to diffuse over the distance x. The diffusion of a molecule in the bulk of a liquid over a length of 10 μm is a million times faster than diffusion over 1 cm. These short diffusion times give transient species a chance to react with other molecules at the interface of the liquid with a gas, another liquid, or a solid. This opens up the opportunity for designing faster chemical reactions systems favoring new reaction paths in microvessels. Not only diffusion but also mixing of fluids happens much faster in the microdomain. Mixing many small amounts of fluid in parallel leads to a much higher efficiency for reaction than mixing big reactor vessels all at once. Based on the latter, it may be possible to develop chemical microreactors where scaling up is perfectly linear, a feat not possible when scaling up chemical reactors in the traditional fashion

Scaling of amperometric detector electrodes was addressed in Chapter 3. We discussed several benefits derived from miniaturization of amperometric detectors, including higher sensitivity and possibility of measuring in higher resistivity solutions. Closely spaced ultramicro-electrodes collect electrogenerated species with a high efficiency, and the high mass transfer rates make it possible to experiment with shorter time scales. The nonlinear diffusion effects in amperometric micro-electrodes leads to improved sensitivities for sensors with electrode dimensions of a size comparable to the diffusion layer thickness. To compensate for the decrease of the overall absolute current level, an array of microelectrodes is employed. Micro-array electrochemical detectors were applied, for example, in liquid chromatography for the detection of carbamate pesticides in river water.[116] The detection limits, in the subnanogram range (50 to 430 pg), represent as much as a 60-fold improvement over other reported liquid chromatography detectors such as fluorescence detection and electrochemical detection with a single macroelectrode. In this early work the electrochemical detector array consisted of a mixture of graphite and Kel-F, the so-called Kelgraf electrode with a micro-array-like structure. More recently, platinum microelectrodes were microfabricated as detector electrodes for capillary electrophoresis.[117] Using an array of microelectrodes in a thin-chamber-like configuration could further enhance sensitivity. Potentiometric sensors, such as the ion-sensitive field effect transistors (ISFET), are scaling independent. A lot more effort went into trying to miniaturize potentiometric devices than amperometric devices although more benefits can be derived from miniaturizing the latter.

# Thermal Actuators

## Bimetallic Actuators

Thermal actuators are very popular actuators in micromachining. Three different thermal mechanisms employed in microvalves are illustrated in Figure 9.21. The thermopneumatic valve (see Figures 9.21A and 10.4A) traps a liquid in a sealed cavity containing a thin-film heating resistor along one side of the cavity, with a flexible diaphragm wall forming the opposite side of the cavity.[118] Upon heating, the fluid in the chamber expands and evaporates, raising the pressure in the sealed cavity and bulging the flexible wall outward. In a normally open valve, this bulging diaphragm closes a nearby orifice. In the case of a normally closed valve, as illustrated in Figure 9.21A, actuation levers a silicon body away from an orifice. Compared to other microvalve technologies, thermopneumatic actuation exerts tremendous force through a long stroke. In the case of a normally open valve, over 20 N through a stroke of 50 µm was demonstrated.[118] This long stroke allows a thermopneumatically actuated valve to control high flow rates and a very wide range of pressures compared to other microvalve technologies. The thermopneumatic valve is compared with various other thermal valves in Table 10.16. The second thermal actuator, illustrated in Figure 9.21B, involves a bimetal (see also Figure 10.40). When the valve is cold, the nickel and silicon relax and the valve is closed. Upon heating, the nickel expands more than the silicon, lifting the silicon body from the valve seat. Size-wise bimetal actuators are smaller than thermopneumatic ones and consume less power, but they lack the longer stroke of the thermopneumatic valves.[119]

## Shape Memory Alloy Actuators

A third thermal actuator is based on shape-memory alloys (see Figure 9.21C). Shape-memory alloys (SMAs) such as NiTi undergo phase transformations from a weak and easily deformable state at low temperatures to a hard and difficult to deform state at higher temperatures. The material is first held in the desired shape and heated to well above the transition temperature. At these temperatures the crystal structure is in the austenite or parent phase. Cooling transforms the material into martensite, the low-temperature phase, which plastically deforms (solder-like) by as much as 10% at relatively low stresses. When the SMA is heated again above its transition temperature, $T_{tr}$, it transforms back to the high-temperature phase (austenite) and reverse to its originally high temperature shape, exerting a substantial force ($>100$ MN/m$^2$) (see Figure 9.22). The austenite phase has a much higher yield point and can sustain stresses greater than 500 MPa without permanent deformation.[120] The temperature interval over which the shape change takes place typically is 10 to 20°C. The lowest energy path to the austenite exactly retraces the atomic movements responsible for the deformation, causing the shape memory. For practical applications, Ni-Ti, Cu-Zn-Al, and Cu-Al-Ni all have been tried. Depending on the type of alloy and the alloy composition, critical temperatures range between −150 and +150°C. Nitinol, a composition transforming near room temperature, traditionally is prepared with 50% Ti and 50% Ni. Increasing the Ni content decreases the transition temperature by about 25°C per 0.2 at.% Ni. Earlier work with shape memory alloys involved bulk materials. The more IC-compatible sputtered films were made in the late 1980s at Bell Labs by Walker et al.[121] and by Busch et al.[120] at TiNi Alloy Corporation. Thin-film nickel-titanium shape memory alloys, up to 20 µm in thickness, have been sputter deposited. These films exhibit memory behavior comparable to bulk material, and the phase transition to and from martensite lays entirely above ambient temperature (see Figure 9.22).[122] When deposited at room temperature, NiTi films are amorphous. Heating to 500°C causes the film to crystallize and acquire their shape-memory property.[122] The bandwidth of the NiTi films is slightly improved, to about 5 Hz, compared to the NiTi wire (about 1 Hz) because of the more effective cooling in the micro-domain.

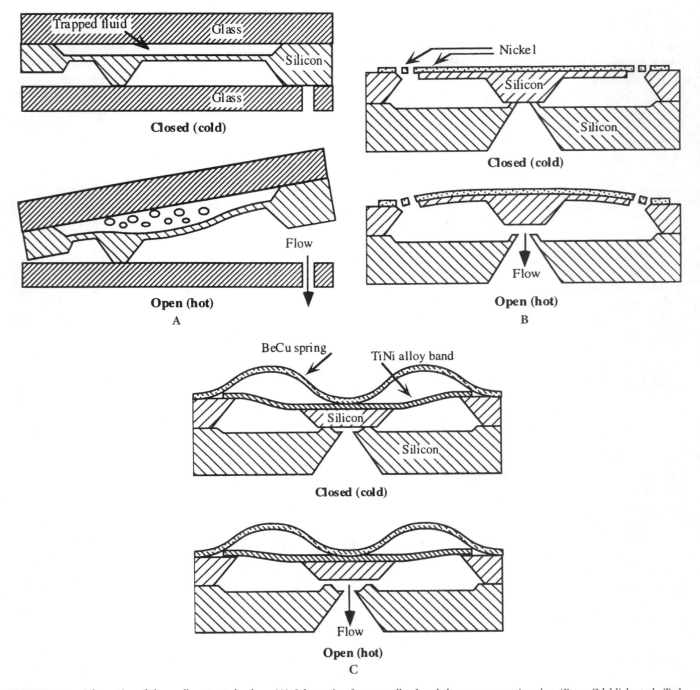

**FIGURE 9.21** Schematics of thermally actuated valves. (A) Schematic of a normally closed thermopneumatic valve. (From Zdeblick et al., Tech. Dig. 1994 Solid-State Sensor and Actuator Workshop, Hilton Head, SC, 251–255, 1994. With permission.) (B) Schematic of a normally closed nickel/silicon bimetal valve. (C) Schematic of a normally closed shape memory alloy valve. (From Barth, P.W., 8th Intl. Conference on Solid-State Sensors and Actuators, Stockholm, Sweden, 276–280, 1995. With permission.)

The energy density and drive conditions for NiTi actuators are compared with magnetic, electrostatic, and piezoactuators in Table 9.2. NiTi outperforms magnetic and electrostatic actuation in terms of energy density and may cause large motions in the range from 10 μm to 1 mm, operational requirements not well matched by electrostatic or piezoelectric technologies. As the required voltage for actuation is low (couple of volts), it becomes compatible with IC technology. The phase change in shape memory alloys is temperature driven, and heat must be removed before the next cycle starts, making the system inefficient, as the cycle rate may be slowed down by the achievable rate of heat transfer. The ultimate efficiency cannot exceed that of the Carnot cycle operating between the same temperature limits.[123] Practical SMA actuators have an efficiency typically less than 10%; for example, 50 J of heat input might be required to obtain 1 J of mechanical energy output.[122] However, in very small devices, heat transfer is rapid and makes this type of actuator more attractive. Another potential disadvantage is the need, in most applications, of mechanical biasing devices to provide a return force such as the BeCu spring, enabling cyclical

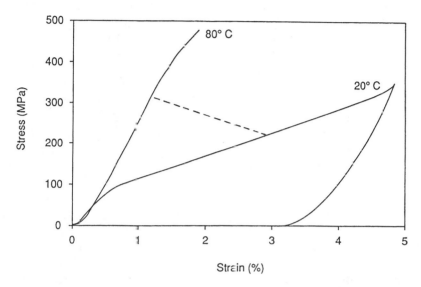

Sample #55C — 16 microns thick

FIGURE 9.22 Stress-strain curve for a TiNi shape memory alloy thin film. The sloping, broken line indicates a load line spring force resisting the actuator, showing a 2% repeatable strain recovery. (From Johnson, A. D., *J. Macromech. Microeng.*, 1, 34–41, 1991. With permission.)

behavior in the normally closed valve sketched in Figure 9.21C (see also Figure 10.4B). The beryllium-copper spring in Figure 9.21C puts bands of SMA in tension over the orifice at low temperature. The SMA bands constitute electrical resistors and are heated by current passage. At high temperature, the SMA element works against the BeCu spring to reach its high-temperature preset form and moves away from the valve seat, opening the valve. At low temperatures the spring easily pushes the silicon body back onto the valve seat, closing the valve. The spring adds somewhat to the size of the SMA actuator and also slows the response time further down (0.5 to 2 Hz). If the high-temperature shape does not change upon cooling the specimen and transforming it to the martensic phase, the phenomenon is referred to as a 'one-way' shape memory effect and springs or other mechanical biasing mechanisms are required. Solutions to avoid mechanical biasing might be within reach: SMAs can, under certain circumstances, remember their low-temperature shape as well, enabling cyclic device operation without the need of a bias.[124] In the latter case one talks about a 'reversible' shape memory effect. The low bandwidth of SMA usually is assumed to be determined by the relatively long cooling thermal time constant. Hunter et al. report an interesting unexplained effect in which very large brief current pulses ($>10^9$ A/m²) imposed during externally shortening and lengthening cycles alter the subsequent NiTi switching properties. The altered NiTi shortens and lengthens very rapidly (within 40 msec) and generates a maximum extrapolated stress of 230 MN/m² and a peak power/mass approaching 50 kW/kg.[7]

Shape-memory plastics also are being developed. Norsorex, for example, is the tradename for a polynorbornene polymer with excellent shape recovery properties at shape memory temperatures of over 35°C. Another polymer is Zeon Shable, a polyester-based polymer blend.[30]

## Other Actuators

A variety of electrochemical and chemical actuators have been demonstrated. In Chapter 10 we review a electrochemical disruptable valve (see Figure 10.31); Hamberg et al.[125] use electrolysis of water to create a pressure on a micromachined membrane. Osmotic pumps use a semipermeable membrane over a pump body to realize a mechanical stroke; the mechanism is clarified in Figure 9.23.[126] A solution in a closed vessel is covered on one side with a semipermeable membrane. A hypertonic solution outside the pump draws solvent from the hypotonic solution inside the pump body, reducing the volume inside and bending the membrane inward. Such an osmotic pump was proposed as an insulin pump for delivering insulin to a patient depending on his changing glucose concentration. In this application, the membrane displacement caused by a high glucose concentration in the patient pushes insulin out of the pump body into the patient's body.[126]

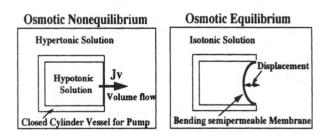

FIGURE 9.23 Mechanism of the osmotic pump. The decrease in volume bends the semipermeable membrane. JV corresponds to the volume flow out of the pump chamber. (From Nagakura et al., 8th Intl. Conference on Solid-State Sensors and Actuators, Stockholm, Sweden, 287–290, 1995. With permission.)

# Integrated Power

As in the case of portable computers, small, light-weight and long lasting energy sources are one of the most urgently needed breakthrough technologies. Microsystems require even smaller power sources, as weight and volume of on-board energy sources are disproportionately large compared to the microsystems they power. The roles of energy storage and energy dissipation in microsystems differ considerably from the world of practical daily experience. Designing microsystems demands close examination of energy budgets and taking advantage of the merits of smallness as well as minimizing its adverse effects.[127]

The specific energy (energy per volume unit) of the power source determines the proper active volume for a given application. If the volume of the packaging is taken into account, energy volume densities of lithium batteries may reach 240 to 360 Wh/l. These lithium-based batteries generate the highest specific energy of any commercially available battery. In these batteries the anode, consisting of high purity lithium, may be combined with many different cathode materials, resulting in different voltages ranging from 1.5 to 3.9 V. Organic solvents in which lithium salts are dissolved and conducting solid polymers function as electrolytes. Beyond button Li batteries 4.8 mm (dia.) by 1.4 mm (high) used in watches and cameras, progress in miniaturizing high-energy density power sources has been limited. Button Li cells are now the best available energy sources for microsystems. Batteries and fuel cell materials deposited with IC technologies on the device substrate itself are in the research stage.[128] Often the thin film materials deposited in constructing those batteries, such as Li, $TiS_2$, $V_2O_5$, etc. prove incompatible with the IC process, and the prospect of integrating them with ICs seems remote. Ultrathin, solid state Li cells, 'energy paper', also start to emerge. Kanebo introduced the polymeric PAS (poly-acenic semiconductor)-based battery in 1993. The polymer PAS film in the battery is only 200 μm thick and has an active surface area of 2200 $m^2$/g. It serves as the anode and the cathode is again lithium based. The voltage is 3.3 V, corresponding to 3 Ni-Cd elements in series. Unfortunately, the energy density, taking the complete, packaged battery into account, is only 5.5 Wh/L.[127] The reversibility, absence of polarity, and extended lifetime of supercapacitors make them an attractive alternative for power in microsystems. Supercapacitors with energy densities of 1.9 Wh/L and slightly higher are available. In supercapacitors an electrical double layer on a very high surface area material such as activated carbon or $IrO_x$ is reversibly charged and discharged. Since it is possible to carbonize photoresist materials, make them porous, and charge them, it seems feasible that ultracapacitors could be integrated on ICs. The overwhelming issue to overcome, just as in the case of a chemical sensor, is packaging. Supercapacitors and batteries incorporate very corrosive and reactive materials, making the challenge even more daunting. All of the above tend to suggest a hybrid implementation as the only possible means of integrating supercapacitors or thin-film batteries with ICs.

Power generation by the alternate heating and cooling of a working fluid or a solid (e.g., shape memory alloys) integrated

on a chip has been attempted for driving a load. The heating in such an engine results from passing a current through a resistor. It would be preferable though to use infrared radiation instead, since no leads need to connect to the chip. It has been projected that a gas-based heat engine of $5 \times 5 \times 5$ $mm^3$ might provide an output of 10 to 100 W/kg. With actuators based on shape memory alloys, an output of up to 1 kW/kg is feasible, with an efficiency ten times lower.[127] Several problems are associated with crafting MEMS engines, including the thermal isolation of heating and cooling sections, minimization of friction, and the difficulty of implementing a flywheel. Some of these problems were successfully addressed by Sniegowski et al.[129,130] who demonstrated a surface micromachined microengine capable of delivering torque to a micromechanism. Angular velocities of 600,000 rpm were registered for the engine driven by an electrostatically comb drive. In an alternative construct, the same engine was also driven by steam.

Given the size of power sources, generating electricity on board or supplying energy from the outside often is preferred. One interesting way to accomplish this, used for several decades in wristwatches, is to use a microgenerator. An eccentrically rotating mass driven by wrist movements supplies energy to a spring. A mechanical watch requires 1 to 2 μW. In order to keep the watch working for 48 hours after it has been removed from the wrist, the loaded spring must contain 4.8 $10^{-5}$ Wh. Given the size of the microgenerator, the system stores about 0.3 Wh/L, more than two orders of magnitude smaller than the specific energy of a button cell for quartz watches. In an automatic quartz watch, the microgenerator drives an electric generator, the electrical energy is then stored in a supercapacitor powering the quartz oscillator, IC, and stepping motor of the watch. The power requirement of a high-quality analog watch is as low as 0.5 μW.[127] As Goemans points out, the possibility of converting motion into electrical energy can be very attractive for cases where battery replacement is unacceptable, kinetic energy is abundantly available, and space is not too limited. He lists biomedical implants, tire pressure monitoring systems, and electronic locks as potential application areas.[127]

Thermo-electric converters may extract energy in applications where heat and temperature difference are available 'for free'. Thin-film thermocouples have been used to power a watch based on the temperature difference between the cool front face of the watch and the warm skin contact. A disadvantage of this approach is the very low efficiency of the conversion.

We already discussed the implementation of a high-voltage, integrated solar cell array by Lee et al.[131] as an electrostatic MEMS power supply (see Chapter 5). The conversion efficiency in that effort was only 0.2% though. Sakakibara et al.[132] were able to generate more than 200 V with a similar solar cell on an area of 1 $cm^2$ and obtained a conversion efficiency of 4.65%. In both cases amorphous silicon was used in a triple-stacked photovoltaic structure generating up to 2.3 V per cell. To obtain a very dense packing of array elements and to make the series connection of the solar cells, the latter group used focused laser beams for patterning electrodes and photovoltaic materials. For future thin-film photovoltaic cells, efficiencies of over 30% are expected. Solar cell technology represents the most MEMS-compatible technology for power integration. Since solar light is only intermittently available, electric storage elements need be implemented as well. Along this line Kimura et al.[133] fabricated a miniature opto-electric transformer consisting of a p-n junction photocell and a multilayer spiral coil transformer. Besides photovoltaic converters for solar light and laser light, microwaves could be used to power microsystems. In the latter case, extremely small receivers and converters would need to be built.

**For URLs on scaling, actuators, and power sources, see Appendix B.**

# References

1. Lebbink, G. K., "Microsystem Technology: Exploring Opportunities", Alphen aan de Rijn/Zaventem, 1994.
2. Trimmer, T., "Micromechanical Systems", Proceedings. Integrated Micro-Motion systems: Micromachining, Control, and Applications (3rd Toyota Conference), Aichi, Japan, October, 1990, p. 1–15.
3. Berry, A. T., Ed. *The Book of Scientific Anecdotes*, Prometheus Books, Buffalo, N.Y., 1993.
4. Thompson, D. W., *On Growth and Form* University Press, Cambridge, MA. 1992.
5. Morowitz, H., J., *Mayonnaise and The Origin of Life*, Berkley Books, New York, 1985.
6. Hayashi, T., "Micromechanism and Their Characteristics", IEEE International Workshop on Micro Electro Mechanical Systems, MEMS '94, Oiso, Japan, 1994, p. 39–44.
7. Hunter, I., W. and Lafontaine, S., "A Comparison of Muscle with Artificial Actuators", Technical Digest of the 1992 Solid State Sensor and Actuator Workshop., Hilton Head Island, SC, 1992, p. 178–185.
8. Trimmer, W. S. N. and Gabriel, K. J., "Design Considerations for a Practical Electrostatic Micro-Motor", *Sensors Actuators*, 11, 189–206, 1987.
9. Trimmer, W. S. N., "Microrobots and Micromechanical Systems", *Sensors Actuators*, 19, 267–287, 1989.
10. Tang, W. C.-K., "Electrostatic Comb Drive for Resonant Sensor and Actuator Applications", Ph.D. Thesis, University of California at Berkeley, 1990.
11. Tang, W. C., T. H. Nguyen, and R. T. Howe, "Laterally Driven Polysilicon Resonant Microstructures", *Sensors Actuators*, 20, 25–32, 1989.
12. Denn, M. M., *Process Fluid Mechanics*, Prentice-Hall, Inc., Englewood Cliffs. New Jersey 07632, 1980.
13. Schmidt, M. A., "Microsensors for the Measurement of Shear Forces in Turbulent Boundary Layers", Ph.D. Thesis, Massachusetts Institute of Technology, 1988.
14. Mohr, J., Bley P., Strohrmann, M., and Wallrabe, U., "Microactuators Fabricated by the LIGA process", *J. Micromech. Microeng.*, 2, 234–241, 1992.
15. Bart, S. F., T. A. Lober, R. T. Howe, J. H. Lang, and M. F. Schlecht, "Design Considerations for Micromachined Electric Actuators", *Sensors Actuators*, 14, 269–292, 1988.
16. Mahadevan, R., "Capacitance calculations for a single-stator, single-rotor electrostatic motor", Proceedings of the IEEE Micro Robots and Teleoperators Workshop, Hyannis, MA, 1987, p. 15/1–8.
17. Mahadevan, R., "Analytical Modelling of Electrostatic Structures", Proceedings. IEEE Micro Electro Mechanical Systems, (MEMS '90), Napa Valley, CA, 1990, p. 120–127.
18. Kumar, S. and D. Cho, "A Perturbation Method for Calculating the Capacitance of Electrostatic Motors", Proceedings. IEEE Micro Electro Mechanical Systems, (MEMS '90), Napa Valley, CA, 1990, p. 27–33.
19. Mehregany, M., P. Nagarkar, S. D. Senturia, and L. J. H., "Operation of Microfabricated Harmonic and Ordinary Side-Drive Motors", Proceedings. IEEE Micro Electro Mechanical Systems, (MEMS '90), Napa Valley, CA, 1990, p. 1–8.
20. Fujita, H. and K. J. Gabriel, "New Opportunities for Micro Actuators", 6th International Conference on Solid-State Sensors and Actuators (Transducers '91), San Francisco, CA, 1991, p. 14–20.
21. Price, R., H., Cunningham, S. J. and Jacobsen, S. C., "Field Analysis for the Electrostatic Eccentric Drive Micromotor ("Wobble Motor")", *J. Electrost.*, 28, 7–38, 1992.
22. Furuhata, T., T. Hirano, L. H. Lane, R. E. Fontana, L. S. Fan, and H. Fujita, "Outer Rotor Surface-Micromachined Wobble Micromotor", Proceedings. IEEE Micro Electro Mechanical Systems, (MEMS '93), Fort Lauderdale, FL, 1993, p. 161–166.
23. Tai, Y., L. Fan, and R. Muller, "IC-Processed Micromotor: Design, Technology, and Testing", Proceedings. IEEE Micro Electro Mechanical Systems, (MEMS '89), Salt Lake City, UT, 1989, p. 1–6.
24. Fujita, H., "Electrostatic and Superconducting Microactuators", Proceedings. Micro System Technologies '90, Berlin, Germany, 1990, p. 818.
25. Ohnstein, T., T. Fukiura, J. Ridley, and V. Bonne, "Micromachined Silicon Microvalve", Proceedings. IEEE Micro Electro Mechanical Systems, (MEMS '90), Napa Valley, CA, 1990, p. 95–98.
26. Petersen, K., "Micromechanical Membrane Switches on Silicon", IBM J. Res. Develop., Report No. 23, p.376, 1979.
27. Schumann, W. O., *Elektrische Durchbruchfeldstarke von Gasen*, Springer, Berlin, 1923.
28. Busch-Vishniac, I. J., "The Case for Magnetically Driven Microactuators", *Sensors and Actuators A*, A33, 207–220, 1992.
29. Bollee, B., "Electrostatic Motors", *Philips Tech. Rev.*, 30, 178–194, 1969.
30. Gandhi, M. V. and Thompson, B. S., *Smart Materials and Structures*, Chapman & Hall, London, 1992.
31. Hueter, T. F. and Bolt, R. H., *Sonics*, John Wiley & Sons, New York, 1955.

32. Pallas-Areny, R. and J. G. Webster, *Sensors and Signal Conditioning*, John Wiley & Sons, New York, 1991.

33. Herbert, J. M., *Ferroelectric Transducers and Sensors*, Gordon and Breach Science Publishers, New York, 1982.

34. Khazan, A. D., *Transducers and Their Elements*, PTR Prentice Hall, Englewood Cliffs, NJ, 1994.

35. Flynn, A. M., L. S. Tavrow, S. F. Bart, R. A. Brooks, D. J. Ehrlich, K. R. Udayakumar, and L. E. Cross, "Piezoelectric Micromotors", *J. Microelectromech. Syst.*, 1, 44–52, 1992.

36. Udayakumar, K. R., S. F. Bart, A. M. Flynn, J. Chen, L. S. Tavrow, L. E. Cross, R. A. Brooks, and D. J. Ehrlich, "Ferroelectric Thin Film Ultrasonic Micromotors", Proceedings. IEEE Micro Electro Mechanical Systems, (MEMS '91), Nara, Japan, 1991, p. 109–113.

37. van Lintel, H. T. G., F. C. M. van der Pol, and S. Bouwstra, "A Piezoelectric Micropump Based on Micromachining of Silicon", *Sensors Actuators*, 15, 153–167, 1988.

38. Smits, J. G., "Piezoelectric Micropump with Three Valves Working Peristaltically", *Sensors and Actuators A*, A21, 203–206, 1990.

39. Shoji, S., S. Nakagawa, and M. Esashi, "Micro-pump and Sample-Injector for Integrated Chemical Analyzing Systems", *Sensors and Actuators A*, A21, 189–192, 1990.

40. Abe, T. and M. L. Reed, "RF-Magnetron Sputtering of Piezoelectric Lead-Zirconate-Titanate Actuator Films Using Composite Targets", IEEE International Workshop on Micro Electro Mechanical Systems, MEMS '94, Oiso, Japan, 1994, p. 164–169.

41. Kolesar, E. S. and C. S. Dyson, "Object Imaging with a Piezoelectric Robotic Tactile Sensor", *J. Microelectromech. Syst.*, 4, 87–96, 1995.

42. AMP, "Piezo Film Sensors", AMP Incorporated, Piezo Film Sensors, P.O.Box 799, Valley Forge, PA 19482, 1993.

43. Nalwa, H. S., Ed. *Ferroelectric Polymers-Chemistry, Physics, and Applications*, Marcel Dekker, Inc., New York, 1995.

44. Hirata, Y., H. Okuyama, S. Ogino, T. Numazawa, and H. Takada, "Piezoelectric Composites for Micro-Ultrasonic Transducers Realized with Deep-Etch X-ray Lithography", Proceedings. IEEE Micro Electro Mechanical Systems, (MEMS '95), Amsterdam, Netherlands, 1995, p. 191–196.

45. Preu, G., A. Wolff, D. Cramer, and U. Bast, "Microstructuring of Piezoelectric Ceramic", Proceedings of the Second European Ceramic Society Conference (2nd ECerS '91), Augsburg, Germany, 1991, p. 2005–2009.

46. Lubitz, K., "Mikrostrukturierung von Piezokeramik", *VDI-Tagungsbericht*, 796, 1989.

47. Muller, R. S., "From IC's to Microstructures: Materials and Technologies", Proceedings of the IEEE Micro Robots and Teleoperators Workshop, Hyannis, MA, 1987, p. 2/1–5.

48. Tjhen, W., T. Tamagawa, C.-P. Ye, C.-C. Hsueh, P. Schiller, and D. Polla, L., "Properties of Piezoelectric Thin Films for Micromechanical Devices and Systems", Proceedings. IEEE Micro Electro Mechanical Systems, (MEMS '91), Nara, Japan, 1991, p. 114–119.

49. Allen, R., "Sensors in Silicon", *High Technology*, September, 43–81, 1984.

50. Holland, R. and E. P. Eernisse, *Design of Resonant Piezoelectric Devices*, The M.I.T. Press, Cambridge, MA, 1969.

51. Rogacheva, N. N., *The Theory of Piezoelectric Shells and Plates*, CRC Press, Boca Raton, FL, 1994.

52. Ye, C., T. Tamagawa, and D. L. Polla, "Pyroelectric $PbTiO_3$ Thin Films for Microsensor Applications", 6th International Conference on Solid-State Sensors and Actuators (Transducers '91), San Francisco, CA, 1991, p. 904–907.

53. Hsieh, H. Y., A. Spetz, and J. N. Zemel, "Wide Range Pyroelectric Anemometers for Gas Flow Measurements", 6th International Conference on Solid-State Sensors and Actuators (Transducers '91), San Francisco, CA, 1991, p. 38–40.

54. Voorthuyzen, J. A. and P. Bergveld, "The PRESSFET: An Integrated Electret-MOSFET Based Pressure Sensor", *Sensors Actuators*, 14, 349–360, 1988.

55. Voorthuyzen, J. A., P. Bergveld, and A. J. Sprenkels, "Semiconductor-Based Electret Sensors for Sound and Presure", *IEEE Trans. Electr. Insul.*, 24, 267–276, 1989.

56. Sprenkels, A. J., R. A. Groothengel, A. J. Verloop, and P. Bergveld, "Development of an Electret Microphone in Silicon", *Sensors Actuators*, 17, 509–512, 1989.

57. Wolffenbuttel, R. F., J. F. L. Goosen, and P. M. Sarro, "Design Considerations for a Permanent-Rotor-Charge Excited Micromotor with an Electrostatic Bearing", *Sensors and Actuators A*, A25–27, 583–590, 1991.

58. Glaser, K. and G. Fuhr, "Electrorotation-The Spin of Cells in Rotating High Frequency Electric Fields", in *Mechanistic Approaches to Interactions of Electric and Electromagnetic Fields with Living Systems*, Blank, M. and Findl, E., Eds., Plenum Press, New York, 1987.

59. Fuhr, G., R. Hagedorn, T. Muller, B. Wagner, and W. Benecke, "Linear Motion of Dielectric Particles and Living Cells in Microfabricated Structures by Traveling Electric Fields", Proceedings. IEEE Micro Electro Mechanical Systems, (MEMS '91), Nara, Japan, 1991, p. 259–264.

60. Moesner, F. M. and T. Higuchi, "Devices for Particle Handling by an AC Electric Field", Proceedings. IEEE Micro Electro Mechanical Systems, (MEMS '95), Amsterdam, Netherlands, 1995, p. 66–71.

61. Melcher, J. R., "Traveling-Wave Induced Electroconvection", *Phys. Fluids*, 9, 1548–1555, 1966.

62. Bart, S. and J. H. Lang, "An Analysis of Electroquasistatic Induction Micromotors", *Sensors Actuators*, 20, 97–106, 1989.

63. Bart, S. F., L. S. Tavrow, M. Mehregany, and J. H. Lang, "Microfabricated Electrohydrodynamic Pumps", *Sensors and Actuators A*, A21, 193–197, 1990.

64. Zipernowsky, "Zipernowsky electrostatic motor", *Electr. World*, 14, 260, 1889.

65. Fujita, H. and A. Omodaka, "Electrostatic Actuators for Micromechatronics", Proceedings of the IEEE Micro Robots and Teleoperators Workshop, Hyannis, MA, 1987, p. 14/1–10.

66. Takeshima, N., K. J. Gabriel, M. Ozaki, J. Takahashi, H. Horiguchi, and H. Fujita, "Electrostatic Parallelogram Actuators", 6th International Conference on Solid-State Sensors and Actuators (Transducers '91), San Francisco, CA, 1991, p. 63–6.

67. Wagner, B., M. Kreutzer, and W. Benecke, "Electromagnetic Microactuators with Multiple Degrees of Freedom", 6th International Conference on Solid-State Sensors and Actuators (Transducers '91), San Francisco, CA, 1991, p. 614–617.

68. Wise, K. D., "Integrated Microelectromechanical Systems: A Perspective on MEMS in the 90s", Proceedings. IEEE Micro Electro Mechanical Systems, (MEMS '91), Nara, Japan, 1991, p. 33–8.

69. Romankiw, L. T., "Thin Film Inductive Heads; From One to Thirty One Turns", Proceedings of the Symposium on Magnetic Materials, Processes, and Devices, Hollywood, FL, 1989, p. 39–53.

70. Cardot, F., J. Gobet, M. Bogdanski, and F. Rudolf, "Fabrication of a Magnetic Transducer Composed of a High-Density Array of Microelectromagnets with On-Chip Electronics", *Sensors and Actuators A*, A 43, 11–16, 1994.

71. Ahn, C. H., Y. J. Kim, and M. G. Allen, "A Fully Integrated Micromachined Toroidal Inductor with a Nickel-Iron Core (The Switched DC/DC Boost Converter Application)", 7th International Conference on Solid-State Sensors and Actuators (Transducers '93), Yokohama, Japan, 1993, p. 70–73.

72. Wagner, B., M. Kreutzer, and W. Benecke, "Permanent Magnet Micromotors on Silicon Substrates", *J. Microelectromech. Syst.*, 2, 23–29, 1993.

73. Kim, Y.-K., M. Katsurai, and H. Fujita, "Fabrication and Testing of a Micro Superconducting Actuator Using the Meissner Effect.", Proceedings. IEEE Micro Electro Mechanical Systems, (MEMS '90), Napa Valley, CA, 1990, p. 61–6.

74. Peirine, R. and I. Busch-Vishniac, "Magnetically Levitated Micro-Machines", Proceedings of the IEEE Micro Robots and Teleoperators Workshop, Hyannis, MA, 1987, p. 19/1–5.

75. Kumar, S., D. Cho, and W. N. Carr, "Experimental Study of Electric Suspension for Microbearings", *J. Microelectromech. Syst.*, 1, 23–30, 1992.

76. Pister, K. S. J., R. S. Fearing, and R. T. Howe, "A Planar Air Levitated Electrostatic Actuator System", Proceedings. IEEE Micro Electro Mechanical Systems, (MEMS '90), Napa Valley, CA, 1990, p. 67–71.

77. Benecke, W., "Silicon-Microactuators: Activation Mechanisms and Scaling Problems", 1991 International Conference on Solid-State Sensors and Actuators, Transducers '91, San Francisco, CA, USA, 1991, p. 46–50.

78. Rombach, P. and W. Langheinrich, "An Integrated Sensor Head in Silicon for Contactless Detection of Torque and Force", *Sensors and Actuators A*, A41–42, 410–416, 1994.

79. Quandt, E. and K. Seemann, "Fabrication of Giant Magnetostrictive Thin Film Actuators", Proceedings. IEEE Micro Electro Mechanical Systems, (MEMS '95), Amsterdam, Netherlands, 1995, p. 273–277.

80. Honda, T., K. I. Arai, and M. Yamaguchi, "Fabrication of Actuators Using Magnetostrictive Thin Films", IEEE International Workshop on Micro Electro Mechanical Systems, MEMS '94, Osio, Japan, 1994, p. 51–56.

81. Fukuda, T., H. Hosokai, H. Ohyama, H. Hashimoto, and F. Arai, "Giant Magnetostrictive Alloy (GMA) Applications to Micro Mobile Robot as a Micro Actuator without Power Supply Cables", Proceedings. IEEE Micro Electro Mechanical Systems, (MEMS '91), Nara, Japan, 1991, p. 210–215.

82. Humphrey, E. F. and D. H. Tarumoto, Eds. *Fluidics*, Fluidic Amplifier Associates, Inc., Boston, MA, 1965.

83. Terry, S. C., J. H. Jerman, and A. J. B., "A Gas Chromatographic Air Analyzer Fabricated on a Silicon Wafer", *IEEE Trans. Electron Devices*, ED-26, 1880–1886, 1979.

84. Bassous, E., H. H. Taub, and L. Kuhn, "Ink Jet Printing Nozzle Arrays Etched in Silicon", Appl. Phys. Lett., 31, 135–137, 1977.

85. Manz, A., H. D.J., F. J.C., V. E., L. H., and W. H.M., "Integrated Electroosmotic Pumps and Flow Manifolds for Total Chemical Analysis Systems", 6th International Conference on Solid-State Sensors and Actuators (Transducers '91), San Francisco, CA, 1991, p. 939–41.

86. Zdeblick, M. J. and J. A. Angell, "A microminiature Electric-to-Fluidic Valve", 4th International Conference on Solid-State Sensors and Actuators (Transducers '87), Tokyo, Japan, 1987, p. 248–253.

87. Northrup, M. A., M. T. Ching, R. M. White, and R. T. Watson, "DNA Amplification with a Microfabricated Reaction Chamber", 7th International Conference on Solid-State Sensors and Actuators (Transducers '93), Yokohama, Japan, 1993, p. 924–926.

88. Gravesen, P., J. Branebjerg, and O. S. Jensen, "Microfluidics: A Review", *J. Micromech. Microeng.*, 3, 168–182, 1993.

89. Harley, J., H. Bau, J. N. Zemel, and V. Dominko, "Fluid Flow in Micron and Submicron Size Channels", Proceedings. IEEE Micro Electro Mechanical Systems, (MEMS '89), Salt Lake City, UT, 1989, p. 25–8.

90. Flik, M. I., B. I. Choi, and K. E. Goodson, "Heat Transfer Regimes in Microstructures", ASME 1991, Micromechanical Sensors, Actuators, and Systems, Atlanta, GA, 1991, p. 31–46.

91. Manz, A., E. Verpoorte, C. S. Effenhauser, N. Burggraf, D. E. Raymond, D. J. Harrison, and H. M. Widmer, "Miniaturization of Separation Techniques Using Planar Chip Technology", *HRC-Journal of High Resolution Chromatography*, 16, 433–436, 1993.

92. Potter, M. C. and J. F. Foss, *Fluid Mechanics*, Great Lakes Press, Okemos, MI, 1982.

93. Allen, T. J. and R. L. Ditsworth, *Fluid Mechanics*, McGraw-Hill, New York, 1972.

94. Solarz, R. W. and W. F. Krupke, "Diode Pumped Lasers Finding Commercial Applications", *Res. & Dev.*, May, 89–90, 1992.

95. Little, W. A., "Applications of Closed-Cycle Cryocoolers to Small Superconducting Devices", Proceedings of the NBS Cryocooler Conference, Boulder, CO, 1978, p. 75–80.

96. Little, W. A., "Microminiature Refrigeration", *Rev. Sci. Instrum.*, 55, 661–680, 1984.

97. Ozisik, M. N., *Basic Heat Transfer*, McGraw-Hill, New York, 1977.

98. Corcoran, P., H. V. Shurmer, and J. W. Gardner, "Integrated Tin Oxide Sensors of Low Power Consumption for Use in Gas and Odour Sensing", *Sensors and Actuators B*, B15–16, 32–37, 1993.

99. Pfahler, J., J. Harley, and H. Bau, "Gas and Liquid Flow in Small Channels", ASME 1991, Micromechanical Sensors, Actuators, and Systems, Atlanta, GA, 1991, p. 49–60.

100. Liu, J. and Y.-C. Tai, "MEMS for Pressure Distribution of Gaseous Flows in Microchannels", Proceedings. IEEE Micro Electro Mechanical Systems, (MEMS '95), Amsterdam, Netherlands, 1995, p. 209–215.

101. Arkilic, E. B., K. S. Breuer, and M. A. Schmit, "Gaseous Flow in Microchannels", in *Applications of Microfabrication to Fluid Mechanics*, Chicago, Ill, 1994, p. 57–66.

102. Nyborg, W. L. M., "Acoustic Streaming", in *Physical Acoustics*, Mason, W. P., Ed., Academic Press, New York, 1965, p. 265–331.

103. Moroney, R. M., R. M. White, and R. T. Howe, "Fluid Motion Produced by Ultrasonic Lamb Waves", IEEE 1990 Ultrasonics Symposium Proceedings, Honolulu, HI, 1990, p. 355–358.

104. Moroney, R. M., R. M. White, and R. T. Howe, "Microtransport Induced by Ultrasonic Waves", *Appl. Phys. Lett.*, 59, 774–776, 1991.

105. Shiokawa, S., Y. Matsui, and T. Ueda, "Study on SAW Streaming and its Application to Fluid Devices", *Japanese Journal of Applied Physics, Part 1*, 29, 137–139, 1990.

106. Miyazaki, S., T. Kawai, and M. Araragi, "A Piezo-Electric Pump Driven by A Flexural Progressive Wave", Proceedings. IEEE Micro Electro Mechanical Systems, (MEMS '91), Nara, Japan, 1991, p. 283–288.

107. Jorgenson, J. W. and E. J. Guthrie, "Liquid Chromatography in Open-Tubular Columns", *J. Chromatography*, 255, 335–348, 1983.

108. Harrison, D. J., Z. Fan, K. Fluri, and K. Seiler, "Integrated Electrophoresis Systems for Biochemical Analyses", Technical Digest of the 1994 Solid State Sensor and Actuator Workshop., Hilton Head Island, SC, 1994, p. 21–24.

109. Jacobson, S. C., R. Hergenroder, A. W. Moore, and J. M. Ramsey, "Electrically Driven Separations on a Microchip", Technical Digest of the 1994 Solid State Sensor and Actuator Workshop., Hilton Head Island, SC, 1994, p. 65–68.

110. Richter, A., A. Plettner, K. Hoffmann, and H. Sandmaier, "Electrohydrodynamic Pumping and Flow Measurement", Proceedings. IEEE Micro Electro Mechanical Systems, (MEMS '91), Nara, Japan, 1991, p. 271.

111. Colgate, E. and H. Matsumoto, "An Investigation of Electro-Wetting-Based Microactuation", *J. Vac. Sci. Technol.*, A8, 3625–3633, 1990.

112. Abaxis, "Piccolo", Product Literature, Abaxis, 1996.

113. Ozaki, K., "Pumping Mechanism Using Periodic Phase Changes of a Fluid", IEEE International Workshop on Micro Electro Mechanical Systems, MEMS '95, Amsterdam, Netherlands, 1995, p. 31–36.

114. Manz, A., N. Graber, and H. M. Widmer, "Miniaturized Total Chemical Analysis Systems; A Novel Concept for Chemical Sensing", *Sensors and Actuators B*, B1, 244–248, 1990.

115. Golay, M. J. E., *Gas Chromatography*, Academic Press, New York, 1958.

116. Anderson, J. L., K. K. Whiten, J. D. Brewster, T.-Y. Ou, and W. K. Nonidez, "Microarray Electrochemical Flow Detectors at High Applied Potentials and Liquid Chromatography with Electrochemical Detection of Carbamate Pesticides in River Water", *Anal. Chem.*, 57, 1366–1373, 1985.

117. Reay, R. J., R. Dadoo, C. W. Storment, R. N. Zare, and G. T. A. Kovacs, "Microfabricated Electrochemical Detector for Capillary Electrophoresis", Technical Digest of the 1994 Solid State Sensor and Actuator Workshop., Hilton Head Island, SC, 1994, p. 61–64.

118. Zdeblick, M. J., R. Anderson, J. Jankowski, B. Kline-Schoder, L. Christel, R. Miles, and W. Weber, "Thermopneumatically Actuated Microvalves and Integrated Electro-Fluidic Circuits", Technical Digest of the 1994 Solid State Sensor and Actuator Workshop., Hilton Head Island, SC, 1994, p. 251–255.

119. Barth, P. W., "Silicon Microvalves for Gas Flow Control", 8th International Conference on Solid-State Sensors and Actuators (Transducers '95), Stockholm, Sweden, June, 1995, p. 276–280.

120. Busch, J. D. and A. D. Johnson, "Shape-Memory Properties in Ni-Ti Sputter-Deposited Film", *J. Appl. Phys.*, 68, 6224–6228, 1990.

121. Walker, J. A., K. J. Gabriel, and M. Mehregany, "Thin-Film Processing of TiNi Shape Memory Alloy", *Sensors and Actuators A*, A21, 243–246, 1990.

122. Johnson, A. D., "Vacuum-deposited TiNi Shape Memory Film: Characterization and Applications in Microdevices", *J. Micromech. Microeng.*, 1, 34–41, 1991.

123. Dario, P., M. Bergamasco, L. Bernardi, and A. Bicchi, "A Shape Memory Alloy Actuating Module for Fine Manipulation", Proceedings of the IEEE Micro Robots and Teleoperators Workshop, Hyannis, MA, 1987, p. 16/1–5.

124. Quandt, E., C. Halene, H. Holleck, K. Feit, M. Kohl, and P. Schlossmacher, "Sputter Deposition of TiNi and Ti-NiPd Films Displaying the Two Way Shape Memory Effect", 8th International Conference on Solid-State Sensors and Actuators (Transducers '95), Stockholm, Sweden, June, 1995, p. 202–205.

125. Hamberg, M. W., C. Neagu, J. G. E. Gardeniers, D. IJntema, J., and M. Elwenspoek, "An Electrochemical Micro Actuator", Proceedings. IEEE Micro Electro Mechanical Systems, (MEMS '95) Amsterdam, Netherlands, 1995, p. 106–110.

126. Nagakura, T., K. Ishihara, T. Furukawa, K. Masuda, and T. Tsuda, "Auto-Regulated Medical Pump Without Energy Supply", 8th International Conference on Solid-State Sensors and Actuators (Transducers '95), Stockholm, Sweden, June, 1995, p. 287–290.

127. Goemans, P. A. F. M., "Microsystems and Energy : The Role of Energy", Microsystem technology: exploring opportunities, 1994.

128. Bates, J. B., G. R. Gruzalski, and C. F. Luck, "Rechargeable Solid State Lithium Microbatteries", Proceedings. IEEE Micro Electro Mechanical Systems, (MEMS '93), Fort Lauderdale, FL, 1993, p. 82–86.

129. Sniegowski, J. and E. Garcia, "Microfabricated Actuators and Their Applications to Optics", Micro-Optics/Micromechanics and Laser Scanning and Shaping, (Proceedings of the SPIE), San Jose, CA, 1995, p. 46–64.

130. Garcia, E. and J. Sniegowski, "Surface Micromachined Microengine", *Sensors and Actuators A*, A48, 203–14, 1995.

131. Lee, J. B., Z. Chen, M. G. Allen, A. Rohatgi, and R. Arya, "A Miniaturized High-Voltage Solar Cell Array as an Electrostatic MEMS Power Supply", *J. Microelectromech. Syst.*, 4, 102–108, 1995.

132. Sakakibara, T., H. Izu, T. Kura, W. Shinohara, H. Iwata, S. Kiyama, and S. Tsuda, "High-Voltage Photovoltaic Micro-Devices Fabricated by a New Laser-Processing", Proceedings. IEEE Micro Electro Mechanical Systems, (MEMS '95), Amsterdam, Netherlands, 1995, p. 282–287.

133. Kimura, M., N. Miyakoshi, and M. Daibou, "A Miniature Opto-Electric Transformer", Proceedings. IEEE Micro Electro Mechanical Systems, (MEMS '91), Nara, Japan, 1991, p. 227–232.

# Microfabrication Applications

## Introduction

To better understand where revenues can be derived from microfabrication applications, we first address definitions for micromachined sensors, actuators, components, and instruments, terms that are familiar to micromachinists but often are confused in marketing reports. We also adapt a generic sensor classification option. A market study using those definitions and classification method and explicitly describing the manufacturing method (traditional or nontraditional) would reveal more than current sensor market studies from which it is hard to define what is comprised in the term 'microfabricated products'. An analysis of the overall market for micromachinery as distilled from a variety of recent surveys follows. In this analysis we distinguish between the nascent market for micromachines, fabricated using nontraditional machining methods (lithography techniques often involving single crystal or polycrystalline silicon), and the larger more mature market for traditional precision engineered products such as optical pickup heads in compact disc players and read-write heads in hard disk drives. The overall market discussion is followed by a series of application reviews: sensors in the automotive industry; sensors and instruments in biomedical applications; gas sensors in industrial, home, and automotive markets; and miniaturization in analytical equipment. These application sections are meant to highlight how the characteristics of specific applications/markets influence the choice of an optimum micromanufacturing technology.

We hope that a better understanding of the different micromachining technologies and their optimized application, acquired in the preceding chapters, will provide guidance in determining which devices and approaches may mature into sizable markets. Because of the pervasive application of miniaturization in a variety of industries, we are limiting our description to a fraction of the existing work. The home page of *Fundamentals of Microfabrication* (http://www.crcpress.com/microfab) will supply a regularly updated list of breakthrough miniaturization examples, listing the precision engineering techniques employed. Appendix D lists references for more detailed market information on sensors.

## Definitions and Classification Methods

Micromachining marketing studies rarely clarify whether the quoted revenues apply to the micromachine alone or include the micromachine with some ancillary electronics, comprise a higher level subsystem, or describe the whole system. Moreover, the precision machining technology employed often remains unspecified. A good market study should start from a clear definition of the products covered. For several of the definitions and the classification method given below we closely followed the recent suggestions by the National Research Council.[1] We also refer to the Appendix E at the end of this book for further familiarization with micromachining related terms.

### Sensors

A sensor element is a device that converts one form of energy into another (e.g., ZnO, a piezoelectric material, converts mechanical energy into electricity) and provides the user with a usable energy output in response to a specific measurable input. The sensor element may incorporate almost any material including plastic, semiconductor, metal, ceramic, etc. Since the early 1980s, the word 'sensor element', unfortunately automatically invokes the notion of a silicon-built device. Measurands may belong to the radiation, thermal, electrical, chemical, mechanical or magnetic field domains. A sensor includes a sensor element or an array of sensor elements with physical packaging and external electrical or optical connections. Synonyms for sensor are transducer or detector. A sensor system includes the sensor and its assorted signal processing hardware (analog or digital). The word transducer sometimes refers to a sensor system, especially in the process control industry. The process control industry also uses the term 'transmitter', i.e., a type of sensor system with a current output (e.g., 4–20 mA), which allows transmission of information over a long distance without loss of accuracy.[2]

In the case of silicon-based sensors, some additional jargon has developed. A Si sensor element is called a 'sensor die' and refers to a micromachined Si chip. It typically sells for $0.50 to $2.00 as a commodity product, although the price tag can rise to $50.00 or more for a high-performance structure sold in smaller quantities. A 'Si-sensor' alludes to a first-level, packaged Si die or sensor element with or without basic electronic circuitry. The former defines an 'integrated Si sensor' where a monolithic component contains the sensor and one or more electronic components amplify and condition (standardize) the sensor output signal. Typical selling prices range from $2.5 for quantities of a million units per year to over $100.00 for complex sensors in smaller unit quantities. At yet a higher level is a 'smart Si sensor', corresponding in function to the above-defined sensor system: a packaged integrated sensor containing some part of the signal processing unit to provide performance enhancement to the user. Signal processing might include autocalibration, interference reduction, compensation for parasitic effects, offset correction, and self-test. The standardization of the sensor signal makes the device bus compatible, enabling efficient communication between the central processor and the sensor. The various silicon-based components defined are schematically represented in Figure 10.1.

The acronym 'MEMS', for micro-electromechanical systems, originally applied to Si micromachined miniaturized electromechanical systems but now refers to any subminiaturized system including chemical sensors and non-silicon structures.

## Hybrid Sensors

Traditionally, hybrid integration in the IC industry means combining thin-film ICs with thick-film bonding technology. Viewed from the Si sensor angle, a hybrid sensor keeps the electronics part separate from the sensor part. The hybrid might then consist of two pieces of Si on the same substrate connected

with a short wire bridge, or it might be a Si sensor mounted in a header plugged into an electronics board. Because the sensor package determines the final size, the hybrid sensors often are not much larger than their Si counterparts. Hybrids usually are simpler to design and can be produced more economically for smaller production volume runs. For chemical and biosensors they often represent the viable option to implement micromachining technology. Prices for Si sensors given in market reports mostly refer to hybrid devices with the Si sensor mounted in a header and the signal conditioning kept on separate chips. As an example, Figure 10.2* represents the bulk Si micromachined piezoresistive accelerometer manufactured by Lucas NovaSensor.

From the precision engineering angle, a hybrid device is one in which parts produced by different technologies are joined and assembled as illustrated in the large aspect ratio magnetic motor in the inset on page 317. In this motor only the components with the smallest features are produced by means of LIGA, whereas the other parts are produced by traditional precision mechanical machining.

## Microstructures or Microcomponents

A microstructure or microcomponent defines a precision machined part that is not a sensor, actuator, or instrument. Rather, it is a device such as a microlens, micromirror, micronozzle, microneedle, etc. that acquires a useful function only

---

\*        Figure 10.2 appears as color plates after page 144.

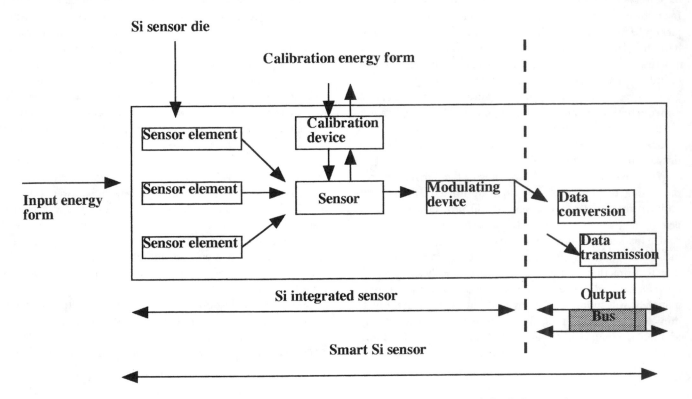

**FIGURE 10.1**   Definition of a Si sensor die, a Si sensor, a Si integrated sensor, and a smart Si sensor. (After Reference 1.)

when combined with other components. Selling prices of Si microstructures in large volumes may range from $0.25 to $100.00.[2] The latter numbers are hard to confirm since most microstructures are delivered to clients as part of large development contracts.

An example of a microstructure built by the author and his team is the scanning tunneling microscope (STM) cantilever with integrated sharp Si tip shown in Figure 10.3. The Si tip is sharpened by consecutive oxidation sharpening (see also Example 4, Combination Wet and Dry Etching, in Chapter 2). While we delivered the first STM tips to the client company under a best effort, research and development contract, eventually these components were sold at a fixed prize. Since an STM tip determines the quality of an STM picture it is a nice example of how a small micromachined component might provide a competitive edge to an STM instrument (~$20K to $60K) manufacturer. In such a case, it is important that the instrument manufacturer controls the micromachining technology or better yet has it in-house.

## Actuators

Interestingly, the definition of a sensor element as a device that converts energy from one form into another also applies to an actuator. A device that converts electrical energy into mechanical energy, by piezoelectricity, more generally is termed an 'output transducer' or an actuator, although it may be considered a sensor by definition. The appropriate use of 'sensor' or 'actuator' is not based on the physics involved but on the intent of the application.[1] If the intent is to measure a change, one will refer to a sensor; if the intent is the change or action itself, one defines an actuator.

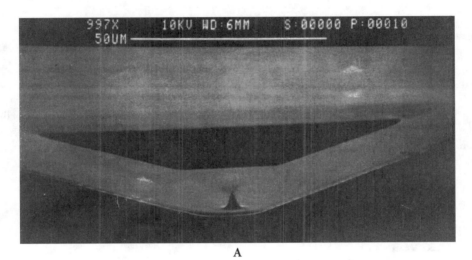

A

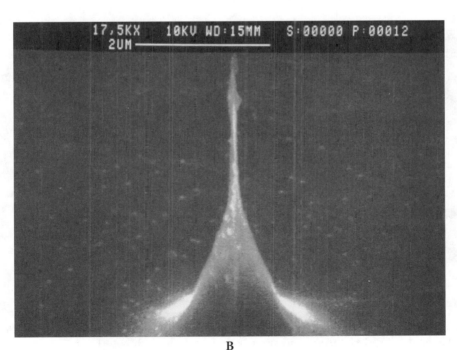

B

FIGURE 10.3   Example of microcomponent made with Si micromachining: single crystal STM tip on single crystal cantilever beam. (A) High aspect ratio Si tip. Photograph taken from a 45° angle. Tip is actually 40% taller than it appears (B). (From *Tips for STM* (TSDC, 1991). With permission.)

Actuators are components that convert energy into an appropriate action often dictated by a sensor control unit. They facilitate actions such as opening a valve, positioning a mirror, etc. Micro-actuators are actuators capable of producing micron-scale motion. Since an actuator 'acts', some power is needed, which often distinguishes an actuator from a sensor. Miniaturization of actuators is less obvious as power does not scale advantageously in the microdomain (power is proportional to the volume, i.e., l³, see Chapter 9). To induce micromotion, actuators do not need to be micron-scale themselves. Busch-Vishniac calls the latter 'micro-actuators' as opposed to microfabricated actuators which are micromachined actuators.[4]

The selling price for actuators in large quantities may range from $5.00 to $200.00. A few Si micromachined actuators have finally reached the commercial market. Examples as shown in Figure 10.4 include: Redwood Microsystems' thermopneumatic valve, Microflow Analytical's shape memory alloy microvalve, and Analog Devices' surface micromachined accelerometer (discussed in detail in Example 1, Analog Devices' Accelerometer in Chapter 5) that describes a commercially available sensor-actuator combination (Figure 5.35) with a self test involving electrostatically actuated interdigitated poly-silicon fingers.

## Microsystem or Micro-Instrument

A microsystem or micro-instrument (Figure 10.5) integrates sensors, actuators, and electronic components on a small footprint, collects and interprets data, makes decisions, and enforces actions upon its environment. As microsystems are combinations of sensors, actuators, and processing units, they are very application specific. They should be distinguished from application-specific ICs (ASICs) which can be grouped into a limited number of classes within which design and production follows well-defined and common steps.[5]

As examples of microsystems a microlaser and optical chemical analysis system are shown in Figure 10.6A and B, respectively. The latter is based on a LIGA microspectrometer already discussed in Chapter 6. None of these microsystems is commercially available. They either remain in the research phase or have been abandoned because chosen manufacturing techniques were not viable commercially. No good figures are available but we suggest that a microsystem may carry a price tag somewhere between that of large instruments ($2,000.00 to $45,000.00) and sensors ($0.50 to $100.00) — say, $200.00 to $500.00.

## Generic Sensor Classification Methods and Their Use

### Classification Methods

Jargon from different disciplines clings to sensors, making discussions difficult. A useful scheme for classifying sensors would make communication between sensor researchers from different disciplines (be it manufactures, immunologists,

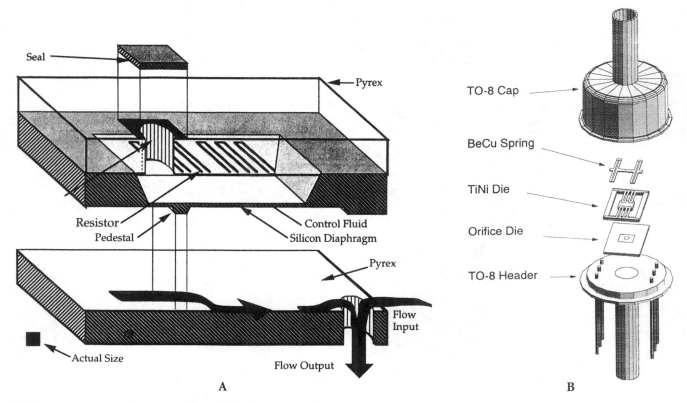

**FIGURE 10.4** Examples of commercially available actuators. (A) Thermopneumatic valve by Redwood Microsystems (Fluistor™). Normally closed microvalve featuring a liquid-filled cavity which flexes a Si diaphragm when heated, forcing the valve cover to lift off the valve seat. (From Zdeblick et al., Technical Digest 1994, Solid State Sensor and Actuator Workshop, Hilton Head Island, SC, 251–255, 1994. With permission.) (B) Microflow Analytical's shape memory alloy (SMA) valve. Normally closed microvalve. A current through the NiTi die lifts it from the orifice die. (Courtesy of Dr. D. Johnson, TiNi Corporation.)

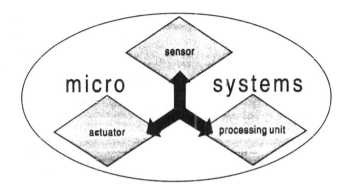

FIGURE 10.5   Scope of microsystems. (From Lebbink, G.K., Ed., *Microsystem Technology: Exploring Opportunities* (Samson Bedrijfs Informatie, 1994). With permission.)

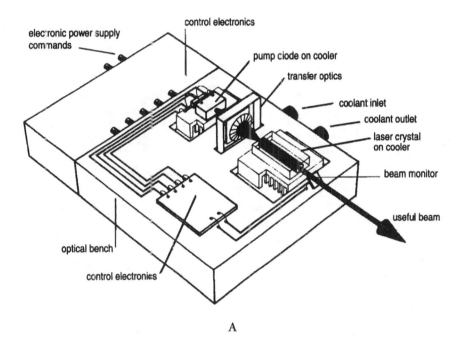

A

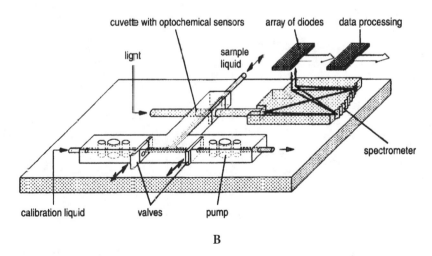

B

FIGURE 10.6   Example of microsystems: (A) microlaser. (B) Optochemical analysis system (KfK). (From Lebbink, G.K., Ed., *Microsystem Technology: Exploring Opportunities* (Samson Bedrijfs Informatie, 1994). With permission.)

mechanical engineers, or market researchers) easier. Several attempts to classify sensors have occurred. We are distilling the most important aspects here.

White[7] describes a sensor classification scheme derived from a Hitachi Research Laboratory communication. He distinguishes ten different measurement domains:

- Acoustic
- Biological
- Chemical
- Electric
- Magnetic
- Mechanical
- Optical
- Radiation
- Thermal
- Other

Middlehoek[8] follows Lion[9] and contracts this table into six signal domains:

- Radiant
- Mechanical
- Thermal
- Electrical
- Magnetic
- Chemical

For simplicity we adopted the latter list here. Based on Göpel et al.[10] and Habekotte,[11] Table 10.1 exemplifies sensing principles in these six signal domains. Since the listed energy domains can be energy input as well as energy output, a 6 × 6 matrix results. We may thus have a mechanical sensor converting mechanical energy into thermal energy, i.e., a thermomechanical sensor (e.g., a friction calorimeter) as well as a thermal sensor converting thermal energy into mechanical energy (i.e., a mechanothermal sensor, e.g., a bimetallic strip). In the literature no logic seems to be applied concerning the use of 'thermo-optical' or 'opto-thermal'. We have taken on the convention that the 'o' goes to the output energy form. Middlehoek et al. further distinguish between self-generating and modulating sensors. A sensor based on a modulating principle requires an auxiliary energy source, e.g., a fiberoptic magnetic-field sensor in which a magnetostrictive jacket is used to convert a magnetic field into an induced strain in the optical fiber; a sensor based on a self-generating principle does not, e.g., a thermocouple in which a change in temperature directly results in an electrical signal that can be measured. The difference between the two is illustrated in Figure 10.7. Since the modulating sensor needs another energy input, one also refers to such sensors as passive, whereas the self-generating sensors are active.

## Apply Specifications List to Sensor Principle Matrix

The sensor principle matrix as listed in Table 10.1 needs to be challenged with a detailed sensor specification list. While discussing the specifications required for measuring a certain measurand, jargon confusion comes about. A specification list (expanded from White's list[7]) should include static and dynamic performance characteristics such as:

Static
- Sensitivity
- Dynamic range of measurand
- Resolution
- Selectivity/Specificity
- Stability (long-term and short-term)
- Response time
- Ambient conditions (e.g., temperature, under or above water, pressure, etc.)
- Operating life
- Shelf life
- Cost, size, weight

**TABLE 10.1**  Example of Transduction Principles

| Input (Primary Signal) | Output (Secondary Signal) | | | | | |
|---|---|---|---|---|---|---|
| | Mechanical (Mechano-) | Thermal (Thermo-) | Electrical (Electro-) | Magnetic (Magneto-) | Radiant (Photo- or Radio-) | Chemical (Chemo-) |
| Mechanical | Acoustics, fluidics | Friction calorimeter, cooling effects | Piezoelectricity, piezoresistivity | Piezo magnetic effect | Photo-elasticity | |
| Thermal | Thermal expansion, bimetallic strip | | Pyroelectricity, Seebeck effect | | Radiant emission | Reaction |
| Electrical | Piezoelectricity, electrometer | Joule heating, Peltier effect | Langmuir probe | | Electroluminescence | Electrolysis |
| Magnetic | Magnetostriction, magnetometer | Thermomagnetic | Magneto resistance | | | |
| Radiant | Radiation pressure | Thermopile, bolometer | Photo-electric, Dember | | | Photoreactions |
| Chemical | Hygrometer | Calorimeter, thermal conductivity | Amperometry, flame ionization, Volta effect | Nuclear magnetic resonance | Chemiluminescence | |

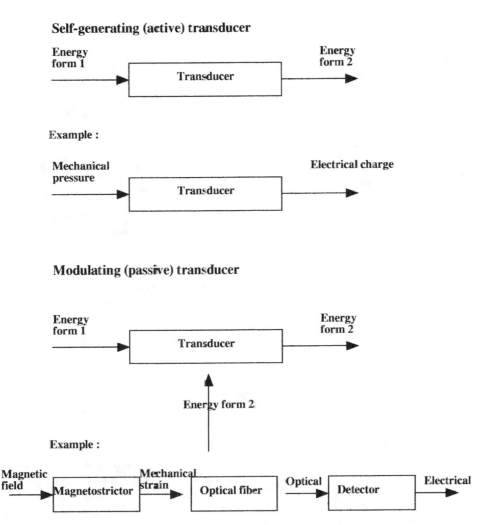

**FIGURE 10.7** Comparison of self-generating (active) transducer and modulating transducer (passive); the piezoelectric sensor represents a self-generating transducer, and the fiberoptic magnetic field sensor represents a modulating transducer. (After Reference 1.)

- Number of devices needed
- Accuracy
- Output format
- Operating voltage, current, power
- Quality, reliability, MTBF (mean time between failures)
- Hysteresis
- Threshold
- Nonlinearity
- Minimum detectable signal

Dynamic
- Dynamic error response
- Hysteresis
- Instability and drift
- Noise
- Dynamic range (operating range)

A good specification list must further detail all application-related intricacies. An example might be a pH sensor for sensing pH in the secondary flow loop of a nuclear reactor. The presence of radiation and water at high temperature and pressure will significantly reduce the sensor principles of use. Given the endlessly fragmented nature of the sensor business, it is impossible to present a generic specification list, but the example clarifies why it is indispensable.

Figure 10.8 depicts the use of the above classification scheme and specification list in sensor design. Illustrated is the search for a sensor to measure glucose in a drop of blood (chemical domain). The specifications — very low cost (<15 cents), disposable, millions of devices per month, easy to read, etc. — lead us to an electrochemical approach (electrochemical conversion option in Table 10.1). For implementation of the preferred sensor materials and manufacturing technology we challenge the electrochemical conversion option with the list of manufacturing options presented in Chapter 7 (Table 7.7) and follow the checklist in Chapter 8 (Table 8.1). Several iterations are needed in the design process. The checklist in Table 8.1 serves as the control gate. At the end of the process we decide on thick-film sensor electrodes on a disposable plastic strip in a continuous process, the option pursued by industry.

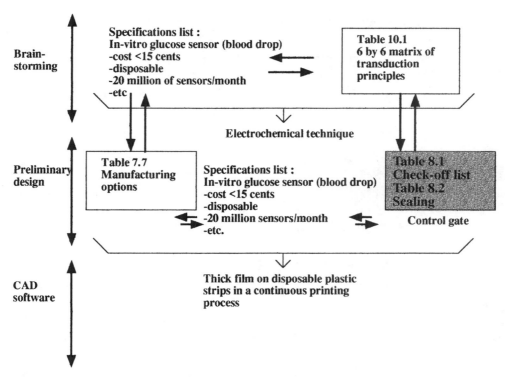

**FIGURE 10.8**  For implementation of the preferred sensor materials and manufacturing technology we challenge the electrochemical conversion option with the list of manufacturing options presented in Chapter 7 (Table 7.7) and follow the checklist in Chapter 8 (Table 8.1).

## Overall Market for Micromachines

### Introduction

Feynman presented two lectures on micromachining: 'There's Plenty of Room at the Bottom', presented at Caltech on December 16, 1959, and 'Infinitesimal Machinery', presented at the Jet Propulsion Labs on February 23, 1983.[12,13] He could not suggest a lot of use for micromachinery, but he wasn't worried; he said: "I'm fascinated but I don't know why." In this section we will investigate applications and markets for micromachined structures in an attempt to go beyond academic curiosity.

### Micromachining vs. IC Precision Engineering

In the previous chapters we emphasized how micromachining represents only one of many precision microfabrication tools available while we outlined the importance of an intelligent choice of the preferred manufacturing option for fabricating a given microdevice rather than using the available manufacturing tool. Currently, revenues of Si micromachined products represent merely 1% of revenues of those in the IC industry. A large part of those revenues stems from automotive pressure sensors[14] (see Table 10.2). Our approach to miniaturization might help generate micromachined products with a larger market share as the machining technology will tend to be more correctly matched to the application. We believe that a broader vision of micromachining will lead to the realization of other 'killer applications' in the same league as the magnetic 'read-write' heads and compact disc pickup heads. These devices constitute triumphs of precision engineering. The magnetic read-write

head alone represents a larger market ($3 billion in 1996) than all of the Si micromachining applications combined (see Table 10.2). Until recently, miniaturization applications not involving Si as a building material and/or made with more traditional machining methods were not given the attention Si micromachines received.

The current lack of commercially successful applications for Si micromachines conflicts with the observation that in the evolution of electronic products, miniaturization only follows as a very natural development initiated by a continuing market pull. With radio transceivers; optical pickup systems for compact disc-players, hard disks, and read-and-write thin-film heads, camcorders, etc., features or functions that take up too much space must be compressed; otherwise, they might not survive. Also outside the electronics area (say, in automotive sensors and biomedical devices), the market pull is towards miniaturization. We conclude that the main reason for the Si

**TABLE 10.2**  Silicon Micromachining vs. IC Technology

|                                                   | Silicon Micromachining     | IC Technology |
|---------------------------------------------------|----------------------------|---------------|
| World market 1995 revenues (million dollars)      | 1316 (±1% of IC market)    | 130,000       |
| World market in automotive sensors                | 414 (32% of total market)  | —             |
| World market in automotive pressure sensing       | 300 (80% of all automotive)| —             |

*Source:*  Based on Sullivan, F.A., *World Emerging Sensor Technologies — High Growth Markets Uncovered*, 1993.[14]

micromachining commercial shortfall so far is that micromachinists are too far removed from a detailed understanding of the applications. They are not alert enough to the market demands and are pushing micromachining technologies they control rather than solving industrial problems with the most suitable micromachining technology. A solution based on a more mature precision engineering technology is as good a solution as one based on, say, surface micromachining. Sensor applications also call for complicated, novel packaging schemes. Packaging has not been addressed much yet; it requires an intimate knowledge of the application, an aspect which is often absent in academic circles.

## A Sign Of the Times

The current predicament of Si micromachining as the potential 'GaAs of the 1980s', i.e., a superior technology with little acceptance from an entrenched and highly standardized industry, has something to do with the intellectual/philosophical climate of research and development today. Bacon (1561–1625) explained how in the advancement of knowledge one is easily misled by what he calls 'idolatry'. Two of the 'idols' he identifies point in the direction of a universal tendency to oversimplify often manifested as the assumption of more order in a given body of phenomena than actually exists, and a tendency to be struck by novelty.[15] These 'idols' perfectly apply in the case of micromachining: little commonality exists between the many different microdevices made possible by micromachining, and the striking visual aspects of Si micromachined devices easily give a sense of novelty to any observer. The appealing visual aspect of Si micromachining drives most of the interest in micromachining shown by the popular press and the efforts of some academics. The above climate contributed to overly optimistic expectations for very fast results with a market size dwarfing the IC industry and an overemphasis on Si as an answer to all miniaturization problems. Subsequently, government and industry funding sources experienced a hangover only curable by major future successes.

Bacon's idols aside, the continuing very large academic involvement in the micromachining field has some other explanations. The IC industry technologically and financially outdistanced universities and forcibly pushed the latter to explore topics requiring less startup expenses and where innovation still seemed likely. It was a perfect fit as micromachining lent itself excellently for numerous Ph.D. topics. In the 1980s, Si micromachining also became a favored filler for cleanroom overcapacity, especially in Europe where the IC industry suffered major setbacks against U.S. and Japanese competition. Today we need to face up to the consequence that the micromachining industry is minuscule compared to the IC industry but has a large number of people involved in its research and development. Middlehoek et al. estimated that in 1994 about 10,000 researchers worldwide were involved in Si sensor research and that $7.5 billion had been spent over the preceding 25 years.[16] Possibly, a more open perspective on tradeoffs between using traditional and nontraditional micromachining technology could have resulted in a much better track record. Another major shortcoming pertained to the lack of interdisciplinary teams. More than any other field, micro-engineering requires generalists rather than specialists. Multidisciplinary work from the design phase on is a must in the development of a successful sensor product. Many of the participating university groups characterize themselves as multidisciplinary but the opposite is true; the lack of multidisciplinary teams obstructed the production of more practical results. It reminds us of a statement by C.P. Snow (1905–1980), who remarked that the separate departments at a university have ceased to communicate with each other. The letters 'uni' in university, he says, lost their meaning as the institution, striving for more and more power as government funds replenish research dollars, has turned into a loose confederation of disconnected mini-states instead of an organization devoted to the joint search for knowledge and truth.

## Overall Micromachining Market Size

Feynman speculated on the use of micromachining for some esoteric devices such as microfabricated light shutters, probing devices for electronic circuits, and adjustable masks. He was also convinced about the future of truly three-dimensional circuitry in computers. In general, he doubted the wide applicability of the technology. Others are very positive and maintain that the microtechnology market will exceed that of microelectronics.[17] We tend to share the more positive outlook, with the understanding that micromachining be defined as broadly as we have in this book.

The market numbers and market projections for micromachining technology presented in Table 10.3, collected from a wide variety of marketing information sources, are very promising. Some of the data may be somewhat misleading, though, as they include big commercial successes in Si sensors such as simple photodiodes, charge coupled devices (CCDs), and Hall elements, for the most part developed before 1970 and almost exclusively based on standard Si manufacturing techniques (see Table 10.4). Unfortunately, for micromachines based on nonstandard processes, the reality is still mainly one of a technology push. Since 1970 only a few large commercial successes such as Si bulk micromachined pressure and acceleration sensors took hold. The more recently announced poly-silicon surface micromachined accelerometers and pressure sensors and projection display micro mirror arrays all constitute very large markets and will brighten the picture considerably in the next few years. Also included in this table are numbers for sensors and instrumentation not involving micromachining. Some sources include those numbers to indicate the potential for Si micromachining. Some of the numbers in Table 10.3 should be considered fiction. We can with some accuracy (see Tables 10.2 and 10.5) deduce that the 1995 Si micromachining market accounted for about $1.3 billion. But to project more than $10 billion by the year 2000 for MEMS, which would mean a tenfold growth in 5 years seems too aggressive. Today, we are hard pressed to find Si micromachined products that could realize this growth projection in just 5 years. Such high numbers become believable only if the definition of micromachining is broadened to include non-silicon technology.

**TABLE 10.3**  Summary of Market Projections for Silicon Sensors and Silicon Microstructures, as well as Total Sensor and Instrumentation Markets Where Silicon Will Have a Growing Impact

| World Market | From | 1995 | 1997 | 1999 | 2000 |
|---|---|---|---|---|---|
| For Si sensors for pressure, flow, and acceleration in all applications | Frost & Sullivan (1993) | 1,316<br>U.S. 54.4%<br>Europe 24.4%<br>Pacific Rim 21.2% | 1,858 | 2,594 | — |
| For Si sensors (pressure, flow, and acceleration) in automotive applications | Frost & Sullivan (1993) | 414 | 621 | 884.6 | — |
| For Si sensors (pressure, flow, and acceleration) and Si microstructures | NovaSensor (1990) | — | — | — | 3,665 |
| For Si micromechanics (sensors, actuators, microstructures, and microsystems) | sgt Sensor Gruppe (1992) | 2,700 | — | — | 11,900<br>5.5 sensors<br>2.3 microstructures<br>2.6 microsystems<br>1.5 microactuators |
| For smart sensors (pressure, temperature, acceleration, displacement, and proximity) and including Si | Frost & Sullivan (1993) | 778 | — | 1,491 | — |
| For all sensors | Sensor Business Digest (1995) | 8,500 | — | — | 13,100 |
| World market for all industrial and scientific instruments | Richard K. Miller & Associates (1994) | 59,000 | — | — | 82,000 |

*Note:*  Revenues expressed in millions of dollars.

**TABLE 10.4**  Commercially Available Si Sensors

| | |
|---|---|
| Radiant signals | Photodiodes, phototransistors, avalanche photodiodes, Schottky photodiode, PIN-diode array, color sensor, CCDs |
| Mechanical signals | Piezoresistive and capacitive pressure sensor, accelerometer, optical and magnetic distance sensor |
| Thermal signals | Thermopile |
| Magnetic signals | Hall element, magnetoresistive magnetic field sensor |
| Chemical signals | ISFET, hydrogen sensor |

*Source:*  Based on Middlehoek.[16]

## Forecast of the World Market for Si Micromachined Sensors

In the following world market projection, micromachined sensors are narrowly defined as three-dimensional devices made by etching micromechanical structures in Si. Since the given numbers pertain to well-defined structures and are projected from existing markets, we tend to give them more credibility.

Table 10.5 shows the revenues and revenue growth rates, along with unit shipments, of the total world market for Si micromachined sensors. This market consists almost exclusively of pressure sensors, accelerometers, and flow sensors. All other micromachining applications, such as Si bulk micromachined valves, surface micromachined projection displays, etc., do not make any dent yet in the total numbers quoted here. For chemical and biological sensors and certainly for micro-instrumentation, the sales of silicon-based technology today are minuscule. Revenue growth projected in Table 10.5 is largely driven by increased applications for Si accelerometers, although some sources predict that both pressure and flow sensors will post above-average growth rates as well.[14,18]

About 3 million Si micromachined devices were shipped worldwide in 1989. The number of units shipped reached 7.7 million in 1992. The compound annual unit growth rate (CAGR) projected for 1989–1999 is 15.9%. In 1999, nearly 38.8 million units are expected to be shipped. Average prices for pressure, accelerometers, and flow sensors are decreasing. The average price of $188.00 for a sensor in 1989 dropped to approximately $119.00 in 1992 and is expected to fall to $67.00 by 1999. These prices represent an average for the three types of Si micromachined sensors with the automotive sensors on the low end and medical and industrial applications on the high end. In the automotive arena, sensors typically should cost less than $10.00. Achieving a unit cost of less than $4.00 would open up many additional applications for automotive pressure sensors, as they would compete more effectively against traditional electromechanical approaches.

In 1989, bulk micromachined pressure sensors accounted for virtually all revenues in Si micromachined sensors. By 1990, bulk micromachined accelerometers had established a strong foothold in the market, and flow sensors were beginning to take up a small share.

TABLE 10.5   Silicon Micromachined Sensor Market: Unit Shipment and Revenue Forecasts, Worldwide 1989–1999[a]

| Year | Units (000) | Revenues ($ Million) | Revenue Growth Rate (%) |
|---|---|---|---|
| 1989 | 3026 | 570.5 | — |
| 1990 | 5741 | 744.6 | 30.5 |
| 1991 | 6844 | 851.7 | 14.4 |
| 1992 | 7760 | 925.4 | 8.6 |
| 1993 | 8816 | 977.1 | 5.6 |
| 1994 | 10,836 | 1116.2 | 14.2 |
| 1995 | 13,980 | 1316.3 | 17.9 |
| 1996 | 18,127 | 1564.4 | 18.9 |
| 1997 | 23,514 | 1857.6 | 18.7 |
| 1998 | 30,355 | 2199.8 | 18.4 |
| 1999 | 38,792 | 2593.8 | 17.9 |

[a]   Compound annual growth rate (1992–1999): 15.9%.

*Note:*   All figures are rounded.

*Source:*   Frost & Sullivan Market Intelligence, 1993.[18]

The pressure sensor market, Frost & Sullivan claim, will continue to lead, but market share dropped steadily to 80% in 1995 and is projected to drop to 71.6% by 1999.[18] The share of pressure sensors will be influenced, in the short run, by the decision of automotive manufacturers to pursue mass air flow (MAF) sensing instead of the more traditional manifold absolute pressure (MAP) sensing for engine control (see section on Automotive Sensor Needs). Accelerometers will show the strongest growth in market share, from over 10% in 1992 to 22.9% in 1999. Flow sensors will slowly increase their market share, from 3% in 1992 to 5.5% by 1999.[8]

The automotive industry has accounted for the largest share of the market for micromachined sensors, e.g., 33.7% of the market in 1989, 35.7% in 1992, and 50% by 1996. Si micromachined sensors in the automotive industry are projected to account for 55.9% of revenues by 1999.[18]

The second largest application of micromachined sensors is found in process control, while medical follows as a close third. It is expected that the share of medical applications will become larger than the process control applications, even though the automotive market will continue to increase its share.

## Why Use Micromachining Technology?

A long list of reasons confirms why micromachining presents opportunities for product innovation. Some of the most obvious reasons are summarized in Table 10.6. Usually not all reasons apply at once. For example, the small dimensions of micromachines are crucial in medical and space applications but lack importance in the automotive industry.

## Market Character

The cost of ICs can be pushed down only when the quantity of parts numbers in the millions per year. For silicon-based sensors

TABLE 10.6   Why Use Micromachining?

- Minimizing of energy and materials consumption during manufacturing
- Redundancy and arrays
- Integration with electronics, simplifying systems (e.g., single point vs. multipoint measurement)
- Reduction of power budget
- Faster devices
- Increased selectivity and sensitivity
- Wider dynamic range
- Exploitation of new effects through the breakdown of continuum theory in the microdomain
- Cost/performance advantages
- Improved reproducibility
- Improved accuracy and reliability

to succeed on the same scale as ICs have, the current manufacturing environment dictates that one must concentrate on mass consumption products such as cars, air conditioners, vacuum cleaners, toys, dishwashers, etc. However, about 50,000 types of sensors exist to measure 100 different physical and chemical parameters;[19] from many sensors only 1000 to 10,000 are needed a year. Notable exceptions are photodiodes, CCDs, Hall elements, and temperature sensors where standard product manufacture is well over a million a year and IC-type prices apply. We can conclude that fragmentation stands out as the main characteristic in the sensor market. This often excludes semiconductor solutions even though they might be technically superior. The cost/performance improvement does not always justify the application of IC technology, and hybrid approaches often are the best approach.

The dynamics of the market for micro-instrumentation is quite different from sensors, as instrumentation might bring about much higher price tags (say, hundreds of dollars rather than $10 or less). In the latter case, smaller number of parts may be offset by a higher profit margin. The situation has been compared to that of ASIC (Application Specific Integrated Circuit) manufacture, leading to the term ASIS for Application Specific Integrated System. We believe this to be a gross misnomer as little integration seems likely for most applications and hybrid approaches will again be the norm. Hybrid technology will be employed not necessarily out of cost considerations but because one monolithic manufacturing technique is unlikely to deliver the bandwidth in feature sizes and material types necessary to cover the various needs common in small instrumentation.

# Automotive Sensor Needs

## Introduction

In a complex electronic/electro-mechanical system such as the modern car, the need for effective, accurate, reliable, and low-cost sensors is pressing. The automotive industry accounts for the largest share of the market for micromachined Si sensors and that share will only increase. Si micromachined sensors will continue to penetrate automotive applications for both safety and performance features. This should fuel strong growth in the latter part of the decade.

From Table 10.7, presenting the evolution of the total world market for automotive Si based sensors, we conclude that the compound annual growth rate (CAGR) for automotive applications is 14% for the period 1989–1999. Revenues were projected to top $500 million in 1996 and reach over $884 million in 1999.

As with Si micromachined sensors in general, pressure and acceleration dominate the present automotive market. We will detail the most recent trends in those two types of automotive Si sensors. A third important automotive sensor, the so-called lambda probe, a solid state oxygen sensor, is dealt with under gas sensors. We do not mean to claim that no other opportunities for micromachining in automotive sensing exist. A list of sensor needs in the automotive and transportation industry in general is presented in Table 10.8. Figure 10.9* shows where some of these sensors are positioned in a car.

## Recent Trends in Micromachined Pressure and Acceleration Sensors for Automotive Applications

### Pressure

*Introduction and Market Perspectives*

Since 1958, Si has been used in the construction of pressure sensors, the most mature Si micromachined devices commercially available today. Costs have come down from $1000 in the 1960s to a few dollars per sensor and less than $.50 for a die. Two main varieties are used in cars: piezoresistive and capacitive. Also, two major manufacturing options are employed: bulk micromachining in single crystal Si (SCS) and surface micromachining in poly-Si. Pressure sensors are available in ranges from fractions of a psi all the way to 15,000 psi. They come as *absolute pressure sensors* (measuring with respect to vacuum, Pref = 0), *gauge pressure sensors* (measuring relative to atmospheric pressure, Pref = 1), and *differential or relative*

**TABLE 10.7** Automotive Si Micromachined Sensor Market: Worldwide Revenue Forecasts, 1989–1999[a]

| Year | Revenues ($ Million) | Revenue Growth Rate (%) |
|------|------|------|
| 1989 | 192.3 | — |
| 1990 | 312.2 | 62.4 |
| 1991 | 355.3 | 13.8 |
| 1992 | 352.9 | (0.7) |
| 1993 | 313.6 | (11.1) |
| 1994 | 343.5 | 9.5 |
| 1995 | 414.2 | 20.6 |
| 1996 | 509.1 | 22.9 |
| 1997 | 620.9 | 22.0 |
| 1998 | 747.1 | 20.3 |
| 1999 | 884.6 | 18.4 |

[a]   Compound annual growth rate (1992–1999): 14.0%.

*Source:*   Frost & Sullivan Market Intelligence, 1993.[18]

**TABLE 10.8**   Examples of Sensors and Other Micromachined Components in Use or Needed in Transportation Context

| Micromachine | Application | Status |
|------|------|------|
| Absolute pressure sensor | Manifold pressure sensing | Micromachined sensor exists and is on the market |
| Accelerometer | Air-bag release | Micromachined products are on the market |
| Air flow sensor | Air/fuel ratio control | Expensive: not micromachined |
| Alcohol sensors | Detection of alcohol level in blood of drivers | No reliable products available |
| Azide sensor | Air bag propellant | Does not exist |
| Battery sensor charge/density | Charge/density of electrolyte | Research |
| Contraband detector | Customs | No good solutions |
| Driver identification | Theft prevention | No good solutions yet |
| Gyros | Navigation | No inexpensive solutions |
| Head-up display | Information on window | Not affordable yet |
| Humidity sensors | Cabin climate | Dissatisfying products available |
| Level sensor | Oil and gas level | Available |
| Light sensors | Turn on the lights | Available |
| NOX sensor | Pollution control | Not available |
| Oil quality monitor | Engine protection | Not available |
| Oxygen Sensor (two will be required soon) | Air/gas ratio | Available, but new technology is needed for more demanding applications |
| Parking sensors | Collision avoidance | Acoustic technology available |
| Road sensors | Roughness, ice, rain | Not available at cost |
| Road/bridge condition sensors | Preventive road maintenance | Not available |
| Rain sensors | Automatic | Japanese technology available (based on piezoelectric detector) |
| Smart windows | Defog, clear view | Not available at low cost |
| Solid state cameras | Looking behind truck or car | Available |
| Spill detector | Truck accidents clearing | No good solutions available |
| Temperature sensors | Inside and outside vehicle | Available |
| Tire pressure sensor | Pressure reading on board | Under development |
| Torque sensors | Work | Under development |

*   Figure 10.9 appears as a color plate after page 144.

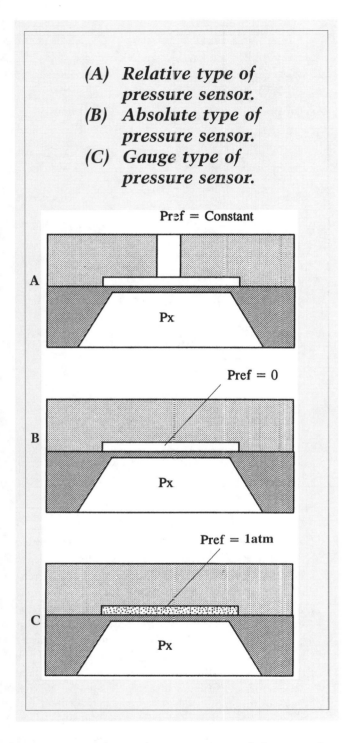

**(A) Relative type of pressure sensor.**
**(B) Absolute type of pressure sensor.**
**(C) Gauge type of pressure sensor.**

Pref = Constant

A    Px

Pref = 0

B    Px

Pref = 1atm

C    Px

*pressure sensors* (measuring one pressure relative to another, Pref = constant).

The capacitive pressure sensor is more sensitive, less temperature dependent, and holds up better in harsh environments, but, in case one relies on bulk micromachining in its manufacture, it must be carved out of a larger piece of Si to attain the same sensitivity as a piezoresistive sensor, making it more expensive. The electronics for a capacitive pressure sensor are complicated and expensive. Most of all, in the case of a capacitive sensor, the small signals necessitate integration of sensor and electronics on the same chip. Stability still poses a major problem with Si pressure sensors and, unless a resonant pressure

sensor is built (see below), accuracy typically is not better than 0.1% FS. With a resonant approach, the accuracy may be improved to 0.01% FS.

Table 10.9 reviews the different pressure sensor applications in a car. We point out where Si micromachining might be competitive with traditional electromechanical designs and project the expected revenues for these sensors up until 1998.

From the different automotive systems reviewed in Table 10.9, engine management takes up the lion's share of pressure sensor technology incorporated, i.e., up to 50%. From the same table we learn that besides the engine oil pressure switch there are at least six types of pressure sensors involved in the power-train system (engine + transmission): manifold absolute pressure sensors (MAP), barometric pressure sensors (BARO), exhaust gas recirculation sensors (EGR), fuel evaporative pressure sensors (FP), transmission fluid pressure sensors (TP), and in-cylinder pressure sensors (CP). In several of those applications, Si sensor technology will play an increasingly important role.

A prime example of the growth of Si pressure sensing devices in this group is the development of the MAP sensor used to control the air-fuel ratio in the engine management system. MAP sensors were the first high-volume Si sensors in the automotive industry, providing engine control by an inferred measurement of the quantity of air being drawn into the engine. This measurement can be combined with other parameters to approximate engine load. In Figure 10.10* we show the General Motors MAP sensor. Notice how the packaging dwarfs the sensor.

The first pressure sensors applied to engine control were introduced in 1973 by General Motors' Delco electronics division for use on the Cadillac Seville. The sensors were an adaptation of a proven electromechanical device originally developed for aerospace applications. In 1979 the first Si pressure sensors were introduced in Delco's MAP and barometric pressure monitoring systems. The Ford MAP sensor, known as SCAP (Si capacitance absolute pressure), was the result of a joint design effort in the early 1980s by Ford and Motorola. Since that time, Si micromachined sensors have become the *de facto* standard for use in engine management systems in the U.S.

Although MAP feeds into a major market, the application is on the decline. Due to a desire for a more direct and accurate measurement technique, MAP sensors are being replaced by mass air flow (MAF) sensors. The major drawback to MAF sensors at this point is their high cost (about $30.00), a significantly higher price than for MAP sensors ($10.00). Several micromachining options are being looked at in an attempt to reduce the cost of a MAF sensor, but none have led to a promising approach yet. In MAF-based systems, the MAF sensor replaces the MAP sensor in conjunction with a barometric sensor needed to correct for altitude. These barometric pressure sensors (BARO) often are identical to MAP sensors. Obviously, losing the MAP sensor application does not mean losing all the pressure sensor applications in engine control. From Table 10.9, we can conclude that besides BARO, EGR, FP, climate control pressure sensors (CLCS), and tire pressure monitoring systems (TPMS) all are excellent opportunities for silicon-based

---

* Figure 10.10 appears as a color plate after page 144.

**TABLE 10.9**   Current and Potential Pressure Sensor Applications

| System | Application | Projected 1998 North America Market (M $) | Use of Silicon Technology | Measuring Range (kPa) | Year Expected or Introduced |
|---|---|---|---|---|---|
| Engine management and transmission, i.e., power train | Engine oil pressure switch ($3.00) and sensor ($6.00) | 41 | Uncertain | | Current, mature application |
| | Manifold absolute pressure ($10.00) (MAP) | 34 | Si: Yes, but might be replaced by MAF | 0–105 | Current application |
| | Barometric pressure sensor (BARO) ($10.00) | 107 | Si: Yes | 50–105 | Current application |
| | Exhaust gas recirculation (EGR) ($10.00) | 36 | Si: Yes | 0–105 | Required by CARB,[a] started in 1989 |
| | Fuel pressure (FP) ($10.00) | 107 | Si: Yes | 0–105 | Required by CARB 1994 |
| | Transmission oil pressure (TP) ($4.00) | 54 | Uncertain | 0–2000 | Current |
| | Cylinder pressure (CP) (?) | ? | Si: No | 10,000 | 1997 or later |
| Suspension/braking and traction control | ABS accumulator pressure ($4.00) | 52 | Uncertain | 17,000 to 20,000 | Current |
| | Tire pressure monitoring systems (TPMS) (?) | 500 | Si: Yes | 500 | 1994–1995 |
| | Active suspension hydraulics ($8.00) | 15.24 | Si: Yes | 20,000 | 1994–1995 |
| In-cabin passenger comfort | Lumbar seat support pressure ($10.00) | — | Si: Yes | | Current |
| | Climate control pressure sensor (CLCS) ($10.00) | 20.9 | Si: Yes | 50–105 | Current |

[a]   CARB = California Air Resource Board.

solutions. Modifications to packaging make similar technology work for all of these types of pressure sensors. A higher accuracy is required for EGR, FP, and BARO than for TPMS and CLCS.

### Pressure Sensing in Automotive Applications: Technology Update

A significant number of different technologies are employed in various automotive pressure sensors. In the following list we attempt to place typical technologies in the stage of their respective product life cycles.

- Mechanical (in decline): Increasing demands for additional information and reliability will lead to the elimination of these devices in all but the least demanding, most cost-sensitive applications.
- Potentiometric/diaphragm (in decline): Demands for increased reliability and information are leading to the elimination of this technology.
- Piezoresistive/capacitive (reaching maturity): Accepted as a reliable, cost-effective product.
- Silicon bulk micromachined piezoresistive/capacitive (growth stage): Miniaturization and signal conditioning potential are likely to see significantly more opportunity

as cost/unit lowers. The preferred technology today is bulk Si micromachining of piezoresistive pressure sensors.

- Poly-silicon surface micromachined piezoresistive sensors with an initial market introduction by SSI Technologies, Inc., in 1995:[21] The poly-silicon surface micromachined pressure sensors are more compatible with regular IC processes than bulk micromachined sensors and one assumes automotive pressure sensors of the future to follow that avenue. For applications where accuracy is not a major factor, poly-silicon represents the least costly solution today. The demands on the long-term accuracy and repeatability and the difficulty in controlling poly-silicon properties in different laboratories might be factors preventing poly-silicon from being implemented quickly. Interesting to note is that Japanese companies are moving slower into poly-silicon applications than their American counterparts as they further explore long-term stability issues a poly-silicon solution might entail.
- Hall effect/diaphragm: Prototype stage (new technology).
- Fiberoptic (infancy): New technology not yet accepted. Opportunity will correspond to the evolution in optical vehicle wiring systems.

- Resonant beam (research and development stage): In resonant beam pressure sensors the principle of operation is simply to measure the change in resonant frequency of a micromachined Si beam as the pressure exerted on the sensor's diaphragm changes.

Because of the promise held by resonant beam applications we detail this technology trend a bit further. Sensors relying on vibrating structures for the measurement of physical properties such as pressure and acceleration have a number of advantages over piezoresistive, piezoelectric, and capacitive methods compared in Table 10.10. All the sensors listed in this table must be used with high precision amplifiers and trimmed analog circuits to accurately sense and condition small changes in the value of electronic components. These sensitive circuits and small signals can be highly susceptible to fluctuations in temperature, electrical noise, mechanical and electrical drift, temperature and pressure cycling hysteresis, ionizing radiation, and magnetic fields. These factors and others result in measurement inaccuracies, effectively degrading overall precision and resolution. As a result, best-case production performance of solid state sensors has generally been limited to a long-term accuracy of about 0.1% or 1000 ppm. Resonating devices, on the other hand, are primarily affected by the mechanical characteristics of the structure, not by its electrical behavior. In addition, since the system itself only measures frequency, the interface to a digital system is vastly simplified. In practice, resonant sensors can attain measurement accuracies at least 10 times higher than conventional sensing methods. The basic building block for a resonant beam structure attached and embedded in a pressure-sensitive membrane is shown in Figure 10.11. The specific structure shown is one by Lucas NovaSensor.[22]

Although this simple device demonstrates that with a resonant sensing scheme a 0.01% or 100 ppm accuracy is attainable, the fabrication process is far from ideal. The most important shortcoming of the pictured device is that the thickness of the pressure membrane is controlled by timing rather than an etch stop, making the design less attractive for manufacturing. Moreover, this manufacturing process involves a time-consuming etch-back step. We suggest that employing silicon-on-insulator (SOI) technology can improve upon this design.

TABLE 10.10  Pressure Sensor Technology Comparison Table

| Parameter | Piezoelectric | Piezoresistive | Capacitive |
|---|---|---|---|
| Self-generating | Yes | No | No |
| Impedance | High | Low | High |
| Signal level | High | Low | Moderate |
| High temperature | Yes | No | No |
| Accuracy (F.S.) | ±1% | ±1% | ±0.2% |
| Static calibration | No | Yes | Yes |
| Rugged, high sensitivity | Yes | No | Yes |
| Damped designs | No | Yes | Yes |
| Cost | High | Low | High |
| Integrated electronics required | Yes | No | Yes |

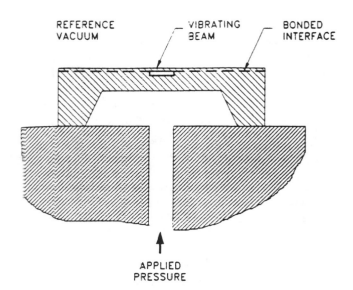

FIGURE 10.11  Cross-sectional sketch of a resonant beam pressure sensor chip, showing diaphragm, outer frame, and Si fusion-bonded beam suspended over a shallow cavity (Lucas NovaSensor). (From Petersen et al., 6th Intl. Conf. on Solid-State Sensors and Actuators, San Francisco, CA, 664–667, 1991. With permission.)

If resonant devices could be made at a cost, say, below $5.00, a major market opportunity would open up. The resonant beam measuring principle is generic and can be used for a wide range of pressure sensors but also is the preferred technology for acceleration and rate sensing (see below).

*Comparing Micromachining Approaches for Pressure Sensing*

The impetus for obtaining higher accuracy in automotive sensors hinges on energy efficiency and/or regulatory mandates, particularly for pressure sensors. This may require a review of the transducing method itself, currently dominated by piezoresistive devices. The significantly higher sensitivity of both capacitive and resonant devices (order of magnitude) as well as their inherent digital compatibility may propel them to the foreground for high accuracy and very low-pressure applications. We believe that this need for increased accuracy favors a better known, crystalline material such as epi-silicon as the mechanical material rather than a poly-silicon approach.

Poly-silicon now holds the cost advantage over an SOI approach. But the long-term poly-silicon material stability still requires demonstrating and the technology will not readily lead to a high accuracy method. Reproducing the same mechanical material properties for poly-silicon in different laboratories has been a major challenge. At one point MEMS engineers were talking about University of California, Berkeley, poly-silicon and University of Wisconsin poly-Si!

Like SOI, bulk micromachining constitutes a more secure option to meet the accuracy demands of future pressure sensor generations but often involves too many glass and/or Si wafers stacked up in one device to allow for an economic solution. Moreover, bulk micromachining is a more expensive technology and less CMOS compatible than poly-Si. SOI might be able to compete better as clever designs could obviate the complicated multiple wafer solutions that characterize today's bulk micro-machining stacks for pressure, acceleration, and gyros (a 3- to 5-wafer stack design is not uncommon). SOI technology will also become more and more available and less expensive as it becomes more commonly used in the IC industry. We believe that the SOI approach will become a competitive technology for low-cost, high-accuracy sensors without the reproducibility and long-term stability issues of the poly-silicon material. More-over, a generic approach based on resonant beams for pressure, acceleration, and rate sensing is rendered feasible this way.

We can summarize that for now an inexpensive solution for a pressure sensor with moderate accuracy will be based on poly-Si. In the long run, a resonant SOI approach will be more suited for a generic strategy to produce higher accuracy devices.

### Generic Design Features

The approach for pressure sensing and, to some extent, for all Si micromachined sensors for large industrial market applications where cost is the main issue must incorporate as many of the following attributes as possible:

- CMOS compatibility because of the increasing demand for digital-compatible, highly integrated Si smart sensors
- Six-inch compatible because of IC trend/availability
- As little fixturing as possible
- Al and IC-line compatible etchants
- Generic, making it easy to implement a different sensing range or different package or even a different sensor type
- Highest accuracy potential (favors capacitative and resonant devices)
- Wafer-level isolation of electrical connections, making it easier to use in harsh media (removing sensor metallization from the media itself)
- Dry etching whenever possible
- Single-sided processing if feasible

### Accelerometers

### Introduction and Market Perspectives

Accelerometers signified the next big Si micromachining entry in the commercial world after the Si pressure sensor. Over the years a variety of accelerometers based on different physical detection schemes such as piezoelectric, piezoresistive (thin- and thick-film), capacitive, tunneling, and resonant members have been attempted in a wide variety of manufacturing schemes. Today, poly-silicon surface micromachined capacitive and single crystal Si bulk micromachined accelerometers quickly are taking over a large share of the automotive acceler-ation market. A major difference between accelerometers and pressure sensors is that less diverse types of accelerometers are needed in the car. Pressure sensors require various ranges and many different packages to accommodate the sensors in their unique environments. Accelerometers, on the other hand, require little difference in packaging; two main types (i.e., high *g* and low *g*) cover all the needs. This explains why the first surface micromachined commercial sensor was an accelerometer rather than a pressure sensor.

In 1991, the Intermodal Surface Transportation Efficiency Act (ISTEA) was signed into law. ISTEA airbag requirements ensured that by 1998 100% of the new vehicle fleet were equipped with airbags. The average U.S. car production from 1989 to 2000 is expected to be 8 million. Chrysler is expected to produce 1.25 million cars, Ford 2.5 million, and General Motors 4.25 million. In 1995, Chrysler already put airbags in 100% of their cars, Ford 40%, and GM 20%.

Until recently airbags sensors used in the U.S. were electrome-chanical types supplied by Breed and TRW. Three electrome-chanical accelerometers discriminated between zones according to the severity of a vehicle crash. Two sensors were used in the crush zone, usually on frame rails behind the front bumpers or on the lower portion of radiator supports. These sensors deter-mined the severity of impact and activation of the airbag. Another sensor was located outside the crush zone, either in or near the passenger compartment. This sensor, often known as the 'safing' sensor, prevents deployment of the airbag should someone inadvertently impact (or tamper with) the front sensors.

Micromachined accelerometers show the potential to com-pletely overtake these electromechanical devices. Si microma-chined accelerometers make single-point sensing possible. Single-point sensing requires only one accelerometer located in the passenger compartment, reducing the number of sensors and associated wiring needed in the system. In addition to reducing the number of sensors required in each system, Si accelerometers are expected to cost approximately $8.00 per unit, significantly less than electromechanical accelerometers at approximately $15.00 per unit. Single-point systems, at the cur-rent level of design, still employ an electromechanical safing sensor. Thus, the savings in dollars from sensors alone, not considering wiring or assembly steps, amount to $22.00 per vehicle, signifying a large incentive for the vehicle manufacturer.

The development of the Si micromachined accelerometer market follows a development cycle similar to the microma-chined pressure sensor market. In the early stages, pressure devices cost on the order of $60.00 to $110.00 depending on the volumes required. By the late 1980s, the price had dropped to less than $10.00 each and every car contained at least one. This same exponential decrease in cost is occurring now with micro-machined accelerometers.

The early bulk micromachined Si accelerometers lacked damping and overload protection until Lucas NovaSensor intro-duced a piezoresistive accelerometer combining a micro-actuator for reliable self-testing and overload protection on the same chip (see Figure 10.2).*[23] In 1991, Analog Devices introduced the first commercial product to use poly-silicon surface micromachining, the so-called AXDL-50, a capacitive accelerometer with electro-static actuator built-in for self-testing (see Figure 5.35).

The operating range of airbag accelerometers is 50 *g*, while the operating range of suspension and anti-lock braking system accelerometers is 0 to 2 *g*, representing the level of noise in the Si accelerometer for many technologies.

Silicon accelerometers in suspension systems face more hur-dles than the Si accelerometers in airbag systems: Low *g* is more

---

*    Figure 10.2 appears as a color plate after page 144.

difficult (signal/noise limit!), signal conditioning is required with each remotely mounted package (one suspension sensor per wheel), and the sensor is exposed to the harsh automotive environment, i.e., more complex packaging. These factors contribute to the estimated cost of $15.00, significantly higher than the airbag accelerometer.

Low *g* accelerometers are also used on antilock braking systems (ABS) featuring traction control. No federal law mandates ABS/traction control, and due to the cost constraints only high-end and midsize cars are expected to utilize the system.

Market projection for automotive accelerometers are shown in Table 10.11.[24] In this table airbags are the largest application. Other market studies for accelerometers in suspension applications for North America show higher numbers than those in Table 10.11, with 1998 revenues at $64 million and a compound annual growth rate for the period 1991 to 1998 forecast at 135.2%.[14] But even the lower prognosis for ABS accelerometers in Table 10.11 seems too high. Indeed, more recent market figures indicate that the market for low *g* in ABS remains small and less secure, as one can invent around the need for an accelerometers in this application.

In 1995 Analog Devices introduced a 5 *g* accelerometer for low *g* automotive applications, pushing the poly-silicon surface micromachining technology they pioneered with their 50-*g* device to its limits.

### Acceleration Measurement Methods: Technology Update

The following briefly describes the major accelerometer technologies available. The selection of accelerometer type depends on such factors as weight, sensitivity, frequency response, dynamic range, temperature requirements, and price. Table 10.12 compares the primary sensing techniques used in accelerometers for automotive applications.

**Piezoelectric Accelerometers** The simplest piezoelectric (PE) accelerometer consists of a disc made of piezoelectric material such as quartz, lead zirconate titanate (PZT), or PVDF piezo-film, electrically contacted by two electrodes on either side (see also piezomaterials in Chapter 9). The electrodes detect the resultant strain in the piezoelectric when movement exerts a force on the seismic mass attached to the piezoelectric. Such PE accelerometers are self-generating, of small size, and cover a wide frequency range (greater than 30,000 Hz). They also

handle accelerations from a fraction of 1 *g* up to 200,000 *g* for shock measurement. Temperature variations affect the output of PE crystals, with quartz and certain newer PE ceramics being less temperature-sensitive than barium titanate or lead zirconate titanate ceramics. The raw, unconditioned output of a PE accelerometer shows very high impedance, and, while it can be treated as either voltage or a charge, most sensors must be used with special charge amplifiers. PE crystals also tend to pick up noise in a noisy electrical environment. Most piezoelectric materials available exhibit very large temperature coefficients over the full automotive temperature range. Acceptable performance can be obtained over 0 to 85°C, making this technology suitable for location in the passenger compartment with its more benign environment.

In addition to their self-generating capacity, small size, and wide frequency range, the piezoelectric accelerometer has the advantage of being rugged, stable, and lacking moving parts prone to breaking down. They are inexpensive and can be mounted in any orientation. However, PE accelerometers are not suitable for very low-frequency and/or low-level accelerations (under 10 Hz) and reportedly are not as effective as the wirestrain-gauge for measuring long-duration shocks (exceeding 10 msec).

The trend leans toward making smaller and lighter PE accelerometers capable of maintaining a low frequency response. Because of their reported higher sensitivity, use of ceramics outranks the use of quartz. A trend toward integrating the electronics into the package prevails. It should be remembered that many applications require DC or very low frequency, excluding the application of piezoelectric devices.

Atochem Sensors (recently taken over by AMP, Inc.), a company selling PVDF piezo-film, reportedly has a suspension accelerometer in development.[25] The piezopolymer is very inexpensive, but for high accuracy devices we are doubtful that this technology could be better than a resonant SOI approach.

**Piezoresistive Accelerometers** Piezoresistive (PR) accelerometers are built by either bonding or diffusing stress-sensitive resistors onto or into the springs suspending the inertial mass at locations of peak stress. This easy-to-implement approach provides DC response, results in low impedance resistors, and works well over a relatively wide temperature range (−50 to +150°C). The drawback of piezoresistive accelerometers

**TABLE 10.11** Market for Automotive Accelerometers ($MM) in the U.S.

| Application | 1991 | 1992 | 1993 | 1994 | 1995 | 1996 | 1997 | 1998 | 1999 | 2000 |
|---|---|---|---|---|---|---|---|---|---|---|
| Airbag | | | | | | | | | | |
| Cars | 20 | 42 | 62 | 58 | 96 | 80 | 80 | 80 | 80 | 80 |
| Vans | 1 | 13 | 27 | 30 | 55 | 47 | 49 | 51 | 53 | 55 |
| Total | 21 | 55 | 89 | 88 | 151 | 127 | 129 | 131 | 133 | 135 |
| | | | | | | | | | | |
| ABS | 0 | 8 | 8 | 16 | 24 | 31 | 50 | 52 | 54 | 56 |
| Suspension | 0 | 6 | 13 | 18 | 26 | 19 | 19 | 20 | 21 | 22 |
| Total | 0 | 14 | 21 | 34 | 50 | 50 | 69 | 72 | 75 | 78 |

*Source:* Based on MIRC 1990.[24]

**TABLE 10.12**  Accelerometer Technology Comparison Table

| Parameter | Piezoelectric | Piezoresistive | Capacitive |
|---|---|---|---|
| DC response | No | Yes | Yes |
| Self-generating | Yes | No | No |
| Impedance | High | Low | High |
| Signal level | High | Low | Moderate |
| High temperature | Yes | No | No |
| Accuracy (F.S.) | ±1% | Bond. = ±2% | ±0.2% |
| | | Diff. = ±1% | |
| Static calibration | No | Yes | Yes |
| Rugged, high sensitivity | Yes | No | Yes |
| Damped designs | No | Yes | Yes |
| Cost | High | Low | High |
| Electronics required | Yes | No | Yes |

is moderate output signal (100 mV full scale for a 10-V bridge excitation) and relatively complex temperature sensitivity and zero terms. They are frequently used in automotive applications such as active-suspension control where good low-frequency performance is required.

The first Si accelerometer, demonstrated in 1976, consisted of a single cantilever carrying piezoresistors near its root. The device was fragile and required the inclusion of a liquid-filled cell for damping. Many improved designs have appeared since. As their output is based on resistance changes independent of frequency, typical microaccelerometers will operate at DC and can be used to measure constant acceleration. This type of single crystal Si micromachined accelerometer has potential for low cost and high reproducibility at large volumes. A further feature endearing them to automotive engineering is their ability to include signal processing and communication functions within the sensor package at little extra cost. Thick-film piezoresistive accelerometers might yet be cheaper. Thick-film piezoresistive accelerometers were introduced as a means to miniaturize circuits without incurring the expense associated with fabrication in silicon. The piezoresistive properties of thick-film inks can be used to form strain sensors. This approach enabled the production of a number of sensors, which carried over into producing accelerometers by 1993. One approach uses an alumina cantilever beam with readout provided by a pair of thick-film piezoresistors printed on each side of the beam optimizing the signal.

**Variable-Capacitance Accelerometers**  In a variable capacitive (VC) accelerometer such as Analog Devices' AXDL-50 (Figure 5.35) the seismic mass is attached to one plate of a parallel plate capacitor. Acceleration changes the distance between the plates, resulting in capacitance changes. If the capacitor is part of a resonant RC network, the change in frequency will be a measure of acceleration.

Advantages of capacitive acceleration sensing include low-temperature sensitivity and simple signal pickup, few moving parts, easy interpretation, reliability, noncontacting design, high sensitivity, and insensitivity to magnetic fields. Moreover, self-testing is easily implemented with a capacitive design, involving

only a few extra processing steps. Disadvantages are limited frequency range, limited accuracy, difficulty in calibrating, and the need for conductive surfaces and a power supply (in contrast to the PE accelerometers). The sensors, when using bulk micromachining manufacturing techniques, typically are larger than piezoresistive devices. The signal processing electronics are difficult because very small parasitics can adversely affect linearity. Micromachined capacitive accelerometers require complex signal conditioning of very low level signals. So low, in fact, that the electronics should be integrated on the sensor for optimum performance. This requires the integration of several technologies on the same chip, inevitably leading to compromises in yield, cost, or the performance characteristics of the sensor itself.

**Resonant Beam Technology**  Electrical readout in an accelerometer may be performed using piezoresistors located in the flexure area, by measuring the change in capacitance of the moving members with respect to the fixed member(s), or by the measurement of frequency change of a resonant beam. This latter method is the least developed, but, as we learned in the Pressure Sensor section, the resonant beam method shows very good potential, as the S/N ratio is better than for the other techniques. Table 10.13 shows major tradeoffs of the three approaches. A common advantage of resonant beam and piezoresistive is that they do not depend on measurements relative to 'fixed' members, as does the capacitive approach. The small signal magnitude ($\Delta C$) of a capacitive sensor means that on-chip electronics are mandatory to avoid errors due to stray capacitance variations. The resonant beam approach encompasses both excitation and readout of the beam resonant frequency.

**TABLE 10.13**  Electrical Readout Approaches

| | Gauge Factor | Signal Magnitude | Temperature Coefficient of Sensitivity |
|---|---|---|---|
| Piezoresistive | $\Delta R/R$ = 2 to 3% | $\Delta R = 100 \ \Omega$ | 2000 ppm/°C |
| Capacitive | $\Delta C/C$ = 10% | $\Delta C = 0.1$ pF | 100 ppm/°C |
| Resonant beam | $\Delta f/f$ = 10% | $\Delta f = 20$ kHz | 100 ppm/°C |

**Tunneling Accelerometer**  New high sensitivity microsensors have been developed using high-resolution position sensors based on electron tunneling. Prototype accelerometers built this way, with a proof mass of the order of 100 mg, had a noise resolution of $10^{-7} \ g/\sqrt{\text{Hz}}$ between 4 and 70 Hz and $6 \times 10^{-7} \ g/\sqrt{\text{Hz}}$ at 400 Hz in a damped situation.[26] These types of sensitivities are not needed for the automotive application.

*Comparing Manufacturing Options for Accelerometers*

Obtaining sufficient sensitivity poses a potential problem when using Si to manufacture low-$g$ accelerometers. Process control and handling considerations limit the minimum flexure thickness for suspending the inertial mass to about 15 μm and, given the relatively low density of Si, obtaining deflection at 1 or 2 $g$ without making the proof mass large becomes difficult.

Realizing a large proof mass directly affects chip size and thus cost, presenting the developer with an unattractive option. To make a large proof mass is especially challenging for poly-silicon accelerometers. The proof mass of a surface micromachined accelerometer may be 0.1 μg whereas a typical value for a bulk micromachined device is 0.1 mg, 1000 times larger. Then again, the formation of a mesa structure (proof mass) from a single crystal Si is more complicated due to rapid and irregular etching of convex corners by orientation-dependent etchants such as EDP (ethylene diamine pyrocatechol) and KOH. Corner compensation schemes must be invoked to avoid irregular corners (see Chapter 4). Despite some of the misgivings expressed above, a surface micromachined ±5 g capacitive accelerometer announced in 1995 by Analog Devices demonstrates that low g poly-silicon sensors and cost-effective integration of all the electronics is possible.

Critics of the capacitive poly-silicon accelerometer still maintain that the technology is not proven, that the integration of electronics is too complex, and that a bulk micromachined sensor requires only 12 mask levels, whereas the poly-silicon device requires 28 mask levels. Along this line, major Japanese companies have chosen single crystal solutions for building accelerometers to avoid the difficulties associated with reproducing poly-silicon material properties and the possible unknown long-term instabilities.

The pros and cons of surface micromachining (see also Chapter 5) can then be summarized as follows:

Pros:
- A surface micromachined device has a built-in support, resulting in cost effectiveness.
- Surface micromachining only requires one-sided processing compatible with conventional IC processes.
- Devices made by surface micromachining are smaller in size, have higher yields, and accommodate more on-chip circuitry.

Cons:
- The out-of-plane dimensions in surface micromachining in general is limited to ~2 μm, hence the light mass of the proof mass resulting in a low sensor sensitivity. To increase the mass, a heavy metal deposit may be added to the proof mass region (e.g., Au), as demonstrated in the LIGA technique.
- The suspended Si structure is more compliant (due to the high ratio of length-to-thickness of the device structure) and hence has a tendency to stick to the support substrate, especially in overdrive situations.

As is clear from the section on pressure sensing, we believe that Si on insulator (SOI) surface micromachining combines the best features of surface micromachining with the best features of single crystalline Si. Using dry etching to etch the proof mass from the epi-material avoids corner compensation issues; the epi-material ensures larger mass and single crystal performance. This approach might overtake both surface and bulk micromachining as SOI wafers become less expensive and more available. A generic design of a resonant beam structure in an SOI wafer may be possible. This design could constitute the basis of a wide family of pressure, acceleration, and rate sensors. In the latter case, a double-ended tuning fork (DETF) (Figure 10.12) is dry etched out of the epi-silicon layer of the SOI wafer and underetched by wet etching. An inexpensive, solid rate, micromachined gyro is the holy grail for micromachinists. An excellent recent review on the status of micromachining in inertial sensors, including DEFT-based gyros, was written recently by Smit.[27]

Also, LIGA accelerometers have been produced but only in a research and development setting.[28] In this case sensor elements are formed by selective electroplated metal (often Ni) via a photolithographically delineated polymer mold. The X-ray lithography adds considerably to the cost, making it difficult to justify LIGA for a price-sensitive automotive application (see Chapter 6).

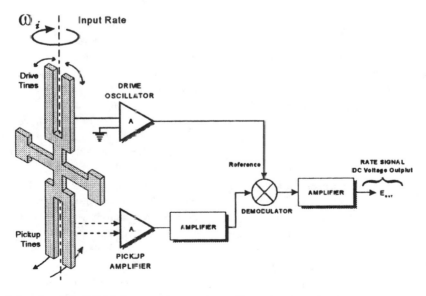

**FIGURE 10.12** Double-ended tuning fork (DETF) for a rate sensor. The DETF structure may be dry etched in the epilayer of an SOI wafer and released by wet etching the oxide layer underneath.[29]

## Conclusions

In the automotive world, Si micromachining is already well entrenched, with both surface and bulk micromachined sensors for pressure and acceleration continuing to take larger market shares away from traditional electromechanical approaches. This trend will become even more pronounced as several new technology tools are now enabling the yet faster development of the next generation Si pressure and accelerometer sensors. These tools include Si fusion bonding (SFB) and all types of CAD/CAM software (see also Chapter 8).[23]

The next challenge in this field for micromachinists is developing an inexpensive Si gyro, reducing the number of layers in a sensor stack, and simplifying packaging by integrating it into the IC process. We suggest that fewer layers and simpler batch packaging will be afforded by using an SOI approach.

# Micromachining in Chemical Sensors and Instrumentation for Biomedical Applications

## Introduction

Biomedical sensors and instrumentation represent a huge opportunity for micromachining in the broad sense of the word, i.e., for micromachining encompassing traditional and nontraditional miniaturization methods. Success in this field is not guaranteed when only considering Si with or without integrated electronics as the machining substrate. The Si avenue has been pursued for more than 25 years with little success to report. Innovation in biomedical sensors has come mainly from the invention of new chemical detection schemes rather than from clever new micromachining methods. Actually, in biomedical sensors, a reversal toward hybrid and continuous manufacturing methods took place rather than moving toward more integration in Si embodiments. In micro-instrumentation, one envisions that analyses will be performed significantly faster, only minute samples will be required, and dead volume in connections eliminated. Despite the many claims, contributions from micromachining are mostly at the component level rather than at the system level. Many machining technologies are being experimented with. Unfortunately, systematic analysis of the different machining options the way we promote in this book has not been applied. When doing such a systematic analysis one comes to the conclusion that micromachining clearly counts as a fragmented set of supporting technologies rather than a monolithic new industry.

As examples of micromachining in biosensors we will detail electrochemical sensors, immunosensors, and lithography in protein patterning. All three play an important role in medical diagnostics applications. These three examples are preceded by a historical perspective on the use of miniaturized biosensors in tabletop and handheld equipment. In Table 10.14 we show the 1994 U.S. market for *in vitro* diagnostics, an application area where micromachined biosensors, if applied wisely, may claim a substantial market.[30]

**TABLE 10.14**   U.S. *In Vitro* Diagnostic Market Overview

| Market Segment | 1994 Estimated Sales (U.S. Sales $ million) |
|---|---|
| Clinical chemistry | 2940 |
| Immunochemistry | 2100 |
| Infectious disease testing | 670 |
| Hematology/cellular analysis | 500 |
| Other[a] | 615 |
| Total | 6825 |

[a]  Includes toxicology, cytology, histopathology, genetic testing, and serology.

*Source:*   Based on the BBI Newsletter.[30]

Concerning miniaturization in biomedical equipment, we will first discuss some of the generic challenges and then review progress in subminiaturization of a few important example instruments: microphysiometers, electrophoresis units, polymerase chain reaction (PCR) chambers, and flow cytometry.

Whereas in the automotive application the market for micromachined products is mature enough to warrant projection of realistic revenue numbers, predicted revenues for micromachined sensors and instrumentation for biomedical applications mostly refer to speculation as commercial successes still remain short of realization. Accordingly, we have opted to discuss general application issues rather than second-guess the size of the market for micromachined sensors and instruments.

## Historical Perspective of Biosensors in Tabletop and Handheld Equipment

During the past 25 years the commercialization of chemical sensors based on microfabrication failed to keep pace with earlier projections. We believe that an overemphasis on integration of chemical sensors with electronic functions on Si hampered the development. We suggest speeding up the progress by applying micromachining tools more correctly. Progress in biosensors based almost exclusively on classical manufacturing techniques has been significant. The fully automated Kodak Ektachem 400 is a tabletop clinical blood analyzer based on planar electrochemical sensor slides, and has been on the market for some time. The Ektachem as well as the Seralyzer from the Ames Division of Miles Laboratory, the Refletron Plus system from Boehringer Mannheim (with a 16-test menu), the Analyst from Dupont, the Nucleus from Nova Biomedical, and other similar instruments cover a wide range of blood chemistries.[31,32] However, none of these products involve Si micromachining. The progress enabling their development mainly was based on the development of solid-state stabilized reagents for so-called 'dry electrochemistry',[33] on miniaturized electrochemical sensors, and to a minor extent on advances in thin-film technology.

It was anticipated that micromachining would play a more prominent role in the development of sensors for the smaller, next-generation, handheld and portable, diagnostic instruments.

In reality, the micromachining and Si sensor content in the more advanced, handheld and portable, biomedical products is surprisingly limited. Some such products for analysis of blood electrolytes and gases on the market include the GEM-Premier from Mallinckrodt Sensor Systems, PPG's similar STATPal, and the i-STAT Portable Clinical Analyzer providing data on sodium, chloride, potassium, urea nitrogen, glucose, and hematocrit in less than 2 min. Another recently introduced handheld instrument is the AccuMeter cholesterol test from Chemtrak, designed to screen lipids to evaluate cardiovascular disease risk and the portable blood analysis systems by Diametrics Medical Inc. (electrochemical) and AVL (optical fluorescence). Abaxis introduced the Piccolo™ for general blood chemistry analysis, a small table-top centrifugal instrument with a spinning reagent disk and optical read-out. Biologix Inc. is developing the electrochemistry based handheld pHacs STAT for blood pH, potassium and oxygen.[30,31] Of all the handheld biomedical analyzers on the market only i-STAT's Portable Clinical Analyzer involves Si in the sensor construction itself, merely due to a historical production mode. Some of these handheld products are shown in Figure 10.13:* i-STAT's instrument with disposable sensor cartridge; Biologix' pH sensor cube and reader (under development); Porton Diagnostics potassium sensor and reader (recently introduced by PD$_x$ in a redesigned configuration with a small table-top reader rather than a handheld instrument).

While the number of new tools available for microfabrication, as learned from Table 7.7, has grown dramatically, few methods have been gainfully applied to biosensor construction. In contrast to mechanical sensors, where the excellent mechanical properties of single crystalline Si often tends to favor Si technology, the choice of the optimum manufacturing technology for chemical sensors is far less evident. For chemical sensors the Si in the sensor often serves no other role than as a passive substrate. Consequently, the following trend emerges: whereas mechanical sensors (pressure, acceleration, temperature, etc.) are moving toward more integration embodied in CMOS-compatible surface micromachining, chemical sensors and biosensors are moving away from integration on Si and toward hybrid technology on plastic or ceramic substrates with silk-screening or drop delivery systems for the application of the organic layers.

An additional point must be made with respect to development of *in vitro* blood glucose sensing with handheld devices such as Medisense's ExacTech blood glucose monitoring system and similar products by Johnson & Johnson (Lifescan) and Wampole Laboratories (Answer).[31] Glucose sensing represents the single largest biosensor market and a significant amount of all research and development in sensors is geared toward improving this technology. A breakthrough in this field came with the 1987 introduction of the amperometric glucose sensor ExacTech by Medisense which proved that small, planar electrochemical probes can compete with colorimetric paper strips. The new Medisense glucose sensor is based on an innovative mediator chemistry and a traditional thick-film manufacturing

process. The sales of mediated amperometric biosensors now represents a dominant 65% of the world biosensor market.[34] A comparison of manufacturing techniques for these types of sensors reveals the difficulty of competing with a Si approach. One often forgets that in biosensors such as glucose sensors one is not competing with an expensive serial manufacturing process, but with a continuous printing type processes (see Chapter 3), making products even cheaper than in batch production. The cost target for a disposable glucose sensor is about $.10, about a factor of 10 less than an unpackaged physical sensor die! Unless a significant performance advantage arises, Si technology cannot compete in this case. It is our belief that, besides new chemistries, continuous processes to manufacture disposable electrochemical sensors will be the ultimate answer to the pressing need for disposable, inexpensive diagnostics.

## Electrochemical Sensors

### Background

Microfabrication of electrochemical sensors, using IC technology, has been extremely challenging, mainly due to process incompatibility issues, packaging problems, failure to incorporate a true reference electrode, and the difficulties involving patterning relatively thick organic layers such as ion-selective membranes and hydrogels.

We have come to the conclusion that for chemical sensing, especially for *in vitro* applications, a modular, hybrid approach rather than an integrated Si approach marks the road to progress. A concept for a microfabricated, modular design for chemical sensor arrays is introduced in this section.

### *In Vitro* Use of Electrochemical Sensors

For blood electrolyte (pH, $Na^+$, $K^+$, $Ca^{2+}$, etc.) and gas analysis ($O_2$ and $CO_2$) with tabletop or handheld instrumentation one wants to develop, inexpensive, multiuse or disposable chemical sensors, up to now attempted via electrochemical technology. Some of the biggest hurdles to overcome when using Si for this application are the construction of a true on-board reference electrode (the Achilles heel of any electrochemical sensor) in the same plane as a FET (field effect transistor) device, the fashioning of a stable internal reference, incompatibility of sensor materials with Si processes, electrolyte shunting the electronics, high fabrication cost, and packaging issues in general.[35] The ion-sensitive field effect transistor (ISFET), combining a thin, ion-sensitive coating on the gate of a FET as shown in Figure 10.14A, only reached limited niche applications after 25 years of research and development. Its research in the U.S. seems at a standstill for now.

Technical and fundamental problems associated with ISFETs were treated in detail in Reference 35 and will only be summarized here. Most sensor experts have come to appreciate that for chemical sensors one should only implement on-board electronics if it is absolutely essential for the integrity of the signal. The extended gate field effect transistor (EGFET) device was conceived to avoid some of the problems associated with the leakage problems of an ISFET.[36] By depositing the ISE membranes a

---

* Figure 10.13 appears as a color plate after page 144.

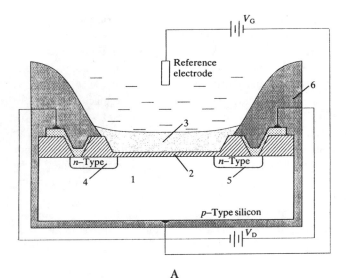

A

B

**FIGURE 10.14** Integration of chemical-sensitive material and FETs: (A) Chemistry layers on the gate of the FET: ion sensitive field effect transistor (ISFET). (1) Si substrate, (2) insulator, (3) chemically sensitive membrane, (4) source, (5) drain, and (6) insulating encapsulant. (B) Chemistry layers extended away from the FET: extended gate field effect transistor (EGFET). The seven small squares at the bottom of the photo are the extended gates. The rectangular pads on top are the contact pads.

distant away on conducting lines extending away from the FET gates, encapsulation was made easier. The EGFET seemed like an elegant solution at the time (Figure 10.14B), but i-STAT, a commercial entity selling blood electrolyte analysis equipment, eliminated all on-board electronics and abandoned the EGFET approach after a lot of research. The Portable Clinical Analyzer from i-STAT uses Si in the disposable chemical sensor cartridge

only as a substrate and a contacting base; electronics are kept in the handheld reader, clearly a more economical approach (Figure 10.13)*. The sensor electrodes, gels, and membranes are all deposited with semiconductor-type processes.[37] Without electronics on-board the continued use of Si as a support has mostly a historical explanation. It should also be noted that the fluidic channels in the disposable i-STAT cartridge are made from molded plastic and not with Si micromachining (Figure 10.13)*. For multi-use electrochemical reference electrodes, thick hydrogel films, with a large capacity to store reference solution, are preferred for stability and, for electrochemical sensor electrodes, thick ion selective membranes are better at preventing pinholes. Consequently, optimized chemical sensor materials end up much thicker than the common IC electronics layers and are more easily prepared with thick-film technology. An additional reason for the slow commercialization of silicon-based electrochemical sensors is that the currently available electrochemical chemical sensing technologies are multi-use, proven, and cost effective. In the U.S., ISFETs currently are manufactured only at the University of Utah, Salt Lake City. Orion Fisher, Corning, Sentron, and Rosemount as well as UniFet, Inc. (San Diego, CA) sell ISFET-based pH sensors in the U.S. but do not produce them in-house. Orion and Rosemount buy their pH ISFETs in research quantities from Neuchatel (Switzerland) and target the food industry where glass pH electrodes cannot be used for food (pH measurement in dough, for example). These commercially available ISFETs are for pH only and feature a simple pH sensitive inorganic membrane on the gate. To modify FETs with polymer membranes for detection of other ions is difficult because of the thickness of the membrane and the need for a stable internal reference electrolyte. Usually it is simpler and less expensive to build electrochemistry on a non-silicon substrate (see Figure 10.17A).

Another problem in the manufacture of Si chemical sensor arrays manifests itself in the decrease in manufacturing yield with increasing number of array elements. When depositing the sensor materials associated with the different array elements, each new layer added to the wafer reduces the yield of the finished product. And each time a new element is added to the array, all of the previous processes need to be reconsidered, which hampers the introduction of new products, especially since the materials involved are almost all nonstandard. To increase the manufacturing yield dramatically, one would be better off fabricating a different wafer with only one type of sensor material and then combining the individual sensors into an array, e.g., with pick-and-place techniques. This modular approach would enable the independent development of different chemistries for different analyte sensing and obviate all compatibility issues. This approach entails a sensor array somewhat larger than an integrated Si approach, but adequate for most *in vitro* applications. Figure 10.15 compares integrated and modular approaches to building chemical sensor arrays. Further below we will expand upon this idea and introduce a modular, hybrid fabrication method based on photoformed glass as well as pick-and-place technology to arrange individual chemical sensor elements in any desirable array (panel) configuration.

---

*    Figure 10.13 appears as a color plate after page 144.

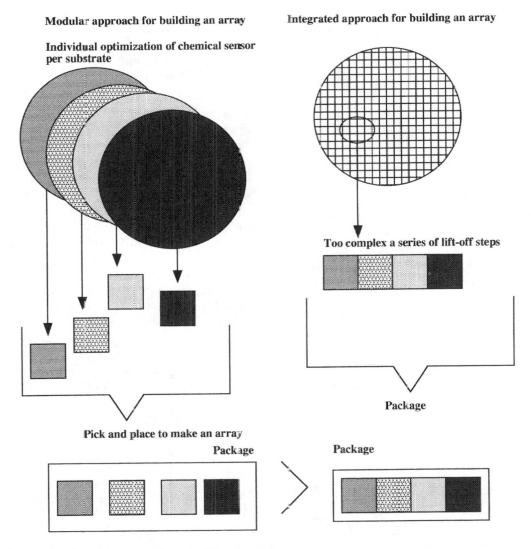

**FIGURE 10.15** Integrated and modular approaches to building chemical sensor arrays.

### *In Vivo* Use of Electrochemical Sensors

Cost of disposable sensors for *in vivo* use is not as big a concern as for *in vitro* disposable sensors; size is of more crucial importance because the smaller the sensor the smaller the intrusion on the patient. It is estimated that a catheter-based pH, $CO_2$, and $O_2$ sensor array may cost up to $250.00 and would serve a market of well over $300 M a year in the U.S. alone. Unfortunately, the warm, wet saline *in vivo* environment perhaps puts forth the most severe environment in which Si sensors might be used. No Si based *in vivo* chemical sensors are commercially available today. Biocompatibility is the single most complex issue facing *in vivo* sensor development and it needs addressing up front in the sensor design. This is in line with our philosophy of starting a sensor design from the package. The package for an electrochemical *in vivo* pH, $CO_2$, and $O_2$ sensor typically is a dual lumen catheter modified to incorporate the sensor structure with the required chemically sensitive materials (see Example 2 in Chapter 4). In building such a sensor the author at first neglected to investigate biocompatibility issues up front and concentrated on the embedded micromachined electrochemical sensors instead.[38] The measurements obtained in saline solutions were encouraging but the funding for continued research dried up before serious biocompatibility testing could be started. The development of the disposable *in vivo* pH, $CO_2$, and $O_2$ probe, the so-called RT-MECSS (i.e., Room Temperature Micro Electronic Chemical Smart Sensor) took 4 years and $6 M. The RT-MECSS puts chemistry and electronics on separate planes while keeping the signal line as short as possible (see Figure 4.70).[39] The latter represents a considerable improvement over both ISFET and EGFET. Despite this novel approach venture capitalists were only interested, and rightly so, in the biocompatibility of the sensor package. During the 1980s, they had seen scores of start-up companies fail over biocompatibility related issues of fiberoptic-based *in vivo* pH, $CO_2$, $O_2$ sensors. In response, jointly with NASA Ames, we have launched an ambitious program addressing the biocompatibility of *in vivo* chemical sensor packages up front before putting more efforts in updated micromachining designs.[40] The experimental catheter used for biocompatibility studies in animals is shown schematically in Figure 10.16A. The dual lumen structure holds a working and a reference chamber. The reference chamber has a hole contacting the external medium, and the working electrode chamber has a hole contacting the ion selective membrane (our design is similar to the dual lumen sensor described by Fogt et al.[41]). The same pH sensitive membrane materials used to cover the outside of the RT-MECSS structure are tested with this probe but no electronics are integrated into the catheter. The purpose of the external 'sleeve' as shown in Figure 10.16A is the incorporation of a reservoir for slow release of an anticoagulant such as herudine or heparin from a polymer matrix to further extend biocompatibility. A typical biocompatibility test sensor for PH (without the anticoagulant sleeve) is compared with a traditional glass pH sensor and a traditional reference electrode in Figure 10.16B. To avoid the need for integrated electronics and to enable biocompatibility measurements of the employed sensor materials in an untethered animal, this catheter is interfaced with a 450-kHz amplitude-modulated (AM) implantable telemetry transmitter which incorporates a high impedance ($>10^{12}$ $\Omega$) preamplifier and allows for nearly continuous measurement of electrochemical performance of the membranes and temperature at a distance of about 2 ft from the animal. *In vivo* and *in vitro* telemetric pH data are now being collected and compared. If biocompatibility results warrant such an endeavor, the RT-MECSS design will be put back into the catheter for nontelemetric *in vivo* applications. We believe that this telemetric approach to study biocompatibility of the exposed chemical sensor materials could be a major help for *in vivo* sensor

development in general. It is a means to test the package before putting in the micromachines.

## A Modular Approach to Building Chemical Sensor Arrays

While Si makes it easy to electrically interconnect the sensor probe and the processing logic to one another, Si makes physically mounting them difficult. Packaging is easier when using an insulating substrate as the conductive sides of a Si chip readily generate a leakage path. Thick-film processes are more suited to the chemical sensor construction and most chemical sensor materials are incompatible with IC processing. The very point of using Si (its standardness) is thus forfeited in an electrochemical environment. Worse, a problem of central importance to the manufacturing yield of these analog components needs addressing since it imposes drastic costs on integrated Si chemical sensor arrays: when the different sensor materials, associated with the independent components are deposited using batch semiconductor (thin-film) techniques, each new layer added to the wafer may interfere with the chemical activity of the previous layers and can cripple the yield of the finished array (see Figure 10.15). As a critical consequence of the integrated array approach, we need to reckon with the largely nonstandard materials, and their modes of deposition need to be

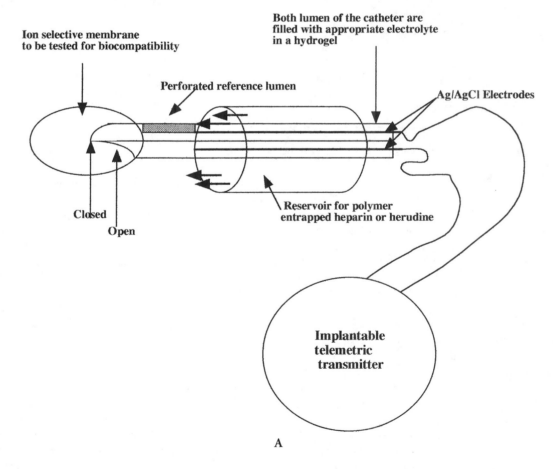

A

**FIGURE 10.16**   (A) Electrochemical test probe for materials biocompatibility. (B) Test probe compared to commercial pH electrode and reference electrode (without the reservoir for heparin or herudine).

reinvented for each new element added to the array. Since it is crucial to subject the sensor probe to as few processing steps as possible, implementing the electronics on a separate die from the bioprobes(s), and implementing it on-board only if the application absolutely makes it essential clearly make for the better approach . It is observed that the integrated electrochemical Si sensors pay a high cost up front (in high design and development cost) as well as for each new unit with an added or different array element produced (in markedly reduced yield).

The success of i-STAT's product introduction renewed interest in making planar electrochemical sensors for *in vitro* applications. In view of the problems encountered in Si technology, Buck et al.'s efforts at the Center for Emerging Cardiovascular Technologies (CECT)[42,43] focus on depositing electrochemical

components on inexpensive non-silicon substrates (ABS, Kapton, etc.). Along the same line, Pace and colleagues at DuPont use thick-film silk-screening to fabricate ion-selective electrodes on A/AgCl contacts on alumina substrates.[44] Flow channels fabricated on the substrates enable a differential measurement between 'identical sensors', with one sensor in the sample stream and one in a calibration stream. This compensates for drift effects that are the same on the two sensors, showing how a system approach can improve individual sensor performance. Mallincrodt's commercial blood analyzer is based on even more traditional technology with Ag/AgCl reference electrodes formed on Ag pegs in a plastic body.

In the author's approach we designed probes of one flavor (e.g., all pH sensors) that are mass produced on large photoformed

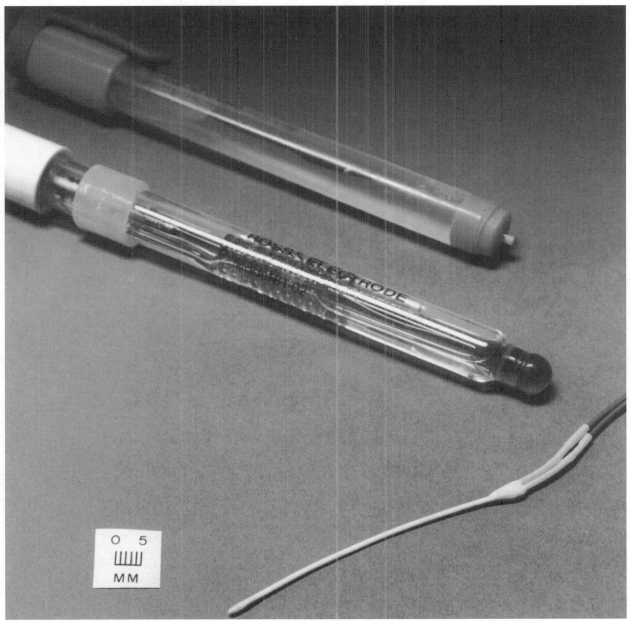

**B**

FIGURE 10.16 (Continued)

glass wafers or polyimide sheets using currently available thick-film technologies (silk-screening and drop-delivery systems) (see Chapter 7 for details on photoformed glass manufacturing). The individual probe concept is sketched in Figure 10.17A. The cup design for individual electrochemical cells, characteristic of the RT-MECSS design (Figure 4.68), is maintained in both the photoformed glass and polyimide approach, as well as the contact scheme with leads coming from the backside. We then rely on pick-and-place techniques to combine the individual sensors of different flavors into an array on a so-called bioplatform (Figure 10.17B). The modularity in manufacturing any desired electrolyte/gas sensing panel by simply putting different sensor element combinations in an array permits the potential intro-duction of a large family of products without the need to reinvent the process steps for each new panel desired. Photoformed glass is available in large panels and its micromachining is being per-fected in Japan in work on flat-panel displays as well as in other research efforts. Also large sheets of polyimide (flex circuit mate-rial, Kapton) can be bought from Dupont (see Chapter 1 on photosensitive and nonphotosensitive polyimides).

### Conclusion

The next generation electrochemical sensors will be planar, thick-film, hybrid devices which will have the cost and fabrication advantages of solid state technology but are likely to still have some of the same limitations as present electrochemical sensors, i.e., repeated calibrations to compensate for long-term drift, and the need for periodic maintenance. An important lesson in reviewing this micromachining application is that there is as much need for further optimization of chemical sensitive layers as for exploration of new manufacturing options. Investigating continuous printing processes for electrochemical sensor manu-facture will be most rewarding in the future. For the intermediate time a modular approach with thick-film deposition processes is advisable. Some of the calibration and cleaning problems will be resolved by implementing plastic molded microfluidics.

## Immunosensors

### Introduction

A most prominent example of a biosensor is an immun-osensor, i.e., a sensor monitoring either an antigen (Ag) or an antibody (Ab) based on the selective binding of antigen-antibody pairs or in a broader sense any sensor based on the pairing of a molecular recognition affinity pair. The ability of natural cells to monitor biochemical events are mimicked in this type of sensor. Therefore, they could constitute the ultimate in sensor selectivity and sensitivity.

Immunoassays have been providing scientists and clinicians with a method to make sensitive and specific measurements of the amount of antigen or antibody present in a sample. Con-ventional immunoassays require various steps such as accurate addition of reagents in a set sequence before a measurable signal is generated. The interest in immunosensing is to adapt these assays into a sensor format and to make measurements by sim-ply bringing the sample in contact with the sensor. Our interest here is to explore how micromachining could help this cause.

### Early Immunosensor Work

Early on immunosensor work was geared toward direct detection of antigens or antibodies with potentiometric tech-niques, i.e., by simply measuring a voltage. Antigen-antibody binding on the surface of a potentiometric-sensing electrode leads to a new charge distribution. In principle this redistribu-tion of charges should be measurable. The effect of direct bind-ing on the potential of a sensing electrode was observed on polymeric membrane-coated electrodes as well as on metal elec-trodes. However, the signal was found to be small, i.e., 1 to 5 mV, making the precision of the measurement poor.[45] Also, ISFETs have been attempted as one special embodiment of potentio-metric devices and several 'ImmunoFETs' were made in labo-ratories around the world.[46] Unfortunately, as Janata et al.[47,48] concluded, the potential changes on potentiometric devices are due to a mixed potential generated from the antigen-antibody binding as well as other charge transfer reactions occurring at the electrode surface/electrolyte interface. To make a potentio-metric sensor work, Janata et al. pointed out that a substrate surface with an infinitely high charge transfer resistance is needed as only then do protein charges exclusively account for the potential generated. Such a perfect polarizable surface has not been identified yet. Not even densely packed Langmuir-Blodget films qualify, as they do not block all current across the interface either.[49] In Figure 10.18A, an ImmunoFET is schemat-ically shown with a membrane, perhaps a Langmuir-Blodgett film, and attached antibody molecules.

An alternative direct measuring immunosensor is the capac-itive affinity sensor developed by Biotronic Systems Corp. (Maryland). The immunosensor is based on the capacitance change associated with the antigen-antibody binding between an insulated pair of interdigitated electrodes on a solid sub-strate. The capacitance change arises because the antibody, a large molecule with small dielectric constant, displaces water with a high dielectric constant. Again, as with the potentiomet-ric sensor, unless a perfect polarizable interface, i.e., a perfect stacking of only the antigen-antibody complexes at the capac-itor surface, is realized, capacitive changes cannot unequivo-cally be associated with the reaction of interest. A top and cross-sectional view of this capacitive sensor is shown in Figure 10.18B.[50]

Langmuir-Blodgett films might be used in another way for immunosensing, similar to the one used by natural cells incor-porating such films. The structure found in most biological membranes is a bilayer lipid membrane (BLM) wherein two layers of lipid molecules are arranged so that their hydrophobic parts interpenetrate, whereas their hydrophilic parts form the two surfaces of the bilayer. A synthetic membrane mimicking a BLM of that nature may be put across an orifice and the con-ductance across the membrane measured. Figure 10.19 displays two possible configurations to accomplish this type of measure-ment. Operation of this type of sensor is based on a selective interaction between an analyte in aqueous solution and recep-tors embedded in the membrane resulting in a sudden measur-able current across the membrane.[51,52] The ionic current modulation can be related to the concentration of the analyte. In principle, any biological complementary pair such as enzyme-substrate, lectin-saccharide, antibody-antigen, etc. could be used. One binding event can modulate the conduction of hundreds of ions, an example of chemical amplification; hence, one expects high sensitivity. Also, the signal is due to a nonequilibrium process that speeds the response time and is not sensitive to interferences encountered in equilibrium processes such as mixed potentials. Although this method clearly offers a way out of the dilemma encountered with direct sensing techniques, it is also one of the more mechanically unstable devices as it is based on a membrane that measures only about 50 Å thick. Developing a membrane surviving longer than 24 hours is a major accomplishment.[53]

The nonavailability of perfectly polarizable electrodes, non-selective protein interactions with the electrode and the influence of pH and ionic strength (through Debye shielding) make the prospect of developing selective and sensitive direct potentiometric or capacitive immunosensors rather dim. These problems have caused a shift toward indirect measuring techniques involving labels and amplification schemes. For example, in an enzyme immunoassay (EIA), immunochemical affinity is relied on for selectivity and chemical amplification of an enzyme label for sensitivity. A major problem introduced through the use of most labeling techniques is that only via a separation step can one differentiate free and bound labeled species as both contribute to the signal (an assay involving a separation step is called a heterogeneous assay). It is especially challenging to develop a sensor where this step is not required; this might mean that a micro-instrument is in order, such as an instrument with some micro-fluidics flushing the unbound species from the sensor surface.

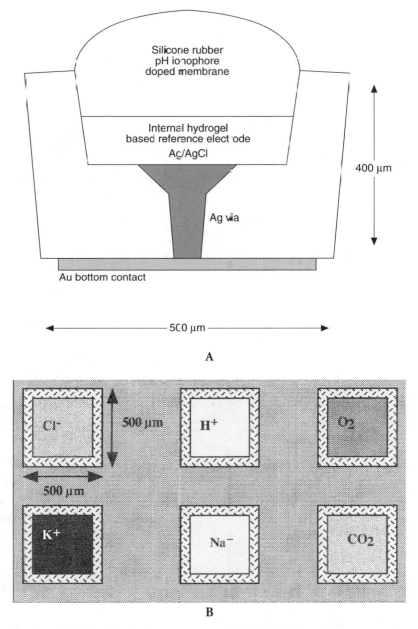

**FIGURE 10.17** Modular approach to building chemical sensor arrays: (A) individual chemical sensor probe, and (B) array on a platform (probe carrier).

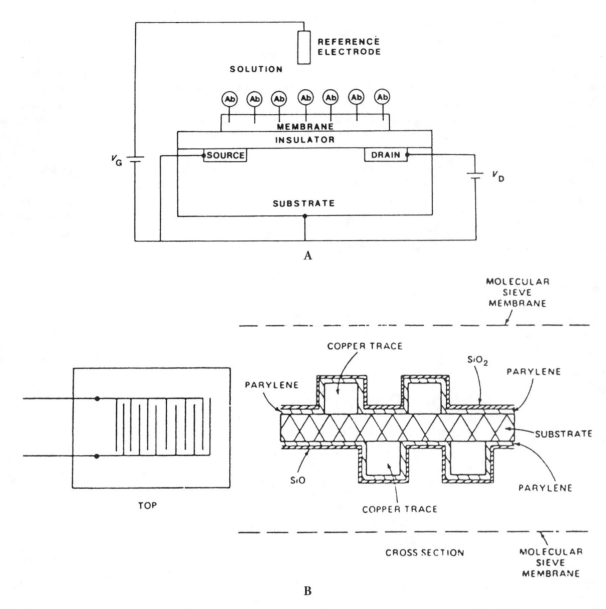

FIGURE 10.18   Examples of micromachined, direct immunosensors. (A) Schematic diagram of an ImmunoFET. (B) Top and cross-sectional views of capacitive immunosensor. (From Newman et al., 2nd Intl. Meeting on Chemical Sensors, Bordeaux, France, 596–598, 1986. With permission.)

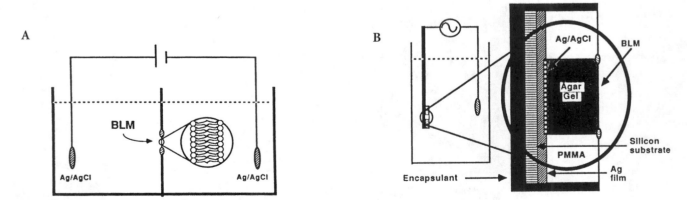

FIGURE 10.19   Two types of implementations of BLM. (A) Standard BLM design showing the bilayer in the aperture between two electrolyte solutions. (B) Sensor produced by forming a BLM on a hydrogel interfaced to a Si substrate via an Ag-AgCl electrode and thin Ag film. (From Reichert et al., *Molecular Engineering of Ultrathin Polymeric Films*, Elsevier, London, 345–376, 1987. With permission.)

Several stumbling blocks still must be overcome in immunosensors, such as assays involving too many steps to qualify as sensors, the deposition of uniform protein coatings, unselective protein adsorption, distinguishing between bound and unbound antigen, and the inherent irreversibility of immunosensors. The latter is based in the large association constants ($K_a$) involved in an antigen-antibody binding reaction (typical $K_a$ values range between $10^5$ and $10^9$ M$^{-1}$). The Ag-Ab association constants are composed of large forward ($k_1$) and small reverse ($k_{-1}$) rates. These kinetic parameters make antibodies very selective for the analyte of interest but also make them quite irreversible.

Problems with potentiometric and capacitive techniques have made optical immunosensing on thin films more popular. In many cases, micromachining could provide inexpensive manufacturing options. But, as Turton wrote in 1987, "Claims for a new immunoassay are as likely aimed at investors as at biochemists."[54] When considering immunosensors one should be even more skeptical about new claims. Such simplified immunoassays mostly still are too complex to qualify as true immunosensors. In most cases, the techniques developed to make individual immunosensors are doomed to remain curiosities only, unless panels for a set of different target analytes can be made.

### Innovation in Immunosensors

We will highlight only a few research developments here to illustrate how new chemistries and micromachining techniques might impact immunosensing. For a more detailed review on immunosensors with commercial potential, we refer to Buerk[55] and Madou et al.[56]

As explained in the introduction, a key issue with immunosensors is their seemingly inherent irreversibility. Barnard et al.[57] introduced an ingenious way to make an assay continuous anyway by using the controlled release of labeled antibodies to maintain sensitivity. They used fluorescein-labeled antibody (F-Ab) and Texas Red-labeled immunoglobulin G antigen (TR-Ag) in a competitive immunoassay based on fluorescence energy transfer. When the antibody and antigen bind, the two fluorophor labels are very close to one another, allowing nonradiative transfer of energy. When the fluorescein molecule is excited by light, it donates energy to the Texas Red fluorophor, enhancing its fluorescent light emission. To circumvent the irreversibility of the immunoassay, these researchers used a controlled-release delivery system capable of sustaining a constant release of fresh immunochemicals as shown in Figure 10.20 (similar to our approach to biocompatibility illustrated in Figure 10.16). Specifically they used ethylene-vinyl acetate as the controlled-release delivery system of reagents into the reaction chamber, i.e., the sensing region of the optical fiber. Unlabeled Ag diffused into the reaction chamber and competed with the Ag-TR for the available binding sites on the F-Ab, and bound species diffused out of the reaction chamber while a constant release of reagents maintained a constant concentration of reagents in the reaction chamber. The fluorescence changes induced by the competition reaction was used to calculate the amount of unlabeled antigen in the sample solution. In cases

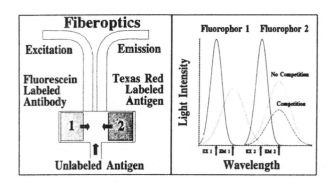

**FIGURE 10.20** A new approach to immunosensing. Reversibility is induced by implementing a polymer-release mechanism. Labeled antibody and antigen are released from polymer reservoirs and fluorescence measured by fiberoptics (left panel). The fluorescence emission from the second fluorophor (Texas Red) is reduced when unlabeled antigen competes with the reaction (right panel). (From Barnard, S.M. and D.R. Walt, *Science*, 251, 927–929, 1991. With permission.)

where response time is not very critical but continuous monitoring is of major importance, this technique seems very promising. A micromachined version of the sensor shown in Figure 10.20 seems imminently feasible. This could involve micromachining a reaction chamber, release polymer reservoirs, and a waste dump, all connected with flow channels and a window in the reaction chamber for the fiberoptics. Plastic injection molding or photoformed glass would be good approaches here.

Examples of optical and electrochemical immunosensors developed by this author are shown in Figures 10.21 and 10.22, respectively. As summarized in Table 10.15, there are pros and

**TABLE 10.15** Comparison of Optical and Electrochemical Immunosensor

|  | Optical | Electrochemical |
|---|---|---|
| Instrument cost | Often expensive | Inexpensive |
| Sensor cost | Fair | Low |
| Turbid solutions | Problematic | Usually no problem |
| Dynamic range | Limited | Wide |
| Selectivity | Good | Fair |
| Sensitivity | Very good | Good |
| Simplicity of method | Often simple | Simple |
| Color | Sometimes problematic | No problem |
| Species size | Better for larger molecules | Better for smaller molecules |
| EMI = electromagnetic interference | No | Yes |
| Resistance to radiation and corrosion | Yes | No |
| Cross-talk | No | Yes |
| Ambient light | Problem (must be modulated) | No problem |
| Response curve | Sigmoidal | Nernstian |

cons to either approach. By providing more details on both methods this table will become clearer.

Several features make electrochemical techniques attractive for immunosensing. Since accuracy is achieved independent of color and turbidity of the solution, rapid measurements are obtained without tedious sample preparation.

Analysis of a whole range of analytes in matrices as complex as whole blood can be done. Electrochemical probes are easy to use as analytical tools measuring ions and neutral molecules, typically in a range of 1 mol/l to 1 μmol/l. Examples of even wider dynamic range exist. Coupled with chemical amplification (enzyme or liposome), detection limits of picomoles/liter have been attained. In addition, the sensor is inexpensive.

Our own one-step electrochemical immunosensor in Figure 10.21 combines the fundamentals of enzyme immunoassay, electrochemistry, and gel filtration.[58] All the components necessary for the assay are incorporated into the sensor; only the sample has to be added for analysis. The base of the sensor consists of working and reference electrodes deposited onto an appropriate substrate. To make the components as cheap as possible, we used thick-film technology on a plastic substrate. Layered on the electrode surface is a mixture of an inactive apo-enzyme (apo-glucose oxidase) and enzyme substrate (glucose in case the enzyme is glucose oxidase) and mediator, reagents necessary to generate an electrical signal. Above the apo-enzyme matrix is a wettable gel containing antibody specific for the

analyte of interest, topped with a layer of analyte labeled with a prosthetic group or cofactor capable of activating the apo-enzyme. A drop of sample placed on the surface of the sensor immediately hydrates the sensor with a fixed volume dictated by the total volume of hydroscopic gel in the device. Sample analyte and cofactor-labeled analyte compete for a limiting amount of immobilized antibody in layer 1. The amount of cofactor-labeled analyte that diffuses to the apo-enzyme layer is directly proportional to the concentration of sample analyte. If the sample does not contain antigen, the labeled antigen remains firmly bound to the antibody layer and will not be able to activate the apo-enzyme. Activation of apo-enzyme by cofactor-labeled analyte results in catalytic breakdown of the enzyme substrate. Reduction or oxidation of one of the reaction products results in a current proportional to the amount of apo-enzyme activated and thus to the concentration of analyte in the sample. The working electrode detects a product of the enzyme reaction with the substrate in an amperometric mode. Using theophylline for analyte and apo-glucose oxidase as inactive enzyme which FAD renders active demonstrates this simple, easy to use, and quantitative immunodiagnostic system. The principle of the new assay is generic and may be applicable to the development of sensors for a variety of analytes with biomedical interest. The detection limit of the FAD/apo-glucose oxidase system is well below $10^{-10}$ $M$, adequate for the analysis of a majority of analytes. This detection limit can be reduced

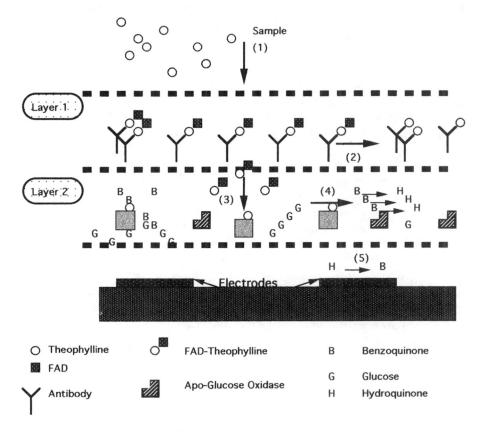

**FIGURE 10.21** Layer configuration of immunosensor using theophylline as example reaction scheme: (1) Theophylline sample diffuses into first layer. (2) Free theophylline displaces prebound FAD-theophilline conjugate. (3) FAD-theophylline diffuses into the second layer and activates apo-glucose oxidase. (4) Benzoquinone consumed and hydroquinone produced. (5) Hydroquinone is oxidized at the metal electrode, leading to an increase in current.

further by optimization of the system. A further challenge is to find ways of reproducibly layering the different organic materials shown in Figure 10.21. This challenge does not concern Si micromachining but rather involves learning how to manufacture and manipulate hydrogel materials.

Using optical techniques one typically has one partner of the immunochemical complementary pair immobilized onto a surface with some optical characteristics modified when binding takes place. In optical sensing the signal is not subject to electrical interference, no reference electrode is required, and simultaneous measurement of two or more analytes is possible by measuring at two or more wavelengths. While classical optical assays rely mostly on absorption spectroscopy, luminescence in the form of fluorescence or phosphorescence (depending on the excited state being single or triplet) is more important for immunosensors.

The optical immunosensor we developed is based on Langmuir-Blodgett technology.[59] As early as 1936, Irving Langmuir and Katherine Blodgett developed, in a most ingenious way, a method to measure film thickness comprising a small fraction of a wavelength of light. It consisted of building multilayers of barium stearate on a slide in a series of steps in a staircase fashion as shown in Figure 10.22A. The method they used is known today as the Langmuir-Blodgett deposition method (see Figure 3.36 in Chapter 3 for a typical set-up). Since then a lot of progress has been made in automating this deposition technology (see Chapter 3). The difference in thickness of two adjacent steps as shown in Figure 10.22A corresponds to a double layer of stearate (approximately 48 Å) and forms an interference-based color gauge which tracks thickness increases. When a film of unknown thickness is coated on the steps and compared with an uncoated slide with steps, the eye can measure differences of thickness by means of color comparison down to about 10 Å. Greater sensitivity is obtained by illuminating films with sodium light rather than white light. Differences of thickness equal to 5 Å have been made plainly visible to the eye this way. This means that adsorbed layers of molecules or atoms having an optical thickness of 5 Å can be made visible to the eye without aid of optical apparatus and can be measured with a probable error of about 2.5 Å. Some preliminary experiments were carried out with this simple step technique in the authors' laboratory. When albumin was adsorbed on one half of the staircase slide and the slide then was exposed to anti-albumin, we noticed a thickness increase by the shift of color. What does this mean from the biosensor point of view? For many years ellipsometers have been used to detect the deposition of antibody onto antigen pre-adsorbed on a flat reflecting surface. In this way, any molecule with a molecular weight higher than 60,000 (e.g., viruses, antibodies, proteins) could be detected. The ellipsometer is a very involved and expensive instrument, whereas the system described above can perform the same task in a simple and inexpensive way. Langmuir-Blodgett deposition, as pointed out earlier, is a difficult manufacturing proposition, though. Fortunately, steps can also be built using etching techniques and employing inorganic materials with optimum refractive indices for maximum interference effect.[59] More recently we found yet a better way to make inter-

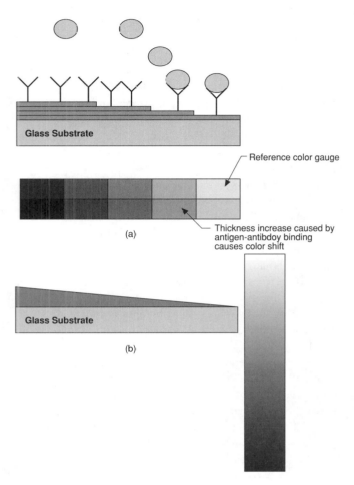

**FIGURE 10.22** Optical immunosensors. (a) Staircase made by Langmuir-Blodgett technology (additive technique) or etching (substractive technique); (b) wedge made by pulling sample out of an etchant (substractive technique). The color version of this same Figure 10.22 appears as a color plate after page 144.

ference slides: a substrate pulled at a uniform rate from an etchant solution forms a continuous optical wedge making for a more elegant and faster manufacturing option than both Langmuir-Blodgett or staircase etching (Figure 10.22B). We believe the wedge approach to be an excellent illustration of how micromachining could benefit immunosensor development, i.e., by making inexpensive, disposable, optical transducers possible. Whereas the wedge approach is more appropriate for the detection of larger molecules, the detection of smaller molecules is easier with the electrochemical approach.

Many other opportunities for innovative micromachining applications in optical immunosensors present themselves. A prime example can be found in surface plasmon resonance (SPR) sensors. Incident plane polarized light at a specific angle can be almost totally absorbed into a thin metal film (e.g., 500 Å Ag or Au) deposited onto a prism, fiber, or waveguide surface due to resonant coupling of the light energy into a free-electron cloud within the metal. This effect is known as surface plasmon resonance and can be extremely sensitive to the refractive index changes surrounding the metal film. By adsorption of species on the metal film, the specific angle of the light needed to create the resonant absorption changes (Figure 10.23).[60] This adsorption could be from gases or the binding of macromolecules onto the metal film. The main limitation is that the sensitivity depends on the molecular weight, or more correctly, the optical thickness of the absorbed layer, meaning that small molecules will not be good candidates. Pharmacia's (Sweden) SPR-based 'immunosensor' is a general purpose, expensive, and rather large instrument incorporating a prism for detection and four parallel flow channels in removable sample chambers. A significant amount of research effort is directed toward trying to miniaturize this large optical system. Comparison of diffraction gratings and prism systems clearly indicates that the diffraction grating is preferable for miniaturization. Garabedian et al.[61] attempted to micromachine the whole surface plasmon sensing system by implementing such a grating, but their resulting micro-instrument is way too complex for manufacturing at this stage. An elegant alternative to both the grating and prism approaches may come from using an optical fiber or a planar optical waveguide. For example, Jorgenson et al.[62] simplified SPR sensing dramatically by using such a fiberoptic approach. As a sensing element, they used a multimode fiber with a section of the cladding removed and a thin layer of highly reflecting metal symmetrically deposited directly on the fiber core (see Figure 10.24A). This design has potential for simplified, robust, remote, and disposable sensing. Unlike traditional SPR measurements, employing a discrete excitation wavelength while modulating the angle of incidence, the SPR fiberoptic sensor

system uses a fixed angle of incidence and modulates the excitation wavelength. The refractive index can then be determined from the resonance spectrum in the reflected or transmitted spectral intensity distribution. In Figure 10.24B the SPR sensing surface is enclosed in a flow-cell. Using micromachined planar waveguides in a flow-channel instead of the set-up shown here might result in a promising commercial product.

In the emerging field of immunosensors, plenty of commercial activity is going on. BioStar Medical Products' optical interference slide is said to determine the presence of bacteria, poisons, hormones, etc. Airborne targets might in the future be detected by very small, handheld, optical attenuation-based devices produced by Diametrix Detectors, Inc. Other interesting new immunosensor approaches under development are Idetek's optical grating technique (see Figure 10.26*), Ohmicron Corp., and AAI-ABTECH's chemiresistor-based immunosensor technology to detect pesticides and Biocircuits Corporation's immunosensitive fluorescent sensor.

### Conclusion

As in the case of electrochemical sensors, progress in immunosensors so far has concentrated on new chemistries rather than micromachining. None of the direct immunosensor techniques have succeeded in carving out a commercial presence. In view of the many technical problems that remain to be solved, a time span of 5 to 8 years will be required to see a remarkable change in this status. Large companies today favor optical over electrochemical techniques and the emphasis has shifted back toward indirect immunosensing approaches as the chances for success are higher.

A sense of innovation resonates in this field as small companies are introducing immunosensors based on techniques borrowed from the IC industry. In contrast to the electrochemical sensors discussed above, optical thin-film immunosensors are quite amenable to IC-type micromachining techniques. In the next few years we will continue to see many innovative approaches being introduced as IC techniques further permeate the biotechnology field. The acceptance of technologies from other disciplines is often slow but always very rewarding. One should not only consider the IC field as a source of innovation in immunosensing but other disciplines as well, such as drug release and atomic force microscopy (AFM). We already saw how, by using drug release technology, a continuous immunoassay was realized. We believe that with AFM enabling us to see individual atoms, one might eventually be able to 'visualize' the specific binding between complementary protein pairs as it happens. This is truly a time when the whole notion of immunomethods could be dramatically altered because of the confluence of so many new emerging technologies impacting the field. Nevertheless, many problems still need to be dealt with, including:

- Long-term stability of enzymes and bioreceptors

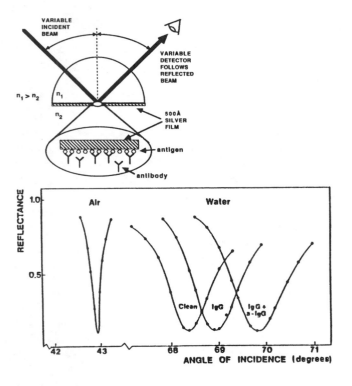

**FIGURE 10.23** Measurements with an SPR sensor based on a prism. (From Sutherland, R.M. and C. Dahne, *Biosensors Fundamentals and Applications*, Turner, A.P.F., Karube, I., and Wilson, G.S., Eds., Oxford University Press, Oxford, 655–678, 1987. With permission.)

---

* Figure 10.26 appears as a color plate after page 144.

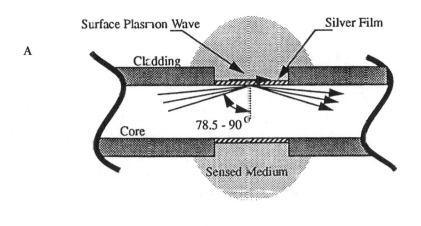

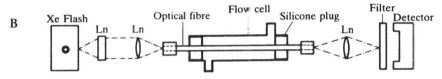

**FIGURE 10.24** Simplified SPR measurements. (A) Illustration of the SPR fiber. (B) Illustration of the SPR fiber enclosed in a flow-cell. (From Jorgenson, R.C. and S.S. Yee, *Sensors and Actuators B*, B12, 213–220, 1993. With permission.)

- Sterilizability and biocompatibility for *in vivo* applications
- Nonspecific adsorption of other species, e.g., proteins
- Immobilization and mediation of enzyme-based sensors
- Reduction of interferences from other substances in blood
- Miniaturization of sensors and sensor arrays for *in vivo* applications
- Variability in manufacturing

## Microlithography Applied to a New Class of Biosensors

### Introduction

Microlithographic techniques can be applied successfully to the field of protein patterning and polymer synthesis. It is easy to imagine arrays with several different enzymes, antibodies, or DNA-probes immobilized precisely onto a transducer surface as a diagnostic panel for clinical diagnosis or for environmental monitoring.[63] Several efforts try to exploit such devices commercially. We will highlight only those efforts that convey new ways of applying lithography to make chemical sensors.

### Grating Technique

The major problem in immunosensors is the prevention of nonspecific protein adsorption competing for detection sites. Proteins, in particular, bind with considerable avidity to a wide range of surfaces. A reference surface in an immunosensor should be a relative reference, i.e., adsorbing all the same proteins except for the protein of interest. The best way to avoid nonspecific binding effects in immunosensors is to make a reference as similar to the sensing surface as possible, i.e., a reference subject to all the same nonspecific protein binding phenomena as the sensor surface itself except for the antigen-antibody of interest. This enables the best possible correction for nonspecific binding. An interesting way of implementing this

idea is to create an optical grating pattern with the antigen-antibody coupling providing the grating structure itself, possibly based upon the loss of antigenicity upon ultraviolet radiation as observed by Panitz and Giaver.[64] Panitz and Giaver found that the antigenic sites on proteins display sensitivity to ultraviolet radiation in air and that a small dose of ultraviolet radiation can destroy the antigenicity of antigens (Figure 10.25). One presumes that the ultraviolet light breaks up chemical bonds in the adsorbed protein layer and simultaneously produces ozone. The broken bonds are targets for the highly reactive ozone, resulting in the progressive oxidation of the protein layer. The protein can actually be completely removed from the surface by too long an irradiation (say, 10 min). With a short radiation time the optical density does not change but antibodies will no longer bind to the antigenetic sites. Recently, this scheme was put to use in the fabrication of an elegant biograting for immunosensing as shown in Figure 10.26*.[65] By using a photomask grating, inactivating alternating bands of antibodies, a biological diffraction grating is created. In this device the antigen-antibody coupling itself constitutes the grating structure, causing diffraction only if the target antigen is present. A CD-type He-Ne laser beam diffracts from the grating with an intensity related to the antigen concentration. The above grating device comes close to embodying an ideal relative reference as the radiated protein bands are almost identical to the active surface except for being incapable of reacting with the target complementary protein.[56]

### Very Large-Scale Immobilized Polymer Synthesis

The above example involves the 'destructive' lithographic patterning of proteins. Nondestructive methods have been developed as well. A well-known example is the one by Fodor et al.,[66] shown in Figure 10.27A. It involves the protection and photodeprotection of an aminosilanated layer on a sensor substrate. Under illumination through a mask, the photoprotection

---

* Figure 10.26 appears as a color plate after page 144.

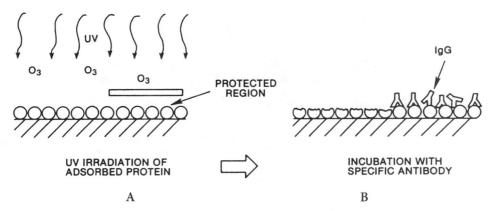

**FIGURE 10.25** Ultraviolet radiation of antigen; opportunities to create patterns of live/dead proteins. (A) A preadsorbed layer of antigen is briefly exposed, in air, to an intense ultraviolet source, resulting in the production of ozone ($O_3$). The combination of ultraviolet damage and ozone results in a partial oxidation of the adsorbed proteins. The area of protein shielded from the irradiation is not oxidized. (B) Following such treatment the protein-coated substrate is incubated with antiserum specific to the adsorbed antigen. The partially oxidized protein layer is no longer antigenic, and IgG molecules are able to bind only to the previously shielded portion of the antigen layer. (After Reference 64.)

groups can be cleaved off where desired, freeing amino groups to react locally to ligand A, making the sensor sensitive to antigen A′ in that spot. A subsequent photodeprotection can free another area to react with ligand B, making the sensor sensitive to antigen B′ in that array element, etc. Affymax, a Palo Alto, CA, company set up around this technology, calls this new type of technology VLSIPS for 'very large-scale immobilized polymer synthesis'. Affymax makes checkerboard arrays of squares measuring 50 μm on a side and using a particular combinatorial masking strategy succeeded, with a 10-step synthesis, in building an array of 1024 peptide-like compounds (a 20-step synthesis would yield 1,048,576 compounds).[66] Since such a system can direct the chemical synthesis and analysis of large numbers of potentially useful compounds on a single small chip (say, a centimeter square), it becomes orders of magnitude quicker and cheaper to identify potential ligands to build new drugs or analyze DNA sequences (e.g., DNA Chip).[67]

A wide range of proteins have been patterned with the above technique, rendered visual by a number of techniques including fluorescence microscopy, atomic force microscopy, and the growth response of cultured cells.[68]

There are also efforts to produce nanopatterned proteins.[69] Many proteins self assemble into molecular lattices, such as the bacterial S protein. Douglas et al.[69] used metal-decorated crystals of the S proteins of *Sulfolobus acidocaldarius* as a protein mask to pattern hexagonal arrays of 10-nm diameter holes into graphite surfaces. Figure 10.27B demonstrates how the self-assembled protein lattice may be used for nanopatterned protein surfaces.

### Conductive Polymer Patterning

In the late 1980s, this author started work on patterning conductive polymers such as polypyrrole and polyaniline with lithography techniques. In one example, arrays of conductive polymer posts were formed on a conductive substrate. Increased ion access to individual conductive polymer posts as compared to access to a uniform film of the same polymer lead to higher electrochemical reversibility which has its application

in polymer battery electrodes, electrochromic devices, enzyme-based biosensors, and micro-electronic or molecular electronic devices. Figure 10.28 depicts the procedure for fabricating three-dimensional arrays of electronically conductive polymers as well as an SEM micrograph of the resulting patterned conductive polymer posts.[70,71]

## Miniaturization in Biomedical Instruments

### Introduction

Micromachining efforts for building microinstrumentation may be seen as a better approach to solve chemistry and biology problems compared to building selective chemical sensors. Some underlying reasons are that the resulting instrument may justify significantly higher costs than disposable chemical sensors and help solve the shortcomings of chemical sensors by enabling calibration, separating of bound and unbound species, cleaning the detectors, filtering out unwanted compounds, and obviating the need for very selective chemical sensors. In Chapter 9, a basic theory of hydrodynamics and diffusion predicts faster and more efficient chromatographic and electrophoretic separations for miniaturized biomedical and analytical equipment. Performance is expected to increase with miniaturization because of the favorable scaling properties of some important instrument processes (for example, heating and cooling are faster while the effect of diffusion is reduced). Micromachining might also allow cofabrication of many interconnected functional instrument blocks. Tasks that are now performed in a series of conventional benchtop instruments could then be combined into one unit, reducing labor and risk of sample contamination. And, because microinstruments could potentially be batch fabricated at low cost, they might be used only once and then thrown away to prevent sample contamination. Potential applications of microinstruments include disposable diagnostic kits for infectious agents, tests for chemical purity assurance, and instruments for biotechnology. As with chemical sensors, results using IC based micromachining methodology after just about 15 years are very limited. In this

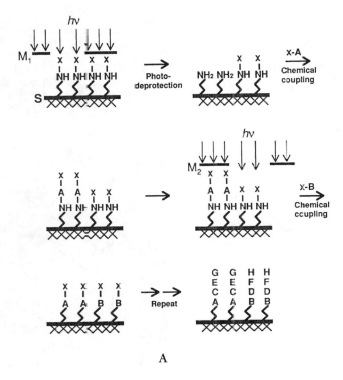

A

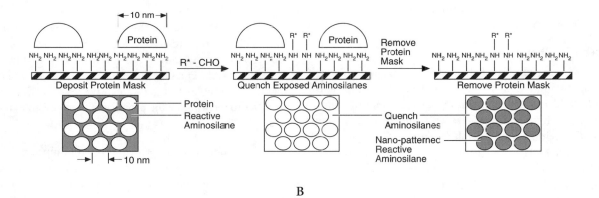

B

**FIGUR 10.27** Micro- and nanopatterned protein surfaces. (A) Photodeprotection patterning scheme. (After Fodor et al.[66]) (B) Nanopatterning of surface chemistry using self-assembled protein mask. (After Douglas et al.[69])

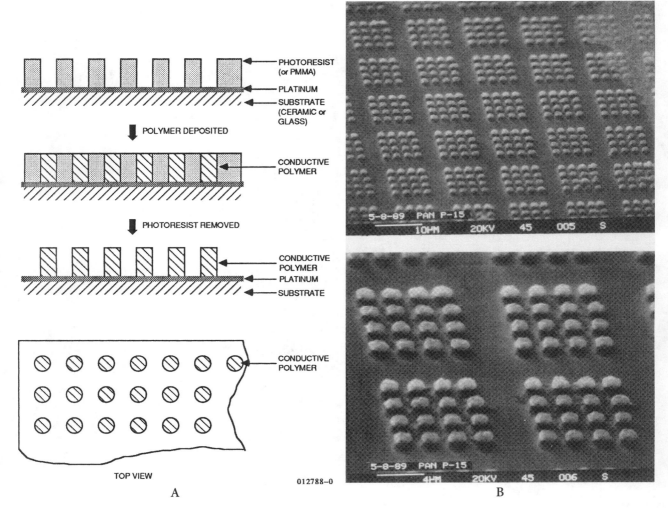

**FIGURE 10.28**  Micropatterning of conductive polymers. (A) Process sequence. (B) SEM micrograph of conductive polymer sub-micron electrate array (polyaniline doped with tosylate).

light we will again analyze the potential paths for progress in applying micromachining more fruitfully.

### Generic Challenges

The generic technical challenges in making a miniature biomedical instrument includes moving sample, wash, and calibration fluid through tiny conduits, controlling the temperature of the fluid, getting the sample into and out of the instrument providing fresh viable reagents, detection, and most of all, a cost-effective manufacturing method. In one extreme one might envision a totally integrated option such as the micro total analysis system ($\mu$-TAS),[72] promoted mainly in Europe and illustrated schematically in Figure 10.29A. More realistically, one might look at a hybrid construct as sketched in Figure 10.29B. For most micro-instruments it indeed makes more sense to envision a disposable cassette incorporating the specific reagents needed for a certain test and a separate reader instrument. An important set of decisions then includes the functions to integrate with the disposable cassette and those to incorporate in the reader, i.e., how to partition the instrument and disposable element.

### Fluid Propulsion

There are various technologies for moving small quantities of fluids from reservoirs to mixing and reaction sites, to detectors, and eventually to waste or to a next instrument. These propulsion techniques may be integrated with the cassette holding the disposables, but the state of the art is such that today it is better to incorporate the propulsion mechanism in the reader instead.

**Micropumps**  In principle, micropumps can be built in the disposable cassette together with microflow sensors and microvalves. Membrane deflection in a micropump displaces fluid through one of two integrated check valves. Relaxation of the membrane pulls fluid through the second check valve to refill the pump chamber. A micropump based on two one-way valves may achieve a precise flow control of the order of 1 $\mu$l/min. Advantages are fast response, high sensitivity, and negligible dead volume. Unfortunately, these pumps require complicated fabrication processes, generate only modest flow rates and low pressures, and consume a large amount of chip area and considerable power. As a consequence of the low

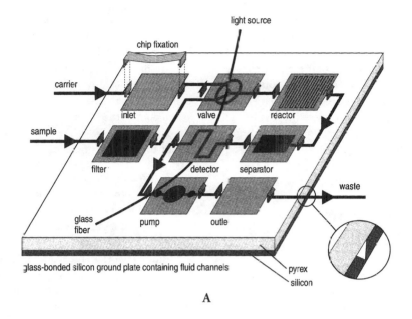

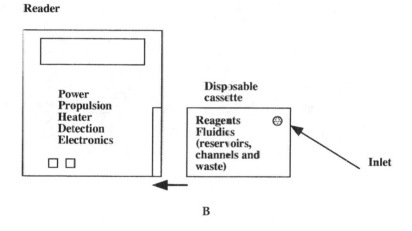

**FIGURE 10.29** Partitioning of functions between instrument and a disposable (A) Configuration for a micro total analysis system (μ-TAS). (From Lebbink, G.K., *Mikroniek*, 3, 70–73, 1994. With permission.) (B) Generic hybrid microinstrument, with inlet, valves, fluid propulsion, detection, waste, and thermal control. Various options exist for partitioning between cassette and reader.

pumping pressure, they often become useless when the device dimensions are reduced and the resistance to fluid flow and required driving pressure increase.

The Si micromachined pump closest to market is Debiotech's micropump shown in Figure 10.30A.* This belt-worn drug delivery pump features a piezoelectric design pioneered at the University of Twente, a generic example of which is shown in Figure 10.30B.[73] The size of most Si micromachined pumps is quite large and one wonders why Si was ever chosen as a substrate. Reasonably, plastic molding technology is now being explored as a viable solution (see also Chapter 6 for discussion on LIGA machined pumps).[74,75]

The Molecular Devices' light-addressable, potentiometric sensor (LAPS) (Figure 10.33), one of the few micromachining-based commercially available biochemical instruments, uses conventional peristaltic pumps to drive reagents through the

chamber holding the micromachined pH sensor.[76] The advantages of this approach are that it relies on well-developed, commercially available components and that a wide range of flow rates are attainable. The LAPS instrument itself is discussed further below.

Micromachined valves, key to mechanical pumping, are a bit more advanced than micropumps and several types are commercially available. Thermally actuated designs are illustrated in Figure 10.4 and Figure 10.40 and their performances are compared in Table 10.16. These electronic technologies are still too complicated, too large, power hungry, and expensive for deployment in disposable cassettes. A more appropriate technology for a cassette may be a 'sacrificial' or 'use-once' valve. Such a valve could ensure device sterility, isolate reagents until they are required, and entrap a vacuum to draw sample into the device (like a vacutainer blood-collection tube). We developed such sacrificial valves, formed from a membrane that covers a micromachined orifice. Thin metal membranes (such as silver) can

---

* Figure 10.30A appears as a color plate after page 144.

be opened by passing a current from the metal via a contacting electrolyte solution to a counterelectrode (U.S. Patent 4,874,500).[77] The valve can be opened with an applied bias as low as 1 to 1.5 V. Alternatively, the membrane may be made from a low-melting-point material such as paraffin wax. Resistive heaters integrated into the orifice could be used to melt and open the valve. Although the 'use-once' microvalve in Figure 10.31 involves an Ag membrane suspended over an orifice, this patented electrochemical valve technology is very generic. The author's current research involves reversible electrochemical valves with polymeric membranes which change their water permeability upon application of large fields.

Some more information on valving technology is provided in the section below on Gas Chromatography. For a more in-depth review of micropumps and microvalves and other microfluidic devices, see also the review by Gravesen et al.[78]

In contrast to diaphragm pumps, the two fluid propulsion technologies that we shall investigate next, acoustic streaming and electro-osmosis use no moving parts. Both of these phenomena produce pumping forces at the walls of a tube. This driving force can be generated along the entire length of a narrow tube to create a distributed pump that moves fluids without an applied pressure. Both modes of fluid transport have been documented, but have not been incorporated in a disposable fluidic element. Both transport mechanisms being surface-force based, scale favorably with decreasing capillary size. In contrast, pressure-based flow relies on a volumetric force, which scales poorly: the pressure per unit length required to maintain a constant linear flow rate in a capillary tube is proportional to the inverse-square of the capillary diameter (see Chapter 9).

**Acoustic Streaming**   Acoustic streaming is a constant (DC) fluid motion induced by an oscillating sound field at a solid-fluid boundary (see also Chapter 9). Transducers do not need to be integrated with the disposable; the cassette with capillary flow channels could simply be laid on top of the acoustic pump network in the reader instrument. Importantly, acoustic streaming is also well suited for mixing reagents.[79] Although it is more complex to implement than electro-osmosis (see below), the insensitivity of acoustic streaming to the chemical nature of the fluids inside the fluidic channels and its ability to mix fluids make it a potentially viable approach. A typical flow

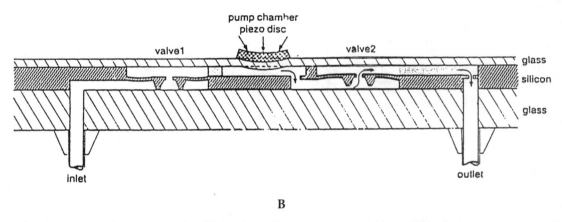

FIGURE 10.30   Micropump technology. (A, the color plate) Debiotech's micropump is a Si micromachined piezopump meant to be a belt-worn drug delivery mechanism. The Oncojet is disposable and may deliver a drug at a rate of 10 to 50 ml/day. The glass-silicon sandwich pump is about 2 × 1 cm in size. (Courtesy Aray Saaman, Debiotech.) (B) Generic design of a piezopump.[73]

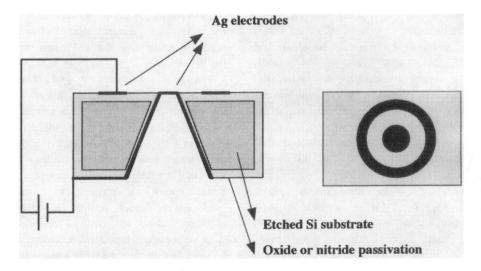

FIGURE 10.31   Disposable electrochemical valve: principle of operation.[77]

rate measured for water in a small metal pipe lying on a piezo-electric plate is 0.02 cc/sec at 40 V, peak to peak.[80]

**Electrophoresis/Electro-osmosis** Electrophoresis moves charged particles in a fluid and is used as a separation method while electro-osmosis drives the fluid near the capillary-walls and is used as a pumping mechanism (see also Chapter 9). The whole fluid column flows when a large electric field is applied to the fluid in a narrow capillary. Typical electro-osmotic flow velocities are on the order of 1 mm/sec with 1200 V/cm applied electric field. For example, in free-flow capillary electrophoresis work by Jorgenson, electro-osmotic flow of 1.7 mm/sec was reported.[81] This is fast enough for most analytical purposes. The technology for generating this flow rate does not involve any moving parts and is easily implemented. All that is needed is an electrode in the reservoirs at each end of the flow channel. Also, if electrophoresis is used for DNA analysis, electro-osmosis becomes an attractive means for fluid transport, since the required electric field is already present. Some disadvantages of electro-osmosis are the required high voltage and direct electrical-to-fluid contact with resulting sensitivity of flow rate to the charge of the capillary wall and to the ionic strength and pH of the solution. It is consequently more difficult to make it into a generic propulsion method compared to acoustic streaming. While the conventional conduits for electro-osmosis are micro-capillary tubes, electro-osmotic pumps have also been micro-machined from planar substrates.[82-86] A variety of other valveless propulsion techniques is under consideration. A very attractive potential candidate is centrifugation of fluids in fluidic channels and reservoirs crafted in compact disc-like plastic substrates.[87] The centrifugal force generated in a compact disc-player-like instrument overcomes, depending on the rotation speed, the capillary forces in the fluidic network and moves fluid in a valveless manner from reaction chamber to reaction chamber and eventually to a waste site. Measurements may be fluorescent or absorption based. Informatics embedded on the same disc, provide test-specific information. Electrohydrodynamic pumping is another recent example of a valveless propulsion approach. In electrohydrodynamic (EHD) pumping, fluid forces are generated by the interaction of electric fields and charges in the fluid (see Chapter 9). This interaction yields a force on the charge which then transfers momentum to the fluid. A major constraint for its use in biological environments is its reduced effectiveness with conducting fluids.[88]

**Vacuum Pressure Reservoir** Fluid transport can also be accomplished by employing chambers within the cassette or reader that are at a higher or lower pressure than the sample and reagent reservoirs. Valves are required to separate the evacuated or pressurized chambers from the rest of the fluid path until fluid motion is required. Opening of a valve initiates the flow of fluid toward an evacuated chamber or away from a pressurized chamber. An advantage of this mechanism over those previously described is that it is a straightforward way to generate a large fluid-motive force. Its drawbacks are that it can only be actuated once, valves may be difficult to implement, and the reliable manufacture of pressurized or evacuated chambers may be challenging. Unless one uses osmotic

pumping by implementing a salt reservoir, external actuation seems the safest implementation method for pressure-gradient actuation.

**Blister Pouch** Another example of external actuation is embodied in the Kodak blister pouch HIV test.[89] This device consists of a plastic pouch containing the reagents for PCR (polymerase chain reaction, see below) and DNA detection. An automatic mechanical roller, similar to the one used in credit card image transfer, is used to squeeze the pouch and move the reagents through the miniature channels in the pouch. This approach meets the goal for a low-cost disposable containing all the reagents and fluidics. A schematic of the blister pouch approach is shown in Figure 10.32. We believe this design to be a good example of clever partitioning; more integrated approaches will have a tough time competing in cost with this design (see also below under PCR).

### Thermal Control

Heating and cooling of fluid within a microinstrument, like fluid propulsion, can be accomplished with elements that are either contained within the cassette or the reader.

**Integrated Heaters** Electrical current passed through resistors integrated on thin membranes can cause a very rapid temperature increase. This is because the resistor has very little thermal mass and is in intimate thermal contact with the membrane and the fluid to be heated (see Chapter 9). Integrated heaters are simple to fabricate on a variety of materials.

**External Temperature Control** Temperature control can also be accomplished with a heating and cooling system that is external to the microsystem. For example, the Kodak pouch uses two external heaters (Figure 10.32). The heaters are cooled with forced air to speed cycling times. Fast heating and cooling are possible with external heaters: Wittwer et al. have demonstrated 30-sec PCR thermal cycling using a hot-air cycler and glass capillary PCR chambers.[79,80] External heaters offer the advantages of simpler (less expensive) micro-instruments and no cassette-to-reader electrical connections. In general, however, external heaters are slower and require more electrical power than integrated heaters.

**Radiative Heating** A third option is to heat fluid within the microinstrument using radiation generated in the reader. The radiation could be infrared, radiowave, or microwave. The radiation could be adsorbed by the fluid or by absorptive structures (such as black ink) placed on the cassette. An advantage of radiative heating is that it avoids electrical contact to the micro-instrument. Radiative heating using the CD player IR laser is an option under investigation at Gamera Bioscience.[87]

### Sample Introduction

The exact mechanism for introducing samples into the micro-instrument depends on the target sample. This is one of the most difficult areas because not only is pretreatment of the sample often required, but one also must transition from a relatively bulky external receptacle structure, which can be

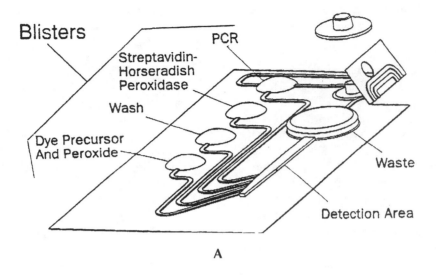

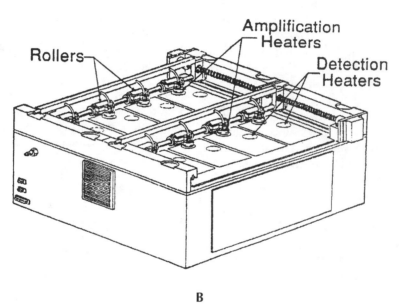

**FIGURE 10.32**  Kodak's blister pouch and propulsion mechanism. (From Findlay, J.B. et al., *Clin. Chem.*, 39, 1927–1933 (1993). With permission.)

handled and accessed manually, to the microfluidic components inside without incurring too much dead space. Very few machining techniques cover the whole dynamic size range required. Because the sample-introduction method may place important constraints on the manufacturing process, a few sample introduction options are considered here.

**Injection Valves**  Traditional sample injection systems such as for flow injection analysis (FIA) rely on some sort of commutating valve, either rotary or sliding, actuated manually or mechanically. Harrow et al.,[92] some years back, compared different commercially available sample injection systems and Shoji et al.[93] were some of the first to micromachine a sample injector consisting of two three-way microvalves and a channel. As pointed out above, these valve systems today are too complex and expensive to consider integrating them in a disposable.

**Micromachined Sample Wells**  Perhaps the simplest way to introduce the sample into the micro-instrument is by

creating wells into which the sample is dropped. For example, most of the substrates with electro-osmotic pumps have plastic tubes glued to them to make sample reservoirs. Similarly, the Biotrack blood coagulation device requires that a drop of sample be placed in a depression on the instrument surface.[94] The sample is then wicked into an internal chamber by capillary action. Problems with sample wells include: the large volume that they require, or if the sample well is small, the need to accurately align the sample with the well. Because the wells are open, one risks sample contamination or the inclusion of air bubbles in the sample. Air bubbles are a major problem in miniature fluidic systems because of the associated surface tension and debubblers, which are typically quite bulky, are often a necessity.

**Injection Ports**  Samples may also be injected directly into the micro-instrument or disposable cassette. An example of this is the Kodak pouch HIV test where an injection septum is provided in the disposable pouch. This reduces the chance of sample contamination but again requires exact alignment of the injection needle with the sample injection port. Furthermore, the manufacture of the cassette pouch is complicated by the need for an injection septum.

**Integrated Sample Collection**   In an integrated sample collection device could be the cassette. For example, the cassette could be manufactured in the form of a blood collection tube (vacutainer), a syringe, a pipette, or a swab. However, this would complicate the cassette manufacture considerably and may make the cassette design specific to one type of sample.

*Manufacturing*

The choice of fluid transport and heating techniques must go hand-in-hand with the choice of cassette manufacturing technique. A thorough review of manufacturing techniques and how to choose from them can be found in Chapters 7 and 8. The principle advantages of nontraditional methods for making micro-instruments are that very intricate structures with very fine features can be produced and that heaters, electro-osmosis electrodes, and acoustic streaming pumps can be readily realized. On the other hand, nontraditional microfabrication is optimized for making devices that are considerably smaller than most biochemical reaction sample volumes. A 100-µl reagent volume is the same size as a Si chip that is 14 mm on a side — huge by microelectronics standards. It is certainly possible to make Si micromachined parts that are as large as the starting wafer (typical 4-in. diameter); however, the costs of the wafer and wafer processing are likely to be much higher than for other manufacturing techniques. Additionally, variation from a previously developed microfabrication sequence can require an expensive and time-consuming process development effort. In the meantime, traditional methods have provided answers to most of the miniaturization needs in diagnostic applications.

The above generic issues in microinstrumentation for biomedical applications constitute prime examples of the need for micromachinists to look at all miniaturization options rather than to be blinded by Si micromachining as the only way to make progress. Furthermore, a hybrid approach is the only reasonable short-term solution for miniaturized total chemical analysis systems.

## Examples of Miniaturization in Biomedical Equipment

### Microphysiometry

In 1988, Hafeman, Parce, and McConnell introduced a new sensing principle for pH based on an electrolyte/insulator/semiconductor arrangement (EIS).[76,95] They called it a light-addressable potentiometric sensor, with the acronym LAPS. A diagram of the system is shown in Figure 10.33A. By modulating the light of an LED shining from the backside of the Si wafer onto the silicon/insulator interface, the flatband potential (i.e., the DC bias resulting in zero electrical field in the Si near the insulator/silicon interface) can be determined from the alternating photocurrent. The flatband potential sensitively reflects the pH at the electrolyte/insulator interface. The presence of protonatable silanol and amine groups on the insulator surface ensures that the flatband potential depends on pH

in a Nernstian fashion.[95] In other words, the LAPS is very similar to a pH ISFET in that the surface charges at an insulator are measured, but in the LAPS configuration no discretely insulated gates are needed and the monitoring is optical rather than electrical.

Based on the LAPS principle, Molecular Devices Corp. (Sunnyvale, CA) introduced the Cytosensor Microphysiometer. The instrument monitors the metabolism of living cells. Real-time response of cells is followed by monitoring the rate of acidification for sudden changes due to addition of, for example, drugs or toxins. The measurement of the rate of pH change instead of absolute changes decreases the problem posed by potential drift. The heart of the device is a microvolume chamber containing living cells in diffusive contact with the LAPS pH sensor (Figure 10.33B). A plunger defines the microchamber size. The system periodically halts the flow of a fresh nutrient solution through the chamber, to allow acidic waste products to build up, and the pH sensitive chip picks up differences in metabolic rate of the cells.[96] The smaller the chamber the larger the pH changes for the same number of cells. Besides the pH sensor the amount of integration on Si is obviously minimal. Molecular Devices has presented research on a more integrated instrument. Currently each channel in the Cytosensor Microphysiometer has an individual Si sensor chip in each of eight separate flow chambers (~ 1 ml/sample). To achieve higher density and throughput, the logical next steps are to place several flow channels on a single chip, each with its own sensor spot, and to reduce the chamber volume further. This was accomplished in the laboratory by wet etching Si to produce an eight-channel flow-through chip as shown in Figure 10.33C. The chamber volume here is reduced to ~1 µl. Pumps, valves, and debubblers remain all off-chip. Further integration would mean producing microfabricated valve arrays (32 valves in an area of 23 × 23 mm are required) and debubbler arrays. Due to the type of actuators and flexible membrane used today, existing microvalves have difficulties reaching such densities. The debubblers, involving conduits covered with a gas permeable, hydrophophic membrane might be a little easier to integrate. The microphysiometer under development measures the mean properties of a few thousands cells, yet a further exciting development would be to measure individual cell metabolism at the bottom of cubic wells containing only one cell.[97]

Molecular Devices also developed the capability for depositing multiple discrete chemistries, or detection sites (say, for immunoassays) on single chips, with up to 1500 test sites possible on a 4-in. Si wafer. By modulating an array of LEDs in a sequential fashion, the LAPS technology can then be used for measurements of discrete assays (Figure 10.33D). The chemical reaction producing protons which are adsorbed onto the chip's surface at each local test site, thus changing the electrical potential at that site, causes a measurable change in the local photocurrent. Enzyme immunoassays, for example with urease as a label, can be followed sensitively in this manner in a low-volume LAPS chamber. The small volume ensures maximum pH change for the reaction product formed when the urease is brought in contact with a substrate solution. The rate of change of the pH vs. time is proportional to the amount of analyte present.

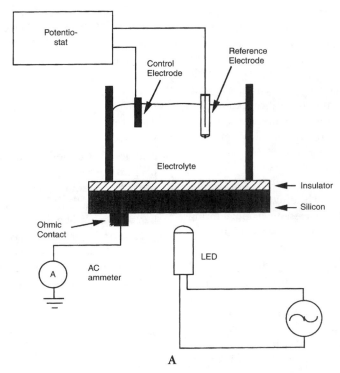

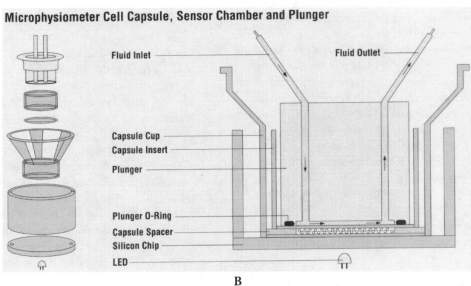

**FIGURE 10.33**   Light-addressable potentiometric sensor (LAPS) for biochemical systems. (A) Diagram of the LAPS sensor. (B) Microphysiometer cell capsule, sensor chamber, and plunger. (C) Three-dimensional view of the assembly of an eight-channel, flow-through chip, with a cover slip, LEDs, and fluid connections. (D) Discrete chemistries produce variations in the local surface potential that can be determined by selective illumination with one or another of the light-emitting diodes. (Figures B and D, from Hafeman, D.G., Parce, J.W., and McConnell, H.M., *Science*, 240, 1182–1185 (1988). With permission.)

For immunoassay applications Molecular Devices developed the Threshold® Immuno-Ligand Assay System. One Si LAPS chip in this instrument has nine active areas (eight sensor sites and one reference site), and the enzyme substrate reaction leads to micro-pH changes within a detection volume of 0.5 μl per site.[98] Given the many steps required before the actual pH detection takes place this $40,000 system hardly qualifies as an 'immunosensor' but the approach to miniaturization taken by Molecular Device does illustrate a very sensible, conservative approach towards miniaturization: starting with one crucial micromachined part and gradually integrating more as technology matures and micromachining offers a cost/performance advantage.

### Electrophoresis

As we saw above, electrically charged particles in an electrical field migrate toward the electrode of the opposite charge in a process called electrophoresis. The physical characterization

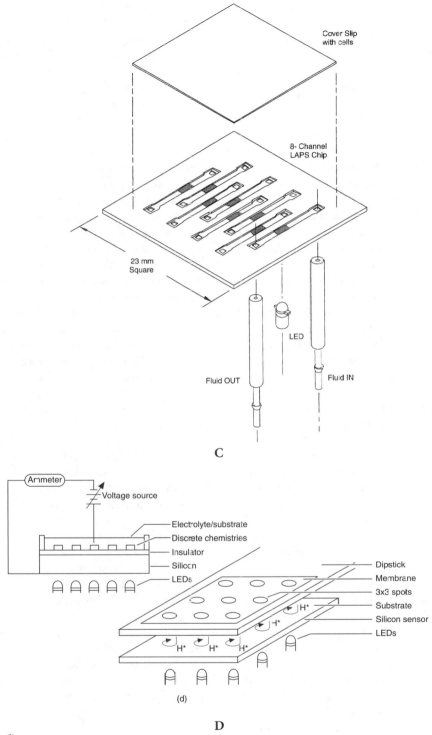

**C**

**D**

FIGURE 10.33 (Continued)

of analytes migrating as an electrophoretic zone is conventionally carried out by comparing the migration distances, i.e., the retention times of unknown with those of known chemicals. For detection of zones the most common approaches involve the use of stains, radiography of radiolabeled analytes, fluorescence or ultraviolet absorption of fluorescent or ultraviolet-absorbing labels, or immunoreaction with specially prepared antisera. Separation methods based on measuring retention times as in electrophoresis, gas chromatography, high-pressure liquid chromatography, etc. rely on separation instead of selective detectors. In principle, with a small enough micromachined instrument, these separation systems may be equivalent to a sensor. Columns in both capillary electrophoresis and high pressure liquid chromatography have already been miniaturized to diameters well below 1 mm with traditional machining technology. The motivation for miniaturization in this case was often the high expense and hazardous nature of the mobile phase and the increased speed of analysis with smaller sample volumes. Micromachined channels may afford a next level of miniaturization.

Electrophoresis is carried out in a variety of experimental set-ups which can be divided into two major categories: free solution in which no stabilizers are added, or with anticonvection stabilizers such as paper or gels. Miniaturization of these two approaches poses different challenges.

In the case of free solution electrophoresis in a capillary, the electro-osmosis force becomes significant and can be used advantageously as a propulsion force for the sample in the electrophoresis unit (see above). Electrophoresis in free solution has certain other attractive features. First, electrophoretic mobilities can be measured without the introduction of complicating factors such as adsorption of analyte to the stabilizer, molecular sieving effects, or tortuosity of migration paths. Thus measurement of the true mobility of an analyte is more straightforward. Second, electrophoretic separation of large particles, including whole cells, is readily accomplished, whereas this is more difficult in the microporous network of a stabilizer. Unfortunately electrophoresis in free solution suffers from a serious problem with convection. The passage of electric current through the electrophoretic medium results in joule heating of the medium. This heat is dissipated only through the boundaries of the electrophoresis channel, with as a natural consequence the evolution of temperature gradients within the electrophoresis medium, leading to convective flows. These flows can easily disrupt a separation by mixing of zones. A second problem is resolution; the method is not good enough to separate DNA fragments differing by one base only. Still, the technique is very amenable to automation, and by carrying free zone electrophoresis out in small bore capillary tubes or etched channels the magnitudes of temperature and density gradients remain rather small.

Gel electrophoresis with its higher resolution is vital to the genetic engineer, as it represents the main way by which nucleic acid fragments may be visualized directly. The method relies on the fact that nucleic acids are polyanionic at neutral pH, i.e., they carry multiple negative charges due to the phosphate groups on the phosphodiester backbone of the nucleic acid strands. This means that the molecules will migrate towards the positive electrode when placed in an electric field. The technique is carried out using a gel matrix, which separates the nucleic acid molecules according to size, i.e., a sieving gel. The sieving gel also stabilizes the separation zones by preventing convection. Two gel types commonly used are agarose and polyacrylamide. Polyacrylamide gels are used to separate small nucleic acid molecules in applications such as DNA sequencing, as the pore size is smaller than that achieved with agarose. The above gel electrophoresis method is not well suited for automation and on-line detection. Also, gel electrophoresis poses more difficult micromachining challenges as it requires that a gel be placed in the small microchannels, which could be impractical. Moreover any nonelectrophoresing polymer would probably still have a significant mobility caused by electro-osmosis, so optimization of this system would involve gaining control over the velocity of the electro-osmotic flow, relative to the electrophoretic flow and its effect on the sieving polymer. Volkmuth et al.[99,100] have started addressing the problem of a mobile sieving gel by testing microlithographically patterned immobile arrays of obstacles in the microchannels instead. Lithographically constructed, immobile obstacle arrays, with their extremely reproducible topography, will make it possible to understand the motion and fractionation of large polymer molecules in complex but well characterized topologies and may ultimately make a micromachined, high resolution gel electrophoresis unit possible.

### Flow Cytometry

Flow cytometry is a method for high-throughput measurements of physical and chemical characteristics of cells and other microscopic biological particles. Typically, in commercial cytometers, optical measurements are made as particles are passing, single-file, through a sensing region. The particles are constrained to such a single line-up by hydrodynamic focusing. In this method, the sample flow is a narrow stream formed as the innermost flow of two concentric fluid flows (see Figure 10.34A). The faster the outer sheath flow, the narrower the inner sample flow is focused. Hydrodynamic focusing requires absolute laminar flow in all positions in the flow cell; from some high speed on, the inner stream will defocus through turbulence. Existing commercial equipment requires frequent adjustment of the optical components, but simpler instrumentation is expected through the introduction of optical fibers replacing lenses for sample illumination and light collection. Shapiro et al. in 1986,[101] as the next step towards more user-friendly equipment, suggested integration of flow channels, optical elements, and signal conditioning electronics. The first such attempts led to complex micromachined sheath flow cells; for example, a five-layer stainless steel/glass laminate[102] and a four-wafer Si micromachined flow chamber[103] were early attempts. More recently, Sobek et al.[104] made a much simpler micromachined flow cytometer, which seems to put us back on the right track in terms of correct use of available micromachining tools. Cost consideration did result in a $30 \times 15 \times 1$-mm chip made from two symmetrically machined HOYA T-4040 synthetic quartz (fused silica) parts. These two parts, bonded at 1000°C, define the ovoid cross-section of the sheath flow channel, the circular cross-section of the sample injector, and the ovoid measuring capillary (Figure 10.34B). A smooth flat surface of the capillary is essential for ensuring good optical measurements in the measuring region. The quartz starting material used here comes in 4-in. wafers just like Si wafers. This is an important point as most processing equipment is built to take Si wafers only. The resulting cytometer can focus a sample stream, introduced into the injector with a syringe, to a 10-μm width and can maintain this focusing to a mean velocity up to at least 10 m/sec.

The next step in this development would be to find the correct package for the cytometer and an interface to the macroworld. Connectors to the sheath flow ports and the sample injection mechanisms are areas ripe for innovative approaches. A complete overall redesign of a cytometer should begin with the package.

### Polymerase Chain Reaction

Polymerase chain reaction is to genes what Gutenberg's printing press was to the written word, but it took the 1995

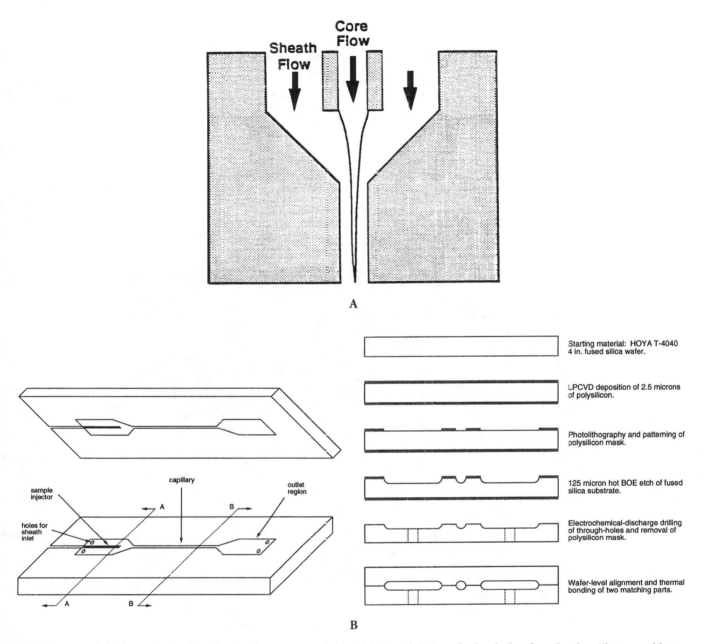

**FIGURE 10.34** (A) Schematic of a flow chamber for cytometry. (B) (Right) Exploded view of a fused silica flow chamber. The top and bottom have matching injection and outlet areas with a capillary joining them. Through-holes in the bottom plate are for sheath inflow outflow. The sample is introduced through the injector into the center of the sheath flow. (Left) The fabrication sequence shown along the cross-section A-A. (From Sobeck et al., Technical Digest of the 1994 Solid-State Sensor and Actuator Workshop, Hilton Head Island, SC, 260–63, 1994. With permission.)

O.J. Simpson trial to make PCR a household word. Through it, a desired DNA sequence of viral, bacterial, plant, or human origin can be amplified hundreds of millions of times in a matter of hours. By amplifying minute amounts of DNA to analyzable quantities, circumventing the need for lengthy cell-culture methods, the technique has had major impact on clinical medicine, genetic disease diagnostics, forensic science, and evolutionary biology.

The PCR is a process based on the polymerase enzyme, which can synthesize a complementary strand to a given DNA strand in a mixture containing the four DNA bases and two primer DNA fragments (each about 20 bases long) flanking the target sequence. The mixture is heated to separate (denature or melt)

the strands of double-helix DNA containing the target sequence and then cooled to allow (1) the primers to find and bind to their complementary sequences on the separate strands and (2) the polymerase to extend the primers into new complementary strands. Repeated heating and cooling cycles multiply the target DNA exponentially, since each new double strand separates to become two templates for further synthesis. In Figure 10.35A* we illustrate how each PCR cycle doubles the amount of DNA.

---

* Figure 10.35A appears as a color plate after page 144.

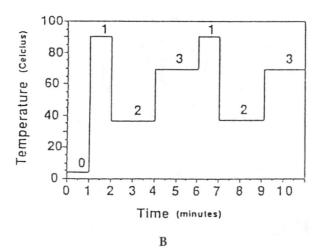

B

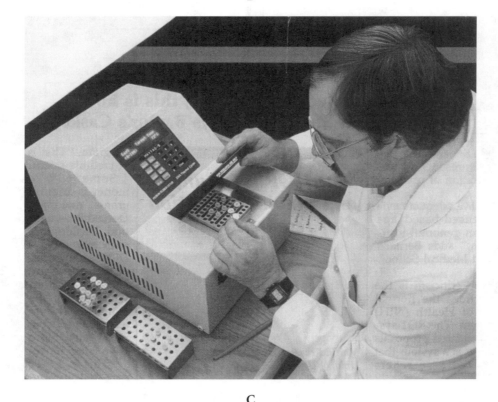

C

**FIGURE 10.35**  PCR: (A) a color plate illustration (appearing after page 144) on how PCR cycle doubles DNA; (B) a typical PCR temperature profile; (C) Commercial PCR instrument and PCR tubes.

A typical temperature profile (Figure 10.35B) for PCR looks as follows:

1. Denature @ 93°C for 15 to 30 sec.
2. Anneal primer @ 55°C for 15 to 30 sec.
3. Extend primers @ 72°C for 30 to 60 sec.

The primer extension step has to be increased by roughly 60 sec/kbase to generate products longer than a few hundred bases. The above are typical instrument times; in fact, the denaturing and annealing steps occur almost instantly but temperature rates in commercial instruments (Figure 10.35C) usually are less

than 1°C/sec when metal blocks or water are used for thermal equilibration and samples are contained in plastic microcentrifuge tubes. Instantaneous temperature changes are not possible because of sample, container, and cycler heat capacities, and extended amplifications times of 2 to 6 hours result. By micromachining thermally isolated, low-thermal-mass PCR chambers, one might be able to mass produce a faster, more energy efficient, and a more specific PCR instrument. Some major attempts to that effect have already been made.

Northrup et al.[105] fashioned Si microchambers with integrated thermally isolated heater elements and found that thermal cycling resulted in amplification without the need for active cooling. This is a direct result of the effective cooling of these small Si structures (see Chapter 9). Northrup et al. found that they could heat the PCR reactants as quickly as 35°C/sec. Further work has shown that integrated heaters on both sides of the reactant volume are required for optimal temperature uniformity. We believe that to make progress in PCR for diagnostic applications the heater/cycle miniaturization is not the most important aspect; miniaturizing a thermal cycler is probably the easier part. Moreover, the heater/cooler should be part of a fixed reader instrument to make an inexpensive disposable possible. Along this line Wilding et al.[106] performed a less integrated experiment with Si/glass microchambers of 4.4 to 8.9 μl inserted in a commercially available heater/cooler structure. The PCR total cycling time was reduced but specificity was found to be inferior to commercial cyclers. This inferior performance probably is linked to the fact that Si is poisonous to the polymerase enzyme. The specificity problem can be avoided by derivatizing the Si surface or by machining PCR chambers from a more enzyme-compatible material. The most glaring problem with the two described Si micromachined PCR approaches is that they do not provide an easy way to apply and remove the DNA samples; the interface with the macroworld is problematic. We suggest that the major problem with microsystems is the making of connectors, receptacles, and packages that link the microstructure to the macroworld. In principle, to avoid a hybrid approach, this requires a manufacturing method with a very wide dynamic range, i.e., from microns to several millimeters or even centimeters. For this reason Si is not necessarily the best approach. The next two examples illustrate viable alternatives. Using a thermal cycler based on recirculating hot air and 10 μl samples in thin glass capillary tubes and with the sample temperature monitored by a miniature thermocouple, Wittwer et al.,[90,91] optimized PCR temperature and cycle times and found that they could improve product specificity significantly while decreasing the required amplification time by an order of magnitude. They found that, for a 536-basepair β-globin fragment from human genomic DNA, optimal denaturation at 92 to 94°C took place in less than 1 sec, annealing for 1 sec or less at 54 to 56°C gave the best product specificity and yield, and only elongation, at 75 to 79°C required longer times. Denaturation yield actually decreased with times greater than 30 sec and nonspecific amplification was minimized with a rapid denaturation to annealing temperature transition. This instrument is still lacking an interface to the real world but seems more amenable to manufacturing since the partitioning of the various parts was executed more sensibly.

Automated amplification coupled with detection systems would revolutionize the medical diagnostic application of PCR. An interface for the μ-PCR and the macroworld needs to be designed first. Sample receptacle, reagents, and all fluidics, mixing, and valving (if required) as well as waste area should be integrated in an inexpensive disposable. Heat cycling and propulsion mechanisms better reside in the permanent reader. An elegant example of progress in this direction comes from the Eastman Kodak Company (see Figure 10.32).[89] (The technology was bought by Johnson and Johnson in 1995.) This company developed a disposable pouch-based PCR instrument with a total amplification time of 43 min and a sensitivity and specificity adequate for the detection of human immunodeficiency virus (HIV). A sample and the PCR cocktail are injected into the amplification compartment through a fill port in the pouch. Pouches are aligned on a fixed instrument with the amplification compartments on the upper amplification heaters and the detector areas on the lower detection heaters (Figure 10.32). On completion of the amplification, the next analysis step is accomplished by an automatic roller that forces detection chemistry fluids and amplified samples from compartments through channels leading to a temperature-controlled detection area. In the detection zone, biotinylated PCR products are captured via hybridization to probes immobilized on latex particles. Multiple probe sites can be accommodated in the detection area. In the prototype instrument, treptavidin-horseradish peroxidase is allowed to interact with those probes, and any that is unbound is removed with the subsequent wash solution. In the presence of peroxicase, leuco dye is oxidized to a visible blue color, indicating a positive hybridization event. Replacing the relative thick walls of traditional, low surface-to-volume, polypropylene reaction tubes by the pouch "blisters", with their relatively high surface-to-volume ratio and very thin walls (100 μm), readily conformable to the shape of the instrument heaters, permits rapid heating and cooling. The above inexpensive and ingenous way to couple PCR and detection is what a Si micromachining approach has to compete with.

# Gas Sensors

## Introduction to the Gas Sensor Market

The world gas sensor market in 1990 was estimated at about $0.9 billion. Solid state gas sensors such as zirconia-based oxygen sensors for automotive, toxic, and combustible gas sensors for the home and chemical and petrochemical industry, and humidity sensors for the energy conservation market (e.g., drying operations) constitute up to 90% of that total.[107] This market today is served by unselective solid-state gas sensor products and there is plenty of room for innovation.[108] The lack of progress in terms of selectivity in solid state gas sensors is making for a very fast growing market for the more selective but more expensive wet electrochemical gas sensors (some with a 30%/year growth rate) and the newly emerging inexpensive optochemical techniques. With the automotive world as the driving force, significant efforts in using microfabricated solid state gas sensors are under development. So far results on micromachine solid-state gas sensors have been promising in terms of thermal budget minimization but discouraging in terms of sensitivity and inexpensive manufacturing methods. Excellent miniaturized wet electrochemical gas sensors are now available, all based on traditional machining methods. We will list the merits and disadvantages of using micromachining and explore or suggest new directions to be taken to reach a better product offering in the nontraditional machined devices.

## Solid State Gas Sensors and Micromachining

The market size for automotive oxygen sensors is difficult to assess because most of them are produced captively by automobile manufacturers. We estimated the total 1990 U.S. market for such sensors to be between $100 million and $120 million.[107] Several factors will make solid state oxygen sensors a very attractive business opportunity in the near future. Government regulations in California will soon be requiring two oxygen sensors per car, and the demand for a sensor measuring oxygen both in lean and rich gas mixtures, i.e., so-called wide-range oxygen sensors is high.

The high operation temperature of zirconia-based oxygen sensors (>300°C) precludes the use of Si technology for their fabrication. In Chapter 7 we discussed alternative planar wide-range zirconia oxygen sensors for automotive applications including one sensor fabricated by using plasma-spray deposition (see Figure 7.21C).[109,110] The use of plasma spraying might enable the batch fabrication of such sensors at a cost below $2.00 rather than the current $12.00 and up. Controlling the plasma spray process to form the porosity gradient in the sensor is still a major challenge.

The 1990 market for small solid state gas sensors and sensor systems in the toxic and combustible gas arena was some $120 million in the U.S. alone. Broken down by product, the U.S. market for these chemical sensors comprises about $25 million for hydrogen sulfide, $20 million for total hydrocarbon monitoring, $12 million each for carbon monoxide and oxygen deficiency, and small percentages for a myriad of other gases. The use of a ceramic tube in a classical Taguchi sensor (see Figure 3.42) maximizes the utilization of the power from the coiled Pt wire heater inside the tube, so that most of the power is used to heat the tin oxide covering the tube. The tin-oxide paste is applied on the outside of the ceramic body over thick-film resistance measuring pads by dip coating and is sintered at high temperature. The sintering stabilizes the intergranular contacts where the sensitivity of the gas sensor resides. The structure is not a design suited for mass production as it involves excessive hand labor. The problems with the thick-film structure are mainly in the area of reproducibility: compressing and sintering a powder, the deposition of the catalyst (e.g., the size of palladium crystallites and how close they are to the intergranular contacts), and the use of binders and other ceramics (e.g., for filtering) are all very difficult to control. It is obvious from Example 1 in Chapter 3 that the application of IC techniques to Taguchi sensors (see Figure 3.43) can improve the state of the art dramatically in terms of the required power budget; a power reduction by a factor of ten is quite common. Unfortunately, as explained in Chapter 3, no thin-film tin-oxide gas sensor can yet compete with the thick-film-based traditional Taguchi gas sensor in terms of sensitivity. If one could make a thin film tin-oxide with the same gas sensitivity as a thick film then micromachining the heater elements and thin-film deposition of the gas-sensitive material would further improve reproducibility and make for an excellent product. For long-term stability relatively low temperature semiconductor catalyst should be used (perhaps operating at temperatures <200°C).

A major obstacle to more rapid progress with thin-film micromachined gas sensors remains that the sensitivity of thin films to gases is not yet very well understood. The currently available devices, already quite small and inexpensive, will only be replaced quickly when newer technology improves upon the existing product in terms of reproducibility, cost, sensitivity, response time, concentration range, specificity, reversibility, stability, and power consumption.

At this time the best course for future progress in the Taguchi-type sensor is to optimize sensitivity of thin-film oxide semiconductors, use planar Si devices (reproducibility and cost), and concentrate on implementing less power-consuming micro-heater elements. Along this line Motorola has completed construction of a pilot line for large-scale production of micromechanical Si thin-film tin oxide gas sensors. Details about the nature of the thin-film tin oxide are scarce at this point.[111] For very low cost and less demanding applications the Motorola sensor might succeed, for more demanding applications and for the short-term, planar thick film structures on ceramic might present a better option.[108]

The market for humidity sensing in the U.S. is about the same size or larger than the total toxic gas sensor market ($120 million). A particularly interesting set of new micromachined gas sensors, including humidity sensors, are based on thin-film-coated chemiresistors and acoustic devices.[35] We will discuss the JPL microhygrometer, further below.

## Electrochemical Gas Sensors

Back in the 1960s, semiconductor devices were heralded as the replacement for electrochemical sensors; however, the electrochemical gas sensor sales are experiencing a growth rate of 30%/year. The latter is probably due to the slow rate of improvement in the selectivity of solid state gas sensors and superior selectivity of electrochemical sensors brought on the market by companies such as City Technology and Neotronics.[107]

A micromachined oxygen sensor was presented in Example 2 of Chapter 4 (Figure 4.66). Here we add an example where the attempt was to make a very fast and sensitive small electrochemical gas sensor. In any type of gas sensor, sensitivity is critically dependent on the number of sites where reacting gas, solid, or liquid phase and catalyst meet, i.e., the number of triple points.[35] In an electrochemical sensor these sites constitute points where metal electrode, gas, and electrolyte come together. Usually the triple points, in a wet electrochemical cell, are covered by a thick layer of electrolyte. This author made faster responding electrochemical gas sensors by innovative methods of exposing the triple points more directly to the gas media. By making the liquid path through which the gas must diffuse to reach the triple points as short as possible or even do away with it completely a much faster sensor was realized. We believe it is possible this way to create gas sensors that combine some of the positive characteristics of wet electrochemical (selective) and solid state gas sensors (fast). The sensor we invented for this purpose is the Back-cell™.[112] Two versions of the Back-cell sensors were investigated: sensors using a porous ceramic substrate and those based on a Si substrate with micromachined pores. The two approaches are compared in Figure 10.36. For electrolyte a thin layer of Nafion or a hydrogel was deposited over the metal electrodes on the front of the porous Si or ceramic slab and the gas to be detected was only allowed to reach the sensing electrode from the back of the porous substrate. The details of the microfabrication procedure were explained elsewhere.[113,114]

The response of this sensor (to 90% of full signal) to a step change in gas concentration (say, of oxygen) from 0 to 100% occurs in less than 300 msec; this is extremely fast compared to other amperometric sensors (typical response time 30 to 90 sec). Gas does not need to diffuse through a thick layer of electrolyte to reach the triple points where they react. In other words, the triple point is almost dry (as water is involved in the reaction, gas still diffuses through a few monolayers of water to react at the metal). The recovery of the sensor back to baseline upon a

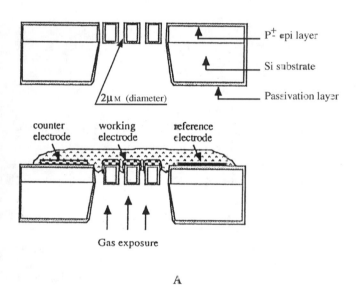

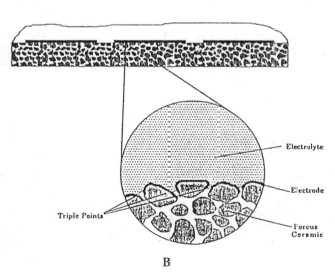

FIGURE 10.36 Back-cell™ gas sensor approaches for fast responding electrochemical sensors: (A) Si embodiment,[113] and (B) porous ceramic embodiment.[114]

step change in oxygen concentration from 100 to 0% is slightly slower than the initial response: 400 msec.

For ease of manufacturing, the ceramic approach is preferable. The major problem with the current device is water management. The fast response of the sensor gets lost when water fills the pores on the back of the metal electrodes. Making the pores hydrophobic is a solution we are exploring now.

## Optochemical Gas Sensor

An optochemical gas sensor approach developed by Quantum Group Inc.[115] is being adopted by First Alert as an inexpensive and selective fire alarm. This sensor comprises a porous, semitransparent substrate into which a self-regenerating chemical sensor reagent selective for CO is impregnated. Reaction with CO changes the color of the reagent and this change is picked up by a diode photo-detector. Oxygen in the air regenerates and clears the substrate. In this fashion the sensor is biomimetic and simulates human response to airborne toxins such as carbon monoxide, mercury, ethylene oxide, volatile organic materials, and hydrogen sulfide. This is a very inexpensive and promising approach and with further fine-tuning of the chemistry of various reactants we might expect a large family of products based on this optochemical approach with a selectivity akin to an electrochemical sensor but at a cost below that of a solid state gas sensor.

The biomimetic sensor has a functional lifetime of over one year and the mimicking of the human response to toxins is achieved by the use of a molecular encapsulant that encapsulates at least one gas component. Work is concentrating on extending the sensor's lifetime and speeding up the regeneration of the sensor. The optochemical gas sensor from Quantum Group Inc. could be miniaturized by using nontraditional manufacturing techniques, but finding selective chemistries for other gases, such as $NO_2$, $O_3$, $H_2S$, etc., is a higher priority.

# Miniaturization of Analytical Equipment

## Introduction

Microprocessors provide instrument developers with unprecedented capabilities for monitoring and controlling instrument functions and timing sequences in addition to their obvious benefits for use in data processing and communications. Taking further advantage of the IC revolution there is a trend to build nonelectronic components of analytical instruments relying on the same type of batch fabrication equipment that made ICs possible. The size of miniaturized instruments is huge, though, compared to batch-fabricated IC components, and the same economy of numbers rarely applies. This leads to other candidate micromachining methods, including traditional methods, now being reevaluated for the purpose of building smaller analytical equipment.

Research in instrumentation miniaturization is especially challenging as science requirements call for advances in performance with simultaneous large reductions in device mass, volume, and cost. Many conventional instruments operate at or near theoretical limits, and reducing the scale (volume and mass) often leads to sharp reduction of performance. This situation calls for new fundamental principles for instrumentation technology development. In other instances, we shall see, miniaturization is making for better instruments.

From the time a picture of a gas chromatograph, the size of a matchbox, was printed on the cover of the April 1983 issue of *Scientific American*[116] (Terry and Angell built a first prototype between 1974 and 1975), there has been a continued effort to micromachine analytical instruments. Few micro-instruments reached the market. They are still relatively large, tabletop or suitcase size, with only a few key components based on micromachining. Some examples are the MTI gas chromatograph, looking significantly less integrated than the device suggested in the 1983 *Scientific American* article (the MTI gas chromatograph column itself is not micromachined, only the inlet and detector are). Other examples include the two biomedical instruments based on Molecular Devices' patented LAPS technology reviewed above. In the LAPS devices, the Si content is a minute contributor to the $40,000 dollar price tag but it is the most essential part. The same is true for micro-instruments such as the STM and AFM proximal probes based on micromachined tips. For the advent of monolithic subminiaturized equipment we will have to wait quite a bit longer. The verdict on the wisdom of using Si micromachining for micro-instrumentation is still out, but looking at the miniaturization job from a precision engineering perspective one becomes aware of the many options. In the following we will analyze what the major hurdles are for further miniaturization of a variety of important analytical instruments.

## Gas Chromatography

Gas chromatography is a technique widely used for the separation and analysis of gaseous samples. A typical gas chromatograph (GC) consists of a carrier gas supply, a sample injection system, a separating fused silica column, an output detector, and a data processing unit, as depicted in the block diagram in Figure 10.37.[117] Separation of the sample vapors is achieved via their differential migration through a capillary column. A precise and reproducible volume of sample vapor is injected at the input of the column by a valve and swept through it by an inert carrier gas. The column is lined with a liquid stationary phase, a substance capable of absorbing and desorbing each of the component vapors depending upon their unique partition coefficient. The migration rate of each vapor along the column depends on the sample injection pressure, the carrier gas velocity, the temperature, and the degree to which the vapor is absorbed by the stationary phase. The column's output is thus a series of vapor peaks separated by regions of pure carrier gas. To detect those peaks, the output gas stream from the column is passed over a detector which measures a particular property of the gas, such as the thermal conductivity, which can be related to the concentration of sample vapor in the carrier gas. The detector produces a signal which is amplified and used in the analysis of the sample mixture; the identity and quantity of each vapor in the mixture can be determined from, respectively, its retention time in the column and the area under its output peak.

Gas chromatography systems tend to be large, fragile, and bulky pieces of laboratory equipment, and miniaturization would present many advantages:

- Integration of the instrument into a process,
- Portable and handheld units,
- Fast response time (1 min or less) in comparison with a large unit with comparable plate numbers (see Chapter 9 on scaling),
- Less calibration and carrier gases needed,
- Low dead volume,
- Higher accuracy.

The first integrated GC, consisting of a long separating column, a sample injection valve, and a thermal conductivity detector (TCD), fabricated on a single Si wafer, was developed at Stanford as early as 1974. The MTI commercial mini-GC, based on that original work, uses a traditional fused silica capillary column but retains the other micromachined components. In Figure 10.38* we show a picture of the commercially available MTI GC with micromachined thermal conductivity

---

\*      Figure 10.38 appears as a color plate after page 144.

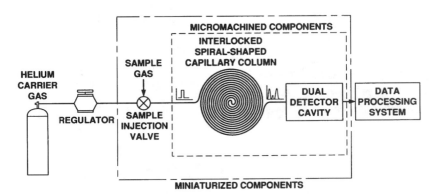

**FIGURE 10.37** Functional block diagram of a gas chromatography (GC) system. (From Kolesar, E.S. and R.R. Reston, Separation and Detection of Toxic Gases with a Silicon Micromachined Gas Chromatography System, 1995. With permission.)

detector and injector. By the time MTI produced their first GC, 100-μm diameter, fused silica GC columns had become commercially available and they proved to be superior to the Si micromachined columns. The reasons behind their superiority must be sought in the perfect symmetry of the monolithic silica columns. The smoothness of the stationary phase coating the inside of a capillary column is a function of the profile symmetry. The uniformity of coating thickness is extremely important to maintain very sharp separation peaks. The micromachined designs described in the literature usually consist of a cover sheet of Pyrex glass which is anodically bonded to a U-shaped, V-groove, or rectangular groove etched in Si.[118] Columns also have been made by joining two U-grooved halves (see Figure 10.39). All these designs have a major potential deficiency; they are bound to accumulate coating on the corners or joints leading to a less perfect separation.[119] Analytes spend a longer time in areas where the coating is thick compared to places where it is thin, leading to a broadening of peaks. Thermal gradients might also be induced by joining dissimilar materials, again causing peak broadening. It also should be realized that the column has to be replaced on a regular basis due to aging and fouling. Therefore, the price of a total integrated system would have to be very low.[120]

Research efforts for making better micromachined GC columns are continuing,[117] but, since existing fused silica columns are perfectly adequate, the focus would be more gainfully put on other GC problem arenas. Specifically, micromachining could be of great immediate use in the arena of better, more reproducible, miniaturized valves and fast heating and cooling

of thermally isolated column and injector for fast temperature-programmed chromatograms.

A GC only works properly if the volume of the injected gas is small compared to the volume of the column. The reproducibility of that amount of gas is crucial. This calls for a miniature, highly accurate sample-injection valve. Hewlett Packard has come up with a thermally-actuated valve intended for a very small, possibly a handheld, GC.[121,122] A schematic of HP's normally closed nickel/silicon bimetal valve is shown in Figure 10.40A. Resistive heating of the nickel/silicon bimetal causes the membrane deformation and opening of the valve. The competing IC Sensor's valve (Figure 10.40B) is also a normally closed bimetallic valve but uses an annular aluminum region on a Si diaphragm with integral diffused resistors instead. By varying the electrical power dissipated in the resistors and thus the temperature of the Si diaphragm, the displacement of a central boss can be controlled.[123] The HP valve exhibits an order-of-magnitude increase in both pressure and flow control capability compared to the IC Sensor's valve. This improvement is made possible by using a torsion bar suspension (see Figure 10.40A) and a nickel-and-silicon bimetal combination, which allows larger valve dimensions and higher stress levels for consequent longer stroke and higher force. For very small structures, this thermal type of actuation becomes an attractive option due to geometric scaling (see Chapter 9). The amount of thermal mass decreases as the volume of heated material decreases, and the thermal loss decreases as the thickness of the support structure for the bimetal decreases, while the force per unit area remains high. The normally closed, thermopneumatic

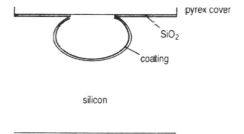

 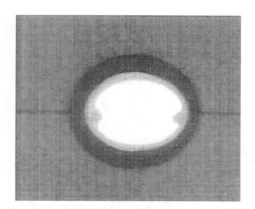

FIGURE 10.39  Profiles of micromachined columns. Coating nonuniformity or defects or on assymetrical tube profiles leads to peak broadening.

valve (the Fluistor™) from Redwood Microsystems was shown already in Figure 10.4A; in this valve a liquid-filled cavity upon heating forces a Si diaphragm to lift off the valve seat (a normally open valve is also available from the same company).[113] Although providing a tremendous force (over 20 N) through a long stroke (between 50 and 150 μm), this valve, is thermally inefficient due to the large volume of heated fluid and is difficult to assemble due to the working fluid in the device; moreover, it is quite slow (~ 400 msec). A fourth alternative is the normally closed shape memory alloy (SMA) valve from Microflow and sold by TiNi Corporation which was shown in Figure 10.4B.[124] Sputtered thin-film NiTi is the SMA element and is deformed in tension through an integral BeCu bias spring pushing it against a valve seat. Passing a small current through the actuator element heats it up and it regains its original shape, lifting it against the spring from the valve seat and allowing gas to flow. These four commercially available thermal valve technologies are compared in Table 10.16.[125] Parameters of interest for valve performance include power consumption, actuation speed, force, stroke, cost, and package size. From the point of view of

performance the HP valve stands out; in terms of manufacturability and IC process compatibility the IC valve has the edge. Redwood Microsystems has gone the furthest in incorporating their valves into higher level subsystems, including pressure and feedback control to create a completely electronically programmable pressure regulator. This company has also paid more attention to applying its valves in biomedical instruments carrying fluids.[6,125]

To build a gas chromatograph that accepts samples with a wide range of gas volatilities (e.g., compounds with vapor pressures at room temperature from 100 to 2 10⁻⁵ mmHg), the injector and column assembly must be heated and cooled quickly and uniformly and the system must be temperature hardened to temperatures of a few hundred degrees. Overton et al., at Louisiana State University's Institute for Environmental Studies[119] accomplish this in the Personal Chromatograph (the PC) by employing low thermal mass chromatographic columns in intimate contact with a heater and an RTD sensor. A typical chromatographic column assembly consists of a 3-m long spiral of a tubular silica glass column (100 μm i.d. and

**TABLE 10.16** Comparison of Four Commercially Available Micromachined Thermal Valves

| Parameters | SMA (Microflow Analytical, Inc.) | Al/silicon (IC Sensors, Inc.) | Ni/silicon (Hewlett Packard) | Thermopneumatic (Redwood Microsystems) |
|---|---|---|---|---|
| Pressure (PSIG) | 80 | 25 | 150 | 100 |
| Flow (sccm, air, or N₂) @ T = 25°C | 6000 | 100 | 1000 | 2000 |
| Power (W) | 0.29 | 0.5 | 1.03 | 2.0 |
| Response time (msec) | 100 | 100 | 200 | 400 |

*Source:* After Reference 125.

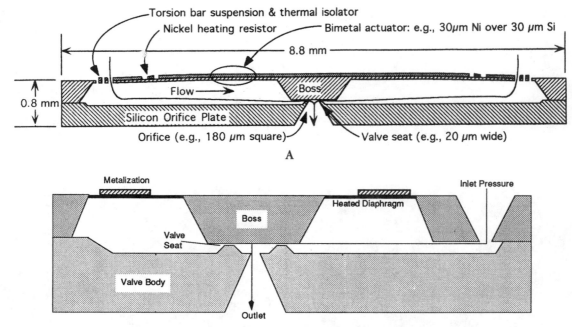

**Cross Section of a Micromachined, Bimetallically-Actuated Diaphragm Valve.**

B

**FIGURE 10.40** Commercially available, micromachined, thermally actuated valves. (A) Hewlett Packard's bimetallic Ni/Si valve. (B) IC Sensor's bimetallic Al/Si valve.

300 μm o.d.) with resistive heater wires (0.25 mm o.d.) and an RTD temperature sensor (0.02 mm o.d.) wound around it and all put together in a 1.1-mm i.d. PTFE tube. The whole column assembly is further housed in an alumina-silica thermal insulator block. For comparison, the more traditional HP 5890 GC oven loses control over the column temperature at programming rates of 1°C/sec, while, through good thermal management, the PC column heaters can be accurately temperature programmed at rates up to 5°C/sec over temperature ranges up to 250°C (see Figure 10.41). A cooling fan activated by the electronic control board hastens the return of the column to its initial temperature. The low thermal mass column ovens thus allow for fast temperature-programmed chromatograms and relatively short turnaround times of about 2 min in analysis of semivolatile analytes. The above results were obtained using traditional machining techniques; a second generation instrument would benefit from a micromachining approach. For example, miniature Joule-Thomson cryogenic refrigerators or thermoelectric coolers for local cooling could replace the large fan. A micromachined Joule-Thomson refrigerator, by MMR, is shown in Figure 10.42 and consists of a heat exchanger, an expansion capillary, and a reservoir.[127,128] These devices today cool only very small structures such as radiation detectors and

laser diodes but we confirmed with MMR that making larger area Joule-Thomson refrigerators of a size to fit underneath a GC column is feasible. A disadvantage of this approach is the need for high pressure, high purity gas to feed the expansion chamber. A thermoelectric cooler as used in the surface acoustic wave hygrometer in Figure 10.51A does not need a high-pressure gas but is quite power hungry and is more difficult to use for cooling large areas.

Besides Overton's very applied and traditional approach and Kolesar's[117] completely integrated nontraditional approach to miniaturization of GCs, a number of other important GC developments are of note in the U.S. For analysis of a variety of organic chlorines and long-lived species in the stratosphere Elkins et al.[129] have developed a four-channel chromatograph with each channel incorporating an electron capture detector. This GC is incorporated on high-altitude aerospace platforms to investigate the impact of chlorofluorocarbons (CFCs) on the ozone hole.[130] Although a bulky traditional instrument today, this research team at the NOAA Aeronomy Lab in Colorado is now investigating means to miniaturize their equipment. This group is buying and testing separate micromachined parts. This seems to be a rather general phenomenon; traditional instrument groups are working on a part-by-part redesign of their instru-

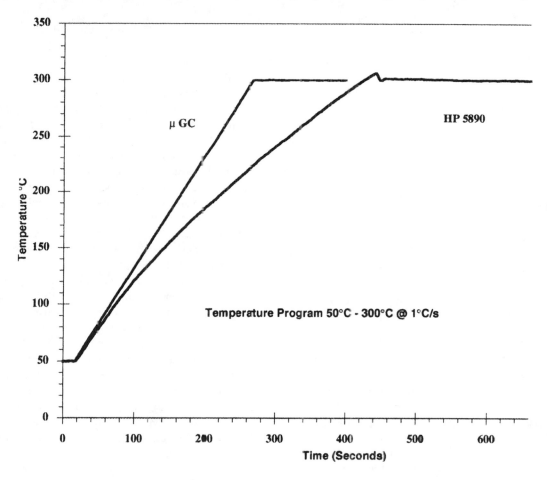

**FIGURE 10.41** Gas chromatograph improvements through better thermal management. Comparison of heating rates of a Hewlett Packard and Personal Chromatograph gas chromatograph; better thermal management enables faster heating while maintaining better control of the column temperature. (Courtesy of Dr. E. Overton, LSU, USA.)

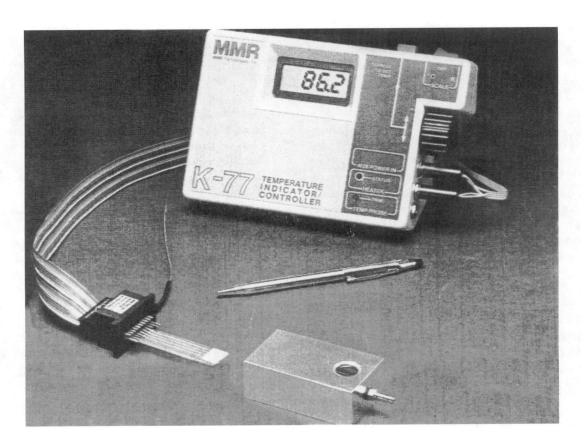

**FIGURE 10.42**    A micromachined Joule-Thomson refrigerator. (Courtesy of MMR Technologies, Mountain View, CA.)

ments, whereas micromachinists are attempting a total instrument redesign. Unfortunately, the latter work often without the benefit of a good instrument group's guidance. An HP development of a handheld GC may be the only effort that combines both excellent micromachining capabilities and instrumentation knowledge. RVM Scientific, Inc., also has built a novel handheld miniaturized gas chromatograph for detection of contraband drugs in cargo containers and chemical weapons treaty verification.[131] There are several small GC products now being introduced to the market. The SnapShot, for example, is a handheld GC from Photovac International Inc. (Deer Park, NY). The instrument has a battery pack providing 4 to 6 hrs of operation and may be used for personal exposure monitoring. Besides MTI, Stanford Research Instruments is offering rugged GCs for field use.[132]

## Mass Spectrometry

Mass spectrometers (MS) are extremely versatile tools which have found, at least for the moment, their greatest use in laboratory science. The detection of the component atomic and molecular masses and their abundance is indeed essential for the understanding of almost any chemical presence and/or process. Lack of broader use is attributed to their large mass, volume, and power requirements. Especially powerful would be a miniaturized gas chromatography-mass spectrometry (GC-MS) combination, with the GC for the isolation of the various compounds of a complex mixture and the MS for further identification. Innovative vacuum systems, smaller GC ovens, new sample inlets, and advanced electronics are now making some of this possible. For example, a GC-

MS system built in a suitcase by Lawrence Livermore National Laboratory in California for real-time field use is shown in Figure 10.43A.[132] Ferran Scientific has successfully implemented a 1-cm scale array of quadrupoles for gas analysis (Figure 10.43B) and a similar MS is under development at Jet Propulsion Laboratory (JPL) in Pasadena, CA. JPL researchers suggest that the Ferran instrument lacks in mechanical precision and hence resolution. In their own work a miniaturized quadrupole mass filter, one machined with traditional (Figure 10.44A), but very precise machining methods and one dramatically smaller in size fabricated employing LIGA (Figure 10.45)* are being pursued. The JPL research team deemed quadrupole and ion-trap mass spectrometers most suited for miniaturization as their operation does not require a permanent magnetic field, and they can be fabricated as parallel arrays of instruments.[133] The array concept is extremely important considering the following simple scaling insights. Miniaturization of an instrument dimension, l, by a factor k leads to a cubic mass and volume reduction ($l^3$ scaling) and an input aperture area reduction, and thus sensitivity reduction of $k^2$. To offset the lost sensitivity, one can construct an array of $k^2$ spectrometers all working in parallel. One ends up with a spectrometer array of sensitivity comparable to its larger cousin but with mass and volume reductions of k. A present quadrupole or magnetic-sector instrument may be of the order of $10^4$ cm$^3$ volume and 10 to 15 kg mass and may require about 20 W power for operation. The JPL traditionally machined microspectrometer, a 4 × 4 array of nine miniature quadrupole mass analyzers, weighs

---

*        Figure 10.46A appears as a color plate after page 144.

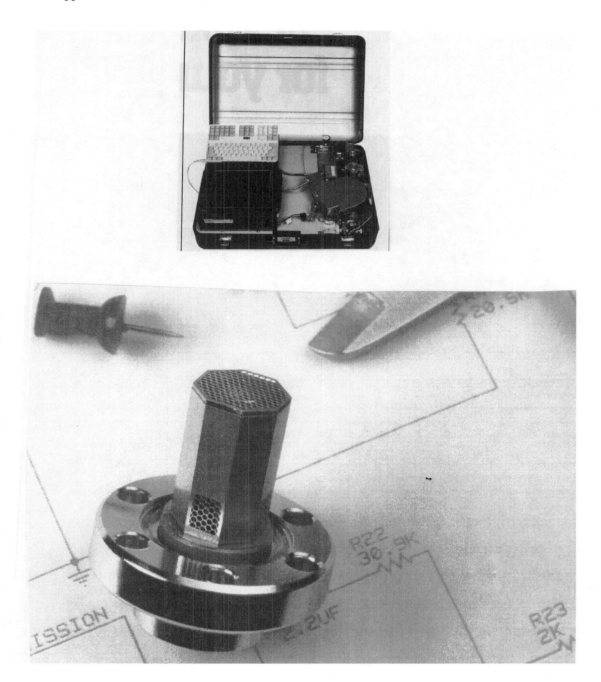

**FIGURE 10.43** (A) A GC-MS system built in a suitcase for real-time field use. (Courtesy of Lawrence Livermore National Laboratory, CA.) (B) A 1-cm scale array of quadrupoles for gas analysis (traditional mechanical precision machining). Built by Ferran Scientific, CA.

less than 0.8 kg mass (including electronics), is 20 cm³ in volume, and draws less than 8 W power. The rod alignment, straightness, and roundness accuracy for a mass range of 1 to 250 amu and a mass resolution of 0.5 amu (FWHM) at 200 amu must be 0.15% (see mass spectrum in Figure 10.44B). Initially the rods were machined at 2-cm length but they can be further reduced to 1.5 or even 1 cm. The alignment was carried out using a machinable ceramic alignment jig to accurately grip the rods. The holes in the jigs are machined using a conventional, high-precision lathe (±0.0002-in. tolerance). With the LIGA technique the attempt is to make the quadrupole rods 1 mm high and 80 μm in diameter but these specifications have not been met yet (see Figure 10.45*). LIGA might provide more

accuracy for the straightness of vertical walls but the tolerances in the x,y directions are still dependent on the aligner and, as we may deduce from Figure 7.1, a 0.15% tolerance on dimensions below 100 μm is quite a challenge.

## Optical Absorption Instruments

The most common analyzers used in analytical chemistry for both gases and liquids are spectroscopic absorption techniques. A considerable amount of effort has been put into

---

* Figure 10.45 appears as a color plate after page 144.

A

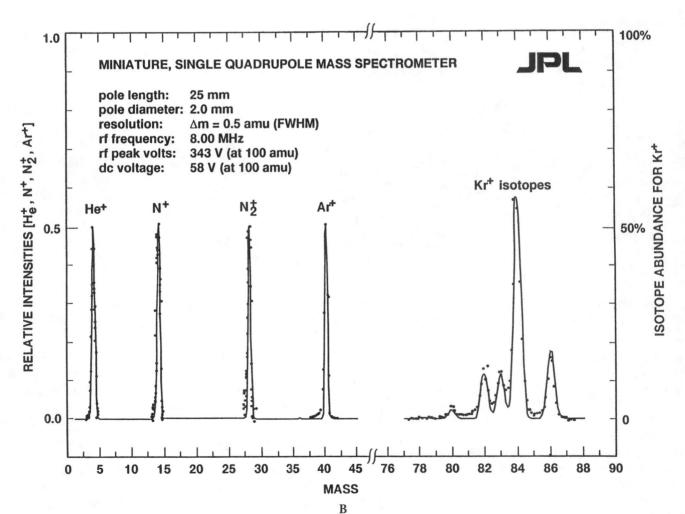

B

FIGURE 10.44   Miniature mass spectrometry. (A) JPL's small quadrupole array made by traditional, but very precise, machining. (B) Spectrum generated from the single quadrupole mass spectrometer in A.

nontraditional machining tools. To illustrate the issues one faces in micromachining this type of equipment relying on nontraditional methods, we will analyze a compact tunable infrared laser absorption spectrometer for measuring gases. An example of such an infrared instrument, the so-called Argus, measuring $NO_2$ and $CH_4$ at the 0.1-ppb level, was built with traditional machining methods at NASA Ames as pictured in Figure 10.46A.* The machine pictured weighs about 23 kg and consumes 63 W.

The principle of gas detection by means of infrared absorption in the Argus instrument is based on Beer's law. To obtain the required sensitivity of 0.1 ppb for $NO_2$ at 4.5 μm and $CH_4$ at 3.3 μm, an optical path length of 18.8 m is required. To accomplish this on a small footprint, a Herriott cell is used (tubular structure in the photograph of Figure 10.46A). The Herriot cell is 26.1 cm long, and a laser beam injected into it

---

\* Figure 10.46A appears as a color plate after page 144.

transverses the cell 72 times for a total optical path of 18.8 m. From the schematic in Figure 10.46B we also notice that a beam splitter provides a second beam that passes through a frequency marker cell (reference cell) to provide wave-number calibration for the second harmonic line fitting procedure. The marker cell contains a gas at low pressure, providing several sharp, Doppler-broadened, spectral lines of accurate, known frequency. Four InSb infrared detectors and two tunable laser diodes are mounted in a 1.8-L $LN_2$ Dewar (circular structure in Figure 10.46A). The accuracy of the instrument is 3%, its precision is 1%, and the response time is 10 sec.

In an attempt to further miniaturize the above spectrometer to a footprint of a few square centimeters using nontraditional machining methods, we set out to replace the Herriott cell by a set of LIGA machined micromirrors positioned in a folded configuration as illustrated in Figure 10.47. The infrared mirrors may be made from solid electroplated gold or by evaporating gold on poly(methylmethacrylate) (PMMA) walls.

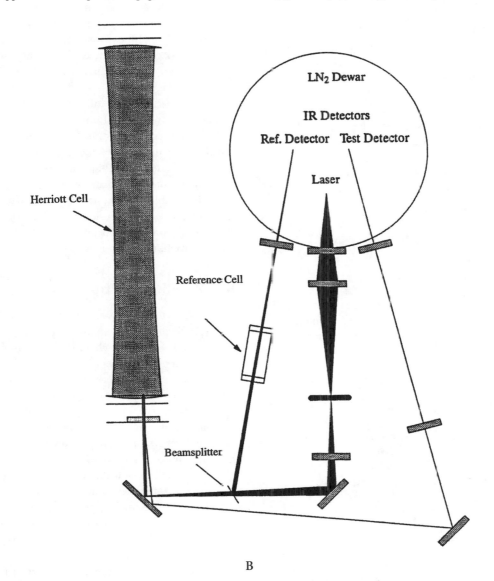

B

**FIGURE 10.46** An example of IR absorption spectrometer instrument, the so-called Argus, made with traditional machining methods at NASA Ames. See color plate 10.46A after page 144.

The vertical height of the mirrors is limited by the spot size of the infrared laser beam. Assuming that the beam has a Gaussian intensity profile, the variation of the beam spot radius, w, as a function of the distance traveled by the beam, z, can be expressed as:

$$w^2(z) = w_0^2 \left[ 1 + \left( \frac{\lambda z}{\pi w_0^2} \right)^2 \right] \qquad 10.1$$

where $w_0$ is the initial spot size. It can be verified that with a 4.5-μm laser (for $CH_4$) an initial spot radius of 100 μm at a distance (z) of 1.6 cm (i.e., the distance between the mirrors) has expanded, by the time the beam is incident on the first mirror, to more than twice its original size. A beam with a spot radius, somewhat larger, say 300 μm expands by less than 3% which could be accommodated by square mirrors of 1000 μm height.[134] Interestingly, scaling in infrared optics works against too much miniaturization; a 300-μm beam is better than a 100-μm beam. Next we will investigate the intensity loss in the instrument due to absorption and scattering by the micromirrors.

Gold was chosen as the preferred material for the micromirrors because of its high reflectivity in the mid-infrared region. Super polished gold surfaces can reflect as much as 98% of an incident infrared beam in the specular direction. However, the current gold micromirrors, electroformed or evaporated on PMMA walls, have an inherent surface roughness which may cause scattering of incident beams. If the surface roughness has a Gaussian distribution about the mean surface level, and the rms roughness is small compared to the wavelength, then the ratio of specular reflectance, $R_s$, to the total reflectance of a reflective surface, $R_0$, is written as:

$$\frac{R_s}{R_0} = e^{-(4\pi\delta\cos\theta_0/\lambda)^2} \qquad 10.2$$

where δ is the rms roughness, and $\theta_0$ is the angle of incidence.[134] Working in the infrared range presents an advantage in this case as the roughness of the mirrors is small compared to the used

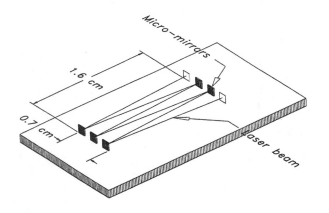

**FIGURE 10.47** Layout of a folded optical system. (From Desta, Y.M., Murphy, M., Madou, M.J., and Hines, J., *Microlithography and Metrology in Micromachining*, (SPIE, Austin, TX, 1995). With permission.)

wavelength. With an angle of incidence close to the normal direction one can typically accommodate up to 100 reflections before serious losses occur. With a 1.6-cm distance between mirrors this means 1.6 m of optical path-length, which may not be adequate for a 0.1-ppb sensitivity but would serve well for applications which require only a few parts per million of sensitivity. The conclusion is that we might well be able to miniaturize the Herriot cell but a miniaturized infrared analyzer of this sort will be suited only for less demanding applications.

To avoid space-hungry liquid nitrogen cooling (round vessel in Figure 10.46A*) one might rely on thermoelectric cooling for the lasers and detectors instead. This is demonstrated in the JPL microhygrometer in Figure 10.51. To transform the two-gas analyzer into a multipurpose gas analyzer, a micromachined broadband light source, such as the heated filament in Figure 3.43*, could replace the two lasers. As discussed in Chapter 6 on LIGA applications, by incorporating a self-focusing grating a set number of important absorption lines may be isolated for sensing a large number of important gases. The problem in this case is to maintain sufficient light intensity at the desired sensing wavelengths.

Another most difficult issue remains the focusing optics of the miniaturized spectrometer. At present, laser beam radiation sources which can produce well-collimated beams in the mid-infrared region are not available; therefore, focusing optics might become necessary for the proper functioning of a miniaturized spectrometer. Although some micromachined Fresnel lenses have been demonstrated (see Figure 5.23B), their multicomponent nature makes them unsuited for use in a LIGA-fabricated, folded optical system. Since the LIGA process is a lithographic method, focusing of light with LIGA-produced optics is possible only in one plane, e.g., with a cylindrical lens. It takes truly three-dimensional lenses to focus light to a point, revealing one of the most important shortcomings of nontraditional machining techniques for optical applications.

## Ion Mobility Spectrometer

In an ion mobility spectrometer (IMS), ions, produced at atmospheric pressure in an ionization cell by a [63]Ni beta source are accelerated in a drift chamber (uniform field of 150 to 250 V/cm), where they are separated according to their mobilities, detected as a current on a Faraday plate, and plotted on a time axis in accordance to their time of arrival (see Figure 10.48). About 6 years ago we set out to micromachine some critical components of an IMS instrument. We recognized that field uniformity, temperature, and pressure could be controlled with micromachined Si sensors. Our aim was to substitute also the [63]Ni ionization source with a less fragmenting ionization source employing a 'softer' field ionization technique (FI). In field ionization one creates mainly positive parent ions ($M^+$) and thus simpler spectra for complex environmental gas mixtures result. Field ionization is well known from field ionization mass spectrometry (FIMS). But mass spectrometry is carried out in vacuum, and considering the extreme electric fields required for FI, one might expect that under atmospheric pressure conditions,

---

\*     Figures 10.46A and 3.43 appear as color plates after page 144.

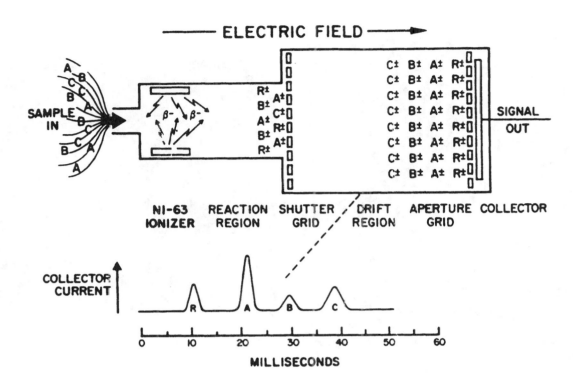

**FIGURE 10.48**  Ion mobility spectrometry with a Ni ionization source.

catastrophic electrical breakdown between the ionizing electrodes would occur. The idea on how to avoid breakdown or sparking at atmospheric pressures is based on the peculiar behavior of micro-electrodes operating at the left side of the Paschen curve (see Figure 2.7). The minimum of the Paschen curve at atmospheric pressure in air is around 6.4 $\mu$m and one notices that on the left side of that minimum it takes much higher voltages before breakdown occurs.[135]

For an ionization source operating on the left side of the Paschen curve we microfabricated an array of microvolcanoes with a typical throat opening of 1 $\mu$m and a volcano rim to gate distance of less than a micrometer. Both the volcano rim and gate electrode may be made from a variety of inert metals. An SEM photomicrograph of a single microvolcano and an array of microvolcanoes are shown in Figure 10.49A (the metal used in this case is Pt). Given the small dimensions of the microvolcanoes, the intense electric fields required for field ionization can be produced with significantly lower voltages than other FI sources. Referring back to the Paschen curve in Figure 2.7, it is clear that the microvolcano sources should operate at atmospheric pressure with voltages up to several hundred volts and perhaps as high as a kilovolt before they induce sparking. In collaboration with Dr. Coggiolla at SRI International we have shown that these sources can indeed resist electrical breakdown at atmospheric pressure and produce usable current levels from positive ions formed from gases such as pyridine, butane, and toluene in a set-up as shown in Figure 10.49B. The fabrication process of the microvolcanoes, which are embedded in a brittle 1-$\mu$m thick Si dioxide layer, has been very problematic and sturdier, more reliable volcano sources are needed to continue this study.

Given the extreme importance of the atmospheric micro-ionizer concept we recently rekindled efforts to build sturdier microvolcano arrays. To understand more about the fundamentals of the gaseous reactions in the high fields of the volcano rims, a fundamental study of pre-avalanche ionization phenomena has been initiated as well. In a separate effort we also have shown that negative ions can be formed at atmospheric pressure on arrays of microtips.[135]

## Innovative Optics and Smart Pixel Arrangements

The area where micromachinists expect the next biggest commercial breakthroughs is in micromachined optical components and systems. Some important realizations in this field are the TI projection display based on the surface micromachined Digital Micromirror Device[136] (see Chapter 5), Bloom's deformable light grating for high resolution displays,[137] and KfK's spectrometer gratings[138] (see Chapter 6).

These technologies will also start showing up in analytical equipment. For example, optical systems need dynamic corrections for atmospheric distortions, thermal effects, and mechanical vibrations. But adaptive optics are typically very expensive, difficult to fabricate, and sensitive to temperature and hysteresis. Micromachining mirrors from Si with integrated drive and sensing presents an attractive solution here.[139]

One of the most established applications for solid state sensors, although not often typified as MEMS, is visual imaging with CCDs (charge-coupled devices). In a CCD imager, each element in a two-dimensional array generates an electrical charge in proportion to the amount of light it receives. Each

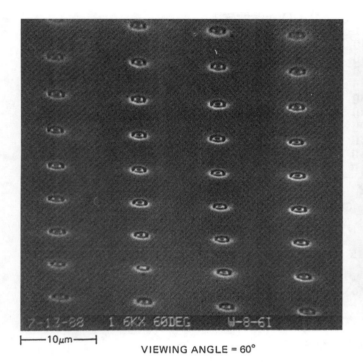

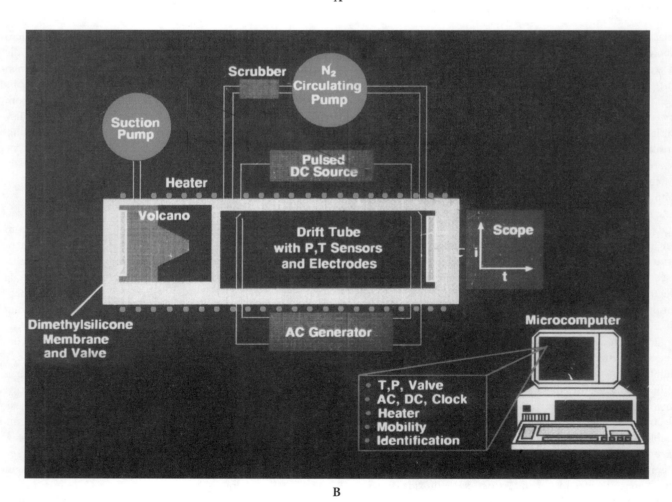

**FIGURE 10.49**   Ion mobility spectrometry with micromachined volcano sources. (A) SEM micrograph of micromachined volcanoes. (B) Set-up for ion mobility spectrometry with micro volcano ionization sources.

charge stored by CCD elements along a row is subsequently transferred to the next element in bucket-brigade fashion as the light input is read-out line by line. The resolution of a CCD is determined by the number of picture elements (pixels) on the CCD. Si as the pixel semiconductor can be used for a wide variety of electromagnetic radiation wavelengths, from gamma rays to infrared. There is a trend now to configure pixels in clever ways to make novel optical sensors feasible.

One example of a smart pixel configuration is embodied in the retina chip shown in Figure 10.50.[140]* This is an integrated circuit chip working like the human retina to select out only the necessary information from a presented image to greatly speed up image processing. The chip features 30 concentric circles with 64 pixels each. The pixels increase in size from $30 \times 30$ $\mu$m for the inner circle to $412 \times 412$ $\mu$m for the outer circle. The circle radius increases exponentially with eccentricity. The center of the chip, called the fovea, is filled with 104 pixels measuring $30 \times 30$ $\mu$m placed on an orthogonal pattern. The total chip area is $11 \times 11$ mm. The chip has been designed for those applications where real time coordination between the sensory perception and the system control is of prime concern. The main application area is active vision, and its potential application is expected in robot navigation, surveillance, recognition, and tracking. The system can cover a wide field of view with a relatively low number of pixels without sacrificing the overall resolution and leads to a significant reduction in required image processing hardware and calculation time. For example, the fast but insensitive large pixels on the rim of the retina chip might pick up a sudden movement in the scenery (peripheral vision), prompting the robot equipped with this 'eye' to redirect itself in the direction of the movement to better focus on the moving object with the more sensitive fovea pixels.

## Microhygrometer

At JPL, as part of a Micro Weather Station development program, a new microhygrometer operating on accurate dewpoint measuring principles was demonstrated. A schematic illustrating the surface acoustic wave hygrometer is shown in Figure 10.51A. This device precisely measures the dewpoint/frostpoint temperature of ambient air. A quartz surface acoustic wave (SAW) microsensor detects the presence of thin layers of water on its surface as its temperature is varied by a miniature thermoelectric cooler. A fast digital feedback loop is used to control the sensor's temperature such that the amount of water on the sensor is held constant. This temperature is the dewpoint or frostpoint. This approach is functionally similar to the conventional chilled mirror hygrometers. However, the SAW moisture sensor provides much higher sensitivity and faster response for small amounts of water than does the optical detectors used in conventional hygrometers. This advantage, coupled with a high performance digital feedback loop and precise measurements of the sensor temperature, enables accurate mea-

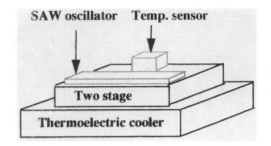

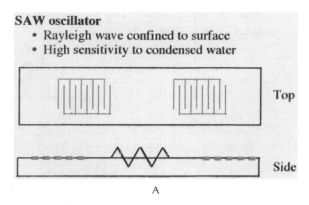

**FIGURE 10.51** Surface acoustic wave (SAW) hygrometer.[141] (A) Schematic of the direct dewpoint sensor (Courtesy Dr. Hoenk, J.P.L., Pasadena, CA.) (B) Photograph of SAW hygrometer. The SAW device (long, rectangular bar) is mounted next to a platinum temperature sensor (square white package), on top of a miniature two-stage thermoelectric cooler. The device size is approximately $1 \times 1$ cm.

surements with a response over an order of magnitude faster than conventional airborne hygrometers. Figure 10.51B** is a photograph of the sensor.[141]

At NASA Ames researchers are embarking on the construction of an expanded array of sensors of the type shown in Figure 10.51. The proposed sensor array combines a set of chemiresistors and a set of acoustic waveplate devices on a thermally isolated hotplate; their measurement modes are, respectively, conductivity changes of chemically sensitive films and mass changes of the same coatings. By controlling the temperature the response time of the sensors can be modulated and the sensitivity of the devices can be increased by operating at low temperature in an integration mode. Metal phthalocyanines (MePcs) and other organics such as 1,4-polybutadiene are used as chemical coatings because of their chemical stability and the fact that these compounds have been shown to exhibit selectivity to several important gases.[35] The specific target gases of the NASA experiment are $O_3$, $H_2O$, NOX, and CO. Once the instrument has been proven, other target gases could be added. Sensitivities in the sub-parts-per-billion range have been demonstrated with both chemiresistors and SAW devices, and identification and quantification of the listed gases would contribute to the atmospheric chemistry studies and form a new type of instrument with considerable market appeal.

---

* Figure 10.50 appears as a color plate after page 144.

** Figure 10.51B appears as a color plate after page 144.

## Other Applications

Miniaturization is contributing to a wide variety of new products across many different industries. We necessarily had to limit our survey, but to keep the reader up to date on a much wider range of miniaturization breakthroughs we refer to the given URL's in Appendix B.

## References

1. "Expanding the Vision of Sensor Materials", National Research Council, 1995.
2. Bryzek, J., K. Petersen, J. R. Mallon, L. Christel, and F. Pourahmadi, *Silicon Sensors and Microstructures*, Novasensor, Fremont, CA, 1990.
3. Editorial, Tips for STM, TSDC (Teknekron), Menlo Park, CA, USA, 1991.
4. Busch-Vishniac, I. J., "The Case for Magnetically Driven Microactuators", *Sensors and Actuators A*, A33, 207–220, 1992.
5. Fluitman, J. H. J., "Microsystem Technology: Exploring Opportunities", Microsystem Technology: Exploring Opportunities, 1994.
6. Zdeblick, M. J., R. Anderson, J. Jankowski, B. Kline-Schoder, L. Christel, R. Miles, and W. Weber, "Thermopneumatically Actuated Microvalves and Integrated Electro-Fluidic Circuits", Technical Digest of the 1994 Solid State Sensor and Actuator Workshop., Hilton Head Island, SC, 1994, p. 251–255.
7. White, R. M., "A Sensor Classification Scheme", *IEEE Trans. Ultrason. Ferroelectr. Freq. Control*, UFFC-34, 124–126, 1987.
8. Middlehoek, S. and S. Audet, *Silicon Sensors*, Academic Press, London, 1989.
9. Lion, K. S., "Transducers-Problems and Prospects", *IEEE Trans. Ind. Electron.*, 16, 2–5, 1969.
10. Gopel, W., J. Hesse, and J. N. Zemel, Eds. *Sensors: A Comprehensive Survey*, Vol. 1, VCH, New York, 1989.
11. Habekotte, E., "De Technologie van Sensoren en Actuatoren in de Werktuigbouw", in *Sensoren and Actuatoren in de Werktuigbouw/Machinebouw*, Centrum voor Micro-Electronica, The Hague, 1993, p. 15–96.
12. Feynman, R., P., "There's Plenty of Room at the Bottom", *J. Microelectromech. Syst.*, Vol.1, 60–66, 1992.
13. Feynman, R., P., "Infinitesemal Machinery", *J. Microelectromech. Syst.*, Vol.2, 4–14, 1993.
14. Sullivan, F. a., "Automotive Sensor Markets-Legislation Guarantees Growth, New Technology Gurantees Battle Royal", Report No. 822–40, Mountain View, CA, 1992.
15. Van Doren, C., *A History of Knowledge*, Ballantine Books, New York, 1991.
16. Middlehoek, S. and U. Dauderstadt, "Haben Mikrosensoren aus Silizium eine Zukunft?", *Technische Rundschau*, July, 102–105, 1994.
17. Gabriel, K., J. Jarvis, and W. Trimmer, "Smal Machines, Large Opportunities: A Report on the Emerging Field of Microdynamics", National Science Foundation, 1989.
18. Sullivan, F. a., "World Emerging Sensor Technologies-High Growth Markets Uncovered", Report No. 915–40, Mountain View, CA, 1993.
19. Collins, A., J., "Problems Associated with Bringing a Sensor Technology to the Market Place", *Sensors and Actuators A*, A31, 77–80, 1992.
20. Sandmaier, H. and H. L. Offereins, "Von "ASIC" zu "ASIS"", in Technische Rundschau, Krull, F., Ed., Technische Rundschau, 86, Bern, 1994, p. 48–52.
21. SSI Technologies, I., "Solid-State Integrated Pressure Sensor", 1995, SSI Technologies, Inc., Janesville, WI.
22. Petersen, K., F. Pourahmadi, J. Brown, P. Parsons, M. Skinner, and J. Tudor, "Resonant Beam Pressure Sensor Fabricated with Silicon Fusion Bonding", 6th International Conference on Solid-State Sensors and Actuators (Transducers '91), San Francisco, CA, 1991, p. 664–667.
23. Bryzek, J., K. Petersen, L. Christel, and F. Pourahmadi, "New Technologies for Silicon Accelerometers Enable Automotive Applications", *SAE Technical Papers Series*, 920474, 25–32, 1992.
24. Corporation, M. I. R., "New and Emerging Markets for Automotive Sensors in North America", MIRC, Mountain View, CA, USA, 1990.
25. AMP, "Piezo Film Sensors", AMP Incorporated, Piezo Film Sensors, P.O.Box 799, Valley Forge, PA 19482, 1993.
26. Rockstad, H. K., T. W. Kenny, J. K. Reynolds, and W. J. Kaiser, "A Miniature High-Sensitivity Broad-Band Accelerometer Based on Electron Tunneling Transducers", *Sensors and Actuators A*, A43, 107–114, 1994.
27. Smit, G. N., "Performance Threshold for the Application of MEMS Inertial Sensors in Space", in *Microengineering Technology for Space Systems (Aerospace Report No. ATR-95(8168)-2*, Helvajian, H., Ed., The Aerospace Corporation, El Segundo, CA, 1995, p. 45–63.
28. Burbaum, C., J. Mohr, P. Bley, and W. Ehrfeld, "Fabrication of Capacitive Acceleration Sensors by the LIGA Technique", *Sensors and Actuators A*, A25, 559–563, 1991.
29. Editorial, "GyroChip", (Commercial brochure), 1995.
30. Editorial, "Microminiature Tools Affecting Medical Products", *The BBI Newsletter*, 17, 88–91, 1994.
31. Roe, J. N., "Biosensor Development", *Pharmaceutical Research*, 9, 835–844, 1992.
32. Editorial, *Sensor Business Digest*, 1993, 1–11.
33. Battaglia, E. A., *US Patent*, in US patent 4,214,968, 1980.
34. Turner, A. P. F., "Mediated Enzyme Electrodes 10 Years On", The Third World Congress on Biosensors, New Orleans, LA, 1994.

35. Madou, M. J. and S. R. Morrison, *Chemical Sensing with Solid State Devices*, Academic Press, New York, 1989.

36. Lauks, I., J. Van der Spiegel, W. Sansen, and M. Steyaert, "Multispecies integrated electrochemical sensor with on-chip CMOS circuitry", 3rd International Conference on Solid-State Sensors and Actuators (Transducers '85), Philadelphia, PA, 1985, p. 122–4.

37. Lauks, I., "Multi-element Thin Film Chemical Microsensors", SPIE International Society for Optical Engineering, Los Angeles, Ca, 1983, p. 138–50.

38. Madou, M. J. and T. Otagawa, "Micron and Submicron Electrochemical Sensors", *AICHE Symposium Series*, 267, 85, 7–14, 1989.

39. Madou, M. J. and T. Otagawa, *Microelectrochemical Sensor and Sensor Array*, in US Patent 4,874,500, 1989.

40. Somps, C., M. Madou, and J. Hines, "Telemetric Ion Selective Electrodes for in Vivo Applications", The Sixth International Meeting on Chemical Sensors, Gaithersburg, Md., 1996, p. No pages numbering provided.

41. Fogt, E., J. Kelley, J. Lund, M. S. Norenberg, and M. Schwinghammer, "Potassium Impregnated Tube Ion-Selective Electrodes (KITISE)", *Sensors and Actuators B*, B1, 267–271, 1990.

42. Cosofret, V. V., M. Erdosy, T. A. Johnson, R. P. Buck, R. B. Ash, and M. R. Neuman, "Microfabricated Sensor Arrays Sensitive to pH and K+ for Ionc Distribution Measurements in the Beating Heart", *Anal. Chem.*, 76, 1647–1653, 1995.

43. Cosofret, V. V., M. Erdosy, T. A. Johnson, D. A. Bellinger, R. P. Buck, R. B. Ash, and M. R. Neuman, "Electroanalytical and Surface Characterization of Encapsulated Implantable Membrane Planar Microsensors", *Anal. Chim. Acta*, 314, 1–11, 1995.

44. Pace, S. J. and J. D. Hamerslag, "Thick Film Biosensors", in *Biosensors and Chemical Sensors*, ACS, Washington, D.C., 1992.

45. Aizawa, M., S. Kato, and S. Suzuki, "Immunoresponsive membrane.I.Membrane Potential Change Associated with an Immunochemical Reaction Between Membrane-Bound Antigen and Free Anti-Body", *J. Membr. Sci.*, 2, 125–132, 1977.

46. Janata, J., "An Immunoelectrode", *J. Am. Chem. Soc.*, 97, 2915–2916, 1975.

47. Janata, J. and G. F. Blackburn, "Immunochemical Potentiometric Sensors", *Ann. N.Y. Acad. Sci.*, 428, 286–292, 1984.

48. Collins, S. and J. Janata, "A Critical Evaluation of the Mechanism of Potential Response of Antigen Polymer Membranes to Corresponding Antiserum", *Anal. Chim. Acta*, 136, 93–99, 1982.

49. Anzai, J.-I. and T. Osa, "Langmuir-Blodgett Membranes in Chemical Sensor Applications", *Selective Electrode Rev.*, 12, 3–34, 1990.

50. Newman, A. L., K. W. Hunter, and W. D. Stanbro, "The Capacitive Affinity Sensor: A New Biosensor", 2nd International Meeting on Chemical Sensors, Bordeaux, France, 1986, p. 596–598.

51. Reichert, W. M., C. J. Bruckner, and J. Joseph, "Langmuir-Blodgett Films and Black Lipid Membranes in Biospecific Surface-Selective Sensors", in *Molecular Engineering of Ultrathin Polymeric Films*, Stroeve, P. and Franses, E., Eds., Elsevier, London, 1987, p. 345–376.

52. Ligler, F. S., T. L. Fare, K. D. Seib, J. W. Smuda, A. Singh, P. Ahl, M. E. Ayers, A. Dalziel, and P. Yager, "Fabrication of Key Components of a Receptor-Based Biosensor", *Medical Instrumentation*, 22, 247–256, 1988.

53. Fendler, H., *Membrane Mimetic Systems*, John Wiley & Sons, 1982.

54. Turton, *Novel Immunoassays*, John Wiley & Sons, New York, 1988.

55. Buerk, D. G., *Biosensors Theory and Applications*, Technomic Publishing Co.,Inc., Lancaster, 1993.

56. Madou, M. J. and J. Joseph, "Immunosensors with Commercial Potential", *Immunomethods*, 3, 134–152, 1993.

57. Barnard, S. M. and D. R. Walt, "Chemical Sensors based on Controlled Release Polymer Systems", *Science*, 251, 927–929, 1991.

58. Joseph, J., A. Jina, and M. Madou, "Separation-Free Electrochemical Immunosensor", Annual Meeting of the AIChE, Anaheim, CA, 1991.

59. Joseph, J. and K. Itoh, *Etching Interference Slides*, in US Patent 5,169,599, 1992.

60. Sutherland, R. M. and C. Dahne, "IRS Devices for Optical Immunoassays", in *Biosensors, Fundamentals and Applications*, Turner, A. P. F., Karube, I. and Wilson, G. S., Eds., Oxford University Press, Oxford, 1987, p. 655–678.

61. Garabedian, R., C. Gonzalez, J. Richards, A. Knoesen, R. Spencer, S. D. Collins, and R. L. Smith, "Microfabricated Surface Plasmon Sensing System", *Sensors and Actuators A*, A43, 202–7, 1995.

62. Jorgenson, R. C. and S. S. Yee, "A Fiber-Optic Chemical Sensor Based on Surface Plasmon Resonance", *Sensors and Actuators B*, B12, 213–220, 1993.

63. Connolly, P., G. R. Moores, W. Monoghan, J. Shen, S. Britland, and P. Clark, "Microelectronic and Nanoelectronic Interfacing Techniques for Biological Systems", *Sensors and Actuators B*, B6, 113–121, 1992.

64. Clementi, E., G. Corongiu, M. H. Sarma, and R. Sarma, H., Eds. *Structure and Motion: Membranes, Nucleic Acids & Proteins*, Adenine Press, Guilderland, N.Y., 1985.

65. Editorial, "Biograting", *R&D Magazine*, 51, 1993.

66. Fodor, P. A. S., J. L. Read, M. C. Pirrung, L. Sryer, A. T. Lu, and D. Solas, "Light-Directed, Spatially Addressable Parallel Chemical Synthesis", *Science*, 251, 767–773, 1991.

67. Eggers, M., "DNA Chip Momentum", *Harc Corollary*, 1993.

68. Britland, S. T., G. R. Moores, P. Clark, and P. Connolly, "Patterning and Cell Adhesion and Movement on Artificial Substrate, a Simple Method", *J. Anat.*, 170, 235–236, 1990.

69. Douglas, K., G. Deavaud, and N. A. Clark, "Transfer of Biologically Derived Nanometer Scale Patterns to Smooth Substrates", *Science*, 257, 642–644, 1992.

70. Madou, M. and T. Otagawa, *Tetrasulfonated Metal Phthalocyaninie Doped Electrically Conducting Electrochromic Poly(Dithiophene) Polymers*, Osaka Gas Company, USA, Patent No. US Patent 5, 151, 224, 1992.

71. Otagawa, T. and M. Madou, *Permanently Doped Polyaniline and Method Thereof*, Osaka Gas Company, USA, Patent No. US Patent 5,002,700, 1991.

72. Lebbink, G. K., "Microsysteem Technology: Een Niet te Missen Ontwikkeling", *Mikroniek*, 3, 70–73, 1994.

73. van Lintel, H. T. G., F. C. M. van der Pol, and S. Bouwstra, "A Piezoelectric Micropump Based on Micromachining of Silicon", *Sensors Actuators*, 15, 153–167, 1988.

74. Bustgens, B., W. Bacher, W. Menz, and W. Schomburg, K., "Micropump Manufactured by Thermoplastic Molding", Proceedings. IEEE Micro Electro Mechanical Systems, (MEMS '94), Oiso, Japan, January, 1994, p. 18–21.

75. Schomburg, W. K., J. Vollmer, B. Bustgens, J. Fahrenberg, H. Hein, and W. Menz, "Microfluidic Components in LIGA Technique", *J. Micromech. Microeng.*, 4, 186–191, 1994.

76. Hafeman, D. G., J. W. Parce, and H. M. McConnell, "Light-Addressable Potentiometric Sensor for Biochemical Systems", *Science*, 240, 1182–1185, 1988.

77. Madou, M. J. and M. Tierney, "Micro-Electrochemical Valves and Methods", in US Patent 5,368,704, 1994.

78. Gravesen, P., J. Branebjerg, and O. S. Jensen, "Microfluidics: A Review", *J. Micromech. Microeng.*, 3, 168–182, 1993.

79. Tsao, T. R., R. M. Moroney, B. A. Martin, and R. M. White, "Electrochemical Detection of Localized Mixing Produced by Ultrasonic Flexural Waves", IEEE 1991 Ultrasonics Symposium Proceedings, Orlando, FL, 1991, p. 937–40.

80. Miyazaki, S., T. Kawai, and M. Araragi, "A Piezo-Electric Pump Driven by A Flexural Progressive Wave", Proceedings. IEEE Micro Electro Mechanical Systems, (MEMS '91), Nara, Japan, 1991, p. 283–288.

81. Jorgenson, J. W. and E. J. Guthrie, "Liquid Chromatography in Open-Tubular Columns", *J. Chromatography*, 255, 335–348, 1983.

82. Seller, K., Fan, Z., H., Fluri, K., and Harrison, D., J., "Electroosmotic Pumping and Valveless Control of Fluid Flow with a Manifold of Capillaries on a Glass Chip", *Anal. Chem.*, 66, 3485–3476, 1994.

83. Jacobson, S. C., Kounty, L. B., Hergenroder, R., Moore, A. W., and Ramsey, J. M., "Microchip Capillary Electrophoresis with an Integrated Postcolumn Reactor", *Anal. Chem.*, 66, 3472–3476, 1994.

84. Manz, A., E. Verpoorte, C. S. Effenhauser, N. Burggraf, D. E. Raymond, D. J. Harrison, and H. M. Widmer, "Miniaturization of Separation Techniques Using Planar Chip Technology", *HRC-Journal of High Resolution Chromatography*, 16, 433–436, 1993.

85. Harrison, D. J., P. G. Glavina, and A. Manz, "Towards Miniaturized Electrophoresis and Chemical Analysis Systems on Silicon-An Alternative to Chemical Sensing", Sensors and Actuators B, B10, 107–116, 1993.

86. Woolley, A., T. and Mathies, R. A., "Ultra-High Speed DNA Fragment Separations Using Micromachined Capillary Array Electrophoresis Chips", *Proceedings of the National Academy of Sciences*, 91, 11348–11352, 1994.

87. Private Communication, Mian, A., *CDX: A Multimedia Platform to Enable Universal Fluidics*, (Commercial brochure), Gamera Bioscience Corporation, 1996.

88. Bart, S. F., L. S. Tavrow, M. Mehregany, and J. H. Lang, "Microfabricated Electrohydrodynamic Pumps", *Sensors and Actuators A*, A21, 193–197, 1990.

89. Findlay, J. B., S. Atwood, M., L. Bergmeyer, J. Chemelli, K. Christy, T. Cummins, W. Donish, T. Ekeze, J. Falvo, and D. a. o. Patterson, "Automated Closed-Vessel System for in vitro Diagnostics Based on Polymerase Chain Reaction", *Clin. Chem.*, 39, 1927–1933, 1993.

90. Wittwer, C. T. and D. J. Garling, "Rapid Cycle DNA Amplification: Time and Temperature Optimization", *BioTechniques*, 10, 76–83, 1991.

91. Wittwer, C. T., G. C. Fillmore, and D. J. Garling, "Minimizing the Time Required for DNA Amplifications by Efficient Heat Transfer to Small Samples", *Analytical Biochemistries*, 186, 328–331, 1990.

92. Harrow, J. J. and J. Janata, "Comparison of Sample Injection Systems for Flow Injection Analysis", *Anal. Chem.*, 55, 2461–2463, 1983.

93. Shoji, S., S. Nakagawa, and M. Esashi, "Micro-pump and Sample-Injector for Integrated Chemical Analyzing Systems", *Sensors and Actuators A*, A21, 189–192, 1990.

94. Hillman, R. S., M. E. Cobb, J. D. Allen, I. Gibbons, V. E. Ostoich, and L. Winfrey, J., *Capillary Flow Device*, Patent No. US Patent 4,963,498, 1990.

95. Madou, M. J., B. H. Loo, K. W. Frese, and S. R. Morrison, "Bulk and Surface Characterization of the Silicon Electrode", *Surf. Sci.*, 108, 135–152, 1981.

96. Bousse, L. and W. Parce, "Aplying Silicon Micromachining to Cellular Metabolism", *IEEE Eng. Med. Biol. Mag.*, 13, 396–401, 1994.

97. Baxter, G. T., L. J. Bousse, T. D. Dawes, J. M. Libby, D. N. Modlin, J. C. Owicki, and J. W. Parce, "Microfabrication in Silicon Microphysiometry", *Clin. Chem.*, 40, 1800–1804, 1994.

98. Briggs, J., V. T. Kung, B. Gomez, K. C. Kasper, P. A. Nagainis, R. S. Masino, L. S. Rice, R. F. Zuk, and V. E. Ghazarossian, "Sub-Femtomole Quantitation of Proteins with Threshold, for the Biopharmaceutical Industry", *BioTechniques*, 9, 598–606, 1990.

99. Volkmuth, W. D. and R. H. Austin, "DNA Electrophoresis in Microlitographic Arrays", *Nature*, 358, 600–602, 1992.

100. Volkmuth, W. D., Duke, T., Wu, M. C., and Austin, R. H., "DNA Electrodiffusion in a 2D array of Posts", *Physical Electrodiffusion in a 2D Array of Posts*, 72, 2117–2120, 1994.

101. Shapiro, H. M. and M. Hercher, "Flow Cytometers Using Optical Waveguides in Place of Lenses for Spectroscopic Illumination and Light Collection", *Cytometry*, 1986, 221–223.

102. Miyake, R., H. Ohki, I. Yamazaki, and R. Yabe, "A Development of Micro Sheath Flow Chamber", Proceedings. IEEE Micro Electro Mechanical Systems, (MEMS '91), Nara, Japan, 1991, p. 259–264.

103. Sobek, D., A. M. Young, M. L. Gray, and S. D. Senturia, "A Microfabricated Flow Chamber for Optical Measurements in Fluids", Proceedings. IEEE Micro Electro Mechanical Systems, (MEMS '93), Fort Lauderdale, FL, 1993, p. 219–224.

104. Sobek, D., S. D. Senturia, and M. L. Gray, "Microfabricated Fused Silica Flow Chambers for Flow Cytometry", Technical Digest of the 1994 Solid State Sensor and Actuator Workshop., Hilton Head Island, SC, 1994, p. 260–63.

105. Northrup, M. A., M. T. Ching, R. M. White, and R. T. Watson, "DNA Amplification with a Microfabricated Reaction Chamber", 7th International Conference on Solid-State Sensors and Actuators (Transducers '93), Yokohama, Japan, 1993, p. 924–926.

106. Wilding, P., M. A. Shoffner, and L. J. Kricka, "PCR in a Silicon Microstructure", *Clin. Chem.*, 40, 1815–1818, 1994.

107. Madou, M., "Solid-State Gas Sensors: World Markets and New Approaches to Gas Sensing", NIST, Report No. NIST Special Publication 865, Gaithersburg, MD, 1994.

108. Madou, M. J., "Compatibility and Incompatibility of Chemical Sensors and Analytical Equipment with Micromachining", Technical Digest of the 1994 Solid State Sensor and Actuator Workshop., Hilton Head Island, SC, 1994, p. 164–171.

109. Oh, S. and M. Madou, "Planar-Type, Gas Diffusion-Controlled Oxygen Sensor Fabricated by the Plasma Spray Method", *Sensors and Actuators B*, B14, 581–582, 1992.

110. Oh, S., "A Planar-Type Sensor for Detection of Oxidizing and Reducing Gases", *Sensors and Actuators B*, B20, 33–41, 1994.

111. Lyle, R. P., L. Marchadier, B. Patissier, and S. Schroeder, "Characterization of Micromachined Chemical Gas Sensors", The Sixth International Meeting on Chemical Sensors, Gaithersburg, Md., USA, 1996, p. No page numbering in proceedings.

112. Madou, M. J. and T. Otagawa, *Fast Response Time Microsensors for Gaseous and Vaporous Species*, in US Patent 4,812,221, 1989.

113. Maseeh, F., M. J. Tierney, W. S. Chu, J. Joseph, L. Kim, and T. Otagawa, "A Novel Silicon Micro Amperometric Gas Sensor", 6th International Conference on Solid-State Sensors and Actuators (Transducers '91), San Francisco, Ca, 1991, p. 359–362.

114. Tierney, M. J., L. Kim, M. Madou, and T. Otagwa, "Microelectrochemical Sensor for Nitrogen Oxides", *Sensors and Actuators B*, B13, 408–11, 1993.

115. Goldstein, M., K., *Biomimetic Sensor That Simulates Human Response to Airborne Toxins*, in US Patent 5,063,164, Quantum Group, Inc., 1991.

116. Angell, J. B., S. C. Terry, and P. W. Barth, "Silicon Micromechanical Devices", *Sci. Am.*, 248, 44–57, 1983.

117. Kolesar, E. S. and R. R. Reston, "Separation and Detection of Toxic Gases with a Silicon Micromachined Gas Chromatography System", 1995.

118. Terry, S. C., J. H. Jerman, and A. J. B., "A Gas Chromatographic Air Analyzer Fabricated on a Silicon Wafer", *IEEE Trans. Electron Devices*, ED-26, 1880–6, 1979.

119. Private Communication, Overton, E. B., *Gas Chromatograph Column Profile*, LSU.

120. Leeuwis H., a. v. V., J., "Integrated Micro Gas Chromatograph", in *Microsystem Technology*, Lebbink, G. K., Ed., Samsom BedrijfsInformatie bv, Alphen aan den Rijn/Zaventem, 1994, p. 90–94.

121. Gordon, G. B. and P. W. Barth, *Thermally-Actuated Microminiature Valve*, in US Patent 5,058,856, Hewlett-Packard Company, Palo Alt, Ca, 1991.

122. Barth, P. W., C. C. Beatty, L. A. Field, J. aracterization of the Silicon Electrode", *Surf. Sci.*, 108, 135–152, 1981.

123. Jerman, H., "Electrically-Activated, Normally-Closed Diaphragm Valves", 6th International Conference on Solid-State Sensors and Actuators (Transducers '91), San Francisco, CA, 1991, p. 1045–1048.

124. Editorial, "Silicon Micromachined Proportional Valve", Microflow Analytical, 1995.

125. Barth, P. W., "Silicon Microvalves for Gas Flow Control", 8th International Conference on Solid-State Sensors and Actuators (Transducers '95), Stockholm, Sweden, June, 1995, p. 276–280.

126. Carney, K. R., E. B. Overton, R. L. Wong, C. F. Jackisch, and C. F. Steele, "Use of a Microchip GC for Ambient Analysis-Measurement of Airborne Compounds: Sampling, Analysis and Data Interpretation", ACS Symposium Series: Sampling and Analysis of Airborne Pollutants, , 1993, p. 21–38.

127. Little, W. A., "Microminiature Refrigeration", *Rev. Sci. Instrum.*, 55, 661–680, 1984.

128. Little, W. A., "Applications of Closed-Cycle Cryocoolers to Small Superconducting Devices", Proceedings of the NBS Cryocooler Conference, Boulder, CO, 1978, p. 75–80.

129. Elkins, J. and D. Fahey, "Airborne Southern Hemisphere Ozone Experiment", NASA Mail Stop 245–5, 1995.

130. Elkins, J. W., T. M. Thompson, T. H. Swanson, J. H. Butler, B. D. Hall, S. O. Cummings, D. A. Fischer, and A. G. Raffo, "Decrease in the Growth Rates of Atmospheric Chlorofluorocarbons 11 and 12", *Nature*, 364, 780–783, 1993.

131. Holland, P. M., R. V. Mustacich, J. Everson, W. F. Foreman, M. Leone, A. H. Sanders, and W. J. Naumann, "Correlated Column Micro Gas Chromatography Instrumentation for the Vapor Detection of Contraband Drugs in Cargo Containers", SPIE, San Diego, CA, 1995.

132. Goldner, H., "'Honey, They Shrunk the GC!' Gas Chromatogrpahs Go Mini", *Res. & Dev.*, June, 45, 1993.

133. Stalder, R. E., S. Boumsellek, T. R. Van Zandt, T. W. Kenny, M. H. Hecht, and F. E. Grunthaner, "Micromachined Array of Electrostatic Energy Analyzers for Charged Particles", *J. Vac. Sci. Technol. A*, Micromachined array of electrostatic energy analyzers for charged particles, 12, 2554–2558, 1993.

134. Desta, Y. M., M. Murphy, M. Madou, and J. Hines, "Integrated Optical Bench for a $CO_2$ Gas Sensor", Microlithography and Metrology in Micromachining, (Proceedings of the SPIE), Austin, TX, 1995, p. 172–177.

135. Madou, M. J. and S. R. Morrison, "High-field Operation of Submicrometer Devices at Atmospheric Pressure", 6th International Conference on Solid-State Sensors and Actuators (Transducers '91), San Francisco, CA, 1991, p. 145–149.

136. Hornbeck, L. J., "Projection Displays and MEMS: Timely Convergence for a Bright Future", Micromachining and Microfabrication Process Technology, (Proceedings of the SPIE), Austin, TX, 1995, p. 2.

137. Apte, R. B., F. S. A. Sandejas, W. C. Banyai, and D. M. Bloom, "Deformable Grating Light Valves for High Resolution Displays", Technical Digest of the 1994 Solid State Sensor and Actuator Workshop., Hilton Head Island, SC, 1994, p. 1–6.

138. Staerk, H., A. Wiessner, C. Muller, and J. Mohr, "Design Considerations and Performance of a Spectro-Streak Apparatus Applying a Planar LIGA Microspectrometer for Time-Resolved Ultrafast Fluorescence Spectrometry", *Rev. Sci. Instrum.*, 67, 2490–2495, 1996.

139. Miller, L. M., E. C. Vote, T. W. Kenny, W. J. Kaiser, M. L. Agronin, R. Bartman, K., and N. Wolff, M., "Adaptive MicroOptics", AVS-1–day Workshop, Micromachining: Technology & Applications, Pasadena, 1993.

140. IMEC, "Silicon Detectors", *IMEC Brochure*, 1994.

141. Hoenk, M. E., T. R. Van Zandt, D. A. McWatters, R. K. Watson, C. Kukkonen III, W. Kaiser, and D. Cheng, *Surface Acoustic Wave Hygrometer Flight Tests on The NASA DC-8 Airborne Laboratory*, JPL, Pasadena, CA, 1995.

# Appendix A
## Metrology Techniques for MEMS

| | Spot Size or Probe Area | Resolution/Magnification | Thickness Range | Applications | Remarks (n = Refractive Index) |
|---|---|---|---|---|---|
| Reflection spectroscopy (visible: 400–900 nm) | 5.5-50 µm | | 40 Å-5µm | Nonabsorbing films | Lim. meas. below 100 Å, n meas. only above 700 Å. Fast. |
| Reflection spectroscopy (UV: 200–400 nm) | 10 µm | | 25 Å-2µm | Transparent very thin films on metals | Lim. meas. below 50 Å, n meas. only above 100 Å |
| Ellipsometry (632.8 nm standard), 790 nm, 830 nm, 1300 nm and 1500 nm (optional) | 12-100 µm | 1-2 µm | 10-1000 Å | Nonabsorbing films | Automated, highest accuracy. Thickness and n meas. |
| Optical microscopes | | Magnification 10 to 1000 | 0.3–5 µm | Universal tool for all dimensions > 0.29 µm | Poor depth of field. |
| Scanning electron microscope (SEM) | Spots down to 10-30 Å | Lateral: -35 Å (standard SEM); -9 Å (field-emission SEM); -150 Å (low-voltage SEM); magnification is 20 to 150,000 | | SEM pictures, CD meas. below 0.8 µm. | Good for surface topography. Large depth of focus. Needs vacuum. Can be destructive. |
| Transmission electron microscope (TEM) | 10 Å | 1–2 Å (resolving power $10^4$ times better than optical microscope). Magnification is 500 to 500,000 | 10 Å-1 µm | Atomic structure defect analysis nanoprobe | Requires thin sample preparation |
| Scanning acoustic microscopy | 1.3 mm (50 MHz) | Lateral:10 nm – 2 mm Depth: 10 nm – 5 mm | Signal penetrates 1 λ | Inspection inside a ceramic packaged IC. Interface inspection, wide range of samples. | Nondestructive. Subsurface information. Needs liquid medium. |
| Stylus Profilometer | Tip size limits bandwidth. | Resolution of better than 1 Å and lateral resolution of ~0.03 mm | | Meas. sharp steps from a few Å to several micrometers. Materials & Process R & D Low volume QC | Meas. surf. profile using a contact stylus. May damage soft samples. |
| Nomarski microscope (differential interference contrast-DIC) | 0.3 µm | | | Separation of images of different layers. Sample roughness inspection. | Pseudo three-dimensional image. |
| Confocal Scanning optical microscope (CSOM) | 0.3 µm | 0.3 µm | < 1 µm | Cross sectioning. Independent layer imaging. | Nondestructive. No vacuum. Diffraction lim. resol. |
| Atomic force microscope (AFM) | Radius of curvature of probe tip: 50-100 Å | Atomic imaging (one atom) | < 1 Å | Surface roughness. Tribology. | Slow scan. Probe tip wear. Nondestructive. |

| | Spot Size or Probe Area | Resolution/Magnification | Thickness Range | Applications | Remarks (n = Refractive Index) |
|---|---|---|---|---|---|
| Total integrated scattering | | With visible light 0.15 µm is possible | | Lithography control, etch structure monitoring and metal grain size determination | Nondestructive, rapid quantitative process monitor |
| Scanning Tunneling Microscope (STM) | 50 - 100 Å | Lateral: 0.1 – 1 nm Depth: 0.1 nm | | Atomic resolution. Measurement of chemical properties. | Sample must be conduct. Slow scan. Probe wear. Cannot meas. surfaces larger than a few µm². |
| Optical Interference Microscope | | Lateral: 0.2 – 2 µm. Depth: 400 µm – 2 cm. | | Cross sectioning. | Lateral resolution limited by diffraction. |
| Near-Field Scanning Optical Microscope | 0. 02 – 4 µm | Lateral: 1 – 200 nm. Depth: 10 nm | | General imaging. Lithography. | No vacuum. |

See also: WWW Directory of Microscopy and Microanalysis Products and Services: http://www.mwrn.com/product/product.htmt
Guide to Microscopy and Microanalysis on the Internet: http://www.mwrn.com/guide.htm

# Appendix B

## Further Reading on the Internet (For Regular Updates, See the Homepage)

## Chapter 1. Lithography

### Electron Beam Lithography

**10 nm Electron Beam Lithography. Nanofabrication Technology at the NSL**
http://www.tc.umn.edu/nlhome/m017/nanolab/research/fab.html#EBL
> Technical note: Nanotechnology is presenting great opportunities for engineering innovative magnetic materials and devices, ultra-high density magnetic storage and micromagnetics.

**Electron Beam Lithography**
http://www.xraylith.wisc.edu/~ebeam/ebeam/ebeam.html
> Technical note

**Electron Beam Lithography**
http://www.imec.be/6/phantoms/domains.html#nano
> Technical note: Current activities in device structuring nanofabrication research are dominated by e-beam lithography for shape definition followed by dry/wet etching and lift-off.

**Opto.forum Archives: Diffractive Optics Foundry Run**
http://iris.eecs.uic.edu/opto.forum/0081.html
> Technical note: Diffractive optics technology. The application of the technology draws on high resolution electron beam lithography and dry etch micromachining techniques

**CO-OP- Honeywell/ARPA CO-OP Workshop**
http://co-op.gmu.edu/workshops/doe/doe_announce.html
> Technical note: HONEYWELL is offering its Diffractive Optics technology on a foundry basis through the ARPA CO-OP program.

**aiss GmbH Company overview / (Electron Beam Lithography)**
http://www.aiss.de/
> Technical note: Our main goal is to provide mask makers or direct writers with tools to enhance both quality and throughput.

**nanoCAD - introduction to the product**
http://www.aiss.de/nanoCAD/introduction.html
> Technical note: nanoCAD provides functions essential for innovative designs from rapidly developing technological branches, e.g. binary optics, micro optics, micro mechanics and micro-/nanosystems.

**Home of PYRAMID/ (Electron Beam Lithography)**
http://www.eng.auburn.edu/~sylee/pyramid.html
> Technical note: In electron-beam and optical lithography, blurring in a written pattern would become more severe as the critical dimension of a circuit is reduced down to 0.1 um and below.

**New Leica VB6 at CNF**
http://www.nnf.cornell.edu/CNF/Changes/VB6.html
> Description of product: The new Leica VB6 E-beam system was installed and a beam was obtained during October, 1996. (We are now in the final stages of getting it ready for users.)

**MTL Research Area: Fabrication Technology**
http://www-mtl.mit.edu/MTL/Report94/SUB/SUE.html
> Description of projects - MTL Research Area: Submicron and Nanometer Structures

**CATS On The Web! Transcription Enterprises**
http://www.te-cats.com/
>  Technical note: CATS is the most advanced and fully featured data preparation software available for semiconductor photomask manufacturing, inspection, and for direct-write-on-wafer.

**Developing Tools for Nanolithography**
http://www.ncsa.uiuc.edu/Pubs/access/93.2/Nano.html
>  Overview article: Nano-li-thog-ra-phy: a technique used for integrated circuit fabrication done on a dwarfed scale. (The prefix nano represents 10-9, or one-billionth of the unit adjoined.)

## Images/Lithography

http://mems.cwru.edu/images/hiratio.gif
>  High-Aspect-Ratio Photolithography (190K).gif

http://mems.cwru.edu/images/hiratio.gif
>  Photograph only / Nanolithography/AFM image

http://www.opt-sci.arizona.edu/stm/spm/osc.html
>  Photograph only: AFM image of 5nm tall Silicon Oxide letters on a Silicon surface.

**MS/PhD theses of CXrL students**
http://www.xraylith.wisc.edu/info/PhdThesis.html
>  Description of project with pictures: Structures and devices made by XRL

**Nanolithography using the Scanning Tunneling Microscope**
http://weber.u.washington.edu/~pearsall/proj1.html
>  Description of project: We are using the scanning tunneling microscope to achieve localized oxidation of silicon. The oxide is used as a mask for transferring the pattern directly on silicon.

**Steve Konksek's Home Page**
http://weber.u.washington.edu/~pearsall/konsek.html
>  Description of project: I am studying the physics and chemistry of direct write nanolithography using the scanning tunneling microscope.

**Annual Report**
http://nanoweb.mit.edu/annual.html

## Masks

**Economical Mask Making using Desktop Publishing**
http://mems.isi.edu/archives/tools/PSMASKMAKER/
>  Technical note on mask making with software.

**Layout editor: L-Edit Documentation**
http://hypatia.dartmouth.edu/levey/ssml/ledit/ledit.html
>  Technical note: L-Edit is a LAYOUT EDITOR from Tanner Research. It is a VLSI design tool, flexible enough to do micromachining design, printed circuitboard layout, and other CAD work.

## Miscellaneous

**PMMA**
http://www.me.umist.ac.uk/historyp/pmma.htm
>  Technical note: PMMA is a transparent, hard, stiff material with excellent uv stability, low water absorption and high abrasion resistance.

## More from the grab bag:

>  **A pictorial intro to lithography.**
>  http://www.physics.udel.edu/~watson/scen102/litho/index.html
>  **A recent update on microlithography.**
>  http://www.spie.org/web/meetings/programs/ml96/ml96 home.html
>  **Also 'Semiconductor International' is a good take-off point.**
>  http://www.semiconductor-intl.com/home.htm

**The Homepage of Microlithography World.**

http://www.pennwell.com/litho.html

**A good historical overview. Comparison of advanced lithography strategies.**

http://www.xraylith.wisc.edu/thesis/1994/xiao/main.html

**A commercial company with process simulation software. Optolith for lithography simulation.**

http://www.silvaco.com/products/athena/athena.html

**A primer by IBM on Lithography.**

http://www.research.ibm.com/litho/

**Dr. Quate's group at Stanford on AFM in Lithography.**

http://snf.stanford.edu/Projects/Quate-Soh/AppliScanning.html

**DARPA's strategy on implementing lithography.**

http://esc.sysplan.com/ETO/ALP/alpplan.html

**An introduction to LIGA from the LIGA Club (UK). For more URL's on LIGA see Chapter 5.**

http://www.dl.ac.uk/LIGA/index.html

**Manipulation of atoms with light .**

http://www.lightforce.harvard.edu/lightforce/

**A somewhat unconventional introduction to ion-beam lithography**

http://iris.microcosm.com/alg/Fabtek95.htm

**An introduction to IfM's lithography capabilities.**

http://www.latech.edu/tech/engr/ifm/litho.html

**An example of 3 nm lithography using e-beam.**

http://www.elec.gla.ac.uk/~dc/3nmlith.html

**A gallery of images on atom lithography from NIST.**

http://www.nist.gov/public_affairs/gallery/lithcp.htm

**A primer on Lithography (including 3D lithography by MicroScape).**

http://www.microscape1.com/lithography.html

**A guided tour of the NNF Lithography capabilities at Cornell.**

http://www.nnf.cornell.edu/NanoLine/NNF/LabMap/GuidedTour.html

**The advanced Lithography Group Home Page.**

http://www.microcosm.com/alg/index.html

**Extreme UV lithography at LLNL.**

ahttp://grace.lbl.gov/interf/kg2.html

**A commercial outfit specializing in stereo-lithography.**

http://www.materialise.com/default.htm

**Parallel Electron Beam Lithography.**

http://trinity.tamu.edu/~rennie/ibl.html

**Talk on phase shift mask generation.**

http://www.aiss.de/me93/me93.html

**Joint European Enterprise in Phase-Shift (JEEPS), Advanced Mask Technology: towards 0.25 micron.**

http://albion.ncl.ac.uk/esp-syn/text/6908.html

**Hitachi's e-beam equipment.**

http://www.hitachi.co.jp/Div/keisokuki/byoue.html

**A pictorial explantion of stereo lithography.**

http://www.cranfield.ac.uk/aero/rapid/SYSTEMS/explanation.html

**A short note about the original work on stereolithography.**

http://hkumea.hku.hk/~mensing/3dsystem.htm

**A bibliography on stereolithography.**

http://www.cs.hut.fi/~ado/rp/bibliography3_14.html

Fundamentals of Microfabrication

**A primer on stereo lithography.**
http://www.cranfield.ac.uk/coa/rapid/PROCEEDING/jacobs1.html
**The case for ion-beam lithography.**
http://iris.microcosm.com/alg/Fabtek96.htm
**Ion beam facilities at LBNL.**
http://www.lbl.gov/Tech-Transfer/Ion-Sources-Bus-Ops.html
**1995 Electron, Ion and Photon Beam Nanofabrication and Technology Conference**
http://www.eecs.umich.edu/~pang/EIPB95Prog.html
**Ion Beam Lithography Art Gallery.**
http://trinity.tamu.edu/~rennie/ibl.html
**Scanning tunneling microscopy at IBM.**
http://www.almaden.ibm.com:80/vis/stm/gallery.html
**A home built AFM.**
http://www.biochem.mpg.de/spm/SPM-AFM.html
**STM work at the Technical University of Florida**
http://pss.fit.edu/scanprob.html
**Direct-write processing. Writing with AFM.**
http://luciano.stanford.edu/~shimbo/afmdwp.html
**Some historical background of scanning probe microscopy from In-Focus (September 1995).**
http://www.ou.edu/research/electron/mirror/in-focus/9509.html
**More links to Scanning Probe Microscopy.**
http://www.ifm.liu.se/Applphys/spm/links.html
**More links to advanced lithography sites.**
http://eto.sysplan.com/ETO/ALP/links.html

# Chapter 2. Pattern Transfer with Dry Etching

## From the grab bag:

**Dry etching at Bristol.**
http://www.tlchm.bris.ac.uk/etching/etchhome.htm
**An introduction to a text book on: "Principles of Plasma Discharges and Materials."**
http://hera.eecs.berkeley.edu/~ajl/Textbook2.html
**Dry etching in Glasgow.**
http://www.elec.gla.ac.uk/groups/dryetch/Welcome.html
**Typical course topics on dry etching.**
http://www.ucce.edu/296.html
**Picture of an ion milling machine.**
http://stud1.tuwien.ac.at/~e8925123/im_e.html
**On etching Si in UV and with $XeF_2$.**
http://www.physik.fu-berlin.de/~twesten/abs152.html
**Doctoral thesis on $XeF_2$ etching. Abstract.**
http://mems.isi.edu/archives/dissertation-abstracts/ucla/ms95-2.html
**Plasma Etching in Semiconductor Fabrication. Table of Content of Morgan's book on the topic.**
http://www.elsevier.nl:80/inca/publications/store/5/0/3/6/3/8/
**Introduction to a simulation program for etching and deposition processes.**
http://www.tmai.com/PRODUCT/donatello.html
**Dry etchers available at the Nanofabrication Lab in Stanford.**
http://www-nanofab.stanford.edu/NanoFab/physical/equipment/etch/dry.html

**Oxford Instruments CAIBE System at the Cornell NNF.**

http://www.nnf.cornell.edu/NanoLine/NNF/LabMap/Oxford.html

**Plasma cleaning at Los Alamos.**

http://www-emtd.lanl.gov/TD/Prevention/PlasmaCleaning.html

**References in Plasma Science and Technology.**

http://www-plasma.umd.edu/references.html

**A white paper on a proposed Plasma Chemistry Database.**

http://www.csn.net/~morgan/white_ paper.html

**Axel Scherer on combining e-beam lithography with dry etching to define features with a 6 nm lateral feature size.**

http://www.systems.caltech.edu/EE/Faculty/Scherer.html

**Low damage selective & non-selective RIE of GaAs/AlGaAs in $SiCl_4$&$SiCl_4$/$SiF_4$ Plasmas.**

http://www.elec.gla.ac.uk/groups/dryetch/saad1 html

**Dry etch with high selectivity in etching $SiO_2$ over $Si_3N_4$.**

http://epswww.epfl.ch/aps/BAPSGEC96/abs/S1C004.html

**Si trench etching at Plasma Quest using a HBr chemistry at 0°C.**

http://www.plasmaquest.com/prodinfo/process.sitrench.htm

**A short intro to RIE from MAT in Berlin, with SEM pictures.**

http://www-mat.ee.tu-berlin.de/research/dryetch/dryetch.htm

**Some French work on dry etching of SOI wafers to craft pressure sensors and accelerometers. Nice schematics and SEM pictures.**

http://emsto.laas.fr/Europe/EUROPRACTICE/Public/Page7.html

**Comparing six different chemistries for dry etching in an ECR of thin film magnetic materials.**

http://www.phys.ufl.edu/~nanoscale/reports/year1/pattern.html

**Dry etching process simulation from the Fraunhofer Institute.**

http://www.isit.fhg.de/english/simulation/master.html

**Pictures and information on Applied Materials Dry Etchers.**

http://www.appliedmaterials.com/html1/prodtech/products-dry.html

**Surface Technology Systems (STS). A UK manufacturer of dry etching systems. With amazing SEM pictures of dry etching work at Stanford.**

http://www.stsystems.com/

**Plasma cleaning systems by Anatech.**

http://www.thomasregister.com:8000/olc/anatech/cleaning.htm

**Barrel etchers for plasma stripping.**

http://mail.cccbi.chester.pa.us/spi/catalog/instruments/etchers1.html

**Dry etching at SINTEF (Oslo).**

http://www.oslo.sintef.no/ecy/7230/micro12.html

**Deep reactive ion etching of Si at Stanford.**

http://snf.stanford.edu/Projects/Kovacs-Klaa/mems.html

**Ion milling at Oak Ridge National Laboratory.**

http://www.ornl.gov/orcmt/bmp/ionmill.html

**Good pictorial introduction to focused ion beam milling.**

http://timonf.aero.org/nlectc-wr/fib.html

**MERIE, Helicon and ECR are being developed at ISRC.**

http://chips.snu.ac.kr/plasma.html

# Chapter 3. Pattern Transfer with Additive Techniques

## From the grab bag:

### Process Equipment (From SPECNet).

Automation & Robotics || CAD || CVD || CAM/CIM || Cleaning Equipment || Cluster Tools || Contamination Control Equipment || Crystal Growing Equipment || Developing || Diffusion Oxidation Annealing Systems || Dry Etch Systems || Epitaxy || Exposure

Sources || Handling Transfer Linear Motion Systems || Inspection Equipment || Ion Implantion || Ion Beam Equipment || Laser Beam Equipment || Lithography || Measurement Equipment || Ovens || Photoresit Application || Physical Vapor Deposition || Power Supplies || Pressure Instrumentation || Recirculators || Sawing–Lapping–Polishing || Software Interfaces || Vacuum Pumps and Accessories || Vacuum Deposition || Wafer Identification || Wet Etch Systems || Other

**A free subscription to Solid State Technology.**

http://www.solid-state.com/sst/subscribe.html

**Silicon as an Element.**

http://chemlab.pc.maricopa.edu/periodic/Si.html

**A simple minded introduction to Silicon. Thank you Silicon.**

http://www2.interpath.net/sbi/AART/opportunities/silicon.html

**A historical document: "First Commercial Silicon Transistor,"**

http://www.ti.com/corp/docs/history/sitrans.htm

**Pathway to the production of high-purity silicon.**

http://www.egg.or.jp/MSIL/english/msilhist0-e.html

**Growth of single crystal Silicon.**

http://www.egg.or.jp/MSIL/english/msilhist4-e.html

**Towards Silicon wafers of 12″ diameter!**

http://www.egg.or.jp/MSIL/english/semicon/12wafer-e.html

**Proposed unified model for both thin and thick oxides.**

http://www.eas.asu.edu/~whidden/pcsi1.html

**Properties of silicon dioxide.**

http://www.ai.mit.edu/people/tk/tks/silicon-dioxide.html

**A list of references on CVD from 1993 to 1996.**

http://www.lci.espci.fr/VO/references/VO-publi-CVD.html

**A very good pictorial introduction to CVD.**

http://chiuserv.ac.nctu.edu.tw/~htchiu/cvd/home.html

**Society of Vacuum Coaters.**

http://www.svc.org/

**Thin Solid Films. Issues content and download articles.**

http://bulb.mit.edu/mumble.sav/journals/00406090/combined-toc.html

**A good introduction to CVD by Lamp Postech.**

http://www-ce.postech.ac.kr/~lamp/cvd.htm

**A good introduction on LPCVD by Lamp Postech.**

http://www-ce.postech.ac.kr/~lamp/cvd/pecvd.htm

**A student seminar on Silicon on Insulator (SOI).**

http://caligari.dartmouth.edu/~cgl/microeng/es194/student/psr/194sem.html

**An introduction to MBE.**

http://ccwf.cc.utexas.edu/~anselm/mbechapter.html

**Argonne National Lab in Microwave Plasma research.**

http://www.anl.gov/LabDB/Current/Ext/M355.html

**CVD at Sandia with photo of set-up.**

http://www.sandia.gov/ttrans/pam/Microelectronics/cvs/CVDScihomepage.html

**Superhard coatings.**

http://www.birl.nwu.edu/birlhtml/birl/pvdts.html

**American Vacuum Society.**

http://www.vacuum.org/

**The Semiconductor Subway.**

http://www-mtl.mit.edu/semisubway.html

**An overview of the Epi-Center Molecular Beam Epitaxy Laboratory at the University of Illinois.**
http://www.mrl.uiuc.edu/~ritley/epi.html

**Tapping mode AFM is used to study epi-Si.**
http://www.di.com/AppNotes/Semi/Main.html

**Why epitaxy in space?**
http://www.svec.uh.edu/epitaxy.html

**Metrology products from Tencor.**
http://www.tencor.com/prodsol.html

**References on all types of plasmas.**
http://www-plasma.umd.edu/references.html

**Molecular Beam Epitaxy Set-up (schematic).**
http://131.155.110.159/images/mbe.html

**Molecular Beam Epitaxy at UT-Austin.**
http://taylor.ece.utexas.edu/projects/ece/mrc/groups/mbe.html

**Chemical Beam Epitaxy (CBE) with a $Si_3N_4$ mask.**
http://www.physik.tu-muenchen.de/tumphy/e/einrichtungen/wsi/E26/research/S_EPI.html

**Spreading and solidification of molten particles in a plasma spray process (movies).**
http://www.hprc.utoronto.ca/HPRC/visualization/examples/plasma/plasma.html

**Surface and Coatings Technology.**
http://bulb.mit.edu/bulb/journals/02578972/combined-toc.html

**Overview of use of plasmas in manufacturing**
http://www.engr.wisc.edu/centers/ercpam/useplasma.html

**A short list of important review works on CVD.**
http://klft.tn.tudelft.nl:8080/~chrisk/chrisk_cvdoverview.html

**Epitaxial lateral overgrowth for integrated silicon smart sensors**
http://muresh.et.tudelft.nl/dimes/1993/main/section1.4.2.13.html

**Process simulator.**
http://www.silvaco.com/products/athena/athena.html

**Publications by Solid State Electronics (1994/1995).**
http://www.ele.kth.se/FTE/fte_publ.html

**Field emitter group at MCNC. Showing Spindt-type field emitters.**
http://www.mcnc.org/HTML/ETD/EMAD/field_emitter/field_emitter.html (Photo of MCNC Field Emmiter)

**A treatise on aerosols with a sketch of a spray pyrolysis set-up**
http://www.physik.rwth-aachen.de/group/iaphys/aerosols/aerosols.html

**Chemical Mechanical Planarization (CMP) supplies from TBW Industries Inc.**
http://www.tbw-inc.com/cmp.html

**A course on chemical mechanical planarization.**
http://www.asms.rpi.edu/cont_ed/courses/cmp.htm

**Laser Assisted Molecular Beam Deposition (LAMBD)**
http://wings.buffalo.edu/academic/department/chem/Fac_Res_Int/Garvey/Instruments/Thin_Film_Setup/index.html

**Info on ion beam assisted deposition.**
http://www.iws.fhg.de/ext/firmen/ionbea.htm

**Ion beam surface treatment at SANDIA.**
http://www.sandia.gov/pulspowr/ppeng/ibest.html

**Focussed Ion Beam in micromachining at Maryland University.**
http://www.ee.umd.edu/Newsletter/vol2_no1/articles/ion.htm

**Ion beam materials processing at Los Alamos National Labs.**
http://fjwsys.lanl.gov/APT/ion_beam.html

A commercial group in micro dispensing technology.

http://www.metronet.com/~dwallace/mfabweb/mfab/compinfo.htm

A schematic of a Langmuir-Blodgett set-up.

http://www.cchem.berkeley.edu/~mmmgrp/LBT.html

The Electrochemical Science and Technology Information Resource (ESTIR) . Ideal for questions on plating, corrosion, etc.

http://www.cmt.anl.gov/estir/meet.htm

Very nice step by step explanation of Through-Hole-Plating.

http://www.databahn.net/galleria/retail/greenckt/stack/volumes/volvi/copplate.htm

Electroplating in the finishing industry.

http://www.finishing.com/index.html

Some background on chemical-mechanical polishing (CMP) for sensors and actuators.

http://www.el.utwente.nl/tdm/mmd/projects/polish/polish.htm

# Chapter 4. Wet Bulk Micromachining

## From the grab bag:

Introduction to Bulk Micromachining with pictures.

http://hypatia.dartmouth.edu/courses/es65/labs/lab2/bulk.html

Mechanical properties of Silicon.

http://www.ai.mit.edu/people/tk/tks/silicon-mechanical.html

Chemical properties of Silicon.

http://www.ai.mit.edu/people/tk/tks/silicon-chemical.html

Thermal properties of Silicon.

http://www.ai.mit.edu/people/tk/tks/silicon-thermal.html

Electrical properties of Silicon.

http://www.ai.mit.edu/people/tk/tks/silicon-electrical.html

Crystallography of Silicon with pictures of the crystal lattice viewed at different angles.

http://www.ai.mit.edu/people/tk/tks/diamond-structure.html

Si as an element in the periodic Table and a photo of a chunk of Si.

http://chemlab.pc.maricopa.edu/periodic/Si.html

An introduction to Microengineering by Danny Banks.

http://www.ee.surrey.ac.uk/Personal/D.Banks/umintro.html

The building of a piezoresistive pressure sensor. A nice introduction with plenty of pictures.

http://olympus.ece.jhu.edu/~tim/papers/sensor.html

An introduction to Si and GaAs micromachining.

http://tima-cmp.imag.fr/tima/mcs/micromachining.html

A pictorial explanation on how binary optic components are micromachined.

http://www.tbe.com/products/optics/tutor/manufac.html

Photo of Si etching set-up and nice background description of Si etching methods.

http://www.esat.kuleuven.ac.be/~lapadatu/Activity/etching.html

A series of contact mode AFM pictures of thin Si membranes etched in CsOH, demonstrating different topographies.

http://casimir.eecs.uic.edu/~serry/SPM/membranes.html

Tunnel epitaxy.

http://hypatia.dartmouth.edu/~cgl/microeng/courses/es194/student/psr/194sem_4b.html

Some background on glass etching.

http://www.brunswickmicro.nb.ca/~glassetch/intro.htm

In the middle of the discussion of $XeF_2$ as isotropic etchant of Si .

http://mems.isi.edu/archives/Discussion Group/1995/19.html

**Microscopic images of LEPI Si (Light-emitting porous silicon)**

http://www.science.uwaterloo.ca/research_groups/confocal/semiapp/semiapp.html

**Introduction to porous Si by Sailor at UCSD.**

http://checfs1.ucsd.edu/Faculty/sailor/PorousSilicon.html

**On overview of porous Si based devices.**

http://lisa.polymtl.ca/LISA-Brochure/Por.Si-Yacouba-p47.html

**Porous Silicon Nanostructures Homepage from Munich's Fraunhofer Institute.**

http://amp.nrl.navy.mil/code6651/porous-nano.html

**Porous Si experimental conditions.**

http://www.chembio.uoguelph.ca/thomas/poster1/pos1pg2.htm

**Detailed procedure for anisotropic etching of Si with pictures.**

http://www.eecs.uic.edu/~peter/eecs449/le/Silicon_Etching.html

**Emerging faces in anisotropically etched crystals by Carlo Sequin.**

http://s2k-ftp.cs.berkeley.edu:8000/tech-reports/csd/csd-91-639/all.ocr

**Simulation of anistropic and isotropic wet etching.**

http://mems.isi.edu/archives/Discussion_Group/1995/98.html

**An Overview Book on Etching of III-V Compounds.**

http://www.elsevier.com/catalogue/SAF/420/09215/09215/422419/422419.html

**Etch rate of TMAH vs. EDP.**

http://hypatia.dartmouth.edu/~cgl/microeng/MEM/fab/processing/etching/anisotropic.html

**Good list of questions for an exam on anistropic etching.**

http://ece-www.colorado.edu/~bart/ecen4375/rep03-96.txt

**Anisotropic etching of 110 Si at LLNL.**

http://www.llnl.gov./IPandC/op96/08/8a-ani.html

**Kirt Williams from BSAC and bubbles formation during Si etching.**

http://mems.isi.edu/archives/Discussion_Group/1996/212.html

**Some Caltech musings on Si etching.**

http://red.caltech.edu/Research/MEMS/etching.txt

**On-line isotropic /anisotropic etch simulation.**

http://red.caltech.edu/Research/MEMS/software.html

# Chapter 5. Surface Micromachining

## Thin Films Mechanical Properties

**MAT Berlin - Research**
http://www-mat.ee.tu-berlin.de/research/research.htm
> Description of research projects on the Characterization of thin films

**Measurement of Thin film Interface Strength**
http://hypatia.dartmouth.edu/levey/ssml/spallation.html
> Measurement of Thin film Interface Strength by a Laser SpallationTechnique

**DepositionTechnologies (including Carbon Film Deposition)**
http://www.anl.gov/LabDB/Current/Ext/M351.html
> Description of thin film deposition technologies

**Atomistic Simulation, Defect Control in Materials**
http://ny.frontiercomm.net/~casa/index.htm
> A program for the simulation of materials and defects

**IBM Journal of R & D - Adhesion science and technology**
http://www.almaden.ibm.com/journal/rd38-4.html#three
> Abstracts of papers on adhesion science and technology

**Probing Nano-Scale Forces with the AFM**
http://www.di.com/AppNotes/ForcCurv/Main.html#matsci

Excellent application note on the Probing of Nano-Scale Forces with the AFM/Mechanical properties of thin films
**Characterization of the Mechanical Properties of Thin Films and Interfaces—**
http://snf.stanford.edu/Projects/W.Nix/Nix.html
Figure showing a typical stress-strain curve for a sputter deposited aluminum film showing several elastic unloading cycles.
**Thin Film Stress Measurement**
http://www.ten.com/products/thinfilmstress.html
Equipment for Thin Film Stress Measurement
**Fatigue Test Structure**
http://mems.mcnc.org/figs/fatigue.gif
Photograph only
**Measurements of Polyimide Mechanical Properties using Resonant String Structures**
http://www.ece.gatech.edu/research/labs/msmsma/Materials/pistring.html
Description of project
**Determination of Young's modulus and residual stress of electroless nickel**
http://mems.cwru.edu/roy/abstracts/usystems96.html
Abstract of project: DETERMINATION OF YOUNG'S MODULUS AND RESIDUAL STRESS OF ELECTROLESS NICKEL USING TEST STRUCTURES FABRICATED IN A NEW SURFACE MICROMACHINING PROCESS
**In Situ Measurement of Young's Modulus and Residual Stress of Thin Electroless**
http://mems.cwru.edu/roy/abstracts/mrs94.html
Abstract of project: IN SITU MEASUREMENT OF YOUNG'S MODULUS AND RESIDUAL STRESS OF THIN ELECTROLESS NICKEL FILMS FOR MEMS APPLICATIONS
**Surface Roughness and Adhesion of Electroless Plated Nickel on Polysilicon**
http://mems.cwru.edu/roy/abstracts/ecs94-2.html
Abstract of project
**Surface micromachining tutorial**
http://www-mtl.mit.edu:8001/htdocs/tutorial.html
Surface micromachining is the process of forming moving structures by placing the structures on initially rigid platforms, then removing the platforms, usually by etching the material away.
**Surface micromachining history**
http://www-mtl.mit.edu:8001/htdocs/history.html
In 1984, Roger T. Howe completed his doctoral thesis at Berkeley on a novel gas sensor which utilized a resonating polysilicon bridge, formed using surface micromachining.
**Research Project Micromachining**
http://wwwetis.et.tudelft.nl/groups-info/pjfrench/ResearchProjects.html
Surface, Epi, and Bulk Micromachining/Research Project Micromachining/Faculty of Electrical Engineering, Electronic Instrumentation Laboratory, Mekelweg 42628, CD Delft., The Netherlands.
**LETI: Microtechnologies (Microsensors and Actuators)**
http://www-dta.cea.fr/wwwcea/leti/uk/miccap.htm
SURFACE MICROMACHINING ON SOI (SILICON OXIDE)/LETI: Microtechnologies (Microsensors and Actuators)
**Integrated Silicon Capacitive R.M.S.-to-DC Converter**
http://wwwetis.et.tudelft.nl/1994/2.12.html
Example of Surface Micromachining /Abstract with Photo
**Surface Micromachining**
http://mail.vdivde-it.de/IT/emsto/Europractice/Bosch/bosche.html
Foundry Service: Surface Micromachining/ Short application note with pictures
**Integrated Silicon Capacitive R.M.S.-to-DC Converter Based on the Surface Micromachining**
http://muresh.et.tudelft.nl/dimes/1993/main/section1.4.2.12.html
Example of surface micromachining.
**CTN: 3-D Surface Micromachining**
http://www.ee.ust.hk/eee/faculty/eenguyen/pers/research/3d.htm
Abstract: Recently a lot of attention has been focused on micro-electromechanical systems, or introducing a similar miniaturization technology to mechanical and electro-mechanical systems.
**Controlled Structure Release for Silicon Surface Micromachining**
http://mems.isi.edu/archives/dissertation-abstracts/ucb/phd93_4.html
Abstract: Silicon surface micromachining is a new technology that uses many of the common microfabrication techniques found in silicon integrated circuit (IC) production.

**Optimizing Sacrificial etching for surface micromachining**
http://wwwetis.et.tudelft.nl/1994/2.20.html
> Abstract with figure: The aim of this project is to optimize the sacrificial etching of phospho-silicate glass (PSG) for surface micromachining applications.

**Three-Layer Polysilicon Surface Micromachining Process**
http://mems.mcnc.org/smumps/Mumps.html
> Excellent example of Surface Micromachining Process with figures and text.

**Multi-Project Chip Activities Demonstrate Enormous Potential of Surface Micromachining**
http://www.analog.com/publications/magazines/accel_news/issue4/multproj.html
> Abstract of project by Analog Devices

**PSG for surface micromachining**
http://muresh.et.tudelft.nl/dimes/1993/main/section1.4.2.17.html
> Abstract: The effect of phosphorus content and PSG processing on the essential characteristics of PSG and upon the stress in overlying micromachined polysilicon layers.

**Design, Fabrication, Position Sensing, and Control of Electrostatic, Surface-Micromachined Microactuator.**
http://mems.isi.edu/archives/dissertation-abstracts/ucb/phd95_2.html
> Abstract: This dissertation describes the design, fabrication, position sensing and control of an electrostatically-driven microactuator.

**MEMS TechNet: Substrate Offerings (for micromachining)**
http://mems.mcnc.org/technet/offering.html
> Announcement: MEMS TechNet is pleased to announce the addition of Surface and Bulk Micromachining Substrates to the directory of available MEMS services offered through TechNet.

**Surface micromachined micromotors**
http://muresh.et.tudelft.nl/dimes/1993/main/section1.4.2.14.html
> Article: Surface micromachining provides the possibility to fabricate mechanical structures on chip, using process steps similar to those used for the fabrication of integrated circuits. <

**Nickel Surface Micromachining**
http://mems.cwru.edu/roy/abstracts/ecs94-1.html
> Abstract

**Polycrystalline silicon carbide for surface micromachining**
http://mems.cwru.edu/roy/abstracts/mems96.html
> Abstract

**Surface Micromachining**
http://hypatia.dartmouth.edu/courses/es65/labs/lab1/surface.html
> Laboratory outline: The Micro-Cantilever. This laboratory presents the fundamental operations in most electromechanical devices created using the techniques of surface micro-machining.

# MUMPs

**Three-Layer Polysilicon Surface Micromachining Process/Overview**
http://mems.mcnc.org/smumps/Mumps.html
> Overview with figures:The Multi-User MEMS Processes or MUMPs is a DARPA-supported program that provides the domestic (USA and Canada) industry, government and academic communities with cost-effective, proof-of-concept surface micromachining fabrication.

**MUMPs Design Guidelines and Rules**
http://mems.mcnc.org/smumps/Mrules.html
> Technical note: The purpose of the design rules is to ensure the greatest possibility of successful fabrication. The rules have evolved through process development, the experience of the MCNC staff, and the experience from previous MUMPs runs.

**SmartMUMPs**
http://mems.mcnc.org/smumps/smart.html
> A How-To Guide to Creating SmartMUMPs: Smart MEMS systems that integrate electronics and electromechanical devices can now be built using SmartMUMPs, a process designed to provide low-cost access to integrated MEMS using the Multi-User MEMS Process.

**Introduction to SmartMUMPs**
http://mems.mcnc.org/smumps/Fcmems.html
> Technical note with figures: The integration of microelectronics with MEMS is a key component to the success of most MEMS devices. By placing electronics closer to sensors and actuators, the device performance is improved.

**Smart MUMPsDesign Handbook**
http://mems.mcnc.org/smumps/SMTOC.html
> Technical note: This document contains revision 4 of the MUMPs Process Overview and Design Rules.


## SIMOX

**SIMOX wafers**
http://www.egg.or.jp/MSIL/english/semicon/simox-e.html
> Technical Information Sheet: SIMOX substrate manufacturing process

**SIMOX/HITEN: Silicon-on-Insulator Page/Links**
http://www.hiten.com/hiten/categories/SOISOS.html
> News and articles on SOI and links to centers working on SOI

**SIMOX/MAT Berlin - Publications in 1995**
http://www-mat.ee.tu-berlin.de/papers/public95/walt1ab.html
> Abstrct: Beta-SiC deposited on SIMOX substrates: characterization of the SiC/SOI system at elevated temperatures

**SIMOX**
http://bulb.mit.edu/mumbledir/bulbit.cgi?EA000052.09244247.EA940026.182.170-174
> Mechanical light modulator fabricated on a silicon chip using SIMOX technology, Wiszniewski W.R., Collins R.E., Pailthorpe B.A.


## Surface Micromachining Applications

**Optimizing Sacrificial etching for surface micromachining**
http://wwwetis.et.tudelft.nl/groups-info/pjfrench/Researchprojects/SacrificialEtch.html
> Abstract with figure

**Surface micromachined micromotors**
http://muresh.et.tudelft.nl/dimes/1993/main/section1.4.2.14.html
> Extended abstract: Surface micromachining provides us with the possibility to fabricate mechanical structures on chip, using process steps similar to those used for the fabrication of integrated circuits.

**Spiral/MCNC/JPEG image 338x258 pixels**
http://mems.mcnc.org/figs/spiral.jpg
> Photograph only

**Hinge/MCNC/GIF image 335x259 pixels**
http://mems.mcnc.org/figs/hingec.gif
> Photograph only

**Accelerometer/MCNC/GIF image 267x196 pixels**
http://mems.mcnc.org/cug/c5.fig.22.gif
> Photograph only

**Accelerometer and 68HC05/Motorola**
http://design-net.com/csic/WHATSNEW/PRESSRLS/PR950912.htm
> Announcement: MOTOROLA'S ACCELEROMETER DESIGN FEATURES SEALED G-CELL IN PLASTIC PACKAGE

**ADXL05/Accelerometer/Allied Electronics**
http://www.allied.avnet.com/allied/adi/adxl05.htmlx
> Product information with figure: Analog Devices, Inc. - ADXL05. Single Chip Accelerometer with Signal Conditioning

**ADXL50/Adobe Document/Analog Products**
http://www.analog.com/adibin/locate?ADXL50
> Product information

**OLYMPUS micro cantilevers for SPM**
http://www.olympus.co.jp/LineUp/Technical/Cantilever/levertopE.html#menu
> Product information. Links to the subject of micro cantilevers and AFM

**Technical data of silicon nitride cantilevers**
http://www.olympus.co.jp/LineUp/Technical/Cantilever/specnitrideE.html
> Product information with photos and figures

**Heart cell contractions measured using a micromachined polysilicon force transducer**
http://synergy.icsl.ucla.edu/htdocs/abstracts/GLin4.html
> Abstract: A microelectromechanical systems (MEMS) force transducer, with a volume less than one cubic millimeter, is being developed to measure forces generated by living, isolated cardiac muscle cells.

**Research Project Micromachining**

http://wwwetis.et.tudelft.nl/groups-info/pjfrench/Researchprojects/epipoly.html

Abstract with photo: Epitaxial poly for surface micromachined smartsensors/The goal of this project is to use the epitaxial reactor to selectively grow both polycrystalline and monocrystalline silicon films in a single deposition step.

**Micromachined adaptive mirrors.**

http://guernsey.et.tudelft.nl/swiss/swiss.html

Excellent article on micro-optics with figures

**Micromachined Movable Platforms as Integrated Optic Devices**

http://www.ece.gatech.edu/research/labs/msmsma/Electroplating/microoptics.html

Abstract with photo: In order to get good contact and light absorption structures, bulk micromachined movable platforms were fabricated on silicon substrates.

## More from the Grab Bag:

**A quick tutorial on surface micromachining.**

http://www-mtl.mit.edu:8001/htdocs/tutorial.html

**An exhibition of MEMS pictures at UCLA.**

http://synergy.icsl.ucla.edu/sems.html

**Fabrication and morphology of polysilicon.**

http://a76.iue.tuwien.ac.at:8001/diss/puchner/diss/node32.html

**Polysilicon grain growth.**

http://a76.iue.tuwien.ac.at:8001/diss/puchner/diss/node33.html

**Extrinsic diffusion.**

http://a76.iue.tuwien.ac.at:8001/diss/puchner/diss/node34.html

**Epitaxial realignment of polysilicon.**

http://a76.iue.tuwien.ac.at:8001/diss/puchner/diss/node35.html

**Nickel Surface Micromachining.**

http://mems6.eciv.cwru.edu/WWW/roy/nickel.html

**An introduction on surface micromachining (Analog Devices).**

http://nitride.eecs.berkeley.edu:8001/htdocs/tech.html

**MEMS Shortcourse Notes (CWRU).**

http://mems6.cwru.edu/shortcourse.html

**Formation, Properties and Applications of Porous Silicon.**

http://mems.isi.edu/archives/dissertation-abstracts/ucb/phd91_1.html_CalTech's

**Micromachining Homepage.**

http://www.cco.caltech.edu/~changliu/microlab.html

**MEMS in Canada.**

http://www.cmc.ca/Fabrication/Micromachining/micromachhp.html

**Materials constants of typical IC materials.**

http://www.ai.mit.edu/people/tk/tks/mcon.html

**An outline of the production of polycrystalline Silicon.**

http://www.egg.or.jp/MSIL/english/msi_hist3-e.html

**Properties of amorphous Silicon.**

http://www.ee.princeton.edu/~ampayne/abstracts_ist/jap75a.html

# Chapter 6. LIGA

## Electrodeposition

**Electroless plated nickel microstructure**

http://mems.cwru.edu/roy/pix/nickel1.gif

Photograph only

**Electroless Nickel Plating (293K)**

http://mems.cwru.edu/images/plating.gif

Photograph only

**Electrodeposition of Metals on Semiconductors**

http://www.phy.bris.ac.uk/research/microstructure/dep.htm

Short description with figure

**Metallic Microstructures Fabricated Using Photosensitive Electroplating Molds**
http://www.ece.gatech.edu/research/labs/msmsma/Electroplating/electroplating.html
> Abstract with figures: A novel process for the inexpensive fabrication of high aspect ratio electroplated microstructures is presented.

# LIGA

**UW-MEMS- Search Engine**
http://mems.engr.wisc.edu/search.html

**UW-MEMS- Image Archive**
http://mems.engr.wisc.edu/images/
> Examples of devices micromachined using Deep X-Ray Lithography (DXRL)/Pictures

**X-Ray Nanolithography**
http://nanoweb.mit.edu/xrnano.html
> Short overview with figure

**MEMS - UW-Madison**
http://mems.engr.wisc.edu/index.html
> MEMS homepage at the university of Winsconsin

**UW-MEMS- Master Index**
http://mems.engr.wisc.edu/master.html
> Master index of MEMS activities at the university of Winsconsin

**UW-MEMS- Research**
http://mems.engr.wisc.edu/research.html
> Current Research projects on MEMS at the university of Winsconsin

**UW-MEMS- Products**
http://mems.engr.wisc.edu/products.html
> Products Using UW-MEMS at the university of Winsconsin

**UW-MEMS- Archives**
http://mems.engr.wisc.edu/archives.html
> UW-MEMS Archives of photos, publications, and patents at the university of Winsconsin

**UW-MEMS- Other MEMS sites**
http://mems.engr.wisc.edu/moreinfo.html
> Links to MEMS centers worldwide

**A Fast Track for Micromachining Beamline**
http://www.src.wisc.edu/www/src_news/94_12/membln.html
> A short overview on micromachining: Micromachining has been used to create a wide range of ultra-miniature optical sensors, mechanicalmotion devices, and mechanical motion sensors.

**X-ray Lithography, University of Wisconsin at Madison**
http://www.xraylith.wisc.edu/
> Description of the Center for X-ray Lithography at the University of Winsconsin and of their experimental activities in the area of the X-ray lithography process development.

**X-ray Lithography/Student/University of Wisconsin at Madison**
http://www.xraylith.wisc.edu/~ocola/
> Homepage with descriptions and links in the area of X-ray lithography

**X-ray Nanolithography**
http://www-mtl.mit.edu/MTL/Report94/SUB/x-ray.html
> Project description: For several years we have been developing the tools and methods of x-ray nanolithography (i.e., sub-100 nm features).

**Nickel Micromotor and Gear Train Formed Using the LIGA Process at the University**
http://144.126.176.216/mems/c1_s1.htm#f1_4
> Description with figure of Nickel micromotor and gear train formed using the LIGA process at the University of Wisconsin

**X-ray Lithography**
http://suntid.bnl.gov:8080/nslsnews/0394/a07.html
> Micromachining at the NSLS: Shows a variety of test objects produced at the NSLS/ Figures

**W. Ehrfeld: Past and Future of LIGA**
http://www.uni-mainz.de/IMM/LNews/LNews_1/we.html
> Overview and history

**Microfabrication Using LIGA**
http://me62.lbl.gov:8001/engin_cap/micro.html
> A short description with example (figure): As modern manufacturing demands increasing precision and miniaturization, new processes are appearing to answer new challenges: A microfabrication process called LIGA.

**LIGA in the UK**
http://www.uni-mainz.de/IMM/LNews/LNews_2/tolfree.html
> Overview: The LIGA technique (X-ray lithography, electrodeposition, and injection/reaction molding or vacuum embossing) is employed to manufacture three-dimensional microstructures satisfying the most rigorous requirements

**S/I joins LIGA research**
http://www.uni-mainz.de/IMM/LNews/LNews_3/si.html
> Description of the LIGA Scanner System

**LIGA news Home Page**
http://www.uni-mainz.de/IMM/LNews/LIGA_home.html
> This is the WWW homepage of LIGA news

**The Microsystems Technology Program**
http://hbksun17.fzk.de:8080/FMT/kom1e.html
> Description of the LIGA project in Karlsruhe

**UW-MEMS- Precision Engineering**
http://mems.engr.wisc.edu/research/precision.html
> Technical note with figures: The basic Deep X-Ray Lithography and Electrodeposition Process has been extended to heights above one centimeter with improved resolution

**Multi-Project LIGA Run - Process Information**
http://mems.mcnc.org/ligaproc.html
> Technical note with figures

**LIGA SEMS**
http://mems.mcnc.org/ligasems.html
> SEM photographs from LIGA based processes

**LIGA at SMSSL**
http://cdr.stanford.edu/DD/SMSSL/NovelMicrostructures/LIGA.html
> Description of a LIGA project: LIGA Fabrication of X-Ray Telescope Grids

**LIGA news Third Issue / September 1995**
http://www.uni-mainz.de/IMM/LNews/LNews_3/cont3.html
> News from "LIGA news": Progress in Industrialising LIGA Technology

**LIGA phantasy**
http://www.uni-mainz.de/IMM/LNews/LNews_4/ligaimag.html
> Examples of microstructures using the LIGA process

**Design tools for LIGA**
http://www.uni-mainz.de/IMM/LNews/LNews_4/dortmund.html
> News of activities from the Physical Design of Micro Systems group at the University of Dortmund

## Molding

**Metallic Microstructures Fabricated Using Photosensitive Electroplating Molds**
http://www.ece.gatech.edu/research/labs/msmsma/Electroplating/electroplating.html
> Abstract with figures

**A Metallic Microaccelerometer Fabricated using Photosensitive Polyimide Electroplating**
http://www.ece.gatech.edu/research/labs/msmsma/Electroplating/accelerometer.html
> Photographs only

**Injection moulding modelling**
http://www.uni-mainz.de/IMM/LNews/LNews_3/microinj.html
> A short article with figures: At present, injection moulding is considered to be the most suitable technique for replication of plastic LIGA parts.

**Etching and Molding of 3DSi-Structures**
http://www.tu-harburg.de/ht/english/aetz.html
> Technical note

## More from the Grab Bag:

**The LIGA NEWS Home Page.**
http://www.uni-mainz.de/IMM/LNews/LIGA_home.html
**The LIGA HomePage at LBL.**
http://grace.lbl.gov/liga/ligapg.html
**LIGA in California.**
http://www.uni-mainz.de/IMM/LNews/LNews_1/califo.html
**LIGA in the UK.**
http://www.uni-mainz.de/IMM/LNews/LNews_2/tolfree.html
**LIGA in Japan.**
http://www.uni-mainz.de/IMM/LNews/LNews_3/umeda.html
**LIGA SEM Picture Gallery.**
http://mems.mcnc.org/ligasems.html
**LIGA at IMT.**
http://hbksun17.fzk.de:8080/IMT/liga.htm
**More LIGA at IMT.**
http://hbksun17.fzk.de:8080/IMT/avt b4.htm
**LEMA. Multi-User LIGA at KfK.**
http://hbksun17.fzk.de:8080/PMT/lemae.html
**Microsystem engineering at KfK.**
http://hbksun17.fzk.de:8080/FZK/Blick/englisch/micro.html
**High aspect ratio micromachining using UV sensitive polymers at Georgia Tech.**
http://www.ece.gatech.edu/research/labs/msmsma/GTMEMS.html
**Multiproject LIGA at MCNC.**
http://mems.mcnc.org/ligaproc.html

# Chapter 7. Comparison of Micromachining Techniques

## Laser Machining

**Resonetics, Inc - Micromachining Technology**
http://www.resonetics.com/
> Homepage of Resonetics, Inc.

**Resonetics, Inc - Illustrated Guide To Applications**
http://www.resonetics.com/applications.html
> Page on materials for laser micromachining by Resonetics, Inc.

**Laser Micromachining**
http://130.225.95.15/mic/staff/mm.htm
> Homepage with examples and pictures

**The Center for Applied Optical Sciences**
http://www.lasersoptrmag.com/links/univers.htm
> University Research Programs with lasers

## Links to the Web on Nanotechnology

**Electronic Nanocomputers**
http://www.mitre.org:80/research/nanotech/electronic.html
> Clicking on the mouse-sensitive boxes in the chart below will guide you to resources and information about many aspects of electronic nanocomputers.

**Future Nanocomputer Technologies**
 http://www.mitre.org:80/research/nanotech/futurenano.html
> Links to Nanotechnology pages on the Web

**Institute for Molecular Manufacturing**
 http://www.imm.org/
> Links to Nanotechnology pages on the Web

**Lycos Related Sites Search: nanotechnology**
 http://a2z.lycos.com/cgi-bin/pursuit?cat=a2z&query=nanotechnology&x=31&y=10
> Links to Nanotechnology pages on the Web

**NanoLink**
 http://sunsite.nus.sg/MEMEX/nanolink.html
> NanoLink - Key Nanotechnology Sites on the Web

**Small is Beautiful: a collection of nanotechnology links**
 http://www.nas.nasa.gov/NAS/Education/nanotech/nanotech.html
> The purpose of this page is to help facilitate access to information on the emerging science of nanotechnology.

**Nanotechnology page at Xerox PARC**
 http://nano.xerox.com/nano
> Introduction to the core concepts of molecular nanotechnology, followed by links to further reading.

**Brad Hein's Nanotechnology Page**
 http://www.public.iastate.edu/~bhein/nanotechnology.html
> Links to lots of publications

**Nanotechnology Information**
 http://www.aeiveos.com/nanotech/
> Links to Nanotechnology pages on the Web

**Chemistry on the Internet: The best of the Web 1995**
 http://hackberry.chem.niu.edu:80/Infobahn/Paper35/
> A selection of high quality chemical information on the Internet, presented as a poster at the ACS Symposium in Chicago on August 21, 1995 as part of the Chemistry on the Infobahn session.

**Molecular Nanotechnology**
 http://www.lucifer.com/~sean/n-mnt.html
> Links to Nanotechnology pages on the Web

**Nanotechnology and Such**
 http://www.rpi.edu/~greeng3/topics/nano.html
> Links to Nanotechnology pages on the Web

**Sean Morgan's Nanotechnology Pages**
 http://www.lucifer.com/~sean/Nano.html
> At this site you can access dozens of links to molecular nanotechnology, scanning probe microscopy, molecular modelling and nanoelectronics.

**DIMES Delft Institute of MicroElectronics and Submicron Technology**
 http://muresh.et.tudelft.nl/dimes/index.html
> Description of projects on Nanotechnology

**Foresight Institute**
 http://www.foresight.org/
> Introduction to the activities of the Foresight Institute on Nanotechnology

**Nanotechnology**
 http://www.ncb.gov.sg/jkj/nt/ntmiracle.html
> Links to Nanotechnology pages on the Web

**Ralph Merkle's Home Page**
 http://merkle.com/merkle
> Links to Nanotechnology pages on the Web

**Welcome to Nanothinc -- The World of Nanotechnology**
 http://nanothinc.com/
> There is a time for some things, and a time for all things; A time for great things, and a time for small things. Miguel de Cervantes . . . the time for nanotechnology is NOW

## Molecular

**Zurich Scientists Position Individual Molecules at Room Temperature**
http://www.zurich.ibm.com/News/Molecule/
      Description with pictures

**The Fourth Foresight Conference on Molecular Nanotechnology**
http://nano.xerox.com/nanotech/nano4.html
      SUMMARY: The conference is now over. This web page now serves as an archival source of information about the 1995 conference.

**March 1996 NASA Ames Computational Molecular Nanotechnology Workshop**
http://www.nas.nasa.gov/NAS/Projects/nanotechnology/workshop/
      Summary of conference

**The Potential of Nanotechnology for Molecular Manufacturing**
http://www.rand.org/publications/MR/MR615/mr615.html
      Overview article

**Strange Molecules With Strange Properties**
http://nanothinc.com/published/ostman/Cyberlife/Cyberlife3.html
      Overview article from the book: Nanotechnology - The Next Revolution by Charles Ostman

**Molecular Machine Theory**
http://ftp.ncifcrf.gov/pub/delila/nano2.ps

**Computational Molecular Nanotechnology at NASA Ames Research Center**
http://www.nas.nasa.gov/NAS/Projects/nanotechnology/
      Description with links to the Web

**Molecular Manufacturing: Adding Positional Control to Chemical Synthesis**
http://nano.xerox.com/nanotech/CDAarticle.html
      Article with figures

**Steps Towards Molecular Manufacturing**
http://www.nanothinc.com/nanosci/supramol/mbb/cda-news/cda-news.html
      Technical note

**Molecular Manufacturing Shortcut Group (MMSG)**
http://www.islandone.org/MMSG/
      Description of the mission of the Molecular Manufacturing Shortcut Group (MMSG), A Chapter of the National Space Society

**USB Buckyball home page**
http://buckminster.physics.sunysb.edu/
      This is the home page for the Fullerene research groups in the Physics Department at SUNY@Stony Brook, on Long Island, New York.

**Information Theory and Molecular Recognition page**
http://www-lmmb.ncifcrf.gov/~toms/
      Technical note on the use of information theory in molecular biology.

**Molecular Assembly Sequence Software (MASS)**
http://www.carol.com/mass.shtml

      Description of Molecular Assembly Sequence Software (MASS), by Carol Shaw, can be used as a high speedviewer for molecular designs.

**Representation Models in Molecular Graphics**
http://scsg9.unige.ch/eng/toc.html
      Overview article

**A Molecular Planetary Gear**
http://nano.xerox.com/nanotech/gearAndCasing.html
      Technical note: Many conventional designs can be scaled down to the molecular size range (if some care is taken in the design).

**Tech info on the molecular planetary gear**
http://nano.xerox.com/nanotech/gearAndCasingTech.html
      Technical note

**Materials and Process Simulation Center Home Page**
http://www.wag.caltech.edu/

# Nanotechnology

**Cyberlife - Micromachines, Nanotech and Processes Available Today**
http://nanothinc.com/published/ostman/Cyberlife/Cyberlife4.html
> Overview article from the book: Nanotechnology - The Next Revolution by Charles Ostman

**Cyberlife - Biomechanics in the Nanorealm**
http://nanothinc.com/published/ostman/Cyberlife/Cyberlife5.html
> Overview article from the book: Nanotechnology - The Next Revolution by Charles Ostman

**Cyberlife - WHAT ARE THE GOALS, AND HOW DO WE GET THERE?**
http://nanothinc.com/published/ostman/Cyberlife/Cyberlife6.html
> Overview article from the book: Nanotechnology - The Next Revolution by Charles Ostman

**Nanotechnology Bibliography**
http://nanotech.rutgers.edu/biblio.html

**NASA Conference on Computing for Nanotechnology**
http://www.nas.nasa.gov/NAS/Projects/nanotechnology/
> Description of the Nanotechnology work at NASA

**Nanotechnology: Myth or Miracle?**
http://www.aeiveos.com/nanotech/ntmiracle.html
> Overview article with links to the Web

**Nanotechnology**
http://athos.rutgers.edu:80/nanotech/
> List of papers on Nanotechnology with links to the Web

**Nanotech in space**
http://www.public.iastate.edu/~bhein/txt/mmsg.txt
> The National Space Society (NSS) position paper on space and molecular nanotechnology.

**Nanosystems Book Summary**
http://nano.xerox.com/nanotech/nanosystems.html
> A review of the book Nanosystems: molecular machinery, manufacturing, and computation by K. Eric Drexler

**Nanotech essay from ORNL**
http://www.ornl.gov/publications/labnotes/jun94/nanotech.html
> Overview article - Thinking nanotech: A group of Lab researchers believes it's not too early, or far out, to start doing R & D on atomic-scale machinery.

**Nanotechnology and Medicine Paper**
http://nano.xerox.com/nanotech/nanotechAndMedicine.html
> Overview article: Nanotechnology, the manufacturing technology of the 21st century should let us build a broad range of complex molecular machines (including, not incidentally, molecularcomputers) economically.

**Nanotechnology and MEMS Information**
http://137.132.166.53:1024/Tom/nanotech.html

**Nanotechnology on the WWW**
http://www.lucifer.com/~sean/Nano.html
> Homepage with numerous links to the Web on Nanotechnology

**Nanotechnology**
http://nano.xerox.com/nano
> Overview article with links to the Web

**NanoNews**
http://www.nanothinc.com/News/1021/space.html
> Overview article: Design and Manufacture of Space-Qualified MEMS Components (The Aerospace Corporation has a project underway which will help find answers to questions about the appropriateness of MEMS in space systems).

**NanoNet: Publications**
http://snf.stanford.edu/NNUN/NanoNet/publications.html
> List of publications that will be included in the National Nanofabrication Users Network (NNUN) that will be network-wide.

**Nanotechnology Mailing Lists**
http://www.lucifer.com/~sean/n-lists.html#MSTNET

**Introduction to Nanotechnology and Nanoelectronics**
http://www.mitre.org:80/research/nanotech/intronano.html
> Tutorial on the subject of Nanotechnology

**Mechanical Nanotechnology**

http://www.mitre.org:80/research/nanotech/mechanical.html

Technical note: Mechanical Nanocomputers

**The Museum of Nanotechnology**

http://www.hotwired.com/wired/scenarios/museum.html

The Museum of Nanotechnology is a fictional piece by Charles Platt with a possible future scenario for a nanotechnological museum/ Pictures

**Nanotechnology could lead to next industrial revolution**

http://www.ornl.gov/Press_Releases/archive/mr950712-00.html

Short overview

**Nanotechnology: the coming revolution in molecular manufacturing**

http://www.foresight.org/NanoRev/index.html

Short overview with links to the Web

**On Nanotechnology**

http://www.foresight.org/NanoRev/index.html

Short overview with links to the web (See Notes and References at the end of page).

**Overview of Nanotechnology. intro.html**

http://nanotech.rutgers.edu/nanotech/intro.html

Overview of Nanotechnology Adapted by J. Storrs Hall from papers by Ralph C. Merkle and K. Eric Drexler.

**Quantum Dots**

http://nanotech.rutgers.edu/nanotech/intro.html

List of preliminary descriptive papers about quantum dots (or "artificial atoms") that can be found on the Web

**The potential of Nanotechnology**

http://www.rand.org/publications/MR/MR615/mr615.html

Overview article

**Response to Trends in Nanotechnology**

http://www.foresight.org/SciAmDebate/SciAmResponse.html

Page on the Foresight Institute vs Scientific American Debate on nanotechnology with links to the Web

**Nanotech: Engines of Hyperbole?**

http://www.wired.com/Etext/1.6/features/nanotech.html

Overview article: You know the hype: Teensy-weensy little robots will someday push atoms around and build wholenew cities, eat pollution, make us immortal. But really, is any of this even remotely possible?

**Nanotechnology in Manufacturing**

http://www.fourmilab.ch/autofile/www/chapter2_84.html

Overview article: What Next? The Coming Revolution In Manufacturing Autodesk Technology Forum Presentation by John Walker May 10th, 1990

**Nanotechnology**

http://metaverse.com/futurequest/106.html

**Nanotechnology Musings**

http://www.cs.washington.edu/homes/pauld/fishnet/issue/01-08.html

**Nanowackology at SherryArt**

http://www.sherryart.com/nano/nanohome.html

Overview article: Understanding Terrifying Science Through Humor

**Nanomagnetics at the NSL**

http://www.tc.umn.edu/nlhome/m017/nanolab/research/nanomag.html

Technical note: Magnetic memory is one of the oldest solid-state memory technologies, yet it is still one of the most widely used in today's data storage.

**Nanoelectronics at the NSL**

http://www.tc.umn.edu/nlhome/m017/nanolab/research/quantum.html

Description of projects

**The Nanoelectronics Nanocomputing Home Page**

http://www.mitre.org:80/research/nanotech/

THE Nanoelectronics Home Page provides the Internet Gateway to nanoelectronics research and development information and resources from around the world.

**NanoTechnology**

http://www.ti.com/research/docs/nano.htm

> Description of the Nano Technology work at Texas Instruments

**The Nanomanipulator Project (UNC Chapel Hill)**

http://www.cs.unc.edu/nano/etc/www/nanopage.html

> Description of project: The nanoManipulator (nM) project is a collaboration between the departments of Computer Science and Physics at the University of North Carolina at Chapel Hill.

**Stanford NanoFabrication Facility Home Page**

http://www-nanonet.stanford.edu/NanoFab/

> Description: Welcome to the Stanford NanoFabrication Facility

**NanoNet Home Page**

http://snf.stanford.edu/NNUN/

> The National Science Foundation has established the National Nanofabrication Users Network (NNUN) as an integrated partnership of user facilities

**Nanostructures - What's Possible...**

http://snf.stanford.edu/NNUN/NanoNet/brochure_Possible.html

> Examples of nanostructures with figures

**...is limited only by your imagination.**

http://snf.stanford.edu/NNUN/NanoNet/brochure_Limited.html

> Photograph only

**Probing Nano-Scale Forces with the AFM**

http://www.di.com/AppNotes/ForcCurv/Main.html#matsci

> Article with figures

**Nanotechnology**

http://130.225.95.15/mic/research/nano/nanoele/research/research.htm

> Examples of nanostructures

**Construct's 3d Nanotech work in progress**

http://www.construct.net/projects/nanotech/conference/

> List of nanostructures projects

**Nanofabrication at Delft**

http://muresh.et.tudelft.nl/dimes/1993/main/chapter1.5.3.html

> Nanofabrication projects at Delft: Development of the technology and instrumentation for the fabrication of structures with dimensions in the nanometer regime.

**Electron and ion beam instrumentation for nanostructure fabrication**

http://muresh.et.tudelft.nl/dimes/1993/main/chapter1.5.6.html

> Nanofabrication project at Delft: Commercial electron beam lithography instruments can produce structures with dimensions down to about 20 nm.

**Fundamental aspects of nanocrystal engineering**

http://muresh.et.tudelft.nl/dimes/1993/main/chapter1.5.7.html

> Nanofabrication projects at Delft: The long term goal of the research program in the work group is the fabrication of nanometer-size structures with atomically sharp boundaries and well-defined shapes.

**Project Nanostructures**

http://sol.physik.tu-berlin.de/htm_grdm/pg3mission.htm

> Nanostructures projects at the TU in Berlin

**Fabrication of Quantum dots/GIF image 613x866 pixels**

http://sol.physik.tu-berlin.de/htm_grdm/img_grdm/makeqdse.gif

> Figures only

**Simulation in Nanostructures**

http://sol.physik.tu-berlin.de/htm_grdm/img_grdm/qdtheory.gif

> Simulation of strain relaxation and electronic levels in arbitrarily shaped nanostructures.

**Animation in Nanostructures**

http://sol.physik.tu-berlin.de/htm_grdm/img_grdm/hp1_5.mpeg

> Animation of strain relaxation and electronic levels in arbitrarily shaped nanostructures.

**Chemicals and process technology**

http://www.uni-mainz.de/IMM/LNews/LNews_3/surtec.html

> Technical note on Process-Technology and Chemicals for Micro-Engineering

**Electroforming**
 http://www.amtx.com/electroforming.html
        Overview paper
**Electroless Nickel Plating (293K).gif**
 http://mems.cwru.edu/images/plating.gif
        Photograph only
**Electroplating for Electronic Components**
 http://www.electroplating.com/
              Homepage on Electroplating Technology for the Electronic Component Industry
**End Point Detection in Semiconductor Processes**
 http://www.hiden.co.uk/endpoint.html
        Technical note with figures
**Focused Ion Beam Milling**
 http://timonf.aero.org/nlectc-wr/fib.html
        Technical note
**ION BEAM ASSISTED DEPOSITION (IBAD)**
 http://www.iws.fhg.de/ext/firmen/ionbea.htm
        Technical note
**Ion Implantation**
 http://www.iws.fhg.de/ext/firmen/ionimp.htm
        Technical note
**Plasma etching for sensor applications**
 http://muresh.et.tudelft.nl/dimes/1993/main/section1.4.2.16.html
              Project description: The goal of this research is to develop plasma etching processes for the fabrication of various micro-
              mechanical structures for sensors and actuators.
**Self Assembly and Nanotechnology/Xerox**
 http://nano.xerox.com/nanotech/nano4/whitesidesAbstract.html
        Overview paper

## More from the Grab Bag:

**The Institute for Micromanufacturing (IfM).**
http://www.latech.edu/tech/engr/ifm/litho.html
**Supplier of Laser Micromachining.**
http://www.resonetics.com/
**Precision manufacturing center (PMC).**
http://www.eng2.uconn.edu/pmc/index.html
**CNC wire EDM supplier.**
http://www.wireedm.com/
**The Potential of Nanotechnology for Molecular Manufacturing (RAND).**
http://www.rand.org/publications/MR/MR615/mr615.html
**Nanotechnology and the next 50 Years (R. Smalley presentation).**
http://cnst.rice.edu/dallas12-96.html
**Molecular Nanotechnology: Research Funding Sources (R. H. Smith).**
http://planet-hawaii.com/nanozine/nanofund.htm
**Nanotechnology. A matter of faith.**
http://www.virtualschool.edu/mon/Bionomics/Nanotechnology.html
**Best links to all nano related web materials. Start here !**
http://www.nas.nasa.gov/NAS/Education/nanotech/nanotech.html
**Molecular engineering.**
http://www.fourmilab.ch/autofile/www/section2_84_15.html

**The science fiction aspect of nanotechnology.**

http://www.hotwired.com/wired/scenarios/museum.html

**An introduction to nanomachining.**

http://www.nanothinc.com/NanoWorld/Introduction/RichardSmithIntro/smith_r1.html

**Design consideration for an assembler.**

http://nano.xerox.com/nanotech/nano4/merklePaper.html

**A picture gallery of STM pictures.**

http://www-i.almaden.ibm.com/vis/stm/atomo.html

**An update by foresight.**

http://nanotech.rutgers.edu/nanotech/papers/update8.html

**Nanofabrication and micromachining links.**

http://www.lucifer.com/~sean/n-mem.html

**More Nano-Links:**

    http://nanotech.rutgers.edu/nanotech/

    http://nano.xerox.com/nanotech/reversible.html

    http://nano.xerox.com/nanotech/feynmanPrize.html

    http://nano.xerox.com/nanotech/nano4.html

    http://nano.xerox.com/nanotech/feynman.html

    http://www.carol.com/mass.shtml

    http://nanothinc.com//

    http://www.wired.com/Etext/1.6/features/nanotech.html

    http://www.ibf.unige.it//

    http://www-im.lcs.mit.edu//

**Precision Machining: Sensors and Controls.**

http://www.me.mtu.edu/research/mfg/mmt/prec/precision.html

**Nanotechnology bibliography.**

http://nanotech.rutgers.edu/nanotech/biblio.html

**Bioelectronics. A simple introduction.**

http://www.circuitworld.com/subpage/jan96.htm

**A roadmap to bioelectronics.**

http://www.pst-elba.it/Brussels/bioelmap.html

**Karube's Lab.**

http://t-rex.bio.rcast.u-tokyo.ac.jp/index2.html

**Usenet group on nanomachining.**

http://sunsite.unc.edu/usenet-i/groups-html/sci.nanotech.html

# Chapter 8. Micromachine Development and Packaging

## CAD

**Microcosm Technologies MEMCAD Modeling Software**

  http://mems.mcnc.org/technet/microcosm.html

      Technical note

**MEMCAD**

  http://arsenio.mit.edu/projects/MEMCAD.html

      Technical note

**MEMSCAD**

  http://symphony.icsl.ucla.edu/

      3D MEMS Simulation

**CAD Tools**
http://nitride.eecs.berkeley.edu:8001/htdocs/cad.html
>Technical note

**MISTIC Homepage**
http://www.eecs.umich.edu/mistic/
>MISTIC is a planar process compiler for thin-film and micromachined devices. This tool automatically generates fabrication process flows starting from a two-dimensional geometrical device description.

**MEMS Tools and Software offered through the Clearinghouse**
http://mems.isi.edu/archives/tools/
>Economical Mask-Making Using Desk-Top Publishing

**Microsystem Group Research Topics/(CAD)**
http://tima-cmp.imag.fr/tima/mcs/topics.html#cadtools
>Technical note: CAD tools for MEMS

## Computing

**Reversible Electronic Logic Article**
http://nano.xerox.com/nanotech/electroTectOnly.html

**Helical Logic Computing**
http://nano.xerox.com/nanotech/helicalIntro.html
>This page is the abstract and introduction of Helical Logic

**Reversible Logic Summary**
http://nano.xerox.com/nanotech/reversible.html
>Article with links to other publications

**Materials and Process Simulation Center (MSC) at the California Institute of Technology**
http://www.wag.caltech.edu/
>Description of center: Purpose - develop the tools necessary for atomic level modeling and simulation of materials and others

**Computational Nanotechnology by Ralph Merkle**
http://nano.xerox.com/nanotech/compNano.html
>Article on Nanotechnology: The major research objectives in molecular nanotechnogy are the design, modeling, and fabrication of molecular machines and molecular devices.

**Computational results at Xerox**
http://nano.xerox.com/nanotech/comp

**D. H. Robertson: Selected Research Projects**
http://chem.iupui.edu/Research/Robertson/Robertson.html
>This page presents some of my research which I thought might lend itself to a WWW presentation. This includes images, graphs and animations - no sound yet.

**Concurrent Computing Laboratory for Materials Simulations, Louisiana State University**
http://www.cclms.lsu.edu/
>Description of projects

## Micromachining Materials

**Colorless and Low Dielectric Polyimide Thin Film Technology**
http://www.rti.org/technology/poly_film.html
>Application note: A need exists for high temperature, flexible polymeric film and coating materials that have high optical transparency and radiation stability for space applications.

**Colorless Polyimide Thin Film Technology**
http://db-www.larc.nasa.gov/tops/Exhibits/Ex_W-619/Ex_W-619.html
>Technical Information Sheet: Colorless polyimide films are being developed for use on large space structures and components such as solar concentrators, reflectors, solar cells and thermal control coating systems.

**PSG for surface micromachining**
http://muresh.et.tudelft.nl/dimes/1993/main/section1.4.2.17.html
>Abstract of project

**Formation, Properties and Applications of Porous Silicon**
http://mems.isi.edu/archives/dissertation-abstracts/ucb/phd91_1.html
>Abstract of project

**Amorphous Silicon: The Dominant Active Matrix Technology**
http://itri.loyola.edu/Dsply Jp/c5_s4.htm
>   Review article with figures

**Micromachines Program**
http://tima-cmp.imag.fr/tima/mcs/micromachining.html
>   Paper on micromachining materials

**Omron Corporation**
http://144.126.176.216/MEMS/C_Omron.htm
>   Description of micromachining research activities at Omron Corporation

**Wide and continuous wavelength tuning...**
http://luciano.stanford.edu/~larson/aplvcsel.html
>   Technical note: Wide and continuous wavelength tuning in a vertical-cavity surface-emitting laser using a micromachined deformable-membrane mirror

**Emerging Smart Materials Systems**
http://world.std.com/~hbstrock/sta/exec.html
>   Technical note - EMERGING SMART MATERIALS SYSTEMS: TECHNOLOGIES, APPLICATIONS and MARKET OPPORTUNITIES

**Piezoresistive and Piezoelectric Micromachined Pendulous Accelerometers**
http://verp.www.media.mit.edu/projects/SmartPen/node30.html
>   Technical note

**MEMS-Based Sensors**
http://itri.loyola.edu/MEMS/C3_S2.htm
>   Description of SENSOR DEVELOPMENT IN JAPAN with figures

**A model of silicon crystal**
http://synergy.icsl.ucla.edu:80/Figures/crystal.ps

**Polycrystalline silicon carbide**
http://mems.cwru.edu/roy/projects/poly sic.html
>   Technical note

**XRD and XTEM investigation of polycrystalline silicon carbide on polysilicon**
http://mems.cwru.edu/publications.html
>   Technical note


## Packaging

**ESI Electronic Packaging Products Guide**
http://www.elcsci.com/mmimenu.htm
>   Technical note on tools for laser micromachining

**Microstructure Micropackaging**
http://mems.isi.edu/archives/dissertation-abstracts/ucb/ms91_3.html
>   Thesis abstract: Two processes are developed to fabricate electrostatic comb drives hermetically encapsulated in a silicon nitride or polysilicon micropackage.

**Micropackaging and interconnection**
http://www.elis.rug.ac.be/brochure/node27.html
>   Abstract with figure

**Micro Packaging**
http://www.micropackaging.com/
>   Homepage of company devoted to provide solutions to the challenges of surface mount device packaging.

**MATERIALS for ELECTRONIC PACKAGING (MEP)**
http://www.smtplus.com/mep.htm
>   OVERVIEW of MATERIALS for ELECTRONIC PACKAGING

**ADVANCED PACKAGING, MICROASSEMBLY, AND TESTING TECHNIQUES**
http://144.126.176.216/mems/d_6.htm
>   Technical note

**KAEDING - RESEARCH PROJECTS**
http://www-leland.stanford.edu/group/mtmc/package.html
>   Technical note: Package-Level Thermal Management using Novel Micromachined Structures

**A Variety of Typical Closed, Rigid Electronic Packages**

http://144.126.176.216/mems/c6_s2.htm#f6_1

>> Photograph of a variety of typical closed, rigid electronic packages: dual in-line package (DIP), pingrid array (PGA), surface mounts, direct die mounting.

**A Method of Integrated Vacuum Packaging**

http://144.126.176.216/mems/c6_s2.htm#f6_3

>> Photograph of method of integrated vacuum packaging.

**PAT Processes Occur in All Procedures and at All Levels**

http://144.126.176.216/mems/c6_s3.htm#f6_8

>> Technical note on Fabrication Processes, Packaging, Assembly and Testing.

**Packaging, Assembly, and Testing. c6_s1**

http://144.126.176.216/mems/c6_s1.htm

>> Short overview: PACKAGING, ASSEMBLY, AND TESTING: INTRODUCTION TO PAT AND ITS RELATIONSHIP TO MEMS (Importance of PAT) by Stephen C. Jacobsen

**Packaging, Assembly, and Testing. c6_s2**

http://144.126.176.216/mems/c6_s2.htm

>> Overview: PAT PROCESSES OCCUR AT ALL FOUR SYSTEMS LEVELS Level 1: Microelectronics

**Packaging, Assembly, and Testing. c6_s3**

http://144.126.176.216/mems/c6_s3.htm

>> Technical note - REVIEW OF PACKAGING, ASSEMBLY, AND TESTING: Package Definitions, Requirements, and Design.

**Packaging, Assembly, and Testing. c6_s4**

http://144.126.176.216/mems/c6_s4.htm

>> Examples of SAMPLE PACKAGES — LEVELS AND PROCESSES

**Packaging, Assembly, and Testing. c6_s5**

http://144.126.176.216/mems/c6_s5.htm

>> SUMMARY AND CONCLUSIONS of technical note on Packaging: General - Review of Observations

**Microassembly Automation Laboratory**

http://www.llnl.gov/eng/ee/erd/maal/maalhome.html

>> Description of the Microassembly Automation Laboratory at Lawrence Livermore National Laboratory (LLNL).

## More from the Grab Bag:

**The MEMCAD Homepage.**

http://arsenio.mit.edu/projects/MEMCAD.html

**The Intellisense Homepage.**

http://www.intellis.com/

**Microcosm Technologies Home Page. A commercial supplier of MEMCAD software.**

http://www.memcad.com/home.html

**A position paper on Structured Design Methods for MEMS by Crary.**

http://red.caltech.edu/NSF_MEMS_Workshop/crary

**The CadStuff page. Excellent starting point for all information on CAD.**

http://tribeca.ios.com/~compvent/cadstuff.html

**The Home page for the MOSIS VLSI Fabrication Service.**

http://www.isi.edu/mosis/

**Design rules for all MOSIS technologies, in Postscript.**

http://info.broker.isi.edu:70/mosis/designrules

**An introduction to PATRAN.**

http://www.engr.utk.edu/software/eng/patran.html SUPREM007.

http://www-ee.stanford.edu/ee/tcad/programs/supremOO7.html

**SUPREMIV.GS (Besides Si this program also models GaAs).**

http://www-ee.stanford.edu/ee/tcad/programs/suprem-IV.GS/Book/Introduction.html

**Garry Fedder gives an overview of the various MEMCAD efforts worldwide.**

http://red.caltech.edu/NSF_MEMS_Workshop/fedder

The University of Michigan MEMS CAD package; 'MISTIC.'

http://www.eecs.umich.edu/mistic/index.html

**Software documentation for Abaqus.**

http://www.cenapad.unicamp.br/CORNELL/UserDoc/Software/Num/abaqus/

**An introduction to the various needed CAD tools for surface micromachining. Includes a description of FASTCAP.**

http://nitride.eecs.berkeley.edu:8001/htdocs/cad.html

**A brief primer on advanced microelectronic packaging.**

http://www.plk.af.mil/PLhome/VT/AMP/packtut1.html

**A detailed description of plastic encapsulated microelectronics.**

http://spezia.eng.umd.edu/general/demos/pem/contents/contents.html

**Dicing and grinding equipment.**

http://www.microtronic.com/lo_index.htm

**Low cost packaging from the DARPA technology office.**

http://esto.sysplan.com/ESTO/El-Packaging/LowCost/Presentation/index.html

**A list of books on packaging (OPAMP Technical Books).**

http://www.opampbooks.com/ELE_PACK/

**An electronic packaging design house (EPD).**

http://www.uscad.com/epd/epd.html

**A listing of companies in the electronic packaging business.**

http://www.ibc.co.il/v4040507.html

**Electronic packaging at IBM.**

http://www.chips.ibm.com/products/packaging/

**An introduction to electronic packaging.**

http://www.mcc.com/projects/pcmcia/toc.html

**A presentation on cryogenic electronic packaging.**

http://esto.sysplan.com/ESTO/El-Packaging/Presentation/CryoPack/slide1.html

**Kurt Petersen on his DARPA project on Silicon Fusion Bonding.**

http://eto.sysplan.com/ETO/MEMs/Prog_Summaries/Petersen.html

**On anodic and frit bonding.**

http://www.ebl.rl.ac.uk/europractice/mc3/crl.html

**Principles of Si to Si and Silicon to glass anodic bonding (with pictures).**

http://www.oslo.sintef.no/avd/31/pera/Sensors.html

**A collection of references on packaging.**

http://www.afep.cornell.edu/Packaging/Resources/References.htm

# Chapter 9. Scaling Laws, Actuators and Power in Miniaturization

## Fluidics

**Addressable Micromachined Jet Arrays**

http://www.ece.gatech.edu/research/labs/msmsma/Microjet/microjet.html

> Technical note

**Andrew Kamholz's homepage**

http://weber.u.washington.edu/~kamholz/research/research.html

> Description of project: Plan to use microelectromechanical systems (MEMS) technology to create a series of devices which perform continuous small-molecule extractions from a particle-laden suspension.

**Microfluidics**

http://www.pharmacia.se/biosensor/introduction/microfluidics.html

**Microfluidics**

http://www.uni-mainz.de/IMM/LNews/LNews_4/cont4.html

> News on latest advances with links

**CFD-designing micro mixers**

http://www.uni-mainz.de/IMM/LNews/LNews_4/aea.html
> Technical note: The Application of CFD to the Design of Microfluidic Mixers

**Microfluidic reactor systems**

http://www.uni-mainz.de/IMM/LNews/LNews_4/mire.html
> Technical note: Microreactors - New Applications for Microfluidic Devices

**Geoff Dolan Project**

http://hypatia.dartmouth.edu/courses/es65/projects/dolan/explain.html
> Description of project on Microfluidic Devices

**Fluid Handling**

http://www.el.utwente.nl/tdm/mmd/projects/fluid.htm
> Description of project on Fluid Handling

**MEMS Fluid Sensing and Control Summaries**

http://esto.sysplan.com/ESTO/MEMS/Prog_Summaries/fluid.html
> MEMS Program Summaries on Fluid Sensing and Control sponsored by DARPA

**Microfluidics**

http://www.pharmacia.se/biosensor/introduction/microfluidics2.html

**Microfluidics International Corp.**

http://guide.nature.com/company/microfluidicsinternationalcorp
> Homepage

**MicroFluidics Center**

http://www.mal.eecs.uic.edu/cover.htm
> Description of projects at the MicroFluidics Center

**Dennis L. Polla, Ph.D.**

http://pro.med.umn.edu/bme/polla_d.html
> Description of research project on the deposition and characterization of piezoelectric and pyroelectric thin films on silicon-based microstructure materials

**MIT Project: Silicon Wafer Bonding Micromachined Devices**

http://goesser.mit.edu/SRC/Schmidt.html
> Description of research project: The objective of this task is to establish a technology platform for fabrication of integrated micromechanical devices, specifically sensors.

**Piezoelectric**

http://verp.www.media.mit.edu/projects/SmartPen/node31.html
> Technical note on PE materials and sensors/Links

**SMART System**

http://www.biotech.pharmacia.se/smartsys/smartsys.htm
> Description of SMART system with links: SMART System is optimized for micropurification and micropore chromatography.


## From the Grab Bag:

**Scaling: Why is Animal Size so Important?**

http://www.cup.org/Titles/31/0521319870.html

**CISM Course on Scaling and Fractality in Continuum Mechanics.**

http://www.polito.it/iniziati/CISM/

**Scaling laws of bending moment.**

http://www.caip.rutgers.edu/vizlab_group_files/RESEARCH/HPCD/SHIP/DIAGSHIP_PAPER/node6.html

**Examining Scaling Laws for Turbulence at low Rayleigh Number via Direct Simulations.**

http://www.mmm.ucar.edu/blt/gac95/node71.html

**Fluid dynamics and Thermodynamics of Micromechanical Devices.**

http://ostrich.usc.edu/rsg/mmd/mmd_index.html

**Theory of Minature Direct Drive Actuators.**

http://found.cs.nyu.edu/rsw/theory.html

**History and tutorial on piezo.**

http://www.piezo.com/info.html

**Piezoelectric Actuators (Sensor Technology Limited).**

http://www.sensortech.ca/actuator.html

**Piezo Actuators (PZT Translators) with subnanometer resolution and microsecond response time.**

http://www.physikinstrumente.com/pages/pztact.htm

**Shape Memory Alloys and Their Applications.**

http://www.uni.uiuc.edu/%7Erichlin/chem.html

**Ferroelectric polymers and pyroelectric sensors.**

http://ap-dec87b.physik.uni-karlsruhe.de/ferropol/fppspub.htm

**Electro-vs piezo actuators. Tutorial.**

http://www.newport.com/tutorials/Electro_vs_Piezo_Actuators.html

**Fluid control. Commercial products from Vickers.**

http://www.vickers-systems.com/products/

**Control of aircraft wings with MEMS.**

http://ho.seas.ucla.edu/research/m3micro.html

**Massively Actuated and Sensed Structures.**

http://caswww.colorado.edu/%7Ergmenon/project.html

**Advanced fluid systems. Electro and magneto rheological fluids. Smart fluids.**

http://www.a-f-s.com/

# Chapter 10. Microfabrication Applications

## Accelerometers

**Single and Multilayer Electroplated Microaccelerometers**

http://www.ece.gatech.edu/research/labs/msmsma/Electroplating/accelerometer.html

     Photographs only

**Accelerometer/GIFimage 640x324 pixels**

http://infoserve.unisa.edu.au/eng/mec/IMAGES/ACCELERO.GIF

     Photograph only

**Hitachi Closed Loop Capacitive Accelerometer**

http://144.126.176.216/mems/c3_s2.htm#f3_7

     Extended abstract with figures

**Nissan Integrated Silicon Accelerometer**

http://144.126.176.216/mems/c3_s2.htm#f3_8

     Extended abstract with figures

**Tohoku University Integrated Capacitive Accelerometer**

http://144.126.176.216/mems/c3_s2.htm#f3_9

     Extended abstract with figures

**Modern Accelerometers**

http://verp.www.media.mit.edu/projects/SmartPen/node27.html

     Description of different implementations of the basic accelerometer: the pendulous accelerometer, the vibrational accelerometer, and the electromagnetic accelerometer.

**Accelerometer Intrinsics**

http://verp.www.media.mit.edu/projects/SmartPen/node15.html

     Description of projects on accelerometers: Accelerometers are the sensors at the very foundation of inertial navigation; they sense inertia.

**Accelerometer/GIFimage 267x196 pixels**

http://mems.mcnc.org/cug/c5.fig.22.gif

     Figure only

**Accelerometer and 68HC05 by Motorola**

http://design-net.com/csic/WHATSNEW/PRESSRLS/PR950912.htm

     Technical note: MOTOROLA'S ACCELEROMETER DESIGN FEATURES SEALED G-CELL IN PLASTIC PACKAGE.

**Accelerometer by Allied Electronics**

http://www.allied.avnet.com/allied/adi/adxl05.htmlx

>> Technical note: Analog Devices, Inc. - ADXL05 Single Chip Accelerometer with Signal Conditioning

**Accelerometer/GIFimage 640x324 pixels**

http://infoserve.unisa.edu.au/eng/mec/IMAGES/ACCELERO.GIF

>> Figure only

**MIT microaccelerometer. GIF image 764x993 pixels**

http://hypatia.dartmouth.edu/courses/es65/images/henning4.gif

>> Photograph only

**CSMT Sensor Technology/ (Microseismometer and Microaccelerometers)**

http://137.79.14.14/seis.html

>> CENTER FOR SPACE MICROELECTRONICS TECHNOLOGY

**ACCELER2.GIF**

http://infoserve.unisa.edu.au/eng/MEC/IMAGES/ACCELER2.GIF

>> Photograph only

**MAT Berlin - Two-Axis Accelerometer**

http://www-mat.ee.tu-berlin.de/research/ac_sensor/ac_sens.htm

>> Description of project: Two-Axis Micromachined Accelerometer for Gesture Recognition

**Northeastern University: Microsensors: Microaccelerometer**

http://www.ece.neu.edu/edsnu/zavracky/mfl/programs/acc/acc.html

>> Overview article: Microaccelerometer

**Top View of a Force-Balanced Accelerometer Chip**

http://144.126.176.216/mems/c5_s2.htm#f5_4

>> Technical note with figure: Top view of a force-balanced accelerometer chip formed by surface micromachining together with CMOS detection electronics realized at UC Berkeley.

**Surface-Micromachined Accelerometer**

http://144.126.176.216/mems/c5_s2.htm#f5_5

>> Technical note with figure: Surface-micromachined accelerometer with on-chip signal processing electronics from Analog Devices.

**Circuit Organization in the Hitachi Force-Balanced Accelerometer**

http://144.126.176.216/mems/c5_s3.htm#f5_12

>> Technical note with figure: Circuit organization in the Hitachi force-balanced accelerometer.

**Integrated Accelerometer - Analog Devices**

http://144.126.176.216/mems/c6_s4.htm#f6_9

>> Technical note with figure: Integrated accelerometer - Analog Devices (USA).

## Actuators

### Microactuators: Overview

**Microactuators.c4_s1**

http://144.126.176.216/mems/c4_s1.htm

>> MICROACTUATORS: INTRODUCTION by Richard S. Muller

**Microactuators.c4_s2**

http://144.126.176.216/mems/c4_s2.htm

>> DEVELOPMENTS IN JAPAN: These and other generalized concepts about microactuators formed a basis for comparing microactuation efforts in Japan and the United States.

**Actuator Arrays**

http://simon.cs.cornell.edu/Info/People/karl/MicroActuators/

>> Description of project: Micro-Electro-Mechanical-Actuator-Arrays - Our actuators are based on microfabricated torsional resonators, rectangular grids etched out of single-crystal silicon and suspended by two rods that act as torsional springs.

**Actuators**

http://www.el.utwente.nl/tdm/mmd/projects/actuat.htm

>> Description of project on actuators at Twente

**Aluminum MEMS Structures**
http://transducers.stanford.edu/stl/Projects/AlumMEMS.html
  Description of project: This project seeks to develop all-aluminum micromachined devices for use as sensors and actuators.

**Deforming Film Actuators - Tohoku University**
http://144.126.176.216/mems/c6_s4.htm#f6_11
  Technical note: Deforming film actuators

**Description of IMMI's BITS Microactuators**
http://www.micromachines.com/summary_paper.html
  IMMI's technical publication related to silicon micromachining: Microactuators for Rigid Disk Drives

**MAT Berlin - Micro Actuator - Design**
http://www-mat.ee.tu-berlin.de/research/actuator/design.htm
  Article on the Design of Surface Micromachined Micro Actuators

**MAT Berlin - Micro Actuator - Fabrication**
http://www-mat.ee.tu-berlin.de/research/actuator/fabricat.htm
  Article on the Fabrication of Surface Micromachined Micro-Actuators

**Microactuator**
http://www.latech.edu/~saravana/phase.html
  Description of project: Thermally Driven phase Change Microactuator

**Microactuators**
http://144.126.176.216/mems/c4_s2.htm#f4_2
  Technical note

**Microactuators**
http://144.126.176.216/mems/c4_s2.htm#f4_1
  Technical note

**Microactuators**
http://144.126.176.216/mems/c4_s2.htm#f4_3
  Technical note

**MICROACTUATORS: TECHNOLOGY, DEVICES, AND IDEAS IN THE UNITED STATES**
http://144.126.176.216/mems/d_4.htm
  Overview

**Microfabricated nickel actuators**
http://mems.cwru.edu/roy/projects/nickel.html
  Description of project: Microfabricated Nickel Actuators

**Organic Thermal Actuators**
http://transducers.stanford.edu/stl/Projects/organic.html
  Description of project - Objectives: Development and demonstration of robust organic-based thermal bimorph actuators that can be used to manipulate small objects for inspection and general positioning in realistic environmental conditions.

**Shape Memory Alloy**
http://tag-www.larc.nasa.gov/tops/tops95/exhibits/mat/mat-20-95/mat02095.html
  Technical note: Design, Fabrication, and Testing of an SMA Actuator

**SMA Actuator/TiNi**
http://lisa.polymtl.ca/LISA-Brochure/TiNi-Mario.html
  Technical note: Magnetron Sputtering of TiNi Shape Memory Alloys for Microactuators

**Survey of Micro-Actuator Technologies**
http://nanothinc.com/nanosci/microtech/mems/ten-actuators/gilbertson.html
  Overview

## Bearings

**Assembled nanoscale bearing**
http://www.aeiveos.com/nanotech/nano/bigBearing.gif
  Figure only

**Disassembled bearing**
http://www.aeiveos.com/nanotech/nano/bigBearingApart.gif
> Figure only

**Bearing (angle view)**
http://www.aeiveos.com/nanotech/nano/planGearAngle.gif
> Figure only

**Bearing (end view)**
http://www.aeiveos.com/nanotech/nano/planGearEnd.gif
> Figure only

**Bearing (side view)**
http://www.aeiveos.com/nanotech/nano/planGearSide.gif
> Figure only

**SMART HYDRODYNAMIC BEARING**
http://weewave.mer.utexas.edu/MED_files/MED_research/Smart_Bearings/smart_bearings.html
> Article on SMART HYDRODYNAMIC BEARING APPLICATIONS OF MICRO-ELECTRO-MECHANICAL SYSTEMS.

**IFM Research Areas**
http://www.latech.edu/tech/engr/ifm/research.htm
> Description of project: Sensor embedded grinding balls and bearings are an example of integrating piezoelectric sensors and data processing microelectronics for on-line measurements in manufacturing processes.

**Closeup of CaMEL Bearing2 Element**
http://mems.mcnc.org/figs/sem10.gif
> Photograph only

**Micromachined Smart Bearings**
http://weewave.mer.utexas.edu/MED_files/MED_research/mems_sum.html#bearings


# Biomedical

**Heart cell contractions measured using a micromachined polysilicon force transducer**
http://synergy.icsl.ucla.edu/htdocs/abstracts/GLin4.html
> Abstract of project: A microelectromechanical systems (MEMS) force transducer, with a volume less than one cubic millimeter, is being developed to measure forces generated by living, isolated cardiac muscle cells.

**Cell Impedance Probe Project**
http://hypatia.dartmouth.edu/courses/es65/projects/cell/explain.html
> Description of project

**Microsystems for the Hearing Impaired.**
http://www.shef.ac.uk/uni/projects/mesu/projects/speaker.html
> Description of a speaker capable of being inserted into the outer ear canal for in-ear hearing aids and monitoring hearing loss in infants.

**Duodeno fiberscope Currently Marketed by Olympus**
http://144.126.176.216/mems/c6_s2.htm#f6_7
> Technical note

**Microfabricated cell culture 'device'. GIF image 816x631 pixels**
http://hypatia.dartmouth.edu/courses/es65/images/henning7.gif
> Photograph only

## Biological Sensors

### Patterning Cultured Neurons
http://transducers.stanford.edu/stl/Projects/ControlledPatt.html
> Description of project: Acoustic microscope image of PC-12 cells, showing a cell's outgrowth turning around a ninety-degree corner in a micromachined guidance channel.

### Thin-Film Microelectrode Arrays for Neural Connectivity Studies
http://transducers.stanford.edu/stl/Projects/Thin-Film2.html
> Description of project: Development of planar, two-dimensional arrays of thin-film iridium microelectrodes for recording from and stimulation of neural tissue in both primary cell line cultures and slice.

**Micro Probes for Neurophysiology**

http://transducers.stanford.edu/stl/Projects/MicroMech.html

> Description of project: Representative microprobe with strain sensitive piezoresistor at base of probe shank.

**Penetrating Cortical Probes**

http://transducers.stanford.edu/stl/Projects/penetrate.html

> Description of project: Development of robust micromachined silicon probes for cortical recording and stimulation.

**Regeneration TypeNeural Interfaces**

http://transducers.stanford.edu/stl/Projects/regen.html

> Description of project: Development of micromachined neural interfaces that become "locked" into regenerating (deliberately cut and re-joined) nerves to form stable electrical links between on-chip microelectrodes and axons in the nerve.

**Hybrid Biosensors**

http://transducers.stanford.edu/stl/Projects/hybrid2.html

> Description of project: Development of hybrid biosensors based on living, cultured cells grown on microelectrode array substrates.

**Controlled Patterning of Cultured Neurons/Figure**

http://transducers.stanford.edu/stl/Projects/ControlledPatt.html

> Description of project: This work seeks to demonstrate the use of micromachined micron-scale physical structures and photolithographically-defined nanometer-scale patterns of organic molecules to direct the outgrowths of neural cells in culture.

**Thin-Film Microelectrode Arrays for Neural Connectivity Studies/Figure**

http://transducers.stanford.edu/stl/Projects/Thin-Film2.html

> Description of project: Development of planar, two-dimensional arrays of thin-film iridium microelectrodes for recording from and stimulation of neural tissue in both primary cell line cultures and slices.

**Injectionmoulding micropump**

http://www.uni-mainz.de/IMM/LNews/LNews_4/imm.html

**Micro-Total-Analysis Systems. &micro; TAS**

http://www.uni-mainz.de/IMM/LNews/LNews_4/manz.html

**BIOMEDICAL APPLICATIONS GROUP - Projects & Programs**

http://www.cnm.es/gab'projpubl.htm

**BIOMEDICAL APPLICATIONS GROUP - HEMOSENSORS**

http://www.cnm.es/gab'Haemo.htm

# DNA

**Nanotechnology in Ned Seeman's Laboratory**

http://seemanlab4.chem.nyu.edu/nanotech.html

> Description of project: Nanotechnological applications of DNA - The attachment of specific sticky ends to a DNA branched junction enables the construction of stick figures, whose edges are double-stranded DNA.

# Cantilevers

**OLYMPUS micro cantilevers for SPM**

http://www.olympus.co.jp/LineUp/Technical/Cantilever/levertopE.html#menu

> Technical note

**Technical data of silicon nitride cantilevers**

http://www.olympus.co.jp/LineUp/Technical/Cantilever/specnitrideE.html

> Technical note: OLYMPUS micro cantilevers for DC (contact) mode AFM and LFM

**AFM Thermomechanical Data Storage**

http://www-leland.stanford.edu/~chui/afmdata.html

> Description of project: Improved Cantilevers for AFM Thermo-mechanical Data Storage

**Cantilevers for MRFM**

http://cdr.stanford.edu/DD/SMSSL/NovelMicrostructures/CantileversForMRFM.html

> Description of project: Single Crystal Cantilevers for Magnetic Resonance Force Microscopy

**Image of cantilevers. JPEG image 1024x960 pixels**

http://hypatia.dartmouth.edu/levey/ssml/20HRES1.jpg

> Photograph only - Stress engineering in micromechanical devices: image of cantilevers

## Choppers

**Optical light chopper. GIF image 616x455 pixels**

http://hypatia.dartmouth.edu/courses/es65/images/henning10.gif

> Photograph only

## Chromatography

**TU-HHHLT, A Micro Gaschromatograph**

http://www.tu-harburg.de/ht/english/microgas.html

> Technical note

**Organization of a Miniature Gas Chromatography System**

http://144.126.176.216/mems/c1_s1.htm#f1_5

> Technical note: Organization of a miniature gas chromatography system, an early example of MEMS. All components except the display and main entry valve were proposed as a single chip in 1972.

## Flowmeters

**Flow Controller/Redwood Mass**

http://www.redwoodmicro.com/flow_cutaway.html

**Microrotor for blood flow. GIF image 552x712 pixels**

http://www.redwoodmicro.com/flow_cutaway.html

> Photograph only

**Flowsensors**

http://www.el.utwente.nl/tdm/mmd/projects/flowsens/flowsens.htm

> Description of project

**Thermal Flow Sensor**

http://www.el.utwente.nl/tdm/mmd/projects/thermflo/thermflo.htm

> Description of project

## Mass Flowmeters

**Monolithic Mass Flowmeter with On-Chip CMOS Interface Circuitry**

http://144.126.176.216/mems/c5_s2.htm#f5_2

> Technical note: Monolithic mass flowmeter with on-chip CMOS interface circuitry.

## Gas Detectors

**Silicon Micromachined Gas Detector - University of Michigan**

http://144.126.176.216/mems/c6_s4.htm#f6_18

> Technical note: a microciliary motion system being developed at the University of Tokyo

## Gears

**Guckel LIGA gears. GIF image 816x654 pixels**

http://hypatia.dartmouth.edu/courses/es65/images/henning1.gif

> Photograph only

**IFM Research Areas**

http://www.latech.edu/tech/engr/ifm/research.htm

> Description of projects: Contrasting microstructures - a portion of a microgear and a naturally occurring radiolarian. These natural microstructures are plentiful and have possible applications such as micro-filters

## Gyroscopes

**Nickel tuning fork gyroscope. GIF image 896x704 pixels**
http://hypatia.dartmouth.edu/courses/es65/images/henning11.gif
> Photogrph only

## Hinges

**Hinge/MCNC/GIF image 335x259 pixels**
http://mems.mcnc.org/figs/hingec.gif
> Photograph only

**Closeup Sideview of Hinge**
http://mems.mcnc.org/figs/hingec.gif
> Photograph only

## Magnetic Devices

**A Planar Variable Reluctance Magnetic Micromotor**
http://www.ece.gatech.edu/research/labs/msmsma/Magnetic/motor.html
> Description of project: A new self-propelled planar variable reluctance magnetic micromotor is realized with a micro-machined nickel-iron rotor and a fully integrated stator

**MicroInductor for High Frequency Applications**
http://www.ece.gatech.edu/research/labs/msmsma/Magnetic/hifinductor.html
> Description of project

**An Integrated Electromagnet for use as a Microrelay Driving Element**
http://www.ece.gatech.edu/research/labs/msmsma/Magnetic/electromagnet.html
> Description of project

**Micromachined Polymer Magnets**
http://www.ece.gatech.edu/research/labs/msmsma/Magnetic/polymermagnet.html
> Description of project

## Membranes

**Thin film gas-selective membranes**
http://www.tu-harburg.de/ht/english/gassel.html
> Description of project: Thin film gas selective membranes are used to enhance the selectivity of semiconductor gas-sensors, which is poor due to their operation principle.

## Micromachines

**Micromachines**
http://www.src.wisc.edu/www/highlights/micromachine.html
> Photograph only: The picture shows a miniature gear train fabricated in Prof. Henry Guckel's group at UW Madison.

**Atomic-scalemachine. GIF image 382x298 pixels**
http://www.ornl.gov/publications/labnotes/jun94/nanomachines.gif
> Nanotechnology is miniaturization to the extreme: atomic-scale machines constructed to perform tasks in molecular environments.

## Micromotors

**MEMS Projects**
http://cms.njit.edu/MEMSFolder/MEMSProjects.html
> Description of projects: Microelectromechanical systems are integrated sensors, actuators, and structural components fabricated using the same materials and the same processes that are used to fabricate microelectronic devices.

**Wisconsin magnetostatic motor. GIF image 712x924 pixels**
http://hypatia.dartmouth.edu/courses/es65/images/henning8.gif
> Photograph only

**Mehregany electrostatic motor. GIF image 676x898 pixels**
  http://hypatia.dartmouth.edu/courses/es65/images/henning6.gif
      Photograph only
**Salient-Pole Micromotor (234K)**
  http://mems.cwru.edu/images/microSPmotor.gif
      Photograph only
**Wobble Micromotor (236K)**
  http://mems.cwru.edu/images/microwobblemotor.gif
      Photograph only
**Outer-Rotor Micromotor (240K)**
  http://mems.cwru.edu/images/microouterrotor.gif
      Photograph only
**Salient-Pole Millimotor (318K)**
  http://mems.cwru.edu/images/milliSPmotor1.gif
      Photograph only
**Micromachined Rotating Gyroscope**
  http://www.shef.ac.uk/uni/projects/mesu/projects/gyro.html
      Description of project: A novel two-axis rotating micro-gyroscope with predicted high resolution and large dynamic range.
**Micro-Electric Generator. generato.html**
  http://www.shef.ac.uk/uni/projects/mesu/projects/generato.html
      Description of project: Miniature device for generating power for microsystems from mechanical vibrations in the surrounding media.
**Wobble Micromotor (236K).gif**
  http://mems.cwru.edu/images/microwobblemotor.gif
      Photograph only
**Outer-Rotor Micromotor (240K).gif**
  http://mems.cwru.edu/images/microouterrotor.gif
      Photograph only
**Salient-Pole Millimotor (318K).gif**
  http://mems.cwru.edu/images/milliSPmotor1.gif
      Photograph only
**Salient-Pole Micromotor (microSPmotor.gif)**
  http://mems.cwru.edu/images/microSPmotor.gif
      Photograph only
**Rotary Side Drive Motor - Top View**
  http://mems.mcnc.org/figs/sem5.gif
      Photograph only
**Rotary Side Drive Motor - Side View**
  http://mems.mcnc.org/figs/sem1.gif
      Photograph only
**Harmonicor Wobble Motor**
  http://mems.mcnc.org/figs/sem4.gif
      Photograph only
**Rotary Comb Drive**
  http://mems.mcnc.org/figs/sem12.gif
      Photograph only
**Rotary Side Drive Motor with Stacked Poly1 and Poly2 Rotor and Stator**
  http://mems.mcnc.org/figs/sem6.gif
      Photograph only
**Closeup of Bearing and Rotor of Stacked Poly Motor**
  http://mems.mcnc.org/figs/sem7.gif
      Photograph only
**Closeup of Stator-Rotor Air Gap of Stacked Poly Motor**
  http://mems.mcnc.org/figs/sem11.gif
      Photograph only

**Poly2 Rotary Side Drive Motor**

http://mems.mcnc.org/figs/sem8.gif

Photograph only

## Micropumps

**Micromembrane pump**

http://www.uni-mainz.de/IMM/LNews/LNews 4/fzk.html

Description of project: a micro membrane pump, made of plastic material, for moving small volumes of gases or liquids

**Micro-Blast**

http://www.el.utwente.nl/tdm/mmd/projects/mublast/mublast.htm

Description of project: The project aims to the development of a working model of a micropump by means of the LIGA technology.

**TUHH, Micromechanical Pumps**

http://www.tu-harburg.de/ht/english/pump.html

Description of project: Micromechanical pumps can versatilely be applied for controlled continuous medication, as micro-dosing system in biochemical laboratories and as a controlled ink transport system for penplotters.

**A Silicon Micromachined Micropump**

http://infoserve.unisa.edu.au/eng/mec/docum/project2.htm#pump

Description of project: The main objectives of this project are to design and fabricate a silicon micromachined micropump having a constant flow, low leakage and high pressure head.

## Microvalves

**Redwood Microsystems**

http://www.redwoodmicro.com/welcome.html

Redwood MicroSystems Home Page

**Silicon Micromachined Prototype Valve**

http://mems.isi.edu/archives/industrypages/microflow/MicroFlow1.html

Technical note: MicroFlow's shape memory alloy actuated silicon microvalve is designed to provide proportional flow or pressure control of gasses.

**Microvalves**

http://www.sma-mems.com/recent.htm

Overview: PROGRESS IN THIN FILM SHAPE MEMORY MICROACTUATORS

**Microvalve.GIF image 233x144 pixels**

http://www.llnl.gov/eng/ee/bitmaps/microvalve.gif

Picture only

**Electromagnetically Driven Microvalve - NTT**

http://144.126.176.216/mems/c6_s4.htm#f6_16

Technical note

## Optical Devices

### Mirrors

**A MEMS-Based Array of Actuated Micromirrors**

http://144.126.176.216/mems/c6_s2.htm#f6_2

Technical note: A MEMS-based array of actuated micromirrors (Texas Instruments).

**Programmable Micro Mirror Array**

http://mems.mcnc.org/figs/pxlarray.gif

Photograph only

**Integrated Silicon Micro Optics in ETIS**

http://guernsey.et.tudelft.nl/index.html

Description of projects

**Micromachined Movable Platforms as Integrated Optic Devices**

http://www.ece.gatech.edu/research/labs/msmsma/Electroplating/microoptics.html

> Description of project: In order to get good contact and light absorption structures, bulk micromachined movable platforms were fabricated on silicon substrates.

**Ultrafast Optics at the NSL**

http://www.tc.umn.edu/nlhome/m017/nanolab/research/optics.html

> Description of projects

**Simple Binary Optical Elements for Aberration Correction in Confocal Microscopy**

http://hypatia.dartmouth.edu/levey/ssml/publications/OpticsLetters.95/OpticsLett.95.html

> Article: Aberration correcting binary optics for deep scanning confocal microscopy

**Fresnel Lens/GIF image 713x401 pixels**

http://144.126.176.216/mems/fhomr_2.gif

> Figure Omron.2 shows a concept system of recognition sensor of less than 1 mm OD. Optical elements such as a microlens and a laser diode are mounted on a silicon cantilevered system. Resolution is expected to be -0.5 mm.

## Power Supplies for MEMS

**A Miniaturized High Voltage Solar Cell Array As An Electrostatic MEMS Power**

http://www.ece.gatech.edu/research/labs/msmsma/MEMSpower/cellarray.html

> Technical note: Additional external power connections or voltage/currentconversion circuitry are commonly used in micro-electromechanical systems (MEMS).

## Relays

**IMMI's Products**

http://www.micromachines.com/products.html

> Technical note: Silicon microrelay fabricated using IMMI's proprietary BITS technology.

**Microrelay contact**

http://mems.cwru.edu/roy/pix/relayc.gif

> Photograph only

**Northeastern: microsensors: microrelay**

http://www.ece.neu.edu/edsnu/zavracky/mfl/programs/relay/relay.html

> Article: Electrostatically actuated micromechanical switches using surface micromachining

**Microrelay contact (relayc.gif)**

http://mems.cwru.edu/roy/pix/relayc.gif

> Photograph only`

**Miniature electromechanical relays**

http://mems.cwru.edu/roy/projects/microrelays.html

> Description of project

**Design, fabrication, and characterization of electrostatic microrelays**

http://mems.cwru.edu/roy/abstracts/spie95.html

> Abstract of project

**Fabrication of Electrostatic Nickel Microrelays by Nickel Surface Micromachining**

http://mems.cwru.edu/roy/abstracts/mems95.html

> Description of project

**MicroRelay**

http://mems.mcnc.org/figs/relay.gif

> Photograph only

**MicroRelay**

http://www.ece.neu.edu/edsnu/zavracky/mfl/programs/relay/relay.html

> Photograph only

## Resonators

**Integrated Resonant Magnetic Field Sensor**

http://wwwetis.et.tudelft.nl/groups-info/abossche/Researchprojects/resonantmagneticsensor.html

> Description of project: Integrated Resonant Magnetic Field Sensor with High Sensitivity and High Dynamic Range

**MIT Micromechanics Group, MEMS Research**
http://umech.mit.edu/mems.html
> Technical note: Folded Beam Resonator

**Surface-Micromachined Micromechanical Resonator Chip**
http://144.126.176.216/mems/c5_s2.htm#f5_3
> Technical note: Surface-micromachined micromechanical resonator chip formed as a high-Q mechanical filter fabricated at the University of California at Berkeley.

**Linear Comb Resonator**
http://mems.mcnc.org/figs/lrsntr91.gif
> Photograph only


# Sensors

**Integrated sensors**
http://muresh.et.tudelft.nl/dimes/1993/main/chapter1.4.2.html
> Description of projects

**Epitaxial poly for surface micromachined smart sensors**
http://wwwetis.et.tudelft.nl/groups-info/pjfrench/Researchprojects/epipoly.html
> Description of project: The goal of this project is to use the epitaxial reactor to selectively grow both polycrystalline and monocrystalline silicon films in a single deposition step.

**Solid-State Microsensors**
http://future.sri.com/TM/about_TM/aboutSSM.html
> Technical note: A solid-state sensor uses semiconductor material for the sensing function and/or a microsensor that is fabricated using the photolithographic, etching, and deposition techniques of the IC industry.

**Integrated Sensors/Research Project Micromachining**
http://wwwetis.et.tudelft.nl/groups-info/pjfrench/Researchprojects.html
> Description of project at the Faculty of Electrical Engineering, Electronic Instrumentation Laboratory, Mekelweg 42628 CD, Delft, The Netherlands.

**Sensor Technology at LLNL**
http://www.llnl.gov/sensor_technology/SensorTech_contents.html
> List of sensors

**Microsensors**
http://reuben.afit.af.mil:8000/~bfreeman/thesis.html
> Photographs and links

**LWIM Presentation at Hilton Head Conference, June 1996**
http://www.janet.ucla.edu/lpe.lwim/hilton.head/
> Wireless Integrated Microsensors: Slide Show

**The Stanford Micro-Calorimeter Sensor**
http://cdr.stanford.edu/~tep/SMCS.html
> Photographs and links

**Microsystems Design Group Home Page**
http://dmtwww.epfl.ch/ims/sysmic/index.html
> Sensors

**Microsystems Design Group Projects Page**
http://dmtwww.epfl.ch/ims/sysmic/projects/index.html
> Description of projects

**MIT Project: Silicon Wafer Bonding Micromachined Devices**
http://www-mtl.mit.edu/SRC/Schmidt.html
> Description of project: The objective of this task is to establish a technology platform for fabrication of integrated micromechanical devices, specifically sensors.

**Sensors**
http://www.el.utwente.nl/tdm/mmd/projects/sensor.htm
> Description of project on sensors

## Chemical Sensors

**MAT Berlin - Basic Modules for Chemical Sensors**

http://www-mat.ee.tu-berlin.de/research/chemsens/chemsens.htm

> Description of projects

**Micromachined Electrophoresis and Fluidic Devices**

http://transducers.stanford.edu/stl/Projects/electro-Rich Ray.html

> Objectives: Develop integrated, on-column AC conductivity-based detector for capillary electrophoresis, including miniaturized lock-in amplifier and digitizer.

**Micromachined Spectrometers**

http://transducers.stanford.edu/stl/Projects/spectro.html

> Descripton of project - Objectives: To develop a broadband, simple and robust micromachined spectrophotometer of identification of gases and liquids, detection of specific emission spectra, colorimetry, etc.

**Electroanalytical System Based on Thin-Film Microelectrode Arrays**

http://transducers.stanford.edu/stl/Projects/electrosys.html

> Description of project - Objectives: Development of systems for detection of metallic impurities (i.e. lead, cadmium, copper, etc.) at parts-per-billion levels in drinking and process waters.

## Fabry Perot Sensors

**Skin Effect Circuit Models**

http://weewave.mer.utexas.edu/MED_files/MED_research/MEMS_chem_snsr/slides/chem_snsr_slds.html

> Fabry Perot Sensors

## Image Sensors

**3D Computational Image Sensors**

http://www.ece.neu.edu/edsnu/Darpa-3D/NU-3D.html

> Description of project: 3D Computational Image Sensors For Advanced Low Power Visual Processing

## Inertial Sensors

**Inertial measurement units**

http://esto.sysplan.com/ESTO/MEMS/Prog_Summaries/IMU.html

## Mechanical Sensors

**Aluminum-MEMS Systems Development**

http://transducers.stanford.edu/stl/Projects/aluminum.html

> Description of project: Development of dry released, CMOS-compatible aluminum micromechanical structures that can be electrostatically deflected and sensed in a variety of transducer applications.

**Micromechanical Force Probes**

http://transducers.stanford.edu/stl/Projects/force.html

> Description of project

**Micromachined Tactile Sensor Arrays**

http://transducers.stanford.edu/stl/Projects/traction.html

> Description of project: A CMOS Compatible Traction Stress Sensor Array for use in High Resolution Tactile Imaging.

**Accelerometers**

http://transducers.stanford.edu/stl/Projects/KenResearch.html

> Description of project: MULTI-AXIS ACCELEROMETER ARRAYS

**MEMS Devices Through Deep Reactive Ion Etching of Single-Crystal Silicon**

http://transducers.stanford.edu/stl/Projects/mems.html

> Description of project: Objectives: Develop novel single-crystal silicon micromechanical devices that take advantage of deep vertical sidewalls for high capacitance (for electrostatic drive and sensing)

## Pressure Sensors

**UW-MEMS- Research - Polysilicon Planar Pressure Transducer**

http://mems.engr.wisc.edu/pt.html

**Optically-Interrogated Fabry-Perot Pressure Transducers**

http://weewave.mer.utexas.edu/MED_files/MED_research/F_P_sensor_folder/FP_pressure_sensor.html

**Fabry-Perot Based Pressure Transducers**

http://weewave.mer.utexas.edu/MED_files/MED_research/mems_sum.html#FPsensors

**Fabry Perot pressure sensors**

http://weewave.mer.utexas.edu/MED_files/MED_research/F_P_sensor_folder/FP_pressure_sensor.html

    Description of project

**Toyota Surface Micromachined Piezoresistive Microdiaphragm**

http://144.126.176.216/mems/c3_s2.htm#f3_1

    Technical note with figure: Schematic and cross-sectional view of Toyota's surface micromachined piezoresistive microdiaphragm pressure sensor

**Toyota 1K-Element Piezoresistive Pressure/Tactile Sensor Array**

http://144.126.176.216/mems/c3_s2.htm#f3_2

    Technical note with figure: Toyota 32 x 32 (1K)-element piezoresistive pressure/tactile sensor array

**Toyota Capacitive Pressure Sensor with CMOS Electronics**

http://144.126.176.216/mems/c3_s2.htm#f3_3

    Technical note Photograph and cross-sectional structure of Toyota capacitive pressure sensor with CMOS electronics

**Nippon densoIntegrated Pressure and Temperature Sensor Chip**

http://144.126.176.216/mems/c3_s2.htm#f3_5

    Technical note with figure: Top view of the Nippondenso integrated pressure and temperature sensor chip

**Yokogawa Resonant Microbeam Pressure Sensor**

http://144.126.176.216/mems/c3_s2.htm#f3_6

    Technical note with figure: Yokogawa resonant microbeam pressure sensor (Ikeda et al. 1990b): construction of the sensor (a) and cross-sectional SEM photograph of the resonator

**Tohoku University Integrated Capacitive Pressure Sensor**

http://144.126.176.216/mems/c3_s2.htm#f3_10

    Technical note with figure: Tohoku University integrated capacitive pressure sensor

**Piezoresistive Pressure Sensor**

http://144.126.176.216/mems/c5_s3.htm#f5_6

    Technical note with figure: Piezoresistive pressure sensor with on-chip bipolar readout electronics reported by Toyota in 1983, one of the earliest examples of an integrated MEMS sensor with integrated electronics.

**Organization of the Toyota Tactile Imager Readout Electronics**

http://144.126.176.216/mems/c5_s3.htm#f5_7

    Technical note with figure: Organization of the Toyota tactile imager readout electronics. The device implements a 32 x 32 element array of pressure sensors with on-chip selection electronics. The overall die size is 10 mm x 10 mm.

**Pressure Sensor Evolution in the Toyota Tactile Imager**

http://144.126.176.216/mems/c5_s3.htm#f5_8

    Technical note with figure: Pressure sensor evolution in the Toyota tactile imager from a bulk micromachined device (left) to a surface-micromachined structure (right). Both devices are vacuum sealed at wafer level.

**Circuit Organization in the Digitally-Compensated Toyota Pressure Sensor**

http://144.126.176.216/mems/c5_s3.htm#f5_9

    Technical note with figure: Circuit organization in the digitally-compensated pressure sensor from Toyoda Machine Works and Toyota.

**Structure and Excitation Scheme for the Yokogawa DPharp Resonant Pressure Sensor**

http://144.126.176.216/mems/c5_s3.htm#f5_10

    Technical note with figure: Structure and excitation scheme used for the Yokogawa DPharp resonant pressure sensor.

**Integrated Pressure Sensor - NOVA**

http://144.126.176.216/mems/c6_s4.htm#f6_10

    Technical note with figure: Integrated pressure sensor - NOVA (USA).

**Micromachined Resonant Pressure Sensor - Yokogawa**

http://144.126.176.216/mems/c6_s4.htm#f6_12

    Technical note with figure: Micromachined resonant pressure sensor - Yokogawa (Japan).

**Capacitive pressure sensor**

http://arsenio.mit.edu/groups/schmidt/pressure.html

    Description of project

### Sensor-Circuit Integration and System Partitioning

**Sensor-Circuit Integration and System Partitioning. c5_s1**
http://144.126.176.216/mems/c5_s1.htm
>SENSOR-CIRCUIT INTEGRATION AND SYSTEM PARTITIONING: INTRODUCTION by Kensall D. Wise

**Sensor-Circuit Integration and System Partitioning. c5_s2**
http://144.126.176.216/mems/c5_s2.htm
>SENSOR-CIRCUIT INTEGRATION IN THE UNITED STATES

**Sensor-Circuit Integration and System Partitioning. c5_s3**
http://144.126.176.216/mems/c5_s3.htm
>SENSOR-CIRCUIT INTEGRATION IN JAPAN

**Sensor-Circuit Integration and System Partitioning. c5_s4**
http://144.126.176.216/mems/c5_s4.htm
>ISSUES IN SENSOR-CIRCUIT INTEGRATION

**Sensor-Circuit Integration and System Partitioning. c5_s5**
http://144.126.176.216/mems/c5_s5.htm
>CONCLUSIONS and REFERENCE

### SOI Sensors

**Northeastern University: Microsensors: SOI**
http://www.ece.neu.edu/edsnu/zavracky/mfl/programs/soi/soi.html
>Silicon-on-insulator Research (SOI) Sensors. Recent advances in silicon-on-insulator(SOI) technology has generated increased interest in their application to high speed CMOS circuits.

### Stress Sensors

**MIT shear stress sensor. GIF image 688x916 pixels**
http://hypatia.dartmouth.edu/courses/es65/images/henning5.gif
>Photograph only

### Thermal Sensors

**Microfabricated Arrays of Thermocouples.**
http://hypatia.dartmouth.edu/levey/ssml/publications/henning/NNF/DartmouthTCMOSFET.html
>Description of project: There is a growing need for embedded sensors, particularly for 'smart maintenance' applications.

**Switched thin film thermocouple arrays. JPEG image 553x415 pixels**
http://hypatia.dartmouth.edu/images/hochwitz/tftc.jpg
>Picture only

**Electrochemically Etched, Thermally Isolated Single-Crystal Silicon Structures**
http://transducers.stanford.edu/stl/Projects/electro-Erno K.html
>Description of project - Objectives: Apply micromachining techniques to realize thermally isolated, single crystal silicon islands, suspended by thin dielectric bridges. These islands are capable of containing active circuitry.

**Organic Thermal Acuators**
http://transducers.stanford.edu/stl/Projects/organic.html
>Description of project - Objectives: Development and demonstration of robust organic-based thermal bimorph actuators that can be used to manipulate small objects for inspection and general positioning in realistic environmental conditions.

### Tunneling Sensors

**Tunneling Sensors Home Page**
http://cdr.stanford.edu/DD/SMSSL/TunnelingSensors/Tunneling.html
>Overview article on tunneling sensors

**Inductive Proximity Sensors**
http://weewave.mer.utexas.edu/MED files/MED_research/mems_sum.html#Proxsensors
>Description of project

**Two-coil inductive proximity sensor**

http://weewave.mer.utexas.edu/MED_files/MED_research/indctv_prox_folder/prox_SPIE_95/prx_SPIE_95_tlk.html

> Description of project

**Yield of manufactured sensors**

http://weewave.mer.utexas.edu/MED_files/MED_research/F_P_sensor_folder/F_P_yld_paper/yield_F_P_sensor.html

> Description of project

## Sliders

**Electromechanical slider. GIF image 652x508 pixels**

http://hypatia.dartmouth.edu/courses/es65/images/henning13.gif

> Photograph only

## Spectrometers

**Northeastern University:Microsensors:Microspectrometer**

http://www.ece.neu.edu/edsnu/zavracky/mfl/programs/spec/microspe.html

> Article: A Fabry-Perot Spectrometer Microspectrometer Fabricated using Surface Micromachining Technology. Surface micromachining techniques are being used to create a miniature, low cost replacement for conventional optical spectrometers.

## Strain Gauges

**Microstrain gauge. GIF image 664x658 pixels**

http://hypatia.dartmouth.edu/courses/es65/images/henning12.gif

> Photograph only

## Structures

**Poly-SiC lateral resonant structure**

http://mems.cwru.edu/roy/pix/sic1.gif

> Photograph only

**Comb-finger section of poly-SiC microstructure**

http://mems.cwru.edu/roy/pix/sic2.gif

> Photograph only

**Electroless plated nickel microstructure**

http://mems.cwru.edu/roy/pix/nickel1.gif

> Photograph only

**Intelligent micromachines**

http://www.sandia.gov/LabNews/LN03-15-96/intell.html

> Description of project: Sandia team produces intelligent micromachines. Tiny new microelectronic machines are small, smart, cheap.

**Large Pop-up Structure**

http://mems.mcnc.org/figs/bboard1.gif

> Photograph only

## Tools

**IFM Research Areas**

http://www.latech.edu/tech/engr/ifm/research.htm

> Description of project: 22-micrometer (0.0087-inch) diameter, two fluted micromilling tool of high speed steel fabricated with focused ion beam micromachining.

## Tweezers

**Microfabricated Tweezers with a Large Gripping Force and a Large Range of Motion**

http://mems.isi.edu/archives/dissertation-abstracts/casewestern/phd94-1.html

> Extended abstract: The objective of this thesis is to gain a basic understanding of bi-metallic microactuation through a study of microfabricated bimetallic microtweezers.

**Microforceps - Berkeley Sensor and Actuator Center**

http://144.126.176.216/mems/c6 s4.htm#f6_13

        Technical note: Microforceps - Berkeley Sensor and Actuator Center (USA).

**Microfabricated Tweezers with a Large Gripping Force and a Large Range of Motion**

http://mems.isi.edu/archives/dissertation-abstracts/casewestern/phd94-1.html

        Description of project: The objective of this thesis is to gain a basic understanding of bi-metallic microactuation through a study of microfabricated bimetallic microtweezers.

**Amicrogripper**

http://synergy.icsl.ucla.edu:80/ftp/pub/micrographs/gripplan.jpg

        Photograph only

**Microtweezer (270K).gif**

http://mems.cwru.edu/images/utweezer1.gif

        Photograph only

**IMMI's Products**

http://www.micromachines.com/products.html

        Description of products: It was once thought that solid-state technology will eventually replace electromechanical relays. What has happened is actually the opposite.

**IMMI's Technology**

http://www.micromachines.com/technology.html

        At UCLA and Caltech, we have developed a complete set of technologies for fabricating electromagnetic microactuators using single-crystal silicon as the structural material and electroplated permalloy with planar copper coils as the actuation mechanism.

**Integrated sensors**

http://muresh.et.tudelft.nl/dimes/1993/main/chapter1.4.2.html

        Long list of smart sensors projects

**Microelectromechanical Systems Development in Japan**

http://144.126.176.216/mems/c1_s1.htm#f1_5

        Review article with figures: Organization of a Miniature Gas Chromatography System.

**Micromachined Movable Platforms as Integrated Optic Devices**

http://www.ece.gatech.edu/research/labs/msmsma/Electroplating/microoptics.html

**Sensor Technology at LLNL**

http://www.llnl.gov/sensor technology/SensorTech contents.html

        Excellent long list of sensors with photos and figures

**Silicon Microstructures and Microactuators for Compact Computer Disk Drives**

http://www.computer.org/conferen/mss95/miu/miu.htm

        Overview of Silicon Micromachining with figures. IMMI's technical publication related to silicon micromachining

**UW-MEMS - Image Archive**

http://mems.engr.wisc.edu/images/

        Examples with Photos: a) Deep X-Ray Lithography (DXRL) b) Polysilicon Surface Micromachining

## More from the grab bag:

**Micromachining as it applies to microrobotics.**

http://wwwipr.ira.uka.de/~wallner/BARMINT/chap2.html

**Japanese Technology Evaluation Center. What some US micromachinists think about the micromachining efforts in Japan.**

http://144.126.176.216/MEMS/toc.htm

**The type of applications sponsored by the Electronics Technology Office at DARPA.**

http://eto.sysplan.com/ETO/MEMs/Prog Summaries/dist nets.html

**The results of a 1995 Xerox study on the long-term benefits of using distributed MEMS to couple computation to the physical world (slides).**

http://www.parc.xerox.com/spl/projects/memsisat/sld001.htm

**Dosing of small amounts of liquids. Drop on demand micropumps.**

http://www.fgb.mw.tu-muenchen.de/mikrosys/e_dosing.htm

**Biosensors and Bioelectronics.**

http://www.elsevier.com/catalogue/SAH/125/05000/05015/405913/405913.html

**Selected papers on fiber optic sensors.**

http://www.spie.org/web/abstracts/oepress/MS108.html

**The Inventors HomePage-Usenet.**

http://sunsite.unc.edu/usenet-i/groups-html/alt.inventors.html

# Appendix C

**TABLE 1**  <100> Silicon Etch Rates in [mm/hr] for Various KOH Concentrations and Etch Temperatures as Calculated from Eq. [A-1] by Setting $E_0 = 0.595$ eV and $k_0 = 2480$ mm/hr · $(mol/L)^{-4.25}$

| | Temperature [°C] | | | | | | | | |
|---|---|---|---|---|---|---|---|---|---|
| % KOH | 20° | 30° | 40° | 50° | 60° | 70° | 80° | 90° | 100° |
| 10 | 1.49 | 3.2 | 6.7 | 13.3 | 25.2 | 46 | 82 | 140 | 233 |
| 15 | 1.56 | 3.4 | 7.0 | 14.0 | 26.5 | 49 | 86 | 147 | 245 |
| 20 | 1.57 | 3.4 | 7.1 | 14.0 | 26.7 | 49 | 86 | 148 | 246 |
| 25 | 1.53 | 3.3 | 6.9 | 13.6 | 25.9 | 47 | 84 | 144 | 239 |
| 30 | 1.44 | 3.1 | 6.5 | 12.8 | 24.4 | 45 | 79 | 135 | 225 |
| 35 | 1.32 | 2.9 | 5.9 | 11.8 | 22.3 | 41 | 72 | 124 | 206 |
| 40 | 1.17 | 2.5 | 5.3 | 10.5 | 19.9 | 36 | 64 | 110 | 184 |
| 45 | 1.01 | 2.2 | 4.6 | 9.0 | 17.1 | 31 | 55 | 95 | 158 |
| 50 | 0.84 | 1.8 | 3.8 | 7.5 | 14.2 | 26 | 46 | 79 | 131 |
| 55 | 0.66 | 1.4 | 3.0 | 5.9 | 11.2 | 21 | 36 | 62 | 104 |
| 60 | 0.50 | 1.1 | 2.2 | 4.4 | 8.4 | 15 | 27 | 47 | 78 |

**TABLE 2**  <100> Silicon Etch Rates in [mm/hr] for Various KOH Concentrations and Etch Temperatures as Calculated from Eq. [A-1] by Setting $E_0 = 0.60$ eV and $k_0 = 4500$ mm/hr · $(mol/L)^{-4.25}$

| | Temperature [°C] | | | | | | | | |
|---|---|---|---|---|---|---|---|---|---|
| % KOH | 20° | 30° | 40° | 50° | 60° | 70° | 80° | 90° | 100° |
| 10 | 2.2 | 4.8 | 10.1 | 20.1 | 38 | 71 | 126 | 216 | 362 |
| 15 | 2.3 | 5.1 | 10.6 | 21.2 | 40 | 74 | 132 | 228 | 381 |
| 20 | 2.3 | 5.1 | 10.7 | 21.3 | 41 | 75 | 133 | 229 | 383 |
| 25 | 2.3 | 5.0 | 10.4 | 20.6 | 39 | 73 | 129 | 222 | 372 |
| 30 | 2.1 | 4.7 | 9.8 | 19.4 | 37 | 68 | 121 | 209 | 350 |
| 35 | 2.0 | 4.3 | 8.9 | 17.8 | 34 | 63 | 111 | 192 | 321 |
| 40 | 1.7 | 3.8 | 8.0 | 15.9 | 30 | 56 | 99 | 171 | 285 |
| 45 | 1.5 | 3.3 | 6.9 | 13.7 | 26 | 48 | 85 | 147 | 246 |
| 50 | 1.2 | 2.7 | 5.7 | 11.3 | 22 | 40 | 71 | 122 | 204 |
| 55 | 1.0 | 2.2 | 4.5 | 9.0 | 17 | 31 | 56 | 96 | 161 |
| 60 | 0.7 | 1.6 | 3.4 | 6.7 | 13 | 24 | 42 | 72 | 121 |

563

**TABLE 3**  Calculated Etch Rates of Thermally Grown Silicon Dioxide in [nm/hr] for Various KOH Concentrations and Etch Temperatures. Calculation was Based on Best Numerical Fit of Experimental Data. The Activation Energy was Taken to be 0.85 eV

| % KOH | Temperature [°C] | | | | | | | | |
|---|---|---|---|---|---|---|---|---|---|
| | 20° | 30° | 40° | 50° | 60° | 70° | 80° | 90° | 100° |
| 10 | 0.40 | 1.22 | 3.5 | 9.2 | 23 | 54 | 123 | 266 | 551 |
| 15 | 0.63 | 1.91 | 5.4 | 14.4 | 36 | 85 | 193 | 416 | 862 |
| 20 | 0.88 | 2.66 | 7.5 | 20.0 | 50 | 118 | 268 | 578 | 1200 |
| 25 | 1.14 | 3.46 | 9.8 | 26.0 | 65 | 154 | 348 | 752 | 1560 |
| 30 | 1.42 | 4.32 | 12.2 | 32.5 | 81 | 193 | 435 | 940 | 1950 |
| 35 | 1.44 | 4.37 | 12.4 | 32.8 | 82 | 195 | 440 | 949 | 1970 |
| 40 | 1.33 | 4.03 | 11.4 | 30.3 | 76 | 180 | 406 | 876 | 1820 |
| 45 | 1.21 | 3.67 | 10.4 | 27.5 | 69 | 163 | 369 | 797 | 1650 |
| 50 | 1.08 | 3.28 | 9.3 | 24.6 | 62 | 146 | 330 | 713 | 1480 |
| 55 | 0.95 | 2.87 | 8.1 | 21.6 | 54 | 128 | 289 | 624 | 1290 |
| 60 | 0.81 | 2.45 | 6.9 | 18.4 | 46 | 109 | 246 | 532 | 1100 |

From Seidel et al., *J. Electrochem. Soc.*, 137, 3612–3626, 1990. With permission.

# Appendix D

## Suggested Further Reading

### Books on Micromachining and Sensors (alphabetical by author):

**B**

Boisde, G. and A. Harmer: *Chemical and Biochemical Sensing With Optical Fibers and Waveguides.* Artech House, Boston, 1996.

Buerk, D. G.: *Biosensors: Theory and Applications.* Lancaster: Technomic Publishing Co., Inc., 1993.

Bushan, B. (Ed.): *Handbook of Micro/Nano Tribology.* Boca Raton, CRC Press, 1995.

Buttgenbach, S.: *Mikromechanik,* Teubner Studienbucher. Stuttgart, 1991 (German).

**C**

Cobbold, R. S.: *Transducers for Biomedical Measurements: Principles and Applications.* New York: John Wiley & Sons, 1974.

Crandall, B. C. (Ed): *Nanotechnology: Molecular Speculations on Global Abundance.* The MIT Press, Cambridge, Massachusetts, 1996.

**E**

Edelman, P. G. and Wang, J. (Eds.): *Biosensors and Chemical Sensors.* Washington, D.C. American Chemical Society, 1992.

Ehrlich, Daniel, J. Tsao, and Jeffrey Y.: *Laser Microfabrication: Thin Film Processes and Lithography.* Boston, Academic Press, 1989.

**F**

Frank, R.: *Understanding Smart Sensors.* Boston. Artech House, 1996.

Fung, C. D., Cheung, P. W., Ko, W. H., and Fleming, D. G.: *Micromachining and Micropackaging of Transducers.* Amsterdam, Elsevier, 1985.

**H**

Hall, E. A. H.: *Biosensors.* Englewood Cliffs, N.J.: Prentice Hall, 1991.

Hellman, H.: *Beyond Your Senses: The New World of Sensors.* New York. Dutton, 1997.

Heuberger, A.: *Mikromechanik.* Springer Verlag. Heidelberg, 1989 (German).

Hoch, H. C. L. W., Jelenski, L., and H. G. Craighead (Eds.): *Nanofabrication and Biosystems.* Cambridge University Press, Cambridge, 1996.

**J**

Janata, J.: *Principles of Chemical Sensors.* New York. Plenum Press, 1989.

**K**

Khazan, Alexander D.: *Transducers and their Elements: Design and Application.* Englewood Cliffs, N.J., PTR Prentice Hall. 1994

Klein Lebbink, G.: *Microsystem Technology: Exploring Opportunities.* Samsom BedrijfsInformatie bv., Alphen aan den Rijn/Zaventem, Belgium, 1994.

**L**

Lambrechts, M. and Sansen, W.: *Biosensors: Microelectrochemical Devices.* Bristol, Institute of Physics Publishing, 1992.

**M**

Madou, M. J. and S. R. Morisson: *Chemical Sensing with Solid State Devices.* New York, Academic Press, 1989.

Mallouk, Thomas E. and Harrison, Daniel J.: *Interfacial Design and Chemical Sensing.* Washington, D.C., American Chemical Society, 1994.

Middelhoek, S. (Ed.): *Sensors.* Amsterdam, Elsevier-Sequoia, 1990.

Miller, Richard K.: *Survey on Silicon Micromachining and Microstructures.* Madison, GA, Future Technology Surveys, 1989.

Muller, R. S., Howe, R. T., Senturia, S. T., Smith, R. L., and White, R. M. (Eds.): *Microsensors.* New York, IEEE, 1991.

**N**

Nakamura, Robert M., Kasahara, Yasushi, and Rechnitz, Garry A. (Eds.): *Immunochemical Assays and Biosensor Technology for the 1990s.* Washington, D.C., American Society for Microbiology, 1992.

Neuman, R. M., Fleming, G. D., Cheung, W. P., and Ko, W. H.: *Physical Sensors for Biomedical Applications*. Boca Raton: CRC Press, 1980.

Norton, Harry N.: *Handbook of Transducers*. Englewood Cliffs, N.J., Prentice Hall 1989.

**R**

Regis, E.: *Nano: The Emerging Science of Nanotechnology: Remaking the World-Molecule by Molecule*. Boston, Little, Brown and Company, 1995.

**S**

Scheller, F. and Schubert, F.: *Biosensors*. Amsterdam, Elsevier, 1992.

Singh, Prithiapal, Sharma, Bhanu, and Tyle, Praveen: *Diagnostics in the Year 2000: Antibody, Biosensor, and Nucleic Acid Technologies*. New York, Van Nostrand Reinhold, 1993.

Slack, P. T. and Kress-Rogers, Erika: *The Chemical Sensing Needs of the Food Industry*. London, Chemical Sensors Club, 1989.

**T**

Tandeske, D.: *Pressure Sensors: Selection and Applications*. New York, Marcel Dekker, 1991.

Togawa, T.: *Biomedical Transducers and Instruments*. Boca Raton, CRC Press, 1997.

Technical Insights, Inc.: *Sensor Technology Sourcebook: Guide to Worldwide Research & Development*. Englewood/Fort Lee, N.J., Technical Insights Inc., 1995.

Thompson, Michael: *Surface-Launched Acoustic Wave Sensors: Chemical Sensing and Thin-Film Characterization*. New York, John Wiley & Sons, 1997.

Turner, A. P. F., Karube, I., and Wilson, G. S.: *Biosensors: Fundamentals and Applications*. Oxford Science Publications, 1987.

**U**

Usmani, A. M. and Akmal, N. (Eds).: *Diagnostic Biosensor Polymers*. Washington, D.C., American Chemical Society (ACS), 1994.

**V**

Valcarcel, M.; Luque de Castro, M. D.: *Flow-Through (Bio)Chemical Sensors*. Amsterdam, Elsevier, 1994.

**W**

Wagner, Gabriele and Guilbault, George G. (Eds.): *Food Biosensor Analysis* (Food Science and Technology; 60). New York, Marcel Dekker, 1994.

Walcher, H.: *Position Sensors*. VDI Verlag, 1985.

Wolffenbuttel (Ed.): *Silicon Sensors and Circuits: On-Chip Compatibility*. London. Chapman & Hall, 1966.

## Journals and Periodicals on Micromachining and Sensors:

1. Sensors and Materials.
   (from 1989 to 1995, 6 issues a year in 1996, 8 issues)
   (Scientific Publishing Division of MYU K.K., Tokyo, Japan).

2. Analytical Chemistry.
   (Semimonthly)
   http://pubs.acs.org/journals/ancham/about.html
   (American Chemical Society, Washington, D.C. 20036, USA).

3. Sensors, The Journal of Machine Perception.
   (Helmers Publishing, Inc., Peterborough, N.H., USA)

4. Micromachine Devices
   (From the Editors of R&D Magazine, Cahners Publishing Co., Newton, MA, USA).

5. Nanobiology.
   (Quarterly) http://www.carfax.co.uk/index.htm
   (Carfax Publishing Company, Abingdon, Oxfordshire, U.K.).

6. Biosensors & Bioelectronics.
   (Monthly) http://www.elsevier.nl/inca/publications/store/4/0/5/9/1/3/405913.pub.shtml
   (Elsevier Advanced Technology, Oxford, England).

7. LIGA.
   (Five issues since January 1995)
   http://www.uni-mainz.de/IMM/LNews/LIGA_home.html
   (IMM Institute of Microtechnology Mainz, Germany).

8. SENSOR Technology.
   (Monthly) http://www.insights.com/sensor_tech.html
   (Technical Insights, Inc., Englewood, NJ, USA). (Bought by John Wiley in July 1996).

9. Sensor Business Digest.
   (Monthly).
   (Vital Information, Foster City, CA, USA).

10. Journal of Micro Electro Mechanical Systems.
    (Quarterly)
    http://www.ieee.org/pub_preview/mems_toc.html
    (A joint IEEE/ASME publication, New York, USA).

11. Journal of the Electrochemical Society (JECS).
    (Monthly) http://ecs.electrochem.org/journal.html
    (Electrochemical Society, Pennington, NJ, USA).

12. Journal of Micromechanics and Microengineering.
    http://www.iop.org/Journals/jm
    (Institute of Physics (IOP), Bristol, UK, Philadelphia, USA).

13. Sensor Review. The International Journal of Sensing for Industry.
    (Quarterly)
    (IFS Publications Ltd., Bedford, England).

14. Nanotechnology.
    (4 issues per year) http://www.iop.org/Journals/na
    (Bristol, UK, Philadelphia, USA: Institute of Physics Publishing (IOP).

## Series on Micromachining and Sensors:

1. *Sensors: a Comprehensive Survey*
   Edited by W. Gopel, J. Hesse and J. N. Zemel
   Volume 1 Fundamentals and General Aspects
   Volume 2 Chemical and Biochemical Sensors-Part I
   Volume 3 Chemical and Biochemical Sensors-Part II
   Volume 4 Thermal Sensors
   Volume 5 Magnetic Sensors
   Volume 6 Optical Sensors
   Volume 7 Mechanical Sensors
   Volume 8 Micro- and Nanosensors/Sensors: Market Trends
   Cumulative Sensor Index: Combined and Optimized index of the eight sensor volumes.
   Sensors Update
   Sensors Update 1, 1996
   Sensors Update 2, 1996
   Sensors Update 3, 1997
   Edited by H. Baltes, W. Gopel and J. Hesse.

2. *Handbook of Sensors and Actuators*
   Amsterdam, Elsevier Science B.V.
   Series Editor: S.Middlehoek, Delft University of Technology, The Netherlands
   Volume 1 Thick Film Sensors (Edited by M.Prudenziati)
   Volume 2 Solid State Magnetic Sensors (by C.S. Roumenin)
   Volume 3 Intelligent Sensors (edited by H. Yamasaki)
   Volume 4 Semiconductor Sensors in Physico-Chemical Studies (edited by L.Yu. Kuprianov)-1996

## Market Studies on Micromachining and Sensors:

1. *Sensor Daten Info*
   SDI Verlag & Consulting, Hedwigstr. 5, D-80636 Munchen,
   Tel. 49(0)89-1232039
   Fax 49(0)89-898139-24
2. *Sensor and Instrumentation Markets, 1994*
   Richard K. Miller & Associates, Inc.
   Published by Richard K. Miller & Associates, Inc.
   Norcross, GA, USA
3. *The European Sensor Market, 1989*
4. *New and Emerging Markets for Automotive Sensors in North America, 1990*
5. *Automotive Sensor Markets, 1992*
6. *World Emerging Sensor Technologies, 1993*
7. *Sensor Market '94: A Strategic Assessment of the International Market for Sensors, 1994*
8. *Rapidly Developing Biomedical Sensor Market; Sensors Play an Increasingly Important Role in Patient Monitoring, 1994*
   3,4,5,6, 7 and 8 from:

MIRC / Frost and Sullivan
Market Intelligence
2525 Charleston Road
Mountain View, CA 94043 USA
Tel. 415-961-9000
Fax. 415-961-5042
54 rue Vandehoven
1150 Brussels, Belgium
Tel. 32(2)762-2781
Fax. 32(2)771-7248

## Important Proceedings/Conferences on Micromachining and Sensors:

### Hilton Head Conferences

1. Technical Digest of the 1984 Solid State Sensor Conference
   Hilton Head, Island,SC, USA
2. Technical Digest of the 1986 Solid State Sensor and Actuator Workshop
   Hilton Head Island, SC, USA
3. Technical Digest of the 1988 Solid State Sensor and Actuator Workshop
   Hilton Head Island, SC, USA
4. Technical Digest of the 1990 Solid State Sensor and Actuator Workshop
5. Technical Digest of the 1992 Solid State Sensor and Actuator Workshop
   Hilton Head, Island, SC, USA
6. Technical Digest of the 1994 Solid State Sensor and Actuator Workshop
   Hilton Head, Island, SC, USA
7. Technical Digest of the 1996 Solid State Sensor and Actuator Workshop
   Hilton Head, Island, SC, USA

### MEMS

Proceedings. IEEE Micro Electro Mechanical Systems, (MEMS '89)
  Salt Lake City, UT, USA
Proceedings. IEEE Micro Electro Mechanical Systems, (MEMS '90)
  Napa Valley, CA, USA
Proceedings. IEEE Micro Electro Mechanical Systems, (MEMS '91)
  Nara, Japan
Proceedings. IEEE Micro Electro Mechanical Systems, (MEMS '92)
  Travemunde, Germany
Proceedings. IEEE Micro Electro Mechanical Systems, (MEMS '93)
  Fort Lauderdale, FL, USA
Proceedings. IEEE Micro Electro Mechanical Systems,

(MEMS '94)
Oiso, Japan

Proceedings. IEEE Micro Electro Mechanical Systems, (MEMS '95)
Amsterdam, The Netherlands

Proceedings. IEEE Micro Electro Mechanical Systems, (MEMS '96)
San Diego, CA, USA

## Chemical Sensors

1st International Meeting on Chemical Sensing, Fukuoka, Japan, 1983

2nd International Meeting on Chemical Sensors, Bordeaux, France, 1986

3rd International Meeting on Chemical Sensors, Cleveland, Ohio,USA,1990

4th International Meeting on Chemical Sensors, Tokyo, Japan, 1992

5th International Meeting on Chemical Sensors, Rome, Italy, 1994

6th International Meeting on Chemical Sensors, Gaithersburg, Md, USA, 1996

## Transducers

1. 1st International Conference on Solid-State Sensors and Actuators

2. 2nd International Conference on Solid-State Sensors and Actuators (Transducers '83)
   Delft, The Netherlands.

3. 3rd International Conference on Solid-State Sensors and Actuators (Transducers '85)
   Philadephia, PA, USA

4. 4th International Conference on Solid-State Sensors and Actuators (Transducers '87)
   Tokyo, Japan

5. 5th International Conference on Solid-State Sensors and Actuators (Transducers '89)

6. 6th International Conference on Solid-State Sensors and Actuators (Transducers '91)
   San Fransisco, CA, USA

7. 7th International Conference on Solid-State Sensors and Actuators (Transducers '93)
   Yokohama, Japan

8. 8th International Conference on Solid-State Sensors and Actuators (Transducers '95)
   Stockholm, Sweden

9. 9th International Conference on Solid-State Sensors and Actuators (Transducers'97)
   Chicago, IL, USA

## Other Conference Proceedings:

- Bandyopadhyay, Promode R., Breuer, Kenneth S., and Blechinger, C. J.: Application of microfabrication to fluid mechanics: presented at 1994 International Mechanical Engineering Congress and Exposition, Chicago, Ill, Nov. 6-11, 1994. New York: American Society of Mechanical Engineers, 1994. FED (Series); vol. 197.

- Buck, Richard P. (Ed.): Biosensor technology: fundamentals and applications: Proceedings of the International Symposium on Biosensors held at the University of North Carolina at Chapel Hill, Sept. 7-9, 1989. New York, M. Dekker, 1990.

- Datta, Madhav, Sheppard, Keith, and Snyder, Dexter D.: Proceedings of the first International Symposium on Electrochemical Microfabrication. Pennington, NJ: Electrochemical Society, 1992. Proceedings/Electrochemical Society; v. 94-3.

- Datta, Madhav; Sheppard, Keith; Dulovic, John O.: Proceedings of the Second International Symposium on Electrochemical Microfabrication. Pennington, NJ: Electrochemical Society, 1995. Proceedings/Electrochemical Society; v. 94-32.

- Electrochemical Society. Electrodeposition Division: Electrochemical microfabrication: 1st International symposium: 180th Meeting. Proceedings - Electrochemcial Society PV; 92-3, 1992.

- Harashima, Fumio (Ed.): Integrated micro-motion systems: micromachining, control, and applications: 3rd Toyota conference: Papers. Amsterdam, New York: Elsevier Science Pub. Co., 1990.

- Howard, R. E. (Ed.): Science and technology of microfabrication: symposium held December 4-5, 1986, Boston, Mass. Pittsburgh, Pa: Materials Research Society, 1987.

- Jet Propulsion Laboratory: Workshop on Microtechnologies and applications to space systems: proceedings of the Workshop held at the Jet Propulsion Laboratory, May 27-28, 1992. Pasadena, Ca: National Aeronautics and Space Administration, Jet Propulsion Laboratory, California Institute of Technology, 1993.

- Langenbeck, Peter (Ed.): Micromachining optical components and precision engineering: ECO1 22-23 Sept. 1988, Hamburg. Conference proceedings. Bellingham, WA: SPIE, 1989. Spie Proceedings series; v. 1015.

- Leistiko, Otto (Ed.): Micromechanics (MME '95). Selected Papers. Philadelphia: Institute of Physics, 1996.

- LEOS 1991. IEEE Lasers and Electro-Optics Society (LEOS) Summer Topical Meeting on Epitaxial Materials and In-situ processing for Optoelectronic devices, July 29-31, 1991 Sheraton Newport Beach, Newport Beach, Ca; and Microfabrication for photonics and optoelectronics, July 31-Aug. 2, 1991. New York: IEEE, 1991.

- Markus, Karen W. (Ed.): Micromachining and microfabrication process technology: 23-24 Oct. 1995, Austin, Texas. Bellingham, WA. USA: SPIE, 1996. Spie Proceedings series; v. 2639.

- Mathewson, Paul R. ard Finley, John (Eds.) Biosensor design and application: Symposium: 201st Annual Meeting. Papers. ACS. Biotechnology Secretariat, 1992.
- Micromachining and Microfabrication process technology. Bellingham, WA. USA: SPIE, The International Society for Optical Engineering, 1995-. Annual.
- Micromechanics. Bristol, Uk.; Philadelphia: Institute of Physics Publishing, 1996-. Selected Papers
- Microlithography and metrology in micromachining sponsored by the National Institute of Standards and Technology; Semiconductor Equipment and Materials International; Society of Photo-optical Instrumentation Engineers. Bellingham, WA. USA: SPIE, The International Society for Optical Engineering, 1995-. Annual
- Middelhoek, Simon, and Van der Spiegel, Jan (Eds.): Sensors and Actuators: state of the art of sensor research and development. Lausanne, Elsevier Sequoia, 1987-.
- National Institute of Standards and Technology; Semiconductor Equipment and Materials International; Society of Photo-optical Instrumentation Engineers: Microlithography and metrology in micromachining. Bellingham, WA. USA: SPIE, The International Society for Optical Engineering, 1995-. Annual
- Pang, Stella W. and Chang, Shih-Cha: Micromachining and Microfabrication process technology II: 14-15 October, 1996, Austin, Texas. Bellingham, WA. USA: SPIE, 1996. Spie Proceedings series; v. 2880.
- Postek, Michael T. (Ed.): Microlithography and metrology in micromachining: 23-24 Oct. 1995, Austin Texas. Bellingham, WA. USA: SPIE, 1996. Spie Proceedings series; v. 2640.
- Postek, Michael T. and Friedrich, Craig (Eds.): Microlithography and metrology in micromachining II: 14-15 Oct. 1996, Austin Texas. Bellingham, WA. USA: SPIE, 1996. Spie Proceedings series; v. 2879.
- Rogers, Kim R., Mulchandani, Ashok, and Zhou, Weichang: Biosensor and chemical sensor technology: process monitoring and control: Symposium: 209th National Meeting: Papers. ACS, Division of Biochemical Technology, 1995.
- Roop, Ray and Chau, Kevin (Eds.): Micromachined devices and components: held as part of the 1995 symposium on micromachining and microfabrication: Papers. Bellingham, WA: SPIE, 1995
- Ryssel, Heiner and Stephani, Dieter (Eds.): Microcircuit Engineering 92: proceedings of the Internaional Conference on Microfabrication, Sept. 21-24, 1992, Erlangen, Germany. Microelectronic engineering, vol. 21. Amsterdam, New York: Elsevier, 1993.
- Usmani, Arthur M. and Akmal, Naim (Eds.): Diagnostic biosensor polymers: developed from a Symposium sponsored by the Division of Industrial and Engineering Chemistry, at the 205th National meeting of the American Chemical Society, Denver, CO, March 28-April2, 1993. ACS, 1994.
- Weck, M.: Micromachining of elements with optical and other submicrometer dimensional and surface specifications. The Hague, The Netherlands: 2-3 April, 1987. Bellingham, WA: SPIE, 1987. Spie Proceedings series; v. 803.

See also the NEXUS bibliography on microsystems (Http://wwwetis.et.tudelft.nl/nexus/by_author.html) and visit Fun.Micro@CRC.com.

# Appendix E
## Glossary*

**Abrasive:** A hard and wear-resistant material (such as a ceramic) used to wear, grind, or cut away other material.

**Accuracy:** The degree of correctness with which the measuring system yields the "true value" of a measured quantity, where the "true value" refers to an accepted standard, such as a standard meter or volt. Typically described in terms of a maximum percentage of deviation expected based on a full-scale reading.

**Affinity:** A thermodynamic measurement of the strength of binding between molecules, say between an antibody and antigen. Each antibody/antigen pair has an association constant, $K_a$, expressed in L/mol.

**Algorithm:** A set of well-defined mathematical rules or operations for solving a problem in a finite number of steps.

**AM 1:** The air mass 1 spectrum of a light source is equivalent to that of sunlight at the earth's surface when the sun is at zenith.

**Ampere (amp) [A]:** Measure of electric current: 1A = 1 coulomb/second.

**Amperometric Sensor:** Amperometric sensors involve a heterogeneous electron transfer as a result of an oxidation/reduction of an electro-active species at a sensing electrode surface. A current is measured at a certain imposed voltage of the sensing electrode with respect to the reference electrode. Analytical information is obtained from the current-concentration relationship at that given applied potential.

**Analyte:** A chemical species targeted for qualitative or quantitative analysis.

**Angstrom [Å]:** Measure of length: 1 Å= $10^{-10}$ m.

**Anisotropic:** Exhibiting different values of a property in different crystallographic directions.

**Anneal:** Heat process used to remove stress, crystallize or render deposited material more uniform.

**Anode:** The electrode in an electrochemical cell or galvanic couple that experiences oxidation, or gives up electrons.

**Arrhenius equation:** The equation representing the rate constant as $k = Ae^{E_a/RT}$ where A represents the product of the collision frequency and a steric factor, and $e^{-E_a/RT}$ is fraction of collisions with sufficient energy to produce a reaction.

**ASIC:** Application-specific integrated circuit; an IC designed for a custom requirement.

**Atomic mass unit (a.m.u.):** A unit of mass used to express relative atomic masses. It is equal to 1/12 of the mass of an atom of the isotope carbon-12 and is equal to $1.66033 \times 10^{-27}$.

**Atomic number (also proton number Z):** The number of protons within the atomic nucleus of a chemical element.

**Atomic weight:** The weighted average mass of the atoms in a naturally occuring element.

**Austenite:** Face-centered cubic iron; also iron and steel alloys that have the FCC crystal structure.

**Avogadro's number:** The number of atoms in exactly 12 grams of pure $^{12}C$, equal to $6.022 \times 10^{12}$.

**Band gap energy ($E_g$):** For semiconductors and insulators, the energies that lie between the valence and conduction bands; for intrinsic materials, electrons are forbidden to have energies within this range.

**Bandwidth:** The range of frequencies over which the measurement system can operate within a specified error range.

**BAW:** Bulk acoustic wave.

**Bilayer lipid membrane (BLM):** The structure found in most biological membranes, in which two layers of lipid molecules are so arranged that their hydrophobic parts interpenetrate, whereas their hydrophilic parts form the two surfaces of the bilayer.

**Binary:** Numbering system based on powers of 2 using only the digits 0 and 1, called "bits".

**Biosensor:** The term "biosensor" is a general designation that denotes either a sensor to detect a biological substance or a sensor which incorporates the use of biological molecules such as antibodies or enzymes. Biosensors are a subcategory of chemical sensors.

**Bipolar-junction transistor:** Transistor with n-type and p-type semiconductors having base-emitter and collector-base junctions.

**Bit:** see binary.

**Body-centered cubic (BCC):** A common crystal structure found in some elemental metals. Within the cubic unit cell, atoms are located at corner and cell center positions.

**Brazing:** A metal joining technique that uses a molten filler metal alloy having a melting temperature greater than about 425°C (800°F).

**Breakdown:** Failure of a material resulting from an electrical overload. The resulting damage may be in the form of thermal damage (melting or burning) or electrical damage (loss of polarization in piezoelectric materials).

**Bus:** Transmission medium for electrical or optical signals that perform a particular function, such as computer control.

---

* See also Lexicon of Semiconductor Terms: http://rel.semi.harris.com/docs/lexicon-old/

**Byte:** A group of eight bits that can represent any of $2^8 = 256$ different entities.

**Calibration:** A process of adapting a sensor output to a known physical quantity to improve sensor output accuracy.

**Capacitance (C):** The charge-storing ability of a capacitor, defined as the magnitude of charge stored on either plate divided by the applied voltage. A 1-F capacitor charged to 1 V contains C of charge (see also capacitor) and 1 C is an amount of charge equal to that of about $6.24 \times 10^{18}$ electrons.

**Capacitor:** Energy storage circuit element having two conductors separated by an insulator.

**Cathode:** The electrode in an electrochemical cell or galvanic couple at which a reduction reaction occurs; in other words the electrode receiving electrons from an external circuit.

**Ceramic:** A nonmetallic material made from clay and hardened by firing at high temperature; it contains minute silicate crystals suspended in a glassy cement.

**Cermet:** A composite material consisting of a combination of ceramic and metallic meterials. The most common cermets are the cemented carbides, composed of an extremely hard ceramic (e.g. WC, TiC), bonded together by a ductile metal such as cobalt or nickel.

**Chip:** A die (unpackaged semiconductor device) cut from a silicon wafer, incorporating semiconductor circuit elements such as a sensor, actuator, resistor, diode, transistor, and/or capacitor.

**Chromatography:** The general name for a series of methods for separating mixtures by employing a system with a mobile phase and a stationary phase.

**CMOS:** Complementary metal oxide semiconductor - integrated circuit containing n-channel and p-channel MOS-FETs.

**Codons:** Organic bases in sets of three that form the genetic code.

**Conductor:** Material such as the metals copper or aluminum that conducts electricity via the motion of electrons.

**Copolymer:** A polymer that consists of two or more dissimilar monomer units in combination in its molecular chains. Also a polymer formed from the polymerization of more than one type of monomer.

**Corrosion:** Deteriorative loss of a metal as a result of dissolution environmental reactions.

**Covalent bond:** A primary interatomic bond that is formed by the sharing of electrons between neighboring atoms.

**Coulomb [C]:** Measure of electrical charge: 1 C is an amount of charge equal to that of about $6.24 \times 10^{18}$ electrons.

**Creep:** The time-dependent permanent deformation that occurs under stress; for most materials it is important only at elevated temperatures.

**Crosslinked polymer:** A polymer in which adjacent linear molecular chains are joined at various positions by covalent bonds.

**Cross-sensitivity:** The influence of one measurand on the sensitivity of a sensor, another measurand.

**Crosstalk:** Electromagnetic noise transmitted between leads or circuits in close proximity to each other.

**Crystal structure:** For crystalline materials, the manner in which atoms or ions are arrrayed in space. It is defined in terms of the unit cell geometry and the atom positions within the unit cell.

**Curie temperature (also Curie point) ($T_c$):** The temperature above which a ferromagnetic or ferrimagnetic material becomes paramagnetic. For iron the Curie point is 760° and for nickel 356°C.

**Current [A]:** Measure of rate of flow of electric charge: a one-ampere current is a flow of 1 C of charge per second.

**Cutoff:** Condition in a diode or bipolar junction transistor in which the potential across a p-n junction prevents current flow.

**Cyclotron:** A type of particle accelerator in which an ion introduced at the center is accelerated in an expanding spiral path by use of alternating electrical fields in the presence of a magnetic field.

**Debye Shielding:** The Debye length in front of a sensing electrode depends on the ionic strength of the electrolyte used. In a 0.001N NaCl the Debye length measures 96.5 Å, while for a 1.0 N solution it is reduced to 3.0 Å. An adsorbed protein can stick out from the surface for as much as 50 to 100 Å. As a result, the charges which could contribute to the surface potential will be shielded in a 1.0 N solution. To make more sensitive measurements a solution of low ionic strength should be used.

**Degradation:** A term used to describe the deteriorative processes that occur with polymeric materials, including swelling, dissolution, and chain scission.

**Deoxyribonucleic acid (DNA):** A huge nucleotide polymer having a double helical structure with complementary bases on two strands. Its major functions are protein synthesis and the storage and transport of genetic information.

**Design:** To plan and delineate with an end in mind and subject to constraints.

**Devitrification:** The process in which a glass (noncrystalline or vitreous solid) transforms to a crystalline solid.

**Denaturation:** The breaking down of the three-dimensional structure of a protein resulting in the loss of its function.

**Dialysis:** A phenomenon in which a semipermeable membrane allows transfer of both solvent molecules and small solute molecules and ions.

**Diamagnetism:** A weak form of induced or nonpermanent magnetism for which the magnetic susceptibility is negative. A type of magnetism associated with paired electrons, that causes a substance to be repelled from the inducing magnetic field.

**Die:** see chip.

**Dielectric:** Any material that is electrically insulating.

**Dielectric constant ($\varepsilon$):** The ratio of the permittivity of a medium to that of a vacuum. Also called the relative dielectric constant or relative permittivity.

**Dielectric displacement:** The magnitude of charge per unit area of capacitor plate.

**Dielectric (breakdown) strength:** The magnitude of an electric field necessary to cause significant current passage through a dielectric material.

**Dimer:** A molecule formed by the joining of two identical monomers.

**Diffusion:** A thermochemical process whereby controlled dopants are introduced into a substrate.

**Diffusion coefficient:** The constant of proportionality between the diffusion flux and the concentration gradient in Fick's first law. Its magnitude is indicative of the rate of atomic diffusion.

**Digital:** Refers to systems employing only quantized (discrete) states to convey information (also see "analog").

**Diode:** two-terminal device that conducts current well in one direction and poorly in the other.

**Dip:** Dual In-line Package - common ceramic or plastic enclosure for an integrated circuit.

**Dipole (electric):** A pair of equal yet opposite electrical charges that are separated by a small distance.

**Dislocation:** A linear crystalline defect around which there is atomic misalignment. Plastic deformation corresponds to the motion of dislocations in response to an applied shear stress. Edge, screw, and mixed dislocations are possible.

**DNA Probes:** A DNA or nucleic acid probe is a short strand of DNA that locates and binds to its complementary sequence in samples containing single strands of DNA or RNA enabling identification of specific sequences. Nucleic acid probe assays exploit the fundamental hybridization reaction that occurs spontaneously between two complementary DNA:DNA or DNA:RNA strands. As in immunoassays, detection of the hybrid requires that the probe be labeled. Various direct and indirect methods have been devised for the detection of the hybrid. Direct labeling involves attaching the label directly to the probe sequence; indirect labeling binds an antibody to the DNA:DNA or DNA:RNA hybrid. As in immunoassays, non-isotopically-labeled probes are preferred over radio-labeled probes primarily because of radiation hazards, disposal problems, and short reagent shelf life. In addition, the factors determining the detection limits of hybridization assays based on labeled probes are similar to those in immunoassays. Therefore, the development of a simple, inexpensive and sensitive direct detection system which eliminates the use of labels is highly desirable.

**DNA Sequencing:** Their are two main classical methods for sequencing DNA: The first method, developed by Allan Maxam and Walter Gilbert, involves chemicals used to cleave the DNA at certain positions, generating a set of fragments that differ by one nucleotide. The second method, developed by Fred Sanger and Alan Coulson, involves enzymatic synthesis of DNA strands that terminate in a modified nucleotide. Analysis of fragments is similar for both methods and involves gel electrophoresis and autoradiography or fluorescence. The enzymatic method has largely replaced the chemical method as the technique of choice, although there are some situations where chemical sequencing can provide data more easily than the enzymatic method.

**Domain:** A region of a ferromagnetic or ferrimagnetic material in which all atomic or ionic magnetic moments are aligned in the same direction.

**Doping:** Process of introducing impurity atoms into a semiconductor to affect its conductivity.

**DRAM:** Dynamic Random Access Memory — memory in which each stored bit must be refreshed periodically.

**Drift:** Gradual departure of the instrument output from the calibrated value. An undesired slow change of the output signal.

**Dynamic characteristics:** A description of an instrument's behavior between the time a measured quantity changes value and the time the instrument obtains a steady response.

**Dynamic error:** The error that occurs when the output does not precisely follow the transient response of the measured quantity.

**Dynamic range:** The ratio of the largest to the smallest values of a range, often expressed in decibels.

**DSP:** Digital Signal Processing; a process by which a sampled and digitized data stream (real-time data such as sound or images) is modified in order to extract relevant information. Also, a digital signal processor.

**Ductility:** A measure of a material's ability to undergo plastic deformation before fracture; expressed as percent elongation (%EL) or percent area reduction (%AR) from a tensile test.

**EDP:** Ethylene diamine pyrocatechol.

**Elastic deformation:** A nonpermanent deformation that totally recovers upon release of an applied stress.

**Elastomer:** A polymeric material that may experience large and reversible elastic deformations.

**Electrical breakdown:** Condition in which, particularly with high electric field, a nominal insulator becomes electrically conducting.

**Electric field [V/m]:** In simplest form, the potential difference between two points divided by the distance between the two.

**Electroluminescence:** In electrical engineering: the emission of visible light by a p-n junction across which a forward-biased voltage is applied. In electrochemistry: emission of light by a molecule which is being reduced or oxidized on a biased electrode. If the exciting cause is a photon, rather than an electron, the process is called photoluminescence.

**Electrolyte:** A solution through which an electric current may be carried by the motion of ions.

**Electrolyte/Insulator/Silicon (EIS):** Structures at the heart of a broad family of potentiometric silicon sensors. The best-known member of the family is the ion-sensitive field effect transistor, known as the ISFET or CHEMFET and the light-addressable potentiometric sensor LAPS . The principle of operation of devices using such structures is as follows. A potential with respect to a reference electrode is generated at the interface between the liquid solution and the insulator. The surface potential ($\psi_0$) is determined by that ionic species which has the fastest exchange rate ($i_o$) with the membrane covering the insulator. If no intentional membrane is deposited on an oxide covered insulator that species will be $H^+$. Surface potential changes in turn change the Si flat-band voltage $V_{FB}$. The flat-band voltage is the potential one needs to apply to the Si in order to have the bands flat throughout the semiconductor. The flat band voltage of an EIS structure has been shown to be given by: $V_{FB} = E_{REF} - \phi^{Si}/q - \psi_0 - Q_{ins}/C_{inss}$ where $V_{FB}$ stands for the flat-band voltage of the structure, $E_{REF}$ for the reference electrode potential, $\Phi^{Si}$ for the work function of silicon, $\psi_0$ for the surface potential at the insulator/electrolyte interface, $Q_{ins}$ for the charge at the insulator/silicon interface and $C_{INS}$ for the insulator capacitance. At least two terms in the above equation are not known with a precision greater than a few hundred millivolts. This is true for $E_{Ref}$ as well as for $Q_{INS}/C_{INS}$ which can vary from device to device by several hundred millivolts. For a given EIS sensor, these inaccurately known quantities are constant, and variations in flat-band voltage can be equated to variations of the surface potential.

**Electromotive force (emf) series:** A series of chemical elements arranged in order of their electromotive force. The electromotive force is the greatest potential difference that can be generated by a particular source of electric current. In practice this potential may be observable only when the source is not supplying current, because of its internal resistance.

**Electron:** Elementary negative particle whose charge is $-1.602 \times 10^{19}$ coulombs.

**Electronegative:** Describing elements that tend to gain electrons and form negative ions. The halogens are typical electronegative elements.

**Electron state (level):** One of a set of discrete, quantized energies that are allowed for electrons. In the atomic case each state is specified by 4 quantum numbers.

**EMI:** Electromagnetic interference.

**Energy [J]:** Capacity for performing work or to cause heat flow. Like work itself, it is measured in Joules.

**Enthalpy (H):** A property of a system equal to E + PV, where E is the internal energy of the system, P is the presssure of the system, and V is the volume of the system. At constant pressure the change in enthalpy equals the energy flow as heat.

**Enzyme:** A large molecule, usually a protein, that catalyzes biological reactions.

**Enzyme immunoassay/EIA:** In an enzyme immunoassay (EIA), an enzyme-labeled antibody or antigen is used for the detection and quantification of the antigen-antibody reaction. In an electrochemical EIA, the enzyme-catalyzed reaction is monitored electrochemically (amperometric, potentiometric, voltammetric or conductometric). In EIA, the antibody-antigen reaction furnishes the needed specificity. The enzyme label provides the sensitivity via chemical amplification.

**Epitaxial or epi:** A single-crystal semiconductor layer grown upon a single-crystal substrate having the same crystallographic characteristics as the substrate material.

**EPROM:** Electrically Programmable Read-Only Memory -— nonvolatile memory device.

**Extrinsic:** Characterizes doped, rather than pure, semiconductor.

**Fab:** For "fabrication", a term referring to the making of semiconductor devices such as microprocessors.

**Face-centered cubic:** A crystal structure found in common elemental metals also FCC. Within the cubic unit cell, atoms are located at all corner and face-centered positions.

**Farad:** The unit of capacitance (see "capacitance").

**Faraday:** A constant representing the charge on one mole of electrons; 96, 485 coulombs.

**FEA:** Finite element analysis.

**Fermi energy:** The energy level in a solid at which the probablity of finding an electron is 1/2. For a metal, the energy corresponding to the highest filled electron state in the valence band at 0 K.

**Ferroelectric material:** A dielectric material such as Rochelle salt and barium titanate with a domain structure containing dipoles (asymmetric distributions of electrical charge) which spontaneously align. Their domain structure makes them analogous to ferromagnetic materials. They exhibit hysteresis and usually the piezoelectric effect.

**Ferromagnetism:** Permanent and large magnetizations found in some metals (e.g. Fe, Ni and Co), resulting from the parallel alignment of neighboring magnetic moments.

**FET:** Field-Effect Transistor — semiconductor device whose insulated gate electrode controls current flow.

**Fiber-optic:** Relates to transmission of information as modulated light in tiny transparent fibers instead of copper wires.

**Fick's first law:** The diffusion flux is proportional to the concentration gradient. This relationship is employed for steady-state diffusion situations.

**Fick's second law:** The time rate of change of concentration is proportional to the second derivative of concentration. This relationship is employed in non-steady-state diffusion situations.

**Filler:** An inert foreign substance added to a polymer to improve or modify its properties.

**Firing:** A high temperature heat treatment that increases the density and strength of a ceramic piece.

**Flat-band potential:** see under Electrolyte/Insulator/Silicon (EIS).

**Flip-flop:** binary device whose outputs change value only in response to an input pulse.

**Fluorescence:** Luminescence (see also under "luminescence") which persists less than a second after the exciting cause has been removed. If the luminescence persists significantly longer it is called phosphorescence.

**FM:** Frequency Modulation - information coding scheme in which the frequency of a steady wave is changed.

**Forward bias:** The conducting bias for a p-n junction rectifier that assures electron flow to the n side of the junction.

**Free energy (G):** A thermodynamic quantity that is a function of the enthalpy (H), the Kelvin temperature (T) and the entropy (S) of a system; G=H-TS. At equilibrium, the free energy is at a minimum. Under certain conditions the change in free energy for a process is equal to the maximum useful work.

**Frequency:** Number of times per second that a quantity representing a signal, such as a voltage, changes state. Also, the number of waves (cycles) per second that pass a given point in space.

**Frequency response:** Two relations between sets of inputs and outputs. One relates frequencies to the output-input amplitude ratio; the other relates frequencies to the phase difference between the output and input.

**Gain:** The ratio of the amplitude of an output to input signal.

**Galvanic corrosion:** The preferential corrosion of the more chemically active of two metals electrically coupled and exposed to an electrolyte.

**Gate:** Circuit whose logical output variables are determined by its inputs.

**Gauss:** The cgs unit used in measuring magnetic induction.

**Gene:** A given segment of the DNA molecule that contains the code for a specific protein.

**Glass:** An amorphous solid obtained when silica is mixed with other compounds, heated above its melting point, and then cooled rapidly.

**Glass transition temperature ($T_g$):** The temperature at which, upon cooling, a noncrystalline ceramic or polymer transforms from a supercooled liquid to a rigid glass.

**Grain boundary:** The interface separating two adjoining grains having different crystallographic orientations.

**Grain growth:** The increase in average grain size of a polycrystalline material: for most materials, an elevated temperature heat treatment is necessary.

**Grain size:** The average grain diameter as determined from a random cross section.

**Green ceramic body:** A ceramic piece, formed as a particulate aggregate, that has been dried but not fired.

**Ground:** To make electrical connection to the earth or to the chassis of a device (verb); the connection point so used (noun).

**GUI:** Graphical User Interface — hardware, software, and firmware that produces the display on modern personal computers.

**Hall effect:** The phenomenon whereby a force is applied to a moving electron or hole by a magnetic field that is applied perpendicular to the direction of motion. The force direction is perpendicular to both the magnetic field and the particle motion directions.

**Hardness:** The measure of a material's resistance to deformation by surface indentation or by abrasion. There are various scales in use to express hardness. The Mohs scale is qualitative and somewhat arbitrary and ranges from 1 on the soft end for talc to 10 for diamond. Quantitative scales are the Rockwell (HR), Brinell (indicated by HB), Knoop (HK) and Vickers (HV). Knoop and Vickers are referred to as microhardness testing methods on the basis of load and indenter size.

**Heat capacity ($C_V$ at constant volume and $C_p$ at constant pressure):** The quantity of heat required to produce a unit temperature rise per mole of material.

**Henry (H):** Unit of inductance (see "inductance"). One henry (H) is the inductance of a closed circuit in which an electromotive force of 1 volt is produced when the electric current in the circuit varies uniformly at the rate of 1 ampere per second.

**Henry's law:** The amount of gas dissolved in a solution is directly proportional to the pressure of the gas above the solution.

**Heme:** An iron complex.

**Hemoglobin:** A biomolecule composed of four myoglobinelike units (proteins plus heme) that can bind and transport four oxygen molecules in the blood.

**HEMT:** High electron mobility transistor.

**Home Page:** a site or "page" on the World Wide Web (WWW).

**Homogeneous and Heterogeneous Assays:** A homogeneous assay does not require a separation step to remove free antigen from bound antigen and relies upon the fact that the function of the label is modified upon binding, leading to a change in signal intensity. Because of high background signal a heterogeneous approach incorporating a separation step of bound and unbound makes the detection limit lower, approaching the values obtained by RIA. The homogeneous assay is less technically demanding.

**http:** Hypertext transfer protocol — transfer protocol used on the WWW.

**Hysteresis:** The difference in the output when a specific input value is approached first with an increasing and then with a decreasing input. This phenomenon occurs in ferroelectric materials and results in irreversible loss of energy through heat dissipation.

**IC:** See "integrated circuit".

**Impedance:** The complex ratio of a force-like quantity (force, pressure, voltage, temperature, or electric field) to a corresponding related velocity-like quantity (velocity, volume velocity, current, heat flow, or magnetic field strength).

**Index of refraction (n):** The ratio of the velocity of light in a vacuum to the velocity in some medium.

**Inductance [in Henry, H]:** That property of an electric circuit which tends to oppose change in current in the circuit. One henry (H) is the inductance of a closed circuit in which an electromotive force of 1 volt is produced when the electric current in the circuit varies uniformly at the rate of 1 ampere per second.

**Inductor:** Energy storage circuit component consisting of a coil of wire and possibly a magnetic material.

**Infrared:** Invisible electromagnetic radiation having a longer wavelength, and lower frequency, than visible red light.

**Inhibitor:** A chemical substance that, when added in relatively low concentrations, slows down a chemical reaction.

**Insertion point (in lithography context):** Adaptation of a new lithography technique is referred to as the insertion point of that technique.

**Insulator:** Material that conducts electricity very poorly.

**Integrated circuit (IC):** Semiconductor circuit, typically on a very small silicon chip, containing microfabricated transistors, diodes, resistors, capacitors, etc.

**Internet:** Worldwide digital communication network in which packets of information travel between senders and recipients.

**Interstitial diffusion:** a diffusion mechanism that causes atomic motion from interstitial site to interstitial site.

**Intrinsic:** Characterizes pure undoped semiconductor; electrical conductivity depends only on temperature and the band gap energy.

**I/O:** Input/output information transfer between computer and peripherals such as keyboard or printer.

**Ionic bond:** A coulombic interatomic bond existing between two adjacent and oppositely charged ions.

**Ionophore:** A macro-organic molecule capable of specifically solubilizing an inorganic ion of suitable size in organic mediums.

**ISE (Ion selective electrode):** Ions in solution are quantified by measuring the change in voltage (i.e. potentiometric) resulting from the distribution of ions (by ion exchange controlled by the ion exchange current $i_o$) between a sensing membrane (the ion selective membrane) and the solution. This potential is measured at zero current with respect to a reference electrode which is also in contact with the solution. The potential measured is proportional to the logarithm of the analyte concentration. The oldest and best known ISE is the pH sensor based on a glass membrane. More recently, polymeric membranes have been formed incorporating ionophores (see ionophore) rendering the membrane specific to certain ions only.

**ISFET:** Ion Sensitive Field Effect Transistor: a logical extension of ISE's. They can be conceptualized by imagining that the lead from an ion-selective electrode, attached via a cable to a FET in the high impedance input stage of a voltmeter, is made shorter until no lead exists and the selective membrane is attached directly to the FET. For an ISFET, the property measured is the lateral conductivity between two opposing doped regions (the source and drain) surrounding the active area. The underlying change is a change in flat-band voltage.

**Isomorphous:** Having the same structure. In the phase diagram sense, isomorphicity means having the same crystal structure or complete solid solubility for all compositions.

**Isothermal:** At a constant temperature. In an isothermal process heat is, if necessary, supplied or removed from the system at just the right rate to maintain constant temperature.

**Isotropic:** Having identical values of a property in all crytallographic directions.

**Kilobyte (kB):** $2^{10}$ (= 1024, or about one thousand) bytes of information.

**Kilohertz (kHz):** One thousand cycles per second (see also "frequency").

**Kinetic molecular theory:** A model that assumes that an ideal gas is composed of tiny particles (molecules) in constant motion.

**Label or marker:** A problem endemic in immunoassays is the absence of a chemical signal created by the antibody-antigen binding, in contrast with an enzyme-substrate binding reaction which produces a chemical reaction product. As a result of this absence, the use of a label or marker is usually required to detect the bound antibody-antigen complex. Several markers have been established for use in immunoassays. Examples of such markers are:

- Particles (e.g. latex, gold particles, erythrocytes)
- Metal and dye sols (e.g. Au, Palanil® Luminous Red G)
- Chemiluminescent and bioluminescent compounds (e.g. Luciferase/luciferins, Luminol and derivatives, Acridinium esters, Peroxidase)
- Electrochemical active species (ions, redox species, ionophores)
- Fluorophores (e.g. Dansyl chloride DANS, Rare earth metal chelates, Umbelliferones)
- Chromophores
- Enzymes (e.g. Alkaline phosphatase, β-D-Galactosidase, Peroxidase), substrates, cofactors
- Liposomes
- Iodine-125, tritium, $^{14}C$, $^{75}Se$, $^{57}Co$.

**Laser:** Light Amplification by the Stimulated Emission of Radiation — quantum device that produces coherent light.

**Laser trimming:** A method for adjusting the value of thin- or thick-film resistors by using a computer-controlled laser system.

**LCD:** Liquid Crystal Display — display device employing light source and electrically alterable optically active thin film.

**LED:** Light-Emitting diode — semiconducting diode that produces visible or infrared radiation.

**Leakage:** The loss of all or parts of a useful agent, as of the electric current that flows through an insulator or the magnetic flux that passes outside useful flux circuits.

**Lewis acid:** An electron-pair acceptor.

**Lewis base:** An electron-pair donor.

**Life (lifetime):** The length of time the sensor can be used before its performance changes.

**Linear coefficient of thermal expansion:** see thermal expansion coefficient, linear.

**Linearity:** The degree to which the calibration curve of a device conforms to a straight line.

**Limit of detection:** The smallest measurable input. This differs from resolution, which defines the smallest measurable change in input. For a temperature measurement, this would provide an indication of the lowest temperature a sensor could generate an output in response to.

**Lipids:** Water-insoluble substances than can be extracted from cells by nonpolar organic solvents.

**Luminescence:** defined as the mission of light from a substance in an electronically excited state. Depending on whether the excited state is singlet or triplet, the emission is called fluorescence (less than one second decay) or phosphorescence (longer than 1 second decay). Depending on the source, molecules get the needed extra energy from different types of luminescence are distinguished: radioluminescence, photoluminescence (in the same category are fluorescence and phosphorescence), chemiluminescence and bioluminescence, electrochemiluminescence, sonochemi-luminescence and thermoluminescence.

**Magnetic field strength (designated by H) [A/m]:** Magnetic field produced by a current, independent of the presence of magnetic material. The units of H are ampere-turns per meter, or just amperes per meter.

**Magnetic flux density or magnetic induction (designated by B):** The magnetic field produced in a substance by an external magnetic field. The units of B are tesla (T). One tesla is the magnetic flux density given by a magnetic flux of 1 weber per square meter. One weber is a magnetic flux that, linking a circuit of 1 turn, would produce in it an electromotive force of 1 volt if it were reduced to zero at a uniform rate in 1 second. Both B and H are field vectors. One henry (H) is the inductance of a closed circuit in which an electromotive force of 1 volt is produced when the electric current in the circuit varies uniformly at the rate of 1 ampere per second. The magnetic field strength and flux density are related according to: $B = \mu H$, where $\mu$ is the permeability (see under permeability).

**Magnetic susceptibility ($\chi_m$)** The proportionality constant between the magnetization M (see under "magnetization") and the magnetic field strength H. The magnetic susceptibility is unitless.

**Magnetization (M):** The total magnetic moment per unit volume of material. Also, a measure of the contribution to the magnetic flux by some material within an H field. The magnitude of M is proportional to the applied field as: $M = \chi_m \times H$, with $\chi_m$ the magnetic susceptibility.

**Magnetostrictive material:** A material that changes dimension in the presence of a magnetic field or generates a magnetic field when mechanically deformed.

**Martensite:** A metastable iron phase supersaturated in carbon that is the product of a diffusionless (a thermal) transformation from austenite.

**Mask:** Pattern on glass, like a photographic negative, for producing integrated-circuit elements on semiconductor wafer.

**MCM:** MultiChip Module; the interconnection of two or more semiconductor chips in a semiconductor-type package.

**Mean:** Numerical average of data values.

**Measurand:** A physical quantity, condition, or property that is to be measured.

**Mechatronics:** The synergistic combination of precision mechanical engineering with electronic control.

**Megabyte (MB):** $2^{20}$ (= 1,048,576, or about one million) bytes of information.

**Megahertz (Mhz):** One million cycles per second (see also "frequency").

**MEMS:** Stood originally for Micro-ElectroMechanical System — microscopic mechanical elements, fabricated on silicon chips by techniques similar to those used in integrated circuit manufacture, for use as sensors, actuators, and other devices. Today almost any miniaturized device (based on Si technology or traditional precision engineering, chemical or mechanical) is referred to as a MEMS device.

**Microphone:** Device that produces voltage or current in response to a sound wave.

**Microprocessor:** Chip containing the logical elements for performing calculations and carrying out stored instructions.

**Microstructure:** In material engineering the structural features of a material such as grain boundaries, grain size and structure, subject to observation under a microscope, selective etching etc. In MEMS microstructure unfortunately is also used to designate a micromachined feature.

**Miller indices:** A set of 3 integers (4 for hexagonal) that designate crystallographic planes, as determined from reciprocals of fractional axial intercepts.

**MIPS:** Millions of Instructions per Second — a measure of computing power.

**Mobility (electron, and hole):** the proportionality constant between the carrier drift velocity and applied electric field.

**Modulus of elasticity (E):** The ratio of stress to strain when deformation is totally elastic. Also the Young's modulus.

**Molding (plastics):** Shaping a plastic material by forcing it, under pressure at a high temperature, into a mold cavity.

**Molarity:** Concentration in a liquid solution (symbol c), in terms of the number of moles of a solute dissolved in $10^6$ mm$^3$ ($10^3$ cm$^3$) of solution in mol l$^{-1}$.

**Molality:** The molality or molal concentration (symbol m) is the amount of substance per unit mass of solvent or mol kg$^{-1}$.

**Monoclonal antibodies:** Produced by injecting animals to elicit a response from lymphocytes to produce antibodies. Lymphocytes which produce antibodies with strong binding capability can be isolated and used to produce only one kind of antibody (monoclonal) on a permanent basis once the lymphocytes are immortalized. This is accomplished by fusing them (combining them genetically) with cancer cells which have the distinction of living indefinitely in a culture. Monoclonal antibodies can be produced repeatedly and collected for use in immunodetection.

**Moore's law:** after Gordon Moore: "The number of transistors per computer chip will double roughly every two years".

**MOSFET:** Metal-Oxide-Semiconductor Field-Effect Transistor — device where gate electrode potential controls current flow

**MUX:** Device for combining several signals or data streams into a single flow.

**NAND:** NOT-AND — logic gate whose output is the negation of that of the AND gate.

**Nernst equation:** An equation relating the potential of an electrochemical cell to the concentrations of the cell components: $E = E^o + RT/zF \ln C_1/C_2$ with z the charge exchanged at the electrode and $C_1$ and $C_2$ concentrations of two electro-active compounds.

**NOR:** NOT-OR — logic gate whose output is the negation of that of the OR gate.

**NOT:** Logic gate whose output is binary 1 when its input is 0, and whose output is a 0 when its input is a 1.

**Nucleation:** The initial stage in a phase transformation, evidenced by the formation of small particles (nuclei) of the new phase, which are capable of growing.

**Nucleotide:** A monomer of the nucleic acids composed of a five-carbon sugar, a nitrogen-containing base, and phosphoric acid.

**n-type:** Characterizes a semiconductor containing predominantly mobile electrons (also see "p-type").

**Ohm ($\Omega$):** Unit of resistance. One ohm is the electrical resistance between two points of a conductor when a constant potential difference of 1 volt, applied to these points, produces a current of 1 ampere in the conductor.

**Ohmmeter:** Tool for measuring electrical resistance.

**Op-amp:** Operational Amplifier — semiconductor amplifier characterized by high gain and high internal resistance.

**OR:** Logic gate whose output is a binary 1 if any of its inputs is a 1; zero otherwise.

**Oscillator:** Circuit that produces an alternating voltage (current) when supplied by a steady (DC) energy source.

**Osmosis:** The flow of solvent into a solution through a semi-permeable membrane.

**Osmotic pressure ($\pi$):** The pressure that must be applied to a solution to stop osmosis: $\pi = MPT$.

**Oxidation:** The removal of one or more electrons from an atom, ion or molecule.

**Package:** Protective enclosure for a chip or a sensor, typically made of plastic or ceramic.

**Paramagnetism:** A relatively weak form of magnetism resulting from the independent alignment of atomic dipoles (magnetic) with an applied magnetic field. Also a type of induced magnetism, associated with unpaired electrons, that causes a substance to be zapped into the inducing magnetic field.

**Permeability [$\mu$]:** From the relation between magnetic induction and magnetic field ($B = \mu \times H$); for free space, $\mu_0 = 1.26 \times 10^{-6}$ H/m.

**Permitivity [$\varepsilon$]:** From the relation between polarization charge and electric field; for free space, $\varepsilon_0 = 8.85 \times 10^{-12}$ F/m.

**Phase shift:** A time difference between the input and output signals.

**Phase transformation:** A change in the number and/or character of the phases that make up the microstructure of an alloy.

**Phonon:** A single quantum of vibrational or elastic energy.

**Phosphorescence:** Luminescence that occurs at times greater than on the order of a second after an electron excitation event (see also "luminescence").

**Photodiode:** Semiconductor diode that produces voltage (current) in response to illumination (see also "phototransitor").

**Photomicrograph:** The picture made with a microscope.

**Phototransistor:** Transistor that, when powered, produces amplified voltage (current) in response to illumination.

**Piezoelectric material:** A ferroelectric material in which an electrical potential difference is created due to mechanical deformation, or conversely, in which the application of a voltage causes dimensional changes in the material.

**Pinhole:** The term pinhole embraces a wide variety of oxide defects and is used in a broad sense today. Listed in this category are cracks caused by thermal contraction after oxidation or by handling, and regions of oxide with low dielectric strength caused by dust particles, inadequate masking, contamination, or poor resist adhesion.

**Pin-out:** Diagram showing for electronic components the relations between connecting pins and internal components.

**Pixel:** Picture Element — smallest element of an image, such as a dot on a computer monitor screen.

**Pitting:** A form of very localized corrosion wherein small pits or holes form, usually in a vertical direction.

**pK value:** A measure of the strength of an acid on a logarithmic scale. The pK value is given by $\log_{10}(1/K_a)$, where $K_a$ is the acid dissociation constant pK values often are used to compare the strengths of different acids.

**Plastic deformation:** Permanent or nonrecoverable deformation, accompanied by permanent atomic displacement.

**Plasticizer:** A low molecular weight polymer additive that enhances flexibility and workability and reduces stiffness and brittleness.

**Point defect:** A crystalline defect associated with one or several atomic sites.

**Poisson's ratio ($\nu$):** For elastic deformation, the negative ratio of lateral and axial strains that result from an applied axial stress.

**Polarization (P):** The total electric dipole moment per unit volume of dielectric material.

**Polyclonal antibodies:** Antibodies produced by an animal's white blood cells (lymphocytes, specifically) in response to an antigen. This response occurs naturally or can purposely be created by injecting an animal, such as a rabbit or goat, with a specific antigen. More than one kind of anti-body is produced since more than one lymphocyte is producing antibodies. This is referred to as "polyclonal". The polyclonal antibodies are isolated from the animal and can be used for detection purposes. Because the antibodies are actually a mixture with different affinities (binding capability) for the antigen of interest, some variability in performance can occur from one test to another or one batch of antibodies to another.

**Polysilicon:** Polycrystalline silicon used as conductor in integrated circuits, and especially FETs.

**Potentiometric Device:** Monitors the voltage between a sensing electrode and a reference electrode. A high input impedance voltmeter is used to minimize current flow. The voltage typically is proportional to the logarithm of the analyte concentration.

**Power [W]:** Product of voltage and current in a component; also, refers to the field of electric energy supply.

**Precision:** The degree of reproducibility among several independent measurements of the same true value under specified conditions.

**Printed circuit board:** PC — selectively metallized insulating sheet for supporting and interconnecting circuit components.

**p-Type semiconductor:** A semiconductor for which the predominant charge carriers responsible for electrical conduction are holes.

**Pyroelectricity:** The property of certain crystals, such as tourmaline, of acquiring opposite electrical charges on opposite faces when heated.

**Quantization:** The concept that energy can occur only in discrete units called quanta.

**Q factor:** A rating, applied to coils, capacitors, and resonant circuits, equal to the reactance divided by the resistance. The ratio of energy stored to energy dissipated per cycle in an electrical or mechanical system.

**RAM:** Random Access Memory — read-write memory with elements accessible in any order.

**Range:** The difference between the minimum and maximum values of sensor output in the intended operating range. Defines the overall operating limits of a sensor.

**Reactance:** Portion of impedance that characterizes non-dissipative, energy storage effects (also see "impedance").

**Reactant:** A starting substance in a chemical reaction. It appears to the left of the arrow in a chemical reaction.

**Recrystallization:** The formation of a new set of strain-free grains within a previously cold-worked material due to an annealing heat treatment.

**Rectifier:** Device that converts bi-directional to one-way current flow.

**Reduction:** The addition of one or more electrons to an atom, ion, or molecule.

**Reflection:** Deflection of a light beam at the interface between 2 media.

**Refraction:** Bending of a light beam when passing from one medium to another, at different velocities of light.

**Refractory:** A metal or ceramic that does not deteriorate rapidly or does not melt when exposed to extremely high temperatures.

**Relative magnetic permeability ($\mu_r$):** The ratio of the magnetic permeability of some medium to that of a vacuum (unitless), or: $\mu_r = \mu/\mu_o$, where $\mu_0$ is the permeability of vacuum, a universal constant, which has a value of $1.257 \times 10^{-6}$ H/m.

**Reliability (life, multi-use vs. single, calibration vs. accuracy drift):** How well a sensor maintains both precision and accuracy over its expected lifetime. Also includes the robustness of the sensor.

**Repeatability:** The exactness with which a measuring instrument repeats indications when it measures the same property under the same conditions.

**Residual stress:** A persisting stress in a material free of external forces or temperature gradients.

**Resistance [$\Omega$–ohm]:** Characteristic of a resistor: in a 1-ohm resistance a current of 1 A produces a voltage drop of 1 V.

**Resistivity ($\rho$):** The reciprocal of electrical conductivity, and a measure of a material's resistance to passing electric current.

**Resistor:** Energy dissipative element consisting of a poor conductor in series with connecting wires.

**Resolution:** The smallest measurable change in input that will produce a small but noticeable change in the output. In the context of chemical separations, defines the completeness of separation.

**Resonant frequency:** The frequency at which a moving member or a circuit has a maximum output for a given input.

**Response time:** The time it takes for the sensor's output to reach its final value. A measure of how quickly the sensor will respond to changes in the environment. In general, this parameter is a measure of the speed of the sensor and must be compared with the speed of the process.

**Reverse bias:** The insulating bias for a p-n junction rectifier; electrons flow into the p side of the junction.

**RF: Radio Frequency** — refers to alternating voltages and currents having frequencies between 9 kHz and 3 MHz.

**rms:** Root mean square.

**ROM:** Read only memory; memory used for permanent, storage of unalterable data; nonvolatile memory.

**Sacrificial anode:** An active metal or alloy that corrodes and protects another metal or alloy to which it is electrically coupled.

**Sacrificial layer:** A thin film that is later removed to release a microstructure from its substrate.

**Scanning electron microscope (SEM):** a microscope producing an image by using reflected electron beams that scan the surface of a specimen.

**Semiconductor:** Nonmetallic material, such as silicon, whose electrical conductivity is moderate and alterable by doping.

**Selectivity:** The ability of a sensor to measure only one metric or, in the case of a chemical sensor, to measure only a single chemical species.

**Sensitivity:** The amount of change in a sensor's output in response to a change at a sensor's input over the sensor's entire range. Provides an indication of a sensor's ability to detect changes. For some sensors, the sensitivity is defined as the input parameter change required to produce a standardized output change.

**SI units:** International System of units based on the metric system and units derived from the metric system.

**Signal-to-noise-ratio:** The ratio of the output signal with an input signal to the output signal with no input signal.

**Single crystal:** A crystalline solid for which the periodic and repeated atomic pattern extends throughout its entirety without interruption.

**Sintering:** Particle coalescence of a powdered aggregate by diffusion that is accomplished by firing at an elevated temperature.

**Slip:** Plastic deformation resulting from dislocation motion; also, the shear displacement of 2 adjacent planes of atoms.

**Slip casting:** A forming technique used to shape ceramic materials. A slip or suspension of solid particles in water is poured into a porous mold. A solid layer forms on the inside wall as water is absorbed by the mold, leaving a shell (or a solid piece) in the shape of the mold.

**Smart sensor:** A sensor in which the electronics that process the output from the sensor, and forms the modifier, are partially or fully integrated on a single chip.

**Solvent:** The component of a solution that dissolves a solute.

**Span:** The difference between the highest and lowest scale values of an instrument.

**Specific heat:** The heat capacity per unit mass of material.

**Specific Modulus (specific stiffness):** The ratio of elastic modulus to specific gravity for a material.

**Specific strength:** The ration of tensile strength to specific gravity for a material.

**Spinning:** Fiber forming process: a multitude of fibers are spun as molten material is forced through many small orifices.

**Squeeze-film damping:** Effect of ambient fluid and spacing on the vertical movement of a structural member with respect to a substrate.

**Stability:** The ability of a sensor to retain specified characteristics after being subjected to designated environmental or electrical test conditions.

**Stabilizer:** A polymer additive that counteracts deteriorative processes.

**Standard atmosphere:** A unit of pressure equal to 760 mm Hg.

**Standard hydrogen electrode (SHE):** A platinum conductor in contact with a 1 M $H^+$ ions and bathed by hydrogen gas at one atmosphere.

**Steady-state diffusion:** The diffusion condition for which there is no net accumulation or depletion of diffusing species. The diffusion flux is independent of time.

**Step response:** The response of a system to an instantaneous jump in the input signal.

**Stiction:** Static friction; adhering of thin micromachined layers to a substrate.

**Stoichiometry:** For ionic compounds, the state of having exactly the ratio of cations to anions specified by the chemical formula. Stoichiometric quantities refers to quantities of reactants mixed in exactly the correct amounts so that all are used up at the same time.

**Strain (symbol ε):** The change in gauge length of a specimen, in the direction of an applied stress, divided by its original gauge length.

**Strain gauge:** An element (wire or foil) that measures a strain based on electrical resistance changes of the gauge that result from a change in length or dimension strain of the wire or foil.

**Stress concentration:** The concentration or amplification of an applied stress at the tip of a notch or small crack.

**Stress corrosion (cracking):** A form of failure resulting from the combined action of a tensile stress and a corrosion environment, occuring at lower stress levels than required when the corrosion environment is absent.

**Superconductivity:** A phenomenon characterized by the disappearance of the electrical resistivity at temperatures approaching 0 K.

**Surface Plasmon:** A collective motion of electrons in the surface of a metal conductor, excited by the impact of light of appropriate wavelength at a particular angle.

**Systematic error:** An error that always occurs in the same direction.

**TAB bonding:** Tape automated bonding; semiconductor packaging technique that uses a tiny lead-frame to connect circuitry on the surface of the chip to a substrate instead of wire bonds.

**Tensile strength (TS):** The maximum engineering stress, in tension, sustainable without fracture; also called "ultimate (tensile) strength".

**Tesla [T]:** Unit of magnetic induction: 1 T = 1 weber/m² (also, 1T = $10^4$ gauss).

**TFT:** Thin film transistor.

**Thermal conductivity (κ):** For steady-state heat flow, the proportionality constant between the heat flux and the temperature gradient. Also, a parameter characterizing the ability of a material to conducting heat.

**Thermal expansion coefficient, linear (α):** The fractional change in length divided by the change in temperature.

**Thermal fatigue:** A type of fatigue failure that introduces the cyclic stresses by fluctuating thermal stresses.

**Thermal shock:** The fracture of a brittle material resulting from stresses introduced by a rapid temperature change.

**Thermal stress:** A residual stress introduced within a body resulting from a change in temperature.

**Thermistor:** A temperature-measuring device, that contains a resistor or semiconductor whose resistance varies with temperature.

**Thermocouple:** A temperature-measuring device, which contains a pair of end-joined dissimilar conductors in which an electromotive force is developed by thermoelectric effects when the joined ends and the free ends of the conductors are a different temperature.

**Thermoplastic polymer:** A substance that when molded to a certain shape under appropriate conditions can later be remelted.

**Thermoset polymer:** A substance that when molded to a certain shape under pressure and high temperatures cannot be softened again or dissolved.

**Threshold:** The smallest input signal that will cause a readable change in the output signal.

**Time constant:** The time it takes for the output change to reach 63% of its final value.

**Transient response:** The response of the sensor to a step change in the measurand.

**Toughness:** A measure of the amount of energy absorbed by a material as it fractures, indicated by the total area under the material's tensile stress-strain curve.

**Transduction (self-generating or modulating):** The conversion of the signal to be measured into another, more easily accessible form. Source of energy for transmission of the sensor signal.

**Transduction mode (direct or indirect):** How the sensor acquires the desired information from the material. In general, this parameter is an indication of the ability of the sensor signal to provide information regarding a material property or state of interest.

**Transformer:** Device using magnetically linked inductors to change AC voltage level.

**Transmission:** Refers to system for carrying electric power at voltages above 100,000 volts.

**Transmission electron microscope (TEM):** A microscope that produces an image by using electron beams to transmit (pass through) the specimen, making examination of internal features at high magnifications possible.

**Transistor:** Semiconductor device used for amplification and switching.

**Tribology:** The science and technology of two interacting surfaces in relative motion and of related subjects and practices. The popular equivalent is friction, wear, and lubrication in surfaces sliding against each other, as in bearing and gears.

**Turn-on-voltage:** Applied voltage required to produce conduction in a diode.

**ULSI:** Ultra large scale integration; a chip with over 1,000,000 components.

**Unit cell:** The basic structural unit of a crystal structure, defined in terms of atom (or ion) positions within a parallelepiped volume.

**URL:** Universal Resource Locator — address of a World Wide Web site.

**Usenet:** Interlinked bulletin boards available via Internet and commercial on-line services.

**UV:** Ultraviolet — characterization of short-wavelength light for exposing photoresist in making semiconductor devices.

**Valence band:** The electron energy band that contains the valence electrons in solid materials.

**Valence electrons:** The electrons in the outermost occupied electron shell, that participate in interatomic bonding.

**van der Waals bond:** A secondary, permanent or induced, interatomic bond between adjacent molecular dipoles.

**Viscoelasticity:** a type of deformation exhibiting the mechanical characteristics of viscous flow and elastic deformation.

**Viscosity (symbol is $\eta$):** The ratio of the magnitude of an applied shear stress to the velocity gradient that it produces; in other words: a measure of a noncrystalline material's resistance to permanent deformation.

**Vitrification:** During firing of a ceramic body, the formation of a liquid phase that becomes a glass-bonding matrix upon cooling.

**VLSI:** Very large scale integration; a chip with 100,000 to 1,000,000 components.

**Volt:** Unit of electrical potential difference (see "voltage").

**Voltage [V]:** Potential difference between two points: energy to move a 1-C charge through a 1-V potential difference is 1-J.

**Wafer:** Semiconductor disk out of which integrated circuits are made (also see "chip", "mask").

**Watt (W):** Unit of power. One watt is the power that, in 1 second, gives rise to an energy of 1 joule.

**Weber:** Unit of magnetic flux. One weber is a magnetic flux that, linking a circuit of 1 turn, would produce in it an electromotive force of 1 volt if it were reduced to zero at a uniform rate in 1 second.

**Weight percent (wt%):** Concentration specification on the basis of weight (or mass) of a particular element relative to the total alloy weight (or mass).

**Whisker:** A very thin, single crystal of high perfection which has an extremely large length-to-diameter ratio. Whiskers are used as the reinforcing phase in some composites.

**World Wide Web (WWW):** Graphical hypertext system linking many Internet computers.

**Yielding:** The onset of plastic deformation.

**Yield strength:** The stress required to produce a very slight yet specified amount of plastic strain; a strain offset of 0.002 is commonly used.

**Zener diode:** Semiconductor diode that has a well-defined turn-on voltage for conduction in the reverse direction.

**Zero offset:** The output of a sensor at zero input for a specified supply voltage or current.

**Zone refining:** A metallurgical process for obtaining a highly pure metal that depends on continuously melting the impure material and recrystallizing the pure metal.

# Index